P9-CKL-862

Biotechnology

Second Edition

Volume 8b

Biotransformations II

Biotechnology

Second Edition

Fundamentals

Volume 1
Biological Fundamentals

Volume 2
Genetic Fundamentals and
Genetic Engineering

Volume 3
Bioprocessing

Volume 4
Measuring, Modelling and Control

Products

Volume 5a
Recombinant Proteins, Monoclonal
Antibodies and Therapeutic Genes

Volume 5b
Genomics and Bioinformatics

Volume 6
Products of Primary Metabolism

Volume 7
Products of Secondary Metabolism

Volumes 8a and b
Biotransformations I and II

Special Topics

Volume 9
Enzymes, Biomass, Food and Feed

Volume 10
Special Processes

Volumes 11a–c
Environmental Processes I–III

Volume 12
Legal, Economic and
Ethical Dimensions

All volumes are also displayed on our Biotech Website:
http://www.wiley-vch.de/books/biotech

A Multi-Volume Comprehensive Treatise

Biotechnology

Second, Completely Revised Edition

Edited by
H.-J. Rehm and G. Reed
in cooperation with
A. Pühler and P. Stadler

Volume 8b

Biotransformations II

Volume Editor:
D. R. Kelly
Industrial Consultant:
J. Peters

Weinheim · New York · Chichester · Brisbane · Singapore · Toronto

Series Editors:
Prof. Dr. H.-J. Rehm
Institut für Mikrobiologie
Universität Münster
Corrensstraße 3
D-48149 Münster
FRG

Dr. G. Reed
1029 N. Jackson St. #501-A
Milwaukee, WI 53202-3226
USA

Prof. Dr. A. Pühler
Biologie VI (Genetik)
Universität Bielefeld
P.O. Box 100131
D-33501 Bielefeld
FRG

Prof. Dr. P. J. W. Stadler
Artemis Pharmaceuticals
Geschäftsführung
Pharmazentrum Köln
Neurather Ring 1
D-51063 Köln
FRG

Volume Editor:
Dr. D. R. Kelly
Department of Chemistry
Cardiff University
P.O. Box 912
Cardiff CF1 3TB
UK

Industrial Consultant:
Dr. J. Peters
Bayer AG
Geschäftsbereich Pharma
TO Biotechnologie
D-42096 Wuppertal
FRG

Library of Congress Card No.: applied for

British Library Cataloguing-in-Publication Data:
A catalogue record for this book is available from the British Library

Die Deutsche Bibliothek – CIP-Einheitsaufnahme

A catalogue record for this book
is available from Der Deutschen Bibliothek
ISBN 3-527-28324-2

Printed on acid-free and chlorine-free paper.

Composition and Printing: Zechner Datenservice und Druck, D-67346 Speyer.
Bookbinding: J. Schäffer, D-67269 Grünstadt.
Printed in the Federal Republic of Germany

Preface

In recognition of the enormous advances in biotechnology in recent years, we are pleased to present this Second Edition of "Biotechnology" relatively soon after the introduction of the First Edition of this multi-volume comprehensive treatise. Since this series was extremely well accepted by the scientific community, we have maintained the overall goal of creating a number of volumes, each devoted to a certain topic, which provide scientists in academia, industry, and public institutions with a well-balanced and comprehensive overview of this growing field. We have fully revised the Second Edition and expanded it from ten to twelve volumes in order to take all recent developments into account.

These twelve volumes are organized into three sections. The first four volumes consider the fundamentals of biotechnology from biological, biochemical, molecular biological, and chemical engineering perspectives. The next four volumes are devoted to products of industrial relevance. Special attention is given here to products derived from genetically engineered microorganisms and mammalian cells. The last four volumes are dedicated to the description of special topics.

The new "Biotechnology" is a reference work, a comprehensive description of the state-of-the-art, and a guide to the original literature. It is specifically directed to microbiologists, biochemists, molecular biologists, bioengineers, chemical engineers, and food and pharmaceutical chemists working in industry, at universities or at public institutions.

A carefully selected and distinguished Scientific Advisory Board stands behind the series. Its members come from key institutions representing scientific input from about twenty countries.

The volume editors and the authors of the individual chapters have been chosen for their recognized expertise and their contributions to the various fields of biotechnology. Their willingness to impart this knowledge to their colleagues forms the basis of "Biotechnology" and is gratefully acknowledged. Moreover, this work could not have been brought to fruition without the foresight and the constant and diligent support of the publisher. We are grateful to VCH for publishing "Biotechnology" with their customary excellence. Special thanks are due to Dr. Hans-Joachim Kraus and Karin Dembowsky, without whose constant efforts the series could not be published. Finally, the editors wish to thank the members of the Scientific Advisory Board for their encouragement, their helpful suggestions, and their constructive criticism.

H.-J. Rehm
G. Reed
A. Pühler
P. Stadler

Scientific Advisory Board

Contents

Contributors

Dr. Paul Bentley
Columbia University
Chemistry Department
Havemeyer Hall
Mail Box 3135
3000 Broadway (119th Street)
New York, NY 10027
USA
Chapter 12

Prof. Dr. G. Michael Blackburn
Department of Chemistry
The University of Sheffield
Sheffield, S3 7HF
UK
Chapter 11

Prof. Dr. Uwe T. Bornscheuer
Institut für Technische Chemie
und Biotechnologie
Ernst-Moritz-Arndt-Universität
Soldtmannstrasse 16
D-17487 Greifswald
Germany
Chapter 6

Dr. Antony D. M. Curtis
Department of Chemistry
Keele University
Keele, Staffordshire, ST5 5BG
UK
Chapter 1

Dr. Sabine Flitsch
Department of Chemistry
University of Edinburgh
West Mains Road
Edinburgh EH9 3JJ
Scotland
Chapter 5

Dr. Ian Fotheringham
NSC Technologies
601 East Kensington Road
Mt. Prospect, IL 60056
USA
Chapter 8

Dr. Arnaud Garçon
Department of Chemistry
The University of Sheffield
Sheffield, S3 7HF
UK
Chapter 11

Dr. Simon A. Jackman
Biological Laboratory
University of Kent
Canterbury, Kent CT2 7NJ
UK
Chapter 3

Dr. Simon Jones
Department of Chemistry
Bedson Building
University of Newcastle upon Tyne
Newcastle upon Tyne, NE1 7RU
UK
and
Cancer Research Institute
Arizona State University
Box 871604
Tempe, AZ 85287-2404
USA
Chapter 4

Dr. David R. Kelly
Department of Chemistry
Cardiff University
P.O. Box 912
Cardiff CF1 3TB
UK
Chapter 2

Dr. Jane McGregor-Jones
6 Cartmel Avenue
Maghull, Merseyside L31 9BJ
UK
and
Cancer Research Institute
Arizona State University
Box 871604
Tempe, AZ 85287-2404
USA
Chapter 10

Dr. Jassem Mahdi
Department of Chemistry
University of Wales
P.O. Box 912
Cardiff, CF1 3TB
UK
Chapter 2

Dr. Jürgen Rabenhorst
Haarmann & Reimer GmbH
Postfach 1253
D-37601 Holzminden
Germany
Chapter 9

Dr. Gary K. Robinson
Biological Laboratory
University of Kent
Canterbury, Kent CT2 7NJ
UK
Chapter 3

Dr. Michael Schedel
Bayer AG
PH-TO Biotechnologie
Biotechnikum Mikrobiologie
Gebäude 226
Friedrich-Ebert-Strasse 217
D-42096 Wuppertal
Germany
Chapter 7

Dr. Jane Stratford
Biological Laboratory
University of Kent
Canterbury, Kent CT2 7NJ
UK
Chapter 3

Dr. Gregory M. Watt
Department of Chemistry
University of Edinburgh
West Mains Road
Edinburgh EH9 3JJ
Scotland
Chapter 5

Introduction

DAVID R. KELLY
Cardiff, UK

The two years since the publication of Volume 8a have been a period of consolidation for the business community. As large chemical and pharmaceutical companies have merged into ever larger companies with increasing focus on their primary mission (the "best of breed" concept), small companies specializing in biotransformations have prospered, by providing expertise on demand. The lack of familiarity of most synthetic chemists with biotransformations, has in fact worked to the advantage of these small companies. Increasing demands for enantiomerically pure materials will provide further fuel for this trend. Many medium sized, fine and speciality chemical manufacturers are adding biotransformations to their repertoire of skills by acquisitions or internal developments. Much of the impetus in the speciality area, is the demand for so called natural or "nature identical" products.

On the other hand the spectre of genetic modification (GM) has undoubtedly acted as a brake on the development of many new products and processes. It remains to be seen what the public and legislators will tolerate. Attempts to stop the growing of GM crops and banish foodstuffs from GM organisms from the human food chain, lie outside the ambit of biotransformations. But regulation of these activities may well have a "knock on effect". For example enzymes derived from GM organisms, are used in the beverage industry for dehazing and clarification. Will these be regulated? Hydrolytic enzymes from GM organisms are used in laundry. Will non-foodstuff uses escape regulation?

On the academic front, enantioselective reactions which were once the sole preserve of enzymes are now being encroached upon by abiotic catalysts. Two good examples are the Baeyer–Villiger reaction and ester formation/hydrolysis. Conversely the first enzymes and catalytic antibodies capable of catalyzing pericyclic processes such as the Diels–Alder reaction are now emerging.

Volume 8b commences with the cornerstone of organic synthesis: carbon–carbon bond formation. Although there is no enzymatic equivalent of the Grignard or organolithium reagent, aldol and benzoin condensations can be performed with exquisite stereochemical control. These reactions are constrained by a limited range of anionic components, but are fairly promiscuous with electrophiles. Lyases (Chapter 2) are involved in a galaxy of reactions, which includes just about every prosthetic group/cofactor known or unknown. In connection with the latter, it has been discovered in the last year, that histidine ammonia-lyase which was long thought to contain a catalytically active dehydroalanine residue, actually contains an unprecedented 4-methylene-imidazol-5-one residue. This result throws doubt on the presence of the dehydroalanine residue in other enzymes, thought to contain it. Chapter 2 covers carboxy-lyases (many of which are involved in carbon–carbon bond formation), hydro-lyases (including the intricate biosynthesis of deoxysugars) and the ammonia-lyases many of which are involved in the biosynthesis of amino acids. Halocompounds (Chapter 3) have a reputation as un-natural compounds which are persistent and damaging to the environment or at best as exotic marine natural products. This reputation is unfounded. The majority of halocompounds in the environment originate from eological and natural sources. The fascinating topic of the biosynthesis of the ten known natural organofluorocompounds provides an insight into an element largely spurned by nature. The cleavage and formation of phosphate bonds (Chapter 4) is neglected in most biotransformations reviews. Yet for most phosphorylated and pyrophosphorylated cofactors the majority of the binding energy to the enzyme originates from this linkage. Whereas, the traditional organic chemist utilizes halosubstituents as nucleofuges, nature utilizes phosphates and their derivatives, because they can be selectively activated by carefully placed protonation sites.

Short polypeptide, RNA, and DNA sequences (<18 residues) are now routinely synthesized using automated machinery. In contrast, the synthesis of even a trisaccharide requires a dedicated team of specialists and although spectacular advances have been made, a general approach to the controlled synthesis of polysaccharides is still lacking. Simultaneous control of stereo- and regioselectivity have confounded conventional chemical approaches. Enzymatic glycosidation (Chapter 5) provides a powerful approach to escape this impasse.

The "Applications" superchapter covers selected industrial scale processes (commissioned by Dr. Jörg Peters) and the use of biotransformation products as intermediates for organic synthesis. Chapter 6 reviews general aspects of industrial biotransformations and banishes some myths about their limitations. The carbohydrate theme of chapter 5 is continued in chapter 7, which describes the development of the microbial oxidation of aminosorbitol, in the synthesis of 1-deoxynojirimycin, a key precursor of the anti-diabetes drug Miglitol®. This is an important example of a synthetic route, which is fast and efficient, because the biotransformation step removes the need to employ protecting groups to enforce regioselectivity.

The majority of amino acids are manufactured by fermentation, although there are also examples of amino acids produced employing isolated enzymes (e.g., L-*tert.* leucine). Developments in molecular biology and genome manipulation have enabled biosynthetic pathways to be optimized to previously unimaginable levels (Chapter 8). This area has been further promoted by the continuing demand for the dipeptide artificial sweetener, Aspartame.

Many of the important natural flavorings or fragrances are extracted in small amounts from exotic plants and the supply is affected by seasonal variations due to the weather, crop pests, and even political factors. Although many flavorings and fragrances can be synthesized via chemical routes, the resulting products cannot be marketed using the term "natural", since EU and US legislation has defined the term "natural": The product must be produced from natural starting materials applying physical, enzymatic, or microbial processes for their final production. These are ideal circumstances for the development of biotransformation routes, described in Chapter 9. For example, potential precursors of vanillin are found in numerous plants and even waste products. Biotransformations can convert these materials with low or even negative value (i.e., disposal costs) into valuable products.

Biotransformations have yielded a wealth of new routes to novel and known compounds. These have provided marvellous opportunities for synthetic organic chemists. Chapter 10 gives brief descriptions of an enormous range of biotransformations and describes applications of the products in the synthesis of natural products.

The final superchapter in Volume 8b provides a glimpse into two areas which are likely to play a major role in future developments in biotransformations. Catalytic antibodies (Chapter 11) are literally artificial enzymes. They can be engineered to catalyze reactions which have no counterparts in the natural world. Sophisticated strategies for raising antibodies and screening techniques for identifying active catalysts, have already yielded catalysts of exquisite selectivity. Intrinsic activity and turnover continue to be limited, but in the best cases they are only a few orders of magnitude lower than conventional enzymes.

The final chapter deals with wholly synthetic enzymes. These may be constructed from the familiar amino acid building blocks of conventional enzymes and may even borrow design features, but the structures are, nevertheless, totally alien. For example, poly-leucine catalyzes the enantioselective epoxidation of enones by hydrogen peroxide and cyclic dipeptides catalyze the enantioselective addition of cyanide to aldehydes. This area will surely benefit from the ever continuing improvements in computer processing power and new software for molecular modeling.

All predictions of the future have one thing in common. They are wrong! But this does not absolve us from trying to discern avenues that are ripe for development. A Third Edition of this volume is likely to contain a wider range of reactions (possibly including pericyclic and photochemical processes), more engineered pathways and more site directed mutagenesis. Gene transfer and heterologous expression is now common place. Are there better recipient organisms than the ubiquitous *E. coli*? Enzymes from thermophiles were touted as the universal panacea for industrial processes at elevated temperatures, but their early promise has not been fulfilled. Why? Enzyme catalyzed reactions in organic solvents are highly desirable for hydrophobic substrates, but are currently thwarted by low activity. Is it possible to design and synthesize an enzyme which is soluble in organic solvents (e.g., a lipase mutant)? These proposals represent reasonable extensions of current technology. However, it is certain that the most spectacular advances will be made in areas, that cannot be predicted at present.

I am pleased to acknowledge Dr. JÖRG PETERS' helpful comments on this Introduction.

Cardiff, March 2000 D. R. Kelly

Carbon–Carbon Bond Formation, Addition, Elimination and Substitution Reactions

1 Carbon–Carbon Bond Formation Using Enzymes

ANTHONY D. M. CURTIS

Keele, UK

1 Introduction

If the formation of carbon to carbon (C–C) bonds is undeniably fundamental to organic chemistry then the ability to induce asymmetry in the constructive process must be the undeniable goal of the modern organic chemist. Research into chemical methods for enantioselective synthesis has yielded many elegant methods for asymmetric C–C bond formation but the concepts of efficiency and atom economy seem to have been ignored in many cases. Enantioselective methods are, of course, of paramount importance to the industrial chemist, not least due to ever-tightening regulations governing single enantiomer pharmaceuticals.

Historically, reviews concerning the utility of enzymes in organic synthesis paid only cursory respect to the potential of enzymes to catalyze the formation of C–C bonds. However, trends in chemistry are not uncommon and there is certainly a trend for such reviews to devote an increasing amount of space to the subject. Indeed, as contemporary microbiological methods enable the study of enzymes from different species this exciting field of research has witnessed an exponential period of development, as reflected by the increasing number of articles devoted to the synthesis of C–C bonds (RACKER, 1961a; OGURA and KOYAMA, 1981; FINDEIS and WHITESIDES, 1984; BUTT and ROBERTS, 1986; JONES, 1986; DAVIES et al., 1989; TURNER, 1989, 1994; SERVI, 1990; CSUK and GLÄNZER, 1991; DRUECKHAMMER et al., 1991; WONG et al., 1991, 1995a; SANTANIELLO et al., 1992; LOOK et al., 1993; FANG and WONG, 1994; WONG and WHITESIDES, 1994; WALDMANN, 1995; WARD, 1995; GIJSEN et al., 1996; HOFFMANN, 1996; FESSNER and WALTER, 1997; TAKAYAMA et al., 1997a, b). The aim of this chapter is to present an overview of this rapidly growing field of research using appropriate examples from the literature which exemplify the versatility of the particular enzymes discussed. As the majority of discoveries in this field have taken place within the past two decades, this survey will focus predominantly on work which has appeared in the primary literature during that period. Readers are directed to the review articles cited above for a historical perspective of developments within this field.

Many C–C bond forming reactions fall within Enzyme Commission subclass 4.1, carbon–carbon lyases. Lyases are enzymes which cleave or form bonds other than by hydrolysis or oxidation. Systematic names for lyases are constructed from the template "substrate group-lyase". Groups are defined at the sub-subclass level, e.g., decarboxylases (4.1.1), aldehydes (4.1.2), and oxoacids (4.1.3). Thus, threonine aldolase (EC 4.1.2.5), which catalyzes the aldol condensation of glycine with acetaldehyde, is systematically named "threonine acetaldehyde-lyase". Reactions which have only been demonstrated in the carbon–carbon bond forming direction or for which this direction is more important are termed synthases. However, such designations have not found favor and trivial names are common. Other carbon–carbon bond forming reactions which appear in the following discussion may be found in classes 2 (transferases), 5 (isomerases), and 6 (ligases).

The laboratory use of enzymes for C–C bond formation has developed in quite distinct areas, often determined by the class of enzyme used. In recent years, a large amount of information concerning the effective use of aldolases in asymmetric synthesis has appeared, perhaps surpassing all of the other enzyme systems capable of forming C–C bonds. In an attempt to provide a balanced chapter, this review will present developments in the use of many enzymes, particularly on their use in the synthesis of both naturally occurring and unnatural bioactive products. Reference will be made to specialist reports in each area where they are available.

2 Aldolases

There are numerous chemical methods for achieving enantioselective aldol additions cited in the literature, some simple and some quite sophisticated. Catalytic processes are, of course, preferable over all others, provided the catalyst is cheap and readily available. However, stereoselective aldol reactions are a normal metabolic process for the vast majority of organisms and, given the skills of the microbiologist, naturally occurring catalysts have now become widely available for use in the laboratory. Several excellent accounts of the use of aldolases (the general name applied to this group of lyases and synthases) in organic synthesis detail the progress made in this area, notably the prolific contributions by the groups of WHITESIDES and WONG (WONG and WHITESIDES, 1994; GIJSEN et al., 1996; FESSNER and WALTER, 1997; TAKAYAMA et al., 1997a, b).

At the time of writing, over 30 aldolases have been identified and isolated, most of which catalyze the reversible stereospecific addition of a ketone to an aldehyde. These enzymes are further classified according to the mechanism by which they operate. Type I aldolases, typically associated with animals and higher plants, form an imine intermediate with the donor ketone in the enzyme active site. Type II aldolases, on the other hand, require a Zn^{2+} cofactor to act as a Lewis acid in the enzyme active site; such aldolases predominate in bacteria and fungi. Type II aldolases tend to be more stable than their type I counterparts.

Aldolases are also characterized by their specificity for particular donor substrates and the products they deliver:

(1) Dihydroxyacetone phosphate (DHAP) dependent lyases yield ketose 1-phosphates upon reaction with an aldehyde acceptor.
(2) Pyruvate dependant lyases and phosphoenolpyruvate (PEP) dependent synthases yield 3-deoxy-2-keto-acids.
(3) 2-Deoxyribose 5-phosphate aldolase (DERA) is an acetaldehyde dependent lyase which yields 2-deoxyaldehydes.
(4) Glycine dependent aldolases yield substituted α-amino acids.

Several aldolases are commercially available and have been widely used. Other aldolases have not yet been subjected to such rigorous synthetic investigations, though their *in vivo* reactions may give some indication as to their utility.

2.1 DHAP Dependent Aldolases

The aldol addition of a ketone donor and an aldehyde acceptor can potentially yield four stereoisomeric products which would require costly chromatographic separations. Fortunately for the synthetic chemist, aldolases exist which can deliver all of the four possible stereoisomers. This is illustrated by the asymmetric aldol addition of dihydroxyacetone phosphate (DHAP) to D-glyceraldehyde 3-phosphate or L-lactaldehyde (Fig. 1). *In vivo*, D-fructose-1,6-diphosphate (FDP) aldolase (EC 4.1.2.13), L-rhamnulose-1-phosphate (Rha 1-P) aldolase (EC 4.1.2.19), L-fuculose-1-phosphate (Fuc 1-P) aldolase (EC 4.1.2.17), and tagatose 1,6-diphosphate (TDP) aldolase (EC 4.1.2.-) form products which differ in the relative conformation of the hydroxy substituents on C-3 and C-4; FDP aldolase gives the (3*S*,4*R*) stereochemistry of L-fructose, Rha 1-P aldolase gives (3*R*,4*S*) as in L-rhamnulose, Fuc 1-P aldolase gives (3*R*,4*R*) as in L-fuculose, and TDP aldolase gives the (3*S*,4*S*) configuration of L-tagatose.

Each of the above aldolases has been isolated from a variety of mammalian and microbial sources. FDP type I aldolase from rabbit muscle (rabbit muscle aldolase, RAMA) has found extensive use in organic synthesis and is commercially available, as are FDP type II aldolase, Rha 1-P type II aldolase, and Fuc 1-P type II aldolase. All four complementary aldolases have been cloned and overexpressed (GARCIA-JUNCEDA et al., 1995).

The four DHAP-dependent aldolases have each been the subject of synthetic investigations. Each aldolase will accommodate a wide range of aldehydic substrates; the substrate specificity of each these enzymes has been elaborated in detail elsewhere (WONG and WHITESIDES, 1994). Aliphatic aldehydes, including those with α-substituents, and monosaccharides are generally suitable acceptor substrates, while aromatic, sterically hindered, and α,β-unsaturated aldehydes are not generally substrates. A recent review notes that over 100 aldehydes have been used as substrates for DHAP dependent aldolases in the synthesis of monosaccharides and the sequential use of an aldolase and an isomerase facilitates the synthesis of hexoses from the ketose aldol prod-

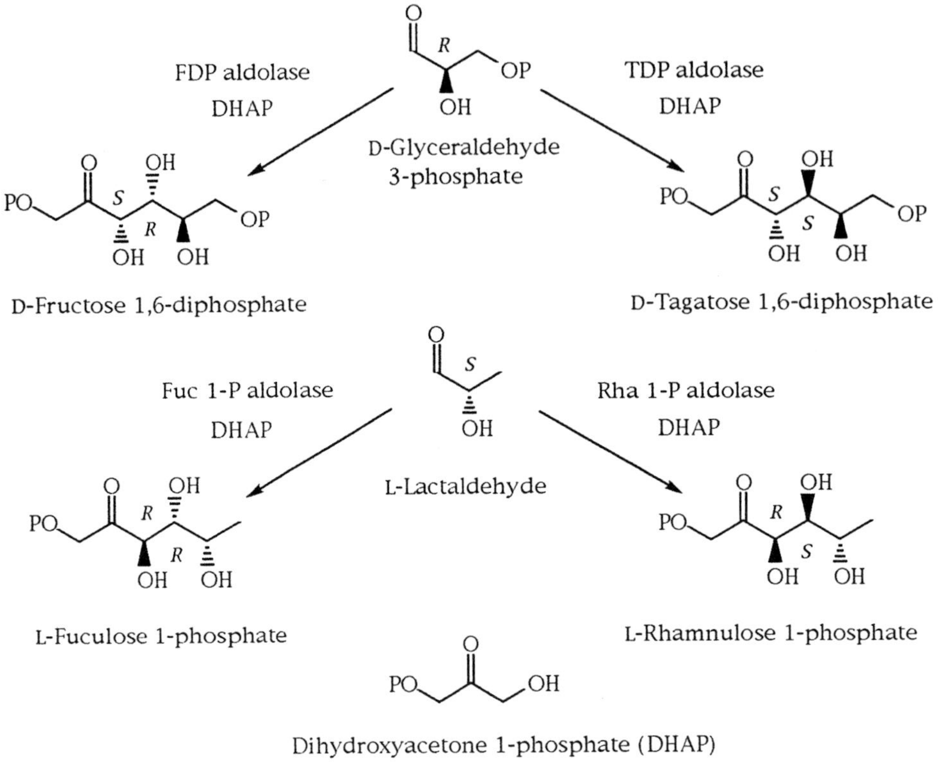

Fig. 1. Four stereo-complementary DHAP dependent aldolases.

ucts (WONG and WHITESIDES, 1994; TAKAYAMA et al., 1997b). In contrast, few analogs of DHAP, for example, (**1**) and (**2**) (Fig. 2), are substrates for FDP aldolase, all being very much less active than the natural ketone substrate (BISCHOFBERGER et al., 1988; BEDNARSKI et al., 1989). Recently, however, the 1-thio-analog of DHAP (**3**) was found to be a substrate and has found use in carbohydrate synthesis (FESSNER and SINERIUS, 1994; DUNCAN and DRUECKHAMMER, 1995).

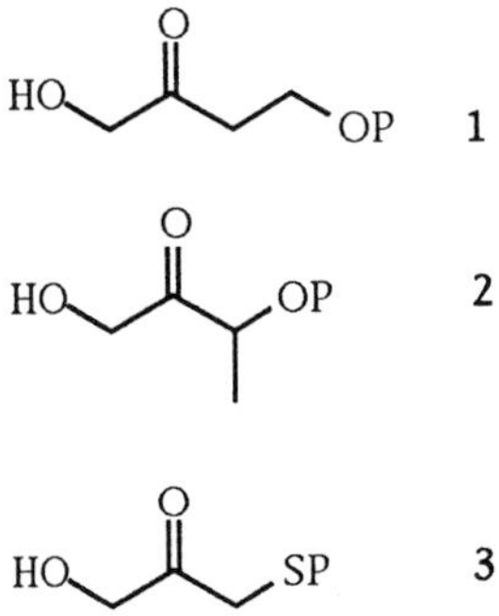

Fig. 2. Dihydroxyacetone phosphate (**3**) (DHAP) analogs.

The aldolases discussed thus far require DHAP as the ketone donor but the prohibitive commercial cost of DHAP may make the synthetic use of such DHAP dependent aldolases rather unattractive. However, many chemical and enzymatic methods have been reported for preparing DHAP in quantities sufficient for synthetic use (WONG and WHITESIDES, 1994; FESSNER and WALTER, 1997). A recent chemical method is able to give multigram quantities of DHAP from dimer (**4**) in 61% overall yield (Fig. 3) (JUNG et al., 1994) and an enzymatic process using glycerol phosphate oxidase gives almost quantitative yields

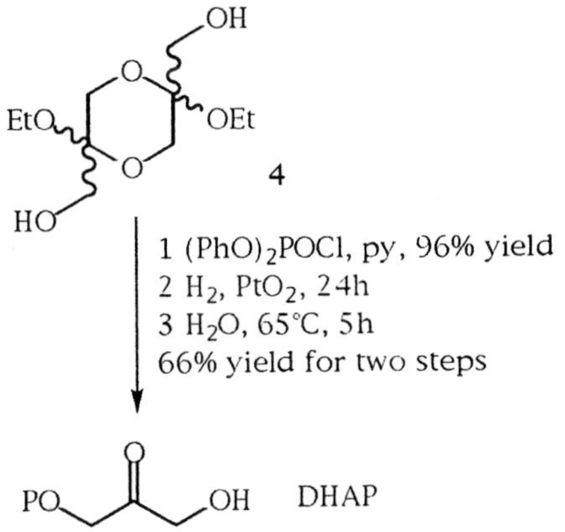

Fig. 3. Synthesis of dihydroxyacetone phosphate (DHAP).

of highly pure DHAP (FESSNER and SINERIUS, 1994).

Of all the aldolases discussed thus far, FDP type I aldolase (rabbit muscle aldolase, RAMA) has proved to be a most useful catalyst. However, the stereochemical outcome of reactions catalyzed by RAMA strongly depends upon the reaction conditions employed. For example, under kinetically controlled conditions D-glyceraldehyde 3-phosphate was preferred by the enzyme over L-glyceraldehyde 3-phosphate with a selectivity of 20:1 (BEDNARSKI et al., 1989). A model for this selectivity has been proposed (LEES and WHITESIDES, 1993). Dynamic resolution is possible in thermodynamically controlled reactions when the product is able to form a six-membered cyclic hemiketal. Compounds with the least number of diaxial interactions predominate (Fig. 4) (DURRWACHTER and WONG, 1988; BEDNARSKI et al., 1989).

FDP aldolase, more particularly RAMA, may be immobilized, used in a dialysis bag, or as the free enzyme in solution. Many examples of its use have appeared, notably in the synthesis of carbohydrates and analogs thereof (WONG and WHITESIDES, 1994; GIJSEN et al., 1996). The aldehyde substrate may be used as a racemic mixture, although the use of single enantiomers avoids the painful separation of diastereomeric product mixtures. Sharpless asymmetric dihydroxylation when used in tandem with an aldolase catalyzed condensation facilitates the synthesis of carbohydrates as single enantiomers (*vide supra*). Both enantiomers of D- and L-fructose were obtained from the dihydroxylation of an α,β-unsaturated aldehyde (not itself a substrate for aldolases), followed by reaction with DHAP catalyzed by the appropriate aldolase (HENDERSON et al., 1994a).

A new route to deoxysugars has become available through the use of the 1-thio-analog

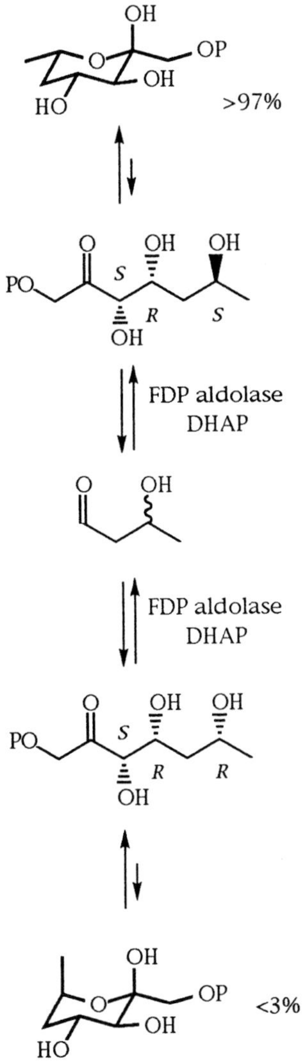

Fig. 4. Resolution of racemic aldehydes using FDP aldolase.

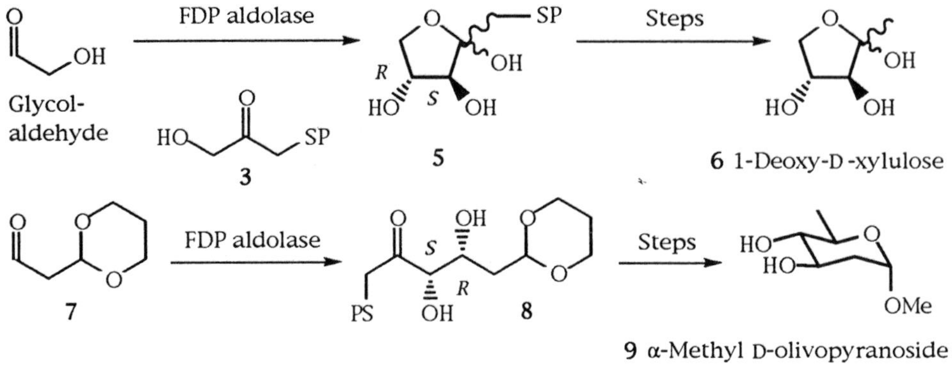

Glycol-aldehyde
FDP aldolase
3
5
Steps
6 1-Deoxy-D-xylulose
7
FDP aldolase
8
Steps
9 α-Methyl D-olivopyranoside

Fig. 5. Applications of the thio-analog of dihydroxyacetone phosphate (**3**) in synthesis.

of DHAP (**3**) (Fig. 5) (DUNCAN and DRUECKHAMMER, 1996). FDP aldolase catalyzed the condensation of (**3**) with glycolaldehyde to give 1-deoxy-1-thio-D-xylulose 1-phosphate (**5**), further elaboration of which yielded 1-deoxy-D-xylulose (**6**). Acetal (**7**) also reacted with (**3**) in the presence of FDP aldolase to deliver the diol (**8**), an intermediate in the synthesis of D-olivose methyl glycoside (**9**).

RAMA catalyzed the addition of DHAP to the semialdehyde (**10**) to give the diol (**11**), which is an intermediate in the synthesis of 3-deoxy-D-*arabino*-heptulosonic acid (**12**) (Fig. 6) (TURNER and WHITESIDES, 1989).

Analogs of so-called "higher sugars" sialic acid and 3-deoxy-D-*manno*-2-octulosonic acid (D-KDO), e.g., compounds (**13**) and (**14**), can be prepared by employing RAMA to catalyze the reaction of DHAP and hexoses or pentoses, respectively (BEDNARSKI et al., 1986). Higher sugars have been synthesized from simpler starting materials but with unexpected results (KIM and LIM, 1996). RAMA catalyzed reaction of the aldehyde (**15**) with DHAP followed by treatment with phosphatase, surprisingly gave a 62% yield of ketose (**16**) (Fig. 7). This has (3*S*,4*S*)-configurations at the newly formed chiral centers whereas RAMA catalysis normally delivers products with (3*S*,4*R*)-configurations, although other workers have noted this change in stereoselectivity (GIJSEN and WONG, 1995b; GIJSEN et al., 1996).

RAMA accepts aldehyde substrates bearing a variety of polar functionalities. 3-Azido-2-hydroxypropanal, either as the racemate or in optically pure form, was used as the substrate in a RAMA catalyzed synthesis of deoxynojirimycin (**17**) from L-3-azido-2-hydroxypropanal and deoxymannojirimycin (**18**) from D-3-

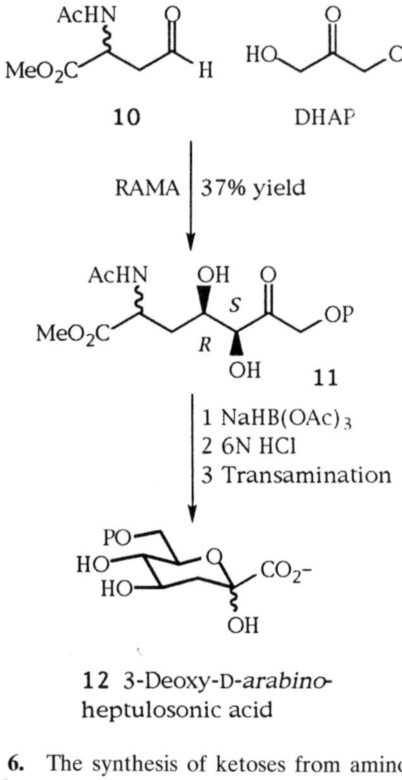

10
DHAP
RAMA | 37% yield
11
1 $NaHB(OAc)_3$
2 6N HCl
3 Transamination
12 3-Deoxy-D-*arabino*-heptulosonic acid

Fig. 6. The synthesis of ketoses from amino acid synthons.

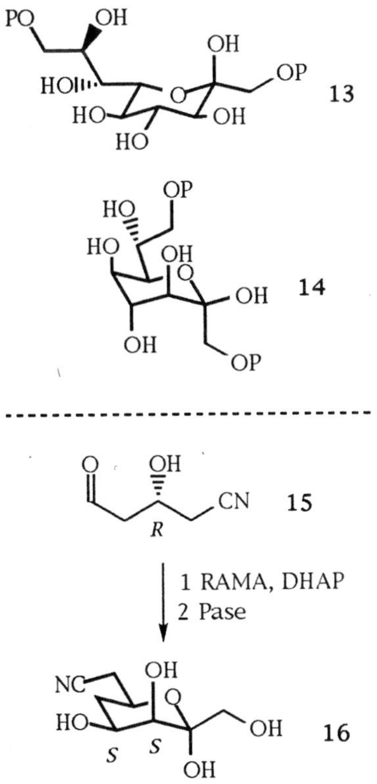

PO OH
OH
OP
HO
O
HO OH
HO
13

OP
HO
HO OH
O
OH 14
OH
OP

O OH
CN 15
R

1 RAMA, DHAP
2 Pase

NC OH
O
HO
OH 16
S S
OH

Fig. 7. RAMA catalyzed synthesis of carbohydrates.

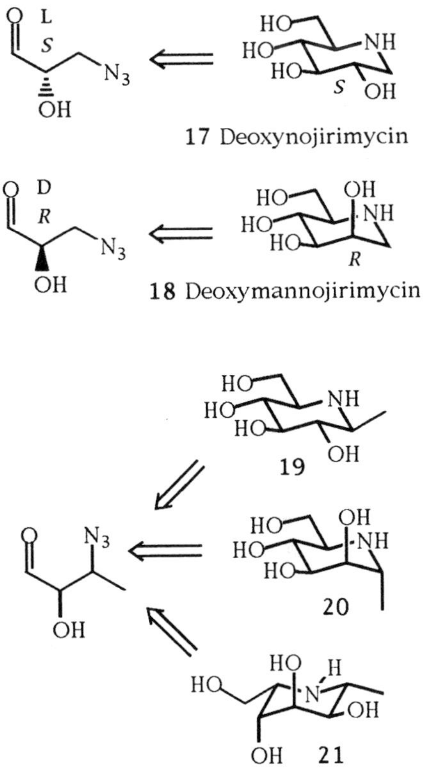

O L
S
N_3
OH
HO
HO NH
HO
S OH
17 Deoxynojirimycin

O D
R
N_3
OH
HO OH
HO NH
HO
R
18 Deoxymannojirimycin

HO
HO NH
HO
19 OH

O N_3
OH

HO OH
HO NH
HO
20

HO H
HO N
OH
OH 21

Fig. 8. Retrosynthetic analysis of the use of azido aldehydes in the RAMA catalyzed synthesis of azasugars.

azido-2-hydroxypropanal (Fig. 8); the same strategy was adopted by both EFFENBERGER and WONG (PEDERSON et al., 1988, 1990; ZIEGLER et al., 1988; VON DER OSTEN et al., 1989) and has recently found use in the preparation of iminocyclitols (MORÍS-VARAS et al., 1996). Analogously, 3-azido-2-hydroxybutanal was used as starting material for the synthesis of several iminoheptitols (**19–21**) (STRAUB et al., 1990).

The synthesis of aza-analogs of 1-deoxy-*N*-acetylglucosamine (**22**) and 1-deoxy-*N*-acetylmannosamine (**23**) employed optically pure 3-azido-2-acetamidopropanal, (*S*)- or (*R*)-(**24**), respectively, as the aldehyde substrate in a RAMA catalyzed reaction with DHAP (Fig. 9) (PEDERSON et al., 1990; KAJIMOTO et al., 1991a). Reduction of aldol products prior to removal of the phosphate group leads to the formation of 6-deoxy azasugars (KAJIMOTO et al., 1991b).

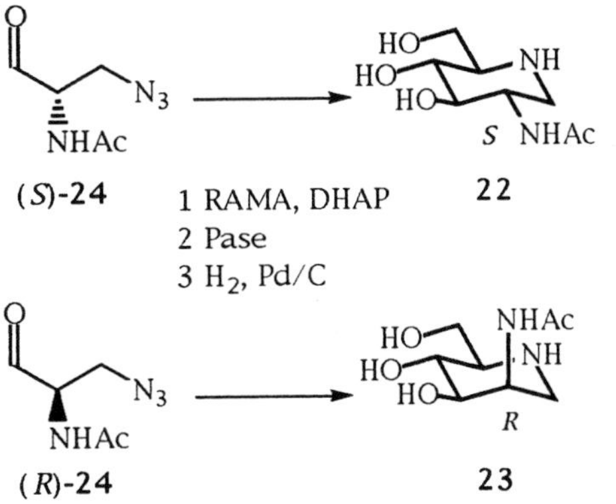

O
N_3
NHAc
(*S*)-24
HO
HO NH
HO
S NHAc
22

1 RAMA, DHAP
2 Pase
3 H_2, Pd/C

O
N_3
NHAc
(*R*)-24
HO NHAc
HO NH
HO
R
23

Fig. 9. Azido aldehydes in the RAMA catalyzed synthesis of azasugars.

1,4-Dideoxy-1,4-imino-D-arabitinol (**25**), a potent α-glucosidase inhibitor found in *Arachniodes standishii* and *Angylocalyx boutiqueanus*, and other polyhydroxylated pyrrolidines, such as heterocycles (**26a**) and (**26b**) can be prepared using the RAMA catalyzed condensation of DHAP and 2-azidoaldehydes (Fig. 10) (HUNG et al., 1991; LIU et al., 1991; KAJIMOTO et al., 1991a; TAKAOKA et al., 1993). Homoaza sugars, e.g., (**27**), may be prepared from 3-azido-2,4-dihydroxyaldehydes by the same methodology (HENDERSON et al., 1994b; HOLT et al., 1994).

Aldolase catalyzed reaction of thioaldehydes with DHAP enables the preparation of a range of thiosugars. 2-Thioglycolaldehyde (**28**) yielded enantiomeric 5-thio-D-*threo*-pentulofuranoses (**29**) and (**30**) when reacted with DHAP in the presence of FDP aldolase or Rha 1-P aldolase, respectively (Fig. 11), while RAMA catalyzed reaction of aldehyde (**31**) yielded thioketose (**32**), elaboration of which gave 1-deoxy-5-thio-D-glucopyranose peracetate (**33**) (Fig. 12). Further thiopyranose compounds are obviously accessible by using FDP aldolase, Fuc 1-P aldolase, or Rha 1-P aldolase

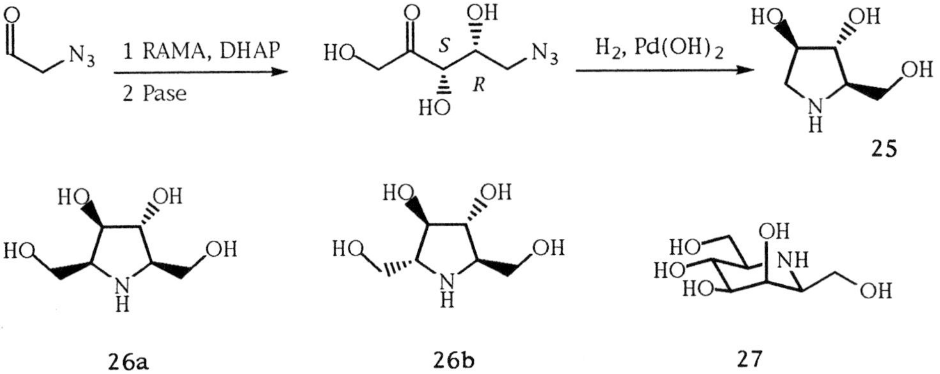

Fig. 10. Azasugars.

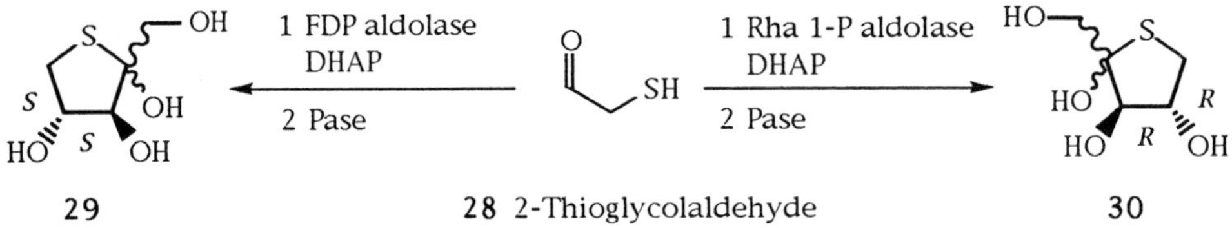

Fig. 11. Aldolase catalyzed synthesis of thiofuranoketoses.

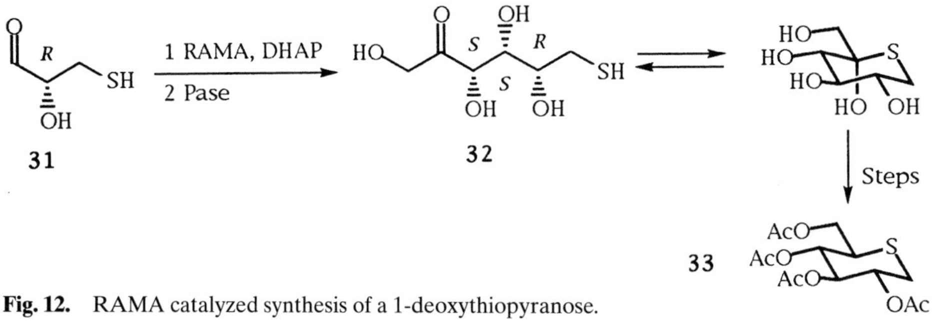

Fig. 12. RAMA catalyzed synthesis of a 1-deoxythiopyranose.

and enantiomers of (**31**) as required (EFFENBERGER et al., 1992a; CHOU et al., 1994).

Cyclitols are interesting targets for enzyme catalyzed synthesis. The RAMA catalyzed condensation of DHAP and chloroacetaldehyde forms the basis of a general synthesis of a range of cyclitols and C-glycosides, for example, (**34**) and (**35**) (Fig. 13) (SCHMID and WHITESIDES, 1990; NICOTRA et al., 1993). Nitrocyclitol (**36**) was prepared by coupling 2-hydroxy-3-nitropropanal and DHAP catalyzed by RAMA, followed by cyclization by an intramolecular Henry reaction in 51% overall yield (CHOU et al., 1995). Similarly cyclopentene

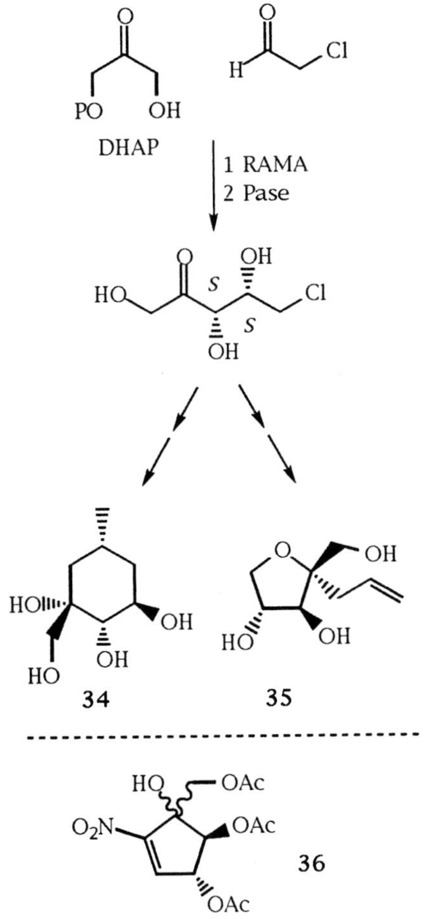

Fig. 13. The synthesis of carbocycles and C-glycosides.

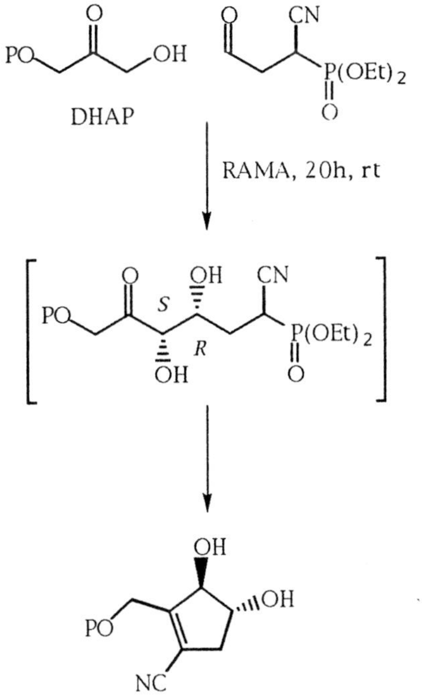

Fig. 14. RAMA catalyzed aldol condensation with *in situ* olefination.

(**37a**) was prepared using a one-pot strategy that required a RAMA catalyzed condensation as the first step, followed by cyclization by Horner-Wadsworth-Emmons olefination at pH 6.1–6.8 (Fig. 14) (GIJSEN and WONG, 1995a).

Unlikely as it may appear, the use of aldolases is not entirely restricted to carbohydrate synthesis. RAMA was used to install the chirality required for the synthesis of (+)-*exo*-brevicomin (**38**) (Fig. 15), seemingly a favored target for constructive biotransformations, from 5-oxohexenal (SCHULTZ et al., 1990). The C-9–C-16 fragment (**39**) of pentamycin has been prepared from L-malic acid using RAMA to catalyze the final elaboration; an alternative route starts from D-glucose, also employing RAMA (MATSUMOTO et al., 1993; SHIMAGAKI et al., 1993). The C-3–C-9 fragment of aspicilin was similarly prepared in a stereoselective

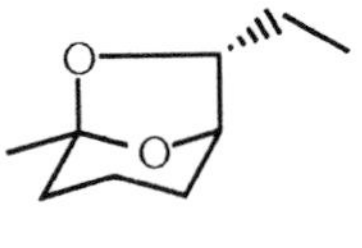

38 (+)-*exo*-Brevicomin

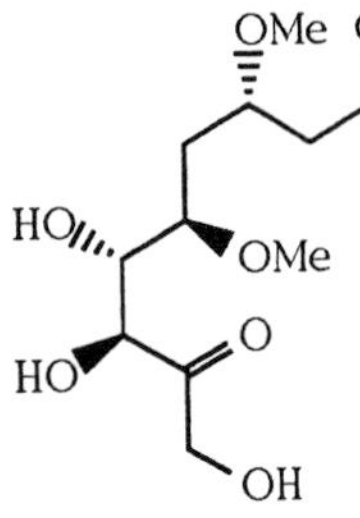

39 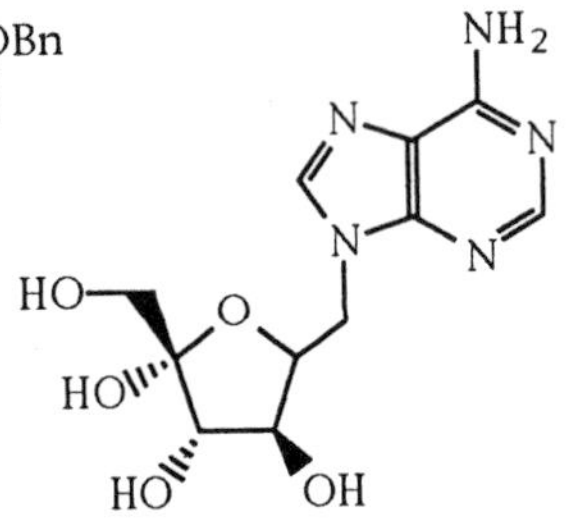 40

Fig. 15. Synthetic targets prepared using RAMA catalyzed aldol condensation.

manner using RAMA (CHENEVERT et al., 1997) and the synthesis of homo-C-nucleoside (**40**) also has a RAMA catalyzed condensation as a key step (LIU and WONG, 1992).

FDP aldolase has been used to effect a bidirectional synthesis strategy in which a series of disaccharide mimics have been prepared from α,ω-dialdehydes (EYRISCH and FESSNER, 1995). For example, ozonolysis of cyclohexene (**41**) and subsequent enzymatic elaboration gave compound (**42**) (Fig. 16). The same "tandem aldolization" methodology was used to convert cyclohexene (**43**) into disaccharide (**44**) (Fig. 17) (FESSNER and WALTER, 1997). The strategy noted above has come under recent scrutiny: in the FDP aldolase catalyzed desymmetrization of a series of dialdehydes in which the two carbonyl groups are remote from each other, only one aldehyde group reacted with the enzyme and the stereochemistry of the product did not appear to have been influenced by any chirality present in the starting materials (KIM et al., 1997).

Rha 1-P aldolase, Fuc 1-P aldolase, and TDP aldolase have been used less widely than FDP aldolase, though this may well change now that each enzyme has been overexpressed (GARCÍA-JUNCEDA et al., 1995). Rha 1-P aldolase and Fuc 1-P aldolase both accept a variety of aldehydic substrates. The products from both enzymes have the *R*-configuration at C-3 but the stereochemistry at C-4 can be less well defined (FESSNER et al., 1991). Both enzymes will effect dynamic resolution of racemic 2-hydroxyaldehydes (FESSNER et al., 1992).

Rha 1-P aldolase and Fuc 1-P aldolase have been used in the synthesis of a number of azasugars following analogous strategies to those

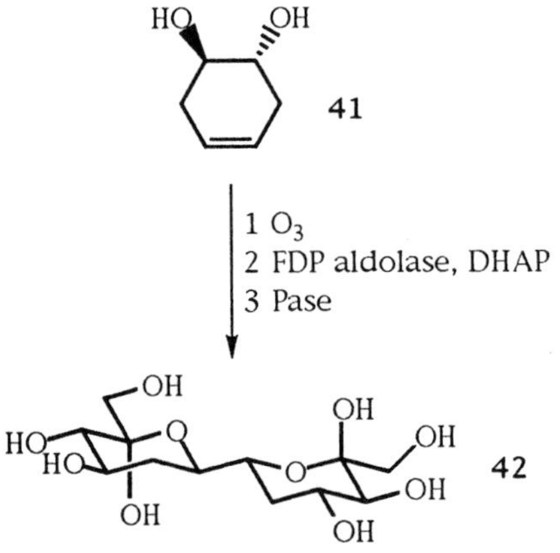

41

1 O_3
2 FDP aldolase, DHAP
3 Pase

42

Fig. 16. FDP aldolase catalyzed aldol condensation of a dialdehyde.

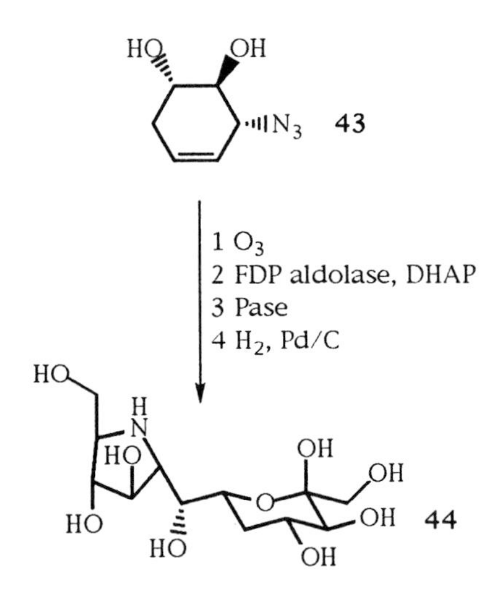

Fig. 17. FDP aldolase catalyzed aldol condensation of an azido dialdehyde, as a route to aminosugars.

outlined above for FDP aldolase (LIU et al., 1991; KAJIMOTO et al., 1991b; LEES and WHITESIDES, 1992). It should be noted that all of these aldolases may be used in combination with other enzymes to extend their synthetic utility. Aldolase catalyzed condensation and subsequent reaction with a corresponding isomerase gives access to aldoses from ketose intermediates (FESSNER and EYRISCH, 1992).

Recent investigations show that TDP aldolase also shows a low specificity for the aldehyde substrate but in all cases examined a diastereomeric mixture of products resulted. Of interest here is the anomalous stereochemical outcome of the reactions; the major product in each case had the (3*S*,4*R*) stereochemistry of D-fructose and not the expected (3*S*,4*S*) stereochemistry of D-tagatose (compare this with analogous observations made with FDP aldolase as noted above). Only D-glyceraldehyde, the natural substrate, delivered the expected configuration (FESSNER and EYRISCH, 1992; EYRISCH et al., 1993). This lack of stereospecificity has severely hindered the use of TDP aldolase in organic synthesis.

2.2 Pyruvate, Phosphoenolpyruvate (PEP) Dependent Aldolases

These particular enzymes, while performing opposite biological functions and requiring different substrates, can be examined together as they can both be used to synthesize ketoacids. However, only two aldolases from this group will be discussed at length, the potential of the other members not yet having been realized.

N-Acetylneuraminic lyase (NeuAc aldolase, pyruvate dependent, EC 4.1.3.3), also known as sialic acid aldolase, and the corresponding synthase (PEP dependent, EC 4.1.3.19) would both appear to deliver the same aldol products but the latter enzyme has not yet been fully characterized (GIJSEN et al., 1996). A type I aldolase, NeuAc aldolase catalyzes *in vivo* the reversible aldol condensation of *N*-acetyl-α-D-mannosamine (ManNAc, (**45**)) and pyruvate to yield the α-anomer of *N*-acetyl-5-amino-3,5-dideoxy-D-*glycero*-D-*galacto*-2-nonulosonic acid (NeuAc, (**46**)) (Fig. 18). The enzyme is relatively stable and can be used in solution, in dialysis tubing, or immobilized. It is specific for pyruvate and an excess of pyruvate is used to favor the formation of aldol products (UCHIDA et al., 1984, 1985; BEDNARSKI et al., 1987; AUGÉ et al., 1984, 1985; AUGÉ and GAUTHERON, 1987; KIM et al., 1988). The enzymes from Clostridia and *Escherichia coli* are commercially available.

A high conversion (>80%) of ManNAc (**45**) to NeuAc (**46**) is possible using the isolated enzyme though the work-up can be tricky (Fig. 18) (KRAGL et al., 1991; LIN et al., 1992; FITZ et al., 1995). A wide variety of aldehydes are tolerated by the enzyme; a recent review states that over 60 aldoses are substrates providing certain structural motifs are displayed, though substrates containing less than four carbons are not tolerated (WONG and WHITESIDES, 1994; FITZ et al., 1995; GIJSEN et al., 1996). The stereochemical outcome of reactions catalyzed by NeuAc aldolase strongly depends upon substrate structure and seems to be under thermodynamic control. Most substrates, like the natural substrate D-ManNAc (**45**), yield products with the *S*-configuration at the new chiral center, arising from attack at the *si*-face of the aldehyde substrate. Some substrates, however, such as L-mannose, L-xylose, and D-altrose, give the thermodynamically more stable products with the *R*-conforma-

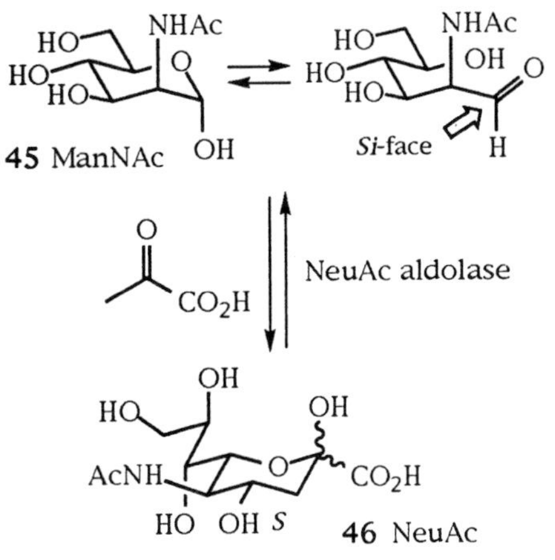

Fig. 18. NeuAc aldolase catalyzed aldol condensation of *N*-acetylmannosamine with pyruvate.

tion, resulting from attack at the *re*-face of the substrate (Lin et al., 1992; Fitz et al., 1995).

Many biologically relevant target compounds have been realized from NeuAc aldolase catalyzed reactions, including a wide variety of NeuAc derivatives. For example, L-NeuAc (**47**) (from L-ManNAc), D-KDO (**48**) (from D-arabinose), L-KDO (**49**) from L-arabinose and L-KDN (**50**) (from L-mannose) were all synthesized using NeuAc aldolase (Fig. 19) (Augé et al., 1990; Lin et al., 1997).

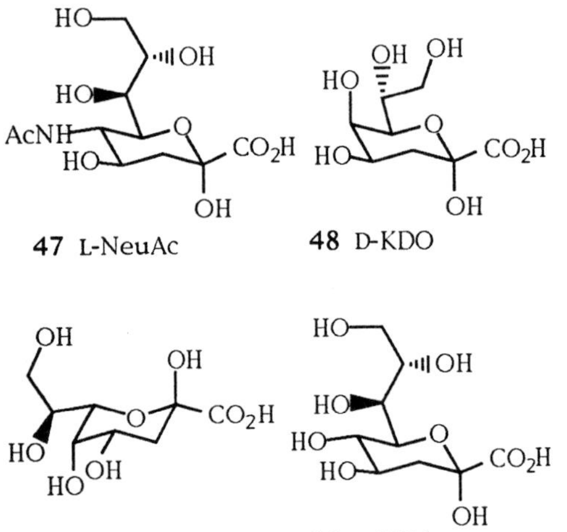

Fig. 19. Higher sugars prepared using NeuAc aldolase catalyzed aldol condensation.

6-*O*-(*N*-Carbobenzyloxyglycinoyl)-*N*-acetyl mannosamine (**51**) underwent a NeuAc aldolase catalyzed transformation to yield the aldol product (**52**), from which the fluorescent sialic acid (**53**) was prepared (Fig. 20) (Fitz and Wong, 1994). Polyacrylamides with α-sialoside and poly-C-sialosides appendages have been synthesized, which inhibit agglutination by the influenza virus (Spaltenstein and Whitesides, 1991; Nagy and Bednarski, 1991; Spevak et al., 1993). Pyrrolidine (**55**), derived from the NeuAc catalyzed reaction of protected D-mannosamine (**54**) followed by catalytic hydrogenation, was converted to 3-(hydroxymethyl)-6-epicastanospermine (**56**) (Fig. 21) (Zhou et al., 1993).

Again, despite performing distinctly different functions *in vivo*, 3-deoxy-D-*manno*-2-octulosonate lyase (2-keto-3-deoxyoctanoate aldolase, KDO aldolase, EC 4.1.2.23) and 3-deoxy-D-*manno*-2-octulosonate 8-phosphate synthase (phospho-2-keto-3-deoxyoctanoate synthase, KDO 8-P aldolase) may be used to catalyze the formation of the D-KDO skeleton; D-KDO 8-P has been prepared on a preparative scale using KDO 8-P aldolase (Bednarski et al., 1988). KDO aldolase isolated from *E. coli*

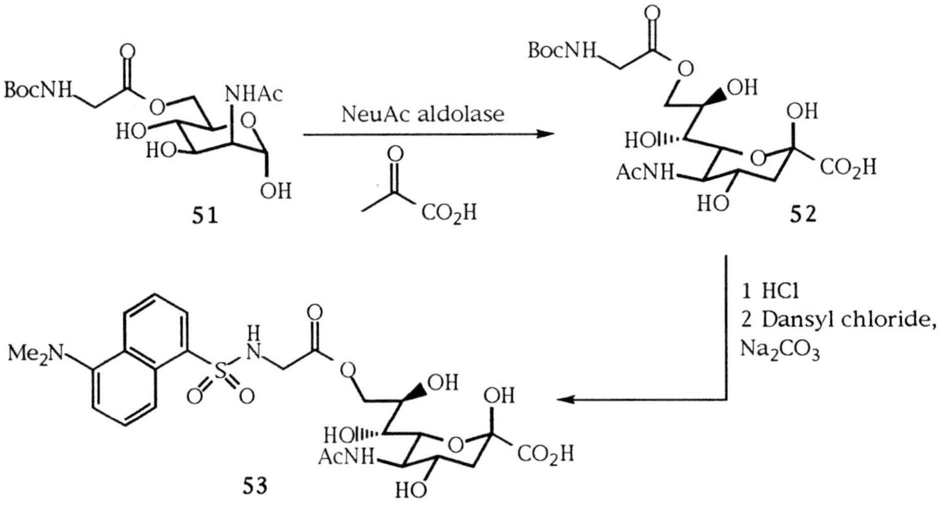

Fig. 20. NeuAc aldolase catalyzed aldol condensation in the synthesis of a fluorescent sugar.

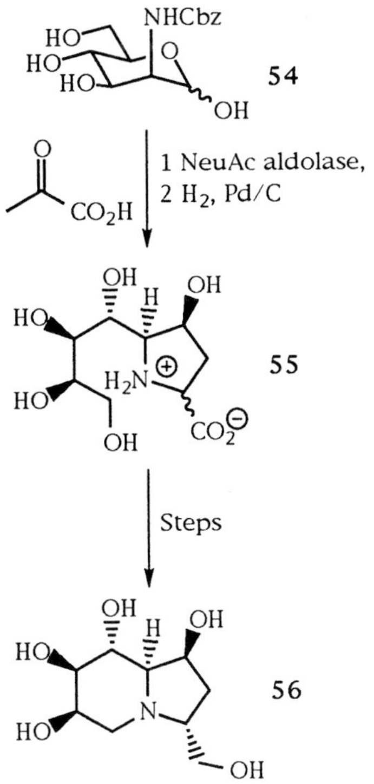

Fig. 21. NeuAc aldolase catalyzed aldol condensation of *N*-carbobenzyloxymannosamine (**54**) with pyruvate in the synthesis of 3-(hydroxymethyl)-6-epicastanospermine (**56**).

will accept a number of different aldose substrates though the rate of reaction is much slower than that for the natural substrate D-arabinose (GHALAMBOR and HEATH, 1966). KDO aldolase from *Aureobacterium barkeri*, strain KDO-37-2, shows a much wider substrate specificity. Substrates require the *R*-configuration at C-3 but the stereochemical requirements for C-2 are more relaxed, the *S*-configuration being preferred for kinetically controlled reactions and the *R*-configuration being favored under thermodynamic conditions. A number of carbohydrate derivatives were prepared during the course of these studies (SUGAI et al., 1993).

The 2-keto-4-hydroxyglutarate (KHG) aldolases from bovine liver and *E. coli* have been isolated, both enzymes having been shown to accept a range of analogs of pyruvate as the ketone donor (ROSSO and ADAMS, 1967; NISHIHARA and DEKKER, 1972; SCHOLTZ and SCHUSTER, 1984; FLOYD et al., 1992).

2-Keto-3-deoxy-6-phosphogluconate (KDPG) aldolase from *Pseudomonas fluorescens* will not accept simple aliphatic aldehydes as substrates: rather, the only requirement appears to be that there must be a polar substituent at C-2 or C-3. KDPG aldolase generates a new chiral center at C-4 with the *S*-configuration (ALLEN et al., 1992).

A number of other aldolases which are dependent on pyruvate or PEP have been isolated but their substrate specificities have not been investigated in any depth (GIJSEN et al., 1996).

2.3 2-Deoxyribose-5-Phosphate Aldolase (DERA)

2-Deoxyribose 5-phosphate aldolase (DERA, EC 4.1.2.4) is unique in that the donor species is an aldehyde. A type I aldolase, DERA catalyzes *in vivo* the reversible addition of acetaldehyde and D-glyceraldehyde 3-phosphate to give 2-deoxyribose 5-phosphate. The enzyme has been obtained from several different organisms but DERA from *E. coli* is available in large quantities as a consequence of cloning and overexpression (BARBAS et al., 1990; CHEN et al., 1992; WONG et al., 1995b). DERA will also accept propanal, acetone, and fluoroacetone as donor moieties though their rates of reaction are considerably slower than that of the natural donor. The enzyme will accept a wide range of aldehyde acceptor substrates and is capable of effecting dynamic resolution of a racemic mixture of aldehydes, D-isomers reacting preferentially to L-isomers, and the new stereocenter created during reaction generally has the *S*-configuration. DERA has been used to prepare a number of carbohydrate analogs from suitably substituted aldehydes, examples being furanose (**57**), pyranose (**58**), thiosugar (**59**), azasugar (**60**), and simple aldehyde (**61**) (Fig. 22).

Conveniently, when acetaldehyde is used as the donor substrate the products from DERA catalyzed reactions are themselves aldehydes and are thus capable of acting as acceptor sub-

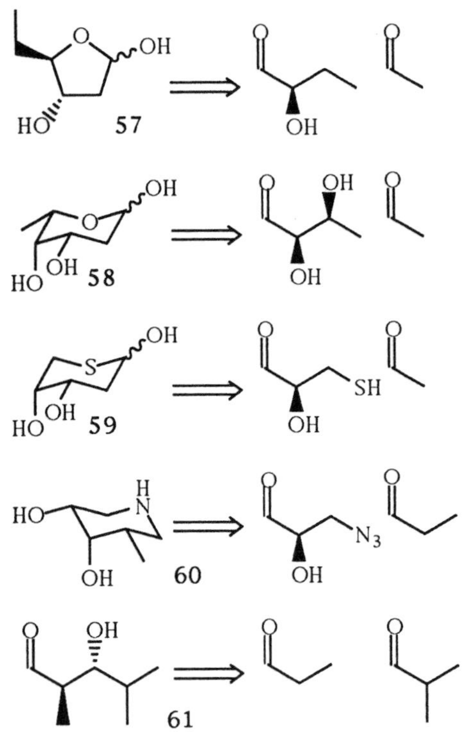

Fig. 22. Retrosynthetic analysis of the use of 2-deoxyribose 5-phosphate aldolase (DERA).

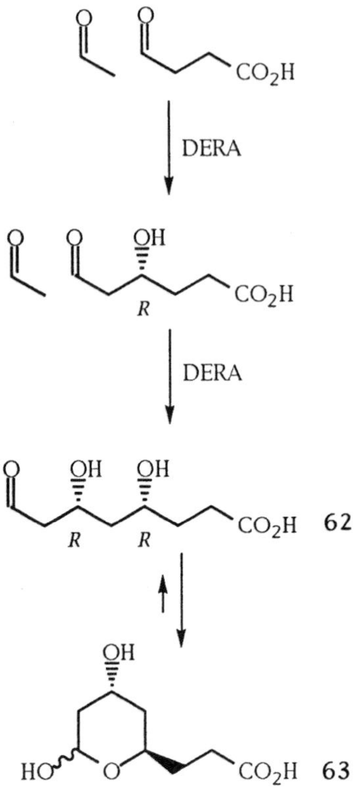

Fig. 23. 2-Deoxyribose 5-phosphate aldolase (DERA) catalyzed acetaldehyde condensations.

strates for further reaction (GIJSEN and WONG, 1994). As observed with a number of α-substituted aldehydes, if the product from the first addition of acetaldehyde is unable to form a cyclic hemiacetal then reaction with a second molecule of acetaldehyde proceeds to yield 2,4-dideoxyhexoses, e.g., (**62**), cyclization of which yields cyclic hemiacetal (**63**) thus preventing further addition (Fig. 23). The best substrate for such sequential additions appears to be succinic semialdehyde, the carboxylic acid group mimicking the phosphate group of D-glyceraldehyde 3-phosphate (WONG et al., 1995b).

In the same vein, DERA has also been used in combination with FDP-aldolase (RAMA) to give a range of 5-deoxyketoses with three axial substituents (**64**) together with other compounds which arise from thermodynamic equilibration (Fig. 24), and in combination

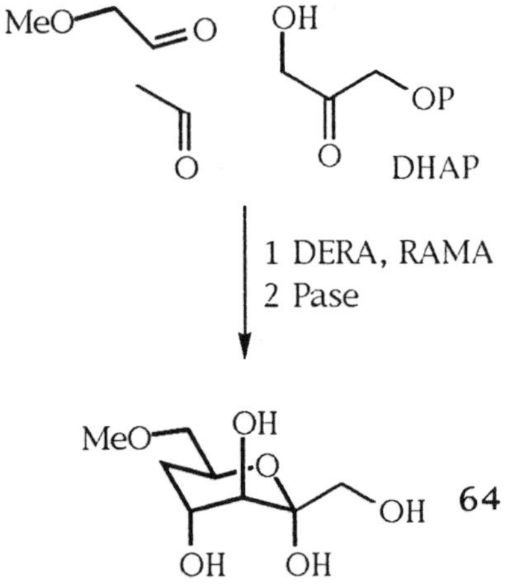

Fig. 24. One-pot multiple enzyme catalyzed synthesis of sugars.

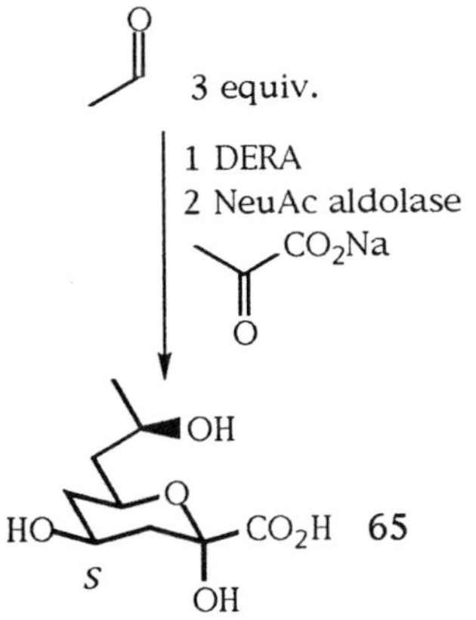

3 equiv.
1 DERA
2 NeuAc aldolase

65

Fig. 25. Two step multiple enzyme catalyzed synthesis of sugars.

with NeuAc-aldolase to yield derivatives of sialic acid (**65**) (55% yield) (Fig. 25) (GIJSEN and WONG, 1995b, c). In the latter reaction the configuration of C-4 is the opposite to that normally observed in NeuAc-aldolase catalyzed reactions and results from equilibrated coupling.

2.4 Glycine Dependent Aldolases

Serine hydroxymethyltransferase (SHMT, serine aldolase) and threonine aldolase (both EC 2.1.2.1) require pyridoxal 5-phosphate and tetrahydrofolate to catalyze the reversible addition of glycine and an aldehyde to deliver β-hydroxy-α-amino acids. Despite their apparent synthetic potential, neither enzyme has received intensive investigative attention. The term "threonine aldolase" is also used to refer to pyridoxal phosphate requiring enzymes (EC 4.1.2.5) which do not require tetrahydrofolate.

In vivo SHMT catalyzes the reversible aldol reaction of glycine with "formaldehyde" (in the form of 5,10-methylenetetrahydrofolate and water) to give L-serine. SHMT can be isolated from numerous organisms and cloning experiments have been successfully carried out (FESSNER and WALTER, 1997). SHMT will react with other aldehyde substrates in the absence of tetrahydrofolate, as demonstrated by synthetic studies carried out using SHMT from rabbit liver and pig liver (LOTZ et al., 1990; SAEED and YOUNG, 1992). An excess of glycine was required to favor formation of the new amino acids. L-Amino acids predominated but long reaction times resulted in equilibration of product stereochemistry, delivering the thermodynamically more stable *threo* isomers.

Threonine aldolase from mammalian tissues catalyzes *in vivo* the liberation of glycine and acetaldehyde from L-threonine, although L-*allo*-threonine appears to be a more active substrate for the enzyme (FESSNER and WALTER, 1997). The enzyme from *Candida humicola* showed tolerance to a variety of aliphatic and aromatic aldehyde acceptors, although α,β-unsaturated aldehydes were not accepted (VASSILEV et al., 1995a). A large excess of glycine was required to drive the preparative reactions and the (2*S*,3*S*):(2*S*,3*R*) diastereoselectivity varied depending upon the aldehyde used. L-Threonine aldolase was used in a synthesis of α-amino-β-hydroxy-γ-butyrolactone (**66**) (Fig. 26) (VASSILEV et al., 1995b). Kinetic and thermodynamic control has been applied to the reac-

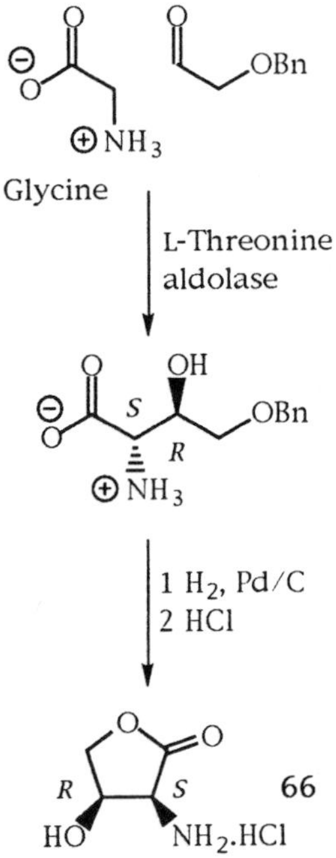

Glycine
L-Threonine aldolase
1 H_2, Pd/C
2 HCl
66

Fig. 26. The synthesis of α-amino-β-hydroxy-γ-butyrolactone (**66**) using L-threonine aldolase.

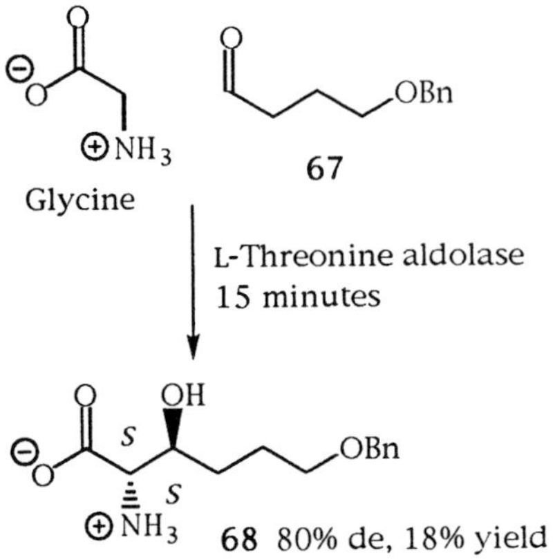

Fig. 27. Kinetically controlled synthesis of amino acids using L-threonine aldolase.

tions of L-threonine aldolase. Enzymic condensation of γ-benzyloxybutanal (**67**) (Fig. 27) and glycine for 15 min gave the (2*S*,3*S*)-amino acid (**68**) as the major product (80% de, 18% yield), whereas over 15 h the thermodynamically more stable (2*S*,3*R*)-diastereoisomer predominated (20% de, 70% yield) (SHIBATA et al., 1996). Recent studies have focused on the use of recombinant D- and L-threonine aldolases (KIMURA et al., 1997).

A threonine aldolase from *Streptomyces amakusaensis* has been purified and preliminary synthetic investigations conducted, including the resolution of several racemic *threo*-phenylserines to deliver enantiomerically pure (2*R*,3*S*)-D-amino acids (HERBERT et al., 1993, 1994).

3 Transaldolase and Transketolase

Transaldolase (EC 2.2.1.2) and transketolase (EC 2.2.1.1) are enzymes found in both the oxidative and reductive pentose phosphate pathway (RACKER, 1961b, c; MOCALI et al., 1985) and an example of each enzyme obtained from bakers' yeast is commercially available.

Transaldolase catalyzes *in vivo* the transfer of a dihydroxyacetone unit between phosphorylated metabolites; for example, the C-1–C-3 aldol unit is transferred from D-sedoheptulose 7-phosphate (**69**) to D-glyceraldehyde 3-phosphate, delivering D-erythrose 4-phosphate (**70**) and D-fructose 6-phosphate (**71**) (Fig. 28). However, despite its availability and potentially low acceptor substrate specificity, transaldolase has not been fully exploited in organic synthesis. This can probably be attributed to the fact that reactions catalyzed by transaldolase have the same stereochemical outcome as is obtained by using fructose 1,6-diphosphate (FDP) aldolase (RAMA).

Transaldolase has been used in a multienzyme system for the conversion of starch to D-fructose. The transaldolase catalyzed transfer of dihydroxyacetone from D-fructose 6-phosphate to D-glyceraldehyde enabled the final step of the synthesis, a step which could not be satisfactorily completed using a phosphatase (MORIDIAN and BENNER, 1992).

Transketolase, on the other hand, has seen rather more widespread use as a reagent in organic synthesis. Transketolase catalyzes *in vivo* the reversible transfer of a nucleophilic hydroxyacetyl unit from a ketose phosphate to

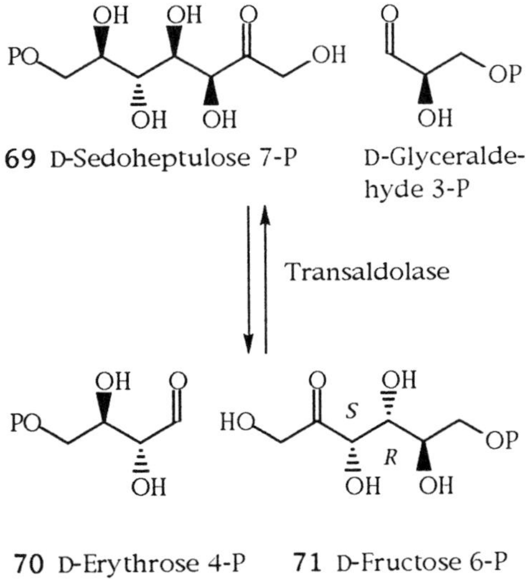

Fig. 28. Transaldolase catalyzed transfer of a dihydroxyacetone unit.

an aldose phosphate, specifically the C-1–C-2 ketol unit of D-xylulose 5-phosphate (**72**) to D-ribose 5-phosphate (**73**), yielding D-sedoheptulose 7-phosphate (**69**) and D-glyceraldehyde 3-phosphate (Fig. 29). D-Erythrose 4-phosphate (**70**) also acts as a ketol donor, yielding D-fructose 6-phosphate. The enzyme requires both thiamine pyrophosphate and Mg^{2+} as cofactors.

β-Hydroxypyruvic acid may also be used as the ketol donor for transketolase, though it is transferred to an aldose acceptor with an activity 4% of that of D-xylulose 5-phosphate (SRERE et al., 1958; BOLTE et al., 1987; HOBBS et al., 1993). The key advantage to using hydroxypyruvic acid, however, is that the transfer process is now irreversible, due to the elimination of carbon dioxide subsequent to transfer of the ketol unit.

Transketolases have been isolated from several different species; the enzymes used most commonly are obtained from bakers' yeast or from spinach leaves (BOLTE et al., 1987; SCHNEIDER et al., 1989; DEMUYNCK et al., 1990a, b; LINDQVIST et al., 1992), although recent reports have discussed the use of transketolase from *E. coli* (HOBBS et al., 1993; HUMPHREY et al., 1995; MORRIS et al., 1996). The substrate specificity of transketolase has been surveyed and its synthetic potential compared to the established chemistry of fructose 1,6-diphosphate (FDP) aldolase (RAMA) (DEMUYNCK et al., 1991; KOBORI et al., 1992; DALMAS and DEMUYNCK, 1993). A wide range of aldehydes will act as ketol acceptors, the rate of reaction being increased if there is an oxygen atom α or β to the carbonyl group, and the new chiral center formed always has the *S*-configuration. Transketolase can also be used to effect a kinetic resolution of racemic 2-hydroxyaldehydes as only the *R*-enantiomer reacts (EFFENBERGER et al., 1992b).

Recent investigations into the synthetic utility of transketolase from *E. coli* have yielded promising results. The use of a genetically modified strain of *E. coli* to provide transketolase of sufficient purity for biotransformations, combined with an efficient synthesis of potassium hydroxypyruvate, enables the gram-scale synthesis of triol (**75**) from aldehyde (**74**) in 76% yield (Fig. 30) (MORRIS et al., 1996).

Several natural product syntheses employing transketolase have been reported. The glycosidase inhibitor 1,4-dideoxy-1,4-imino-D-arabinitol (**25**) was prepared by using the transketolase catalyzed condensation of racemic aldehyde (**76**) with lithium hydroxypyruvate and kinetic resolution as the key step. Condensation gave the azido compound (**77**) in 71% effective yield, with further chemical

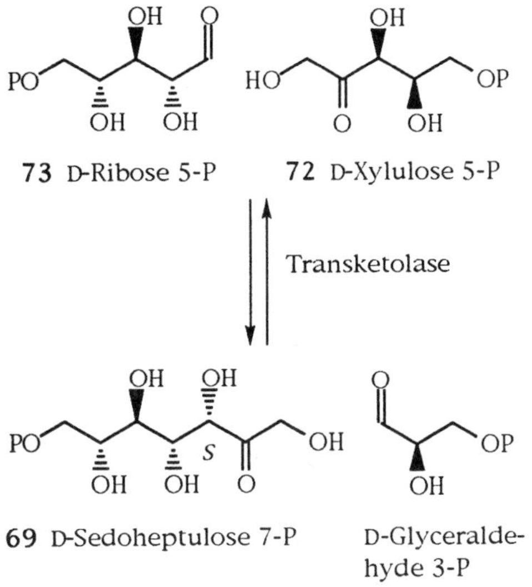

Fig. 29. Transketolase catalyzed transfer of a hydroxyacetyl unit.

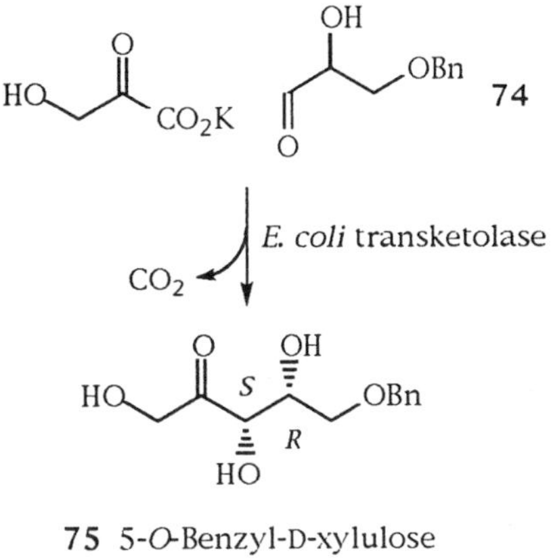

Fig. 30. Transketolase catalyzed condensation of β-hydroxypyruvate.

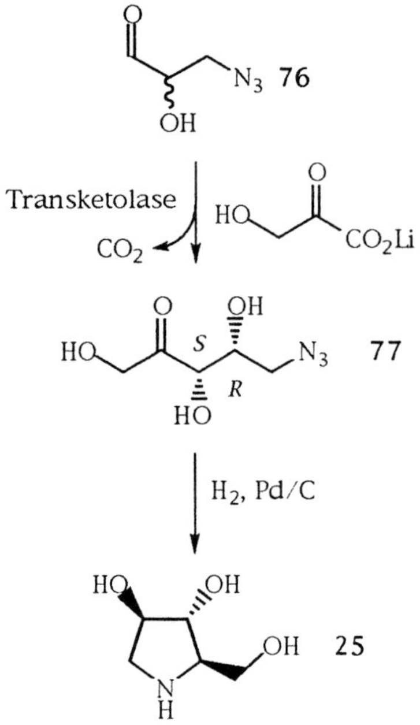

Fig. 31. The synthesis of 1,4-dideoxy-1,4-imino-D-arabitinol (**25**) using transketolase.

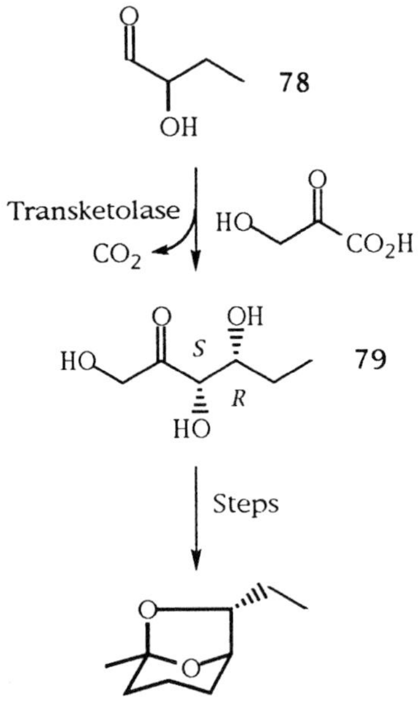

Fig. 32. The synthesis of a bark beetle pheromone using transketolase.

elaboration delivering the pyrrolidine derivative (**25**) in good overall yield (Fig. 31) (ZIEGLER et al., 1988).

(+)-*exo*-Brevicomin (**38**) represents an ideal target as it possesses the stereochemical motif obtained from a transketolase catalyzed transformation. Taking advantage of the enzymatic kinetic resolution, intermediate (**79**) was obtained in 45% yield by reacting 2-hydroxybutyraldehyde (**78**) with hydroxypyruvic acid in the presence of transketolase and cofactors. Ketone (**79**) may also be obtained from the RAMA catalyzed condensation of propanal and DHAP, followed by enzymic dephosphorylation. Further chemical operations delivered (+)-*exo*-brevicomin (**38**) (Fig. 32) (MYLES et al., 1991).

Yeast transketolase was used to prepare triol (**81**) from 3-cyano-2-hydroxyaldehyde (**80**) as the key step in a synthesis of fagomine (**82**) (Fig. 33) (EFFENBERGER and NULL, 1992). The same catalyst delivered 5-thio-D-*threo*-2-pentulofuranose (**29**) from the thiol-substituted aldehyde (**31**) (Fig. 34) (EFFENBERGER et al., 1992a) which is also accessible from the FDP aldolase condensation of DHAP and 2-thioglycolaldehyde (**28**) (Fig. 11) (EFFENBERGER et al., 1992a; CHOU et al., 1994).

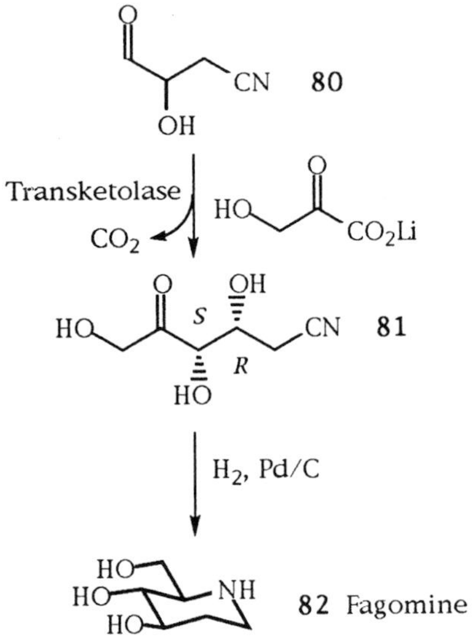

Fig. 33. The synthesis of fagomine (**82**) using transketolase.

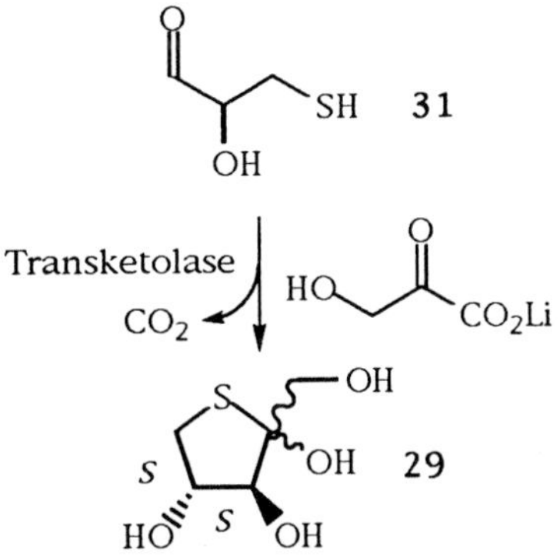

Fig. 34. The synthesis of 5-thio-D-*threo*-2-pentulofuranose (**29**) using transketolase.

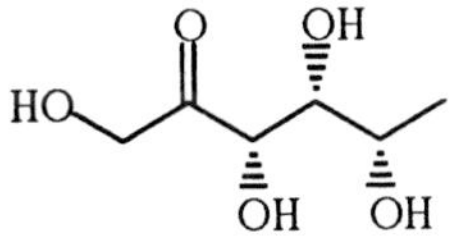

Fig. 35. Deoxysugars prepared using transketolase.

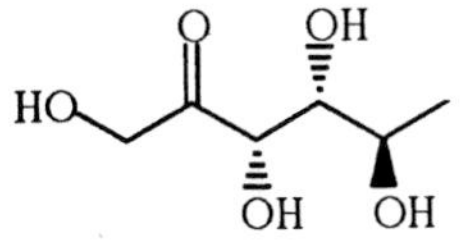

Transketolase catalyzed condensation of a racemic mixture of *erythro*- and *threo*-2,3-dihydroxybutanals with hydroxypyruvate was specific for the (2*R*)-stereoisomers, and yielded exclusively 6-deoxy-D-fructose (**84**) and 6-deoxy-L-sorbose (**85**) (Fig. 35) (HECQUET et al., 1994a, 1996). Similarly the triols (**86**) and (**87**) were prepared from the corresponding dithiane precursors. The corresponding aldehydes are intermediates for the synthesis of fagomine (**82**) and 1,4-dideoxy-1,4-imino-D-arabitinol (**25**), respectively (Fig. 36) (HECQUET et al., 1994b).

4 Pyruvate Decarboxylase

The capability of bakers' yeast to catalyze the acyloin condensation was first observed over 70 ago and has since been examined in several review articles (FUGANTI and GRASSELLI, 1985; DAVIES et al., 1989; SERVI, 1990; CSUK and GLÄNZER, 1991; WARD, 1995; HOFFMANN, 1996; FESSNER and WALTER, 1997).

In this classic biotransformation, a C–C bond is formed by reaction of an aldehyde, e.g., benzaldehyde, and an acetaldehyde equivalent to yield an α-hydroxyketone (**88**); with bakers' yeast the acetaldehyde equivalent adds to the *si*-face of the aldehydic carbonyl group (Fig. 37) (FUGANTI et al., 1984). The ketol product (**88**) is usually accompanied by varying amounts of the corresponding diol (**89**), derived from subsequent enzymatic reduction on the *re*-face of the carbonyl, and the major by-product is benzyl alcohol. The reduction step has been studied in some detail (CSUK and GLÄNZER, 1991).

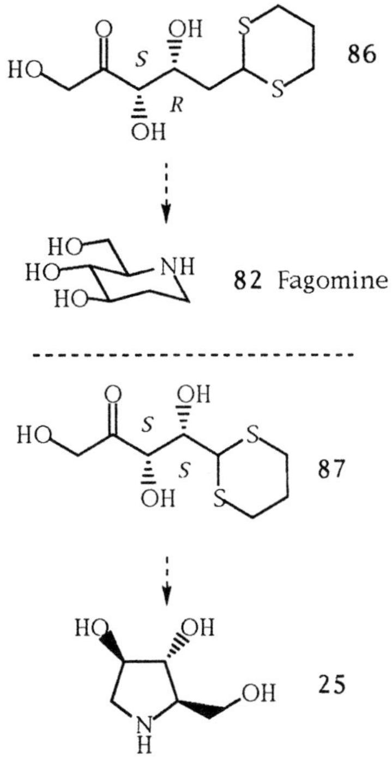

Fig. 36. The synthesis of fagomine (**82**) and 1,4-dideoxy-1,4-imino-D-arabitinol (**25**) from intermediates prepared using transketolase.

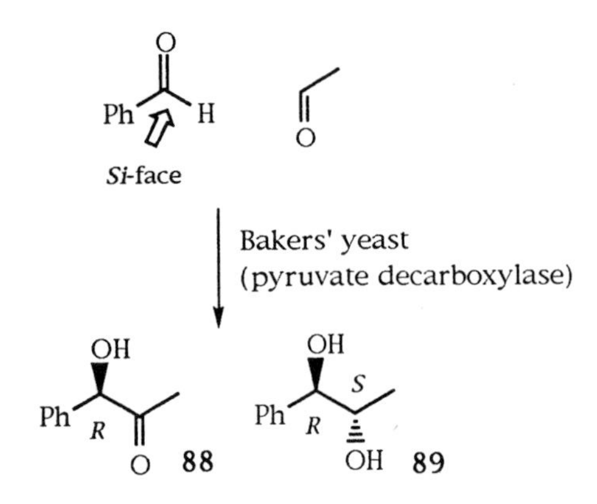

Fig. 37. Pyruvate decarboxylase mediated acyloin condensation.

The acyloin condensation is catalyzed by pyruvate decarboxylase in the presence of thiamine pyrophosphate and Mg^{2+}. Decarboxylation of pyruvic acid supplies the acetaldehyde equivalent and the condensation is highly enantioselective. The caveat is that the biotransformation usually gives low yields of the desired condensation products. However, both reactants and biocatalyst are cheap and the enzymic condensation can deliver chiral compounds which may not be readily obtained from the chiral pool.

The main reason why this particular biotransformation continues to be of interest is that the venerable condensation of benzaldehyde with pyruvic acid to give L-3-hydroxy-3-phenylpropan-2-one (phenylacetyl carbinol, PAC) (**88**) is used in the industrial synthesis of L-(−)-ephedrine and D-pseudoephedrine (Rose, 1961). In fact, the vast majority of recent publications concerning the enzymatic acyloin condensation discuss the production of PAC. Despite this, all these reports have indications for the laboratory-scale production of α-hydroxyketones.

It has been reported that the judicious addition of acetone to the reaction mixture suppresses diol formation, by undergoing reduction and inhibiting further reduction of PAC (Shimazu, 1950). Moreover, the use of carbohydrate-free yeast partially surpresses the formation of several by-products (Glänzer et al., 1987).

During their investigations using isolated yeast pyruvate decarboxylase, Crout et al. (1991) observed that aldehydes actually inhibited the enzyme. Indeed, at a concentration of 0.05 mol L^{-1} benzaldehyde completely inhibits the decarboxylation reaction, thus explaining why the large-scale production of PAC is disfavored if the concentration of benzaldehyde in the reactor is too high. Various polymeric supports have been investigated for the immobilization of yeast cells during PAC production (Nikolova and Ward, 1994). In a fed-batch process, immobilized *Candida utilis* led to a PAC yield of 15.2 g L^{-1} (Shin and Rogers, 1995), while the use of isolated pyruvate decarboxylase from C. *utilis* delivered an optimum yield of 28.6 g L^{-1} (Shin and Rogers, 1996). Most recently, the large-scale production of PAC has been optimized further to give a "cleaner" process and higher yields of PAC (Oliver et al., 1997).

An alternative enzyme, benzylformate decarboxylase isolated from *Pseudomonas putida* ATCC 12633, has potential for use in the production of PAC and other α-hydroxyketones by condensing benzylformate with acetaldehyde (Wilcocks et al., 1992; Wilcocks and Ward, 1992). Benzylformate decarboxylase from *Acinetobacter calcoaceticus* will also deliver PAC from benzylformate and acetaldehyde in an optimized yield of 8.4 g L^{-1} (Prosen and Ward, 1994).

The yeast catalyzed acyloin condensation is not restricted to the use of benzaldehyde and pyruvic acid, although pyruvate decarboxylase does appear to show some substrate specificity. Variously substituted cinnamaldehydes have been used extensively as substrates for pyruvate decarboxylase to provide chiral compounds since the seminal observation that diol (**91a**) may be isolated from a fermenting mixture containing cinnamaldehyde (**90a**) (Fuganti and Grasselli, 1977). For example, aldehydes (**90a–d**) gave the corresponding diols (**91a–d**) in 30%, 50%, 30%, and 10% yield, respectively (Fig. 38). Larger substituents α to the carbonyl group inhibited the condensation but addition of acetaldehyde to the reaction mixture was reported to dramatically increase the yield of condensation products (Fuganti et al., 1979). Diols (**93a, b**) were obtained in 15% and 10%, respectively, from

R¹ O
R² 90
Bakers' yeast (pyruvate decarboxylase)
a R^1 = H, R^2 = H
b R^1 = H, R^2 = Br
c R^1 = H, R^2 = Me
d R^1 = Me, R^2 = H
R¹ OH S R R² OH 91

Fig. 38. Bakers' yeast mediated acyloin condensation of cinnamaldehyde to give *erythro*-diols (**91**).

2- and 3-methyl-5-phenylpenta-2,4-dien-1-al (**92a, b**) (Fig. 39) (FUGANTI and GRASSELLI, 1978). Recently, the condensation of cinnamaldehyde to give (2*S*,3*R*)-5-phenylpent-4-en-2,3-diol in the presence of bakers' yeast was optimized using statistical methods to modify the experimental design (EBERT et al., 1994).

α-Ketoacids other than pyruvic acid can participate in the condensation with benzaldehyde to yield the corresponding diols in >95% ee, though the range of such alternative donors appears extremely limited (FUGANTI et al., 1988).

R¹ O
R² 92
Bakers' yeast (pyruvate decarboxylase)
a R^1 = H, R^2 = Me
b R^1 = Me, R^2 = H
R¹ OH S R R² OH 93

Fig. 39. Bakers' yeast mediated acyloin condensation of 2- and 3-methyl-5-phenylpenta-2,4-dien-1-al (**92a, b**).

CROUT et al. provided firm experimental evidence that pyruvate decarboxylase is indeed responsible for α-hydroxyketone formation by carrying out a series of biotransformations with the commercially available enzyme (CROUT et al., 1991; KREN et al., 1993). Pyruvate decarboxylase was subsequently the subject of a molecular modeling study, based upon the published crystal structure of the enzyme from *Saccharomyces uvarum* (DYDA et al., 1993; LOBELL and CROUT, 1996).

Other species have been shown to have pyruvate decarboxylase activity. The enzyme from *Zymomonas mobilis* has been isolated and purified; while by-products resulting from reduction were suppressed, the purified enzyme does not give higher yields than purified decarboxylases from other species (BRINGER-MEYER and SAHM, 1988). Interestingly, the stereoselectivity of pyruvate decarboxylase from brewers' yeast differed from that obtained from Z. *mobilis* or from wheat germ in reactions with acetaldehyde (Fig. 40) (CHEN

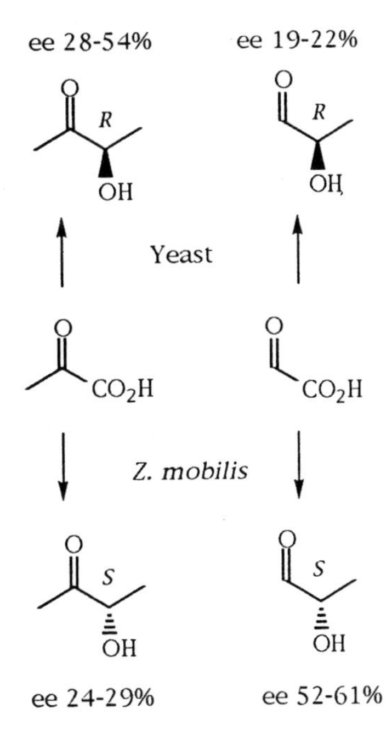

Fig. 40. Enantiocomplementary acyloin condensations.

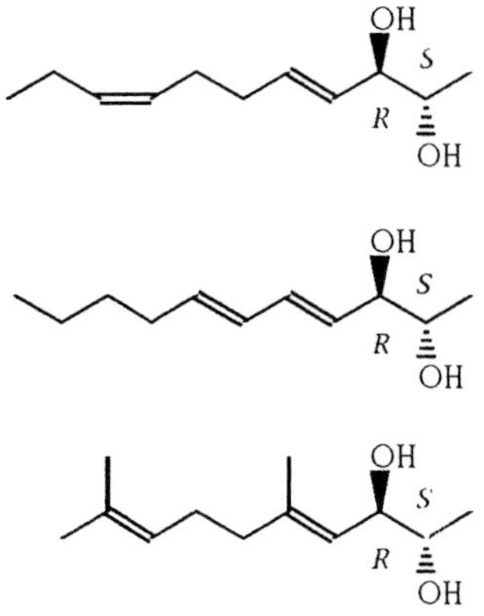

Fig. 41. *Mucor circinelloides* CBS 39468 mediated acyloin condensation and reduction products.

and JORDAN, 1984; CROUT et al., 1986; ABRAHAM and STUMPF, 1987; BORNEMANN et al., 1993).

Several acyclic unsaturated aldehydes underwent a condensation reaction in the presence of fermenting *Mucor circinelloides* CBS 39468; (2*S*,3*R*)-diols (**94–96**) (Fig. 41) were obtained from the corresponding aldehydes, presumably the reduction products of initially formed α-hydroxyketones (STUMPF and KIESLICH, 1991).

Saccharomyces fermentati and *S. delbrueckii* promoted the acyloin condensation between benzaldehyde and pyruvic acid, giving product yields which were higher than with other microorganisms, but did not offer any advantage over *Saccharomyces cerevisiae* when chal-

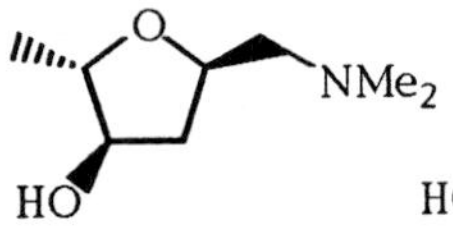

97 D-(-)-*allo*-Muscarine 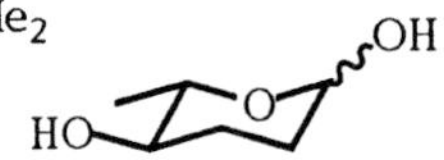 **98** L-Amicetose 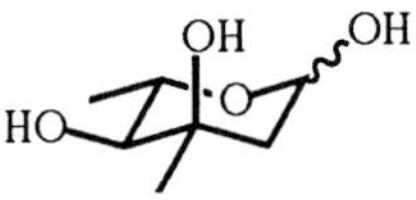 **99** L-Mycarose 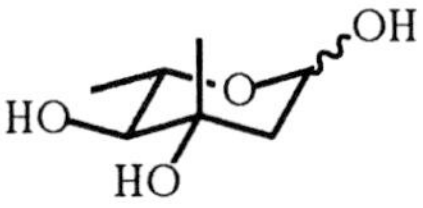 **100** L-Olivomycose

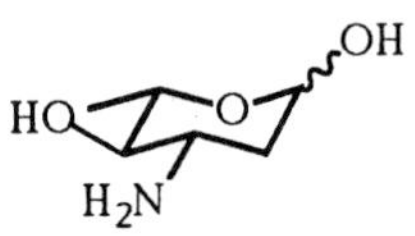

101 L-Acosamine 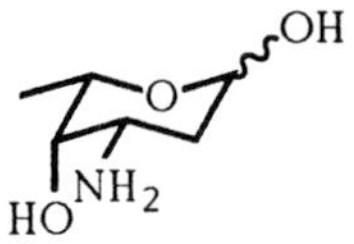 **102** L-Daunosamine 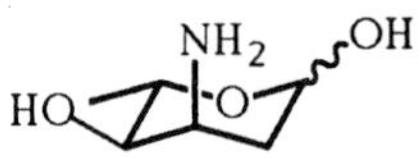 **103** L-Ristosamine 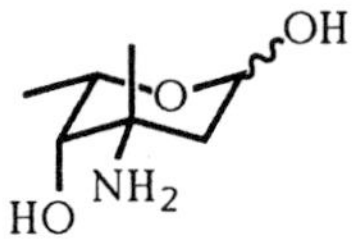 **104** L-Vancosamine

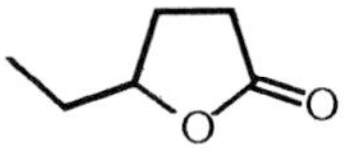

105 4-Hexanolide **107** (-)-Frontalin **38** (+)-*exo*-Brevicomin

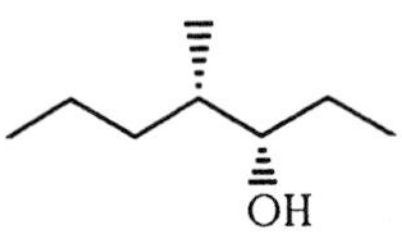

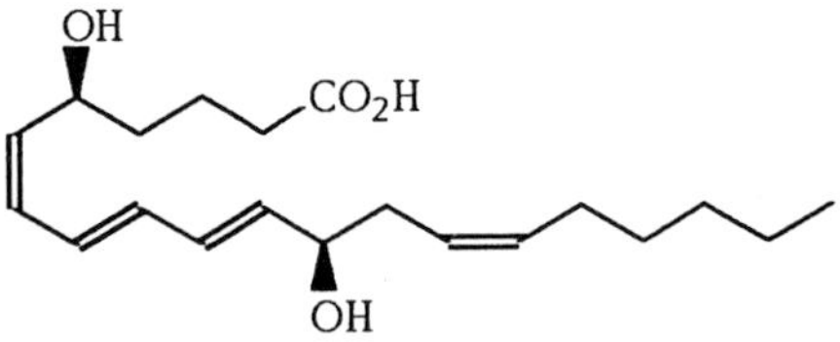

106 (3*S*,4*S*)-4-Methyl-3-heptanol

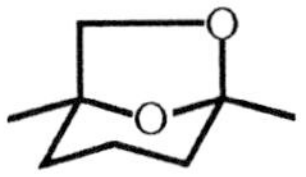

108 LTB_4 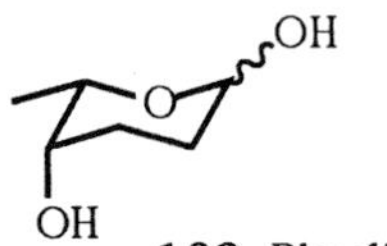 **109** Rhodinose

Fig. 42. Natural products synthesized using the biotransformation products of cinnamaldehydes.

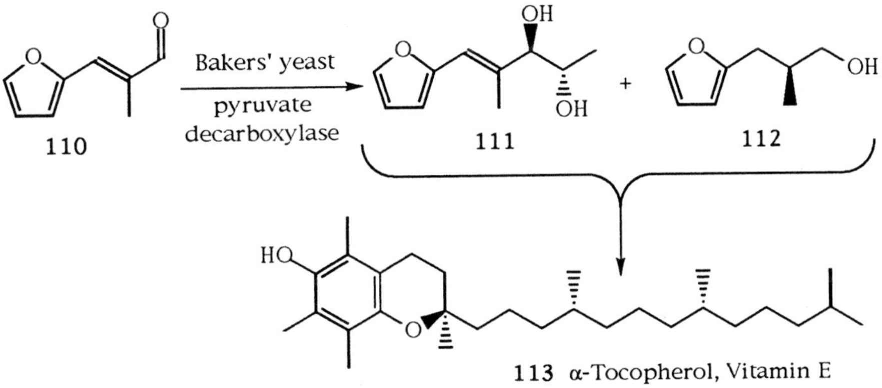

Fig. 43. Biocatalytic synthesis of α-tocopherol (**113**).

lenged with other aldehydes or oxoacids (CARDILLO et al., 1991). The yeast catalyzed acyloin condensation has been used as a key step in many syntheses of natural products, the FUGANTI group having eminently demonstrated that biotransformation of cinnamaldehyde derivatives can provide useful chiral starting materials (FUGANTI and GRASSELLI, 1977, 1985). The following are just some of the targets that have been prepared: D-(−)-*allo*-muscarine (**97**) (FRONZA et al., 1978), L-amicetose (**98**), L-mycarose (**99**), and L-olivomycose (**100**) (FUGANTI and GRASSELLI, 1978; FUGANTI et al., 1979), *N*-acyl derivatives of aminohexoses including L-acosamine (**101**), L-daunosamine (**102**), L-ristosamine (**103**) and L-vancosamine (**104**) (FRONZA et al., 1980, 1981, 1982, 1985, 1987; FUGANTI et al., 1983a), (+)- and (−)-4-hexanolide (**105**) (BERNARDI et al., 1980; FUGANTI et al., 1983b), (3*S*,4*S*)-4-methyl-3-heptanol (**106**), a pheromone from *Scolytus multistriatus* (FUGANTI et al., 1982), (−)-frontalin (**107**) (FUGANTI et al., 1983c), (+)-*exo*-brevicomin (**38**) (BERNARDI et al., 1981; FUGANTI et al., 1983b), LTB_4 (**108**) (FUGANTI et al., 1983d), and (+)- and (−)-rhodinose (**109**) (SERVI, 1985) (Fig. 42).

Following an earlier synthesis, both the diol (**111**) and by-product alcohol (**112**), obtained in 20% and 10% yield, respectively, from reaction of α-methyl-β-(2-furyl)acrolein (**110**) in fermenting yeast, were used in a synthesis of α-tocopherol (vitamin E, (**113**)) (Fig. 43) (FUGANTI and GRASSELLI, 1979, 1982).

The synthesis of solerone (5-oxo-4-hexanolide, (**116**)) was achieved using pyruvate decarboxylase from bakers' yeast in a key step. Ethyl 4-oxobutanoate (**114**), a novel substrate for the enzymic acyloin condensation, yielded ethyl 4-hydroxy-5-oxohexanoate (**115**), which was readily converted into solerone (**116**) by acid (Fig. 44) (HARING et al., 1997).

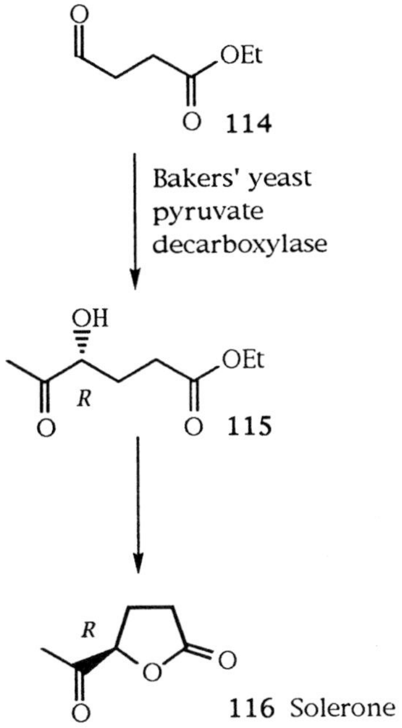

Fig. 44. Biocatalytic synthesis of solerone (**116**).

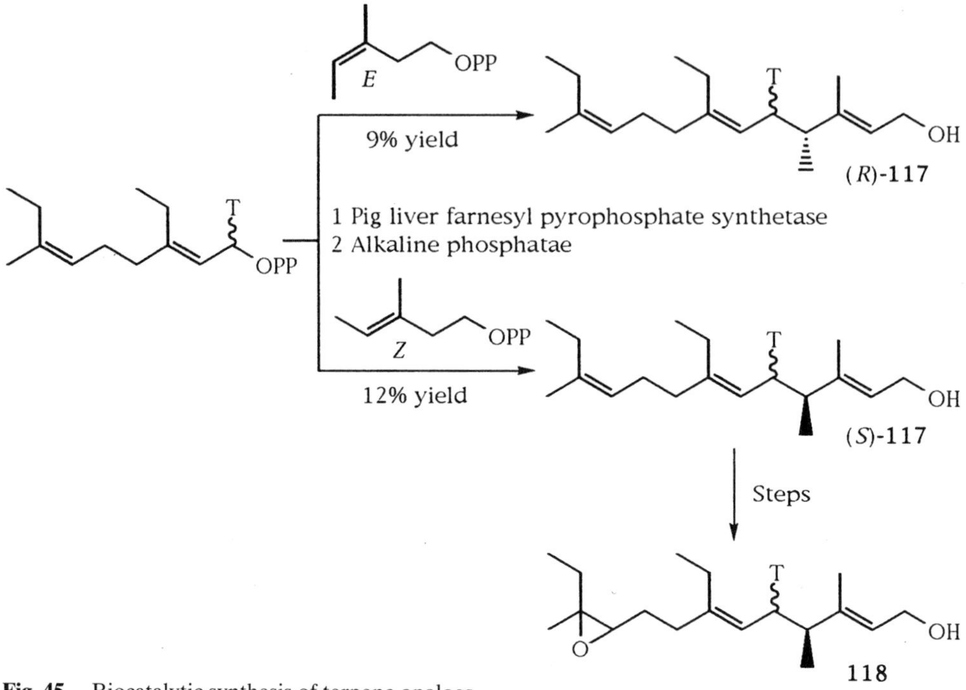

OPP
E
OPP
9% yield
T
OH
(*R*)-**117**
T
OPP
1 Pig liver farnesyl pyrophosphate synthetase
2 Alkaline phosphatae
OPP
Z
12% yield
T
OH
(*S*)-**117**
Steps
T
O
OH
118

Fig. 45. Biocatalytic synthesis of terpene analogs.

5 Enzymes from Terpenoid and Steroid Biosynthesis

Examples of the laboratory use of enzymes from the biosynthetic pathways which lead to terpenes and steroids are relatively sparse, despite abundant microscale biosynthetic studies. Pig liver farnesyl pyrophosphate synthetase has been used to prepare both enantiomers of 4-methyldihomofarnesol (**117**). (*S*)-(**117**) was used in the synthesis of juvenile hormone (**118**) (Fig. 45) (KOYAMA et al., 1985, 1987).

Several enzymes from the biosynthesis of steroids are of use in organic synthesis. Pig liver 2,3-oxidosqualene cyclase has long been known to catalyze the cyclization of 2,3-oxidosqualene (**119**) and derivatives substituted at the Δ18–19 double bond, to yield the expected steroidal derivatives, in this case lanosterol (**120**) (Fig. 46) (DJERASSI, 1981; GOAD, 1981).

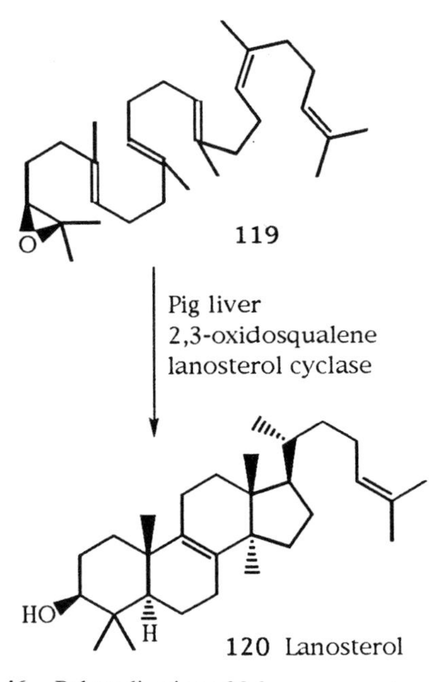

Fig. 46. Polycyclization of 2,3-epoxy-squalene.

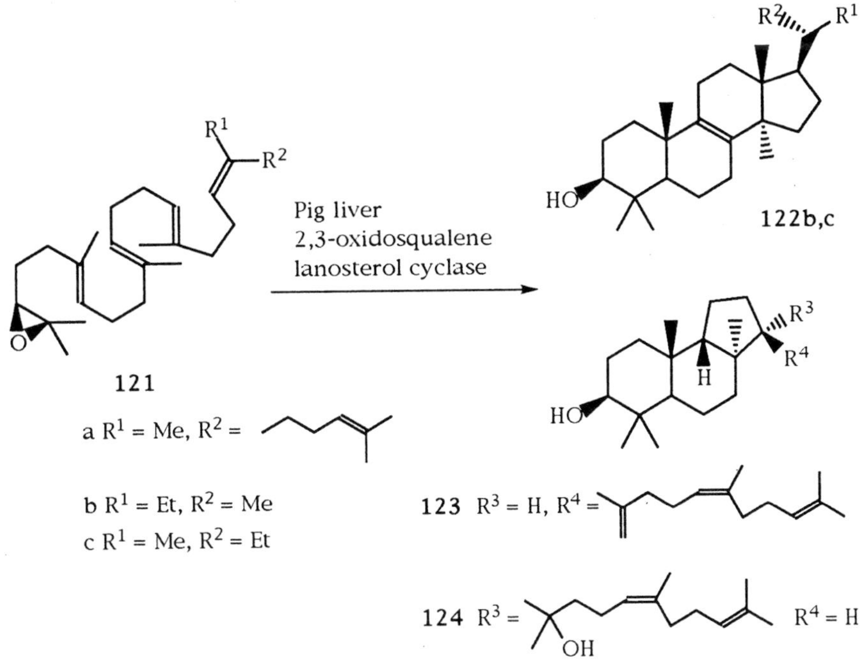

Fig. 47. Polycyclization of 2,3-epoxy-squalene analogs (**121**).

More recently, the enzyme has been used to prepare steroids (**122b, c**) or tricyclic compounds (**123**) and (**124**) from the corresponding oxidosqualene analogs (**121a–c**) which possess the unnatural *Z*-configuration at the Δ18–19 double bond (Fig. 47) (HÉRIN et al., 1979; KRIEF et al., 1987).

Bakers' yeast has been used as a source of sterol cyclase, allowing the efficient cyclization of squalene oxide derivatives to the expected steroidal compounds. The cyclization is accompanied by a kinetic resolution and is curiously promoted by ultrasonication; whether ultrasound increases cyclase activity or simply distributes the substrate more efficiently is debatable (BUJONS et al., 1988; MEDINA and KYLER, 1988; MEDINA et al., 1989; XIAO and PRESTWICH, 1991). Interestingly, in the transformation of racemic compound (**125**) to optically pure steroid (**126**), the vinyl group rearranges from C-8 to C-14 in the same way as a hydrogen does in the natural substrate (**121a**) (Fig. 48).

A comparison has been made between pig liver 2,3-oxidosqualene cyclase and the ultra-

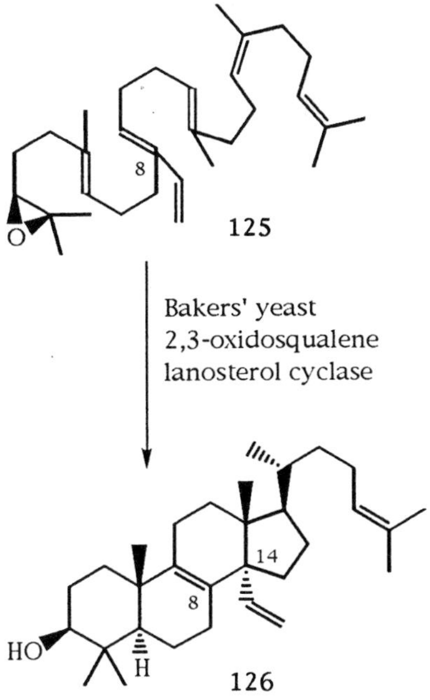

Fig. 48. Epoxy-polyene cyclization.

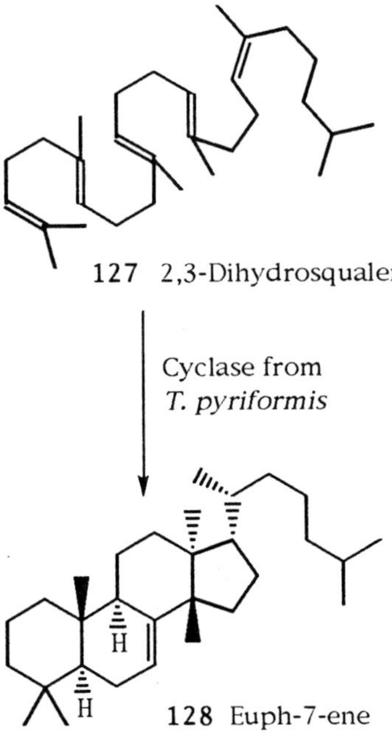

127 2,3-Dihydrosqualene

Cyclase from
T. pyriformis

128 Euph-7-ene

Fig. 49. Polyene cyclization.

sonicated bakers' yeast preparation in catalysis of the cyclizations mentioned above (KRIEF et al., 1991). A protozoal cyclase from *Tetrahymena pyriformis* has been used to synthesize the unnatural steroid euph-7-ene (**128**) from 2,3-dihydrosqualene (**127**) (Fig. 49) (ABE and ROHMER, 1991).

6 Other Enzyme Catalyzed C–C Bond Forming Reactions

It is well established that enzyme catalyzed pericyclic reactions form part of many biosynthetic pathways. The Diels–Alder reaction, undoubtedly one of the most powerful synthetic transformations available to the organic chemist, has long been speculated as being a key reaction in the biosynthesis of many natural products but an enzyme responsible for facilitating such a reaction, a so-called "Diels–Alder-ase", could not be identified (LASCHAT, 1996). In fact, it may readily be concluded that a Diels–Alder-ase does not exist, based upon the difficulty encountered in isolating such an enzyme. Several reports of biocatalytic Diels–Alder reactions have appeared in the literature. For example, in the presence of bakers' yeast the reaction of maleic acid and cyclopentadiene in water gave exclusively the *exo*-cycloadduct (**129**), whereas in the absence of yeast the *endo*-cycloadduct (**130**) predominated (Fig. 50) (RAMA RAO et al., 1990). Whether this effect may be attributed to a specific enzyme is, however, uncertain.

The long-awaited breakthrough came when OIKAWA, ICHIHARA, and coworkers detected enzymic activity for a Diels–Alder reaction in a cell-free extract from *Alternaria solani*, a pathogenic fungus which causes early blight disease in potatoes and tomatoes (OIKAWA et al., 1995; ICHIHARA and OIKAWA, 1997). Following on from previous work in which they proposed that solanapyrones (**131–134**), phytotoxins produced by *A. solani*, were biosynthesized via a [4+2] cycloaddition prosolanapyrone III (**135**) was treated with the cell-free extract at 30°C for 10 min to yield cyclized products (**131**) and (**132**) with an *endo*:*exo* ra-

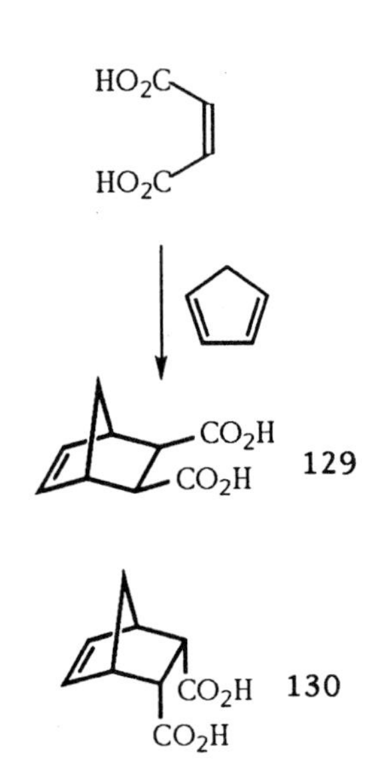

Fig. 50. Bakers' yeast mediated Diels–Alder reaction.

tio of 47:53 (Fig. 51). When compound (**135**) was reacted in the absence of the cell-free extract, the uncatalyzed reaction delivered products (**131**) and (**132**) with an *endo*:*exo* ratio of 97:3; a similar *endo*:*exo* ratio was obtained when a denatured cell-free extract was used as the reaction medium. Overall, the enzyme catalyzed Diels–Alder reaction proceeds with a conversion of 15% and an *endo*:*exo* ratio of 13:87 (this translates to an ee of >92% in favor of solanapyrone (**132**)). Such enzymic cycloadditions are particularly interesting from a mechanistic point of view: usually, the *endo* cycloadduct predominates when Diels–Alder reactions are carried out in the absence of catalyst or if chemical methods of catalysis are used. It must be noted that, at this stage, no single enzyme was isolated to which these results could be attributed.

Since the above report, the Japanese group has offered a second example of a biocatalyzed Diels–Alder reaction (OIKAWA et al., 1997). Pyrone (**136**) was converted to macrophomic acid (**137**) by reaction with phosphoenolpyruvate in a cell-free extract from *Macrophoma commelinae*; the proposed mechanism involves an initial [4+2] cycloaddition, followed by loss of carbon dioxide to

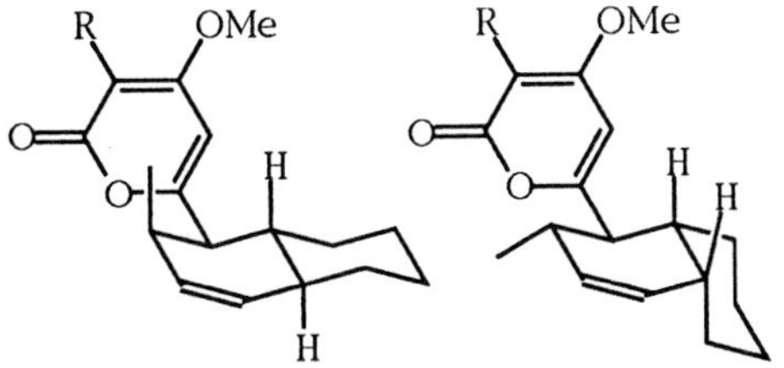

131 R = CHO
133 R = CH_2OH

132 R = CHO
134 R = CH_2OH

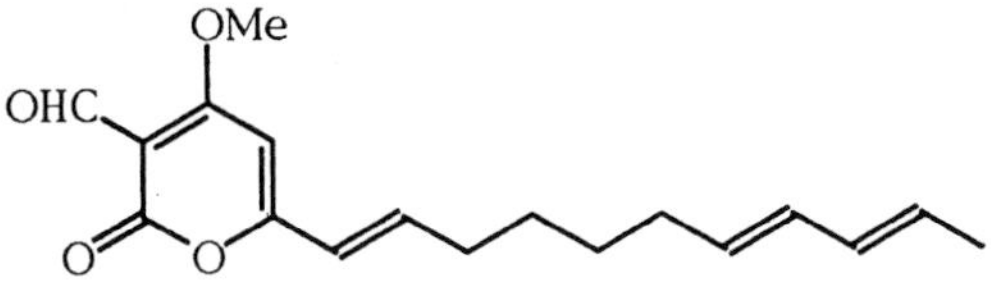

135 Prosolanapyrone III

Fig. 51. Intramolecular Diels–Alder reaction.

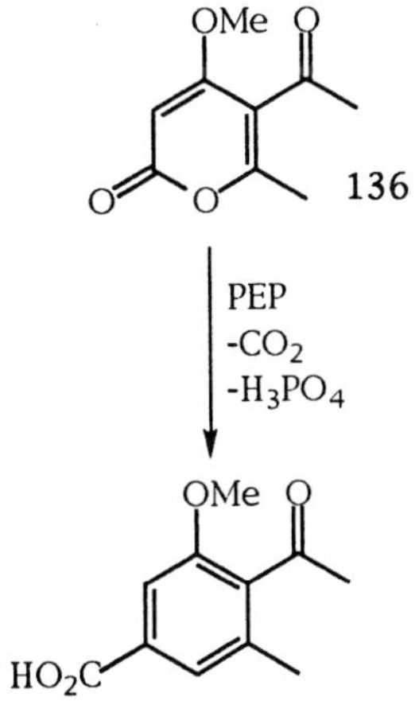

136

PEP
$-CO_2$
$-H_3PO_4$

137 Macrophomic acid

Fig. 52. Diels–Alder reaction of phosphoenolpyruvate?

deliver the product (**137**) (Fig. 52). However, the implied involvement of a Diels–Alder-ase in this reaction is not certain as an alternative biosynthetic route may be envisaged.

C–C bond formation has been achieved using certain enzymes from the biosynthesis of amino acids. A variety of β-substituted α-amino acid derivatives were prepared from β-chloroalanine by using *E. coli* tryptophan synthase and tyrosine phenol lyase as reaction catalysts (NAGASAWA and YAMADA, 1986).

Reactions using yeast have delivered some rather unexpected products. The cyclic hemiacetal (**139**) was isolated from a condensation of cinnamaldehyde in the presence of fermenting yeast with added acetaldehyde (BERTOLLI et al., 1981). It is probably formed by Michael addition of the enolate of ketol (**138**) to cinnamaldehyde, followed by hemiacetal formation (Fig. 53).

In another apparent Michael addition, α,β-unsaturated ketones and esters react with 2,2,2-trifluoroethanol in the presence of fermenting bakers' yeast to give products in high enantiomeric excess, but unknown absolute stereochemistry. For example, ethyl acrylate (**140**) reacts to give lactone (**142**) in 47% yield and 79% ee via the addition product (**141**) (Fig. 54) (KITAZUME and ISHIKAWA, 1984).

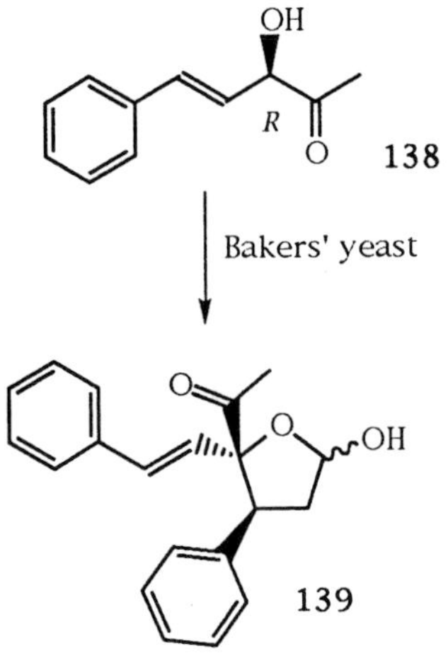

Fig. 53. Bakers' yeast mediated Michael addition to cinnamaldehyde.

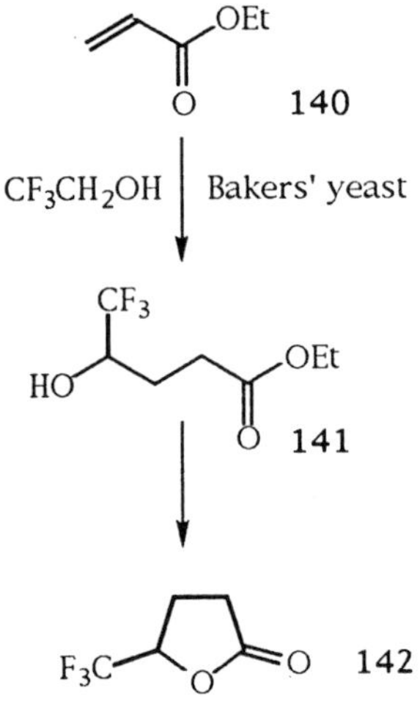

Fig. 54. Bakers' yeast mediated Michael addition.

A strange enzymatic alkylation reaction was observed during the reduction of 3-oxobutyronitrile (**143**) using fermenting bakers' yeast in ethanol (ITOH et al., 1989). The biotransformation did not deliver the expected 3-hydroxybutyronitrile but gave, instead, the diastereomeric (3*S*)-2-ethyl-3-hydroxybutyronitriles (**144**) and (**145**) in a *syn*:*anti* ratio of 2:1 (Fig. 55) each of >99% ee. The alkylated keto

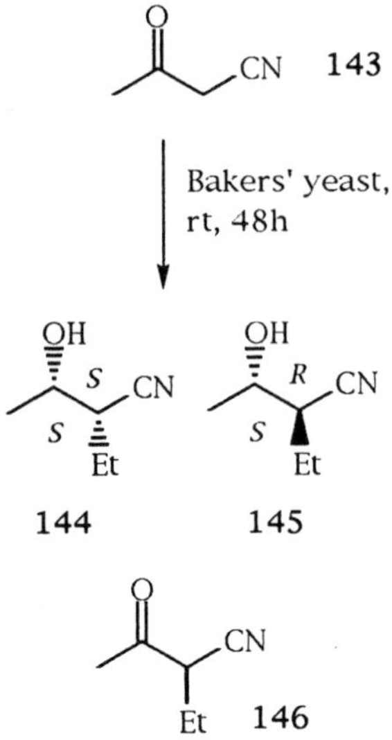

Fig. 55. An unusual bakers' yeast mediated alkylation.

nitrile intermediate (**146**) could be obtained if the reaction if was terminated before it went to completion. A mechanism to account for these observations was not proposed but alkylation appears to precede the reductive process. One possibility is C-acetylation by acetyl-CoA, followed by reduction, elimination, and further reduction of the alkene bond.

7 Conclusion

C–C bond formation is fast becoming the sole subject of many review articles in the primary chemistry literature, a fact which bears witness to the steady increase in the number of reports concerning the use of enzymes to perform such a biotransformation. The majority of organic chemists will thus recognize the use of, for example, bakers' yeast as a laboratory reagent but, unfortunately, do not seek to use it (likewise other enzymic systems), in the belief that it is inconvenient and somewhat messy. However, as noted above, even lowly bakers' yeast can accomplish asymmetric C–C bond formation reactions. While this present work provides an account of the state of the art, a healthy combination of breakthroughs in modern microbiology and a growing awareness among chemists of the potential benefits of using enzymes will give future reviewers much to write about.

8 References

ABE, I., ROHMER, M. (1991), Enzymatic cyclization of 2,3-dihydrosqualene into euph-7-ene by a cell-free system from the protozoan *Tetrahymena pyriformis*, *J. Chem. Soc., Chem. Commun.*, 902–903.

ABRAHAM, W.-R., STUMPF, B. (1987), Enzymatic acyloin condensation of acyclic aldehydes, *Z. Naturforsch.* **42C**, 559–566.

ALLEN, S. T., HEINTZELMAN, G. R., TOONE, E. J. (1992), Pyruvate aldolases as reagents for stereospecific aldol condensation, *J. Org. Chem.* **57**, 426–427.

AUGÉ, C., GAUTHERON, C. (1987), The use of an immobilized aldolase in the 1st synthesis of a natural deaminated neuraminic acid, *J. Chem. Soc., Chem. Commun.*, 859–860.

AUGÉ, C., DAVID, S., GAUTHERON, C. (1984), Synthesis with immobilized enzyme of the most important sialic acid, *Tetrahedron Lett.* **25**, 4663–4664.

AUGÉ, C., DAVID, S., GAUTHERON, C., VEYRIÈRES, A. (1985), Synthesis with an immobilized enzyme of *N*-acetyl-9-*O*-acetylneuraminic acid, a sugar reported as a component of embryonic and tumor antigens, *Tetrahedron Lett.* **26**, 2439–2440.

AUGÉ, C., GAUTHERON, C., DAVID, S. (1990), Sialyl aldolase in organic synthesis: from the trout egg acid, 3-deoxy-D-glycero-D-galacto-2-nonulosonic acid (KDN), to branched-chain higher ketoses as possible new chirons, *Tetrahedron* **46**, 201–214.

BARBAS, C. F. I., WANG, Y.-F., WONG, C.-H. (1990), Deoxyribose- 5-phosphate aldolase as a synthetic catalyst, *J. Am. Chem. Soc.* **112**, 2013–2014.

BEDNARSKI, M. D., WALDMANN, H. J., WHITESIDES, G. M. (1986), Aldolase-catalyzed synthesis of complex C-8 and C-9 monosaccharides, *Tetrahedron Lett.* **27**, 5807–5810.

BEDNARSKI, M. D., CHENAULT, H. K., SIMON, E. S., WHITESIDES, G. M. (1987), Membrane-enclosed enzymatic catalysis (MEEC) – a useful, practical new method for the manipulation of enzymes in organic synthesis, *J. Am. Chem. Soc.* **109**, 1283–1285.

BEDNARSKI, M. D., CRANS, D. C., DICOSIMO, R., SIMON, E. S., STEIN, P. D., WHITESIDES, G. M. (1988), Synthesis of 3-deoxy-D-manno-2-octulosonate-8-phosphate (KDO-8-P) from D-arabinose – generation of D-arabinose-5-phosphate using hexokinase, *Tetrahedron Lett.* **29**, 427–430.

BEDNARSKI, M. D., SIMON, E. S., BISCHOFBERGER, N., FESSNER, W.-D., KIM, M.-J. et al. (1989), Rabbit muscle aldolase as a catalyst in organic synthesis, *J. Am. Chem. Soc.* **111**, 627–635.

BERNARDI, R., FUGANTI, C., GRASSELLI, P., MARINONI, G. (1980), Synthesis of the enantiomeric forms of 4-hexanolide (γ-caprolactone) from the optically active 5-phenyl-4-pentene-2,3-diol prepared from cinnamaldehyde and bakers' yeast, *Synthesis*, 50–52.

BERNARDI, R., FUGANTI, C., GRASSELLI, P. (1981), On the steric course of addition of Grignard reagents on to α,β-dialkoxy *erythro* and *threo* chiral aldehydes. Synthesis of (+)- and (−)-*exo* and *endo*-brevicomin, *Tetrahedron Lett.* **22**, 4021–4024.

BERTOLLI, G., FRONZA, G., FUGANTI, C., GRASSELLI, P., MOJORI, L., SPREAFICO, F. (1981), On the α-ketols obtained from α,β-unsaturated aromatic aldehydes and bakers' yeast, *Tetrahedron Lett.* **22**, 965–968.

BISCHOFBERGER, N., WALDMANN, H., SAITO, T., SIMON, E. S., LEES, W. et al. (1988), Synthesis of analogs of 1,3-dihydroxyacetone phosphate and glyceraldehyde-3-phosphate for use in studies of fructose-1,6-diphosphate aldolase, *J. Org. Chem.* **53**, 3457–3465.

BOLTE, J., DEMUYNCK, C., SAMAKI, H. (1987), Utilization of enzymes in organic chemistry: transketolase catalyzed synthesis of ketoses, *Tetrahedron Lett.* **28**, 5525–5528.

BORNEMANN, S., CROUT, D. H. G., DALTON, H., HUTCHINSON, D. W., DEAN, G. et al. (1993), Stereochemistry of the formation of lactaldehyde and acetoin produced by the pyruvate decarboxylases of yeast *(Saccharomyces* sp.) and *Zymomonas mobilis* – different Boltzmann distributions between bound forms of the electrophile, acetaldehyde in the enzymatic reactions, *J. Chem. Soc., Perkin Trans. I*, 309–311.

BRINGER-MEYER, S., SAHM, H. (1988), Metabolic shifts in *Zymomonas mobilis* in response to growth conditions, *FEMS Microbiol. Rev.* **54**, 131–142.

BUJONS, J., GUAJARDO, R., KYLER, K. S. (1988), Enantioselective enzymatic sterol synthesis by ultrasonically stimulated bakers' yeast, *J. Am. Chem. Soc.* **110**, 604–606.

BUTT, S., ROBERTS, S. M. (1986), Recent advances in the use of enzyme-catalyzed reactions in organic research – the synthesis of biologically-active natural products and analogs, *Nat. Prod. Rep.* **3**, 489–503.

CARDILLO, R., SERVI, S., TINTI, C. (1991), Biotransformation of unsaturated aldehydes by microorganisms with pyruvate decarboxylase activity, *Appl. Microbiol. Biotechnol.* **36**, 300–303.

CHEN, G. C., JORDAN, F. (1984), Brewers' yeast pyruvate decarboxylase produces acetoin from acetaldehyde – a novel tool to study the mechanism of steps subsequent to carbon dioxide loss, *Biochemistry* **23**, 3576–3582.

CHEN, L., DUMAS, D. P., WONG, C.-H. (1992), Deoxyribose-5-phosphate aldolase as a catalyst in asymmetric aldol condensation, *J. Am. Chem Soc.* **114**, 741–748.

CHENEVERT, R., LAVOIE, M., DASSER, M. (1997), Use of aldolases in the synthesis of non-carbohydrate natural products. Stereoselective synthesis of aspicilin C-3–C-9 fragment, *Can. J. Chem.* **75**, 68–73.

CHOU, W.-C., CHEN, L., FANG, J.-M., WONG, C.-H. (1994), A new route to deoxythiosugars based on aldolases, *J. Am.Chem. Soc.* **116**, 6191–6194.

CHOU, W.-C., FOTSCH, C., WONG, C.-H. (1995), Synthesis of nitrocyclitols based on enzymatic aldol reaction and intramolecular nitroaldol reaction, *J. Org. Chem.* **60**, 2916–2917.

CROUT, D. H. G., LITTLECHILD, J., MORREY, S. M. (1986), Acetoin metabolism – stereochemistry of the acetoin produced by the pyruvate decarboxylase of wheat-germ and by the α-acetolactate decarboxylase of *Klebsiella aerogenes*, *J. Chem. Soc., Perkin Trans. I*, 105–108.

CROUT, D. H. G., DALTON, H., HUTCHISON, D. W., MIYAGOSHI, M. (1991), Studies on pyruvate decarboxylase: acyloin formation from aliphatic, aromatic and heterocyclic aldehydes, *J. Chem. Soc., Perkin Trans. I*, 1329–1334.

CSUK, R., GLÄNZER, B. I. (1991), Bakers' yeast-mediated transformations in organic chemistry, *Chem. Rev.* **91**, 49–97.

DALMAS, V., DEMUYNCK, C. (1993), Study of the specificity of a spinach transketolase on achiral substrates, *Tetrahedron: Asymmetry* **4**, 2383–2388.

DAVIES, H. G., GREEN, R. H., KELLY, D. R., ROBERTS, S. M. (1989), *Biotransformations in Preparative Organic Chemistry*, pp. 221–231. London: Academic Press.

DEMUYNCK, C., FISSON, F., BENNANI-BAITI, I., SAMAKI, H., MANI, J.-C. (1990a), Immunoaffinity purification of transketolases from yeast and spinach leaves, *Agric. Biol. Chem.* **54**, 3073–3078.

DEMUYNCK, C., BOLTE, J., HECQUET, L., SAMAKI, H. (1990b), Enzymes as reagents in organic chemistry – transketolase-catalyzed synthesis of D-[1,2-$^{13}C_2$]xylulose, *Carbohydr. Res.* **206**, 79–85.

DEMUYNCK, C., BOLTE, J., HECQUET, L., DALMAS, V. (1991), Enzyme-catalyzed synthesis of carbohydrates: synthetic potential of transketolase, *Tetrahedron Lett.* **32**, 5085–5088.

DJERASSI, C. (1981), Recent studies in the marine sterol field, *Pure Appl. Chem.* **53**, 873–890.

DRUECKHAMMER, D. G., HENNEN, W. J., PEDERSON, R. L., BARBAS III, C. F., GAUTHERON, C. M. et al. (1991), Enzyme catalysis in synthetic carbohydrate chemistry, *Synthesis*, 499–525.

DUNCAN, R., DRUECKHAMMER, D. G. (1995), A pseudoisomerization route to aldose sugars using aldolase catalysis, *J. Org. Chem.* **60**, 7394–7395.

DUNCAN, R., DRUECKHAMMER, D. G. (1996), Preparation of deoxy sugars via aldolase-catalyzed synthesis of 1-deoxy-thioketoses, *J. Org. Chem.* **61**, 438–439.

DURRWACHTER, J. R., WONG, C.-H. (1988), Fructose 1,6-diphosphate aldolase catalyzed stereoselective synthesis of C-alkyl and N-containing sugars – thermodynamically controlled C–C bond formations, *J. Org. Chem.* **53**, 4175–4181.

DYDA, F., FUREY, W., SWAMINATHAN, S., SAX, M., FARRENKOPF, B., JORDAN, F. (1993), Catalytic centers in the thiamine diphosphate dependent enzyme pyruvate decarboxylase at 2.4 Å resolution, *Biochemistry* **32**, 6165–6170.

EBERT, C., GARDOSSI, L., GIANFERRARA, T., LINDA, P., MORANDINI, C. (1994), A multivariate reexamination of experimental condition effects on acyloin-type condensation mediated by *Saccharomyces cerevisiae*, *Biocatalysis* **10**, 15–23.

EFFENBERGER, F., NULL, V. (1992), Enzyme-catalyzed reactions. 13. A new, efficient synthesis of fagomine, *Liebigs Ann. Chem.*, 1211–1212.

EFFENBERGER, F., STRAUB, A., NULL, V. (1992), Enzyme-catalyzed reactions. 14. Stereoselective synthesis of thiosugars from achiral starting compounds by enzymes, *Liebigs Ann. Chem.*, 1297–1301.

EFFENBERGER, F., NULL, V., ZIEGLER, T. (1992), Preparation of optically pure L-2-hydroxyaldehydes with yeast transketolase, *Tetrahedron Lett.* **33**, 5157–5160.

EYRISCH, O., FESSNER, W.-D. (1995), Disaccharide mimetics by enzymatic tandem aldol reaction, *Angew. Chem.* (Int. Edn. Engl.) **34**, 1639–1641.

EYRISCH, O., SINERIUS, G., FESSNER, W.-D. (1993), Facile enzymatic *de novo* synthesis and NMR spectroscopic characterization of D-tagatose 1,6-bisphosphate, *Carbohydr. Res.* **238**, 287–306.

FANG, J.-M., WONG, C.-H. (1994), Enzymes in organic synthesis: alteration of reversible reactions to irreversible processes, *Synlett*, 393–402.

FESSNER, W.-D., EYRISCH, O. (1992), Enzymes in organic synthesis. 2. One-pot synthesis of tagatose 1,6-bisphosphate by diastereoselective aldol addition, *Angew. Chem.* (Int. Edn. Engl.) **31**, 56–58.

FESSNER, W.-D., SINERIUS, G. (1994), Enzymes in organic synthesis. 7. Synthesis of dihydroxyacetone phosphate (and isosteric analogs) by enzymatic oxidation – sugars from glycerol, *Angew. Chem.* (Int. Edn. Engl.) **33**, 209–212.

FESSNER, W.-D., WALTER, C. (1997), Enzymatic C–C bond formation in asymmetric synthesis, *Top. Curr. Chem.* **184**, 97–194.

FESSNER, W.-D., SINERIUS, G., SCHNEIDER, A., DREYER, M., SCHULZ, G. E. et al. (1991), Enzymes in organic synthesis. 1. Diastereoselective enzymatic aldol additions – L-rhamnulose and L-fuculose 1-phosphate aldolases from *Escherichia coli*, *Angew. Chem.* (Int. Edn. Engl.) **30**, 555–558.

FESSNER, W.-D., BADIA, J., EYRISCH, O., SCHNEIDER, A., SINERIUS, G. (1992), Enzymes in organic synthesis. 5. Enzymatic synthesis of rare ketose 1-phosphates, *Tetrahedron Lett.* **33**, 5231–5234.

FINDEIS, M. A., WHITESIDES, G. M. (1984), Enzymic methods in organic synthesis, *Ann. Rep. Med. Chem.* **19**, 263–272.

FITZ, W., WONG, C.-H. (1994), Combined use of subtilisin and *N*-acetylneuraminic acid aldolase for the synthesis of a fluorescent sialic acid, *J. Org. Chem.* **59**, 8279–8280.

FITZ, W., SCHWARK, J.-R., WONG, C.-H. (1995), Aldotetroses and C(3)-modified aldohexoses as substrates for *N*-acetylneuraminic acid aldolase – a model for the explanation of the normal and the inversed stereoselectivity, *J. Org. Chem.* **60**, 3663–3670.

FLOYD, N. C., LIEBSTER, M. H., TURNER, N. J. (1992), A simple strategy for obtaining both enantiomers from an aldolase reaction – preparation of L-4-hydroxy-2-ketoglutarate and D-4-hydroxy-2-ketoglutarate, *J. Chem. Soc., Perkin Trans. I*, 1085–1086.

FRONZA, G., FUGANTI, C., GRASSELLI, P. (1978), Noncarbohydrate-based enantioselective synthesis of D-(−)-*allo*-muscarine, *Tetrahedron Lett.* **19**, 3941–3942.

FRONZA, G., FUGANTI, C., GRASSELLI, P. (1980), Synthesis of *N*-trifluoroacetyl-L-acosamine and L-daunosamine, *J. Chem. Soc., Chem. Commun.*, 442–444.

FRONZA, G., FUGANTI, C., GRASSELLI, P., PEDROCCHI-FANTONI, G. (1981), Synthesis of the *N*-benzoyl derivatives of L-*arabino*, L-*xylo*, and L-*lyxo* (L-vancosamine) isomers of 2,3,6-trideoxy-3-*C*-methyl-3-aminohexose from a non-carbohydrate precursor, *Tetrahedron Lett.* **22**, 5073–5076.

FRONZA, G., FUGANTI, C., GRASSELLI, P., MAJORI, L., PEDROCCHI-FANTONI, G., SPREAFICO, F. (1982), Synthesis of enantiomerically pure forms of *N*-acyl derivatives of *C*-methyl analogs of the aminodeoxy sugar L-acosamine from noncarbohydrate precursors, *J. Org. Chem.* **47**, 3289–3296.

FRONZA, G., FUGANTI, C., GRASSELLI, P., PEDROCCHI-FANTONI, G. (1985), Carbohydrate-like chiral synthons. Synthesis of the *N*-trifluoroacetyl derivatives of 4-amino-2,4,6-trideoxy-L-*lyxo*, 4-amino-2,4,6-trideoxy-L-*arabino*, and 4-amino-2,4,6-trideoxy-L-*ribo*-hexose from the (2*S*,3*R*) 2,3-diol formed from cinnamaldehyde in fermenting bakers' yeast, *Carbohydr. Res.* **136**, 115–124.

FRONZA, G., FUGANTI, C., PEDROCCHI-FANTONI, G., SERVI, S. (1987), Stereochemistry and synthetic applications of the products of yeast reduction of 3-hydroxy-3-methyl-5-phenylpent-4-en-2-one, *J. Org. Chem.* **52**, 1141–1144.

FUGANTI, C., GRASSELLI, P. (1977), Transformations of non-conventional substrates by fermenting bakers' yeast: production of optically active methyl-diols from aldehydes, *Chem. Ind.* (London), 983.

FUGANTI, C., GRASSELLI, P. (1978), Stereospecific synthesis from non-carbohydrate precursors of the deoxy- and methyl-branched deoxy-sugars L-amicetose, L-mycarose and L-olivomycose, *J. Chem. Soc., Chem. Commun.*, 299–300.

FUGANTI, C., GRASSELLI, P. (1979), Efficient stereoselective synthesis of natural α-tocopherol (vitamin E), *J. Chem. Soc., Chem. Commun.*, 995–996.

FUGANTI, C., GRASSELLI, P. (1982), Synthesis of the C_{14} chromanyl moiety of natural α-tocopherol (vitamin E), *J. Chem. Soc., Chem. Commun.*, 205–206.

FUGANTI, C., GRASSELLI, P. (1985), Stereochemistry and synthetic applications of products of fermentation of α,β-unsaturated aromatic aldehydes by bakers' yeast, in: *Enzymes in Organic Synthesis, Ciba Foundation Symposium III* (PORTER, R., CLARK, S., Eds.), pp. 112–127. London: Pitman.

FUGANTI, C., GRASSELLI, P., MARINONI, G. (1979), Further studies on the transformations of unsaturated aldehydes by fermenting bakers' yeast: a facile synthesis of L-olivomycose, *Tetrahedron Lett.* **20**, 1161–1164.

FUGANTI, C., GRASSELLI, P., SERVI, S., ZIROTTI, C. (1982), Synthesis of the enantiomeric forms of *cis* and *trans* 1-benzyloxy-2,3-epoxybutane and of (3*S*,4*S*) 4-methyl-3-heptanol, *Tetrahedron Lett.* **23**, 4269–4272.

FUGANTI, C., GRASSELLI, P., PEDROCCHI-FANTONI, G. (1983a), Stereospecific synthesis of *N*-benzoyl-L-daunosamine and L-ristosamine, *J. Org. Chem.* **48**, 909–910.

FUGANTI, C., GRASSELLI, P., PEDROCCHI-FANTONI, G., SERVI, S., ZIROTTI, C. (1983b), Carbohydrate-like chiral synthons. Preparation of (*R*) γ-hexanolide, (5*R*,6*S*,7*S*) 6,7-isopropylidendioxy-δ-octanolide and (+)-*exo*-brevicomin from (2*S*,3*S*,4*R*) 2,3-isopropylidendioxy-4-benzyloxyhept-6-ene), *Tetrahedron Lett.* **24**, 3753–3756.

FUGANTI, C., GRASSELLI, P., SERVI, S. (1983c), Synthesis of (−)-frontalin from the (2*S*,3*R*)-diol prepared from α-methylcinnamaldehyde and fermenting bakers' yeast, *J. Chem. Soc., Perkin Trans. I*, 241–244.

FUGANTI, C., SERVI, S., ZIROTTI, C. (1983d), Non-carbohydrate based synthesis of natural LTB_4, *Tetrahedron Lett.* **24**, 5285–5288.

FUGANTI, C., GRASSELLI, P., SERVI, S., SPREAFICO, F., ZIROTTI, C. (1984), On the steric course of bakers' yeast reduction of α-hydroxyketones, *J. Org. Chem.* **49**, 4087–4089.

FUGANTI, C., GRASSELLI, P., POLI, G., SERVI, S., ZORZELLA, A. (1988), Decarboxylative incorporation of α-oxybutyrate and α-oxovalerate into (*R*)-α-hydroxyethyl ketones and normal-propyl ketones on reaction with aromatic and α,β-unsaturated aldehydes in bakers' yeast, *J. Chem. Soc., Chem. Commun.*, 1619–1621.

GARCIA-JUNCEDA, E., SHEN, G.-J., SUGAI, T., WONG, C.-H. (1995), A new strategy for the cloning, overexpression and one-step purification of 3 DHAP-dependant aldolases – rhamnulose 1-phosphate aldolase, fuculose 1-phosphate aldolase and tagatose 1,6-diphosphate aldolase, *Bioorg. Med. Chem.* **3**, 945–953.

GHALAMBOR, M. A., HEATH, E. C. (1966), The biosynthesis of cell wall lipopolysaccharide in *Escherichia coli*. V. Purification and properties of 3-deoxy-D-*manno*-octulosonate aldolase, *J. Biol. Chem.* **241**, 3222–3227.

GIJSEN, H. J. M., WONG, C.-H. (1994), Unprecedented asymmetric aldol reactions with three aldehyde substrates catalyzed by 2-deoxyribose-5-phosphate aldolase, *J. Am. Chem. Soc.* **116**, 8422–8423.

GIJSEN, H. J. M., WONG, C.-H. (1995a), Synthesis of a cyclitol via a tandem enzymatic aldol – intramolecular Horner-Wadsworth-Emmons reaction, *Tetrahedron Lett.* **36**, 7057–7060.

GIJSEN, H. J. M., WONG, C.-H. (1995b), Sequential one-pot aldol reactions catalyzed by 2-deoxyribose-5-phosphate aldolase and fructose-1,6-diphosphate aldolase, *J. Am. Chem. Soc.* **117**, 2947–2948.

GIJSEN, H. J. M., WONG, C.-H. (1995c), Sequential three- and four-substrate aldol reactions catalyzed by aldolases, *J. Am. Chem. Soc.* **117**, 7585–7591.

GIJSEN, H. J. M., QIAO, L., FITZ, W., WONG, C.-H. (1996), Recent advances in the chemoenzymatic synthesis of carbohydrates and carbohydrate mimetics, *Chem. Rev.* **96**, 443-473.

GLÄNZER, B. I., FABER, K., GRIENGL, H. (1987), Enantioselective hydrolyses by bakers' yeast. 3. Microbial resolution of alkynyl esters using lyophilized yeast, *Tetrahedron* **43**, 5791–5796.

GOAD, L. J. (1981), Sterol biosynthesis and metabolism in marine invertebrates, *Pure Appl. Chem.* **53**, 837–852.

HARING, D., SCHREIER, P., HERDERICH, M. (1997), Rationalizing the origin of solerone (5-oxo-4-hexanolide): biomimetic synthesis and identification of key metabolites in sherry wine, *J. Agric. Food Chem.* **45**, 369–372.

HECQUET, L., BOLTE, J., DEMUYNCK, C. (1994a), Chemoenzymatic synthesis of 6-deoxy-D-fructose and 6-deoxy-L-sorbose using transketolase, *Tetrahedron* **50**, 8677–8684.

HECQUET, L., LEMAIRE, M., BOLTE, J., DEMUYNCK, C. (1994b), Chemo-enzymatic synthesis of precursors of fagomine and 1,4-dideoxy-1,4-imino-D-arabitinol, *Tetrahedron Lett.* **35**, 8791–8794.

HECQUET, L., BOLTE, J., DEMUYNCK, C. (1996), Enzymatic synthesis of "natural-labeled" 6-deoxy-L-sorbose precursor of an important food flavor, *Tetrahedron* **52**, 8223–8232.

HENDERSON, I., SHARPLESS, K. B., WONG, C.-H. (1994a), Synthesis of carbohydrates via tandem use of the osmium-catalyzed asymmetric dihydroxylation and enzyme-catalyzed aldol addition reactions, *J. Am. Chem. Soc.* **116**, 558–561.

HENDERSON, I., LASLO, K., WONG, C.-H. (1994b), Chemoenzymatic synthesis of homoazasugars, *Tetrahedron Lett.* **35**, 359–362.

HERBERT, R. B., WILKINSON, B., ELLAMES, G. J., KUNEC, E. K. (1993), Stereospecific lysis of a range of β-hydroxy-α-amino acids catalyzed by a novel aldolase from *Streptomyces amakusaensis*, *J. Chem. Soc., Chem. Commun.*, 205–206.

HERBERT, R. B., WILKINSON, B., ELLAMES, G. J. (1994), Preparation of (2*R*, 3*S*) β-hydroxy-α-amino acids by use of a novel *Streptomyces* aldolase as a resolving agent for racemic material, *Can. J. Chem.* **72**, 114-117.

HÉRIN, M., SANDRA, P., KRIEF, A. (1979), Stereospecific enzyme cyclization of a synthetic 2,3-oxidosqualene analog bearing an 18*Z* carbon–carbon double bond, *Tetrahedron Lett.* **33**, 3103–3106.

HOBBS, G. R., LILLY, M. D., TURNER, N. J., WARD, J. M., WILLETS, A. J., WOODLEY, J. M. (1993), Enzyme-catalyzed carbon–carbon bond formation: use of transketolase from *Escherichia coli*, *J. Chem. Soc., Perkin Trans. I*, 165–166.

HOFFMANN, N. (1996), Bakers' yeast – a live reagent for synthesis in organic chemistry, *Chemie in Unserer Zeit* **30**, 201–203.

HOLT, K. E., LEEPER, F. J., HANDA, S. (1994), Synthesis of β-1-homonorjirimycin and β-1-homomannorjirimycin using the enzyme aldolase, *J. Chem. Soc., Perkin Trans. I*, 231–234.

HUMPHREY, A. J., TURNER, N. J., MCCAGUE, R., TAYLOR, S. J. C. (1995), Synthesis of enantiomerically pure α-hydroxyaldehydes from the corresponding α-hydroxycarboxylic acids: novel substrates for *Escherichia coli* transketolase, *J. Chem. Soc., Chem. Commun.*, 2475–2476.

HUNG, R. R., STRAUB, J. A., WHITESIDES, G. M. (1991), α-Aminoaldehyde equivalents as substrates for rabbit muscle aldolase – synthesis of 1,4-dideoxy-D-arabitinol and 2(*R*),5(*R*)-bis(hydroxymethyl)-3(*R*),4(*R*)-dihydroxypyrrolidine, *J. Org. Chem.* **56**, 3849–3855.

ICHIHARA, A., OIKAWA, H. (1997), Biosynthesis of phytotoxins from *Alternaria solani*, *Biosci. Biotech. Biochem.* **61**, 12–18.

ITOH, T., TAKAGI, Y., FUJISAWA, T. (1989), A novel carbon–carbon bond formation in the course of bakers' yeast reduction of cyanoacetone, *Tetrahedron Lett.* **30**, 3811–3812.

JONES, J. B. (1986), Enzymes in organic synthesis, *Tetrahedron* **42**, 3351–3403.

JUNG, S.-H., JEONG, J.-H., MILLER, P., WONG, C.-H. (1994), An efficient multigram-scale preparation of dihydroxyacetone phosphate, *J. Org. Chem.* **59**, 7182–7184.

KAJIMOTO, T., CHEN, L., LIU, K. K.-C., WONG, C.-H. (1991a), Palladium-mediated stereocontrolled reductive amination of azido sugars prepared from enzymatic aldol condensation – a general approach to the synthesis of deoxy azasugars, *J. Am. Chem. Soc.* **113**, 6678–6680.

KAJIMOTO, T., LIU, K. K.-C., PEDERSON, R. L., ZHONG, Z., ICHIKAWA, Y. et al. (1991b), Enzyme-catalyzed aldol condensation for asymmetric synthesis of azasugars – synthesis, evaluation, and modeling of glycosidase inhibitors, *J. Am. Chem. Soc.* **113**, 6187–6196.

KIM, M.-J., LIM, I. T. (1996), Synthesis of unsaturated C_8–C_9 sugars by enzymatic chain elongation, *Synlett*, 138–140.

KIM, M.-J., HENNEN, W. J., SWEERS, H. M., WONG, C.-H. (1988), Enzymes in carbohydrate synthesis – *N*-acetylneuraminic acid aldolase catalyzed reactions and preparation of *N*-acetyl-2-deoxy-D-neuraminic acid-derivatives, *J. Am. Chem. Soc.* **110**, 6481–6486.

KIM, M.-J., LIM, I. T., KIM, H.-J., WONG, C.-H. (1997), Enzymatic single aldol reactions of remote dialdehydes, *Tetrahedron: Asymmetry* **8**, 1507–1509.

KIMURA, T., VASSILEV, V. P., SHEN, G.-J., WONG, C.-H. (1997), Enzymatic synthesis of β-hydroxy-α-amino acids based on recombinant D- and L-threonine aldolases, *J. Am. Chem. Soc.* **119**, 11734–11742.

KITAZUME, T., ISHIKAWA, N. (1984), Induction of center of chirality into fluoro compounds by microbial transformation of 2,2,2-trifluoroethanol, *Chem. Lett.*, 1815–1818.

KOBORI, Y., MYLES, D. C., WHITESIDES, G. M. (1992), Substrate specificity and carbohydrate synthesis using transketolase, *J. Org. Chem.* **57**, 5899–5907.

KOYAMA, T., MATSUBARA, M., OGURA, K. (1985), Isoprenoid enzyme systems of silkworm. II. Formation of the juvenile hormone skeletons by farnesyl pyrophosphate synthetase II, *J. Biochem. Jpn.* **98**, 457–463.

KOYAMA, T., OGURA, K., BAKER, F. C., JAMIESON, G. C., SCHOOLEY, D. A. (1987), Synthesis and absolute configuration of 4-methyl juvenile hormone I (4-MEJH I) by a biogenetic approach – a combination of enzymatic synthesis and biotransformation, *J. Am. Chem. Soc.* **109**, 2853–2854.

KRAGL, U., GYGAX, D., GHISALBA, O., WANDREY, C. (1991), Enzymatic 2-step synthesis of *N*-acetylneuraminic acid in the enzyme membrane reactor, *Angew. Chem.* (Int. Edn. Engl.) **30**, 827–828.

KREN, V., CROUT, D. H. G., DALTON, H., HUTCHINSON, D. W., KÖNIG, W. et al. (1993), Pyruvate decarboxylase: a new enzyme for the production of acyloins by biotransformation, *J. Chem. Soc., Chem. Commun.*, 341–343.

KRIEF, A., SCHAUDER, J.-R., GUITTET, E., HERVÉ DU PENHOAT, C., LALLEMAND, J.-Y. (1987), About the mechanism of sterol biosynthesis, *J. Am. Chem. Soc.* **109**, 7910–7911.

KRIEF, A., PASAU, P., QUÉRÉ, L. (1991), Comparison of the behavior of oxidosqualene cyclases from pig liver and yeast to toward epoxysqualene analogs possessing a Δ18–19 *Z* or *E* C, C double bond, *Bioorg. Med. Chem. Lett.* **1**, 365–368.

LASCHAT, S. (1996), Pericyclic reactions in biological systems – does nature know about the Diels–Alder reaction? *Angew. Chem.* (Int. Edn. Engl.) **35**, 289–291.

LEES, W. J., WHITESIDES, G. M. (1992), The enzymatic synthesis of 1,5-dideoxy-1,5-diimino-D-talitol and 1-deoxygalactostatin using fuculose 1-phosphate aldolase, *Bioorg. Chem.* **20**, 173–179.

LEES, W. J., WHITESIDES, G. M. (1993), Diastereoselectivity (enantioselectivity) of aldol condensations catalyzed by rabbit muscle aldolase at C-2 of RCHOHCHO if R has an appropriately placed negatively charged group, *J. Org. Chem.* **58**, 1887–1894.

LIN, C.-H., SUGAI, T., HALCOMB, R. L., ICHIKAWA, Y., WONG, C.-H. (1992), Unusual stereoselectivity in sialic acid aldolase-catalyzed aldol condensations – synthesis of both enantiomers of high-carbon monosaccharides, *J. Am. Chem. Soc.* **114**, 10138–10145.

LIN, C.-C., LIN, C.-H., WONG, C.-H. (1997), Sialic acid aldolase-catalyzed condensation of pyruvate and *N*-substituted mannosamine: a useful method for the synthesis of *N*-substituted sialic acids, *Tetrahedron Lett.* **38**, 2649–2652.

LINDQVIST, Y., SCHNEIDER, G., ERMLER, U., SUNDSTROM, M. (1992), 3-Dimensional structure of transketolase, a thiamine diphosphate dependent enzyme, at 2.5 Å resolution, *EMBO J.* **11**, 2373–2379.

LIU, K. K.-C., WONG, C.-H. (1992), A new strategy for the synthesis of nucleoside analogs based on enzyme-catalyzed aldol reactions, *J. Org. Chem.* **57**, 4789–4791.

LIU, K. K.-C., KAJIMOTO, T., CHEN, L., ZHONG, Z., ICHIKAWA, Y., WONG, C.-H. (1991), Use of dihydroxyacetone phosphate dependent aldolases in the synthesis of deoxyazasugars, *J. Org. Chem.* **56**, 6280–6289.

LOBELL, M., CROUT, D. H. G. (1996), Pyruvate decarboxylase: a molecular modeling study of pyruvate decarboxylation and acyloin formation, *J. Am. Chem. Soc.* **118**, 1867–1873.

LOOK, G. C., FOTSCH, C. H., WONG, C.-H. (1993), Enzyme-catalyzed organic synthesis – practical routes to azasugars and their analogs for use as glycoprocessing inhibitors, *Acc. Chem. Res.* **26**, 182–190.

LOTZ, B. T., GASPARSKI, C. M., PETERSON, K., MILLER, M. J. (1990), Substrate specificity studies of aldolase enzymes for use in organic synthesis, *J. Chem. Soc., Chem. Commun.,* 1107–1109.

MATSUMOTO, K., SHIMAGAKI, M., NAKATA, T., OISHI, T. (1993), Synthesis of acyclic polyol derivatives via enzyme-mediated aldol reaction, *Tetrahedron Lett.* **34**, 4935–4938.

MEDINA, J. C., KYLER, K. S. (1988), Enzymatic cyclization of hydroxylated surrogate squalenoids with bakers' yeast, *J. Am. Chem. Soc.* **110**, 4818–4821.

MEDINA, J. C., GUAJARDO, R., KYLER, K. S. (1989), Vinyl group rearrangement in the enzymatic cyclization of squalenoids – synthesis of 30-oxysterols, *J. Am. Chem. Soc.* **111**, 2310–2311.

MOCALI, A., ALDINUCCI, D., PAOLETTI, F. (1985), Preparative enzymic synthesis and isolation of D-threo-2-pentulose 5-phosphate (D-xylulose 5-phosphate), *Carbohydr. Res.* **143**, 288–293.

MORADIAN, A., BENNER, S. A. (1992), A biomimetic biotechnological process for converting starch to fructose – thermodynamic and evolutionary considerations in applied enzymology, *J. Am. Chem. Soc.* **114**, 6980–6987.

MORÍS-VARAS, F., QIAN, X. N., WONG, C.-H. (1996), Enzymatic/chemical synthesis and biological evaluation of 7-membered iminocyclitols, *J. Am. Chem. Soc.* **118**, 7647–7652.

MORRIS, K. G., SMITH, M. E. B., TURNER, N. J., LILLY, M. D., MITRA, R. K., WOODLEY, J. M. (1996), Transketolase from *Escherichia coli:* a practical procedure for using the biocatalyst for asymmetric carbon–carbon bond synthesis, *Tetrahedron: Asymmetry* **7**, 2185–2188.

MYLES, D. C., ANDRULIS III, P. J., WHITESIDES, G. M. (1991), A transketolase-based synthesis of (+)-*exo*-brevicomin, *Tetrahedron Lett.* **32**, 4835–4838.

NAGASAWA, T., YAMADA, H. (1986), Enzymatic transformations of 3-chloroalanine into useful aminoacids, *Appl. Biochem. Biotechnol.* **13**, 147–165.

NAGY, J. O., BEDNARSKI, M. D. (1991), The chemical-enzymatic synthesis of a carbon glycoside of *N*-acetylneuraminic acid, *Tetrahedron Lett.* **32**, 3953–3956.

NICOTRA, F., PANZA, L., RUSSO, G., VERANI, A. (1993), Simple and stereoselective chemoenzymatic synthesis of an α-*C*-mannoside, *Tetrahedron: Asymmetry* **4**, 1203–1204.

NIKOLOVA, P., WARD, O. P. (1994), Effect of support matrix on ratio of product to by-product formation in L-phenylacetyl carbinol synthesis, *Biotechnol. Lett.* **16**, 7–10.

NISHIHARA, H., DEKKER, E. E. (1972), Purification, substrate specificity and binding, β-decarboxylase activity, and other properties of *Escherichia coli* 2-keto-4-hydroxyglutarate aldolase, *J. Biol. Chem.* **247**, 5079–5087.

OGURA, K., KOYAMA, T. (1981), Applications of enzymes to organic synthesis – on the enzymatic reactions of C–C bond formation, *J. Synth. Org. Chem. Jpn.* **39**, 459–466.

OIKAWA, H., KATAYAMA, K., SUZUKI, Y., ICHIHARA, A. (1995), Enzymatic activity catalyzing *exo*-selective Diels–Alder reaction in solanapyrone biosynthesis, *J. Chem. Soc., Chem. Commun.,* 1321–1322.

OIKAWA, H., YAGI, K., WATANABE, K., HONMA, M., ICHIHARA, A. (1997), Biosynthesis of macrophomic: plausible involvement of intermolecular Diels–Alder reaction, *J. Chem. Soc., Chem. Commun.,* 97–98.

OLIVER, A. L., RODDICK, F. A., ANDERSON, B. N. (1997), Cleaner production of phenylacetyl carbinol by yeast through productivity improvements and waste minimization, *Pure Appl. Chem.* **69**, 2371–2385.

PEDERSON, R. L., KIM, M.-J., WONG, C.-H. (1988), A combined chemical and enzymatic procedure for the synthesis of 1-deoxynojirimycin and 1-deoxymannojirimycin, *Tetrahedron Lett.* **29**, 4645–4648.

PEDERSON, R. L., LIU, K. K.-C., RUTAN, J. F., CHEN, L., WONG, C.-H. (1990), Enzymes in organic synthesis – synthesis of highly enantiomerically pure 1,2-epoxy aldehydes, epoxy alcohols, thiirane, aziridine, and glyceraldehyde-3-phosphate, *J. Org. Chem.* **55**, 4897–4901.

PROSEN, E., WARD, O. P. (1994), Optimization of reaction conditions for production of *S*-(−)-2-hydroxypropiophenone by *Acinetobacter calcoaceticus*, *J. Ind. Microbiol.* **13**, 287–291.

RACKER, E. (1961a), Cleavage and synthesis of carbon to carbon bonds, in: *The Enzymes* (BOYER, P. D., LARDY, H., MYRBACK, K., Eds.), pp. 305–317. New York: Academic Press.

RACKER, E. (1961b), Transketolase, in: *The Enzymes* (BOYER, P. D., LARDY, H., MYRBACK, K., Eds.), pp. 407–412. New York: Academic Press.

RACKER, E. (1961c), Transaldolase, in: *The Enzymes* (BOYER, P. D., LARDY, H., MYRBACK, K., Eds.), pp. 397–406. New York: Academic Press.

RAMA RAO, A., SRINIVASAN, T. N., BHANUMATHI, N. (1990), A stereoselective biocatalytic Diels–Alder reaction, *Tetrahedron Lett.* **31**, 5959–5960.

ROSE, A. H. (1961), *Industrial Microbiology*. London: Butterworths.

ROSSO, R. G., ADAMS, E. (1967), 4-Hydroxy-2-ketoglutarate aldolase of rat liver – purification, binding of substrates and kinetic properties, *J. Biol. Chem.* **242**, 5524–5534.

SAEED, A., YOUNG, D. W. (1992), Synthesis of L-β- hydroxyaminoacids using serine hydroxymethyltransferase, *Tetrahedron* **48**, 2507–2514.

SANTANIELLO, E., FERRABOSCHI, P., GRISENTI, P., MANZOCCHI, A. (1992), The biocatalytic approach to the preparation of enantiomerically pure chiral building blocks, *Chem. Rev.* **92**, 1071–1140.

SCHMID, W., WHITESIDES, G. M. (1990), A new approach to cyclitols based on rabbit muscle aldolase (RAMA), *J. Am. Chem. Soc.* **112**, 9670–9671.

SCHNEIDER, G., SUNDSTROM, M., LIDQVIST, Y. (1989), Preliminary crystallographic data for transketolase from yeast, *J. Biol. Chem.* **264**, 21 619–21 620.

SCHOLTZ, J. M., SCHUSTER, S. M. (1984), Substrates of hydroxyketoglutarate aldolase, *Bioorg. Chem.* **12**, 229–234.

SCHULTZ, M., WALDMANN, H., VOGT, W., KUNZ, H. (1990), Stereospecific C–C bond formation with rabbit muscle aldolase – a chemoenzymatic synthesis of (+)-*exo*-brevicomin, *Tetrahedron Lett.* **31**, 867–868.

SERVI, S. (1985), 2,2,5-Trimethyl-1,3-dioxolane-4-carbaldehyde as a chiral synthon: synthesis of the two enantiomers of methyl 2,3,6-trideoxy-α-L-*threo*-hex-2-enopyranoside, key intermediate in the synthesis of daunosamine. and of (+)- and (−)-rhodinose, *J. Org. Chem.* **50**, 5865–5867.

SERVI, S. (1990), Bakers' yeast as a reagent in organic synthesis, *Synthesis,* 1–25.

SHIBATA, K., SHINGU, K., VASSILEV, V. P., NISHIDE, K., FUJITA, T. et al. (1996), Kinetic and thermodynamic control of L-threonine aldolase catalyzed reaction and its application to the synthesis of mycestericin D, *Tetrahedron Lett.* **37**, 2791–2794.

SHIMAGAKI, M., MUNESHIMA, H., KUBOTA, M., OISHI, T. (1993), Chemoenzymatic carbon–carbon bond formation leading to noncarbohydrate derivative – stereoselective synthesis of pentamycin C-11–C-16 fragment, *Chem. Pharm. Bull.* **41**, 282–286.

SHIMAZU, Y. (1950), Formation and decomposition of D-(−)-phenylacetylcarbinol by yeast, *J. Chem. Soc. Jpn.* **71**, 503–505 (*Chem. Abstr.* **45**, 9004).

SHIN, H. S., ROGERS, P. L. (1995), Biotransformation of benzaldehyde to L-phenylacetyl carbinol, an intermediate in L-ephedrine production, by immobilized *Candida utilis, Appl. Microbiol. Biotechnol.* **44**, 7–14.

SHIN, H. S., ROGERS, P. L. (1996), Production of L-phenylacetylcarbinol (L-PAC) from benzaldehyde using partially purified pyruvate decarboxylase (PDC), *Biotechnol. Bioeng.* **49**, 52–62.

SPALTENSTEIN, A., WHITESIDES, G. M. (1991), Polyacrylamides bearing pendant α-sialoside groups strongly inhibit agglutination of erythrocytes by influenza virus, *J. Am. Chem. Soc.* **113**, 686–687.

SPEVAK, W., NAGY, J. O., CHARYCH, D. H., SCHAEFER, M. E., GILBERT, J. H., BEDNARSKI, M. D. (1993), Polymerized liposomes containing C-glycosides of sialic acid – potent inhibitors of influenza virus *in vitro* infectivity, *J. Am. Chem. Soc.* **115**, 1146–1147.

SRERE, P., COOPER, J. R., TABACHNICK, M., RACKER, E. (1958), Oxidative pentose phosphate cycle. I. Preparation of substrates and enzymes, *Arch. Biochem. Biophys.* **74**, 295–305.

STRAUB, A., EFFENBERGER, F., FISCHER, P. (1990), Enzyme-catalyzed reactions. 4. Aldolase-catalyzed C–C bond formation for stereoselective synthesis of nitrogen-containing carbohydrates, *J. Org. Chem.* **55**, 3926–3932.

STUMPF, B., KIESLICH, K. (1991), Acyloin condensation of acyclic unsaturated aldehydes by *Mucor* species, *Appl. Microbiol. Biotechnol.* **34**, 598–603.

SUGAI, T., SHEN, G.-J., ICHIKAWA, Y., WONG, C.-H. (1993), Synthesis of 3-deoxy-D-*manno*-2-octulosonic acid (KDO) and its analogs based on KDO aldolase-catalyzed reactions, *J. Am. Chem. Soc.* **115**, 413–421.

TAKAOKA, Y., KAJIMOTO, T., WONG, C.-H. (1993), Inhibition of *N*-acetylglucosaminyltransfer enzymes – chemical-enzymatic synthesis of new 5-membered acetamido azasugars, *J. Org. Chem.* **58**, 4809–4812.

TAKAYAMA, S., MCGARVEY, G. J., WONG, C.-H. (1997a), Microbial aldolases and transketolases: new biocatalytic approaches to simple and complex sugars, *Ann. Rev. Microbiol.* **51**, 285–310.

TAKAYAMA, S., MCGARVEY, G. J., WONG, C.-H. (1997b), Enzymes in organic synthesis: recent developments in aldol reactions and glycosylations, *Chem. Soc. Rev.* **26**, 407–415.

TURNER, N. J. (1989), Recent advances in the use of enzyme-catalyzed reactions in organic synthesis, *Nat. Prod. Rep.* **6**, 625–644.

TURNER, N. J. (1994), Recent advances in the use of enzyme-catalyzed reactions in organic synthesis, *Nat. Prod. Rep.* **11**, 1–15.

TURNER, N. J., WHITESIDES, G. M. (1989), A combined chemical-enzymatic synthesis of 2-deoxy-D-*arabino*-heptulosonic acid 7-phosphate, *J. Am. Chem. Soc.* **111**, 624–627.

UCHIDA, Y., TSUKADA, Y., SUGIMORI, T. (1984), Purification and properties of *N*-acetylneuraminate lyase from *Escherichia coli*, *J. Biochem.* **96**, 507–522.

UCHIDA, Y., TSUKADA, Y., SUGIMORI, T. (1985), Distribution of *N*-acetylneuraminic lyase in bacteria and its production by *Escherichia coli*, *Agric. Biol. Chem.* **49**, 181–187.

VON DER OSTEN, C. H., SINSKEY, A. J., BARBAS, C. F., PEDERSON, R. L., WANG, Y.-F., WONG, C.-H. (1989), Use of a recombinant bacterial fructose-1,6-diphosphate aldolase in aldol reactions – preparative syntheses of 1-deoxynojirimycin, 1-deoxymannojirimycin, 1,4-dideoxy-1,4-imino-D-arabinitol, and fagomine, *J. Am. Chem. Soc.* **111**, 3924–3927.

VASSILEV, V. P., UCHIYAMA, T., KAJIMOTO, T., WONG, C.-H. (1995a), L-Threonine aldolase in organic synthesis: preparation of novel β-hydroxy-α-amino acids, *Tetrahedron Lett.* **36**, 4081–4084.

VASSILEV, V. P., UCHIYAMA, T., KAJIMOTO, T., WONG, C.-H. (1995b), An efficient chemo-enzymatic synthesis of α-amino-β-hydroxy-γ-butyrolactones, *Tetrahedron Lett.* **36**, 5063–5064.

WALDMANN, H. (1995), Enzymatic carbon–carbon bond formation, in: *Organic Synthesis Highlights II* (WALDMANN, H., Ed.), pp. 147–155. Weinheim: VCH.

WARD, O. P. (1995), Application of bakers' yeast in bioorganic synthesis, *Can. J. Bot.* **73**, S1043–S1048.

WILCOCKS, R., WARD, O. P. (1992), Factors affecting 2-hydroxypropiophenone formation by benzylformate decarboxylase from *Pseudomonas putida*, *Biotechnol. Bioeng.* **39**, 1058–1063.

WILCOCKS, R., WARD, O. P., COLLINS, S., DEWDNEY, N. J., HONG, Y. P., PROSEN, E. (1992), Acyloin formation by benzylformate decarboxylase from *Pseudomonas putida*, *Appl. Environ. Microbiol.* **58**, 1699–1704.

WONG, C.-H., WHITESIDES, G. M. (1994), *Enzymes in Organic Chemistry, Tetrahedron Organic Series* 12, pp. 195–251. Oxford: Pergamon Press.

WONG, C.-H., SHENG, G. J., PEDERSON, R. L., WANG, Y. F., HENNEN, W. J. (1991), Enzymatic catalysis in organic synthesis, *Methods Enzymol.* **202**, 591–620.

WONG, C.-H., HALCOMB, R. L., ICHIKAWA, Y., KAJIMOTO, T. (1995a), Enzymes in organic synthesis: application to the problems of carbohydrate recognition (part 1), *Angew. Chem.* (Int. Edn. Engl.) **34**, 412–432.

WONG, C.-H., GARCÍA-JUNCEDA, E., CHEN, L., BLANCO, O., GIJSEN, H. J. M., STEENSMA, D. H. (1995b), Recombinant 2-deoxyribose-5-phosphate aldolase in organic synthesis: use of sequential two-substrate and three-substrate aldol reactions, *J. Am. Chem. Soc.* **117**, 3333–3339.

XIAO, X.-Y., PRESTWICH, G. D. (1991), Enzymatic cyclizations of 26-hydroxy-2,3-oxidosqualenes and 29-hydroxy-2,3-oxidosqualenes give 19-hydroxylanosterols and 21-hydroxylanosterols, *Tetrahedron Lett.* **32**, 6843–6846.

ZHOU, P., SALLEH, H. M., HONEK, J. F. (1993), Facile chemoenzymatic synthesis of 3-(hydroxymethyl)-6-epicastanospermine, *J. Org. Chem.* **58**, 264–266.

ZIEGLER, T., STRAUB, A., EFFENBERGER, F. (1988), Enzyme-catalyzed reactions. 3. Enzyme-catalyzed synthesis of 1-deoxymannonorjirimycin, 1-deoxynorjirimycin and 1,4-dideoxy-1,4-imino-D-arabinitol, *Angew. Chem.* (Int. Edn. Engl.) **27**, 716–717.

2 Lyases

JASSEM G. MAHDI
DAVID R. KELLY
Cardiff, Wales, UK

1 Introduction

Lyases are defined as *enzymes cleaving C–C, C–O, C–N and other bonds by other means than by hydrolysis or oxidation* (Anonymous, 1984, 1992). In most cases the products are unsaturated molecule (frequently an alkene) plus an X–H species. The lyases are a disparate group of enzymes, catalysing a wide range of processes with few features in common, beyond their assignment to the same enzyme commission group. Lyases utilize an extraordinary range of prosthetic groups which include thiamine pyrophosphate (JORDAN, 1999; LINDQVIST and SCHNEIDER, 1993), iron sulfur clusters (FLINT and ALLEN, 1996), NAD(P)(H), pyridoxal 5′-phosphate (JANSONIUS, 1998; JOHN, 1995; HAYASHI, 1995; ALEXANDER et al., 1994), biotin (TOH et al., 1993) and backbone dehydroalanine (GULZAR et al., 1997), or 4-methylidene-imidazole-5-one residues (SCHWEDE et al., 1999).

They have been rarely (if ever) reviewed together and indeed one major biotransformations text does not even mention them as a single group in the index (DRAUZ and WALDMANN, 1995). The fundamental biochemistry of lyases (including protein crystallography) is a subject of frenetic activity at present. Several lyases are used in industrial processes, such as the production of amino acids, however, laboratory applications are very rare. It is hoped that this review will bring this subject to a wider audience and promote exploitation of these fascinating enzymes.

2 Conventions

Each lyase is discussed in a separate section, according to Enzyme Commission class numbers. Each section is presented in the sequence: definition of the biotransformation, background, enzymology and enzyme structure studies, stereochemistry, synthetic applications, technological aspects. The last of these subsections covers large scale reactions and methodology leading to manufacturing processes. In shorter sections these headings are not shown explicitly. The conventions for referring to figures, references and compounds used throughout this, and the previous volume, are adhered to in this chapter. However, if these are marked "cf." the attribution refers to a similar, related or parallel process, rather than that directly under discussion at the time. Tab. 1 summarizes the principal classes of lyases.

3 Enzyme Commission Classification

Lyases are enzymes that cleave C–C, C–O, C–N and other bonds by elimination to form multiple bonds or rings. In many cases this involves cleavage of C–X and C–H bonds on adjacent carbon atoms by an E1 or E2 mechanism. Reactions which involve hydrolysis (but not hydration) or oxidation (e.g., desaturases) are excluded. Lyases are assigned to Enzyme Commission class 4. The subclass (1–6 and 99) is assigned according to the bond broken and the subsubclass according to the group eliminated. As usual for enzyme commission codes, the subsubsub class is a unique designator for each enzyme. For example phenylalanine ammonia-lyase is assigned to EC 4.3.1.5. This indicates a carbon–nitrogen lyase (EC 4.3), which eliminates ammonia from the substrate (EC 4.3.1).

The systematic names for lyases are constructed according to the pattern *substrate group-lyase*, e.g., phenylalanine ammonia-lyase. The hyphen is an important component of the systematic name because it distinguishes this class of enzymes from others with similar names, e.g., hydro-lyases and hydrolases. However, this rule is relaxed for recommended names, e.g., aldolase, decarboxylase and dehydratase.

In some cases elimination is followed immediately by addition and hence overall the reaction is apparently a substitution. If there is evidence of an unsaturated intermediate the enzyme is nevertheless classified as a lyase.

The intramolecular lyases are a small group of enzymes which are classified as isomerases

Tab. 1. Representative Examples of Lyase Catalyzed Reactions

EC Code Systematic Names	Reaction
4.1 Carbon–carbon lyases	
4.1.1 Carboxy-lyases	**1** Pyruvate → Pyruvate decarboxylase, EC 4.1.1.1 → **2** Acetaldehyde + CO_2
	26 Oxalate → Oxalate decarboxylase, EC 4.1.1.2 → **27** Formate + CO_2
	28 Oxaloacetate → Oxaloacetate decarboxylase, EC 4.1.1.3 → **1** Pyruvate + CO_2
	40 Acetoacetate → **41** Acetoacetate decarboxylase, EC 4.1.1.4 → **39** Acetone + CO_2
	65 (*S*)-Acetolactate → Acetolactate decarboxylase, EC 4.1.1.5 → **5** (***R***)-Acetoin + CO_2
	78 Benzylformate → Benzoylformate lyase, EC 4.1.1.7 → **4** Benzaldehyde + CO_2
	137 L-Histidine → Histidine decarboxylase, EC 4.1.1.22 → **138** Histamine + CO_2

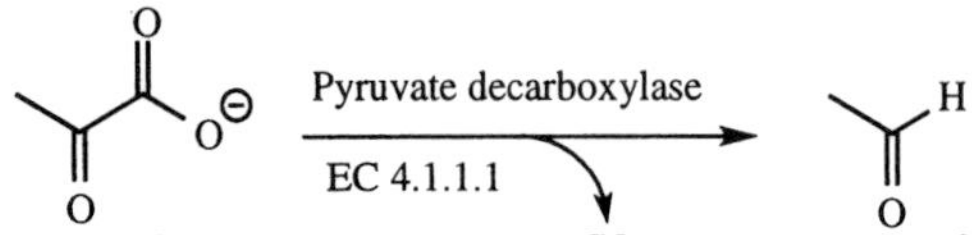

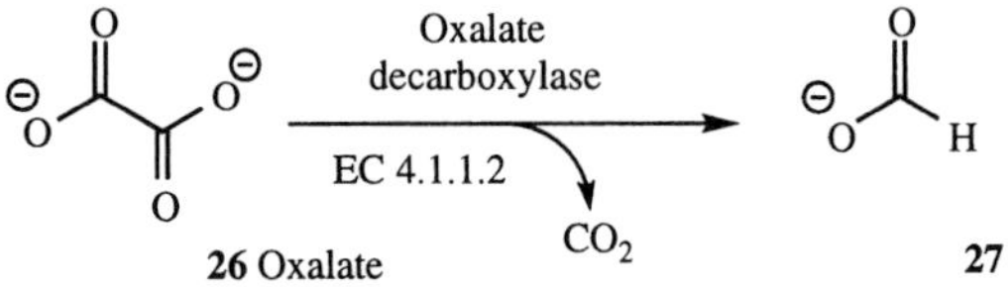

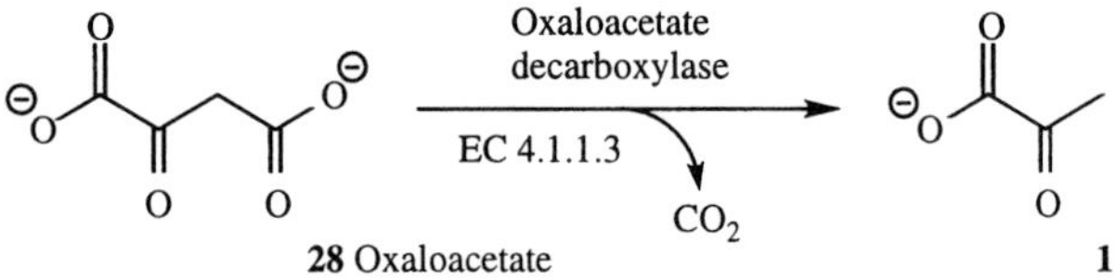

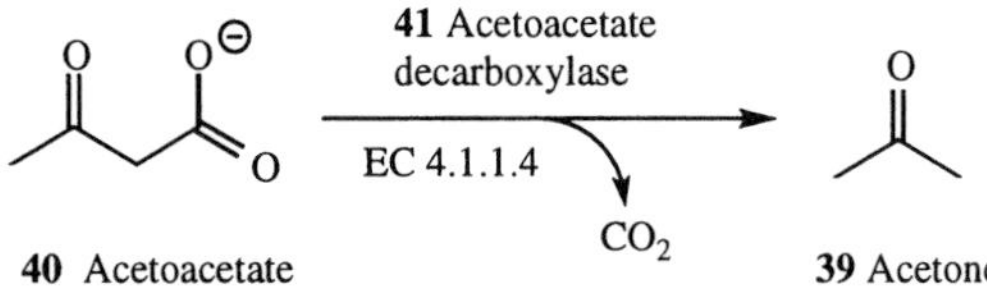

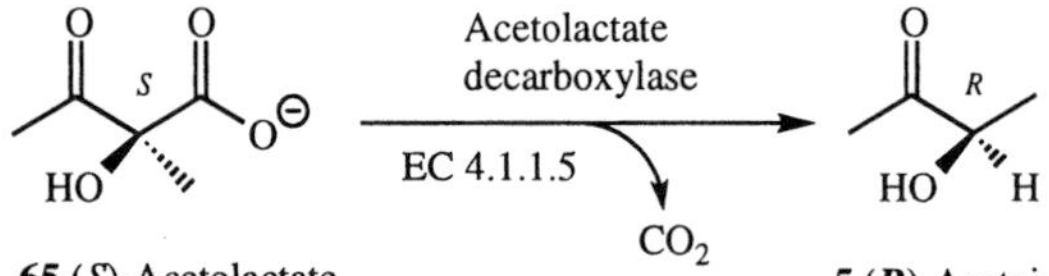

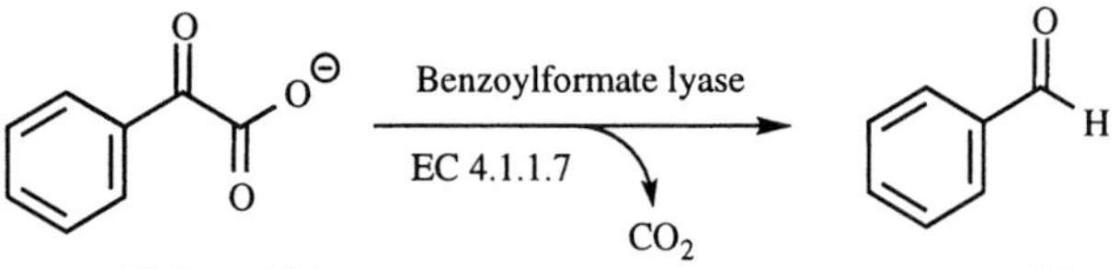

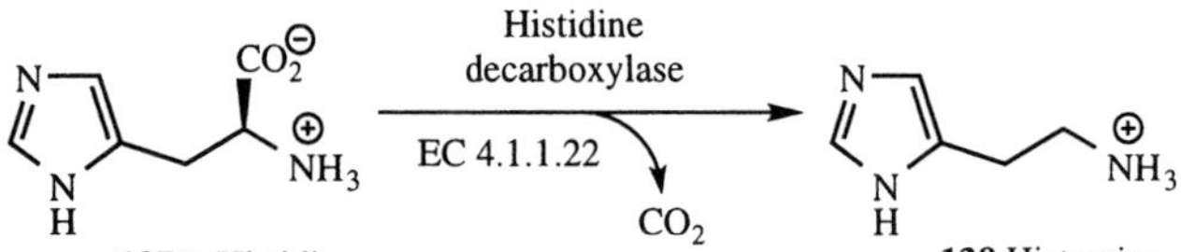

Tab. 1 (continued)

EC Code Systematic Names		Reaction
4.1.2	Aldehyde-lyases	**139** R = $CO_2^{\ominus}$ → (Orotidine 5'-phosphate decarboxylase, EC 4.1.1.23; − CO_2) **140** R = H
		141 a R = H; L-Tyrosine; b R = OH; L-DOPA → (Tyrosine decarboxylase, EC 4.1.1.25; − CO_2) **142** a R = H; Tyramine; b R = OH; Dopamine
		143 a R = H; L-Tryptophan; b R = OH; 5-Hydroxy-L-tryptophan → (Aromatic L-amino acid decarboxylase, EC 4.1.1.28; − CO_2) **144** a R = H; Tryptamine; b R = OH; Serotonin
4.1.3	Oxo-acid-lyases	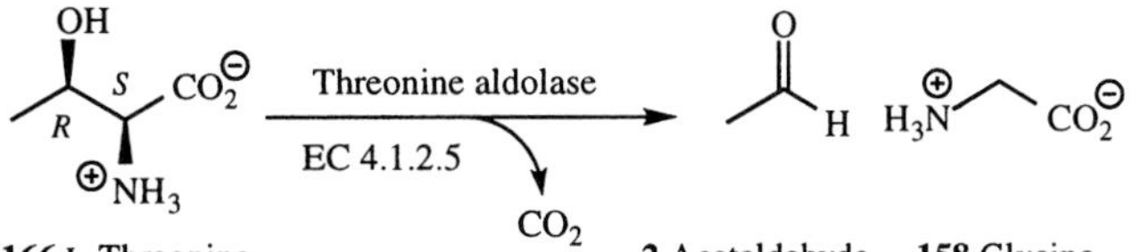 **166** L-Threonine → (Threonine aldolase, EC 4.1.2.5; − CO_2) **2** Acetaldehyde + **158** Glycine
4.1.4	Other carbon–carbon lyases	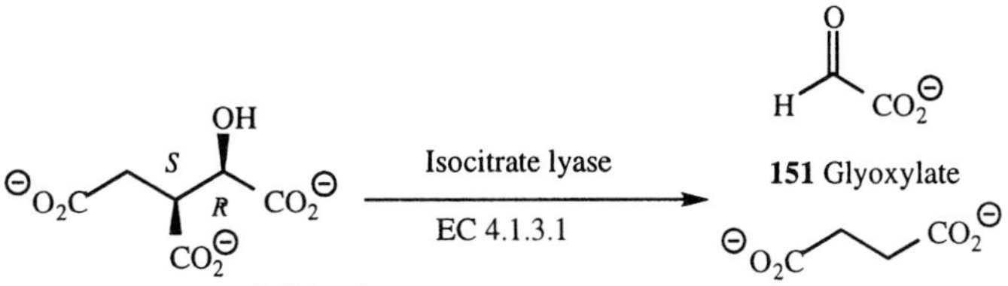 **168** Isocitrate → (Isocitrate lyase, EC 4.1.3.1) **151** Glyoxylate + **198** Succinate

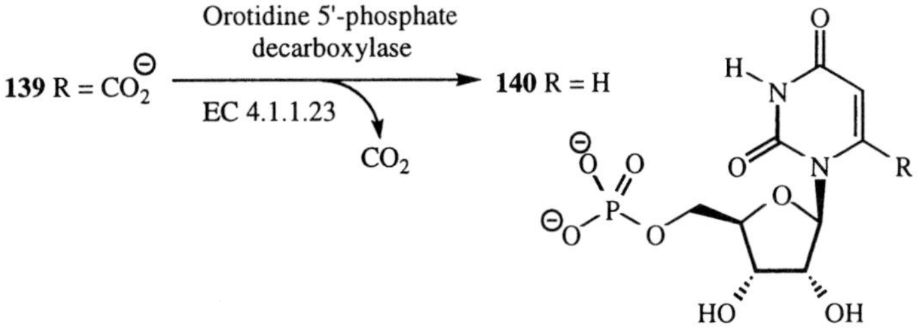

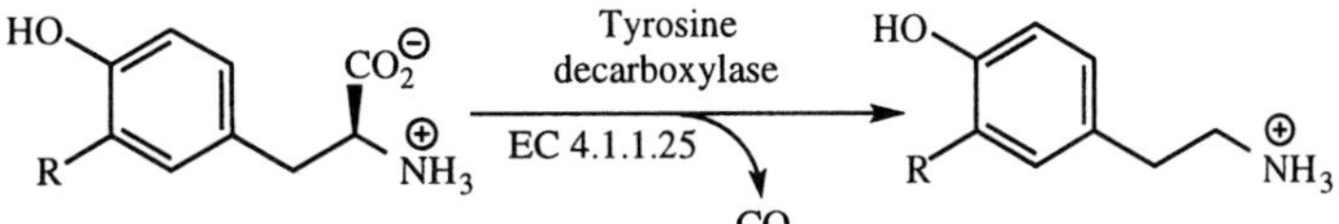

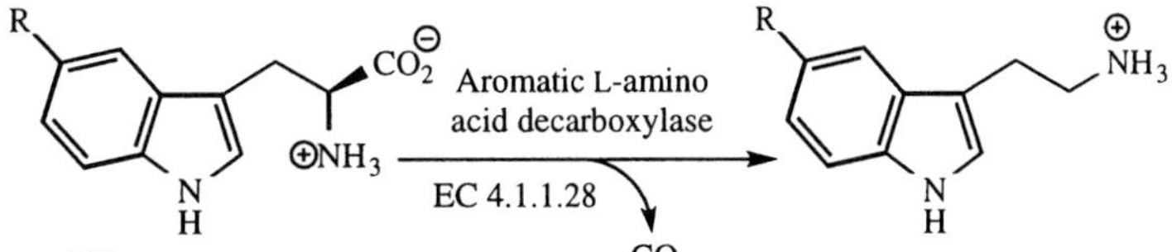

Tab. 1 (continued)

EC Code Systematic Names	Reaction
4.2 Carbon–oxygen lyases 4.2.1 Hydrolyases	CO_2 → $HCO_3^{\ominus}$ (Carbonic anhydrase, EC 4.2.1.1, $^{\ominus}OH$) **196** Fumarate → **197** L-(*S*)-(-)-Malate (Fumarase, EC 4.2.1.2, H_2O) **257** 3-Dehydroquinate → **258** 3-Dehydroshikimate (3-Dehydroquinate dehydratase, EC 4.2.1.10, H_2O)
4.2.2 Acting on polysaccharides	**321** → **322** + **323** (Pectin and pectate lyases, EC 4.2..2.2 & EC 4.2.2.10, H_2O) **a** R' = $Ca^{2\oplus}$ **b** R' = Me

Tab. 1 (continued)

EC Code Systematic Names	Reaction
4.3 Carbon–nitrogen lyases	
4.3.1 Ammonia lyases	**196** Fumarate → Aspartase, EC 4.3.1.1, $NH_4^{\oplus}$ → **101** L-(*S*)-(+)-Aspartate
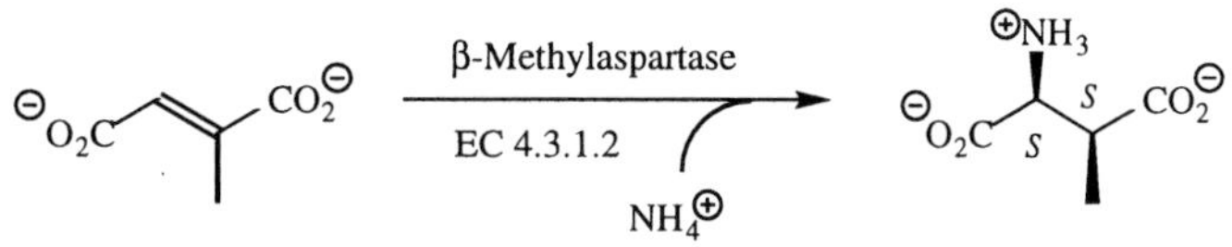	**335** Mesaconate → β-Methylaspartase, EC 4.3.1.2, $NH_4^{\oplus}$ → **336** L-*threo*-(2*S*,3*S*)-3-Methylaspartate
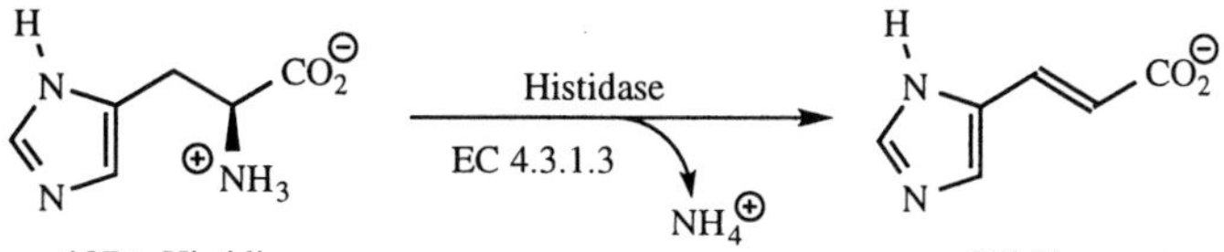	**137** L-Histidine → Histidase, EC 4.3.1.3 → **368** Urocanate + $NH_4^{\oplus}$
	388 L-Phenylalanine → Phenylalanine ammonia-lyase, EC 4.3.1.5 → **389** *trans*-Cinnamate + $NH_4^{\oplus}$

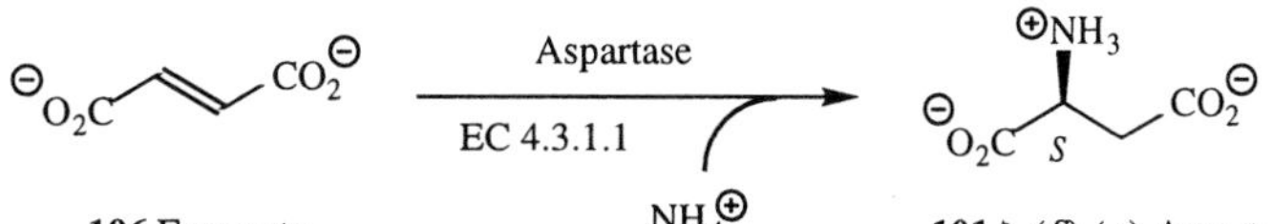

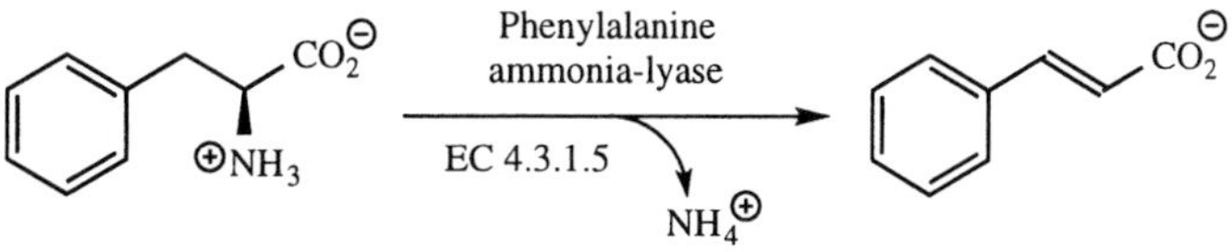

(EC 5). They catalyze reactions in which a group is eliminated from one part of a molecule to leave a multiple bond while remaining covalently attached to the molecule.

A remarkable number of lyases are involved in central metabolism in the citric acid (Krebs cycle), glycolysis and neoglucogenesis. These are summarized in Fig. 1.

4 Examples of Lyases

4.1 Carbon–Carbon Lyases (EC 4.1)

Many of the enzymes under this heading are covered in Chapter 1, this volume: Carbon–Carbon Bond Formation Using Enzymes.

4.1.1 Carboxy-Lyases (EC 4.1.1.X)

This subsubclass contains enzymes which catalyze decarboxylation (IDING et al., 1998; CROUT et al., 1994). Most carboxy-lyases utilize thiamine pyrophosphate (JORDAN, 1999; LINDQUIST and SCHNEIDER, 1993), pyridoxal phosphate or biotin (TOH et al., 1993) as cofactors.

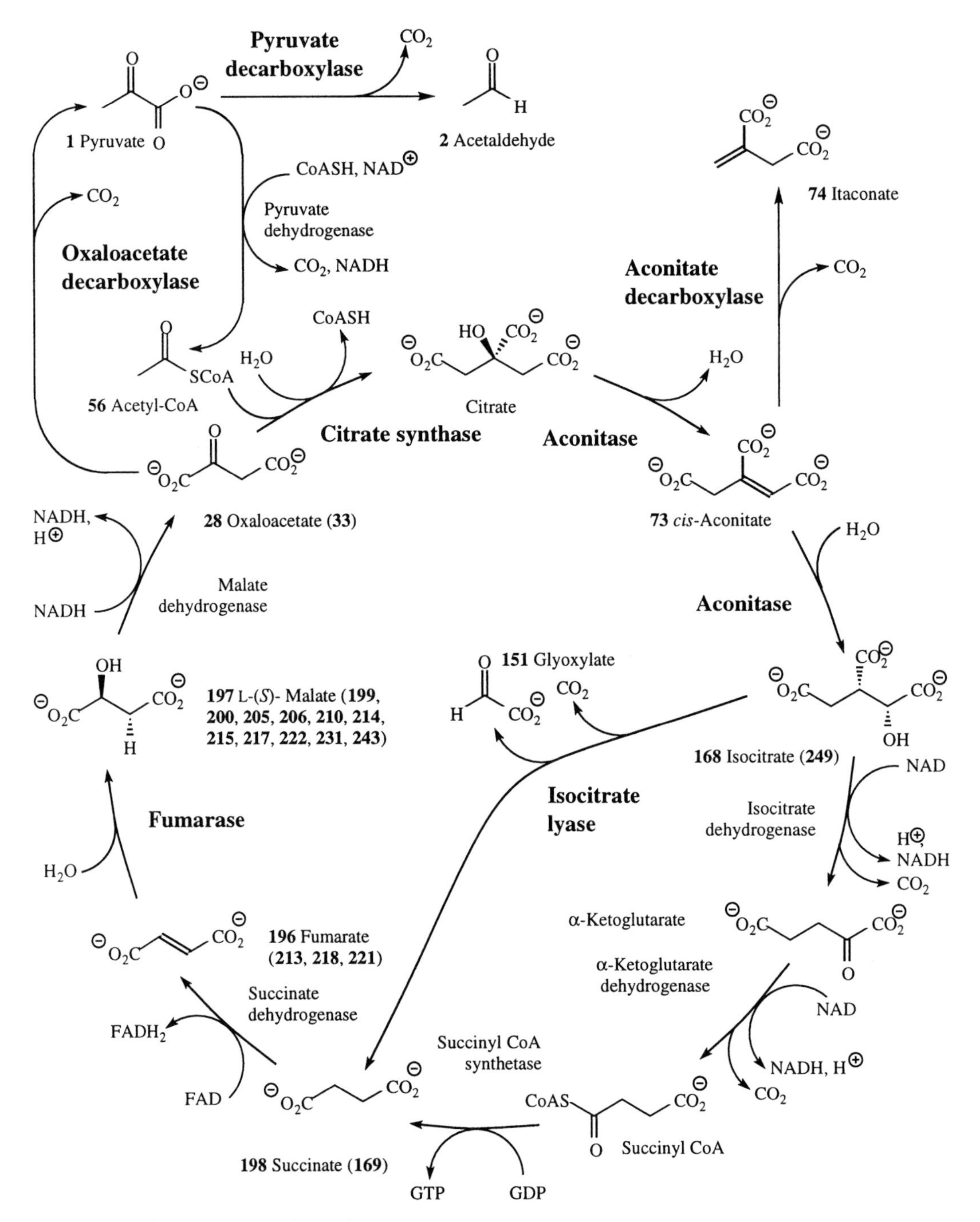

Fig. 1. Lyases in the citric acid (Krebs) cycle and related biosynthetic sequences. Bold numbers in brackets refer to labeled analogs.

4.1.1.1 Pyruvate Decarboxylase (EC 4.1.1.1)

Introduction

Pyruvate decarboxylase catalyzes the decarboxylation of pyruvate (**1**) to acetaldehyde (**2**) and acyloin type condensation reactions (Fig. 2) (OBA and URITANI, 1982). Synthetic applications of pyruvate decarboxylase (PDC) are described in Chapter 1 of this volume and have been reviewed recently elsewhere (SCHORKEN and SPRENGER, 1998; FESSNER, 1998; FESSNER and WALTER, 1997). This section will describe mechanistic aspects and biotechnological applications. PDC is widely distributed in plants and fungi, but is rare in prokaryotes and absent in animals (CANDY and DUGGLEBY, 1998; VAN WAARDE, 1991). The major metabolic role of PDC is as a component of the anaerobic pathway from pyruvate to ethanol. Pyruvate (**1**, Fig. 2) undergoes decarboxylation to give acetaldehyde (**2**) which is reduced by alcohol dehydrogenase to ethanol. The NAD^+ produced by the reductive step is then utilized in cellular metabolism (PRONK et al., 1996). This pathway is of course the basis of the brewing industry and much effort has been made to enhance its productivity for both beverage production and for ethanol for fuel (INGRAM et al., 1999; DENG and COLEMAN, 1999; SCOPES, 1997). The operation of this pathway is essential for the survival of crop plants which are subjected to flooding, such as rice (TADEGE et al., 1999; SU and LIN, 1996; SACHS et al., 1996; PESCHKE and SACHS, 1993).

Mechanism and Enzyme Structure

Pyruvate decarboxylase is of considerable synthetic importance because the acetaldehyde α-anion equivalent (**3**) can be intercepted by exogenous aldehydes to give acyloins (e.g., **5**, **6**). Surprisingly, the reaction with acetaldehyde (**2**), reproducibly gives 3-hydroxybutanon-2-one (acetoin) with a low enantiomeric excess. Whole yeast and the purified PDC from yeast both give (*R*)-3-hydroxybutanon-2-one (**5**) of about 50% enantiomeric excess, whereas *Zymomonas mobilis* PDC gives the opposite enantiomer, but with essentially the same enantiomeric excess (BORNEMANN et al., 1993). Wheat germ PDC converts pyruvate to racemic 3-hydroxybutanon-2-one, whereas in the presence of exogenous acetaldehyde, the (*S*)-enantiomer is produced with a low enantiomeric excess (CROUT et al., 1986).

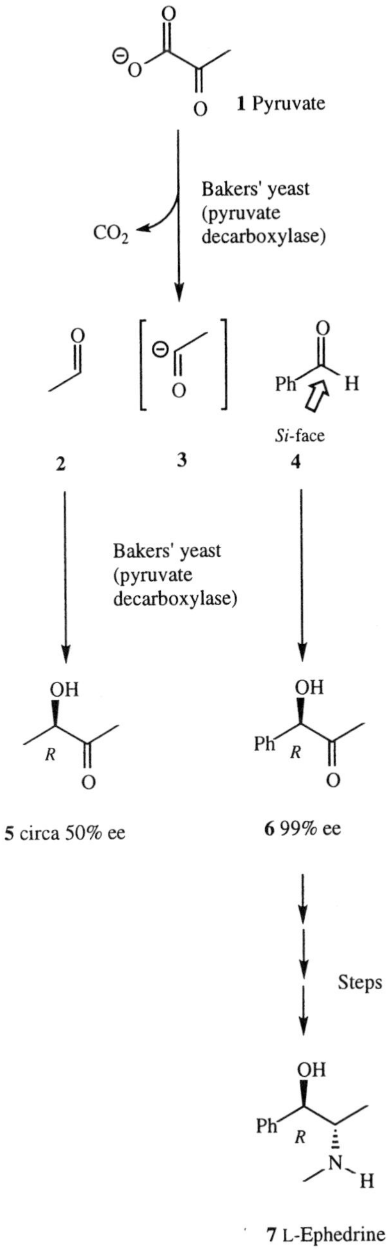

Fig. 2. Pyruvate decarboxylase mediated decarboxylation and acyloin condensation.

None of these reactions are of appreciable practical value, however, interception by benzaldehyde (**4**) is highly stereoselective and the (*R*)-phenylacetyl carbinol (**5**) so formed has been used in the manufacture of L-ephedrine (**7**) since 1932 (Knoll process; SHUKLA and KULKARNI 1999; NIKOLOVA and WARD, 1991). Similarly high enantiomeric excesses have been achieved with other aromatic and heteroaromatic aldehydes using both yeast PDC (KREN et al., 1993, CROUT et al., 1991b; CARDILLO et al., 1991) and *Zymomonas mobilis* PDC (BORNEMANN et al., 1996). Moreover, aliphatic aldehydes (C_2–C_{12}) are also good substrates (MONTGOMERY et al., 1992).

The active site of PDC contains thiamine pyrophosphate (TPP, **8**, Fig. 3), which is also known as thiamin diphosphate (ThDP). Recent advances in the study of TPP linked enzymes are reported in a collection of articles in volume 1385 of *Biochimica et Biophysica Acta* (Various, 1998). TPP (**8**) mediates both decarboxylation and the acyloin condensation (SPRENGER and POHL, 1999; SCHORKEN and SPRENGER, 1998). Magnesium ions or calcium ions are required for activity, as part of the pyrophosphate binding site (VACCARO et al., 1995). Fundamental chemical aspects of the mechanism are well understood and have been engineered into abiotic chemical models (IMPERIALI et al., 1999; KNIGHT and LEEPER, 1998; cf. KLUGER et al., 1995) and catalytic antibodies (TANAKA et al., 1999). PDC catalyzes the rate of decarboxylation of pyruvate by a factor of about 10^{12} above the spontaneous rate, which is equivalent to transition state stabilization of some 70 kJ mol^{-1} (HUHTA et al., 1992).

TPP (**8**) is bound tightly, but non-covalently to PDC (Fig. 4) and is converted to the ylide (**9**) by intramolecular deprotonation (HUBNER et al 1998) of C-2 of the thiazolium ring, by the amino group of the aminopyrimidine substituent (KERN et al., 1997; FRIEDEMANN and NEEF, 1998), which is involved in a proton relay, with a glutamate side chain (KILLENBERG-JABS et al., 1997). The thiazolium ring, the aminopyrimidine ring and the intervening methylene group adopt an unusual V conformation which brings C-2 of the thiazolium ring and the amino substituent into close proximity. This arrangement increases the rate of deprotonation by at least a factor of 10^4, relative to that observed in solution. The ylide (**9**) (SCHELLENBERGER, 1998) undergoes nucleophilic addition to the ketone group of pyruvate (**1**) to give the zwitterion (**10**), which evolves carbon dioxide. The product (**11**) acts as an enamine/vinyl sulfide rather than an enol and hence undergoes either protonation (**12**) or alkylation (**13**) adjacent to the oxygen to give an imminium salt. Cleavage of the C-2 thiazolium bond and restitution of the carbonyl group yields either acetaldehyde (**2**) or the acyloin product (**6**). The stereochemistry of the intermediates has been inferred from molecular modeling studies (LOBELL and CROUT, 1996b). A complete catalytic decarboxylation cycle requires the consumption of one proton, which is appended to C-1 of acetaldehyde. Single turnover experiments with [2-^{3}H]-TPP, demonstrate that 43% of the tritium label (*H, Fig. 5) is incorporated into [1-^{3}H]-acetaldehyde and 54% into [^{3}H]-water (HARRIS and WASHABAUGH, 1995a, b). It is beyond doubt that the ylide (**9**) exists at the active site at some stage, however, it has not been detected by ^{13}C-NMR (SCHELLENBERGER, 1998; LI and JORDAN, 1999). It is, therefore, reasonable to speculate, that protonation of the ylide is part of a relay which brings the "consumed proton" into the active site and that ylide formation is linked with binding of pyruvate as substrate or allosterically. Further aspects of Fig. 5 are discussed below.

S. cerevisiae, PDC consists of the products of the genes *PDC1*, *PDC5* and *PDC6*. The predominant isozyme monomer is the gene product of *PDC1*, which is 80% identical with the

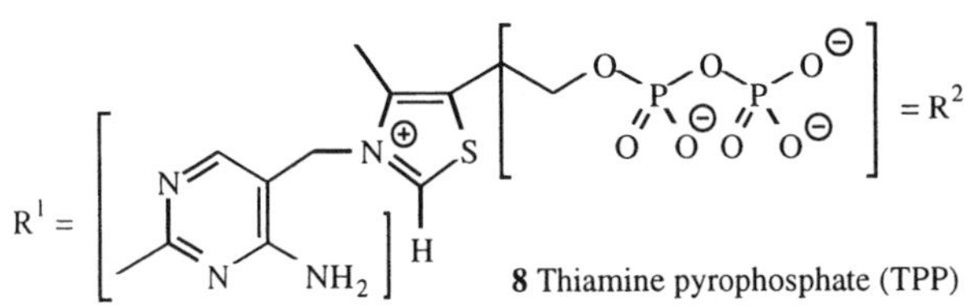

R^1 = ; = R^2

8 Thiamine pyrophosphate (TPP)

Fig. 3. The structure of thiamine pyrophosphate, R^1 and R^2 are abbreviations used in the following two figures.

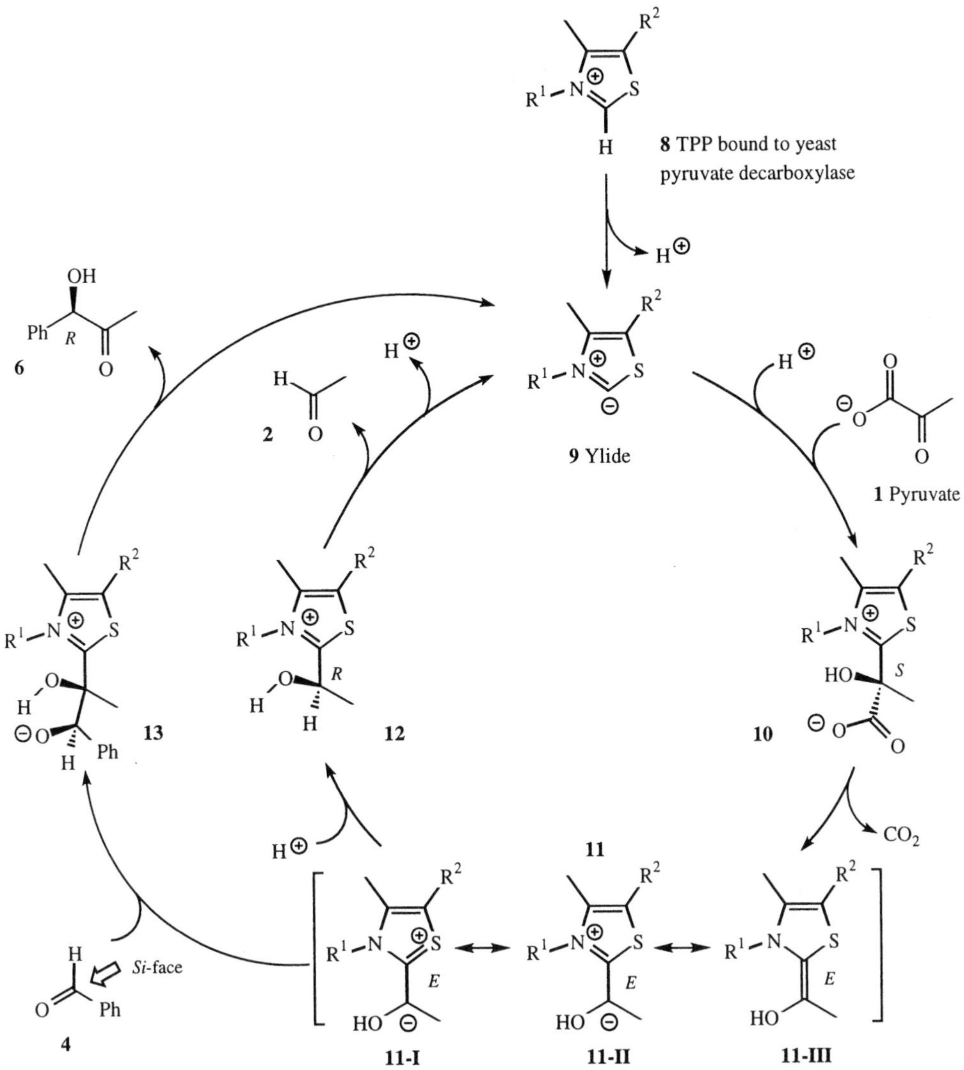

Fig. 4. "Minimal mechanism" for the pyruvate decarboxylase (PDC) catalyzed decarboxylation of pyruvate. Stereochemical assignments are based on the molecular modeling studies of LOBELL and CROUT (1996b) (cf. Fig. 3, for R^1 and R^2).

PDC5 gene product. Normally these monomers associate into mixed tetramers (α_4, $\alpha_3\beta_1$, $\alpha_2\beta_2$ etc.; FARRENKOPF and JORDAN, 1992) which can be separated by ion-exchange chromatography, however, by using a non-*PDC5* expressing mutant, a homotetrameric pdc1 (MW 240,000 Da) has been isolated directly (KILLENBERG-JABS et al., 1996). The expression of *PDC5* is repressed by thiamine but *PDC1* is unaffected, hence under thiamine limiting conditions both are expressed (MULLER et al., 1999). *PDC6* is only expressed at very low levels and appears to be metabolically unimportant under normal circumstances (HOHMANN, 1991a, b). *PDC2* is a regulatory gene which controls basal expression of *PDC1* (HOHMANN, 1993). pdc1 and pdc5 both have four cysteine residues per monomeric unit,

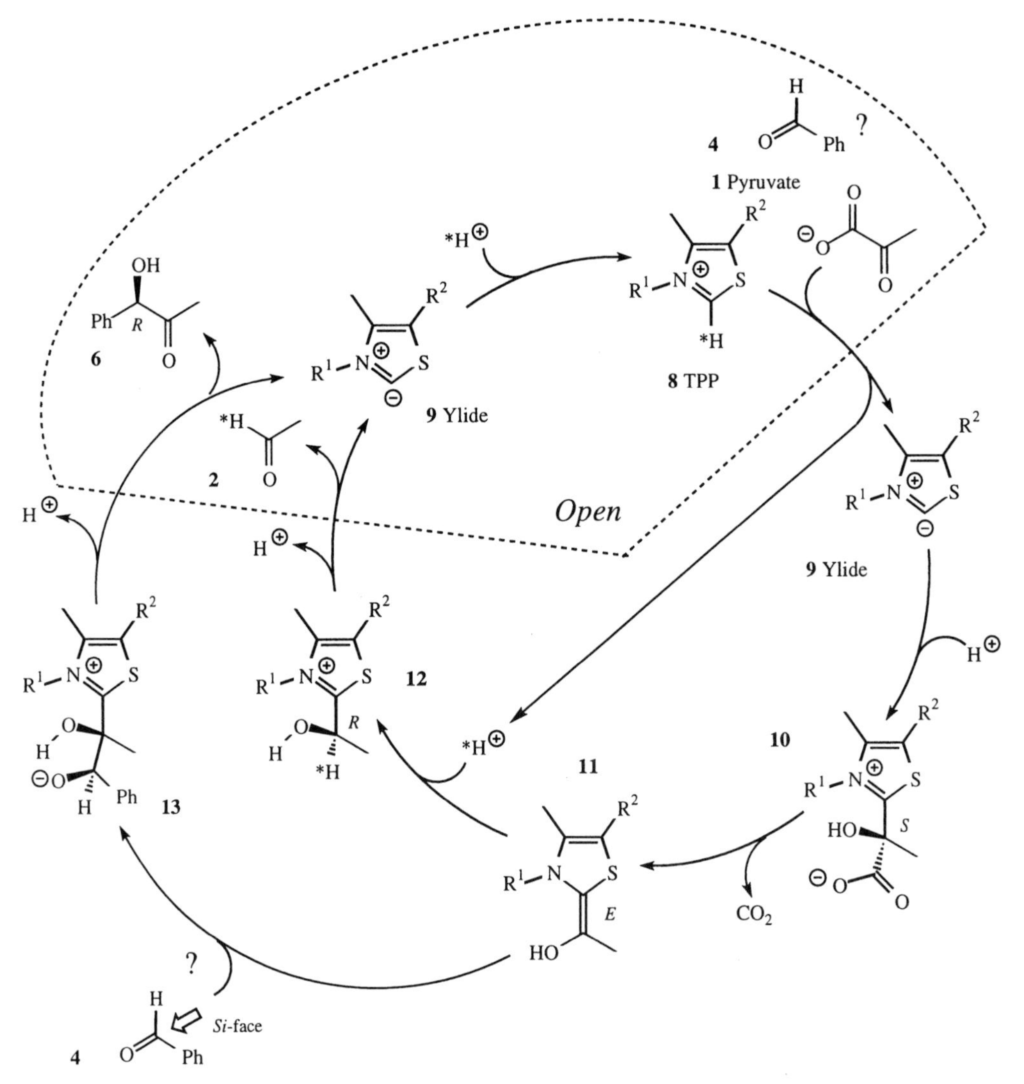

Fig. 5. Complete catalytic sequence for the pyruvate decarboxylase (PDC) catalyzed decarboxylation of pyruvate, incorporating stereochemical assignments (LOBELL and CROUT, 1996b), [2-^{3}H]-TPP labeling studies, (HARRIS and WASHABAUGH, 1995a, b) and opening and closing of the active site (ALVAREZ et al., 1995) (cf. Fig. 3, for R^1 and R^2).

whereas pdc6 only has one (Cys221; BABURINA et al., 1996). This residue which is located on the surface of the protein at an interface seems to be involved in allosteric activation by pyruvate and other carbonyl containing compounds, which triggers conformational changes (LI and JORDAN, 1999; JORDAN et al., 1998; BABURINA et al. 1998a, b).

PDC undergoes slow allosteric activation by pyruvate (**1**) or substrate surrogates such as pyruvamide or ketomalonate. This may take several seconds to minutes and causes an approximately ten fold increase in activity. Each allosteric binding of a pyruvate molecule suffices to activate several thousand catalytic cycles (ZENG et al., 1993; BABURINA et al.,1994). It has been proposed that addition of the cysteine-221 thiol group (**14**) to pyruvate (**1**), give a hemithioketal (**15**) triggers the initial slow allosteric activation of PDC (Fig. 6). However, recent results suggest that pyruvate binding at the regulatory site, without thiol addition causes slow initial activation, and that hemithioketal formation occurs during the catalytic cycle, rather than prior to it. In this model, thiol addition opens the active site for admission of pyruvate and elimination closes it for the decarboxylation, protonation and acetaldehyde formation steps. Finally, addition again enables release of acetaldehyde and carbon dioxide (ALVAREZ et al., 1995; Fig. 5). A closed conformation for the decarboxylation/protonation stages of the cycle is consistant with the need to sequester the highly basic ylide (**9**) (pK_a 17.5 in solution), away from solvent and the [2-^{3}H]-TPP labeling study, which indicates minimal hydrogen exchange with the solvent (cf. the *waterproof* active site, LOBELL and CROUT, 1996a).

The X-ray crystal structures of wild type *Saccharomyces cerevisiae* PDC (from *PDC1*, α4), PDC from *PDC1* expressed in *E. coli* and *S. uvarium* are PDC, have been determined (LU et al., 1997; ARJUNAN et al., 1996; DYDA et al., 1990, 1993). Both enzymes have almost identical sequences and are tetrameric. Each monomeric unit consisting of a single peptide chain, is tightly associated into a dimer with the TPPs situated at the interface. Two dimeric pairs associate to give the tetramer (KONIG, 1998; KONIG et al., 1998). Each monomeric unit consists of three domains: the *N*-terminal, α-domain which binds the pyrimidine group of TPP, the central β-regulatory domain and the *C*-terminal γ-domain which binds the pyrophosphate group of TPP. The X-ray crystal structure of *Zymomonas mobilis* is also a tetramer (DOBRITZSCH et al., 1998), but tight packing of the subunits and the absence of appropriate cysteine residues precludes allosteric activation by pyruvate and the enzyme is locked in an activated conformation (SUN et al., 1995; FUREY et al., 1998; KONIG 1998).

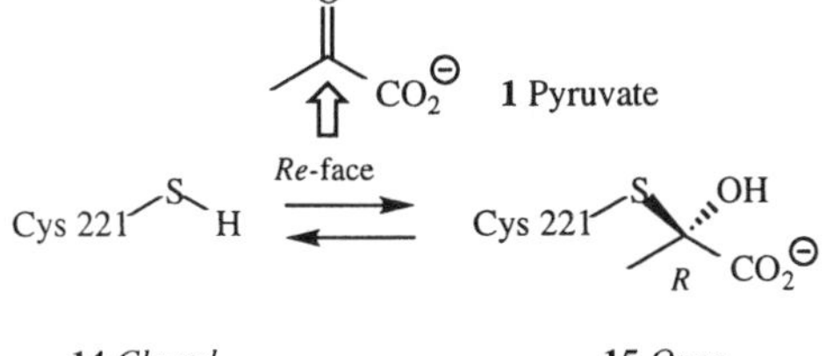

Fig. 6. It has been proposed that addition of cysteine-221 to pyruvate opens a lid which enables another molecule of pyruvate to enter the active site (ALVAREZ et al., 1995; LOBELL and CROUT, 1996b).

Biotransformations

The Knoll process for the manufacture of (*R*)-phenylacetyl carbinol (**5**) from benzaldehyde (**4**) (Fig. 7) using yeast continues to be the most efficient route to this material, nevertheless, there are still significant problems to be solved (OLIVER et al., 1997). A high concentration of benzaldehyde is essential for high productivity in batch processes, but it is toxic to yeast and reduced to benzyl alcohol (NIKOLOVA and WARD, 1991; WARD, 1995) which complicates the work-up. Increasing the supply of benzaldehyde when PDC activity is maximal and reducing the supply when alcohol dehydrogenase activity is maximal minimizes these effects (OLIVER et al., 1997). Immobilized and semi-continuous production gives similar benefits (TRIPATHI et al., 1997). Reduction of L-phenylacetyl carbinol (**6**) to (2*R*,3*S*)-phenylpropan-2,3-diol (**16**) is also a significant side reaction (MOCHIZUKI et al., 1995), although the former can be isolated by bisulfite adduct formation (SHUKLA and KULKARNI, 1999).

In principle, use of purified PDC has the potential to avoid all these problems, but thus far this has not been achieved on a practical scale, moreover benzaldehyde also inhibits PDC (CHOW et al., 1995). Partially purified PDC from *Candida utilis*, was used to produce L-phenylacetylcarbinol (**6**) at a final concentration of 28.6 g L^{-1} (190.6 mM) (SHIN and ROGERS, 1996a). When this enzyme was immobilized in spherical polyacrylamide beads, higher concentrations of benzaldehyde were tolerated (300 mM) and the half life of the enzyme was increased to 32 days (SHIN and ROGERS, 1996b). Similarly the thermal stability of brewers' yeast PDC can be improved by covalent conjugation with amylose via a glycylglycine spacer (OHBA et al., 1995).

Z. mobilis is an anaerobic gram-negative bacterium which is widely used in tropical regions for the fermentation of alcoholic beverages (DOELLE et al., 1993). Unlike most other

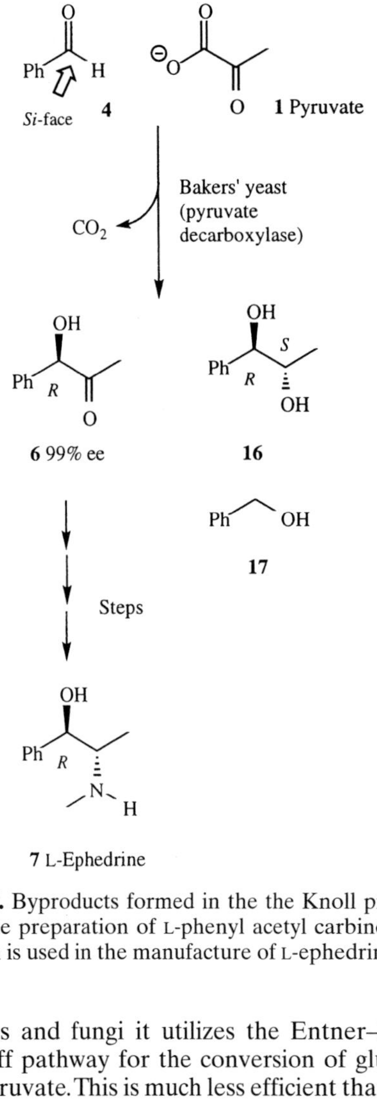

Fig. 7. Byproducts formed in the the Knoll process for the preparation of L-phenyl acetyl carbinol (**6**), which is used in the manufacture of L-ephedrine (**7**).

plants and fungi it utilizes the Entner–Doudoroff pathway for the conversion of glucose to pyruvate. This is much less efficient than glycolysis and hence large amounts of ethanol accumulate and PDC is one of the most abundant proteins produced by the organism (GUNASEKARAN and RAJ, 1999). Unlike yeast PDC, *Z. mobilis* PDC maintains the tetrameric form above pH 8, when catalytic action is lost due to dissociation of TPP and magnesium ions. It also has normal Michaelis–Menten kinetics (K_m pyruvate 0.3–4.4 mM), whereas all other PDCs to date have sigmoidal kinetics (CANDY and DUGGLEBY, 1998; POHL et al., 1995). Overall these factors render *Z. mobilis* PDC a more effective pyruvate decarboxylation catalyst (1.7–10x) than yeast PDC, particularly at low pyruvate concentrations (SUN et al., 1995).

The *E. coli* strain DH1 has been transformed with the plasmid pLOI295, harboring the gene encoding PDC from *Z. mobilis* ATCC 31821. 27% of the soluble cell protein consisted of active recombinant *Z. mobilis* PDC, which was purified to homogenerity. The enzyme was somewhat less active than yeast PDC, with benzaldehyde (Tab. 2, entry 1), and other aromatic aldehydes (Tab. 2, entries 2–9), however, the relative reactivities were similar. Heteroaromatic aldehydes (Tab. 2, entries 10–12) and propanal were poor substrates (BORNEMANN et al., 1996). Cinnamaldehydes are good substrates for yeast mediated acyloin formation. In virtually all cases the initally formed acyloins are stereospecifically reduced to the corresponding diols. These substrates are covered in detail in the chapter on C–C bond formation.

Fluoropyruvate (**20**) undergoes decarboxylation by yeast PDC to give acetate (**21**), rather than fluoroacetaldehyde (**22**) (Fig. 8). Decarboxylation proceeds normally (cf. Figs. 4, 5) to give the enol (**23**) (Fig. 9), which undergoes elimination of fluoride to give the imminium alkene (**24**), which presumably undergoes hydration (**25**) and cleavage of the C-2 thiazolium bond to give acetate (**21**) and the ylide (**9**) (GISH et al., 1988). Fluoropyruvate (**20**) also acts irreversibly at the allosteric site (SUN et al., 1997) and hence it would be of great benefit to investigate the kinetics of decarboxylation of fluoropyruvate by *Z. mobilis* PDC. The

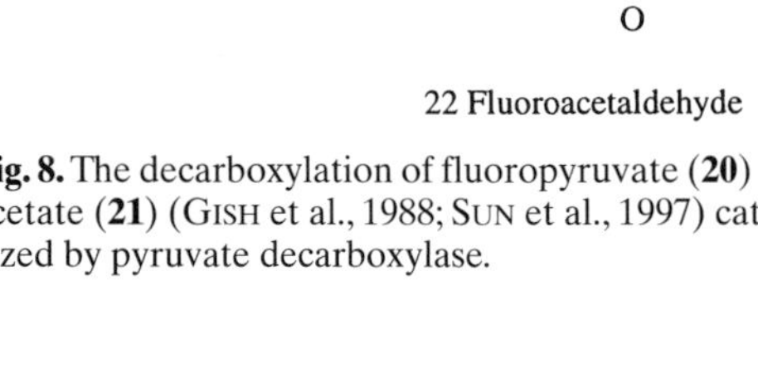

Fig. 8. The decarboxylation of fluoropyruvate (**20**) to acetate (**21**) (GISH et al., 1988; SUN et al., 1997) catalyzed by pyruvate decarboxylase.

Tab. 2. Acyloin Condensation Catalyzed by Whole Yeast, Yeast Pyruvate Decarboxlase (KREN et al., 1993) and *Z. mobilis* Pyruvate Decarboxylase (BORNEMANN et al., 1996)

18 → **19**: **1** Sodium pyruvate; Yeast or pyruvate decarboxylase

Entry	Ar	Yeast	Yeast PDC[a]		*Z. mobilis* PDC	
		% ee	Yield	% ee	Yield	% ee
1	Phenyl	97	–	99	15	98
2	2-fluorophenyl	87	60	99	60	98
3	3-fluorophenyl	95	25	99	30	97
4	4-fluorophenyl	97	31	99	35	98
5	2-chlorophenyl	81	32	98	5	98
6	3-chlorophenyl	86	25	99	20	98
7	4-chlorophenyl	77	–	99	35	98
8	2,3-difluorophenyl	97	40	99	–	–
9	2,6-difluorophenyl	87	–	92	–	–
10	2-furyl	–	–	–	low	80-90
11	3-furyl	–	–	–	low	80-90
12	3-thienyl	–	–	–	0	–

practical impact of this result is that pyruvates bearing potential nucleofuges are unsuitable substrates for alkylation.

4.1.1.2 Oxalate Decarboxylase (Oxalate Carboxy-Lyase, EC 4.1.1.2)

Oxalate decarboxylase catalyzes the decarboxylation of oxalate (**26**) to formate (**27**) (Fig. 10). Oxalate occurs as a byproduct of photorespiration in plants (spinach is a rich source) and is produced by many microorganisms and fungi. Several species of *Penicillium* and *Aspergillus* convert sugars into oxalic acid in high yields. Hyperoxaluria is a major risk factor in human urinary stone disease (SHARMA et al., 1993). The enzyme is readily available and is used in kits for determining oxalate in urine (SANTA-MARIA et al., 1993). The biotechnological potential of this enzyme remains to be exploited, however, differential binding of carboxylate substituted biomimetic dyes indicates that substrates other than oxalate may be accepted (LABROU and CLONIS, 1995). Further aspects of oxalate decarboxylation are covered in Sect. 4.1.1.8, Oxalyl-CoA Decarboxylase.

4.1.1.3 Oxaloacetate Decarboxylase (Oxaloacetate Carboxy-Lyase, EC 4.1.1.3)

Oxaloacetate plays a central role in primary metabolism as an intermediate in the citric acid (Krebs) cycle (Fig. 1) and as the "starting material" for gluconeogenesis. All amino acids except for leucine and lysine can be converted to oxaloacetate. The majority are converted to oxaloacetate via citric acid cycle intermediates. Aspartate and asparagine are converted via other pathways and the remainder are converted via pyruvate and carboxylation to oxa-

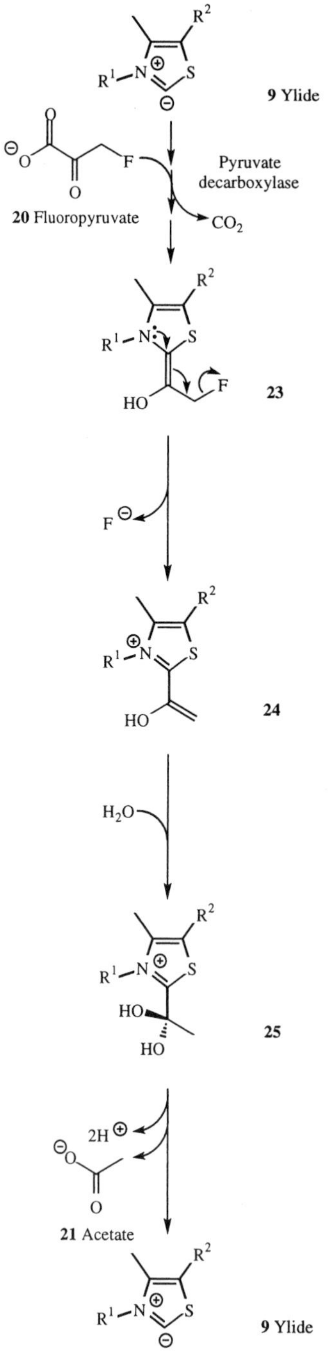

Fig. 9. The mechanism for the decarboxylation of fluoropyruvate (**20**) to acetate (**21**) by pyruvate decarboxylase (GISH et al., 1988; SUN et al., 1997); cf. Figs. 3, 4, 5 for R^1 and R^2 the sequence **11** ∅ **12** ∅ **9**.

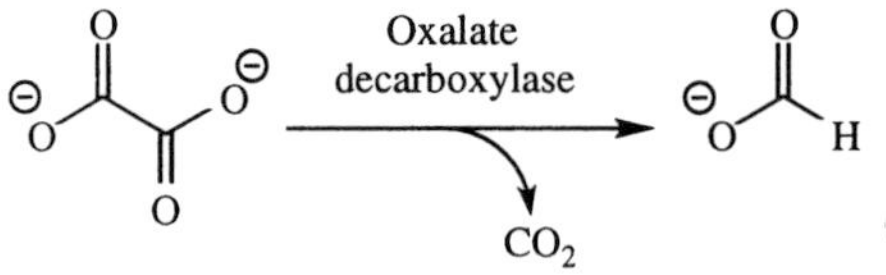

Fig. 10. The decarboxylation of oxalate (**26**) to formate (**27**) catalyzed by oxalate decarboxylase.

loacetate. There are two classes of enzymes which are capable of the interconversion of carbon dioxide and pyruvate (**1**) with oxaloacetate (**28**) (Fig. 11). Pyruvate carboxylases (EC 6.4.1.1) are ligases and hence carboxylation (which is thermodynamically disfavored), is coupled to the hydrolysis of ATP to ADP. Whereas oxaloacetate decarboxylases (OAD, oxaloacetate carboxy-lyase, EC 4.1.1.3), catalyzes the reverse and thermodynamically more favorable reaction. The latter enzymes are not able to catalyze the net carboxylation of pyruvate.

There are four classes of enzymes that catalyze the decarboxylation of oxaloacetate. The most important are the OADs that require magnesium(II) or manganese(II) ions and are consequently sensitive to EDTA (e.g., *Micrococcus lysodeikticus* OAD et al., 1999; *Corynebacterium glutamicum* OAD et al., 1995) and the biotinylated protein (TOH et al., 1993) located on the cytoplasmic membrane, which

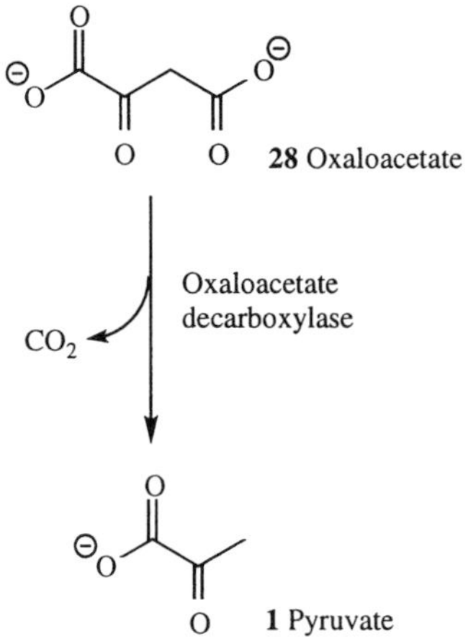

Fig. 11. The decarboxylation of oxaloacetate (**28**) to pyruvate (**1**) catalyzed by oxaloacetate decarboxylase.

acts as a sodium ion pump (DIMROTH, 1994; DIMROTH and THOMER, 1993). In addition L-malate dehydrogenase (L-malic enzyme, EC 1.1.1.38; EDENS et al., 1997; SPAMINATO and ANDREO, 1995) and pyruvate kinase in muscle, both have oxaloacetate decarboxylating activity, but oxaloacetate is not a substrate *in vivo* (KRAUTWURST et al., 1998).

The mechanism of the OAD catalyzed decarboxylation of oxaloacetate (**28**) involves the formation of a magnesium(II) or manganese(II) α-ketoacid complex (**29**) (Fig. 12), which undergoes retro-aldol cleavage to give the magnesium enolate (**30**) plus carbon dioxide. This mechanism is quite different to that of other β-ketoacid decarboxylases (e.g., acetoacetate decarboxylase) which involve imine formation by the ketone to initiate decarboxylation. Imine based OADs, have been prepared by rationale peptide design, but the rates of reaction are much slower than those of comparable enzymes (ALLERT et al., 1998; JOHNSSON et al., 1993).

Oxaloacetate (**28**) enolizes rapidly in solution and hence special precautions are required to elucidate the stereochemistry of decarboxylation. This was achieved in a remarkable sequence catalyzed by five enzymes. Incubation of (2*S*)-[2-^{3}H,3-^{3}H$_2$]-aspartic acid (**31**) with L-aspartase in deuterium oxide gave stereospecifically labeled (2*S*,3*S*)-[2-^{3}H,3-^{2}H,^{3}H]-aspartic acid (**32**) via fumarate (Fig. 13). Transamination yielded (3*S*)-[3-^{2}H,^{3}H]oxaloacetate (**33**), which was immediately decarboxylated *in situ* with the OAD from *Pseudomonas putida* to give (3*S*)-[3-^{1}H,^{2}H,^{3}H]-pyruvate (**34**). To prevent enolization and loss of the radiolabels, this was reduced with L-lactate dehydrogenase to (2*S*,3*S*)-[3-^{1}H,^{2}H,^{3}H]-lactate (**35**), with which was isolated by anion exchange chromatography.

Oxidative decarboxylation with L-lactate oxidase gave (2*S*)-[2-^{1}H,^{2}H,^{3}H]-acetate (**36**) (Fig. 14). The chirality of which was determined by CORNFORTH's method (CORNFORTH et al., 1969; LUTHY et al., 1969), which utilizes a further four enzymes. Decarboxylation by OAD proceeds with inversion of stereochemistry, whereas the corresponding reaction with the biotin dependent pyruvate carboxylases and malic enzyme all proceed with retention of configuration (PICCIRILLI et al., 1987).

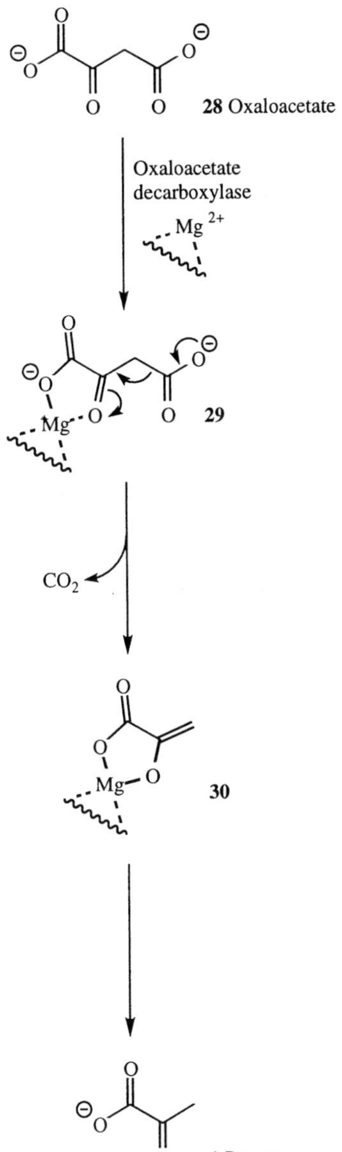

Fig. 12. Mechanism for the decarboxylation of oxaloacetate (**28**) to pyruvate (**1**) catalyzed by oxaloacetate decarboxylase. Magnesium(II) bound to OAD is shown without ligands for the purposes of simplicity only. It is most likely that water or carboxylate ligands are displaced when oxaloacetate is bound.

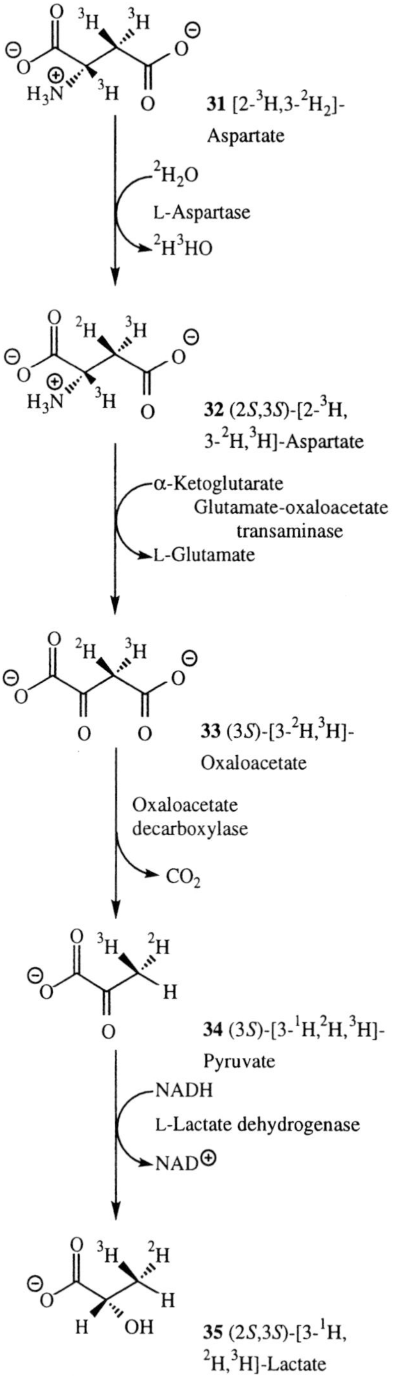

31 [2-^{3}H,3-^{2}H$_2$]-Aspartate

^2H_2O
L-Aspartase
$^2H^3HO$

32 (2*S*,3*S*)-[2-^{3}H,3-^{2}H,^{3}H]-Aspartate

α-Ketoglutarate
Glutamate-oxaloacetate transaminase
L-Glutamate

33 (3*S*)-[3-^{2}H,^{3}H]-Oxaloacetate

Oxaloacetate decarboxylase
CO_2

34 (3*S*)-[3-^{1}H,^{2}H,^{3}H]-Pyruvate

NADH
L-Lactate dehydrogenase
$NAD^{\oplus}$

35 (2*S*,3*S*)-[3-^{1}H,^{2}H,^{3}H]-Lactate

Fig. 13. Stereoselectivity of the decarboxylation of labeled oxaloacetate (**33**) catalyzed by *Pseudomonas putida* OAD.

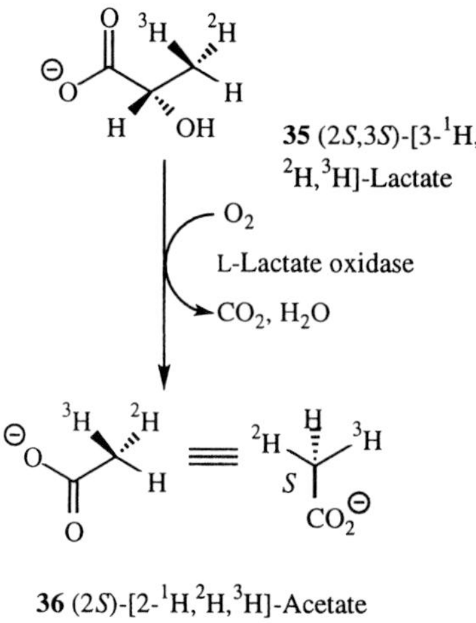

35 (2*S*,3*S*)-[3-^{1}H,^{2}H,^{3}H]-Lactate

O_2
L-Lactate oxidase
CO_2, H_2O

36 (2*S*)-[2-^{1}H,^{2}H,^{3}H]-Acetate

Fig. 14. Oxidative decarboxylation of lactate (**35**) yields acetate (**36**) of known stereochemistry.

4.1.1.4 Acetoacetate Decarboxylase (Acetoacetate Carboxy-Lyase, EC 4.1.1.4)

Acetoacetate decarboxylase (AAD) catalyzes the decarboxylation of acetoacetate to acetone, a process which parallels the abiotic decarboxylation of acetoacetic acid (**37a**) (Fig. 15). The salts and esters of β-ketoacids are generally stable to decarboxylation, whereas the corresponding acids decarboxylate spontaneously. Consequently, this reaction is most often employed during the acid catalyzed hydrolysis of β-ketoesters (**37c**).The mechanism of the abiotic reaction is fairly well understood (HUANG et al., 1997; BACH and CANEPA, 1996). Intramolecular proton transfer yields the enolic form of the ketone (**38a**), plus carbon dioxide, in a process which is in effect a retro-aldol reaction. The obligatory nature of the proton transfer can be inferred from the corresponding reaction of trimethylsilyl β-ketoesters (**37b**), which gives isolable trimethysilyl enol ethers (**38b**). It might be anticipated that the enzyme catalyzed process would simply involve protonation of the β-ketoacid salts which are the predominant form at physiological pH. However, early experiments, demonstrated that exchange of the ketonic oxygen was an obligatory step in the decarboxylation of acetoacetate by AAD (HAMILTON and WESTHEIMER, 1959), which is not consistant with a simple protonation mechanism. It is now well established, that reaction of acetoacetate (**40**) (Fig. 16) with a lysine residue of AAD (**41**), gives a protonated imine (Schiff's base) derivative (**42**), which undergoes decarboxylation to give the enamine (**43**), which in turn is hydrolyzed to acetone (**39**). The pH optimum for *Clostridium acetobutylicum* AAD is 5.95, whereas the pK_a for the ε-amino group of lysine in solution is 10.5, hence the effective pK_a at the active site must be reduced by a factor of at least 20,000 to enable the ε-amino group of lysine to act as an effective nucleophile (WESTHEIMER, 1995). *Clostridium acetobutylicum* AAD bears adjacent lysine residues at the active site. Lysine-115 acts as the nucleophile, whereas the ε-ammonium group of lysine-116 serves to reduce the pK_a of the adjacent residue by electrostatic effects (HIGHBARGER et al., 1996). This reduction in pK_a has also been exploited in the design of 1,3-diaminonorbornane decarboxylation catalysts (OGINO et al., 1996). Catalytic antibodies which are capable of catalysing aldol reactions and decarboxylation, also bear deeply buried lysine residues which are essential for catalytic activity (BARBAS III et al., 1997; BJORNESTEDT et al., 1996)

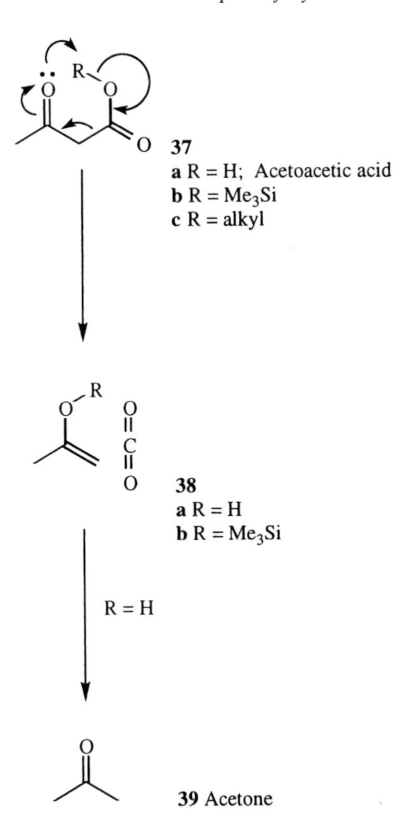

Fig. 15. Abiotic mechanism for the decarboxylation of acetoacetic acid (**37a**).

The decarboxylation of (2*R*) and (2*S*)-[2-^{3}H]-acetoacetate in deuterium oxide catalyzed by AAD gives racemic [2-^{1}H,^{2}H,^{3}H]-acetate. A series of control experiments demonstrated unambigously that this was a consequence of the decarboxylation step and was not due to hydrogen exchange reactions with the solvent (ROZZELL Jr and BENNER, 1984). In contrast the homolog, racemic 2-methyl-3-oxobutyrate (**44**) (**45**) (Fig. 17) is decarboxylated enantioselectively. Treatment of the crude reaction mixture with sodium borohydride gives a mixture of diastereomeric alcohols (**46**) (**48**) which both have the (2*S*)-configuration and when the reaction is run in deuterium oxide, (3*R*)-[3-^{2}H]-2-butanone (**47**) is produced, indicating that decarboxylation occurs with retention of stereochemistry (BENNER et al., 1981). Similarly, racemic 2-oxocyclohexanecarboxylate (**49**) (**50**) (Fig. 18) undergoes enantioselective decarboxylation as demonstrated by recovery of the (1*S*)-diastereomeric alcohols (**51**) (**53**) after sodium cyanoborohydride reduction. Decarboxylation of the deuterio-derivative (**54**) in water gives (2*S*)-[2-^{2}H]-cyclohexanone (**55**) (Fig. 19), again with retention of stereochemistry (BENNER and MORTON, 1981). The biotransformations potential of AAD has thus far only been investigated with a few "unnatural substrates". The ability to decarboxylate substrates such as 2-oxocyclohexanecarboxylate (**49**) (**50**), which are structurally far removed from the natural substrate (**40**) suggests it may be applicable to a diverse range of β-ketoacids.

In the first half of the twentieth century, acetone was manufactured by fermentation, but this technology was supplanted in the 1950s by more cost effective abiotic processes based on petrochemicals. With increasing concerns about enviromental impacts and the need for

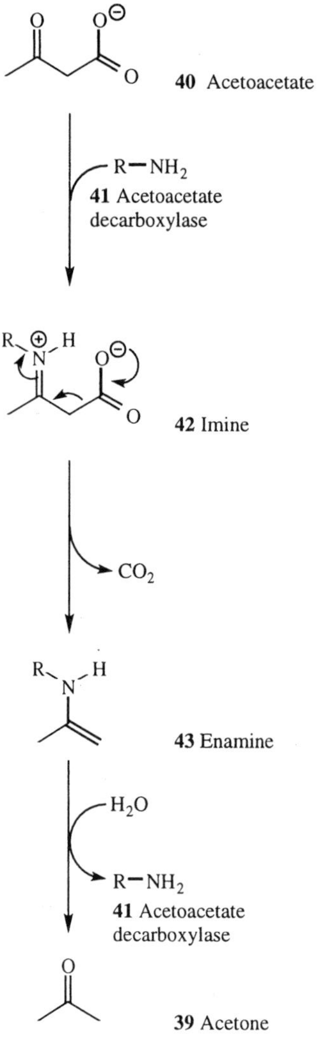

Fig. 16. Mechanism for the decarboxylation of acetoacetate (**40**) by acetoacetate decarboxylase.

the development of renewable resources, the fermentation process is now being reexamined (Reviews: DUERRE, 1998; GIRBAL and SOUCAILLE, 1998; SANTANGELO and DUERRE, 1996; WOODS, 1995; DUERRE et al., 1992). The original organism used for the starch (corn mash) based fermentation was *Clostridium acetobutylicum*, however, later strains were grown on sucrose (molasses) and this has led to some genetic drift such that 39 cultures from various collections can now be divided into four

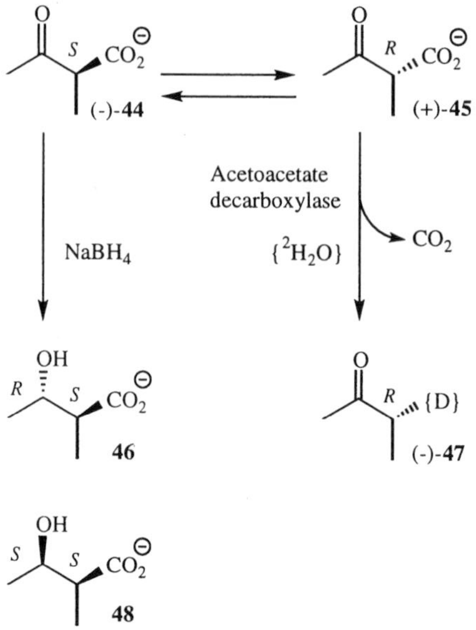

Fig. 17. Enantioselective decarboxylation of racemic 2-methyl-3-oxobutyrate (**44**) (**45**).

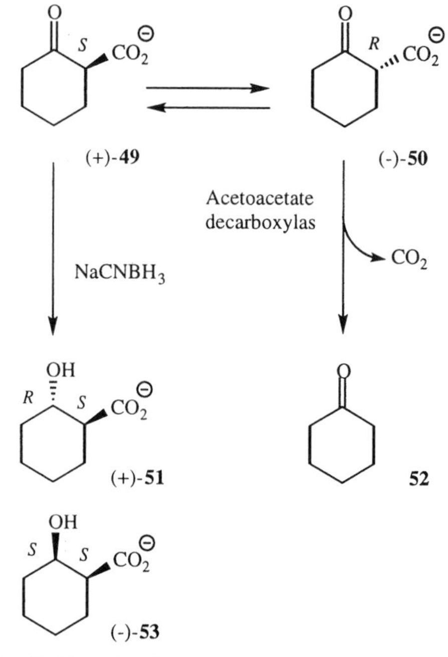

Fig. 18. Enantioselective decarboxylation of racemic 2-oxocyclohexanecarboxylate (**49**) (**50**).

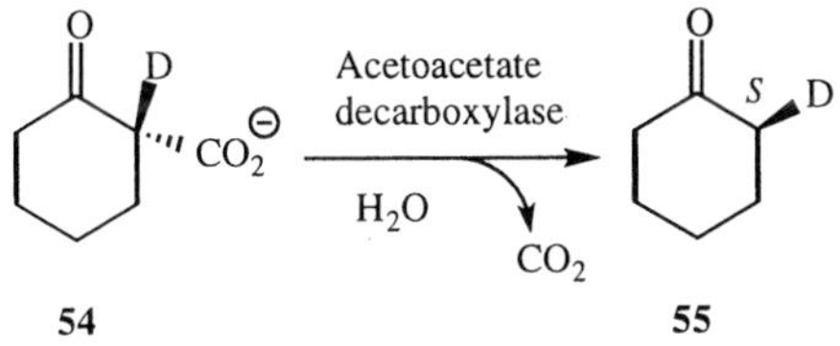

Fig. 19. Decarboxylation of the deuterio-derivative (**54**) in water gives (2*S*)-[2-^{2}H]-cyclohexanone (**55**) with retention of stereochemistry.

groups based on DNA–DNA reassociation (JONES and KEIS, 1995; JOHNSON and CHEN, 1995; JOHNSON et al., 1997; TOTH et al., 1999). The two most important groups are those which are related to *C. acetobutylicum*, and *C. beijerincki*; in addition two cultures (NRRL B643 and NCP 262) have 94% relatedness and *C. saccharoperbutylacetonicum* forms a separate group on its own (JOHNSON and CHEN 1995).

The so called "solvent producing microorganisms" yield a mixture of acetone (**39**), butanol (**63**), plus less valuable metabolites such as ethanol, butanoate (**61**) and isopropanol (**64**)), which are produced by a sequence of steps which resemble fatty acid metabolism (Fig. 20; BENNETT and RUDOLPH, 1995). The early steps in the sequence appear to be fairly similar in all organisms examined thus far (STIM-HERNDON et al., 1995), however, a variety of aldehyde and alcohol dehydrogenases are employed (CHEN, 1995; REID and FEWSON, 1994).

In *C. acetobutylicum* ATCC-824, the adc operon (encoding acetoacetate decarboxylase; PETERSEN and BENNETT, 1990; GERISCHER and DUERRE, 1990), is contiguous with structural genes for acetoacetyl-coenzyme A:acetate/butyrate:coenzyme A transferase (ctfB and ctfA), an alcohol/aldehyde dehydrogenase (aad) (FISCHER et al., 1993; PETERSEN et al., 1993), plus a repressor (solR) (NAIR et al., 1999). These four genes form the *sol* operon, which is induced or derepressed before solventogenesis (SAUER and DUERRE, 1995) and does not operate in strains which are unable to produce "solvents" (STIM-HERNDON et al., 1996). Two putative promoter sites are located upstream of the sol operon (FISCHER et al., 1993; DUERRE et al., 1995).

The primary limitation in the production of "solvents" by *C. acetobutylicum* and related species is product inhibition (acetate, butyrate and butanol), this is multifactorial and synergistic. Acetone and ethanol are not inhibitors except at higher levels (YANG and TSAO, 1994). However, the addition of acetate to MP2 media increases "solvent" production and induces the expression of the *sol* operon in *C. beijerincki* (CHEN and BLASCHEK, 1999a, b)

In the initial exponentional growth phase of the fermentation, acetate and butyrate production predominates and without external control the pH typically drops to 4.3. The fermentation shifts to butanol, ethanol and acetone production, shortly before the stationary phase (GIRBAL et al, 1995; GRUPE and GOTTSCHALK, 1992). Both nitrogen and phosphate limited fermentation favor acetate and butyrate formation, but not "solvent" formation. A two stage continuous fermentation, with a nitrogen limited first stage fermentation maximized acid production, which was followed by a second fermentation with supplemental glucose and nitrogen sources which maximized butanol production (0.4 g L^{-1} h^{-1}; LAI and TRAXLER 1994). Acetone production without concommitant butanol production is an important objective. Plasmid pFNK6 contains a synthetic operon (ace) in which the three homologous acetone formation genes (adc, ctfA and ctfB) are transcribed from the adc promoter. *C. acetobutylicum* ATCC 824 (pFNK6) gave the highest level of expression of AAD and CoA-transferase at pH 4.5, but the highest levels of "solvents" were produced at pH 5.5. Acetone, butanol and ethanol were produced at 95%, 37% and 90% higher concentrations those produced by the plasmid free strain (MERMELSTEIN et al., 1993). In the first study of the expression of clostridial genes in a nonclostridial host, a synthetic operon (ace4) composed of four *C. acetobutylicum* ATCC 824 genes (adc, ctfA, ctfB and th1) under the control of the th1 promoter was constructed and introduced into three strains of *E. coli* on the vector pACT. *E. coli* ER2275(pACT) and ATCC 11303(pACT) both produced 40mM acetone in shake cultures and addition of acetate and glucose further increased the titres. In a bioreactor *E. coli* ATCC 11303(pACT) at pH 4.5, 125–154 mM acetone accumulated, which is equal to or higher than that achieved by wild *C. acetobutylicum* (BERMEJO et al., 1998).

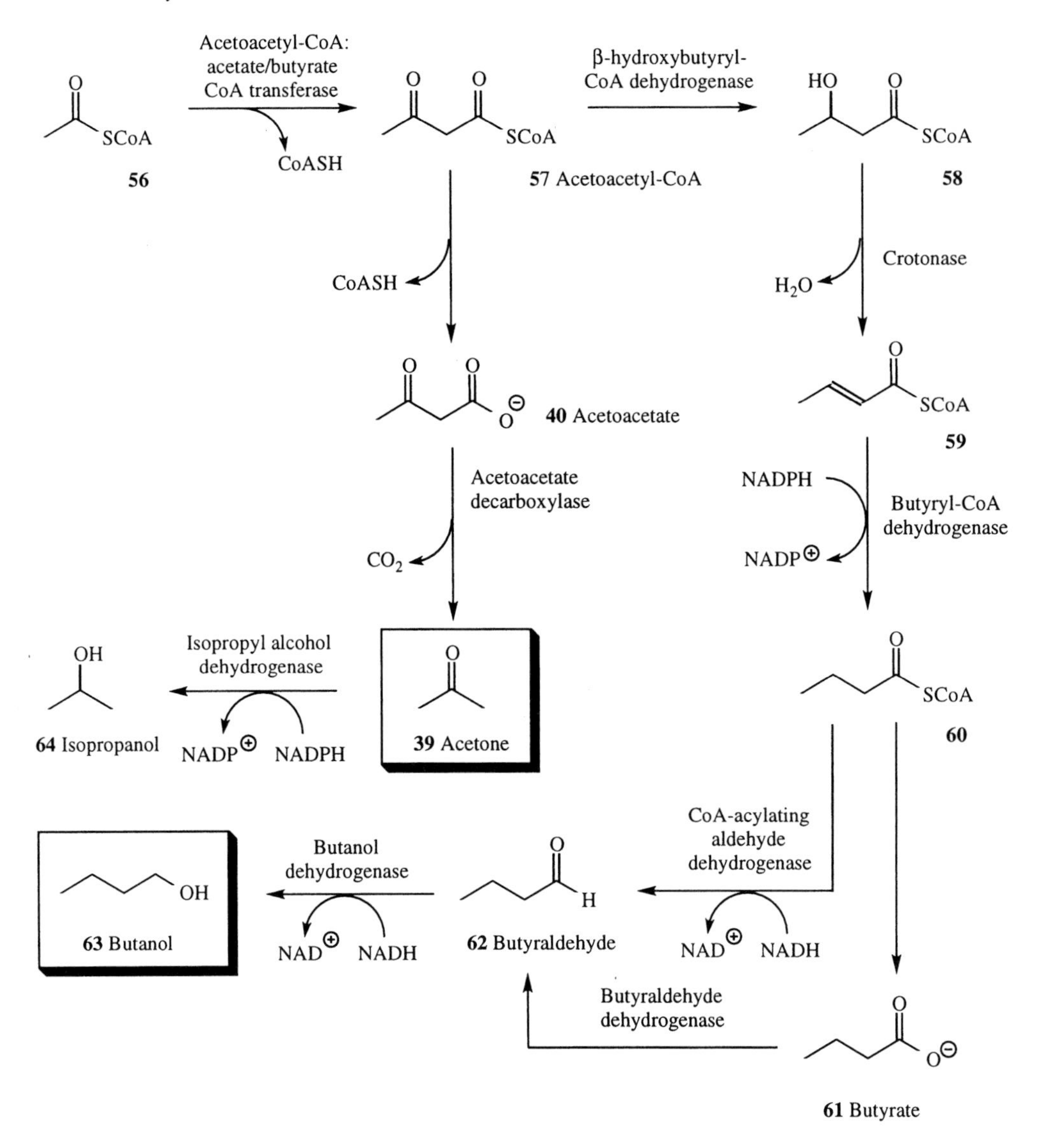

Fig. 20. The production of acetone (**39**) and butanol (**63**) by *C. acetobutylicum* and related species. Isopropanol (**64**) is most frequently formed by *C. beijerincki* (CoASH = coenzyme A).

4.1.1.5 Acetolactate Decarboxylase (2-Hydroxy-2-Methyl-3-Oxobutyrate Carboxyl-Lyase, EC 4.1.1.5)

Acetolactate decarboxylases from *Klebsiella aerogenes* (formerly *Aerobacter aerogenes*) catalyzes the decarboxylation of (*S*)-α-acetolactate (**65**) to give (*R*)-acetoin (**5**) (Fig. 21). (2*S*)-α-Acetolactate (**65**) is decarboxylated with inversion of configuration (CROUT et al., 1984; ARMSTRONG et al., 1983) and the carboxylate group is substituted by a proton from the solvent (DRAKE et al., 1987). Surprisingly, (2*R*)-α-acetolactate (**66**) is also decarboxylated to (*R*)-acetoin (**5**), but at a slower rate (CROUT et al., 1986).

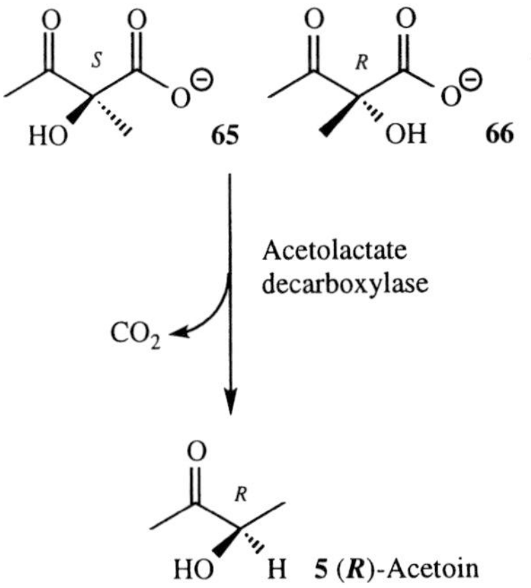

Fig. 21. Decarboxylation of α-acetolactate [(2-hydroxy-2-methyl-3-oxobutanoate] (**65**) (**66**) to (*R*)-acetoin (3-hydroxy-2-butanone) (**5**), catalyzed by acetolactate decarboxylase.

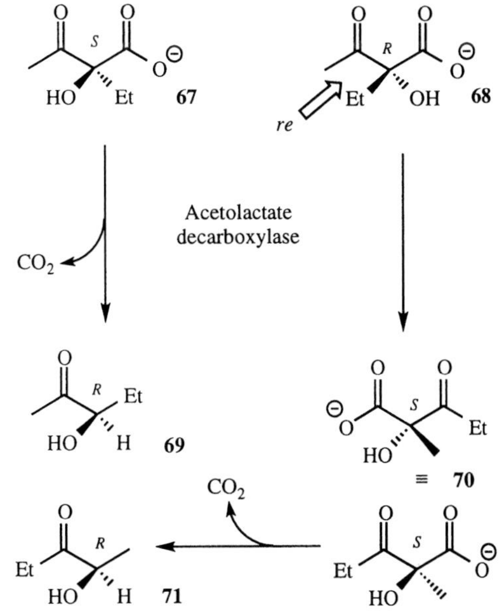

Fig. 22. Decarboxylation of (2*S*)-a-acetohydroxybutyrate [(2*S*)-2-ethyl-2-hydroxy-3-oxobutanoate] (**67**), and tertiary ketol rearrangement of (2*R*)-α-acetohydroxybutyrate (**68**), to (2*S*)-2-hydroxy-2-methyl-3-oxopentanoate (**70**) and decarboxylation catalyzed by acetolactate decarboxylase.

A clue to the mechanism was provided by the decarboxylation of racemic α-acetohydroxybutyrate (**67**) (**68**) with acetolactate decarboxylase from a *Bacillus* sp. (Fig. 22). The (*S*)-enantiomer (**67**) was decarboxylated rapidly to give the ketol (**69**), which was followed by slow decarboxylation of the (*R*)-enantiomer (**68**) to give the isomeric ketol (**71**). Both ketols were formed with high enantiomeric excesses. This apparently anomolous result can be rationalized by a stereospecific enzyme catalyzed tertiary ketol rearrangement of (**68**) with transfer of the carboxylate group to the *re*-face of the ketone to give the α-propiolactate (**70**). This compound which is a homolog of (*S*)-α-acetolactate (**65**) then undergoes decarboxylation by the "normal" pathway. When (*R*)-α-acetolactate (**66**) is subjected to the tertiary ketol rearrangement, the enantiomeric (*S*)-α-acetolactate (**65**) is formed and hence this also explains the ability of this enzyme to transform both enantiomers of α-acetolactate (CROUT and RATHBONE, 1988). An abiotic base catalyzed tertiary ketol rearrangement (CROUT and HEDGECOCK, 1979) was excluded by control experiments.

α-Acetolactate (**65**) is produced biosynthetically by the condensation of two molecules of pyruvate (**1**) with concomittant decarboxylation, catalyzed by acetolactate synthase [acetolactate pyruvate-lyase (carboxylating); EC 4.1.3.18]. The enzyme is TPP dependent and hence the mechanism is very similar to that of pyruvate decarboxylase described in Sect. 4.1.1.1 (CARROLL et al., 1999; CHIPMAN et al., 1998; HILL et al., 1997) (Fig. 23). α-Acetolactate (**65**) is the biological precursor of diacetyl (**72**), which is a desirable aroma in some fermented diary products, but an off-flavor in beer. Acetoin has no flavor impact on beer (HUGENHOLTZ, 1993). The conversion of α-acetolactate (**65**) to diacetyl (**72**) is an oxidative decarboxylation. Plausibly the anionic intermediate of decarboxylation of α-acetolactate (**65**) is intercepted by oxygen (BOUMERDASSI et al., 1996) or peroxide acting as electrophiles. This process apparently does not occur during decarboxylation by acetolactate decarboxylase. α-Acetolactate (**65**) undoubtedly undergoes spontaneous, extracellular oxidative decarboxylation (SEREBRENNIKOV et al., 1999), but the evidence is accumulating for a parallel intracellular enzyme (RONDAGS et al., 1997, 1998; JORDAN et al., 1996) or quinone cata-

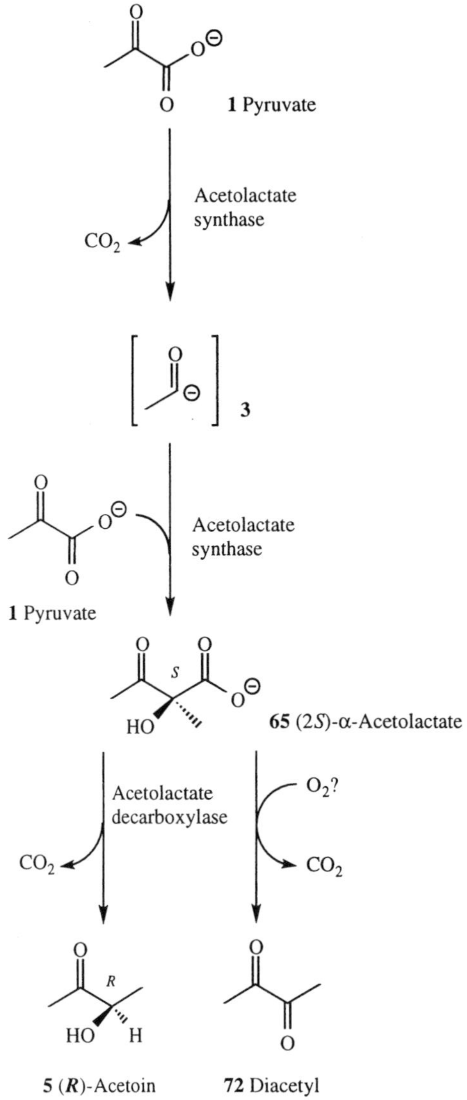

Fig. 23. The biosynthesis of (2*S*)-α-acetolactate [(*S*)-2-hydroxy-2-methyl-3-oxobutanoate] (**65**) from pyruvate (**1**), catalyzed by acetolactate synthase. Non-oxidative decarboxylation to give (*R*)-acetoin (**5**), is catalyzed by acetolactate decarboxylase. Oxidative decarboxylation to diacetyl (**72**) occurs spontaneously, but may also be catalyzed by an unidentified

lyzed process (PARK et al., 1995). Conversion of acetoin to diacetyl or *vice versa* does not seem to be a significant pathway in most microorganisms.

The conversion of α-acetolactate (**65**) to acetoin (**5**) reduces the "pool" of α-acetolactate, available for conversion to diacetyl, hence by regulating the levels of acetolactate decarboxylase, it should be possible to control the levels of diacetyl (RENAULT et al., 1998; BENITEZ et al., 1996; KOK, 1996). However, selection for *Lactobacillus rhamnosus* mutants with low levels of acetolactate decarboxylase activity had no affect on diacetyl levels (MONNET and CORRIEU, 1998; PHALIP et al., 1994). Acetolactate accumulates, but is not converted to diacetyl unless the redox potential is high (MONNET et al., 1994). In contrast incorporation of the acetolactate decarboxylase gene from *Acetobacter aceti* into the genome of brewers' yeast successfully reduced the levels of diacetyl (TAKAHASHI et al., 1995; YAMANO et al., 1994, 1995). Moreover the free enzyme from *Bacillus subtilis* containing the structural gene for ALDC production originating from a *Bacillus brevis* strain is now used directly in beer brewing to reduce diacetyl levels (DE BOER et al., 1993; GERMAN, 1992).

There is considerable unrealized potential for the use of both acetolactate synthase and acetolactate decarboxylase in synthesis (CROUT et al., 1994)

4.1.1.6 Aconitate Decarboxylase (cis-Aconitate Carboxy-Lyase, EC 4.1.1.6)

Aconitate decarboxylase from *Aspergillus terreus* catalyzes the decarboxylation of *cis*-aconitate (**73**) to itaconate (**74**) (Fig. 24). There has been considerable interest in the production of itaconate (**74**) by various strains of *A. terreus* for use as a monomer for plastics manufacture. However, the enzymology has largely been neglected, because of the instability of the enzyme (BENTLEY and THIESSEN, 1957). The generally accepted mechanism for the decarboxylation involves an allylic shift of the alkene bond with protonation on the (2-*re*,3-*se*) face of *cis*-aconitate (**73**) (RANZI et al., 1981). However, recent work in which the enzyme was immobilized in a silicate matrix, has indicated that the initial decarboxylation yields citraconate (**75**), which is subsequently isomer-

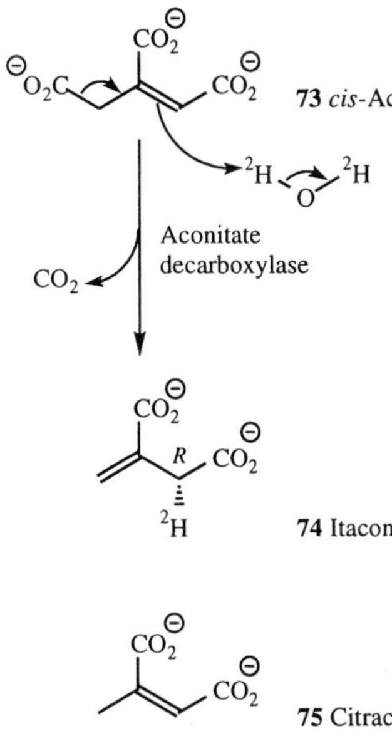

73 *cis*-Aconitate

Aconitate decarboxylase

74 Itaconate

75 Citraconate

Fig. 24. Decarboxylation of cis-aconitate (**73**) to itaconate (**74**) catalyzed by aconitate decarboxylase.

ized to itaconte (**74**) by a previously unknown enzyme (BRESSLER and BRAUN, 1996). This mechanistically unlikely result remains to be confirmed.

cis-Aconitate is a citric acid (Krebs) cycle intermediate (Fig. 1) which is produced predominantly in the mitochondria, whereas aconitate decarboxylase is located exclusively in the cytosol (JAKLITSCH et al., 1991). Much work has been devoted to increasing the productivity and efficiency of production of itaconate (**74**) from carbohydrate foodstocks (BONNARME et al., 1995). Selection on itaconic acid concentration gradient plates yielded *A. terreus* TN-484, which gave 82 g L^{-1} itaconic acid from 160 g L^{-1} glucose (51% efficiency) in a shaking flask (YAHIRO et al., 1995). Although this is a good level of productivity and efficency, the challenge is to achieve the same results with cheaper foodstocks. Corn starch gave the best results for a range of starchy materials with *A. terreus* NRRL 1960, but the productivity was only 18 g L^{-1} and the efficiency 34% (PETRUCCIOLI et al., 1999). Pretreatment of the the corn starch with nitric acid enabled more than 60 g L^{-1} itaconic acid to be produced from 140 g L^{-1} corn starch (50% efficiency) using *A. terreus* TN-484; a result which is almost equivalent to that achieved with glucose (YAHIRO et al., 1997). Protoplast fusion between *A. terreus* and *A. usamii* (a glucoamylase producer), produced fusant mutants which were subcultured to confirm stability. The F-112 mutant maximally produced 35.9 g L^{-1} of itaconic acid from 120 g L^{-1} soluble starch (KIRIMURA et al., 1997). The other major factor in itaconate production is aeration. Interuption of aeration for five minutes caused the cessation of itaconate production and this was only restored after a further 24 hours aeration. Moreover when protein synthesis was also inhibited after interuption of aeration no further itaconate production occurred (GYAMERAH et al., 1995; BONNARME et al., 1995). Given these results it is not surprising that air lift reactors have been favored for large scale conversions (PARK et al., 1994; OKABE et al., 1993) On a small scale a 10 mm air lift reactor has been monitored directly by ^{31}P-NMR at 161 MHz (LYNSTRAD and GRASDALEN 1993).

4.1.1.7 Benzoylformate Decarboxylase (Benzoylformate Carboxy-Lyase, EC 4.1.1.7)

Benzoylformate decarboxylase (BFD) is a rare enzyme which catalyzes the decarboxylation of benzoylformate (phenyl glyoxalate; **78**) to benzaldehyde (**4**) (Fig. 25). The most intensely investigated example is the BFD from the mandelate pathway of *Pseudomonas putida*.

Phenylacetate (**76**) is a common intermediate in the microbial metabolism of various aromatic compounds, including phenylalanine. In *Pseudomonas putida*, phenylacetate is degraded by benzylic hydroxylation to mandelate (**77**), oxidation to benzoylformate (**78**) (LEHOU and MITRA, 1999), decarboxylation to benzaldehyde (**4**) catalyzed by BFD and oxidation to give benzoate (**79**), which is oxidized further by ring fission(TSOU et al., 1990). Some other organisms utilize nominally similar pathways, but with a different complement of enzymes. For example, the denitrifying bacterium *Thauera aromatica* utilizes a membrane bound

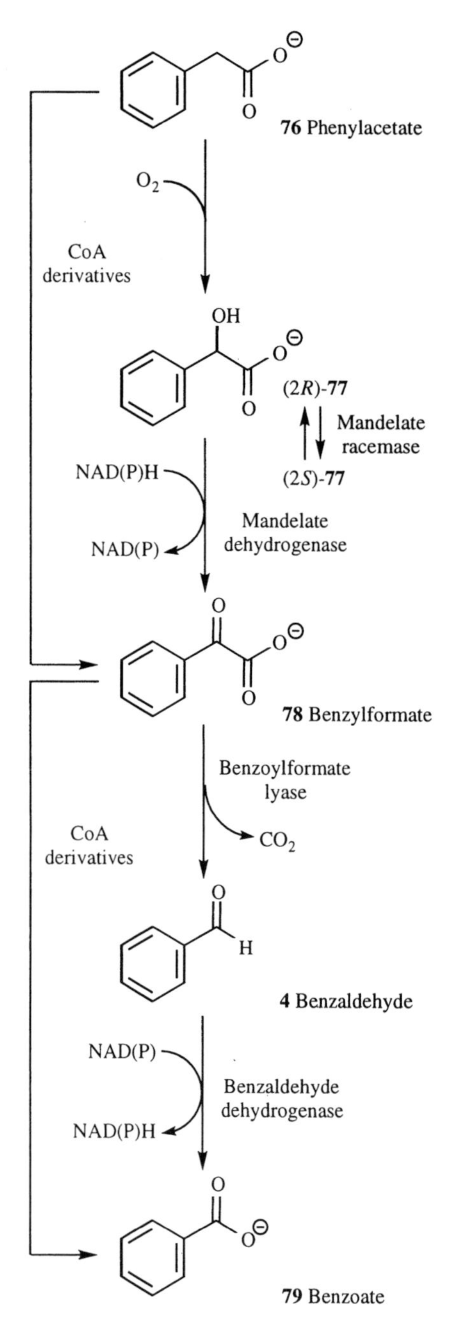

Fig. 25. Generalized pathway for the catabolism of phenylacetate (**76**).

molybdenium–iron–sulfur protein, which converts phenylacetyl-CoA directly to benzylformate (**78**) in the absence of dioxygen (RHEE and FUCHS, 1999). Another denitrifying bacterium *Azocarus evensii*, catalyzes the oxidative decarboxylation of benzylformate (**78**) to benzoyl-CoA with a six subunit iron–sulfur enzyme (HIRSCH et al., 1998).

BFD from *Pseudomonas putida* is a tetrameric TPP (thiamine pyrophosphate) dependent decarboxylase composed of 57 kDa monomers (HASSON et al., 1995, 1998). As with other TPP dependent decarboxylases the intermediate enamine can be intercepted by exogenous aldehydes to give acyloins (cf. pyruvate decarboxylase, Sect. 4.1.1.1). The acyloin condensation with acetaldehyde (**2**) gives (*S*)-(–)-2-hydroxypropiophenone (**80**) with up to 92% enantiomeric excess (Fig. 26). Thus far, only *p*-substituted benzylformates have been shown to act as substrates and pyruvate; α-ketoglutarate and α-ketobutryrate are not substrates (REYNOLDS et al., 1988; WEISS et al., 1998). The fairly limited synthetic applications of BFD have been reviewed (IDING et al., 1998; SCHORKEN and SPRENGER, 1998; SPRENGER and POHL, 1999). [*p*-(Bromomethyl)benzoyl]formate (**81a**) reacts with BFD at about 1% of the rate of benzylformate (**78**) and causes inhibition which can be reversed by the addition of TPP. The "enamine" (**82**) which is formed by the "normal" mechanistic sequence, undergoes elimination of bromide to give a transient quinone-dimethide (**83**), which tautomerizes to the acyl derivative and undergoes slow hydrolysis via the hydrate (**84**) to give *p*-methylbenzoate (**85**) (Fig. 27). This mechanism of this reaction parallels the conversion of 2-fluoropyruvate (**20**) to acetate (**21**) (Fig. 9) by pyruvate decarboxylase, however, [*p*-(fluoromethyl)benzoyl]formate (**81b**) undergoes decarboxylation without elimination of fluoride, which suggests that the intervening phenyl group imposes an energy barrier which impedes elimination (DIRMAIER et al., 1986; REYNOLDS et al., 1998).

The production of (*S*)-(–)-2-hydroxypropiophenone from benzylformate (**78**) and acetaldehyde (**2**) by *Acinetobacter calcoaceticus*, has been optimized at pH 6.0 and 30 °C. Maximal productivity was 8.4 g L^{-1} after two hours. In one hour, 6.95 g L^{-1} of (*S*)-(–)-2-hydroxypropiophenone (**80**) was formed which corresponds to a productivity of 267 mg per g of dry cells per hour (PROSEN and WARD, 1994).

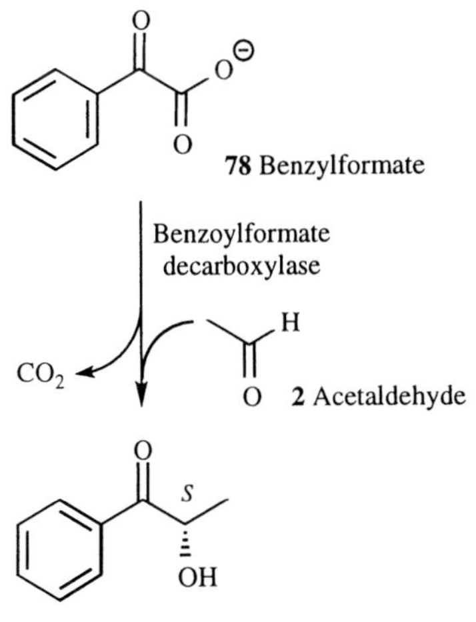

Fig. 26. Acyloin condensation of benzylformate (**78**) and acetaldehyde (**2**) catalyzed by benzylformate decarboxylase.

Whole cell and cell extracts of *Pseudomonas putida* ATCC 12633 also catalyze this transformation, but benzaldehyde was a major byproduct together with small amounts of benzyl alcohol (Willcocks et al., 1992, Willcocks and Ward 1992).

4.1.1.8 Oxalyl-CoA Decarboxylase (Oxalyl-CoA Carboxy-Lyase, EC 4.1.1.8)

Oxalyl-CoA decarboxylase (OCD) catalyzes the decarboxylation of oxalyl-CoA (**86**) to formyl-CoA (**87**; Fig. 28) and is TPP dependent. This offers the prospect that the "enamine" intermedaite (cf. **11**, Fig. 4), which is functionally equivalent to a formate anion (**88**) could be trapped by electrophiles, such as aldehydes (cf. Sect. 4.1.1.1, pyruvate decarboxylase). The most heavily investigated OCD is that from *Oxalobacter formigenes*, which is a gram-negative anaerobe found in the gut, which is able to grow using oxalate as the sole energy and carbon source, if small amounts of acetate are also available (Cornick and Allison, 1996; Cornick et al., 1996). In humans the absence of *O. formigenes*, in the gut is a major

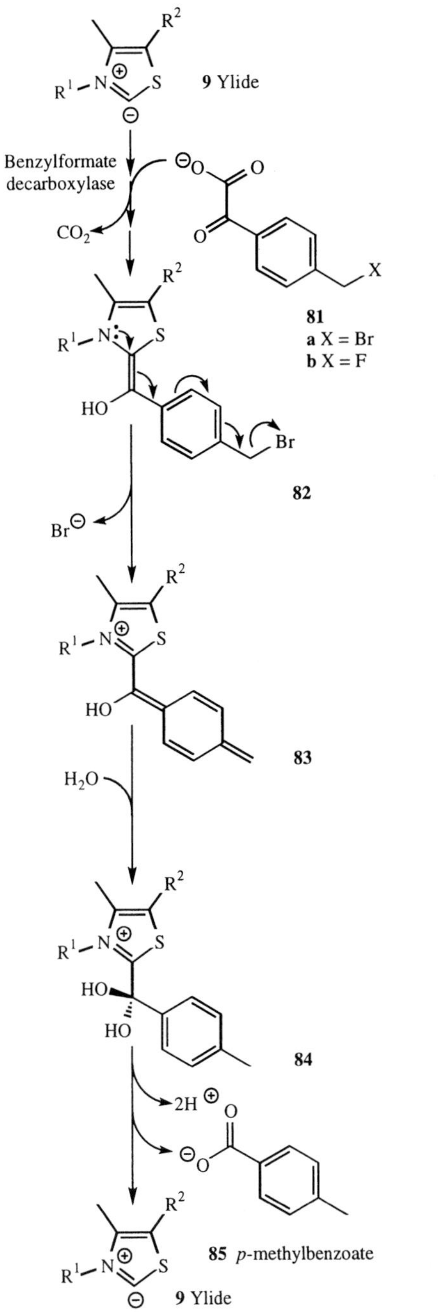

Fig. 27. The mechanism for the decarboxylation of [*p*-(bromomethyl)benzoyl]formate (**81a**) to *p*-methylbenzoate (**85**) by benzylformate decarboxylase (BFD) cf. Figs. 3, 4, 5 for R^1 and R^2 and the sequence **9** → **10** → **11** → **12** → **9** and Figs. 8, 9 for the corresponding reaction of fluoropyruvate (**20**).

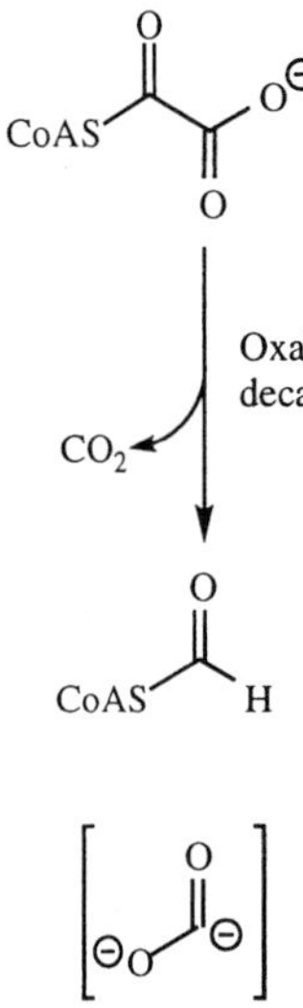

86 Oxalyl-CoA

Oxalyl-CoA decarboxylase

CO_2

87 Formyl-CoA

88 Formate umpolung

Fig. 28. Oxalyl-CoA decarboxylase.

risk factor for hyperoxaluria and calcium oxalate kidney stone disease (SIDHU et al., 1998, 1999).

OCD in *O. formigene* is located in the soluble cytoplasmic fraction (BAETZ and ALLISON, 1992) and is linked to a transport system (OxlT; RUAN et al., 1992), which exchanges extracellular oxalate for intracellular formate. Decarboxylation of oxalyl-CoA (**86**) to formyl-CoA (**87**), followed by hydrolysis consumes a proton and hence transport of formate out of the cell develops an internally negative membrane potential. In effect, formate is acting as a proton carrier and hence transport plus decarboxylation performs the function of an indirect proton pump (FU and MALONEY, 1998). However, there is no obligatory relationship between decarboxylation and the transport system, because functional OCD was produced when a PCR fragment containing the gene (oxc) was overexpressed in *E. coli* (BAETZ and PECK, 1992; cf. SIDHU et al., 1997; LUNG et al., 1994).

It is interesting to speculate if the formate anion equivalent (**88**) is involved in the biosynthesis of common cellular constituents such as Citric acid (Krebs) cycle intermediates (Fig. 1) and amino acids. ^{14}C labeling studies indicate that some 54 % of the total cell carbon is derived from oxalate and at least 7% from acetate. Carbonate is assimilated but not formate. Apparently, there are multiple pathways for the assimilation of oxalate (CORNICK and ALLISON, 1996; CORNICK et al., 1996). The strictly anerobic, obligatory oxalotrophic bacterium, *Bacillus oxalophilus*, utilizes tartronate semialdehyde synthase (EC 4.1.1.47; Sect. 4.1.1.24) or the serine pathway to assimilate oxalate and OCD and formate dehydrogenase for oxidation (ZAITSEV et al., 1993).

4.1.1.9 Malonyl-CoA Decarboxylase (Malonyl-CoA Carboxy-Lyase, EC 4.1.1.9) and Related Malonate Decarboxylating Enzymes

Malonyl-CoA decarboxylase (MCD) catalyzes the decarboxylation of malonyl-CoA (**90**) to acetyl-CoA (**56**; Fig. 29). Some forms of this enzyme are also able to act as malonate-CoA transferase (EC 2.8.3.3), which catalyzes the transfer of coenzyme A from acetyl-CoA (**56**) to malonate (**89**). MCD is not able to catalyze the net carboxylation of acetyl-CoA. This requires the biotin dependent enzyme, acetyl-CoA carboxylase, in which the thermodynamically disfavored carboxylation is linked to the hydrolysis of ATP to ADP (EC 6.4.1.2).

The mechanism for the abiotic decarboxylation of malonic acid (**92**) proceeds via a six-membered transition state (Fig. 30; HUANG et al., 1997) in which a proton is transferred from the carboxylate group undergoing decarboxylation to the incipient carboxylate group of acetate (**21**) in a concerted sequence of bond migrations (GOPOLAN, 1999). This mechanism is essentially identical to that of the decarboxylation of acetoacetate (**37a**; Fig. 15). Despite the relative ease of the abiotic reaction and the potential to catalyze it by protonation and augmented proton transfer, there do not appear to be any decarboxylases in which malonate undergoes direct decarboxylation. The enzymatic decarboxylation of malonate apparently requires obligatory formation of the corresponding thioester to facilitate the decarboxylation. The ability of the sulfur atom of the thioester to acts as an electron sink and hence fa-

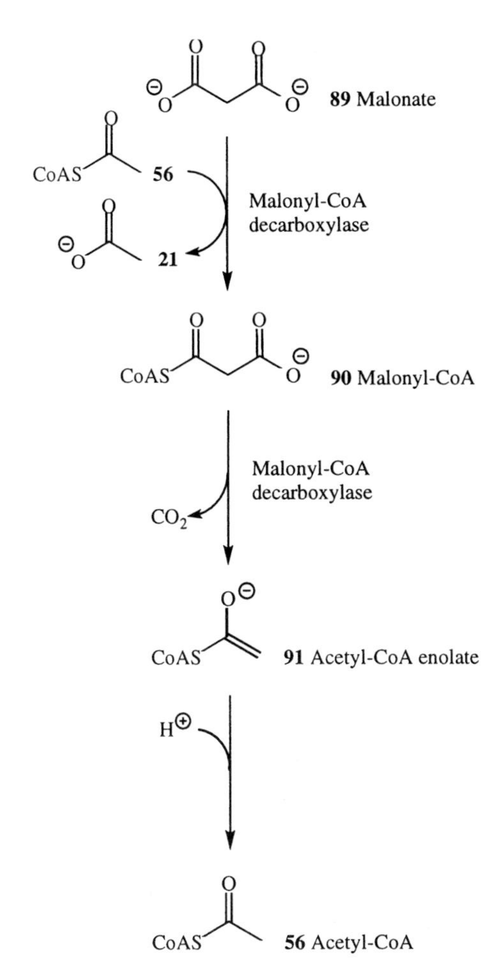

Fig. 29. Malonyl-CoA decarboxylase catalyzes the decarboxylation of malonyl-CoA (**90**) to acetyl-CoA enolate (**91**) which undergoes protonation to give acetyl-CoA (**56**). Some forms of the enzyme are also able to catalyze metathesis of malonate (**89**) and acetyl-CoA (**56**) with malonyl-CoA (**90**) and acetate (**21**).

cilitate the reaction is well established (BACH and CANEPA, 1996), nevertheless, the absence of enzymes able to catalyze the direct decarboxylation of malonate is surprising.

MCDs have been found in mammalian, avian and plant tissues, plus various microorganisms (KOLATTUKUDY et al., 1981). In humans, MCD deficiency results in mild mental retardation, seizures, metabloic acidosis and excretion of malonate in the urine (GAO et al., 1999b; FITZPATRICK et al., 1999; SACKSTEDER et al., 1999). In mammals, MCD and acetyl-CoA carboxylase regulate the levels of malonyl-CoA in muscular tissues as part of a complex system regulating glucose and fatty acid metabolism (GOODWIN and TAEGTMEYER, 1999; ALAM and SAGGERSON, 1998).

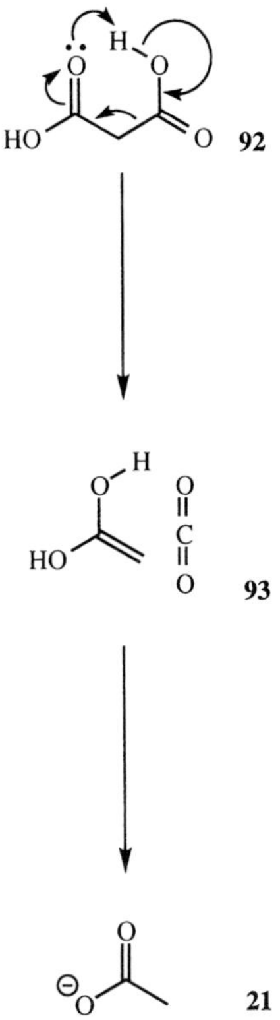

Fig. 30. Abiotic mechanism for the decarboxylation of malonic acid (**37a**).

MCD activity has been described for *Mycobacterium tuberculosis*, *Pseudomonas ovalis* (TAKAMURA and KITAYAMA, 1981), *Pseudomonas fluorescens* (KIM et al., 1979), *Pseudomonas putida* (CHOHNAN et al., 1999), *Sporomusa malonica*, *Klebsiella oxytoca* and *Rhododbacter capsulatus* (DEHNING and SCHINK, 1994), *Citrobacter divs.* (JANSSEN and HARFOOT, 1992), *Acinetobacter calcoaceticus* (KIM and BYUN, 1994), *Malonomonas rubra* (DEHNING and SCHINK, 1989) and *Klebsiella pneumoniae*.

There is some controversy over the exact nature of the MCD activity in microorganisms, which has arisen from work with two species that utilize malonate as sole energy source (DIMROTH and HILBI, 1997). *Malonomomas rubra* is a microaerotolerant fermenting bacterium which can grow by decarboxylation of malonate to acetate under anaerobic conditions. It is phylogenetically related to the sulfur reducing bacteria (KOLB et al., 1998). *Klebsiella pneumoniae* is able to grow aerobically on malonate, or anaerobically if supplemented with yeast extract. Both of these species are able to decarboxylate malonate without the intermediacy of malonyl-CoA. In each case, malonate is decarboxylated as the thioester of an acyl-carrier protein (ACP), which bears an acetylated 2′-(5″-phosphoribosyl)-3′-dephosphocoenzyme A prosthetic group (**94b**) (Fig. 31) (BERG et al., 1996; SCHMID et al., 1996). This group was previously identified as a component of *K. pneumoniae* citrate (*pro*-3*S*)-lyase (ROBINSON et al., 1976). The *K. pneumoniae* malonate decarboxylase consists of four subunits (α–δ) involved in the catalytic cycle (Fig. 32), plus a further enzyme (*MdcH*) which transfers malonyl-CoA (**90**) to the thiol form of the ACP (**94a**) (HOENKE and DIMROTH 1999). Decarboxylation then yields the acetyl derivative (**94b**), which undergoes *trans*-acylation with malonate (**89**) to commence the catalytic cycle (HOENKE et al., 1997). In this sequence, there is no means by which the energy released by decarboxylation ($DG°' = -17.4$ kJ mol^{-1}) can be "harvested" and it is lost as heat. This is in any case, unnecessary because more than sufficient energy is created by the aerobic oxidation of acetate to carbon dioxide and water in the citric acid cycle and respiratory chain.

The decarboxylation system of the anerobe *M. rubra* consists of water soluble cytoplasmic components (some of which resemble those of *K. pneumoniae*) plus membrane bound components which are involved in energy harvesting. Acetate and ATP are required for the priming step in which ACP-SH (**94a**) is converted to the acetyl form (**94b**) (Fig. 33). This undergoes metathesis with malonate to give malonyl-ACP (**94c**) and decarboxylation is linked to the transfer of carbon dioxide to *MadF*, the biotin carrier protein (BCP) (**96**) with regeneration of acetyl-ACP (**94b**). The carboxybiotin protein (**95**) is believed to diffuse to the membrane, where it is decarboxylated by *MadB* in an exogonic reaction which is linked to sodium ion pumping across the membrane. The Na^+ gradient so formed is then exploited for ATP synthesis by the F1F0 ATP synthases. This process has been termed decarboxylative phosphorylation and is responsible for all ATP synthesis in several anaerobic bacteria (DIMROTH and SCHINK, 1998).

It is claimed that the apparent MCD activity of *Acinetobacter calcoaceticus*, *Citrobacter divs.*, *K. oxytoca*, and some *Pseudomonas* spp.

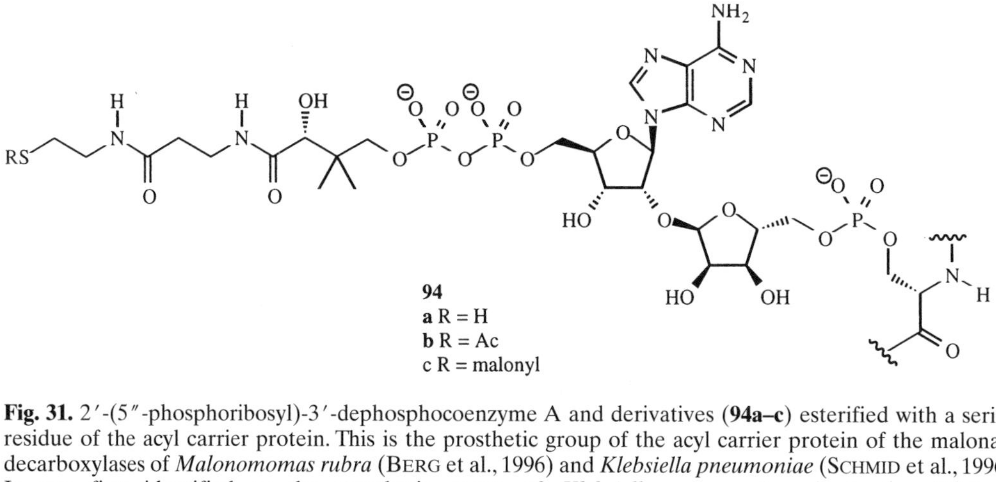

94
a R = H
b R = Ac
c R = malonyl

Fig. 31. 2′-(5″-phosphoribosyl)-3′-dephosphocoenzyme A and derivatives (**94a–c**) esterified with a serine residue of the acyl carrier protein. This is the prosthetic group of the acyl carrier protein of the malonate decarboxylases of *Malonomomas rubra* (BERG et al., 1996) and *Klebsiella pneumoniae* (SCHMID et al., 1996). It was first identified as the prosthetic group of *Klebsiella pneumoniae* citrate (*pro*-3*S*)-lyase (ROBINSON et al., 1976).

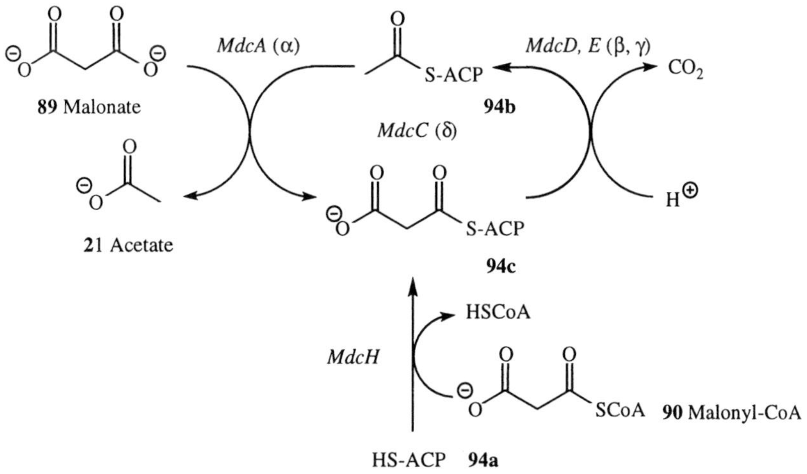

Fig. 32. Proposed reaction mechanism for malonate decarboxylation by *Klebsiella pneumoniae*. The gene cluster cosists of eight consecutive genes which have been termed *mdcABCDEFGH* and the divergently orientated *mdcR* gene. The anticipated functions of the gene products are as follows (functional subunits of the enzyme are shown in brackets): *MdcA*, acetyl-S-acyl carrier protein: malonate acyl carrier protein transferase (α); *MdcB*, involved in the synthesis and attachment of the prosthetic group; *MdcC*, acyl carrier protein (ACP)(δ); *MdcD,E*, malonyl-S-acyl carrier protein decarboxylase (β, γ); *MdcF*, membrane protein possibly involved in malonate transport; *MdcG*, involved in the synthesis and attachment of the prosthetic group; *MdcH*, malonyl-CoA:acyl carrier protein-SH transacylase; *MdcR* is a protein of the LysR regulator family. It is probably a transcriptional regulator of the *mdc* genes (HOENKE et al., 1997).

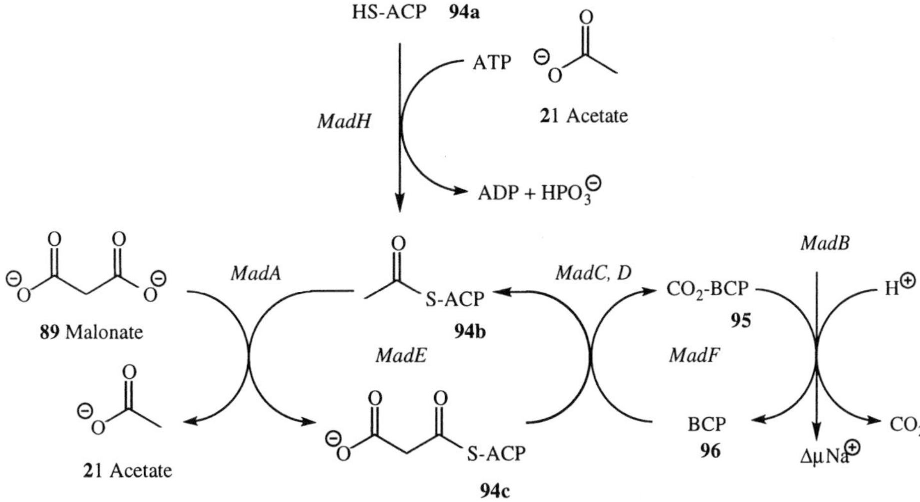

Fig. 33. Proposed reaction mechanism for malonate decarboxylation by *Malonomonas rubra*. The gene cluster contains 14 consecutive genes which have been termed *madYZGBAECDHKFLMN*. The functions of the gene products are as follows: *MadA*, acetyl-S-acyl carrier protein: malonate acyl carrier protein-SH transferase; *MadB*, carboxybiotin decarboxylase; *MadC, D* carboxy-transferase subunits; *MadE* acyl carrier protein (ACP); *MadF*, biotin carrier protein (BCP); *MadG* may be involved in the biosynthesis of MadH; *MadH*, deacetyl acyl carrier protein: acetate ligase; *MadL* and *MadM* are membrane proteins which could function as a malonate carrier. The function of *madY, Z, K* and *N* is currently unknown (BERG et al., 1997).

is due to the use of assays in which malonate is present in addition to malonyl- and acetyl-CoA. This remains to be established definitively, however, there are striking parallels between some of the subunits identified in the MCDs of these species and those from *M. rubra* and *K. pneumoniae* (DIMROTH and HILBI, 1997).

The stereochemistry of decarboxylation of (*R*)-[1-^{13}C, 2-^{3}H]-malonate by the biotin dependent "MCD" of *M. rubra* (MICKLEFIELD et al., 1995) and the biotin independent "MCDs" of *P. fluorescens* and *Acinetobacter calcoaceticus* (HANDA et al., 1999) in deuterium oxide have been determined. In every case the carboxylate group was replaced by hydrogen with retention of configuration. Methods for the synthesis of chiral malonates used in this study are described in the section on fumarases (Sect. 4.2.1.2).

The decarboxylation of malonate to acetate has few biotechnological merits other than for the removal of malonate from situations in which it may be viewed as a contaminant. Nevertheless the organisms discussed above produce large amounts of the decarboxylase system and there is great potential for their use with "unnatural substrates".

There has been considerable interest in the decarboxylation of 2,2-disubstituted malonates, which can potentially give chiral carboxylic acids (cf. MUSSONS et al., 1999; for related abiotic catalysts). OHTA's group screened microorganisms from soil samples on phenylmalonic acid. This study yielded *Alcaligenes bronchisepticus* KU 1201, which is capable of asymmetric decarboxylation of arylmalonates with good to excellent enantiomeric excesses (Tab. 3) (MIYAMOTO and OHTA, 1990, 1991).

The gene for the arylmalonate decarboxylase was identified and expressed in *E. coli* (MIYAMOTO and OHTA, 1992). Remarkably, the amino acid sequence had no significant homologies with known proteins and the thiol nucleophile was furnished by a cysteine residue rather than coenzyme A (KAWASAKI et al., 1995). Hydrophobic interactions with the aromatic ring were shown to be especially important for substrate recognition (KAWASAKI et al., 1996a, 1997; OHTA, 1997). Whole cell mediated decarboxylation gave good results with α-fluoro-α-phenylmalonates, particularly when electron withdrawing substituents were present on the the aromatic ring (MIYAMOTO et al, 1992). The generally poorer results with all substrates bearing aromatic rings with electron donating substituents (e.g., OMe) is probably only partially a function of slower decarboxylation, but is also due to the presence of alternative degradation pathways (MIYAMOTO and OHTA, 1990, 1991. The purified recombinant enzyme quantitatively converts α-fluoro-α-phenylmalonic acid (**99**) to enantiomerically pure (*R*)-α-fluorophenylacetic acid (**100**) (FUKUYAMA et al., 1999).

4.1.1.10 Aspartate α-Decarboxylase (L-Aspartate 1-Carboxy-Lyase, EC 4.1.1.11)

Aspartate α-decarboxylase (AαD) catalyzes the decarboxylation of L-aspartate (**101**) to β-alanine (**102**) (Fig. 35) (WILLIAMSON, 1985). β-Alanine (**102**) is a biosynthetic precursor of pantothenate in bacteria (DUSCH et al., 1999) and is a degradation product of uracil in plants and fungi (GANI and YOUNG, 1983).

AαD is a tetramer in which each monomer consists of a six stranded double ψ β-barrel capped by small α-helices (ALBERT et al., 1998). This motif is found in several protein superfamilies and includes enzymes such as DMSO reductase, formate dehydrogenase and the aspartic proteinases (CASTILLO et al., 1999). In the X-ray crystal structure catalytic pyruvoyl groups were found in three subunits and an ester in the fourth. The enzyme is initially produced as an inactive proenzyme (π-protein) (**103**; Fig. 36), which undergoes *N,O*-acyl migration between Gly-24 and Ser-25 to give the amino ester (**104**). β-elimination of the β-protein (**105**) and hydrolysis yields the catalytically active α-subunit with a terminal pyruvoyl residue (**107**). Purified recombinant enzyme consisted predominantly of π-protein (**103**). Incubation of this at 50 °C for 49 hours gave a fully processed tetrameric enzyme containing three subunits bearing pyruvoyl groups. Unchanged serine was found at the terminus of some of the α-subunits (**106**), in the tetramer, but the exact proportion could not be determined (RAMJEE et al., 1997).

Tab. 3. Decarboxylation of α-Aryl-α-Methylmalonates (MIYAMOTO and OHTA, 1990)

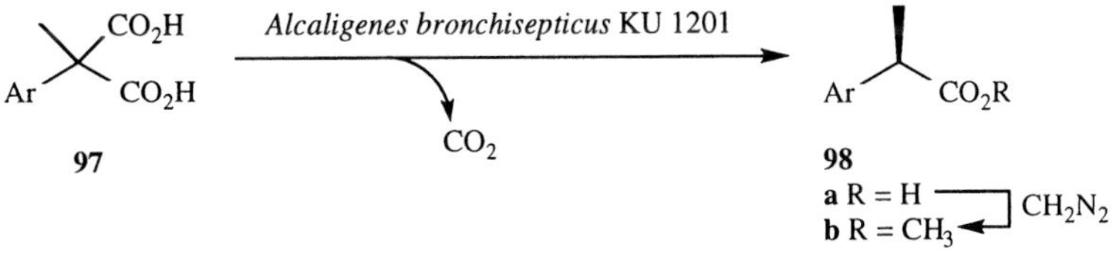

Ar–C(CO_2H)(CO_2H) **97** —*Alcaligenes bronchisepticus* KU 1201 ($-CO_2$)→ Ar–CH(CH_3)–CO_2R **98**

a R = H
b R = CH_3 (CH_2N_2)

No.	Substrate	Substrate Concentration	Yield %	ee %	Configuration %
1	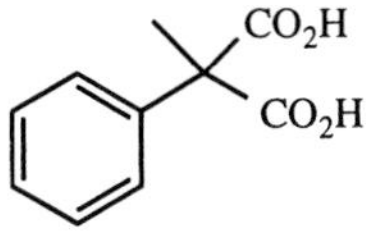 CO_2H, CO_2H	0.3	87	91	*R*
2		0.4	93	96	*R*
3		0.5	90	98	*R*
4	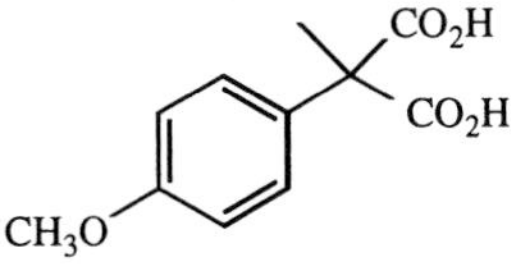 CH_3O, CO_2H, CO_2H	0.1	48	99	*R*
5		0.2	35	97	*R*
6	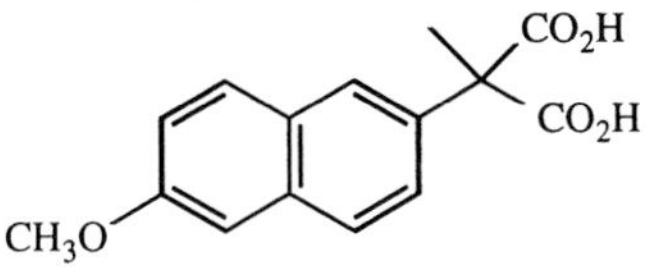 CH_3O, CO_2H, CO_2H	0.3	95	>95	*R*
7		0.5	96	>95	*R*
8	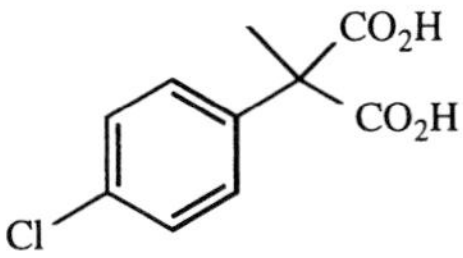 Cl, CO_2H, CO_2H	0.3	95	98	*R*
9		0.5	85	97	*R*
10	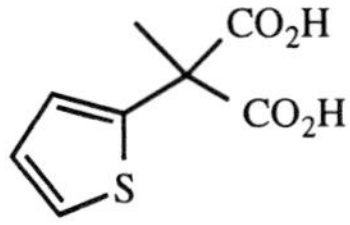 S, CO_2H, CO_2H	0.3	98	95	*S*
11		0.5	97	91	*S*

It has been known for over a hundred years that α-ketoacids catalyze the decarboxylation of α-amino acids, nevertheless some aspects of the reaction are still controversial. In principle, two zwitterionic imine rotomers (**108**) (**109**) can be formed. The "*syn*" rotomer (**108**) benefits by an intramolecular hydrogen bond (BACH and CANEPA, 1997), whereas the "*anti*"rotomer is able to undergo cyclization to the uncharged oxazolidin-5-one (**110**), which is an intermediate in the abiotic reaction. Both rotomers and the oxazolidin-5-one are capable of eliminating carbon dioxide to give the azomethine ylide (**111 to 125**) which has been trapped in abiotic model studies (GRIGG et al., 1989). AαD has not yet found applications in biotransformations, however, developments such as the the X-ray crystal structure determination and the isolation of large amounts of enzyme, should encourage its exploitation. Histidine decarboxylase, *S*-adenosylmethionine decarboxylase and phosphatidylserine decarboxylase all bear catalytic pyruvoyl groups and hence AαD provides an exemplar of this class of enzymes.

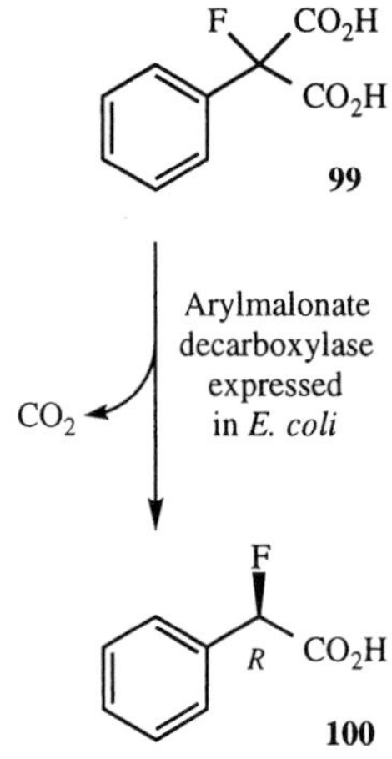

Fig. 34. *Alcaligenes bronchisepticus* KU 1201 arylmalonate decarboxylase expressed in *E. coli* (MIYAMOTO and OHTA, 1992) catalyzes the decarboxylation of α-fluoro-α-phenylmalonic acid (**99**) to enantiomerically pure (R)-α-fluorophenylacetic acid (**100**) in quantitative yield! (FUKUYAMA et al., 1999).

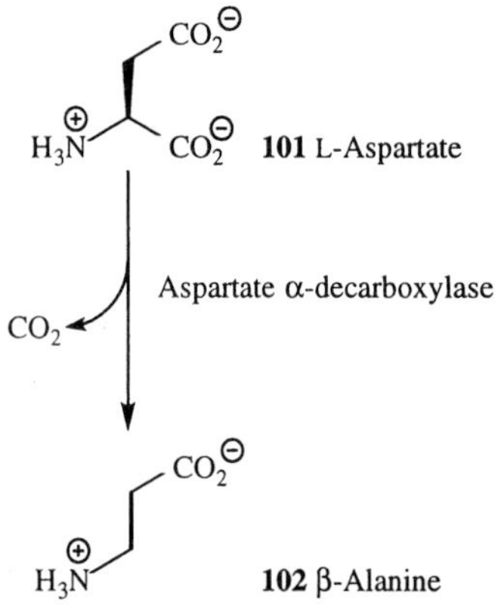

Fig. 35. L-Aspartate α-decarboxylase catalyzes the decarboxylation of L-aspartate (**101**) to β-alanine (**102**).

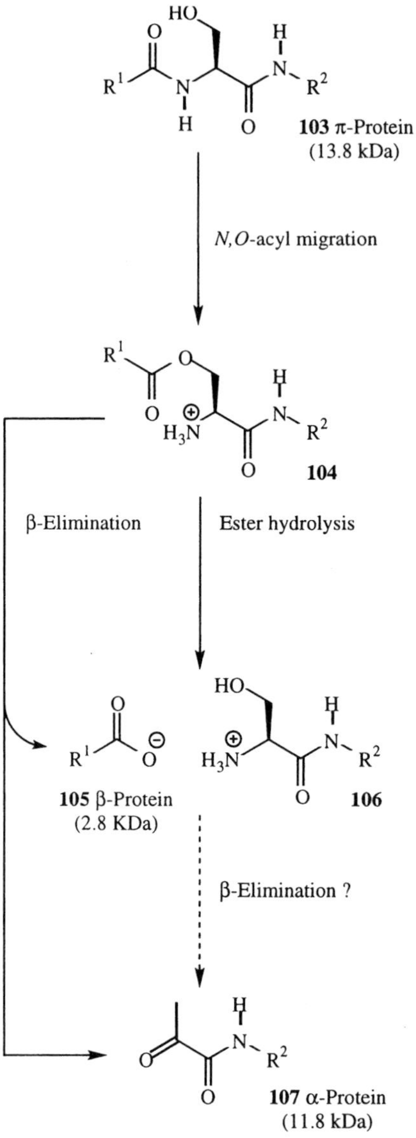

Fig. 36. Processing of α-aspartate decarboxylase (π-protein) (**103**) to catalytically active α-protein.

4.1.1.11 Aspartate 4-Decarboxylase (L-Aspartate β-Carboxylase, L-aspartate 4-Carboxy-Lyase, Desulfinase, EC 4.1.1.12)

Aspartate 4-decarboxylase catalyzes the decarboxylation of L-aspartate (**101**) to L-alanine (**111**) (Fig. 38) (CHANG et al., 1982). This reaction has been used in various forms for the manufacture of L-alanine, which is difficult to produce by fermentation. The process has been optimized with resting cells of *Pseudomonas dacunhae* grown at pH 7.0–7.5 on glutamate (CALIK et al., 1997) and immobilized on κ-carrageenan (CALIK et al., 1999). By using

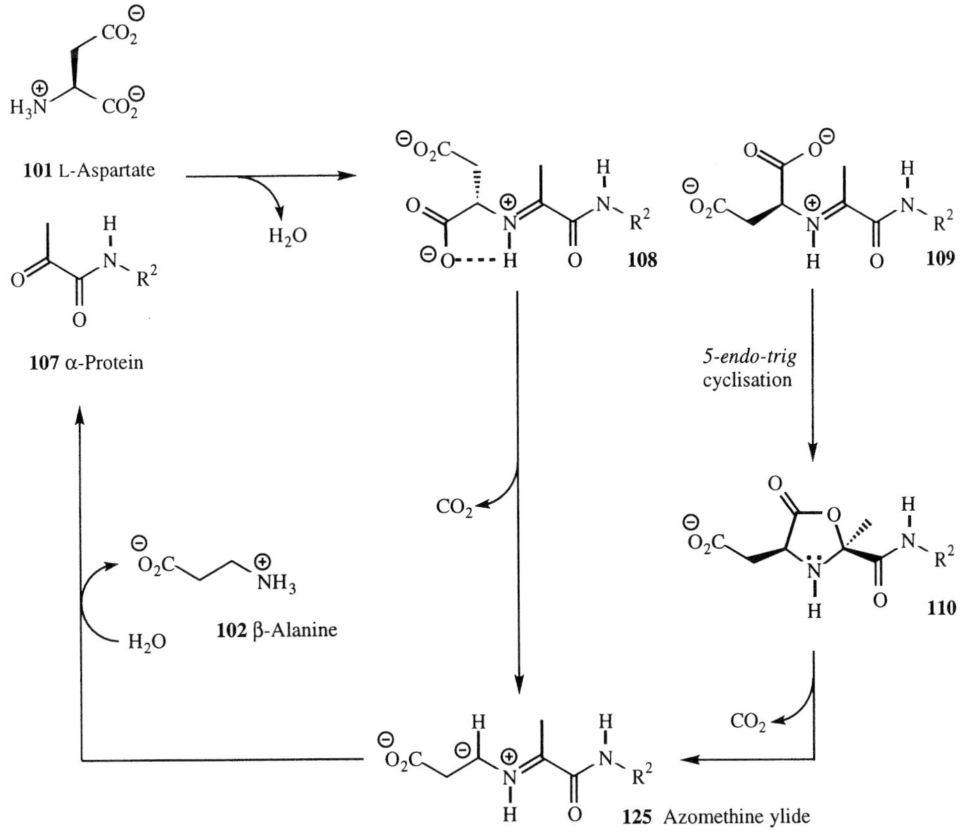

Fig. 37. Mechanism of decarboxylation by α-aspartate decarboxylase (**107**).

D,L-aspartate as substrate both L-alanine and D-aspartate were produced using whole cells immobilized in carrageenan or enzyme supported on acylonitrile activated silica gel (ABELYAN, 1999; ABELYAN et al., 1991; SENUMA et al., 1989). Coimmobilized aspartase and cells with aspartate decarboxylase activity, converted fumarate to aspartate which was decarboxylation *in situ* (ABELYAN and AFRIKYAN, 1997). Remarkably the combined catalyst was stable for 450–460 days. Aspartate 4-decarboxylase contains a pyridoxal 5′-phosphate prosthetic group, which is discussed in detail in Sect. 4.1.1.13, L-Glutamate Decarboxylase.

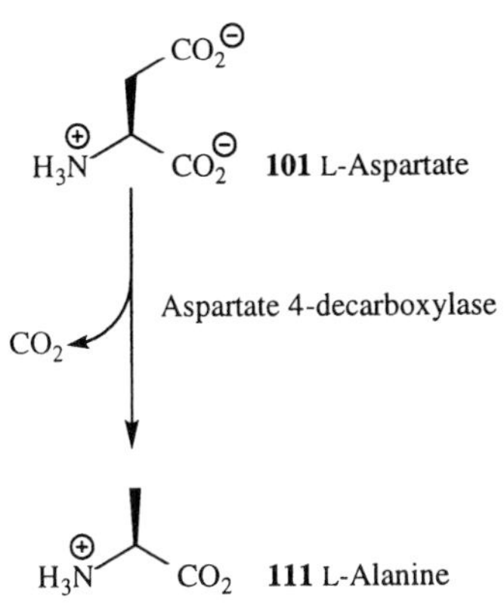

Fig. 38. Aspartate-4-decarboxylase.

All α-amino acid decarboxylases and most other amino acid decarboxylases, rely on imine formation between the amino group of the amino acid and the carbonyl group of a pyruvoyl (**108**) (**109**), pyridoxal 5′-phosphate or some other group to promote decarboxylation. Imine formation is also the first step in transamination catalyzed by aminotransferases. Thus by impeding the hydrogen shift step in transamination and promoting the protonation required for decarboxylation, it should be possible to switch an enzyme from transamination to decarboxylation. This has been achieved with an *E. coli* aspartate aminotransferase triple mutant (Y225R/R292K/R386A), which catalyzes the 4-decarboxylation of aspartate eight times as fast as transamination. This is a greater than 25 million fold increase in decarboxylation relative to the wild type enzyme (GRABER et al., 1999)

4.1.1.12 Valine Decarboxylase (L-Valine Decarboxylase, EC 4.1.1.14)

Valine decarboxylase catalyzes the decarboxylation of L-valine (**112**) to 2-methylpropanamine (**113**) (Fig. 39) and also acts on L-leucine, norvaline and isoleucine. It has a pyridoxal 5′-phosphate (PLP) (**116**) prosthetic group, but seems to have been largely overlooked since its discovery (SUTTON and KING, 1962).

4.1.1.13 Glutamate Decarboxylase (L-Glutamate 1-Carboxy-Lyase, EC 4.1.1.15)

Glutamate decarboxylase (GAD) catalyzes the decarboxylation of L-glutamate (**114**) to 4-aminobutanoate, which is more widely known as γ-aminobutyric acid (GABA) (**115**) (Fig. 40). GABA plays a role in pH regulation, nitrogen storage and defence in plants (SHELP et al., 1999) and as a neurotransmitter in mammals (WAAGEPETERSEN et al., 1999; SCHWARTZ et al., 1999). GAD also decarboxylates L-aspartate (**101**) to β-alanine (**102**).

The prosthetic group of GAD is nominally pyridoxal 5′-phosphate (PLP) (**116a**), but in common with many other PLP dependent enzymes, the usual form of the holoenzyme contains the cofactor bonded to the peptide chain via an aldimine (**116b**) formed with an ε-amino group of a lysine residue (reviews, JANSONIUS, 1998; JOHN, 1995; HAYASHI, 1995; ALEXANDER et al., 1994).

The mechanism and stereochemistry of reactions catalyzed by *E. coli* GAD have been determined by detailed isotope studies (Fig. 41). The "normal" decarboxylation pathway is initiated by imine (**117**) formation between the PLP–GAD aldimine derivative (**116b**) and glutamate (**114**). Decarboxylation yields a quinoid type intermediate (**118**) which can reprotonate at two sites. Protonation of the former C-α-center occurs on the same face as the C-4′-*Si* side of the quinoid intermedaite and hence the carboxylate group is replaced by the incoming proton with retention of configuration.The imine (**119**) so formed, undergoes metathesis to give the PLP–GAD aldimine derivative (**116b**) and GABA (**115**). Decarboxylation may also occur concommitantly with

$H_3N^{\oplus}$ $CO_2^{\ominus}$ **112** L-Valine

Valine decarboxylase

CO_2

$H_3N^{\oplus}$ **113** 2-Methylpropanamine

Fig. 39. Valine decarboxylase.

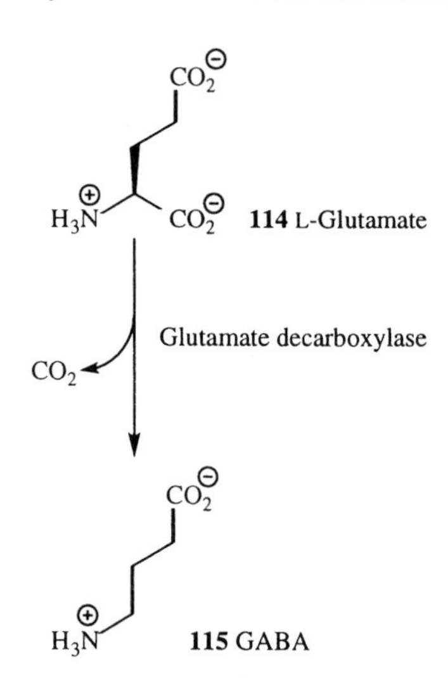

Fig. 40. L-Glutamate decarboxylase.

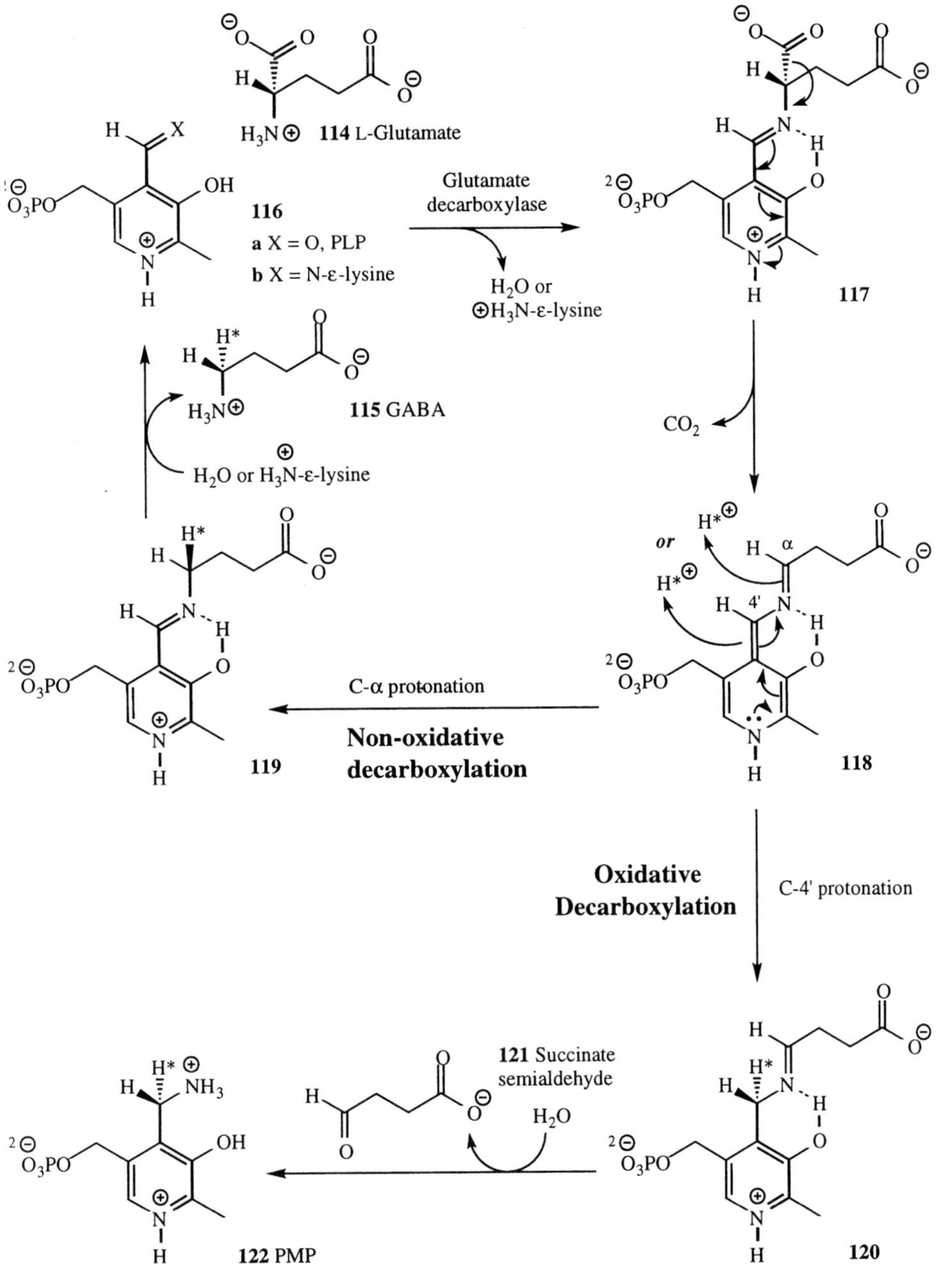

114 L-Glutamate

116
a X = O, PLP
b X = N-ε-lysine

Glutamate decarboxylase

H_2O or $^{\oplus}H_3N$-ε-lysine

117

115 GABA

H_2O or $H_3\overset{\oplus}{N}$-ε-lysine

CO_2

or

118

C-α protonation

Non-oxidative decarboxylation

119

Oxidative Decarboxylation

C-4' protonation

121 Succinate semialdehyde

H_2O

122 PMP

120

Fig. 41. The mechanism of decarboxylation by L-glutamate decarboxylase. Abbreviations: GABA, γ-aminobutyric acid (**115**); PLP, pyridoxal 5′-phosphate (**116**); PMP, pyridoxamine 5′-phosphate (**122**). In this and other diagrams showing PMP derivatives, the phenolic hydroxyl group–imine interaction [(**117**)–(**120**)] is shown as a hydrogen bond (*dotted line*) for the purposes of simplicity. On the basis of solution phase pK_a's, a protonated imine–phenoxide interaction would be a more accurate depiction.

transamination (cf. Khristoforov et al., 1995). C-4′-*Si*-face protonation of the quinoid intermedaite (**119**) gives the imine (**120**), which hydrolyzes to succinate semialdehyde (**121**) and stereospecifically labeled PMP (**122**) in a single turnover process (Tilley et al., 1992, 1994). A similar pathway occurs with α-methylglutamate (**123**) (Fig. 42), which is oxidatively deaminated to levulinate (**124**) in a multi-turnover catalytic process which requires dioxygen (Bertoldi et al., 1999).

PLP dependent aminotransferases have a Ser-X-Ala-Lys sequence in the active site, whereas decarboxylases have a Ser-X-His-Lys sequence. The lysine residue is responsible for aldimine formation and C-4′-*Si*-face protonation of the quinoid intermedate (**118**), whereas the histidine binds the α-carboxylate groups and protonates at C-α (Gani, 1991, Smith et al., 1991). Human GAD consists of two isoforms (GAD65 and GAD67; Schwartz et al., 1999), which closely resemble aspartate transaminase and ornithine decarboxylase (Qu et al., 1998).

GAD has been used extensively for the the stereospecific synthesis of GABA hydrogen isotopomers (**115**) (Young, 1991; Battersby et al., 1982b) and for the synthesis of D-glutamate. L-glutamate produced by fermentation was racemized with recombinant glutamate racemase from *Lactobacillus brevis* ATCC-8287. The L-glutamate was then converted to GABA using GAD from *E. coli* ATCC-11246 which enabled the unreactive D-glutamate to be isolated (Yagashi and Ozaki, 1998).

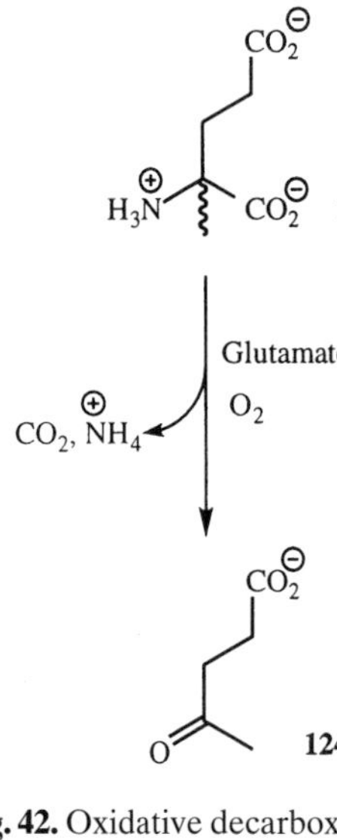

Fig. 42. Oxidative decarboxylation of α-methylglutamate (**123**) by glutamate decarboxylase.

4.1.1.14 Ornithine Decarboxylase (EC 4.1.1.17)

Ornithine decarboxylase (ODC) is a PLP dependent enzyme, which catalyzes the decarboxylation of L-ornithine (**127**) to putrescine (**128**) (1,4-diaminobutane) (Fig. 43). It also decarboxylates L-lysine, but more slowly (Swanson et al., 1998). ODC has been heavily investigated because putrescine (**128**) and the next highest homolog, cadaverine, (which is derived from L-lysine) are precursors of polyamines (e.g., spermideine and spermine; Morgan, 1999), which are accumulated in cancer cells and stimulate proto-oncogene expression (Tabib and Bachrach, 1999). ODC is the rate limiting enzyme in polyamine biosynthesis and its activity is down regulated by increasing polyamine levels which accelerate degradation (Toth and Coffino, 1999; Di Gangi et al., 1987). ODC is widely distributed and well established methods are available for its isolation from *E. coli* (Morris and Boeker, 1983), *Physarum polycephalum* (Mitchell, 1983), yeast (Tyagi et al., 1983), germinated barley seeds (Kyriakidis et al., 1983), mouse kidney (Seely and Pegg, 1983), rat liver (Hayashi and Kameji, 1983)

X-ray crystal structures have also been determined for mouse ODC (Kern et al., 1999), *Lactobacillus*-30a ODC (L30a OrnDC; Momany et al., 1995) and *Trypanosoma brucei* ODC as the native enzyme and as a complex (Grishin et al., 1999) with the suicide inhibitor α-difluoromethylornithine (Seely et al., 1983). *Trypanosoma brucei* is the causative agent of African sleeping sickness and its ODC is a recognized drug target (McCann et al., 1983). These are large enzymes with disparate, complex, multimeric structures.The active site lies at the interface of a dimers and PLP is bound as an aldimine with the ε-amino group of a lysine residue (e.g., Lys-60 for *Trypanosoma brucei* ODC; Osterman et al., 1999, 1997) as is

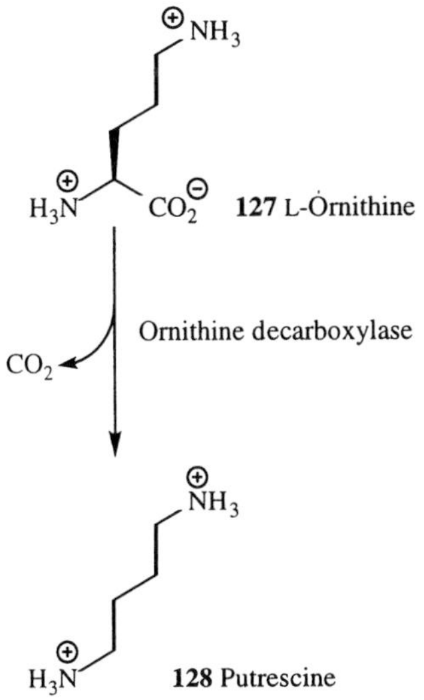

127 L-Ornithine

Ornithine decarboxylase

CO_2

128 Putrescine

Fig. 43. Ornithine decarboxylase (ODC) catalyzed decarboxylation of L-ornithine (**127**) to putrescine (**128**).

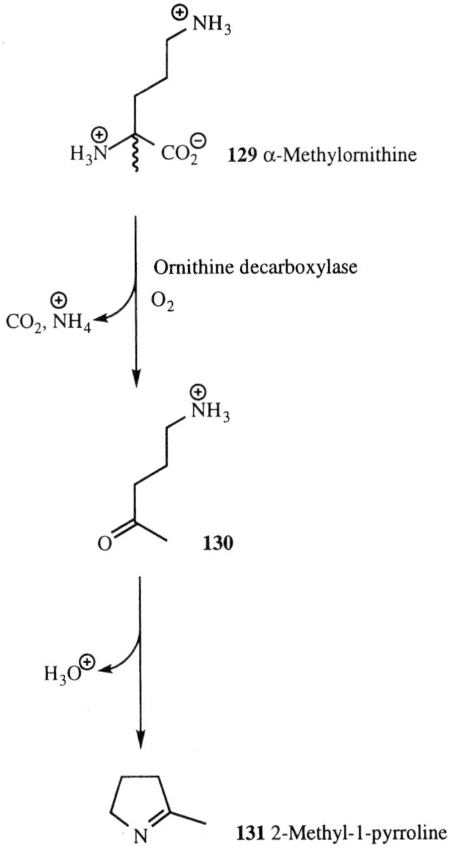

found for GAD and most other α-amino acid decarboxylases.

L-Ornithine (**127**) undergoes decarboxylation by ODC, such that the carboxylate group is replaced by hydrogen with retention of configuration (ORR and GOULD, 1982) This reaction has been used for the stereospecific synthesis of putrescine hydrogen isotopomers (YOUNG, 1991). α-Methylornithine (**129**) (Fig. 44) undergoes oxidative decarboxylation with ODC in the presence of dioxygen to give the ammonium ketone (**130**) which cyclizes to 2-methyl-1-pyrroline (**131**) (BERTOLDI et al., 1999).

4.1.1.15 Lysine Decarboxylase (L-Lysine Carboxy-Lyase, EC 4.1.1.18)

Lysine decarboxylase is a PLP dependent enzyme, which catalyzes the decarboxylation of L-lysine (**132**) to cadaverine (**133**) (Fig. 45) in a reaction which is homologous with that of ornithine decarboxylase (Fig. 43). Most forms of both enzymes are capable of catalysing both reactions, to some degree, however, unusually

129 α-Methylornithine

Ornithine decarboxylase

O_2

CO_2, NH_4^+

130

H_3O^+

131 2-Methyl-1-pyrroline

Fig. 44. Ornithine decarboxylase catalyzed oxidative decarboxylation of α-methylornithine (**129**) to 2-methyl-1-pyrroline (**131**).

Selenomonas ruminantium lysine decarboxylase, catalyzes both reactions with similar kinetics parameters (TAKATSUKA et al., 1999a, b). Most workers have utilized lysine decarboxylase from *Bacillus cadaveris* or the constitutive enzyme from *E. coli* (BOEKER and FISCHER, 1983), but a second form has been identified recently (LEMONNIER and LANE, 1998; YAMAMOTO et al., 1997). Lysine decarboxylases decarboxylates L-lysine (**132**) such that the carboxylate group is replaced by hydrogen with retention of configuration (ORR et al., 1982; BATTERSBY et al., 1982a). This reaction has been used for the stereospecific synthesis of cadaverine hydrogen isotopomers (YOUNG, 1991).

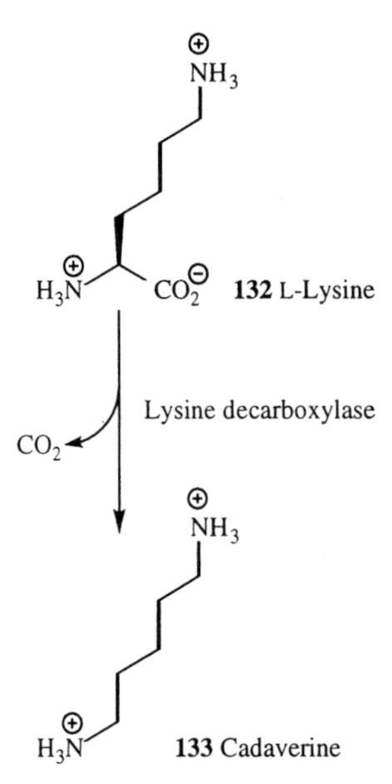

Fig. 45. Lysine decarboxylase (ODC) catalyzed decarboxylation of L-lysine (**127**) to cadaverine (**128**) (1,5-diaminopentane).

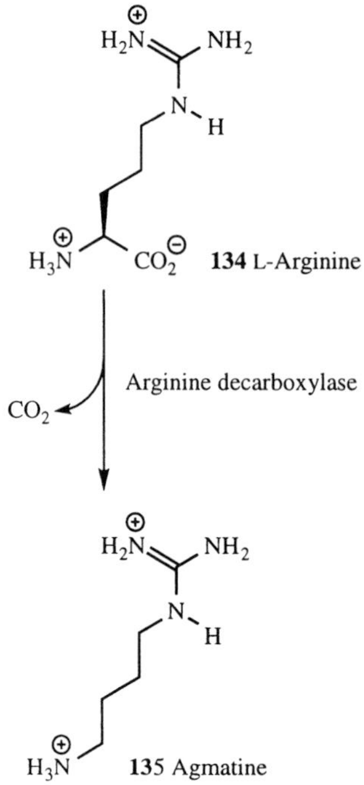

134 L-Arginine

Arginine decarboxylase

CO_2

135 Agmatine

Fig. 46. Arginine decarboxylase (ADC) catalyzed decarboxylation of L-arginine (**134**) to agmatine (**135**) (4-aminobutylguanidine).

4.1.1.16 Arginine Decarboxylase (L-Arginine Carboxy-Lyase, EC 4.1.1.19)

Arginine decarboxylase catalyzes the decarboxylation of L-arginine (**134**) to agmatine (**135**) (4-aminobutylguanidine) (Fig. 46). It is a PLP dependent enzyme which decarboxylates L-arginine such that the carboxylate group is replaced by hydrogen with retention of configuration (ORR et al., 1982). As with the other basic amino acid decarboxylases, it has been used to prepare agmatine hydrogen isotopomers (YOUNG, 1991; RICHARDS and SPENSER, 1983).

4.1.1.17 Diaminopimelate Decarboxylase (meso-2,6-Diaminoheptanedioate Carboxy-Lyase, EC 4.1.1.20)

Diaminopimelate decarboxylase catalyzes the decarboxylation of *meso*-2,6-diaminopimelate (**136**) to L-lysine (**132**) (Fig. 47), which is the last step in the biosynthesis of L-lysine from L-aspartate and pyruvate (SCAPIN and BLANCHARD, 1998; CHATTERJEE, 1998). *meso*-2,6-Diaminopimelate (**136**) is a cross-linking constituent of the peptidoglycans of virtually all gram-negative and some gram-positive bacteria. L-Lysine is an essential dietary amino acid for mammals, because they lack the diaminopimelate pathway, hence inhibitors of the enzymes of this pathway could have antimicrobial or herbicidal properties with low mammalian toxicity (SONG et al., 1994).

Incubation of *meso*-2,6-diaminopimelate (**136**) with diaminopimelate decarboxylase in deuterium oxide yields (2*S*,6*R*)-[6-$^2H_1$]-lysine and hence the decarboxylation occurs such that the incoming hydrogen replaces the carboxylate group with inversion of stereochemistry (ASADA et al., 1981; KELLARD et al., 1985). This result is unusual because PLP dependent enzymes, such as diaminopimelate decarboxylase generally catalyze decarboxyla-

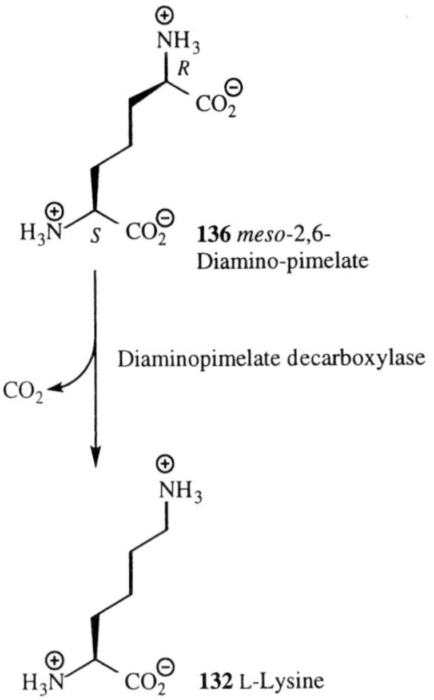

136 *meso*-2,6-Diamino-pimelate

Diaminopimelate decarboxylase

132 L-Lysine

Fig. 47. Diaminopimelate decarboxylase catalyzed decarboxylation of *meso*-2,6-diaminopimelate (*meso*-2,6-diaminoheptanedioate) (**136**) to L-lysine (**132**).

tion with retention of configuration. However, the center undergoing decarboxylation has the (*R*)-configuration rather the (*S*)-configuration which is typical of most natural amino acids, consequently with a similar mode of binding for the two enantiomers the hydrogen is delivered from the same side. Stereospecific synthetic routes to *meso*-2,6-diaminopimelate (**136**) and other stereoisomers (WILLIAMS and YUAN, 1992) have been reported and decarboxylation has been used to prepare a range of hydrogen isotopomers (ASADA et al., 1981; KELLARD et al., 1985; YOUNG, 1991), but other synthetic opportunities remain to be exploited.

4.1.1.18 Histidine Decarboxylase (Histidine Carboxy-Lyase, EC 4.1.1.22)

Histidine decarboxylase (HDC) catalyzes the decarboxylation of L-histidine (**137**) to histamine (**138**) (Fig. 48). Most forms of the enzyme have a pyruvoyl group at the active site, however, PLP dependent forms are present in *Klebsiella planticola*, *Enterobacter aerogenes* (KAMATH et al., 1991) and *Morganella morganii* (SNELL and GUIRARD, 1986).

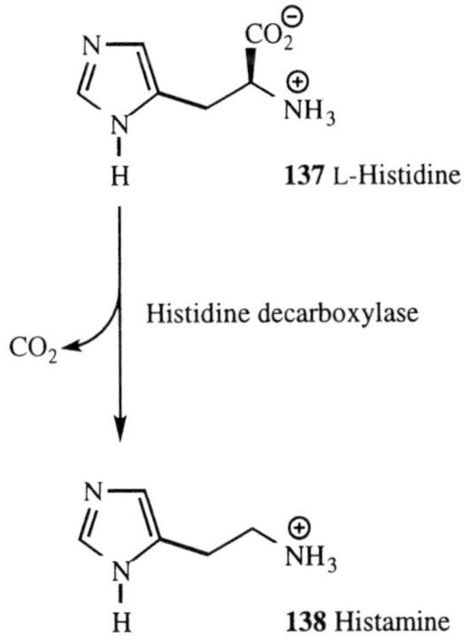

137 L-Histidine

Histidine decarboxylase

138 Histamine

Fig. 48. The histidine decarboxylase catalyzed decarboxylation of L-histidine (**137**) to histamine (**138**).

Pyruvoyl-dependent HDCs are produced by *Lactobacillus buchneri*, *Clostridium perfringens*, *Micrococcus* sp.n., *Leuconostoc oenos* 9204 (COTON et al., 1998) and *Lactobacillus* 30a (SNELL, 1986) All of these enzymes show high sequence homology and have probably evolved from a common ancestral protein. *Lactobacillus* 30a HDC has been studied extensively and the X-ray crystal structure determined (PARKS et al., 1985; GALLAGHER et al., 1993). *Lactobacillus* 30a HDC is formed from hexameric prohistidine decarboxylase (π_6; SNELL and HUYNH, 1986), by autoproteolysis between Ser-81 and Ser-82 and elimination to give a "*N*"-terminal pyruvoyl group derived from Ser-82 (the α-chain; cf. Fig. 36.), plus the remainder of the peptide chain with a normal terminal carboxylate group (the β-chain). The α- and β-chains are tightly bound together and three of these subunits form a trimer, which binds to a second trimeric unit to give a dumbbell shaped hexamer; (ab)6 (GALLAGHER et al., 1993).Each pyruvoyl group lies in an active site at the interface between two subunits in a trimer (GELFMAN et al., 1991; PISHKO and RO-

BERTUS, 1993). Decarboxylation of L-histidine (**137**) by HDC occurs with retention of configuration (BATTERSBY et al., 1980; BATTERSBY, 1985; SANTANIELLO et al., 1981; YOUNG, 1991).

4.1.1.19 Orotidine-5′-Phosphate Decarboxylase (ODCase, Orotidine 5′-Monophosphate Decarboxylase [OMP], Orotidine-5′-Phosphate Carboxy-Lyase, EC 4.1.1.23)

Orotidine-5′-phosphate decarboxylase (OPD) catalyzes the decarboxylation of orotidine 5′-phosphate (**139**) to uridine 5′-phosphate (**140**) (Fig. 49), which is the final step in *de novo* pyrimidine biosynthesis (GAO et al., 1999a). This reaction is especially notable because OPD catalysis increases the rate by a factor of 10^{17} over the spontaneous rate, which is higher than that of any other enzyme described thus far. To appreciate the magnitude of catalysis; consider that orotic acid is estimated to have a half life for decarboxylation of some 78 million years, in neutral aqueous solution at room temperature (MILLER et al., 1998). The precise origin of the catalytic efficiency is currently an enigma (EHRLICH et al., 1999).

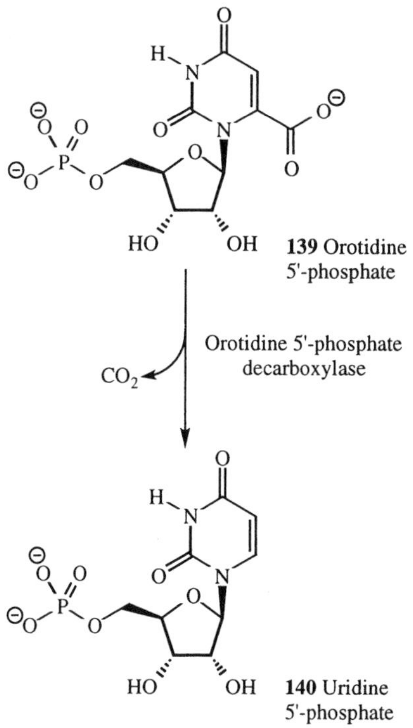

139 Orotidine 5'-phosphate

Orotidine 5'-phosphate decarboxylase

CO_2

140 Uridine 5'-phosphate

Fig. 49. The orotidine 5′-phosphate decarboxylase (ODC) catalyzed decarboxylation of orotidine 5′-phosphate (**139**) to uridine 5′-phosphate (**140**).

4.1.1.20 Tyrosine Decarboxylase (L-Tyrosine Carboxy-Lyase, EC 4.1.1.25)

Tyrosine decarboxylase (TDC) catalyzes the decarboxylation of L-tyrosine (**141a**) to tyramine (**142a**) and L-DOPA (**141b**) to dopamine (**142b**).(Fig. 50). TDC is a common plant enzyme involved in the synthesis of numerous secondary metabolites. Tyramine (**142a**) is a desirable flavoring in small amounts a range of food products including wine, cheese and sausages (MORENO-ARRIBAS and LONVAUD-FUNEL, 1999), and (together with dopamine) a biosynthetic precursor of numerous alkaloids in

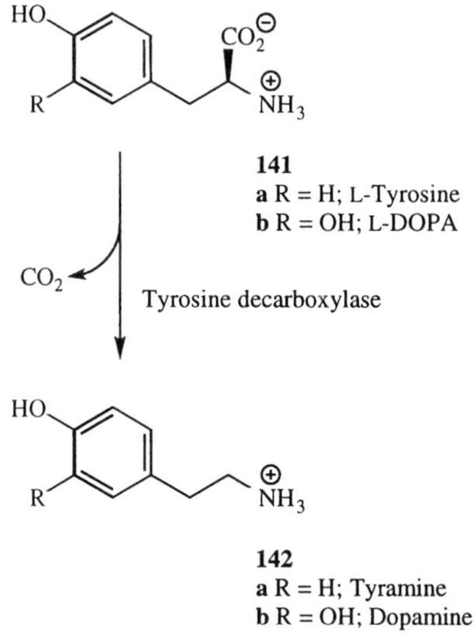

141
a R = H; L-Tyrosine
b R = OH; L-DOPA

CO_2

Tyrosine decarboxylase

142
a R = H; Tyramine
b R = OH; Dopamine

Fig. 50. The tyrosine decarboxylase catalyzed decarboxylation of L-tyrosine (**141a**) to tyramine (**142a**) and L-DOPA (3,4-dihydroxyphenylalanine) (**141b**) to dopamine (**142b**).

plants (Facchini, 1998). TDC is a PLP dependent enzyme and catalyzes decarboxylation such that the departing carboxylate group is replaced by hydrogen with retention of configuration. This reaction has been used to prepare tyramine hydrogen isotopomers (Belleau et al., 1960; Belleau and Burba, 1960; Young, 1991).

TDC has been utilized in a multi-enzyme process for the production of dopamine (**142b**) from catechol. Condensation of catechol, pyruvate and ammonia catalyzed by tyrosine phenol-lyase gave L-DOPA (**141b**),which was decarboxylated with rat liver L-DOPA decarboxylase or *Streptococcus faecalis* TDC. The L-DOPA decarboxylase was inhibited by L-DOPA concentrations greater than 1 mM, whereas TDC was unaffected up to 40 mM, consequently TDC was selected for further development. Decarboxylation was carried out in an enzyme reactor at 37 °C for 12 h, with pH control by addition of hydrochloric acid and supplements of PLP to compensate for decarboxylative transamination. At 100 mM L-DOPA (**141b**), conversion to dopamine (**142b**) was quantitative. When the decarboxylation was run with *in situ* synthesis of L-DOPA, the productivity was much lower than with two consecutive "single pot" reactions. (Lee et al., 1999a, b). The same process has also been run with whole cells expressing the relevent enzymes. *Erwinia herbicola* cells with phenol tyrosine-lyase activity and *Streptococcus faecalis* cells with TDC activity (Anderson et al., 1987) were coimmobilized in glutaraldehyde crosslinked porcine gelatin beads and placed in a packed bed reactor together with catechol, pyruvate and ammonia, Dopamine was produced but the conversion was low (Anderson et al., 1992a, b). These processes are discussed in greater detail in the section describing phenol tyrosine lyase (Sect. 4.1.4.2).

4.1.1.21 Aromatic L-Amino Acid Decarboxylase (DOPA Decarboxylase, Tryptophan Decarboxylase, Hydroxytryptophan Decarboxylase, EC 4.1.1.28)

Aromatic L-amino acid decarboxylase (AAAD) catalyzes the decarboxylation of L-tryptophan (**143a**) to tryptamine (**144a**); 5-hydroxy-L-tryptophan (**143b**) to serotonin (**144b**) (Fig. 51) and L-DOPA (3,4-dihydroxyphenylalanine) (**141b**) to dopamine (**142b**) (Fig. 50). Prior to 1984 the enzyme commission recognized DOPA decarboxylase (EC 4.1.1.26) and tryptophan decarboxylase (EC 4.1.1.27). However, these distinctions are not warranted by the selectivity of the enzymes and they have now been incorporated into the AAAD subsubsubclass. AAADs also generally, have some phenylalanine decarboxylase (EC 4.1.1.53) activity, and conversely, phenylalanine decarboxylases generally have some activity with L-DOPA (**141b**), tryptophan (**143a**) and 5-hydroxy-L-tryptophan (**143b**). AAADs are also

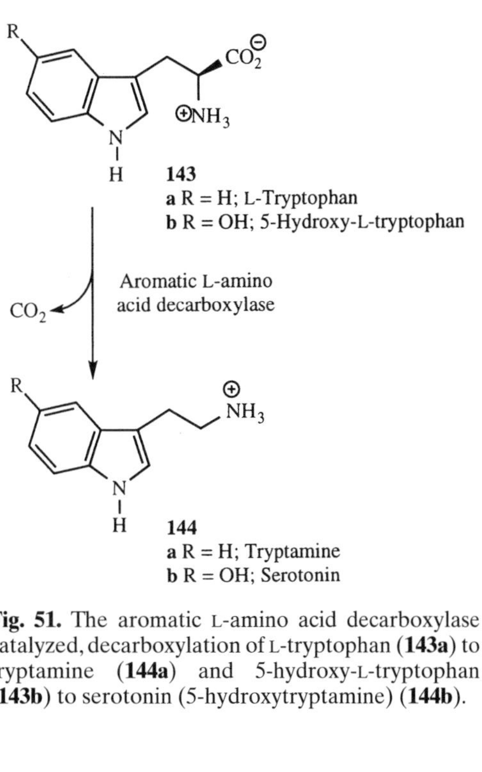

Fig. 51. The aromatic L-amino acid decarboxylase catalyzed, decarboxylation of L-tryptophan (**143a**) to tryptamine (**144a**) and 5-hydroxy-L-tryptophan (**143b**) to serotonin (5-hydroxytryptamine) (**144b**).

frequently described as L-DOPA decarboxylases. Typically, the ratio of activity of an AAAD between L-DOPA (**141b**) and 5-hydroxy-L-tryptophan (**143b**) is 5:1 (JEBAI et al., 1997). The plurality of substrates decarboxylated by AAADs suggests that a mixtures of enzymes or isozymes may be responsible for the activity, but at least with the well characterized enzymes from pig kidney and rat liver the vast majority of the data indicates that activity is due to a single protein.

AAAD is present in fungi (NIEDEND et al., 1999), plants, insects, fish (NAGAI et al., 1996) and mammals (ZHU and JUORIO, 1995; POUPON et al., 1999). In mammals, AAAD clearly plays a role in the biosynthesis of the neurotransmitters; serotonin (**144b**) and L-DOPA (**141b**), but other roles have been suggested including the prevention of apoptosis (BERRY et al., 1996; ZHU and JUORIO, 1995).

There are reliable methods for the extraction of pig kidney AAAD (VOLTATTORNI et al., 1987; DOMINICI et al., 1993), but instability and low specific activity limit the availability. Expression of the gene in *E. coli* gives protein with twice the specific activity of the extracted enzyme and each dimer binds two molecules of PLP rather than one as with the extracted enzyme. It appears that the latter contains PLP bound covalently in an non-catalytically active form (MOORE et al., 1996) and the differences between the two forms are apparent in recently reported preliminary crystal structure data (MALASHKEVICH, 1999). Extraction of rat liver or kidney AAAD (SOURKES, 1987) has also been supplanted by expression of the gene in *E. coli,* but in this case the recombinant protein is kinetically identical to the extracted protein (JEBAI et al., 1997).

As with all other PLP dependent, L-amino acid α-decarboxylases, decarboxylation occurs such that the carboxylate group is replaced by hydrogen with retention of configuration (e.g., 5-hydroxy-L-tryptophan, (**143b**) BATTERSBY et al., 1990). Despite the ability of AAAD to accomodate a wide ring off substrates it has barely been exploited for biotransformations, presumably because of its limited availability until recently. α-Methyl-L-DOPA (**145**) undergoes decarboxylation as expected to give α-methyldopamine (**146**) (Fig. 52), this is oxidatively deaminated to give 3,4-dihydroxyphenylacetone (**147**), which irreversibly inactivates the enzyme by formation of a ternary adduct with PLP and an active site lysine residue (BERTOLDI et al., 1998).

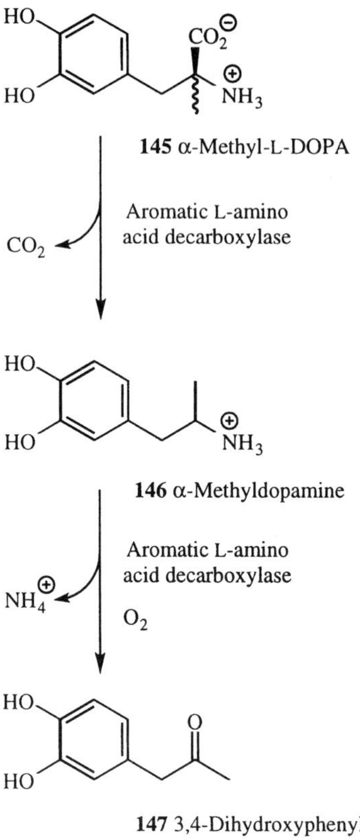

Fig. 52. The oxidative deamination of α-methyl-DOPA (**145**) produces 3,4-dihydroxyphenylacetone (**147**), which deactivates AAAD.

4.1.1.22 Phosphoenolpyruvate Carboxylases and Carboxykinases (EC 4.1.1.31, 32, 38, 49)

There are four lyases which catalyze the carboxylation of phosphoenolpyruvate (PEP) (**148**) (Fig. 53) to oxaloacetate (**28**) and the reverse reaction (Tab. 4). PEP is an excellent phosphate donor which is frequently used for recycling ADP to ATP. An efficient mole scale chemical synthesis has been reported (HIRSCHBEIN et al,1982).

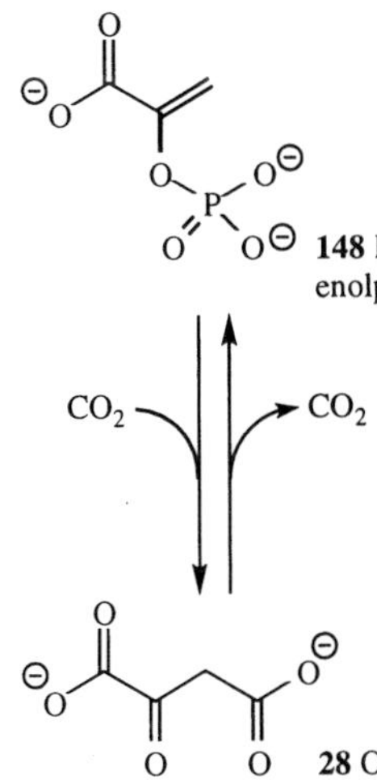

148 Phosphoenolpyruvate

CO_2 CO_2

28 Oxaloacetate

Fig. 53. The carboxylation of phosphoenolpyruvate (**148**) to give oxaloacetate (**28**) catalyzed by phosphoenolpyruvate carboxylases (PEP-CL) and carboxykinases (PEP-CK). Cofactors are not shown, see Fig. 54.

Tab. 4. Phosphoenolpyruvate Carboxylases and Carboxykinases

EC No.	Name Phosphoenolpyruvate-	Phosphate Donor	Other Products
4.1.1.31	-carboxylase	Orthophosphate	H_2O
4.1.1.32	-carboxykinase	GTP	GDP
4.1.1.38	-carboxylase	Pyrophosphate	Orthophosphate
4.1.1.49	-carboxykinase	ATP	ADP

PEP-carboxylases (CL) are present in plants, algae and bacteria, but the properties vary widely depending on the source (YANO et al., 1995). PEP-CL is especially important in C_4 plants, where it catalyzes the initial step of photosynthetic carbon fixation (Hatch-Slack pathway; LEPINIEC et al., 1994). PEP-CL will accept phosphate donors other than PEP, such as 2-phosphoenolbutyrate and phosphoenol-3-fluoropyruvate, but dephosphorylation without carboxylation is the predominant reaction (GONZALEZ and ANDREO, 1988).

The carboxykinases (CK) are assigned on the basis of their ability to utilize GTP or ATP, but both nucleotides or others may show some activity with a given enzyme. An extreme example is provided by the PEP-CK from the halophile *Vibrio costicola*, which is able to utilize four nucleotides with the following V_{max}/K_M values ($M^{-1}s^{-1}$) ATP (6.8), GTP (1.30), CTP (0.87), ITP (0.66). Decarboxylation is fully reversible and hence it may be used to incorporate isotopic labels (e.g., ^{13}C) into oxaloacetate (**28**) (SALVARREY et al., 1995; TARI et al., 1996).

4.1.1.23 Phenylpyruvate Decarboxylase (Phenylpyruvate Carboxy-Lyase, EC 4.1.1.43)

Phenylpyruvate decarboxylase catalyzes the decarboxylation of phenylpyruvate (**149**) to phenylacetaldehyde (**150**) (Fig. 54) and also has some activity with indole pyruvate (see Sect. 4.1.1.26). This reaction is the second step in the catabolism of phenylalanine, which commences with transamination to form phenylpyruvate. Moreover the decarboxylation is homologous to benzoylformate decarboxylation which is a later step in the same pathway (Sect. 4.1.1.7) (SCHNEIDER et al., 1997; BARROWMAN and FEWSON, 1985). This reaction is virtually unexplored, but there is the prospect that trapping with an aldehyde could be used to form an acyloin as observed with benzylformate decarboxylase (Fig. 26).

4.1.1.24 Tartronate Semi-Aldehyde Synthase [Glyoxylate Carboxy-Lyase (dimerizing), EC 4.1.1.47]

Tartronate semi-aldehyde synthase catalyzes the decarboxylation of glyoxylate (**151**) and condensation with another molecule of glyox-

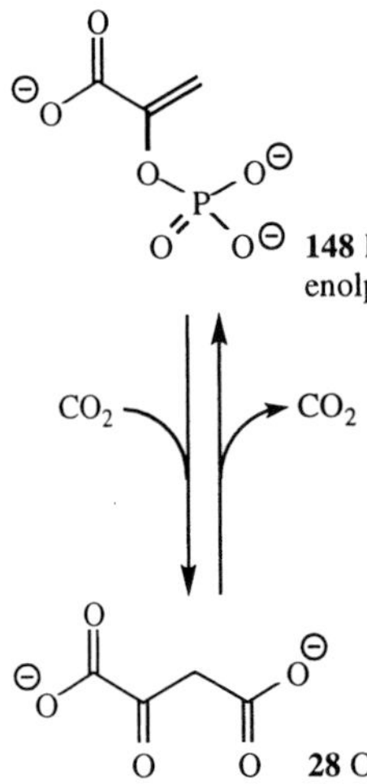

148 Phosphoenolpyruvate

28 Oxaloacetate

Fig. 54. The phenylpyruvate decarboxylase catalyzed decarboxylation of phenylpyruvate (**28**) to phenylacetaldehyde (**148**).

ylate to give tartronate semi-aldehyde (**152**) (Fig. 55), in a reaction analogous to the acyloin condensation of pyruvate (Fig. 2). The enzyme is reported to be a flavoprotein, although a TPP prosthetic group would be expected. This reaction forms part of the tartronate pathway, which is utilized by many organisms for the metabolism of glyoxylate. The strictly anerobic, obligatory oxalotrophic bacterium, *Bacillus oxalophilus*, utilizes tartronate semi-aldehyde

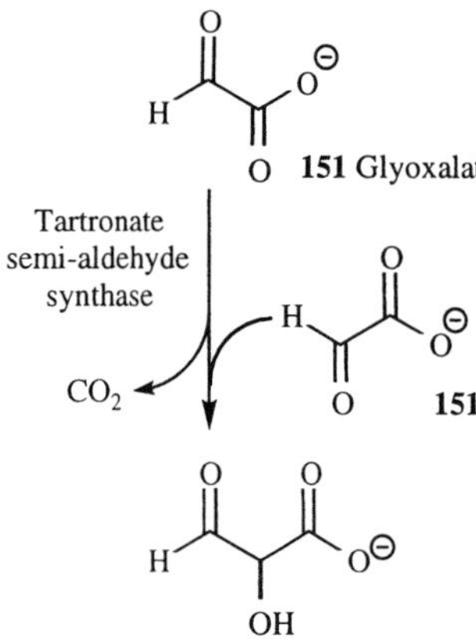

151 Glyoxalate

151 Glyoxalate

152 Tartronate semi-aldehyde

Fig. 55. The acyloin condensation of glyoxalate (**151**) to give tartronate semi-aldehyde (**152**), catalyzed by tartronate semi-aldehyde synthase.

synthase or the serine pathway to assimilate oxalate and oxalyl-CoA decarboxylase and formate dehydrogenase for oxidation of oxalate (ZAITSEV et al., 1993). Glycine fed to *Pseudomonas fluoresens*, is converted to glyoxylate, and proccessed to the antibiotic obafluorin via the tartronate semi-aldehyde pathway (HERBERT and KNAGGS, 1992).

4.1.1.25 Dialkylglycine Decarboxylase [2,2-Dialkylglycine Decarboxylase (Pyruvate), EC 4.1.1.64]

Dialkylglycine decarboxylase (DAGD) from *Pseudomonas cepacia* catalyzes the oxidative decarboxylation of dialkyl glycines (**153**) to ketones (**39**) (Fig. 56) and the PMP (**122**) form of the enzyme. Transamination with α-ketoacids such as pyruvate (**1**) regenerates the "internal imine" (**116b**) and an amino acid such as L-alanine (**111**). Thus this enzyme catalyzes two classical PLP-dependent reactions (cf. Fig. 41), with different substrates.

DAGD is a tetramer which resembles apartate aminotransferase. In solution, it consists of two conformers of differing activity (ZHOU and TONEY, 1999). The equilibrium between the two conformers is perturbed by alkali metal ions. Li^+ and Na^+, act as inhibitors, whereas K^+ and Rb^+ are activators. Similarly the enzyme adopts two different conformational forms in the crystalline state, depending on the alkali metal ion present (TONEY et al., 1995; HOHENESTER, et al., 1994). It is generally presumed that the conformers observed in the crystal structures correspond to those in solution. However, crystal structures of DAGD bound to other inhibitors (e.g., 1-aminocyclopropane carboxylate (**162**)), show only the "active" K^+/Rb^+ conformer (MALASHKEVICH, 1999).

As with other PLP-dependent decarboxylases, non-oxidative decarboxylation may also occur with DAGD. The ratio between non-oxidative and oxidative decarboxylation is determined by the propensity of the enzyme to protonate the quinoid intermediate (**118**; Fig. 41) at C-α and C-4$'$, which in turn is a function of

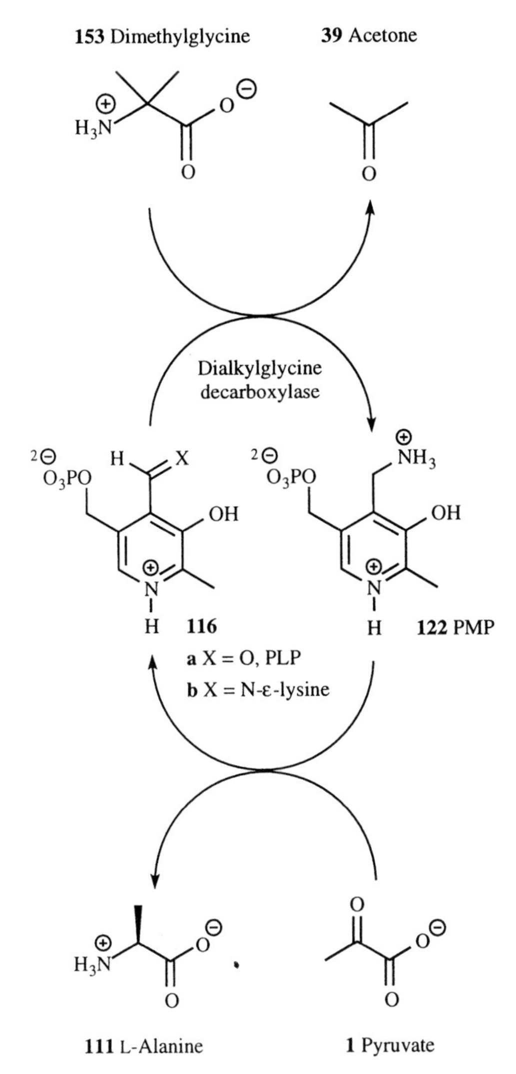

Fig. 56. Dialkylglycine decarboxylase (DAGD) catalyzed oxidative decarboxylation of dimethylglycine (**153**) to acetone (**39**). Pyruvate (**1**) is the preferred substrate for transaminating the PMP-form of the enzyme to the "internal" imine (**116b**).

the fit of the substrate to the active site. Dimethylaminoglycine (**153**) is the best substrate for DAGD (Tab. 5, entry 1). Larger alkyl groups can be accommodated (entries 2 and 3), but replacement of an α-alkyl group by hydrogen (entries 4–7) or carboxylate (entries 9, 10) results in appreciable non-oxidative decarboxylation (SUN et al., 1998a, b). L-Alanine (**111**) is more reactive than D-alanine (**156**) (entries 4,5), which has low and comparable reactivity to L-phenylglycine (**157**) (entry 6). D-Phenylglycine is not a substrate.

Most extraordinarily, 1-aminocyclopropane carboxylate (**162**), forms an imine with the PLP group, but does not undergo decarboxylation. The X-ray crystal structure of this adduct shows the aldimine bond is out of the plane of the pyridinium ring (cf.**117**, Fig. 41) and thus is unable to act as an electron sink for the pair of electrons produced by decarboxylation (MALASHKEVICH, 1999).

This enzyme has tremendous potential for the resolution of α-methyl and other α-alkyl amino acids, which are poor substrates with amidases and proteases.

4.1.1.26 Indolepyruvate Decarboxylase (EC 4.1.1.72)

Indolepyruvate decarboxylase (IDPC) catalyzes the decarboxylation of indole-3-pyruvate (**163**) to indole-3-acetaldehyde (**164**) (Fig. 57). Phenylpyruvate decarboxylase (EC 4.1.1.43) also catalyzes this reaction. In bacteria, L-tryptophan (**143a**) is converted to indole-3-acetic acid (**165**) by two pathways; the indole-3-acetamide pathway (EMANUELE and FITZPATRICK, 1995) and the indole-3-pyruvate pathway (MANULIS et al., 1998). The latter commences with the transamination of L-tryptophan (**143a**) to indole-3-pyruvate (**163**) (KOGA et al., 1994; GAO et al., 1997), followed by decarboxylation to indole-3-acetaldehyde (**164**) and oxidation to indole-3-acetic acid (**165**). In yeast, a similar pathway is followed except that indole-3-acetaldehyde (**164**) is reduced to indole-3-ethanol (tryptophol, IRAQUI et al., 1999; FURUKAWA et al., 1996). The biosynthesis of indole-3-acetic acid (**165**) in plants has been extensively investigated because of its importance as a plant growth regulator, nevertheless important aspects of this pathway remain to be elucidated (KAWAGUCHI et al., 1996). Consequently, much of the discussion that follows will deal with bacterial species, particularly those that are plant pathogens or are root associated. Many of these organisms overproduce indole-3-acetic acid (**165**) in culture when fed supplementary L-trytophan (**143a**) and

Tab. 5. Decarboxylation and Transamination Catalyzed by Dialkylglycine Decarboxylase

Entry	Substrate	Decarboxylation		Non-Oxidative Decarboxylation	Transamination	
		k_{cat} (s^{-1})	k_{cat}/K_M ($M^{-1}s^{-1}$)	(%)	k_{cat} (s^{-1})	k_{cat}/K_M ($M^{-1}s^{-1}$)
1	**153**	25 (1)	7.2 (0.5) $\cdot 10^3$	< 0.001	na	na
2	**154**	0.98 (0.03)	2.2 (0.2) $\cdot 10^2$	< 0.01	na	na
3	**155**	2.3 (0.2)	43 (3)	< 0.01	na	na
4	**111**	3.3 (0.3) $\cdot 10^{-2}$	4.7 (0.7)	9.8 (3.0)	21 (2)	3.5 (0.6) · 103
5	**156**	0.20 (0.02)	28 (5)	3.7 (0.2)	< 10-5	nd
6	**157**	0.20 (0.03)	24 (5)	5.8 (0.8)	nd	nd
7	**158**	1.2 (0.1) $\cdot 10^{-4}$	5.2 (0.9) $\cdot 10^{-4}$	< 1.5	2.5 (0.3) · 10–4	1.1 (0.2) · 10–3
8	**159**	na	na	na	7.2 (0.5) $\cdot 10^{-3}$	7.7 (1.0) $\cdot 10^{-2}$
9	**160**	7.98 (0.8) $\cdot 10^{-3}$	1.1 (0.2)	56 (9)	na	na
10	**161**	25 (6) $\cdot 10^{-3}$	0.11 (0.03)	94 (6)	nd	nd

Estimated errors are shown in brackets (SUN et al., 1998a, b)

153 Dimethylglycine **154** 1-Aminocyclopentane carboxylate **155** 1-Aminocyclohexane carboxylate **111** L-Alanine **156** D-Alanine **157** L-Phenylglycine

158 Glycine **159** *iso*-Propylamine **160** Methylaminomalonate **161** Aminomalonate **162** 1-Aminocyclopropane carboxylate

consequently they have the potential to biotransform other exogenous substances (BRANDL and LINDLOW, 1998).

The gene for IDPC has been identified in the root associated bacterium, *Azospirillum brasilense* (BROEK et al., 1999; ZIMMER et al.,

1998), the plant gall inducer, *Erwinia herbicola* (BRANDL and LINDOW, 1996, 1997), seven species of *Enterobacteriaceae* (ZIMMER et al., 1994) and *Enterobacter cloacae* (KOGA et al., 1991) and IDPC activity has been detected in *Bradyrhizobium elkanii* (MINAMISAWA et al., 1996). The IDPC gene from the plant associated bacterium *Enterobacter cloacae*, has been expressed in *E. coli* and the enzyme characterized. It is a tetramer which is TPP and magnesium dependent. IDPC is inactive with indole-3-lactate and oxaloacetate (**33**), but does decarboxylate pyruvate (**1**) (19% activity relative to indolepyruvate) (KOGA, 1995; KOGA et al., 1992). Although there have been no synthetic applications of this enzyme thus far, the activity with pyruvate suggests the enzyme should be fairly substrate permissive and clearly there are excellent prospects for acyloin formation. Whole organism multi-enzyme processes may also be possible. The aminotransferase from *Candida maltosa* accepts a range of fluoro- and methyl-tryptophans (BODE and BIRNBAUM, 1991). Indolepyruvate ferredoxin oxidoreductase catalyzes the oxidative decarboxylation of arylpyruvates (SIDDIQUI et al., 1997, 1998).

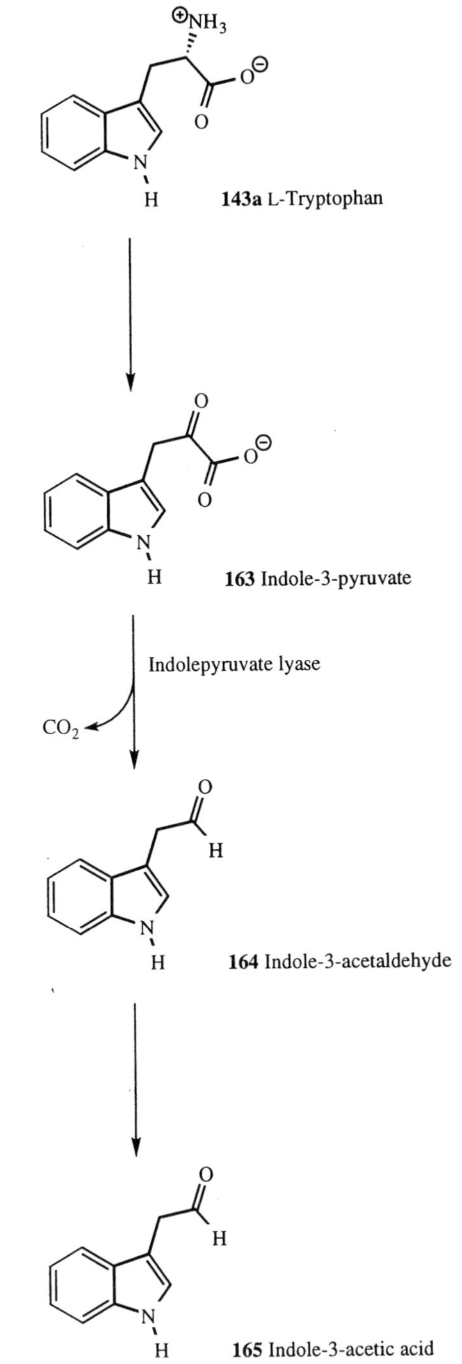

143a L-Tryptophan

163 Indole-3-pyruvate

Indolepyruvate lyase

CO_2

164 Indole-3-acetaldehyde

165 Indole-3-acetic acid

Fig. 57. The indole-3-pyruvate pathway for the conversion of L-tryptophan (**143a**) to indole-3-acetic acid (**165**).

4.1.2 Aldehyde-Lyases (EC 4.1.2)

Aldehyde lyases are enzymes which catalyze the cleavage of alcohols to carbanions and aldehydes. The vast majority are aldolases, involved in carbohydrate metabolism. Two examples (involving non-carbohydrates) are provided here for illustrative purposes.

4.1.2.1 Threonine Aldolase (L-Threonine Acetaldehyde-Lyase, EC 4.1.2.5)

Threonine aldolase is a PLP-dependent enzyme which catalyzes the retro-aldol cleavage of L-threonine (**166**) to acetaldehyde (**2**) and glycine (**158**) (Fig. 58). A similar reaction occurs with serine, but the enzyme is both PLP and tetrahydrofolate dependent and is classified as serine (or glycine) hydroxymethyltransferase (EC 2.1.2.1). This enzyme also acts

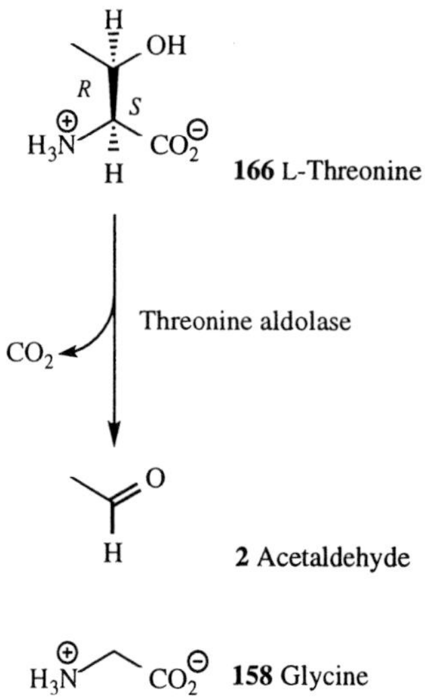

Fig. 58. The threonine aldolase catalyzed retro-aldol cleavage of L-threonine (**166**) to acetaldehyde (**2**) and glycine (**158**).

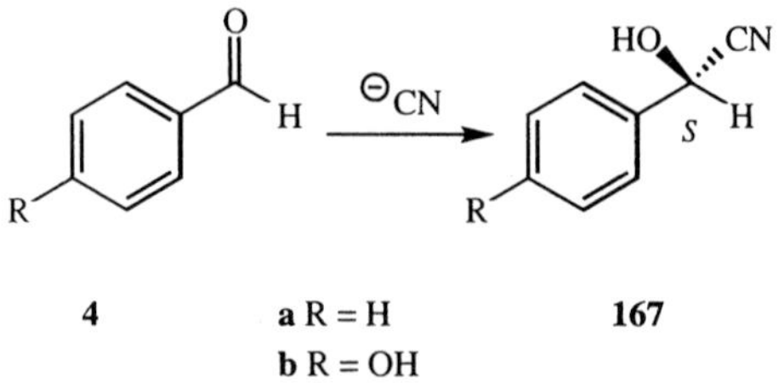

on L-threonine (**166**) (OGAWA and FUJIOKA, 1999). Perhaps surprisingly, enzymes have been isolated that can transform all four stereoisomers of threonine. In most cases, recognition of the α-amino acid center (i.e., D- or L-) is fairly specific, but recognition of the stereochemistry of the carbon bearing the hydroxyl group is more relaxed (WADA et al., 1998; KATAOKA et al., 1997). These enzymes are discussed in detail in Chapter 1, this volume: Carbon–Carbon Bond Formation Using Enzymes.

4.1.2.2 Mandelonitrile Lyase (EC 4.1.2.10) and Hydroxymandelonitrile Lyase (EC 4.1.2.11)

Mandelonitrile lyase and hydroxymandelonitrile lyase catalyzes the addition of cyanide to benzaldehyde (**4a**) or *p*-hydroxybenzaldehyde (**4b**) to give mandelonitrile (**167a**) or (*p*-hydroxy)mandelonitrile (**167b**) respectively (Fig. 59). These enzymes are oftern termed oxynitrilases. They occur predominantly in plants (WAJANT and EFFENBERGER, 1996), where catalyze the formation and elimination

Fig. 59. The mandelonitrile lyase or hydroxymandelonitrile lyase catalyzed addition of cyanide to benzaldehyde (**4a**) or *p*-hydroxybenzaldehyde (**4b**) to give mandelonitrile (**167a**) or (*p*-hydroxy)mandelonitrile (**167b**).

of cyanogenic glycosides which are utilized for protection against predators (WAGNER et al., 1996). The synthetic potential of this reaction has been reviewed in Chapter 6 of Volume 8a of the *Biotechnology* series (BUNCH, 1998) and elsewhere (POCHLAUER, 1998; FESSNER, 1998; FESSNER and WALTER, 1997; EFFENBERGER, 1999).

4.1.3 Oxo-Acid-Lyases (EC 4.1.3)

Oxo-acid-lyases catalyze the cleavage of α-hydroxy carboxylic acids to α-carbonyl carboxylic acids and carbanions in a retro-aldol type reaction. This subsubclass includes several enzymes from the citric acid (Krebs) cycle and related pathways.

4.1.3.1 Isocitrate Lyase (Isocitrate Glyoxylate-Lyase, Isocitrase, EC 4.1.3.1)

Plants (but not animals) are able to convert acetyl-CoA (**56**) to oxaloacetate (**33**) via the glyoxylate pathway. The first step in this pathway is the cleavage of isocitrate (**168**) to glyoxylate (**151**) and succinate (**169**) catalyzed by isocitrate lyase, which can be rationalized as a retro-aldol reaction (Fig. 60) (REHMAN and MCFADDEN, 1996). When the reaction is carried out in deuterium oxide, (2*S*)-[2-$^2H_1$]-succinate (**169**) is produced, hence C–C bond cleavage occurs with inversion of configuration (DUCROCQ et al., 1984).

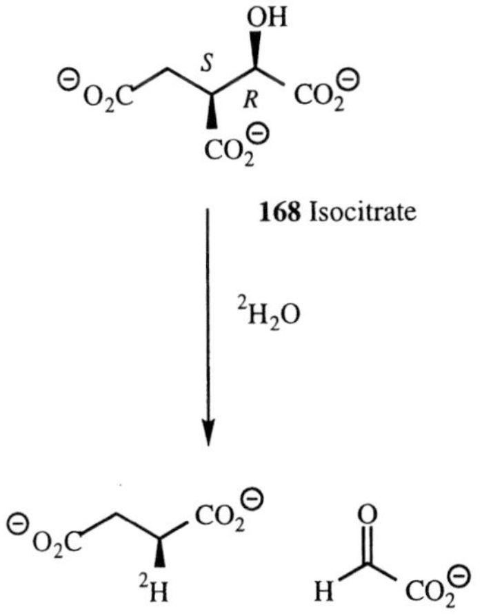

168 Isocitrate

^2H_2O

169 (2*S*)-[2-$^2H_1$]-Succinate 151 Glyoxylate

Fig. 60. The isocitrate lyase catalyzed retro-aldol cleavage of isocitrate to succinate and glyoxylate in deuterium oxide.

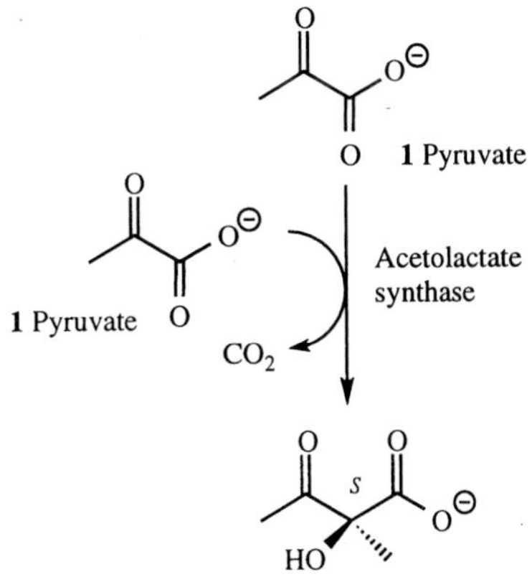

1 Pyruvate

1 Pyruvate

Acetolactate synthase

CO_2

65 (2*S*)-α-Acetolactate

Fig. 61. Acetolactate synthase catalyzes the decarboxylative aldol condensation of two molecules of pyruvate (**1**), to give (2*S*)-α-acetolactate (**65**).

4.1.3.2 Acetolactate Synthase Acetolactate Pyruvate-Lyase (carboxylating), EC 4.1.3.18]

α-Acetolactate (**65**) is produced biosynthetically by the condensation of two molecules of pyruvate (**1**) with concommittant decarboxylation, catalyzed by acetolactate synthase (Fig. 61). The enzyme is TPP dependent and hence the mechanism is very similar to that of pyruvate decarboxylase described in Sect. 4.1.1.1 (Carroll et al., 1999; Chipman et al., 1998; Hill et al., 1997) (Fig. 23).

4.1.4 Other Carbon–Carbon Lyases

4.1.4.1 Tryptophanase [L-Tryptophan Indole-Lyase (deaminating), EC 4.1.99.1]

Tryptophanase is a PLP-dependent enzyme, which catalyzes the cleavage of L-tryptophan (**143a**) to indole (**170**), pyruvate (**1**) and ammonia (Fig. 62). It also catalyzes the net reverse reaction, as well as α,β-elimination, β-replacement and α-hydrogen exchange with various L-amino acids. The β-subunit of tryptophan synthase or desmolase (EC 4.2.1.20) catalyzes similar reactions and the coupling of indole with serine to give L-tryptophan. However, this is a bifunctional enzyme which utilizes

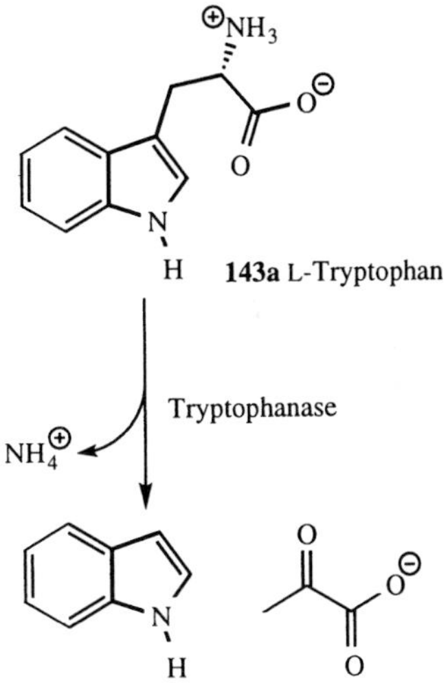

143a L-Tryptophan

Tryptophanase

NH_4^+

170 Indole 1 Pyruvate

Fig. 62. Tryptophanase catalyzed cleavage of L-tryptophan (**143a**) to indole (**170**), pyruvate (**1**) and ammonia.

the α-subunit to cleave 3-indolyl-D-glycerol 3′-phosphate to indole and glyceraldehyde 3′-phosphate (WOEHL and DUNN, 1999). Tryptophanase is unable to accomplish this latter reaction.

Tryptophanase is a tetramer, consisting of identical units (52,000 Da) each of which contains one molecule of PLP. It is activated by potassium ions and inactivated by cooling, which causes release of the PLP and dissociation into dimers (EREZ et al., 1998). The X-ray crystal structure resembles that of aspartate aminotransferase and tyrosine phenol-lyase, with the active site at a dimer interface (ISUPOV et al., 1998).

The mechanism of cleavage proceeds via formation of an "external" aldimine (**171**) (Fig. 63), between the substrate and the tryptophanase bound PLP group (**116b**). Proton transfer between the α-carbon atom and protonation of the indole gives the quinoid intermediate (**172**), which is poised for cleavage of the exocyclic indole C–C bond to give the aminoacrylate (**173**). This undergoes hydrolysis to complete the cycle and yield the initial form of the enzyme (**116b**), pyruvate (**1**) and ammonium ions (IKUSHIRO et al., 1998; PHILLIPS, 1989). The transfer of hydrogen from the C-a center to the 3-position of the indole ring involves a base with pK_a = 7.6 and deprotonation/protonation of the indolic nitrogen involves a base with pK_a = 6.0. The mechanism of action of tryptophanase has a number of similarities to that of tyrosine phenol-lyase and the section

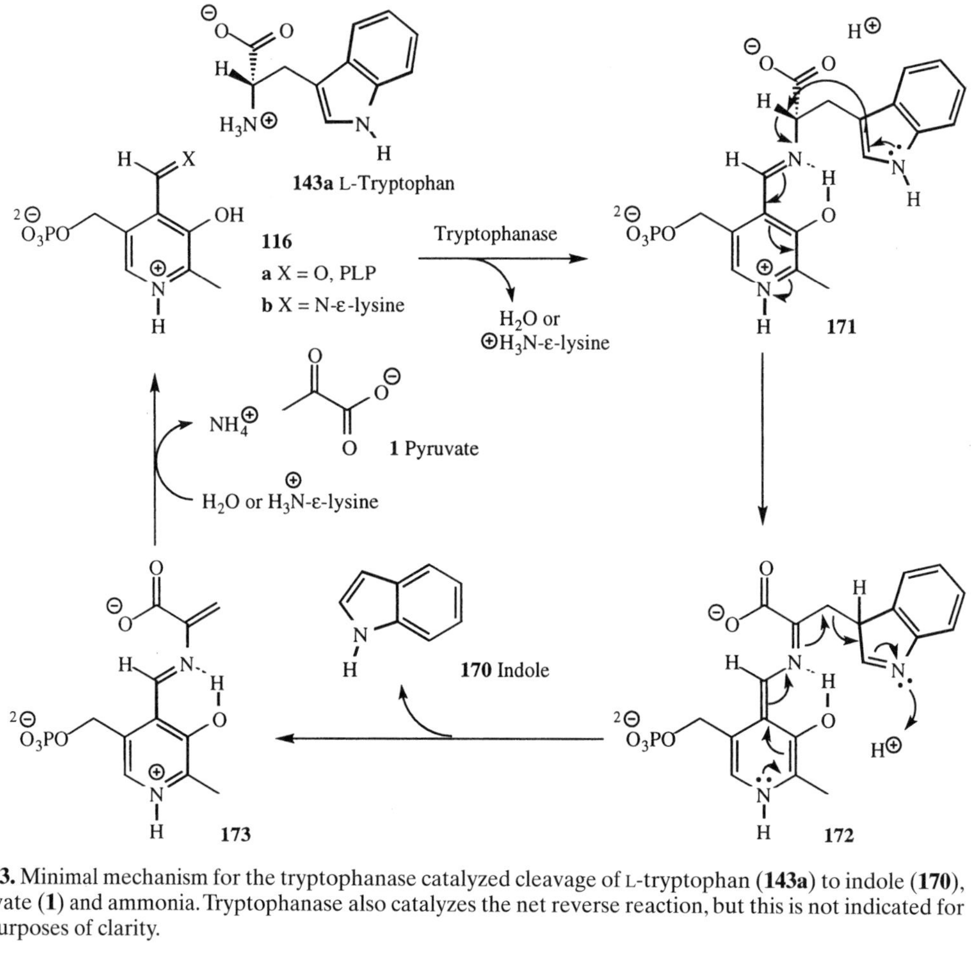

Fig. 63. Minimal mechanism for the tryptophanase catalyzed cleavage of L-tryptophan (**143a**) to indole (**170**), pyruvate (**1**) and ammonia. Tryptophanase also catalyzes the net reverse reaction, but this is not indicated for the purposes of clarity.

on the latter (Sect. 4.1.4.2) should be consulted for further details.

Currently, tryptophan is resolved industrially by the enzyme catalyzed hydrolysis of racemic *N*-carbamoyl or hydantoinyl derivatives, but more direct routes would be desirable. The *Enterobacter aerogenes* tryptophanase gene was expressed in the pyruvate producing *E. coli* strain W1485lip2. After 32 hours in culture the 28.6 g L^{-1} of pyruvate had been produced. Addition of ammonia and indole initiated L-tryptophan production which reached 23.7 g L^{-1} after a further 36 hours (KAWASAKI et al., 1996b).

The synthesis of compounds containing ^{11}C is particularly demanding because of the short-half life ($t_{1/2}$ = 20.34 minutes) and the limited availability of costly labeled precursors. These and a number of other problems were solved in a neat synthesis of [3-^{11}C]-5-hydroxy-L-tryptophan (**178**). Racemic [3-^{11}C]-alanine (**174**) (**175**), 5-hydroxyindole (**177**) plus ammonium salts and coenzymes were passed through a column packed with immobilized alanine racemase, D-amino acid oxidase and tryptophanase. The alanine racemase ensures that the [3-^{11}C]-L-alanine (**175**) is equilibrated with [3-^{11}C]-D-alanine (**174**), which is oxidatively decarboxylated to [3-^{11}C]-pyruvate (**176**) and ammonium ions. Both of these then react with 5-hydroxyindole (**177**) catalyzed by tryptophanase to give [3-^{11}C]-5-hydroxy-L-tryptophan (**178**) (Fig. 64) in 60% yield (IKEMOTO et al., 1999; AXELSSON et al., 1992). Tryptophanase from *E. coli* has been assayed with a broad range of "tryptophans" in which the benzene ring has been replaced by pyridine or thiophene. 4-Aza, 5-aza, 6-aza and 7-aza-L-tryptophan (**143a**) (Fig. 65) were all very slow substrates (*kcat* <1% of L-tryptophan), whereas β-indazoyl-L-alanine (**179**) was a better substrate and the thiophene analogs approached the reactivity of L-tryptophan (SLOAN and PHILLIPS, 1996; PHILLIPS, 1989).

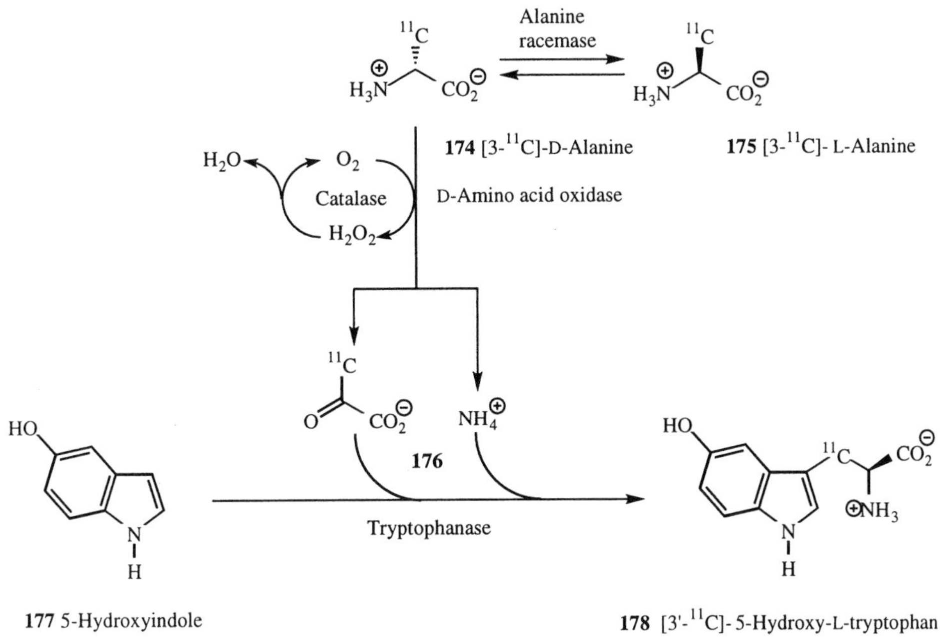

174 [3-^{11}C]-D-Alanine; **175** [3-^{11}C]-L-Alanine; Alanine racemase; Catalase; D-Amino acid oxidase; **176**; Tryptophanase; **177** 5-Hydroxyindole; **178** [3'-^{11}C]-5-Hydroxy-L-tryptophan

Fig. 64. Multi-enzyme system for the preparation of [3-^{11}C]-5-hydroxy-L-tryptophan (**178**).

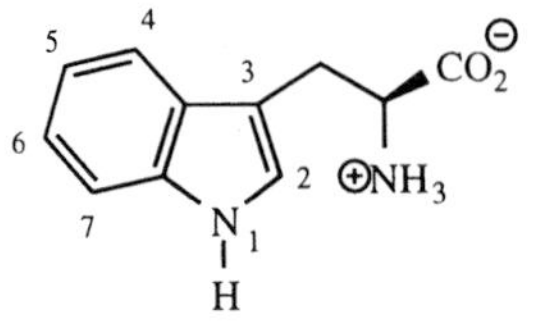

143a L-Tryptophan

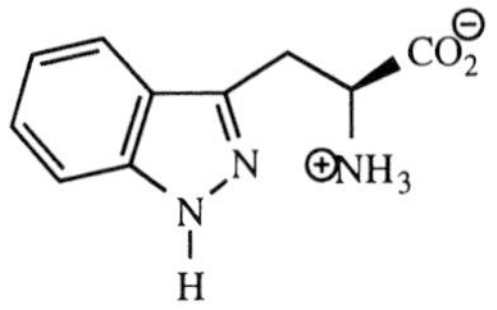

179 β-Indazoyl-L-alanine

Fig. 65. L-Tryptophan substitution and analogs.

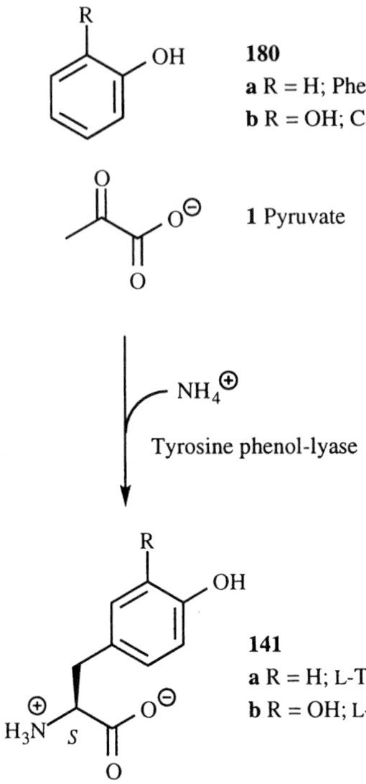

180
a R = H; Phenol
b R = OH; Catechol

1 Pyruvate

$NH_4^{\oplus}$
Tyrosine phenol-lyase

141
a R = H; L-Tyrosine
b R = OH; L-DOPA

Fig. 66. Tyrosine phenol-lyase catalyzes the synthesis of L-tyrosine (**141a**) from phenol (**180a**), ammonia and pyruvate (**1**). By using catechol (**180b**) in place of phenol L-DOPA (**141b**, 3-(3,4-dihydroxyphenyl)-L-alanine) can also be prepared.

4.1.4.2 L-Tyrosine Phenol-Lyase (deaminating) (TPL, β-Tyrosinase, EC 4.1.99.2)

Tyrosine phenol-lyase (TPL, β-tyrosinase) is a PLP dependent enzyme, which catalyzes the racemization, α,β-elimination and β-substitution of L-tyrosine (**141a**) and other amino acids such as L- or D-serine (**184a**), *S*-alkyl L-cysteines (**184c,d**) and β-chloro-L-alanine (**184e**) (SUNDARARAJU et al., 1997). It also catalyzes the net synthesis of L-tyrosine from pyruvate (**1**), ammonia and phenol (**180a**, Fig. 66) at high concentrations of ammonium pyruvate. This latter reaction has attracted considerable attention because TPL accepts a range of aromatic substrates in place of phenol (Tab. 6).

Tab. 6. Products from Tyrosine Phenol-Lyase (TPL) Catalyzed Reactions

Product	Reference
L-Tyrosine (**141a**)	cf. SUNDARARJU et al., 1997
L-[β-^{11}C]-tyrosine	AXELSSON et al., 1992
L-DOPA (**141b**)	LEE et al., 1999a; SUZUKI et al., 1992
L-[β-^{11}C]-DOPA	IKEMOTO et al., 1999;
Dopamine (**142b**)	LEE et al., 1999a, ANDERSOŃ et al., 1992a, b
(2*S*,3*R*)-β-methyltyrosine	KIM and COLE, 1999
Fluorinated tyrosines	PHILLIPS et al., 1997; VON TERSCH et al., 1996; FALEEV et al., 1995, 1996; URBAN and VON TERSCH, 1999
2-Chloro-L-tyrosine	FALEEV et al., 1995:
2-Methyl- and 3-methyl-L-tyrosine	FALEEV et al., 1995:
2-Azido-L-tyrosine	HEBEL et al, 1992
1-Amino-2-(4-hydroxyphenyl)ethyl phosphinic acid	KHOMUTOV et al., 1997

For example, substitution of catechol (**180b**) enables the synthesis of L-DOPA (**141b**), which is used in the treatment of Parkinson's disease (LEE et al., 1999a).

TPL is found primarily in enterobacteria plus a few arthropods and plays a role in the biosynthesis of shikimates (DEWICK 1998). TPL's have been identified in *Erwinia herbicola* (and expressed in *E. coli*, FOOR et al., 1993), *Citrobacter freundii* (POLAK and BRZESKI, 1990), *Citrobacter intermedius* (NAGASAWA, 1981) and *Symbiobacterium sp.* (LEE et al., 1997). X-ray crystal structures have been determined for the TPLs from *Erwina herbicola* (PLETNEV, 1997, 1996) and that from *Citrobacter freundii* with bound 3-(4′-hydroxyphenyl)propionic acid (SUNDARARAJU et al., 1997) and without (ANTSON et al., 1992, 1993).

The mechanism of action of tyrosine phenol-lyase is similar to that of tryptophanase. Aldimine formation between the "internal" PLP-aldimine (**116**) (Fig. 67) and L-tyrosine (**141a**) yields an aldimine which is deprotonated by "base-1" (pK_a = 7.6) to give the quinomethide (**181**). Concurrent deprotonation of the phenolic hydroxyl group by "base-2" (pK_a = 8.0) and C-4 protonation of the phenol by "base-1-H^+" gives the quinone (**182**), which undergoes protonation by "base-2-H^+" and rate limiting bond cleavage to phenol (**180a**), plus the pyruvate enamine derivative of PLP (**173**) (cf. Fig. 63). Hydrolysis restores the resting state of the enzyme (AXELSSON et al., 1992). The elimination and substitution reactions of serine derivatives bearing a β-nucleofuge (**184**) may be rationalized by an abbreviated form of this mechanism. In this case aldimine (**185**) (Fig. 68), formation and deprotonation results in β-elimination to give the the pyruvate enamine derivative of PLP (**173**) directly (CHEN and PHILLIPS, 1993). The identity of base-1 remains uncertain despite the X-ray crystal structure data. In *Citrobacter freundii* TPL it is probably Lys-257 or a water molecule bound to Lys-256. Base-2 is almost certainly Arg-381 (SUNDARARAJU et al., 1997).

TPL has been identified in several species (mainly Enterobacteriaceae), grown on media containing L-tyrosine (**180a**), but TPLs are generally inhibited by catechol (**180b**) which limits their application to the synthesis of L-DOPA (**141b**). A thermostable TPL, which is tolerant of catechol, was identified in a thermophile, *Symbiobacterium* sp. SC-1, however, this only grows in coculture with *Bacillus* sp. SK-1, which makes large scale cultivation extremely difficult. The gene for TPL was identified and overexpressed in *E. coli* (15% of soluble proteins) and isolated in 48% yield (LEE et al., 1996, 1997). The enzyme is a tetramer (MW 202 kDa), consisting of four identical subunits, each of which contains one molecule of pyridoxal 5′-phosphate (PLP). Remarkably, the enzyme retains full activity at 70 °C for at least 30 minutes (LEE et al., 1999b). *Symbiobacterium thermophilum* is another obligate, symbiotic, thermophile that only grows in coculture with a specific thermophilic, *Bacillus sp.*, strain S. It produces both tryptophanase and TPL (SUZUKI et al., 1992). Expression of the gene for TPL in *E. coli* yielded a 375 times increase in TPL production relative to that of *S. thermophilum* (HIRAHARA et al., 1993). The amino acid sequences of the TPLs from the two *Symbiobacterium* sp. differ at only three positions (LEE et al., 1997).

Symbiobacterium sp. SC-1 TPL has been utilized in a multi-enzyme process for the production of dopamine (**142b**) from catechol. Condensation of catechol (**180b**), pyruvate (**1**) and ammonium salts catalyzed by TPL gave L-DOPA (**141b**) (Fig. 69), which was decarboxylated with rat liver L-DOPA decarboxylase or *Streptococcus faecalis* tyrosine decarboxylase. The L-DOPA decarboxylase was inhibited by L-DOPA concentrations greater than 1 mM, whereas tyrosine decarboxylase was unaffected up to 40 mM, consequently tyrosine decarboxylase was selected for further development. Decarboxylation was carried out in an enzyme reactor at 37 °C for 12 hours, with pH control by addition of hydrochloric acid and supplements of PLP to compensate for decarboxylative transamination. At 100 mM L-DOPA (**141b**), conversion to dopamine (**142b**) was quantitative. When the decarboxylation was run with *in situ* synthesis of L-DOPA, the productivity was much lower than with two consecutive "single pot" reactions. (LEE et al., 1999a, b). *Erwinia herbicola* TPL and *Streptococcus faecalis* tyrosine decarboxylase have been investigated as components of a multi-enzyme system. Kinetic studies were run at pH 7.1 which is a compromise value dictated by

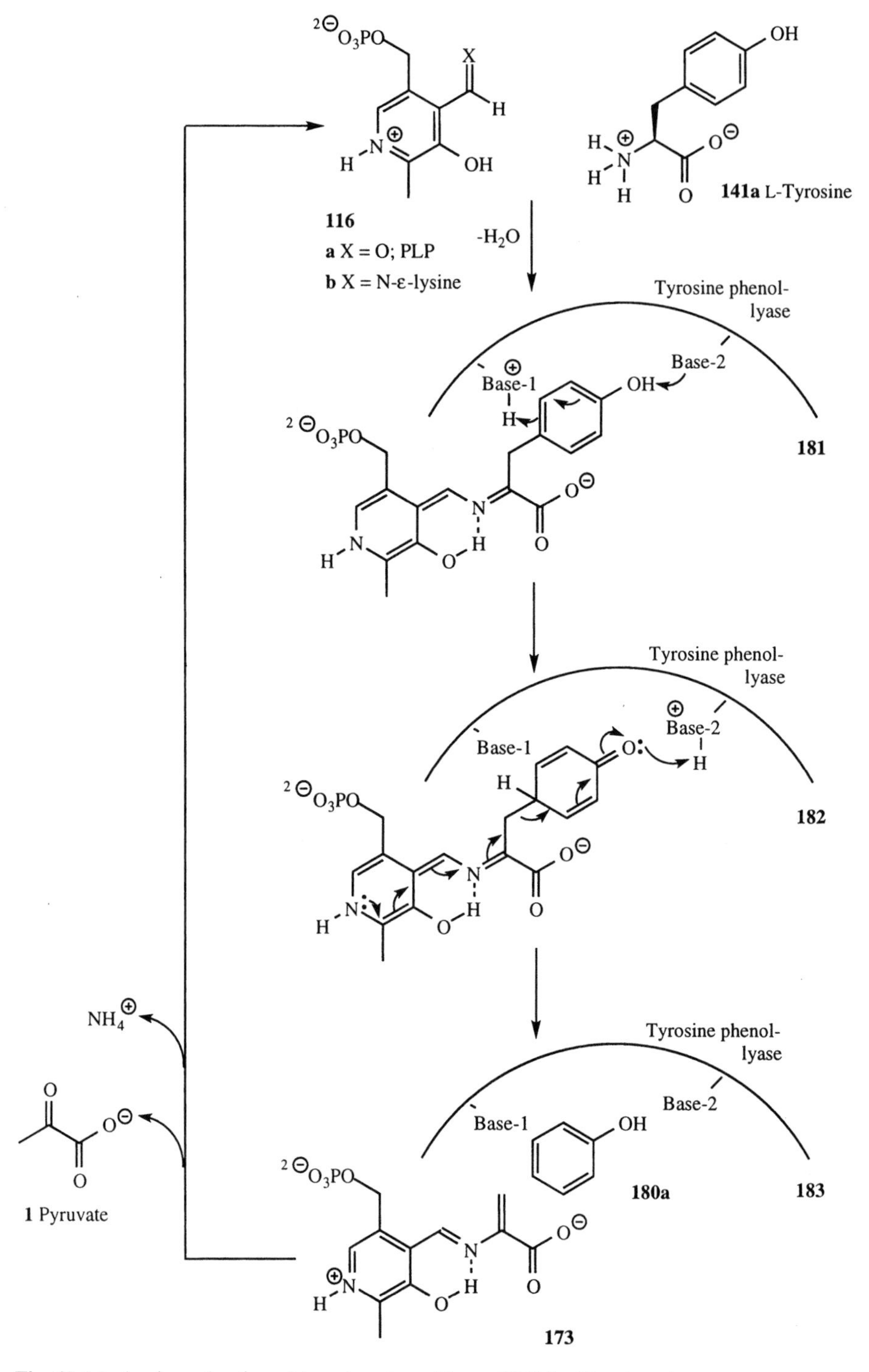

Fig. 67. Mechanism of action of tyrosine phenol-lyase (TPL) with L-tyrosine.

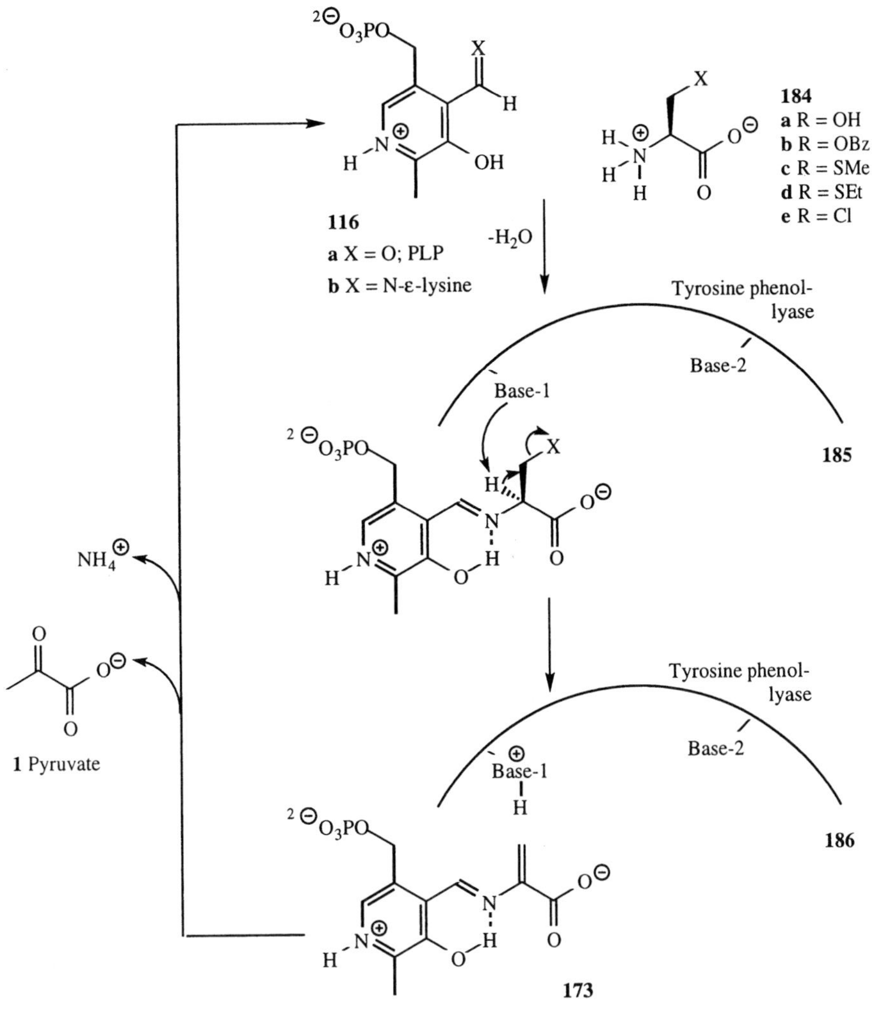

Fig. 68. Mechanism of action of tyrosine phenol-lyase (TPL) with β-substituted L-alanines (**184**).

differing values for the pH optima. The TPL followed pseudo 1st order kinetics with catechol, pyruvate and ammonium salts, whereas, the tyrosine decarboxylase was inhibited by dopamine and combinations of pyruvate and catechol (ANDERSON et al., 1992a). The same process has also been run with whole cells expressing the relevent enzymes. *Erwinia herbicola* cells with phenol tyrosine-lyase activity and *Streptococcus faecalis* cells with tyrosine decarboxylase activity (ANDERSON et al., 1987) were co-immobilized in glutaraldehyde cross-linked porcine gelatin beads (mean diameter 2.8 mm). The beads were placed in a packed bed reactor together with catechol, pyruvate and ammonia. Dopamine was produced, but the conversion was low (ANDERSON et al., 1992b). Whole *Erwinia herbicola* cells were microencapsulated in alginate-poly-L-lysine-alginate membraned microcapsules, and tested for TPL activity in a rotary shaker incubator. An agitation rate of at least 240 rpm was required to ensure that the microencapsulated cells achieved the activity of the free cells (LLOYD-GEORGE and CHANG, 1993, 1995).

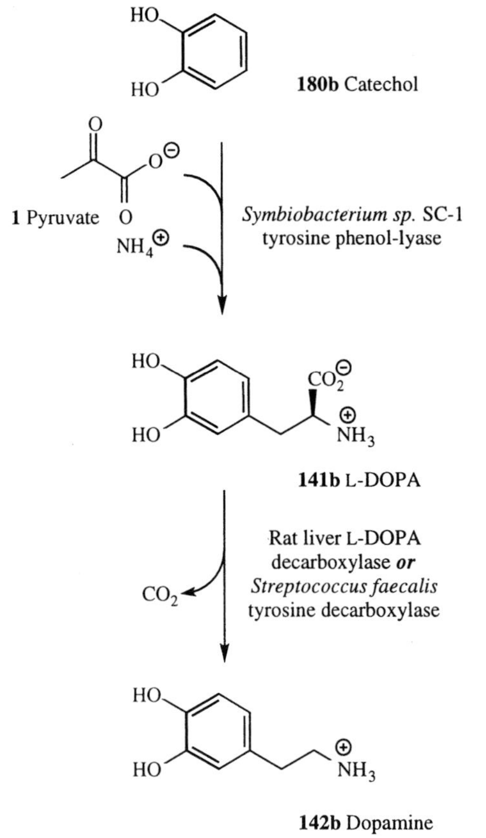

Fig. 69. The multi-enzyme synthesis of dopamine (**142b**) (LEE et al., 1999).

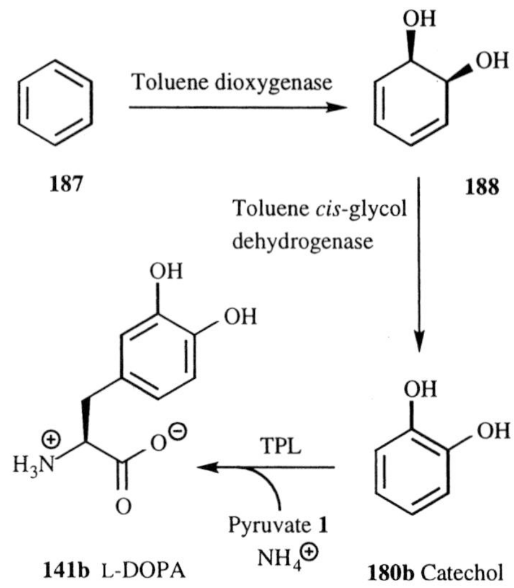

In a neat approach, toluene dioxygenase, toluene *cis*-glycol dehydrogenase and TPL from *Citrobacter freundii* (BAI and SOMERVILLE, 1998) have been coexpressed in *E. coli*. The maximum concentration of L-DOPA (**141b**) (Fig. 70) achieved in 4 hours was only 3 mM due to the toxicity of benzene (**187**). However, with *Pseudomonas aeruginosa* 14 mM L-DOPA was achieved in 9 hours (PARK et al., 1998).

TPL have been suggested as a means for removing phenol (**166a**) from waste water by conversion to L-tyrosine (**141a**) which is insoluble. Optimal concentrations of substrates were 60 mM phenol, 0.1 M pyruvate and 0.4 M ammonia and the reaction was effective up to 70 °C, pH 6.5–9.0. Intact or acetone dried cells

Fig. 70. Hybrid pathway for the synthesis of L-DOPA (**141b**), expressed in *E. coli* and *Pseudomonas aeruginosa* (PARK et al., 1998).

expressing TPL reduced 100 mM phenol to 8 mM within 24 hours at 37 °C (LEE et al., 1996).

4.2 Carbon–Oxygen Lyases (EC 4.2)

Carbon–oxygen lyases catalyze the cleavage of a carbon–oxygen bond, with the formation of unsaturated products. The vast majority of enzymes in this subclass are hydro-lyases (EC 4.2.1) which catalyze the elimination of water. Elimination of sugars from carbohydrates (EC 4.2.2) and a few other (mostly complex) cases (EC 4.2.99) complete the subclass

4.2.1 Hydro-Lyases (Hydratases and Dehydratases, EC 4.2.1)

Hydro-lyases are a group of enzymes that catalyze cleavage of carbon–oxygen bonds with the elimination of water and typically (but not exclusively) the formation of an alkene. In most cases the elimination of water occurs adjacent to a carboxylic acid, ketone or ester (**189**) (Fig. 71). Deprotonation occurs adjacent to the carbonyl group to give an enolate (**190**) which undergoes β-elimination to give an α,β-unsaturated carbonyl compound (**191**). This is the common E1cB mechanism (monomolecular elimination from the conjugate base) or in Guthrie-IUPAC nomenclature $A_{xh}D_H + D_N$ (GUTHRIE and JENCKS, 1989). When there is a hydroxyl group adjacent to the carbonyl group (as commonly occurs in sugars), tautomerization of the enol (**191**, X = OH) occurs to give the 2-ketocarboxylic ester (**192**).

There are two stereochemical classes of hydratase-dehydratase enzymes. Those that catalyze the addition of water to α,β-unsaturated thioesters have *syn*-addition-elimination stereochemistry, whereas those that catalyze the addition of water to α,β-unsaturated carboxylates have *anti*-stereochemistry. In a limited investigation of the spontaneous reaction, addition of deuterium oxide to α,β-unsaturated thioesters was moderately selective for *anti*- over *syn*-addition (4.3 : 1), whereas α,β-unsaturated esters only showed a minute preference (1.3 : 1) (MOHRIG et al., 1995). Thus the stereochemical preferences of the enzyme catalyzed reaction do no reflect the innate stereochemical preferences of the solution phase reaction.

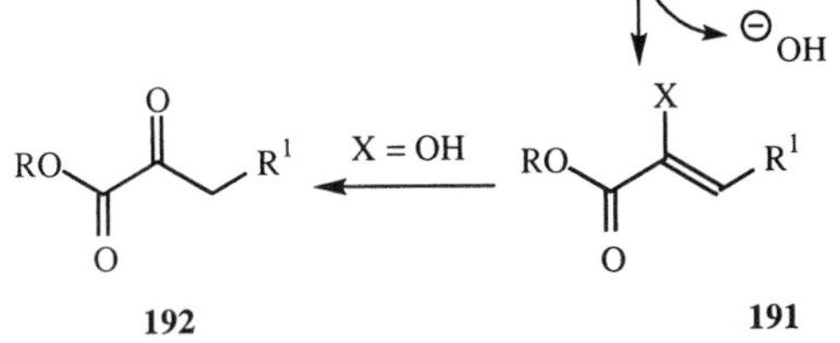

Fig. 71. Mechanisms for the elimination of β-hydroxyl-esters (**189**) to α,β-unsaturated esters (**191**) and α-ketoesters (**192**).

4.2.1.1 Carbonic Anhydrase (Carbonate Dehydratase, EC 4.2.1.1)

Carbonic anhydrase (CA) is a zinc dependent metalloprotein which catalyzes the hydration of carbon dioxide to carbonate. It is a ubiquitous enzyme present in animals, plants, algae and some bacteria and is frequently present as several different forms in the same organism. There are at least 14 different forms in human tissues (THATCHER et al., 1998) and much effort has been devoted to efforts to develop inhibitors, principially for eye conditions (DOYON and JAIN, 1999; KEHAYOVA et al., 1999; WOODRELL et al., 1999). Three evolutionarily distinct classes (α, β, γ) have been defined and were originally thought to be characteristic of animals, plants and bacteria, however, as more have been discovered, it has emerged that they are distributed indiscriminately (LINDSKOG, 1997; HUANG et al., 1998; ALBER et al., 1999). Extraction of bovine CA has been recommended as an undergraduate experiment (BERING et al., 1998).

The CA catalyzed hydration of carbon dioxide is one of the fastest known enzyme reactions. Typical kinetics parameters are K_M 1.2 · 10^{-2} M, k_{cat} 1.0 · 106 s^{-1}, V_{Max}/K_M 8.3 · 10^7 $M^{-1}s^{-1}$. Although K_M is unremarkable, the value for k_{cat} is extremely high and only exceeded by a few enzymes (e.g., catalase). The active site of CA consists a shallow depression containing a zinc ion bound to three histidines and a hypernucleophilic hydroxide ligand (**193**) (Fig. 72). This adds to carbon dioxide to generate bicarbonate (**194**), which is displaced by water (**195**) and deprotonated in the rate limiting step to regenerate the hydroxide ligand (**193**) (QIAN et al., 1999; EARNHARDT et al., 1999; MUGURUMA, 1999). CA also catalyzes the hydration of aldehydes and ketones, and the hydrolysis of esters, phosphates (SHERIDAN and ALLEN, 1981) and cyanamide (BRIGANTI et al., 1999). There is considerable unrealized potential for the use of CA as an esterase. Bovine CA III increases the rate of decarboxylation of amino acids by L-arginine, L-lysine or L-ornithine decarboxylases used in a biosensor (BOTRE and MAZZEI, 1999).

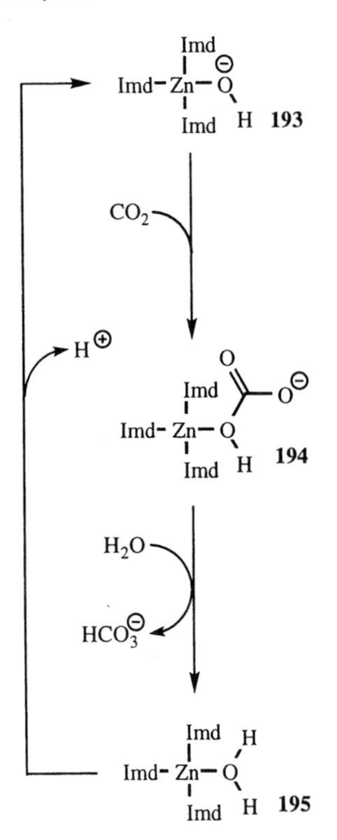

Fig. 72. Minimal mechanisms for the hydration of carbon dioxide by carbonic anhydrase. Zinc is present as Zn(II). Hydroxide, bicarbonate and water ligands are depicted without charge transfer to the zinc atom.

196 Fumarate

Fumarase

197 L-(*S*)-(-)-Malate

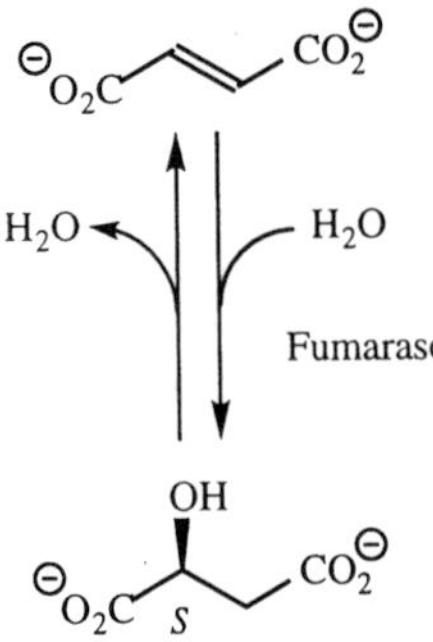

Fig. 73. Fumarase catalyzes the interconversion of fumarate (**196**) and L-malate (**197**).

4.2.1.2 Fumarase (EC 4.2.1.2)

Introduction

Fumarase catalyzes the addition of water to the alkenic bond of fumarate (**196**) to give L-malate (**197**, Fig. 73). It is one of a number of related enzymes which catalyze additions to the alkene bond of fumarate, such as aspartase (EC 4.3.1.1, ammonium, SHI et al., 1997), and the amidine lyases; arginosuccinase (EC 4.3.2.1, L-arginine, COHEN-KUPEIC et al., 1999; TURNER et al., 1997) and adenylsuccinase (EC 4.3.2.2, AMP, LEE et al., 1999). Many of these enzymes have high sequence homology, even though they catalyze different reactions or none at all (WOODS et al., 1986, 1988a). For example δ1-crystallin from duck lens is catalytically inactive (PIATIGORSKY and WISTOW, 1991; VALLEE et al., 1999), even though it shares 90% sequence homology with arginosuccinase (SIMPSON et al., 1994; cf. WU et al., 1998).

Fumarase is a remarkably efficient enzyme. It has been estimated that the rate enhancement relative to spontaneous hydration of fumarate is some $3.5 \cdot 10^{15}$ fold which equates to a transition state stabilization of -30 kcal mol^{-1}. On this basis it is the second most efficient enzyme known (BEARNE and WOLFENDEN, 1995) after orotidine 5′-monophosphate decarboxylase (-32 kcal mol^{-1}, EHRLICH et al., 1999).

Fumarase is widely distributed in animals, plants (BEHAL and OLIVER, 1997; KOBAYASHI et al., 1981), yeast (KERUCHENKO et al., 1992), fungi and bacteria (BATTAT et al., 1991) and archaebacteria (COLOMBO et al., 1994). In humans, mutations of fumarase (COUGHLIN et al., 1998) are associated with diseases such as fumaric aciduria and progressive encephalopathy (DE VIVO, 1993; BOURGEN et al., 1994).

In *E. coli*, fumarase activity arises from three distinct genes, *fumA*, *fumB* and *fumC* (BELL et al., 1989). The gene products of *fumA* and *fumB* are fumarase A (FUMA) and fumarase B (FUMB), which are examples of class I fumarases. These are iron dependent (4Fe-4S), heat labile, dimeric enzymes (MW 120 kDa, UEDA et al., 1991), whereas *fumC* yields fumarase C (FUMC) a class II fumarase. These are iron-independent, heat stable, tetrameric enzymes (MW 200 kDa), which include yeast fumarase, *Bacillus subtilis* fumarase (*citG*) and mammalian fumarases. Class II fumarases have high sequence homology and have been extensively studied, mechanistically and kinetically (Woods et al., 1988b).

Class I Fumarases

In *E. coli*, fumarase A and fumarase C are expressed under aerobic cell growth conditions, whereas fumarase B is more abundant during anaerobic growth (Tseng, 1997; Woods and Guest, 1987). Fumarase A (and C) act in the citric acid cycle during aerobic metabolism, whereas fumarase B enables fumarate to act as an anaerobic electron acceptor in the reductive sequence leading from oxaloacetate (**28**) to succinate (**198**) (Fig. 74) (Woods et al., 1988b). There is some evidence

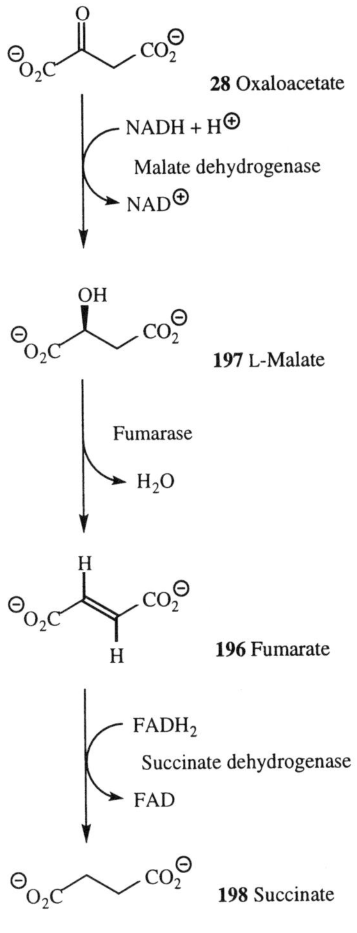

28 Oxaloacetate

NADH + $H^{\oplus}$

Malate dehydrogenase

$NAD^{\oplus}$

197 L-Malate

Fumarase

H_2O

196 Fumarate

$FADH_2$

Succinate dehydrogenase

FAD

198 Succinate

Fig. 74. The role of fumarase in anaerobic metabolism.

that fumarase C may act as a back-up to fumarase A under conditions of iron deprivation (Hassett et al., 1997). The amino acid sequences of fumarase A and fumarase B have 90% sequence homology to each other, but almost no homology to the class II fumarases. Each monomeric unit of fumarase A and B contains a [4Fe-4S] cluster as do *Euglena gracilis* fumarase (Shibata et al., 1985; Rikin and Schwartzbach, 1989) and *Rhodobacter capsulatus* fumarase. Other hydrolyases which contain a non-redox active [4Fe-4S] cluster (review, Flint and Allen, 1996) include aconitase (Gruer at al., 1997; Beinert et al., 1996), dihydroxy-acid dehydratase (Flint et al., 1996; Limberg and Thiem, 1996), isopropylmalate isomerase, phosphogluconate dehydratase, tartrate dehydratase, maleate dehydratase, lactyl-CoA dehydratase, 2-hydroxyglutaryl-CoA dehydratase and *Peptostreptococcus asaccharolyticus* serine dehydratase (Hofmeister et al., 1994). There is little or no sequence homology between aconitase and fumarase A, although both bear the dipeptide moiety Cys–Pro which is often found in the sequences of iron–sulfur proteins. Both are oxidized by oxygen, but fumarase A is more susceptible to further oxidation and "reactivation" by iron and a reductant takes longer (Flint et al., 1992b). Fumarase A, fumarase B, aconitase and *E. coli* di-hydroxy-acid dehydratase are inactivated by superoxide (O_2^-) with release of iron and by hyperbaric oxygen (Flint et al., 1993; Ueda et al., 1991).

The addition of water to fumarate catalyzed by fumarase A is stereospecific (Fig. 75). The hydroxyl group and the hydrogen undergo *trans*-addition with the incoming hydrogen added to the (3*R*)-position (**199**). Prolonged incubation does not result in the formation of [^{2}H]-fumarate, which testifies to the stereochemical fidelity of the reaction (Flint et al., 1994). There is no hydrogen isotope effect and hence hydrogen abstraction is not the rate determining step.

Fumarase A catalyzes the transfer of ^{18}O from (2*S*)-[2-$^{18}O_1$]-malate (**200**) (Figs. 76, 77) and ^{2}H from (2*S*,3*R*)-[3-$^2H_1$]-malate (**206**) to acetylene dicarboxylate (**202**) to give ^{18}O and ^{2}H labeled oxaloacetate enol (**203**) (**208**) (Flint et al., 1992a; Fell et al., 1997), which isomerizes to the keto-form of oxaloacetate

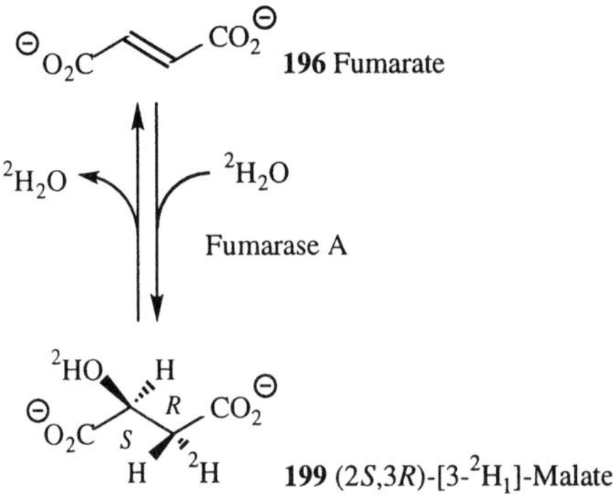

Fig. 75. Fumarase A, catalyzed addition of the elements of water occurs stereospecifically in the *trans*-orientation.

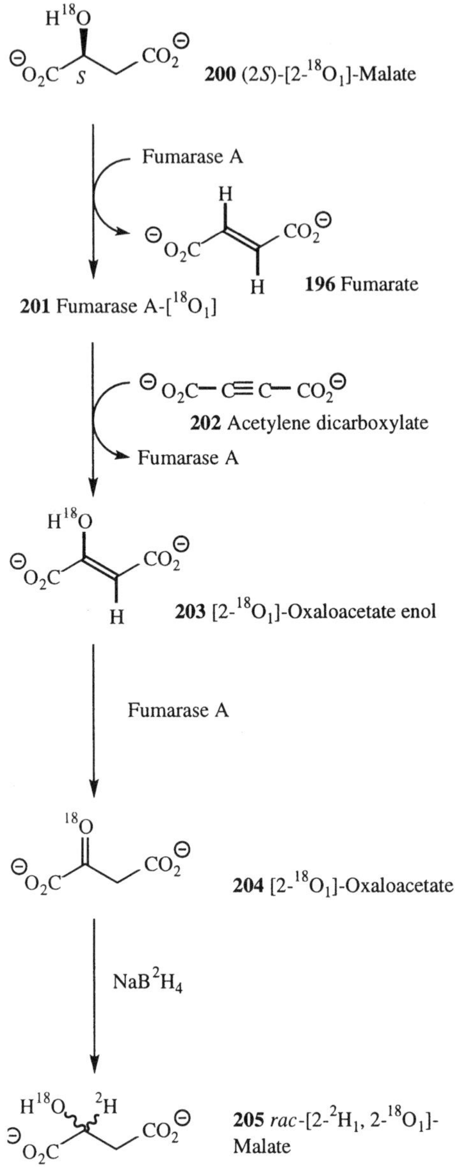

Fig 76. The hydration of acetylene dicarboxylate catalyzed by fumarase A-[$^{18}O_1$] (**201**) from *E. coli.*

(**204**) (**209**). Further oxygen or hydrogen exchange was prevented by reduction with sodium borodeuteride to the malates (**205**) (**210**). It was calculated that at an infinite concentration of acetylene dicarboxylate (**202**), some 33% of the ^{18}O and 100% of the ^{2}H would be transfered. This indicates that these atoms are strongly bound at the active site, presumably with the oxygen acting as a ligand to the [4Fe-4S]-cluster and the hydrogen bound to a basic residue (**211**) (Fig. 78. FLINT and MCKAY, 1994). Moreover fumarase also catalyzes the isomerization of oxaloacetate enol (**208**) to the ketone form (**209**), at the same active site which is responsible for hydration–dehydration. The reaction occurs with retention of the enolic oxygen and is stereoselective with the hydrogen delivered to the 3-*pro*-(*S*)-position (FLINT, 1993).

Fumarase A has high substrate specificity which is typical of fumarases in general. The only substrates which have reactivity which is comparable to fumarate (relative rate = 100, 10 mM substrate) are L-malate (41), acetylene dicarboxylate (97) and (2*S*,3*S*)-tartrate (11). 2-Fluorofumarate (13) is a reasonably reactive substrate. It undergoes addition of the hydroxy group at the 2-position to give a 2,2-fluorohydrin which spontaneously eliminates hydrogen fluoride to give oxaloacetate. This reaction is identical to that catalyzed by the class II fumarases and is described in detail in the subsequent section.

In summary, class I fumarases are rare, unstable enzymes bearing a [4Fe-4S] cluster which has a non-redox role.The stereochemistry of hydration of fumarate is identical to that of the class II fumarases

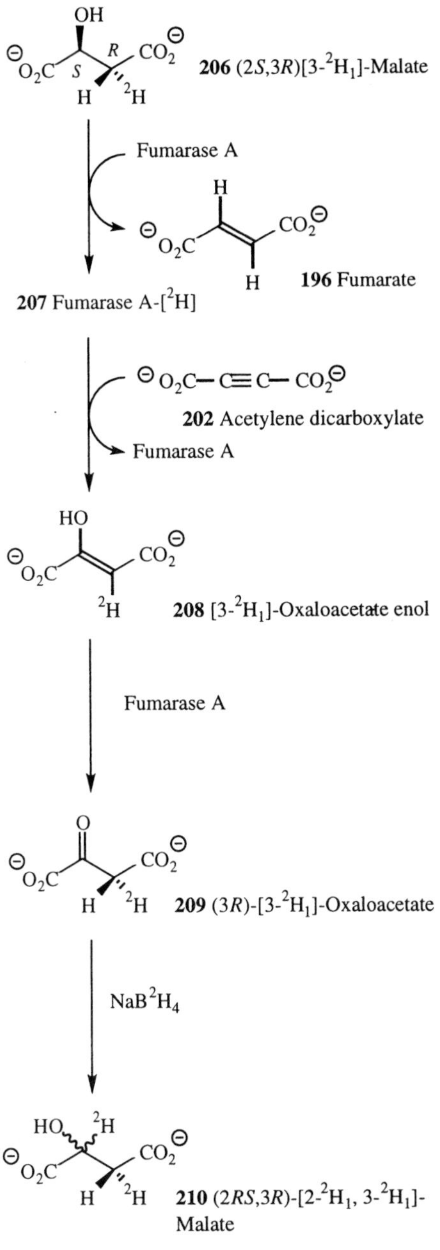

206 (2*S*,3*R*)[3-$^2H_1$]-Malate

Fumarase A

196 Fumarate

207 Fumarase A-[^2H]

202 Acetylene dicarboxylate

Fumarase A

208 [3-$^2H_1$]-Oxaloacetate enol

Fumarase A

209 (3*R*)-[3-$^2H_1$]-Oxaloacetate

NaB^2H_4

210 (2*RS*,3*R*)-[2-$^2H_1$, 3-$^2H_1$]-Malate

Fig 77. The hydration of acetylene dicarboxylate catalyzed by fumarase A-[$^2H_1$] (**207**) from *E. coli*.

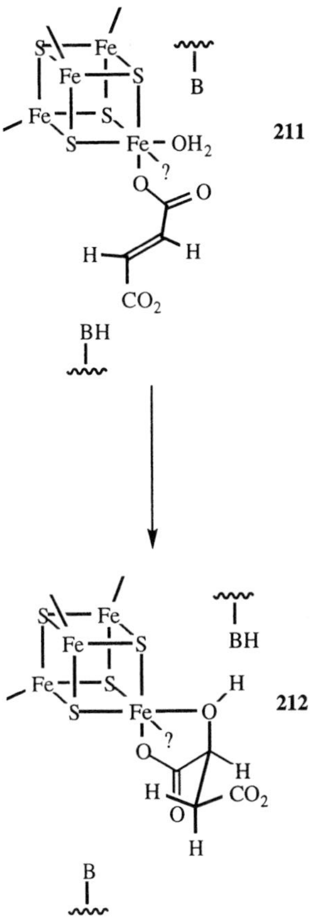

211

212

Fig. 78. Outline mechanism for the hydration of fumarate (**196**) by a [4Fe-4S]-cluster.

Class II Fumarases

In *S. cerevisiae* both cytoplasmic and mitochondrial fumarases (class II) are encoded by a single nuclear gene FUM1. The mitochondrial enzyme is produced with a 23-residue translocation sequence, which is responsible for directing it to the mitochondria and is removed during importation into the mitochondria (KNOX et al., 1998). Similar systems are employed in humans, pigs (O'HARE and DOONAN, 1985), rats (SUZUKI et al., 1989; TUBOI et al., 1990) and mice, which apparently only utilize class II fumarases. In *S. cerevisiae*, some 30% of the enzyme activity is associated with the mitochondrial fraction and the remaining 70% with the post-ribosomal fraction, although the specific activity of the former is some 3–4 fold higher. (WU and TZAGOLOFF, 1987).

Stereochemistry

Early work demonstrated that the addition of deuterium oxide to fumarate catalyzed by fumarase was stereospecific (FISHER et al., 1955). Initially the evidence suggested that the addition occurrred with *syn*-stereochemistry (FARRAR et al., 1957; ALBERTY and BENDER, 1959). However, the synthesis of stereospecifically labeled standards, clearly indicated *anti*-stereochemistry with the hydrogen atom in the (3*R*)-position (GAWRON and FONDY, 1959; ANET, 1960; GAWRON et al., 1961), which was confirmed by a neutron diffraction study (BAU et al., 1983). This stereochemical outcome is identical to that of the class I fumarases (Fig. 75).

Enzymology and Structures

Kinetic analysis of fumarase hydration is complex. At low substrate concentrations, classical Michaelis–Menten kinetics are followed, at substrate concentrations greater than five times K_M substrate activation occurs and above 0.1 M inhibition occurs. Although the conversion of substrate is fast, recycling of the enzyme to the active state is slow and is catalyzed by anions, including the substrates (ROSE, 1997, 1998; BEHAL and OLIVER, 1997). Thus, although there is no deuterium isotope effect in the hydration reaction, k_{cat} is reduced if deuterium oxide is used in place of water as solvent. X-ray crystal structures for *E. coli* (WEAVER and BANASZAK, 1996; WEAVER et al., 1995) and yeast (WEAVER et al., 1998; KERUCHENKO et al., 1992) fumarases have revealed a binding site for anions, structurally close to the active site, which acts allosterically and is apparently responsible for the unusual kinetics noted above. Mutation of a histidine residue to asaparagine in the active site (H188N), resulted in a large decrease in specific activity, but the same change at the second site (H129N) had essentially no effect, however, both mutations reduced the affinity for carboxylic acids at the respective sites (WEAVER et al., 1997). It may be that the asparagine residue at the second site in the mutant acts as a surrogate for the ligand's carboxylate group and hence compensates for the lack of normal allosteric action. The pig liver enzyme also has a second site, as deduced from a kinetics study (BEECKMANS and VAN DRIESSCHE, 1998). A preliminary report of crystallization and diffraction data for this enzyme was reported, however, thus far there has been no crystal structure (SACCHETTINI et al., 1986).

The extreme stereochemical fidelity of fumarase catalysis has enabled it to be used in the synthesis of stereospecifically labeled malates (review, YOUNG, 1991). As has been noted previously, (2*S*,3*R*)-[3-$^2H_1$]-malate (**199**) can be prepared by fumarase catalyzed addition of deuterium oxide to fumarate (**196**) (AXELSSON et al., 1994) (Fig. 75). The converse reaction in which deuterated fumarate (**213**) is hydrated by water gives (2*S*,3*S*)-[2,3-$^2H_2$]-malate (**214**) (Fig. 79) (ENGLARD and HANSON 1969; AXELSSON et al., 1994). Alternatively, this can be prepared by malate dehydrogenase catalyzed stereospecific reduction of fully deuterated oxaloacetate (**216**), with (4*R*)-[4-$^2H_1$]-NADH, to give fully deuterated malate (**215**) and exchange of the (3-*pro-R*)-deuteron with hydrogen ions catalyzed by fumarase (CLIFFORD et al., 1972; YOUNG, 1991). In contrast stereospecific reactions of the dideuteromalate (**217**) gave a mixture of products (Fig. 80). Fumarase catalyzed dehydration in tritiated water stereospecifically gave the monodeuterofumarate (**216**). However, this can bind in the active site in two different orientations, thus hydration was non-selective and a mixture of isotopomers (**219**) (**220**) were formed (LENZ et al., 1976; YOUNG, 1991).

Malonate (**89**) (Fig. 29) is proprochiral, ie. at least two groups attached to the central carbon atom must be labeled to render it chiral and hence amenable to stereochemical studies of decarboxylation or homologation (FLOSS et al., 1984). This has been achieved in an extremely neat way by the fumarase catalyzed hydration of [1,4-$^{13}C_2$]-fumarate (**221**) to give labeled malate (**222**), which can be stored for prolonged periods (Fig. 81). Rapid cleavage with permanganate, at pH 10, gives chiral malonate (**223**), with minimal opportunity for exchange of the acidic hydrogens (HUANG et al., 1986; JORDAN et al., 1986). This methodology has been used in studies of decarboxylation (MICKLEFIELD et al., 1995; HANDA et al., 1999), fatty acid (JORDAN et al., 1986; JORDAN and SPENCER, 1991), 6-methylsalicyclic acid (SPENCER and JORDAN, 1990; JORDAN and SPENCER, 1990, SPENCER and JORDAN, 1992b) and orsel-

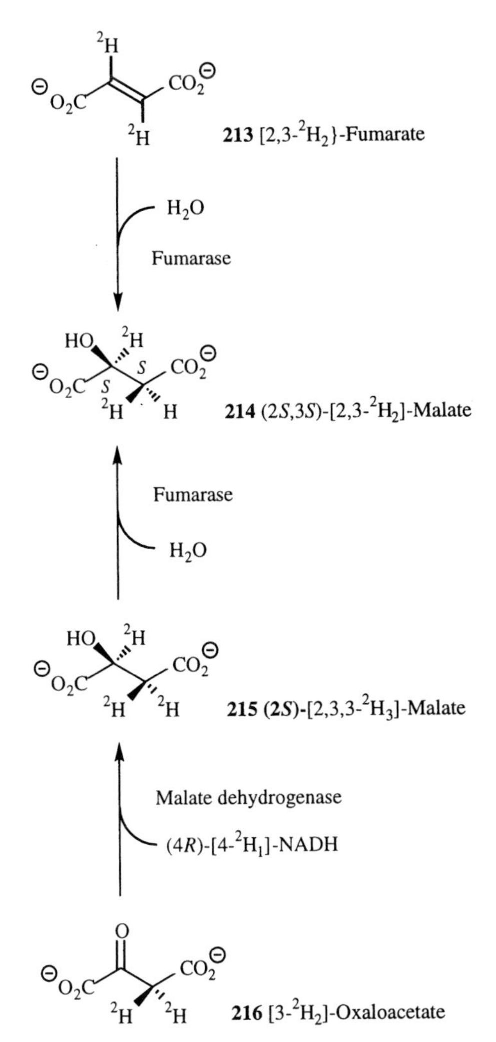

Fig. 79. Two routes to (2*S*,3*S*)-[2,3-$^2H_2$]-malate (**214**) (ENGLARD and HANSON, 1969; CLIFFORD et al., 1972; YOUNG, 1991).

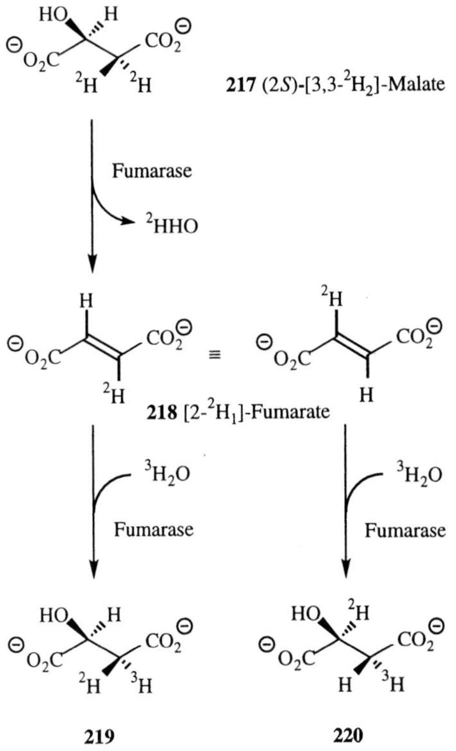

linic acid biosynthesis (SPENSER and JORDAN, 1992a).

Non-Natural Substrates

Fumarases are highly specific for fumarate (**196**) and L-malate (**197**). A limited range of other unsaturated dicarboxylic acids undergo hydration, but with the exception of 2-fluorofumaric acids and 2,3-difluorofumaric acid, the

217 (2*S*)-[3,3-$^2H_2$]-Malate; Fumarase; ^{2}HHO; **218** [2-$^2H_1$]-Fumarate; ^3H_2O; Fumarase; **219**; **220**

Fig. 80. Stereospecific dehydration–hydration yields a mixture of isotopomers (**219**) (**220**) (LENZ et al., 1976).

reaction rates are much slower. Pig liver fumarase catalyzed hydration of 2-chlorofumarate (**224**), yields enantiomerically pure L-*threo*-chloromalate (>99.5% ee) (**225**) (Fig. 82). Unfortunately, the reaction has an unfavorable equilibrium constant (K_{eq} = 6.2), which limits the theoretical yield to 86%. However, using PAN gel immobilized fumarase, 63.7 g of chlorofumaric acid was converted to 54.5 g (76% yield) of L-*threo*-chloromalic acid, in two reaction/isolation cycles, over two weeks. The recovered immobilized fumarase retained 60% of the initial activity (FINDEIS and WHITESIDES, 1987). Interestingly, the corresponding fluorofumarate (**228**) reacts with the opposite regiochemistry to give an unstable fluoroalcohol (**229**), which spontaneously eliminates hydrogen fluoride to give the tritiated oxaloacetate (**230**) (Fig. 83). The stereochemistry of the tritium label was determined by reduction

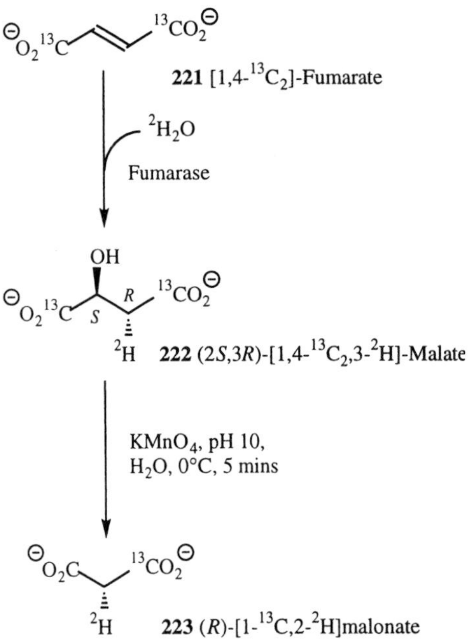

221 [1,4-$^{13}C_2$]-Fumarate

^2H_2O
Fumarase

222 (2*S*,3*R*)-[1,4-$^{13}C_2$,3-^{2}H]-Malate

$KMnO_4$, pH 10,
H_2O, 0°C, 5 mins

223 (*R*)-[1-^{13}C,2-^{2}H]malonate

Fig. 81. The synthesis of chiral malonate (**223**) (HUANG et al., 1986).

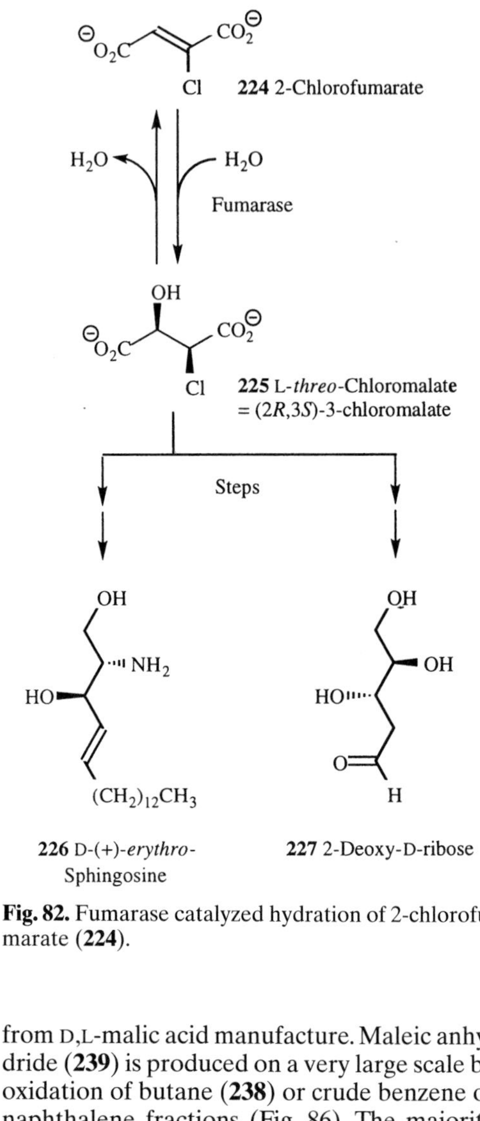

224 2-Chlorofumarate

H_2O H_2O
Fumarase

225 L-*threo*-Chloromalate
= (2*R*,3*S*)-3-chloromalate

Steps

226 D-(+)-*erythro*-Sphingosine

227 2-Deoxy-D-ribose

Fig. 82. Fumarase catalyzed hydration of 2-chlorofumarate (**224**).

with malate dehydrogenase and treatment with fumarase, which yielded unlabeled fumarate (**196**). Consequently the tritium label must have been located at the (3*R*)-position. (MARLETTA et al., 1982). The hydration of 2,3-difluorofumaric acid (**233**) also proceeds with elimination of hydrogen fluoride, but in this case *in situ* reduction yields L-*threo*-fluoromalic acid (**235**) which was isolated as the dimethyl ester in 49% yield (Fig. 84) (FINDEIS and WHITESIDES, 1987). Fumarase binds the nitronates (**237a,b**) more tightly than fumarate (**196**), whereas the nitrocompounds (**236a,b**) are barely inhibitory. Similarly the nitonates (**237b,c,d**) binds to aspartase, more strongly than aspartate. Although, all of the analogs are inhibitors but not substrates, the fact that they bind so well indicates there are prospects for designing unnatural substrates (PORTER and BRIGHT, 1980).

Biotechnology

Fumaric acid has been manufactured by fermentation of glucose by *Rhizopus nigricans* but nowadays most is produced as a byproduct from D,L-malic acid manufacture. Maleic anhydride (**239**) is produced on a very large scale by oxidation of butane (**238**) or crude benzene or naphthalene fractions (Fig. 86). The majority of this is used in the production of plastics and paints and detergents. Hydrolysis of maleic anhydride gives maleic acid (**240**), fumaric acid (**241**) or D,L-malic acid (**242**). All of these acids are used as acidulants (preservatives) for canned fruit, preserves, soft drinks, cosmetics and pharmaceuticals. However, there is considerable (and increasing) demand for L-malic acid, which is the naturally occurring enantiomer (apple acid). The current world production of L-malic acid is about 500 tonnes per year, most of which is used in food and phar-

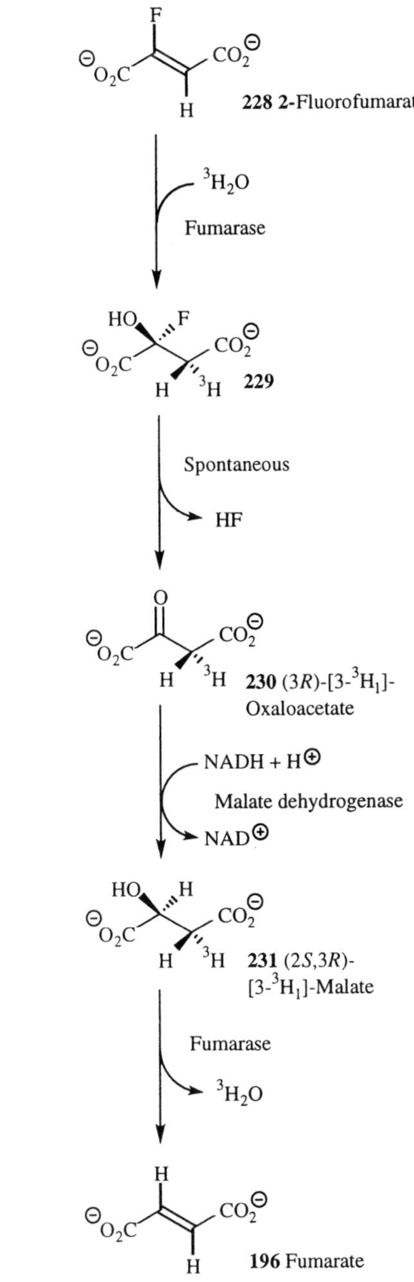

228 2-Fluorofumarate

^3H_2O Fumarase

229

Spontaneous

HF

230 (3*R*)-[3-$^3H_1$]-Oxaloacetate

NADH + $H^{\oplus}$

Malate dehydrogenase

$NAD^{\oplus}$

231 (2*S*,3*R*)-[3-$^3H_1$]-Malate

Fumarase

^3H_2O

196 Fumarate

Fig. 83. Fumarase catalyzed hydration of 2-fluorofumarate (**226**).

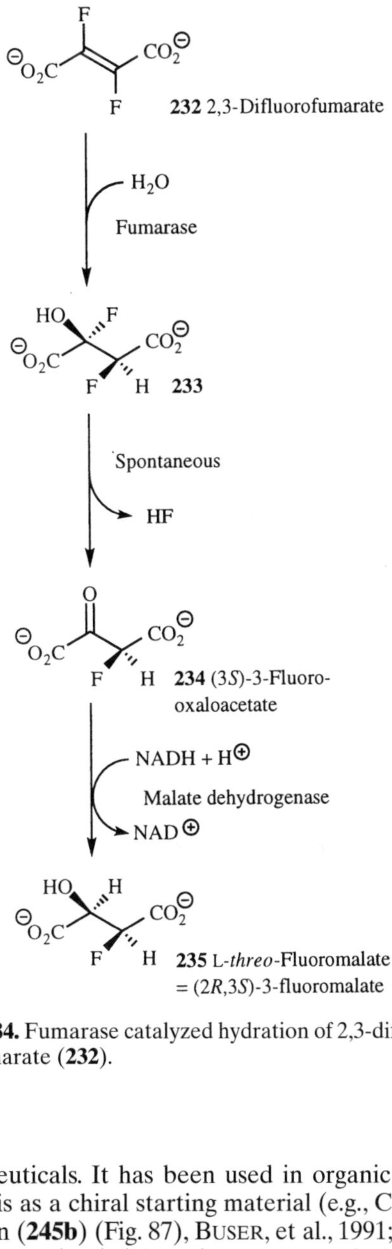

232 2,3-Difluorofumarate

H_2O Fumarase

233

Spontaneous

HF

234 (3*S*)-3-Fluoro-oxaloacetate

NADH + $H^{\oplus}$

Malate dehydrogenase

$NAD^{\oplus}$

235 L-*threo*-Fluoromalate = (2*R*,3*S*)-3-fluoromalate

Fig. 84. Fumarase catalyzed hydration of 2,3-difluorofumarate (**232**).

maceuticals. It has been used in organic synthesis as a chiral starting material (e.g., Clozylacon (**245b**) (Fig. 87), BUSER, et al., 1991; Pyrrolam A (**247**) (Fig. 88), HUANG et al., 1999), however, such applications are rare because it is less versatile than the tartaric acids, which have a C_2-axis of symmetry (GAWRONSKI and GAWRONSKI, 1999).

Most of this demand for L-malate (**197**) is satisfied by the fumarase catalyzed hydration of fumarate (**196**) (Fig. 75). This has been ac-

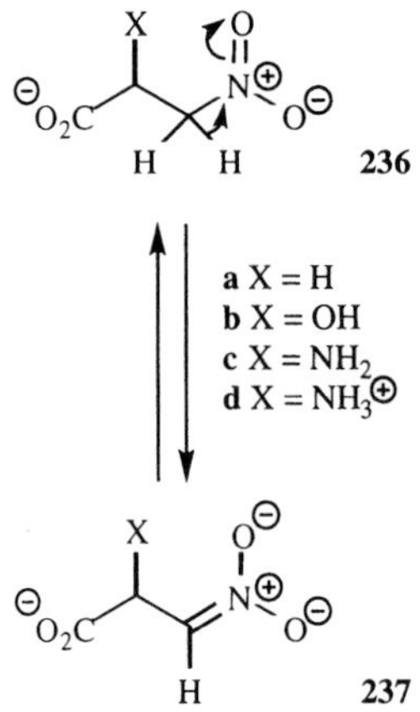

236

a X = H
b X = OH
c X = NH_2
d X = $NH_3^{\oplus}$

237

Fig. 85. Fumarase inhibitors (PORTER and BRIGHT, 1980).

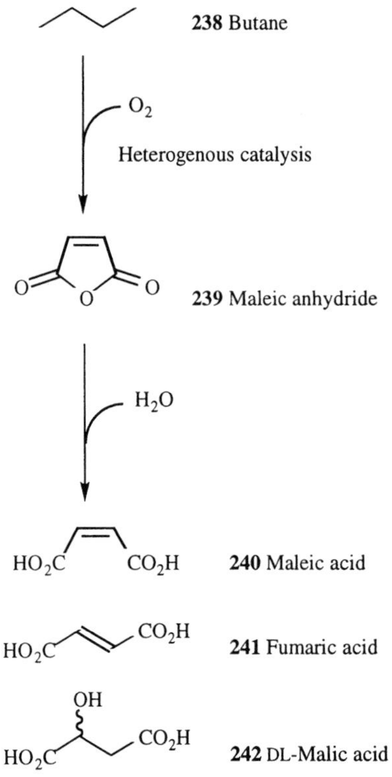

238 Butane

O_2
Heterogenous catalysis

239 Maleic anhydride

H_2O

240 Maleic acid

241 Fumaric acid

242 DL-Malic acid

Fig. 86. The manufacture of C_4-dicarboxylic acids.

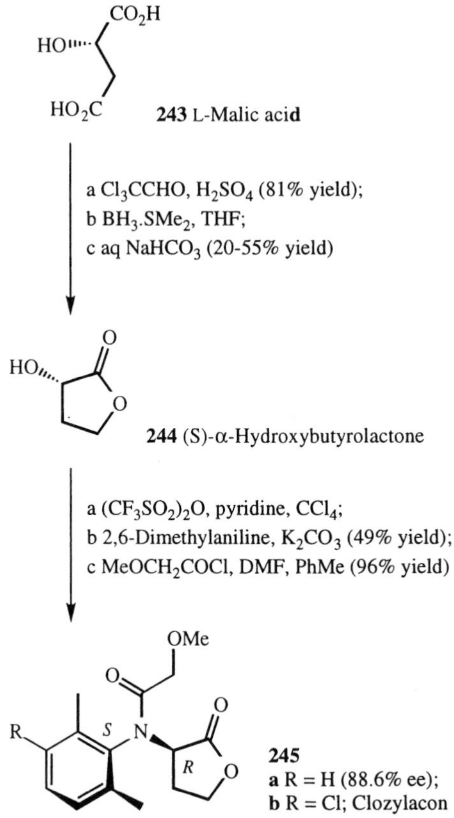

243 L-Malic acid

a Cl_3CCHO, H_2SO_4 (81% yield);
b $BH_3.SMe_2$, THF;
c aq $NaHCO_3$ (20-55% yield)

244 (S)-α-Hydroxybutyrolactone

a $(CF_3SO_2)_2O$, pyridine, CCl_4;
b 2,6-Dimethylaniline, K_2CO_3 (49% yield);
c $MeOCH_2COCl$, DMF, PhMe (96% yield)

245
a R = H (88.6% ee);
b R = Cl; Clozylacon

Fig. 87. Synthesis of the fungicide CGA 80000, Clozylacon (**245b**) (Ciba-Geigy) from L-malic acid (**243**) (BUSER, et al., 1991).

complished with immobilized whole cells of *Brevebacterium ammoniagenes* or laterly *B. flavum* (CHIBATA, et al., 1987b; WANG, et al., 1998). Various immoblized gels have been applied in the continuous process such as polyacrylamide and κ-carrageenan (CHIBATA, et al., 1995, 1987a). Fumarase activity (unit mL^{-1} gel) and relative productivity (%) of cell immobilized *Brevibacterium flavum* increased respectively from 10.2 and 273 on polyacrylamide to 15.0 and 897 on κ-carrageenan. The immobilized *B. flavum* on either gel showed higher fumarase activity and relative productivity compared to *B. ammoniagenes* cells on polyacrylamide gel (TOSA et al., 1982; TANAKA et al., 1983a, 1983b, 1983c, 1984). Many other organisms have been investigated for im-

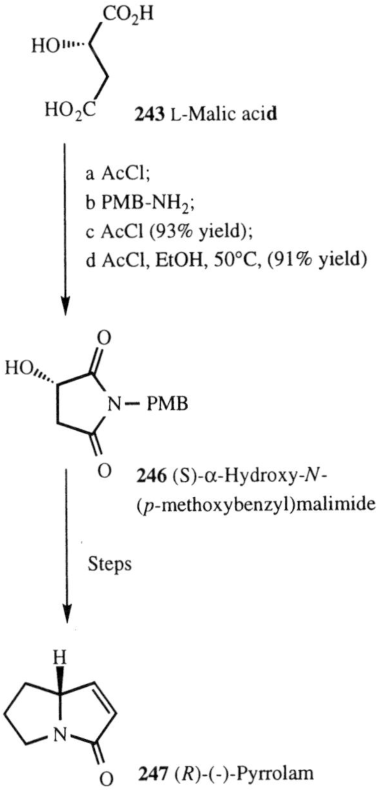

CO_2H
HO
HO_2C **243** L-Malic acid

a AcCl;
b PMB-NH_2;
c AcCl (93% yield);
d AcCl, EtOH, 50°C, (91% yield)

HO
O
N – PMB
O **246** (S)-α-Hydroxy-*N*-(*p*-methoxybenzyl)malimide

Steps

H
N
O **247** (*R*)-(-)-Pyrrolam

Fig. 88. Synthesis of the pyrrolizidine alkaloid, (–)-(*R*)-pyrrolam A (**247**) from L-malic acid (**243**) (HUANG et al., 1999) using the method of LOUWRIER et al. (1996) for the preparation of the malimide (**246**).

proved fumarase activity including *Corynebacterium equi*, *Escherichia coli*, *Microbacterium flavum*, *Proteus vulgaris*, *Pichia farinosa*, *Leuconostos brevis* (MIALL, 1978; DUFFY, 1980) and yeast (WANG et al., 1998) but the highest levels of process intensification (higher space-time yields) will require the use of enzymes (CHIBATA, et al., 1995).

B. flavum immobilized in κ-carrageenan and chinese gallotannin has a operational half life of 310 days at 37 °C. Although the half life is excellent by any measure, productivity could be increased by a higher working temperatures, moreover this would be an advantage for poorly soluble magnesium or calcium fumarate salts. *Thermus thermophilus* fumarase showed optimum activity of 85 °C and over 80% of the activity remained after a 24-h incubation at 90 °C. Furthermore, the enzyme showed good chemostability; 50% of the initial activity was detected in assay mixtures containing 0.8% M guanidine hydrochloride (MIZOBATA, et al., 1998). The fumC genes from another *T. thermophilus* strain (KOSAGE et al., 1998) and *Sulfolobus solfataricus* (COLOMBO, et al., 1994) have been expressed in *E. coli*. Both retain full activity at 70 °C and in the latter case this is despite an 11 amino acid *C*-terminal deletion.

4.2.1.3 Aconitase (Aconitate Hydratase, Citrate/Isocitrate Hydro-Lyase, EC 4.2.1.3) and Citrate Dehydratase (EC 4.2.1.4)

Aconitase catalyzes the dehydration of citrate (**248**) to *cis*-aconitate (**73**) and hydration to isocitrate (**249**) (Fig. 89). The prosthetic group is a [4Fe-4S] cluster and hence the fundamental chemistry, resembles that of class I fumarases (BEINERT et al., 1996; GRUER et al., 1997). Aconitase has significant sequence homology with iron regulatory factor or iron-responsive element binding protein and indeed the major difference between them seems to be the presence or absence of the [4Fe-4S] cluster (BEINERT and KENNEDY, 1993). Citrate dehydratase was reported to dehydrate citrate to *cis*-aconitate, but not isocitrate to *cis*-aconitrate.

The conversion of citrate (**248**) to isocitrate (**249**) catalyzed by aconitase is a key step in the citric acid (Krebs) cycle (Figs. 1, 89). The first step is the *trans*-elimination of the hydroxyl group and the (*pro-R*)-hydrogen of the (*pro-R*)-methylene carboxylate group of citrate (**248**) to give *cis*-aconitate (**73**). In the reverse direction, this can be regarded as the addition of a proton to the *re*-face of C-2 and a hydroxyl group to the *re*-face of C-3. Hydration of *cis*-aconitate (**73**) proceeds by *trans*-addition of hydroxyl to the *re*-face of C-2 and a proton to the *re*-face of C-3. Thus each addition or elimination of a proton or a hydroxyl from the two carbons centres, occurs on the same side for

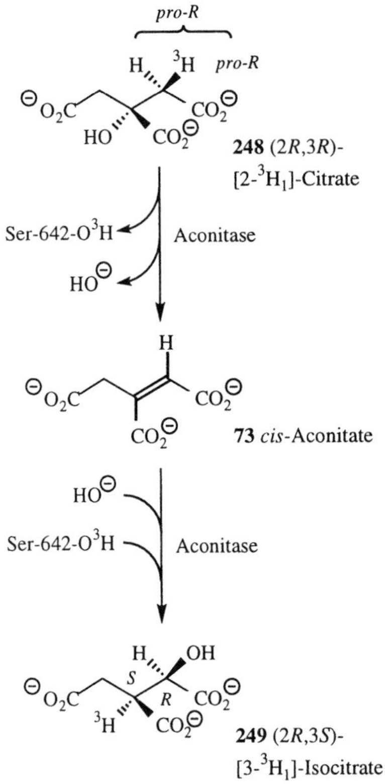

Fig. 89. The aconitase catalyzed isomerization of citrate (**248**) to isocitrate (**249**). For the purposes of brevity, no distinction is made in the text, between labeled and unlabeled citrate and isocitrate.

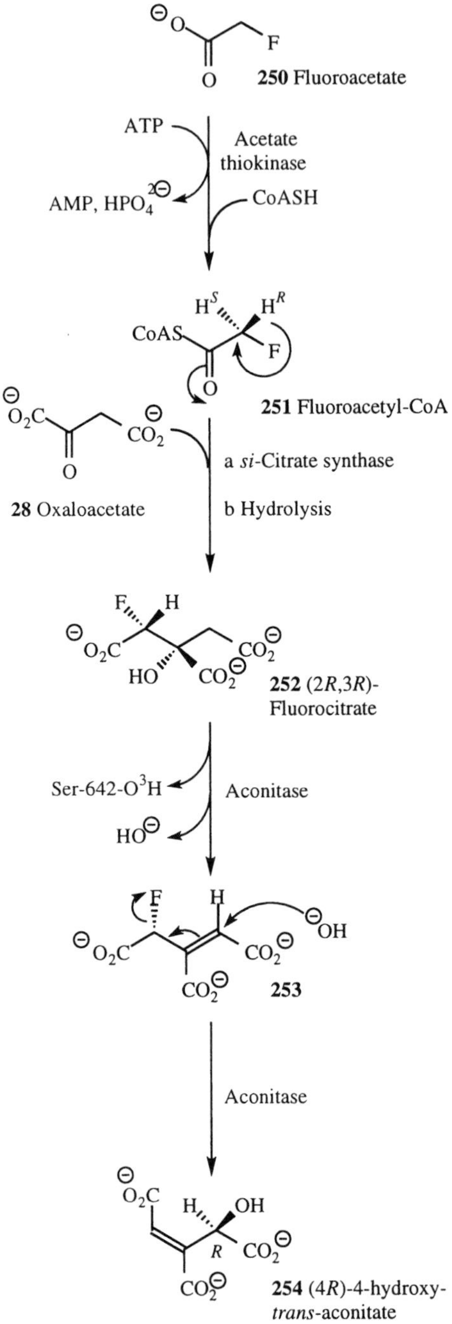

Fig. 90. The biosynthesis of fluorocitrate (**252**) from fluoroacetate and oxaloacetate (**28**) and the lethal biosynthesis of (4*R*)-4-hydroxy-*trans*-aconitate (**254**) catalyzed by aconitase.

each carbon centre. Remarkably, the proton (tritium) abstracted from C-2 of citrate (probably by Serine-642) is delivered back to the opposite face of *cis*-aconitate to give H-3 of isocitrate, whereas the hydroxyl group is exchanged with solvent. Hence either *cis*-aconitate must flip in the active site so that the retained proton can be transferred to the opposite face or less likely the proton must be transferred from one face to the other. This is all the more extraordinary, because it occurs without dissociation of *cis*-aconitate (**73**) from aconitase.

Fluoroacetate (**250**) (Fig. 90) is one of the most toxic small molecules known (LD_{50} rat 0.2 mg kg^{-1}). It is converted *in vivo* to

(2*R*,3*R*)-fluorocitrate (**252**), which acts as both a competitive and semi-irreversible inhibitor of aconitase. Aconitase catalyzed elimination proceeds normally (**253**), but rehydration occurs with S_N2' displacement of fluoride, to give (4*R*)-4-hydroxy-*trans*-aconitate (**254**) which binds very strongly. It cannot be displaced with a 20-fold molar excess of citrate (**248**), but with a 10^6 fold molar excess of isocitrate (**249**) lag kinetics are observed indicating displacement. X-ray crystal structures have been determined for aconitase binding *trans*-aconitate, nitrocitrate (LAUBLE et al., 1994) and (4*R*)-4-hydroxy-*trans*-aconitate (**254**) (LAUBLE et al., 1996). Further aspects of fluoroacetate and fluorocitrate metabolism are discussed in Chapter 3, this volume: Halocompounds.

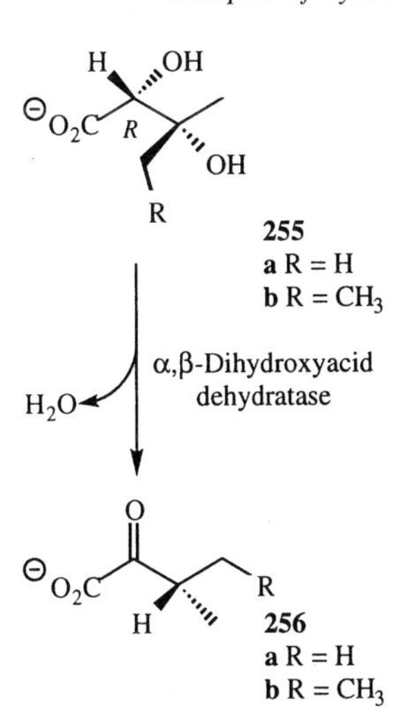

Fig. 91. The *α,β*-dihydroxyacid dehydratase catalyzed dehydration of (2*R*)-2,3-dihydroxy-3-methylbutanoate (**255a**) and (2*P*,3*P*)-2,3-dihydroxy-3-methylpentanoate (**255b**) to 2-oxo-3-methylbutanoate (**256a**) and (3*S*)-2-oxo-3-methylpentanoate (**256b**), respectively.

4.2.1.4 *α,β*-Dihydroxy Acid Dehydratase (2,3-Dihydroxy-Acid Hydro-Lyase, EC 4.2.1.9)

α,β-Dihydroxyacid dehydratase (DHAD) catalyzes the dehydration of (2*R*)-2,3-dihydroxy-3-methylbutanoate (**255a**) (Fig. 91) to 2-oxo-3-methylbutanoate (**256a**) and (2*R*,3*R*)-2,3-dihydroxy-3-methylpentanoate (**255b**) to (3*S*)-2-oxo-3-methylpentanoate (**256b**), These are the penultimate intermediates in the biosynthesis of L-valine and L-isoleucine, respectively. Unusually the last four steps in these pathways are all catalyzed by a common set of enzymes. DHAD has an absolute requirement for the (2*R*)-configuration, but substrates with both the (3*R*)- and (3*S*)-configurations are dehydrated stereospecifically. Thus (2*R*,3*S*)-2,3-dihydroxy-3-methylpentanoate undergoes dehydration to give the enantiomer of the "normal" product (*ent*-**256b**), which is a precursor of L-alloleucine.

DHAD has been found in bacteria, yeast, algae and higher plants (PIRRUNG et al., 1991). *E. coli* DHAD contains a [4Fe-4S] cluster prosthetic group (FLINT et al., 1996), whereas spinach DHAD contains a [2Fe-2S] cluster (FLINT et al., 1993; FLINT and EMPTAGE, 1988). Most substrate studies have been performed with *Salmonella typhimurium* (by CROUT's group) and spinach DHAD (by PIRRUNG's group) and in both cases it has been demonstrated that the *α*-proton is lost to the solvent during the course of dehydration.

Substrate homolog rate study results (Tab. 7) demonstrate that substituents larger than methyl are better accommodated at the R^1 position than the R^2 position (**255e**,**f** and **g**,**h**), however, if only one methyl group is present, rates of dehydration are faster when the methyl group is located in the R^2 position than the R^1 position (**255i**,**j**). One interpretation, is that the binding site for R^2 is tight and responsible for steering the stereochemistry of elimination, whereas the R^1 binding site is loose and accommodates larger groups (ARMSTRONG et al., 1985; cf. LIMBERG and THIEM, 1996). Comparable result have been obtained with spinach DHAD (PIRRUNG et al., 1991).

4.2.1.5 3-Dehydroquinate Dehydratase (EC 4.2.1.10)

3-Dehydoquinate dehydratase (DHQD) catalyzes the dehydration of 3-dehydoquinate (**257**) to give 3-dehydroshikimate (**258**) (Fig. 92). This is the central step in the catabolic qui-

Tab. 7. The *Salmonella typhimurium* α,β-Dihydroxyacid Dehydratase Catalyzed Dehydration of Substrate Homologs (ARMSTRONG et al., 1985)

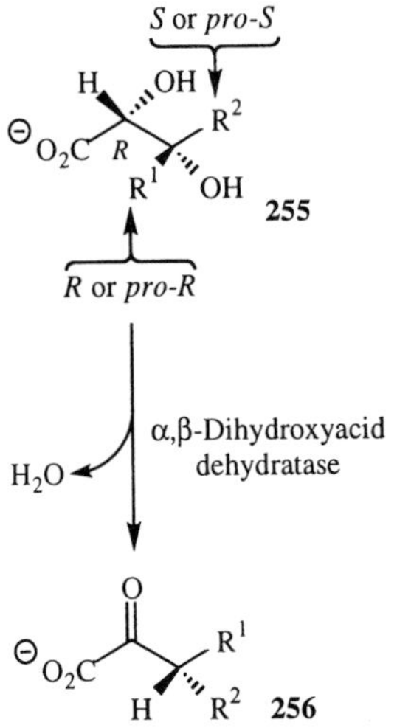

255	Substituents R^1	R^2	Relative rate %
a	Me	Me	100
b	Et	Me	44–46
c	Me	Et	~48
d	Et	Et	19–20
e	Pr	Me	7–9
f	Me	Pr	2–3
g	Bu	Me	1–2
h	Me	Bu	0–1
i	Me	H	7–10
j	H	Me	45–47
k	H	H	3–4

nate pathway and the biosynthetic shikimate pathway (MANN, 1987, MANN et al, 1994). DHQDs are divided into two types. The type I enzymes are heat labile, anabolic enzymes which catalyze *syn*-elimination, via a imine derived from lysine, whereas the type II enzymes are heat stable and catalyze an *anti*-elimination via an unknown mechanism (FLOROVA et al., 1998).

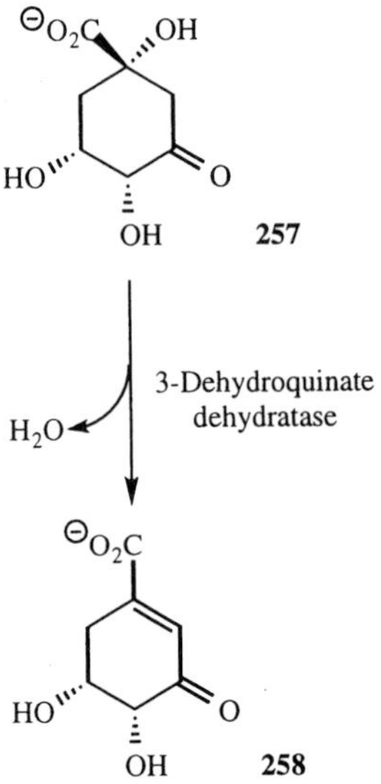

Fig. 92. The 3-dehydroquinate dehydratase catalyzed dehydration of 3-dehydoquinate (**257**) to give 3-dehydroshikimate (**258**).

4.2.1.6 Enolase (2-Phospho-D-Glycerate Hydro-Lyase, EC 4.2.1.11)

Enolase catalyzes the dehydration of 2-phospho-D-glycerate (**259**) to phosphoenolpyruvate (PEP) (**148**) (Fig. 93). Enolase lends its name to a superfamily of enzymes which catalyze deprotonation adjacent to carboxylate groups. The active sites are located in β-barrel (TIM barrel) domains and contain Mg^{2+}, bound to a conserved sequence. Examples include yeast enolase, muconate lactonising enzyme, mandelate racemase, D-gluconate dehydratase from *E. coli* and D-glucarate dehydratases from several eubacteria. The enolase superfamily is divided into three subfamiles, mandelate racemase, muconate lactonizing enzyme (WILLIAMS et al., 1992) and enolase. All members of the enolase *subfamily* of the enolase superfamily bear a conserved lysine, which abstracts protons from centres with (*R*)-stereochemistry, adjacent to a carboxylate group (BABBITT et al., 1995, 1996). This nomenclature is discussed further in connection with

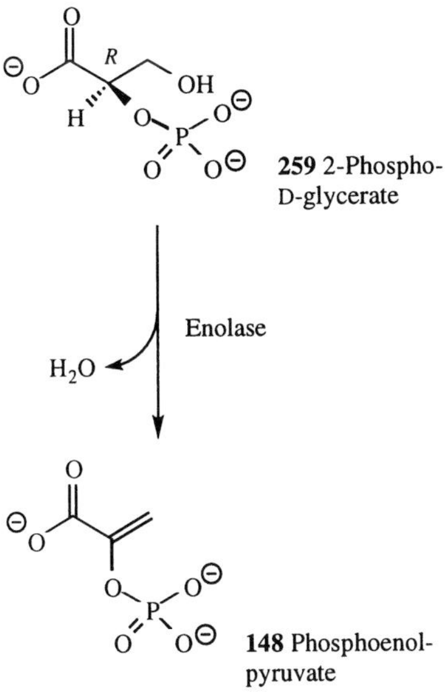

259 2-Phospho-D-glycerate

Enolase

H_2O

148 Phosphoenolpyruvate

Fig. 93. The enolase catalyzed dehydration of 2-phospho-D-glycerate (**259**) to phosphoenolpyruvate (PEP) (**148**).

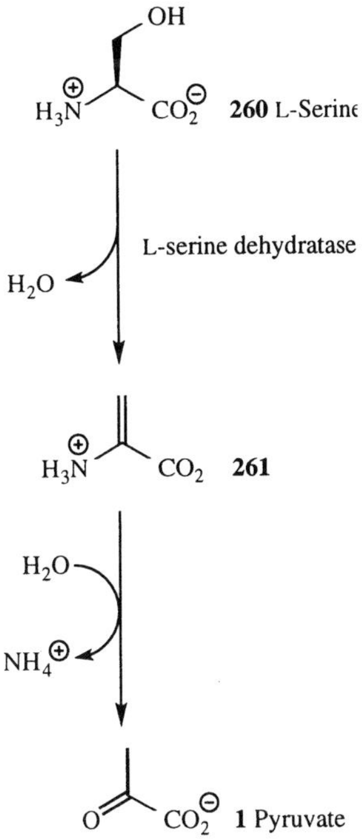

260 L-Serine

L-serine dehydratase

H_2O

261

H_2O

NH_4^+

1 Pyruvate

Fig. 94. The L-serine dehydratase catalyzed dehydration of L-serine (**259** to **260**) to pyruvate (**1**).

166 L-Threonine

L-Threonine dehydratase

H_2O

H_2O

NH_4^+

262 2-Oxobutanoate

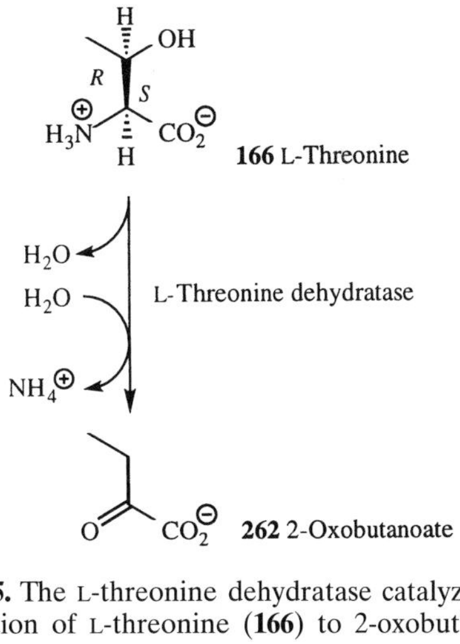

Fig. 95. The L-threonine dehydratase catalyzed dehydration of L-threonine (**166**) to 2-oxobutanoate (**262**).

D-gluconate dehydratase and D-glucarate dehydratase (Sect. 4.3.1.1).

4.2.1.7 L-Serine Dehydratase (EC 4.2.1.13), D-Serine Dehydratase (EC 4.2.1.14) and L-Threonine Dehydratase (EC 4.2.1.15)

L-Serine dehydratase catalyzes the dehydration of L-serine (**259 to 260**) to pyruvate (**1**) (Fig. 94) and L-threonine dehydratase catalyzes the dehydration of L-threonine (**166**) to 2-oxobutanoate (**262**) (Fig. 95). Both enzymes also catalyze the other reaction more slowly than with the "native" substrate. Similarly D-serine dehydratase also has some activity with D-threonine. Most forms of these enzymes are PLP-dependent, but no bacterial L-serine dehydratase has been shown to be PLP dependent. The L-serine dehydratase from *Peptostreptococcus asaccharolyticus* serine dehydratase contains a [4Fe-4S] cluster (HOFMEISTER et al., 1994; GRABOWSKI et al., 1993) and hence is related to fumarase and α,β-dihydroxyacid dehydratase.

4.2.1.8 Imidazoleglycerol Phosphate Dehydratase (EC 4.2.1.19)

Imidazoleglycerol phosphate dehydratase catalyzes the dehydration of imidazoleglycerol phosphate (**263**) to imidazoleacetol phosphate (**264**) (Fig. 96), which is a late step in the biosynthesis of L-histidine in plants and microorganisms. L-Histidine is an essential nutrient for animals, because they lack the requisite biosynthetic pathway. The elimination is fairly unusual because the proton lost is essentially non-acidic, moreover no co-factor has been identified other than Mn^{2+} ions. Elimination proceeds with inversion of configuration at C-3 and the proton is derived from the solvent (PARKER et al., 1995).

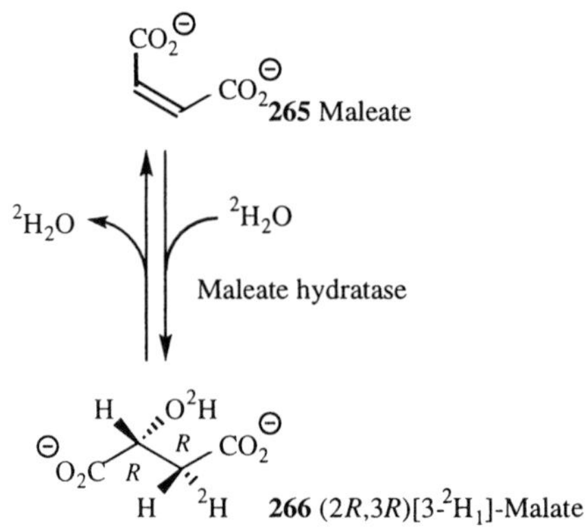

4.2.1.9 Maleate Hydratase [(*R*)-Malate Hydro-Lyase, EC 4.2.1.31]

Maleate hydratase catalyzes the addition of water to maleate (**265**) to give D-malate (**266**) (Fig. 97). 2-Methyl- and 2,3-dimethylmaleate are also substrates. The stereochemistry of hydration has remarkable similarities to that of fumarase. Both enzymes catalyze the *trans*-addition of the elements of water to the alkene bond, with the hydrogen introduced into the (3*R*)-position of malate. Consequently fumarase gives (2*S*)-L-malate (**197**) (Fig. 73) whereas maleate dehydratase gives the enantiomer, (2*R*)-D-malate (**265 to 266**).

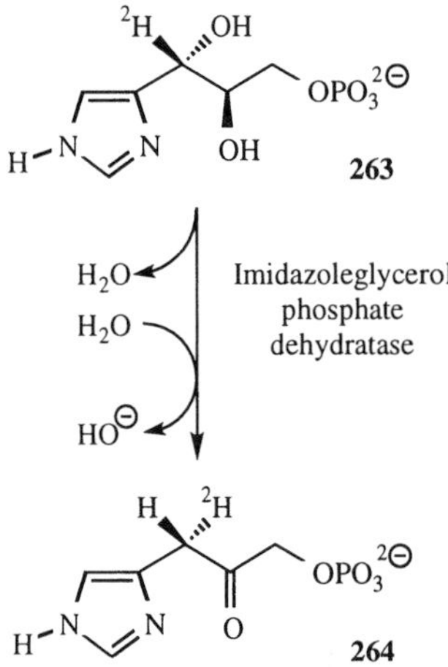

Fig. 96. The imidazoleglycerol phosphate dehydratase catalyzed dehydration of imidazoleglycerol phosphate (**263**) to imidazoleacetol phosphate (**264**).

Fig. 97. Stereochemistry of the hydration of maleate (**265**) to D-malate (**266**) catalyzed by maleate hydratase. For the purposes of brevity, no distinction is made in the text, between labeled and unlabeled D-malate.

Maleate hydratase has been purified from rabbit kidneys, *Arthrobacter* sp. strain MCI2612 and *Pseudomonas pseudoalcaligenes* (VAN DER WERF et al., 1992). The rabbit kidney and *Arthrobacter* sp. strain MCI2612 forms contain an iron–sulfur cluster (DREYER, 1985), whereas that from *Pseudomonas pseudoalcaligenes* does not (VAN DER WERF et al., 1993).

L-Malate (**197**) is a valuable bulk commodity (see Sect. 4.2.1.2, Fumarase), whereas D-malate (**266**) has only found a few minor applications in organic synthesis. Nevertheless potentially, the most efficient and cheapest way to prepare it, is by maleate hydratase catalyzed hydration of maleate. Kinetic analysis of D-malate production by permeabilized *Pseudomonas pseudoaligenes* has shown that the reaction suffers from product inhibition and biocatalyst inactivation. Optimal production occurred at pH 8 and 35 °C, but this was accompanied by complete biocatalyst deactivation over 11 hours (MICHIELSEN et al., 1998). Clearly, considerable development is required to make this a practical method.

4.3 Carbohydrate Lyases (EC 4.2.X.Y)

The lyases are involved in the metabolism of carbohydrates may be separated into two classes: the hydro-lyases which catalyze the elimination of water and the dissacharide and higher lyases which catalyze the the cleavage of glycosidic bonds

4.3.1 Carbohydrate Hydro-Lyases (EC 4.2.1.X)

4.3.1.1 Monosaccharide Carboxylic Acid Dehydratases

The monosaccharide carboxylic acid dehydratases (Tab. 8), catalyze the α,β-elimination of water to give enols which isomerize to the corresponding ketone (e.g., Fig. 98). The stereochemistry at C-2 and C-3 is lost during the course of the elimination, consequently, D-altronate (**274**), D-mannonate (**276**), D-gluconate (**280**) and D-glucosaminate (**279**) all give the same product; 2-dehydro-3-deoxy-D-gluconate (**275**) (Fig. 99). In general enzyme catalyzed eliminations which involve α,β-unsaturated esters, proceed with *anti*-stereochemistry (MOHRIG et al., 1995; PALMER et al., 1997), hence the enol (**289**) derived from D-mannonate (**276**), will be the (*E*)-rather than the (*Z*)-stereoisomer formed by the other substrates.

Galactonate dehydratase is a member of the enolase superfamily of enzymes, which includes yeast enolase, muconate lactonizing enzyme and mandelate racemase. The active sites are located in β-barrel (TIM barrel) domains and contain Mg^{2+}, bound to a conserved sequence. The prototype enzyme is mandelate racemase from *Pseudomonas putida* which has been structurally characterized. Lysine-166 acts as a (*S*)-selective base (LANDRO et al., 1994) and histidine-297 as a (*R*)-selective base. All members of the mandelate racemase subgroup of the enolase family have either one of both of these residues (BABBITT et al., 1995, 1996).

C-2 of D-galactonate (**272**) has the (*R*)-configuration (Fig. 100). Galactonate dehydratase

$^{\ominus}O_2C$ ... OH, OH, OH, OH **270** D-Arabinonate

D-Arabinonate dehydratase

H_2O

OH OH OH **288** Enol

O OH OH **271** 2-Dehydro-3-deoxy-D-arabinonate

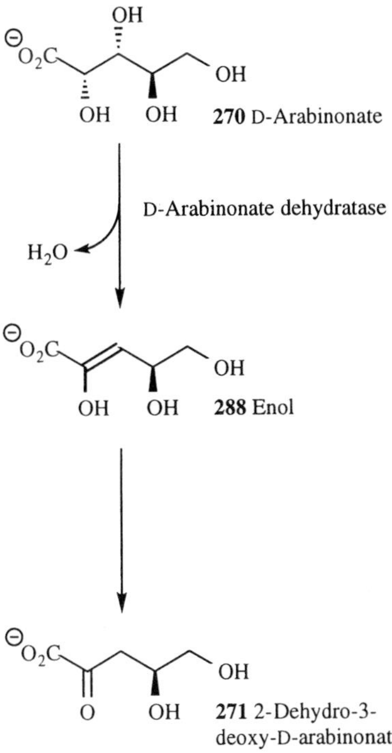

Fig. 98. D-Arabinonate dehydratase catalyzes the α,β-elimination of water from D-arabinonate (**270**) to give an enol (**288**) which tautomerizes to 2-dehydro-3-deoxy-D-arabinonate (**271**).

from *E. coli* bears histidine-285 and aspartate-258, which act as the base and histidine-185 which acts as a general acid catalyst to promote elimination (WIECZOREK et al., 1999). Protonation of the enol, occurs such that hydrogen is incorporated in the (3-*pro-S*)-position (PALMER et al., 1997). Other members of this group include L-rhamnonate dehydratase from *E. coli* and D-glucarate dehydratases from several eubacteria (HUBBARD et al., 1998).

Glucarate dehydratase from *Pseudomonas putida* acts on D-glucarate (**281**) and L-idarate (**294**) (epimers at C-5) to give exclusively 5-dehydro-4-deoxy-D-glutarate (**297**) (Fig. 101). The two substrates are interconverted via the enol (**295**), which partitions between epimerization (reprotonation) and dehydration in a ratio of 1:4. Early experiments suggested that

Tab. 8. Monosaccharide Carboxylic Acid Dehydratases (EC 4.2.1.X)

X	Substrate, Product		Reference
5	OH $^{\ominus}O_2C$ OH OH OH **270** D-Arabinonate	$^{\ominus}O_2C$ OH O OH **271** 2-Dehydro-3-deoxy-D-arabinonate	
6	OH OH $^{\ominus}O_2C$ OH OH OH **272** D-Galactonate	O OH $^{\ominus}O_2C$ OH OH **273** 2-Dehydro-3-deoxy-D-galactonate	DAHMS et al., 1982; WIECZOREK et al., 1999
7	OH OH $^{\ominus}O_2C$ OH OH OH **274** D-Altronate	O OH $^{\ominus}O_2C$ OH OH **275** 2-Dehydo-3-deoxy-D-gluconate	ROBERT-BAUDOUY et al., 1982; DREYER, 1987
8	OH OH $^{\ominus}O_2C$ OH HO HO **276** D-Mannonate	O OH $^{\ominus}O_2C$ OH OH **275** 2-Dehydo-3-deoxy-D-gluconate	ROBERT-BAUDOUY et al., 1982; DREYER, 1987
25	OH $^{\ominus}O_2C$ OH OH OH **277** L-Arabinonate	$^{\ominus}O_2C$ OH O OH **278** 2-Dehydro-3-deoxy-L-arabinonate	
26	$^{\oplus}NH_3$ OH $^{\ominus}O_2C$ OH HO HO **279** D-Glucosaminate	O OH $^{\ominus}O_2C$ OH OH **275** 2-Dehydo-3-deoxy-D-gluconate	IWAMOTO et al., 1995
39	OH OH $^{\ominus}O_2C$ OH HO HO **280** D-Gluconate	O OH $^{\ominus}O_2C$ OH OH **275** 2-Dehydo-3-deoxy-D-gluconate	GOTTSCHALK and BENDER, 1982

Tab. 8. (continued)

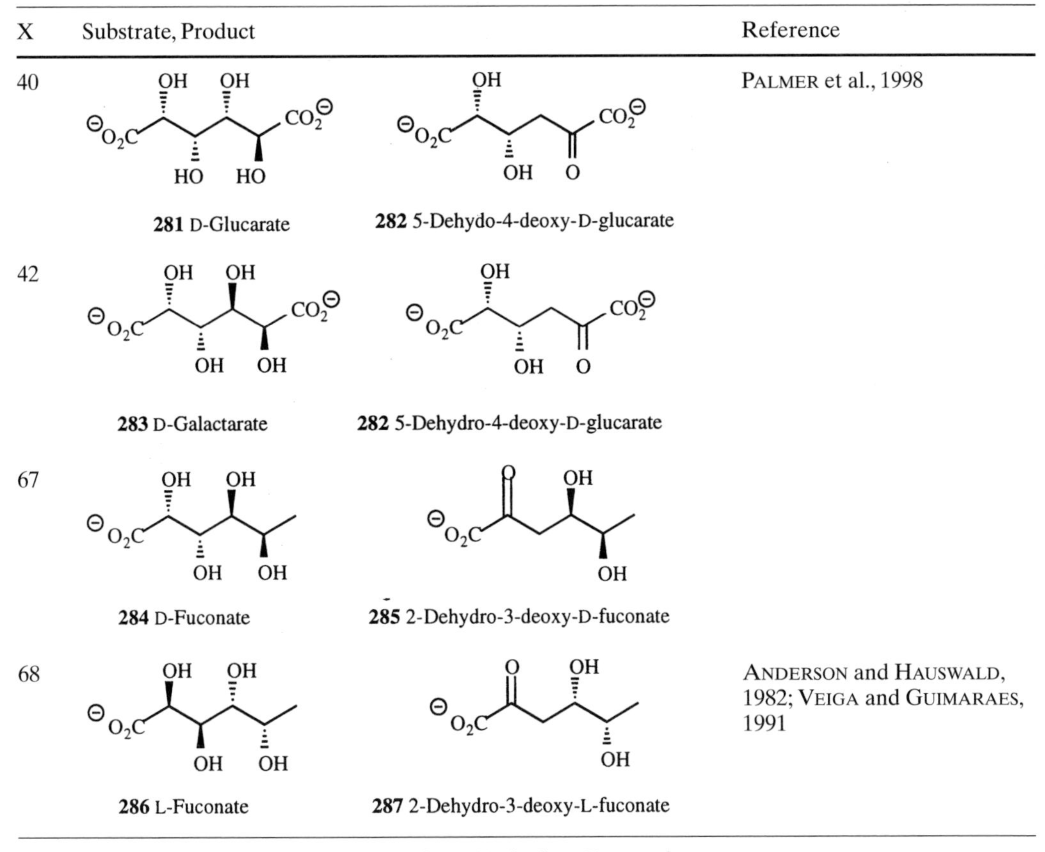

X	Substrate, Product		Reference
40	**281** D-Glucarate	**282** 5-Dehydo-4-deoxy-D-glucarate	PALMER et al., 1998
42	**283** D-Galactarate	**282** 5-Dehydro-4-deoxy-D-glucarate	
67	**284** D-Fuconate	**285** 2-Dehydro-3-deoxy-D-fuconate	
68	**286** L-Fuconate	**287** 2-Dehydro-3-deoxy-L-fuconate	ANDERSON and HAUSWALD, 1982; VEIGA and GUIMARAES, 1991

All 1982 references in this table are taken from *Methods in Enzymology*

the dehydration of D-glucarate (**281**) was only partially regioselective, based on the partial scrambling of a radiochemical label from C-1 to C-6. However, epimerization of D-glucarate to D-idarate which has C_2-symmetry renders C-1 and C-6 equivalent (PALMER and GERLT, 1996; PALMER et al., 1998). Elimination of the 4-hydroxyl group gives an α-hydroxyacrylate (**296**) which is stereospecifically reprotonated by water. The hydrogen is incorporated exclusively with (4-*pro-S*) stereochemistry (PALMER et al., 1997). Given that glutarate dehydratase is capable of deprotonating both D-glucurate (**281**) and L-idarate (**294**), the model for the enolase family predicts that two bases should be present. An X-ray structure determination enabled the assignment of the (*S*)-specific base as lysine-213 (which is activated by lysine-211) and the (*R*)-specific base as histidine-297 and aspartate-270 (GULICK, *et al.*, 1998).

4.3.1.2 3,6-Dideoxysugars

3,6-dideoxyhexoses are found, (with a few exceptions) exclusively in the cell wall lipopolysaccharides of gram-negative bacteria of the family, *Enterobacteriaceae,* where they act as the dominant *O*-antigenic determinants (JOHNSON and LIU, 1998; KIRSCHNING et al., 1997; LIU and THORSON, 1994). There are five 3,6-dideoxysugars known in nature: D-abequose (**298**), L-ascarylose (**299**), L-colitose (**300**), D-paratose (**301**) and D-tyvelose (**302**) (Fig. 102). Many organisms produce most or all of these simultaneously and all except for col-

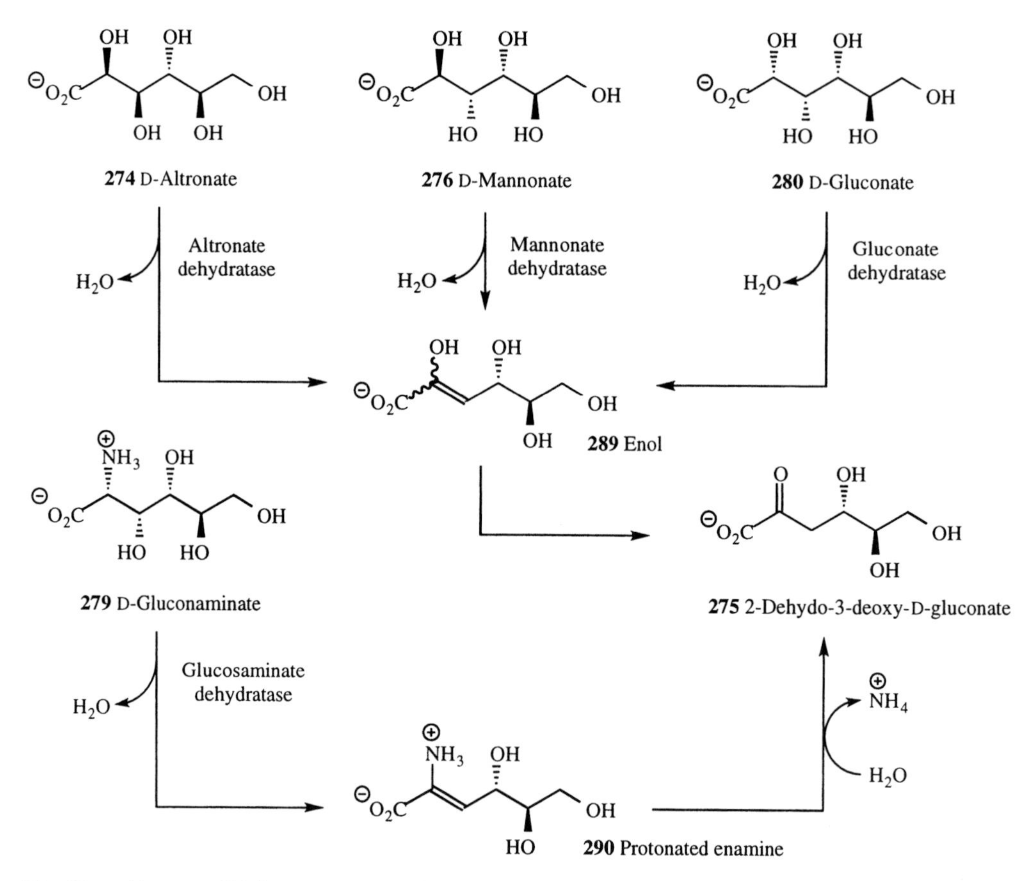

Fig. 99. D-Altronate (**274**), D-mannonate (**276**) and D-gluconate (**280**) undergo α,β-elimination of water to give an enol (**289**) which tautomerizes to 2-dehydro-3-deoxy-D-gluconate (**275**). Dehydration of D-glucosaminate (**279**) also gives the same product.

itose (**300**), are produced biosynthetically by via a complex pathway commencing with CDP-D-glucose (**303c**). Colitose (**300**) is formed biosynthetically from CDP-D-mannose.

The biosynthesis of CDP-ascarylose (**299c**) from D-glucose (**303a**) by *Yersinia pseudotuberculosis* has been investigated in considerable detail (Fig. 103; review HALLIS and LIU, 1999b). The pathway employs two dehydratases (one with a "hidden" redox function), plus two oxidoreductases and an epimerase to deliver a product in which two stereocentres have been retained, two inverted and two reduced (Fig. 103). Three overlapping clones containing the entire gene cluster required for CDP-ascarylose biosynthesis have been identified (THORSON et al., 1994b). Each enzyme has a trivial, but conveniently short designation which is used in all the figures, except Fig. 103 where the full name is shown for reference purposes. The full enzyme commission designation is shown in the text.

Biosynthesis commences with D-glucose (**303a**) which is converted sequentially into the 1-phosphate (**303b**) and the 1-*O*-cytidylyl (**303c**) derivatives catalyzed by α-D-glucose-1-phosphate cytidylyltransferase (E_P) (THORSON et al., 1994a).

CDP-D-glucose 4,6-dehydratase (Eod) [EC 4.2.1.45] catalyzes the transformation of CDP-D-glucose (**303c**) to CDP-6-deoxy-L-*threo*-D-*glycero*-4-hexulose (**304c**). [4-$^2H_1$]-CDP-D-glucose (**306**) (Fig. 104) has been used to investigate the pathway. It consists of three distinct steps; oxidation of the 4-hydroxy group to a

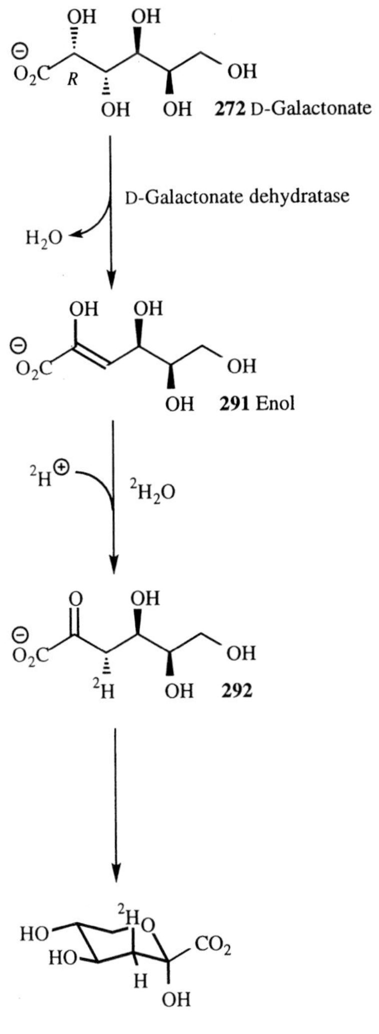

272 D-Galactonate

D-Galactonate dehydratase

291 Enol

292

293 (3*S*)-[3-$^2H_1$]-2-Dehydro-3-deoxy-D-galactonate hemiketal

Fig. 100. D-Galactonate dehydratase catalyzes the α,β-elimination of water from D-galactonate (**272**) to give an enol (**291**) which tautomerizes to 2-dehydro-3-deoxy-D-galactonate (**273**) (Tab. 8). In deuterium oxide, (3*S*)-[3-$^2H_1$]-2-dehydro-3-deoxy-D-galactonate (**292**) is formed which cyclizes to the pyranosyl hemiketal (**293**).

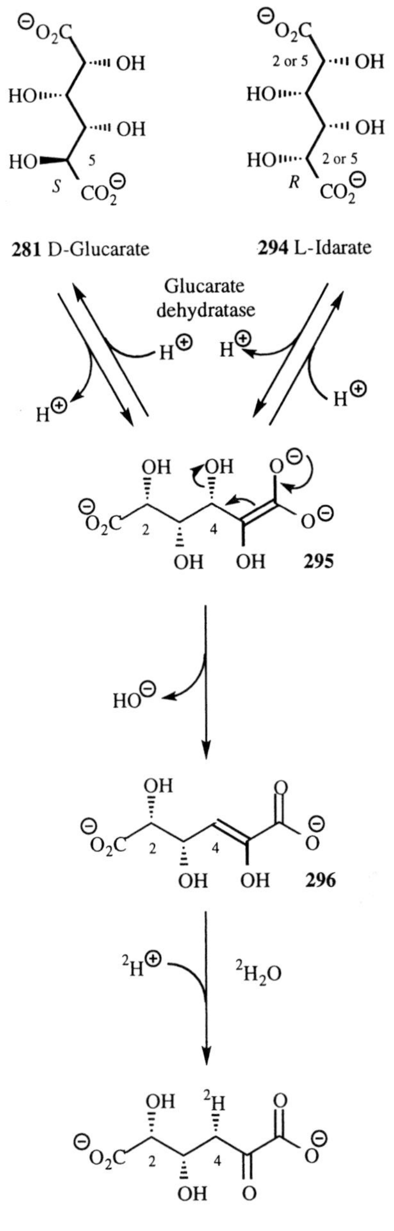

281 D-Glucarate

294 L-Idarate

Glucarate dehydratase

295

296

297 (4*S*)-4-$^2H_1$]-5-Dehydro-4-deoxy-D-glucarate

Fig. 101. Glucurate dehydratase catalyzes the dehydration of D-glucurate (**281**) and L-idarate (**294**).

ketone (**307**) and reduction of NAD to NADH, deprotonation and β-elimination of the 6-hydroxyl group (**308**) and conjugate addition of the hydride ion which originated from C-4 to C-6 to give the 6-deoxy-sugar (**309**). The replacement of the 6-hydroxyl

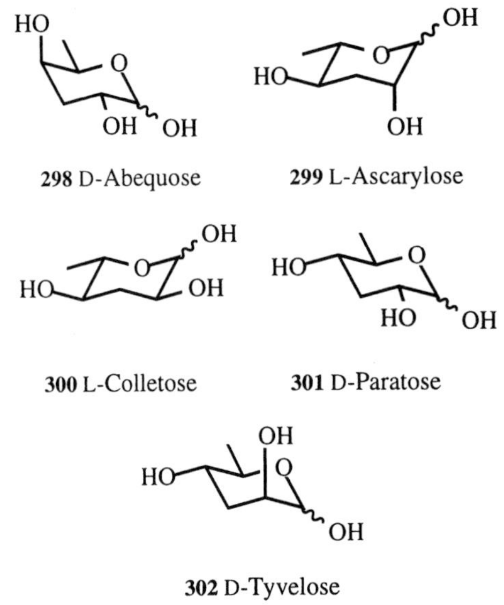

298 D-Abequose

299 L-Ascarylose

300 L-Colletose

301 D-Paratose

302 D-Tyvelose

Fig. 102. Naturally occurring 3,6-dideoxyhexoses.

group by hydride occurs with inversion of configuration (YU et al., 1992; RUSSELL and YU, 1991). This stereochemical course is also followed by the GDP-D-mannose dehydratase from the soil organism (ATCC 19241) (OTHS et al., 1990) and the TDP-D-glucose 4,6-dehydratase from *E. coli* (WANG and GABRIEL, 1970; SNIPES et al, 1977). The basic features of the mechanism are nicely confirmed by the behavior of the difluoro-analog substrate (**310**) (Fig. 105). After traversing the reaction sequence shown above (Fig. 100), the monofluoro-derivative (**311**) so formed undergoes a further β-elimination of fluoride to give the alkene (**308**), which is the normal intermediate in the sequence. However, there is no further hydride to donate and instead addition of a nucleophile located on the protein backbone occurs to give an isolatable covalent derivative (**312**) (CHANG et al., 1998).

The facial selectivity of hydride donation from NADH is rather difficult to determine for this class of enzymes, because NADH is only present during the catalytic cycle. Nevertheless it has been demonstrated that the product (**304c**) is reduced by [4*S*-$^2H_1$]-NADH with incorporation of label when incubated with apo-E_{od} and no label is transferred from [4*R*-$^2H_1$]-NADH (Fig. 106). It is conceivable that

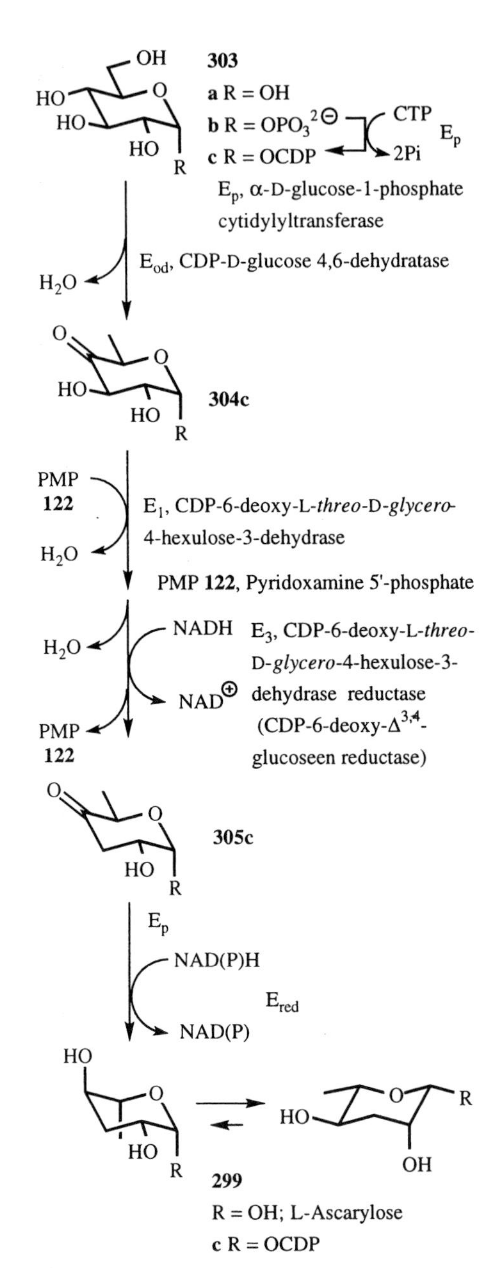

Fig. 103. The biosynthesis of CDP-ascarylose (**299c**) from D-glucose (**303a**) by *Yersinia pseudotuberculosis* (**122**, PMP, pyridoxamine 5′-phosphate).

NADH hydride donation to the α,β-unsaturated ketone (**308**) might occur from the other face, but this is implausible (HALLIS and LUI, 1998). (*pro-S*) Hydride donation has also been

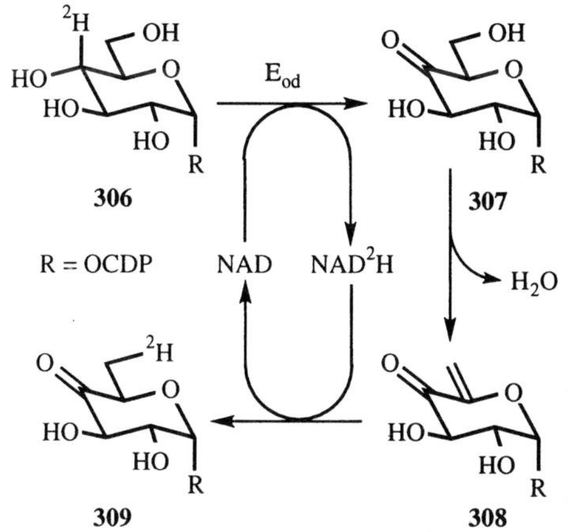

Fig. 104. The isomerization of [4-$^2H_1$]-CDP-D-glucose (**306**) to [6-$^2H_1$]-CDP-6-deoxy-L-*threo*-D-glycero-4-hexulose (**309**) by CDP-D-glucose 4,6-dehydratase (E_{od}) from *Yersinia pseudotuberculosis*.

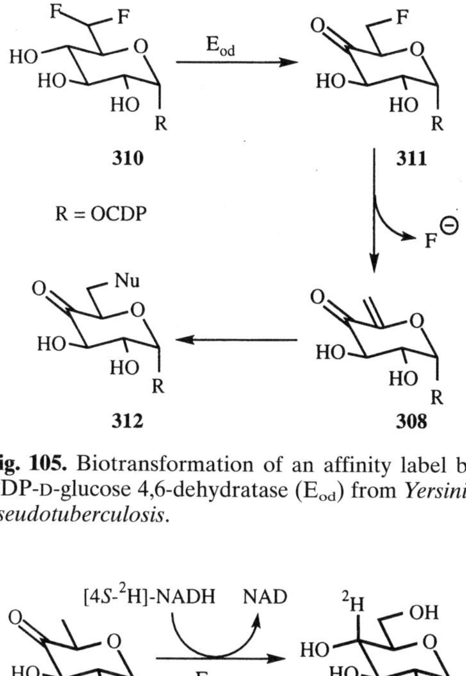

Fig. 105. Biotransformation of an affinity label by CDP-D-glucose 4,6-dehydratase (E_{od}) from *Yersinia pseudotuberculosis*.

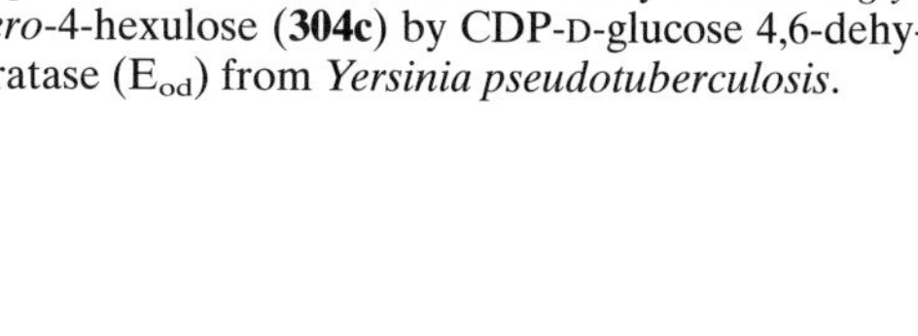

Fig. 106. Reduction of CDP-6-deoxy-L-*threo*-D-*glycero*-4-hexulose (**304c**) by CDP-D-glucose 4,6-dehydratase (E_{od}) from *Yersinia pseudotuberculosis*.

demonstrated for TDP-D-glucose 4,6-dehydratase (WANG and GABRIEL, 1970), L-*myo*-inositol-1-phosphate synthase (BRYUN and JENNESS, 1981; LOEWUS et al., 1980) and UDP-D-glucose-4-epimerase (NELSESTUEN and KIRKWOOD, 1971), which all utilize NAD and act on sugars.

Most of the 4,6-dehydratases belong to a rare group of enzymes which contain a tightly bound NAD(H). As demonstrated above, this is responsible for the redox process and is retained throughout the catalytic cycle as a prosthetic group, rather than a co-substrate (THOMPSON, et al., 1992; NAUNDORF and KLAFFKE, 1996). However, CDP-D-glucose 4,6-dehydratase is a dimer of identical sub-units (42,500 Da), which only binds one equivalent of NAD, despite the presence of two binding sites and there is a large anti-co-operative effect for binding a second molecule of NAD. NADH is bound much more strongly and is released when the substrate binds indicating that this may be the normal "resting" state of the enzyme *in vivo* (HE et al., 1996).

The deoxygenation at C-3 is catalyzed by CDP-6-deoxy-L-*threo*-D-*glycero*-4-hexulose-3-dehydrase (E_1) a pyridoxamine 5′-phosphate (PMP)-linked, [2Fe-2S] containing protein and CDP-6-deoxy-L-*threo*-D-*glycero*-4-hexulose-3-dehydrase reductase (formerly designated CDP-6-deoxy-$\Delta^{3,4}$-glucoseen reductase) (E_3) a [2Fe-2S]-containing flavoprotein (LO et al., 1994). In the first step CDP-6-deoxy-L-*threo*-D-*glycero*-4-hexulose (**304c**) (Fig. 107) undergoes condensation with PMP (**122**) to form the corresponding imine (**313**) which undergoes 1,4-elimination of water to give the protonated 2-aza-1,4-diene (**314**). Both steps are reversible and the elimination is stereospecific; only the (4-*pro-S*)-hydrogen of PMP undergoes exchange (WEIGEL et al., 1992). The reaction stops at this stage, unless CDP-6-deoxy-L-*threo*-D-*glycero*-4-hexulose-3-dehydrase reductase is present. This forms a tight complex with the dehydrase (CHEN et al., 1996) and mediates a reduction in which electrons are transferred within the reductase from NADH to FAD to the [2Fe-2S]-center and from there to the [2Fe-2S]-center of the dehydrase. The details of the transfer of electrons from there to the PMP-sugar adduct are somewhat speculative. It is likely to involve stepwise addition of

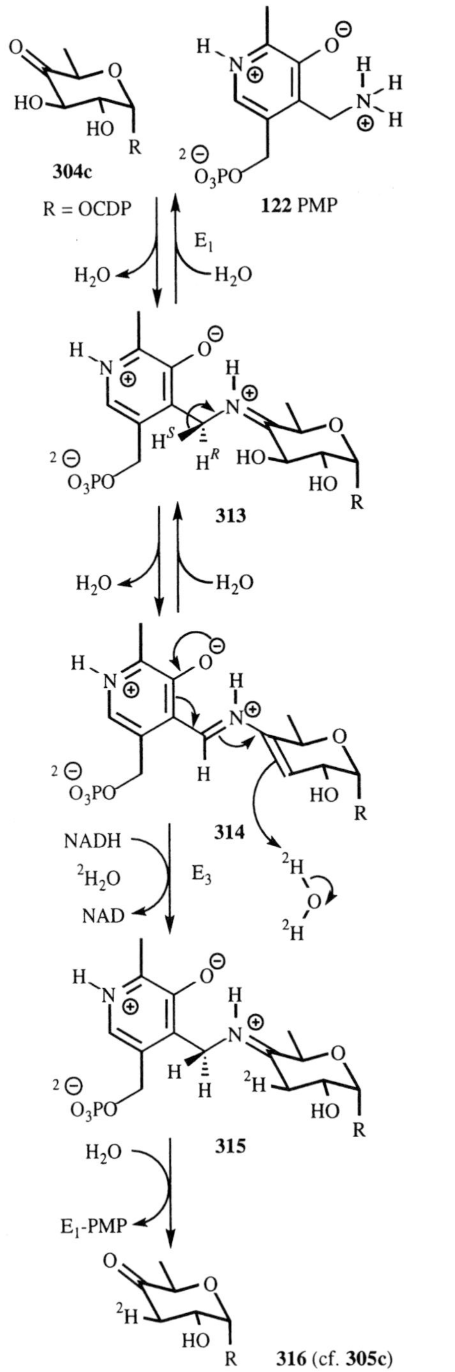

Fig. 107. The E_1, CDP-6-deoxy-L-*threo*-D-*glycero*-4-hexulose-3-dehydrase catalyzed reduction (R = OCDP).

single electrons and this is supported by the detection of an ESR signal for a phenoxy radical (JOHNSON et al., 1996). One attractive possibility is that electron transfer occurs to an *o*-quinomethide formed by protonation at C-3 of the sugar (PIEPER et al., 1997). This occurs stereospecifically with incorporation of a proton from water exclusively at the C-3 equatorial position (**316**) (PIEPER et al., 1995).

Reduction steps

CDP-3,6-dideoxy-D-*glycero*-D-*glycero*-4-hexulose is the common intermedaite for D-abequose (**298**) (LINDQUIST et al., 1994), L-ascarylose (**299**), D-paratose (**301**) (HALLIS et al., 1998) and D-tyvelose (**302**) (HALLIS and LUI, 1999a) by reduction and/or epimerization (Fig. 108). The gene encoding CDP-paratose synthase (rfbS) in *Salmonella typhi* has been identified, sequenced, amplified (PCR) and cloned into a pET-24(+) vector. Expression and purification of the enzyme enabled it to be fully charactorized. This homodimeric oxidoreductase transfers the *pro-S* hydrogen from NADH, and has a 10-fold preference for NADPH over NADH. (HALLIS *et al.*, 1998).

4.3.2 Lyases which Act on Polysaccharides (EC 4.2.2)

4.3.2.1 Pectin and Pectate Lyases

Pectin, pectins or pectic substances are generic names used to describe a group of heterogenous acidic polysaccharides found widely in primary plant cell walls and intercellular layers. They make up for example, 30% of the dry weight of apples, but are usually present in much smaller amounts in woody tissues. The principal sugar residue is D-galacturonic acid which may be partially esterified (pectinic acids) or present as the free acid or calcium salt (pectic acids) (Fig. 109). The pectinic acids are freely soluble in water and are good gelling agents, which are used as pharmaceutical excipients and as gelling agents for fruit jellies (PILNIK and VORAGEN, 1992). Pectic acids are much less soluble and usually require ligands for the counterions of the carboxylates (e.g., EDTA) to solubilize them. The most common

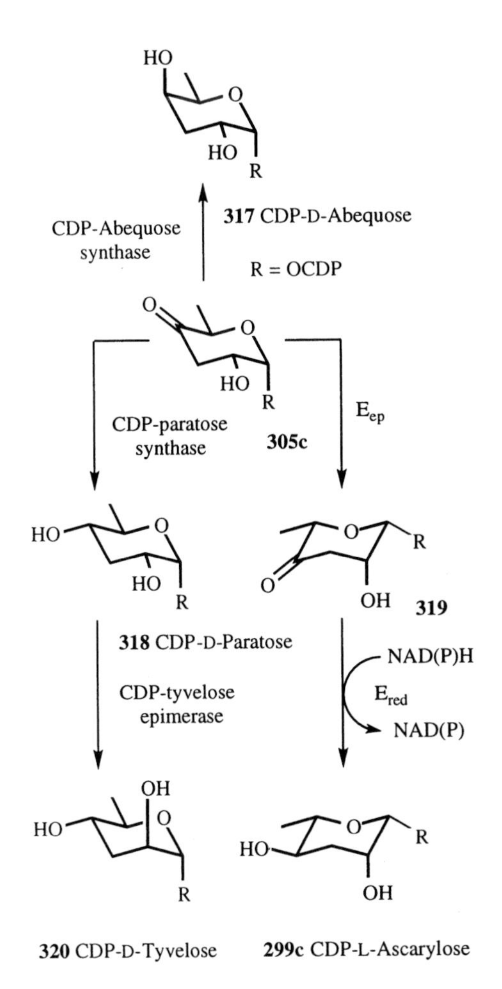

Fig. 108. The E_1, CDP-6-deoxy-L-*threo*-D-*glycero*-4-hexulose-3-dehydrase catalyzed reduction (R = OCDP). CDP-D-tyvelose 2-epimerase (HALLIS and LUI, 1999a), CDP-paratose synthase (HALLIS et al., 1998).

pectins are heteropolysaccharides containing neutral and acidic sugars, but there are also more rare homopolysaccharides such as D-galactans, D-galacturonans and L-arabinans (KENNEDY and WHITE, 1990). Pectinic and pectic acids are cleaved by different classes of enzymes (Tab. 9).

Pectolytic enzymes are essential in the food industry for the clarification of fruit juices and wines. For example, in freshly pressed apple juice, pectins act as solubilising agents for otherwise insoluble cell debris. Hydrolysis of the pectin causes floculation of the insoluble components. Pectolytic enzymes are also useful for dehulling, extractive, pressing (sugar beet and olives) and fermentation (cocoa) processes because they weaken the cell walls and increase yields (ALKORTA et al., 1998). Commercial pectolytic enzymes are almost invariably mixtures which also contain cellulases, hemicellulases and proteases and only part of the pectolytic activity is due to lyases (NAIDU et al., 1998).

Enzymes that degrade pectic substances are frequently the first cell wall degrading enzymes produced by plant pathogens. Depolymerization of pectins not only provides nutrients for the invading organism, but also exposes other cell wall components to degradation. Pectate and pectin lyase activities have been detected in the white rot fungus *Trametes trogii* (LEVIN and FORCHIASSIN, 1998), *Colletotrichum gloeosporiodes* (ORTEGA, 1996), *Botryis cinerea* (MOVAHEDRI and HEALE, 1990; KAPAT et al., 1998a, b), *Aspergillus niger* (VAN DEN HOMBERGH et al., 1997), *Pseudocercosporella herpotrichoides* infecting wheat plants (MBWAGA et al., 1997), *Amycolata* sp. (BRUHLMANN,

Tab. 9. Lyases (EC 4.2.2.X) Acting on Poly-D-Galacturonates (**321**)

X	Name	Regioselectivity	Occurrence	Conditions
2	Pectate lyase	endo	bacteria and pathogenic fungi	Ca^{2+}, pH 9–10, long chains
6	Oligogalacturonide lyase	exo	*Erwinia carotovorans* and *E. aroideae*	–
9	Exopolygalacturonate lyase	exo	rare, *Clostridium*, *Streptomyces*, *Erwinia*, *Fusarium*	Ca^{2+}, pH 9–10
10	Pectin lyase	endo	common in fungi	pH 5–6, long chains

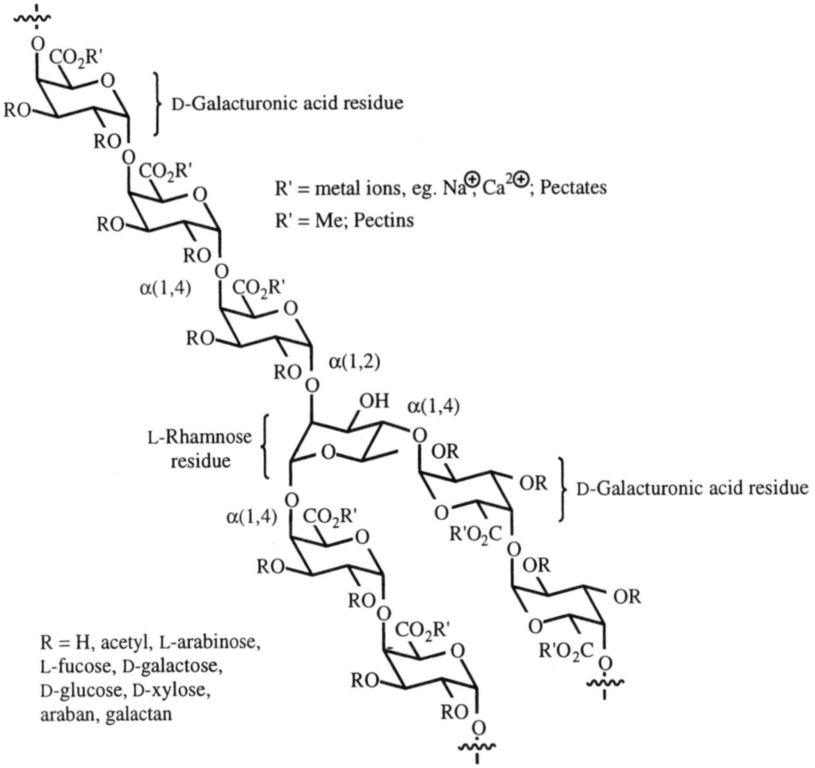

Fig. 109. Generalized structure for pectic substances. The distance between "branching" rhamnose residues depends on the source, e.g., in citrus pectin there are typically 25 intervening D-galacturonic acid residues and 16 in tomato pectin.

1995), *Pseudomonas marginalis* N6031 (NIKAIDOU et al., 1995) and a variety of organisms on rottern cocoyams (UGWUANYI and OBETA, 1997)

4.3.2.2 Pectate Lyases [Poly(1,4-α-D-Galacturonide)Lyase, Pectate Transeliminase, EC 4.2.2.2, formerly EC 4.2.99.3]

Pectate lyases catalyzes the cleavage of pectate (**321a**) to give D-galacturonate (**332a**) and 4-deoxy-α-D-gluc-4-enuronate (**323a**) terminated oligomers. The cleavage reaction is a *trans*-α,β-elimination, operates at pH 8.0–10 and has an absolute requirement for Ca^{2+} counterions. As with the pectin lyases, attack is random, *endo*-selective and occurs preferentially with long chain pectates. Care must be taken when using pectate lyases to get obtain material free from esterases, which will enable cleavage of pectins.

Pectate lyases have been identified in many bacteria and pathogenic fungi. The X-ray crystal structures for pectate lyase C (PelC) and pectate lyase E (PelE) from *Erwinia chrysthanemi* and from *Bacillus subtilis* (BsPel) (PICKERSGILL et al., 1994) are very similar to each other and that of *Aspergillus niger* pectin lyase (MAYANS et al., 1997). The structures are composed of all parallel β-strands wound into an unusual right-handed parallel β-helix (JENKINS et al., 1998; PICKERSGILL et al., 1994 and cf. 1998). The active/calcium binding site is located in a cleft between this feature and two T3-loops, which also contains a highly basic region which is presumably responsible for deprotonation adjacaent to the carboxylate group.

The genes for *Butyrivibrio fibrisolvens endo-β*-1,4-glucanase plus the *Erwinia* pectate lyase and polygalacturonase were each inserted between a yeast expression secretion cassette and yeast gene terminator and cloned into yeast centromeric shuttle vectors. Coexpression in *Saccharomyces cerevisiae* successfully resulted in secreted activity for all three enzymes (VAN RENSBURG et al., 1994).

4.3.2.3 Pectin Lyases [Poly(Methoxygalacturonide)Lyase, EC 4.2.2.10]

Pectin lyase only acts on poly(methyl D-galacturonate esters) (**321b**) (Fig. 110) and has no activity with the corresponding acids. The main source of this enzyme is *Aspergillus* sp. which releases pectin lyases A and B types when grown on pectin (DELGADO, *et. al.*, 1993; VAN DEN HOMBERGH *et al.*, 1997). Pectin lyases are *endo*-selective for cleavage of poly(methyl D-galacturonate esters) (**321b**) at random positions and show very little activity with di- and trisaccharides. pH optima range from 5.5 to 8.6 and are generally lower than those for the pectate lyases. However, pectin lyase A activity drops above pH 7 due to conformational changes at the active site triggered by the proximity of aspartate residues (MAYANS *et al.*, 1997).

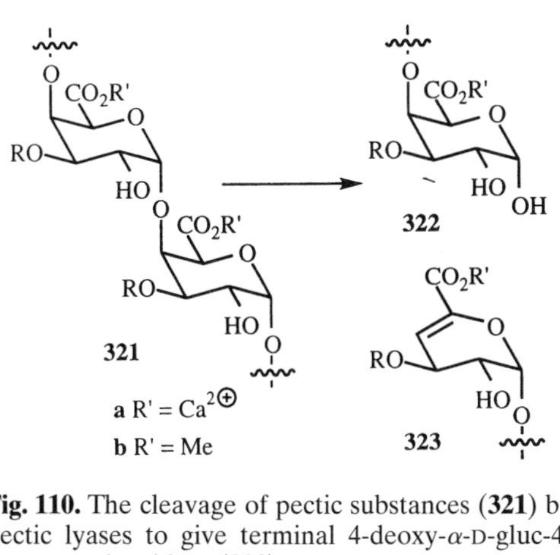

321 a R' = $Ca^{2\oplus}$ b R' = Me; **322**; **323**

Fig. 110. The cleavage of pectic substances (**321**) by pectic lyases to give terminal 4-deoxy-α-D-gluc-4-enuronosyl residues (**323**).

4.4 Carbon–Nitrogen Lyases (EC 4.3)

4.4.1 Ammonia Lyases (EC 4.3.1.X)

The ammonia lyases have achieved considerable industrial importance because of their use in the manufacture of amino acids. This area has been spurred by the development of Aspartame (**325**), (Nutrasweet®, α-L-aspartyl-L-phenylalanine methyl ester, Fig. 111), which is an artificial sweetener that is some 200 times sweeter than sucrose. Unlike many other artificial sweetness its has no bitter after taste and it is used extensively in soft drinks, confectionery and many other processed foods. The patent for its use ran out in the 1980s and hence for producers to remain competitive they must develop new, cheaper and patentable ways to manufacture it (OYAMA, 1992).

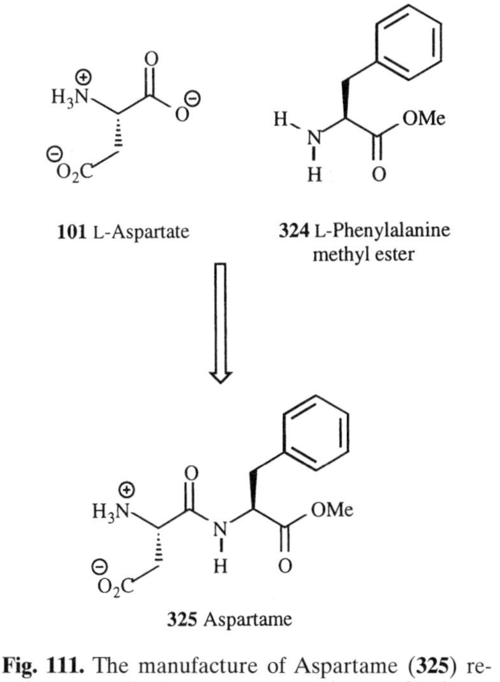

101 L-Aspartate; **324** L-Phenylalanine methyl ester; **325** Aspartame

Fig. 111. The manufacture of Aspartame (**325**) requires cost effective routes to L-aspartate (**101**) and L-phenylalanine methyl ester (**324**) (OYAMA, 1992).

4.4.1.1 Aspartase (L-Aspartase, Aspartase Ammonia-Lyase; EC 4.3.1.1)

Aspartase catalyzes the addition of ammonia to the alkene bond fumarate (**196**) to give L-aspartate (**101**) (Fig. 112). The reaction is for all practical purposes, absolutely specific for fumarate and L-aspartate (FALZONE et al., 1988), but hydroxylamine can substitute for ammonia (KARSTEN and VIOLA, 1991). The equilibrium constant for the reaction heavily favors aspartate formation over the reverse process (K_{eq} $5 \cdot 10^{-3}$, DG° 3.2 kcal mol^{-1}). Carbon–nitrogen bond cleavage is at least partially rate limiting, but carbon–hydrogen bond cleavage is not. This can be rationalized as a conventional conjugate addition (Michael) reaction to give a stabilized carbanion which is rapidly protonated. In the reverse reaction this equates to efficient deprotonation adjacent to the carboxylate group of L-aspartate followed by slow β-elimination (PORTER and BRIGHT, 1980; NUIRY et al., 1984)

Aspartase was discovered by Harden (1901) in *Bacillus coli communis* over one hundred years ago, and in 1932 it became the first microbial enzyme to be used in the synthesis of an amino acid. It is widely distributed in bacteria as well as in fungi, higher plants and animals but not in mammals. The enzyme has been purified and characterized from microorganism including *Bacillus fluorescens*, *B. subtilis*, *B. strearothermophilus*, *Pseudomonas fluorescens*, *P. trefolii*, *Escherichia feundii*, *E. coli*, *Bacterium cadaveris*, *Brevibacterium ammoniagenes* and *Brevibacterium flavum* (DRAUZ and WALDMANN, 1995). L-Aspartase is postulated to play a significant role in the mineralization of soils, by the release of ammonium ions (SENWO and TABATABAI, 1996, 1999).

Aspartase is one of a number of related enzymes which catalyze additions to the alkene bond of fumarate, such as class II fumarases (EC 4.3.1.2, water, WOODS et al., 1986), and the amidine lyases; arginosuccinase (EC 4.3.2.1, L-arginine, COHEN-KUPEIC et al., 1999; TURNER et al., 1997) and adenylsuccinase (EC 4.3.2.2, AMP, LEE et al, 1999). These enzymes have significant sequence homology (WOODS et al., 1988a) and catalyze the addition-elimination reactions with identical stereochemistry, i.e. *trans*, with the hydrogen added to the (3-*pro-R*-) position (Fig. 113) (GAWRON and FONDY, 1959). The reaction is fully reversible and hence can be used to prepare hydrogen isotopomers of aspartate (**328**) from, for example pertritiated-aspartate (**327**) (Fig. 114) (PICCIRILLI et al., 1987; cf. YOUNG, 1991). Large scale syntheses of of ^{15}N-labeled L-aspartate (SOOKKHEO et al., 1998) and (2*S*,3*S*)-[2,3-$^2H_2$]- and (2*S*,3*R*)-[3-^{2}H]-aspartate (LEE et al., 1989) have been achieved using immobilized aspartase.

Asparatase from *E. coli* (and most other sources) is a tetramer of identical 50 kDa subunits (50 kDa). It is in a pH dependent equilib-

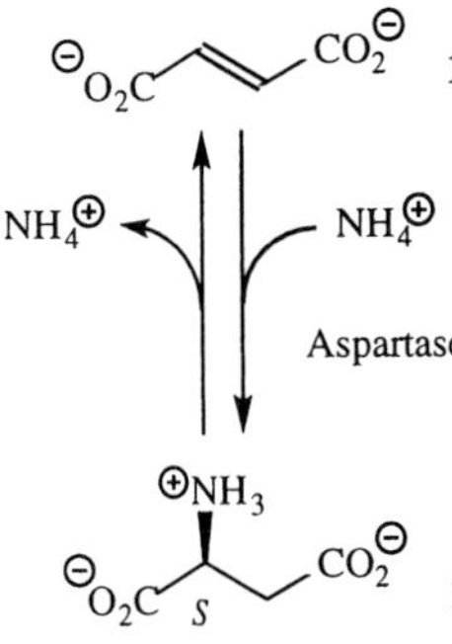

196 Fumarate

NH_4^+ NH_4^+ Aspartase

101 L-(*S*)-(+)-Aspartate

Fig. 112. The aspartase catalyzed the addition of ammonia to fumarate (**196**) to give L-aspartate (**101**).

196 Fumarate

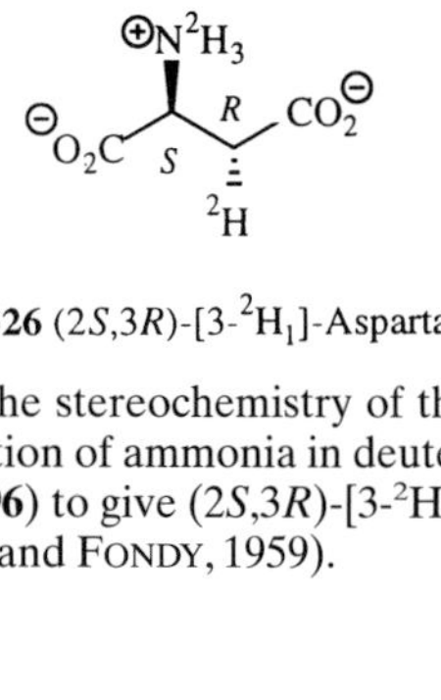

$N^2H_4^+$ $N^2H_4^+$ D_2O Aspartase

326 (2*S*,3*R*)-[3-$^2H_1$]-Aspartate

Fig. 113. The stereochemistry of the aspartase catalyzed addition of ammonia in deuterium oxide to fumarate (**196**) to give (2*S*,3*R*)-[3-$^2H_1$]-aspartate (**326**) (GAWRON and FONDY, 1959).

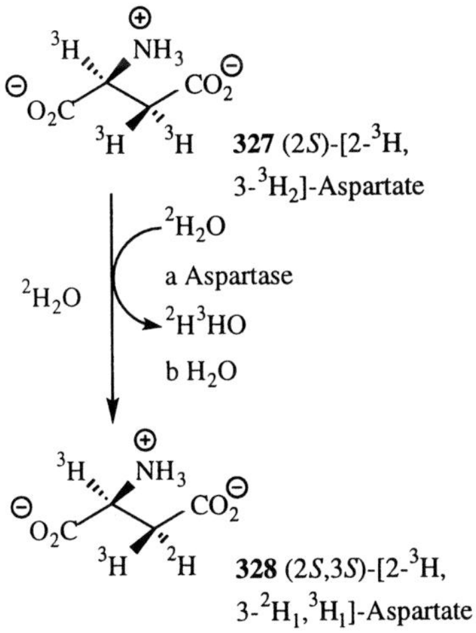

Fig. 114. Stereospecific exchange of the (3-*pro-R*)-tritium of (2*S*)- [2-^{3}H,3-$^3H_2$]-aspartate (**327**) to give (2*S*,3*S*)-[2-^{3}H,3-$^2H_1$,$^3H_1$]-aspartate (**328**) (PICCIRILLI et al., 1987).

rium between two forms. At pH 7.5 and above, divalent metal ions (e.g., Mg^{2+}) are required for activity, and in the amination direction a lag time is observed which is eliminated by the addition of L-aspartate (**101**) or structural analogs. At lower pH values there is no requirement for metal ions and no lag time (KARSTEN and VIOLA, 1991). In contrast, the aspartase from *Bacillus* sp. YM55-1 is insensitive to magnesium ions over the complete pH range (KAWATA, et al., 1999) and the aspartase from *H. alvei* is activated by magnesium ions at all pHs (YOON et al., 1995; cf. NUIRY et al., 1984). In contrast to the absolute substrate specificity, aspartase is activated by a range of metal ions. Transition metal ions are bound better than alkali earth metals, but they are inhibitory above 100 µM, whereas there is no significant inhibition by alkali earth metals. Electron paramagnetic resonance measurements indicated that at higher pH, Mn^{2+} is bound more tightly (FALZONE et al., 1988).

The X-ray crystal structure of aspartate ammonia-lyase from *E. coli* has been determined to 2.8 Å. A putative active site was identified but the precise residues involved in catalysis could not be assigned (SHI et al., 1993, 1997). A succession of candidate residues have been postulated as active site acids and bases, but site specific mutagenic studies have excluded all of these, thus far (JAYASEKERA et al., 1997). For example, L-aspartate β-semialdehyde (**329**), undergoes deamination by aspartase to give fumaric β-semialdehyde (**330**), which reacts with cysteine-273 to give a catalytically inactive covalent adduct with the putative structure (**331**) (Fig. 115) (SCHINDLER and VIOLA, 1994). Inhibition of inactivation by fumarate and divalent metal ions, indicates that inactivation by L-aspartate β-semialdehyde (**329**), occurs at the active site. However, mutagenesis of Cysteine-273, yielded mutants (C273A, C273S) which were still catalytically active, but less sensitive to inactivation (GIORGIANNI et al., 1995). Similarly, treatment of aspartase with *N*-ethylmaleimide caused loss of activity due to modification of cysteines-140 or 430, but mutagenesis (C430W) increased activity (MURASE et al., 1991). Analysis of conserved residues and comparisons with the structure of fumarase, provided a further list of

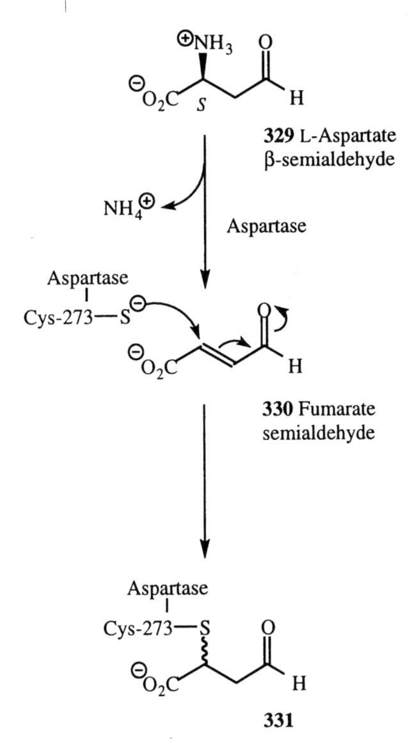

Fig. 115. Covalent modification of aspartase by fumarate semi-aldehyde (330) formed by aspartase catalyzed deamination of L-aspartate semi-aldehyde (**329**).

candidates for mutagenesis (C389S, C389A, H123L), but these caused only minor reductions in activity. However, these studies resulted in the identification of lysine-327 as a binding site for one of the carboxylate groups of the substrate and the other is probably bound by Arginine-29 (SARIBAS et al., 1994; JAYASEKERA et al., 1997). In summary, the role of the active site residues in aspartase is subtle and quite different to that of fumarase. Their full identification will probably require an X-ray crystal structure with a substrate analog.

Biotechnology

The major use for L-aspartate (**101**) is an important food additive. It is also used in many medicines, as a precursor of the artifical sweetner, Aspartame (**325**; Fig. 112), and as a chiral synthon. L-Glutamate (**114**) and L-aspartate (**101**) are the most important amino donors in transamination in nature. L-Aspartate is the most common donor in the manufacture of amino acids by transamination (Fig. 116), but there are a few transaminases which do not accept it as an amino donor. However, by exploiting a L-glutamate (**114**)/α-ketoglutarate (**334**) couple, it can be used with virtually all transaminases. Moreover the unfavorable equilibrium constant (K_{eq} = circa 1) can be displaced by decarboxylation of α-ketoglutarate (**334**) to pyruvate (**1**) (STIRLING, 1992).

L-Aspartate has been produced in batch culture since 1960, using free, intact *E. coli* cells with high aspartase activity. In 1973, a continuous process was introduced which used *E. coli* cells entrapped in polyacrylamide gel, packed into a column. Such columns have a half-life of 120 days at 37 °C. Immobilization in k-carrageenan, hardened with glutaraldehyde and hexamethylene diamine extended the half life to almost two years and this system was introduced into production in 1978. Using a 1,000 L column packed with hardened k-carrageenan immobilized *E. coli* some 3.4 tonnes of L-aspartate can be produced per day and 100 tonnes per month. *E. coli* EAPc7 produces seven times higher aspartase activity than the parent strain (*E. coli* ATCC 11303), but also has fumarase activity. This can be abolished by incubating the culture broth with 50 mM, L-aspartate at 45 °C for 1 hour. After immobilization, the productivity of L-aspartate treated

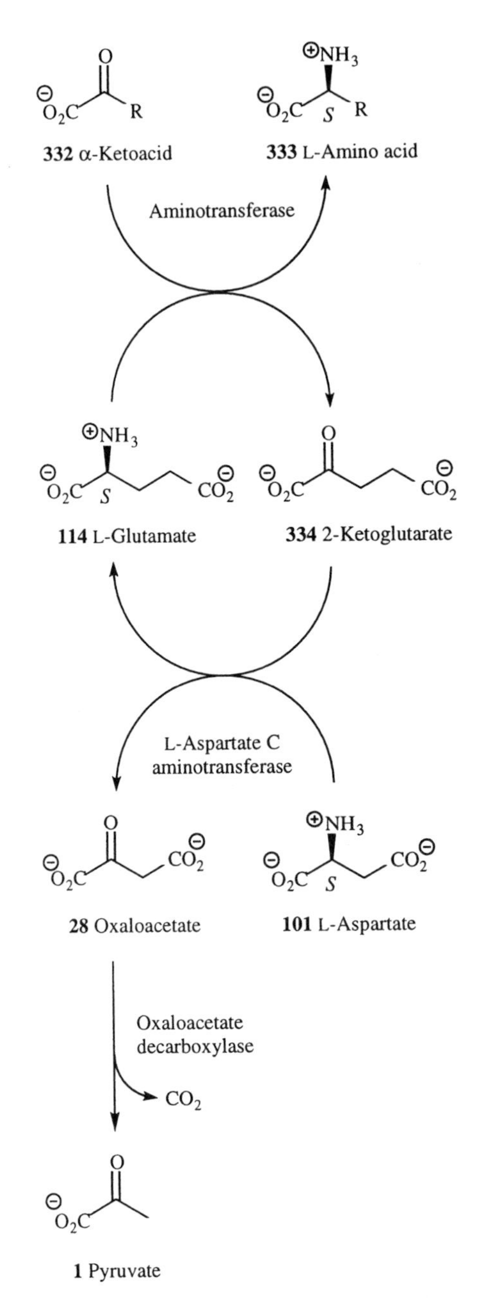

Fig. 116. L-aspartate (**101**) is an amino donor for transamination with most a-ketoacids (**332**). When it cannot act as a direct donor, the amino group can be donated via L-glutamate (**114**). The equilibrium constant is typically about 1, but may be displaced towards product (**333**) formation by *in situ* decarboxylation of oxaloacetate (**28**) to give pyruvate (**1**), which is a less reactive substrate.

E. coli EAPc7 was some six times the parent strain and the half life was 126 days. This system was commercialized in 1982 (CHIBATA, et al., 1995). The activity of L-aspartase from *E. coli* has also been enhanced 2 to 3 fold by site-directed mutagenesis (C430W) (MURASE et al., 1991), limited *C*-terminal proteolysis and acetylation of amino groups by acetic anhydride (MURASE and YUMOTO, 1993).

The amination of fumarate to give aspartate is exothermic, which means that on a large scale, an expensive heat exchanger system is required to minimize thermal degradation of the enzyme. Consequently, there has been much interest in aspartases from thermostable organisms, which will tolerate operation at higher temperatures. A thermostable aspartase has been purified from the thermophile, *Bacillus* sp. YM55-1, which has an activity 3 to 4 times higher than those from *Escherichia coli* and *Pseudomonas fluorescens* respectively at 30 °C (KAWATA, et al., 1999).

4.4.1.2 β-Methylaspartase (Methylaspartase Ammonia-Lyase, EC 4.3.1.2)

β-Methylaspartase catalyzes the addition of ammonia to mesaconate ((*E*)-2-methylfumarate) (**335**) to give L-*threo*-(2*S*,3*S*)-3-methylaspartate (**336**) (Fig. 117). The stereochemistry of addition is *anti*. Ammonia adds to the *si*-face of C-2 and hydrogen to the *re*-face of C-3. This reaction is the second step in the catabolism of L-glutamate (**114**) under anaerobic conditions (KATO and ASANO, 1997). In the prior step, L-glutamate (**114**) undergoes an extraordinary rearrangement to give L-*threo*-3-methylaspartate (**336**), which is catalyzed by coenzyme B_{12}-dependent glutamate mutase (EC 5.4.99.1).

The β-methylaspartase from the obligate anaerobe *Clostridium tetanomorphum* has been extensively investigated. Intimate details of the mechanism have been deduced in a long series of studies (GANI et al., 1999) and the gene has been cloned, sequenced and expressed in *E. coli* (GODA et al., 1992). The preliminary data on enzymes from other sources, suggests that they are fairly similar. β-methylaspartases from the facultative anaerobes, *Cit-*

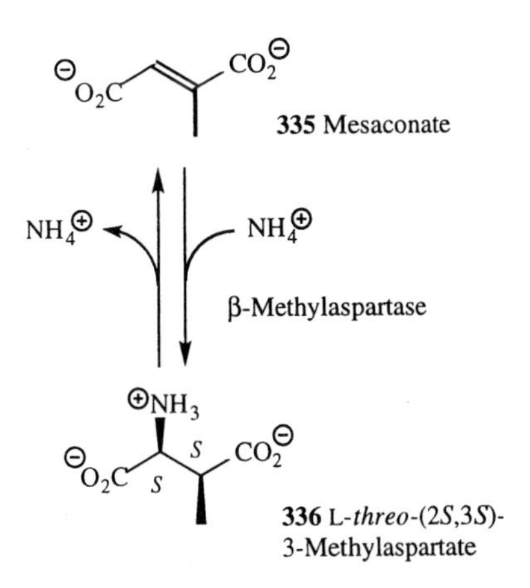

Fig. 117. The β-methylaspartase catalyzed addition of ammonia to mesaconate ((*E*)-2-methylfumarate) (**335**) to give L-*threo*-(2*S*,3*S*)-3-methylaspartate (**336**).

robacter sp. strain YG-0505 and *Morganella morganii* strain YG-0601 (KATO and ASANO, 1995a), *Citrobacter amalonaticus* strain YG-1002 (ASANO and KATO, 1994a, b; KATO and ASANO, 1995b) have been purified and crystallized. The gene for β-methylaspartase in *Citrobacter amalonaticus* strain YG-1002, has been cloned, sequenced and expressed in *E. coli,* but synthetic applications have barely been explored (KATO and ASANO, 1998).

The remainder of this section will deal exclusively with the β-methylaspartase from the *Clostridium tetanomorphum*, unless indicated otherwise. The gene has been cloned, sequenced and expressed in *E. coli.* The open reading frame, codes for a polypeptide of 413 amino acids (45,539 Da), which would exist as a homodimer in solution (GODA et al., 1992). The deduced sequence for *Citrobacter amalonaticus* YG-1002 β-methylaspartase is identical in length and has 62.5% identity (KATO and ASANO, 1998).

It has long been assummed (on the basis of substantial circumstantial evidence) that β-methylasparatase, histidine ammonia-lyase (HAL) and phenylalanine ammonia-lyase (PAL) all bear a dehydroalanine group at the active site, which acts as an electrophilic "trap" for ammonia during the catalytic cycle (GULZAR et al., 1995). However, a recent X-ray structure of catalytically active *Pseudomonas putida* HAL has revealed a 4-methylene-imidazole-5-one group at the active site (cf. Fig. 123; SCHWEDE et al., 1999). It is not clear at the time of writing, if this is a general or unique phenomenon. Nevertheless, the gross mechanisms of action of dehydroalanine and 4-methylene-imidazole-5-one group, are sufficiently similar, that a discussion of the chemistry of the former, provides a fair approximation to that of the latter. Both of the β-methylaspartases bear the Ser-Gly-Asp sequence (Tab. 10) which is conserved in all putative dehydroalanine bearing proteins, however, none of the other conserved residues found in HAL's or PAL's are found in the β-methylaspartases. Albeit that the conserved glycine in the –5 position may be a conservative replacement for proline.

β-Methylaspartase has attracted much attention because (unlike aspartase), it accepts a fair range of alternative alkenic substrates (Tab. 11) and nucleophiles (Tab. 12) (BOTTING et al, 1988a). Substitution of the methyl group of mesaconate (**335**, entry 2) by alkyl groups (entries 3–5) reduces the rate, but nevertheless adequate yields of products can be obtained. Amination of the halo-derivatives (entries 7–10), is complicated by enzyme deactivation and only the chloro-fumarate (entry 8) gives usuable yields of products. β-Methylaspartase is much more tolerant of variation of the nucleophile. Alkyl amines (entries 1–8), hydrazines (9–13), hydroxylamines (15–17) and methoxylamines (entries 18–20) are all accepted to some extent. Overall, structural deviation from the natural substrates reduces the rate in an additive way. The cyclopropyl methyl- and bromo-derivatives (**341a**, **b**) are a potent competive inhibitors (K_i 20 and 630 μmol L^{-1} and respectively) (Fig. 118; BADIANI et al., 1996). It would be of great interest to discover if the amino derivative (**341c**) would yield the corresponding cyclopropene and if this reaction could be extended to higher cyclic homologs.

β-Methylaspartase catalyzes the exchange of the C-3 proton of L-*threo*-3-methylaspartate (**336**) with solvent, faster than deamination (BOTTING et al., 1987). This was originally interpreted as indicating a carbanion intermediate, but $^{15}N/^{14}N$-$^{2}H/^{1}H$ isotope fractionation

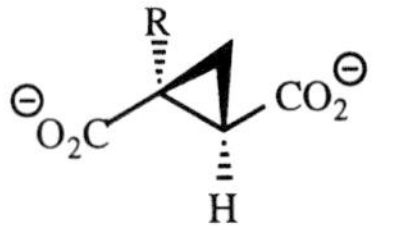

Fig. 118. Cyclopropyl inhibitors (**341**) of β-methylasparatase (BADIANI et al., 1996).

Tab. 10. Amino Acid Sequences of Enzymes Bearing Dehydroalanine/4-Methylene-Imidazole-5-One Precursor Motifs (entry 1, GODA et al., 1992; entry 2, KATO and ASANO, 1998; entries 3-15, LANGER et al., 1994b)

Entry	Enzyme	Species	Range	Residues 5	4	3	2	1	0	1	2	3	4	5	6
1	BMA	*Clostridium tetanomorphum*	168–179	P	V	F	A	Q	**S**	**G**	**D**	D	R	Y	D
2	BMA	*Citrobacter amalonaticus* YG-1002	168–179	P	L	F	G	Q	**S**	**G**	**D**	D	R	Y	I
3	HAL	*P. putida*	138–149	**G**	S	V	G	A	**S**	**G**	**D**	L	A	**P**	**L**
4	HAL	*S. griseus*	142–153	**G**	S	L	G	C	**S**	**G**	**D**	L	A	**P**	**L**
5	HAL	*B. subtilis*	137–148	**G**	S	L	G	A	**S**	**G**	**D**	L	A	**P**	**L**
6	HAL	*R. norvegicus*	250–261	**G**	T	V	G	A	**S**	**G**	**D**	L	A	**P**	**L**
7	HAL	*M. musculus*	250–251	**G**	T	V	G	A	**S**	**G**	**D**	L	A	**P**	**L**
8	PAL	*R. toruloides*	205–216	**G**	T	I	S	A	**S**	**G**	**D**	L	S	**P**	**L**
9	PAL	*R. rubra*	211–222	**G**	T	I	S	A	**S**	**G**	**D**	L	S	**P**	**L**
10	PAL	*O. sativa*	184–195	**G**	T	I	T	A	**S**	**G**	**D**	L	A	**P**	**L**
11	PAL	*P. crispum*	197–208	**G**	T	I	T	A	**S**	**G**	**D**	L	P	**V**	**L**
12	PAL	*G. max*	194–205	**G**	T	I	T	A	**S**	**G**	**D**	L	P	**V**	**L**
13	PAL	*I. batates*	187–198	**G**	T	I	T	A	**S**	**G**	**D**	L	P	**V**	**L**
14	PAL	*L. esculentum*	204–315	**G**	T	I	T	A	**S**	**G**	**D**	L	P	**V**	**L**
15	PAL	*P. vulgaris*	193–204	**G**	T	I	T	A	**S**	**G**	**D**	L	P	**V**	**L**

BMA, β-methylaspartase; HAL, histidine ammonia-lyase; PAL, phenylalanine ammonia-lyase

Tab. 11. The β-Methylaspartase Catalyzed Addition of Ammonia to Substituted Fumarates (AKHTAR et al., 1986, 1987a, b)

$NH_4^{\oplus}$, $Mg^{2\oplus}$, $K^{\oplus}$, β-Methylaspartase

338 → **339**

Entry	R	Yield %	Rate, Comments
1	H	90	very fast, (**196** → **101**)
2	Me	61	fast (**335** → **336**)
3	Et	60	moderately fast
4	^{n}Pr	49	very slow
5	^{i}Pr	54	very slow
6	^{n}Bu	0	none
7	F	–	very slow
8	Cl	60	moderately fast
9	Br	–	moderately fast, inhibitor
10	I	–	not a substrate

measurements demonstrate that elimination is concerted and that product release from the enzyme occurs in a slower step. The C-3 proton exchange reaction has a primary deuterium kinetic isotope of circa 1.6 in the range pH 6.5–9.0, but this is reduced to 1 at high and low concentrations of potassium ions. Magnesium ions are also essential for activity (BOTTING et al., 1988b, 1989; BOTTING and GANI, 1992).

The elimination of ammonia from the 3-epimer of the natural substrate i.e., L-*erythro*-3-methylaspartate (**342**), occurs at about 1% of the rate of the natural substrate (Fig. 119). Isotope studies have demonstrated that epimerization does not occur prior to elimination and hence the process must be a *syn*-elimination. In the addition direction, the *threo*-stereoisomer (**336**) is formed first and then the *erythro*-stereoisomer is formed very slowly (ARCHER et al., 1993a, b; ARCHER and GANI, 1993; BEAR et al., 1998).

The mechanism of action of β-methylaspartase has undergone a series of refinements as the evidence has accumulated (Fig. 120). The current model for the catalytic cycle encompasses 16 intermediates and 23 kinetic parameters. The resting state of the enzyme contains a conjugate addition adduct of the dehydroalanine residue (most likely with cysteine = B_2) and no bound metal ions (**343**). Binding of magnesium ions (**344**), promotes addition of L-*threo*-3-methylaspartate (**336**) and binding of potassium ions promotes the β-elimination of B_2 (**346**) to reveal the dehydrhydroalanine residue (347). Conjugate addition of the amino group of L-*threo*-3-methylaspartate (**336**), gives the adduct (**349**) which undergoes con-

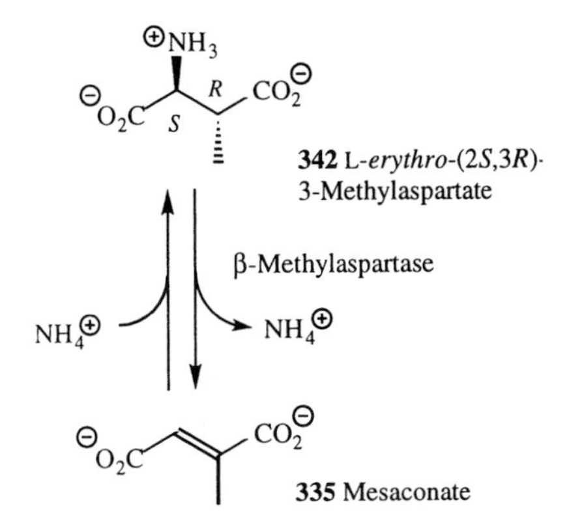

Fig. 119. The β-methylaspartase catalyzed *syn*-elimination of ammonia from L-*erythro*-(2*S*,3*R*)-3-methylaspartate to give (**342**) mesaconate ((*E*)-2-methyl-

Tab. 12. The β-Methylaspartase Catalyzed Addition of Amines to Substituted Fumarates (GULZAR et al., 1994, 1997; BOTTING et al., 1988a)

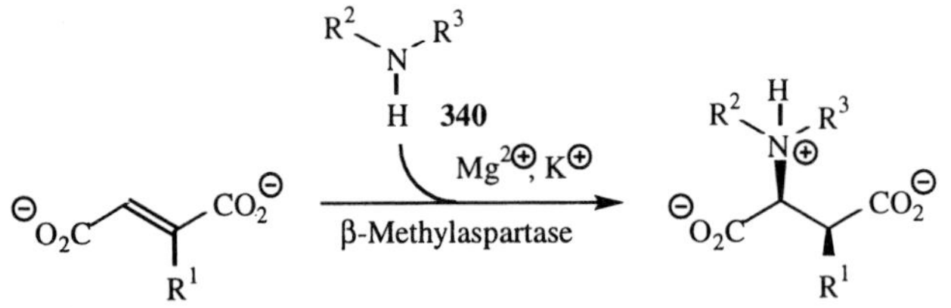

Entry	Substituents Alkene **338** R^1	Amine **340** R^2	R^3	Conversion %	Yield %
1	H	Me	H	55	45
2	Me	Me	H	54	40
3	Et	Me	H	60	35
4	^{n}Pr	Me	H	NR	–
5	^{i}Pr	Me	H	NR	–
6	H	Me	Me	70	38
7	Me	Me	Me	NR	–
8	H	Et	H	5	–
9	H	NH_2	H	89	42
10	Me	NH_2	H	91	41
11	Et	NH_2	H	90	57
12	^{n}Pr	NH_2	H	90	31
13	^{i}Pr	NH_2	H	90	33
14	Cl	NH_2	H	NR	–
15	H	OH	H	90	28
16	Me	OH	H	90	19
17	Et	OH	H	60	12
18	H	OMe	H	80	31
19	Me	OMe	H	70	34
20	Et	OMe	H	NR	–

NR, no reaction

certed elimination (**350**). A sequence of proton transfers then promotes the β-elimination of ammonia (352) and the readdition of B2 to the dehydroalanine residue (**354**), which is followed by product (336) and cofactor release (POLLARD et al., 1999; GANI et al., 1999; BOTTING et al., 1992).

4.4.1.3 Histidase (Histidine Ammonia-Lyase, EC 4.3.1.3)

Histidase catalyzes the addition of ammonia to uroconate (**368**) to give L-histidine (**137**) (Fig. 121). Unlike the β-methylasparatase and phenylalanine ammonia-lyase, histidase is found in mammals (TAYLOR, et al., 1990) and in fact uroconate was first discovered in dog urine in 1874. Histidase is found in plants and numerous microorganisms, e.g., *Pseudomonas testosteroni*, *Bacillus subtilis* and *Streptomyces griseus* (WU, et al., 1992). Several organisms are capable of growth by utilising exogenous histidine. The majority of researchers have studied the enzyme from *Pseudomonas* ATCC 11299, which has been described at various times as *P. putida*, *P. fluorescens*, but is now classified as *P. acidovorans* (CONSEVAGE and PHILLIPS, 1985). We will follow the consensus usauge which is *P. putida*.

Histidase is the first enzyme in the major catabolism of histidine. A deficiency of histidase

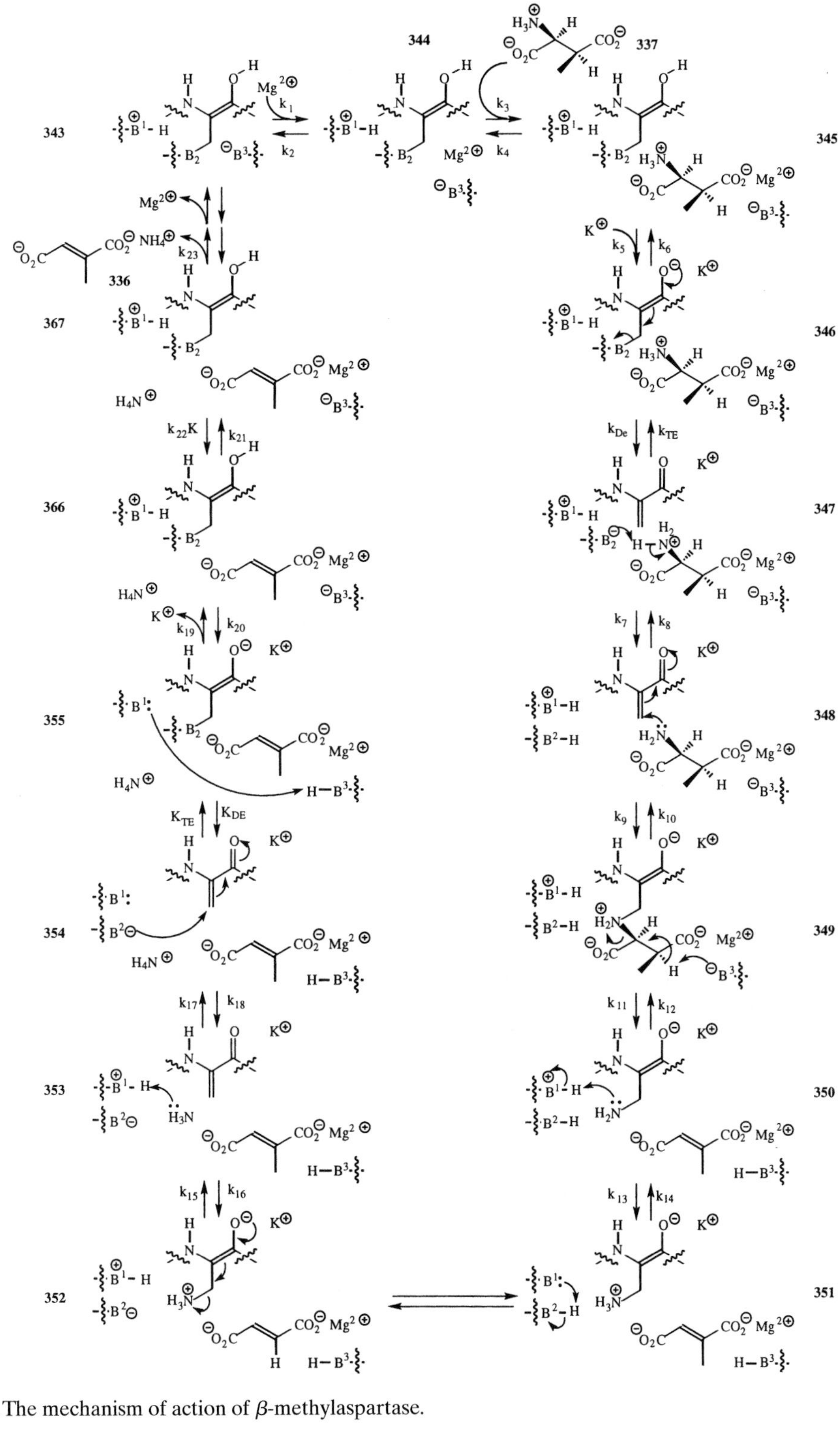

Fig. 120. The mechanism of action of β-methylaspartase.

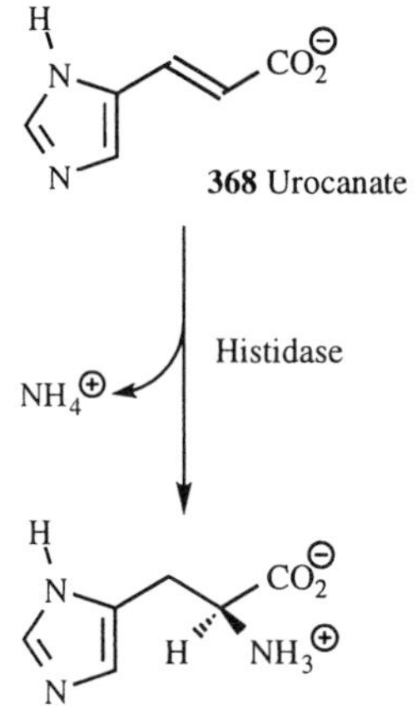

Fig. 121. The histidase catalyzed addition of ammonia to urocanate (**368**) to give L-histidine (**137**).

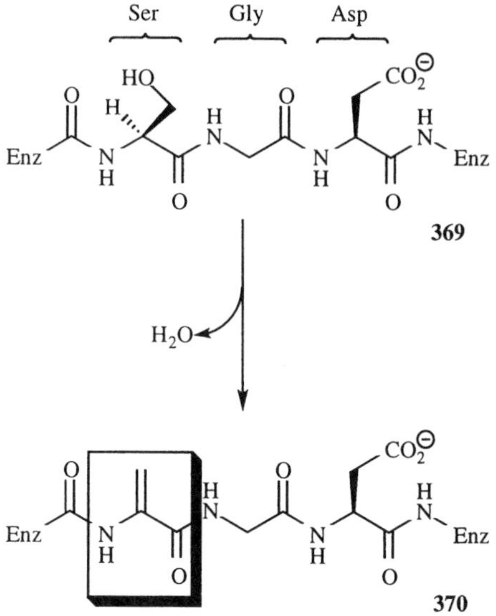

Fig. 122. Formation of dehydroalanine residues in peptides.

activity in humans causes accumulation of histidine in body fluids (histidinemia, TAYLOR, et al., 1991). The amino acid sequence is 93% conserved with rat and mouse sequences including four *N*-glycosylation consensus sites. Histidase activity has been detected primarily in the liver and epidermis as well as in kidney, adrenal gland and skin. (SUCHI et al., 1993; SANO, et al., 1997).

It has long been assumed that the catalytic activity of histidase results from the presence of a dehydroalanine residue (**370**), formed by β-elimination of the hydroxyl group of a serine residue (**369**) (Fig. 122). This would plausibly act as a Michael acceptor for the conjugate addition of ammonia and or the amino group of histidine. The conjugate addition adduct acts in effect as an immobilization device, around which the other components of the reaction can be optimally orientated to promote catalysis (cf. Fig. 120). Putative dehydroalanine residues are found in a conserved sequence; GXXXXSGDLXPL in histidase and phenylalanine lyase, whereas β-methylaspartase only has the key SGD motif (Tab. 10). There are no X-ray crystal structures of enzymes bearing a dehydroalanine residue. Consequently identification of such residues, hangs on the detection of the conserved sequence and labeling/inhibition reactions with nucleophiles such as [^{14}C]-cyanide (CONSEVAGE and PHILLIPS, 1985), [^{14}C]-nitromethane, cysteine (HERNANDEZ et al., 1993), phenyl hydrazine and sodium bisulfite. Double inhibition experiments and protection by L-histidine established that these reagents react at the same catalytically significant site. The adduct formed with [^{3}H]-sodium borohydride and [^{14}C]- cyanide were shown to be [3-^{3}H]-L-alanine and [4-^{14}C]-aspartate respectively (HANSON and HAVIR, 1972; CONSEVAGE and PHILLIPS, 1985).

This work enabled the serine precursor of the putative dehydroalanine to be assigned as serine-143 in *P. putida* histidase (HERNANDEZ and PHILLIPS, 1994; LANGER, et al., 1994a) and serine-254 in rat histidase (TAYLOR and MCINNES, 1994)

Mutation of the key serine residue residue to alanine abolishes activity (TAYLOR and MCINNES, 1994; LANGER et al., 1994b; HERNANDEZ and PHILLIPS, 1994), whereas mutation to cysteine, yields active enzyme, suggesting that the catalysis of post-translational is sufficiently robust to tolerate other nucleofuges (LANGER et al., 1994a). There is also some evidence that the thiol group of a cysteine residue close to the putative dehydroalanine residue may undergo conjugate addition to give a protected or resting state of the enzyme (CON-

SEVAGE and PHILLIPS, 1990). The activity of histidase is increased by 2–10 fold by Mn^{2+}, Fe^{2+}, Cd^{2+} and Zn^{2+} and this correlates with the excellent coordinating properties of dehydroamino acids towards metal ions (JEZOWSKA-BOJCZUC and KOZLOWSKI 1991; 1994). Furthermore, theoretical calculations showed that all dehydroamino acids except dehydroalanine tend to bent a peptide chain towards a turn conformation which has a strong impact on the co-ordination ability of the dehydropeptide ligand (JEZOWSKA-BOJCZUC, et al., 1996).

Methyl- and phenyl-glyoxal inhibit histidase by reaction with arginine residues. Alignment of four histidase and twelve phenylalanine lyase sequences revealed a single completely conserved arginine residue and another conserved in 15 of the sequences, but not *P. putida* histidase (WHITE and KENDRICK, 1993).

The preceding edifice of data and deductions has been shaken by the publication of the X-ray crystal structure of histidase, which shows that dehydroalanine is not present at the active site, but a previously unknown 4-methylidene-imidazole-5-one residue (**372**) (SCHWEDE, et al., 1999), which is formed by an intramolecular condensation and elimination of water. The authors postulate intramolecular nucleophilic attack to give the amidine (**371**), followed by α,β-elimination. One objection to this proposal, is that the nucleophile is an amidic nitrogen, which is only weakly nucleophilic. Conversely, elimination of water first to give dehydroalanine (**370**), reduces the conjugation between the amidic nitrogen and the carbonyl group and changes the conformational preferences of the peptide chain, so as to promote condensation. The 4-methylidene-imidazole-5-one residue (**372**) has a curious conformational feature. The substituent attached to the "amidic" nitrogen is out of the plane of the ring (O–C–N–C_α is 54°) and hence the nitrogen lone pair is sp3-hydridized. This renders the carbonyl, ketone like, and hence more susceptible to nucleophilic addition. In this conformation, the adduct is able utilize canonical forms (**373**) and (**374**) (Fig. 124), but not the aromatic 6π-electron canonical form (**375**). It is conceivable that addition triggers a conformational change which renders the nitrogen substituent co-planar and hence enable this canonical form to be utilized. The presence of a 4-methylidene-imidazole-5-one residue (**372**), rather than a dehydroalanine residue (**370**), equally explains all of the chemistry previously attributed to the latter (SCHWEDE, et al., 1999). In addition, it explains the former puzzling observation that Edman degradation stops two residues before the serine (HERNANDEZ et al., 1993). It remains to be seen if the 4-methylidene-imidazole-5-one residue (**372**) of histidase is a unique abberation or the first member of a new family of enzymes.

The mechanism of interconversion of L-histidine (**137**) and urocanate (**368**) provides a number of puzzles. The most of pressing of which is, how is the non-acidic proton removed in the presence of an ammonium group. The stereochemistry of addition-elimination is stereospecifically *trans* with exchange of the (3′-*pro-R*)-hydrogen (YOUNG, 1991). Exchange of the proton at C-5 with solvent protons provides a clue to the mechanism. This occurs much more rapidly with labeled urocanate (**388a,b to 368a**) than L-histidine (**376 to 137a**) and is not an obligatory step in the mechanism. However, an analogous reaction of an electrophile with L-histidine (**376**), provides an imminium ion (**377**), which is ideally placed to facilitate loss of the 3′-proton. The enamine (**378**) so formed can then eject ammonia to give another imminium ion (**379**), which is restored to neutrality by loss of the original electrophile. Alternatively, release of the electrophile from the enamine (**378**) yields urocanate (**288a to 368**) directly. This mechanism was originally postulated with a proton as an electrophile (FURUTA et al., 1990, 1992), then dehydroalanine (LANGER et al., 1995) and finally with a 4-methylidene-imidazole-5-one residue (**372**) (Fig. 126) (SCHWEDE, et al., 1999).

Mutants of histidase that lack the serine residue (e.g., S143A), which is the progenitor of the 4-methylidene-imidazole-5-one moiety (**372**) (Fig. 123), retain a minute amount of activity with L-histidine (**137**) and K_m is essentially unchanged. More surpisingly, the kinetic parameters for 5-nitro-L-histidine (**385 to 383**) are virtually identical for the wild type and mutant! This can be rationalized by assumming that the nitro group plays the role of the prosthetic moiety in the enzyme. Thus the

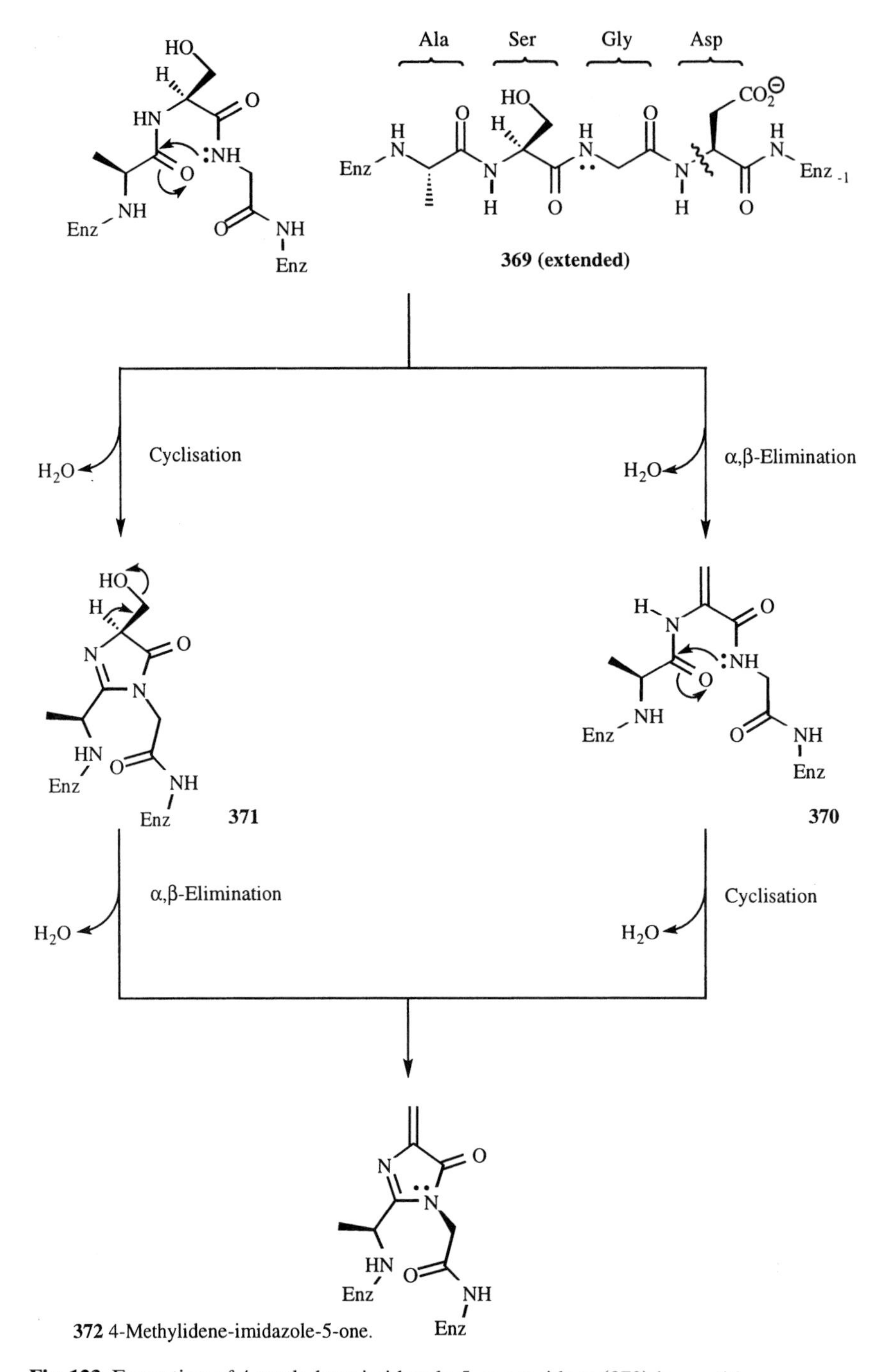

Fig. 123. Formation of 4-methylene-imidazole-5-one residues (**372**) in peptides.

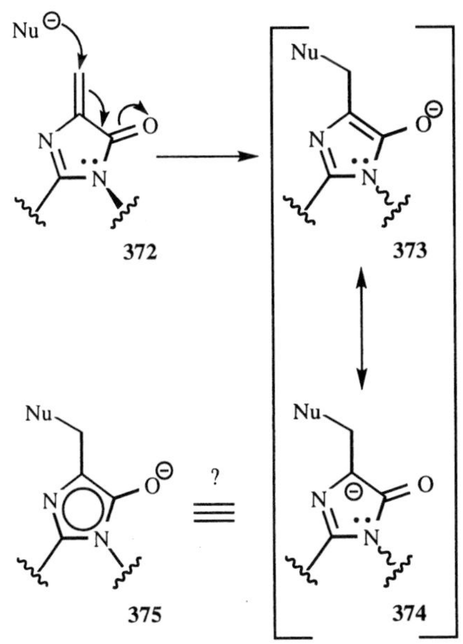

Fig. 124. Canonical forms for the addition adducts of 4-methylene-imidazole-5-one residues (**372**) in peptides. The stereochemistry of the nitrogen substituent controls access to the aromatic form (**375**).

canonical form (**384**) is ideally set up for loss of the 3′-proton, and the elimination of the ammonia is driven by the enamine (**385**) as before (LANGER et al., 1997a; LANGER et al., 1995).

Histidase has not found large scale practical applications as yet, although it has been used for the preparation of (3*R*)-[3-$^3H_1$]-histidine by exchange with tritium oxide (YOUNG, 1991). Urocanic acid acts as a hydroxyl radical scavenger in skin and like histamine (**138**) it is an immunosuppressant (KAMMEYER et al, 1999).

4.4.1.4 Phenylalanine Ammonia-Lyase [PAL, EC 4.3.1.5]

Phenylalanine ammonia-lyase (PAL) catalyzes the deamination of L-phenylalanine (**388**) to give *trans*-cinnamate (**389**) (Fig. 128). PAL is widely distributed in higher plants, some fungi, yeasts and a single prokaryote, *Streptomyces*, but is not found in true bacteria or animals. PAL is of prodigous importance in plants, because some 30–45% of all plant

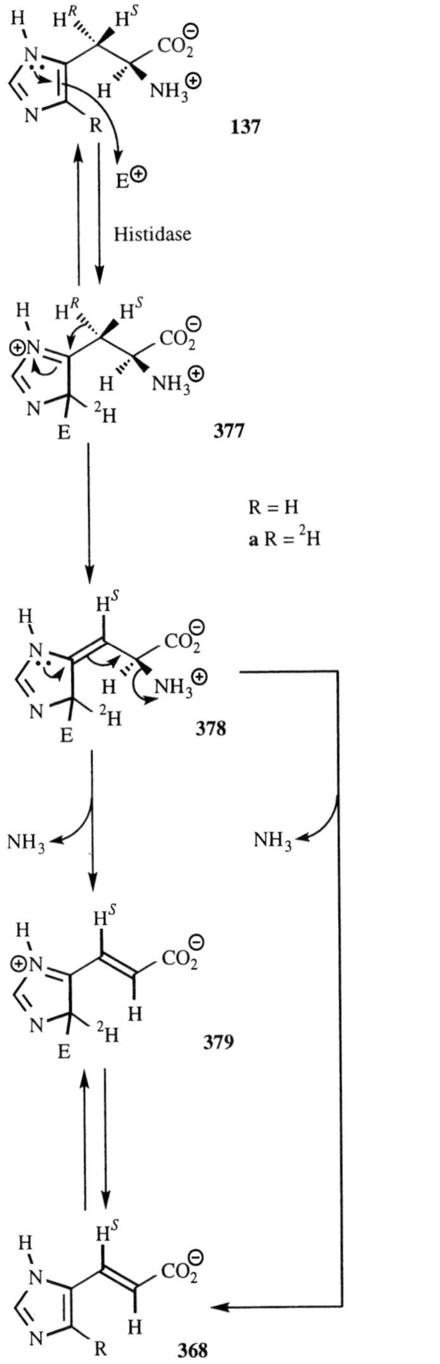

Fig. 125. Generalized mechanism for the histidase catalyzed deamination of labeled L-histidine (**376** to **137a**) and C-5 proton exchange.

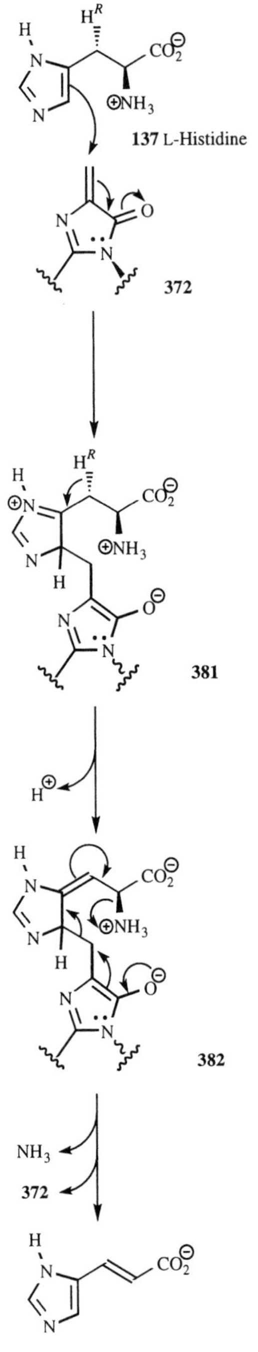

Fig. 126. Mechanism for the histidase catalyzed deamination of L-histidine (**376** to **137**) to urocanate (**368**).

organic material is derived from L-phenylalanine (**388**) and to a lesser extent L-tyrosine through the cinnamate pathway (JONES, 1984). The ammonium ions released in the formation of cinnamate are recycled by conversion of glutamate to glutamine catalyzed by glutamine synthetase (RAZAL et al., 1996). The ultimate products of the cinnamate pathway are lignins, lignans, flavanoids, benzoic acids, coumarins and suberins (MANN, 1987, MANN et al, 1994) and phenols (NICHOLSON and HAMMERSCHMIDT, 1992). PAL levels are increased in plants, in response to environmental stress and pathogen infection (REGLINSKI, et al., 1998; ORCZYK et al., 1996; KIM et al., 1996). In microorganisms, PAL acts as the first step in the catabolism of the exogenous phenylalanine as a carbon source. The enzyme has been purified from numerous sources, including potato tubers, maize, sunflower hypocotyls (JORRIN et al., 1988), *S. pararoseus* and the "smut" fungal pathogens, *Ustilago hordei* (HANSON and HAVIR, 1972) and *U. maydis* (KIM et al., 1996) and two forms from leaf mustard, *Brassica juncea* var. *integrifolia* (LIM et al., 1997, 1998).

L-Phenylalanine is an essential nutrient for human nutrition and is used widely in the food and pharmaceutical industries. Currently, most L-phenylalanine is manufactured by fermentation using an overproducing strain of *E. coli*, cultivated on glucose. The demand for L-phenylalanine methyl ester (**324**) for the manufacture of the artifical sweetner Aspartame (**325**) (Fig. 111) assures continuing efforts to develop improved methods of manufacture (OYAMA, 1992; KINOSHITA and NAKAYAMA, 1978; DUFFY, 1980).

PAL is a member of the putative dehydroalanine family, which also contains β-methylaspartase and histidase. It has been shown recently that the active site of histidase does not contain dehydroalanine (**370**, Fig. 122), but a 4-methylidene-imidazole-5-one (**372**) (Fig. 123) moiety instead (SCHWEDE, et al., 1999). The high sequence homology between PAL and histidase (Tab. 10) suggests that it almost certainly contains the same functional residue. Labeling studies indicated that serine-202 is the progenitor residue in parsly PAL (*Petroselinum crispum* L.). The *PAL1* gene was expressed in *E. coli* and when serine-202 was mutated to alanine, virtually all activity was lost

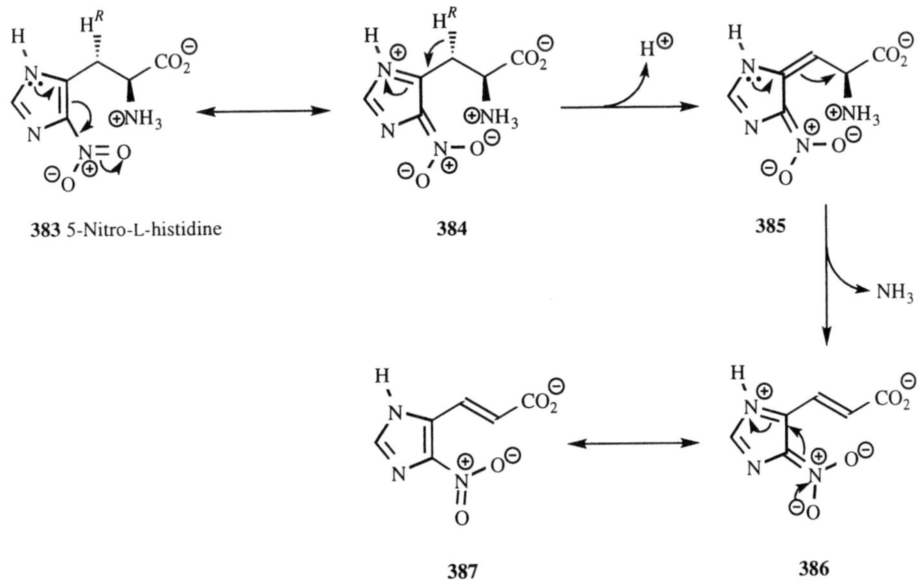

383 5-Nitro-L-histidine

384

385

386

387

Fig. 127. The mechanism for deamination of 5-nitro-L-histidine (**383**), with histidase mutants (e.g., S143A), which lack the 4-methylidene-imidazole-5-one residue (**372**).

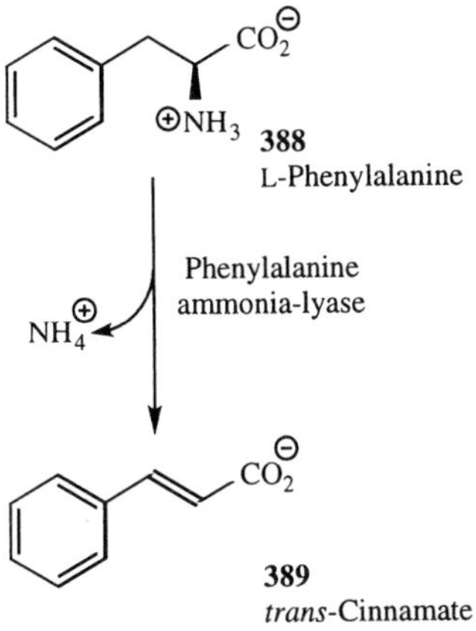

388
L-Phenylalanine

Phenylalanine ammonia-lyase

389
trans-Cinnamate

Fig. 128. The phenylalanine ammonia-lyase (PAL) catalyzed deamination of phenylalanine (**388**) to give *trans*-cinnamate (**389**).

(SCHUSTER and RETEY, 1995). However, activity was retained with 4-nitro-phenylalanine, which parallels the results achieved with 5-nitrohistidine (**383**) and histidase (LANGER et al., 1997a, 1995). Mutation to cysteine (which is able to undergo elimination), yielded a mutant which had activity, identical to the wild type (LANGER et al., 1997b).

Potato tuber PAL catalyzed deamination of L-phenylalanine (**388**), to give *trans*-cinnamate (**389**) is stereospecific. The (3-*pro-S*)-proton is lost, hence elimination occurs with *trans*-stereochemistry (HANSON, et al., 1971; WIGHTMAN, et al., 1972) (Fig. 129). This is exactly as observed with β-methylaspartase and histidase. Intriguingly, PAL from *Rhodotorula glutinis* (and potato tubers) also catalyzes the deamination of D-phenylalanine to *trans*-cinnamate (**389**), albeit at 0.02% of the rate of L-phenylalanine (**388**). *Rhodotorula glutinis* PAL catalyzed deamination of D-phenylalanine results in preferential (but not exclusive) loss of the (3-*pro-R*)-proton, which is consistant with *trans*-stereochemistry for the overall elimination (EASTON and HUTTON, 1994).

Deamination of [$^2H_5$]-L-phenylalanine deuterated in the aromatic ring shows a kinetic isotope effect of 1.09, relative to L-phenylalanine (**388**) (Fig. 129) (GLOGE et al., 1998). This can be rationalized by postulating a Friedels–Craft type reaction with an electrophile [presumably 4-methylidene-imidazole-5-one (**372**)] to give the conjugated carbonium ion

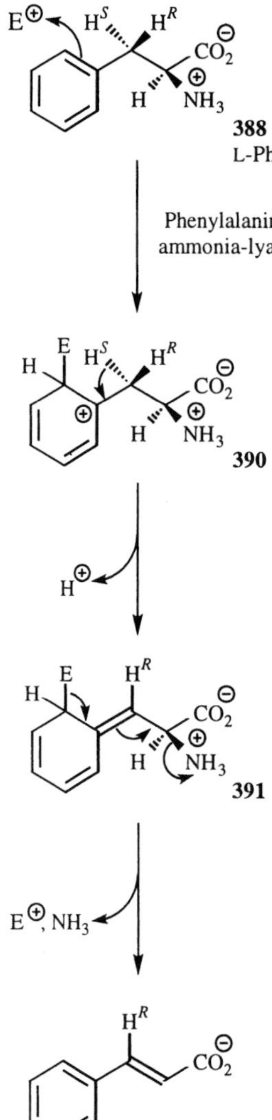

388
L-Phenylalanine

Phenylalanine ammonia-lyase

390

391

389
trans-Cinnamate

Fig. 129. The mechanism for the phenylalanine ammonia-lyase (PAL) catalyzed deamination of phenylalanine (**388a** to **388**) to give *trans*-cinnamate (**389b** to **389**).

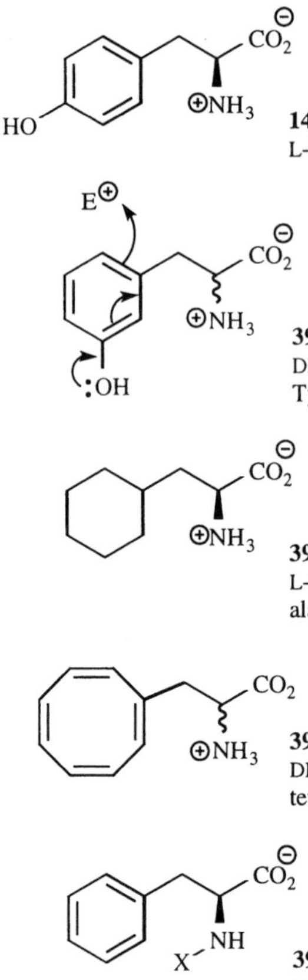

141a
L-Tyrosine

392
DL-*meta*-Tyrosine

393
L-Cyclohexylalanine

394
DL-Cyclooctatetraenealanine

395
a R = OH
b R = NH_2

Fig. 130. Alternative substrates and inhibitors of phenylalanine ammonia-lyase (PAL).

(**390**). Deprotonation yields the triene (**391**), which undergoes 1,4-elimination to urocanate (**389**). L-tyrosine (**141a**) is a poor substrate for PAL, whereas racemic 3-hydroxyphenylalanine (**392**) (D,L-*meta*-tyrosine) is a slightly better substrate than L-phenylalanine, despite the presence of the D-enantiomer (Fig. 130). The 3-hydroxy-group is capable of promoting the Friedels–Craft reaction, whereas the 4-hydroxyl group of L-tyrosine is not (SCHUSTER and RETEY, 1995). Cyclohexylalanine (**393**) does not undergo deamination with PAL, but it is a good inhibitor, which indicates binding is not impaired, Cyclooctatetraenyl alanine (**394**) would not be expected to be able to participate in the Friedels–Craft reaction, but both binding and turnover were impaired and hence it is not possible to make a wholly electronic argument for the lack of reactivity. Substrates in which the amino substituent has been replaced or substituted (**395**) are good inhibitors (MUNIER and BOMPEIX, 1985; JONES and NORTHCOTE, 1984).

The 2-, 3- and 4-pyridylalanines (**397**) (Fig. 131) were prepared by PAL catalyzed amination of the corresponding cinnamates (**396**) using a high concentration of aqueous ammonia, partially neutralized with carbon dioxide. In the reverse direction, the 2-, 3- and 4-pyridylalanines (**397**) are also good substrates for PAL, with higher K_M's and similar or slightly higher V_{max}'s than L-phenylalanine. From the standpoint of classical solution phase chemistry, this result is unexpected. Pyridines undergo Friedel–Craft reactions more slowly than benzene. But this is because under solution phase conditions they are present as the protonated form, in which the pyridinium group acts as an electron withdrawing group. The enzyme catalyzed reaction occurs without protonation, and hence under these circumstances the pyridyl nitrogen acts as an activating group (GLOGE et al., 1998).

The pH optima for PAL is circa 8.7 (depending on source). Activity drops off quickly at lower pH, but higher pH is sometimes beneficial. Most PALs are deactivated by thiols, but only a few by metals ions, e.g., Ag^+, Cu^{2+}, Hg^{2+} and Zn^{2+} (KIM, et al., 1996; LIM, et al., 1997; 1998). The thermodynamically preferred mode of action of PAL is deamination of L-phenylalanine (**388**) to give *trans*-cinnamate (**389**). However, in the presence of high concentrations of ammonium salts the reverse reaction is also feasible, particularly if the amino acid precipitates from solution. This is the direction of interest for the vast majority of industrial processes. Ammonia concentrations up to 8 M have been utilized for L-phenyalanine production (TAKAC, et al., 1995), but 5 M suffices for most purposes (YAMADA et al., 1981; EVANS, et al., 1987; GLOGE, 1998).

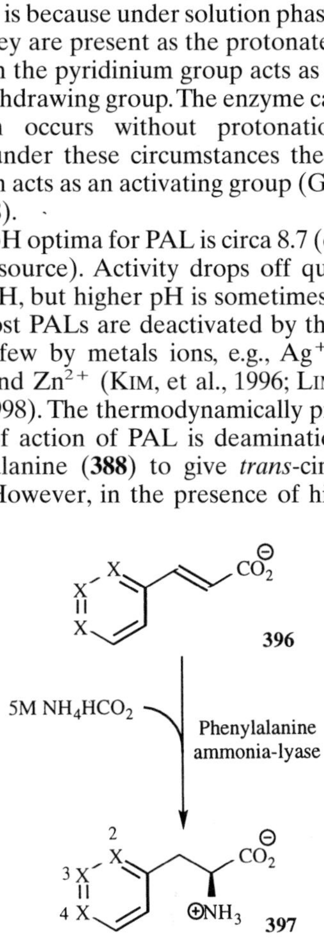

396

5M NH_4HCO_2

Phenylalanine ammonia-lyase

397

2-Pyridyl, 63% yield
3-Pyridyl, 59% yield
4-Pyridyl, 75% yield

Fig. 131. Pyridyl cinnamates (**395**) are good substrates for phenylalanine ammonia-lyase (PAL) catalyzed amination.

trans-Cinnamic acid is very cheap because it is easily prepared by aldol condensation of benzaldehyde and a host of acetic acid enolate equivalents (e.g., diethyl malonate). Unfortunately, *trans*-cinnamate (**389**) is a competitive inhibitor of of most forms of PAL (*Ustilago maydis* PAL is an exception, KIM et al., 1996), which limits the reactor concentration and hence productivity. β-cyclodextrin (**398**) (Fig. 132) has been advocated as a sequestrant for *trans*-cinnamate in the deamination direction, but it remains to be seen if it is beneficial for amination (EASTON et al., 1995).

Despite the widespread abundance of PAL, industrial processes have almost exclusively used PAL from *Rhodotorula glutinis*, because the organism is easily cultivated, PAL levels are high and the enzyme is easily purified in high yield (KANE and FISKE, 1985; D'CUNHA et al., 1996b). *Sporidiobolus pararoseus*, *Rhodosporidium toruloides* (WATANABE et al., 1992), *Rhodotorula rubra* (EVANS et al., 1987), *Rhodotorula graminis* and *Cladosporium cladosporioides* are unproven alternatives (TAKAC, et al., 1995, DRAUZ and WALDMANN, 1995).

The production of L-phenylalanine (**388**) by whole *Rhodotorula glutinis* cells has been optimized at pH 10.5, 30 °C, 8 M ammonia and a loading of 2 g of substrate per g of dry cells. Using a fed batch operation, the maximum concentration of L-phenylalanine reached 12.57 g L^{-1} (70.19 mM) from 14.8 g L^{-1} (100 mM) cinnamate (**389**). Sodium glutamate and penicillin increased PAL activity (TAKAC et al., 1995).

Thus far the unfavorable equilibrium and inhibition by cinnamate have made the production of L-phenylalanine using PAL, uncompetitive compared to manufacture by fermentation. However, PAL also catalyzes the amination of *trans*-methyl cinnamate (which is cheaper than cinnamic acid) at 90% of the rate of the natural substrate. The product L-phenyl-

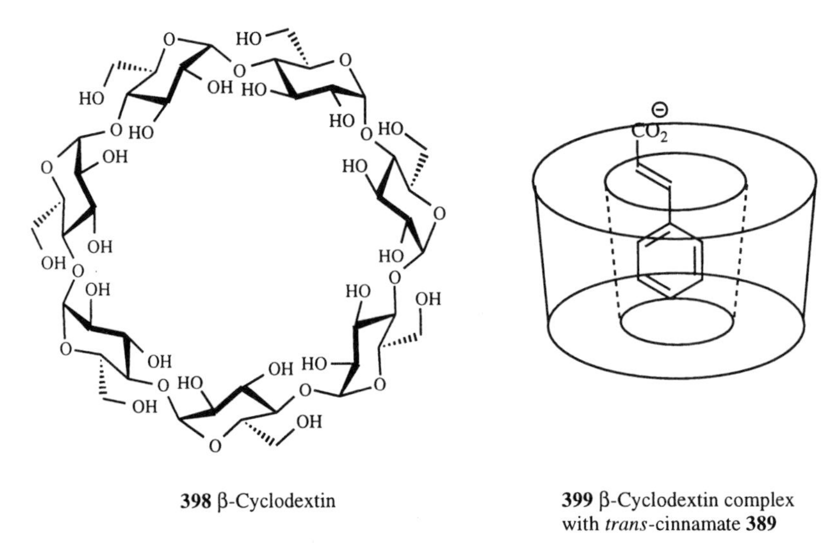

398 β-Cyclodextin

399 β-Cyclodextin complex with *trans*-cinnamate **389**

Fig. 132. β-Cyclodextrin is a cyclic hepta-saccharide , which preferentailly binds *trans*-cinnamate (**389**) over L-phenylalanine (**388**), to give the complex (**399**) (EASTON et al., 1995).

alanine methyl ester is more valuable than L-phenylalanine, because it is a direct precursor of Aspartame (**325**). The amination was run with whole *Rhodotorula glutinis* cells, in a two phase heptane: water system to overcome the low solubility of *trans*-methyl cinnamate in water. The conditions were optimized at pH 9.0, 30 °C, 16–18 h, 0.1 M substrate and 1M ammonium sulfate and 70% conversion was achieved (D'CUNHA et al., 1994). But the activity of the cells declined rapidly during the bioconversion and they were not resuable. By adding small amounts of magnesium ions and 10% glycerol, the cells could be recycled up to nine times with a total yield of 92g L^{-1} (D'-CUNHA et al., 1996a). PAL from the red basidomyces yeast *Rhodosporidium toruloides* has been purified (REES and JONES, 1996) and used in a single phase octanol: water mixture. At least 2% water was required for activity. The best result was obtained with freshly lyophilized PAL, in a 96.4:3.6, v/v, mixture which retained, up to 17% of the activity in aqueous media (REES and JONES, 1997). The use of enzymes in organic solvents offers the prospect of using hydrophobic substrates at higher concentrations than would otherwise be possible in aqueous media (GUPTA, 1992; TRAMPER, et al., 1992)

4.5 Other Carbon–Nitrogen Lyases

The gram-negative bacterium DSM 9103 is able to grow on ethylenediaminetetra-acetic acid (EDTA) (**400**) (Fig. 133) as the sole source of carbon and nitrogen. The initial step in the degradation is stepwise removal of the acetate groups as glyoxylate by an oxidative process (WITSCHEL et al., 1997). The same species also grows on (*S,S*)-ethylenediaminedisuccinate (**401**) and cell free extracts, without added cofactors, degrade this substrate. An enzyme was purified 41-fold that catalyzed reversible formation of fumarate (**196**) and *N*-(2-aminoethyl) aspartic acid (**402**). (*S,S*)- and (*S,R*)-ethylenediaminedisuccinate were accepted as substrates, but the (*R,R*)-stereoisomer was unchanged in both cell free and purified enzyme assays (WITSCHEL and EGLI, 1997). This data clearly indicates a stereospecific lyase system.

4.6 Carbon–Halide Lyases (EC 4.5.1.X)

This subclass was originally conceived for the enzyme which catalyzes the elimination of hydrogen chloride from DDT (DDT-dehy-

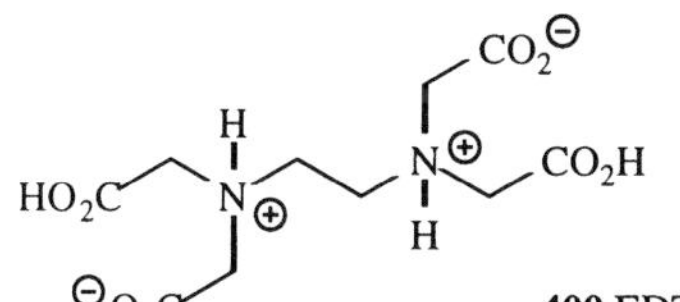

Fig. 133. The degradation of (*S,S*)-ethylenediaminedisuccinate (**401**).

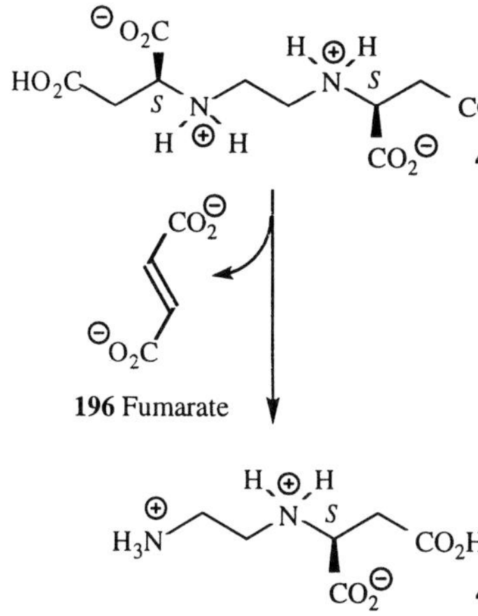

drochlorinase, EC 4.5.1.1). The other early member of this class is 3-chloro-D-alanine dehydrochlorinase (EC 4.5.1.2), which is a pyridoxal 5′-phosphate dependent protein, that also catalyzes β-replacement reactions. Further aspects of these enzymes are covered in detail in Chapter 3, this volume and in reviews (FETZNER, 1998; FETZNER and LINGRENS, 1994).

4.7 Other Lyases

4.7.1 Alkyl Mercury-Lyase (EC 4.99.1.2)

Alkylmercury lyase catalyzes the cleavage of alkylmercury bonds in the presence of a thiol to give a hydrocarbon and mercury (II) as a mercaptide (GUPTA et al., 1999). Plants and microorganisms are extraordinarily adept at handling materials that are toxic to higher organisms (BALDI, 1997). Toxic alkyl mercury compounds are transformed either by alkyl mercury-lyase or by a reductase which produces metallic mercury. There are considerable prospects for the use of such microorganisms and enzymes in organometallic chemistry, but at present the main use of these enzymes is for land remediation (GHOSH et al., 1996; HEATON et al., 1998).

5 References

ABELYAN, V. A. (1999), Simultaneous production of D-aspartic acid and L-alanine, *Appl. Biochem. Microbiol.* **35**, 152–156.

ABELYAN, V. A., AFRIKYAN, E. G. (1997), One step synthesis of L-alanine from fumaric acid, *Appl. Biochem. Microbiol.* **33**, 554–556.

ABELYAN, V. A., BAGDASARYAN, S. N., AFRIYAN, E. G. (1991), Intracellular L-aspartate-β-decarboxylase of *Pseudomonas sp.* and *Alcalignes sp.* and its immobilization, *Biochemistry (Moscow)* **56**, 899–905.

AKHTAR, M., COHEN, M. A., GANI, D. (1986), Enzymatic synthesis of 3-halogenoaspartic acids using β-methylaspartase – inhibition by 3-bromoaspartic acid, *J. Chem. Soc., Chem. Commun.* 1290–1291.

AKHTAR, M., COHEN, M. A., GANI, D. (1987a), Stereochemical course of the enzymatic amination of chlorofumaric acid and bromo-fumaric acid by 3-methylaspartate ammonia-lyase, *Tetrahedron Lett.* **28**, 2413–2416.

AKHTAR, M., BOTTING, N. P., COHEN, M. A., GANI, D. (1987b), Enantiospecific synthesis of 3-substituted aspartic acids via enzymic amination of substituted fumaric acids, *Tetrahedron* **43**, 5899–5908.

ALAM, N., SAFFERSON, E. D. (1998), Malonyl-CoA and the regulation of fatty acid oxidation in soleus muscle, *Biochem. J.* **334**, 233–241.

ALBER, B. E., COLANGELO, C. M., DONG, J., STALHANDSKE, C. M. V., BAIRD, T. T., et al. (1999), Kinetic and spectroscopic characterization of the γ-carbonic anhydrase from the methanoarchaeon *Methanosarcina thermophila*, *Biochemistry* **38**, 13119–13128.

ALBERT, A., DHANARAJ, V., GENSCHEL, U., KHAN, G. L., RAMJEE, M. K. et al. (1998), Crystal structure of aspartate decarboxylase at 2.2 Å resolution provides evidence for an ester in protein self processing, *Nature Struct. Biol.* **5**, 289–293.

ALBERTY, R. A., BENDER, P. (1959), Proton magnetic resonance spectra of malic acid and its salts in deuterium oxide, *J. Am. Chem. Soc.* **81**, 542–546.

ALLERT, M., KJELLSTRAND, M., BROO, K., NILSSON, A., BALTZER, L. (1998), A designed folded polypeptide model system that catalyzes the decarboxylation of oxaloacetate, *J. Chem. Soc., Perkin Trans II*, 2271–2274.

ANET, F. A. L. (1960), The configuration of deuterio-L-malic acid produced enzymatically. Synthesis of *threo*-3-deuterio-DL-malic acid. *J. Am. Chem. Soc.* **82**, 994–995.

ALEXANDER, F. W., SANDMEIER, E., MEHTA, P. K., CHRISTEN, P. (1994), Evolutionary relationships among pyridoxal-5′-phosphate – dependent enzymes – regiospecific α-family, β-family and γ-family, *Eur. J. Biochem.* **219**, 953–960.

ALKORTA, I., GARBISU, C. LLAMA, M. J., SERRA, J. L. (1998), Industrial applications of pectic enzymes: a review, *Process Biochem.* **33**, 21–28.

ALVAREZ, F. J., ERMER, J., HUBNER, G., SCHELLENBERGER, A., SCHOWEN, R. L. (1995), The linkage of catalysis and regulation in enzyme action, solvent isotope effects as probes of protonic sites in the yeast pyruvate decarboxylase mechanism, *J. Am. Chem. Soc.* **117**, 1678–1683.

ANDERSON, R. L., HAUSWALD, C. L. (1982), D-galactonate/D-fuconate dehydratase, *Methods Enzymol.* **90**, 299–302.

ANDERSON, W. A., MOO-YOUNG, M., LEGGE, R. L. (1987a), Optimization of growth conditions for the induction of tyrosine decarboxylase in *Streptococcus faecalis*, *Biotechnol. Lett.* **9**, 685–690.

ANDERSON, W. A., MOO-YOUNG, M., LEGGE, R. L. (1992a), Development of a multienzyme reactor for dopamine synthesis. 1. Enzymology and kinetics, *Biotechnol. Bioeng.* **39**, 781–789.

ANDERSON, W. A., MOO-YOUNG, M., LEGGE, R. L. (1992b), Development of a multienzyme reactor for dopamine synthesis. 2. Reactor engineering and simulation, *Biotechnol. Bioeng.* **40**, 388–395.

Anonymous (1984, 1992), *Enzyme Nomenclature: Recommendations of the Nomenclature Committee of the International Union of Biochemistry on the Nomenclature and Classification of Enzyme-Catalyzed Reactions.* Orlando, FL: Academic Press.

ANTSON, A. A., STROKOPYTOV, B. V., MURSHUDOV, G. N., ISUPOV, M. N., HARUTYUNYAN, E. H. et al. (1992), The polypeptide-chain fold in tyrosine phenol-lyase, a pyridoxal-5′-phosphate-dependent enzyme, *FEBS Lett.* **302**, 256–260.

ANTSON, A. A., DEMIDKINA, T. V., GOLLNICK, P., DAUTER, Z., VONTERSCH, R. et al. (1993) 3-Dimensional structure of tyrosine phenol-lyase, *Biochemistry* **32**, 4195–4206.

ARCHER, C. H., GANI, D. (1993), Kinetics and mechanism of *syn*-elimination of ammonia from (2*S*,3*R*)-3-methylaspartic acid by methylaspartase, *J. Chem. Soc., Chem. Commun.* 140–142.

ARCHER, C. H., THOMAS, N. R., GANI, D. (1993a), Syntheses of (2*S*,3*R*)-3-methylaspartic and (2*S*,3*R*)[3-H^2]-3-methylaspartic acids – slow substrates for a *syn*-elimination reaction catalyzed by methylaspartase, *Tetrahedron Asymmetry* **4**, 1141–1152.

ARCHER, C. H., THOMAS, N. R., GANI, D. (1993b), Kinetics and mechanism of *syn*-elimination of ammonia from (2*S*,3*R*)-3-methylaspartic acid by methylaspartase, *J. Chem. Soc., Chem. Commun.*, 140–142.

ARMSTRONG, F. B., LIPSCOMB, E. L., CROUT, D. H. G., MORGAN, P. J. (1985), Structure–activity studies with the α,β-dihydroxyacid dehydratase of *Salmonella typhimurium*, *J. Chem. Soc. Perkin Trans. I*, 691–696.

ARMSTRONG, F. B., HEDGECOCK, C. J. R., REARY, J. B., WHITEHOUSE, D., CROUT, D. H. G. (1974), Stereochemistry of the reductoisomerase and α,β-dihydroxyacid dehydatase steps in valine and isoleucine biosynthesis, *J. Chem. Soc., Chem. Commun.*, 351–352.

ARMSTRONG, F. B., LIPSCOMB, E. L., CROUT, D. H. G., MITCHELL, M. B., PRAKASH, S. R. (1983) Biosynthesis of valine and isoleucine: synthesis and biological activity of (2*S*)-α-acetolactic acid (2-hydroxy-2-methyl-3-oxobutanoic acid) and (2*R*)- and (3*S*)-α-acetohydroxybutyric acid (2-ethyl-2-hydroxy-3-oxobutanoic acid), *J. Chem. Soc., Perkin. Trans.1*, 1197–1201.

ARJUNAN, P., UMLAND, T., DYDA, F., SWAMINATHAN, S., FUREY, W. et al. (1996), Crystal structure of the thiamin diphosphate-dependent enzyme pyruvate decarboxylase from the yeast *Saccharomyces cerevisiae* at 2. 3 angstrom resolution, *J. Mol. Biol.* **256**, 590–600.

ASADA, Y., TANIZAWA, K., SAWADA, S., SUZUKI, T., MISINO, H., SODA, K. (1981), Stereochemistry of *meso*-α,ε-diaminopimelate decarboxylase reaction – The 1st evidence for pyridoxal 5′-phosphate dependent decarboxylation with inversion of stereochemistry, *Biochemistry* **20**, 6881–6886.

ASANO, Y., KATO, Y. (1994a), Crystalline 3-methylaspartase from a facultative anaerobe, *Escherichia coli* strain YG1002, *FEMS Microbiol. Lett.* **118**, 255–258.

ASANO, Y., KATO, Y. (1994b), Occurrence of 3-methylaspartase ammonia lyase in facultative anaerobes and their application to synthesis of 3-substituted (*S*)-aspartic acids, *Biosci. Biotech. Biochem.* **58**, 223–224.

AXELSSON, B. S., BJURLING, P., MATSSON, O., LANGSTROM, B. (1992), $^{11}C/^{14}C$ kinetic isotope effects in enzyme mechanism studies, *J. Am. Chem. Soc.* **114**, 1502–1503.

AXELSSON, B. S., O'TOOLE, K. J., SPENCER, P. A.,

YOUNG, D. W. (1994), Versatile synthesis of stereospecifically labeled D-amino acids via labeled aziridines – Preparation of (2*R*, 3*S*)-[3-H-$^2H_1$]-serine and (2*R*, 3*R*)-[2, 3-$^2H_2$]-serine; (2*S*, 2′*S*, 3*S*, 3′*S*-[3, 3′-$^2H_2$]- and (2*S*, 2′*S*, 3*R*, 3′*R*)-[2, 2′, 3, 3′-$^2H_4$]-cystine and; (2*S*, 3*S*)-[3-$^2H_1$]- and (2*S*, 3*R*)-[2,3-$^2H_2$]-β-chloroalanine, *J. Chem. Soc., Perkin Trans 1.*, 807–815.

BABBITT, P. C., MRACHKO, G. T., HASSON, M. S., HUISMAN, G. W., KOLTER, R. et al. (1995), Functionally diverse enzyme superfamily that abstracts the α-protons of carboxylic acids, *Science* **267**, No. 5201, 1159–1161.

BABBITT, P. C., HASSON, M. S., WEDEKIND, J. E., PALMER, D. R. J., BARRETT, W. C. et al. (1996), The enolase family: a general strategy for enzyme catalyzed abstraction of the α-protons of carboxylic acids, *Biochemistry* **35**, 16489–16501.

BABURINA, I., GAO, Y. H., HU, Z. X., JORDAN, F., HOHMANN, S., FUREY, W. (1994), Substrate activation of brewers yeast pyruvate decarboxylase is abolished by mutation of cysteine-221 to serine, *Biochemistry* **33**, 5630–5635.

BABURINA, I., MOORE, D. J., VOLKOV, A., KAHYAOGLU, A., JORDAN, F., MENDELSOHN, R. (1996), Three of four cysteines, including that responsible for substrate activation, are ionized at pH 6. 0 in yeast pyruvate decarboxylase: Evidence from Fourier transform infrared and isoelectric focusing studies, *Biochemistry* **35**, 10249–10255.

BABURINA, I., LI, H. J., BENNION, B., FUREY, W., JORDAN, F. (1998a), Interdomain information transfer during substrate activation of yeast pyruvate decarboxylase: The interaction between cysteine 221 and histidine 92, *Biochemistry* **37**, 1235–1244.

BABURINA, I., DIKDAN, G., GUO, F. S., TOUS, G. I., ROOT, B., JORDAN, F. (1998b), Reactivity at the substrate activation site of yeast pyruvate decarboxylase: Inhibition by distortion of domain interactions, *Biochemistry* **37**, 1245–1255.

BACH, R. D., CANEPA, C. (1996), Electronic factors influencing the decarboxylation of β-ketoacids. A model enzyme study, *J. Org. Chem.* **61**, 6346–6353.

BACH, R. D., CANEPA, C. (1997), Theoretical model for pyruvoyl dependent enzymatic decarboxylation of α-amino acids, *J. Am. Chem. Soc.* **119**, 11725–11733.

BACH, R. D., CANEPA, C., GLUKHOVTSEV, M. N. (1999), Influence of electrostatic effects on activation barriers in enzymatic reactions: Pyridoxal 5′-phosphate dependent decarboxylation of α-amino acids, *J. Am. Chem. Soc.* **121**, 6542–6555.

BADIANI, K., LIGHTFOOT, P., GANI, D. (1996), Synthesis and mode of action of 1-substituted *trans*-cyclopropane 1,2-dicarboxylic acids: inhibitors of the methylaspartase reaction, *J. Chem. Soc., Chem. Commun.* 675–677.

BAETZ, A. L., ALLISON, M. J. (1992), Localization of oxalyl-coenzyme A decarboxylase and formyl-coenzyme A transferase in *Oxalobacter formigenes* cells, *Systemat. Appl. Microbiol.* **15**, 167–171.

BAI, Q., SOMERVILLE, R. L. (1998), Integration host factor and cyclic AMP receptor protein are required for TyrR-mediated activation of TPL in *Citrobacter freundii*, *J. Bacteriol.* **180**, 6173–6186.

BALDI, F. (1997), Microbial transformation of mercury species and their importance in the biogeochemical cycle of mercury, *Metal Ions Biol. Systems* **34**, 213–257.

BARBAS III, C. F., HEINE, A., ZHONG, G. F., HOFFMANN, T., GRAMATIKOVA, S. et al. (1997), Immune versus natural selection: Antibody aldolases with enzymatic rates but broader scope, *Science* **278**, No. 5346, 2085–2092.

BARROWMAN, M. M., FEWSON, C. A. (1985), Phenylglyoxylate decarboxylase and phenylpyruvate decarboylase from *Acinetobacter calcoaceticus*, *Curr. Microbiol.* **12**, 235–239.

BATTAT, E., PELEG, Y., BERCOVITZ, A., ROKEM, J. S., GOLDBERG, I. (1991), Optimization of L-malic acid production by *Aspergillus flavus* in a stirred fermenter, *Biotechnol. Bioeng.* **37**, 1108–1116.

BATTERSBY, A. R. (1985), Enzymic synthesis of labelled chiral substances, in: *Enzymes in Organic Synthesis, Ciba Foundation Symposium 111* (PORTER, R., CLARK, S., Eds.), pp. 22–30. London: Pitman Press.

BATTERSBY, A. R., NICOLETTI, M., STAUNTON, J., VLEGGAAR, R. (1980), Studies of enzyme mediated reactions. Part 13. Stereochemical course of the formation of histamine by decarboxylation of (2*S*)-histidine with enzymes from *Clostridium welchii* and *Lactobacillus* 30a, *J. Chem. Soc., Perkin Trans. I.*, 43–51.

BATTERSBY, A. R., MURPHY, R., STAUNTON, J. (1982a), Studies of enzyme mediated reactions. 14. Stereochemical course of the formation of cadaverine by decarboxylation of (2*S*)-lysine with lysine decarboxylase (EC 4.1.1.18) from *Bacillus cadaveris*, *J. Chem. Soc., Perkin I*, 449–453.

BATTERSBY, A. R., STAUNTON, J., TIPPETT, J. (1982b), Studies of enzyme mediated reactions. 15. Stereochemical course of the formation of γ-aminobutyric acid (GABA) by decarboxylation of (2*S*)-glutamic acid with glutamate decarboxylase from *Escherichia coli*, *J. Chem. Soc., Perkin I*, 455–459.

BATTERSBY, A. R., SCOTT, A., STAUNTON, J. (1990), Studies of enzyme mediated reactions. 16. Stereochemical course of the formation of 5-hydroxytryptamine (serotonin) by decarboxylation of (2*S*)-5-hydroxytryptophan with the aromatic L-amino acid decarboxylase (EC 4.1.1.28) from hog kidney, *Tetrahedron* **46**, 4685–4696.

BAU, R., BREWER, I., CHIANG, M. Y., FUJITA, S., HOFFMAN, J. et al. (1983), Absolute configuration of a chiral CHD group via neutron diffraction – conformation of the absolute stereochemistry of the enzymatic formation of malic acid, *Biophys. Res. Commun.* **115**, 1048–1052.

BEAR, M. M., MONNE, C., ROBIE, D., CAMPION, G., LANGLOIS, V. et al. (1998), Synthesis and polymerization of benzyl (3*R*,4*R*)-3-methylmalolactonate via enzymatic preparation of the chiral precursor, *Chirality* **10**, 727–733.

BEARNE, S. L., WOLFENDEN, R. (1995), Enzymatic hydration of an olefin: The burden borne by fumarase, *J. Am. Chem. Soc.* **117**, 9588–9589.

BEECKMAN, S., VAN DRIESSCHE, E. (1998), Pig heart fumarase contains two distinct substrate binding sites differing in affinity, *J. Biol. Chem.* **273**, 31661–31669.

BEHAL, R. T., OLIVER, D. J. (1997), Biochemical and molecular characterization of fumarase from plants: purification and characterization of the enzyme cloning, sequencing, and expression of the gene, *Arch. Biochem. Biophys.* **348**, 65–74.

BEINERT, H., KENNEDY, M. C. (1993), Aconitase, a two faced protein-enzyme and iron regulatory factor, *FASEB J.*, **7**, 1442–1449.

BEINERT, H., KENNEDY, M. C., STOUT, C. D. (1996) Aconitase as iron–sulfur protein, enzyme and iron regulatory protein, *Chem. Rev.* **96**, 2335–2373.

BELL, P. J., ANDREWS, S. C., SIVAK, M. N., GUEST, J. R. (1989), Nucleotide sequence of the FNR regulated fumarase gene (fumB) of *Escherichia coli* K-12, *J. Bacteriol.* **171**, 3494–3503.

BELLEAU, B., BURBA, J. (1960), The stereochemistry of the enzymatic decarboxylation of amino acids, *J. Am. Chem. Soc.* **82**, 5751–5752.

BELLEAU, B., FANG, M., BURBA, J., MORAN, J. (1960), The absolute optical specificity of monoamine oxidase, *J. Am. Chem. Soc.* **82**, 5752–5753.

BENITEZ, T., GASENT-RAMIREZ, J. M, CASTREJON, F., CODON, A. C. (1996), Development of new (yeast) strains for the food industry, *Biotechnol. Progr.* **12**, 149–163.

BENNER, S. A., MORTON, T. H. (1981), Analogs for acetoacetate as an enzyme substrate. Stereochemical preferences, *J. Am. Chem. Soc.* **103**, 991–993.

BENNER, S. A., ROZZELL, J. D., MORTON, T. H. (1981), Stereospecificity and stereochemical infidelity of acetolactate decarboxylase (AAD), *J. Am. Chem. Soc.* **103**, 993–994.

BENNETT, G. N., RUDOLPH, F. B. (1995), The central metabolic pathway from acetyl-CoA to butyryl-CoA in *Clostridium acetobutylicum*, *FEMS Microbiol. Rev.* **17**, 241–249.

BENTLEY, R., THIESSEN, C. P. (1957), Biosynthesis of itaconic acid in *Aspergillus terreus*. III. The properties and reaction mechanism of *cis*-aconitic acid decarboxylase, *J. Biol. Chem.* **226**, 703–.

BERG, M., HILBI, H., DIMROTH, P. (1996), The acyl carrier protein of malonate decarboxylase of *Malonomonas rubra* contains 2′-(5″-phosphoribosyl)-3′-dephosphocoenzyme A as a prosthetic group, *Biochemistry* **35**, 4689–4696.

BERG, M., HILBI, H., DIMROTH, P. (1997), Sequence of a gene cluster from *Malonomonas rubra* encoding components of the malonate decarboxylase Na^+ pump and evidence for their function, *Eur. J. Biochem.* **245**, 103–115.

BERMEJO, L. L., WELKER, N. E., PAPOUTSAKIS, E. T. (1998), Expression of *Clostridium acetobutylicum* ATCC 824 genes in *Escherichia coli* for acetone production and acetate detoxification, *Appl. Environ. Microbiol.* **64**, 1079–1085.

BERING, C. L., KUHNS, J. J., ROWLETT, R. (1998), Purification of bovine carbonic anhydrase by affinity chromatography – an undergraduate experiment, *J. Chem. Ed.* **75**, 1021–1024.

BERRY, M. D., JURIO, A. V., LI, X. M., BOULTON, A. A., (1996), Aromatic L-amino acid decarboxylase: a neglected and misunderstood enzyme, *Neurochem. Res.* **21**, 1075–1087.

BERTOLDI, M., DOMINICI, P., MOORE, P. S., MARAS, B., VOLTATTORNI, C. B. (1998), reaction of DOPA decarboxylase leads to an oxidative deamination producing 3,4-dihydroxyphenylacetone, *Biochemistry* **37**, 6552–6561.

BERTOLDI, M., CARBONE, V., VOLTATTORNI, C. B. (1999), Ornithine and glutamate decarboxylases catalyze an oxidative deamination of their α-methyl substrates, *Biochem. J.* **342**, 509–512.

BJORNESTEDT, R., ZHONG, G. F., LERNER, R. A., BARBAS, C. F. (1996), Copying nature's mechanism for the decarboxylation of β-keto acids into catalytic antibodies by reactive immunization, *J. Am. Chem. Soc.* **118**, 11720–11724.

BODE, R., BIRNBAUM, D. (1991), Enzymatic production of indolepyruvate and some of its methyl and fluoro-derivatives, *Acta Biotechnol.* **11**, 387-393.

BOEKER, E. A., FISCHER, E. H. (1983), Lysine decarboxylase (*Escherichia coli-b*), *Methods Enzymol.* **94**, 180–184.

BONNARME, P., GILLET, B., SEPULCHRE, A. M., ROLE, C., BELOEIL, J. C., DUCROCQ, C. (1995), Itaconate biosynthesis in *Aspergillus terreus*, *J. Bacteriol.* **177**, 3573–3578.

BORNEMANN, S., CROUT, D. H. G., DALTON, H., HUTCHINSON, D. W., DEAN, G. et al. (1993), Stereochemistry of the formation of lactaldehyde and acetoin produced by the pyruvate decarboxylases of yeast (*Saccharomyces* sp.) and *Zymomonas mobilis* – different Boltzmann distribuons between bound forms of the electrophile, acetaldehyde, in the 2 enzymatic-reactions, *J. Chem. Soc., Perkin Trans. 1*, 309–311.

BORNEMANN, S., CROUT, D. H. G., DALTON, H., KREN, V., LOBELL, M. et al. (1996), Stereospecific formation of *R*-aromatic acyloins by *Zymomonas mobilis* pyruvate decarboxylase, *J. Chem. Soc., Perkin Trans. 1*, 425–430.

BOTRE, F. MAZZEI, F. (1999), Interactions between carbonic anhydrase and some decarboxylating enzymes as studied by a new bioelectrochemical approach, *Bioelectrochem. Bioenerget.* **48**, 463–467.

BOTTING, N. P., AKHTAR, M., COHEN, M. A., GANI, D. (1987), Mechanism of the elimination of ammonia from 3-substituted aspartic acids by 3-methylaspartase, *J. Chem. Soc., Chem. Commun.* 1371–1373.

BOTTING, N. P., AKHTAR, M., COHEN, M. A., GANI, D. (1988a), Substrate specificity of the 3-methylaspartate ammonia-lyase reaction – Observation of differential relative reaction rates for substrate product pairs, *Biochemistry* **27**, 2953–2955.

BOTTING, N. P., COHEN, M. A., AKHTAR, M., GANI, D. (1988b), Primary deuterium isotope effects for the 3-methylaspartase catalyzed deamination of (2*S*)-aspartic acid, (2*S*,3*S*)-3-methylaspartic acid and (2*S*,3*S*)-3-ethylaspartic acid, *Biochemistry* **27**, 2956–2959.

BOTTING, N. P., JACKSON, A. A., GANI, D. (1989), 15N-isotope and double isotope fractionation studies of the mechanism of 3-methylaspartase – concerted elimination of ammonia from (2*S*,3*S*)-3-methylaspartic acid, *J. Chem. Soc., Chem. Commun.* 1583–1585.

BOTTING, N. P., GANI, D. (1992), Mechanism of C-3 hydrogen-exchange and the elimination of ammonia in the 3-methylaspartate ammonia-lyase reaction, *Biochemistry* **31**, 1509–1520.

BOTTING, N. P., AKHTAR, M., ARCHER, C. H., COHEN, M. A., THOMAS, N. R. et al. (1992), in: *Molecular Recognition: Chemical and Biochemical Problems II* (ROBERTS, S. M., Ed.) Special Publication No. 111. Cambridge: RSC.

BOUMERDASSI, H., DESMAZEAUD, M., MONNET, C., BOQUIEN, C. Y., CORRIEU, G. (1996), Improvement of diacetyl production by *Lactococcus lactis* ssp *lactis* CNRZ 483 through oxygen control, *J. Dairy Sci.* **79**, 775–781.

BOURGEN, T., CHRETIEN, D., POGGI-BACH, J., DOONAN, S., RABIER, D. et al. (1994), Mutation of the fumarase gene in two siblings with progressive encephalopathy and fumarase deficiency, *J. Clin. Invest.* **93**, 2514–2518.

BOYNTON, Z. L., BENNETT, G. N., RUDOLPH, F. B. (1996), Cloning sequencing and expression of clustered genes encoding β-hydroxybutyryl-coenzyme A (CoA) dehydrogenase, crotonase and butyryl-CoA dehydrogenase from *Clostridium acetobutylicum* ATCC 824, *J. Bacteriol.* **178**, 3015–3024.

BRANDL, M. T., LINDOW, S. E. (1996), Cloning and characterization of a locus encoding an indolepyruvate decarboxylase involved in indole-3-acetic acid synthesis in *Erwinia herbicola*, *Appl. Environ. Microbiol.* **62**, 4121–4128.

BRANDL, M.T., LINDOW, S. E. (1997), Environmental signals modulate the expression of an indole-3-acetic acid biosynthetic gene in *Erwinia herbicola*, *Mol. Plant Microbe Interact.* **10**, 499–505.

BRANDL, M. T., LINDOW, S. E. (1998), Contribution of indole-3-acetic acid production to the epiphytic fitness of *Erwinia herbicola*, *Appl. Environ. Microbiol.* **64**, 3256–3263.

BRESSLER, E., BRUAN, S. (1996), Use of gel entrapment techniques for the resolution of itaconic acid acid biosynthetic pathway in *Aspergillus terreus*, *J. Sol Gel Sci. Technol.* **7**, 129–133.

BRIGANTI, F., MANGANI, S., SCOZZAFAVA, A., VERNAGLIONE, G., SUPURAN, C. T. (1999), Carbonic anhydrase catalyzes cyanamide hydration to urea: is it mimicking the physiological reaction, *J. Biol. Inorg. Chem.* **4**, 528–536.

BROEK, A. V., LAMBRECHT, M., EGGERMONT, K., VANDERLEYDEN, J. (1999), Auxins upregulate expression of the indole-3-pyruvate decarboxylase gene in *Azospirillum brasilense*, *J. Bacteriol.* **181**, 1338–1342.

BRUHLMANN, F. (1995), Purification and characterization of an extracellular pectate lyase from an *Amycolata* sp., *Appl. Environ. Microbiol.* **61**, 3580–3585.

BRYUN, S. M., JENNESS, R. (1981), Stereospecificity of L-*myo*-inositol-1-phosphate synthase for nicotinamide adenine dinucleotide, *Biochemistry* **20**, 5174–5177.

BUNCH, A. (1998), Nitriles, in: *Biotechnology* Vol. 8a, *Biotransformations* (REHM, H.-J., REED, G., PÜHLER, A., STADLER, P. J. W., Eds.), pp. 277–326. Weinheim: WILEY-VCH.

BUSER, H. P., PUGIN, B., SPINDLER, F., SUTTER, M. (1991), Two enantioselective syntheses of a precursor of the biologically most active isomer of CGA 80000 (Clozylacon), *Tetrahedron* **47**, 5709–5716.

CALIK, G., VURAL, H., CALIK, P., OZDAMAR, T. H. (1997), Bioprocess parameters and oxygen transfer effects in the growth of *Pseudomonas dacunhae* for L-alanine production, *Chem. Eng. J.* **65**, 109–116.

CALIK, G., SAVASCI, H., CALIK, P., OZDAMAR, T. H. (1999), Growth and -carrageenan immobilization of *Pseudomonas dacunhae* for L-alanine production, *Enzyme Microb. Technol.* **24**, 67–74.

CANDY, J. M., DUGGLEBY, R. G. (1998), Structure and properties of pyruvate decarboxylase and site-directed mutagenesis of the *Zymomonas mobilis* enzyme, *Biochim. Biophys. Acta.* **1385**, 323–338.

CARDILLO, R., SERVI, S., TINTI, C. (1991), Biotransformation of unsaturated aldehydes by microorganisms with pyruvate decarboxylase activity, *Appl. Microbiol. Biotechnol.* **36**, 300–303.

CARROLL, N. M., ROSS, R. P., KELLY, S. M., PRICE, N. C., SHEEHAN, D., COGAN, T. M. (1999), Characterization of recombinant acetolactate synthase from *Leuconostoc lactis*, *Enzyme Microb. Technol.* **25**, 61–67.

CASTILLO, R. M., MIZUGUCHI, K., DHANARAJ, V., ALBERT, A., BLUNDELL, T. L., MURZIN, A. G. (1999) A six stranded double psi β-barrel is shared by several protein superfamilies, *Structure with Folding & Design* **7**, 227–236.

CHANG, C. C., LAGHAI, A., O'LEARY, M. H., FLOSS, H. G. (1982), Some stereochemical features of aspartate β-decarboxylase, *J. Biol. Chem.* **257**, 3564–3569.

CHANG, C.-W. T., CHEN, X. H., LIU, H.-W. (1998), CDP-6-deoxy-6,6-difluoro-D-glucose: A mechanism based inhibitor for CDP-D-glucose 4,6-dehydratase, *J. Am. Chem. Soc.* **120**, 9698–9699.

CHATTERJEE, M. (1998), Lysine production by *Brevebacterium linenes* and its mutants: Activities and regulation of enzymes of the lysine biosynthetic pathway, *Folia Microbiol.* **43**, 141–146.

CHEN, J. S. (1995), Alcohol dehydrogenase – multiplicity and relatedness in the solvent producing Clostridia, *FEMS Microbiol. Rev.* **17**, 263–273.

CHEN, C. K., BLASCHEK, H. P. (1999a), Effect of acetate on molecular and physiological aspects of *Clostridium beijerinckii* NCIMB 8052, solvent production and strain degeneracy, *Appl. Environ. Microbiol.* **65**, 499–505.

CHEN, C. K., BLASCHEK, H. P. (1999b), Examination of the physiological and molecular factors involved in the enhanced solvent production by *Clostridium beijerinckii* BA101, *Appl. Environ. Microbiol.* **65**, 2269–2271.

CHEN, H., PHILLIPS, R. S. (1993), Binding of phenol and analogs to alanine complexes of tyrosine phenol-lyase from *Citrobacter freundii*: Implications for the mechanism of α,β-elimination and alanine racemization, *Biochemistry* **32**, 11591–11599.

CHEN, X. M. H., PLOUX, O., LIU, H.-W. (1996) Biosynthesis of 3,6-dideoxyhexoses: *In vivo* and *in vitro* evidence for protein-protein interaction between CDP-6-deoxy-L-*threo*-D-*glycero*-4-hexulose 3-dehydratase (E(1)) and its reductase (E(3)), *Biochemistry* **35**, 16412–16420.

CHIBATA, I., TOSA, T., SATO, T., TANAKA, T. (1987a), Immobilization of cells in carragenan, *Methods Enzymol.* **135**, 189–198.

CHIBATA, I., TOSA, T., YAMAMOTO, K., TANAKA, T. (1987b), Production of L-malic acid by immobilized microbial cells, *Methods Enzymol.* **136**, 455–463.

CHIBATA, I, TOSA, T., SHIBATANI, T. (1995), The industrial production of optically active compounds by immobilized biocatalysis, in: *Chirality in Industry* (COLLINS, A. N., SHELDRAKE, G. N., CROSBY J., Eds.), pp. 352–370. Chichester: John Wiley & Sons.

CHIPMAN, D., BARAK, Z., SCHLOSS, J. V. (1998), Biosynthesis of 2-aceto-2-hydroxy-acids: acetolactate synthases and acetohydroxyacid synthases, *Biochim. Biophys. Acta.* **1385**, 401–419.

CHOHNAN, S., KURUSU, Y., NISHIHARA, H., TAKAMURA, Y. (1999), Cloning and characterization of mdc gnes encoding malonate decarboxylase of *Pseudomonas putida*, *FEMS Microbiol. Lett.* **174**, 311–319.

CHOW, Y. S., SHIA, H. S., ADESINA, A. A., ROGERS, P. L. (1995), A kinetic model for the deactivation of pyruvate decarboxylase (pdc) by benzaldehyde, *Biotechnol. Lett.* **17**, 1201–1206.

CLIFFORD, K. H., CORNFORTH, J. W., DONNINGER, C., MALLABY, R. (1972), Stereochemical course of the decarboxylation of *S*-malate on malic enzyme, *Eur. J. Biochem.* **26**, 401–406.

COHEN-KUPEIC, R., KUPIEC, M., SANDBECK, K., LEIGH, J. A. (1999), Functional conservation between the argininosuccinate lyase of the archaeon *Methanococcus maripaludis* and the corresponding bacterial and eukaryal genes, *FEMS Microbiol. Lett.* **173**, 231–238.

COLOMBO, S., GRISA, M., TORTORA, P., VANONI, M. (1994), Molecular cloning, nucleotide sequence and expression of a *Sulfolobus solfataricus* gene encoding a class II fumarase, *FEBS Lett.* **337**, 93–98 (correction 1994, **340**, 151–153).

CONSEVAGE, M. W., PHILLIPS, A. T. (1985), Presence and quantity of dehydroalanine in histidine ammonia-lyase from *Pseudomonas putida*, *Biochemistry* **24**, 301–308.

CONSEVAGE, M. W., PHILLIPS, A. T. (1990), Sequence analysis of the huth-gene encoding histidine ammonia-lyase in *Pseudomonas putida, J. Bacteriol.* **172**, 2224–2229.

CORNFORTH, J. W., REDMOND, J. W., EGGERER, H., BUCKEL, W., GUTSCHOW, C. (1969), Asymmetric methyl groups and the mechanism of malate synthase, *Nature* **221**, 1212–1213.

CORNICK, N. A., ALLISON, M. J. (1996), Assimilation of oxalate, acetate and CO_2 by *Oxalobacter formigenes*, *Can J. Microbiol.* **42**, 1081–1086.

CORNICK, N. A., YAN, B., BANK, S., ALLISON, M. J. (1996), Biosynthesis of amino acids by *Oxalobacter formigenes.* Analysis using C-13-NMR, *Can J. Biochem.* **42**, 1219–1224.

COTON, E., ROLLAN, G. C., LONVAUD-FUNEL, A. (1998), Histidine decarboxylase of *Leuconostoc oenos* 9204: Purification, kinetic properties, cloning and nucleotide sequence of the *hdc* gene, *J. Appl. Microbiol.* **84**, 143–151.

COUGHLIN, E. M., CHRISTENSEN, E., KUNZ, P. L., KRISHNAMOORTHY, K. S., WALKER, V. et al. (1998), Molecular analysis and prenatal diagnosis of human fumarase deficiency, *Mol. Genet. Metab.* **63**, 254–262.

CROUT, D. H. G., HEDGECOCK, C. J. R. (1979), The base catalyzed rearrangement of α-acetolactate (2-hydroxy-2-methyl-3-oxobutanoate): a novel carboxylate ion migration in a tertiary ketol rearrangement, *J. Chem. Soc., Perkin. Trans.* 1 1982–1989.

CROUT, D. H. G., RATHBONE, D. L. (1988), Biotransformations with acetolactate decarboxylase: unusual conversions of both substrate enatiomers into products of high optical purity, *J. Chem. Soc., Chem. Commun.* 98–99.

CROUT, D. H. G., LITTLECHILD, J., MITCHELL, M. B., MORREY, S. M. (1984) Stereochemistry of the decarboxylation of α-acetolactate (2-hydroxy-2-methyl-3-oxobutanoate) by the acetolactate decarboxylase of *Klebsiella aerogenes*, *J. Chem. Soc., Perkin. Trans.* 1 2271–2276.

CROUT, D. H. G., LITTLECHILD, J., MORREY, S. M. (1986) Acetoin metabolism: stereochemistry of the acetoin produced by the pyruvate decarboxylase of wheat germ and by the acetolactate decarboxylase of *Klebsiella aerogenes*, *J. Chem. Soc., Perkin. Trans. 1*, 105–108.

CROUT, D. H. G., LEE, E. R., RATHBONE, D. L. (1990), Absolute-configuration of the product of the acetolactate synthase reaction by a novel method of analysis using acetolactate decarboxylase, *J. Chem. Soc., Perkin Trans.* 1 1367–1369.

CROUT, D. H. G., MCINTYRE, C. R., ALCOCK, N. W. (1991a), Stereoelectronic control of the tertiary ketol rearrangement – implications for the mechanism of the reaction catalyzed by the enzymes of branched-chain amino-acid-metabolism, reductoisomerase and acetolactate decarboxylase, *J. Chem. Soc., Perkin Trans.* 2, 53–62.

CROUT, D. H. G., DALTON, H., HUTCHINSON, D. W., MIYAGOSHI, M. (1991b), Studies on pyruvate decarboxylase – acyloin formation from aliphatic, aromatic and heterocyclic aldehydes, *J. Chem. Soc., Perkin Trans.* 1, 1329–1334.

CROUT, D. H. G, DAVIES, S., HEATH, R. J., MILES, C. O., RATHBONE, D. R.et al. (1994), Applications of hydrolytic and decarboxylating enzymes in biotransformations, *Biocatalysis* **9**, 1–30.

DAHMS, A. S., DONALD, A. (1982), D-*xylo*-Aldonate dehydratase, *Methods Enzymol.* **90**, 302–305.

DAHMS, A. S., SIBLEY, D., HUISMAN, W., DONALD, A. (1982), D-Galactonate dehydratase, *Methods Enzymol.* **90**, 294–298.

D'CUNHA, G. B., SATYANARAYAN, V., NAIR, P. M. (1994), Novel direct synthesis of L-phenylalanine methyl ester by using *Rhodotorula glutinis* phenylalanine ammonia lyase in an organic-aqueous biphasic system, *Enzyme Microb. Technol.* **16**, 318–322.

D'CUNHA, G. B., SATYANARAYAN, V., NAIR, P. M. (1996a), Stabilization of phenylalanine ammonia lyase containing *Rhodotorula glutinis* cells for the continuous synthesis of L-phenylalanine methyl ester, *Enzyme Microb. Technol.* **19**, 421–427.

D'CUNHA, G. B., SATYANARAYAN, V., NAIR, P. M. (1996b), Purification of phenylalanine ammonia lyase from *Rhodotorula glutinis*, *Phytochemistry* **42**, 17–20.

DE BOER, A. S., MARSHALL, R., BROADMEADOW, A., HAZELDEN, K. (1993), Toxicological evaluation of acetolactate decarboxylase, *J. Food. Protect.* **56**, 510–517.

DEHNING, I., SCHINK, B. (1989), *Malonomonas rubra* gen. nov. sp. nov., a microaerotolerant bacterium growing by decarboxylation of malonate, *Arch. Microbiol.* **151**, 427–433.

DEHNING, I., SCHINK, B. (1994), Anaerobic degradation of malaonate via maloyl-CoA by *Sporomusa malonica*, *Klebsiella oxytoca* and *Rhododbacter capsulatus*, *A. Van Leeuwenhoek Int. J. Gen. Mol. Microbiol.* 66, 343–350.

DELGADO, L., TREJO, B. A., HUITRON, C. AGUILAR, G. (1993), Pectin lyase from *Aspergillus* sp. CH-Y-1043. *Appl. Microbiol. Biotechnol.* **39**, 515–519.

DENG, M. D., COLEMAN, J. R. (1999), Ethanol synthesis by genetic engineering in cyanobacteria, *Appl. Environ. Microbiol.* **65**, 523–528.

DE VIVO (1993), The expanding clinical spectrum of mitochondrial diseases, *Brain Dev.* **15**, 1–22.

DEWICK, P. M. (1998), The biosynthesis of shikimate metabolites, *Nat. Prod. Rep.* **15**, 17–58.

DI GANGI, J. J., SEYZADEH, M., DAVIS, R. H. (1987), Ornithine decarboxylase from *Neurospora crassa*, purification, characterization and regulation by inactivation, *J. Biol. Chem.* **262**, 7889–7893.

DIMROTH, P., (1994) Bacterial sodium ion coupled energetics *A. Van Leeuwenhoek Int. J. Gen. Mol. Microbiol.* **65**, 381–395.

DIMROTH, P., HILBI, H. (1997), Enzymatic and genetic basis for bacterial growth on malonate, *Mol. Microbiol.* **25**, 3–10.

DIMROTH, P., SCHINK, B. (1998), Energy conservation in the decarboxylation of dicarboxylic acids by fermenting bacteria, *Arch. Microbiol.* **170**, 69–77.

DIMROTH, P., THOMER, A. (1993), On the mechanism of sodium ion translocation by oxaloacetate decarboxylase of *Klebsiella pneumoniae*, *Biochemistry* **32**, 1734–1739.

DIRMAIER, L. J., GARCIA, G. A., KOZARICH, J. W., KENYON, G. L., (1986), Inhibition of benzylformate decarboxylase by [*p*-(bromomethyl)benzoyl]formate – enzyme catalyzed modification of thiamine pyrophosphate by halide elimination and tautomerization, *J. Am. Chem. Soc.* **108**, 3149–3150.

DOBRITZSCH, D., KONIG, S., SCHNEIDER, G., LU, G. G. (1998), High resolution crystal structure of pyruvate decarboxylase from *Zymomonas mobilis* – implications for substrate activation in pyruvate decarboxylases, *J. Biol. Chem.* **273**, 20196–20204.

DOELLE, H. W., KIRK, L., CRITTENDEN, R., TOH, H., DOELLE, M.B. (1993), *Zymomonas mobilis* – science and industrial application, *Crit. Rev. Biotechnol.* **13**, 57–98.

DOMINICI, P., MOORE, P. S., VOLTATTORNI, C. B. (1993), Modified purification of L-aromatic amino acid decarboxylase from pig kidney, *Protein Expr. Purif.* **4**, 345–347.

DOYON, J. B., JAIN, A. (1999), The pattern of fluorine substitution affects binding affinity in a small library of fluoroaromatic inhibitors for carbonic anhydrase, *Org. Lett.* **1**, 183–185.

DRAUZ, K., WALDMANN, H. (1995), *Enzyme Catalysis in Organic Synthesis, A Comprehensive Handbook*, pp. 481–483, 1050. Weinheim: WILEY-VCH.

DRAKE, A. F., SILIGARDI, G., CROUT, D. H. G., RATHBONE, D. L. (1987), Applications of vibrational infrared circular dichoism to biological problems: Stereochemistry of the proton exchange in acetoin (3-hydroxybutan-2-one) catalyzed by acetolactate dehydrogenase, *J. Chem. Soc., Chem. Commun.* 1834–1835.

DREYER, J. L. (1985), Isolation and biochemical characterization of maleic-acid dehydratase, an iron containing hydro-lyase, *Eur. J. Biochem.* **150**, 145–154.

DREYER, J. L. (1987), The role of iron in the activation of mannonic and altronic acid hydratases, two Fe-requiring hydro-lyases, *Eur. J. Biochem.* **166**, 623–630.

DUCROCQ, C., FRAISSE, D., TABET, J. C., AZERAD, R. (1984), An enzymatic method for the determination of enantiomeric composition and absolute configuration of deuterated or tritiated succinic acid, *Anal. Biochem.* **141**, 418–422.

DUERRE, P. (1998), New insights and novel developments in *Clostridial* acetone/butanol/isopropanol fermentation, *Appl. Microbiol. Biotechnol.* **49**, 639–648.

DUERRE, P., BAHL, H., GOTTSCHALK, G. (1992), Acetone–butanol fermentation – Basis of a modern biotechnological process, *Chem. Ing. Tech.* **64**, 491–498.

DUERRE, P., FISCHER, R. J., KUHN, A., LORENZ, K., SCHREIBER, W. et al. (1995), Solventogenic genes of *Clostridium acetobutylicum* – catalytic properties, genetic organization and transcriptional regulation, *FEMS Microbiol. Rev.* **17**, 251–262.

DUFFY, J. I. (1980), Chemicals by enzymatic and microbial processes, *Chemical Technol. Rev.* **161**, 107–109.

DUSCH, N., PUEHLER, A., KALINOWSKI, J. (1999), Expression of the *Corynebacterium glutamicum panD* gene encoding L-aspartate-α-decarboxylase leads to pantothenate overproduction in *Escherichia coli*, *Appl. Environ. Microbiol.* **65**, 1530–1539.

DYDA, F., FUREY, W., SWAMINATHAN, S., SAX, M., FARRENKOPF, B., JORDAN, F. (1990), Preliminary crystallographic data for the thiamin diphosphate-dependent enzyme pyruvate decarboxylase from brewers' yeast, *J. Biol. Chem.* **265**, 17413–17415.

DYDA, F., FUREY, W., SWAMINATHAN, S., SAX, M., FARRENKOPF, B., JORDAN, F. (1993), Catalytic centers in the thiamin diphosphate dependent enzyme, pyruvate decarboxylase at 2.4-angstrom resolution, *Biochemistry* **32**, 6165–6170.

EARNHARDT, J. N., TU C., SILVERMAN, D. N. (1999), Intermolecular proton transfer in catalysis by carbonic anhydrase, *Can. J. Chem.* **77**, 726–732.

EASTON, C. J., HUTTON, C. A. (1994), Synthesis of each stereoisomer of [3-$^2H_1$]phenylalanine and evaluation of the stereochemical course of the reaction of (*R*)-phenylalanine with (*S*)-phenylalanine ammonia lyase, *J. Chem. Soc., Perkin Trans.* 1, 3545–3548.

EASTON, C. J., HARPER, J. B., LINCOLN, S. (1995), Use of cyclodextrins to limit product inhibition of (*S*)-phenylalanine ammonia lyase, *J. Chem. Soc., Perkin Trans.* 1, 2525–2526.

EDENS, W. A. A., URBAUER, J. L., CLELAND, W. W. (1997) Determination of the chemical mechanism of malic enzyme by isotope effects, *Biochemistry* **36**, 1141–1147.

EFFENBERGER, F. (1999), Enzyme catalyzed preparation and synthetic applications of optically active cyanohydrins, *Chimica* **53**, 3–10.

EHRLICH, J. I., HWANG, C.-H., COOK, P. F., BLANCHARD, J. S. (1999), Evidence for a stepwise mechanism for OMP decarboxylase, *J. Am. Chem. Soc.* **121**, 6966–6967.

EMANUELE, J. J. FITZPATRICK, P. F. (1995), Mechanistic studies of the flavoprotein tryptophan 2-monooxygenase. 1. Kinetic mechanism, *Biochemistry* **34**, 3710–3715.

ENGLARD, S., HANSON, K. R. (1969), Stereospecifically labeled citric acid ccyle intermediates, *Methods Enzymol.* **13**, 567–601.

EREZ, T., GDALEVSKY, G. Y., TORCHINSKY, Y. M., PHILLIPS, R. S., PAROLA, A. H. (1998), Cold inactivation and dissociation into dimers of *Escherichia coli* tryptophanase and its W330F mutant form, *Biochim. Biophys. Acta.* **1384**, 365–372.

EVANS, C. T., HANNA, K., PAYNE, C., CONRAD, D. W., MISAWA, M. (1987), Biotransformation of *trans* cinnamic acid to L-phenylalanine: Optimization of reaction conditions using whole yeast cells, *Enzyme Microb. Technol.* **9**, 417–421.

FACCHINI, P. J. (1998), Temporal correlation of tyramine metabolism with alkaloids and amide biosynthesis in elicited opium poppy cultures, *Phytochemistry* **49**, 481–490.

FALEEV, N. G., SPIRINA, S. N., PERYSHKOVA, O. E., SAPOROVSKAYA, M. B., TSYRYAPKIN, V. A., BELIKOV, V. M. (1995), The synthesis of L-tyrosine and its aromatic ring substituted analogs by microbial cells containing tyrosine phenol-lyase, *Appl. Biochem. Microbiol.* **31**, 155–160.

FALEEV, N. G., SPIRINA, S. N., IVOILOV, V. S., DEMIDKINA, T. V., PHILLIPS, R. S. (1996), The catalytic mechanism of tyrosine phenol-lyase from *Erwinia herbicola*: The effect of substrate structure on pH-dependence of kinetic parameters in the reactions with ring-substituted tyrosines, *Zt. Naturforsch. C* **51**, 363–370.

FALZONE, C. J., KARSTEN, W. E., CONLEY, J. D., VIOLA, R.E. (1988). L-Aspartase from *Escherichia coli*: Substrate specificity and role of divalent metal ions, *Biochemistry* **27**, 9089–9093.

FARRAR, T. C., GUTOWSKY, H. S., ALBERTY, R. A., MILLER, W. G. (1957), The mechanism of the stereospecific enzymatic hydration of fumarate to L-malate, *J. Am. Chem. Soc.* **79**, 3978–3980.

FARRENKOPF, B. C., JORDAN, F. (1992), Resolution of brewers' yeast pyruvate decarboxylase into multiple isoforms with similar subunit structure and activity using high-performance liquid-chromatography, *Protein Expr. Purif.* **3**, 101–107.

FELL, L. M., FRANCIS, J. T., HOLMES, J. L., TERLOUW, J. K. (1997), The intriguing behavior of (ionized) oxalacetic acid investigated by tandem mass spectrometry, *Int. J. Mass Spectrom.* **165**, 179–194.

FESSNER, W. D. (1998), Enzyme mediated C–C bond formation, *Curr. Opin. Chem. Biol.* **2**, 85–97.

FESSNER, W.-D., WALTER, C. (1997) Enzymatic C–C bond formation in asymmetric synthesis. *Topics Curr. Chem.* **184**, 97–194.

FETZNER, S. (1998), Bacterial dehalogenation, *Appl. Microbiol. Biotechnol.* **50**, 633–657.

FETZNER, S., LINGRENS, F. (1994), Bacteriol dehydrogenases – biochemistry, genetics and biotechnological applications, *Microbiol. Rev.* **58**, 641–685.

FINDEIS, M. A., WHITESIDES, G. M. (1987) Fumarase catalyzed synthesis of L-*threo*-chloromalic acid and its conversion to 2-deoxy-D-ribose and D-*erythro*-sphingosine, *J. Org. Chem.* **52**, 2838–2848.

FISHER, H. L., FRIEDEN, C., MCKEE, J. S. M., ALBERTY, R. A. (1955), Concerning the stereospecificity of the fumarase reaction and the demonstration of a new intermediate, *J. Am. Chem. Soc.* **77**, 4436.

FISHER, R. J., HELMS, J., DUERRE, P. (1993), Cloning, sequencing and molecular analysis of the sol operon of *Clostridium acetobutylicum*, a chromosomal locus involved in solventogenesis, *J. Bacteriol.* **175**, 6959–6969.

FITZPATRICK, D. R., HILL, A., TOLMIE, J. L., THORBURN, D. R., CHRISTODOULOU, L. (1999), The molecular basis of malonyl-CoA decarboxylase deficiency, *Am. J. Human Genet.* **65**, 318–326.

FLINT, D. H. (1993), *Escherichia coli* fumarase A catalyzes the isomerization of enol and keto oxalacetic acid, *Biochemistry* **32**, 799–805.

FLINT, D. H. (1994), Initial kinetic and mechanistic characterization of *Escherichia coli* fumarase A, *Arch. Biochem. Biophys.* **311**, 509–516.

FLINT, D. H., ALLEN, R. M. (1996), Iron–sulfur proteins with non-redox functions, *Chem. Rev.* **96**, 2315–2334.

FLINT, D. H., EMPTAGE, M. H. (1988), Dihydroxy acid dehydratase from spinach contains a [2Fe-2S] cluster, *J. Biol. Chem.* **263**, 3558–3564.

FLINT, D. H, MCKAY, R. G. (1994) *Escherichia coli* fumarase A catalyzed transfer of O-18 from C-2 and H-2 from C-3 of malate to acetylene-dicarboxylate to form O-18-labeled and H-2-labeled oxalacetate, *J. Am. Chem. Soc.* **116**, 5534–5539.

FLINT, D. H., NUDELMAN, A., CALABRESE, J. C., GOTTLIEB, H. E. (1992a), Enol oxalacetic acid exists in the *Z*-form in the crystalline state and in solution, *J. Org. Chem.* **57**, 7270–7274.

FLINT, D. H., EMPTAGE, M. H., GUEST, J. R. (1992b), Fumarase A from *Escherichia coli*: Purification and characterization as an iron-sulfur cluster containing enzyme, *Biochemistry* **31**, 10331–10337.

FLINT, D. H., TUMINELLO, J. F., EMPTAGE, M. H. (1993), The inactivation of Fe-S cluster containing hydro-lyases by superoxide, *J. Biol. Chem.* **268**, 22369–22376.

FLINT, D. H., TUMINELLO, J. F., MILLER, T. J. (1996), Studies on the synthesis of the Fe-S cluster of dihydroxy-acid dehydratase in *Escherichia coli* crude extract – Isolation of *O*-acetylserine sulfhydrylases A and B and β-cystathione based on their ability to mobilize sulfur from cysteine and to participate in Fe-S cluster synthesis, *J. Biol. Chem.* **271**, 16053–16067.

FLOROVA, G., DENOYA, C. D., MORGENSTERN, M. R., SKINNER, D. D., REYNOLDS, K. A. (1998), Cloning, expression, and characterization of a Type II 3-dehydroquinate dehydratase gene from *Streptomyces hygroscopicus*, *Arch. Biochem. Biophys.* **350**, 298–306.

FLOSS, H. G., TSAI, M. D., WOODARD, R. W. (1984), Stereochemistry of biological reactions at proprochiral centres, *Topics Stereochem.* **15**, 253–321.

FOOR, F., MORIN, N., BOSTIAN, K. A. (1993), Production of L-dihydroxyphenylalanine in *Escherichia coli* with the tyrosine phenol-lyase gene cloned from *Erwinia herbicola*, *Appl. Environ. Microbiol.* **59**, 3070–3075.

FRIEDEMANN, R., NEEF, H. (1998), Theoretical studies on the electronic and energetic properties of the aminopyrimidine part of thiamin diphosphate, *Biochim. Biophys. Acta.* **1385**, 245–250.

FU, D. X., MALONEY, P. C. (1998), Structure–function relationships in OxlT, the oxalate–formate transporter of *Oxalobacter formigenes* – topological features of transmembrane helix 11 as visualized by site directed fluorescent labeling, *J. Biol. Chem.* **273**, 17962–17967.

FUKUYAMA, Y., MATIOSHI, K., IWASAKI, M., TAKIZAWA, E., MIYAZAKI, M. et al. (1999), Preparative scale synthesis of (*R*)-α-fluorophenylacetic acid, *Biosci. Biotech. Biochem.* **63**, 1664–1666.

FUREY, W., ARJUNAN, P., CHEN, L., SAX, M., GUO, F., JORDAN, F. (1998), Structure–function relationships and flexible tetramer assembly in pyruvate decarboxylase revealed by analysis of crystal structures, *Biochim. Biophys. Acta.* **1385**, 253–270.

FURUKAWA, T., KOGA, J., ADACHI, T., KISHI, K., SYONO, K. (1996), Efficient conversion of L-tryptophan to indole-3-acetic acid and/or tryptophol by some species of *Rhizoctonia*, *Plant Cell Physiol.* **37**, 899–905.

FURUTA, T., TAKAHASHI, H., KASUYA, Y., (1990), Evidence for a carbanion intermediate in the elimination of ammonia from L-histidine catalyzed by histidine ammonia-lyase, *J. Am. Chem. Soc.* **112**, 3633–3636.

FURUTA, T., TAKAHASHI, H., SHIBASAKI, H., KASUYA, Y. (1992), Reversible stepwise mechanism involving a carbanion intermediate in the elimination of ammonia from L-histidine catalyzed by histidine ammonia-lyase, *J. Biol. Chem.* **267**, 12600–12605.

GALLAGHER, T., ROZWARSKI, D. A., ERNST, S. R., HACKETT, M. L. (1993), Refined structure of the pyruvoyl dependent decarboxylase from *Lactobacillus* 30a, *J. Mol. Biol.* **230**, 516–528.

GANI, D. (1991), A structural and mechanistic comparison of pyridoxal 5′-phosphate dependent decarboxylase and transaminase enzymes, *Phil. Trans. Royal Soc. B.* **332**, 131–139.

GANI, D., YOUNG, D. W. (1983) Stereochemistry of the dihydouracil dehydrogenase reaction in the metabolism of uracil to β-alanine, *J. Chem. Soc., Chem. Commun.*, 576–578.

GANI, D., ARCHER, C. H., BOTTING, N. P., POLLARD, J. R. (1999), The 3-methylapartase reaction probed using H-2 and N-15 isotope effects for three substrates: A flip from a concerted to a carbocationic mechanism upon changing the C-3 stereochemistry in the substrate from *R* to *S*, *Bioorg. Med. Chem.* **7**, 977–990.

GAO, S., OH, D. H., BROADBENT, J. R., JOHNSON, M. E., WEIMER, B. C, STEELE, J. L. (1997), Aromatic amino acid catabolism by *Lactococci*, *Lait* **77**, 371–381.

GAO, G., NARA, T., NAKAJIMA-SHIMADA, J., AOKI, T. (1999a), Novel organization and sequences of five genes encoding all six enzymes for *de novo* pyrimidine biosynthesis in *Trypanosoma cruzi*, *J. Mol. Biol.* **285**, 149–161.

GAO, J. M., WABER, L., BENNETT, M. J., GIBSON, K. M., COHEN, J. C. (1999b), Cloning and mutational analysis of human malonyl-coenzyme A decarboxylase, *J. Lipid Res.* **40**, 178–182.

GAWRON, O., FONDY, T. P. (1959), Stereochemistry of the fumarase and aspartase catalyzed reactions and of the Krebs cycle from fumaric acid to *d*-isocitric acid, *J. Am. Chem. Soc.* **81**, 6333–6334.

GAWRON, O., GLAID, A. J., FONDY, T. P. (1961), Stereochemistry of Krebs' cycle hydrations and related reactions, *J. Am. Chem. Soc.* **83**, 3634–3640.

GAWRONSKI, J., GAWRONSKI (1999), *Tartaric and Malic Acids in Organic Synthesis: A Source Book Of Building Blocks, Ligands, Auxilaries and Resolving Agents.* Chichester: John Wiley & Sons.

GELFMAN, C. M., COPELAND, W. C., ROBERTUS, J. D. (1991), Site directed alteration of 4 active site residues of a pyruvoyl-dependent histidine decarboxylase, *Biochemistry* **30**, 1057–1062.

GERISCHNER, U., DUERRE, P. (1990), Cloning, sequencing, and molecular analysis of the acetoacetate decarboxylase gene region from *Clostridium acetobutylicum*, *J. Bacteriol.* **172**, 6907–6918.

GERMAN, A. (1992), Use of α-acetolactate decarboxylase from a recombinant strain of *Bacillus subtilis* for the production of beer and alcohol, *Bull. Acad. Natl. Med.* **176**, 1175–1176.

GHOSH, S., SADHUKHAN, P. C., CHAUDHURI, J., GHOSH, D. K., MANDAL, A. (1996), Volatilization of mercury by immobilized mercury resistant bacterial cells, *J. Appl. Bacteriol.* **81**, 104–108.

GIORGIANNI, F., BERANOVA, S., WESDEMIOTIS, C., VIOLA, R. E. (1995), Elimination of the sensitivity of L-aspartase to activate site directed inactivation without alteration of catalytic activity. *Biochemistry* **34**, 3529–3535.

GIRBAL, J., SAINT-AMANS, I. V. S., SOUCAILLE, P. (1995), How neutral red modified carbon and electron flow in *Clostridium acetobutylicum* grown in chemostats at neutrel pH, *FEMS Microbiol. Lett.* **16**, 151–162.

GIRBAL, L., SOUCAILLE, P. (1998), Regulation of solvent production in *Clostridium acetobutylicum*, *TIBS* **16**, 11–16.

GISH, G., SMYTH, T., KLUGER, R. (1988), Thiamin diphosphate catalysis. Mechanistic divergence as a probe of substrate activation of pyruvate decarboxylase, *J. Am. Chem. Soc.* **110**, 6230–6234.

GLOGE, A., LANGER, B., POPPE, L., RETEY, J. (1998), The behavior of substrate analogs and secondary deuterium isotope effects in the phenylalanine ammonia-lyase reaction, *Arch. Biochim. Biophys.* **359**, 1–7.

GODA, S. K., MINTON, N. P., BOTTING, N. P., GANI, D. (1992), Cloning, sequencing, and expression in *Escherichia coli* of the *Clostridium tetanomorphum* gene encoding β-methylaspartase and characterization of the recombinant protein, *Biochemistry* **31**, 10747–10756.

GOODWIN, G. W., TAEGTMEYER, H., (1999), Regulation of fatty acid oxidation of the heart by MCD and ACC during contractile stimulation, *Am. J. Physiol. Endocrinol. Metabol.* **277**, E772–E777.

GONZALEZ, D. H., ANDREO, C. S. (1988), Carboxylation and dephosphorylation of phosphoenol-3-fluoropyruvate by maize leaf phosphoenolpyruvate carboxylase, *Biochem. J.* **253**, 217–222.

GOPALAN, R. (1999), Kinetics of decarboxylation of ethylmalonic acid: A hundred fold reactivity of the acid molecule over its mono-anion, *Asian J. Chem.* **11**, 1474–1482.

GOTTSCHALK, G., BENDER, R. (1982), D-gluconate dehydratase from *Clostridium pasteurianum*, *Methods Enzymol.* **90**, 283–287.

GRABER, R., KASPER, P., MALASHKEVICH, V. N., STROP, P., GEHRING, H. et al. (1999), Conversion of aspartate aminotransferase into L-aspartate β-decarboxylase by a triple active site mutation, *J. Biol. Chem.* **274**, 31203–31208.

GRABOWSKI, R., HOFMEISTER, A. E. M., BUCKEL, W. (1993), Bacterial L-serine dehydratases – a new family of enzymes containing iron–sulfur clusters, *Biochem. Sci.* **18**, 297–300.

GRIGG, R., HENDERSON, D., HUDSON, A. J. (1989) Decarboxylation of α-amino acids by pyruvic acid and its derivatives: evidence for azomethine ylides in *in vitro* analogs of pyruvoyl enzymic processes, *Tetrahedron Lett.* **30**, 2841–2844.

GRISHIN, N. V., OSTERMAN, A. L., BROOKS, H. B., PHILLIPS, M. A., GOLDSMITH, E. J. (1999), X-ray structure of ornithine decarboxylase from *Trypanosoma brucei*: The native structure and the structure in complex with α-difluoromethylornithine, *Biochemistry* **38**, 15174–15184.

GRUER, M. J., ARTYMIUK, P. J., GUEST, J. R. (1997), The aconitase family: Three structural variations on a common theme, *Trends. Biochem. Sci.* **22**, 3–6.

GRUPE, H., GOTTSCHALK, G. (1992), Physiological events in *Clostridium acetobutylicum* during the shift from acidogenesis to solventogenesis in continuous culture and presentation of a model for shift induction, *Appl. Environ. Microbiol.* **58**, 3896–3902.

GULICK, A. M., PALMER, D. R. J., BABBITT, P. C., GERLT, J. A., RAYMENT, I. (1998), Evolution of enzymatic activities in the enolase superfamily: Crystal structure of D-glucarate dehydratase from *Pseudomonas putida*, *Biochemistry* **37**, 14358–14368.

GULZAR, M. S., AKHTAR, M., GANI, D. (1994), Enantiospecific conjugate addition of *N*-nucleophiles to substituted fumaric acids using methylaspartase, *J. Chem. Soc., Chem. Commun.* 1994, 1601–1602.

GULZAR, M. S., MORRIS, K. B., GANI, D. (1995), Control of the regioselectivity of *N*-nucleophile addition to *N*- carbonyl protected dehydroalanines – a model for the ammonialyase enzymes, *J. Chem. Soc., Chem. Commun.* 1995, 1061–1062.

GULZAR, M. S., AKHTAR, M., GANI, D. (1997), Preparation of *N*-substituted aspartic acids via enantiospecific conjugate addition of *N*-nucleophiles to fumaric acids using methylaspartase: Synthetic utility and mechanistic implications, *J. Chem. Soc., Perkin Trans.*1:, 649–655.

GUNASEKARAN, P., RAJ, K. C. (1999), Ethanol fermentation technology – *Zymomonas mobilis*, *Curr. Sci.* **77**, 56–68.

GUPTA, M. N. (1992), Enzyme function in organic solvents. *Eur. J. Biochem.* **203**, 25–32.

GUPTA, A., PHUNG, L. T., CHAKRAVARTY, L., SILVER, S. (1999), Mercury resistance in *Bacillus cereus* RC607: Transcriptional organization and two new open reading frames, *J. Bacteriol.* **181**, 7080–7086.

GUTHRIE, R. D., JENCKS, W. P. (1989), IUPAC recommendations for the representation of reaction mechanisms, *Acc. Chem. Res.* **22**, 343–349.

GYAMERAH, M. H. (1995), Oxygen requirement and energy relations of itaconic acid fermentation by *Aspergillus terreus* NRRL 1960, *Appl. Microbiol. Biotechnol.* **44**, 20–26.

HALLIS, T. M., LIU, H. W. (1998), Stereospecificity of hydride transfer for the catalytically recycled NAD^+ in CDP-D-glucose 4,6-dehydratase, *Tetrahedron* **54**, 15975–15982.

HALLIS, T. M., LIU, H.-W. (1999a), Mechanistic studies of the biosynthesis of tyvelose: Purification and characterization of CDP-D-tyvelose 2-epimerase, *J. Am. Chem. Soc.* **121**, 6765–6766.

HALLIS, T. M., LIU, H.-W. (1999b), Learning nature's strategy for making deoxy sugars: Pathways, mechanisms and combinatorial applications, *Acc. Chem. Res.* **32**, 579–598.

HALLIS, T. M., LEI, Y., QUE, N. L. S., LIU, H., (1998), Mechanistic studies of the biosynthesis of paratose: purification and characterization of CDP-paratose synthase, *Biochemistry* **37**, 4935–4945.

HAMILTON, G. A., WESTHEIMER, F. H. (1959), On the mechanism of the enzymatic decarboxylation of acetoacetate, *J. Am. Chem. Soc.* **81**, 6332–6333.

HANDA, S., KOO, J. H., KIM, Y. S., FLOSS, H. G. (1999), Stereochemical course of biotin-indepenent malonate decarboxylase catalysis, *Arch. Biochem. Biophys.* **370**, 93–96.

HANSON, K. R., HAVIR, E. A. (1972), The enzymic elimination of ammonia, in: *The Enzymes* 3rd Edn. Vol. 7 (BOYER, S. D., Ed.), pp. 75–166. New York: Academic Press.

HANSON, K. R., WIGHTMAN, R. H., STAUNTON, J., BATTERSBY, A. R. (1971), Stereochemical course of the elimination catalyzed by L-phenylalanine ammonia-lyase and the configuration of 2-benzamidocinnamic azlactone, *J. Chem. Soc., Chem. Commun.*, 185–186.

HARDEN, A. (1901), The chemical action of *Bacillus coli communis* and similar organisms on carbohydrates and allied compounds. *J. Chem. Soc.* **79**, 611–628.

HASSON, M. S., MUSCATE, A., HENEHAN, G. T. M., GUIDINGER, P. F., PETSKO, G. A. et al. (1995), Purification and crystallization of benzoylformate decarboxylase, *Protein Sci.* **4**, 955–959.

HASSON, M. S., MUSCATE, A., MCLEISH, M. J., POLOVNIKOVA, L. S., GERLT, J. A. et al. (1998), The crystal structure of benzoylformate decarboxylase at 1. 6 angstrom resolution: Diversity of catalytic residues in thiamin diphosphate-dependent enzymes, *Biochemistry* **37**, 9918–9930.

HARRIS, T. K., WASHABAUGH, M. W. (1995a), Distribution of the thiamin diphosphate C-2-proton during catalysis of acetaldehyde formation by brewers-yeast pyruvate decarboxylase, *Biochemistry* **34**, 13994–14000.

HARRIS, T. K., WASHABAUGH, M. W. (1995b), Solvent-derived protons in catalysis by brewers-yeast pyruvate decarboxylase, *Biochemistry* **34**, 14001–14011.

HASSETT, D. J., HOWELL, M. L., SOKOL, P. A., VASIL, M. L., DEAN, G. E. (1997), Fumarase C activity is elevated in response to iron deprivation and in mucoid, alginate producing *Pseudomonas aeruginosa*: Cloning and characterization of *fumC* and purification of native FumC, *J. Bacteriol.* **179**, 1442–1451.

HAYASHI, H. (1995), Pyridoxal enzymes – mechanistic diversity and uniformity, *J. Biochem.* **118**, 463–473.

HAYASHI, S., KAMEJI, T. (1983), Ornithine decarboxylase (rat-liver), *Methods Enzymol.* **94**, 154–158.

HE, X. M, THORSON, J. S., LIU, H. W. (1996), Probing the coenzyme and substrate binding events of CDP-D-glucose 4,6-dehydratase: Mechanistic implications, *Biochemistry* **35**, 4721–4731.

HEATON, A. C. P., RUGH, C. L., WANG, N. J., MEAGHER, R. B. (1998), Phytoremediation of mercury and methymercury polluted soils using genetically engineeered plants, *J. Soil Contamin.* **7**, 497–509.

HEBEL, D., FURLANO, D. C., PHILLIPS, R. S., KOUSHIK, S., CREVELING, C. R., KIRK, K. L. (1992), An enzymatic-synthesis of 2-azido-L-tyrosine, *Bioorg. Med. Chem. Lett.* **2**, 41–44.

HERBERT, R. B., KNAGGS, A. R. (1992), Biosynthesis of the antibiotic obafluorin from para-aminophenylalanine and glycine (glyoxylate), *J. Chem. Soc., Perkin Trans. I*, 109–113.

HERNANDEZ, D., PHILLIPS, A. T. (1994), Ser-143 is an essential active-site residue in histidine ammonia-lyase of Pseudomonas putida. *Biochem. Biophys. Res. Commun.* **201**, 1433–1438.

HERNANDEZ, D., STROH, J. G., PHILLIPS, A. T. (1993), Identification of Ser^{143} as the site of modification in the active-site of histidine ammonia-lyase. *Arch. Biochem.. Biophys.* **307**, 126–132.

HIGHBARGER, L. A., GERLT, J. A., KENYON, G. L. (1996), Mechanism of the reaction catalyzed by acetoacetate decarboxylase. Importance of lysine 116 in determining the pK_a of active site lysine 115, *Biochemistry* **35**, 41–46.

HILL, C. M., PANG, S. S., DUGGLEBY, R. G. (1997), Purification of *Escherichia coli* acetohydroxyacid synthase isoenzyme II and reconstitution of active enzyme from its individual pure subunits, *Biochem. J.* **327**, 891–898.

HIRAHARA, T., HORINOUCHI, S., BEPPU, T. (1993), Cloning, nucleotide-sequence, and overexpression in *Escherichia coli* of the β-tyrosinase gene from an obligately symbiotic thermophile, *Symbiobacterium thermophilum*, *Appl. Microbiol. Biotechnol.* **39**, 341–346.

HIRSCH, W., SCHAGGER, H., FUCHS, G. (1998), Phenyl glyoxylate: NAD^+ oxidoreducatse (CoA benzoylating), a new enzyme of anaerobic phenylalanine metabolism in the denitrifying bacterium, *Azoarcus evansii*, *Eur. J. Biochem.* **251**, 907–915.

HIRSCHBEIN, B. L., MAZENOD, F. P., WHITESIDES, G. M. (1982), Synthesis of phosphoenolpyruvate and its use in adenosine triphosphate cofactor regeneration, *J. Org. Chem.* **47**, 3765–3766.

HOENKE, S., DIMROTH, P. (1999), Formation of catalytically active *S*-acetyl(malonate decarboxylase) requires malonyl-coenzyme A:acyl carrier protein-SH transacylase as auxilary enzyme, *Eur J. Biochem.* **259**, 181–187 (correction 1999, **259**, 961).

HOENKE, S., SCHMID, M., DIMROTH, P. (1997), Sequence of a gene cluster from *Klebsiella pneumoniae* encoding malonate decarboxylase and expression of the enzyme in *Escherichia coli*, *Eur. J. Biochem.* **246**, 530–538.

HOFMEISTER, A. E. F., ALBRACHT, S. P. J., BUCKEL, A, W. (1994) Iron sulfur containing L-serine dehydratase from *Peptostreptococcus asaccharolyticus*: correlation of the cluster type with enzymatic activity, *FEBS Lett.* **351**, 416–418.

HOHENESTER, E., KELLER, J. W., JANSONIUS, J. N. (1994), An alkali metal ion size dependent switch in the active site structure of dialkylglycine decarboxylase, *Biochemistry* **33**, 13561–13570.

HOHMANN, S. (1991a), PDC6, a weakly expressed pyruvate decarboxylase from yeast, is activated when fused spontaneously under the control of the *PDC1* promoter, *Curr. Genet.* 1991, **20**, 373–378.

HOHMANN, S. (1991b), Characterization of *PDC6*, a 3rd structural gene for pyruvate decarboxylase in *Saccharomyces cerevisiae*, *J. Bacteriol.* **173**, 7963–7969.

HOHMANN, S. (1993), Characterization of *PDC2*, a gene necessary for high-level expression of pyruvate decarboxylase structural genes in *saccharomyces-cerevisiae*, *Mol. Gen. Genet.* **241**, 657–666.

HUANG, S., BEALE, J. M., KELLER, P. J., FLOSS, H. G. (1986), Synthesis of (*R*)-[1-$^{13}C_1$,2-$^2H_1$)]malonate and (*S*)-[1-$^{13}C_1$,2-$^2H_1$)]malonate and its stereochemical analysis by NMR spectroscopy, *J. Am. Chem. Soc.* **108**, 1100–1101.

HUANG, C. L., WU, C. C., LIEN, M. H. (1997), *Ab initio* studies of decarboxylations of β-keto carboxylic acids, $XCOCH_2COOH$ (X = H, OH, and CH_3), *J. Phys. Chem. A* **101**, 7867–7873.

HUANG, S., XUE, Y., SAUER-ERIKSSON, E., CHIRICA, L., LINDSKOG, S., JONSSON, B.-H. (1998), Crystal structure of carbonic anhydrase from *Neisseria gonorrhoeae* and its complex with the inhibitor acetazolamide, *J. Mol. Biol.* **283**, 301–310.

HUANG, P. Q., CHEN, Q. F., CHEN, C. L., ZHANG, H. K. (1999), Asymmetric synthesis of (–)-(*R*)-pyrrolam A starting from (*S*)-malic acid, *Tetrahedron: Asymmetry* **10**, 3827–3832.

HUBBARD, B. K., KOCH, M., PALMER, D. R. J., BABBITT, P. C., GERLT, J. A. (1998) Evolution of enzymatic activities in the enolase superfamily: Characterization of D-glucarate/galactarate catabolic pathway in *Escherichia coli*, *Biochemistry* **37**, 14369–14375.

HUBNER, G., TITTMANN, K., KILLENBERG-JABS, M., SCHAFFNER, J., SPINKA, M. et al. (1998), Activation of thiamin diphosphate in enzymes, *Biochim. Biophys. Acta.* **1385**, 221–228.

HUGENHOLTZ, J. (1993), Citrate metabolism in lactic acid bacteria, *FEMS Microbiol. Rev.* **12**, 165–178.

HUHTA, D. W., HECKENTHALER, T., ALVAREZ, F. J., ERMER, J., HUBNER, G. et al. (1992), The catalytic power of pyruvate decarboxylase – a stochastic model for the molecular evolution of enzymes, *Acta Chem. Scand.* **46**, 778–788.

IDING, H., SIEGERT, P., MESCH, K., POHL, M. (1998), Application of α-keto acid decarboxylases in biotransformations, *Biochim. Biophys. Acta* **1385**, 307–322.

IKEMOTO, M., SASAKI, M., HARADAHIRA, T., YADA, T., OMURA, H. et al. (1999), Synthesis of L-[β-^{11}C]amino acids using immobilized enzymes, *Appl. Rad. Isotopes* **50**, 715–721.

IKUSHIRO, H., HAYASHI, H., KAWATA, Y., KAGAMIYAMA, H. (1998). Analysis of the pH and ligand induced spectral transitions of tryptophanase: Activation of the coenzyme at the early stages of the catalytic cycle, *Biochemistry* **37**, 3043–3052.

IMPERIALI, B., MCDONNELL, K. A., SHOGREN-KNAAK, M. (1999), Design and construction of novel peptides and proteins by tailored incorporation of coenzyme functionality, *Topics Curr. Chem.* **202**, 1–38.

INGRAM, L. O., ALDRICH, H. C., BORGES, A. C. C., CAUSEY, T. B., MARTINEZ, A. et al. (1999), Enteric bacterial catalysts for fuel ethanol production, *Biotechnol. Prog.* **15**, 855–866.

IRAQUI, I., VISSERS, S., ANDRE, B., URRESTARAZU, A. (1999); Transcriptional induction by aromatic amino acids in *Saccharomyces cerevisiae, Mol. Cell. Biol.* **19**, 3360–3371.

ISHII, S., MIZUGUCHI, H., NISHINO, J., HAYASHI, H., KAGAMIYAMA, H. (1996), Functionally important residues of aromatic L-amino acid decarboxylase probed by sequence alignment and site directed mutagenesis, *J. Biochem.* **120**, 369–376.

ISUPOV, M. N., ANTSON, A. A., DODSON, E. J., DODSON, G. G., DEMENTIEVA, I. S. et al. (1998), Crystal structure of tryptophanase, *J. Mol. Biol.* **276**, 603–623.

IWAMOTO, R., TANIKI, H., KOISHI, J., NAKURA, S. (1995), D-Glucosamine aldolase activity of D-glucosaminate dehydratase from *Pseudomonas fluorescens* and its requirement for Mn^{2+}, *Biosci. Biotech. Biochem.* **59**, 408–411.

JAKLITSCH, W. M., KUBICEK, C. P., SCRUTTON, M. C. (1991), The subcellular organization of itaconate biosynthesis in *Aspergillus terreus*, *J. Gen. Microbiol.* **137**, 533–539.

JANSONIUS, J. N. (1998), Structure, evolution and action of vitamin B_6-dependent enzymes, *Curr. Opin. Struct. Biol.* **8**, 759–769.

JANSSEN, P. H., HARFOOT, C. G. (1992), Anaerobic malonate decarboxylation by *Citrobacter divs.*, *Arch. Microbiol.* **157**, 471–474.

JAYASEKERA, M. M. K., SHI, W. X., FARBER, G. K., VIOLA, R. E. (1997b), Evaluation of functionally important amino acids in L-aspartate ammonia-lyase from *Escherichia coli*, *Biochemistry* **36**, 9145–9150.

JEBAI, F., HANOUN, N., HAMON, M., THIBAULT, J., PELTRE, G. et al. (1997), Expression, purification and characterization of rat aromatic L-amino acid decarboxylase in *Escherichia coli*, *Protein Expr. Purif.* **11**, 185–194.

JENKINS, J., MAYANS, O., PICKERSGILL, R. (1998), Structure and evolution of parallel β-helix proteins, *J. Struct. Biol.* **122**, 236–246.

JETTEN, M. S. M., SINSKEY, A. J. (1995), Purification and properties of oxaloacetate decarboxylase from *Corynebacterium glutamicum*, *A. Van Leeuwenhoek Int. J. Gen. Mol. Microbiol.* **67**, 221–227.

JEZOWSKA-BOJCZUK, M., KOZLOWSKI, H. (1991), Coordination of copper(II) ions by α,β-dehydrodipeptides. Potentiometric and spectroscopic study, *Polyhedron* **10**, 2331–2335.

JEZOWSKA-BOJCZUK, M., KOZLOWSKI, H. (1994), Unusual binding ability of α,β-dehydrodipeptides towards metal ions, *Polyhedron* **13**, 2683–2687.

JEZOWSKA-BOJCZUK, M., VARNAGY, K., SOVAGO, I., PIETRZYNSKI, G., DYBA, M. et al. (1996), Co-ordination of copper(II) ions by prolyl-α,β-dehydroamino acids: comparative studies and general considerations, *J. Chem. Soc., Dalton Trans.* 3265–3268.

JOHN, R. A. (1995) Pyridoxal phosphate dependent enzymes, *Biochim. Biophys. Acta.* **1248**, 81–96.

JOHNSON, D. A. CHEN, J. S. (1995), Taxonomic relationships among strains of *Clostridium acetobutylicum* and other phenotypically related organisms, *FEMS Microbiol. Rev.* **17**, 233–240.

JOHNSON, D. A., LIU, H.-W. (1998) Mechanisms and pathways from recent deoxysugar biosynthesis research, *Curr. Opin. Chem. Biol.* **2**, 642.

JOHNSON, D. A., GASSNER, G. T., BANDARIAN, V., RU-

ZICKA, F. J., BALLOU, D. P. et al. (1996), Kinetic characterization of an organic radical in the ascarylose biosynthetic pathway, *Biochemistry* **35**, 15846–15856.

JOHNSON, J. L., TOTH, J., SANTIWATANAKUL, S., CHEN, J. S. (1997), Cultures of "*Clostridium acetobutylicum*" from various culture collections comprise *Clostridium acetobutylicum*, *Clostridium beijerinckii* and two other distinct types based on DNA–DNA reassociation, *J. Systemat. Bacteriol.* **47**, 420–424.

JOHNSSON, K., ALLEMANN, R. K., WIDMER, H., BENNER, S. A. (1993), Synthesis, structure and activity of artifical, rationally designed catalytic polypeptides, *Nature* **365**, No. 6446, 530–532.

JONES, D. H. (1984), Phenylalanine ammonia-lyase: Regulation of its induction, and its role in plant development, *Phytochemistry* **23**, 1349–1359.

JONES, D. T., KEIS, S. (1995), Origins and relationships of industrial solvent producing clostridial strains, *FEMS Microbiol. Rev.* **17**, 223–232.

JONES, D. H., NORTHCOTE, D. H. (1984), Stability of the complex formed between French bean *Phaseolus vulgaris*) phenylalanine lyase and its transition state analog, *Arch. Biochem. Biophys.* **235**, 167–177.

JORDAN, F. (1999), Interplay of organic and biological chemistry in understanding coenzyme mechanisms: example of thiamin diphosphate-dependent decarboxylations of 2-oxo acids, *FEBS Lett.* **457**, 298–301.

JORDAN, P. M., SPENCER, J. B. (1990), Stereospecific manipulation of hydrogen atoms with opposite absolute orientations during the biosynthesis of the polyketide 6-methylsalicyclic acid from chiral malonates in *Penicillin patulum, J. Chem. Soc., Chem. Commun.*, 238–242.

JORDAN, P. M., SPENCER, J. B. (1991), Mechanistic and stereochemical investigation of fatty acid and polyketide biosynthesis using chiral malonates, *Tetrahedron* **47**, 6015–6028.

JORDAN, P. M., SPENCER, J. B., CORINA, D. L. (1986), The synthesis of chiral malonates and determination of their chirality using fatty acid synthase and mass spectrometry, *J. Chem. Soc., Chem. Commun.*, 911–913.

JORDAN, K.N., O'DONOGHUE, M., CONDON, S., COGAN, T. M. (1996). Formation of diacetyl by cell-free extracts of *Leuconostoc lactis, FEMS Microbiol. Lett.* **143**, 291–297.

JORDAN, F., NEMERIA, N., GUO, F. S., BABURINA, I., GAO, Y. H. et al. (1998), Regulation of thiamin diphosphate-dependent 2-oxo acid decarboxylases by substrate and thiamin diphosphate. Mg(II) – evidence for tertiary and quaternary interactions, *Biochim. Biophys. Acta.* **1385**, 287–306.

JORRIN, J., LOPEZ-VALBUENA, R., TENA, M. (1988), Purification and properties of phenylalanine ammonia-lyase from sunflower (*Helianthus annuus* L.) hypocotyls, *Biochim. Biophys. Acta* **964**, 73–82.

KAMATH, A. V., VAALER, G. L., SNELL, E. E. (1991), Pyridoxal phosphate dependent histidine decarboxylases – Cloning sequencing and expression of genes from *Klebsiella planticola* and *Enterobacter aerogenes* and properties of the overexpressed enzymes, *J. Biol. Chem.* **266**, 9432–9437.

KAMMEYER, A, EGGELTE, T. A., BOS, J. D., TEUNISSEN, M. B. M. (1999), Urocanic acid isomers are good hydroxyl radical scavengers: a comparative study with structural analogs and with uric acid, *Biochim. Biophys. Acta.* **1428**, 117–120.

KANE, J. F., FISKE, M. J. (1985), Regulation of phenylalanine ammonia lyase in *Rhodotorula glutinis, J. Bacteriol.* **161**, 963–966.

KAPAT, A., ZIMAND, G., ELAD, Y. (1998), Effect of two isolates of *Trichoderma harzianum* on the activity of hydrolytic enzymes produced by *Botrytis cinerea*, *Physiol. Mol. Plant Pathol.* **52**, 127–137.

KAPAT, A., ZIMAND, G., ELAD, Y. (1998), Biosynthesis of pathogenicity hydrolytic enzymes by *Botrytis cinerea* during infection of bean leaves and *in vitro*, *Mycol. Res.* **102**, 1017–1024.

KARSTEN, W. E., VIOLA, R. E. (1991), Kinetic studies of L-aspartase from *Escherichia-coli*: pH dependent activity changes. *Arch. Biochem. Biophys.* **287**, 60–67.

KATAOKA, M., IKEMI, M., MORIKAWA, Y., MIYOSHI, T., NISHI, K. et al. (1997), Isolation and characterization of D-threonine aldolase, a pyridoxal-5′-phosphate dependent enzyme from *Arthrobacter* sp. DK-38, *Eur. J. Biochem.* **248**, 385–393.

KATO, Y., ASANO, Y. (1995a), Purification and properties of crystalline 3-methylaspartase from two facultative anaerobes, *Citrobacter* sp. strain YG-0505 and *Morganella morganii* strain YG-0601, *Biosci. Biotech. Biochem.* **59**, 93–99.

KATO, Y., ASANO, Y. (1995b), 3-Methylaspartase ammonia-lyase from a facultative anaerobe strain YG-1002, *Appl. Microbiol. Biotechnol.* **43**, 901–907.

KATO, Y., ASANO, Y. (1997), 3-Methylaspartase ammonia-lyase as a marker enzyme of the mesaconate pathway for (*S*)-glutamate fermentation in Enterobacteriaceae, *Arch. Microbiol.* **168**, 457–463.

KATO, Y., ASANO, Y. (1998), Cloning, nucleotide sequencing and expression of the 3-methylaspartase ammonia lyase gene from *Citrobacter amalonaticus* strain YG-1002, *Appl. Microbiol. Biotechnol.* **50**, 468–474.

KAWAGUCHI, M., SYONO, K. (1996), The excessive production of indole-3-acetic acid and its significance in studies of the biosynthesis of this regulator of plant growth and development, *Plant Cell Physiol.* **37**, 1043–1048.

KAWASAKI, T., WATANABE, M., OHTA, H. (1995), A

novel enzymatic decarboxylation proceeds via a thiol ester intermediate, *Bull Chem. Soc. Jpn.* **68**, 2017–2020.

KAWASAKI, T., HORIMAI, E., OHTA, H. (1996a), On the conformation of the substrate binding to the active site during the course of enzymatic decarboxylation, *Bull. Chem. Soc. Jpn.* **69**, 3591–3594.

KAWASAKI, K., YOKOTA, A., TOMITA, F. (1996b), L-Tryptophan production by a pyruvic acid producing, *Escherichia coli* strain carrying the *Enterobacter aerogenes* tryptophanase gene, *J. Ferment. Bioeng.* **82**, 604–606.

KAWASAKI, T., SAITO, K., OHTA, H. (1997), The mode of substrate recognition of arylmalonate decarboxylases, *Chem. Lett.* 351–352.

KAWATA, Y, TAMURA, K., YANO, S., MIZOBATA, T., NAGAI, J. et al. (1999), Purification and characterization of thermostable aspartase from *Bacillus* sp. YM55-1, *Arch. Biochem. Biophys.* **366**, 40–46.

KEHAYOVA, P. D., BOKINSKY, G. E., HUBER, J. D., JAIN, A. (1999), A caged hydrophobic inhibitor of carbonic anyhydrase, *Org. Lett.*, **1**, 187–188.

KELLARD, J. G., PALCIC, M. M., PICKARD, M. A., VEDERAS, J. C. (1985), Stereochemistry of lysine formation by *meso*-diaminopimelate decarboxylase from wheat-germ, *Biochemistry* **24**, 3263–3267.

KENNEDY, J. F., WHITE, C. A. (1990) The plant, algal, and microbial polysaccharides, in: *Carbohydrate Chemistry* (KENNEDY, J. F., Ed.), pp. 220–261. Oxford: Oxford University Press.

KERN, D., KERN, G., NEEF, H., TITMANN, K, KILLENBERG-JABS, M. et al. (1997), How thiamine diphosphate is activated in enzymes, *Science* **275**, 67–70.

KERN, A. D., OLIVEIRA, M. A., COFFINO, P., HACKERT, M. L. (1999), Structure of mammalian ornithine decarboxylase at 1.6 Å resolution: stereochemical implications of PLP-dependent amino acid decarboxylases, *Structure with Folding & Design* **7**, 567–581.

KERUCHENKO, J. S., KERUCHENKO, I. D., GLADILIN, K. L., ZAITSEV, V. N., CHIRGADZE, N. Y. (1992), Purification, characterization and preliminary X-ray study of fumarase from *Saccharomyces cerevisiae*, *Biochem. Biophys. Acta* **1122**, 85–92.

KHOMUTOV, R. M., FALEEV, N. G., BELYANKIN, A. V., KHURS, E. N., KHOMUTOV, A. R. et al. (1997), 1-Amino-2-(4-hydroxyphenyl)ethyl phosphinic acid – a novel substrate of tyrosine phenol lyase, *Bioorganicheskaya Khimiya* **23**, 919–921.

KHRISTOFOROV, R. R., SUKHAREVA, B. S., DIXON, H. B. F., SPARKES, M. J., KRASNOV, V. P., BUKRINA, I. M. (1995), The interaction of glutamate decarboxylase from *Escherichia coli* with substrate analogs modified at C-3 and C-4, *Biochem. Mol. Biol. Int.* **36**, 77–85.

KILLENBERG-JABS, M., KONIG, S., HOHMANN, S., HUBNER, G. (1996), Purification and characterization of the pyruvate decarboxylase from a haploid strain of *Saccharomyces cerevisiae*, *Biol. Chem. Hoppe-Seyler* **377**, 313–317.

KILLENBERG-JABS, M., KONIG, S., EBERHARDT, I., HOHMANN, S., HUBNER, G. (1997), Role of Glu51 for cofactor binding and catalytic activity in pyruvate decarboxylase from yeast studied by site-directed mutagenesis, *Biochemistry* **36**, 1900–1905.

KIM, Y. S., BYUN, H. S. (1994), Purification and properties of a novel type of malonate decarboxylase from *Acinetobacter calcoaceticus, J. Biol. Chem.* **269**, 29636–29641.

KIM, K., COLE, P. A. (1999), Synthesis of (2*S*,3*R*)-β-methyltyrosine catalyzed by tyrosine phenol-lyase, *Bioorg. Med. Chem. Lett*, **9**, 1205–1208.

KIM, Y. S., KOLATTUKUDY, P. E., BOOS, A. (1979), Malonyl-CoA decarboxylase from *Mycobacterium tuberculosis* and *Pseudomonas fluorescens*, *Arch. Biochem. Biophys.* **196**, 543–551.

KIM, S. H., KRONSTAD, J. W., ELLIS, B. E. (1996), Purification and characterization of phenylalanine ammonia-lyase from *Ustilago maydis*, *Phytochemistry* **43**, 351–357.

KIM, M. S., LEE, W. K., KIM, H. Y., KIM, C., RYU, Y. W. (1998), Effect of environmental factors on flavonol glycoside production and phenylalanine ammonia-lyase activity in cell suspension cultures of *Ginkgo biloba. J. Microb. Biotechnol.* **8**, 237–244.

KINOSHITA, S., NAKAYAMA, K. (1978). Amino acids, in: *Economic Microbiology: Primary Products of Metabolism* Vol. 2 (ROSE, A. H., Ed.),pp. 210-256. London: Academic Press.

KIRIMURA, K., SATO, T., NAKANISHI, N., TERADA, M., USAMI, S. (1997), Breeding of starch utilizing and itaconic acid producing koji molds by interspecific protoplast fusion between *Aspergillus terreus* and *Aspergillus usamii*, *Appl. Microbiol. Biotechnol.* **47**, 127–131.

KIRSCHNING, A., BECHTHOLD, A. F. W., ROHR, J. (1997), Chemical and biochemical aspects of deoxysugars and deoxysugar oligosaccharides, *Topics Curr. Chem.* **188**, 1–84.

KLUGER, R., LAM, J. F., PEZACKI, J. P., YANG, C.-H. (1995), Diverting thiamin from catalysis to destruction. Mechanism of fragmentation of *N*(1′)-methyl-2-(1-hydroxybenzyl)thiamin, *J. Am. Chem. Soc.* **117**, 11383–11389.

KNIGHT, R. L.,LEEPER, F. J. (1998), Comparison of chiral thiazolium and triazolium salts as asymmetric catalysts for the benzoin condensation, *J. Chem. Soc., Perkin Trans. 1.*, 1891–1893.

KNOX, C., SASS, E., NEUPERT, W., PINES, O. (1998), Import into mitochondria, folding and retrograde movement of fumarase in yeast, *J. Biol. Chem.* **273**, 25587–25593.

KOBAYASHI, K., YAMANISHI, T., TUBOI, S. (1981), Physiochemical, catalytic and immunochemical properties of fumarases crystallized separately from mitochondrial and cystolic fractions of rat liver, *J. Biochem.* **89**, 1923–1931.

KOGA, J. (1995), Structure and function of indolepyruvate decarboxylase, a key enzyme in indole-3-acetic-acid biosynthesis, *Biochim. Biophys. Acta.* **1249**, 1–13.

KOGA, J., ADACHI, T., HIDAKA, H. (1991), Molecular cloning of the gene for indolepyruvate decarboxylase from *Enterobacter cloacae*, *Mol. Gen. Genet.* **226**, 10–16.

KOGA, J., ADACHI, T., HIDAKA, H. (1992), Purification and characterization of indolepyruvate decarboxylase – a novel enzyme for indole-3-acetic-acid biosynthesis in *Enterobacter cloacae*, *J. Biol. Chem.* **267**, 15823–15828.

KOGA, J., SYONO, K., ICHIKAWA, T., ADACHI, T. (1994), Involvement of L-tryptophan aminotransferase in indole-3-acetic-acid biosynthesis in *Enterobacter cloacae*, *Biochim. Biophys. Acta.* **1209**, 241–247.

KOK, J. (1996) Inducible gene expression and environmentally regulated genes in lactic acid bacteria, *A. Van Leeuwenhoek Int. J. Gen. Mol. Microbiol.* **70**, 129–145.

KOLATTUKUDY, P. E., POULOSE, A. J., KIM, Y. S. (1981), Malonyl-CoA from avian, mammalian and microbial sources, *Methods Enzymol.* **71**, 150–163.

KOLB, S., SEELINGER, S., SPRINGER, N., LUDWIG, W., SCHINK, B. (1998), The fermenting bacterium *Malonomonas rubra* is phylogenetically related to the sulfur reducing bacteria and contains a c-type cytochrome similar to those of sulfur and sulfate reducers, *System. Appl. Microbiol.* **21**, 340–345..

KONIG, S. (1998), Subunit structure, function and organization of pyruvate decarboxylases from various organisms, *Biochim. Biophys. Acta.* **1385**, 271–286.

KONIG, S., SVERGUN, D. I., VOLKOV, V. V., FEIGIN, L. A., KOCH, M. H. J. (1998), Small-angle X-ray solution-scattering studies on ligand-induced subunit interactions of the thiamine diphosphate dependent enzyme pyruvate decarboxylase from different organisms, *Biochemistry* **37**, 5329–5334.

KOSUGE, T., UMEHARA, K., HOSHINO, T. (1998), Cloning and sequence analysis of fumarase and superoxide dismutase genes from an extreme thermophile *Thermus thermophilus* HB27, *J. Ferment. Bioeng.* **86**, 125–129.

KRAUTWURST, H., BAZAES, S., GONZALAE, F. D., JABALQUINTO, A. M., FREY, P. A., CARDEMIL, E. (1998), The strongly conserved lysine of *Saccharomyces cerevisiae* phosphoenolpyruvate carboxykinase is essential for phosphoryl transfer, *Biochemistry* **37**, 6295–6302.

KREN, V., CROUT, D. H. G., DALTON, H., HUTCHINSON, D. W., KONIG, W. et al. (1993), Pyruvate decarboxylase – a new enzyme for the production of acyloins by biotransformation, *J. Chem. Soc., Chem. Commun.*, 1993, 341–343.

KYRIAKIDIS, D. A., PANAGIOTIDIS, C. A., GEORGATSOS, J. G. (1983), Ornithine decarboxylase (germinated barley-seeds), *Methods Enzymol.* **94**, 162–166.

LABROU, N. E., CLONIS, Y. D. (1995), Biomimetic dye-ligands for oxalate recognizing enzymes – studies with oxlate oxidase and oxalate decarboxylase, *J. Biotechnol.* **40**, 59–70.

LABROU, N. E., CLONIS, Y. D. (1999), Oxaloacetate decarboxylase from *Pseudomonas stutzeri*: Purification and characterization, *Arch. Biochem. Biophys.* **365**, 17–24.

LAI, M. C., TRAXLER, R. W. (1994), A coupled two stage continuous fermentation for solvent production by *Clostridium acetobutylicum*, *Enzyme Microb. Technol.* **16**, 1021–1025.

LANDRO, J. A., GERLT, J. A., KOZARICH, J. W., KOO, C. W., SHAH, V. (1994), The role of lysine 166 in the mechanism of mandelate racemase from *Pseudomonas putida* – mechanistic and crystallographic evidence for stereospecific alkylation by (*R*)-α-phenylglycidate *Biochemistry* **33**, 635–643.

LANGER, M., LIEBER, A., RETEY, J. (1994a), Histidine ammonia-lyase mutant-S143C is post-translationally converted into fully active wild-type enzyme – evidence for serine[143] to be the precursor of active-site dehydroalanine, *Biochemistry* **33**, 14034–14038.

LANGER, M., RECK, G., REED, J., RETEY, J. (1994b), Identification of serine143 as the most likely precursor of dehydroalanine in the active-site of histidine ammonia-lyase – a study of the overexpressed enzyme by site-directed mutagenesis, *Biochemistry* **33**, 6462–6467.

LANGER, M., PAULING, A., RETEY, J. (1995), The role of dehydroalanine in catalysis by histidine ammonia lyase, *Angew. Chem. Int. Edn. Engl.* **34**, 1464–1465.

LANGER, B., STARCK, J., LANGER, M., RETEY, J. (1997a), Formation of the Michaelis complex without involvement of the prosthetic group dehydroalanine of histidine ammonialyase, *Bioorg. Med. Chem. Lett.* **7**, 1077–1082.

LANGER, B., ROTHER, D., RETEY, J. (1997b), Identification of essential amino acids in phenylalanine ammonia-lyase by site-directed mutagenesis, *Biochemistry* **36**, 10867–10871.

LAUBLE, H., KENNEDY, M. C., BEINERT, H., STOUT, C. D. (1994), Crystal structures of aconitase with *trans*-aconitate and nitrocitrate bound, *J. Mol. Biol.* **237**, 437–451.

LAUBLE, H., KENNEDY, M. C., EMPTAGE, M. H., BEINERT, H., STOUT, C. D. (1996), The reaction of fluorocitrate with aconitase and the crystal structure of the enzyme-inhibitor complex, *Proc. Natl. Acad. Sci. USA.* **93**, 13699–13703.

LEE, K.-L., RAMALINGAM, K., SON, J.-K., WOODARD, R. W. (1989), A highly efficient and large scale synthesis of (2*S*,3*S*)-[2,3-$^2H_2$]-and (2*S*,3*R*)-[3-^2H]-aspartic acids via an immobilized aspartase containing microbial cell system, *J. Org. Chem.* **54**, 3195–3198.

LEE, S. G., RO, H. S., HONG, S. P., KIM, E. H., SUNG, M. H. (1996a), Production of L-DOPA by thermostable tyrosine phenol-lyase of a thermophilic *Symbiobacterium* species overexpressed in recombinant *Escherichia coli*, *J. Microbiol. Biotechnol.* **6**, 98–102.

LEE, S.-G., HONG, S. P., SUNG, M. H. (1996b), Removal and bioconversion of phenol in wastewater by a thermostable β-tyrosinase, *Enzyme Microb. Technol.* **19**, 374–377.

LEE, S.-G., HONG, S.-P., CHOI, Y.-H., CHUNG, Y.-J., SUNG, M.-H. (1997), Thermostable tyrosine phenol-lyase of *Symbiobacterium* sp. SC-1: gene cloning, structure determination and overproduction in *Escherichia coli*, *Protein Expr. Purif.* **11**, 263–270.

LEE, S.-G., HONG, S.-P., SUNG, M.-H. (1999a), Development of an enzymatic system for the production of dopamine from catechol, pyruvate and ammonia, *Enzyme Microb. Technol.* **25**, 298–302.

LEE, S.-G., HONG, S.-P., KWAK, M. S., ESAKI, N., SUNG, M.-H. (1999b), Characterization of thermostable tyrosine phenol-lyase from an symbiotic obligatory thermophile, *Symbiobacterium* sp. SC-1, *J. Biochem. Mol. Biol.* **32**, 480–485.

LEE, T. T., WORBY, C., BAO, Z. Q., DIXON, J. E., COLEMAN, R. F. (1999c), His^{68} and His^{141} are critical contributors to the intersubunit catalytic site of adenylosuccinate lyase of *Bacillus subtilis*, *Biochemistry* **38**, 22–32.

LEHOU, I. E., MITRA, B. (1999), (*S*)-Mandelate dehydrogenase from *Pseudomonas putida*: Mechanistic studies with alternative substrates and pH and kinetic isotope effects, *Biochemistry* **38**, 5836–5848.

LEMONNIER, M., LANE, D. (1998), Expression of the second lysine decarboxylase gene of *Escherichia coli*, *Microbiology (UK)* **144**, 751–760.

LENZ, H., WUNDERWALD, P., EGGERER, H. (1976), Partial purification and some properties of oxalacetase, *Eur. J. Biochem.* **65**, 225–236.

LEPINIEC, L., VIDAL, J., CHOLLET, R., GADAL, P., CRETIN, C. (1994), Phosphoenolpyruvate carboxylase – Structure, regulation and evolution, *Plant Sci.* **99**, 111–124.

LEVIN, L., FORCHIASSIN, F. (1998), Culture conditions for the production of pectinolytic enzymes by the white-rot fungus *Trametes trogii* on a laboratory scale, *Acta Biotechnol.* **18**, 157–166.

LI, H., JORDAN, F. (1999), Effects of substitution of tryptophan 412 in the substrate activation pathway of yeast pyruvate decarboxylase, *Biochemistry* **38**, 10004–10012.

LIM, H. W., PARK, S. S., LIM, C. J. (1997), Purification and properties of phenylalanine lyase from leaf mustard, *Molecules Cells* **7**, 715–720.

LIM, H. W., SA, J. H., PARK, S. S., LIM, C. J. (1998), A second form of phenylalanine ammonia lyase from leaf mustard, *Molecules Cells* **8**, 343–349.

LIMBERG, G., THIEM, J. (1996), Synthesis of modified aldonic acids and studies of their substrate efficiency for dihydroxy acid dehydratase, *Aust. J. Chem.* **49**, 349–356.

LINDQUIST, Y., SCHNEIDER, G. (1993), Thiamin diphosphate dependent enzymes – transketolase, pyruvate oxidase and pyruvate decarboxylase, *Curr. Opin. Struct. Biol.* **3**, 896–901.

LINDQUIST, L., SCHWEDA, K. H., REEVES, P. R., LINDBERG, A. A. (1994), *In vitro* synthesis of CDP-D-abequose using *Salmonella* enzymes of cloned *rfb* genes: Production of CDP-6-deoxy-D-*xylo*-4-hexulose, CDP-3,6-dideoxy-D-*xylo*-4-hexulose and CDP-3,6-dideoxy-D-galactose and isolation by HPLC, *Eur. J. Biochem.* **225**, 863–872.

LINDSKOG, S. (1997), Structure and mechanism of carbonic anhydrase, *Pharmacol. Ther.* **74**, 1–20.

LIU, H. W., THORSON, J. S. (1994), Pathways and mechanisms in the biogenesis of novel deoxysugars by bacteria, *Ann. Rev. Microbiol.* **48**, 223–256.

LLOYD-GEORGE, I., CHANG, T. M. S. (1993), Free and microencapsulated *Erwinia herbicola* for the production of tyrosine, *Biomat. Artif. Cells Immobil. Biotechnol.* **21**, 323–333.

LLOYD-GEORGE, I., CHANG, T. M. S. (1995), Characterization of free and alginate-polylysine-alginate microencapsulated *Erwinia herbicola* for the conversion of ammonia, pyruvate, and phenol into L-tyrosine, *Biotechnol. Bioeng.* **48**, 706–714.

LO, S. F., MILLER, V. P., LEI, Y. Y., THORSON, J. S., LIU, H.-W., SCHOTTEL, J. L. (1994), CDP-6-deoxy-$\Delta^{3,4}$-glucoseen reductase from *Yersinia pseudotuberculosis*, *J. Bacteriol.* **176**, 460–468.

LOBELL, M., CROUT, D. H. G. (1996a), New insight into the pyruvate decarboxylase-catalyzed formation of lactaldehyde from H-D exchange experiments: A "water proof" active site, *J. Chem. Soc., Perkin Trans. 1*, 1577–1581.

LOBELL, M., CROUT, D. H. G. (1996b), Pyruvate decarboxylase: A molecular modeling study of pyruvate decarboxylation and acyloin formation, *J. Am. Chem. Soc.* **118**, 1867–1873.

LOEWUS, M. W., LOEWUS, F. A., BRILLINGER, G. U., OTSUKA, H., FLOSS, H.G. (1980), Stereochemistry of the L-*myo*-inositol-1-phosphate synthase reaction, *J. Biol. Chem.* **255**, 1710–1712.

LOUWRIER, S., OSTENDORF, M., BOOM, A., HIEMSTRA, H., SPECKAMP, W. N. (1996), Studies towards the synthesis of (+)-ptilomycalin A; stereoselective *N*-acyliminium ion coupling reactions to enantiopure C-2 substituted lactams, *Tetrahedron* **52**, 2603–2628.

LU, G. G., DOBRITZSCH, D., KONIG, S., SCHNEIDER, G. (1997), Novel tetramer assembly of pyruvate decarboxylase from brewers' yeast observed in a new crystal form, *FEBS Lett.* **403,** 249–253.

LUNG, H. Y., BAETZ, A. L., PECK, A. B. (1994), Molecular cloning, DNA sequence and gene expression of the oxalyl-coenzyme A decarboxylase gene, *oxc* from the bacterium *Oxalobacter formigenes*, *J. Bacteriol.* **176**, 2468–2472.

LUTHY, J., RETEY, J., ARIGONI, D. (1969), Preparation and detection of chiral methyl groups, *Nature* **221**, 1213–1215.

LYNSTRAD, M., GRASDALEN, H. (1993), A new air lift bioreactor used in P-31-NMR studies of itaconic acid producing *Aspergillus terreus*, *J. Biochem. Biophys. Methods* **27**, 105–116.

MALASHKEVICH, V. N., BURKHARD, P., DOMINICI, P., MOORE, P. S., VOLTATTORNI, C. B., JANSONIUS, J. N. (1999), Preliminary X-ray analysis of a new crystal form of recombinant pig kidney DOPA decarboxylase, *Acta Crystallographica D* **55**, 568–570.

MALASHKEVICH, V. N., STROP, P., KELLER, J. W., JANSONIUS, J. N., TONEY, M. D. (1999), Crystal structures of dialkylglycine decarboxylase inhibitor complexes, *J. Mol. Biol.* **294**, 193–200.

MANN, J. (1987), *Secondary Metabolism*, pp. 173–190. Oxford: Oxford Science Publications.

MANN, J., DAVIDSON, R. S., HOBBS, J. B., BANTHORPE, D. V., HARBOURNE, J. B. (1994), *Natural Products: Their Chemistry and Biological Significance*, pp. 372–375. Longman Scientific and Technical.

MANULIS, S., HAVIVCHESNER, A., BRANDL, M. T., LINDOW, S. E., BARASH, I. (1998), Differential involvement of indole-3-acetic acid biosynthetic pathways in pathogenicity and epiphytic fitness of *Erwinia herbicola* pv, *gypsophilae*, *Mol. Plant Microbe Interact.* **11**, 634–642.

MARLETTA, M. A., CHEUNG, Y. F., WALSH, C. (1982), Stereochemical studies on the hydration of monofluorofumarate and 2, 3-difluorofumarate by fumarase, *Biochemistry* **21**, 2637–2644.

MAYANS, O., SCOTT, M., CONNERTON, I., GRAVESEN, T., BENEN, J. et al. (1997), Two crystal structures of pectin lyase A from *Aspergillus* reveal a pH driven conformational change and striking divergence in the substrate-binding clefts of pectin and pectate lyases, *Structure* **5**, 677–689.

MBWAGA, A. M., MENKE, G., GROSSMANN, F. (1997), Investigations on the activity of cell wall-degrading enzymes in young wheat plants after infection with *Pseudocercosporella herpotrichoides* (Fron) Deighton, *J. Phytopathol.* **145**, 123–130.

MCCANN, P. P., BACCHI, C. J., HANSON, W. L., NATHAN, H. C., HUTNER, S. H., SJOERDSMA, A. (1983), Methods for the study of the treatment of protozoan diseases by inhibitors of ornithine decarboxylase, *Methods Enzymol.* **94**, 209–213.

MERMELSTEIN, L. D., PAPOUTSAKIS, E. T., PETERSEN, D. J., BENNETT, G. N. (1993), Metabolic engineering of *Clostridium acetobutylicum* ATCC-824 for increased solvent production by enhancement of acetone formation enzyme activities using a synthetic acetone operon, *Biotechnol. Bioeng.* **42**, 1053–1060.

MIALL, L. M., (1978), Organic acids, in: *Economic Microbiology. Primary Products of Metabolism* (-ROSE, A. H., Ed.), pp. 1–55. London: Academic Press.

MICHIELSEN, M. J. F., MEIJER, E. A., WIJFFELS, R. H., TRAMPER, J., BEEFTINK, H. H. (1998), Kinetics of D-malate production by permeabilized *Pseudomonas pseudoaligenes*, *Enzyme Microb. Technol.* **22**, 621–628.

MICKLEFIELD, J., HARRIS, K. J., GROGER, S., MOCEK, U., HILBI, H. et al. (1995), Stereochemical course of malonate decarboxylation in *Malonomonas rubra*, *J. Am. Chem. Soc.* **117**, 1153–1154.

MILLER, B. G., TRAUT, T. W., WOLFENDEN, R. (1998), Effects of substrate binding determinants in the transition state for orotidine 5′-monophosphate decarboxylase, *Bioorg. Chem.* **26**, 283–288.

MINAMISAWA, K., OGAWA, K., FUKUHARA, H., KOGA, J. (1996), Indolepyruvate pathway for indole-3-acetic acid biosynthesis in *Bradyrhizobium elkanii*, *Plant Cell Physiol.* **37**, 449–453.

MITCHELL, J. L. A. (1983), Ornithine decarboxylase and the ornithine decarboxylase-modifying protein of *Physarum polycephalum*, *Methods Enzymol.* **94**, 140–146.

MIYAMOTO, K., OHTA, H. (1990), Enzyme mediated asymmetric decarboxylation of disubstituted malonic acids, *J. Am. Chem. Soc.* **112**, 4077–4078.

MIYAMOTO, K., OHTA, H. (1991), Asymmetric decarboxylation of disubstituted malonic acid by *Alcaligenes bronchisepticus* KU 1201, *Biocatalysis* **5**, 49–60.

MIYAMOTO, K., OHTA, H. (1992), Purification and properties of a novel arylmalonate decarboxylase from *Alcaligenes bronchisepticus* KU 1201, *Eur. J. Biochem.* **210**, 475–481.

MIYAMOTO, K., TSUCHIYA, S., OHTA, H. (1992), Microbial asymmetric decarboxylation of fluorine containing arylmalonic acid derivatives, *J. Fluorine Chem.* **59**, 225–232.

MIZOBATA, T., FUJIOKA, T., YAMASAKI, F., HIDAKA, M. (1998), Purification and characterization of a thermostable class II fumarase from *Thermus thermophilus*, *Arch. Biochem. Biophys.* **355**, 49–55.

MOCHIZUKI, N., HIRAMATSU, S., SUGAI, T., OHTA, H., MORITA, H., ITOKAWA, H. (1995), Reductive conversion of carbonyl compounds by yeast. 2. Improved conditions for the production and characterization of 1-arylpropane-1, 2-diols and related compounds, *Biosci. Biotechnol. Biochem.* **59**, 2282–2291.

Mohrig, J. R., Moerke, K. A., Cloutier, D. L., Lane, B. D., Person, E. C., Onasch, T. B. (1995), Importance of historical contingency in the stereochemistry of hydratase-dehydratase enzymes, *Science* **269**, No. 5223, 527–529.

Momany, C., Ernst, S., Ghosh, R., Chang, N. L., Hackert, M. L. (1995), Crystallographic structure of a PLP dependent ornithine decarboxylase from *Lactobacillus* 30a to 3.0 Å resolution, *J. Mol. Biol.* **252**, 643–655.

Monnet, C., Corrieu, G. (1998), Selection and properties of *Lactobacillus* mutants producing α-acetolactate, *J. Dairy Sci.* **81**, 2096–2102.

Monnet, C, Schmitt, P, Divies, C. (1994), Diacetyl production in milk by an α-acetolactic acid accumulating strain of *Lactococcus lactis* ssp. *lactis biovar diacetylactis*, *J. Dairy Sci.* **77**, 2916–2924.

Montgomery, J. A., Rosiers, C. D., Brunengraber, H. (1992), Biosynthesis and characterization of 3-hydroxyalkan-2-ones and 2,3-alkanediols – potential products of aldehyde metabolism, *Biol. Mass. Spectr.* **21**, 242–248.

Moore, P. S., Dominici, P., Voltattorni, C. B. (1996) Cloning and expression of pig kidney DOPA decarboxylase: Comparison of the naturally occurring and recombinant enzymes, *Biochem. J.* **315**, 249–256.

Morgan, D. M. L. (1999), Polyamines, an overview, *Mol. Biotechnol.* **11**, 229–250.

Moreno-Arribas, V., Lonvaud-Funel, A. (1999), Tyrosine decarboxylase activity of *Lactobacillus brevis* IOEB-9809 isolated from wine and *L. brevis* ATCC-367, *FEMS Microbiol. Lett.* **180**, 55–60.

Morris, D. R., Boeker, E. A. (1983), Biosynthetic and biodegradative ornithine and arginine decarboxylases from *Escherichia coli*, *Methods Enzymol.* **94**, 125–134.

Movahedi, S., Heale, J. B. (1990), The roles of aspartic proteinase and endo-pectin lyase enzymes in the primary stages of infection and pathogenesis of various host tissues by different isolates of *Botrytis cinerea pers ex pers*, *Physiol. Mol Plant. Pathol.* **36**, 303-324.

Muguruma, C. (1999), *Ab initio* MO study on the catalytic mechanism in the active site of carbonic anhydrase, *THEOCHEM, J. Mol. Struct.* **462**, 439–452.

Muller, E. H., Richards, E. J., Norbeck, J., Byrne, K. L., Karlsson, K. A. et al. (1999), Thiamine repression and pyruvate decarboxylase autoregulation independently control the expression of the *Saccharomyces cerevisiae* PDC5 gene, *FEBS Lett.* **449**, 245–250.

Munier, R. L., Bompeix, G. (1985), Inhibition of the L-phenylalanine ammonia-lyase from *Rhodotorula glutinis* with *N*-substituted-phenylalanines, *C. R.* III **300**, 203–206.

Murase, S., Takagi, J. S., Higashi, Y., Imaishi, H., Yumoto, N., Tokushige (1991), Activation of aspartase by side-directed mutagenesis, *Biochem. Biophys. Res. Commun.* **177**, 414–419.

Murase, S., Yumoto, N. (1993), Characterization of 3 types of aspartase activated by site- directed mutagenesis, limited proteolysis, and acetylation. *J. Biochem.* **114**, 735–739.

Mussons, M. L., Raposo, C., de la Torre, M. F., Moran, J. R., Caballero, M. C. (1999), Dibenz[c,h]acridine receptors for dibutylmalonic acid, *Tetrahedron*, **55**, 4077–4094.

Nagai, T., Hamada, M., Kai, N., Tanoue, Y., Nagayama, F. (1996), Distribution of aromatic L-amino acid decarboxylase in tissues of skipjack tuna using L-DOPA as the substrate, *J. Fish. Biol.* **48**, 1014–1017.

Nagasawa, T., Utagawa, T., Goto, J., Kim, C. J., Tani, Y. et al. (1981), Syntheses of L-tyrosine-related amino-acids by tyrosine phenol- lyase of *Citrobacter intermedius*, *Eur. J. Biochem.* **117**, 33–40.

Naidu, G. S. N., Panda, T. (1998), Production of pectolytic enzymes – a review, *Bioproc. Eng.* **19**, 355–361.

Nair, R. V., Green, E. M., Watson, D. E., Bennett, G. N., Papoutsakis, E. T. (1999), Regulation of the sol locus genes for butanol and acetone formation in *Clostridium acetobutylicum* ATCC 824 by a putative transcriptional represssor, *J. Bacteriol.* **181**, 319–330.

Naundorf, A., Klaffke, W. (1996), Substrate specificity of native dTDP-D-glucose-4,6-dehydratase: Chemo-enzymatic syntheses of artificial and naturally occurring deoxy sugars, *Carb. Res.* **285**, 141–150.

Nelsestuen, G. L., Kirkwood, S. (1971), The mechanism of action of the enzyme uridine diphosphoglucose 4-epimerase, *J. Biol. Chem.* **246**, 7533.

Nicholson, R. L., Hammerschmidt, R. (1992), Phenolic compounds and their role in disease resistance, *Ann. Rev. Phytopathol.* **30**, 369–389.

Niedend, B. R., Parker, S. R., Stierle, D. B., Stierle, A. A. (1999), First fungal aromatic L-amino acid decarboxylase from a paclitaxel producing *Penicillium raistrickii*, *Mycologia* **91**, 619–626.

Nikaidou, N., Naganuma, T., Kamio, Y., Izaki, K. (1995), Production, purification, and properties of a pectin lyase from *Pseudomonas marginalis* N6301, *Biosci. Biotechnol. Biochem.* **59**, 323–324.

Nikolova, P., Ward, O. P. (1991), Production of L-phenylacetyl carbinol by biotransformation – product and by-product formation and activities of the key enzymes in wild-type and ADH isoenzyme mutants of *Saccharomyces cerevisiae*, *Biotechnol. Bioeng.* **38**, 493–498.

Nuiry, I. I., Hermes, J. D., Weiss, P. M., Chen, C-Y., Cook, P. F. (1984), Kinetic mechanism and location of rate-determinig steps for aspartase from *Hafnia alvei*, *Biochemistry* **23**, 5168–5175.

Oba, K., Uritani, I. (1982), Pyruvate decarboxylase from sweet-potato roots, *Methods Enzymol.* **90**, 528–532.

Ogawa, H., Fujioka, M. (1999), Is threonine aldolase identical to serine hydroxymethyltransferase?, *Seikagaki* **71**, 1145–1152.

Ogino, K., Tamiya, H., Kimura, Y., Azuma, H., Tagaki, W. (1996), *cis*-Diamines as active catalysts fro the decarboxylation of oxalacetate, *J. Chem. Soc., Perkin Trans II*, 979–984.

O'Hare, M. C., Doonan, S. (1985), Purification and structural comparisons of the cytosolic and mitochondrial isoenzymes of fumarase from pig liver, *Biochim. Biophys. Acta* **827**, 127–134.

Ohba, H., Yasuda, S., Hirosue, H., Yamasaki, N. (1995), Improvement of the thermostability of pyruvate decarboxylase by modification with an amylose derivative, *Biosci. Biotech. Biochem.* **59**, 1581–1583.

Ohta, H. (1997), Enzyme mediated enantioselective protonation of enolates in aqueous medium, *Bull. Chem. Soc. Jpn.* **70**, 2895–2911.

Okabe, M., Ohta, N., Park, Y. S. (1993), Itaconic acid production in an airlift bioreactor using a modified draft tube, *J. Ferment. Bioeng.* **76**, 117–122.

Oliver, A. L., Roddick, F. A., Anderson, B. N. (1997), Cleaner production of phenylacetylcarbinol by yeast through productivity improvements and waste minimization, *Pure Appl. Chem.* **69**, 2371–2385.

Orczyk, A., Hipskind, J., de Neergaard, E., Goldbrough, P., Nicholson (1996), Stimulation of phenylalanine ammonia lyase in Sorghum in response to innoculation with *Bipolaris maydis*, *Physiol. Mol. Plant. Pathol.* **48**, 55–64.

Orr, G. R., Gould, S. J. (1982), Stereochemistry of the bacterial ornithine, lysine and arginine decarboxylase reactions, *Tetrahedron Lett.* **23**, 3139–3142.

Ortega, J. (1996), Pectolytic enzymes produced by the phytopathogenic fungus *Colletotrichum gloeosporioides*, *Texas J. Sci.* **48**, 123–128.

Osterman, A. L., Brooks, H. B., Rizo, J., Phillips, M. A. (1997), Role of Arg-277 in the binding of pyridoxal 5′-phosphate to *Trypanosoma brucei* ornithine decarboxylase, *Biochemistry* **36**, 4558–4567.

Osterman, A. L., Brooks, H. B., Jackson, L., Abbott, J. J., Phillips, M. A. (1999), Lysine-69 plays a key role in catalysis by ornithine decarboxylase through acceleration of the Schiff base formation, decarboxylation and product release steps, *Biochemistry* **38**, 11814–11826.

Oths, P. J., Mayer, R. M., Floss, H. G. (1990) Stereochemistry and mechanism of the GDP-mannose dehydratase reaction, *Carb. Res.* **198**, 91–100.

Oyama, K. (1992), The industrial production of aspartame, in: *Chirality in Industry* (Collins, A. N., Sheldrake, G. N., Crosby J., Eds.), pp. 237-247. Chichester: John Wiley & Sons.

Palmer, D. R. J., Gerlt, J. A. (1996), Evolution of enzymatic activities: multiple pathways for generating and partitioning a common enolic intermedaite by glucarate dehydratase from *Pseudomonas putida*, *J. Am. Chem. Soc.* **118**, 10323–10324.

Palmer, D. R. J., Wieczorek, S. J., Hubbard, B. K., Mrachko, G. T., Gerlt, J. A. (1997), Importance of mechanistic imperatives in enzyme-catalyzed β-elimination reactions: Stereochemical consequences of the dehydration reactions catalyzed by D-galactonate dehydratase from *Escherichia coli* and D-glucarate dehydratase from *Pseudomonas putida*, *J. Am. Chem. Soc.* **119**, 9580–9581.

Palmer, D. R. J., Hubbard, B. K., Gerlt, J. A. (1998), Evolution of enzymatic activities in the enolase superfamily: Partitioning of reactive intermediates by (*D*)-glucarate dehydratase from *Pseudomonas putida*, *Biochemistry* **37**, 14350–14357.

Park, Y. S., Itida, M., Ohta, N., Okabe, M. (1994), Itaconic acid production using an airlift reactor in repeated batch culture of *Aspergillus terreus*, *J. Ferment. Bioeng.* **77**, 329–333.

Park, S. H., Xing, R. Y., Whitman, W. (1995), Nonenzymatic acetolactate oxidation to diacetyl by flavin, nicotinamide and quinone coenzymes, *Biochim. Biophys. Acta.* **1245**, 366–370.

Park, H.S., Lee, J. Y., Kim, H. S. (1998) Production of L-DOPA (3-(3,4-dihydroxyphenyl)-L-alanine) from benzene using a hybrid pathway, *Biotechnol. Bioeng.* **58**, 339–343.

Parker, A. R., Moore, J. A., Schwab, J. M., Davisson, V. J. (1995), *Escherichia coli* imidazoleglycerol phosphate dehydratase: Spectroscopic characterization of the enzyme product and the steric course of the reaction, *J. Am. Chem. Soc.* **117**, 10605–10613.

Parks, E. H., Ernst, S. R., Hamlin, R., Xuong, N. H. (1985), Structure determination of histidine decarboxylase from Lactobacillus 30a at 3.0 Å resolution, *J. Mol. Biol.* **182**, 455–465.

Peschke, V. M., Sachs, M. M. (1993), Multiple pyruvate decarboxylase genes in maize are induced by hypoxia, *Mol. Gen. Genet.* **240**, 206–212.

Petersen, D. J., Bennett, G. N. (1990), Purification of acetoacetate decarboxylase from *Clostridium acetobutylicium* ATCC-824 and cloning of the acetoacetate decarboxylase gene in *Escherichia coli*, *Appl. Environ. Microbiol.* **56**, 3491–3498.

Petersen, D. J., van der Leyden, J., Bennett, G. N. (1993), Sequence and arrangement of genes encoding the enzymes of the acetone production pathway of *Clostridium acetobutylicum* ATCC-824, *Gene* **123**, 93–97.

PETRUCCIOLI, M., PULCI, V., FEDERICI, F. (1999), Itaconic acid production by *Aspergillus terreus* on raw starchy materials, *Lett. Appl. Microbiol.* **28**, 309–312.

PHALIP, V., MONNET, C., SCHMITT, P., RENAULT, P., GODON, J. J., DIVIES, C. (1994), Purification and properties of the α-acetolactate decarboxylase from *Lactococcus lactis* ssp. *lactis* NCDO-2118, *FEBS Lett.* **351**, 95–99.

PHILLIPS, R. S. (1989), Mechanism of tryptophan indole-lyase: Insights from pre-stady state kinetics and substrate and solvent isotope effects, *J. Am. Chem. Soc.* **111**, 727–730.

PHILLIPS, R. S., VON TERSCH, R. L., SECUNDO, F. (1997), Effects of tyrosine ring fluorination on rates and equilibria of formation of intermediates in the reactions of carbon-carbon lyases, *Eur. J. Biochem.* **244**, 658–663.

PIATIGORSKY, J., WISTOW, G. (1991) The recruitment of crystallins: new functions precede gene duplication, *Science* **252**, 1078–1079.

PICCIRILLI, J. A., ROZZELL, Jr., J. D., BENNER, S. A. (1987), The stereospecificity of oxaloacetate decarboxylase: A stereochemical imperative?, *J. Am. Chem. Soc.* **109**, 8084–8085.

PICKERSGILL, R., JENKINS, J., HARRIS, G., NASSER, W., ROBERT-BAUDOUY, J. (1994), The structure of *Bacillus subtilis* pectate lyase in complex with calcium *Struct. Biol.* **1**, 717–723.

PICKERSGILL, R., SMITH, D., WORBOYS, K., JENKINS, J. (1998), Crystal structure of polygalacturonase from *Erwinia carotovora* ssp. *carotovora*, *J. Biol. Chem.* **273**, 24660–24664.

PIEPER, P. A., GUO, Z., LIU, W.-W. (1995), Mechanistic studies of the biosynthesis of 3,6-dideoxy sugars: stereochemical analysis of C-3 deoxygenation, *J. Am. Chem. Soc.* **117**, 5158–5159.

PIEPER, P. A., YANG, D. Y., ZHOU, H. Q., LIU, H.-W. (1997), 3-deoxy-3-fluoropyridoxamine 5′-phosphate: Synthesis and chemical and biological properties of a coenzyme B_6 analog, *J. Am. Chem. Soc.* **119**, 1809–1817.

PILNIK, W., VORAGEN, A. G. J. (1992), Gelling agents (pectins) from plants for the food industry, in: *Advances in Plant Cell Biochemistry and Biotechnology* Vol. 1, (MORRISON, I., Ed.), pp. 219–270. London: JAI Press.

PIRRUNG, M. C., HOLMES, C. P., HOROWITZ, D. M., NUNN, D. S. (1991), Mechanism and stereochemistry of α,β-dihydroxyacid dehydratase, *J. Am. Chem. Soc.* **113**, 1020–1025.

PISHKO, E. J., ROBERTUS, J. D. (1993), Site directed alteration of 3 active site residues of a pyruvoyl-dependent histidine decarboxylase, *Biochemistry* **32**, 4943–4948.

PLETNEV, S. V., ANTSON, A. A., SINITSYNA, N. I., DAUTER, Z., ISUPOV, M. N. et al. (1997), Crystallographic study of tyrosine phenol-lyase from *Erwinia herbicola*, *Crystallogr. Rep.* **42**, 809–819.

PLETNEV, S. V., ISUPOV, M. N., DAUTER, Z., WILSON, K. S., FALEEV, N. G. et al. (1996), Purification and crystals of tyrosine phenol-lyase from *Erwinia herbicola*, *Biochem. Mol. Biol. Int.* **38**, 37–42.

POCHLAUER, P. (1998), Syntheses of homochiral cyanohydrins in an industrial environment: hydroxy nitrile lyases offer new options, *Chimica Oggi (Chemistry Today)* **16**, 15–19.

POHL, M., MESCH, K., RODENBROCK, A., KULA, M. R. (1995), Stability investigations on the pyruvate decarboxylase from *Zymomonas mobilis*, *Biotechnol. Appl. Biochem.* **22**, 95–105.

POLAK, J., BRZESKI, H. (1990), Isolation of the tyrosine phenol-lyase gene from *Citrobacter freundii*, *Biotechnol. Lett.* **12**, 805–810.

POLLARD, J. R., RICHARDSON, S., AKHTAR, M., LASRY, P., NEAL, T. et al. (1999), Mechanism of 3-methylaspartase probed using deuterium and solvent isotope effects and active-site directed reagents: Identification of an essential cysteine residue, *Bioorg. Med. Chem.* **7**, 949–975.

PORTER, D. T. J., BRIGHT, H. J. (1980), 3-Carbanionic substrate analogs bind very tightly to fumarase and aspartase, *J. Biol. Chem.* **255**, 4772–4780.

POUPON, A., JEBAI, F., LABESSE, G., GROS, F., THIBAULT, J. et al. (1999), Structure modeling and site directed mutagenesis of the rat aromatic L-amino acid pyridoxal 5′-phosphate dependent decarboxylase: A functional study, *Proteins Struct. Funct. Genet.* **37**, 191–203.

PRONK, J. T., STEENSMA, H. Y., VAN DIJKEN, J. P. (1996), Pyruvate metabolism in *Saccharomyces cerevisiae*, *Yeast* **12**, 1607–1633.

PROSEN, E., WARD, O. P. (1994), Optimization of the reaction conditions for production of (*S*)-(–)-2-hydroxypropiophenone by *Acinetobacter calcoaceticus*, *J. Ind. Microbiol.* **13**, 287–291.

QIAN, M. Z., EARNHARDT, J. N., WADHWA, N. R., TU, C. K., LAIPIS, P. J., SILVERMAN, D. N. (1999), Proton transfer to residues of basic pK_a during catalysis by carbonic anhydrase, *Biochim. Biophys. Acta* **1434**, 1–5.

QU, K. B., MARTIN, D. L., LAWRENCE, C. E. (1998), Motifs and structural fold of the cofactor binding site of human glutamate decarboxylase, *Protein Sci.* **7**, 1092–1105.

RAMJEE, M. K., GENSCHEL, U., ABELL, C., SMITH, A. G. (1997), *Escherichia coli* L-aspartate-α-decarboxylase: preprotein processing and observation of reaction intermedaites by electrospray mass spectrometry, *Biochem. J.* **323**, 661–669.

RANZI, B. M., RONCHETTI, F., RUSSO, G., TOMA, L. (1981), Stereochemistry of the introduction of the hydrogen atom at C-2 of *cis*-aconitic acid during its transformation into itaconic acid in *Aspergillus terreus*: a 2H N. M. R. approach, *J. Chem. Soc., Chem. Commun.*, 1981, 1050–1051.

RAZAL, R. A., ELLIS, S., SINGH, S., LEWIST, N. G.,

TOWERS, G. H. N. (1996), Nitrogen recycling in phenylpropanoid metabolism, *Phytochemistry* **41**, 31–35.

REID, M. F., FEWSON, C. A. (1994), Molecular characterization of microbial alcohol dehydrogenases, *Crit. Rev. Microbiol.* **20**, 13–56.

REES, D. G., JONES, D. H. (1996), Stability of L-phenylalanine ammonia-lyase in aqueous solution and as the solid state in air and organic solvents, *Enzyme Microb. Technol.* **19**, 282–288.

REES, D. G., JONES, D. H. (1997), Activity of L-phenylalanine ammonia-lyase in organic solvents, *Biochem. Biophys. Acta* **1338**, 121–126.

REGLINSKI, T., STAVELY, F. J. L., TAYLOR, J. L. (1998), Induction of phenylalanine ammonia lyase activity and control of *Sphaeropsis sapinea* infection in *Pinus radiata* by 5-chlorosalicylic acid. *Eur. J. Forestry Pathol.* **28**, 153–158.

REHMAN, A., MCFADDEN, B. A. (1996), The consequences of replacing histidine 356 in isocitrate lyase from *Escherichia coli*, *Arch. Biochem. Biophys.* **336**, 309–315.

RENAULT, P., CALERO, S., DELORME, C., DROUAULT, S., GOUPIL-FEUILLERAT, N. et al. (1998) From genome to industrial application, *Lait* **78**, 39–52.

REYNOLDS, L. J., GARCIA, G. A., KOZARICH, J. W., KENYON, G. L. (1988), Differential reactivity in the processing of *p*-[(halomethyl)benzoyl]formates by benzoylformate decarboxylase, a thiamin pyrophosphate dependent enzyme, *Biochemistry* **27**, 5530–5538.

RHEE, S. K., FUCHS, G. (1999), Phenyl-CoA:acceptor oxidoreductase, a membrane bound molybdenum-iron-sulfur enzyme involved in the anaerobic metabolism of phenylalanine in the denitriying bacterium, *Thauera aromatica*, *Eur. J. Biochem.* **262**, 507–515.

RICHARDS, J. C., SPENSER, I. D. (1983), 2H NMR spectroscopy as a probe of the stereochemistry of enzymic reactions at prochiral centres, *Tetrahedron* **39**, 3549–3568.

RIKIN, A., SCHWARTZBACH, S. D. (1989), Translational regulation of the synthesis of *Euglena* fumarase by light and ethanol, *Plant Physiol.* **90**, 63–69.

ROBERT-BAUDOUV, J., JIMENO-ABENDANO, J., STOEBER, F. (1982), D-Mannonate and D-altronate dehydratases of *Escherichia coli*-K-12, *Methods Enzymol.* **90**, 288–294.

ROBINSON, J. B., SINGH, M., SRERE, P. A. (1976), Structure of the prosthetic group of *Klebsiella aerogenes* citrate (*pro*-3*S*)-lyase, *Proc. Natl. Acad. Sci. USA* **73**, 1872–1876.

RONDAGS, E., GERMAIN, P., MARC, I. (1997), Quantification of extracellular α-acetolactate oxidative decarboxylation in diacetyl production by an α-acetolactate overproducing strain of *Lactococcus lactis* ssp. *lactis* bv. *diacetylactis*, *Biotechnol. Lett.* **21**, 303–307.

RONDAGS, E., GERMAIN, P., MARC, I., (1998), Kinetic studies on α-acetolactic acid extra and intracellular oxidative decarboxylation to diacetyl by *Lactococcus lactis* ssp. *lactis* var. *diacetylactis* SD 933, *Lait* **78**, 135–143.

ROSE, A. H. (1997), Restructuring the active site of fumarase for the fumarate to malate reaction, *Biochemistry* **36**, 12346–12354.

ROSE, A. H. (1998), How fumarase recycles after the malate fumarate reaction. Insights into the the reaction mechanism, *Biochemistry* **37**, 17651–17658.

ROZZELL Jr., J. D., BENNER, S. A. (1984), Stereochemical imperative in enzymatic decarboxylations. Stereochemical course of the decarboxylation catalyzed by acetoacetate decarboxylase, *J. Am. Chem. Soc.* **106**, 4937–4941.

RUAN, Z. S., ANANTHARAM, V., CRAWFORD, I. T., AMBUDKAR, S. V., RHEE, S. Y. et al. (1992), Identification, purification and reconstitution of OxlT, the oxalate–formate antiport protein of *Oxalobacter formigenes*, *J. Biol. Chem.* **267**, 10537–10543.

RUSSELL, R. N., YU, Y. (1991), Stereochemical and mechanistic studies of the CDP-D-glucose oxidoreductase isolated from *Yersinia pseudotuberculosis*, *J. Am. Chem. Soc.* **113**, 7777–7778.

SACCHETTINI, J. C., MEININGER, T., RODERICK, S., BANAZAK, L. J. (1986), Purification, crystallization, and preliminary X-ray data for porcine fumarase, *J. Biol. Chem.* **261**, 15183–15185.

SACHS, M. M., SUBBAIAH, C. C., SAAB, I. N. (1996), Anaerobic gene expression and flooding tolerance in maize, *J. Exp. Bot.* **47**, 1–15.

SACKSTEDER, K. A., MORELL, J. C., WANDERS, R. J. A., MATALON, R., GOULD, S. J. (1999), MCD encodes peroxisomal and cytoplasmic forms of malonyl-CoA decarboxylase and is mutated in malonyl-CoA decarboxylase deficiency, *J. Biol. Chem.* **274**, 24461–24468.

SALVARREY, M. S., CAZZULO, J. J., CANNATA, J. J. B. (1995), Effects of divalent cations and nucleotides on the (CO_2)-^{14}C-oxaloacetate exchange catalyzed by the phosphoenol pyruvate carboxykinase from the moderate halophile, *Vibrio costicola*, *Biochem. Mol. Biol. Int.* **36**, 1225–1234.

SANO, H, TADA, T, MORIYAMA, A., OGAWA, H., ASAI, K. et al. (1997), Isolation of a rat histidase cDNA sequence and expression in *Escherichia coli* – Evidence of extrahepatic/epidermal distribution, *Eur. J. Biochem.* **250**, 212–221.

SANTANGELO, J. D., DUERRE, P. (1996), Microbial production of acetone and butanol: Can history be repeated? *Chimica Oggi (Chemistry Today)* **14**, 29–35.

SANTA-MARIA, J. R., COLL, R., FUENTESPINA, E, (1993), Comparative study of two commerical kits for determining oxalate in urine, *Clin. Biochem.* **26**, 93–96.

SANTANIELLO, E., MANZOCCHI, A., BIONDI, P. A.

(1981), On the stereochemistry of the decarboxylation of (2*S*)-histidine catalyzed by histidine decarboxylase from *Clostridium welchii* (EC 4.1.1.22), *J. Chem. Soc., Perkin Trans. I*, 307–309.

SARIBAS, A. S., SCHINDLER, J. F., VIOLA, R. E. (1994), Mutagenic investigation of conserved functional amino acids in *Escherichia coli* L-aspartase. *J. Biol. Chem.* **269**, 6313–6319.

SAUER, U., DUERRE, P. (1995), Differential induction of genes related to solvent formation during the shift from acidogenesis to solventogenesis in continuous culture of *Clostridium acetobutylicum*, *FEMS Microbiol. Lett.* **125**, 115–120.

SCAPIN, G., BLANCHARD, J. S. (1998), Enzymology of bacterial lysine biosynthesis, *Adv. Enzymol.* **72**, 279–327.

SCHELLENBERGER, A. (1998), Sixty years of thiamin diphosphate biochemistry, *Biochim. Biophys. Acta.* **1358**, 177–186.

SCHINDLER, J. F., VIOLA, R. E. (1994), Mechanism-based inactivation of L-aspartase from *E. coli*, *Biochemistry* **33**, 9365–9370.

SCHMID, M., BERG, M., HILBI, H., DIMROTH, P. (1996), Malonate decarboxylase of *Klebsiella pneumoniae* catalyzes the turnover of acetyl and malonyl thioester residues on a coenzyme-A like prosthetic group, *Eur. J. Biochem.* **237**, 221–228.

SCHNEIDER, S., MOHAMED, M. E., FUCHS, G. (1997), Anaerobic metabolism of L-phenylalanine via benzoyl-CoA in the denitrifying bacterium *Thauera aromatica*, *Arch. Microbiol.* **168**, 310–320.

SCHORKEN, U., SPRENGER, G. A. (1998), Thiamin-dependent enzymes as catalysts in chemoenzymatic syntheses, *Biochim. Biophys. Acta.* **1385**, 229–243.

SCHUSTER, B., RETEY, J. (1994), Serine-202 is the putative precursor of the active-site dehydroalanine of phenylalanine ammonia-lyase – Site directed mutagenesis studies on the enzyme from parsley (*Petroselinum crispum* L.), *FEBS Lett.* **349**, 252–254.

SCHUSTER, B., RETEY, J. (1995), The mechanism of action of phenylalanine ammonia-lyase – the role of prosthetic dehydroalanine, *Proc. Natl. Acad. Sci. USA* **92**, 8433–8437.

SCHWARTZ, H. L., CHANDONIA, J. M., KASH, S. F., KANAANI, J., TUNNELL, E. et al. (1999), High resolution autoreactive epitope mapping and structural modeling of the 65 kDa form of human glutamic acid decarboxylase, *J. Mol. Biol.* **287**, 983–999.

SCHWEDE, T. F., RETEY, J., SCHULZ, G. E. (1999), Crystal structure of histidine ammonia lyase revealing a novel polypeptide modification as the catalytic electrophile, *Biochemistry* **38**, 5355–5361.

SCOPES, R. K. (1997), Ethanol from biomass: The potential use of thermophilic organisms in fermentation, *Australasian Biotechnol.* **7**, 296–299.

SEELY, J. E., PEGG, A. E. (1983), Ornithine decarboxylase (mouse kidney), *Methods Enzymol.* **94**, 158–161.

SEELY, J. E., POSO, H., PEGG, A. E. (1983), Labeling and quantitation of ornithine decarboxylase protein by reaction with α-[5-^{14}C]difluoromethylornithine, *Methods Enzymol.* **94**, 206–209.

SENUMA, M., OTSUKI, O., SAKATA, N., FURUI, M., TOSA, T. (1989) Industrial production of D-aspartic acid and L-alanine from DL-aspartic acid using a pressurized column reactor containing immobilized *Pseudomonas dacunhae* cells, *J. Ferment. Bioeng.* **67**, 233–237.

SENWO, Z. N., TABATABAI, M. A. (1996), Aspartase activity of soils. *Soil Sci. Soc. Am. J.* **60**, 1416–1422.

SENWO, Z. N., TABATABAI, M. A. (1999), Aspartase activity in soils: effects of trace elements and relationships to other amidohydrolases, *Soil Biol. Biochem.* **31**, 213–219.

SEREBRENNIKOV, V. M., KISRIEVA, Y. S., ZAGUSTINA, N. A., SEMINIKHINA, V. F., BEZBORODOV, A. M. (1999), The ability of diacetyl-producing lactic acid bacteria of the genus *Lactococcus* to release α-acetolactic acid into the medium, *Appl. Biochem. Microbiol.* **35**, 612–620.

SHARMA, S., NATH, R., THIND, S. K. (1993), Recent advances in the measurement of oxalate in biological materials, *Scanning Microscopy* **7**, 431–441.

SHELP, B. J., BOWN, A. W., MCLEAN, M. D. (1999), Metabolism and functions of γ-aminobutyric acid, *Trends Plant Sci.* **4**, 446–452.

SHEN, B. W., HENNIG, M., HOHENESTER, E., JANSONIUS, J. N., SCHIRMER, T. (1998), Crystal structure of human recombinant ornithine aminotransferase, *J. Mol. Biol.* **277**, 81–102.

SHERIDAN, R. P., ALLEN, L. C. (1981), The active site electrostatic potential of human carbonic anhydrase, *J. Am. Chem. Soc.* **103**, 1544–1550.

SHI, W., KIDD, R., GIORGIANNI, F., SCHINDLER, J. F., VIOLA, R. E., FARBER, G. K. (1993), Crystallization and preliminary X-ray studies of L-aspartase from *Escherichia coli*, *J. Mol. Biol.* **234**, 1248–1249.

SHI, W. X., DUNBAR, J., JAYASEKERA, M. M. K., VIOLA, R. E., FARBER, G. K. (1997), The structure of L-aspartate ammonia-lyase from *Escherichia coli*, *Biochemistry* **36**, 9136–9144.

SHIBATA, H., GARDINER, W. E., SCHWARTZBACH, S. D. (1985), Purification and characterization and immunological properties of fumarase form *Euglena gracilis* var. *bacillaris*, *J. Bacteriol.* **164**, 762–768.

SHIN, H. S., ROGERS, P. L. (1996a), Production of L-phenylacetylcarbinol (L-PAC) from benzaldehyde using partially purified pyruvate decarboxylase (PDC), *Biotechnol. Bioeng.* **49**, 52–62.

SHIN, H. S., ROGERS, P. L. (1996b), Kinetic evaluation

of biotransformation of benzaldehyde to L-phenylacetylcarbinol by immobilized pyruvate decarboxylase from *Candida utilis*, *Biotechnol. Bioeng.* **49**, 429–436.

SHUKLA, V. B., KULKARNI, P. R. (1999), Downstream processing of biotransformation broth for recovery and purification of L-phenyl acetyl carbinol (L-PAC), *J. Sci. Ind. Res.* **58**, 591–593.

SIDDIQUI, M. A., FUJIWARA, S., IMANAKA, T. (1997), Indolepyruvate ferredoxin oxidoreductase from *Pyrococcus* sp. K0D1 possesses a mosaic: Structure showing features of various oxidoreductases, *Mol. Gen. Genet.* **254**, 433–439.

SIDDIQUI, M. A., FUJIWARA, S., TAKAGI, M., IMANAKA, T. (1998), *In vitro* heat effect on heterooligomeric subunit assembly of thermostable indolepyruvate ferredoxin oxidoreductase, *FEBS Lett.* **434**, 372–376.

SIDHU, H., ALLISON, M., PECK, A. B. (1997), Identification and classification of *Oxalobacter formigenes* strains using oligonucleotide probes and primers, *J. Clin. Microbiol.* **35**, 350–353.

SIDHU, H., HOPPE, B., HESSE, A., TENBROCK, K., BROMME, S. et al. (1998), Absence of *Oxalobacter formigenes* in cystic fibrosis patients: a risk factor for hyperoxaluria, *Lancet* **352**, No. 9133, 1026–1029.

SIDHU, H., SCHMIDT, M. E., CORNELIUS, J. G., THAMILSELVAN, S., KHAN, S. R. et al. (1999), Direct correlation between hyperoxaluria/oxalate stone disease and the absence oif the gastrointestinal tract dwelling bacterium *Oxalobacter formigenes*, *J. Am.. Soc. Nephrol.* **10**, S334–S340.

SIMPSON, A., BATEMAN, O., DRIESSEN, H., LINDLEY, P., MOSS, D. et al. (1994), The structure of avian eye lens δ-crystallin reveals a new fold for a superfamily of oligomeric enzymes, *Nature Struct. Biol.* **1**, 724–733.

SLOAN, M. J., PHILLIPS, R. S. (1996), Effects of α-deuteration and of aza and thia analogs of L-tryptophan on formation of intermedaites in the reaction of *Escherichia coli* tryptophan indole lyase, *Biochemistry* **35**, 16165–16173.

SMITH, D. M., THOMAS, N. R., GANI, D. (1991), A comparison of the pyridoxal 5′-phosphate dependent decarboxylase and transaminase enzymes at a molecular level, *Experientia* **47**, 11–12.

SONG, Y. H., NIEDERER, D., LANEBELL, P. M., LAM, L. K. P., CRAWLEY, S. et al. (1994), Stereospecific synthesis of phosphonate analogs of diaminopimelic acid (DAP) their interaction with DAP enzymes and antibacterial activity of peptide derivatives, *J. Org. Chem.* **59**, 5784–5793.

SOOKKHEO, B., PHUTAKUL, S., CHEN, S. T., WANG, K. T. (1998), Aspartase catalyzed preparative scale synthesis of ^{15}N-aspartic acid, *J. Chin. Chem. Soc.* **45**, 525–528.

SNELL, E. E. (1986), Pyruvoyl-dependent histidine-decarboxylase from *Lactobacillus*-30a – purification and properties, *Methods Enzymol.* **122**, 128–135.

SNELL, E. E. (1993), From bacterial nutrition to enzyme structure – A personal odyssey, *Ann. Rev. Biochem.* **62**, 1–27.

SNELL, E. E., GUIRARD, B. M. (1986), Pyridoxal phosphate-dependent histidine decarboxylase from *Morganella am*-15, *Methods Enzymol.* **122**, 139–143.

SNELL, E. E., HUYNH, Q. K. (1986), Prohistidine decarboxylase from *Lactobacillus*-30a, *Methods Enzymol.* **122**, 135–138.

SNIPES, C. E., BRILLINGER, G.-U., SELLERS, L., MASCARO, L., FLOSS, H. G. (1977), Stereochemistry of the dTDP-glucose oxidoreductase reaction, *J. Biol. Chem.* **252**, 8113–8117.

SOURKES, T. L. (1987), Aromatic-L-amino acid decarboxylase, *Methods Enzymol.* **142**, 170–178.

SPAMPINATO, C. P., ANDREO, C. S. (1995), Kinetic mechanism of NADP-malic enzyme from maize leaves, *Photosynth. Res.* **43**, 1–9.

SPENCER, J. B., JORDAN, P. M. (1990), Use of chiral malonates to determine the absolute configuration of the hydrogen atoms eliminated during the formation of 6-methylsalicyclic acid by 6-methylsalicyclic acid synthase from *Penicillin patulum, J. Chem. Soc., Chem. Commun.*, 1704–1706.

SPENCER, J. B., JORDAN, P. M. (1992a), Stereospecific elimination of hydrogen atoms with opposite absolute orientations during the biosynthesis of orsellinic acid from chiral malonates in *Penicillin cyclopium, J. Chem. Soc., Chem. Commun.*, 646–648.

SPENCER, J. B., JORDAN, P. M. (1992b), Investigation of the mechanism and steric course of the reaction catalyzed by 6-methylsalicylic acid synthase from *Penicillium patulum* using (*R*)-[1-C-13,2-H-2]malonate and (*S*)-[1-C-13,2-H-2]malonate, *Biochemistry* **31**, 9107–9116.

SPRENGER, G. A., POHL, M. (1999), Synthetic potential of thiamin diphosphate-dependent enzymes, *J. Mol. Catal. B Enzymatic* **6**, 145–159.

STIM-HERNDON, K. P., PETERSEN, D. J., BENNETT, G. N. (1995), Characterization of an acetyl-CoA C-acetyltransferase (thiolase) gene from *Clostridium acetobutylicum* ATCC-824, *Gene* **154**, 81–85.

STIM-HERNDON, K. P., NAIR, R., PAPOUTSAKIS, E. T., BENNETT, G. N. (1996), Analysis of degenerate variants of *Clostridium acetobutylicum* ATCC-824, *Anaerobe* **2**, 11–18.

STIRLING, D. I. (1992), The use of aminotransferases for the production of chiral amino acids and amines, in: *Chirality in Industry* (COLLINS, A. N., SHELDRAKE, G. N., CROSBY J., Eds.), pp. 208–222. Chichester: John Wiley & Sons.

SU, P. H., LIN, C. H. (1996), Metabolic responses of luffa roots to long-term flooding, *J. Plant Physiol.* **148**, 735–740.

SUCHI, M., HARADA, N., WADA, Y., TAKAGI, Y. (1993), Molecular cloning of a cDNA-encoding human histidase, *Biochem. Biophys. Acta* **1216**, 293–295.

SUN, S. X., DUGGLEBY, R. G., SCHOWEN, R. L. (1995), Linkage of catalysis and regulation in enzyme action – carbon-isotope effects, solvent isotope effects, and proton inventories for the unregulated pyruvate decarboxylase of *Zymomonas mobilis*, *J. Am. Chem. Soc.* **117**, 7317–7322.

SUN, S. X., SMITH, G. S., O'LEARY, M. H., SCHOWEN, R. L. (1997), The linkage of catalysis and regulation in enzyme action. Fluoropyruvate as a probe of regulation in pyruvate decarboxylases, *J. Am. Chem. Soc.* **119**, 1507–1515.

SUN, S. X., ZABINSKI, R. F., TONEY, M. D. (1998a), Reactions of alternate substrates demonstrate stereoelectronic control of reactivity in dialkylglycine decarboxylase, *Biochemistry* **37**, 3865–3875.

SUN, S. X., BAGDASSARIAN, C. K., TONEY, M. D. (1998b), Pre-steady-state kinetic analysis of the reactions of alternative substrates with dialkylglycine decarboxylase, *Biochemistry* **37**, 3876–3885.

SUNDARARAJU, B., ANTSON, A. A., PHILLIPS, R. S., DEMIDKINA, T. V., BARBOLINA, M. V. et al. (1997), The crystal structure of *Citrobacter freundii* tyrosine phenol-lyase complexed with 3-(4′-hydroxyphenyl)propionic acid, together with site directed mutagenesis and kinetic analysis, demonstrates that arginine 381 is required for substrate specificity, *Biochemistry* **36**, 6502–6510.

SUTTON, C. R., KING, H. G (1962), Inhibition of leucine decarboxylase by thiol-binding reagents, *Arch. Biochem. Biophys.* **96**, 360–370.

SUZUKI, T., SATO, M., YOSHIDA, T., TUBOI, S. (1989), Rat liver mitochondrial and cystolic fumarases with identical amino acids sequences are encoded from a single gene, *J. Biol. Chem.* **264**, 2581–2586.

SUZUKI, S., HIRAHARA, T., SHIM, J. K., HORINOUCHI, S., BEPPU, T. (1992), Purification and properties of thermostable β-tyrosinase from an obligately symbiotic thermophile, *Symbiobacterium thermophilum*, *Biosci. Biotech. Biochem.* **56**, 84–89.

SWANSON, T., BROOKS, H. B., OSTERMAN, A. L., O'LEARY, M. H., PHILLIPS, M. A. (1998), Carbon-13 isotope effect studies of *Trypanosoma brucei* ornithine decarboxylase, *Biochemistry* **37**, 14943–14947.

TABIB, A., BACHRACH, U. (1999), Role of polyamines in mediating malignant transformation and oncogene expression, *Int. J. Biochem. Cell Biol.* **31**, 1289–1295.

TADEGE, M., DUPUIS, I., KUHLEMEIER, C. (1999), Ethanolic fermentation: new functions for an old pathway, *Trends Plant Sci.* **4**, 320–325.

TAKAC, S., AKAY, BULENT, OZDAMAR, T. H. (1995), Bioconversion of *trans*-cinnamic acid to L-phenylalanine by L-phenylalanine ammonia-lyase of *Rhodotorula glutinis*: parameters and kinetics, *Enzyme Microb. Technol.* **17**, 445–452.

TAKAHASHI, R., KAWASAKI, M., SONE, H., YAMANO, S. (1995), Genetic modification of brewers' yeast to produce acetolactate decarboxylase and the safety aspects of the beer brewed by the transformed yeast, *ACS Symp. Ser.* **605**, 171–180.

TAKAMURA, Y., KITAYAMA, Y. (1981), Purification and some properties of the malonate decarboxylase from *Pseudomonas ovalis*: an oligomeric enzyme with bifunctional properties, *Biochem. Int.* **3**, 483–491.

TAKATSUKA, Y., ONODA, M., SUGIYAMA, T., TOMITA, T., KAMINO, Y. (1999a), Novel characteristics of *Selenomonas ruminantium* lysine decarboxylase capable of decarboxylating both L-lysine and L-ornithine, *Biosci. Biotech. Biochem.* **63**, 1063–1069.

TAKATSUKA, Y., TOMITA, T., KAMINO, Y. (1999b), Identification of the amino acids residues conferring specificity upon *Selenomonas ruminantium* lysine decarboxylase, *Biosci. Biotech. Biochem.* **63**, 1843–1846.

TANAKA, I., TOSA, T., CHIBATA, I. (1983a), Stabilization of fumarase activity of *Brevibacterium flavum* cells by immobilization with κ-carrageenan, *Appl. Biochem. Biotechnol.* **8**, 31–38.

TANAKA, I., TOSA, T., CHIBATA, I. (1983b), Reasons for the high-stability of fumarase activity of *Brevibacterium flavum* cells immobilized with κ-carrageenan gel, *Appl. Biochem. Biotechnol.* **8**, 39–54.

TANAKA, I., TOSA, T., CHIBATA, I. (1983c), Effect of growth-phase on stability of fumarase activity of *Brevibacterium flavum* cells immobilized with κ-carrageenan, *Agric. Biol. Chem.* **47**, 1289–1296.

TANAKA, I., TOSA, T., CHIBATA, I. (1984), Stability of fumarase activity of *Brevibacterium flavum* immobilized with κ-carrageenan and chinese gallotannin, *Appl. Microbiol. Biotechnol.* **19**, 85–90.

TANAKA, F., LERNER, R. A., BARBAS, C. F. (1999), Thiazolium-dependent catalytic antibodies produced using a covalent modification strategy, *Chem. Commun.* **15**, 1383–1384.

TARI, L. W., MATTE, A., PUGAZHENTHI, U., GOLDIE, H., DELBAERE, L. T. J. (1996), Snapshot of an enzyme reaction intermediate in the structure ofthe ATP-Mg^{2+}-oxalate ternary complex of *Escherichia coli* PEP carboxykinase, *Nature Struct. Biol.* **3**, 355–363.

TAYLOR, R. G., LAMBERT, M. A., SEXSMITH, E., SADLER, S. J., RAY, P. N. et al. (1990), Cloning and expression of rat histidase, *J. Biol. Chem.* **265**, 18192–18199.

TAYLOR, R. G., LEVY, H. L., MCINNES, R. R. (1991), Histidase and histidinemia – clinical and molecular considerations, *Mol. Biol. Med.* **8**, 101–116.

TAYLOR, R. G., MCINNES, R. R. (1994), Site-directed mutagenesis of conserved serines in rat histidase-identification of serine-254 as an essential active-site residue, *J. Biol. Chem.* **269**, 27473–27477.

THATCHER, B. J., DOHERTY, A. E., ORVISKY, E., MARTIN, B. M., HENKIN, R. I. (1998), Gustin from human parotid saliva is carbonic anhydrase VI, *Biochem. Biophys. Res. Commun.* **250**, 635–641.

THOMPSON, M. W., STROHL, W. R., FLOSS, H. G. (1992) Purification and characterization of TDP-D-glucose 4,6-dehydratase from anthracycline-producing streptomycetes, *J. Gen. Microbiol.* **138**, 779–786.

THORSON, J. S., KELLY, T. M., LIU, H.-W. (1994a), Cloning, sequencing and overexpression in *Escherichia coli* of the α-D-glucose-1-phosphate cytidylyltransferase gene isolated from *Yersinia pseudotuberculosis*, *J. Bacteriol.* **176**, 1840–1849.

THORSON, J. S., LO, S. F., PLOUX, O., HE, X. M., LIU, H. W. (1994b), Studies of the biosynthesis of 3,6-dideoxyhexoses – molecular cloning and characterization of the ASC (ascrylose) region from *Yersinia pseudotuberculosis*, *J. Bacteriol.* **176**, 5483–5493.

TILLEY, K., AKHTAR, M., GANI, D. (1992) The stereochemical course of reactions catalyzed by *Escherichia coli* glutamic acid decarboxylase, *J. Chem. Soc., Chem. Commun.*, 68–71.

TILLEY, K., AKHTAR, M., GANI, D. (1994), The stereochemical course of decarboxylation, transamination and elimination reactions catalyzed by *Escherichia coli* glutamic acid decarboxylase, *J. Chem. Soc., Perkin. Trans.*1 1994, 3079–3087.

TOH, H., KONDO, H., TANABE, T. (1993), Molecular evolution of biotin-dependent carboxylases, *Eur. J. Biochem.* **215**, 687–696.

TONEY, M. D., HOHENESTER, E., KELLER, J. W., JANSONIUS, J. N. (1995), Structural and mechanistic analysis of 2 refined crystal structures of the pyridoxal phosphate dependent enzyme, dialkylglycine decarboxylase, *J. Mol. Biol.* **245**, 151–179.

TORSELL, K. B. G. (1983), *Natural Product Chemistry: A Mechanistic and Biosynthetic Approach to Secondary Metabolism*, pp. 80-88. Chichester: John Wiley & Sons.

TOSA, T., TANAKA, I., CHIBATA, I. (1982), Stabilization of fumarase activity of *Brevibacterium flavum* cells by immobilization with κ-carrageenan and polyethyleneimine, *Enzyme Eng.* **6**, 237–238.

TOTH, C., COFFINO, P. (1999), Regulated degradation of yeast ornithine decarboxylase, *J. Biol. Chem.* **274**, 25921–25926.

TOTH, J., ISMAIEL, A. A., CHEN, J. S. (1999), The ald gene, encoding a coenzyme A-acylating aldehyde dehydrogenase, distinguishes *Clostridium beijerinckii* and two other solvent producing clostridia from *Clostridium acetobutylicum*, *Appl. Environ. Microbiol.* **65**, 4973–4980.

TRAMPER, J. VERMUE, M. H., BEEFTINK, H. H., VON STOCKAR, U. (Eds.) (1992), *Biocatalysis in Nonconventional Media*. Amsterdam: Elsevier.

TRIPATHI, C. M., AGARWAL, S. C., BASU, S. K. (1997), Production of L-phenylacetylcarbinol by fermentation, *J. Ferment. Bioeng.* **84**, 487–492.

TSENG, C. P. (1997), Regulation of fumarase (*fumB*) gene expression in *Escherichia coli* in response to oxygen, iron and heme availability: role of the *arcA*, *fur*, and *hemA* gene products, *FEMS Microbiol. Lett.* **157**, 67–72.

TSOU, A. Y., RANSOM, S. C., GERLT, J. A., BEUCHTER, D. D., BABBITT, P. C., KENYON, G. L. (1990), Mandelate pathway of *Pseudomonas putida* – sequence relationships involving mandelate racemase, (*S*)-mandelate dehydrogenase and benzylformate decarboxylase and expression of benzylformate decarboxylase in *Escherichia coli*, *Biochemistry* **29**, 9856–9862.

TUBOI, S., SUZUKI, T., SATO, M., YOSHIDA, T. (1990), Rat liver mitochondrial and cytosolic fumarases with identical amino acid sequences are encoded from a single mRNA with two alternative in-phase AUG initiation sites, *Adv. Enzyme Regul.* **30**, 289–304.

TURNER, M. A., SIMPSON, A., MCINNES, R. R., HOWELL, P. L. (1997) Human arginosuccinate lyase: a structural basis for intragenic combination, *Proc. Natl. Acad. Sci. USA.* **94**, 9063–9068.

TYAGI, A. K., TABOR, C. W., TABOR, H. (1983), Ornithine decarboxylase (*Saccharomyces cerevisiae*), *Methods Enzymol.* **94**, 135–139.

UEDA, Y., YUMOTO, N., TOKUSHIGE, M., FUKUI, K., OHYA-NISHIGUCHI, H. (1991), Purification and characterization of two types of fumarases from *Escherichia coli*, *J. Biochem.* **109**, 728–733.

UGWUANYI, J. O., OBETA, J. A. N. (1997), Some pectinolytic and cellulolytic enzyme activities of fungi causing rots of cocoyams, *J. Sci. Food Agr.* **73**, 432–436.

URBAN, J. J., VON TERSCH, R. L. (1999), A computational study of charge delocalization and ring fluoro substituent effects in 4-fluoromethylphenoxides, *J. Org. Chem.* **64**, 3409–3416.

VACCARO, J. A., CRANE, E. J., HARRIS, T. K., WASHABAUGH, M. W. (1995), Mechanism of reconstitution of brewers' yeast pyruvate decarboxylase with thiamin diphosphate and magnesium, *Biochemistry* **34**, 12636–12644.

VALLEE, F., TURNER, M. A., LINDLEY, P. L., HOWELL, P. L. (1999), Crystal structure of an inactive duck δ-II-crystallin mutant with bound arginosuccinate, *Biochemistry* **38**, 2425–2434.

VAN DEN HOMBERGH, J. P. T. W., FRAISSINET-TACHET, L., VAN DE VONDERVOORT, P. J. I., VISSER, J. (1997), Production of the homologous pectin lyase B protein in six genetically defined protease deficient *Aspergillus niger* mutant strains, *Curr. Genet.* **32**, 73–81.

VAN DER WERF, M., VAN DEN TWEEL, W. J. J., HARTMANS, S. (1992), Screening for microorganisms producing D-malate from maleate, *Appl. Environ. Microbiol.* **58**, 2854–2860.

VAN DER WERF, M. J., VAN DEN TWEEL, W. J. J., HARTMANS, S. (1993), Purification and characterization of maleate hydratase from *Pseudomonas pseudoalcaligenes*, *Appl. Environ. Microbiol.* **59**, 2823–2829.

VAN RENSBURG, P., VANZYL, W. H., PRETORIUS, I. S. (1994), Expression of the *Butyrivibrio fibrisolvens endo-β*-1,4-glucanase gene together with the *Erwinia* pectate lyase and polygalacturonase genes in *Saccharomyces cerevisiae*, *Curr. Genet.* **27**, 17–22.

VAN WAARDE, A. (1991) Alcoholic fermentation in multicellular organisms, *Physiol. Zool.* **64**, 895–920.

Various (1998), *Biochim. Biophys. Acta* **1385**, 175–419.

VEIGA, L. A., GUIMARAES, M. F. (1991), L-Fuconate dehydratase – purification and properties of the enzyme from *Pullularia pullulans*, *Arquivos de Biologia e Tecnologia* **34**, 536–553.

VOLTATTORNI, C. B., GIARTOSIO, A., TURANO, C. (1987), Aromatic-L-amino acid decarboxylase from pig-kidney, *Methods Enzymol.* **142**, 179–187.

VON TERSCH, R. L., SECUNDO, F., PHILLIPS, R. S., NEWTON, M. G. (1996), Preparation of fluorinated amino acids with tyrosine phenol lyase – Effects of fluorination on reaction kinetics and mechanism of tyrosine phenol lyase and tyrosine protein kinase Csk, *ACS Symp. Ser.* **639**, 95–104.

WAAGEPETERSEN, H. S., SONNEWALD, U., SCHOUSBOE, A. (1999), The GABA paradox: Multiple roles as metabolite, neurotransmitter and neurodifferentitation agent, *J. Neurochem.* **73**, 1335–1342.

WADA, M., SAKAMOTO, M., KATAOKA, M., LIU, J. Q., YAMADA, H., SHIMIZU, S. (1998), Distribution of threonine aldolase activity with different stereospecificities in aerobic bacteria, *Biosci. Biotech. Biochem.* **62**, 1586–1588.

WAGNER, U. G., HASSLACHER, M., GRIENGL, K., SCHWAB, H., KRATKY, C. (1996), Mechanism of cyanogenesis: The crystal structure of hydroxynitrile lyase from *Hevea brasiliensis*, *Structure* **4**, 811–822.

WAJANT, H., EFFENBERGER, F. (1996), Hydroxynitrile lyases of higher plants, *Biol. Chem. Hoppe-Seyler* **377**, 611–617.

WANG, S.-F., GABRIEL, O. (1970), Biological mechanisms involved in the formation of deoxysugars, *J. Biol. Chem.* **245**, 8–14.

WANG, X. H., GONG, C. S., TSAO, G. T. (1998), Production of L-malic acid during biocatalysis employing wild-type and respiratory deficient yeast, *Appl. Biochem. Biotechnol.* **70-2**, 845–852.

WARD, O. P., (1995), Application of bakers' yeast in bioorganic synthesis, *Can. J. Bot.* **73**, S1043–S1048.

WATANABE, S. K., HERNANDEZ-VELAZCO, G., ITURBE-CHINAS, F., LOPEZ-MUNGIA, A. (1992), Phenylalanine ammonia lyase from *Sporidiobolus pararoseus* and *Rhodosporidium toruloides*: Application for phenylalanine and tyrosine deamination, *World J. Microbiol. Biotechnol.* **8**, 406–410.

WEAVER, T., BANASZAK, L. (1996), Crystallographic studies of the catalytic and a second site in fumarase C from *Escherichia coli*, *Biochemistry*, **35**, 13955–13965.

WEAVER, T., LEES, M., BANASZAK, L. (1997) Mutations of fumarase that distinguish between the active site and a nearby dicarboxylic acid binding site, *Protein Sci.* **6**, 834–842.

WEAVER, T. M., LEVITT, D. G., DONNELLY, M. I., STEVENS, P. P. W., BANASZAK, L., (1995) The multisubunit active-site of fumarase-C from *Escherichia coli*, *Nature Struct. Biol.* **2**, 654–662.

WEAVER, T., LEES, M., ZAITSEV, V., ZAITSEVA, I., DUKE, E. et al. (1998) Crystal structures of native and recombinant yeast fumarase, *J. Mol. Biol.* **280**, 431–442.

WEIGEL, T. M, MILLER, V. P., LIU, H.-W. (1992), Mechanistic and stereochemical studies of a unique dehydration catalyzed by CDP-4-keto-6-deoxy-D-glucose-3-dehydrase – a pyridoxamine 5′-phosphate dependent enzyme isolated from *Yersinia pseudotuberculosis*, *Biochemistry* **31**, 2140–2147.

WEISS, P. M., GARCIA, G. A., KENYON, G. L., CLELAND, W. W., COOK, P. F. (1988), Kinetics and mechanism of benzoylformate decarboxylase using C-13 and solvent deuterium effects on benzylformate and benzylformate analogs, *Biochemistry* **27**, 2197–2205.

WESTHEIMER, F. H. (1995), Coincidences, decarboxylation and electrostatic effects, *Tetrahedron* **51**, 3–20.

WHITE, P. J., KENDRICK, K. E. (1993), Inactivation of histidine ammonia-lyase from *Streptomyces griseus* by dicarbonyl reagents, *Biochim. Biophys. Acta* **1163**, 273–279.

WIECZOREK, S. J., KALIDOVA, K. A., CLIFTON, J. G., RINGE, D., PETSKO, G. A., GERLT, J. A. (1999), Evolution of enzyme activities in the enolase superfamily: Identification of a "new" general acid catalyst in the active site of D-galactonate dehydratase from *Escherichia coli*, *J. Am. Chem. Soc.* **121**, 4540–4541.

WIGHTMAN, R. H., STAUNTON, J., BATTERSBY, A. R. (1972), Studies of enzyme-mediated reactions. Part 1. Syntheses of deuterium or tritium-labeled (3*S*)- and (3*R*)-phenylalanines: Stereochemical course of the elimination catalyzed by L-phenylalanine ammonia-lyase, *J. Chem. Soc., Perkin Trans. I*, 2355–2364.

WILCOCKS, R., WARD, O. P. (1992), Fasctors affecting

2-hydroxypropiophenone fromation by benzoyl formate decarboxylase from *Pseudomonas putida*, *Biotechnol. Bioeng.* **39**, 1058–1063.

WILCOCKS, R., WARD, O. P., COLLINS, S., DEWDNEY, N. J., HONG, Y. P., PROSEN, E. (1992), Acyloin formation by benzylformate decarboxylase from *Pseudomonas putida*, *Appl. Environ. Microbiol.* **58**, 1699–1704.

WILLIAMS, R. M., YUAN, C. G. (1992), Asymmetric synthesis of 2,6-diaminopimelic acids, *J. Org. Chem.* **57**, 6519–6527.

WILLIAMS, S. E., WOOLRIDGE, E. M., RANSOM, S. C., LANDRO, J. A., BABBITT, P. C., KOZARICH, J. W. (1992), 3-Carboxy-*cis*,*cis*-muconate lactonizing enzyme from *Pseudomonas putida* is homologous to the class-II fumarase family: A new reaction in the evolution of a mechanistic motif, *Biochemistry* **31**, 9768–9776.

WILLIAMSON, J. M. (1985), L-Aspartate-α-decarboxylase, *Methods Enzymol.* **113**, 589–595.

WITSCHEL, M., EGLI, T. (1997), Purification and characterization of a lyase from the EDTA-degrading bacterial strain DSM 9103 that catalyzes the splitting of [*S*,*S*]-ethylenediaminedisuccinate, a structural isomer of EDTA, *Biodegradation* **8**, 419–428.

WITSCHEL, M., NAGEL, S., EGLI, T. (1997), Identification and characterization of the two enzyme system catalysing oxidation of EDTA in the EDTA-degrading bacterial strain DSM 9103, *J. Bacteriol.* **179**, 6937–6943.

WOEHL, A., DUNN, M. F. (1999), Mechanisms of monovalent cation action in enzyme catalysis: The tryptophan synthase α-,β-, and α,β-reactions, *Biochemistry* **38**, 7131–7141.

WOODRELL, C. D., KEHAYOVA, P. D., JAIN, A. (1999), Photochemically-triggered decarboxylation/deamination of *o*-nitrodimethoxyphenylglycine, *Org. Lett.* **1**, 619–621.

WOODS, D. R. (1995), The genetic engineering of microbial solvent production, *Trends Biotechnol.* **13**, 29–264.

WOODS, S. A., GUEST, J. R. (1987), Differential roles of the *Escherichia coli* fumarases and FNR-dependent expression of fumarase B and aspartase. *FEMS Microbiol. Lett.* **48**, 219–224.

WOODS, S. A., MILES, J. S., ROBERTS, R. E., GUEST, J. R. (1986), Structural and functional relationships between fumarase and aspartase nucleotide sequences of the fumarase (FumC) and aspartase (-AspA) genes of *Escherichia coli*-K12, *Biochem. J.* **237**, 547–557.

WOODS, S. A., MILES, J. S., GUEST, J. R. (1988a), Sequence homologies between argininosuccinase, aspartase and fumarase – a family of structurally-related enzymes, *FEMS Microbiol. Lett.* **51**, 181–186.

WOODS, S., SCHWARTZBACH, S. D., GUEST, J. R. (1988b), Two biochemically distinct classes of fumarase in *Escherichia coli*, *Biochem. Biophys. Acta.* **954**, 14–26.

WU, M., TZAGOLOFF, A. (1987), Mitochondrial and cytoplasmic fumarases in *Saccharomyces cerevisiae* are encoded by a single nuclear gene FUM1, *J. Biol. Chem.* **262**, 12275–12282.

WU, P. C., KROENING, T. A., WHITE, P. J., KENDRICK, K. E. (1992), Histidine ammonia-lyase from *Streptomyces griseus*, *Gene* **115**, 19–25.

WU, C.-Y., LEE, H.-W., WU, S.-H., CHEN, S.-T., CHIOU, S.-H., CHANG, G.-G. (1998), Chemical mechanism of the endogenous argininosuccinate lyase activity of duck lens δ2-crystallin, *Biochem. J.* **333**, 327–334.

YAGASHI, M., OZAKI, A. (1998), Industrial biotransformations for the production of D-amino acids, *J. Mol. Catal. B* **4**, 1–11.

YAHIRO, K., TAKAHAMA, T., PARK, Y. S., OKABE, M. (1995), Breeding of *Aspergillus terreus* TN-484 for itaconic acid production with high yield, *J. Ferment. Bioeng.* **79**, 506–508.

YAHIRO, K., SHIBATA, S., JIA, S. R., PARK, Y., OKABE, M. (1997), Efficient itaconic acid production from raw corn starch, *J. Fermet. Bioeng.* **84**, 375–377.

YAMADA, S., NABE, K., IZUO, N., NAKAMICHI, K., CHIBATA, I. (1981), Production of L-phenylalanine from *trans*-cinnamic acid with *Rhodotorula glutinis*: containing L-phenylalanine ammonia-lyase activity, *Appl. Environ. Microbiol.* **42**, 773–778.

YAMAMOTO, Y., MIWA, Y., MIYOSHI, K., FURUYAMA, J., OHMORI, H. (1997), The *Escherichia coli IdcC* gene encodes another lysine decarboxylase, probably a constitutive enzyme, *Genes Genet. Systems* **72**, 167–172.

YAMANO, S., TOMIZUKA, K., TANAKA, J., INOUE, T. (1994), High-level expression of α-acetolactate decarboxylase gene from *Acetobacter aceti* ssp. *xylinum* in brewing, *J. Biotechnol.* **37**, 45–48.

YAMANO, S, TOMIZUKA, K, SONE, H, IMURA, M, TAKEUCHI, T. et al. (1995), Brewing performance of a brewers' yeast having α- acetolactate decarboxylase from *Acetobacter aceti* ssp. *xylinum*, *J. Biotechnol.* **39**, 21–26.

YANG, X. P., TSAO, G. T. (1994), Mathmatical modelling of inhibition kinetics in acetone-butanol fermentation by *Clostridium acetobutylicum*, *Biotechnol. Prog.* **10**, 532–538.

YANO, M., TERADA, K., UMIJI, K., IZUI, K. (1995), Catalytic role of an arginine residue in the highly conserved and unique sequence of phosphoenolpyruvate carboxylase, *J. Biochem.* **117**, 1196–1200.

YOON, M. Y., THAYER-COOK, K. A., BERDIS, A. J., KARSTEN, W. E., SCHNACKERZ, K. D., COOK, P. F. (1995), Acid-base chemical mechanism of aspartase from *Hafnia alvei*, *Arch. Biochem. Biophys.* **320**, 115–122.

YOUNG, D. W. (1991), Use of enzymes in the synthesis of stereospecifically labeled compounds, in: *Isotopes in the Physical and Biomedical Sciences, Volume 1, Labeled Compounds (Part B)* (BUNCEL, E., JONES, J. R., Eds.), pp. 341–427. Amsterdam: Elsevier.

YU, Y., RUSSELL, R. N., THORSON, J. S., LIU, L. D., LIU, H.W. (1992), Mechanistic studies of the biosynthesis of 3,6-dideoxyhexoses in *Yersinia pseudotuberculosis* – purification and stereochemical analysis of CDP-D-glucose oxidoreductase, *J. Biol. Chem.* **267**, 5868–5875.

YUMOTO, N., TOKUSHIGE, M. (1988), Characterization of multiple fumarase proteins in *Escherichia coli*, *Biochem. Biophys. Res. Commun.* **153**, 1236–1243.

ZAITSEV, G. M., GOVORUKHINA, N. I., LASKOVNEVA, O. V., TROTSENKO, Y. A. (1993), Properties of the new obligatory oxalotrophic bacterium *Bacillus oxalophilus*, *Microbiology* **62**, 378–382.

ZENG, X. P., FARRENKOPF, B., HOHMANN, S., DYDA, F., FUREY, W., JORDAN, F. (1993), Role of cysteines in the activation and inactivation of brewers' yeast pyruvate decarboxylase investigated with a *PDC1–PDC6* fusion protein, *Biochemistry* **32**, 2704–2709.

ZHOU, X. Z., TONEY, M. D. (1999), pH studies on the mechanism of the pyridoxal phosphate dependent dialkylglycine decarboxylase, *Biochemistry* **38**, 311–320.

ZHU, M. Y., JUORIO, A. V. (1995), Aromatic amino acid decarboxylase – Biological characterization and functional role, *Gen. Pharmacol.* **26**, 681–696.

ZIMMER, W., HUNDESHAGEN, B., NIEDERAU, E. (1994), Demonstration of the indolepyruvate decarboxylase gene homolog in different auxin-producing species of the Enterobacteriaceae, *Can J. Microbiol.* **40**, 1072–1076.

ZIMMER, W., WESCHE, M., TIMMERMANS, L. (1998), Identification and isolation of the indole-3-pyruvate decarboxylase gene from *Azospirillum brasilense* Sp7: Sequencing and functional analysis of the gene locus, *Curr. Microbiol.* **36**, 327–331.

3 Halocompounds

GARY K. ROBINSON
SIMON A. JACKMAN
JANE STRATFORD
Canterbury, UK

1 Introduction

1.1 Introductory Remarks

Haloorganic compounds are ubiquitous throughout all environmental matrices. Their levels range from ppb to saturation levels in water, soil, and air. The anthropogenic origin of many of these compounds is not in question and has been the subject of many reports which have influenced environmental legislation throughout the world. Aside from these anthropogenic inputs, it is becoming clear that the natural formation of haloorganics plays a major role in the global haloorganic balance. The burden of halocompound transformation and turnover falls on the major trophic level, the microorganisms. It has been suggested by PRIES et al. (1994) that the lack of halocompound biodegradation is due to biochemical rather than thermodynamic constraints, i.e., the fidelity of compound recognition by uptake proteins, regulatory proteins, or catabolic enzymes is insufficient for their transformation and degradation. Indeed, it is known that both oxidative conversion of chlorinated compounds with oxygen as the electron acceptor and reductive degradation to methanes or alkanes should yield sufficient energy to support growth (DOLFING et al., 1993).

Despite the apparent recalcitrance of many halogenated compounds it is clear that, providing environmental conditions are favorable, microbial populations may adapt to use these compounds as novel carbon and energy sources. The genetic mechanisms of adaptation, especially relating to haloaromatic transformation have recently been reviewed (VAN DER MEER, 1994). The author makes the clear distinction between *vertical evolution* and *horizontal evolution*. The former is brought about by alterations in the DNA sequence which arise due to the accumulation of mutations. These may arise due to the presence of the halogenated substrate itself or be accelerated in the laboratory by the battery of mutagenic protocols available. Horizontal evolution occurs when DNA sequences are exchanged between cells (intercellular movement) or between plasmid and chromosomal DNA (intermolecular movement). Horizontal movement of genetic information may be caused by recombination or other mechanisms of gene exchange, e.g., conjugation, transduction, or transformation. It is clear that both vertical and horizontal evolution have a role to play in the development of catalysts which may be of use in organic syntheses but techniques which facilitate this will not be discussed here.

The present review will deal exclusively with the *bio*synthesis and *bio*transformation of haloorganic compounds of potential interest to those practitioners developing new routes of organic synthesis. Additionally, it will illustrate the diversity of substrates able to be transformed by *biological* material, whether microbial, plant, or animal. It should be stated from the outset that microorganisms play a central role in the transformation of halogenated compounds. As a consequence the majority of currently available information and the obvious industry preference for microbial systems dictate that halotransformation by microorganisms forms the foundation of any review on the subject. However, where appropriate, halotransformation or synthesis by non-microbial sources will be discussed.

1.2 Distribution and Function

The ubiquity of biological materials containing a covalently bound halogen was thought unlikely only a quarter of a century ago. Indeed, FOWDEN (1968) undertook the first published review of naturally occurring organohalogens and stated that:

"present information suggests that organic compounds containing covalently bound halogens are found only infrequently in living organisms."

Today, approximately 2000 halogenated chemicals, predominantly chlorinated, are known to be discharged into the biosphere by biological systems and natural processes (GRIBBLE, 1992, 1994; VAN PEE, 1996). It is interesting to note that public perception and awareness of the incidence of halogenated compounds in the natural environment is often concentrated, and rightly so, on anthropogenic inputs and their associated recalcitrance. Such concerns are heightened by uninformed statements such as that released by the Scientific Advisory Board to the International Joint Commission on the Great Lakes which stated that:

"There is something nonbiological about halogenated organics (excluding iodinated compounds). ... Chemicals [that] do not occur naturally... are often persistent, since there are often no natural biological processes to metabolize or deactivate them."

The ignorance of this statement, made in 1989, is clear when one considers the biological contribution to atmospheric pollution by volatile hydrocarbons. Estimates of the industrial production of chloride and fluoride as halohydrocarbons (1982–1984) put the release as $2.28 \cdot 10^{12}$ g and $0.273 \cdot 10^{12}$ g, respectively, into the environment. However, the biological contribution, as chloromethane from oceans and burning vegetation is put at $1.4–3.5 \cdot 10^{12}$ g (HARDMAN, 1991). The occurrence of organohalogens in nature is usually quantified in terms of the bulk parameter AOX (adsorbable organic halogen). This parameter has been shown to be at least 300 times greater than that which can be accounted for purely by anthropogenic sources (ASPLUND and GRIMVALL, 1991). AOX production has been shown to take place during leaf litter degradation and therefore the organisms responsible for degradation of leaf litter, primarily basidiomycetes, have been targeted (ASPLUND, 1995). Basidiomycetes produce a cross section of halometabolites including chloromethane (HARPER, 1985) plus chlorinated anisoles (DE JONG et al., 1994), orcinol derivatives (**1**) (OKAMOTO et al., 1993), hydroquinones (**2**) (HAUTZEL et al., 1990) and diphenyl ethers (**3–8**) (Fig. 1) (TAKAHASHI, 1993; OHTA et al., 1995). Recently, it as been shown that the ability to produce AOX is fairly ubiquitous, with 25% of 191 fungal strains examined showing moderate (0.5–5.0 mg L^{-1}) to high (5–67 mg L^{-1}) AOX production (VERHAGEN et al., 1996).

Today, it is recognized that a diverse array of organohalogens is produced by biological systems. These range from the large volume

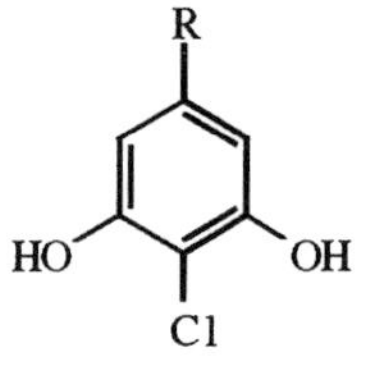

1
a R = Me
b R = CHO

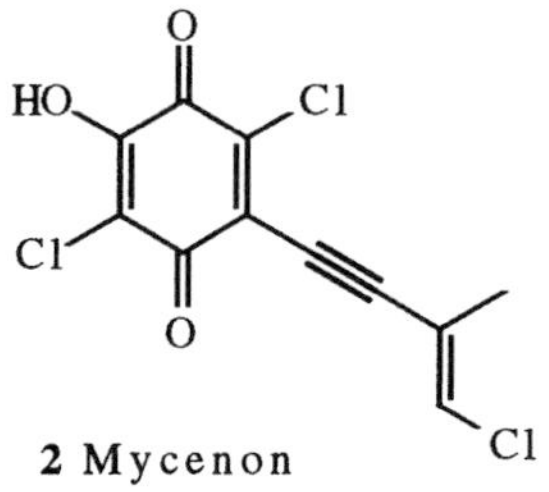

2 Mycenon

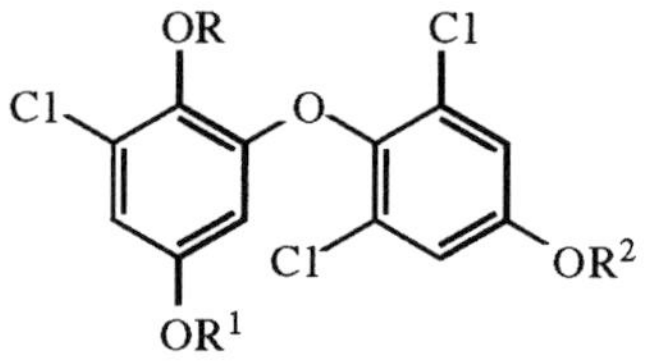

Russuphelins D-F
3 R = R^1 = Me, R^2 = H; D
4 R = Me, R^1 = H, R^2 = Me; E
5 R = H, R^1 = R^2 = Me; F

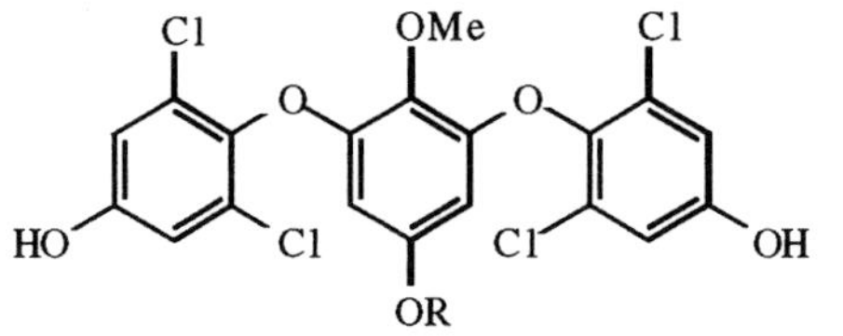

6 R = Me; Russuphelin A
7 R = H; Russuphelin B

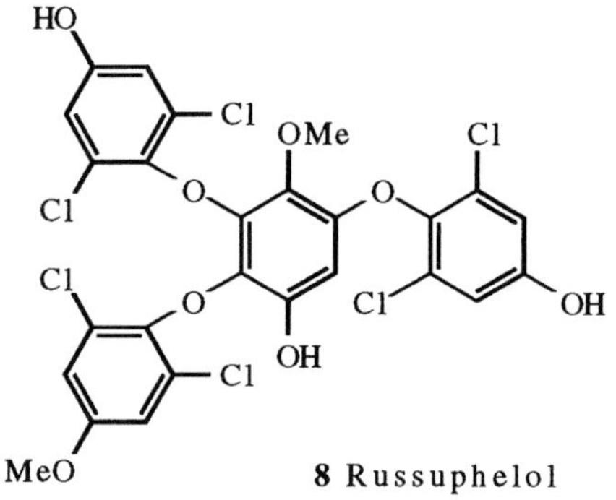

8 Russuphelol

Fig. 1. Chlorinated aromatics from basidiomycetes.

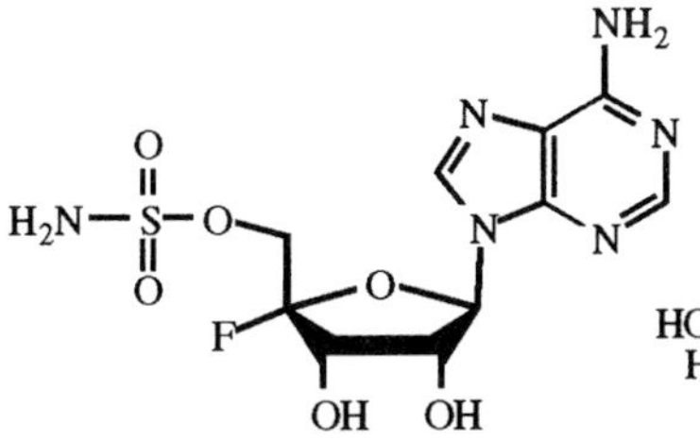

9 Nucleocidin 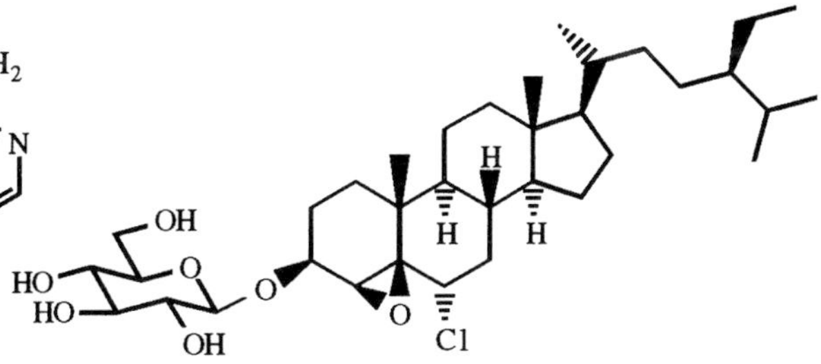 **10** Blattellastanoside A

Fig. 2. Naturally occurring fluorinated nucleoside and cockroach aggregation pheromone.

but simple halogenated alkanes through to the complex secondary metabolites such as nucleocidin (**9**) and the cockroach aggregation pheromone (**10**) (Fig. 2). The simplest halogenated alkane, chloromethane, is produced by a wide cross section of biological systems including marine algae (WUOSMAA and HAGER, 1990), wood rotting fungi (HARPER, 1985), phytoplankton (GSCHWEND et al., 1985), and the pencil cedar (ISIDOROV, 1990). As well as species diversity it has been shown that product diversity can occur within one species. *Asparagopsis taxiformis*, an edible seaweed favored by Hawaiians, produces nearly 100 chlorinated, brominated, and iodinated compounds (MOORE, 1977).

It is clear that naturally occurring organohalogens, while diverse and somewhat ubiquitous, do not exert the same environmental pressure as the man-made organohalogens. These products, which are often used in relatively large amounts, necessitate the evolution of new detoxification mechanisms. In the majority of cases, the nature, position, and number of halosubstituents mirrors that found in the natural organohalogens. However, it is clear that manufactured organohalogens present a different set of problems to biosynthesized organohalogens.

The diversity of organohalogens which are routinely used may best be illustrated by those classes of compounds which are directly applied to the environment, namely the pesticides. The current Pesticide Manual (TOMLIN, 1995) contains approximately 1500 pesticides of which approximately 50% are halogenated. The distribution of halogenation is outlined in Fig. 3.

Although some of these pesticides have been superseded, the data illustrate two contrasting requirements of interest in the present review:

(1) Haloorganics are favored targets for chemosynthesis because they possess a wide spectrum of biological activities.
(2) Haloorganics often possess increased recalcitrance contributing to their increased persistence in the environment.

Microbial biodegradation is considered to be the most important route of breakdown for

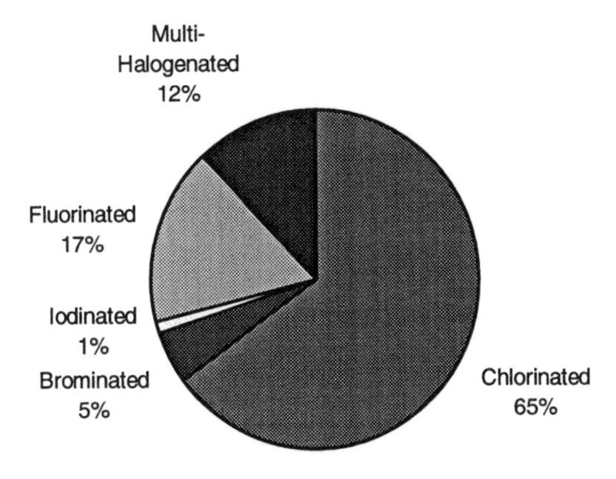

Fig. 3. Distribution of halogenated pesticides (data adapted from TOMLIN, 1995).

the majority of organic chemicals deposited in soil and water and the addition of xenophores such as halosubstituents retards this process (HOWARD et al., 1992). In the pesticides outlined above over 100 compounds possess two or more different halosubstituents. These compounds are obviously more recalcitrant than the non-halogenated homologs and require the development of specific detoxification mechanisms. The pressure to biodegrade pesticides is largely borne by microorganisms and the likelihood of transformation may be increased if the level of application is high. Obviously, abiotic mechanisms contribute to pesticide degradation but 52 of the halogenated pesticides outlined above are applied at levels $>50 \text{ t a}^{-1}$ in at least 1 EU member country (FIELDING, 1992) Consequently, the burden of transformation of these compounds falls upon biological systems and the mechanisms of transformation will be outlined herein.

1.3 Reasons for Synthesis by the Host

Both chlorine and bromine are ubiquitous elements in terrestrial and marine environments. The earth contains 0.031% Cl and 0.00016% Br (w/w) (cf. 0.00005% I) in minerals and 19 g L^{-1} Cl, 65 mg L^{-1} Br, and 0.05 mg L^{-1} I in seawater. The world's oceans compose over 70% of the earth's surface and over 90% of the volume of its crust. Consequently, the richest source of halometabolites are the marine algae which, in a survey by HAGER (1982), were shown to contain up to 20% of extractable halogenated compounds.

The marine habitat is complex, as evidenced by the wide variations in pressure, salinity, and temperature. Therefore, marine microorganisms have developed unique metabolic and physiological capabilities, offering the potential for the production of metabolites unknown in terrestrial environments. Although the present article stresses the synthesis of halometabolites, such synthetic versatility is shown in the production of other, non-halogenated compounds and the reader is directed to other reviews for further information (FENICAL, 1993; JENSEN and FENICAL, 1994).

Whatever the number and diversity of halocompounds the question which needs to be addressed is why invest metabolic energy in the synthesis of halocompounds? NEIDLEMAN and GEIGERT (1986) speculate that two major reasons exist:

(1) Halointermediates are key elements in biosynthesis.
NEIDLEMAN (1975) stated that:
"Microorganisms, ... have discovered the fact, as have organic chemists, that halogenated intermediates, on pathways to non-halogenated end products are useful devices."

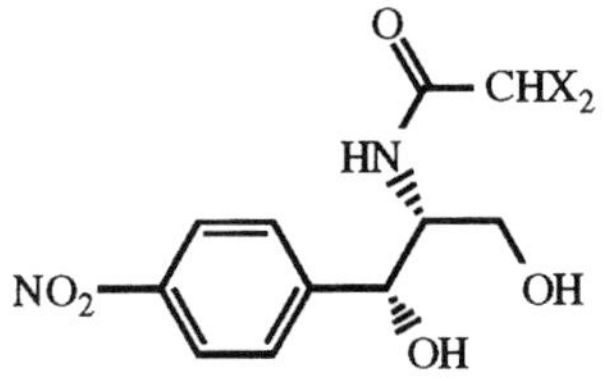

11
a X = H; Corynecin I
b X = Cl; Chloroamphenicol

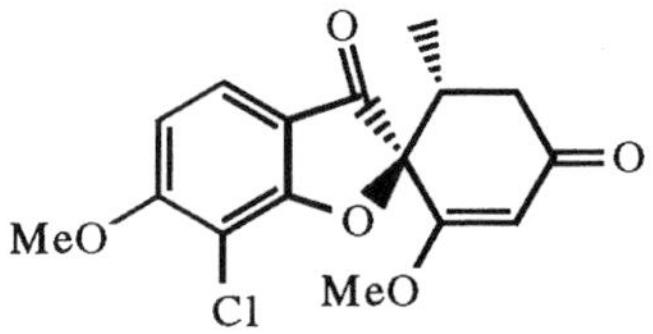

12 Griseofulvin

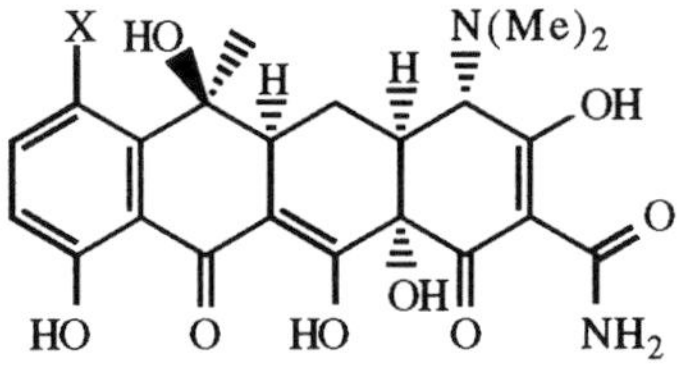

13
a X = H; Tetracycline
b X = Cl; Chlorotetracycline

Fig. 4. Common antibiotics possessing a halosubstituent.

Subsequent to this comment it has been shown that halogenated intermediates are synthesized by a variety of synthetic pathways involving unsaturated carboxylic acids, cyclization of isoprenoids, ring contractions, and methyl migration to yield rearranged terpenoid derivatives (FENICAL, 1979, 1982). Many of these can be rationalized by the intervention of halonium ion intermediates.

(2) Halogenation enhances the bioactive efficacy of most compounds.

As stated at the beginning of this introduction the presence of halogen substituents in pesticides is commonplace. Why this should be so is evidenced by the wide variety of natural organohalogens which possess antibacterial, antifungal, and antitumor activities. Chloramphenicol (**11b**), griseofulvin (**12**), and chlorotetracycline (**13b**) are all common antibiotics that contain chlorine (Fig. 4). However, it does not necessarily follow that the efficacy of these compounds is related to the type, number, and position of the halogen atoms. The bioactivity of bromo-analogs is usually very similar or less than their normally produced chlorinated counterparts. 7-Bromotetracycline has the same level of activity as 7-chlorotetracycline (**13b**) (PETTY, 1961) whereas pyrrolnitrin (**15**, Fig. 5) is more efficacious as a chlorinated, rather than a brominated derivative (VAN PEE et al., 1983). The absence of the halogen atom may also cause different effects with respect to antibiotic activity. Corynecin I (**11a**, Fig. 4), the dechlorinated analog of chloramphenicol (**11b**), possesses only a fraction of the antibiotic activity of its chlorinated congener (SUZUKI et al., 1972) whereas actinobolin (**14a**) and bactinobolin (**14b**) have similar antibiotic activity (Fig. 5).

Halocompounds (e.g., fluoroacetate (**19**, Fig. 13)) and halogenating enzymes are used for defensive purposes by a variety of organisms. This subject is beyond the scope of the present article but has been reviewed (NEIDLEMAN and GEIGERT, 1986).

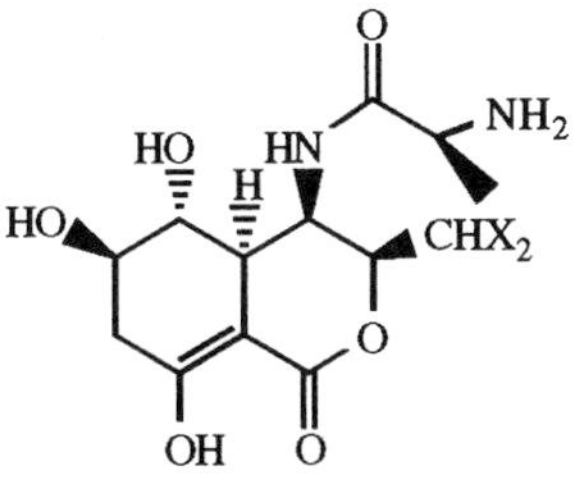

14
a X = H; Actinobolin
b X = Cl; Bactinobolin

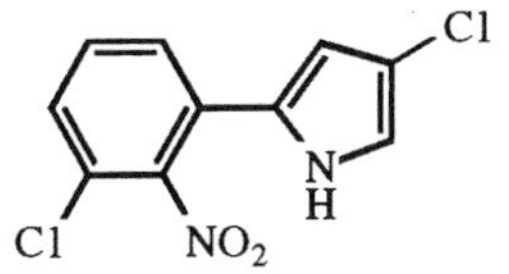

15 Pyrrolnitrin

Fig. 5. Microbial antibiotics possessing a halosubstituent.

2 Synthetic Importance of Halocompound Transforming Enzymes

2.1 Halogenation

A small number of functional groups occupy a dominant position with regard to organic synthesis. All of these groups, C=C (olefinic), C=O (carbonyl), C≡N (cyano), C−OH (alcohol), COOH (carboxyl), NH_2 (amino), NO_2 (nitro), and C≡C (acetylenic) will have been discussed within this volume as they are central to organic syntheses due to their ease of interconversion. Halides are examples of much less central or peripheral groups which are required in the target or simply provide activation or control in the synthetic route. The advantages of biohalogenation are no different to those proposed for other enzyme catalyzed reactions:

Efficient catalysts – usually expressed as turnover number, it has been shown to be approximately 10^5 for chloro- and bromoperoxidases (HAGER et al., 1966; MANTHEY and HAGER, 1981).

Environmental acceptability – traditional halogenation and dehalogenation methodologies use harsh reagents (Friedel-Crafts catalysts, Lewis acids, and heavy metal catalysts) with high environmental impact. Furthermore, unlike heavy metal catalysts, enzymes are completely biodegradable.

Mild reaction conditions – enzymes catalyze reactions in the range, of pH 5–8, typically 5 using haloperoxidases, and at temperatures of 20–40 °C. This minimizes problems of undesired side reactions, a major drawback to many chemosynthetic methodologies.

Wide substrate specificity – enzymes may act on a broad range of substrates giving a range of variously substituted products. The central group of enzymes assumed to catalyze halocompound synthesis, the haloperoxidases, are unusual in that their primary synthetic utility is the halide independent reactions (ZAKS and DODDS, 1995).

Several general assay methods which permit the rapid screening for haloperoxidases (the main halogenating enzyme) have been employed including the use of monochlorodimedone (MCD) (**16a**) and phenol red (**17a**) as substrates (Fig. 6). However, it is unlikely that novel synthetic activity will always be evident when employing limited substrates for discovery. What are required (although it is difficult to see how they may be devised), are specific screens aimed at elucidating enantiospecific halogenations. When considering potential substrates for halogenation there are few structural limits providing that they are relatively electron rich, e.g., pyridine and pyrimidine are fairly unreactive towards haloperoxidase unless an electron donating substituent is present (ITOH et al., 1987b). Additionally, olefinic groups are fairly unreactive unless coupled to an electron rich system, e.g., styrene.

The scope of reactions catalyzed by the haloperoxidases, exemplified largely by the chloroperoxidases, is broad and can be subdivided into halide dependent and halide independent reactions. Halide independent asymmetric oxidations were recently studied by ZAKS and DODDS (1995) who suggested the halide independent reactions, epoxidation of alkenes, formation of sulfoxides, aldehydes and nitroso compounds and N-dealkylations, were catalyzed by a neutral form of the enzyme whereas the halide dependent reactions, halogenation of β-diketones and halohydration, were catalyzed by an acidic form of the enzyme.

The stereoselective haloperoxidase catalyzed epoxidation of alkenes can be highly enantioselective (COLONNA et al., 1993; ALLAIN et al., 1993) and does not require an allylic alcohol (cf. Sharpless epoxidation). Although the abiotic enantioselective epoxidation of alkenes which do not bear an allylic alcohol group continues to be improved, the reaction is currently only highly stereoselective with cyclic alkenes (VELDE and JACOBSEN, 1995).

The oxidation of sulfides to sulfoxides or sulfones has been performed with many types of enzyme. With haloperoxidases in the absence of halide ion the reaction is enantioselective and an oxygen from hydrogen peroxide is incorporated into the sulfoxide. Whereas in the presence of halide the rate is increased, the reaction is not enantioselective, and there is no oxygen incorporation from peroxide. This clearly indicates that in the presence of halide a freely diffusing species such as hypohalous

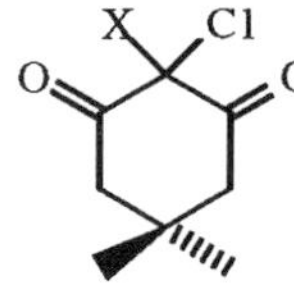

X Cl O O

16
a R = H; monochlorodimedone
b R = Br

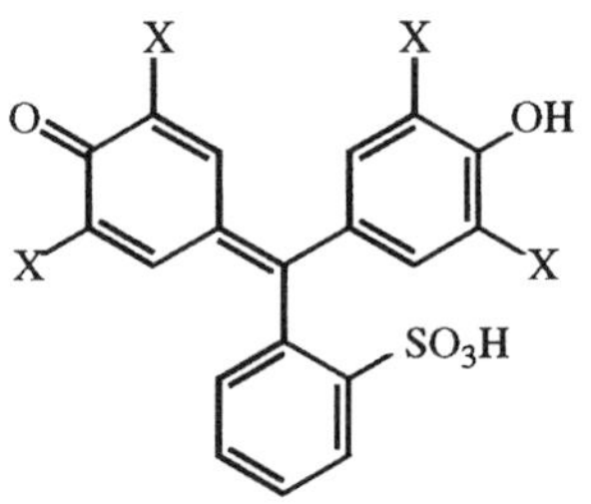

X X O OH X X SO_3H

17
a R = H; Phenol red
b R = Br; Bromophenol blue

Fig. 6. Common substrates for screening haloperoxidases.

acid reacts nonenzymatically with the sulfide (or other substrate molecule) (PASTA et al., 1994).

2.2 Dehalogenation

Dehalogenation is a process normally associated with the detoxification or amelioration of a recalcitrant halocompound. There are several methods of dehalogenating but this review will concentrate on those catalyzed by specific enzymes. Dehalogenation which arises due to facilitated abiotic mechanisms, e.g., spontaneous elimination of a halide ion after ring cleavage of a haloaromatic compound, will not be considered. All of the enzymatic mechanisms require at least one halogen atom, which is removed and replaced by another atom, functional group, or is eliminated. In the case of reductive dehalogenation this will result in a molecule which is deactivated and, unless selective removal was being sought, would be of limited synthetic use. All other dehalogenation mechanisms, oxygenolytic, hydrolytic, thiolytic, epoxide formation, dehydrohalogenation, and hydration result in substitution of the halogen atom with a reactive carbonyl or alcohol group, with the exception of dehydrohalogenation, an elimination reaction resulting in double bond formation.

3 Current Developments in Dehalogenase and Haloperoxidase Biochemistry and Genetics

Biotransformations have come to the forefront of biotechnology over the last 10 years due to the legislative drive towards single enantiomer preparations in the agrochemical, flavor and fragrance, and pharmaceutical markets (CANNARSA, 1996). The primary reasons for this have been the selectivity of enzymes and the potential to produce and manipulate the whole cells and enzymes which catalyze the reactions of interest. In the halogenation and dehalogenation field the impetus has been twofold:

(1) There has been a drive towards understanding the degradative pathways involved in the transformation and mineralization of organohalogens in the environment. This has led to the development of biochemical and molecular techniques which have facilitated our understanding of dehalogenation.

(2) There has been an attempt to better understand how single enantiomer halocompounds are formed in nature. It has been said that "bacterial nonheme haloperoxidases are difficult to isolate and are available in small quantities" (WENG et al., 1991). Only now is progress being made in this field but it still remains in its infancy, especially when ascribing function to haloperoxidases *in vivo*.

Both synthesis and degradation have relied on the development of molecular techniques and a slow increase in the amount of sequence data, both DNA and protein, available for manipulation and exploitation. Additionally, these approaches have caused us to re-evaluate the roles that enzymes such as haloperoxidase may play in halocompound synthesis.

In only a few cases has the genetics and biochemical mechanism for introducing chlorine been examined in detail. Indeed, following the studies of chlorination in the fungus *Caldariomyces fumago*, which confirmed that caldariomycin (**18**, Fig. 7) was chlorinated via a chloroperoxidase catalyzed reaction, it was anticipated this would be a generic mechanism. While haloperoxidase catalyzed incorporation of halide is still generally accepted as the primary mode of halogenation, it has still proved

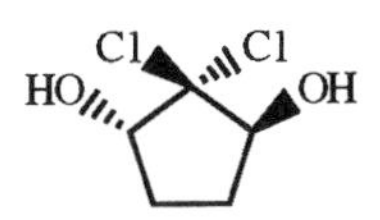

18 Caldariomycin

Fig. 7. Caldariomycin.

difficult studying the biochemistry and genetics of halogenated secondary metabolite formation. The bacterial non heme-type haloperoxidases differ greatly from the heme-type and vanadium containing haloperoxidases. Moreover, it has been shown that the bacterial haloperoxidases form a totally distinct enzyme family for which the catalytic mechanism is still poorly understood (see Sect. 5.1.4).

Chloramphenicol (**11b**, Fig. 4), probably the best known chlorinated antibiotic is believed to be chlorinated due to the action of a chloroperoxidase (NEIDLEMAN and GEIGERT, 1986). It is produced by *Streptomyces venezuelae*, from which bromoperoxidases but not chloroperoxidases have been isolated. It is well known that chloroperoxidases will catalyze the incorporation of both chloride and bromide whereas bromoperoxidases will only, due to thermodynamic constraints, catalyze bromide incorporation. Presently, using molecular techniques, there is no report of a chloroperoxidase being probed and found in an organism known to produce a chlorometabolite. To the authors knowledge, sequence information is only available on a very limited number of haloperoxidases as follows (information taken from the European Bioinformatics Institute WWW site, *http://www. ebi.ac.uk/*):

Chloroperoxidases – sequenced from *Caldariomyces fumago* (NUELL et al., 1988), *Curvularia inequalis* (SIMONS et al., 1995), *Pseudomonas pyrrocina* (WOLFFRAMM et al., 1993), and *Streptomyces lividans* (BANTLEON et al., 1994). Another sequence from *Pseudomonas fluorexcens* is available but unpublished (PELLETIER et al., 1995).

Bromoperoxidases – sequenced from *Streptomyces venezuelae* (FACEY et al., 1996) and *Streptomyces aureofaciens* (PFEIFER et al., 1992; PELLETIER et al., 1994).

Additionally, the role of the haloperoxidases in the synthesis of halometabolites is not unequivocally established, indeed it maybe better argued that they have no role to play in the synthesis of halometabolites. FACEY et al. (1996) have recently shown that a bromoperoxidase–catalase gene from *Streptomyces venezuelae* is not required for chlorination in chloramphenicol (**11b**, Fig. 4) synthesis. However, the non-heme chloroperoxidase from *Pseudomonas pyrrocina* has been shown to be involved in the synthesis of pyrrolnitrin (**15**, Fig. 5) (PELLETIER et al., 1995).

As mentioned previously, there are seven mechanisms for dehalogenating compounds. The biochemistry and genetics of these systems have been thoroughly reviewed by FETZNER and LINGENS (1994) and JANSSEN et al. (1994) and only a cursory glance and update will be provided here. The naming of dehalogenases, particularly those involved in haloaliphatic metabolism, has been confused and this has only started to be addressed. SLATER, BULL, and HARDMAN (1995) have attempted to categorize the various dehalogenating enzymes on the basis of their mechanism and substrate range with further subdivisions to take into account molecular descriptors such as protein and DNA sequence homology. Ultimately this approach will lead to complete descriptions of the genetic and evolutionary relationships between classes and a classification of the different enzyme mechanisms which are undoubtedly involved in dehalogenation by reference to significant tertiary structures.

Currently, the classification of dehalogenases is largely dependent on mechanism and can best be summarized as follows:

(1) **Reductive dehalogenation** – largely associated with anaerobic environments, but not well characterized (Fig. 8) (DOLFING and BUERSKENS, 1995).

(2) **Oxygenolytic dehalogenation** – arylhalide dehalogenation, usually, involves temporary loss of aromaticity and labilization of the halogenated molecule by the introduction of oxygen. This is not

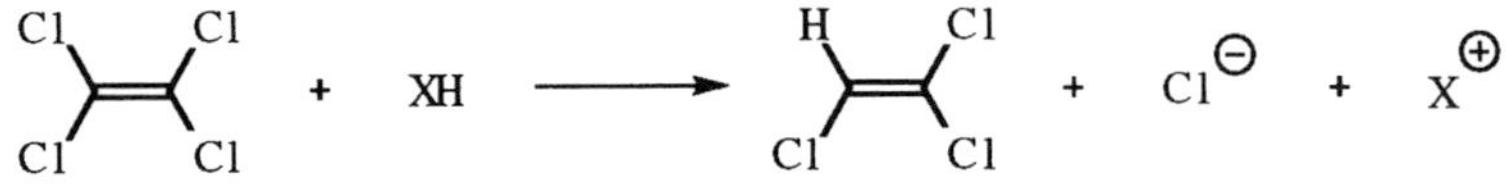

$$Cl_2C{=}CCl_2 + XH \longrightarrow HClC{=}CCl_2 + Cl^{\ominus} + X^{\oplus}$$

Fig. 8. Reductive dehalogenation.

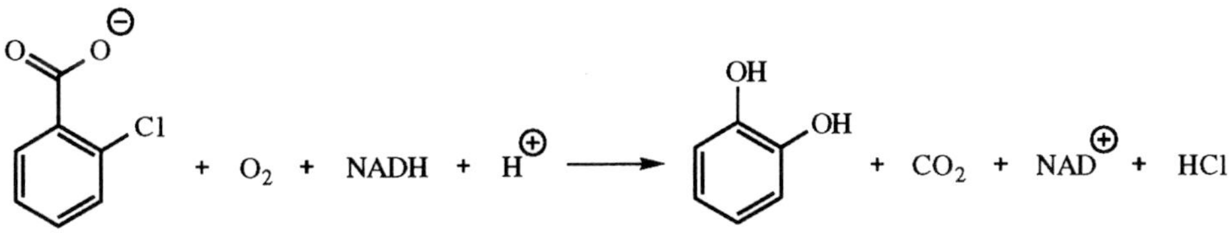

Fig. 9. Oxygenolytic dehalogenation.

considered a true dehalogenase in the context of the present review but the genetics have been extensively investigated and reviewed due to their importance in biodegradation (Fig. 9) (FETZNER and LINGENS, 1994).

(3) **Dehalogenation via epoxide formation** – dehalogenation of haloalcohols to give epoxides (Fig. 10) is frequently followed by hydrolysis of the epoxide to a diol, catalyzed by an epoxide hydrolase. The dehalogenation is catalyzed by lyases which have been variously termed halohydrin epoxidases (CASTRO and BARTNICKI, 1968), haloalcohol dehalogenases and haloalcohol halogen-halide lyases (VAN DEN WIJNGAARD et al., 1989, 1991), and halohydrin hydrogen-halide lyases (NAGASAWA et al., 1992). Recently, two halohydrin hydrogen-halide lyases from *Corynebacterium* sp. have been cloned, sequenced, and expressed in *E. coli* (YU et al., 1994).

(4) **Dehydrohalogenation** – dehalogenation via dehydrohalogenation is not a common mechanism (Fig. 11). The most often cited, and possibly unique example, involves the elimination of hydrogen chloride from γ-hexachlorocyclohexane (lindane) (**82**), to give sequentially: γ-pentachlorocyclohexene (**83**) and 1,3,4,6-tetrachloro-1,4-cyclohexadiene (**84**). The enzyme, termed dechlorinase, is encoded on the chromosome in *Pseudomonas* (or *Sphingomonas*) *paucimobilis*. The gene (*linA*) has been cloned, sequenced, and expressed in *E. coli* (see Fig. 32) (IMAI et al., 1991).

Amino acids with good leaving groups at the β-position such as β-chloroalanine and serine-O-sulfate undergo elimination via imines formed with pyridoxal phosphate, followed by conjugate nucleophilic addition and hydrolysis of the imine to give overall a substitution product (CONTESTABILE and JOHN, 1996; NAGASAWA and YAMADA, 1986).

(5) **Hydrolytic dehalogenation** – the most common dehalogenation mechanisms are hydrolytic, and these are fairly ubiquitous throughout the substrate range (Fig. 12). Whether thiolysis, hydration, or other hydrolytic reactions, they all fall within the classification system put forward by SLATER, BULL, and HARDMAN (1995). All enzymes catalyzing this type of reaction may be classified into 3 groups: hydrolytic dehalogenases, haloalcohol dehalogenas-

Fig. 10. Dehalogenation via epoxide formation.

Fig. 11. Dehydrohalogenation.

Fig. 12. Hydrolytic dehalogenation.

es, and cofactor dependent dehalogenases, which are further subdivided into 3 subgroups and 8 classes. The basis for these divisions is primarily mechanistic but is supported and endorsed by extensive DNA and amino acid sequence information.

4 Mechanistic Aspects of Biohalogenation

The development of our understanding of the nature of biological halogenation has progressed along traditional investigative lines. Beginning with the identification of halogenated compounds from different species and the possible elucidation of their function, it has progressed to consider the biosynthetic pathways involved. From this point, the individual enzymes involved in each transformation are identified, and it is then the role of the biotechnologist to seek to apply these enzyme processes to the production of compounds of interest and value. A plethora of literature is now available concerning halogenated metabolites identified in bacterial and algal species. While it would appear that most halogenations proceed through a single ubiquitous route, namely that catalyzed by the haloperoxidases, the development of standard assays for these enzymes has failed to detect novel catalysts. Only the identification of novel natural products can give some early indication as to their route of biosynthesis.

In this discussion of biohalogenation, we will therefore summarize the broad range of organohalogens found in nature, drawing particular attention to those which have enhanced our understanding of the process and those that have led to technological advances in this area. This forms a very suitable backdrop from which to enter into the more detailed analysis of biosynthetic routes and enzymology. Once again, following the same thread of investigation, this section will be rounded off with a summary of the developments of biohalogenation as a technology in its application to the manufacture of compounds that are of commercial significance.

4.1 Natural Organohalogens

4.1.1 Chlorine and Bromine

Tab. 1 is a survey of natural organohalides, collated according to biogenic origins and structural type. No attempt has been made to give an exhaustive list of organohalogens as new metabolites are being discovered all the time. Compounds of pharmaceutical or agrochemical value are indicated, but the reader is referred to cited original references for more details on the structures of interest.

The most abundant halogenated organics are alkane derivatives produced either through natural combustive processes or as metabolic products in marine algae. Combustion of plants, wood, soil, and minerals containing chloride ions leads to the formation of organochlorine compounds, but a larger contributor to these sinks is the action of volcanoes, with halocarbons such as chloromethane being produced at the rate of $5 \cdot 10^6$ t a^{-1} (RASMUSSEN et al., 1980), cf. 26000 t from anthropogenic inputs (HARPER, 1985). Natural volcanic activity has also been shown to produce more complex highly toxic organohalogens such as chlorofluorocarbons (CFCs), dioxins, and certain polychlorinated biphenyl (PCB) congeners. A considerable proportion of simple halogenated alkanes are produced in the oceans by marine algae, together with smaller quantities of more complex halogenated metabolites (MOORE, 1977; WOOLARD et al., 1976; FENICAL, 1974).

4.1.2 Fluorine

Very few organofluorines have been isolated from biological material to date, with none of these coming from the animal kingdom or marine organisms. It is thought that the main reason for the scarcity of fluorine containing compounds is the high heat of hydration of the fluoride ion which severely restricts its participation in biochemical processes (Tab. 2). In addition, fluorine has a major electronic effect upon compounds in which it is incorporated, restricting their further metabolism.

Tab. 1. Natural Organohalogens

Compounds	Source	Properties/Other Details	Reference
Terpenes			
– mixed chlorinated and brominated	aquatic red algae – *Laurentia* and *Plocamium*	wide range of mono- and polychlorinated monoterpenes,	1
		sesquiterpenes – cytotoxic and antitumor agents	1
– chlorinated	soft corals	diterpenes and a tetraterpene	1
	marine sponges – *Acanthella*	terpenes with rare isonitrile functionality	1
	ferns and related species	10 sesquiterpene indanones	1
	Streptomyces (Australian)	naphthalenequinone	
	basidiomycete *Armillaria ostoyae*	5 chlorinated aryl-sesquiterpene metabolites	
– brominated	green algae – *Neomaris annulata*	herbicides	1
	Cymopolia barbata	cymbarbatol – antimutagenic	1
Nonterpenes			
– chlorinated and brominated	marine red algae – *Laurencia*	large variety including 6 cyclic ethers, some of which are allenes	1
– chlorinated	*Pseudomonas* sp.	bactobolin B	1
	Gluconobacter	enacycloxin II (dichloropolyenic antibiotic)	1
	fungus – *Chaetomium globosum*	4 metabolites incl. chaetovirdin and a unique shikimate derivative	1
	terrestrial plants – e.g., *Rehmannia glutinosa*	15 iridoid derivatives	2
	blue-green alga – *Nostoc linckia*	4 paracyclophanes	1
– brominated	marine sponge – *Haliclona* sp.	2 bromotetrahydropyrans	1
Amino acids and peptides			
– chlorinated	marine sponge – *Dysidea herbacea*	4 trichloromethyl metabolites including dysidin and dysidenin (could produce $CHCl{=}CH_3Cl$ by elimination)	1
	Streptomyces and *Pseudomonas* spp.	amino acids and peptides many of which are potent antibacterials	1
	S. griseosporus	γ-chloronorvaline	1
	P. syringae	4-chlorothreonine	
	blue-green alga – *Anabaena* sp.	cardioactive cyclic peptide puwainaphycin containing 3-amino-14-chloro-2-hydroxy-4-methylpalmitic acid	1
Alkaloids	rarely contain halogens		
– chlorinated	terrestrial plants – *Sinomenium acutum*	acutumine and acutumidine	1
	Melodinus celastroides	2-bis-indole alkaloids	1
	microorganisms – *Streptomyces* sp.	clazamycins A and B	

Tab. 1. Natural Organohalogens (Continued)

Compounds	Source	Properties/Other Details	Reference
Steroids	rarely contain halogens		
– chlorinated	plants – *Withania somnifera* and others	withanolides (e.g., physanolactone) and associated compounds (e.g., jaborochlorodiol)	1
Fatty acids, prostaglandins, and lipids			
– chlorinated	edible jellyfish and white sea jellyfish	6 fatty acid chlorohydrins	1
	octocoral *Telesto riisei* and stolonifer *Clavularia viridis*	8 novel chlorinated prostaglandins, several with antitumour activities	
	fresh water algae, e.g., *Ochromonas danica*	chlorosulfolipids	1
– brominated	seeds of *Eremostachys molucelloides*	9,10-dibromo- and 9,10,12,13-tetrabromosteric acids	1
	marine sponges – *Xestospongia* and *Petrosia*	series of novel bromo acids	1
Pyrroles	reactivity leads to a large number in nature		
– chlorinated	bacterial – *Pseudomonas aeruginosa*	pyroluteorin – antibiotic	1
	Pseudomonas pyrrocinia	pyrrolnitrin	3
	Streptomyces sp.	pyrrolomycin B and an optically active pyrrolomycin	1
	Actinomyces sp.	pyrrolomycin A	1
	Actinosporangium vitaminophilum	10 related pyrrolomycins	1
– brominated	marine bacterium – *Chromobacterium* sp.	polybrominated pyrroles	?
	marine sponges, e.g., *Hymeniacidon*	mono- and dibromophakellin	4
	Agelas sceptrum	sceptrin (contains cyclobutane)	1
	Hymeniacidon sp.	3 bromopyrrolopyrimidine	1
	Astrosclera willeyara	dibromoageliferin	5
Indoles			
– chlorinated	terrestrial plants, e.g., *Pisum sativum*	4-chloroindole ester and carboxcylic acid	
	blue-green alga – *Hapalosiphon fontinalis*	12 chlorinated isonitriles	1
	sponges, e.g., *Batzella* sp.	6 chlorinated metabolites with tryptamine nitrogen attached to C-4 indole position	1
	microbe *Penicillium crustosum*	3,6-chloroindole metabolites – very high structural complexity	1
– brominated	mollusks – *Dicathais* and *Murex* sp.	Tyrian Purple (indigo derivative)	1
	acorn worm – *Glossobalanus* sp.	2 dibromoindoles	1
	red alga – *Laurencia brongniartii*	4 polybromoindoles	1
	marine *Pseudomonas* sp.	6-bromoindole-3-carboxaldehyde	1
	sponge – *Pleroma manoui*	bromoester and hydroxyketone	1
	Iotrochota sp.	indole acrylate	1
	sponges	5- and 6-bromo and 5,6-dibromotryptamines,	6
		6-bromohypaphorine	7

Tab. 1. Natural Organohalogens (Continued)

Compounds	Source	Properties/Other Details	Reference
	tunicates and sponges	dimeric tryptamines and brominated bis-indoles	1
	sponges and related marine organisms	brominated indole hydantoins, cyclic peptides incorporating 2-bromotryptophan	1
	blue-green algae – *Rivularia firma*	6 novel polybrominated diindoles	1
Carbazoles	comparatively rare in comparison to unhalogenated carbazoles		
– chlorinated	blue-green alga – *Hyella caespitosa*	chlorohyellazole	
	bovine urine	3-chlorocarbazole	1
Carbolines			
– brominated	tunicates, e.g., *Eudistoma glaucus*	25 different brominated β-carbolines	1
Indolocarbazoles			
– chlorinated	microbes and fermentation broths	5 halogenated indolo[2,3-α]carbazoles	1
	e.g., *Nocardia aerocolonigenes*	e.g., rebeccamycin – anticancer activity	1
Quinolines			
– chlorinated	*Streptomyces nitrosporeus*	virantmycin	1
	young corn roots – *Zea mays*	glycosyl metabolite	1
– brominated	marine bryozoan – *Flustra foliacea*	bromoquinoline	1
Furans and benzofurans			
– chlorinated	plant – *Gilmaniella humicola*	benzofuran mycorrhizinol	1
– brominated	sponge – *Dendrilla* sp.	2-bromofuran	1
Thiophenes			
– chlorinated	plants, e.g., *Pterocaulon*	5 chlorinated thiophenes	1
Nucleic acids			
– chlorinated	*Actinomyces* sp.	2′-chloropentostatin	1
– brominated	sponge – *Echinodictyum* sp.	novel brominated purine	1
Miscellaneous heterocycles			
– chlorinated	plants and fungi	chlorinated xanthones	1
	plants	isocoumarins	1
– brominated	red alga – *Ptilonia australasica*	unusual polybrominated pyrines	1

Tab. 1. Natural Organohalogens (Continued)

Compounds	Source	Properties/Other Details	Reference
Aromatic compounds			
– chlorinated and brominated	deep sea gorgonian	halogenated azulenes	1
– chlorinated	basidiomycetes (8 genera)	chlorinated anisyl metabolites	8
	basidiomycetes (6 genera)	e.g., drosopholin A	9
	Mississippi salt marsh "needlerush"	1,2,3,4-tetrachlorobenzene	
Phenols and phenolic ethers			
– chlorinated and brominated	sea organisms, e.g., mollusk *Buccinum undatum*	halogenated tyrosines, e.g., 3-chloro-5-bromotyrosine – stabilization of structural proteins	1
	marine sponges	mixed bromochlorodiphenyl ethers	1
– chlorinated	soil *Penicillium* sp.	range of dichlorobiphenyls	
– brominated	acorn worm	2,6-dibromophenol – defense mechanism	1
	sponges	metabolites derived from brominated tyrosine	1
Anthraquinones			
– chlorinated	fungus – *Dermocybe*	5-chlorodermolutein and 5-chlorodermorubin	1
Dioxins and related compounds			
– chlorinated	soil and water microbes using horseradish peroxidase	PCDDs and PCDFs from chlorophenols	1

1: Gribble (1992); 2: Kitagawa et al. (1995); 3: Elander et al. (1968); 4: Kobayashi et al. (1991); 5: Williams and Faulkner (1996); 6: Searle and Molinski (1994); 7: Kondo et al. (1994); 8: Field et al. (1995); 9: Spinnler et al. (1994).

Tab. 2. Comparison of the Physicochemical Properties of the Halides (data taken from HARPER and O'HAGAN, 1994)

	Bond Dissociation Energy CH_3-X [$kcal \cdot mol^{-1}$]	Bond Length (C−X) [Å]	Hydration Energy, X^- [$kcal \cdot mol^{-1}$]	Electronegativity (Pauling Scale)
Halide				
F	110	1.39	117	4.0
Cl	85	1.78	84	3.0
Br	71	1.93	78	2.8
I	57	2.14	68	2.5
H	99	1.09	—	2.2

Of the ten known organofluorine metabolites (Fig. 13), six are modified carboxylic acids (**19–24**) from the plant *Dichapetalum toxicarum*. While most of the compounds are elaborated by plants, the production of fluorothreonine (**27**) and nucleocidin (**9**) from streptomycetes suggests that bacteria may also be good sources for fluorinated structures (MEYER and O'HAGAN, 1992a).

Although this review deals exclusively with organohalogens, it is of passing interest to note the peculiarly high accumulation of the inorganic fluoride K_2SiF_6 in the marine organism *Halichondria moorei* to 10% of dry weight from seawater containing 1.3 ppm fluoride (GREGSON et al., 1979). This salt is a potent anti-inflammatory and may act as part of the defense mechanism of the organism.

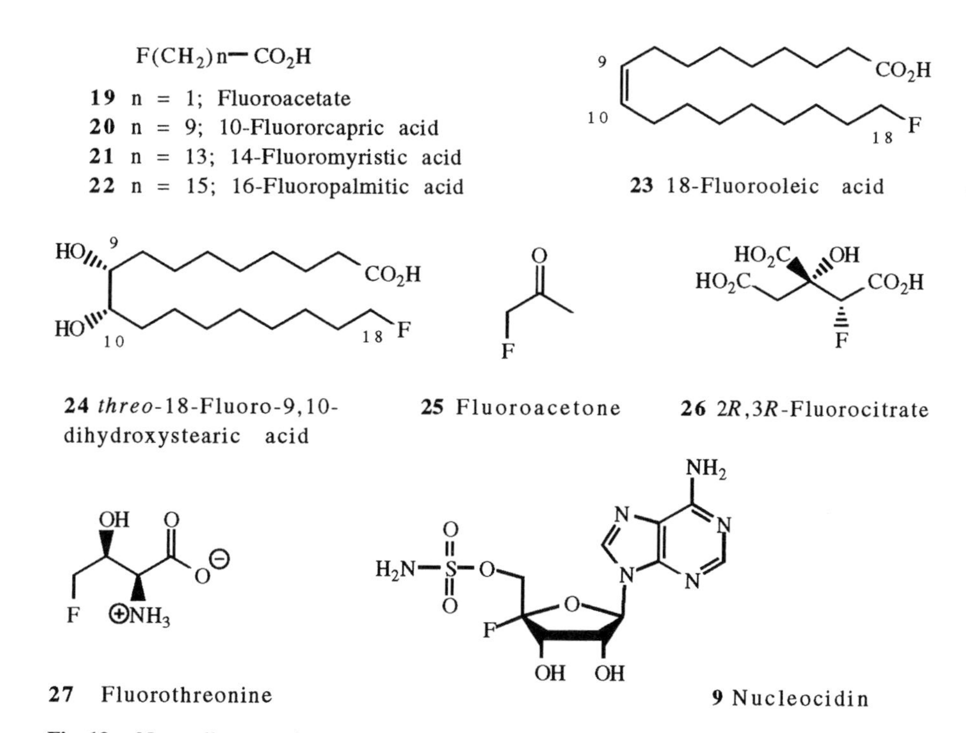

Fig. 13. Naturally occurring organofluorine compounds.

4.1.3 Fluoroacetate

Fluoroacetate (**19**) is the most ubiquitous fluorine metabolite (HARPER and O'HAGAN, 1994). It was first identified in the southern African plant *Dichapetalum cymosum*, and subsequently in over 40 other plant species from Africa, Australia, and South America. The richest single source is *D. braunii* seeds which contain up to 8000 ppm of dry weight, but no ω-fluorofatty acids (O'HAGAN et al., 1993).

Fluoroacetate (**19**) is also produced by the bacterium *Streptomyces cattleya* together with fluorothreonine (**27**). The mechanism of fluoroacetate biosynthesis has not been elucidated, but some possible routes are described in Sect. 5.3. Fluoroacetate is toxic to all cells, due to the lethal synthesis of 2*R*,3*R*-fluorocitrate (**26**). This is a competitive inhibitor of aconitrase (citric acid cycle) and probably binds irreversibly to mitochondrial citrate transporters (MEYER and O'HAGAN, 1992a).

4.1.4 ω-Fluorofatty Acids

Seeds from *Dichapetalum toxicarum* metabolize fluoroacetate to ω-fluorofatty acids, presumably as the starter unit for fatty acid synthetase(s). The major product is 18-fluorooleic acid (**23**), with minor quantities of the capric (**20**), myristic (**21**), and palmitic (**22**) analogs plus the dihydroxylated stearate (**24**) (HARPER et al., 1990). All of the ω-fluorofatty acids are very toxic to cells.

4.1.5 Fluoroacetone

Homogenates of plants such as *Acacia georginae* convert inorganic fluoride into volatile organofluorine compounds, the major component of which is fluoroacetone (**25**) (PETERS and SHORTHOUSE, 1971). It is believed that this pathway is an aberration of fatty acid synthesis. Fluoroacetone (**25**) has also been detected in livers of rats perfused with fluoroacetate (MIURA et al., 1956).

4.1.6 Nucleocidin

Nucleocidin (**9**) was isolated from *Streptomyces calvus*, found in an Indian soil sample. Plausibly, it could be formed by oxidation of the ribose ring of adenosine to the corresponding lactol, followed by substitution of the lactol hydroxyl group by fluoride and sulfamoylation (MAGUIRE et al., 1993).

4.1.7 Fluorothreonine

The second bacterial organofluoride, (2*S*,3*S*)-fluorothreonine (**27**) was discovered during attempts to optimize the production of thienomycin by *Streptomyces cattleya*. Fluoroacetate was also identified. The source of fluoride was traced to casein (containing 0.7% inorganic fluoride) used in the media. No fluorothreonine was produced when it was absent from the media and production could be restored by 2 mM potassium fluoride supplements (SANADA et al., 1986). The stereochemistry of fluorothreonine (**27**) is the same as natural (2*S*,3*R*)-threonine (i.e., L-threonine) (AMIN et al., 1997), although the Cahn-Ingold-Prelog assignment of C-3 is different because of the fluorine substitution.

4.1.8 Iodine

While iodine is less abundant than chlorine or bromine, a significant number of iodinated metabolites have been isolated to date, although they are rare compared to bromo and chloro natural products. However iodomethane has been detected in the oceans and measurements suggest up to $4 \cdot 10^6$ t a^{-1} is produced by marine algae and kelp. The tunicate *Didemnum* sp. produces an aromatic diiodide (**28**) to a concentration of 0.4% dry weight (Fig. 14) (SESIN and IRELAND, 1984), and the marine sponge *Geodia* sp. produces a cyclic peptide containing either a 2-iodo- or 2-bromophenol moiety (**29, 30**) (CHAN et al., 1987). The iodo analog (**29**) is also produced by the sponge *Pseudaxinyssa* sp. together with the chloroanalog (**31**) plus the corresponding nor-methyl compounds (**32–34**) (DE SILVA et al., 1990).

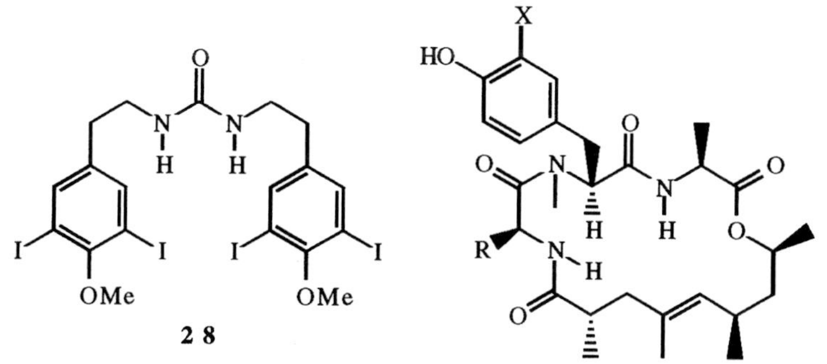

Geodiamolides A-F

R = Me

29 X = I; A

30 X = Br; B

31 X = Cl; C

R = H

32 X = I; D

33 X = Br; E

34 X = Cl; F

28

Fig. 14. Organoiodine compounds isolated from marine organisms.

The calicheamicins e.g. (**35–41**) are extremely potent DNA cleaving agents (enediyne antibiotics, SMITH and NICOLAOU, 1996) which are produced by *Micromonospora echinospora* ssp. *calichensis*. During attempts to optimize the fermentation conditions for the bromo-compounds (**37, 39**), sodium iodide was added to the media and the iodo compounds (**35, 36, 38, 40, 41**) were discovered (Fig. 15) (LEE et al., 1991, 1992).

In higher organisms, thyroxine (**42**) within the thyroid of mammals is the most well-known and studied iodocompound. Details of the action of the peroxidase will be dealt with below.

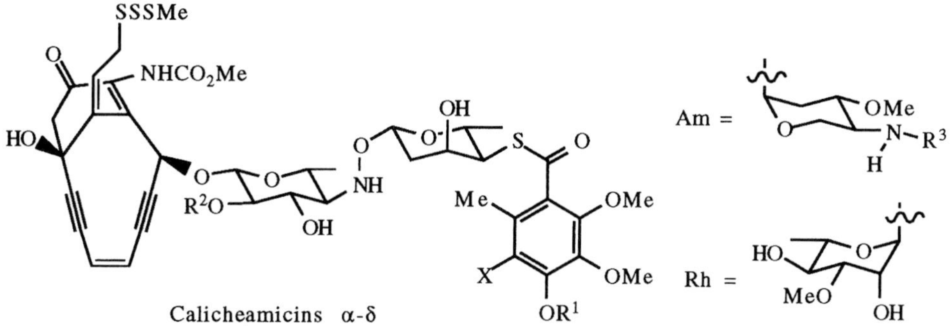

Calicheamicins α-δ

	X	R^1	R^2	R^3	
35	I	H	Am	Et	α_2^I
36	I	Rh	H		α_3^I
37	Br	Rh	Am	^iPr	β_1^{Br}
38	I	Rh	Am	^iPr	β_1^I
39	Br	Rh	Am	Et	γ_1^{Br}
40	I	Rh	Am	Et	γ_1^I
41	I	Rh	Am	Me	δ_1^I

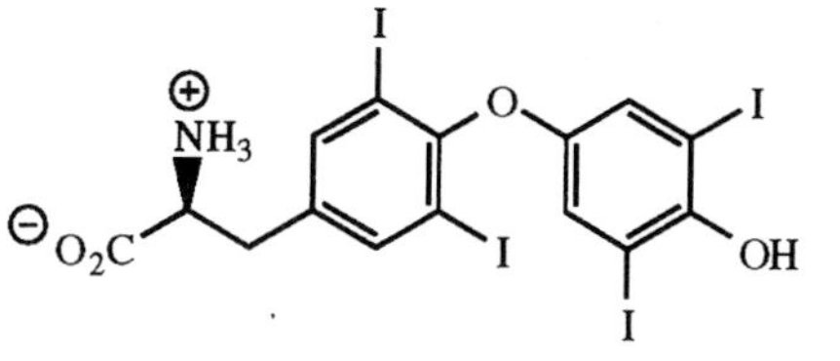

42 Thyroxine

Fig. 15. Organoiodine compounds.

5 Enzymology of Biohalogenation

5.1 Haloperoxidases

5.1.1 Basic Mechanism

Haloperoxidases catalyze halogenation by species which are "synthetically equivalent" to halonium ions (X^+). These are produced by oxidation of halide (X^-) by a peroxide (Fig. 16). Under normal aqueous conditions for an enzymatic reaction, the halonium ion will have hydroxyl as the counter ion and hence will be a hypohalous acid (HOX). The halide can be iodide, bromide, or chloride. The oxidant potential of the halides increase in the sequence $I^- < Br^- < Cl^- < F^-$. Peroxide is not able to provide the oxidation potential for fluorination and, therefore, there are no fluoroperoxidases.

The reaction in Fig. 16 is shown with hydrogen peroxide as the oxidant, however any peroxide (including one bound at an enzyme active site) can participate in the reaction. If the hypohalous acid is bound to an active site (i.e., Enz-OBr) it may react with organic substrates with useful regio- or stereoselectivity, however, if it diffuses into solution it will have identical reactivity to that of "chemically generated" hypohalous acid. The majority of current evidence suggests that free hypohalous acid is produced in most if not all cases, although results from a vanadium bromoperoxidase (TSCHIRRET-GUTH and BUTLER, 1994) and some non-heme haloperoxidases (BONGS and VAN PEE, 1994; FRANSSEN, 1994) indicate substrate binding. Haloperoxidase reactions in which free hypohalous acid is produced have little advantage over conventional chemical methods.

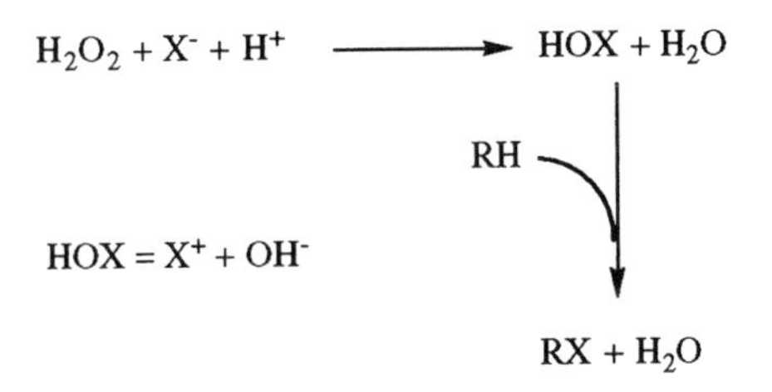

Fig. 16. Haloperoxidase mechanism.

The reactivity of a halonium ion (X^+) increases as the stability of the counterion increases. As noted above, in water, this will usually be hydroxyl to give a hypohalous acid (XOH). As the halide concentration increases, elemental halogen (X_2) or trihalide (X_3^-) will be formed which are more reactive. Under the normal conditions of haloperoxidase reactions such species will halogenate electron rich aromatics or undergo addition to double bonds. Iodine has the lowest oxidation potential and hence iodination is the most favored reaction, with bromination and chlorination requiring increasing levels of activation of the site for halogenation. Consequently, each iodoperoxidase is only able to use iodide as substrate, whereas a bromoperoxidase can brominate or iodinate and a chloroperoxidase can use all three halides. If halide concentrations are too low to give an appreciable reaction with the peroxide, this may also react, e.g., with alkenes to give epoxides.

Haloperoxidases are widespread among plant, bacterial, and animal species with the largest range having been found in algal and bacterial isolates (VAN PEE, 1996). Attempts to find useful haloperoxidase activity have spawned ventures into exotic locations. A team from the Scripps Oceanographic Institute studied marine red and green algae and found 50 bromoperoxidases (HEWSON and HAGER, 1980). The Cetus Corporation searched fungal populations in Death Valley and found 80 chloroperoxidases (HUNTER et al., 1984). The latter search yielded some potentially useful enzymes which will be discussed in more detail below.

5.1.2 Natural Sources of Haloperoxidases

Little is known of the role of haloperoxidases in nature, though it is thought that their primary function may be as part of the defense mechanism of the cell (FRANSSEN and VAN DER PLAS, 1992). Several algal metabolites are toxic to predatory organisms, in addition mammalian myeloperoxidase (LI et al., 1994) and eosinophil peroxidase both produce hypochlorous acid which kills target cells for leukocytes, including microorganisms such as *Staphylococcus aureus* (McCormick et al., 1994). Lactoperoxidase is able to produce hypobromous acid which oxidizes thiocyanate to the antimicrobial hypothiocyanate. Having highlighted this defensive role, there is such a diversity of halogenated metabolites, especially in algal and bacterial species that other roles for these compounds in cellular processes cannot be ruled out.

5.1.3 Haloperoxidase Classes

The haloperoxidases may be divided into four distinct categories according to the prosthetic group. In order of their relative abundance in the organisms studied to date, they are as follows:

(1) Heme (heme-thiolate) containing haloperoxidases;
(2) Vanadium haloperoxidases (Vilter, 1995; Butler and Walker, 1993);
(3) Non-heme non-metal haloperoxidases;
(4) Flavin/heme haloperoxidases.

Early studies indicated that all haloperoxidases contained a heme prosthetic group, with the iron atom ligated to a cysteine thiolate in an arrangement similar to that of the cytochrome P450s. In 1984 the first vanadium containing non-heme haloperoxidase was obtained from the brown alga, *Ascophyllum nodosum* (pigweed) (Vilter, 1984) and subsequently non-heme non-metal haloperoxidases and flavin/heme haloperoxidases were discovered. A summary of haloperoxidase producing organisms and isolated haloperoxidases is shown in Tab. 3. The heme chloroperoxidase of *Caldariomyces fumago* (which produces caldariomycin (**18**, Fig. 7)) is the most extensively studied as well as the most stable haloperoxidase in use at the present time. It can be produced at concentrations up to 280 mg L^{-1} in batch culture (Carmichael and Pickard, 1989; Pickard et al., 1991) but the enzyme displays no significant stereoselectivity. Preliminary X-ray crystal structure data has been reported (Sundaramoorthy et al., 1995).

5.1.4 Enzyme Mechanisms

5.1.4.1 Heme Haloperoxidase

The basic mechanism can be summarized as follows. Hydrogen peroxide reacts with the native enzyme producing "Compound I" in which the ferriprotoporphyrin IX prosthetic group is two oxidizing equivalents above the native state (Fig. 17, Step 1). Compound I is unstable and is able to react with alkenes (epoxidation), sulfides, or halide ions to give Compound EOX (Step 2). This is also very unstable and decomposes to yield the oxidized halide (Step 3) which is the halogenating agent for the organic substrate (Step 4). The haloperoxidase is inactivated by reaction with hypohalous acid and consequently the reaction rate slows down as the reaction progresses (Wagenknecht and Wogan, 1997).

The first heme haloperoxidase to be purified and characterized was the bromoperoxidase of *Penicillus capitatus*, a homodimer of 97 kDa subunits each containing one heme (Manthey and Hager, 1981). Recently, a chloroperoxidase and a bromoperoxidase were purified from *Streptomyces toyocaensis* (Marshall and Wright, 1996).

5.1.4.2 Vanadium Haloperoxidase

Non-heme vanadium haloperoxidases (Vilter, 1995; Butler and Walker, 1993) are generally more stable than heme enzymes to higher temperatures and organic solvent systems (De Boer et al., 1987) but the enzymes isolated to date generally have a lower specific activity than the heme enzymes.

All vanadium bromoperoxidases are similar in structure and amino acid composition, being acidic in nature with 65000 kDa subunits and 0.4 vanadium atoms/subunit. In the catalytic mechanism, hydrogen peroxide binds to the vanadium to form a vanadium–peroxo complex to which the halide ion binds. The resulting complex breaks down with the release of the oxidized halide species (HOX or X_2). This enzyme is also able to catalyze chlorination although at a slower rate than for bromination. The detailed role of the vanadium as an elec-

Tab. 3. Haloperoxidase Producing Cells and Isolated Enzymes

Source	Substrate(s)	Notes	Reference
Chloroperoxidase			
Caldariomyces fumago ATCC 16373, 58814 and 11925 (Woronichin) – heme containing	phenol ether; Cl, Br phenols; Cl, Br, I anilines; Cl, Br pyrazoles, pyridines; Cl nucleic bases; Cl, Br, I	isolated enzyme from Sigma cloned and expressed in *E. coli* and *Streptomyces lividans*	1
Streptomyces aureofaciens – vanadium	phenylpyrroles; Cl, Br nikkomycin; Br obscurolide; Br phenol; Br		1
Pseudomonas pyrrocinia – vanadium	indole; Cl, Br phenylpyrroles; Cl, Br		1
Notomastus lobatus – flavin/heme	phenols; Cl, Br		1
Curvularia inequalis		high stability to peroxide, purified and characterized	2
Eosinophil peroxidase		human gene cloned	3
Bromoperoxidase			
Pseudomonas aurofaciens ATCC 43051	pyrrol to pyrrolnitrin		
Ascophyllum nodosum – vanadium	phenols; Br, I phenol red; Br		5
Corallina pilulifera – vanadium	nucleic bases; Br, I phenols; Br		1
Streptomyces phaeochromogenes			
Penicillus captitatus (green alga) – heme			
Phanaerochaete chrysosporium ligninase – heme			4
Rhodophyta			
Chlorophyta			
Phaeophyta			
Milk, saliva, and tears			
Iodoperoxidase			
Lactoperoxidase from bovine milk – heme	phenols; I estrone; I	purified enzyme – Sigma, Calbiochem. immobilized on Sephadex beads – Sigma immobilized with glucose oxidase on enzymobeads – Bio-Rad bovine and human cDNAs cloned	1
Horseradish root		purified enzyme – Sigma, Calbiochem.	
Thyroid peroxidase	phenols; I		1
Turnip root			
Phaeophyta			

1: GRIBBLE (1992); 2: VANSCHIJNDEL et al. (1993); 3: ROMANO et al. (1992); 4: SCHNEIDER et al. (1996); 5: VAN PEE (1990).

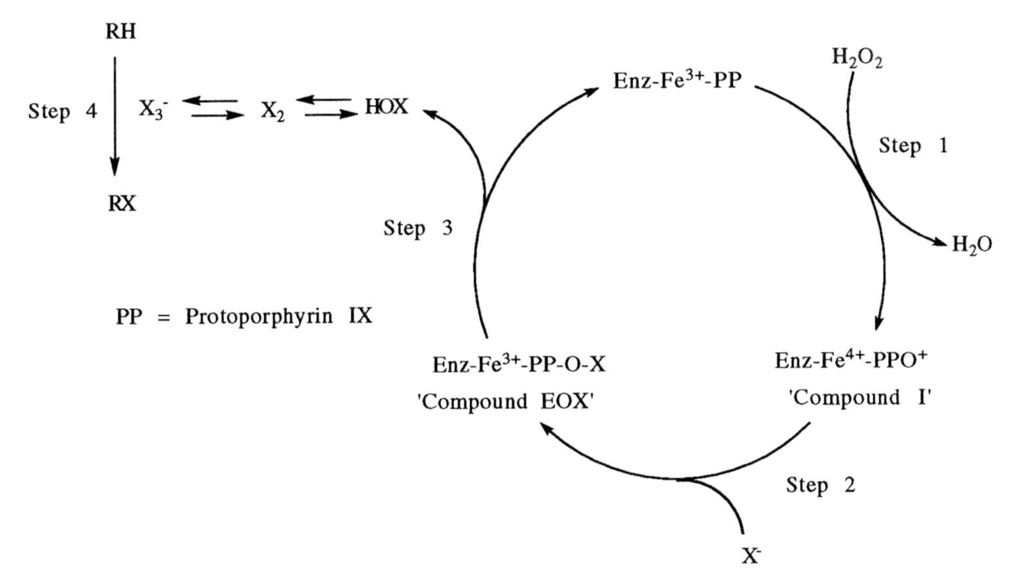

Fig. 17. Mechanism of heme haloperoxidases.

tron transfer catalyst or Lewis base is not known.

Of the non-heme vanadium-dependent haloperoxidases, the chloroperoxidase from *Curvularia inequalis* has been purified (SIMONS et al., 1995), and an X-ray structure for this enzyme is now available (MESSERSCHMIDT et al., 1996). The bromoperoxidases from the brown alga *Ascophyllum nodosum* (WEYAND et al., 1996; WEVER et al., 1985) and the alga *Corallina pilulifera* (ITOH et al., 1988) have also been well studied.

Less well studied but still well characterized are the bromoperoxidases from *Pseudomonas pyrrocinia* (WIESNER et al., 1985). *Pseudomonas aureofaciens* (VAN PEE and LINGENS, 1985a), *Streptomyces phaeochromogenes* (VAN PEE and LINGENS, 1985b), *Streptomyces aureofaciens* (VAN PEE et al., 1987), *Murex trunculis* (JANNUN and COE, 1987), *Xanthoria parietina* (PLAT et al., 1987), *Corallina officinalis* (SHEFFIELD et al., 1993), *Pseudomonas putida* (ITOH et al., 1994), and the chloroperoxidase from *Serratia marcescens* (BURD et al., 1995).

5.1.4.3 Non-Heme Non-Metal Haloperoxidase

The X-ray crystal structure of bromoperoxidase A2 from *Streptomyces aureofaciens* (ATCC 10762) shows a typical α/β hydrolase fold with an active site containing a catalytic triad consisting of Ser 98, Asp 228, and His 257 (HECHT et al., 1994). The non-heme non-metal haloperoxidase have substantial sequence homology with serine hydrolases, including the consensus motif Gly-X-Ser-X-Gly (PELLETIER and ALTENBUCHNER, 1995). Given this similarity, it is perhaps not surprising to find that the esterase from *Pseudomonas fluorescens* also acts as a bromoperoxidase as demonstrated by the bromination of monochlorodimedone (MCD) (**16a**) and phenol red (**17a**) (see Fig. 6). Site directed mutagenesis and inhibitor studies indicate the catalytic triad is essential for halogenation activity (PELLETIER et al., 1995). In an alternative proposal for the mechanism, it has been suggested that a polypeptide backbone *S*-halogenated methionine, may be the active halogenating agent (HAAG et al., 1991).

5.1.4.4 Flavin/Heme Haloperoxidase

This eight subunit enzyme was first isolated from the marine worm *Notomastus lobatus* (CHEN et al., 1991). The flavin moieties are contained within four identical polypeptide chains and two pairs of nonidentical subunits contain the heme groups. The ratio for optimal activity is 1 heme:1 flavin. *N. lobatus* contains a plethora of bromophenols, and it is therefore likely that the enzyme prefers bromide over chloride.

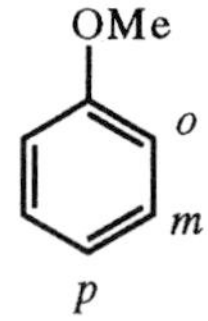

5.1.5 Reactions of the Haloperoxidases

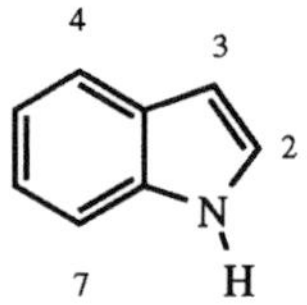

There are comprehensive reviews of haloperoxidase reactions (FRANSSEN, 1994; FRANSSEN and VAN DER PLAS, 1992; DRAUZ and WALDMANN, 1995) and detailed protocols for the halogenation of different substrates (NEIDLEMAN and GEIGERT, 1986).

5.1.5.1 Phenols, Anilines, and other Aromatics

The major substrate requirement for successful haloperoxidase catalyzed halogenation of aromatic compounds is activation of the ring structure. As a result, phenols, phenol ethers, anilines, and electron rich heterocyclic aromatics make good substrates. The transformation of phenol red (**17a**) to bromophenol blue (**17b**) (see Fig. 6) is often used as an indicator of haloperoxidase activity.

While the majority of aromatic halogenations shows no regioselectivity, it is worth noting that *Caldariomyces fumago* chloroperoxidase displays a preference for *para*-bromination of anisole (**43**, Fig. 18) (PICKARD et al., 1991). This enzyme has been used for large scale oxidation of indole (**44**) to oxindole (**45**) (SEELBACH et al., 1997). It has been claimed that *Pseudomonas pyroccinia* chloroperoxidase catalyzed chlorination of indole (**44**) gives 7-chloroindole (WIESNER et al., 1986), however reinvestigation (BONG, unpublished results) has shown that the product is 3-chloroindole as would be expected on the basis of electron density.

OMe o m p

4 3 2 N H 7

43 Anisole **44** Indole

O N H

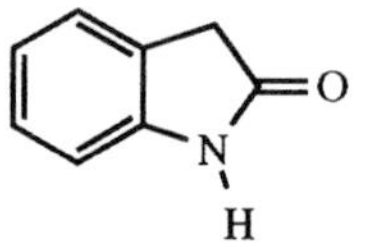

45 Oxindole

Fig. 18. See text.

5.1.5.2 Alkenes

Alkenes (**46**) are halogenated to α,β-halohydrins (**48**) by haloperoxidases (Fig. 19). If the halide concentration is sufficiently high, the intermediate cyclic halonium ion (**47**) will react with halide ion to give a dihalide (**49**). Substrates ranging from simple alkenes to bi-

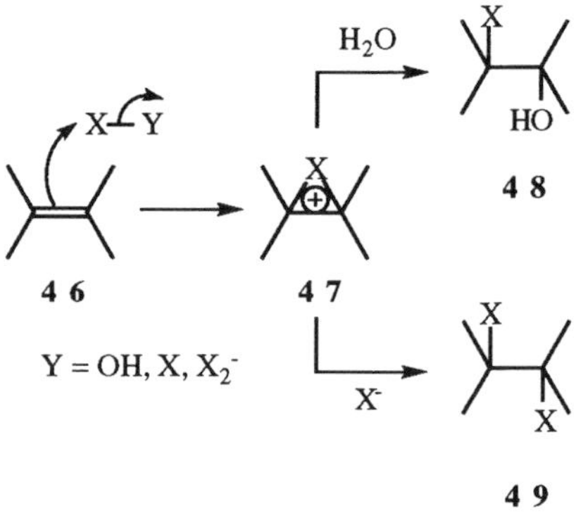

X–Y

H_2O

X HO

48

X ⊕

46 **47**

$Y = OH, X, X_2^-$

X^-

X X

49

Fig. 19. Mechanism of the halogenation of alkenes.

cyclic alkenes and steroids all undergo halogenation (COUGHLIN et al., 1993). The presence of certain functional groups leads to different products to the halohydrins, e.g., haloacetones are produced from unsaturated carboxylic acids (GEIGERT et al., 1983a).

The enzymatic halogenation of propylene was one of the few attempts made to utilize haloperoxidases in a commercial synthesis. Propylene is converted to propylene halohydrin by a haloperoxidase. Halohydrin epoxidase converts the propylene halohydrin to propylene oxide, regenerating the halide. Unfortunately this process was not economical and the pressure to move to halogen independent manufacturing processes ensured the method did not reach commercial reality (NEIDLEMAN, 1980).

5.1.5.3 Alkynes

Haloperoxidases catalyze the halogenation of alkynes (**50**, Fig. 20) to α-haloketones (**53**). As with alkenes, at higher halide ion concentrations dihalides (**54**) can be formed and the use of a mixture of different halide ions can yield heterogeneous dihalides (GEIGERT et al., 1983b).

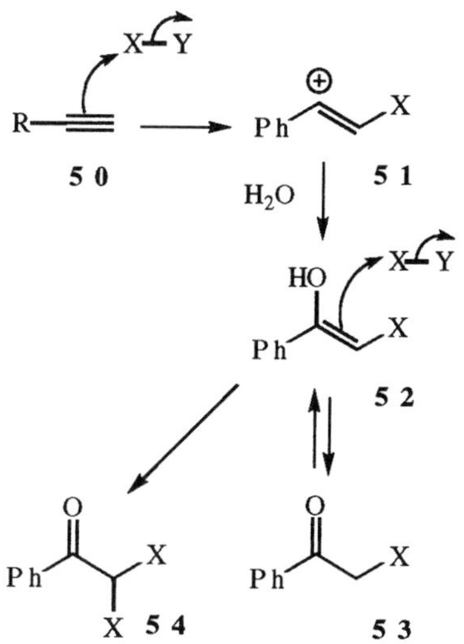

X–Y
R
50
Ph
X
51
H_2O
HO
X–Y
Ph
X
52
O
Ph
X
X
54
O
Ph
X
53

Fig. 20. See text.

5.1.5.4 Cyclopropanes

The cyclopropanes are converted to α,γ-halohydrins by haloperoxidases according to the same principles as for alkenes. One example of this reaction is the conversion of methylcyclopropane (**55**, Fig. 21) to 4-chloro-2-hydroxybutane (**56**) by *C. fumago* haloperoxidase (GEIGERT et al., 1983b).

5.1.5.5 β-Diketones

A wide range of diketones have been halogenated using haloperoxidases, with the major determinant of reactivity being the enol content of the substrate. The simple ketone 2-heptanone, with a low enol content, has low reactivity, whereas the diketone monochlorodimedone (MCD) (**16a**, Fig. 6), which is predominantly enol, has a very high reactivity. This reactivity, together with a marked decrease in UV absorbance upon halogenation has led to its use as a substrate for quantifying haloperoxidase activity.

5.1.5.6 Nitrogen Atoms

The reaction of haloperoxidases with amine compounds yields haloamines, most of which are unstable and rapidly decompose by deamination or decarboxylation, freeing the halide (GRISHAM et al., 1984).

5.1.5.7 Sulfur Atoms

The reaction of haloperoxidases with thiols produces the sulfenyl halide (—SX) which reacts with excess thiol to yield the disulfide (—S—S—) or with a nucleophile such as

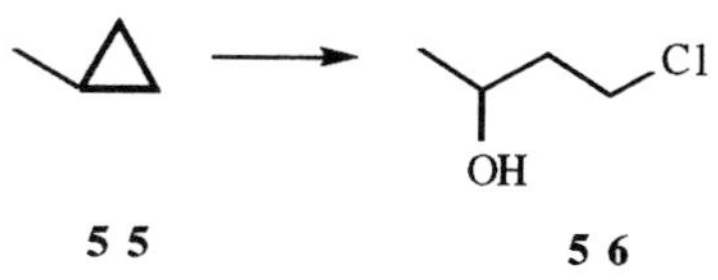

Cl
OH
55
56

Fig. 21. Cyclopropane cleavage by *Caldariomyces fumago* haloperoxidase.

hydroxyl anion to yield the sulfenic acid (−SOH), which can be oxidized to the sulfinic ($-SO_2H$) and sulphonic acid derivatives ($-SO_3H$). In common with other haloperoxidase mediated reactions, the reduction or removal of halide ions favors competitive peroxidase activity with, for example, *C. fumago* chloroperoxidase oxidizing dimethylsulfoxide to its sulfone (GEIGERT et al., 1983c).

5.1.5.8 Inorganic Halogens

The chloroperoxidase from *C. fumago* is able to oxidize iodine to iodate and to dismutate chlorine dioxide to chlorate (THOMAS and HAGER, 1968).

5.1.5.9 Stereo- and Regioselectivity

Several groups have attempted to find regio- or stereoselective biotransformations catalyzed by haloperoxidases. Bromohydrin formation from five cinnamyl substrates by the chloroperoxidase from *C. fumago*, showed no enantioselectivity, but some subtle differences in diastereoselectivity (COUGHLIN et al., 1993). Similarly, the *Corallina pilulifera* bromoperoxidase (non-heme) gave the same regiospecific transformations of anisoles as obtained with sodium hypobromite (NaOBr) and there was no enantioselectivity for bromohydrin formation from a variety of unsaturated substrates (ITOH et al., 1988).

C. fumago chloroperoxidase was used in the regioselective bromohydration of saccharide glycals (**57**, Fig. 22), to produce 2-deoxy-2-bromosaccharides (**58**) (FU et al., 1992; LIU and WONG, 1992). These have both biological activity and potential uses as synthons. The mild, specific halogenation of both sugars and peptides is a potentially rich source of applications for these enzymes.

There remain some puzzles, however, in both the enzymology and natural products chemistry. Most notably, the presence of multiple haloperoxidases in *Streptomyces aureofaciens*, none of which appear to be present in the biosynthetic pathway for 7-chlorotetracycline (**13b**, Fig. 4) which is abundant in these cells. It is an extraordinary fact, that despite the plethora of complex natural products containing halogens, thus far no enzyme has been isolated which is involved in their halogenation. It is inconceivable that complex polyhalocompounds such as aplysiaterpenoid A (**59**, Fig. 23) (cited by FRANSSEN, 1994) could be produced by halogenation using currently known haloperoxidases.

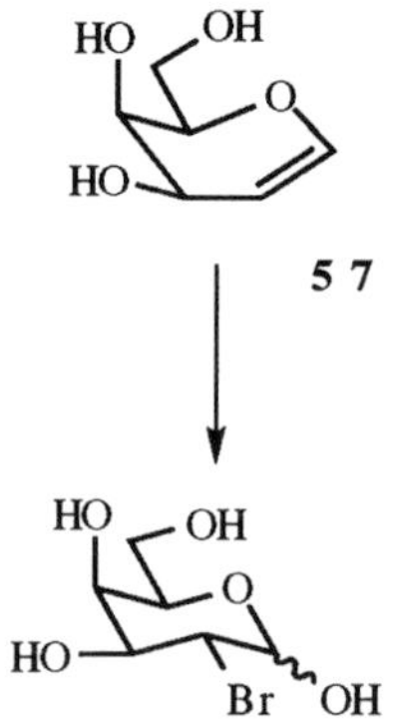

Fig. 22. See text.

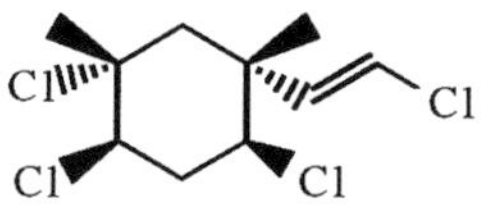

Fig. 23. Aplysiaterpenoid A.

5.2 Halide Methyltransferase

The study of the biosynthesis of chloromethane by the species *Hymenochaetae*, a family of white-rot fungi, has demonstrated a novel biohalogenation independent of that carried out by haloperoxidases (HARPER and HAMILTON, 1988). The mechanism appears to operate via a novel methyltransferase system whereby halide ions are directly methylated by a methyl donor such as S-adenosylmethionine. Normally, in the metabolism of these fungi, the chloromethane goes on to methylate aromatic acids such as benzoate and furoate, producing the corresponding esters. This reaction is also possible when chlorine is replaced by bromine or iodine, but fluoromethane is neither a substrate nor an inhibitor of the enzyme.

Methyl halide formation can be uncoupled from methylation under certain conditions, and it is then that the halomethane is emitted into the atmosphere. Although these emissions were originally thought to come mainly from fungal sources, an *in vivo* halide methyltransferase activity has been detected in a number of species of higher plants (SAINI et al., 1995) and the enzyme responsible for this reaction has been purified and characterized (ATTIEH et al., 1995). It is conceivable that these enzymes might be engineered to transfer alkyl groups more complex than methyl.

5.3 The Enzymes of Fluorine Metabolism

The biosynthesis of the majority of natural organofluorine compounds can be rationalized as anabolites of fluoroacetate (**19**). The terminally fluorinated capric (**20**), myristic (**21**), and palmitic acids (**22**) apparently result from fatty acid biosynthesis initiated by fluoroacetate. There is also presumably a 18-fluorostearic homolog which undergoes Δ9-desaturation to give 18-fluorooleic acid (**23**), which in turn undergoes *cis*-dihydroxylation to give the dihydroxylated stearate (**24**) (see Fig. 13) (HARPER et al., 1990).

Fluorocitrate (**26**) is formed by the action of citrate synthase on fluoroacetate (**19**) and oxaloacetate (**60**) in the "first step" of the Krebs cycle (Fig. 24). Removal of the *pro-R* proton from fluoroacetate (**19**), followed by attack of the *si*-face of oxaloacetate gives stereospecifically (2*R*,3*R*)-fluorocitrate (**26**).

Fluoroacetate (**19**) and fluorothreonine (**27**) are both produced by *Streptomyces cattleya* and labeling studies seem to indicate that they have a common origin. Feeding with [2-^{13}C]-glycine (**61**) (Fig. 25) gave fluoroacetate (**62**) containing ^{13}C in both carbons and fluorothreonine (**63**) with the label at C-3 and C-4. Feeding with [3-^{13}C]-pyruvate (**64**) resulted in incorporation into the fluoromethyl groups of both metabolites (**65**, **66**). The results are accommodated by a pathway (Fig. 25) in which C-2 of glycine (**67**) contributes both C-2 and C-3 of pyruvate (**69**) which in turn is incorporated intact into the fluorinated metabolites (**19**, **27**) (HAMILTON et al., 1997).

The importance of a three carbon intermediate such as pyruvate (**69**) was demonstrated by incorporation experiments with glycerol (Fig. 26) (TAMURA et al., 1995). When (2*R*)-[1-$^{2}H_2$]- and (2*S*)-[1-$^{2}H_2$] glycerol were fed independently to *S. cattleya* only the 2*R*-stereoisomer (**70**) yielded fluoroacetate (**71**) and fluorothreonine (**72**) containing two deuterium atoms. Glycerol kinase only phosphorylates glycerol (**73**) at the *pro-R*-hydroxylmethyl group, direct displacement of the phosphate group (**74**) (or of a derivative) by fluoride is consistent with the results (NIESCHALK et al., 1997). One derivative which might undergo this process is phosphoglycolate (**76**) to give 3-fluoropyruvate (**77**). This reaction has been demonstrated in reverse in several pseudomonads where a haloacetate halidohydrolase enzyme has been identified (WALKER and LIEN, 1981). The conversion of 3-fluoropyruvate (**77**) to fluoroacetate (**19**) also occurs in *Dichapetalum cymosum* (MEYER and O'HAGAN, 1992b, c).

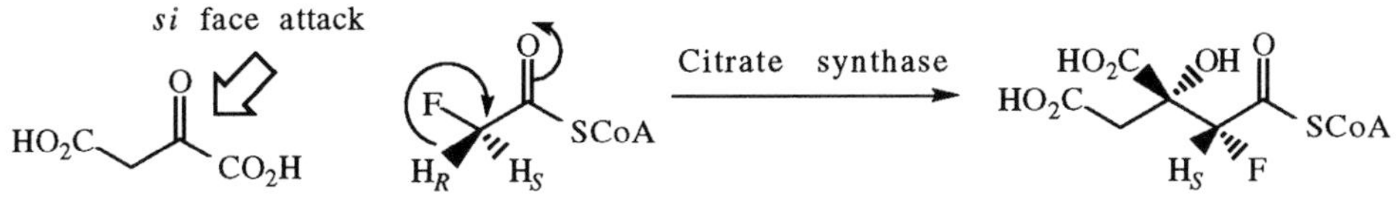

Fig. 24. See text.

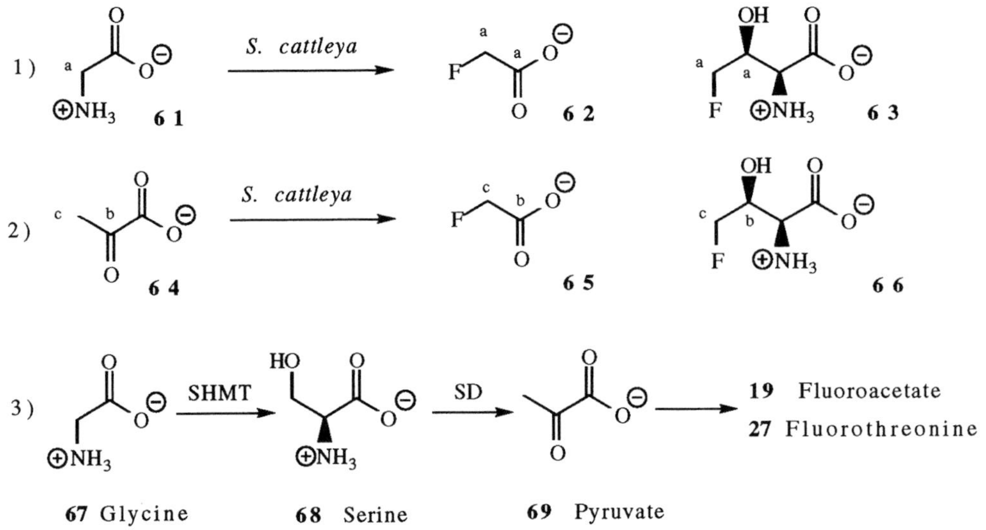

Fig. 25. Biosynthesis of fluoroacetate (**19**) and fluorothreonine (**27**).

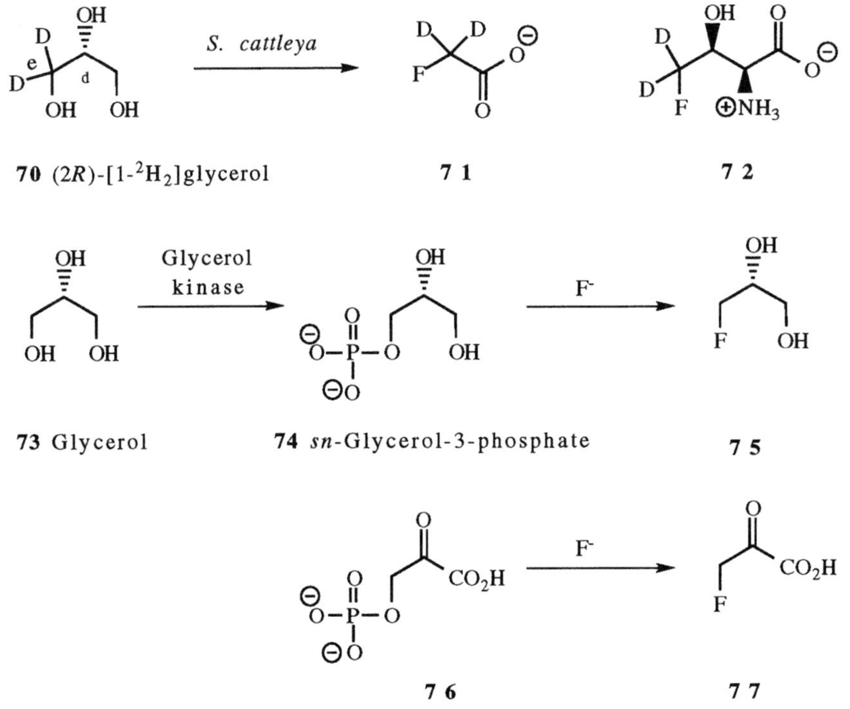

Fig. 26. The role of glycerol in fluoroacetate biosynthesis (**19**).

The labeling data above do not exclude elimination–addition mechanisms which have been discussed in detail by HARPER and O'HAGAN (1994). In one proposal, phosphoglycolic acid (**76**) forms an imine (**79a**) with pyridoxamine phosphate (**78**) (Fig. 27). Loss of a proton (**80a**) and elimination of the terminal phosphate gives a substituted acrylate (**81**) which undergoes nucleophilic addition of fluoride. Reversal of the steps (**80b**, **79b**, **78**) gives fluoropyruvate (**77**). This mechanism, first proposed by MEAD and SEGAL (1973), has been supported by several metabolic studies in *D. cymosum*, but it has yet to be verified by *in vivo* studies of fluoride metabolism in these plants.

The caterpillar *Sindris albimaculatus* accumulates fluoroacetate and can degrade it to carbon dioxide and fluoride. Attempts to force this system to operate in reverse with the enzme generating fluoroacetate have proved unsuccessful.

The possibilities of ethylene and chlorinated organics as precursors for fluoroacetate conversion have also been discussed by HARPER and O'HAGAN (1994), but the evidence for these is sparse. In one final suggestion, based on studies by FLAVIN et al. (1957), PETERS has proposed the presence of a fluorokinase, possibly an existing pyruvate kinase, which catalyzes the ATP- and CO_2-dependent incorporation of fluoride into phosphate (PETERS and SHORTHOUSE, 1964). Once again, he could not confirm this hypothesis in homogenates of plant cells and this lack of detailed metabolic proof leaves the whole question of the mechanism of fluorination open to the suggestions summarized above.

5.4 Advances Towards the Commercial Application of Biohalogenation

Despite the lack of overall success in the application of enzymes to halogenation reactions, several biotechnological approaches are making some headway into developing enzyme systems and applications that enable small steps in the development of processes. Certainly the advent of molecular biology has led to the cloning of chloroperoxidases from human eosinophils, *Caldariomyces fumago*, *Curvularia inequalis*, *Pseudomonas pyrrocinia*, and *Streptomyces aureofaciens* (Sect. 3). With the rapid advancement in expression technologies these should enable the relatively cheap production of haloperoxidases from many different species such that the cost of enzyme will not limit its application. There has also been much investigation of the stability of haloperoxidases under conditions of high temperature and in the presence of organic solvents (DE BOER et al., 1987). Both of these conditions are potentially useful for optimizing manufactur-

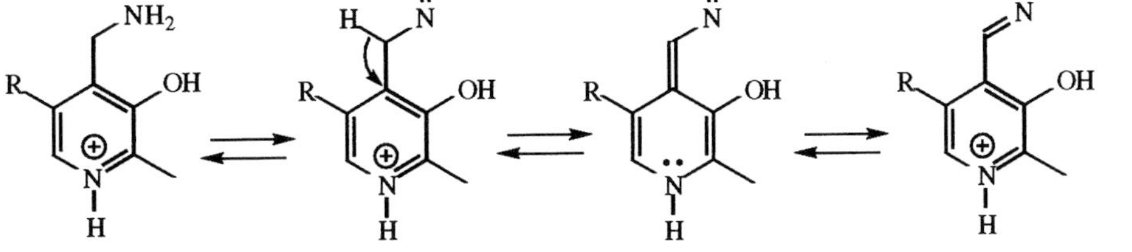

76 X = OPO_3^{2-}; Phosphoglycolic acid
77 X = F; Fluoropyruvate

78-81 R = $CH_2OPO_3^{2-}$
a X = OPO_3^{2-}
b X = F

78 **79ab** **80ab** **81**

Fig. 27. See text.

ing processes as they may be more cost-effective at higher temperatures, and the use of organic solvents may also increase the substrate repertoire that is available to the enzymes. As mentioned earlier (Sect. 5.1.1), the Cetus Corporation isolated several fungal chloroperoxidases from Death Valley, with particular emphasis on those with good resistance to high concentrations of hydrogen peroxide and hypochlorous acid. The chloroperoxidase from *Curvularia inequalis* lost no activity after incubation for 25 h in the presence of 200 mM hydrogen peroxide whereas that from *Caldariomyces fumago* was inactivated within 2 min. A similarly increased stability was observed in the presence of hypochlorous acid.

One further technique applied to haloperoxidases has been immobilization and chemical modification. These techniques have been used with many enzyme and whole cell systems because of the advantages of separating the catalyst from substrates and products, potential increases in stability of immobilized enzymes, and also potential to use enzymes at wider pH and temperature ranges. ITOH et al. (1987a) studied the immobilization of bromoperoxidase from *Corallina pilulifera* onto a variety of matrices utilizing the techniques of covalent binding, adsorption, and entrapment. They determined that ionic adsorption to DEAE-cellulofine and encapsulation in ENT-2000 matrix were the immobilization systems of choice for continuing process development. SHEFFIELD et al. (1994) extended this work with the enzyme from *Corallina officinalis* demonstrating high thermal stability and tolerance to organic solvents when the enzyme was immobilized on cellulose acetate. There are currently two commercially available immobilized enzyme products (see Tab. 3).

Unfortunately, while the promised market for enzymatic halogenation remains extremely good with so many halogenated organics being used as pharmaceuticals, agrochemicals, and fine chemicals, the nature of the reaction process in all enzymes studied to date does not allow for selectivity and therefore we await either the discovery of suitable enzymes from as yet unexplored sources or further developments in protein engineering such that specificity can be conferred upon existing enzymes. The lack of a binding pocket for organic substrates makes the latter approach very difficult to conceive without either the presence of additional molecules to protect certain groups or the incorporation of binding domains through protein engineering.

6 Dehalogenations

6.1 Introduction

Although organohalogens may arise naturally, many are anthropogenic inputs. These tend to be relatively resistant to degradation and, therefore, have become known as major recalcitrant pollutants. However, many of these compounds can be degraded or transformed in the environment by a range of biotic and abiotic processes. The ability of microorganisms to utilize halogenated compounds may have arisen by selective pressure and adaptation created by the presence of such compounds in their environment. Therefore an important relationship has been formed between the environmental exposure of microorganisms and halogenated organics, especially in terms of biodegradation, biotransformation, and biosynthesis. Many naturally occurring microorganisms have been isolated that degrade halogenated compounds to some degree. Some of them have been shown to work in axenic cultures but often, where more complex molecules or mixtures are involved, mixed cultures are required.

The removal of the halogen substituent is the key step in the degradation of halogenated compounds. This requires cleavage of the carbon–halogen bond, a very difficult reaction to achieve as exemplified by the bond dissociation energies given in Tab. 2.

During haloorganic degradation the dehalogenation step is key and maybe rate limiting. It may occur early or late in metabolism. There are five well characterized mechanisms identified for dehalogenation: reductive, oxidative, dehydrogenative, epoxide formation, and hydrolytic (encompassing thiolytic and hydration reactions).

6.2 Aliphatic Dehalogenations

Haloaliphatics encompass the haloalcohols, haloacids, haloalkanes, and their unsaturated derivatives. Chlorinated aliphatics are widely used as solvents, herbicides, and a variety of specialist applications, e.g., tetra- and trichloroethane are used in dry cleaning. Many chlorinated aliphatics are poorly degraded, mainly due to biochemical, rather than thermodynamic factors (PRIES et al., 1994). Their biochemical inertness is caused by an absence of metabolic routes from which energy and anabolic intermediates may be derived. In general, the degree of recalcitrance increases as the degree of chlorination increases and, therefore, dehalogenation is a critical step.

Twenty seven percent of the priority pollutants listed by the US EPA are haloaliphatics. Of these, nearly two thirds are haloalkanes and the rest comprises unsaturated derivatives and activated derivatives. It should be stated that metabolism of haloalkanes and haloaliphatic acids may overlap. If the halogen position on the alkane chain is not terminal, dehalogenation is likely to be preceded by oxidation of the terminal methyl group to yield the 2- or 3-haloaliphatic acids (YOKOTA et al., 1986). Therefore, the examples chosen to illustrate specific dehalogenations are to guide the reader through the myriad of possible reactions without giving an exhaustive list of examples.

Haloalkanes are formed by both natural and anthropogenic routes and are found in all environmental matrices. Haloalkanes can be metabolized in a variety of ways: as a carbon and energy source; cometabolised or partially degraded by microrganisms who are unable to utilize them as a growth source. A range of mono- and dichlorinated species ranging from chloro- or dichloromethane up to halogenated octadecane, can be degraded by a limited range of microorganisms. An extensive review of haloalkane metabolism by microorganisms is available (BELKIN, 1992).

6.2.1 Reductive Dehalogenation

Reductive dehalogenation is a two-electron transfer reaction which involves the release of halogen as a halogenide ion and its replacement by hydrogen. Haloalkane reduction occurs on a range of short chain (C_1 and C_2) chlorinated alkanes by methanogenic, denitrifying, and sulfate reducing organisms in reducing environments in nature (CURRAGH et al., 1994). The precise enzymatic nature of the transformation is not well defined but there is evidence that reduced porphyrin and corrin complexes are able to dehalogenate efficiently. The rate of reductive dechlorination, especially in anaerobic environments with highly halogenated substrates, is slow and may take several months.

Tri- or tetrachlorinated species may be reductively dehalogenated by strictly anaerobic bacteria causing sequential chloride release to produce dichloro- and monochloroderivatives, ultimately yielding methane. Alternatively anaerobic substitutive dehalogenation may totally degrade tri- or tetrachloromethane to form carbon dioxide, although this is possibly due to a nonenzymatic process. Both of these reactions have previously been shown in the same organism and the true catalyst has yet to be unequivocally verified (EGLI et al., 1988, 1990). The list of haloaliphatic compounds which have been shown to be reductively dechlorinated is not extensive but has been reviewed by MOHN and TIEDJE (1992) and FETZNER and LINGENS (1994). It is clear that apart from the haloaromatic transforming strain *Desulfomonile tiedje* DCB-1, the tetrachloroethylene transforming strain PER-K23 (see Fig. 8) is the only pure culture available for further studies into the energy production during reductive dehalogenation (HOLLIGER et al., 1993).

6.2.2 Oxidative Dehalogenation

Cleavage of a covalent carbon–halogen bond may proceed through polymerization of free radical intermediates (Fig. 28) or via oxi-

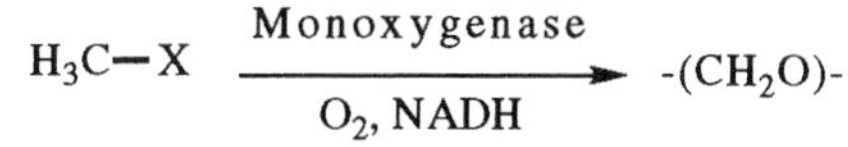

$$H_3C{-}X \xrightarrow[O_2,\ NADH]{\text{Monoxygenase}} \text{-}(CH_2O)\text{-}$$

Fig. 28. See text.

dative hydroxylation at the carbon atom adjacent to the halogen (Fig. 29). The former occurs due to the action of peroxidase, while the latter is oxygenase mediated. Such oxygenases are broadly distributed in nature and may be either mono- or dioxygenases. The action of oxygenase on haloalkanes leads to the formation of the corresponding aldehyde, while further dehydrogenation may yield the acid (Fig. 29). Such a reaction pathway has been shown for 1-chlorobutane grown *Corynebacterium* sp. strain m15-3 (YOKOTA et al., 1986, 1987) and for the ammonia monooygenase (RASCHE et al., 1990).

Oxidative dehalogenation of haloalkanes is usually mediated by a relatively nonspecific monooxygenase, which catalyzes the formation of halohydrins or halogenated epoxides: these spontaneously decompose to aldehydes, releasing the halide ion (CURRAGH et al., 1994). Generally oxidative dehalogenation only occurs with short chain halocarbons comprised of 1–4 carbons: e.g., monohaloethanes and *n*-alkanes are dehalogenated by the ammonia monooxygenase of the nitrifying bacterium *Nitrosomonas europaea*, and the methane oxygenase found in methanogenic microorganisms can dehalogenate C_1 and C_2 compounds.

In general, oxygenase catalyzed dehalogenation reactions of short chain haloalkanes are due to multifunctional enzymes with broad substrate specificity (methane monooxygenase) or involve enzymes from aromatic degradative pathways (toluene dioxygenase). Oxidation can lead to dehalogenation as a result of the formation of labile products, not true dehalogenation. There are no reports of oxygenolytic dehalogenases specific for halogenated alkanes (FETZNER and LINGENS, 1994).

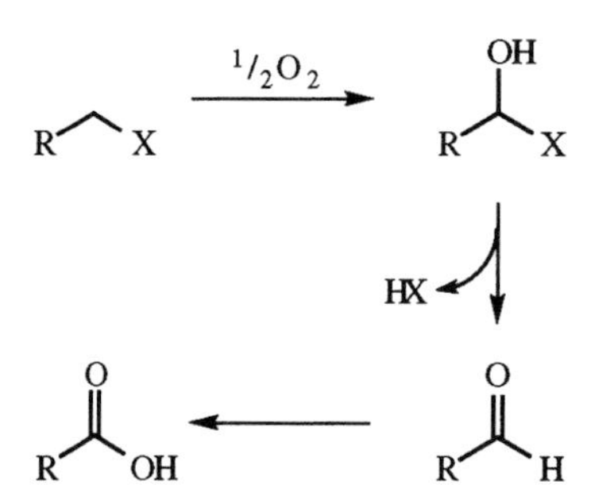

Fig. 29. See text.

6.2.3 Hydrolytic Dehalogenation

Hydrolytic dehalogenation is perhaps the commonest route for the dehalogenation and transformation of haloaliphatic compounds. As outlined above, the hydrolytic transformation of a haloalkane may subsequently give rise to a haloalcohol and a haloacid before dehalogenation occurs. Therefore, it is possible to have multiple dehalogenating activities in one organism. Several organisms able to aerobically degrade dichloroethane have been isolated (KEUNING et al., 1985). The initial step is the hydrolytic removal of one chlorine to form the haloalcohol, chloroethanol. The enzyme responsible has since been cloned and sequenced (JANSSEN et al., 1989) and the structure elucidated (FRANKEN et al., 1991; VERSCHUEREN et al., 1993a, b). Following the formation of the chloroethanol, further oxidation via the aldehyde and acid occurs. It is then at this stage that the second chlorine is removed to yield glycolate. Several enzymes able to catalyze the dehalogenation of the haloacid have been purified (KLAGES et al., 1983; MOTOSUGI et al., 1982) and some have been cloned and sequenced (SCHNEIDER et al., 1991).

During hydrolytic dehalogenation of haloalkanes nucleophilic substitution occurs, in which the halogen atom is replaced by a hydroxyl ion to give the corresponding alcohol (Fig. 30). These reactions are catalyzed by a halidohydrolase-type dehalogenase. The haloalkane dehalogenases can be divided into two classes depending on their substrate specificity. Those which have a fairly restricted range of substrate specificities are gram-negative strains, best exemplified by *Xanthobacter autotrophicus* GJ10 (JANSSEN et al., 1989) for which a detailed mechanism has been determined from X-ray crystal structures. An aspartate residue is alkylated by the haloalkane to give

$$\mathrm{R{\frown}X} \xrightarrow{\ominus\mathrm{OH}} \mathrm{R{\frown}OH}\ \ \mathrm{X}^{\ominus}$$

Fig. 30. See text.

an ester and the displaced halide ion is bound to the ring NHs of two tryptophan residues. A histidine/aspartate pair assist the hydrolysis of the aspartate ester to give the alcohol (Verschueren et al., 1993a, b). The second class of haloalkane dehalogenases is represented by those enzymes isolated from a group of closely related gram-positive organisms. *Rhodococcus erythropolis* strain Y2 (Sallis et al., 1990), *Arthrobacter* sp. strain HA1 (Scholtz et al., 1987), and *Corynebacterium* sp. strain m15-3 (Yokata et al., 1987) all demonstrate much broader substrate specificities and are suggested to be grouped together using the system of Slater et al. (1995).

The haloalcohol dehalogenases are a distinct group of enzymes which result in the formation of epoxides. These are dealt with in a separate section but are still hydrolytic enzymes acting on haloaliphatic compounds (see Sect. 6.2.5).

Of greatest interest and by far the most studied systems are those catalyzing the transformation of the haloaliphatic acids – the haloalkanoic acids, the haloacids, and the haloacetates. The enzymes catalyzing the transformation are broadly divided into two groups of halidohydrolases, namely the haloacetate dehalogenases (EC 3.8.1.3) and the 2-haloacid dehalogenases (EC 3.8.1.2). The division of these enzymes was based on their substrate range, haloacetate dehalogenases acting exclusively on haloacetates and 2-haloacid dehalogenases acting on haloacetates and other short chain fatty acids (C_2–C_4). The two groups of enzymes appear to have non-overlapping activity spectra (Janssen et al., 1985). Both types of halidohydrolases preferentially dehalogenate in the order F>Cl>I>Br. The mechanism of displacement proceeds through a S_N2-type reaction and has been found to differ, depending on the enzyme. Slater et al. (1995) categorized the enzymes into four classes, based on their enantiospecificity and ability to invert the substrate–product configuration. Class 1L and 1D catalyze the inversion of the product configuration compared with the original substrate and are D (Class 1D) or L (Class 1L) specific. Class 2I and 2R are able to dehalogenate both the D and L-enantiomers with either retention (Class 2R) or inversion (2I) of product configuration compared with the original product configuration.

Within these classifications, 1L would appear the most common, having been shown to be present in strains of *Pseudomonas* (Schneider et al., 1991; Kawasaki et al., 1994; Nardi-Dei et al., 1994; Jones et al., 1992; Murdiyatmo et al., 1992), *Moraxella* (Kawasaki et al., 1992) and *Xanthobacter* (van der Ploeg et al., 1991).

A final group of haloaliphatic transforming enzymes are best classified as those glutathione (GSH) dependent dehalogenases. Stucki et al. (1981) isolated a *Hyphomicrobium* species able to dechlorinate dichloromethane which was GSH dependent. The mechanism of these enzymes involves a nucleophilic substitution by GSH, yielding an S-chloromethyl glutathione conjugate which is dehalogenated when it rapidly undergoes hydrolysis in an aqueous environment. The thiohemiacetal thus formed is then in equilibrium with GSH and formaldehyde (Kohler-Staub and Leisinger, 1985). The enzyme has only been shown to work with dichloromethane and a restricted substrate range.

6.2.4 Dehydrohalogenation

Dehydrohalogenation occurs with simultaneous removal of the halogen atom and the hydrogen atom on the neighboring carbon (Fig. 31). This results in the formation of alkenes from simple halogenated hydrocarbons. The major substrates which appear to be transformed by this route are the lindane (**82**, Fig. 32) pesticides, a halogenated cyclohexane, and some steps in the transformation of dichlorodiphenyltrichloroethane (DDT), a haloaromatic pesticide.

The transformation of lindane (γ-hexachlorocyclohexane, γ-HCH) (**82**) by a dehydro-

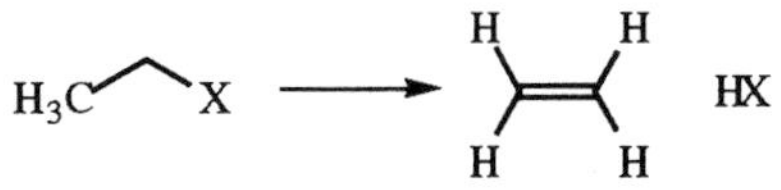

Fig. 31. Dehydrohalogenation.

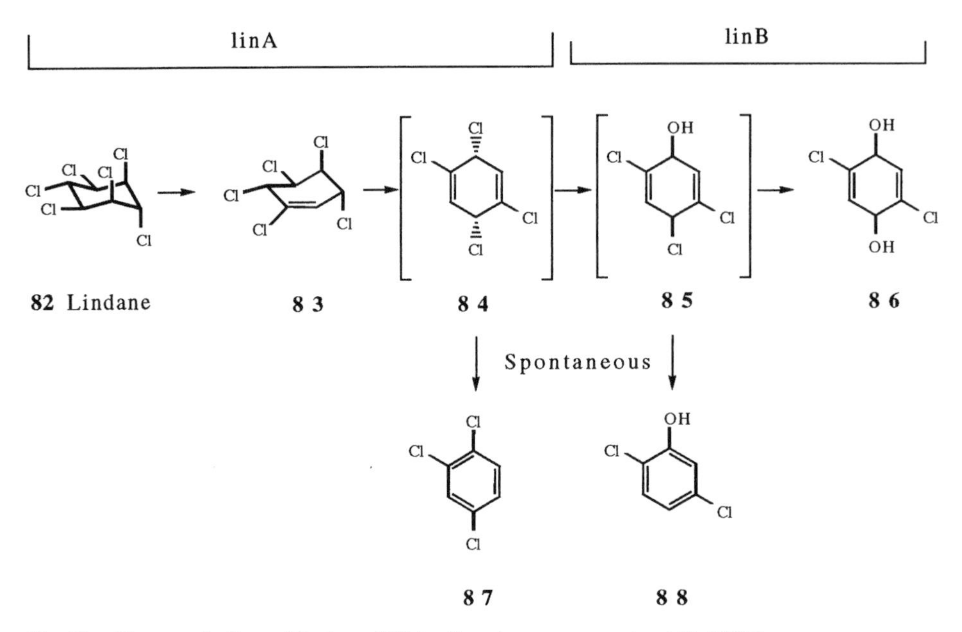

Fig. 32. The metabolism of lindane (**82**) by *Pseudomonas paucimobilis* UT26.

chlorinase is probably the best studied of this limited group of enzymes. Lindane (**82**) was first shown to be aerobically transformed by Tu (1976) and Haider (1979). In *Pseudomonas paucimobilis* UT26, the proposed pathway of lindane (**82**) degradation involves dehydrochlorination to yield γ-pentachlorocyclohexene (**83**) and 1,3,4,6-tetrachloro-1,4-cyclohexadiene (**84**) which may decompose to the dead-end product 1,2,4-trichlorobenzene (**87**) or be acted upon by the hydrolytic dehalogenase LinB (Fig. 32) to give the dichlorodihydroxycyclohexadiene (**86**). Imai et al. (1991) have cloned and sequenced the gene encoding the initial dehydrohalogenase LinA. It has no sequence homology with any bacterial or eukaryotic glutathione S-transferase and did not show DDT dechlorinase activity in the presence of glutathione.

Apart from the restricted range of dehydrochlorinases which have been shown, it is also clear that LinA from *Pseudomonas paucimobilis* UT26 has a very limited substrate range, only converting α-HCH, γ-HCH (**82**), δ-HCH, α-pentachlorocyclohexene, and γ-pentachlorocyclohexene (**83**) (Nagata et al., 1993). Other, less well characterized enzyme systems responsible for the transformation of lindane (**82**) are suggested to be extrachromosomally located and present in *Pseudomonas ovalis* CFT1 and *Pseudomonas tralucida* CFT4 (Johri et al., 1991).

A dehydrohalogenase or dechlorinase, isolated from the house fly (*Musca domestica*), (Lipke and Kearns, 1959) is a glutathione S-transferase (Ishida, 1968) which catalyzes the monodehydrochlorination of DDT.

β-Chloroalanine and serine-O-sulfate form imines (**80**) with pyridoxal phosphate (see Fig. 27). These readily eliminate to give 2-amino acrylates (**81**), which in turn undergo conjugate addition of nucleophiles. Hydrolysis of the imine yields amino acids with a new β-substituent (Contestabile and John, 1996). Using tryptophan synthase or tyrosine phenol lyase this has been developed into a powerful method for the synthesis of β-substituted alanines (Nagasawa and Yamada, 1986).

6.2.5 Epoxide Formation

Epoxidation results from the simultaneous removal of the halogen atom and the hydro-

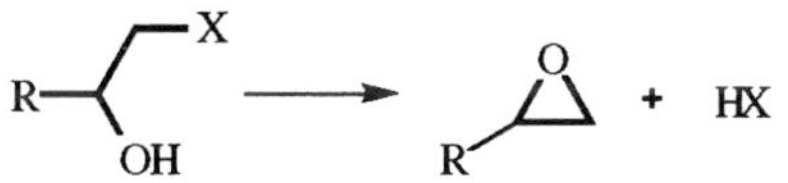

$$R-CH(OH)-CH_2X \longrightarrow R\text{-epoxide} + HX$$

Fig. 33. See text.

gen atom from the adjacent hydroxyl group on the neighboring carbon (Fig. 33).

Although the reaction was first shown nearly 30 years ago (CASTRO and BARTNICKI, 1968) very few studies have extended this finding. VAN DEN WIJNGAARD et al. (1989) have isolated the strains AD1 (*Pseudomonas* sp.), AD2 (*Arthrobacter* sp.) and AD3 (coryneform) and shown that a variety of vicinal haloalcohols may be degraded. The formation of epoxides from a range of haloalcohols (C_2–C_3) has been shown for the enzyme purified from AD2 (VAN DEN WIJNGAARD et al., 1991). The enzyme requires no cofactors or oxygen and probably acts as a simple acid–base catalysis. The haloalcohol dehalogenases from *Corynebacterium* sp. strain N-1074 are the best studied. In the wild-type strain, two haloalcohol transforming enzymes and two epichlorohydrin transforming enzymes have been isolated. The haloalcohol transforming enzymes catalyzed the transformation of halohydrins into epoxides and the reverse reaction. However, they were shown to differ in their enantioselectivity for 1,3-dichloro-2-propanol conversion as well as their substrate range, subunit composition, and immunological reactivity (NAKAMURA et al., 1992). As outlined above for the other hydrolytic enzymes, SLATER et al. (1995) have attempted to classify the haloalcohol dehalogenases on the basis of their substrate specificities. It is clear that the number of examples limits this exercise but it may be significant that the N-terminal amino acid sequence of the enzyme from *Corynebacterium* sp. strain N-1074 is similar to that of *Arthrobacter* strain AD2 (NAGASAWA et al., 1992).

6.3 Aromatic Dehalogenation

Chlorinated aromatic compounds have been widely used as solvents, herbicides, fungicides, and insecticides. The aromatic moiety may range from the the monocyclic (chlorobenzenes) to the polycyclic (polychlorinated biphenyls PCBs). Each may be modified to give a range of haloaromatic pesticides (e.g., 2,4-D, dichlobenil) and other anthropogenic inputs such as dioxins. Natural haloaromatics range from the esoteric 1,2,3,4-tetrachlorobenzene (MILES et al., 1973) to the EPA priority pollutant 2,4-dichlorophenol *produced* by a *Penicillium* sp. (ANDO et al., 1970).

As outlined previously for chlorinated aliphatics (Sect. 6.2), the removal of the halogen substituent is a key step in the degradation of halogenated aromatic compounds. This may occur either as an early reaction, catalyzed by a true dehalogenase, or at a later stage in metabolism, often resulting from the chemical decomposition of unstable primary products of an unassociated enzyme reaction.

Generally, initial halogen removal occurs via reductive, oxygenolytic, or hydrolytic mechanisms while later removal may be a fortuitous event caused by a separate abiotic reaction after the loss of aromaticity. There is an extensive discussion of the microbial metabolism of haloaromatics (REINEKE and KNACKMUSS, 1988).

6.3.1 Reductive Dehalogenation

Under a variety of redox conditions the reductive dehalogenation of a wide cross section of haloaromatics has been demonstrated and discussed extensively (MOHN and TIEDJE, 1992; WACKETT and SCHANKE, 1992). Halobenzenes (RAMANAND et al., 1993), halobenzoates (HAGGBLOM et al., 1993), haloanilines (KUHN et al., 1990), and halocatechols (NEILSON et al., 1987) are only a few of the compounds which have been shown to be reductively dechlorinated. Pentachlorophenol (**89**), which was widely used as a herbicide and wood preservative, undergoes reductive dehalogenation with sequential removal of the chloride substitutions to give 3-chlorophenol (**94**) as the final product (Fig. 34).

It is also evident from the majority of studies performed to date that very few axenic cultures able to perform reductive dehalogenation have been studied. *Desulfonomile tiedje* DCB-1 is one of the few organisms which have been shown to perform reductive dechlorina-

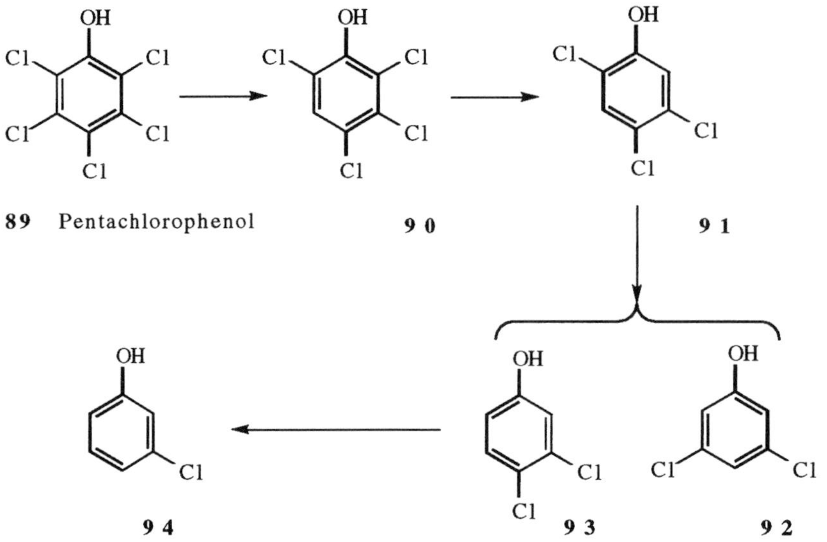

89 Pentachlorophenol **90** **91** **94** **93** **92**

Fig. 34. Degradation of pentachlorophenol.

tion and has become a benchmark organism for these studies. SHELTON and TIEDJE (1984) showed that this organism can grow and derive energy from the dechlorination of 3-chlorobenzoate. The rate of dehalogenation depends upon the number and position of the halogen substitutions, since the presence of the carbonyl group favors removal at *meta* > *ortho* > *para* positions.

Due to the global environmental concerns many studies have been made on the reductive dechlorination of polychlorinated biphenyls. Using Hudson River sediments BROWN et al. (1987) showed that there was an alteration in the PCB congener distribution over a long period of time; highly chlorinated congeners were removed and there was an increase in the lower chlorinated species. It has been shown that the pattern of dechlorination is specific for the particular sediment from which it is derived with dechlorination occurring specifically at *meta* and *para* chloro substituents. A consortium devoid of sediment has never been shown to anaerobically dechlorinate PCBs. It has been demonstrated that as the degree of chlorination increases, so does the probability of dechlorination, and as the number of mono- and dichlorinated congeners increases, so the role of the anaerobes decreases (ABRAMOWICZ, 1990).

6.3.2 Oxidative Dehalogenation

The commonest mechanisms in (halo)aromatic degradation is the introduction of hydroxyl groups to facilitate ring cleavage. Both monooxygenases and dioxygenases may play a role in haloaromatic dehalogenation. In monocyclic aromatic metabolism, the key ring fission substrates (which may be halogenated) are, in order of importance: catechol, protocatechuate, and gentisate. Subsequently, the aromatic ring is cleaved and the halogenated molecule is labilized by the further incorporation of oxygen. Halide removal is thereafter spontaneous and is associated with normal aromatic catabolism.

6.3.2.1 Dioxygenase Catalyzed Dehalogenation

Oxygenolytic attack on halogenated aromatic hydrocarbons may cause cleavage of the chlorine atom fortuitously when both atoms of molecular oxygen are incorporated into the aromatic nucleus. These reactions may be catalyzed by an aromatic dioxygenase and require two adjacent carbons, only one of which is halosubstituted. Of the limited substrates and dioxygenases which fulfill these criteria the metabolism of 2-chlorobenzoate is the best studied (FETZNER et al., 1989, 1992). The 2-halobenzoate-1,2-dioxygenase from *Pseudomonas cepacia* 2CBS, which catalyzes the formation of catechol from 2-halobenzoates, is a multicomponent enzyme which is structurally related to benzoate 1,2-dioxygenase and the TOL plasmid encoded toluate 1,2-dioxygenase. Despite this structural similarity, neither the benzoate nor the toluate 1,2-dioxygenases are able to convert the *ortho* substituted benzoates. Furthermore, although the enzyme from *P. cepacia* 2CBS showed broad substrate specificity, it demonstrated a preference for 2-halosubstituted benzoates, justifying its inclusion and terminology as a dehalogenase. Other oxygenolytic haloaromatic dehalogenases have also been shown to be multicomponent enzymes. Further 2-chlorobenzoate-1,2-dioxygenases have been identified in *P. putida* CLB250 (ENGESSER and SCHULTE, 1989) and *P. aeruginosa* JB2 (HICKEY and FOCHT, 1990) and shown to be plasmid encoded in strain *P. putida* P111 (BRENNER et al., 1993). Dihydroxylation of aromatics is covered in detail in Chapter 10.

Of significant environmental concern, and reliant upon oxygenolytic dehalogenation, is the degradation of PCBs. The aerobic degradation of lightly chlorinated PCBs has been shown to involve an upper pathway in which, e.g., biphenyl (**95**) is dihydroxylated (**96**), rearomatized (**97**), cleaved to a ketoacid (**98**), and then further cleaved to benzoic acid (**99**) (Fig. 35).

This upper pathway has a broad substrate specificity and has been demonstrated in a variety of microbial taxa including *Achromobacter* (AHMED and FOCHT, 1973), *Alcaligenes*, and *Acinetobacter* (FURUKAWA et al., 1978). The action of the four enzyme system results in oxygenolytic cleavage of one of the biphenyl rings to produce (chlorinated) benzoates. Since the early work on microbial degradation many studies have been undertaken. These have been summarized and some general observations have been made (ROBINSON and LENN, 1994):

- Degradation decreases as the extent of chlorination increases.
- *Ortho* substituted biphenyls have a lower degradation rate.
- Ring cleavage always occurs in the least or non-chlorinated ring.
- Degradation decreases if both rings are chlorinated (4′-substituted biphenyls result in the formation of a yellow *meta* cleavage product).

In summary, many dioxygenases are able to catalyze the incorporation of dioxygen into a (halo)aromatic compound and may cause dehalogenation. We have attempted to highlight

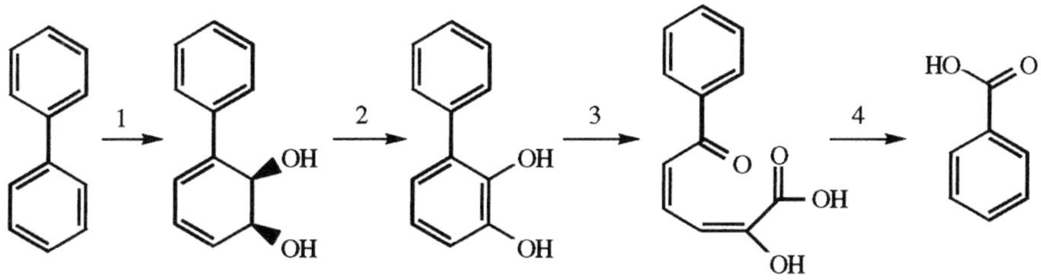

95 96 97 98 99

1 Biphenyl-2,3-dioxygenase
2 Dihydrodiol dehydrogenase
3 2,3-Dihydrobiphenyl-1,2-dioxygenase
4 2-Hydroxy-6-oxo-6-phenylhexa-2,4-dienoic acid hydrolase

Fig. 35. PCB upper pathway.

some of those enzymes which cause dehalogenation due to relaxed substrate specificity and not those enzymes simply associated with spontaneous elimination of halide after cleavage of the aromatic ring.

6.3.2.2 Monooxygenase Catalyzed Dehalogenation

Despite having a high degree of chlorination some microorganisms have been shown to be able to degrade pentachlorophenol (**89**) under aerobic conditions (Fig. 36) (RADEHAUS and SCHMIDT, 1992; SCHENK et al., 1990; TOPP and HANSON, 1990). The initial reaction has been shown to be an NADPH dependent oxygenation to form tetrachlorohydroquinone (**100**). The purified enzyme from *Flavobacterium* sp. strain ATCC 39723 was shown to catalyze the incorporation of $^{18}O_2$ (XUN et al., 1992b). Pentachlorophenol-4-hydroxylase from *Flavobacterium* sp. strain ATCC 39723, has been shown to be a flavoprotein monooxygenase catalyzing the *para* hydroxylation of a broad range of substituted phenols. The requirement for NADPH was dependent upon the leaving group with 1 mole electron donating groups (H and NH_2) and electron withdrawing groups (F, Cl, Br, NO_2, and CN) requiring 1 mole or 2 mole NADPH, respectively (XUN et al., 1992a). Although other isolates able to degrade halophenols have been isolated (GOLOVLEVA et al., 1992; KIYOHARA et al., 1992), the monooxygenase catalyzed dehalogenation would appear to be restricted to the metabolism of more highly halogenated compounds, with the less halogenated aromatics being dehalogenated via the dioxygenase or post cleavage routes outlined above.

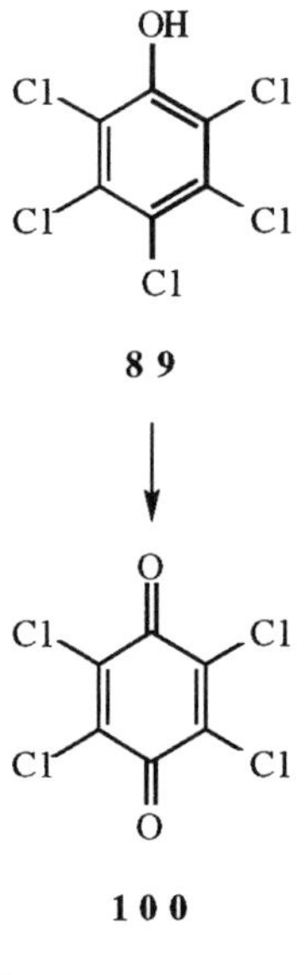

Fig. 36. See text.

6.3.3 Hydrolytic Dehalogenation

The replacement of a halogen with a hydroxyl from water is rare in comparison with the oxidative dehalogenations previously outlined. The delocalization of the π-electrons contributes to the stability of the aromatic ring system but it does not preclude the direct hydrolysis of the carbon–halogen bond. Various s-triazines have been shown to be hydrolytically transformed (COOK and HUTTER, 1986) but the central focus of hydrolytic haloaromatic dehalogenation has been the initial step in the transformation of 4-chlorobenzoate. It has been suggested that 3-chlorobenzoate may also be metabolized by a similar route (JOHNSTON et al., 1972) but this is the only reported example of 3-chlorobenzoate transformation not involving a dioxygenase.

The halidohydrolase responsible for catalyzing the dechlorination of 4-chlorobenzoate (**101**) has been extensively characterized in *Pseudomonas* sp. strain CBS3 (Fig. 37). The multicomponent enzyme system comprises an initial 4-chlorobenzoate-CoA ligase which adenylates the carboxyl group in an ATP dependent reaction. The AMP moiety is then replaced by coenzyme A resulting in the formation of a thioester (**102**). The second enzyme, 4-chlorobenzoyl-CoA dehalogenase, is then able to catalyze the nucleophilic attack at C-4, to yield 4-hydroxybenzoyl-CoA (**103**). Finally, hydrolysis of the thioester yields 4-hydroxybenzoate (**104**) (ELSNER et al., 1991; SCHOLTEN et al., 1991; CHANG et al., 1992; LOFFLER et al., 1992).

Other analogous pathways have been shown in *Arthrobacter* sp. strain SU (SCHMITZ

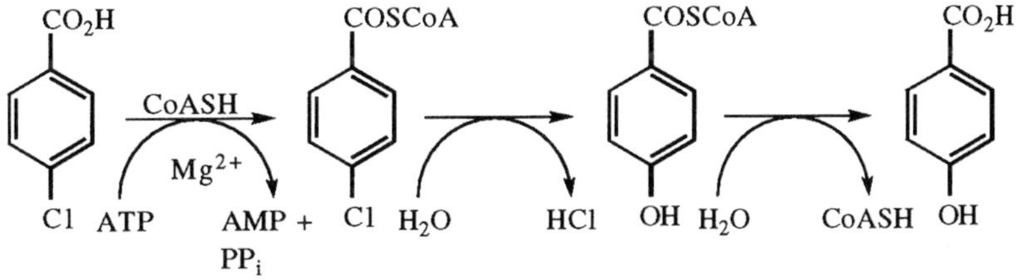

101 **102** **103** **104**

Fig. 37. Dechlorination of 4-chlorobenzoate to 4-hydroxybenzoate by *Pseudomonas* sp. strain CBS3.

et al., 1992) and *Acinetobacter* strain 4-CB1 (COPLEY and CROOKS, 1992).

6.3.4 Dehydrohalogenation

Dehydrohalogenation of an aromatic yields a benzyne which rapidly undergoes addition reactions. Although this reaction can be achieved under extreme abiotic conditions (sodium hydroxide, 300 °C), there is no known enzyme catalyzed example. The corresponding reaction of the alicycle lindane (**82**, Fig. 32) and in the aliphatic moiety of DDT are well known (see Sect. 6.2.4).

6.4 Advances Towards the Commercial Application of Dehalogenases

Dehalogenases have a key role to play in the degradation of persistent haloorganic compounds. Compounds such as lindane (**82**, Fig. 32) DDT, and PCBs have been considered above but it is well known that a large number of commonly isolated microorganisms, such as *Pseudomonas* and *Alcaligenes*, are able to dehalogenate a wide variety of halogenated compounds. Inevitably, the restricted range of selective conditions employed in the laboratory has, up to the present time, resulted in very few well characterized anaerobes being isolated. There can be little doubt that these complex consortia are key catalysts in the turnover of haloaromatic compounds in the environment.

Microorganisms able to dehalogenate specific compounds are finding some uses in the commercial sector. (*S*)-2-Chloropropionic acid is a key chiral synthon required for the synthesis of a range of pharmaceuticals and agrochemicals. Initial attempts to resolve racemic 2-chloropropionic acid with hydrolytic enzymes (CAMBOU and KLIBANOV, 1984) were superseded by the use of an (*R*)-specific dehalogenase enzyme from *Pseudomonas putida* (Fig. 38). Thus from racemic chloropropionic acid (**105**, **106**) the (*R*)-enantiomer (**105**) is converted to (*S*)-lactate (**107**) with inversion of configuration leaving the (*S*)-chloropropionate (**106**) behind (ee 96–98%). This system has been commercialized by Zeneca BioMolecules with the enzyme being produced at >1000 t a^{-1} at Chilton, Teeside.

The bioremediation of halogenated organics in effluent treatment has attracted much research interest. However, the application of a biocatalyst for the removal of halogenated organics in a product stream is novel and may not require any significant alteration to the manufacturing process. Carbury Herne Ltd., a company based in Canterbury and Cardiff, has collaborated with Hercules Inc., a multination-

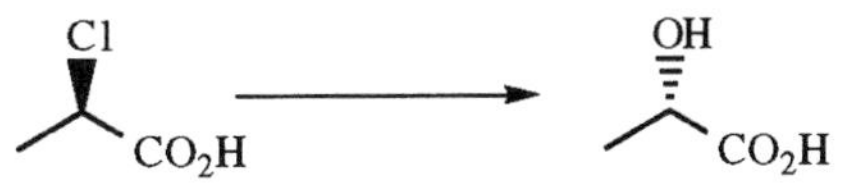

105 (*R*)-2-chloropropionate **107** (*S*)-Lactate

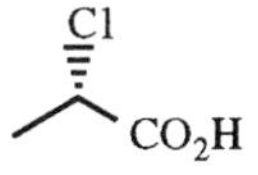

106 (*S*)-2-chloropropionate

Fig. 38. See text.

al speciality chemical company based in Wilmington, Delaware, USA, and they have successfully developed such a process. Products like paper towels, tissues, and packaging materials need a measure of wet strength if they are to perform properly. This property is provided by a polymer which is produced when epichlorohydrin is reacted with an aqueous solution of a poly(aminoamide). Unfortunately, traces of the haloalcohols 1,3-dichloropropan-2-ol (DCP) and 3-chloropropanediol (CPD) are produced as by-products of this reaction. These compounds are contaminants and have no beneficial effects on the finished product. Using a well characterized microbial consortia, *Agrobacterium tumefaciens* NCIMB 40313 and *Arthrobacter histidinolovorans* NCIMB 40274, these molecules are mineralized via glycidol and glucose. This process, incorporating a dehalogenase step, has now been adopted at the production scale at two manufacturing sites producing several thousand tonnes polymer per annum.

7 Final Comments

In this review we have attempted to present an up-to-date appraisal of halogenated molecules: their synthesis and transformation by biological systems. Obviously, there will be omissions that may have enhanced the article but several facts are clear:

- The principal route to halocompound synthesis is that catalyzed by haloperoxidases. However, this class of enzymes does not appear to catalyze the selective (regio-, stereo-, and enantio-) incorporation of halides which obviously occur in nature. It is therefore likely that other, novel biohalogenation mechanisms remain to be discovered.
- Of the reactions that the haloperoxidases are able to catalyze, it is the halide independent reactions that have demonstrated excellent synthetic utility.
- Apart from a limited range of haloperoxidases, no enzymes able to catalyze either the formation or dehalogenation of halocompounds are commercially available at the current time.

It is clear that far more effort has been invested in understanding the effects and persistence of halocompounds in the workplace and the environment than elaborating novel routes for their synthesis. Each of these activities are clearly overlapping and will require continued research and discovery if fundamental understanding and commercial exploitation are to occur concurrently.

8 References

ABRAMOWICZ, D. A. (1990), Aerobic and anaerobic biodegradation of PCBs: A review, *Crit. Rev. Biotech.* **10**, 241–251.

AHMED, M., FOCHT D. D. (1973), Degradation of polychlorinated biphenyls by two species of *Achromobacter*, *Can. J. Microbiol.* **19**, 47–52.

ALLAIN, E. J., HAGER, L. P., DENG, L., JACOBSEN, E. N. (1993), Highly enantioselective epoxidation of disubstituted alkenes with hydrogen peroxide catalyzed by chloroperoxidase, *J. Am. Chem. Soc.* **115**, 4415–4416.

AMIN, R. A., HARPER, D. B., MOLONEY, J. M., MURPHY, C. D., HOWARD, J. K. A., O'HAGAN, D. (1997), A short and stereoselective synthesis of the fluorinated natural product (2*S*,3*S*)-4-fluorothreonine, *J. C. S. Chem. Commun.* 1997, 1471–1472.

ANDO, K., KATO, A., SUZUKI, S. (1970), Isolation of 2,4-dichlorophenol from a soil fungus and its biological significance, *Biochem. Biophys. Res. Commun.* **39**, 1104–1107.

ASPLUND, G. (1995), Origin and occurrence of halogenated matter in soil, in: *Naturally-Produced Organohalogens* (GRIMVALL, A., LEER E. W. B., Eds.), pp. 35–48. Dordrecht: Kluwer.

ASPLUND, G., GRIMVALL, A. (1991), Organohalogens in nature, more widespread than previously assumed, *Environ. Sci. Technol.* **25**, 1346–1350.

ATTIEH, J. M., HANSON, A. D., SAINI, H. S. (1995), Purification and characterization of a novel methyltransferase responsible for biosynthesis of halomethanes and methane thiol in *Brassica oleracea*, *J. Biol. Chem.* **270**, 9250–9257.

BANTLEON, R., ALTENBUCHNER, J., VAN PEE, K.-H. (1994), Chloroperoxidase from *Streptomyces lividans* – isolation and characterization of the enzyme and the corresponding gene, *J. Bacteriol.* **176**, 2339–2347.

BELKIN, S. (1992), Biodegradation of haloalkanes, *Biodegradation* **3**, 299–313.

BONGS, G., VAN PEE, K. H. (1994), Enzymatic chlorination using nonheme haloperoxidases, *Enzyme Microb. Technol.* **16**, 53–60.

BRENNER, V., HERNANDEZ, B. S., FOCHT, D. D. (1993), Variation in chlorobenzoate catabolism by *Pseudomonas putida* P111 as a consequence of genetic alteration, *Appl. Environ. Microbiol.* **59**, 2790–2794.

BROWN, J. F., WAGNER, R. E., FENG, H., BEDARD, D. L., BRENNAN, M. J., CARNAHAN, J. C., RAY, R. J. (1987), Environmental dechlorination of PCBs, *Environ. Toxicol. Chem.* **6**, 579–593.

BURD, W., YOURKEVICH, O., VOSKOBOEV, A. J., VAN PEE, K. H. (1995), Purification and properties of a nonheme chloroperoxidase from *Serratia marcescens*, *FEMS Microbiol. Lett.* **129**, 255–260.

BUTLER, A., WALKER, J. V. (1993), Marine haloperoxidases, *Chem. Rev.* **93**, 1937–1944.

CAMBOU, B., KLIBANOV, A. M. (1984), Lipase-catalyzed production of optically-active acids via asymmetric hydrolysis of esters – effect of the alcohol moiety, *Appl. Biochem. Biotechnol.* **9**, 255–260.

CANNARSA, M. J. (1996), Single enantiomer drugs: new strategies and directions, *Chem. Ind.*, 374–378.

CARMICHAEL, R. D., PICKARD, M. A. (1989), Continuous and batch-production of chloroperoxidase by mycelial pellets of *Caldariomyces fumago* in an airlift fermenter, *Appl. Environ. Microbiol.* **55**, 17–20.

CASTRO, C. E., BARTNICKI, E. W. (1968), Biodehalogenation. Epoxidation of halohydrins, epoxide opening, and transdehalogenation by a *Flavobacterium* species, *Biochemistry* **7**, 3213–3218.

CHAN, W. R., TINTO, W. F., MANCHAND, P. S., TODARO, L. J. (1987), Stereostructures of geodiamolide-A and geodiamolide-B, novel cyclodepsipeptides from the marine sponge *Geodia* sp., *J. Org. Chem.* **52**, 3091–3093.

CHANG, K. H., LIANG, P. H., BECK, W., SCHOLTEN, J. D., DUNAWAY-MARINO, D. (1992), Isolation and characterization of the three polypeptide components of 4-chlorobenzoate dehalogenase from *Pseudomonas* sp. CBS3, *Biochemistry* **31**, 5605–5610.

CHEN, Y. P., LINCOLN, D. E., WOODIN, S. A., LOVELL, C. R. (1991), Purification and properties of a unique flavin-containing chloroperoxidase from the capitellid polychaete *Notomastus lobatus*, *J. Biol. Chem.* **266**, 23909–23915.

COLONNA, S., GAGGERO, N., CASELLA, L., CARREA, G., PASTA, P. (1993), Enantioselective epoxidation of styrene derivatives by chloroperoxidase catalysts, *Tetrahedron: Asymmetry* **4**, 1325–1330.

CONTESTABILE, R., JOHN, R. A. (1996), The mechanism of high-yielding chiral syntheses catalyzed by wild-type and mutant forms of aspartate-aminotransferase, *Eur. J. Biochem.* **240**, 150–155.

COOK, A. M., HUTTER, R. (1986), Ring dechlorination of deethylsimazine by hydrolases from *Rhodococcus corallinus*, *FEMS Microbiol. Lett.* **34**, 335–338.

COPLEY, S. D., CROOKS, G. P. (1992), Enzymatic dehalogenation of 4-chlorobenzoyl coenzyme A in *Acinetobacter* sp. strain 4-CB1, *Appl. Environ. Microbiol.* **58**, 1385–1387.

COUGHLIN, P., ROBERTS, S., RUSH, C., WILLETTS, A. (1993), Biotransformation of alkenes by haloperoxidases – regiospecific bromohydrin formation from cinnamyl substrate, *Biotechnol. Lett.* **15**, 907–912.

CURRAGH, H., FLYNN, O., LARKIN, M. J., STAFFORD, T. M., HAMILTON, J. T. G., HARPER, D. B. (1994), Haloalkane degradation and assimilation by *Rhodococcus rhodochrous* NCIMB 13064, *Microbiology* **140**, 1433–1442.

DE BOER, E., PLAT, H., TROMP, M. G. M., WEVER, R., FRANSSEN, M. C. R., VAN DER PLAS, H. C., MEIJER, E. M., SCHOEMAKER, H. E. (1987), Vanadium containing bromoperoxidase – an example of an oxidoreductase with high operational stability in aqueous and organic media, *Biotechnol. Bioeng.* **30**, 607–610.

DE JONG, E., FIELD, J. A., SPINNLER, H. E., WIJNNBERG, J. B. P. A., DE BONT, J. A. M. (1994), Significant biogenesis of chlorinated aromatics by fungi in natural environments, *Appl. Environ. Microbiol.* **60**, 264–270.

DE SILVA, E. D., ANDERSEN, R. J., ALLEN, T. M. (1990), Geodiamolide-C to geodiamolide-F, new cytotoxic cyclodepsipeptides from the marine sponge *Pseudaxinyssa* sp., *Tetrahedron Lett.* **31**, 489–492.

DOLFING, J., BUERSKENS, J. E. M. (1995), The microbial logic and environmental significance of reductive dehalogenation, *Adv. Microb. Ecol.* **14**, 143–206.

DOLFING, J., VAN DER WIJNGAARD, A. J., JANSSEN, D. B. (1993), Microbiological aspects of the removal of chlorinated hydrocarbons from air, *Biodegradation* **4**, 261–282.

DRAUZ, K., WALDMANN, H. (Eds.) (1995), *Enzyme Catalysis in Organic Synthesis*, pp. 783–791. Weinheim: VCH.

EGLI, C., TSCHAN, T., SCHOLTZ, R., COOK, A. M., LEISINGER, T. (1988), Transformation of tetrachloromethane to dichloromethane and carbon dioxide by *Acetobacterium woodii*, *Appl. Environ. Microbiol.* **54**, 2819–2834.

EGLI, C., STROMEYER, S., COOK, A. M., LEISINGER, T. (1990), Transformation of tetra- and trichloromethane to CO_2 by anaerobic bacteria is a non-enzymatic process, *FEMS Microbiol. Lett.* **68**, 207–212.

ELANDER, R. P., MABE, J. A., HAMILL, R. H., GORMAN, M. (1968), Metabolism of tryptophans by *Pseudomonas aureofaciens*. II Production of pyrrolnitrin by selected *Pseudomonas* species, *Appl. Microbiol.* **16**, 753–758.

ELSNER, A., MULLER, R., LINGENS, F. (1991), Separate cloning and expression analysis of two protein components of 4-chlorobenzoate dehalogenase from *Pseudomonas* sp. CBS3, *J. Gen. Microbiol.* **137**, 477–481.

ENGESSER, K.-H., SCHULTE, P. (1989), Degradation of 2-bromo, 2-chloro and 2-fluorobenzoate by *Pseudomonas putida* CLB250, *FEMS Microbiol. Lett.* **60**, 143–148.

FACEY, S. J., GROSS, F., VINING, L. C., YANG, K., VAN PEE, K.-H. (1996), Cloning, sequencing and disruption of a bromoperoxidase-catalase gene in *Streptomyces venezuelae:* evidence that it is not required for chlorination in chloramphenicol biosynthesis, *Microbiology* **142**, 657–665.

FENICAL, W. (1974), Polyhaloketones from the red seaweed *Asparagopsis taxiformis*, *Tetrahedron Lett.* **51/52**, 4463–4466.

FENICAL, W. (1979), Molecular aspects of halogen-based biosynthesis of marine natural products, *Recent Adv. Phytochem.* **13**, 219–239.

FENICAL, W. (1982), Natural products chemistry in the marine environment, *Science* **215**, 923–928.

FENICAL, W. (1993), Chemical studies of marine bacteria – developing a new resource, *Chem. Rev.* **93**, 1673–1682.

FETZNER, S., LINGENS, F. (1994), Bacterial dehalogenases: Biochemistry, genetics and biotechnological applications, *Microbiol. Rev.* **58**, 641–685.

FETZNER, S., MULLER, R., LINGENS, F. (1989), Degradation of 2-chlorobenzoate by *Pseudomonas cepacia* 2CBS, *Biol. Chem. Hoppe-Seyler* **370**, 1173–1182.

FETZNER, S., MULLER, R., LINGENS, F. (1992), Purification and some properties of 2-halobenzoate 1,2-dioxygenase, a two-component enzyme system from *Pseudomonas cepacia* 2CBS, *J. Bacteriol.* **174**, 279–290.

FIELD, J. A., VERHAGEN, F. J. M., DE JONG, E. (1995), Natural organohalogen production by basidiomycetes, *Trends Biotechnol.* **13**, 451–456.

FIELDING, M. (1992), in: *Water Pollution Research Reports: Pesticides in Ground and Drinking Water Report No. 27* (FIELDING, M., Ed.), pp. 5–11 (publication of the Commission of the European Communities, Directorate General for Science, Research and Development).

FLAVIN, M., CASTRO-MENDOZA, H., OCHOA, S. (1957), Metabolism of propionic acid in animal tissues. II Propionyl coenzyme A carboxylation system, *J. Biol. Chem.* **229**, 981.

FOWDEN, L. (1968), The occurrence and metabolism of carbon–halogen compounds, *Proc. R. Soc. London* B **171**, 5–18.

FRANKEN, S. M., ROZEBOOM, H. J., KALK, K. H., DIJKSTRA, B. W. (1991), Crystal structure of haloalkane dehalogenase: an enzyme to detoxify halogenated alkanes, *EMBO J.* **10**, 1297–12302.

FRANSSEN, M. C. R. (1994), Haloperoxidases: useful catalysts for halogenation and oxidation reactions, *Catalysis Today* **22**, 441–457.

FRANSSEN, M. C. R., VAN DER PLAS, H. C. B. (1992), Haloperoxidases – their properties and their use in organic synthesis, *Adv. Appl. Microbiol.* **37**, 41–99.

FU, H., KONDO, H., ICHIKAWA, Y., LOOK, G. C., WONG, C.-H. (1992), Chloroperoxidase-catalyzed asymmetric synthesis – enantioselective reactions of chiral hydroperoxidase with sulfides and bromohydration of glycals, *J. Org. Chem.* **57**, 7265–7270.

FURUKAWA, K., MATSUMURA, F., TONOMURA, K. (1978), *Alcaligenes* and *Acinetobacter* strains capable of degrading polychlorinated biphenyls, *Agric. Biol. Chem.* **42**, 543–548.

GEIGERT, J., NEIDLEMAN, S. L., DALIETOS, D. J., DEWITT, S. K. (1983a), Haloperoxidases – enzymatic synthesis of α,β-halohydrins from gaseous alkenes, *Appl. Environ. Microbiol.* **45**, 366–374.

GEIGERT, J., NEIDLEMAN, S. L., DALIETOS, D. J. (1983b), Novel haloperoxidase substrates – alkynes and cyclopropanes, *J. Biol. Chem.* **258**, 2273–2277.

GEIGERT, J., DEWITT, S. K., NEIDLEMAN, S. L., DALIETOS, D. J., MORELAND, M. (1983c), DMSO is a substrate for chloroperoxidase, *Biochem. Biophys. Res. Commun.* **116**, 82–85.

GOLOVLEVA, L. A., ZABORINA, O., PERTSOVA, R., BASKUNOV, B., SCHURUKHIN, Y., KUZMIN, S. (1992), Degradation of polychlorinated phenols by *Streptomyces rochei* 303, *Biodegradation* **2**, 201–208.

GREGSON, R. P., BALDO, B. A., THOMAS, P. G., QUINN, R. J., BERGQUIST, P. R., STEPHENS, J. R., HORNE, A. R. (1979), Fluorine is a major constituent of the marine sponge *Halichondria moorei*, *Science* **206**, 1108–1109.

GRIBBLE, G. W. (1992), Naturally occurring organohalogen compounds – A survey, *J. Nat. Prod.* **55**, 1353–1395.

GRIBBLE, G. W. (1994), Natural organohalogens, *J. Chem. Educ.* **71**, 907–911.

GRISHAM, M. B., JEFFERSON, M. M., MELTON, D. F., THOMAS, E. L. (1984), Chlorination of endogenous amines by isolated neutrophils – ammonia-dependent bactericidal, cyto-toxic, and cytolytic activities of the chloramines, *J. Biol. Chem.* **259**, 10404–10413.

GSCHWEND, P. M., MACFARLAND, J. K., NEWMAN, K. A. (1985), Volatile halogenated organic compounds released to seawater from temperate marine macroalgae, *Science* **227**, 1033–1035.

HAAG, T., LINGENS, F., VAN PEE, K.-H. (1991), A metal-ion-independent and cofactor-independent enzymatic redox reaction – halogenation by bacterial nonheme haloperoxidases, *Angew. Chem.* (Int. Edn. Engl.) **30**, 1487–1488.

HAGER, L. P. (1982), Mother nature likes some halogenated compounds, *Basic Life Sci.* **19**, 415–429.

HAGER, L. P., MORRIS, D. R., BROWN, F. S., EBERWEIN, H. (1966), Chloroperoxidase. II Utilization of halogen anions, *J. Biol. Chem.* **241**, 1769–1776.

HAGGBLOM, M. M., RIVERA, M. D., YOUNG, L. Y. (1993), Influence of alternative electron acceptors on the anaerobic biodegradability of chlorinated phenols and benzoic acids, *Appl. Environ. Microbiol.* **59**, 1162–1167.

HAIDER, K. (1979), Degradation and metabolization of lindane and other hexachlorocyclohexane isomers by anaerobic and aerobic soil microorganisms, *Z. Naturforsch. (Sect. C)* **34**, 1066–1069.

HAMILTON, J. T. G., AMIN, M. R., HARPER, D. B., O'HAGAN, D. (1997), Biosynthesis of fluoroacetate and 4-fluorothreonine by *Streptomyces cattleya*. Glycine and pyruvate as precursors, *J. C. S. Chem. Commun.* 797–798.

HARDMAN, D. J. (1991), Biotransformation of halogenated compounds, *Crit. Rev. Biotech.* **11**, 1–40.

HARPER, D. B. (1985), Halomethane from halide ion – a highly efficient fungal conversion of environmental significance, *Nature* **315**, 55–57.

HARPER, D. B., HAMILTON, J. T. G. (1988), Biosynthesis of chloromethane in *Phellinus pomaceus*, *J. Gen. Microbiol.* **134**, 2831–2839.

HARPER, D. B., O'HAGAN, D. (1994), The fluorinated natural products, *Nat. Prod. Rep.* **11**, 123–133.

HARPER, D. B., HAMILTON, J. T. G., O'HAGAN, D. (1990), Identification of *threo*-18-fluoro-9,10-dihydroxystearic acid – A novel ω-fluorinated fatty-acid from *Dichapetalum toxicarium* seeds, *Tetrahedron Lett.* **31**, 7661–7662.

HAUTZEL, R., ANKE, H., SHELDRICK, W. S. (1990), Mycenon, a new metabolite from a *Mycena* species TA-87202 (Basidomycetes) as an inhibitor of isocitrate lyase, *J. Antibiot.* **43**, 1240–1244.

HECHT, T. J., SOBEK, H., HAAG, T., PFEIFER, O., VAN PEE, K. H. (1994), The metal ion free oxidoreductase from *Streptomyces aureofaciens* has an α/β hydrolase fold, *Nat. Struct. Biol.* **1**, 523–537.

HEWSON, W. D., HAGER, L. P. (1980), Bromoperoxidases and halogenated metabolites in marine algae, *J. Phycol.* **16**, 340–345.

HICKEY, W. J., FOCHT, D. D. (1990), Degradation of mono-, di- and trihalogenated benzoic acids by *Pseudomonas aeruginosa* JB2, *Appl. Environ. Microbiol.* **56**, 3842–3850.

HOLLIGER, C., SCHRAA, G., STAMS, A. J. M., ZEHNDER, A. J. B. (1993), A highly purified enrichment culture couples the reductive dechlorination of tetrachloroethene to growth, *Appl. Environ. Microbiol.* **59**, 2991–2997.

HOWARD, P. H., BOETHLING, R. S., STITELER, W. M., MEYLAN, W. M., HUEBER, A. E., BEAUMAN, J. A., LAROSCHE, M. E. (1992), Predictive model for aerobic biodegradability developed from a file of evaluated biodegradation data, *Environ. Toxicol. Chem.* **11**, 593–603.

HUNTER, J. C., FONDA, M., SOTOS, L., TOSO, B., BELT, A. (1984), Ecological approaches to isolation, *Dev. Ind. Microbiol.* **25**, 247–266.

IMAI, R., NAGATA, Y., FUKADA, M., TAGAKI, M., YANO, M. (1991), Molecular cloning of a *Pseudomonas paucimobilis* gene encoding a 17-kilodalton polypeptide that eliminates HCl from γ-hexachlorocyclohexane, *J. Bacteriol.* **173**, 6811–6819.

ISHIDA, M. (1968), Comparative studies on BHC metabolizing enzymes, DDT dehydrochlorinase and glutathione S-transferase, *Agric. Biol. Chem.* **32**, 947–955.

ISIDOROV, V. A. (1990), *Organic Chemistry of the Earths Atmosphere.* Berlin: Springer-Verlag.

ITOH, N., CHENG, L. Y., IZUMI, Y., YAMADA, H. (1987a), Immobilized bromoperoxidase of *Corallina pilulifera* as a multifunctional halogenating biocatalyst, *J. Biotechnol.* **5**, 29–38.

ITOH, N., IZUMI, Y., YAMADA, H. (1987b), Haloperoxidase-catalyzed halogenation of nitrogen-containing aromatic heterocycles represented by nucleic bases, *Biochemistry* **26**, 282–289.

ITOH, N., HASAN, A. K. M. Q., IZUMI, Y., YAMADA, H. (1988), Substrate-specificity, regiospecificity and stereospecificity of halogenation reactions catalyzed by non-heme-type bromoperoxidase of *Corallina pilulifera*, *Eur. J. Biochem.* **172**, 477–484.

ITOH, N., MORINAGA, N., KONZAI, T. (1994), Purification and characterization of a novel metal-containing nonheme bromoperoxidase from *Pseudomonas putida*, *Biochim. Biophys. Acta* **1207**, 208–216.

JANNUN, R., COE, E. L. (1987), Bromoperoxidase from the marine snail *Murex trunculis*, *Comp. Biochem. Physiol.* **88**, 917–922.

JANSSEN, D. B., SCHEPER, A., DIJKHUIZEN, L., WITHOLT, B. (1985), Degradation of halogenated aliphatic compounds by *Xanthobacter autotrophicus* GJ10, *Appl. Environ. Microbiol.* **49**, 673–677.

JANSSEN, D. B., FRIES, F., VAN DER PLOEG, J., KAZEMIER, B., TEPSTRA, P., WITHOLT, B. (1989), Cloning of 1,2-dichloroethane degradation genes of *Xanthobacter autotrophicus* GJ10 and expression and sequencing of the *dhla* gene, *J. Bacteriol.* **171**, 6791–6799.

JANSSEN, D. B., PRIES, F., VAN DER PLOEG, J. R. (1994), Genetics and biochemistry of dehalogenating enzymes, *Ann. Rev. Microbiol.* **48**, 163–191.

JENSEN, P. R., FENICAL, W. (1994), Strategies for dicovery of secondary metabolites from marine bacteria – Ecological perpectives, *Ann. Rev. Microbiol.* **48**, 559–584.

JOHNSTON, H. W., BRIGGS, G. G., ALEXANDER, M. (1972), Metabolism of 3-chlorobenzoic acid by a pseudomonad, *Soil Biol. Biochem.* **4**, 187–190.

JOHRI, S., QUAZI, G. N., CHOPRA, C. L. (1991), Evidence of plasmid mediated dechlorinase activity in *Pseudomonas* sp., *J. Biotechnol.* **20**, 73–82.

JONES, D. H. A., BARTH, P. T., BYROM, D., THOMAS, C. M. (1992), Nucleotide sequence of the structural gene encoding a 2-haloalkanoic acid dehalogenase of *Pseudomonas putida* strain AJ1 and purification of the encoded protein, *J. Gen. Microbiol.* **138**, 675–683.

KAWASAKI, H., TSUDA, K., MATSUHITA, I., TONOMURA, K. (1992), Lack of homology between two haloacetate dehalogenase genes encoded on a plasmid from *Moraxella* species strain B, *J. Gen. Microbiol.* **138**, 1317–1323.

KAWASAKI, H., TOYAMA, T., MAEDA, T., NISHINO, H., TONOMURA, K. (1994), Cloning and sequence analysis of a plasmid-encoded 2-haloacid dehalogenase gene from *Pseudomons putida* no. 109, *Biosci. Biotech. Biochem.* **58**, 160–163.

KEUNING, S., JANSSEN, D. B., WITHOLT, B. (1985), Purification and characterization of hydrolytic haloalkane dehalogenase from *Xanthobacter autotrophicus* GJ10, *J. Bacteriol.* **163**, 635–639.

KITAGAWA, I., FUKUDA, Y., TANIYAMA, T., YOSHIKAWA, M. (1995), Chemical studies on crude drug processing. 8. On the constituents of *Rehmanniae radix*. 2. Absolute stereostructures of rehmaglutin-C and glutinoside isolated from Chinese *Rehmanniae radix*, the dried root of *Rehmannia glutinosa libosch*, *Chem. Pharm. Bull.* **43**, 1096–1100.

KIYOHARA, H., HATTA, T., OGAWA, Y., KAKUDA, T., YOKOYAMA, H., TAKIZAWA, N. (1992), Isolation of *Pseudomonas pickettii* strains that degrade 2,4,6-trichlorophenol and their dechlorination of chlorophenols, *Appl. Environ. Microbiol.* **58**, 1276–1283.

KLAGES, U., KRAUSS, S., LINGENS, F. (1983), 2-Haloacid dehalogenase from a 4-chlorobenzoate-degrading *Pseudomonas* spec. CBS 3, *Hoppe-Seyler's Z. Physiol. Chem.* **364**, 529–535.

KOBAYASHI, J., KANDA, F., ISHIBASHI, M., SHIGEMORI, H. (1991), Manzacidins AC, novel tetrahydropyrimidine alkaloids from the Okinawan marine sponge *Hymeniacidon* sp., *J. Org. Chem.* **56**, 4574–4576.

KOHLER-STAUB, D., LEISINGER, T. (1985), Dichloromethane dehalogenase of *Hyphomicrobium* species strain DM2, *J. Bacteriol.* **162**, 676–681.

KONDO, K., NISHI, J., ISHIBASHI, M., KOBAYASHI, J. (1994), 2 New tryptophan-derived alkaloids from the Okinawan marine sponge *Aplysina* sp., *J. Nat. Prod. (Lloydia)* **57**, 1008–1011.

KUHN, E. P., TOWNSEND, G. T., SUFLITA, J. M. (1990), Effect of sulfate and organic carbon supplements on reductive dehalogenation of chloroanilines in anaerobic aquifer slurries, *Appl. Environ. Microbiol.* **56**, 2630–2637.

LEE, M. D., ELLESTAD, G. A., BORDERS, D. B. (1991), Calicheamicins – discovery, structure, chemistry and interaction with DNA, *Acc. Chem. Res.* **24**, 235–243.

LEE, M. D., DUNNE, T. S., CHANG, C. C., SIEGEL, M. M., MORTON, G. O., ELLESTAD, G. A., MCGAHREN, W. J., BORDERS, D. B. (1992), Calicheamicins, a novel family of antitumor antibiotics. 4. Structure elucidation of calicheamicins β_1^{Br}, γ_1^{Br}, α_2^{I}, α_3^{I}, β_1^{I}, γ_1^{I}, δ_1^{I}, *J. Am. Chem. Soc.* **114**, 985–997.

LI, J. Z., SHARMA, R., DILEEPAN, K. N., SAVIN, V. J. (1994), Polymorphonuclear leukocytes increase glomerular albumin permeability via hypohalous acid, *Kidney Int.* **46**, 1025–1030.

LIPKE, H., KEARNS, C. W. (1959), DDT dehydrochlorinase. I Isolation, chemical properties, and spectrophotometric assay, *J. Biol. Chem.* **234**, 2123–2128.

LIPKE, H., KEARNS, C. W. (1959), DDT dehydrochlorinase. II Substrate and cofactor specificity, *J. Biol. Chem.* **234**, 2129–2132.

LIU, K. K.-C., WONG, C.-H. (1992), Enzymatic halohydration of glycals, *J. Org. Chem.* **57**, 3748–3750.

LOFFLER, F., MULLER, R., LINGENS, F. (1992), Purification and properties of 4-chlorobenzoate-coenzyme A ligase from *Pseudomonas* sp. CBS3, *Hoppe-Seyler's Z. Physiol. Chem.* **373**, 1001–1007.

MAGUIRE, A. R., MENG, W.-D., ROBERTS, S. M., WILLETTS, A. J. (1993), Synthetic approaches towards nucleocidin and selected analogs – anti-HIV activity in 4′-fluorinated nucleoside derivatives, *J. Chem. Soc. Perkin Trans.* **1**, 1795.

MANTHEY, J. A., HAGER, L. P. (1981), Purification and properties of bromoperoxidase from *Penicillus capitatus*, *J. Biol. Chem.* **256**, 1232–1238.

MARSHALL, G. C., WRIGHT, G. D. (1996), Purification and characterization of 2 haloperoxidases from the glycopeptide antibiotic producer *Streptomyces toyocaensis* NRRL-15009, *Biochem. Biophys. Res. Commun.* **219**, 580–583.

MCCORMICK, M. L., ROEDER, T. L., RAILSBACK, M. A., BRITIGAN, B. E. (1994), Eosinophil peroxidase-dependent hydroxyl radical generation by human eosinophils, *J. Biol. Chem.* **269**, 27914–27919.

MEAD, R. J., SEGAL, W. (1973), Formation of β-cyanoalanine and pyruvate by *Accacia georginae*, *Phytochemistry* **12**, 1977–1981.

MESSERSCHMIDT, A., WEVER, R. (1996), X-ray structure of a vanadium-containing enzyme – chloroperoxidase from the fungus *Curvularia inaequalis*, *Proc. Natl. Acad. Sci. USA* **93**, 392–396.

MEYER, M., O'HAGAN, D. (1992a), Rare fluorinated natural products, *Chem. Brit.* **28**, 785–788.

MEYER, J. J. M., O'HAGAN, D. (1992b), Conversion of fluoropyruvate to fluoroacetate by *Dichapetalum cymosum*, *Phytochemistry* **31**, 499–501.

MEYER, J. J. M., O'HAGAN, D. (1992c), Conversion of 3-fluoropyruvate to fluoroacetate by cell-free extracts of *Dichapetalum cymosum*, *Phytochemistry* **31**, 2699–2701.

MILES, D. H., MODY, N. V., MINYARD, J. P., HEDIN, P. A. (1973), Constituents of marsh grass. Survey of the essential oils in *Juncus raemerianus*, *Phytochemistry* **12**, 1399.

MIURA, K., OTSUKA, S., HONDA, K. (1956), Abnormal metabolism in animals poisoned with sodium fluoroacetate, *Bull. Agric. Chem. Soc. Jpn.* **20**, 219–222.

MOHN, W. W., TIEDJE, J. M. (1992), Microbial reductive dehalogenation, *Microbiol. Rev.* **56**, 482–507.

MOORE, R. E. (1977), Volatile compounds from marine algae, *Acc. Chem. Res.* **10**, 40–47.

MOTOSUGI, K., ESAKI, N., SODA, K. (1982), Purification and properties of 2-haloacid dehalogenase from *Pseudomonas putida*, *Agric. Biol. Chem.* **46**, 837–838.

MULLER, R. (1992), Bacterial degradation of xenobiotics, in: *Microbial Control of Pollution, SGM Symposium 48* (FRY, J. C., GADD, G. M., HERBERT, R. A., JONES, C. W., WATSON-CRAIK, I. A., Eds.), pp. 35–57. Cambridge: Cambridge University Press.

MURDIYATMO, U., ASMARA, W., TSANG, J. S. H., BAINES, A. J., BULL, A. T., HARDMAN, D. J. (1992), Molecular biology of the 2-haloacid halidohydrolase from *Pseudomonas cepacia* MBA4, *Biochem. J.* **284**, 87–93.

NAGASAWA, T., YAMADA, H. (1986), Enzymatic transformations of 3-chloroalanine into useful aminoacids, *Appl. Biochem. Biotechnol.* **13**, 147–165.

NAGASAWA, T., NAKAMURA, T., YU, F., WATANABE, I., YAMADA, H. (1992), Purification and characterization of halohydrin hydrogen-halide lyase from a recombinant *Escherichia coli* containing the gene from a *Corynebacterium* species, *Appl. Microbiol. Biotech.* **36**, 478–482.

NAGATA, Y., HATTA, T., IMAI, R., KIMBARA, K., FUKUDA, M., YANO, K., TAKAGI, M. (1993), Purification and characterization of γ-hexachlorocyclohexane (γ-HCH) dehydrochlorinase (LinA) from *Pseudomonas paucimobilis*, *Biosci. Biotechnol. Biochem.* **57**, 1582–1583.

NAKAMURA, T., NAGASAWA, T., YU, F., WATANABE, J., YAMADA, H. (1992), Resolution and some properties of enzymes involved in enantioselective transformation of 1,3-dichloro-2-propanol to (*R*)-3-chloro-1,2-propanediol by *Corynebacterium* sp. strain N-1074, *J. Bacteriol.* **174**, 7613–7619.

NARDI-DEI, V., KURIHARA, T., OKAMURA, O., LIU, J. Q., KOSHIKAWA, H., OZAKI, H., TERASHIMA, Y., ESAKI, N., SODA, K. (1994), Comparative studies of genes encoding thermostable 1-2-halo acid dehalogenase from *Pseudomonas* species YL, other dehalogenases, and two related hypothetical proteins from *Escherichia coli*, *Appl. Environ. Microbiol.* **60**, 3375–3380.

NEIDLEMAN, S. L. (1975), Microbial halogenation, *CRC Crit. Rev. Microbiol.* **5**, 333–358.

NEIDLEMAN, S. L. (1980), Use of enzymes as catalysts for alkene oxide production, *Hydrocarbon Process. (Intl. Edn.)* **59**, 135–138.

NEIDLEMAN, S. L., GEIGERT, J. (1986), *Biohalogenation: Principles, Basic Roles and Applications*. New York: Halsted Press.

NEILSON, A. H., ALLARD, A. S., LINDGREN, C., REMBERGER, M. (1987), Transformation of chloroguaiacols, chloroveratroles and chlorocatechols by stable consortia of anaerobic bacteria, *Appl. Environ. Microbiol.* **53**, 949–954.

NIESCHALK, J., HAMILTON, J. T. G., MURPHY, C. D., HARPER, D. B., O'HAGAN, D. (1997), Biosynthesis of fluoroacetate and 4-fluorothreonine by *Streptomyces cattleya*. The stereochemical processing of glycerol, *J. C. S. Chem Commun.* 799–800.

NUELL, M. J., FANG, G.-H., AXLEY, M. J., KENISBERG, P., HAGER, L. P. (1988), Isolation and nucleotide-sequence of the chloroperoxidase gene from *Caldariomyces fumago*, *J. Bacteriol.* **170**, 1007–1011.

O'HAGAN, D., PERRY, R., LOCK, J. M., MEYER, J. J. M., DASARADHI, L., HAMILTON, J. T. G., HARPE, D. B. (1993), High levels of monofluoroacetate in *Dichapetalum braunii*, *Phytochemistry* **33**, 1043–1045.

OHTA, T., TAKAHASHI, A., MATSUDA, M., KAMO, S., AGATSUMA, T., ENDO, T., NAZOE, S. (1995), Russuphelol, a novel optically active chlorohydroquinone tetramer from the mushroom *Russula subnigricans*, *Tetrahedron Lett.* **36**, 5223–5226.

OKAMOTO, K., SHIMADA, A., SHIRAI, R., SAKAMOTO, H., YOSHIDA, S., OJIMA, F., ISHIGURO, Y., SAKAI, T., KAWAGISHI, H. (1993), Antimicrobial chlorinated orcinol derivatives from mycelia of *Hericium erinaceum*, *Phytochemistry* **34**, 1445–1446.

PASTA, P., CARREA, G., COLONNA, S., GAGGERO, N. (1994), Effects of chloride on the kinetics and stereochemistry of chloroperoxidase catalyzed oxidation of sulfides, *Biochim. Biophys. Acta* **1209**, 203–208.

PELLETIER, I., ALTENBUCHNER, J. (1995), A bacterial esterase is homologous with nonheme haloperoxidases and displays brominating activity, *Microbiology* **141**, 459–468.

PELLETIER, I., PFEIFER, O., ALTENBUCHNER, J., VAN PEE, K.-H. (1994), Cloning of a second nonheme bromoperoxidase gene from *Streptomyces aureofaciens* ATCC 10762. Sequence analysis, expression in *Streptomyces lividans* and enzyme purification, *Microbiology* **140**, 509–516.

PELLETIER, I., ALTENBUCHNER, J., MATTES, R. (1995), A catalytic triad is required by the non-heme haloperoxidases to perform halogenation, *Biochim. Biophys. Acta.* **1250**, 149–157.

PETERS, R. A., SHORTHOUSE, M. (1964), Fluoride metabolism in plants of *Accacia georginae*, *Biochem. J.* **93**, P20.

PETERS, R. A., SHORTHOUSE, M. (1971), Identification of a volatile constituent formed by homogenates of *Accacia georginae* exposed to fluoride, *Nature* **231**, 123–124.

PETTY, M. A. (1961), An introduction to the origin and biochemistry of microbial halometabolites, *Bacteriol. Rev.* **25**, 111–130.

PFEIFER, O., PELLETIER, I., ALTENBUCHNER, J., VAN PEE, K.-H. (1992), Molecular cloning and sequencing of a non-heme bromoperoxidase gene from *Streptomyces aureofaciens* ATCC 10762, *J. Gen. Microbiol.* **138**, 1123–1131.

PICKARD, M. A., KADIMA, T. A., CARMICHEAL, R. D., (1991), Chloroperoxidase, a peroxidase with potential, *J. Ind. Microbiol.* **7**, 235–241.

PLAT, H., KREN, B. E., WEVER, R. (1987), The bromoperoxidase from the lichen *Xanthoria parietina* is a novel vanadium enzyme, *Biochem. J.* **248**, 277–279.

PRIES, F., VAN DER PLOEG, J. R., DOLFING, J., JANSSEN, D. B. (1994), Degradation of halogenated aliphatic compounds: The role of adaptation, *FEMS Microbiol. Rev.* **15**, 279–295.

RADEHAUS, P. M., SCHMIDT, S. K. (1992), Characterization of a novel *Pseudomonas* sp. that mineralizes high concentrations of pentachlorophenol, *Appl. Environ. Microbiol.* **58**, 2879–2885.

RAMANAND, K., BALBA, M. T., DUFFY, J. (1993), Reductive dehalogenation of chlorinated benzenes and toluenes under methanogenic conditions, *Appl. Environ. Microbiol.* **59**, 3266–3272.

RASCHE, M. E., HICKS, R. E., HYMAN, M. R., ARP, D. J. (1990), Oxidation of monohalogenated ethanes and N-chlorinated alkanes by whole cells of *Nitrosomonas europaea*, *J. Bacteriol.* **172**, 5368–5373.

RASMUSSEN, R. A., RASMUSSEN, L. E., KHALI, M. A. K., DALUGE, R. W. (1980), Concentration distribution of methyl chloride in the atmosphere, *J. Geophys. Res.* **85**, 7350–7356.

REINEKE, W., KNACKMUSS, H.-J. (1988), Microbial degradation of haloaromatics, *Ann. Rev. Microbiol.* **42**, 263–287.

ROBINSON, G., LENN, M. J. (1994), The bioremediation of polychlorinated biphenyls (PCBs): Problems and perspectives, *Biotechnol. Gen. Eng. Rev.* **12**, 139–188.

ROMANO, M., MELO, C., BARALLE, F., DRI, P. (1992), cDNA of myeloperoxidase (MPO) and eosinophil peroxidase (EPO) from human blood mononuclear-cells differentiated into granulocyte precursors in liquid cultures, *J. Immunol. Methods* **154**, 265–267.

SAINI, H. S., ATTICH, J. M., HANSON, A. D. (1995), Biosynthesis of halomethanes and methanethiol by higher plants via a novel methyltransferase reaction, *Plant Cell Environ.* **18**, 1027–1033.

SALLIS, P. J., ARMFIELD, S. J., BULL, A. T., HARDMAN, D. J. (1990), Isolation and characterisation of a haloalkane halidohydrolase from *Rhodococcus erythropolis* strain Y2, *J. Gen. Microbiol.* **136**, 115–120.

SANADA, M., MIYANO, T., IWADARE, S., WILLIAMSON, J. M., ARISON, B. H., SMITH, J. L., DOUGLAS, A. W., LIESCH, J. M., INAMINE, E. (1986), Biosynthesis of fluorothreonine and fluoroacetic acid by the thienamycin producer, *Streptomyces cattleya*, *J. Antibiot.* **39**, 259–265.

SCHENK, T., MULLER, R., LINGENS, A. (1990), Mechanism of enzymatic dehalogenation of pentachlorophenol by *Arthrobacter* sp. strain ATCC 33790, *J. Bacteriol.* **172**, 7272–7274.

SCHMITZ, A., GARTEMANN, K. H., FIELDER, J., GRUND, E., EICHENLAUB, R. (1992), Cloning and sequence analysis of genes for dehalogenation of 4-chlorobenzoate from *Arthrobacter* sp. strain SU, *Appl. Environ. Microbiol.* **58**, 4068–4071.

SCHNEIDER, B., MULLER, R., FRANK, R., LINGENS, F. (1991), Complete nucleotide sequences and comparison of the structural genes of two 2-haloalkanoic acid dehalogenases from *Pseudomonas* sp. strain CBS 3, *J. Bacteriol.* **173**, 1530–1535.

SCHNEIDER, H., BARTH, W., BOHME, H. J. (1996), Cloning and characterization of another lignin peroxidase gene from the white-rot fungus *Phanaerochaete chrysosporium*, *Biol. Chem. Hoppe-Seyler* **377**, 399–402.

SCHOLTEN, J. D., CHANG, K. H., BABBIT, P. C., CHAREST, H., SYLVESTRE, M., DUNAWAY-MARINO, D. (1991), Novel enzymatic hydrolytic dehalogenation of a chlorinated aromatic, *Science* **253**, 182–185.

SCHOLTZ, R., LEISINGER, T., SUTER, F., COOK, A. M. (1987), Characterization of 1-chlorohexane halidohydrolase, a dehalogenase of wide substrate range from an *Arthrobacter* species, *J. Bacteriol.* **169**, 5016–5021.

Science Advisory Board of the International Joint Commission of the Great Lakes (1989), 1989 Report. Int. J. Commission, Hamilton, Ontario, October 1989.

SEARLE, P. A., MOLINSKI, T. F. (1994), 5 New alkaloids from the tropical ascidian, *Lissoclinum* sp. – lissoclinotoxin-A is chiral, *J. Org. Chem.* **59**, 6600–6605.

SEELBACH, K., VAN DEURZEN, M. P. L., VAN RANTWIJK, F., SHELDON, B. A., KRAGL, U. (1997), Improvement of the total turnover number and space-time yield for chloroperoxidase catalyzed oxidation, *Biotechnol. Bioeng.* **55**, 283–288.

SESIN, D. F., IRELAND, C. M. (1984), Iodinated products from a didemnid tunicate, *Tetrahedron Lett.* **25**, 403–404.

SHEFFIELD, D. J., HARRY, T., SMITH, A. J., ROGERS, L. J. (1993), Purification and characterization of the vanadium bromoperoxidase from the macroalga *Corallina officinalis*, *Phytochemistry* **32**, 21–26.

SHEFFIELD, D. J., HARRY, T., SMITH, A. J., ROGERS, L. J. (1994), Immobilization of bromoperoxidase from *Corallina officinalis*, *Biotechnol. Techniques* **8**, 579–582.

SHELTON, D. R., TIEDJE, J. M. (1984), Isolation and partial characterization of bacteria in an anaerobic consortium that mineralizes 3-chlorobenzoic acid, *Appl. Environ. Microbiol.* **48**, 840–848.

SIMONS, B. H., BARNETT, P., VOLLENBROEK, E. G. M., DEKKER, H. L., MUIJSERS, A. O., MESSERSCHMIDT, A., WEVER, R. (1995), Primary structure and characterization of the vanadium chloroperoxidase from the fungus *Curvularia inaequalis*, *Eur. J. Biochem.* **229**, 566–574.

SLATER, J. H., BULL, A. T., HARDMAN, D. J. (1995), Microbial dehalogenation, *Biodegradation* **6**, 181–189.

SMITH, A. L., NICOLAOU, K. C. (1996), The enediyne antibiotics, *J. Med. Chem.* **39**, 2103–2117.

SPINNLER, H.-E., DE JONG, E., MAUVAIS, G., SEMON, E., LE QUERE, J.-L. (1994), Production of halogenated compounds by *Bjerkandera adusta*, *Appl. Microbiol. Biotechnol.* **42**, 212–221.

STUCKI, G. R., GALLI, R., EBERSOLD, H. R., LEISINGER, T. (1981), Dehalogenation of dichloromethane by cell extracts of *Hyphomicrobium* DM2, *Arch. Microbiol.* **130**, 881–886.

SUNDARAMOORTHY, M., MAURO, J. M., SULLIVAN, A. M., TERNER, J., POULOS, T. L. (1995), Preliminary crystallographic analysis of chloroperoxidase from *Caldariomyces fumago*, *Acta Crystallogr.* **51**, 842–844.

SUZUKI, T., HONDA, H., KATSUMATA, R. (1972), Production of antibacterial compounds analogous to chloramphenicol by an *n*-paraffin-grown bacterium, *Agr. Biol. Chem.* **36**, 2223–2228.

TAKAHASHI, A., AGATSUMA, T., OHTA, T., NUNOZOWA, T., ENDO, T., NOZOE, S. (1993), Russuphelins B, C, D, E and F, new cytotoxic substances from the mushroom *Russula subnigricans* HONGO, *Chem. Pharm. Bull.* **41**, 1726–1729.

TAMURA, T., WADA, M., ESAKI, N., SODA, K. (1995), Synthesis of fluoroacetate from fluoride, glycerol, and β-hydroxypyruvate by *Streptomyces cattleya*, *J. Bacteriol.* **177**, 2265–2269.

THOMAS, J. A., HAGER, L. P. (1968), The peroxidation of molecular iodine to iodate by chloroperoxidase, *Biochem. Biophys. Res. Commun.* **32**, 770–775.

TOMLIN, C. (1995), *The Pesticide Manual UK* (10th Edn.). R. Soc. Chem., Crop. Protection Publication (TOMLIN, C., Ed.).

TOPP, E., HANSON, R. S. (1990), Degradation of pentachlorophenol by a *Flavobacterium* species grown in continuous culture under various nutrient limitations, *Appl. Environ. Microbiol.* **56**, 541–544.

TSCHIRRET-GUTH, R. A., BUTLER, A. (1994), Evidence for organic substrate binding to vanadium bromoperoxidase, *J. Am. Chem. Soc.* **116**, 411–412.

TU, C. M. (1976), Utilization and degradation of lindane by soil microorganisms, *Arch. Microbiol.* **108**, 259–263.

VAN DEN WIJNGAARD, A. J., JANSSEN, D. B., WITHOLT, B. (1989), Degradation of epichlorohydrin and halohydrins by bacterial cultures isolated from freshwater sediments, *J. Gen. Microbiol.* **135**, 2199–2208.

VAN DEN WIJNGAARD, A. J., REUVEKAMP, P. T. W., JANSSEN, D. B. (1991), Purification and characterization of haloalcohol dehalogenase from *Arthrobacter* sp. strain AD2, *J. Bacteriol.* **173**, 124–129.

VAN DER MEER, J. R. (1994), Genetic adaptation of bacteria to chlorinated aromatic compounds, *FEMS Microbiol. Rev.* **15**, 239–249.

VAN DER PLOEG, J. R., HALL, G., JANSSEN, D. B. (1991), Characterization of the haloacid dehalogenase from *Xanthobacter autotrophicus* GJ10 and sequencing of the *dhl B* gene, *J. Bacteriol.* **173**, 7925–7933.

VAN PEE, K.-H. (1990), Bacterial haloperoxidases and their role in secondary metabolism, *Biotechnol. Adv.* **8**, 185–205.

VAN PEE, K.-H. (1996), Biosynthesis of halogenated metabolites by bacteria, *Ann. Rev. Microbiol.* **50**, 375–399.

VAN PEE, K.-H., LINGENS, F. (1985a), Purification of bromoperoxidase from *Pseudomonas aureofaciens*, *J. Bacteriol.* **161**, 1171–1175.

VAN PEE, K.-H., LINGENS, F. (1985b), Purification and molecular and catalytic properties of bromoperoxidase from *Streptomyces phaeochromogenes*, *J. Gen. Microbiol.* **131**, 1911–1916.

VAN PEE, K.-H., SALCHER, O., FISCHER, P., BOKEL, M., LINGENS, F. (1983), The biosynthesis of brominated pyrrolnitrin derivatives by *Pseudomonas aureofaciens*, *J. Antibiot.* **36**, 1735–1742.

VAN PEE, K.-H., SURY, G., LINGENS, F. (1987), Purification and properties of a non-heme bromoperoxidase from *Streptomyces aureofaciens*, *Biol. Chem. Hoppe-Seyler* **368**, 1225–1232.

VANSCHIJNDEL, J. W. P. M., VOLLENBROEK, E. G. M., WEVER, R. (1993), The chloroperoxidase from the fungus *Curvularia inequalis* – a novel vanadium enzyme, *Biochim. Biophys. Acta* **1161**, 249–256.

VELDE, S. L. V., JACOBSEN, E. N. (1995), Kinetic resolution of racemic chromenes via asymmetric epoxidation – synthesis of (+)-Teretifolione-B, *J. Org. Chem.* **60**, 5380–5381.

VERHAGEN, F. J. M., SWARTS, H. J., KUYPER, T. W., WIJNBERG, J. B. P. A., FIELD, J. A. (1996), The ubiquity of natural adsorbable organic halogen production among basidiomycetes, *Appl. Microbiol. Biotechnol.* **45**, 710–718.

VERSCHUEREN, K. H. G., FRANKEN, S. M., ROZEBOOM, H. J., KALK, K. H., DIJKSTRA, B. W. (1993a), Refined X-ray structures of haloalkane dehalogenase at pH 6.2 and pH 8.2 and implications for the reaction mechanism, *J. Mol. Biol.* **232**, 856–872.

VERSCHUEREN, K. H. G., FRANKEN, S. M., ROZEBOOM, H. J., KALK, K. H., DIJKSTRA, B. W. (1993b), Crystallographic analysis of the catalytic mechanism of haloalkane dehalogenase, *Nature* **363**, 693–698.

VILTER, H. (1984), Peroxidases from phaeophycae. A vanadium(V) dependent peroxidase from *Ascophyllum nodosum* (L) 5, *Phytochemistry* **23**, 1387–1390.

VILTER, H. (1995), Vanadium dependent haloperoxidases, *Met. Ions Biol. Syst.* **31**, 325–362.

WACKETT, L. P., SCHANKE, C. A. (1992), Mechanisms of reductive dehalogenation by transition metal cofactors found in anaerobic bacteria, in: *Metal Ions in Biological Systems*, Vol. 28: Degradation of Environmental Pollutants by Microorganisms and their Metalloenzymes (SIGEL, H., SIGEL, A., Eds.), pp. 329–356.

WAGENKNECHT, H. A., WOGAN, W. D. (1997), Identification of intermediates in the catalytic cycle of chloroperoxidase, *Chem. Biol.* **4**, 367–372.

WALKER, J. R. L., LIEN, B. C. (1981), Metabolism of fluoroacetate by a soil *Pseudomonas* sp. and *Fusarium solani*, *Soil Biol. Biochem.* **13**, 231–235.

WENG, M., PFEIFER, C., KRAUSS, S., LINGENS, F., VAN PEE, K. H. (1991), Purification, characterization and comparison of 2 nonheme bromoperoxidases from *Streptomyces aureofaciens* ATCC 10762, *J. Gen. Microbiol.* **137**, 2539–2546.

WEVER, R., PLAT, H., DEBOER, E. (1985), Isolation procedure and some properties of the bromoperoxidase from the seaweed *Ascophyllum nodosum*, *Biochim. Biophys. Acta* **830**, 181–186.

WEYAND, M., HECHT, H. J., VILTER, H., SCHOMBURG, D. (1996), Crystallization and preliminary X-ray analysis of a vanadium dependent peroxidase from *Ascophyllum nodosum*, *Acta. Crystallogr. D* **52**, 864–865.

WIESNER, W., VAN PEE, K.-H., LINGENS, F. (1985), Purification and properties of bromoperoxidase from *Pseudomonas pyrrocinia*, *Biol. Chem. Hoppe-Seyler* **366**, 1085–1091.

WIESNER, W., VAN PEE, K.-H., LINGENS, F. (1986), Detection of a new chloroperoxidase in *Pseudomonas pyrrocinia*, *FEBS Lett.* **209**, 321–324.

WILLIAMS, D. H., FAULKNER, D. J. (1996), N-Methylated ageliferins from the sponge *Astrosclera willeyana* from Pohnpei, *Tetrahedron* **52**, 5381–5390.

WOLFFRAMM, C., LINGENS, F., MUTZEL, R., VAN PEE, K.-H. (1993), Chloroperoxidase-encoding gene from *Pseudomonas pyrrocina* – sequence, expression in heterologous hosts and purification of the enzyme, *Gene* **130**, 131–135.

WOOLARD, F. X., MOORE, R. E., ROLLER, P. P. (1976), Halogenated acetamides, but-3-en-2-ols and isopropanols from *Asparagopsis taxiformis* (Delile) Trev., *Tetrahedron* **32**, 2843–2846.

WUOSMAA, A. M., HAGER, L. P. (1990), Methyl-chloride transferase – a carbocation route for biosynthesis of halometabolites, *Science* **249**, 160.

XUN, L., TOPP, E., ORSER, C. S. (1992), Diverse substrate range of a *Flavobacterium* pentachlorophenol hydroxylase and reaction stoichiometries, *J. Bacteriol.* **174**, 2898–2902.

XUN, L., TOPP, E., ORSER, C. S. (1992), Confirmation of oxidative dehalogenation of pentachlorophenol by a *Flavobacterium* pentachlorophenol hydroxylase, *J. Bacteriol.* **174**, 5745–5747.

YOKOTA, T., FUSE, H., OMORI, T., MINODA, Y. (1986), Microbial dehalogenation of haloalkanes mediated by oxygenase or halohydrolase, *Agric. Biol. Chem.* **5**, 453–460.

YOKOTA, T., OMORI, T., KODAM, T. (1987), Purification and properties of haloalkane dehalogenase from *Corynebacterium* sp. strain m15-3, *J. Bacteriol.* **169**, 4049–4054.

YU, F., NAKAMURA, T., MIZUNASHI, W., WATANABE, W. (1994), Cloning of 2 halohydrin hydrogen-halide lyase genes of *Corynebacterium* sp. strain N-1074 and structural comparison of the genes and gene-products, *Biosci. Biotechnol. Biochem.* **58**, 1451–1457.

ZAKS, A., DODDS, D. R. (1995), Chloroperoxidase-catalyzed asymmetric oxidations: Substrate specificity and mechanistic study, *J. Am. Chem. Soc.* **117**, 10419–10424.

4 Phosphorylation

SIMON JONES

Tempe, AZ, USA

1 Phosphoryl Transfer in General

1.1 Introduction

The formation and cleavage of the phosphate bond is one of the most important reactions in biological systems. The release of energy from the hydrolysis of a P–O bond (602 kJ mol^{-1}; 144 kcal mol^{-1}) (BERKOWITZ, 1959) is an indication of the free energy available to "power" biological processes, i.e., to couple endogonic processes to the exogonic cleavage of phosphate bonds. The most common way this is achieved, is by transfer of a pyrophosphate group from adenosine triphosphate (ATP) (**1**) (Fig. 1) to an aliphatic alcohol. The alkyl pyrophosphate so formed can then undergo nucleophilic displacement to form a new bond. A typical example is the coupling of dimethylallyl pyrophosphate and isopentenyl pyrophosphate to give geranyl pyrophosphate, the key step in monoterpene biosynthesis. Using the language of synthetic organic chemistry, formation of the pyrophosphate converts the alcohol into a *good leaving group* (nucleofuge). ATP is in effect *Nature's tosyl chloride!*

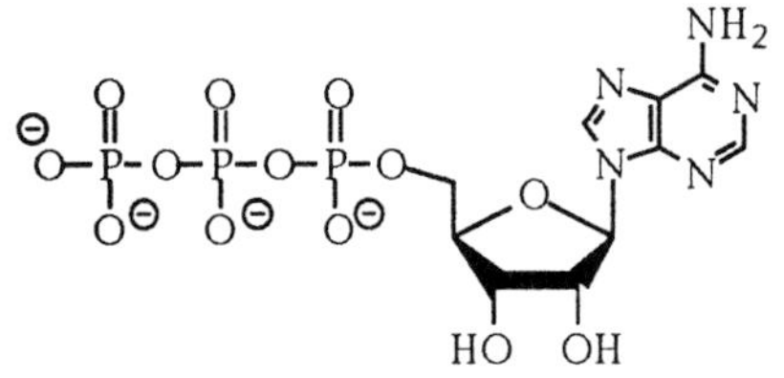

1 Adenosine triphosphate (ATP)

Fig. 1. Adenosine triphosphate (ATP) (**1**).

The importance of nucleotide phosphate metabolism can be judged from the daily energy requirements of an adult woman, 1500–1800 kcal (6300–7500 kJ). This corresponds to the energy of hydrolysis of 200 mol (ca. 150 kg) of ATP to ADP and phosphate, which must be generated by recycling less than 70 g of ATP. The major pathway for the production of ATP is the exogonic oxidation of glucose to carbon dioxide and water (VOET and VOET, 1995).

The key feature that distinguishes phosphate esters from other conceivable derivatives is that they are kinetically highly stable in aqueous solution. For example, the free energy of hydrolysis of acetic anhydride, acetyl phosphate, and inorganic pyrophosphate are all of the same magnitude, whereas the half lives in water are a few seconds, several hours, and years, respectively. Moreover the presence of several oxygens is ideal for facilitating binding to enzymes and for activation of cleavage by protonation (Fig. 2) (KEIKINHEIMO et al., 1996). The predominant counterion *in vivo* is Mg^{2+}.

Phosphate esters have biological roles other than energy metabolism. For example, phosphodiesters provide the backbone of both DNA and RNA, the carriers of the genetic code, while phosphorylation of key enzymes and proteins regulates their activity (KREBS and BEAU, 1979; GIBSON et al., 1990).

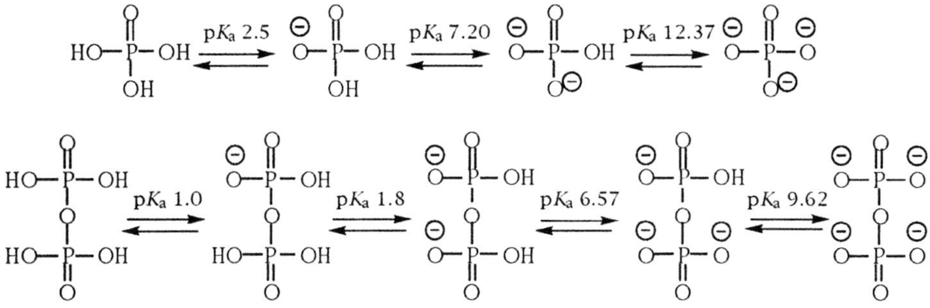

Fig. 2. pK_a values for phosphate and pyrophosphate.

Since many important biological molecules and drugs contain phosphate groups, there is a need for efficient and specific synthetic methods. Chemical synthesis often proceeds in high yield, but currently there is a dearth of stereo- or enantioselective methods. Consequently this must be achieved indirectly via protecting groups. The use of biological systems, whether whole cell or isolated enzymes, can allow the stereoselective and often enantioselective introduction of phosphate groups in comparable yields to those obtained by chemical methods and without the need for protecting groups.

This chapter aims to provide an insight into the use of biological systems for both the formation and cleavage of phosphate esters for synthetic purposes. A range of examples of phosphate forming reactions have been included in tabular form to illustrate the wide applicability of the techniques (see Tab. 3). Other complex enzyme systems involving phosphates in multi-sequence reactions, such as aldolase reactions (ALLEN et al., 1992), are not considered here.

1.2 Enzyme-Catalyzed Phosphate Transfer

When we talk of an enzyme catalyzing the phosphorylation of an alcohol or similar nucleophilic substrate, we should bear in mind that specifically it is an enzyme catalyzed phosphoryl transfer from a phosphate source or cofactor (such as ATP (**1**)) to the nucleophile. Therefore, there are two considerations to be made when examining phosphoryl transfer, the enzyme and the source of phosphate group.

1.2.1 The Choice and Action of the Enzyme

The enzymes that catalyze phosphate transfer are categorized according to the bonds made or cleaved. There are two general groups of enzymes: those that accept phosphate monoesters as substrates and those which accept phosphate diesters or pyrophosphates (Fig. 3) (KNOWLES, 1980). DRAUZ and WALDMANN (1995) correlated this classification with the classes of enzymes specified by the Nomenclature Committee of the International Union of Biochemistry (1984) (Tab. 1). The

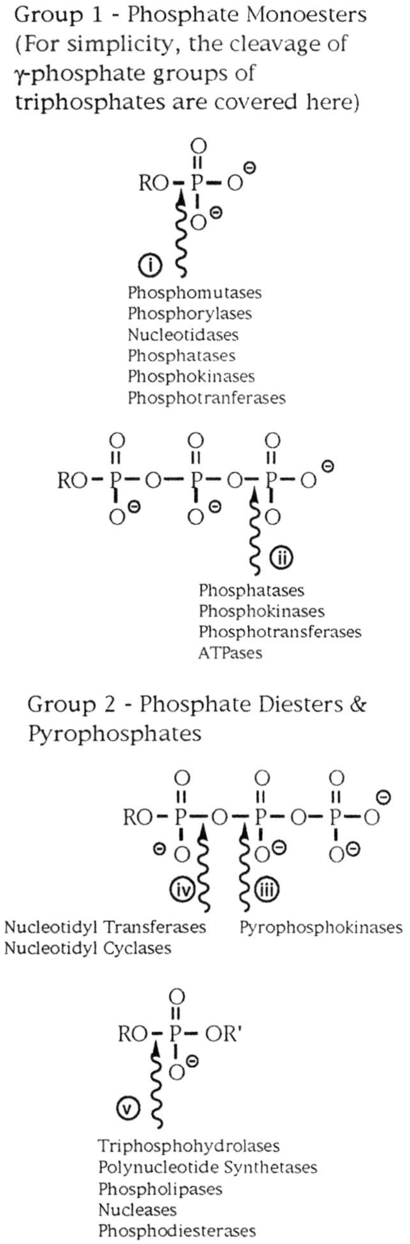

Group 1 - Phosphate Monoesters
(For simplicity, the cleavage of
γ-phosphate groups of
triphosphates are covered here)

(i)
Phosphomutases
Phosphorylases
Nucleotidases
Phosphatases
Phosphokinases
Phosphotranferases

(ii)
Phosphatases
Phosphokinases
Phosphotransferases
ATPases

Group 2 - Phosphate Diesters &
Pyrophosphates

(iv) Nucleotidyl Transferases
Nucleotidyl Cyclases

(iii) Pyrophosphokinases

(v)
Triphosphohydrolases
Polynucleotide Synthetases
Phospholipases
Nucleases
Phosphodiesterases

Fig. 3. The two groups of phosphoryl transfer reactions and enzyme classes.

Tab. 1. Table of Enzymes that Fall into Groups 1 and 2

Cleavage Site	Functional Class	IUB Class & Name	Role
(i)	Phosphomutase	2.7.5. Phosphomutase 5.4.2. Intramolecular phosphotransferase	intramolecular transfer of a phosphoryl group
(i)	Phosphorylase	2.4.1. Hexosyl transferases 2.4.2. Pentosyl transferases	P–O bond formation from phosphorolytic C-hetero-atom cleavage
(i)	Nucleotidases	3.1.3. Phosphoric acid hydrolases 3.1.4. Phosphoric diester hydrolases	transfer of phosphoryl from a nucleoside to water
(i) (ii)	Phosphatases	3.1.3. Phosphoric ester hydrolases 3.6.1. Hydrolases acting on acid anhydrides in phosphorus containing anhydrides	transfer of phosphoryl from a monoester to water
(i) (ii)	Phosphokinase	2.7.1. Phosphotransferases with an alcohol group as acceptor 2.7.2. Phosphotransferases with a carboxyl group as acceptor	transfer of phosphoryl from a nucleoside triphosphate to an acceptor other than water
(i) (ii)	Phospho-transferase	2.7.4. Phosphotransferases with a phosphate group as acceptor	transfer of phosphoryl from a molecule other than a nu-cleoside triphosphate to an acceptor other than water
(ii)	ATPase	3.6.1.3. ATPases	phosphatases which couple ATP cleavage with other processes
(iii)	Pyrophospho-kinase	2.7.6. Diphosphotransferases	transfer of pyrophosphate from ATP to an acceptor other than water
(iv)	Nucleotidyl Transferase	2.7.7. Nucleotidyl Transferases	transfer of nucleotidyl groups
(iv)	Nucleotidyl Cyclase	4.6.1. Phosphorus oxygen lyases	cyclization of a nucleoside triphosphate by formation of a pyrophosphate
(v)	Triphospho-hydrolase	3.1.5. Triphosphoric monoester hydrolases	transfer of triphosphate from a nucleoside triphos-phate to water
(v)	Polynucleotide synthase	6.5.1. Ligases forming phosphoric ester bonds	links two polynucleotides to produce a polynucleotide chain
(v)	Phospholipase	3.1.4. Phosphoric diester hydrolases	hydrolytic cleavage of phosphoglycerides
(v)	Nuclease	3.1.4. Phosphoric diester hydrolases 3.1. Endo- and exonucleases	transfer of a phosphopoly-nucleotide to water
(v)	Phospho-diesterase	2.7.8. Transferases for other substituted phosphate groups 3.1.4. Phosphoric diester hydrolases	transfer of a phosphomono-ester from a phosphodiester, other than a polynucleotide, to water

range of commercially available enzymes is somewhat less than this. On occasions it may be worthwhile to extract and purify a given enzyme, however, this is not practical in general. Consequently the following sections emphasize enzymes which are commercially available or easily extractable. For each example a reference to *Methods in Enzymology* is given which provides *recipes* for purification and assay.

1.2.1.1 Acetate Kinase

Acetate kinase (EC 2.7.2.1) has been used extensively for phosphorylation of adenosine diphosphate (ADP) (**3**) (Fig. 4) and will also accept GDP, UDP, and CDP (HAYNIE and WHITESIDES, 1990; CRANS et al., 1987). The mechanism of phosphoryl transfer has been extensively studied (SPECTOR, 1980) and proceeds with net inversion at phosphorous. The enzyme accepts acetyl phosphate (**8**) (Tab. 2), propionyl phosphate, and carbamoyl phosphate, although the latter two cause reduced enzyme activity (30% and 18%, respectively) but phosphoenolpyruvate (**4**) is not accepted. Spontaneous hydrolysis of acetyl phosphate (**8**) restricts the use of acetate kinase for extended periods. The X-ray crystal structure of acetate kinase shows an "actin fold" which is discussed in the section on glycerol kinase (Sect. 1.2.1.2) (BUSS et al., 1997). Immobilized acetate kinase from *E. coli* has high activity, but is less stable than the much more expensive enzyme from the thermophile, *Bacillus stearothermophilus* (TANG and JOHANSSON, 1997).

1.2.1.2 Glycerol Kinase

Glycerol kinase (EC 2.7.1.30) is usually isolated from rat liver (THOMER and PAULUS, 1973) and commercial enzymes with high activity extracted from *E. coli* (KEE et al., 1988) and a *Cellulomonas* sp. Rat liver glycerol kinase is most stable at pH 5 and readily monophosphorylates glycerol (**15**) (Fig. 6), L-glyceraldehyde (**21**) (Fig. 9), or dihydroxyacetone (LIN, 1977; CRANS and WHITESIDES, 1985a, b). ATP (**1**) is the best cofactor, although different forms of the enzyme can accept uridine triphosphate (UTP), guanosine triphosphate (GTP), and cytidine triphosphate (CTP). ADP (**3**) has been shown to inhibit glycerol kinase noncompetitively at low concentrations, and competitively at higher concentrations. Acetate kinase and glycerol kinase are members of a structural super family which includes sugar kinases, hexokinase, heat shock protein 70, and actin. These are characterized by a two domain structure (actin fold) with the topography $\beta\beta\beta\alpha\beta\alpha\beta\alpha$. The nucleotide binding site lies between the two domains which are joined by a putative hinge. Nucleotide hydrolysis eliminates some of the interdomain bridging interactions which enables formation of an open conformation. The large conformational changes involved in these processes may be involved in regulation (HURLEY, 1996; KABSCH and HOLMES, 1995).

1.2.1.3 Pyruvate Kinase

The most common source of pyruvate kinase is from rabbit muscle from which it can be isolated in crystalline form. It accepts guanosine diphosphate (GDP) with approximately the same tolerance as ATP (**1**) (KAYNE, 1973). However other nucleoside diphosphates (NDPs), such as cytidine diphosphate (CDP), are not so readily accepted, with reaction rates approximately 30% less than that of ADP (**3**). The enzyme shows a very high degree of specificity towards the phosphate donor with only phosphoenolpyruvate (**4**) being accepted. The reaction rate is highly dependent on the concentration of monovalent cations, in particular potassium and ammonium. One drawback is that it is inhibited by ATP (**1**), thus necessitating keeping ATP (**1**) levels low during reaction.

1.2.1.4 Creatine Kinase and Arginine Kinase

These enzymes are both ATP:guanidino phosphotransferases which have high amino acid homology (GROSS et al., 1995; MUHLEBACH, 1994) although they are found in two different biological classes (WATTS, 1973;

MORRISON, 1973). Creatine kinase (EC 2.7.3.2) is isolated from vertebrate muscle (usually from rabbit muscle) and is stable below 50 °C between pH 6.5 and 9.5. It has narrow substrate specificity accepting only a small range of guanidine based amino acids. Arginine kinase (EC 2.7.3.3) is found in most invertebrates and has similar stability and substrate properties to creatine kinase (DUMAS and CAMONIS, 1993; STRONG and ELLINGTON, 1995).

1.2.1.5 Adenylate Kinase

Adenylate kinase (myokinase, EC 2.7.4.3) catalyzes the disproportionation of ATP (**1**) and adenosine monophosphate (**2**) (AMP) to ADP (**3**) (Fig. 4). The enzyme is widespread throughout organisms, with higher concentrations located in regions of high energy turnover, such as muscle (NODA, 1973; KREJCOVA and HORSKA, 1997). Most other nucleotides and inorganic triphosphate can be utilized, but at reduced reaction rates. A divalent cation (usually magnesium or manganese) is needed for the reaction to proceed (WILD et al., 1997).

1.2.1.6 Nucleoside Kinase

Nucleoside kinases catalyze the phosphoryl transfer from a nucleoside triphosphate (NTP) to a nucleoside, giving a nucleoside monophosphate (NMP). There are many different classes of these enzymes depending upon the nucleoside being phosphorylated (ANDERSON, 1973). These enzymes are often specific for the nucleoside. Thus, AMP (**2**) will only be formed from adenosine utilizing adenosine kinase (EC 2.7.1.20). There is however, more flexibility with the NTP donor in most cases; adenosine kinase accepts most other NTPs.

$$\underset{1}{\text{ATP}} + \underset{2}{\text{AMP}} \xrightleftharpoons[]{\text{Adenylate kinase}} 2\ \underset{3}{\text{ADP}}$$

Fig. 4. Formation of ADP (**3**) from ATP (**1**) and AMP (**2**).

1.2.1.7 Hexokinase

Hexokinases (EC 2.7.1.1) have been used extensively for the conversion of glucose to glucose-6-phosphate (**13**) (CHENAULT et al., 1997). There are two main sources of hexokinase: from yeast and from mammals (COLOWICK, 1973). Both sources have certain aspects in common, including strong inhibition by *N*-acetyl glucosamine derivatives (WILLSON et al., 1997; MAITY and JARORI, 1997), broad substrate specificity towards the sugar acceptor, and conversely narrow specificity towards the NTP used as the donor (ATP (**1**) is the best).

1.2.1.8 Alkaline Phosphatase

The two mains sources of alkaline phosphatase (EC 3.1.3.1) are bovine intestines (FERNLEY, 1971) and *E. coli* (REID and WILSON, 1971). As its name suggests, the enzyme is stable at basic pH (typically pH 9.8). Bovine alkaline phosphatase has very high activity, is cheap, but has low stereospecificity (NATCHEV, 1987). Consequently it is used as a general purpose workhorse for the cleavage of monophosphates to alcohols. It will also catalyze transphosphorylation and some enzymes also cleave pyrophosphates. The X-ray crystal structure has been determined (MURPHY et al., 1997; cf. KEIKINHEIMO et al., 1996).

1.2.1.9 Phosphodiesterase

Phosphodiesterase enzymes are often called venom exonucleases since they are present in all snake venoms. One of the more common enzymes used is snake venom phosphodiesterase (phosphodiesterase I, EC 3.1.4.1) (LASKOWSKI, 1971), which is a 5′-exonuclease, i.e., it specifically cleaves oligonucleotides to give 5′-mononucleosides. This enzyme together with bovine intestinal phosphodiesterase has high activity with non-nucleotide substrates. Although snake venom phosphodiesterases are known which act as 3′-exonuclease, spleen phosphodiesterase is used more commonly (phosphodiesterase II, EC 3.1.16.1) (BERNARDI and BERNARDI, 1971).

Tab. 2. A Comparison of the Relative Phosphorylation Potentials of Phosphoryl Transfer Reagents

Cofactor	Structure	Phosphorylation Potential [kJ mol^{-1}] ([kcal mol^{-1}])[a]
Phosphoenolpyruvate **4**	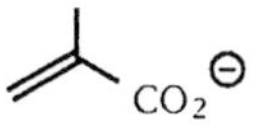 $CO_2^{\ominus}$	−61.9 (−14.8)[b]
Methoxycarbonyl Phosphate **5**	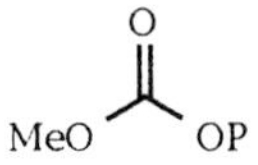 MeO–C(=O)–OP	−51.9 (−12.4)[c]
Carbamyl Phosphate **6**	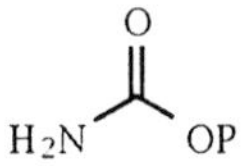 H_2N–C(=O)–OP	−51.4 (−12.3)[d]
1,3-Diphosphoglycerate **7**	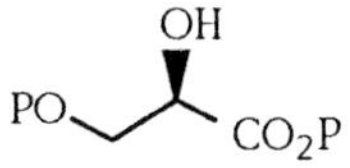 PO, OH, CO_2P	−49.4 (−11.8)[e]
Acetyl Phosphate **8**	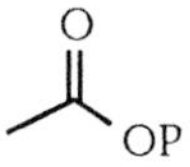 O, OP	−43.1 (−10.3)[b]
Phosphocreatine **9**	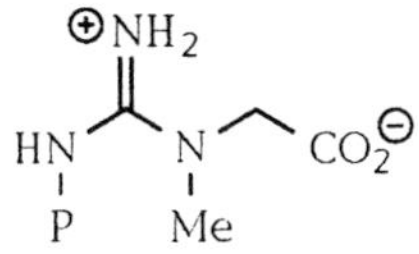 $\oplus NH_2$, HN, P, N, Me, $CO_2^{\ominus}$	−43.1 (−10.3)[e]
ATP **1** → ADP + P_i		−30.5 (− 7.3)[e]
ATP **1** → AMP + PP_i		−41.8 (−10.0)[e]
Arginine Phosphate **10**	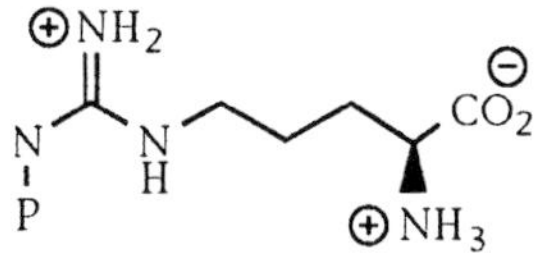 $\oplus NH_2$, N, P, NH, $CO_2^{\ominus}$, $\oplus NH_3$	−32.2 (− 7.7)[e]
Pyrophosphate **11**	PPi	−19.2 (− 4.6)[e]
Glucose-1-phosphate **12**	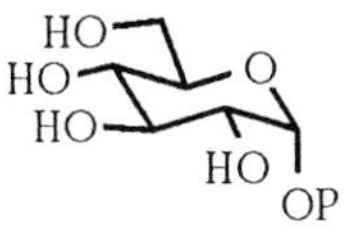 HO, HO, HO, O, HO, OP	−20.9 (− 5.00)[e]
Glucose-6-phosphate **13**	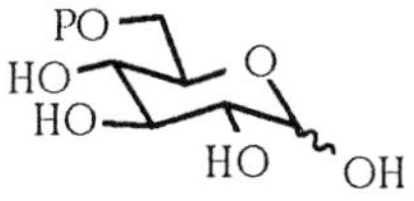 PO, HO, HO, O, HO, OH	−13.8 (− 3.3)[e]
Glycerol-1-phosphate **14**	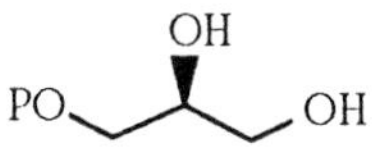 PO, OH, OH	− 9.2 (− 2.2)[e]

[a] Values are for pH 7.0. ATP hydrolysis is highly dependent on magnesium ion concentration; [b] BOLTE and WHITESIDES (1984); [c] KAZLAUSKAS and WHITESIDES (1985); [d] SHIH and WHITESIDES (1977); [e] LEHNINGER (1975).

1.2.2 The Source of Phosphate Group

Phosphoryl transfer necessarily means that a source of phosphate is needed. This is rarely inorganic phosphate and almost invariably a complex cofactor. A comparison of the free energies of hydrolysis of a number of phosphorylating cofactors gives an indication of the phosphorylation potential of that species (Tab. 2). The majority of these phosphorylating agents have some biological role, although the most widely used in nature are the NTPs. ATP (**1**) acts predominantly in central pathways of energy metabolism, GTP is used to drive protein synthesis, CTP phospholipid biosynthesis, and UTP glycosidation. Those agents with a large phosphorylation potential are the primary sources of energy in biological processes. Phosphoenolpyruvate (**4**) and 1,3-diphosphoglycerate (**7**) are both formed during glycolysis and are used as a primary source of phosphate. Phosphocreatine (**9**) and phosphoarginine (**10**) are found in muscle tissue in vertebrates and invertebrates, respectively. These compounds act as an energy "store", which is converted back to ATP (**1**) for use when primary energy sources (such as phosphoenolpyruvate (**4**)) are not available. ATP (**1**) itself then serves as a means to transfer the phosphoryl groups to and from storage, and thus is recognized as a cofactor by many different enzymes. Although better phosphoryl transfer agents are available (cf. phosphoenolpyruvate (**4**)), ATP (**1**) is essential for these processes.

1.2.3 Cofactor Regeneration

The most commonly used NTP as a cofactor for phosphoryl transfer is ATP (**1**). The active form of ATP (**1**) has been found to be the $MgATP^{2-}$ species, which necessitates addition of a source of magnesium ions to the reaction mixture. However, the high cost of ATP (**1**) precludes the use of stoichiometric quantities *in vitro*. Since most enzymes need ATP (**1**) as a cofactor, regeneration of AMP (**2**), or more commonly ADP (**3**) formed after the first phosphoryl transfer is necessary. ATP (**1**) regeneration can be achieved by chemical means or by enzymatic methods. In general, the enzymatic route is preferred. Most commonly a phosphorylating agent with a high phosphorylation potential is used to enable efficient cofactor regeneration. The differing combinations of enzyme and phosphate source are described later.

2 Formation of P–O Bonds

Reactions which involve the synthesis of a P–O bond all require a source of phosphate. This section is divided according to the cofactors used, concluding with a summary of various P–O bond forming reactions (Tab. 3).

2.1 Inorganic Phosphate (P_i) as the Phosphate Source

Glucose-1-phosphate (**12**) has been prepared by the action of phosphorylase-a (EC 2.4.1.1) (Fig. 5) on soluble starch or dextrin in the presence of inorganic phosphate and the glucose-1-phosphate (**12**) formed transformed to glucose-6-phosphate (**13**) using phosphoglucomutase (EC 2.7.5.1) (WONG and WHITESIDES, 1981).

An excellent study of the use of calf intestinal alkaline phosphatase with sodium pyrophosphate as the phosphate source has been made. A variety of diols were used but most of the work was concentrated on the formation of glycerol-1-phosphate (**14**) from glycerol (**15**) (Fig. 6). The regioselectivity of this reaction was found to be greater than 90% in favor of the primary hydroxyl group. However, the reaction was not enantioselective (PRADINES et al., 1988).

2.2 Nucleoside Triphosphates (NTPs) as the Phosphate Source

The use of NTPs is one of the most effective and widespread enzymatic methods of phosphoryl transfer. As noted earlier, regeneration

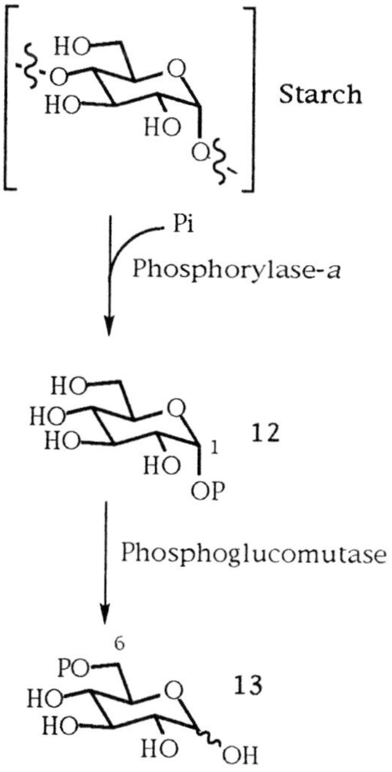

Fig. 5. Synthesis of glucose-6-phosphate (**13**).

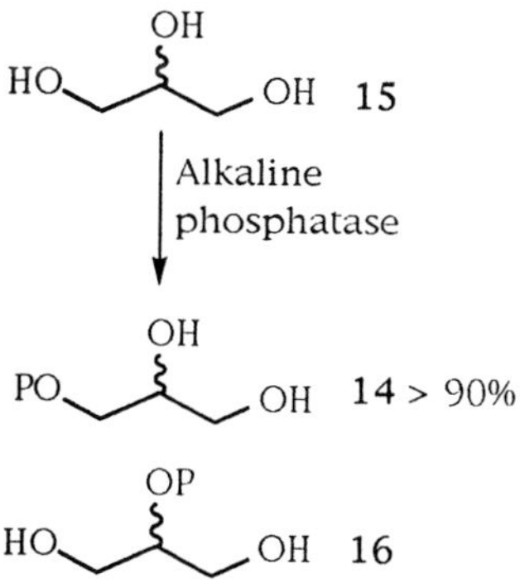

of NTPs is necessary to reduce the cost of the reaction. Several systems exist for this process (HEIDLAS et al., 1992).

2.2.1 Phosphoenolpyruvate/ Pyruvate Kinase

Phosphoenolpyruvate (PEP) (**4**) is an attractive phosphorylating agent for ATP (**1**) regeneration since it has a high phosphorylation potential (-61.9 kJ mol^{-1}). A drawback is the high commercial cost and problematic synthesis of the potassium salt (HIRSCHBEIN et al., 1982), although WHITESIDES has developed an efficient enzymatic method (Fig. 7) of generating PEP (**4**) *in situ* from the relatively inexpensive D-($-$)-3-phosphoglyceric acid (**17**) for the synthesis of NTPs from NMPs (SIMON et al., 1989, 1990). The use of this phosphoryl transfer system is widespread and has broad applicability. For example, a whole range of ring fluorinated hexose sugars have been transformed to their corresponding 6-phosphates, as well as

Fig. 6. Selective phosphorylation of glycerol using alkaline phosphatase.

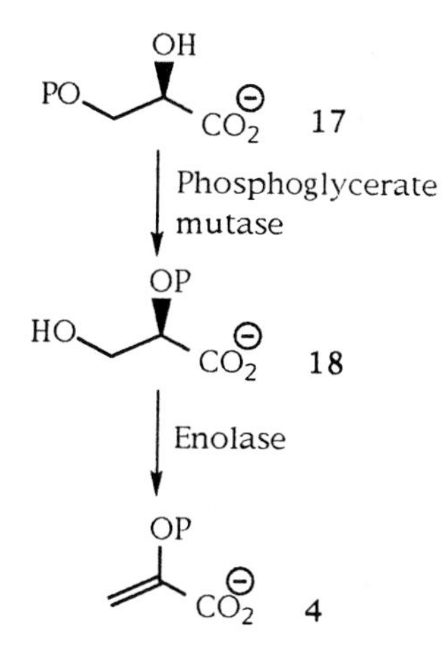

Fig. 7. Enzymatic synthesis of phosphoenolpyruvate (**4**).

those analogs containing sulfur or nitrogen within the pyranose ring (DRUECKHAMMER and WONG, 1985). The reaction is also highly stereospecific. One of the more important uses of enzymatic phosphate synthesis is the preparation of chiral phosphorothioates which are used extensively for the elucidation of biological mechanisms. The synthesis and uses of phosphorothioates have been extensively reviewed (LESNIKOWSKI, 1993; ECKSTEIN, 1983, 1985). As an illustrative example, the 5′-monothiophosphate (**19**) was transformed exclusively into one diastereoisomer of the "pseudo" triphosphate (**20**) (Fig. 8) (MORAN and WHITESIDES, 1984).

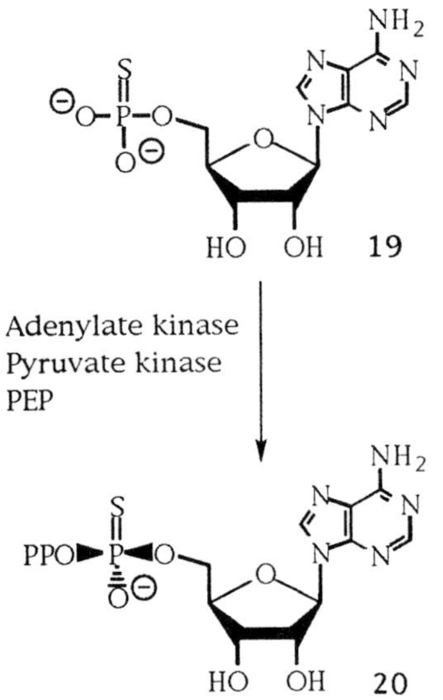

19

Adenylate kinase
Pyruvate kinase
PEP

20

Fig. 8. Stereospecific phosphorylation of a monothiophosphate (**19**).

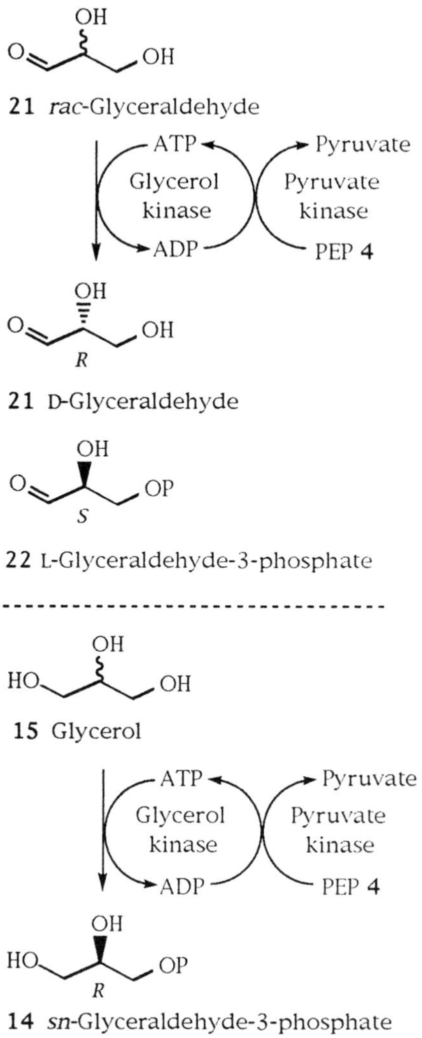

Glycerol kinase and ATP (**1**) have been used to perform kinetic resolutions of *rac*-glyceraldehyde (**21**) (Fig. 9) (WONG and WHITESIDES, 1983). ATP (**1**) was regenerated using PEP (**4**) and pyruvate kinase. Enantioselective phosphorylation of glycerol (**15**) has also been achieved with glycerol kinase (CRANS and WHITESIDES, 1985a, b) and the acetyl phosphate (**8**)/acetate kinase system (RIOS-MERCADILLO and WHITESIDES, 1979). The structural requirements of glycerol kinase were evaluated in a survey of 66 glycerol analogs. This indicated that one terminal hydroxyl group and either a hydroxyl or methyl group at the secondary position was necessary for transformation (CRANS and WHITESIDES, 1985a).

The pyruvate kinase/PEP (**4**) regeneration system has also been exploited in multienzyme sequences as illustrated by the regeneration of UTP from uridine diphosphate (UDP) for the *in situ* synthesis of UDP-glucose (WONG et al., 1982), for use in Leloir glycosidation.

Hydrolysis of the anomeric linkage to adenine is a major pathway for the degradation of adenosine derivatives in solution. Moreover, when bound to most enzymes the oxygen of the ribose ring of adenosine nucleotides does

21 *rac*-Glyceraldehyde

21 D-Glyceraldehyde

22 L-Glyceraldehyde-3-phosphate

15 Glycerol

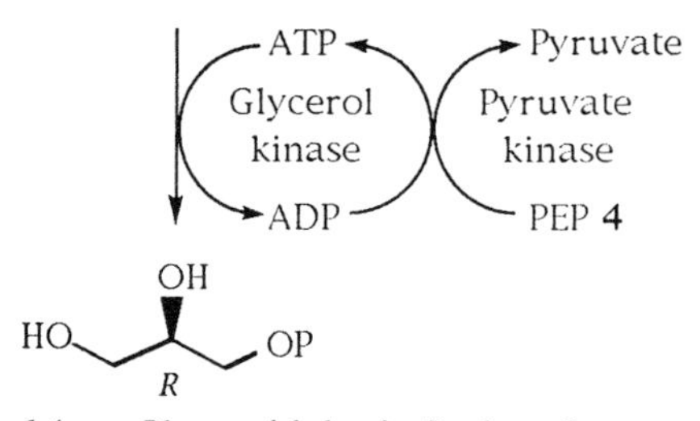

14 *sn*-Glyceraldehyde-3-phosphate

Fig. 9. Specific phosphorylation of glyceraldehyde *rac*-(**21**) and glycerol (**15**).

not participate in substantial hydrogen bonding. Consequently, the carbosugar analog of ATP (**23**) (Fig. 10) should be substantially more stable and should bind to enzymes equally well. Fortuitously the carbosugar analog of adenosine is a microbial natural product, aristeromycin (**23a**). This was converted to the 6′-monophosphate (**23b**) using a chemical method and then to the diphosphate (**23c**) and triphosphate (**23d**) catalyzed by adenylate

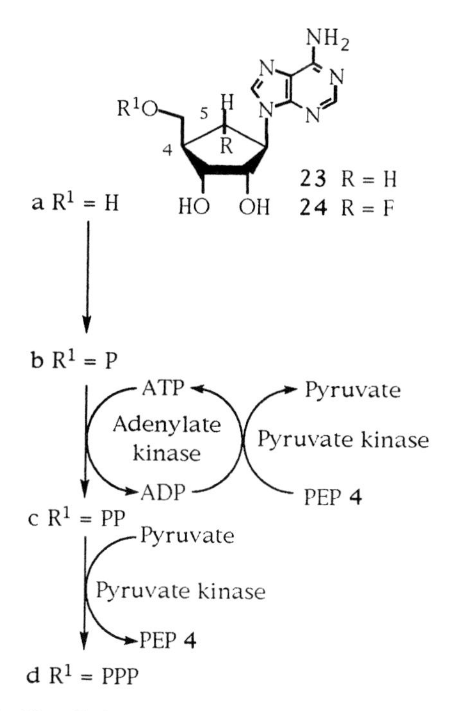

Fig. 10. Carbosugar nucleoside triphosphates used in phosphoryl transfer reactions.

kinase and pyruvate kinase, respectively (cf. SIMON et al., 1990). An identical sequence was used to prepare the fluoroanalog (**24d**). Both carbosugar nucleotides (**23d**) and (**24d**) functioned in place of ATP (**1**) in the synthesis of glucose-6-phosphate (**13**) from glucose catalyzed by hexokinase and *sn*-glycerol-3-phosphate (**14**) from glycerol (**15**) using glycerol kinase (LEGRAND and ROBERTS, 1993).

2.2.2 Acetyl Phosphate/ Acetate Kinase

Acetyl phosphate (**8**) is a good phosphorylating agent. It has a reasonable phosphorylation potential and has been used extensively with acetate kinase in many enzyme reactions due to its relative inexpense. One drawback is that acetate kinase is prone to slight inhibition by acetate ions, which are formed during the reaction. Also, acetyl phosphate (**8**) is not as stable in solution as most other cofactors.

Acetyl phosphate (**8**) can be prepared in large quantities as its diammonium salt from either phosphoric acid, ketene, and ammonia (WHITESIDES et al., 1975), or acetic anhydride, phosphoric acid, and ammonia (LEWIS et al., 1979). The latter is easier to use, since it does not involve the preparation and handling of ketene. In both cases, the ammonium ion present reacts with magnesium ions during the phosphorylation reaction to give a precipitate of $Mg(NH_4)PO_4$, which causes essential magnesium ions to be removed from the reaction mixture. Reaction of acetyl phosphate (**8**) with minute quantities of ammonia present can also cause problems. These can be circumvented by using the potassium or sodium salts which can be prepared either by ion exchange chromatography of the diammonium salt, or by direct preparation (CRANS and WHITESIDES, 1983).

2.2.3 Methoxycarbonyl Phosphate/ Acetate Kinase or Carbamate Kinase

Both acetyl phosphate (**8**) and PEP (**4**) have inherent advantages and disadvantages. Methoxycarbonyl phosphate (**5**) has been developed (KAZLAUSKAS and WHITESIDES, 1985) as an easily synthesized alternative to acetyl phosphate (**8**), with comparable phosphorylating potential to PEP (**4**). Methoxycarbonyl phosphate (**5**) also has the advantage in not containing ammonium groups, thus negating the problems associated with diammonium acetyl phosphate. Following phosphoryl transfer, the by-product of the reaction, methyl carbonate, decomposes to give methanol and carbon dioxide, thus making the workup easier. It is readily prepared in an analogous way to acetyl phosphate with acetic anhydride and phosphoric acid, but decomposes much more rapidly in aqueous solution. One of two enzymes can be used to catalyze phosphoryl transfer to ATP (**1**): acetate kinase or carbamate kinase. The former has been used with methoxycarbonyl phosphate (**5**), creatine kinase, and ATP (**1**) to generate creatine phosphate (**9**) from creatine.

2.2.4 Other Systems

Extensive work has been carried out on the chemoselective formation of ATP (**1**) from ADP (**3**) using the macrocycle (**25a**) (Fig. 11) (HOSSEINI and LEHN, 1985, 1987, 1988, 1991; HOSSEINI et al., 1983). In the presence of acetyl phosphate (**8**), magnesium ions, and a hexokinase, the complex (**25b**) is formed which is able to facilitate a turnover of ATP (**1**) sufficient to catalyze the formation of glucose-6-phosphate (**13**) from glucose by hexokinase (FENNIRI and LEHN, 1993).

The versatility of the use of enzymes can be illustrated in the use of the glycolytic pathway to act as a means of ATP (**1**) regeneration (WEI and GOUX, 1992). With inorganic phosphate and glucose as the "fuel" for the reaction, creatine was converted by creatine kinase to creatine phosphate (**9**) in the presence of ATP (**1**) and the required enzymes for glucose metabolism.

2.3 Coupling to Form Phosphodiesters

Phosphodiesters play an important role in biological functions. DNA polymers contain phosphodiester links as do coenzyme A and NAD(P)(H). There is a wide range of enzymes that catalyze phosphodiester formation by coupling of two components. Usually one of these is a NTP and thus use of a phosphate source or cofactor is not necessary.

2.3.1 Coupling with ATP

Tyrosine residues in peptides have been specifically phosphorylated at the phenolic position in a two step process. Initially a tyrosine 5′-adenosine phosphodiester was formed using glutamine synthetase adenylyltransferase (EC 2.7.7.42), which was selectively dephosphorylated using micrococcal nuclease to give the desired phospho-tyrosine (GIBSON et al., 1990). In a similar manner (MARTIN and DRUECKHAMMER, 1992), pantetheine 4′-phosphate (**26**) was coupled with ATP (**1**) using dephospho-CoA-pyrophosphorylase (EC 2.7.7.3) and inorganic pyrophosphate (Fig. 12) to give 3′-dephospho-coenzyme A (**27a**). The 3′-hydroxyl was specifically phosphorylated by ATP using dephospho-CoA-kinase (EC 2.7.1.24).

2.3.2 Coupling with CTP

N-Acetyl-neuraminic acid (**28**) (Fig. 13), is a sialic acid which is a component of glycoproteins and lipids. It was coupled with CTP

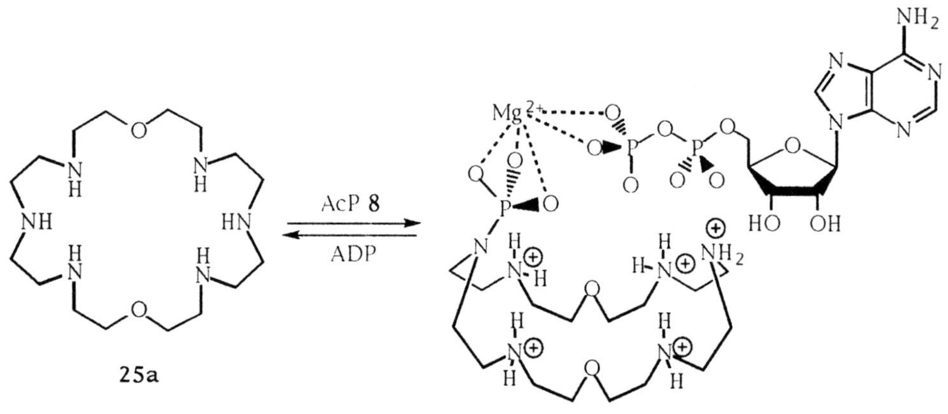

Fig. 11. Macrocyclic amino-crown ether used to catalyze formation of ATP (**1**) from ADP (**3**).

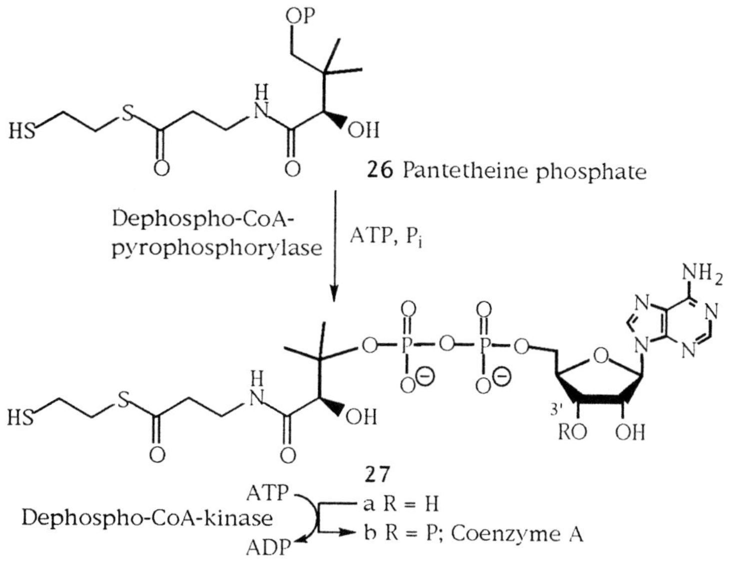

Fig. 12. Formation of Coenzyme A (**27b**).

using cytidine-5′-monophospho-*N*-acetylneuramic acid synthase as a prelude to enzymatic glycosidation. CTP was generated by phosphorylation of CDP with PEP (**4**), catalyzed by pyruvate kinase. CDP was prepared *in situ* by disproportionation of stoichiometric CMP with CTP catalyzed by adenylate kinase (SIMON et al., 1988a).

2.3.3 Coupling with UTP

UDP-glucose has been prepared by coupling of UTP with glucose-6-phosphate (**13**) in the presence of UDP-glucose pyrophosphorylase, inorganic phosphate, and phosphoglucomutase. The unusual feature of this reaction is the source of UTP. Yeast RNA was cleaved into oligonucleotides by nuclease P_1 which were further degraded to the nucleoside pyrophosphates using polynucleotide phosphorylase. Pyruvate kinase mediated phosphorylation by PEP (**4**) gave a mixture of NTPs corresponding to ATP (24%), CTP (18%), UTP (28%), and GTP (30%). The mixture was used crude in the coupling with glucose-6-phosphate (**13**) (WONG et al., 1983a).

This system has also been used with acetate kinase and acetyl phosphate (**8**) in place of pyruvate kinase and PEP (**4**) (HAYNIE and WHITESIDES, 1990).

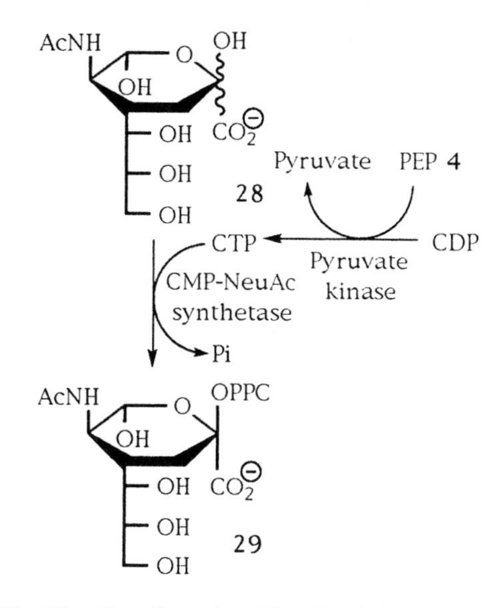

Fig. 13. Coupling of cytidine-5′-triphosphate and *N*-acetylneuramic acid (**28**).

Tab. 3. Survey of P–O Bond Forming Reactions

Substrate	Product	Enzyme	Cofactor	Phosphate Source	Cofactor Regeneration	Reference
D-Arabinose	D-arabinose-5-phosphate	hexokinase	ATP	PEP	pyruvate kinase	BEDNARSKI et al. (1988)
D-Glucose	D-glucose-6-phosphate	hexokinase	ATP	PEP	pyruvate kinase	HIRSCHBEIN et al. (1982)
		hexokinase	ATP	AcP	acetate kinase	CRANS and WHITESIDES (1983)
D-Fructose	D-fructose-6-phosphate	hexokinase	ATP	AcP	acetate kinase	WONG and WHITESIDES (1983)
D-Fructose	D-fructose-1,6-diphosphate	hexokinase and phosphofructokinase	ATP	—	—	WONG et al. (1983b)
D-Ribose	D-ribose-5-phosphate	ribokinase	ATP	PEP	pyruvate kinase	GROSS et al. (1983)
	5-phospho-D-ribosyl-α-1-pyrophosphate (PRPP)	PRPP synthase and ribokinase	ATP	PEP P_i	pyruvate kinase and adenylate kinase	GROSS et al. (1983)
D-Ribose-5-phosphate	5-phospho-D-ribosyl-α-1-pyrophosphate (PRPP)	PRPP synthase	ATP	PEP P_i	pyruvate kinase and adenylate kinase	GROSS et al. (1983)
D-Ribulose-5-phosphate	D-ribulose-1,5-bisphosphate	phosphoribulose kinase	ATP	AcP	acetate kinase	WONG et al. (1980)
		phosphoribulose kinase	ATP	PEP	pyruvate kinase	GROSS et al. (1983)
2′-Deoxy AMP	2′-deoxy ATP	adenylate kinase	ATP	PEP	pyruvate kinase	LADNER and WHITESIDES (1985)
Adenine	ATP	adenylate kinase and adenosine kinase	ATP	AcP	acetate kinase	BAUGHN et al. (1978)
UMP	UTP[a]	adenylate kinase	ATP	PEP	pyruvate kinase	SIMON et al. (1988b)
Deoxycytidine	deoxycytidine triphosphate	deoxycytidine kinase	ATP	—	—	KESSEL (1968)
CMP	CTP[a]	adenylate kinase	CTP	PEP	pyruvate kinase	SIMON et al. (1988b)
		adenylate kinase	ATP	PEP	pyruvate kinase	SIMON et al. (1989)
		nucleoside monophosphokinase	ATP	PEP	pyruvate kinase	AUGÉ and GAUTHERON (1988)

Tab. 3. Continued

Substrate	Product	Enzyme	Cofactor	Phosphate Source	Cofactor Regeneration	Reference
GMP	GTP	guanylate kinase	ATP	PEP	pyruvate kinase	SIMON et al. (1988b)
Dihydroxy-acetone	dihydroxy-acetone phosphate	glycerol kinase	ATP	PEP	pyruvate kinase	WONG and WHITESIDES (1983)
		glycerol kinase	ATP	AcP	acetate kinase	WONG and WHITESIDES (1983)
Creatine	creatine-6-phosphate	creatine kinase	ATP	AcP	acetate kinase	SHIH and WHITESIDES (1977)
Arginine	arginine phosphate	arginine kinase	ATP	PEP	pyruvate kinase	BOLTE and WHITESIDES (1984)
allo-Hydroxy-L-lysine	*O*-phospho-*allo*-hydroxy-L-lysine	hydroxy-lysine kinase	GTP	—	—	HILES and HENDERSON (1972)
Hydroxy-L-lysine	*O*-phospho-hydroxy-L-lysine	hydroxy-lysine kinase	GTP	—	—	HILES and HENDERSON (1972)
L-Serine	L-serine-phosphate	L-serine transferase	P_i	—	—	CAGEN and FRIEDMANN (1972)
L-Threonine	L-threonine-phosphate	L-serine transferase	P_i	—	—	CAGEN and FRIEDMANN (1972)

[a] Needed to initially generate the corresponding NDP.

2.3.4 Coupling with Other Phosphate Sources

Phospholipase D has been used to couple various species with dimyristoyl-L-α-phosphatidyl choline in order to synthesize possible phospholipid inhibitors (WANG et al., 1993). These include (*R*)- and (*S*)-proline and serinol. The reaction was found to be selective for primary alcohols but nitrogen and sulfur nucleophiles were not accepted.

3 Cleavage of P–O Bonds

In marked contrast to the formation of P–O bonds through the specific phosphorylation of functional groups, dephosphorylation is a technique which has become a routine tool in the analysis of nucleic acid oligomers, while remaining relatively unused in synthetic work. This is largely due to the usual ease of cleavage of P–O bonds by chemical methods, and is reflected in the literature pertaining to enzyme catalyzed dephosphorylation (DAVIES et al., 1989; FABER, 1995).

3.1 Hydrolysis of Nucleic Acid Derivatives

The extensive use of phosphodiesterase and alkaline phosphatase enzymes for routine analysis of nucleotides excludes a complete review of this area. Brief descriptions of the mode of action of phosphodiesterase and alkaline phosphatase have already been made. These enzymes are usually used together to completely digest the oligomer of ribonucleic acid to either a 5′- or 3′-nucleoside depending upon the phosphodiesterase chosen. The role of alkaline phosphatase is usually to cleave any phosphate monoesters present, while the phosphodiesterase (or for that matter, any nuclease) cleaves the phosphodiester link. For example, the 2′-phospho-trimer (**30**) (Fig. 14) was first dephosphorylated at the 2′-position by calf intestinal alkaline phosphatase and further digested with nuclease P_1 to give adenosine, 5′-monophospho-adenosine and 5′-monophospho-uridine (SEKINE et al., 1993). Analysis of the digestion products is usually carried out by HPLC of the crude reaction mixture. This method can prove to be a versatile tool, especially when modified phosphodiesters are employed as "markers" for specific groups. The thymidine oligomer (**31**) (Fig. 15), modified with one methyl phosphonate linkage, upon digestion with a 5′-snake venom phosphodiesterase gave thymidine, 5′-monophospho-thymidine, and the methylphosphonate dimer (AGRAWAL and GOODCHILD, 1987).

Extensive mechanistic studies have been carried out on both phosphodiesterase and alkaline phosphatase enzymes using substrates synthesized by specific enzyme phosphorylations. Chiral thio-adenosine derivatives (such

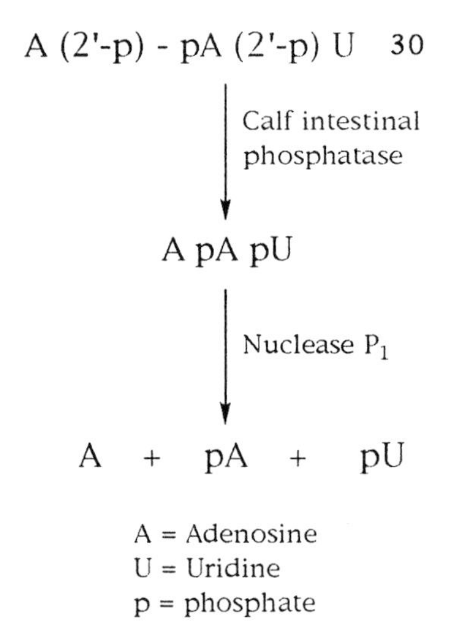

Fig. 14. Typical hydrolysis of phosphate esters using nuclease and phosphatase enzymes.

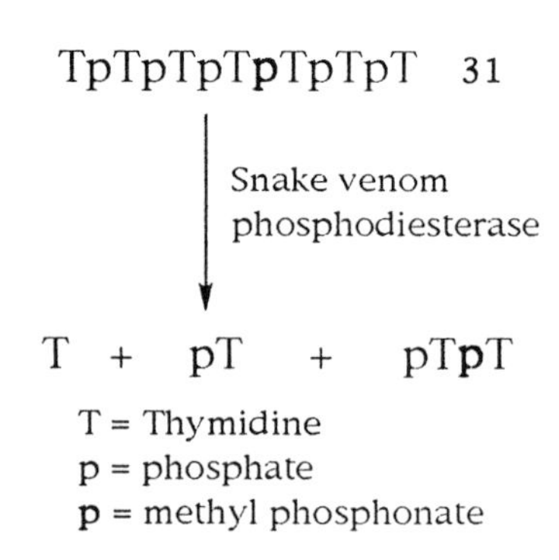

Fig. 15. An example of the use of phosphonates to inhibit cleavage of phosphate diesters.

as (**19**)) have been used in determining the mechanism of action of purple acid phosphatase. It was found that the enzyme catalyzes direct transfer of the phospho group to water (MUELLER et al., 1993). Isotopically labeled uridine (3′-5′) adenosine phosphodiesters have also been used in this way with various enzymes (SEELA et al., 1983). Spleen phosphodiesterase gave the 3′-uridine-monophosphate and adenosine, while nuclease P_1 gave uridine and 5′-adenosine-monophosphate.

Enzymes that catalyze more specific dephosphorylations are also known, although not used as widely as those already mentioned. For example, ATPase catalyzes the hydrolysis of ATP (**1**) to ADP (**3**) (WATT et al., 1984).

3.2 Hydrolysis of Substrates Other than Nucleic Acid Derivatives

The hydrolysis of phosphate esters is chemically a trivial step. The use of enzymes for this process has therefore received little consideration other than for elucidation of enzyme mechanisms, such as in the hydrolysis of *p*-nitrophenol phosphates (NEUMANN, 1968; WILSON et al., 1964), or for more subtle processes where standard techniques are too harsh, as with the hydrolysis of polyprenyl pyrophosphates (FUJII et al., 1982). The epoxy-farnesyl pyrophosphate (**32**) (Fig. 16), for example, was dephosphorylated with alkaline phosphatase (KOYAMA et al., 1987, 1990).

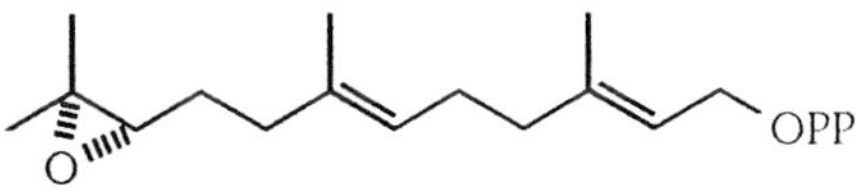

Fig. 16. Epoxy-farnesyl pyrophosphate (**32**).

Carbohydrates have also been used as substrates, although the reaction is not of great synthetic use since the phosphorylated species are usually the more valuable (WONG and WHITESIDES, 1983).

NATCHEV has used the difference in reactivity of phosphodiesterase and alkaline phosphatase enzymes in the cleavage of various

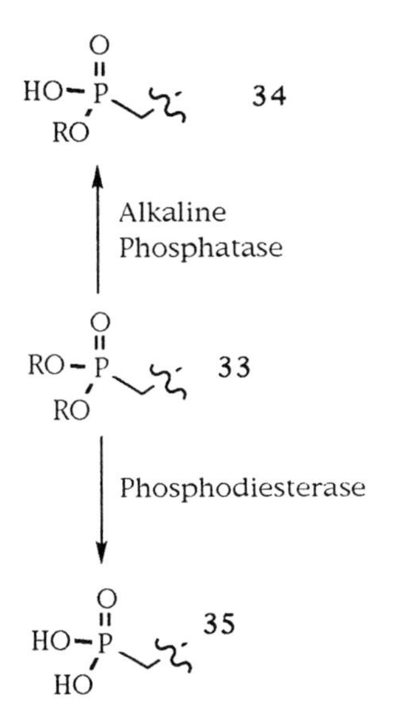

Fig. 17. Selectivities in the hydrolysis of phosphonate esters.

phosphonate diesters (**33**) (Fig. 17) (NATCHEV, 1988a, b). Phosphodiesterase was found to cause cleavage of both ester groups, while alkaline phosphatase cleaved only one of the esters. The reaction was also found to proceed well with phosphoramidite derivatives (**36**) (Fig. 18) (NATCHEV, 1987).

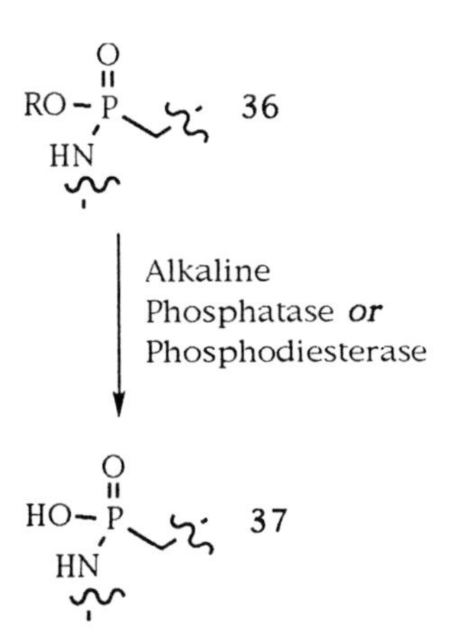

Fig. 18. Hydrolysis of phosphoramidate esters.

4 References

AGRAWAL, S., GOODCHILD, J. (1987), Oligodeoxynucleoside methylphosphonates: Synthesis and enzymic degradation, *Tetrahedron Lett.* **28**, 3539–3542.

ALLEN, S. T., HEINTZELMAN, G. R., TOONE, E. J. (1992), Pyruvate aldolases as reagents for stereospecific aldol condensation, *J. Org. Chem.* **57**, 426–427.

ANDERSON, E. P. (1973), Nucleoside and nucleotide kinases, in: *The Enzymes*, 3rd Edn., IX (BOYER, P. D., Ed.), pp. 49–96. New York: Academic Press.

AUGÉ, C., GAUTHERON, C. (1988), An efficient synthesis of cytidine monophospho-sialic acids with four immobilized enzymes, *Tetrahedron Lett.* **29**, 789–790.

BAUGHN, R. L., ADALSTEINSSON, Ö., WHITESIDES, G. M. (1978), Large-scale enzyme-catalyzed synthesis of ATP from adenosine and acetyl phosphate. Regeneration of ATP from AMP, *J. Am. Chem. Soc.* **100**, 304–306.

BEDNARSKI, M. D., CRANS, D. C., DICOSINIO, R., SIMON, E. S., STEIN, P. D., WHITESIDES, G. M., SCHNEIDER, M. J. (1988), Synthesis of 3-deoxy-D-manno-2-octulosonate-8-phosphate (KDO-8-P) from D-arabinose: Generation of D-arabinose-5-phosphate using hexokinase, *Tetrahedron Lett.* **29**, 427–430.

BERKOWITZ, J. (1959), Correlation scheme for diatomic oxides, *J. Chem. Phys.* **30**, 858–860.

BERNARDI, A., BERNARDI, G. (1971), Spleen acid exonuclease, in: *The Enzymes*, 3rd Edn., IV (BOYER, P. D., Ed.), pp. 329–336. New York: Academic Press.

BOLTE, J., WHITESIDES, G. M. (1984), Enzymatic synthesis of arginine phosphate with coupled ATP cofactor regeneration, *Bioorg. Chem.* 12, 170–175.

BUSS, K. A., INGRAM-SMITH, C., FERRY, J. G., SANDERS, D. A., HASSON, M. S. (1997), Crystallization of acetate kinase from *Methanosarcina thermophila* and prediction of its fold, *Protein Sci.* **6**, 2659–2662.

CAGEN, L. M., FRIEDMANN, H. C. (1972), Enzymatic phosphorylation of serine, *J. Biol. Chem.* **247**, 3382–3392.

CHENAULT, H. K., MANDES, R. F., HORNBERGER, K. R. (1997), Synthetic utility of yeast hexokinase. Substrate specificity, cofactor regeneration, and product isolation, *J. Org. Chem.* **62**, 331–336.

COLOWICK, S. P. (1973), The hexokinases, in: *The Enzymes*, 3rd Edn., IX (BOYER, P. D., Ed.), pp. 1–48. New York: Academic Press.

CRANS, D. C., WHITESIDES, G. M. (1983), A convenient synthesis of disodium acetyl phosphate for use in *in situ* ATP cofactor regeneration, *J. Org. Chem.* **48**, 3130–3132.

CRANS, D. C., WHITESIDES, G. M. (1985a), Glycerol kinase: substrate specificity, *J. Am. Chem. Soc.* **107**, 7008–7018.

CRANS, D. C., WHITESIDES, G. M. (1985b), Glycerol kinase: synthesis of dihydroxyacetone phosphate, *sn*-glycerol-3-phosphate, and chiral analogs, *J. Am. Chem. Soc.* **107**, 7019–7027.

CRANS, D. C., KAZLAUSKAS, R. J., HIRSCHBEIN, B. L., WONG, C.-H., ABRIL, O., WHITESIDES, G. M. (1987), Enzymatic regeneration of adenosine 5′-triphosphate – acetyl phosphate, phosphoenol pyruvate, methoxycarbonyl phosphate, dihydroxyacetone phosphate, 5-phospho-α-D-ribosyl pyrophosphate, uridine-5′-diphosphateglucose, *Methods Enzymol.* **136**, 263–280.

DAVIES, H. G., GREEN, H. G., KELLY, D. R., ROBERTS, S. M. (1989), *Biotransformations in Preparative Organic Chemistry: The Use of Isolated Enzymes and Whole Cell Systems in Synthesis*, pp. 75–86. San Diego, CA: Academic Press.

DRAUZ, K., WALDMANN, H. (1995), *Enzyme Catalysis in Organic Synthesis, A Comprehensive Handbook*, pp. 505–545. Weinheim: VCH.

DRUECKHAMMER, D. G., WONG, C. H. (1985), Chemoenzymatic syntheses of fluoro sugar phosphates and analogs, *J. Org. Chem.* **50**, 5912–5913.

DUMAS, C., CAMONIS, J. (1993), Cloning and sequence analysis of the cDNA for arginine kinase of lobster muscle, *J. Biol. Chem.* **268**, 21 599–21 605.

ECKSTEIN, F. (1983), Phosphorothioate analogs of nucleotides – Tools for the investigation of biochemical processes, *Angew. Chem.* (Int. Edn. Engl.) **22**, 423–439.

ECKSTEIN, F. (1985), Nucleoside phosphorothioates, *Annu. Rev. Biochem.* **54**, 367–402.

FABER, K. (1995), *Biotransformations in Organic Chemistry*, pp. 105–107. Heidelberg, New York: Springer-Verlag.

FENNIRI, H., LEHN, J.-M. (1993), Coupling of supramolecular synthesis of ATP with ATP-consuming enzyme systems, *J. Chem. Soc., Chem. Commun.*, 1819–1821.

FERNLEY, H. N. (1971), Mammalian alkaline phosphatases, in: *The Enzymes*, 3rd Edn., IV (BOYER, P. D., Ed.), pp. 417–447. New York: Academic Press.

FUJII, H., KOYAMA, T., OGURA, K. (1982), Efficient enzymatic hydrolysis of polyprenyl pyrophosphates, *Biochim. Biophys. Acta* **712**, 716–718.

GIBSON, B. W., HINES, W., YU, Z., KENYON, G. L., MCNEMAR, L., VILLAFRANCA, J. J. (1990), Enzymatic synthesis of phosphotyrosine-containing peptides via adenylated intermediates, *J. Am. Chem. Soc.* **112**, 8523–8528.

GROSS, A., ABRIL, O., LEWIS, J. M., GERESH, S., WHITESIDES, G. M. (1983), Practical synthesis of 5-phospho-D-ribosyl-α-1-pyrophosphate (PRPP): Enzymatic routes from ribose 5-phosphate or ribose, *J. Am. Chem. Soc.* **105**, 7428–7435.

GROSS, M., LUSTIG, A., WALLIMANN, T., FURTER, R. (1995), Multiple-state equilibrium unfolding of guanidino kinases, *Biochemistry* **34**, 10350–10357.

HAYNIE, S. L., WHITESIDES, G. M. (1990), Preparation of a mixture of nucleoside triphosphates suitable for use in synthesis of nucleotide phosphate sugars from ribonucleic-acid using nuclease-p1, a mixture of nucleoside monophosphokinases and acetate kinase, *Appl. Biochem. Biotechnol.* **23**, 205–220.

HEIDLAS, J. E., WILLIAMS, K. W., WHITESIDES, G. M. (1992), Nucleoside phosphate sugars – synthesis on practical scales for use as reagents in the enzymatic preparation of oligosaccharides and glycoconjugates, *Acc. Chem. Res.* **25**, 307–314.

HILES, R. A., HENDERSON, M. L. (1972), The partial purification and properties of hydroxylysine kinase from rat liver, *J. Biol. Chem.* **247**, 646–651.

HIRSCHBEIN, B. L., MAZENOD, F. P., WHITESIDES, G. M. (1982), Synthesis of phosphoenolpyruvate and its use in adenosine triphosphate cofactor regeneration, *J. Org. Chem.* **47**, 3765–3766.

HOSSEINNI, M. W., LEHN, J.-M. (1985), Cocatalysis: Pyrophosphate synthesis from acetylphosphate catalyzed by a macrocyclic polyamine, *J. Chem. Soc., Chem. Commun.*, 1155–1157.

HOSSEINNI, M. W., LEHN, J.-M. (1987), Binding of AMP, ADP, and ATP nucleotides by polyammonium macrocycles, *Helv. Chim. Acta* **70**, 1312–1319.

HOSSEINNI, M. W., LEHN, J.-M. (1988), Supramolecular catalysis: Substrate phosphorylations and adenosine triphosphate synthesis with acetyl phosphate catalyzed by a macrocyclic polyamine, *J. Chem. Soc., Chem. Commun.*, 397–399.

HOSSEINNI, M. W., LEHN, J.-M. (1991), Supramolecular catalysis of adenosine triphosphate synthesis in aqueous solution mediated by a macrocyclic polyamine and divalent metal cations, *J. Chem. Soc., Chem. Commun.*, 451–453.

HOSSEINNI, M. W., LEHN, J.-M., MERTES, M. P. (1983), Efficient molecular catalysis of ATP-hydrolysis by protonated macrocyclic polyamines, *Helv. Chim. Acta* **66**, 2454–2466.

HURLEY, J. H. (1996), The sugar kinase heat-shock-protein-70, actin super family – implications of conserved structure for mechanism, *Ann. Rev. Biophys. Biomol. Struct.* **25**, 137.

KABSCH, W., HOLMES, K. C. (1995), Protein motifs. 2. The actin fold, *FASEB J.* **9**, 167–174.

KAYNE, F. J. (1973), Pyruvate kinase, in: *The Enzymes*, 3rd Edn., VIII (BOYER, P. D., Ed.), pp. 353–382. New York: Academic Press.

KAZLAUSKAS, R. J., WHITESIDES, G. M. (1985), Synthesis of methoxycarbonyl phosphate, a new reagent having high phosphoryl donor potential for use in ATP cofactor regeneration, *J. Org. Chem.* **50**, 1069–1076.

KEE, Y., LEE, Y. S., CHUNG, C. H., WAXMAN, L., GOLDBERG, A. L. (1988), Improved methods for purification and assay of glycerol kinase from *Escherichia coli*, *J. Chromatogr. Biomed. Appl.* **428**, 345–351.

KEIKINHEIMO, P., LEHTONEN, J., BAYKOV, A., LAHTI, R., COOPERMAN, B. S., GOLDMAN, A. (1996), The structural basis for pyrophosphatase catalysis, *Structure* **4**, 1491–1508.

KESSEL, D. (1968), Properties of deoxycytidine kinase partially purified from L1210 cells, *J. Biol. Chem.* **243**, 4739–4744.

KNOWLES, J. R. (1980), Enzyme-catalyzed phosphoryl transfer reactions, *Annu. Rev. Biochem.* **49**, 877–919.

KOYAMA, T., OGURA, K., BAKER, F. C., JAMIESON, G. C., SCHOOLERY, D. A. (1987), Synthesis and absolute configuration of 4-methyl juvenile hormone 1 (4-MeJH 1) by a biogenetic approach: a combination of enzymatic synthesis and biotransformation, *J. Am. Chem. Soc.* **109**, 2853–2854.

KOYAMA, T., INOUE, H., OHNUMA, S., OGURA, K. (1990), Prenyltransferase reaction involving enantiomeric discrimination. Enzymatic synthesis of (*S*)-10,11-epoxyfarnesol from racemic 6,7-epoxygeranyl diphosphate and isopentenyl diphosphate, *Tetrahedron Lett.* **31**, 4189–4190.

KREBS, E. G., BEAU, J. A. (1979), Phosphorylation – dephosphorylation of enzymes, *Annu. Rev. Biochem.* **48**, 923–959.

KREJCOVA, R., HORSKA, K. (1997), Adenylate kinases, *Chem. Listy* **91**, 179–188.

LADNER, W. E., WHITESIDES, G. M. (1985), Enzymatic synthesis of deoxy ATP using DNA as starting material, *J. Org. Chem.* **50**, 1076–1079.

LASKOWSKI, M., Sr. (1971), Venom exonuclease, in: *The Enzymes*, 3rd Edn., IV (BOYER, P. D., Ed.), pp. 313–328. New York: Academic Press.

LEGRAND, D. M., ROBERTS, S. M. (1993), Carbocyclic adenosine phosphates as nucleotide mimics in some enzyme-catalyzed reactions, *J. Chem. Soc., Chem. Commun.*, 1284–1285.

LEHNINGER, A. L. (1975), *Biochemistry*, 2nd Edn., p. 398. New York: Worth Publishers.

LESNIKOWSKI, Z. J. (1993), Stereocontrolled synthesis of P-chiral analogs of oligonucleotides, *Bioorg. Chem.* **21**, 127–155.

LEWIS, J. M., HAYNIE, S. L., WHITESIDES, G. M. (1979), An improved synthesis of diammonium acetyl phosphate, *J. Org. Chem.* **44**, 864–865.

LIN, E. C. C. (1977), Glycerol utilization and its regulation in mammals, *Annu. Rev. Biochem.* **46**, 765–795.

MAITY, H., JARORI, G. K. (1997), Yeast hexokinase PII-bound nucleotide conformation at the active site, *Eur. J. Biochem.* **250**, 539–548.

MARTIN, D. P., DRUECKHAMMER, D. G. (1992), Combined chemical and enzymatic synthesis of coenzyme A analogs, *J. Am. Chem. Soc.* **114**, 7287–7288.

MORAN, J. R., WHITESIDES, G. M. (1984), A practical enzymatic synthesis of (S_p)-adenosine 5′-O-(1-thiotriphosphate)((S_p)-ATP-α-*S*), *J. Org. Chem.* **49**, 704–706.

MORRISON, J. F. (1973), Arginine kinase and other invertebrate guanidino kinases, in: *The Enzymes*, 3rd Edn., VIII (BOYER, P. D., Ed.), pp. 457–486. New York: Academic Press.

MUELLER, E. G., CROWDER, M. W., AVERILL, B. A., KNOWLES, J. R. (1993), Purple acid phosphatase: A diiron enzyme that catalyzes a direct phospho group transfer to water, *J. Am. Chem. Soc.* **115**, 2974–2975.

MUHLEBACH, S. M., GROSS, M., WIRZ, T., WALLIMANN, T., PERRIARD, J. C., WYSS, M. (1994), Sequence homology and structure predictions of the creatine kinase isoenzymes, *Mol. Cell. Biochem.* **133**, 245–262.

MURPHY, J. E., STEC, B., MA, L., KANTROWITZ, E. R. (1997), Trapping and visualization of a covalent enzyme–phosphate intermediate, *Nature Struct. Biol.* **4**, 618–622.

NATCHEV, I. A. (1987), Phosphono and phosphino analogs and derivatives of the natural aminocarboxylic acids and peptides. 1. Synthesis and enzyme–substrate interactions of N-phosphono-, and N-phosphinomethylated cyclic amides, *Synthesis*, 1079–1084.

NATCHEV, I. A. (1988a), Organophosphorous analogs and derivatives of the natural L-amino carboxylic acids and peptides. III. Synthesis and enzyme–substrate interactions of D-, DL-, and L-5-dihydroxyphosphinyl-3,4-didehydronorvaline and their cyclic analogs and derivatives, *Bull. Chem. Soc. Jpn.* **61**, 3711–3715.

NATCHEV, I. A. (1988b), Total synthesis, enzyme–substrate interactions and herbicidal activity of plumbemicin A and B (N-1409), *Tetrahedron* **44**, 1511–1592.

NEUMANN, H. (1968), Substrate selectivity in the action of alkaline and acid phosphatases, *J. Biol. Chem.* **243**, 4671–4676.

NODA, L. (1973), Adenylate kinase, in: *The Enzymes*, 3rd Edn., VIII (BOYER, P. D., Ed.), pp. 279–305. New York: Academic Press.

Nomenclature Commitee of the International Union of Biochemistry (1984), *Enzyme Nomenclature*. Orlando, FL: Academic Press.

PRADINES, A., KLAÉBÉ, A., PÉRIÉ, J., PAUL, F., MONSAN, P. (1988), Enzymatic synthesis of phosphoric monoesters with alkaline phosphatase in reverse hydrolysis conditions, *Tetrahedron* **44**, 6373–6386.

REID, T. W., WILSON, I. B. (1971), *E. coli* alkaline phosphatase, in: *The Enzymes*, 3rd Edn., IV (BOYER, P. D., Ed.), pp. 373–415. New York: Academic Press.

RIOS-MERCADILLO, V. M., WHITESIDES, G. M. (1979), Enzymatic synthesis of *sn*-glycerol 3-phosphate, *J. Am. Chem. Soc.* **101**, 5828–5829.

SEELA, F., OTT, J., POTTER, B. V. L. (1983), Oxygen chiral phosphate in uridylyl (3′-5′) adenosine by oxidatio of a phosphite intermediate: Synthesis and absolute configuration, *J. Am. Chem. Soc.* **105**, 5879–5886.

SEKINE, M., IIMURA, S., FURUSAWA, K. (1993), Synthesis of a new class of 2′-phosphorylated oligoribonucleotides capable of conversion to oligoribonucleotides, *J. Org. Chem.* **58**, 3204–3208.

SHIH, Y. S., WHITESIDES, G. M. (1977), Large-scale ATP-requiring enzymatic phosphorylation of creatine can be driven by enzymatic ATP regeneration, *J. Org. Chem.* **42**, 4165–4166.

SIMON, E. S., BEDNARSKI, M. D., WHITESIDES, G. M. (1988a), Synthesis of CMP-NeuAc from N-acetylglucosamine – generation of CTP from CMP using adenylate kinase, *J. Am. Chem. Soc.* **110**, 7159–7163.

SIMON, E. S., BEDNARSKI, M. D., WHITESIDES, G. M. (1988b), Generation of cytidine 5′-triphosphate using adenylate kinase, *Tetrahedron Lett.* **29**, 1123–1126.

SIMON, E. S., GRABOWSKI, S., WHITESIDES, G. M. (1989), Preparation of phosphoenolpyruvate from D-(−)-3-phosphoglyceric acid for use in regeneration of ATP, *J. Am. Chem. Soc.* **111**, 8920–8921.

SIMON, E. S., GRABOWSKI, S., WHITESIDES, G. M. (1990), Covenient synthesis of cytidine 5′-triphosphate, guanosine 5′-triphosphate, and uridine 5′-triphosphate and their use in the preparation of UDP-glucose, UDP-glucuronic acid, and GDP-mannose, *J. Org. Chem.* **55**, 1834–1841.

SPECTOR, L. B. (1980), Acetate kinase: a triple-displacement enzyme, *Proc. Natl. Acad. Sci. USA* **77**, 2626–2630.

STRONG, S. J., ELLINGTON, W. R. (1995), Isolation and sequence analysis of the gene for arginine kinase from the Chelicerate arthropod, *Limulus polyphemus* – insights into catalytically important residues, *Biochim. Biophys. Acta* **1246**, 197–200.

TANG, X. J., JOHANSSON, G. (1997), A bioelectrochemical method for the determination of acetate with immobilized acetate kinase, *Analyt. Lett.* **30**, 2469–2483.

THOMER, J. W., PAULUS, H. (1973), Glycerol and glycerate kinases, in: *The Enzymes*, 3rd Edn., VIII (BOYER, P. D., Ed.), pp. 487–508. New York: Academic Press.

VOET, D., VOET, J. (1995), *Biochemistry*, pp. 428–434; 593–598. Chichester: John Wiley & Sons.

WANG, P., SCHUSTER, M., WANG, Y. F., WONG, C. H. (1993), Synthesis of phospholipid-inhibitor conjugates by enzymatic transphosphatidylation with phospholipase D, *J. Am. Chem. Soc.* **115**, 10487–10491.

WATT, D. R., FINDEIS, M. A., RIOS-MERCADILLO, V. M., AUGÉ, J., WHITESIDES, G. M. (1984), An efficient chemical and enzymatic synthesis of nicotinamide adenine dinucleotide (NAD^+), *J. Am. Chem. Soc.* **106**, 234–239.

WATTS, D. C. (1973), Creatine kinase (Adenosine 5′-triphosphate-creatine phosphotransferase), in: *The Enzymes*, 3rd Edn., VIII (BOYER, P. D., Ed.), pp. 383–455. New York: Academic Press.

WEI, L. L., GOUX, W. J. (1992), ATP cofactor regeneration via the glycolytic pathway, *Bioorg. Chem.* **20**, 62–66.

WHITESIDES, G. M., SIEGEL, M., GARRETT, P. (1975), Large-scale synthesis of diammonium acetyl phosphate, *J. Org. Chem.* **40**, 2516–2519.

WILD, K., GRAFMULLER, R., WAGNER, E., SCHULZ, G. E. (1997), Structure, catalysis and supramolecular assembly of adenylate kinase, *Eur. J. Biochem.* **250**, 326–331.

WILLSON, M., ALRIC, I., PERIE, J., SANEJOUAND, Y. H. (1997), Yeast hexokinase inhibitors designed from 3-D enzyme structure rebuilding, *J. Enzyme Inhib.* **12**, 101.

WILSON, I. B., DAYAN, J., CYR, K. (1964), Some properties of alkaline phosphatase from *Escherichia coli*. Transphosphorylation, *J. Biol. Chem.* **239**, 4182–4185.

WONG, C. H., WHITESIDES, G. M. (1981), Enzyme-catalyzed organic synthesis: NAD(P)H cofactor regeneration by using glucose 6-phosphate and the glucose 6-phosphate dehydrogenase from *Leuconostoc mesenteroides*, *J. Am. Chem. Soc.* **103**, 4890–4899.

WONG, C. H., WHITESIDES, G. M. (1983), Synthesis of sugars by aldolase-catalyzed condensation reactions, *J. Org. Chem.* **48**, 3199–3205.

WONG, C. H., MCCURRY, S. D., WHITESIDES, G. M. (1980), Practical enzymatic syntheses of ribulose 1,5-bisphosphate and ribose 5-phosphate, *J. Am. Chem. Soc.* **102**, 7938–7939.

WONG, C. H., HAYNIE, S. L., WHITESIDES, G. M. (1982), Enzyme-catalyzed synthesis of N-acetyllactosamine with *in situ* regeneration of uridine 5′-diphosphate glucose and uridine 5′-diphosphate galactose, *J. Org. Chem.* **47**, 5416–5418.

WONG, C. H., HAYNIE, S. L., WHITESIDES, G. M. (1983a), Preparation of a mixture of nucleoside triphosphates from yeast RNA: Use in enzymatic synthesis requiring nucleoside triphosphate regeneration and conversion to nucleoside diphosphate sugars, *J. Am. Chem. Soc.* **105**, 115–117.

WONG, C. H., MAZENOD, F. P., WHITESIDES, G. M. (1983b), Chemical and enzymatic syntheses of 6-deoxyhexoses. Conversion to 2,5 dimethyl-4-hydroxy-2,3-dihydrofuran-3-one (furaneol) and analogs, *J. Org. Chem.* **48**, 3493–3497.

5 Enzymes in Carbohydrate Chemistry: Formation of Glycosidic Linkages

SABINE L. FLITSCH

GREGORY M. WATT

Edinburgh, UK

1 Introduction

The glycosidic linkage is used in Nature as a way of reversibly attaching carbohydrates to other biological materials and it forms the chains of oligo- and polysaccharides. Next to the peptide and phosphodiester linkages of nucleic acids, it plays a pivotal role in many biological events ranging from provision of mechanical stability in cellulose and energy storage in amylose or glycogen to mediation of highly specific inter- and intracellular recognition events. Although carbohydrates are highly functionalized molecules, conjugation nearly always takes place through the anomeric center as outlined in Fig. 1. The two partners of the glycosidic linkage are generally called glycosyl donor (e.g., **1**) and glycosyl acceptor (**2**), which is frequently, but not always an alcohol.

The intrinsic chemistry of monosaccharides provides for a startling level of structural complexity of their polymers, which renders them ideally suited as molecules for unique biological specificity. Whereas nucleic acids and polypeptide chains are usually joined in a linear, divalent fashion, the glycoside linkage alone can result in two stereoisomers, the α- and β-anomers (e.g., **4** and **3**). Further structural diversity can be obtained if the glycosyl acceptor (ROH) is multivalent, as in the case of saccharides. Thus, between only 2 different hexoses, 20 different glycosidic linkages can be obtained.

The potential for forming chemically similar, isomeric glycosidic linkages has provided a much greater challenge to synthetic chemistry than peptides and oligonucleotides. Whereas automated synthesizers are now available for the synthesis of the latter two, most oligosaccharide synthesis requires lengthy and laborious synthetic routes and purification protocols. Recent progress in this area has been reviewed elsewhere (PAULSEN, 1982; SCHMIDT, 1991; SCHMIDT and KINZY, 1994; BOONS, 1996; FLITSCH and WATT, 1997).

An alternative route to obtaining glycoconjugates would be by using biological *in vivo* methods, which again have found wide applications for proteins and larger oligonucleotides. However, *in vivo* synthesis has met with limited success because glycoconjugates generally occur in heterogeneous mixtures in small quantities in biological sources and thus isolation from natural sources is not practical in most cases. The manipulation of the biosynthetic pathways within the cell by genetic engineering techniques is very difficult, because each glycosidic linkage is formed by an individual enzyme (glycosyltransferases) which in turn underlie complex control mechanisms.

One of the most successful methods for the practical synthesis of defined oligosaccharides and glycoconjugates has, therefore, been the *in vitro* enzymatic synthesis, using isolated enzymes as biocatalysts. The enzymes used in biosynthesis are the glycosyltransferases (and glycosidases), which have been shown to be amenable to use *in vitro*. Both types of enzymes will be discussed in detail.

2 Glycosidases

It has been known for a long time that glycosidases, which normally catalyze the hydrolysis of glycosidic bonds (i.e., the backward reaction in Fig. 1), can be used "in reverse" for the syn-

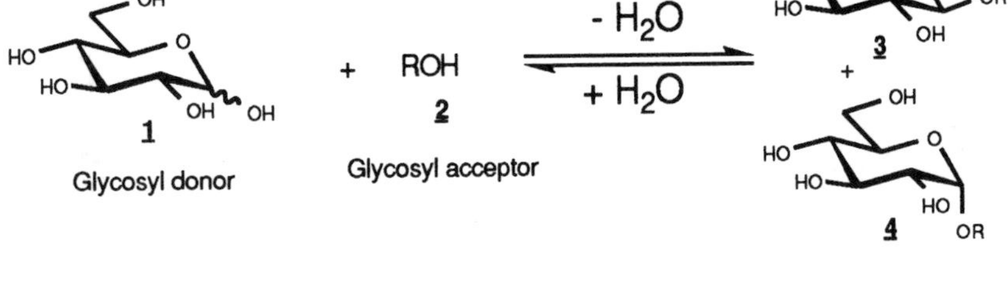

Fig. 1

thesis of glycosidic linkages. Systematic studies using glycosidases for the synthesis of defined saccharides started in the early 1980s, and there are now a large number of examples of glycosidases that are useful in synthesis.

2.1 Availability of Enzymes

Glycosidases can be obtained from a large variety of biological sources, ranging from thermophiles, bacteria and fungi to mammalian tissue. They are generally stable and soluble enzymes. Since they are widely used as analytical tools for biochemical research, many glycosidases are commercially available, although not all glycosidases are suitable for synthesis. Glucuronidases, e.g., have so far failed to yield any glucuronides. Glycosidases are generally highly specific for a particular linkage (α vs. β) and for a particular glycosyl donor. However, they tend to accept a structurally diverse range of glycosyl acceptors, which makes them useful as general catalysts for glycosylation. At the same time, when the acceptor has several hydroxyl groups, glycosidases might have poorer regioselectivity, and can result in mixtures of isomers that need to be separated.

2.2 Kinetic vs. Thermodynamic Glycosylation

In the presence of glycosidases the thermodynamic equilibium of the reaction shown in Fig. 1 lies generally strongly to the left in aqueous medium, favoring hydrolyis of the glycosides (**3**) and (**4**). If glycosidases are used for the formation of glycosidic linkages, reaction conditions need to be chosen which take this into account. In principle, there are two possible approaches: the equilibrium can be shifted towards the glycosides in Fig. 1 by decreasing the water concentration and/or increasing glycosyl donor and/or alcohol concentration. This can be either achieved by using a large excess of alcohol (**2**) or sugar (**1**) or by working in organic solvents. All these approaches have been used.

Alternatively, one can take advantage of the fact that many glycosidase acceptor substrates have a much higher activity compared to water than would be expected from the difference in concentration. Thus, the enzymebound activated intermediate (represented as an oxonium ion in Fig. 2) reacts much faster with an alcohol (R′OH) to give (**7**) and (**8**) than it does with water. In order to encourage this kinetic effect over thermodynamic equilibration the glyco-

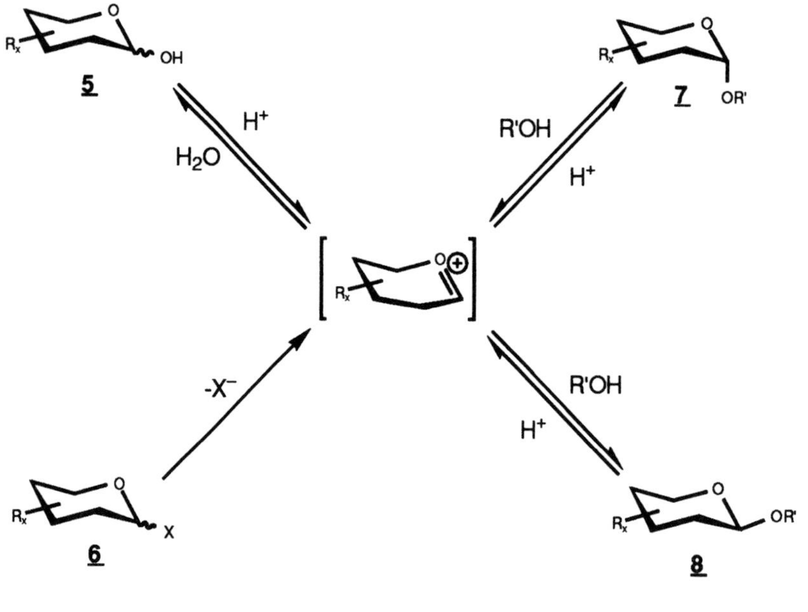

Fig. 2

syl donor needs to be a good substrate for the enzyme, which necessitates the use of activated glycosyl donors, such as glycosyl fluorides or aryl glycosides. This allows the reaction to proceed rapidly before equilibration to (**5**) can take place. Kinetic glycosylations (or "transglycosylations") can, therefore, be performed in aqueous medium with a moderate excess of donor or acceptor. A drawback, however, is the requirement for multistep synthesis of the activated glycosyl donor.

2.3 Glycosyl Donors

Glycosidases that have been used for synthesis generally catalyze the transglycosylation step with very high stereospecificity, in most cases with retention of the anomeric configuration. As a result, the activated glycosyl donor (**6**) needs to have the right anomeric configuration for the particular glycosidase used and will get selectively converted to only one of the two possible anomers (**7** or **8**). For example, a β-galactosidase would require a β-galactosyl donor and would catalyze selectively the formation of the β-galactosidic linkage.

Glycosidases tend also to be highly specific for their donor sugar, although more catholic glycosidases are now being reported, which have not been systematically studied in enzymatic synthesis.

The activated glycosyl donor (**6**) needs to be a good substrate for the enzyme, but needs also to be stable in aqueous medium during the reaction. For these reasons, the most common activated glycosyl donors are, therefore, glycosyl fluorides, phenyl glycosides, nitro-, dinitro-, and chloro-phenyl glycosides. A 1,2-oxazoline derivative of *N*-acetylglucosamine has recently been used as a donor for chitinase (KOBAYASHI et al., 1997). In addition, cheaply available disaccharides such as lactose (for β-galactosidase) and cellobiose (for β-glucosidase) have been used as glycosyl donors. Most of these donors are commercially available. The choice of glycosyl donor depends on the enzyme system used and can markedly influence yield and regioselectivity of the reaction, an example of which is given in Sect. 2.4.3.

2.4 Acceptor Substrates for Transglycosylations

2.4.1 Simple Alcohols as Acceptor Substrates

Alkyl glycosides have found wide applications such as detergents in research and in industry and are, therefore, good targets for enzymatic synthesis using glycosidases, provided that inexpensive starting materials can be employed. Allyl-, benzyl-, and trimethylsilylethyl glycosides can be used as synthetic intermediates carrying a temporary protecting group at the glycosidic center. One of the earlier applications of transglycosylations for the synthesis of simple glycosides was reported by NILSSON,

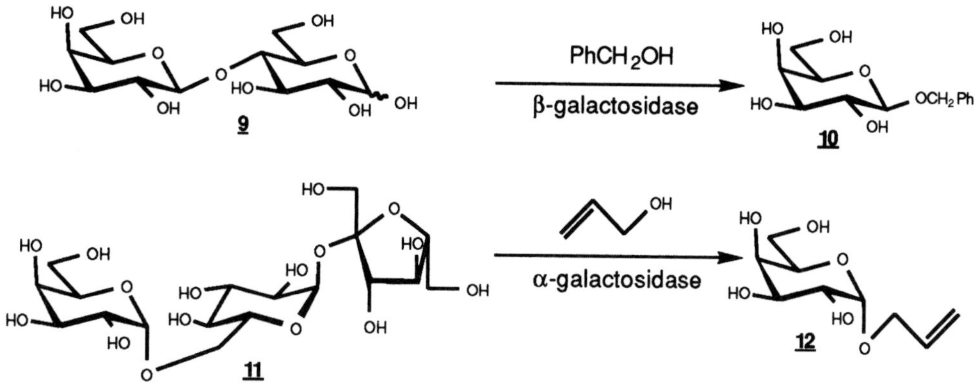

Fig. 3

who found that a mixture of 0.1 M lactose (**9**) and 0.2 M benzyl alcohol in phosphate buffer with 5 mg *β*-galactosidase yielded 6% of benzyl *β*-glucoside (**10**) (NILSSON, 1988) (Fig. 3). Similarly, allyl alcohol was glycosylated with raffinose (**11**) and *α*-galactosidase to give selectively the allyl *α*-galactoside (**12**).

The transglycosylation method is economic for these galactosylations, because lactose and raffinose are inexpensive. Where synthetic activated glycosyl donors such as *para*-nitrophenyl glycosides need to be used, this method becomes too expensive for the synthesis of simple glycosides. Here reverse glycosylation employing organic solvents or the respective alcohol as solvent is much more practical and recent examples will be discussed later on.

2.4.2 Diastereoselective Glycosylation

The diastereoselectivity of transglycosylation with glycosidases has been investigated with more complex prochiral or racemic acceptor substrates such as (**13a, b**), (**15**) and (**17**) (Fig. 4). In the case of the cyclic *meso*-diols (**13a, b**) (n=1, 2); (GAIS et al., 1988) good diastereoselectivities (75% de and 96% de) were obtained when the reactions were performed in 50% (v/v) aqueous acetone. Interestingly, no diastereoselectivity was observed without addition of the co-solvent.

In other cases reported so far, the selectivities have been much lower. For example, for

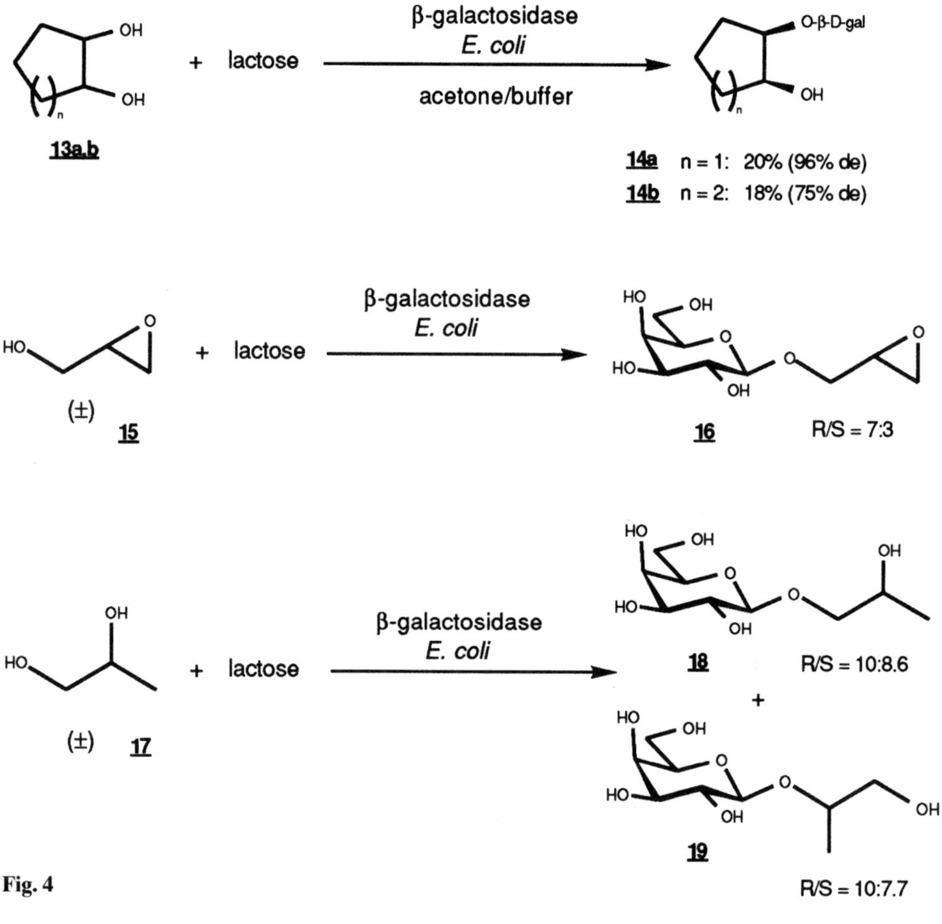

Fig. 4

the galactosylation product of 2,3-epoxypropanol (**15**) R/S values for the aglycon of (**16**) were only 7/3 (BJORKLING and GODTFREDSEN, 1988). Galactosylation of 1,2-propanediol (**17**) resulted in a mixture of regio- and stereoisomers with poor diastereoselectivity (CROUT and MACMANUS, 1990).

2.4.3 Glycosides as Acceptors

A much more challenging application of glycosidases than alkyl glycosides is the use of saccharides themselves as acceptor substrates, which results in the synthesis of oligosaccharides. This provides not only a high degree of stereoselectivity of reaction but also the potential glycosylation of one of several hydroxyl groups. Thus, 8 different β-galactosides can be formed by galactosylation of methyl galactoside (**21**) with *para*-nitrophenyl galactoside (**20**), as a result of galactosylation of the 4 different hydroxyl groups each of the acceptor (**21**) and the donor (**20**) itself. However, with the right choice of reaction conditions, substrates and enzymes, reasonable regioselectivities of reactions can be obtained, which has allowed the use of glycosidases for the synthesis of small oligosaccharides.

The competition of the donor substrate with acceptor can be overcome by using a large excess of acceptor substrate, although this can

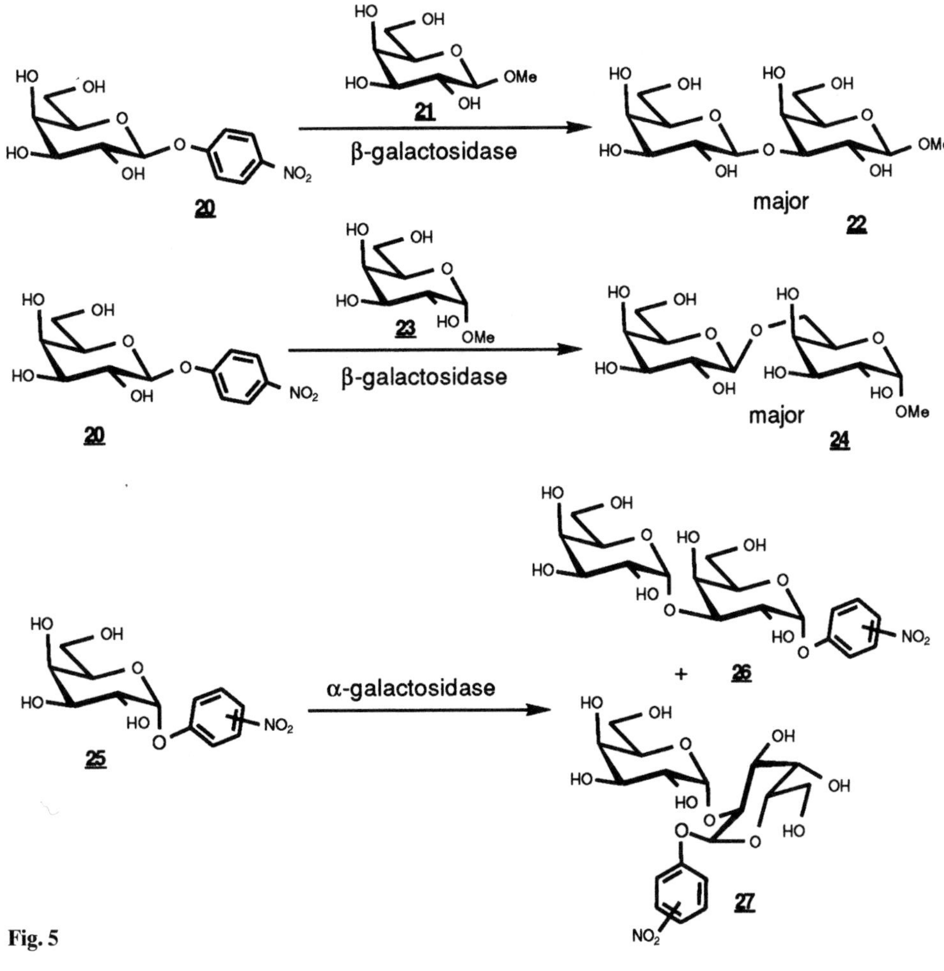

Fig. 5

present problems if the acceptor is valuable or difficult to separate from product.

A bias of regioselectivity is generally observed in glycosidase-catalyzed reactions, often with preferential formation of the 1-6 linkage. However, there have been many reports of good selectivities for other acceptor hydroxyl groups. How the regioselectivity of disaccharide formation can be influenced by apparently subtle changes in glycosyl donor or acceptor structure has been shown by NILSSON (NILSSON, 1987) (Fig. 5). For example, the 1-3 β-glucosidic linkage was mainly formed (**22**), when the β-methylglycoside (**21**) was the acceptor and *para*-nitrophenyl β-glucoside (**20**) the donor, whereas the 1-6 linked disaccharide (**24**) was the main product using methyl α-glycoside (**23**) as the acceptor substrate.

Even more pronounced selectivity was observed with nitrophenyl glycosides (**25**) which act both as acceptors and donors to give the (1-3) or (1-2) linked disaccharides (**26**) and (**27**) in different ratios, depending on whether *ortho*- or *para*-nitrophenyl glycosides are used. With the *ortho*-isomer of (**25**) the ratio of (**26**) to (**27**) was less than 1:6 whereas with the *para*-isomer of (**25**) it was reversed (8:1).

Glycosidases can not only be used for disaccharide synthesis, but also for the assembly of more complex oligosaccharides. Using a β-*N*-

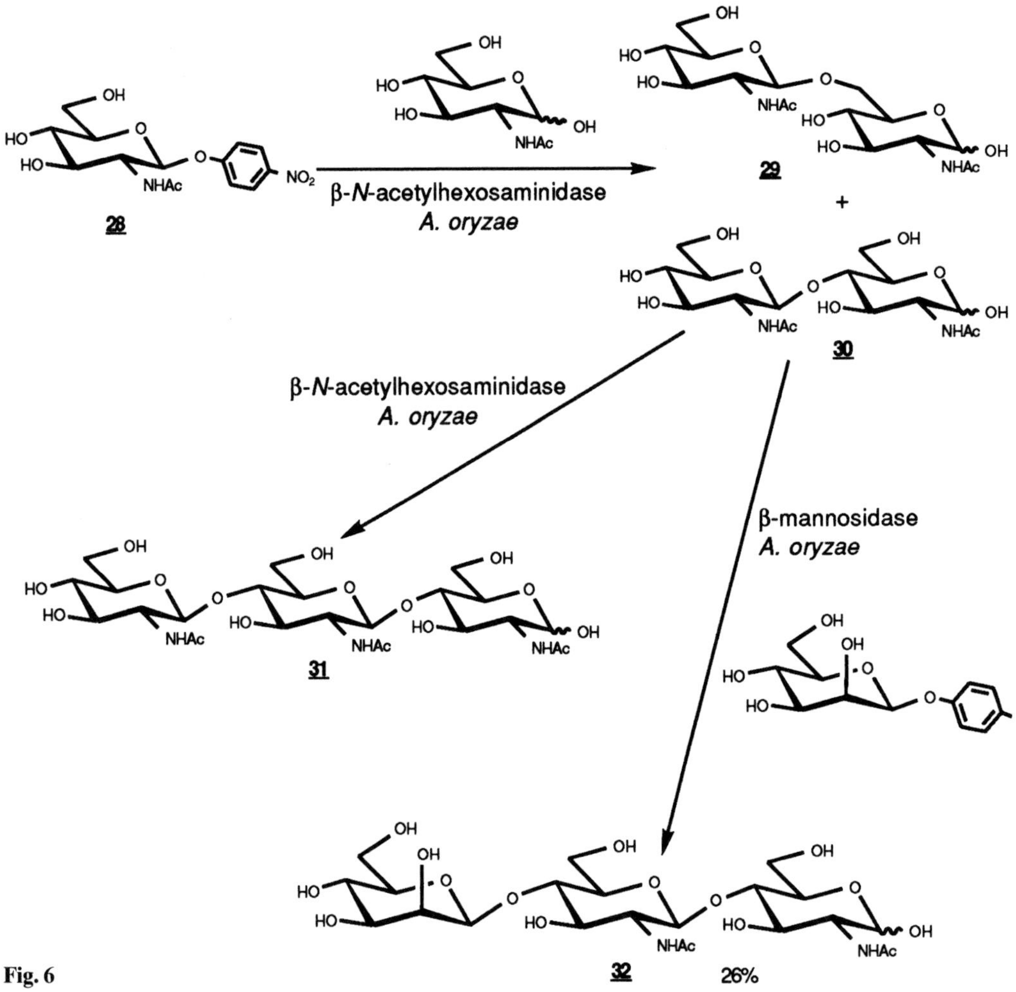

Fig. 6

acetylhexosaminidase from *Aspergillus oryzae*, CROUT and coworkers have synthesized chitooligosaccharides starting from *N*-acetyl glucosamine (Fig. 6). The first transglycosylation using *para*-nitrophenyl GlcNAc (**28**) proceeded with good selectivity for the 1-4 isomer (**30**), although some 1-6 isomer (**29**) was formed. The latter could be removed by selective hydrolysis using the 1-6 specific β-*N*-acetylhexosaminidase from Jack bean. Further elongation to higher chito-oligosaccharides such as (**31**) (with (**30**) acting as the donor and acceptor went with remarkable selectivity for the 1-4 linkage, with no detection of the 1-6 product (SINGH et al., 1995). Equally, glycosylation with a β-mannosidase from *A. oryzae* and *para*-nitrophenyl Man gave the trisaccharide (**32**) in 26% yield based on the donor as the only regioisomer (SINGH et al., 1996).

NILSSON has reported recently on the two-step synthesis of a trisaccharide derivative (**35**) which is related to antigens involved in hyperacute rejection of xenotransplants (Fig. 7). Both the thioethyl group and the 2-*N*-trichloroethoxycarbonyl group allow the further straightforward chemical elaboration of the trisaccharide either for analog synthesis or conjugation to polymers. Both reactions were performed with excess of glycosyl donor and gave 20% and 12% yields based on the acceptor substrates (**33**) and (**34**), respectively (NILSSON, 1997).

2.4.4 Synthesis of Glycoconjugates

Other than saccharides, glycosidases are also able to glycosylate many other classes of natural products such as shown in Fig. 8. OOI and colleagues have used galactosidase from *A. oryzae* for the synthesis of cardiac glycosides such as (**37**) from the precursor (**36**) in 26% yield. These glycosides are chemically unstable and difficult to obtain by conventional chemical synthesis which involves harsh glycosylation conditions and deprotection strategies (OOI et al., 1984).

An important class of glycoconjugates are glycopeptides and glycoproteins. Many advances have been made in the chemical synthesis of glycopeptides, which require large amounts of glycosylated amino acid building

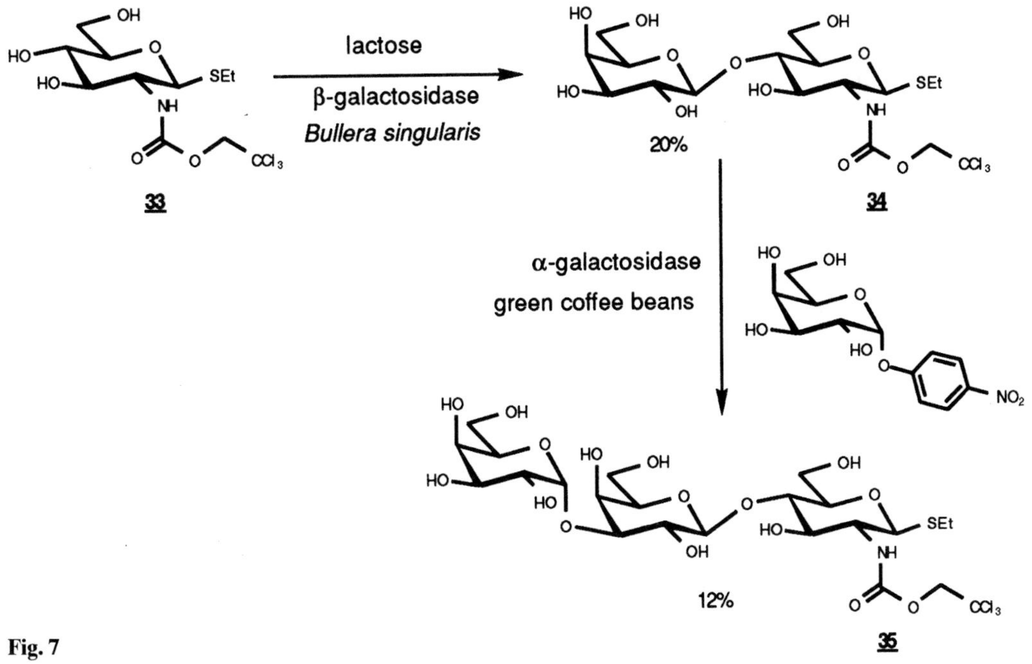

lactose
β-galactosidase
Bullera singularis
20%
α-galactosidase
green coffee beans
12%

Fig. 7

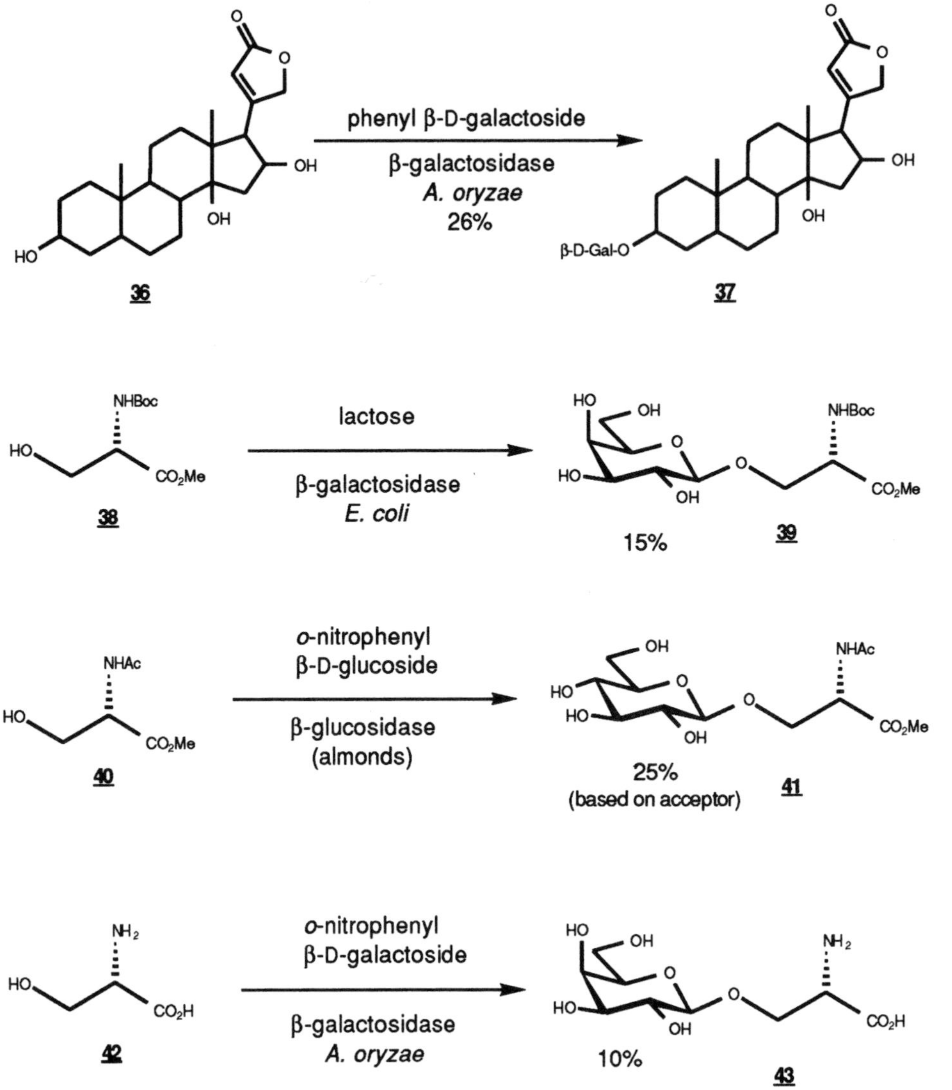

Fig. 8

blocks, such as glycosyl serine derivatives. Such building blocks are accessible by enzymatic synthesis of glycosyl serine derivatives using transglycosylation methodology. CANTACUZENE and colleagues have reported the synthesis of various protected galactosyl serine derivatives such as (**39**) from lactose and the semi-protected serine (**38**) in 15% yield based on lactose (CANTACUZENE et al., 1991).

The need for adding excess of acceptor substrate in order to reduce competition by the donor substrate can be a problem if the acceptor is expensive. For example, the yield of (**39**) based on (**38**) as a starting material is only 8%. This can be overcome (TURNER and WEBBERLEY, 1991) by using excess donor instead, with slow addition of the glycosyl donor during the reaction (with a syringe pump). Using this method, the glycosyl serine derivative (**41**) could be prepared from (**40**) in 25% yield. Alternatively, for the synthesis of glycosyl serine derivatives, the less expensive serine (**42**) itself was used in large excess (30-fold) to generate galactosyl serine (**43**) directly (SAUERBREI and THIEM, 1992; HEIDELBERG and THIEM, 1997; NILSSON et al., 1997).

2.5 Glycosylation by Reverse Hydrolysis

As discussed in Sect. 2.2, the alternative to transglycosylation is the use of glycosidases in the reverse hydrolysis, which has the advantage that no expensive activated glycosyl donors are needed. Reverse hydrolysis is achieved by reducing the water/glycosyl donor ratio such that the reaction is driven towards formation of the glycosidic linkage, rather than hydrolysis. This can either be achieved by using a very large excess of acceptor in aqueous systems, or by using the enzyme in organic solvent mixtures, thus reducing the water activity. The use of organic solvents has the advantage that purification is easier, although not all glycosidases can tolerate organic solvents.

2.5.1 Aqueous Systems

Di- and trisaccharides can be prepared by using very high concentrations of acceptor and donor. This makes use of the high solubility of sugars in water and of the good stability of enzymes under high sugar concentrations. For examples incubating a solution of 30% *N*-acetyl glucosamine and 10% galactose with β-galactosidase from *Escherichia coli* the 1-4 and 1-6 isomers (**44**) and (**45**) could be obtained in overall yield of 7.6% (AJISAKA et al., 1988; AJISAKA and FUJIMOTO, 1989) (Fig. 9). Purification of the disaccharides was possible by activated carbon chromatography, since the affinity for disaccharides is much higher than the affinity for monosaccharides. In addition, the overall yield could be improved to 16% by using a continuous process involving the circulation of

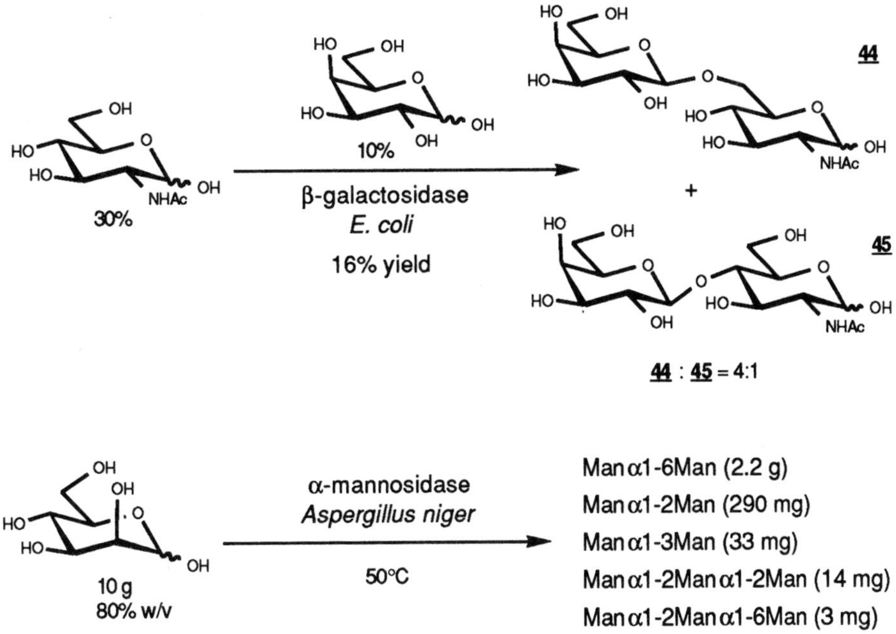

Fig. 9

the reaction mixture through columns of immobilized galactosidase and activated carbon in series.

Mannobioses and mannotrioses were isolated upon incubation of mannose (80% W/V) with an α-mannosidase from *A. niger* (AJISAKA et al., 1995) (Fig. 9).

2.5.2 Use of Organic Solvents

The use of organic co-solvents has been investigated for both the kinetic and thermodynamic glycosylation with almond β-glucosidase using *tert*-butanol or acetonitrile as co-solvents in order to improve yields. *tert*-Butanol was particularly useful because it does not act as an acceptor substrate, presumably because of steric hindrance, but the enzyme remained stable in up to 95%, with an optimum of activity at about 90% butanol in water giving up to 20% yield. Total loss of activity was observed in 100% solvent (VIC and THOMAS, 1992). The yield could be improved by using the acceptor alcohol itself as a solvent, in which case allyl and benzyl glucosides could be synthesized in 40% and 62% yields (VIC and CROUT, 1995). This methodology could also be used for diols such as 1,6-hexane diol using reverse hydrolysis, such that, e.g., the glucoside (**46**) was obtained in good yield (61%) from glucose (Fig. 10) (VIC et al., 1996).

Reverse hydrolysis reactions normally proceed very slowly, but can be accelerated using higher temperatures if the enzyme can tolerate such incubation conditions. The β-glucosidase from almonds, e.g. (Fig. 10) gives higher yields at 50 °C (VIC and CROUT, 1994). Recently, thermophile and thermostable glycosidases have been isolated which can be used at even higher temperatures or under focused microwave irradiation (GELO-PUJIC et al., 1997) to give products in good yields. Thus, glucosides (**48**) and (**49**) were produced in up to 77% and 20%, respectively, from (**47**) using crude homogenate from *Sulfolobus solfataricus* at 110 °C in 2 h. Commercial recombinant thermophilic enzyme libraries are now becoming available, which give a number of biocatalysts for screening. For example, the glycosidase CLONEZYME™ library developed by Recombinant Biocatalysis Inc. has been used for optimizing the synthesis of *N*-acetyl lactosamine (LI and WANG, 1997).

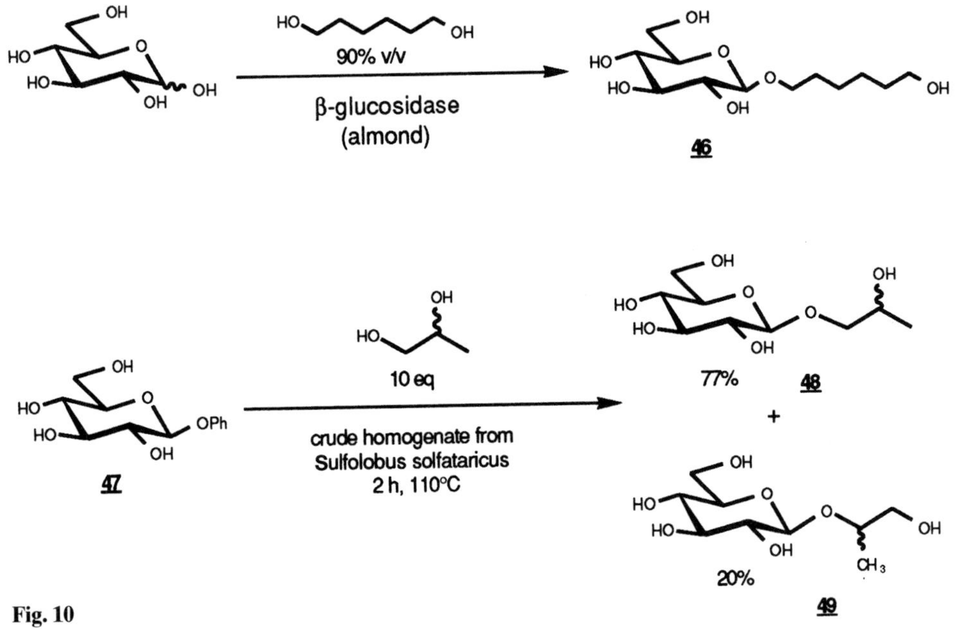

HO OH
90% v/v
β-glucosidase
(almond)
46
47
10 eq
crude homogenate from
Sulfolobus solfataricus
2 h, 110°C
77% 48
+
20% 49

Fig. 10

2.6 Transfer of Disaccharides Using Endoglycosidases

Most glycosidases that have been used as biocatalysts are exoglycosidases, which hydrolyze or synthesize terminal glycosidic linkages. However, a number of endoglycosidases are known and even available, which would allow assembly of larger oligosaccharides from natural or synthetic building blocks. An interesting and useful example is the application of cellulase for such a "block synthesis" approach to synthetic cellulose polysaccharides. KOBAYASHI and colleagues have reported that incubation of β-D-cellobiosyl fluoride (**50**) with cellulase from *Trichoderma viride* in a mixed solvent of acetonitrile/acetate buffer (5:1) yielded predominantly water-soluble and insoluble cello-oligosaccharides of DP larger than 22 and smaller than 8, respectively, depending on reaction conditions (KOBAYASHI et al., 1991) (Fig. 11). More recently, glucanohydrolases have been used to make tetra- and higher saccharides by using suitable acceptor saccharides. For example, a 1,3-1,4-β-D-glucan 4-glucanohydrolase from *Bacillus licheniformis* can catalyze the formation of the tetrasaccharide Glcβ1-3Glcβ1-4Glcβ1-3Glcβ1-OMe from Glcβ1-3Glcβ1-F and Glcβ1-3Glcβ1-OMe in 40% yield. By using a 7-fold excess of acceptor substrate, autocondensation could be suppressed (VILADOT et al., 1997).

2.7 Selective Hydrolysis for the Synthesis of Glycosides

Regioselective glycosylation can also be achieved by selective hydrolysis of multiglycosylated substrates. For example, the potent

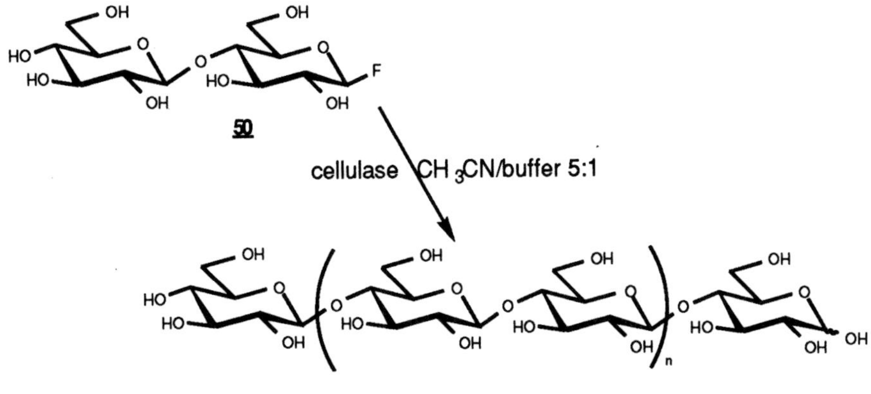

Fig. 11

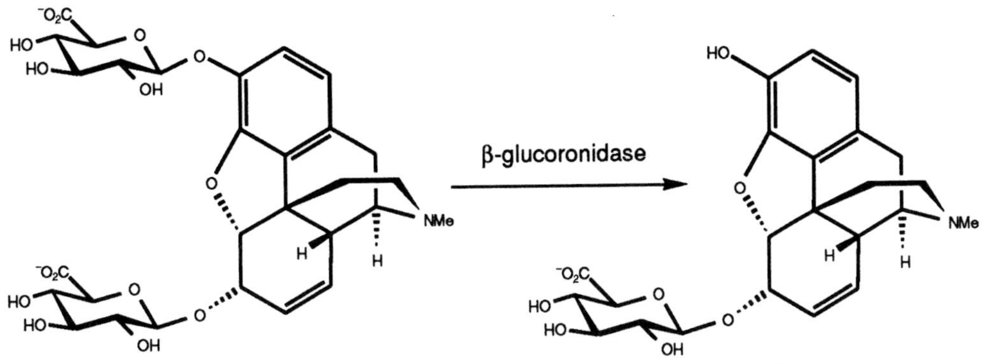

Fig. 12

analgesic morphine-6-glucoronide (**52**) was obtained by selective hydrolysis of the readily available diglucuronide (**51**) using glucuronidase as a catalyst (Fig. 12) (BROWN et al., 1995).

3 Glycosyltransferases

Glycosyltransferases are the enzymes that are naturally responsible for the biosynthesis of glycosides. They catalyze the transfer of a saccharide from a glycosyl donor to a glycosyl acceptor with either inversion or retention of the anomeric center as outlined in Fig. 13. Most of the transferases discussed here use glycosyl phosphates as donors, in particular the sugar nucleotides shown in Fig. 14. However, the glycosyl donor might also be a glycoside itself, such as a di- or polysaccharide. Since glycosyltransferases have evolved to catalyze the formation of a very specific glycosidic linkage in a biosynthetic pathway, they tend to be highly selective both for acceptor and donor substrates as well as highly regio- and stereoselective in the formation of the glycosidic bond.

3.1 Availability of Enzymes

Although a large number of amino acid sequences for glycosyltransferases are now known (FIELD and WAINWRIGHT, 1995), the number of enzymes commercially available is still small compared to glycosidases, and they tend to be relatively expensive enzymes. 5 glycosyltransferases are currently sold through catalogs, namely β-1,4-galactosyltransferase, α-2,3-sialyltransferase, α-2,6-sialyltransferase, α-1,2-mannosyltransferase, and α-1,3-fucosyltransferase V. However, protocols for the isolation of transferases from natural sources are well established (PALCIC, 1994) and cloning and overexpression techniques have greatly increased the number of enzymes that are available for biotransformations in recent years (GIJSEN et al., 1996). Although most of the glycosyltransferases come from higher organisms, there are some examples of successful heterologous overexpression in prokaryotes (WATT et al., 1997; WANG et al., 1993). PALCIC has recently compiled a useful list of transferases that have been used for preparative-scale biotransformations (PALCIC, 1994).

3.2 Availability of Glycosyl Donors

Despite the variety of different glycosyltransferases that are found in Nature, only a very small number of common glycosyl donors are used by the majority of transferases. Their structures are shown in Fig. 14. These are all commercially available, but expensive, if reactions need to be done beyond a 10–50 mg scale. In addition, glycosyltransferase reactions often suffer from product inhibition when stoichiometric amounts of glycosyl donor are used.

Both problems can be overcome by *in situ* regeneration of the sugar nucleotide cofactor such that concentrations of the nucleotide substrates and products are kept low at all times during the biotransformation (ICHIKAWA et al., 1994). Such sugar nucleotide regeneration uses the biosynthetic enzymes *in vitro* and was first developed by WHITESIDES and colleagues (WONG et al., 1982) in the early 1980s. Since then many improvements have been reported, such that glycosyltransferases can now be used to produce complex oligosaccharides on a kilogram scale (ICHIKAWA et al., 1994). The first reported regeneration protocol was for UDP-Glc and UDP-Gal as outlined in Fig. 15. Both

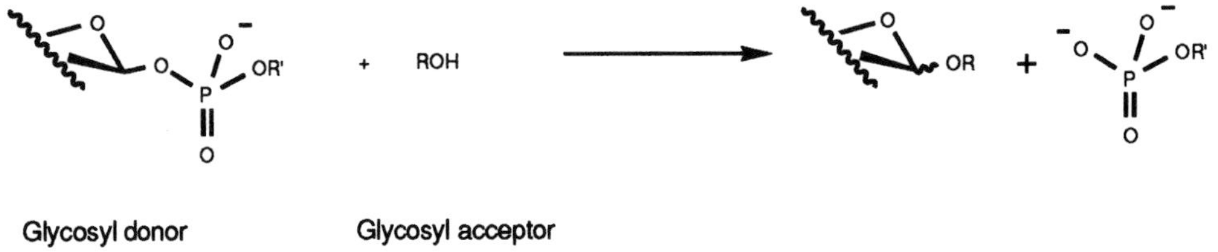

Fig. 13

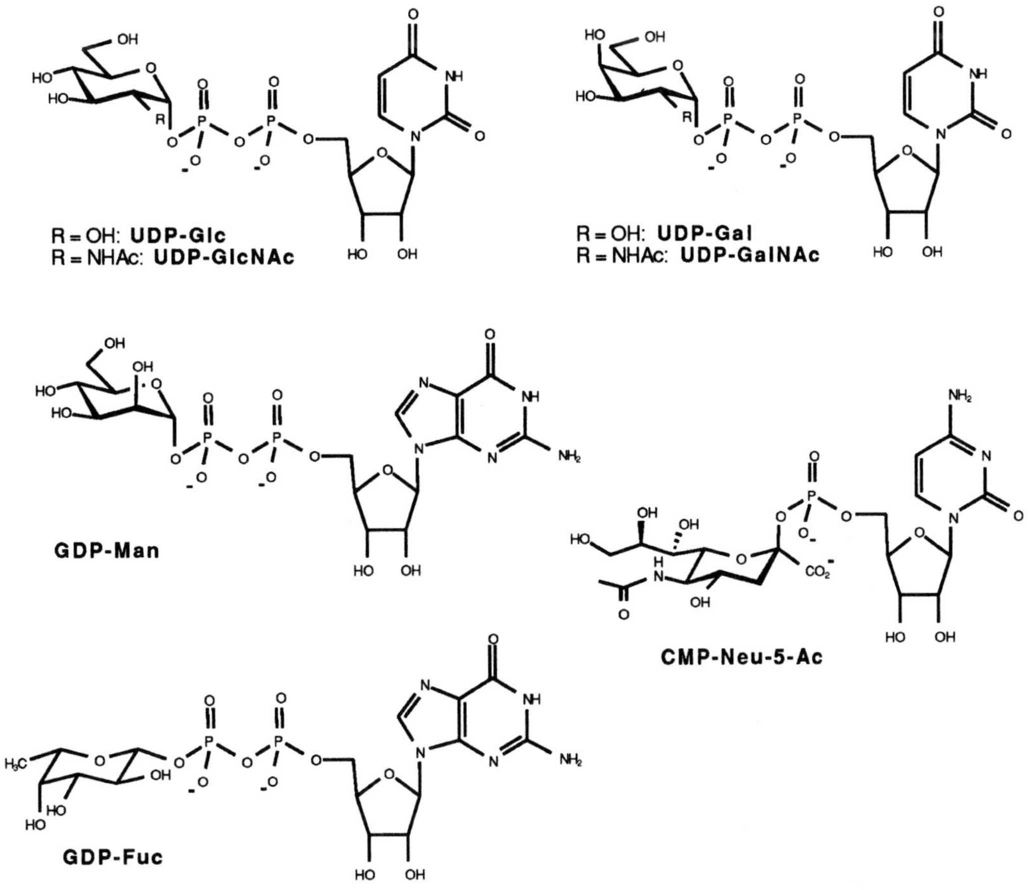

R = OH: **UDP-Glc**
R = NHAc: **UDP-GlcNAc**

R = OH: **UDP-Gal**
R = NHAc: **UDP-GalNAc**

GDP-Man

CMP-Neu-5-Ac

GDP-Fuc

Fig. 14

were generated from glucose-1-phosphate (**53**) and substoichiometric amounts of UTP. Synthesis of UDP-Glc is achieved by UDP-Glc pyrophosphorylase, with the by-product inorganic phosphate (PPi) being hydrolyzed to phosphate using inorganic pyrophosphatase to avoid enzyme inhibition. UDP-Glc can then be used directly in the reaction or be converted to UDP-Gal using UDP-Gal 4-epimerase. The latter equilibrium favors formation of UDP-Glc, which can lead to complications if the subsequent galactosyltransferase can accept UDP-Glc as a substrate. UDP generated from the glycosyltransfer reaction can be regenerated to UTP by using pyruvate kinase and phosphoenolpyruvate as the phosphorylating agent (Ichikawa et al., 1994). With such cofactor regeneration, allyl-*N*-acetyl lactosamine (**55**) was synthesized in 5 d on a 1.7 g scale in 50% yield from (**54**) using equimolar amounts of (**54**), glucose-1-phosphate, and phosphoenolpyruvate and 0.025 equivalents of UDP.

When the monosaccharide starting material is expensive, as in the case of sialic acid, the regeneration system can include synthesis of the monosaccharide itself, as illustrated in Fig. 16 for the regeneration of CMP-NeuAc, the common donor substrate for sialyltransferases (Ichikawa et al., 1994). Synthesis of sialic acid (NeuAc) is achieved from *N*-acetyl mannosamine (**56**) and pyruvate by using a sialyl aldolase. The monosaccharide is then activated to CMP-NeuAc by the action of CMP-NeuAc

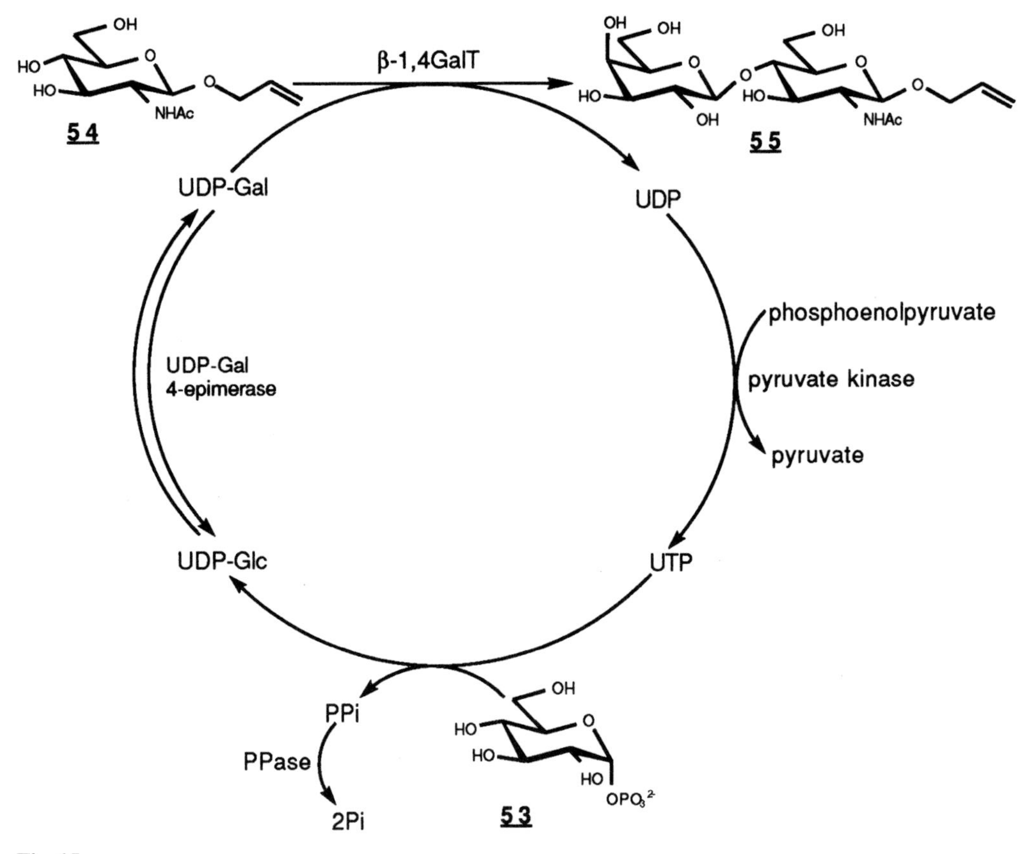

Fig. 15

synthase using CTP. After transfer of the sialic acid to a suitable acceptor substrate by sialyltransferase, the CMP is recycled to CTP in two steps first using ATP and nucleoside monophosphate kinase to generate CDP and then pyruvate kinase to generate CTP. The same pyruvate kinase can also be used to recycle ATP needed for CDP formation.

3.3 Examples of Oligosaccharide Syntheses Using Glycosyltransferases

Glycosyltransferases are generally highly selective for both their glycosyl donor and glycosyl acceptor substrates and catalyze the formation of one specific glycosidic linkage with very high stereo- and regioselectivity. This makes them particularly useful as biocatalysts for the synthesis of much larger and complex oligosaccharides, where regioisomers that would be generated by glycosidases are difficult to separate.

First examples of the use of glycosyltransferases appeared in the early eighties (WONG et al., 1982) (AUGE et al., 1984) using β-1,4-galactosyltransferase from bovine colostrum, which has become the best studied glycosyltransferase and is now commercially available. Di- and trisaccharides (Galβ1-4GlcNAc and Galβ1-4-GlcNAcβ1-6/3Gal) were prepared on a millimole scale using immobilized enzymes and regeneration systems for the sugar nucleotide cofactors. This work has been extended over the past 15 years, perhaps driven by the discovery that oligosaccharide ligands such as the

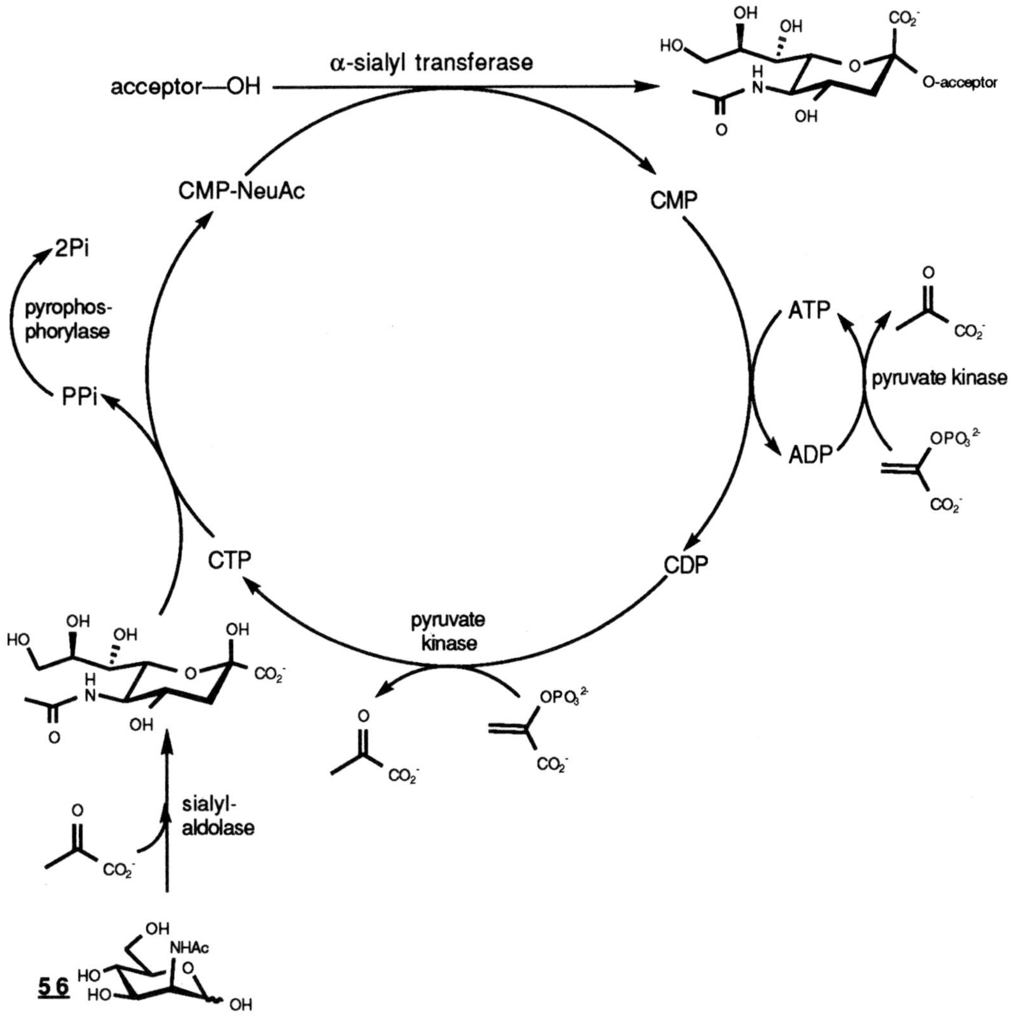

Fig. 16

blood group antigens sialyl Lewis x and sialyl Lewis a are involved in cell adhesion between activated endothelial cells and neutrophils and have potential therapeutic applications in inflammation and cancer metastasis (FEIZI, 1993; LASKY, 1994; PAREKH and EDGE, 1994). Thus, the sialyl Lewis x structure (**60**) (Fig. 17) can be assembled in only three steps from (**57**) using a galactosyltransferase to form (**58**), followed by sialylation to (**59**) and final fucosylation (ICHIKAWA et al., 1992). These reactions have been performed on up to 0.5–1 kg scale and have also been recently used for bivalent sialyl Lewis x structure (DEFREES et al., 1995).

Using similar enzymes, the sialylated undecasaccharide – asparagine conjugate (**62**), an important side chain of glycoproteins, was prepared from chemically synthesized (**61**) in 86% yield by enzymatic transfer of saccharide units in a one-pot reaction (UNVERZAGT, 1996).

Besides focus on galactosyl-, sialyl-, and fucosyltransferases, advances have been made in studying glycosyltransferases involved in the synthesis of the core structures of oligosaccha-

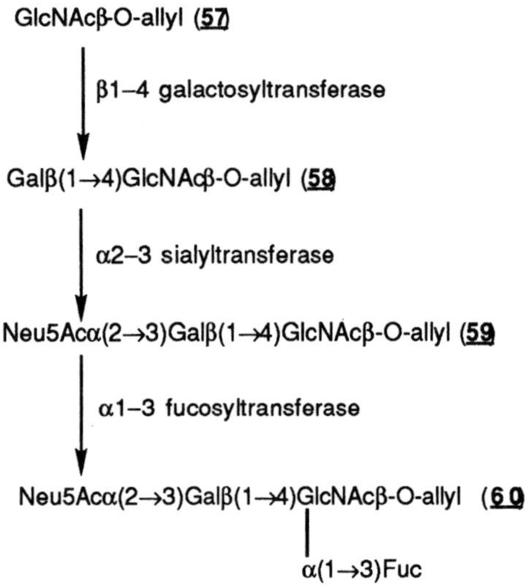

GlcNAcβ-O-allyl (57)

β1–4 galactosyltransferase

Galβ(1→4)GlcNAcβ-O-allyl (58)

α2–3 sialyltransferase

Neu5Acα(2→3)Galβ(1→4)GlcNAcβ-O-allyl (59)

α1–3 fucosyltransferase

Neu5Acα(2→3)Galβ(1→4)GlcNAcβ-O-allyl (60)
α(1→3)Fuc

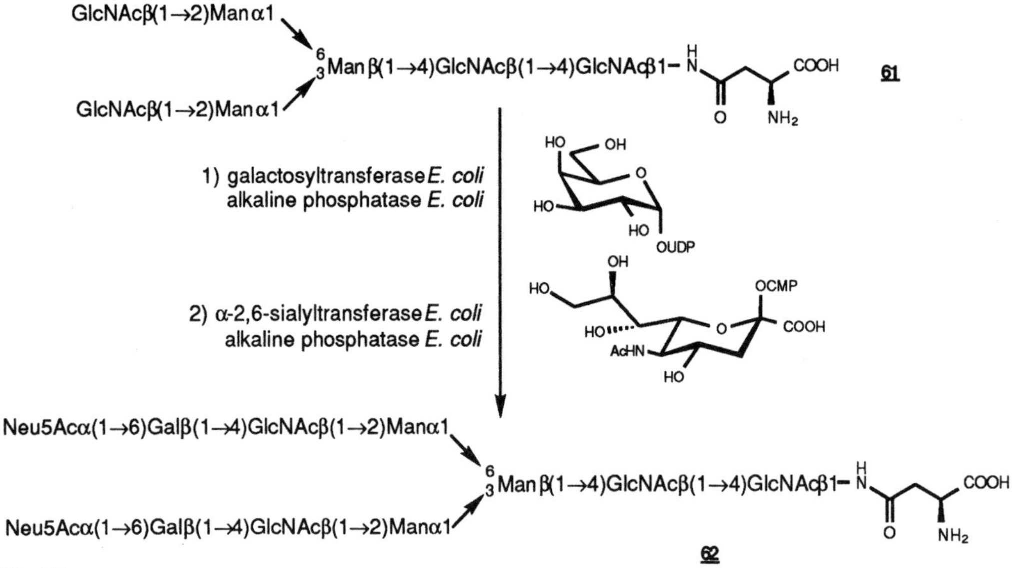

GlcNAcβ(1→2)Manα1 → 6; GlcNAcβ(1→2)Manα1 → 3 Manβ(1→4)GlcNAcβ(1→4)GlcNAcβ1–NH–Asn (COOH, NH2, O) 61

1) galactosyltransferase *E. coli*
alkaline phosphatase *E. coli*
(UDP-Gal: HO, OH, O, HO, HO, OUDP)

2) α-2,6-sialyltransferase *E. coli*
alkaline phosphatase *E. coli*
(CMP-Neu5Ac: OH, HO, OCMP, HO, O, COOH, AcHN, HO)

Neu5Acα(1→6)Galβ(1→4)GlcNAcβ(1→2)Manα1 → 6; Neu5Acα(1→6)Galβ(1→4)GlcNAcβ(1→2)Manα1 → 3 Manβ(1→4)GlcNAcβ(1→4)GlcNAcβ1–NH–Asn (COOH, NH2, O) 62

Fig. 17

ride side chains of glycoproteins. A *β*-1,4-mannosyltransferase from yeast involved in the biosynthesis of the conserved trisaccharide core of glycoproteins was overexpressed in *E. coli* as a soluble His10-fusion protein. Absorption of the enzyme on an immobilized metal (nickel) affinity column provided the mannosyltransferase directly from crude cell extracts as a functionally pure biocatalyst, which was used in the synthesis of the glycolipid (**64**) from synthetic (**63a**) (Fig. 18; WATT et al., 1997). An α-1,2-mannosyltransferase from yeast has also been overexpressed in *E. coli* (WANG et al., 1993).

Most asparagine-linked oligosaccharides in proteins contain a common pentasaccharide core but are very heterogeneous in further elaboration of the oligosaccharide structure (KORNFELD and KORNFELD, 1985). Heterogeneity is first introduced by the biosynthesis of different *N*-acetyl glucosaminidic linkages to the pentasaccharide chore, each being catalyzed by a different GlcNAc-transferase. These enzymes have been investigated by SCHACHTER and colleagues (BROCKHAUSEN et al., 1988, 1992) and have been assigned as GlcNAc-transferases I–VI according to the linkage they form, as shown in Fig. 19. All of these can be isolated in at least milliunit amounts, which is sufficient for milligram synthesis. The GlcNAc-transferases transfer GlcNAc not only to the natural glycoprotein or peptide side chain, but also to shorter oligosaccharide acceptors, which are synthetically available. Thus trisaccharide (**65**) is a substrate for GlcNAc-transferases I and II and (**66**) for GlcNAc-transferase V (Fig. 19) (KAUR et al., 1991; PALCIC, 1994).

Glucuronosyltransferases, which transfer glucuronic acid from UDP-glucuronic acid to acceptor substrates are very important enzymes in the metabolism of endogenous and exogenous lipophilic compounds, since glucuronidation enhances water solubility and thus excretability from the cellular milieu. These enzymes are, therefore, of great interest in drug metabolism, and a liver transferase system has been used for the synthesis of phenol and lipophilic steroid glucuronates in good yields (65% and 30%) on a milligram scale (GYGAX et al., 1991). When crude liver homogenate was used all the necessary enzymes required to convert glucose-1-phosphate to UDP-glucuronic acid were present in the preparation, such that the much less expensive glucose-1-phosphate could be employed as a donor substrate. On the other hand, the liver enzymes and UDP-glucuronic acid are now commercially available.

3.4 Synthesis of Glycoconjugates

One of the greatest advantages of using glycosyltransferases is that complex glycoconjugates such as lipids, peptides, proteins and even cell surfaces can be glycosylated. Thus, oligosaccharides can be assembled directly on the glycoconjugate in a biomimetic fashion, which is difficult with glycosidases because of the high acceptor concentrations required and poor regioselectivity. Equally, chemical synthetic methods of glycosidic bond formation often fail on macromolecules, consequently, the oligosaccharide must be assembled before attachment to the macromolecule. Glycosyl-

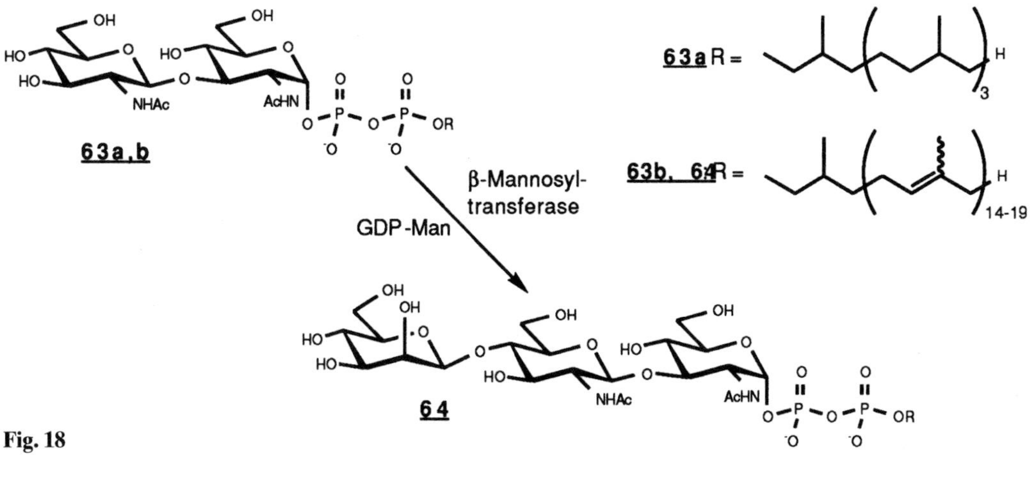

Fig. 18

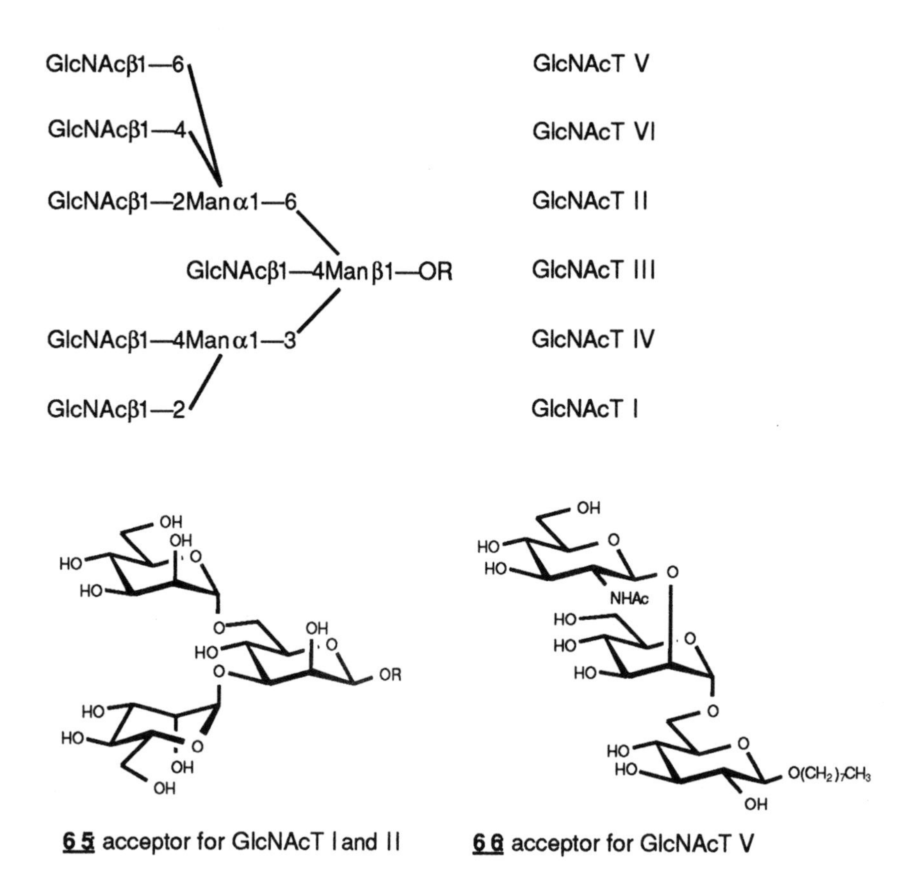

Fig. 19

transferases are, therefore, important tools for the biochemist for modifying the glycosylation of cell surfaces, proteins, and lipids and for investigating structure–function relationships. A few examples are given in the following sections.

3.4.1 Synthesis of Glycolipids

Glycolipids are ubiquitous membrane components of cells and are involved in a wide range of biological processes, which makes them important synthetic targets. Examples of the enzymatic synthesis of glycolipid intermediates in glycoprotein biosynthesis were discussed earlier (Fig. 18; WATT et al., 1997; IMPERIALI and HENRICKSON, 1995).

Another important class of glycolipids are glycosphingolipids such as gangliosides, which are commonly found in eukaryotic cells and participate in a number of cell recognition events (KARLSSON, 1989; HAKOMORI, 1990; ZELLER and MARCHASE, 1992). These lipids contain ceramide as their membrane-associated hydrophobic anchor and complex oligosaccharide headgroups, which contain sialic acids in the case of gangliosides. The biosynthetic pathways of glycosphingolipids appear similar to that of other glycoconjugates, in that the oligosaccharide headgroup is assembled by specific glycosyltransferases from the reducing end (BOUHOURS and BOUHOURS, 1991). Although glycosidic linkages in glycolipids are quite similar to those in glycoproteins, it appears that distinctly separate glycolipid and glycoprotein transferases are used in biosynthesis (MELKERSON-WATSON and SWEELEY, 1991; ITO et al., 1993). Indeed, glycoceramides themselves are often very poor substrates for the glycoprotein glycosyltransferases discussed earlier. On the other hand, the glycolipid-specific transferases are often not available on a sufficient scale for application in millimole glycolipid synthesis (PREUSS et al., 1993).

Several practical solutions have been found to overcome these problems. ZEHAVI and colleagues demonstrated that enzymatic glycosphingolipid synthesis was possible on solid support using bovine milk galactosyltransferase (ZEHAVI et al., 1990). More recently, this approach was extended to sialyltransferases, with transfer of the oligosaccharide product from solid support onto ceramide using ceramide glycanase (NISHIMURA and YAMADA, 1997). Alternatively, more water-soluble chemical precursors of ceramide such as azidosphingosine derivatives were good acceptor substrates for glycosyltransferases, which led to efficient chemo-enzymatic routes of ganglioside GM3 and analogs (GUILBERT et al., 1992; GUILBERT and FLITSCH, 1994).

3.4.2 Glycopeptides and Glycoproteins

Oligosaccharides are linked to peptides and proteins through two main types of linkage: the so-called "*O*-linkage" where the hydroxyl groups of serine and threonine provide sites for glycosylation and the "*N*-linkage" where glycosylation occurs through the amide of asparagine (SCHACHTER, 1991; IMPERIALI and HENRICKSON, 1995). Three main types of enzymatic methods have been reported for the synthesis of glycopeptides and glycoproteins: firstly, the group of WONG has used subtilisin-catalyzed glycopeptide condensation for the synthesis of ribonuclease glycoforms (WITTE et al., 1997). Secondly, enzymes have been used to catalyze the formation of the glycosidic bonds that attach the oligosaccharide to the peptide side chain. Thirdly, glycosyltransferases can extend an existing mono- or oligosaccharide that is already attached to the protein by further glycosylation.

The most common monosaccharide linked to the serine or threonine side chain in *O*-glycan core structures is an α-GalNAc residue (SCHACHTER, 1991). Biosynthesis of the *O*-glycan then occurs by successive addition of individual monosaccharide units from the corresponding activated sugar nucleotides. The first glycosyltransferase in this pathway, peptide GalNAc-transferase, has been isolated and is available for the glycosylation of peptides (CLAUSEN and BENNETT, 1996) (YANG et al., 1992). Considerable effort has gone into studying the relationship between acceptor peptide sequence and enzyme activity. Although the enzyme might glycosylate different threonine and serine residues within a given polypeptide sequence at different rates, so far, no peptide motif for *O*-glycosylation has been identified (CLAUSEN and BENNETT, 1996), and hence specific glycosylation remains elusive.

The formation of *N*-glycans occurs by a very different biosynthetic process. *N*-glycosylation is very selective for a tripeptide motif including the asparagine that is to be glycosylated (Asn-Xaa-Thr/Ser, where Xaa cannot be proline). Furthermore, monosaccharide units are not assembled sucessively on the polypeptide, but are first biosynthesized on a phospholipid (dolichol) and then transferred from the lipid to the nascent polypeptide chain by an oligosaccharide transferase (KORNFELD and KORNFELD, 1985; IMPERIALI and HENRICKSON, 1995). The natural substrate for the oligosaccharyltransferase is a tetradecasaccharide lipid consisting of to GlcNAc, 9 mannose and 2 glucose units, but the enzyme also accepts smaller substrates down to the disaccharide-lipid (GlcNAcβ1-4GlcNAc-lipid). A crude enzyme-containing extract has been used to glycosylate peptides containing the Asn-Xaa-Thr/Ser motif (LEE and COWARD, 1993), but glycosylation of full-length proteins appears to be more difficult (LIU et al., 1994; XU and COWARD, 1997; IMPERIALI, 1997). Thus, the biomimetic approach for the synthesis of *N*-glycosylated proteins using the biosynthetic glycosyltransferases, despite their availability, is still difficult.

An interesting alternative non-biomimetic way of generating *N*-glycans has involved the use of endo-β-*N*-acetyl-glucosaminidase (Endo-A) from *Arthrobacter protophormiae* (FAN et al., 1995), which transfers Man9GlcNAc from Man9GlcNAc2Asn onto GlcNAc-peptide to form Man9GlcNAc2-peptide. This enzyme can also be used to make *N*-glycopeptide analogs with *C*-glycosidic linkages to the peptide (WANG et al., 1997).

Galactosyltransferases, sialyltransferases, and fucosyltransferases previously described can also be used to extend oligosaccharides on peptides and proteins (UNVERZAGT et al., 1994; UNVERZAGT, 1996; SEITZ and WONG, 1997). The disadvantage of all these methods is that they require initial attachment of a single monosaccharide specifically onto the polypeptide chain, which can be achieved either chemically or by truncating the natural oligosaccharide back to a single GlcNAc residue, using appropriate glycosidases such as Endo-H. Further development of these methods should ultimately lead to the synthesis of pure "glycoforms" of proteins, which remains a major challenge to scientists (BILL and FLITSCH, 1996; VERBERT, 1996; STANLEY, 1992).

3.4.3 Cell Surface Glycosylation

An interesting application carrying the use of glycosyltransferases even further into biological systems is the glycosylation of cell surfaces (PAULSEN and RODGERS, 1987). Thus desialylated erythrocytes were treated with sialyltransferases of different specificity resulting in samples of erythrocytes with different surface linkages (e.g., Neu-5-Acα2-6Galβ1-4GlcNAc-R, Neu-5-Acα2-3Galβ1-(3)4GlcNAc-R, Neu-5-Acα2-3Galβ1-3GalNAc-Thr/Ser,). These preparations were used to test the specificity of sialo-oligosaccharide binding proteins such as influenza virus hemagglutinins by measuring its ability to agglutinate the resialylated cells.

3.5 Solid Phase Synthesis

Solid-phase synthesis is a method well established in peptide and nucleic acid chemistry, and there is a considerable interest in applying it to carbohydrate chemistry. Advantages are the ease of purification and the potential for construction of combinatorial libraries. However, development of solid-phase methods has been slow in this area, because of the inherent difficulties of regio- and stereo-control in the chemical formation of glycosidic linkages.

The usefulness of solid-phase synthesis in enzymatic glycolipid synthesis, where one of the major problems is the insolubility of the acceptor lipid in aqueous media, has already been discussed (ZEHAVI et al., 1990; NISHIMURA and YAMADA, 1997). Several examples of oligosaccharide and glycopeptide synthesis using galactosyl- and sialyltransferases have been reported since 1994, such as the synthesis of a sialyl Lewis x glycopeptide (SCHUSTER et al., 1994). One of the practical problems enzyme-catalyzed reactions on solid support has been the choice of enzyme-compatible polymeric support, which has to render support-bound substrates available to the enzyme in aqueous media. The commonly used polystyrene resins do not swell in water and are, therefore, generally unsuitable. Among the resins used successfully are controlled pore glass (HALCOMB et al., 1994), polyethylene glycol polyacrylamide co-polymers (MELDAL et al., 1994), polyacrylamide (KOEPPER, 1994), and sepharose (BLIXT and NORBERG, 1997). Soluble polyacrylamide supports can also be used in combination with glycosyltransferases (NISHIMURA et al., 1994; YAMADA and NISHIMURA, 1995; WIEMANN et al., 1994). Various linkers have been used for oligosaccharides on solid support which can be cleaved with chymotrypsin (YAMADA and NISHIMURA, 1995; SCHUSTER et al., 1994), by hydrazinolysis (HALCOMB et al., 1994), by hydrogenation (NISHIMURA et al., 1994), by photolysis (WIEMANN et al., 1994; KOEPPER, 1994), and by reductive cleavage of disulfide bonds (BLIXT and NORBERG, 1997).

3.6 Acceptor–Donor Analogs

Glycosyltransferases can show remarkable specificity for both acceptor and donor substrates, often beyond even the terminal non-reducing saccharide of the acceptor. For example, an α-2,3-sialyltransferase isolated from porcine submaxillary gland was highly specific for acceptor substrates terminating with the Galβ1-3GalNAc sequence found in *O*-linked oligosaccharides, but did not glycosylate those terminating in Galβ1-4Glc or Galβ1-4GlcNAc, found in *N*-linked chains (REARICK et al., 1979). Such findings have led to the general belief that glycosyltransferases are highly specific and cannot be used for the synthesis of analogs. However, when unnatural, chemically synthesized donor and acceptor analogs were further investigated, a number of unnatural glycosidic linkages could be formed on a milligram scale although the relative rates of reaction were often much reduced (WONG et al., 1995).

The best studied system is the β-1,4-galactosyltransferase from bovine colostrum (Fig. 20). Instead of UDP-Gal as a donor substrate, UDP-Glc, UDP-GalNAc, UDP-GlcNH$_2$, and

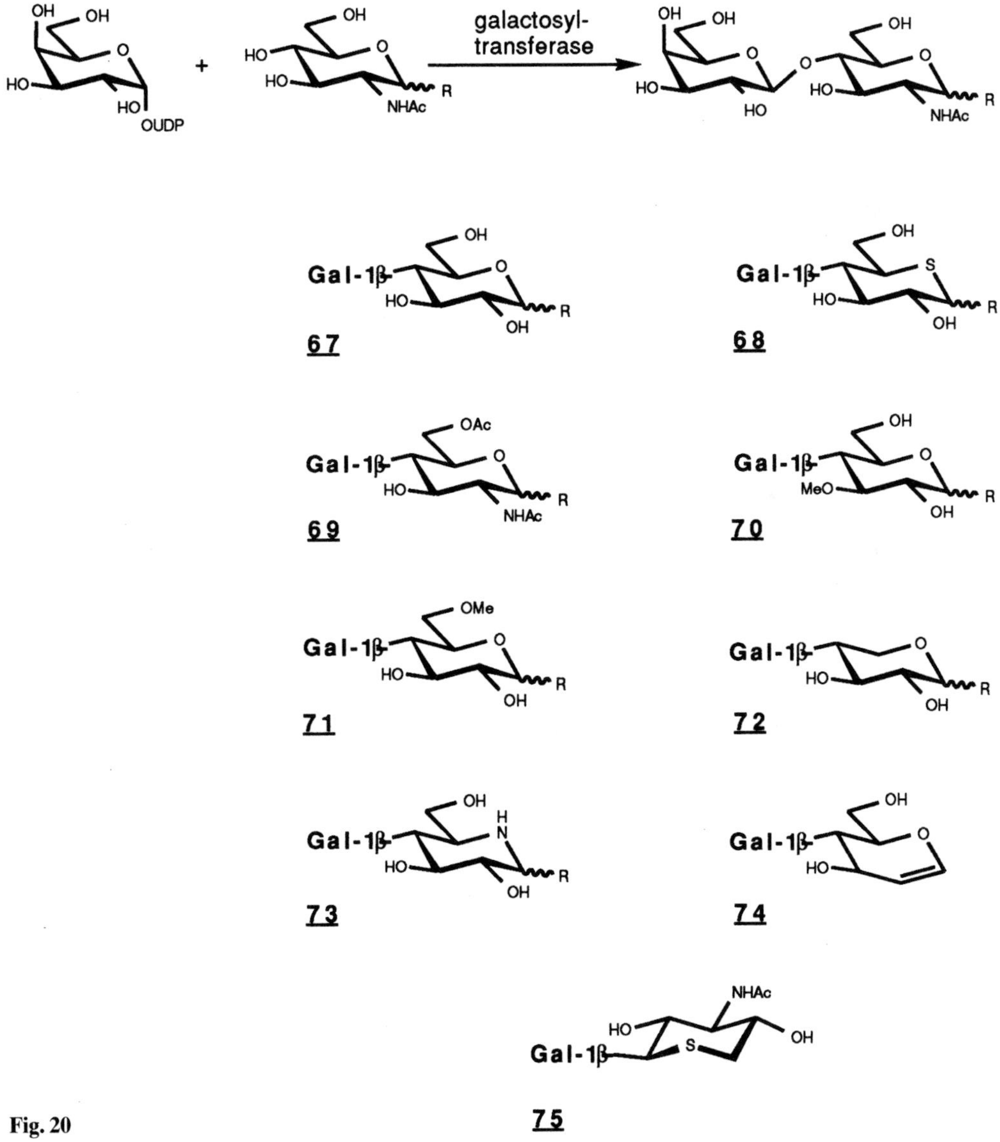

OH OH
O
HO
HO
OUDP
+
OH
O
HO
HO
NHAc
R
galactosyl-
transferase
OH OH
O
HO
HO
O
OH
O
HO
NHAc
R

Gal-1β- HO OH O OH R
67

Gal-1β- HO OH S OH R
68

Gal-1β- HO OAc O NHAc R
69

Gal-1β- MeO OH O OH R
70

Gal-1β- HO OMe O OH R
71

Gal-1β- HO O OH R
72

Gal-1β- HO OH H N OH R
73

Gal-1β- HO OH O
74

Gal-1β- HO NHAc S OH
75

Fig. 20

all the 2,3,4,6-deoxy- and 6-deoxy-6-fluoro-UDP-Gal analogs were successfully transferred to acceptor substrates (PALCIC and HINDSGAUL, 1991; KAJIHARA et al., 1995). An even wider range of acceptor substrates has been used to generate a number of non-natural disaccharides such as (**67–75**) in Fig. 20. Substitutions in the 3- and 6-positions of the acceptor substrate appear to be well tolerated, whereas any modifications in the 4-position, not surprisingly, leads to loss of activity. The 2-position appears also to be sensitive to change, such that a 2-hydroxyl group as in glucose can be tolerated (more so in the presence of a coenzyme, lactalbumin). Glucal is also a substrate leading to (**74**), but mannose (the C-2-epimer) is not a substrate. The enzyme is also very sensitive to negative charge, glucuronic acid and glucose-1-phosphates are not accepted (WONG et al., 1995). When using acceptor analogs with lower binding constants to the enzyme the high regioselectivity of the enzyme might potentially be lost and it is, therefore, important to check authentity of the glycosidic linkage. This has mainly been done using NMR methods, in particular NOE effects, and looking at deshielding effects of ring protons (WONG et al., 1991; NISHIDA et al., 1993a, b). An interesting case of changed regioselectivity of the galactosyltransferase has been reported by THIEM and colleagues (NISHIDA et al., 1993a, b) for acceptor substrates containing 3-*N*-acetyl substituents such as *N*-acetyl 5-thio-gentosamine which led to the formation of 1,1-*O*-linked disaccharide (**75**). This work has suggested that perhaps the *N*-acetyl group of the acceptors *N*-acetyl glucosamine and *N*-acetyl 5-thio-gentosamine is a key substituent in determining regioselectivity.

Various deoxy-UDP-GlcNAc analogs have been studied as substrate donors for *N*-acetyl glucosaminyltransferases I and II (GnT-I and GnT-II) from human milk. Whereas all were accepted by GnT-I, deoxygenation of HO-3 completely abolished ability to act as a donor (SRIVASTAVA et al., 1990).

Analog studies for sialyltransferases have been limited to 9-substituted CMP-Neu-5-Ac (ITO et al., 1993) and to various disaccharide acceptor analogs where the 2-NHAc group has been replaced by a number of substituents such as azido, *O*-pivaloyl and other *N*-acyl substituents (BAISCH et al., 1996a, b). Sialyltransferase have also been used for the synthesis of Lewis x analogs (ICHIKAWA et al., 1992) bivalent (DEFREES et al., 1995) and clustered sialosides (SABESAN et al., 1991).

The substrate specificity of the Lewis x and a fucosyltransferases (α-1,3 and α-1,4) which transfers fucose to the internal GlcNAc residue in Galβ1-3/4GlcNAc seems to be particularly broad (Fig. 21). Apart from deoxygenated acceptor substrates (GOSSELIN and PALCIC, 1996), they can be either sialylated (WONG et al., 1995) or sulfated (AUGE et al., 1997). Within the donor substrate, a large substituent can be tolerated at the C-6 position, which has been used to transfer carbon-backbone elongated GDP-fucose derivatives and even oligosaccharides linked via a spacer to the C-6 position of GDP-fucose in one step to appropriate acceptor substrates (VOGEL et al., 1997).

3.7 Whole-Cell Glycosylation

As an alternative approach to *in vitro* synthesis of oligosaccharides and glycoconjugates is the use of whole-cell systems. In particular in the area of glycoprotein synthesis, the approach of "glycosylation engineering" by heterologous expression of specific glycosyltransferases such that the glycosylation pattern of the glycoproteins is changed, is an area of intense interest and has been reviewed elsewhere (STANLEY, 1992; VERBERT, 1996; BILL and FLITSCH, 1996).

In addition, recombinant whole cells, which had originally been developed as expression systems for glycosyltransferases, have been used directly as catalysts for the synthesis of oligosaccharides (HERRMANN et al., 1994; HERRMANN et al., 1995a). The advantage of such an approach is that the glycosyltransferases do not need to be isolated, although purification of the product from the crude cell mixtures is more difficult. Thus, *E. coli* XL1-Blue cells, harboring a plasmid expressing an α-1,2-mannosyltransferase expressed at a level of about 1 Unit per liter, were incubated with GDP-mannose and various mannose acceptor substrates (mannose, α-methyl-mannoside, Cbz-αManThr-OMe or Boc-Tyr-αManThr-Val-OMe) to give the corresponding α-1-2-

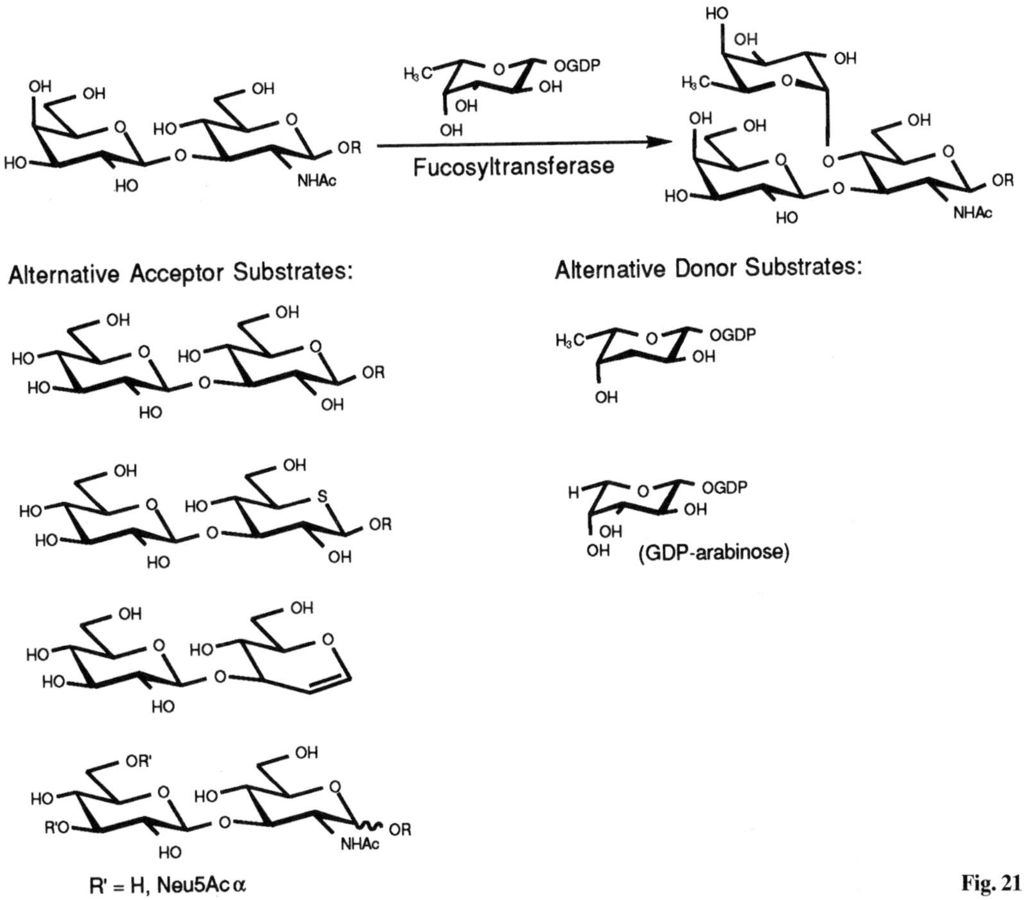

Fig. 21

mannosides in 42–75% isolated yields (HERRMANN et al., 1994). Similarly, a human β-1,4-galactosyltransferase expressed in yeast was used to synthesize *N*-acetyl lactosamine by incubation with UDP-Gal and *N*-acetyl glucosamine (HERRMANN et al., 1995a).

3.8 Glycosyltransferases Using Non-Nucleotide Donor Substrates

A number of glycosyltransferases involved in the biosynthetic formation of glycosidic linkages do not use sugar nucleotides as shown in Fig. 14 as donor substrates, but more simple glycosyl phosphates or glycosides. In contrast to the previously discussed transferases, which are sometimes called "Leloir glycosyltransferases" these "Non-Leloir glycosyltransferases" are involved both in oligosaccharide and polysaccharide synthesis to catalyze mostly reversible reactions. *In vitro* they have been used as biocatalysts for the synthesis of sucrose from glucose-1-phosphate and fructose using sucrose phosphorylase, of trehalose from glucose and glucose-1-phosphate, and for various polysaccharides (WONG et al., 1995).

Another important group of targets are cyclodextrins, which have been synthesized enzymatically using cyclodextrin glycosyltransferases from bacteria (BORNAGHI et al., 1996). Although polysaccharides such as starch and related malto-oligosaccharides are the natural glycosyl donor substrates, it has been shown that α-maltosyl fluoride (Glcα1-4Glcα1-F) can be incorporated into cyclodextrins. The substrate specificity of this enzyme system has been investigated, with the aim to synthesize new derivatives of cyclodextrins for supramolecular studies. It was found that modifications at C-6 of the α-maltosyl fluoride were tolerated by the system and that cyclothiomaltins could be prepared by using 4-thio-α-maltosyl fluoride as the activated disaccharide.

4 Combination of Glycosidases and Glycosyltransferases

Glycosyltransferases and glycosidases can of course be used in consecutive glycosylation reactions. This is particularly advantageous when a highly acceptor-specific transferase is used subsequently to a glycosidase-catalyzed step, which produces regioisomers. Thus, NILSSON has reported the synthesis of sialylated trisaccharides by firstly using a β-galactosidase from *E. coli* to obtain a mixture of β1-3- and β1-4-linked Gal-GlcNAc-OMe disaccharides. Since the β-D-galactoside 3-α-sialyltransferase from porcine submaxillary glands has a higher selectivity for Galβ1-3GlcNAc subsequent sialylation of the product mixture yielded selectively Neu-5-Acα2-3 Galβ1-3-GlcNAc-OMe (NILSSON, 1989).

A further avantage of using a combination of both enzymes is that the glycosidase-catalyzed reaction is reversible, but the transferase reaction is irreversible, such that yields can be improved by using one-pot synthesis (KREN and THIEM, 1995; HERRMANN et al., 1993).

5 Practical Aspects of Synthesis and Analysis

5.1 Isolation and Purification of Enzymes for Biocatalysis

Many of the glycosidases discussed in this chapter are commercially available and inexpensive and particularly the thermophilic enzyme libraries might provide a wide range of useful glycosidation biocatalysts in the future (LI and WANG, 1997). The number of glycosyltransferases available is still very limited and these enzymes need to be purified before use, but a number of reliable purification protocols from mammalian tissue have been published (PALCIC, 1994; SADLER et al., 1982). Affinity chromatography using immobilized nucleosides such as CDP-agarose for sialyltransferases has been successful (PAULSON et al., 1977). The amount of transferases that can be isolated from natural sources variies greatly. A α-2,6-sialyltransferase from porcine liver contains 40 Units per kg (AUGE et al., 1990) whereas 250 mL of human milk yielded about 20 milliunits of an α-1,3/4-fucosyltransferase (PALCIC, 1994).

Where glycosyltransferases are usually only found in very low abundance in natural sources, considerable effort has been spent on developing efficient overexpression systems. So far, a lot of the mammalian glycosyltransferases have required mammalian expression systems, such as those reported for sialyltransferases (GILLESPIE et al., 1992), fucosyltransferases (DEVRIES et al., 1995; DEVRIES et al., 1993; WONG et al., 1992; BAISCH et al., 1996a, b) and GlcNAc-transferases (D'AGOSTARO et al., 1995). More efficient and higher yielding expression was achieved with insect cells (BAISCH et al., 1996a, b) and yeast (MALISSARD et al., 1996; HERRMANN et al., 1995b; BORSIG et al., 1995). The yeast system provided 200 Units of the β-1,4-galactosyltransferase from a 290 L fermentation and 47 Units of α-2,6-sialyltransferase from a 150 L bioreactor.

E. coli expression at a level of about 1 Unit per liter in shake-flask cultures has been achieved for two mannosyltransferases α-1,2- (WANG et al., 1993) and β-1,4- (WATT et al., 1997). The β-1,4-mannosyltransferase yielded much better conversion, when an *N*-terminal hydrophobic peptide sequence was deleted, resulting in a more soluble enzyme.

5.2 Biocatalytic Conversions

Biocatalytic conversions with glycosidases and transferases have been reported for immobilized and soluble enzymes. Immobilization has the advantage that recovery of expensive enzyme and purification of products is easier, and some enzymes are more stable when immobilized (AUGE et al., 1984, 1986, 1990; WATT et al., 1997). However, most glycosidase- and transferase-catalyzed conversions have been performed with soluble enzyme preparations (PALCIC, 1994).

Glycosyltransferase-catalyzed reactions often suffer from product inhibition, in particular by the released nucleotide phosphates. A simple way of removing these products is by further hydrolysis using alkaline phosphatase (UNVERZAGT et al., 1990). Alternatively, cofactor recycling systems discussed earlier keep the nucleotide phosphate concentration low, such that product inhibition is not a problem.

Although the majority of transformations using glycosyltransferases have been reported on a milligram scale, scale-up to kilogram quantities of sialyl Lewis x has been achieved by Cytel Co., California, for tests as a drug candidate for the treatment of reperfusion tissue injury (WONG et al., 1995). Scale-up should in future also be facilitated by the use of enzyme membrane reactors, which allow the use of soluble enzymes with tight control on cofactor and product concentrations. Such a system has been used for the synthesis of sialic acid on a 15 g scale (KRAGL et al., 1991).

5.3 Analysis of Products

Since glycosyltransferases are frequently only available in small quantities, initial analysis of potential substrates and optimization of reaction conditions can make convenient use of radiolabeled sugar–nucleotide donor substrates, which are labeled in the sugar and are commercially available. The assays then involve a separation protocol in which labeled glycosylated product can be effectively separated from unreacted donor substrate and any side products, such as those resulting from hydrolysis of substrates. If the acceptor substrate is a lipid, simple partition into organic solvent can be used for separation (WILSON et al., 1995). Otherwise, paper electrophoresis, thin layer chromatography, or other chromatographic methods need to be employed which can be complicated and lengthy. A general method for the assay of glycosyltransferase activity makes use of synthetic glycoside acceptors attached to hydrophobic aglycones (such as $(CH_2)_8COOMe$ glycosides) which allow rapid adsorption of the radiolabeled product on to reverse-phase C-18 cartridges (PALCIC et al., 1988).

Alternative assay methods relying on non-radioactive methods have also been published (KREN and THIEM, 1997). If glycoproteins or glycopeptides are synthesized, assays based on specific lectins are very useful (CLARK et al., 1990).

A major problem with using conventional HPLC techniques for analysis of non-radioactive oligosaccharide products is the lack of suitable chromophores for UV detection. Thus, refractive index detection (YAMASHITA et al., 1982) or pulsed amperometry (WILLENBROCK et al., 1991) can be used, which are, however, not always compatible with all solvent systems and presence of cofactors. Very recently, quantitative electrospray mass spectrometry was used for screening potential inhibitors against galactosyltransferase by measuring the amounts of benzyl *N*-acetyllactosamine formed in the presence of inhibitors. This promises to be a very fast method for enzyme assays in the future (WU et al., 1997).

5.4 Chromatographic Methods for the Isolation of Products

The reaction products of the glycosylation reactions described here are mostly highly water-soluble compounds, which need to be isolated from unreacted starting materials and other side products in addition to enzymes, cofactors, and buffers after the biotransformation. In particular, glycosidase-catalyzed glycosylations are frequently contaminated with large amounts of starting materials, which are used in excess and are present due to overall low yields. Hence purification using inexpensive chromatography such as activated carbon chromatography (AJISAKA et al., 1995) or carbon-celite chromatography (SINGH et al., 1995) has been useful as a purification step for such crude mixtures.

Since glycosyltransferase-catalyzed reactions tend to be better yielding and highly selective, separation of isomeric mixtures of oligosaccharides is not required and the purification from enzymes, buffer, salts, and starting material is often straightforward. HPLC separation, gel filtration, ion exchange chromatography or gel permeation chromatography (YAMASHITA et al., 1982) are the most widely used techniques for oligosaccharide isolation, in particular on a milligram scale.

6 Conclusions

Both natural and unnatural oligosaccharides and glycoconjugates are accessible by synthetic enzymatic methods using either glycosidases or glycosyltransferases. Both enzyme systems are highly stereospecific and often substrate-specific, such that each particular glycosidic linkage requires its own biocatalyst. So far, there is no known general biocatalyst for formation of glycosidic linkages and, given the high degree of stereo- and regioselectivity required, such a catalyst would have limited use, given that oligosaccharide biosynthesis does not involve a template as is available for oligonucleotide and polypeptide biosynthesis. A great deal of current research is, therefore, focused on finding glycosylation catalysts with new stereo-, regio-, and substrate selectivities, and on then investigating their substrate specificity, such that eventually perhaps a complete set of catalysts for each glycosidic linkage will become available. It should be kept in mind that natural sources might provide only limited activities, since of the thousands of possible oligosaccharide isomers (if one assumes every possible permutation of stereo- and regioselectivity) only a small subset actually occur in natural systems. A very interesting area of research is, therefore, the generation of new biocatalysts using either rational site-specific (SETO et al., 1997) or random mutagenesis techniques combined with directed evolution (MOORE and ARNOLD, 1996).

7 References

AJISAKA, K., FUJIMOTO, H. (1989), Regioselective synthesis of trisaccharides by use of a reversed hydrolysis activity of α- and β-galactosidase, *Carbohydr. Res.* **185**, 139–146.

AJISAKA, K., FUJIMOTO, H. et al. (1988), Enzymic synthesis of disaccharides by use of the reversed hydrolysis activity of β-galactosidases, *Carbohydr. Res.* **180**, 35–42.

AJISAKA, K., MATSUO, I. et al. (1995), Enzymatic synthesis of mannobiosides and mannotriosides by reverse hydrolysis using α-mannosidase from *Aspergillus niger*, *Carbohydr. Res.* **270**, 123–130.

AUGE, C., DAVID, S. et al. (1984), Synthesis with immobilized enzymes of two trisaccharides, one of them active as the determinant of a stage antigen, *Tetrahedron Lett.* **25**, 1467–1470.

AUGE, C., MATHIEU, C. et al. (1986), The use of an immobilized cyclic multi-enzyme system to synthesize branched penta- und hexa-saccharides associated with blood-group I epitopes, *Carbohydr. Res.* **151**, 147–156.

AUGE, C., FERNANDEZ-FERNANDEZ, R. et al. (1990), The use of immobilized glycosyltransferases in the synthesis of sialo-oligosaccharides, *Carbohydr. Res.* **200**, 257–268.

AUGE, C., DAGRON, F. et al. (1997), Synthesis of sulfated derivatives as sialyl Lewis a and sialyl Lewis x analogs, in: *Carbohydrate Mimics: Concepts and Methods* (CHAPLEUR, Y., Ed.), pp. 365–383. Weinheim: WILEY-VCH Verlag.

BAISCH, G., OHRLEIN, R. et al. (1996a), Enzymatic fucosylation of non-natural trisaccharides with cloned fucosyltransferase VI. *Bioorg. Med. Chem. Lett.* **6**, 759–762.

BAISCH, G., OHRLEIN, R. et al. (1996b), Enzymatic α(2-3)-sialylation of non-natural oligosaccharides with cloned sialyltransferase, *Bioorg. Med. Chem. Lett.* **6**, 755–758.

BILL, R. M., FLITSCH, S. L. (1996), Chemical and biological approaches to glycoprotein synthesis, *Chem. Biol.* **3**, 145–149.

BJORKLING, F., GODTFREDSEN, S. E. (1988), New enzyme catalyzed synthesis of monoacyl galactoglycerides, *Tetrahedron* **44**, 2957–2962.

BLIXT, O., NORBERG, T. (1997), Solid-phase enzymatic synthesis of a Lewis a trisaccharide using an acceptor reversibly bound to sepharose, *J. Carbohydr. Chem.* **16**, 143–154.

BOONS, G.-J. (1996), Strategies in oligosaccharide synthesis, *Tetrahedron* **52**, 1095–1121.

BORNAGHI, L., UTILLE, J.-P. et al. (1996), Enzymic synthesis of cyclothiomaltins, *J. Chem. Soc., Chem. Commun.*, 2541–2542.

BORSIG, L., IVANOV, S. X. et al. (1995), Scaled-up expression of human α-2,6(*N*)sialyltransferase in

Saccharomyces cerevisiae, *Biochim. Biophys. Res. Commun.* **210**, 14–20.

Bouhours, D., Bouhours, J.-F. (1991), Genetic polymorphism of rat liver gangliosides, *J. Biol. Chem.* **266**, 12944–12948.

Brockhausen, I., Carver, J. P. et al. (1988), Control of glycoprotein biosynthesis. The use of oligosaccharide substrates and HPLC to study the sequential pathway for *N*-acetyl glucosaminyltransferases I, II, III, IV, V, and VI in the biosynthesis of highly branched *N*-glycans by hen oviduct membranes, *Biochem. Cell Biol.* **66**, 1134–1151.

Brockhausen, I., Moller, G. et al. (1992), Control of glycoprotein-synthesis – characterization of (1-]4)-*N*-acetyl-β-D-glucosaminyltransferases acting on the α-D-(1-]3) *O*-linked and α-D-(1-]6)-linked arms of *N*-linked oligosaccharides, *Carbohydr. Res.* **236**, 281–299.

Brown, R. T., Carter, N. E. et al. (1995), Synthesis of morphine-5-glucuronide via a highly selective enzyme-catalyzed hydrolysis reaction, *Tetrahedron Lett.* **36**, 1117–1120.

Cantacuzene, D., Attal, S. et al. (1991), Stereospecific chemoenzymatic synthesis of galactopyranosyl-L-serine, *Bioorg. Med. Chem. Lett.* **1**, 197–200.

Clark, R. S., Banerjee, S. et al. (1990), Yeast oligosaccharyltransferase: glycosylation of peptide substrates and chemical characterization of the glycopeptide product, *J. Org. Chem.* **55**, 6275–6285.

Clausen, H., Bennett, E. P. (1996), A family of UDP-GalNAc: polypeptide *N*-acetylgalactosaminyl-transferases control the initiation of mucin-type *O*-linked glycosylation, *Glycobiology* **6**, 635–646.

Crout, D. H. G., MacManus, D. A. (1990), Enzymatic synthesis of glycosides using the β-galactosidase of *Escherichia coli*: regio- and stereo-chemical studies, *J. Chem. Soc., Perkin Trans. I*, 1865–1868.

D'Agostaro, G. A. F., Zingoni, A. et al. (1995), Molecular cloning and expression of cDNA encoding the rat UDP-*N*-acetyl glucosaminyltransferase II, *J. Biol. Chem.* **270**, 15211–15221.

DeFrees, S. A., Kosch, W. et al. (1995), Ligand recognition by E-selectin: synthesis, inhibitory activity, and conformational analysis of bivalent sialyl Lewis x analogs, *J. Am. Chem. Soc.* **117**, 66–79.

Devries, T., Norberg, T., Lonn, H., Van den Eynden, D. H. (1993), The use of human milk fucosyltransferase in the synthesis of tumor-associated trimeric X determinants. *Eur. J. Biochem.* **216**, 769–777.

Devries, T., Srnka, C. et al. (1995), Acceptor specificity of different length constructs of human recombinant α-1,3/4-fucosyltransferases – replacement of the stem region and the transmembrane domain of fucosyltransferase-V by protein-A results in an enzyme with GDP-fucosyl hydrolysing activity, *J. Biol. Chem.* **270**, 8712–8722.

Fan, J.-Q., Takegawa, K. et al. (1995), Enhanced transglycosylation activity of *Arthrobacter protophormiae endo-β-N*-acetyl glucosaminidase in media containing organic solvents, *J. Biol. Chem.* **270**, 17723–17729.

Feizi, T. (1993), Oligosaccharides that mediate mammalian cell–cell adhesion, *Curr. Opin. Struct. Biol.* **3**, 701–710.

Field, M. C., Wainwright, L. J. (1995), Molecular cloning of eukaryotic glycoprotein and glycolipid glycosyltransferases: a survey, *Glycobiology* **5**, 463–472.

Flitsch, S. L., Watt, G. M. (1997), *Chemical Synthesis of Glycoprotein Glycans. Glycopeptides and Related Compounds* (Large, D. C., Warren, C. D., Eds.), pp. 207–244. New York: Marcel Dekker.

Gais, H.-J., Zeissler, A. et al. (1988), Diastereoselective D-galactopyranosyl transfer to *meso*-diols catalyzed by β-galactosidases, *Tetrahedron Lett.* **29**, 5743–5744.

Gelo-Pujic, M., Guibe-Jampel, E. et al. (1997), Enzymatic glycosidation in dry media under microwave irradiation, *J. Chem. Soc., Perkin Trans. I*, 1001–1002.

Gijsen, H. J. M., Qiao, L. et al. (1996), Recent advances in the chemo-enzymatic synthesis of carbohydrates and carbohydrate mimetics, *Chem. Rev.* **96**, 443–473.

Gillespie, W., Kelm, S. et al. (1992), Cloning and expression of the Galb1-3GalNAc α-2,3-sialyltransferase, *J. Biol. Chem.* **267**, 21004–21010.

Gosselin, S., Palcic, M. M. (1996), Acceptor hydroxyl group mapping for human milk α1-3 and α1-3/4 fucosyltransferases, *Bioorg. Med. Chem.* **4**, 2023–2028.

Guilbert, B., Flitsch, S. L. (1994), A short chemoenzymic route to glycosphingolipids using soluble glycosyltransferases, *J. Chem. Soc., Perkin Trans. I*, 1181.

Guilbert, B., Khan, T. H. et al. (1992), Chemo-enzymatic synthesis of a glycosphingolipid, *J. Chem. Soc., Chem. Commun.*, 1526–1527.

Gygax, D., Spies, P. et al. (1991), Enzymatic synthesis of β-D-glucuronides with *in situ* regeneration of uridine-5′-diphosphoglucuronic acid, *Tetrahedron* **47**, 5119–5123.

Hakomori, S.-I. (1990), Bifunctional role of glycosphingolipids, *J. Biol. Chem.* **265**, 18713–18716.

Halcomb, R. L., Huang, H. et al. (1994), Solution- and solid-phase synthesis of inhibitors of *H. pylori* attachment and E-selectin-mediated leukocyte adhesion, *J. Am. Chem. Soc.* **116**, 11315–11322.

Heidelberg, T., Thiem, J. (1997), Synthesis of novel amino acid glycoside conjugates, *Carbohydr. Res.* **301**, 145–153.

Herrmann, G., Ichikawa, Y. et al. (1993), A new multi-enzyme system for a one-pot synthesis of sialyl oligosaccharides: combined use of β-galactosidase and α(2,6)sialyltransferase coupled with regeneration *in situ* of CMP-sialyc acid, *Tetrahedron Lett.* **34**, 3091–3094.

HERRMANN, G., WANG, P. et al. (1994), Recombinant whole cells as catalysts for the enzymatic synthesis of oligosaccharides and glycopeptides, *Angew. Chem.* (Int. Edn. Engl.) **33**, 1241–1242.

HERRMANN, G. F., ELLING, L. et al. (1995a), Use of transformed whole yeast cells expressing β-1,4-galactosyltransferase for the synthesis of *N*-acetyl lactosamine, *Bioorg. Med. Chem. Lett.* **5**, 673–676.

HERRMANN, G. F., KREZDORN, C. et al. (1995b), Large-scale production of a soluble human β-1,4-galactosyltransferase using a *Saccharomyces cerevisiae* expression system, *Protein Expr. Purific.* **6**, 72–78.

ICHIKAWA, Y., LIN, Y.-C. et al. (1992), Chemical-enzymatic synthesis and conformational analysis of sialyl Lewis-x and derivatives, *J. Am. Chem. Soc.* **114**, 9283–9290.

ICHIKAWA, Y., WANG, R. et al. (1994), Regeneration of sugar nucleotide for enzymatic oligosaccharide synthesis, *Methods Enzymol.* **247**, 107–127.

IMPERIALI, B. (1997), Protein glycosylation: the clash of the titans, *Acc. Chem. Res.* **30**, 452–459.

IMPERIALI, B., HENRICKSON, T. L. (1995), Asparagine-linked glycosylation: specificity and function of oligosaccharyltransferase, *Bioorg. Med. Chem. Lett.* **3**, 1565–1578.

ITO, Y., GAUDINO, J. et al. (1993), Synthesis of bioactive sialosides, *Pure Appl. Chem.* **65**, 753–762.

KAJIHARA, Y., ENDO, T. et al. (1995), Enzymic transfer of 6-modified D-galactosyl residues: synthesis of biantennary penta- and heptasaccharides having two 6-deoxy-D-galactose residues at the non-reducing end and evaluation of 6-deoxy-D-galactosyl transfer to glycoprotein using bovine β-(1,4)-galactosyltransferase and UDP-6-deoxy-D-galactose, *Carbohydr. Res.* **269**, 273–294.

KARLSSON, K.-A. (1989), Animal glycosphingolipids as membrane attachment sites for bacteria, *Annu. Rev. Biochem.* **58**, 309–350.

KAUR, K. J., ALTON, G. et al. (1991), Use of *N*-acetyl glucosaminyltransferases I and II in the preparative synthesis of oligosaccharides, *Carbohydr. Res.* **210**, 145–153.

KOBAYASHI, S., KASHIWA, K. et al. (1991), Novel method for polysaccharide synthesis using an enzyme: the first *in vitro* synthesis of cellulose via a nonbiosynthetic path utilizing cellulase as catalyst, *J. Am. Chem. Soc.* **113**, 3079–3084.

KOBAYASHI, S., KIYOSADA, T. et al. (1997), A novel method for synthesis of chitobiose via enzymatic glycosylation using a sugar oxazoline as glycosyl donor, *Tetrahedron Lett.* **38**, 2111–2112.

KOEPPER, S. (1994), Polymer-supported enzymic synthesis on a preparative scale, *Carbohydr. Res.* **265**, 161–166.

KORNFELD, R., KORNFELD, S. (1985), Assembly of asparagine-linked oligosaccharides, *Ann. Rev. Biochem.* **54**, 631–664.

KRAGL, U., GYRAX, D. et al. (1991), Enzymatic two-step synthesis of *N*-acetyl neuraminic acid in the enzyme membrane reactor, *Angew. Chem.* (Int. Edn. Engl.) **30**, 827–828.

KREN, V., THIEM, J. (1995), A multienzyme system for one-pot synthesis of sialyl T-antigen, *Angew. Chem.* (Int. Edn. Engl.) **34**, 893–895.

KREN, V., THIEM, J. (1997), Simple and non-radioactive method for determination of sialyltransferase activity, *Biotechnol. Tech.* **11**, 323–326.

LASKY, L. A. (1994), A sweet success, *Nature Struct. Biol.* **1**, 139–141.

LEE, J., COWARD, J. K. (1993), Oligosaccharyltransferase: synthesis and use of deuterium-labeled peptide substrates as mimetic probes, *Biochemistry* **32**, 6794–6801.

LI, J., WANG, P. G. (1997), Chemical and enzymatic synthesis of glycoconjugates. 2. High yielding regioselective synthesis of *N*-acetyl lactosamine by use of recombinant thermophilic glycosidase library, *Tetrahedron Lett.* **38**, 7967–7970.

LIU, Y.-L., HOOPS, G. C. et al. (1994), A comparison of proteins and peptides as substrates for microsomal and solubilized oligosaccharyltransferase, *Bioorg. Med. Chem.* **2**, 1133–1141.

MALISSARD, M., BORSIG, L. et al. (1996), Recombinant soluble β-1,4-galactosyltransferases expressed in *Saccharomyces cerevisiae* – purification, characterization and comparison with human enzyme, *Eur. J. Biochem.* **239**, 340–348.

MELDAL, M., AUZANNEAU, F.-I. et al. (1994), A PEGA resin for use in the solid-phase chemical-enzymatic synthesis of glycopeptides, *J. Chem. Soc., Chem. Commun.*, 1849–1850.

MELKERSON-WATSON, L. J., SWEELEY, C. C. (1991), Purification to apparent homogeneity by immunoaffinity chromatography and partial characterization of the GM3 ganglioside-forming enzyme, CMP-sialic acid:lactosylceramide α2,3-sialyltransferase (SAT-1), from rat liver Golgi, *J. Biol. Chem.* **266**, 4448–4457.

MOORE, J. C., ARNOLD, F. H. (1996), Directed evolution of a *para*-nitrobenzyl esterase for aqueous-organic solvents, *Nature Biotechnology* **14**, 458–467.

NILSSON, K. G. I. (1987), A simple strategy for changing the regioselectivity of glycosidase-catalyzed formation of disaccharides, *Carbohydr. Res.* **167**, 95–103.

NILSSON, K. G. I. (1988), A simple strategy for changing the regioselectivity of glycosidase-catalyzed formation of disaccharides: Part II. Enzymic synthesis *in situ* of various acceptor glycosides, *Carbohydr. Res.* **180**, 53–59.

NILSSON, K. G. I. (1989), Enzymic synthesis of di- and trisaccharide glycosides, using glycosidases and β-

D-galactoside 3-α-sialyltransferase, *Carbohydr. Res.* **188**, 9–17.

Nilsson, K. G. I. (1997), Glycosidase-catalyzes synthesis of di- and trisaccharide derivatives related to antigens involved in the hyperacute rejection of xenotransplants, *Tetrahedron Lett.* **38**, 133–136.

Nilsson, K. G. I., Ljunger, G. et al. (1997), Glycosidase-catalyzed synthesis of glycosylated amino acids: Synthesis of GalNAc α-Ser and GlcNAc β-Ser derivatives, *Biotechn. Lett.* **19**, 889–892.

Nishida, Y., Wiemann, T. et al. (1993a), A new type of galactosyltransferase reaction: transfer of galactose to the anomeric position of *N*-acetyllactosamine, *J. Am. Chem. Soc.* **115**, 2536–2537.

Nishida, Y., Wiemann, T. et al. (1993b), Extension of the β-Gal1,1,-transfer to *N*-acetyl-5-thio-gentosamine by galactosyltransferase, *Tetrahedron Lett.* **34**, 2905–2906.

Nishimura, S.-I., Matsuoka, K. et al. (1994), Chemoenzymatic oligosaccharide synthesis on a soluble polymeric carrier, *Tetrahedron Lett.* **35**, 5657–5660.

Nishimura, S., Yamada, K. (1997), Transfer of ganglioside GM3 oligosaccharide from a water soluble polymer to ceramide by ceramide glycanase. A novel approach for the chemical-enzymatic synthesis of glycosphingolipids, *J. Am. Chem. Soc.* **119**, 10555–10556.

Ooi, Y., Hashimoto, T. et al. (1984), Enzymic synthesis of chemically unstable cardiac glycosides by β-galactosidase from *Aspergillus oryzae*, *Tetrahedron Lett.* **25**, 2241–2244.

Palcic, M. M. (1994), Glycosyltransferases in glycobiology, *Methods Enzymol.* **230**, 300–360.

Palcic, M. M., Heerze, L. D. et al. (1988), The use of hydrophobic synthetic glycosides as acceptors in glycosyltransferase assays, *Glycoconjugate J.* **5**, 49–63.

Palcic, M. M., Hindsgaul, O. (1991), Flexibility in the donor substrate specificity of β1,4-galactosyltransferase: application in the synthesis of complex carbohydrates, *Glycobiology* **1**, 205–209.

Parekh, R. B., Edge, C. J. (1994), Selectins – glycoprotein targets for therapeutic intervention in inflammation, *TIBTECH* **12**, 339–345.

Paulsen, H. (1982), Advances in selective chemical syntheses of complex oligosaccharides, *Angew. Chem.* (Int. Edn. Engl.) **21**, 155–173.

Paulsen, J. C., Rodgers, G. N. (1987), Resialylated erythrocytes for assessment of the specificity of sialooligosaccharide binding proteins, *Methods Enzymol.* **138**, 162–168.

Paulson, J. C., Beranek, W. E. et al. (1977), Purification of a sialyltransferase from bovine colostrum by affinity chromatography on CDP-agarose, *J. Biol. Chem.* **252**, 2356–2362.

Preuss, U., Gu, X. et al. (1993), Purification and characterization of CMP-*N*-acetylneuraminic acid:lactosylceramide (α2-3) sialyltransferase (GM3-synthase) from rat brain, *J. Biol. Chem.* **268**, 26273–26278.

Rearick, J. I., Sadler, J. E. et al. (1979), Enzymatic characterization of β-D-galactoside α2-3 sialyltransferase from porcine submaxillary gland, *J. Biol. Chem.* **254**, 4444–4451.

Sabesan, S., Duus, J. et al. (1991), Synthesis of cluster sialoside inhibitors for influenza virus, *J. Am. Chem. Soc.* **113**, 5865–5866.

Sadler, J. E., Beyer, T. A. et al. (1982), Purification of mammalian glycosyltransferases, *Methods Enzymol.* **83**, 458–514.

Sauerbrei, B., Thiem, J. (1992), Galactosylation and glucosylation by use of β-galactosidase, *Tetrahedron Lett.* **33**, 201–204.

Schachter, H. (1991), Enzymes associated with glycosylation, *Curr. Opin. Struct. Biol.* **1**, 755–765.

Schmidt, R. R. (1991), Synthesis of glycosides. Comprehensive organic synthesis, **6**, 33–63.

Schmidt, R. R., Kinzy, W. (1994), Anomeric-oxygen activation for glycoside synthesis. The trichloroacetimidate method, *Adv. Carbohydr. Chem. Biochem.* **50**, 21–123.

Schuster, M., Wang, P. et al. (1994), Solid-phase chemical-enzymatic synthesis of glycopeptides and oligosaccharides, *J. Am. Chem. Soc.* **116**, 1135–1136.

Seitz, O., Wong, C.-H. (1997), Chemoenzymatic solution- and solid-phase synthesis of *O*-glycopeptides of the mucin domain of MAdCAM-1. A general route to *O*-LacNAc, *O*-sialyl-LacNAc and *O*-sialyl-Lewis-X peptides, *J. Am. Chem. Soc.* **119**, 8766–8776.

Seto, N. O. L., Palcic, M. M. et al. (1997), Sequential interchange of four amino acids from blood group B to blood group A glycosyltransferase boosts catalytic activity and progressively modifies substrate recognition in human recombinant enzymes, *J. Biol. Chem.* **272**, 14133–14138.

Singh, S., Packwood, J. et al. (1995), Glycosidase-catalyzed oligosaccharide synthesis: preparation of *N*-acetyl chitooligosaccharides using the β-*N*-acetyl hexosaminidase from *Aspergillus oryzae*, *Carbohydr. Res.* **279**, 293–305.

Singh, S., Scigelova, M. et al. (1996), Glycosidase-catalyzed synthesis of oligosaccharides: a two-step synthesis of the core trisaccharide of *N*-linked glycoproteins using the β-*N*-acetyl hexosaminidase and the β-mannosidase from *Aspergillus oryzae*, *Chem. Commun.*, 993–994.

Srivastava, G., Alton, G. et al. (1990), Combined chemical-enzymic synthesis of deoxygenated oligosaccharide analogs: transfer of deoxygenated D-GlcpNAc residues from their UDP-GlcpNAc derivatives using *N*-acetyl glucosaminyltransferase I, *Carbohydr. Res.* **207**, 259–276.

Stanley, P. (1992), Glycosylation engineering, *Glycobiology* **2**, 99–107.

TURNER, N. J., WEBBERLEY, M. C. (1991), Stereospecific attachment of carbohydrates to amino acid derivatives using β-glucosidase and β-xylosidase. *J. Chem. Soc., Chem. Commun.*, 1349–1350.

UNVERZAGT, C. (1996), Chemoenzymatic synthesis of a sialylated undecasaccharide–asparagine conjugate, *Angew. Chem.* (Int. Edn. Engl.) **35**, 2350–2353.

UNVERZAGT, C., KUNZ, H. et al. (1990), High-efficiency synthesis of sialyloligosaccharides and sialoglycopeptides, *J. Am. Chem. Soc.* **112**, 9308–9309.

UNVERZAGT, C., KELM, S. et al. (1994), Chemical and enzymatic synthesis of multivalent sialoglycopeptides, *Carbohydr. Res.* **251**, 285–301.

VERBERT, A. (1996), The fascinating challenge to mimic nature in producing recombinant glycoproteins, *Chimica Oggi*, 9–14.

VIC, G., THOMAS, D. (1992), Enzyme-catalyzed synthesis of alkyl β-D-glucosides in organic media, *Tetrahedron Lett.* **33**, 4567–4570.

VIC, G., CROUT, D. H. G. (1994), Synthesis of glucosidic derivatives with a spacer arm by reverse hydrolysis using almond β-D-glucosidase, *Tetrahedron: Asymmetry* **5**, 2513–2516.

VIC, G., CROUT, D. H. G. (1995), Synthesis of allyl and benzyl β-D-glucopyranosides and allyl β-D-galactopyranoside from D-glucose or D-galactose and the corresponding alcohol using almond β-D-glucosidase, *Carbohydr. Res.* **279**, 315–319.

VIC, G., HASTINGS, J. J. et al. (1996), Glycosidase-catalyzed synthesis of glycosides by an improved procedure for reverse hydrolysis: Application to the chemoenzymatic synthesis of galactopyranosyl-(1-4)-*O*-α-galactopyranoside derivatives, *Tetrahedron: Asymmetry* **7**, 1973–1984.

VILADOT, J.-L., MOREAU, V. et al. (1997), Transglycosylation activity of *Bacillus* 1,3-1,4-β-D-glucan 4-glucanohydrolases. Enzymic synthesis of alternate 1,3-1,4-β-D-gluco-oligosaccharides, *J. Chem. Soc., Perkin Trans.* **I 16**, 2383–2387.

VOGEL, C., BERGEMANN, C. et al. (1997), Synthesis of carbon-backbone-elongated GDP-L-fucose derivatives as substrates for fucosyltransferase-catalyzed reactions, *Liebigs Ann. Rec.* **3**, 601–612.

WANG, P., SHEN, G.-J. et al. (1993), Enzymes in oligosaccharide synthesis: active-domain overproduction, specificity study, and synthetic use of an α-1,2-mannosyltransferase with regeneration of GDP-mannose, *J. Org. Chem.* **58**, 3985–3990.

WANG, L. X., TANG, M. et al. (1997), Combined chemical and enzymatic synthesis of a *C*-glycopeptide and its inhibitory activity towards glycoamidases, *J. Am. Chem. Soc.* **119**, 11137–11146.

WATT, G. M., REVERS, L. et al. (1997), Efficient enzymatic synthesis of the core trisaccharide of *N*-glycans using a recombinant β-mannosyltransferase. *Angew. Chem.* (Int. Edn. Engl.) **36**, 2354–2356.

WIEMANN, T., TAUBKEN, N. et al. (1994), Enzymic synthesis of *N*-acetyl lactosamine on a soluble light-sensitive polymer, *Carbohydr. Res.* **257**, C1–C6.

WILLENBROCK, F. W., NEVILLE, D. C. A. et al. (1991), The use of HPLC-pulsed amperometry for the characterization an assay of glycosidases and glycosyltransferases, *Glycobiology* **1**, 223–227.

WILSON, I. B. H., WEBBERLEY, M. C., REVERS, L., FLITSCH, S. L. (1995), Dolichol is not a necessary moiety in lipid-linked oligosaccharide substrates for the mannosyltransferases involved in *N*-linked oligosaccharide biosynthesis, *Biochem. J.* **310**, 909–916.

WITTE, K., SEARS, P. et al. (1997), Enzymatic glycoprotein synthesis: Preparation of ribonuclease glycoforms via enzymatic glycopeptide condensation and glycosylation, *J. Am. Chem. Soc.* **119**, 2114–2118.

WONG, C.-H., HAYNIE, S. L. et al. (1982), Enzyme-catalyzed synthesis of *N*-acetyl lactosamine with *in situ* regeneration of uridine 5′-diphosphate glucose and uridine 5′-diphosphate galactose, *J. Org. Chem.* **47**, 5416–5418.

WONG, C.-H., ICHIKAWA, Y. et al. (1991), Probing the acceptor specificity of β-1,4-galactosyltransferase for the development of enzymatic synthesis of novel oligosaccharides, *J. Am. Chem. Soc.* **113**, 8137–8145.

WONG, C., DUMAS, D. et al. (1992), Specificity inhibition and synthetic utility of a recombinant human α-1,3-fucosyl-transferase. *J. Am. Chem. Soc.* **114**, 7321–7322.

WONG, C., HALCOMB, R. L. et al. (1995), (Part 2) Enzymes in organic synthesis: Application to the problems of carbohydrate recognition, *Angew. Chem.* (Int. Edn. Engl.) **34**, 521–546.

WU, J., TAKAYAMA, S. et al. (1997), Quantitative electrospray mass spectrometry for the rapid assay of enzyme inhibitors, *Chem. Biol.* **4**, 653–657.

XU, T., COWARD, J. K. (1997), C-13 and N-15-labeled peptide substrates as mechanistic probes of oligosaccharyltransferase, *Biochemistry* **36**, 14683–14689.

YAMADA, K., NISHIMURA, S.-I. (1995), An efficient synthesis of sialoglycoconjugates on a peptidase-sensitive polymer support, *Tetrahedron Lett.* **36**, 9493–9496.

YAMASHITA, K., MIZUOCHI, T. et al. (1982), Analysis of oligosaccharides by gel filtration, *Methods Enzymol.* **83**, 105–124.

YANG, W., ABERNETHY, J. L. et al. (1992), Purification and characterization of a UDP-GalNAc-polypeptide *N*-acetyl galactosaminyltransferase specific for the glycosylation of threonine residues, *J. Biol. Chem.* **267**, 12709–12716.

ZEHAVI, U., HERCHMAN, M. et al. (1990), Enzymic glycosphingolipid synthesis on polymer supports. II. Synthesis of lactosyl ceramide, *Glycoconjugate J.* **7**, 229–234.

ZELLER, C. B., MARCHASE, R. B. (1992), Gangliosides as modulators of cell function, *Am. J. Physiol.* **262**, C1341–1355.

Applications

6 Industrial Biotransformations

UWE T. BORNSCHEUER

Greifswald, Germany

1 Introduction

In the last two decades, the application of biocatalysts in industry has considerably increased. This is due to a number of reasons, of which the most important ones might be the better availability of enzymes in large quantities (mainly due to the progress in genetic engineering) and an increasing demand for enantiomerically pure compounds for pharmaceuticals/agrochemicals due to changes in legislation. Furthermore, organic chemists more and more accept biocatalysts as a useful alternative to chemical catalysts. The recent development in the field of directed evolution (STEMMER, 1994; ARNOLD, 1998; BORNSCHEUER, 1998; REETZ and JAEGER, 1999) might further increase the availability of suitable biocatalysts.

1.1 Basic Considerations

The successful application of biotransformations in an industrial environment depends on a number of factors, such as

- availability (and price) of suitable enzymes or microorganisms,
- up- and downstream processing,
- competition with (in-house) chemical methods/established processes,
- time requirements for process development,
- wastewater treatment, solvent disposal,
- legislation/approval of the process or product,
- public perception.

All these factors determine whether a biocatalytic route is superior to a chemical one and the final decision has to be made case-by-case. ROZZELL (1999) pointed out five commonly held ideas (myths) about enzymes (too expensive, too unstable, productivity is low, redox factors cannot be recycled, they do not catalyze industrially interesting reactions). Indeed, the examples shown in his publication as well as those shown here, clearly demonstrate that these myths are rather misconceptions.

In the following survey, some examples for the industrial use of a wide range of biocatalysts are given. They are organized by product not by enzyme or microorganism used, because often several biocatalytic routes have been developed for the same target. Most examples are taken from the recent literature. For further industrialized processes readers are referred to a number of excellent books (COLLINS et al., 1992, 1997; SHELDON, 1993).

It should be pointed out that it is very difficult to provide a complete overview, because information about commercialized processes is hard to get. Moreover, even if details are revealed in the (patent) literature, processes might have changed in the meantime.

2 Enantiomerically Pure Pharmaceutical Intermediates

The worldwide sales of single-enantiomer drugs are increasing dramatically, e.g., from 1996 to 1997 by 21% to almost 90 billion US$. Of the top 100 drugs, 50 are marketed as single enantiomers (STINSON, 1998). As a consequence, the need for enantioselective technologies is also increasing, and companies as well as academia are making considerable efforts to develop efficient synthetic routes including biocatalysis and biotransformations.

A considerable proportion of successful examples for the enzymatic synthesis of optically pure intermediates is based on the action of lipases. This is not surprising keeping in mind that lipases probably represent the most easy-to-use enzymes for organic synthesis. They do not require any cofactors, a wide range of enzymes from various sources are commercially available, and they are active and stable in organic solvents. In contrast to many other enzymes, they accept a huge range of non-natural substrates, which are often converted with high stereoselectivity. Structures of 12 lipases have been elucidated, and this might make the design of tailor-made biocatalysts feasible. Properties and applications of lipases are documented in several thousand publications, and readers are referred to a number of recent reviews (Jaeger and Reetz, 1998; Schmid and Verger, 1998), book sections (Kazlauskas and Bornscheuer, 1998; Bornscheuer and Kazlauskas, 1999), or books (Woolley and Petersen, 1994) for further reading.

2.1 Diltiazem

Both DSM-Andeno (Netherlands) and Tanabe Pharmaceutical (Osaka, Japan) in collaboration with Sepracor (Marlborough, MA) have commercialized lipase-catalyzed resolutions of (+)-(2*S*,3*R*)-*trans*-3-(4-methoxy-phenyl)-glycidic acid methylester (MPGM), a key precursor to diltiazem (Hulshof and Roskam, 1989; Matsumae et al., 1993, 1994; Furui et al., 1996). The DSM-Andeno process uses lipase from *Rhizomucor miehei* (RML), while the Tanabe process uses a lipase secreted by *Serratia marcescens* Sr41 8000. In both cases the lipase catalyzed hydrolysis of the unwanted enantiomer with high enantioselectivity ($E > 100$). The resulting acid spontaneously decomposes to an aldehyde (Fig. 1).

In the Tanabe process, a membrane reactor and crystallizer combine hydrolysis, separation, and crystallization of (+)-(2*R*,3*S*)-MPGM. Toluene dissolves the racemic substrate in the crystallizer and carries it to the membrane containing immobilized lipase. The lipase catalyzes hydrolysis of the unwanted (−)-MPGM to the acid, which then passes through the membrane into an aqueous phase. Spontaneous decarboxylation of the acid yields an aldehyde which reacts with the bisulfite in the aqueous phase. In the absence of bisulfite, this aldehyde deactivates the lipase. The desired (+)-MPGM remains in the toluene phase and circulates back to the crystallizer where it crystallizes. Lipase activity drops significantly after eight runs and the membrane must be recharged with additional lipase. Although the researchers detected no lipase-catalyzed hydrolysis of (−)-MPGM, chemical hydrolysis lowered the apparent enantioselectivity to $E = 135$ under typical reaction conditions. The yield of crystalline (+)-(2*R*,3*S*)-MPGM is >43% with 100% chemical and enantiomeric purity.

2.2 Amines

BASF produces enantiomerically-pure amines using a *Pseudomonas* lipase-catalyzed acylation (Balkenhohl et al., 1997). A key part of the commercialization of this process was the discovery that methoxyacetate esters reacted much faster than simple esters. Activated esters are not suitable due to a competing uncatalyzed acylation (Fig. 2). In the case of (*R*) or (*S*)-1-phenylethylamine, classical

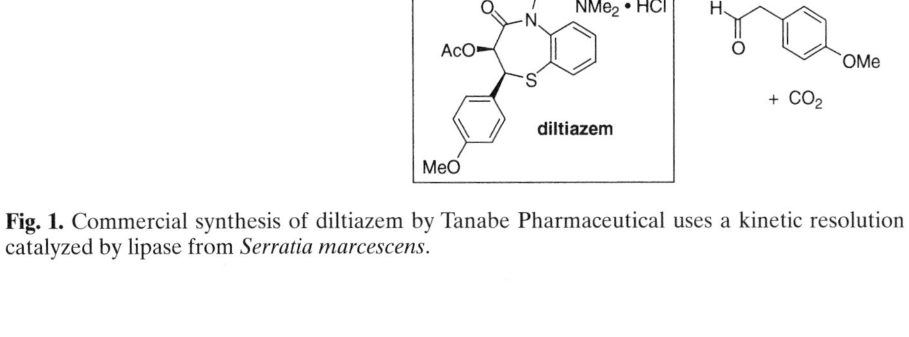

Fig. 1. Commercial synthesis of diltiazem by Tanabe Pharmaceutical uses a kinetic resolution catalyzed by lipase from *Serratia marcescens*.

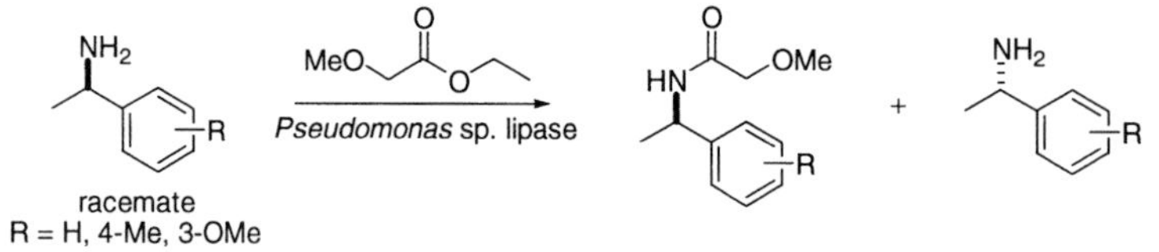

Fig. 2. Large-scale resolution of amines involves a *Pseudomonas* sp. lipase and methoxyacetic acid derivatives.

resolution via diastereomeric salts yielded at best an enantiomeric ratio of 98:2, whereas the biocatalytic product contains less than 0.25% of the opposite enantiomer. In a similar manner a wide range of other amines is resolved. The process is run in a continuous manner using immobilized lipase in a fixed-bed reactor and without additional solvents with an annual production capacity of >2,000 metric tons.

2.3 6-Aminopenicillanic Acid

Penicillin G acylase (PGA, penicillin amidase; for reviews, see BALDARO et al., 1992; TURNER, 1998) catalyzes hydrolysis of the phenylacetyl group in penicillin G (benzylpenicillin) to give 6-aminopenicillanic acid (6-APA) (Fig. 3).

Penicillin G acylase also cleaves the side chain in penicillin V, where the phenylacetyl group is replaced by a phenoxyacetyl group. The commercially available enzymes are derived from *E.coli* strains. Both penicillin G and V are readily available from fermentation, so penicillin manufacturers carry out a PGA-catalyzed hydrolysis to make 6-APA on a scale of approximately 5,000 metric tons per year (MATSUMOTO, 1992). They use 6-APA to prepare semi-synthetic penicillins such as ampicillin, where a D-phenylglycine is linked to the free amino group of 6-APA, or amoxicillin, where a D-4-hydroxyphenyl glycine is linked.

2.4 7-Aminocephalosporanic Acid

The enzymatic production of 7-aminocephalosporanic acid (7-ACA) represents an impressive example, where the biocatalytic route was highly superior to the well-established chemical synthesis. Whereas an enzymatic route to the closely related 6-aminopenicillanic acid (6-APA), a key precursor for the synthesis of semi-synthetic penicillins, was well established in the early 1970s and relies on the action of penicillin G amidase (see Fig. 3), no enzymatic method could be found for 7-ACA. The biocatalytic route was developed by Hoechst researchers and is currently conducted at Biochemie GmbH (Austria). Two enzymes are involved in the transformation of cephalosporin C to 7-ACA (Fig. 4) (CONLON et al., 1995; BAYER and WULLBRANDT, 1999). In the first step, an immobilized D-amino acid oxidase (EC 1.4.3.3) converts cephalosporin C to α-keto-adipinyl-7-ACA. Hydrogen peroxide generated during this step is directly consumed in the decarboxylation to yield glutaryl 7-ACA. Finally, glutaric acid is cleaved by the action of an immobilized glutaryl amidase (EC 3.5.1.3) yielding 7-ACA. The commercial success was mainly based on ecological reasons. In contrast to the chemical route, no toxic substances are produced and the wastewater is readily biodegradable. Thus the amount of waste to be treated was reduced from 31 met-

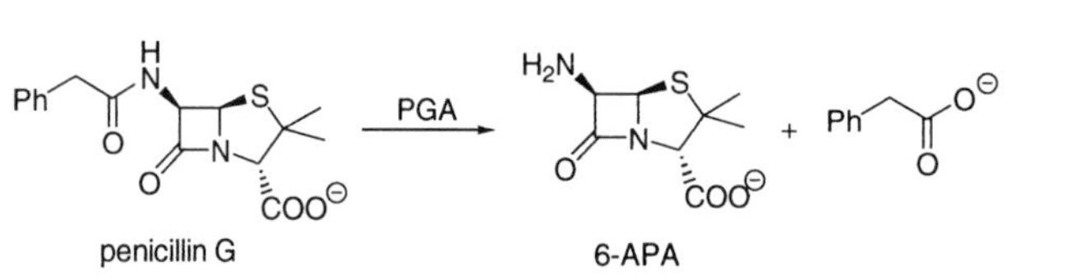

Fig. 3. Commercial process for the production of 6-APA from penicillin G using penicillin G acylase.

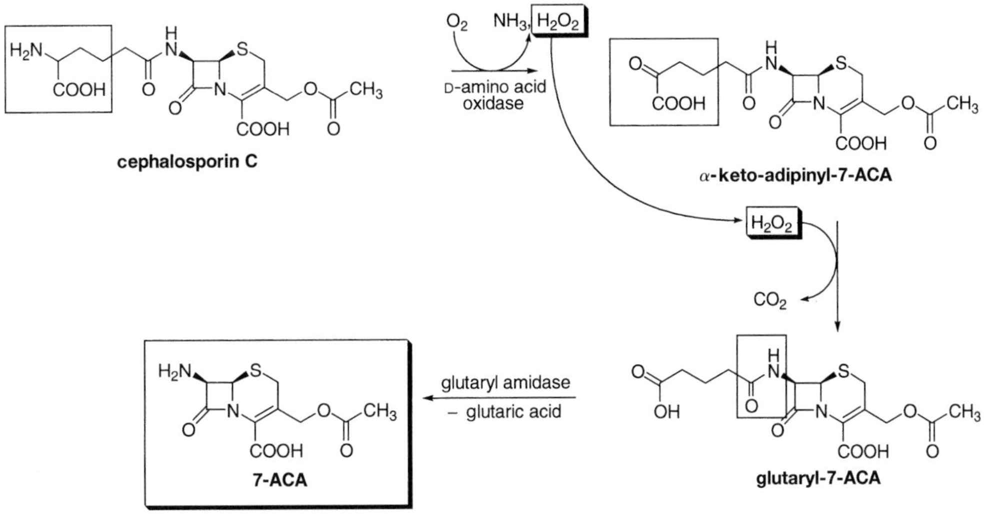

Fig. 4. Biocatalytic route to 7-amino cephalosporanic acid.

ric tons (chemical synthesis) to 0.3 metric tons (enzymatic route) in the production of 1 metric ton of 7-ACA. The cost share for environmental protection on the total process costs was reduced from 21% to only 1% (BAYER and WULLBRANDT, 1999).

2.5 Naproxen

Although a wide range of esterases (EC 3.1.1.1) has been described in the literature (for reviews, see PHYTIAN, 1998; BORNSCHEUER and KAZLAUSKAS, 1999), their industrial application is very limited. This is mostly due to the restricted availability of esterases compared to lipases (with the exception of pig liver esterase). Another exception is carboxylesterase NP, which was originally isolated from *Bacillus subtilis* (strain Thail-8) and was cloned and expressed in *B. subtilis* (QUAX and BROEKHUIZEN, 1994). The esterase shows very high activity and stereoselectivity towards 2-arylpropionic acids, which are, e.g., used in the synthesis of (*S*)-naproxen – (+)-(*S*)-2-(6-methoxy-2-naphthyl) propionic acid – a nonsteroidal anti-inflammatory drug (Fig. 5). (*S*)-naproxen is ca. 150-times more effective than (*R*)-naproxen, the latter also might promote unwanted gastrointestinal disorders. Due to its selectivity toward Naproxen, this esterase is abbreviated in most references as carboxylesterase NP. It has a molecular weight of 32 kDa, a pH optimum between pH 8.5–10.5, and a temperature optimum between 35–55 °C. Carboxylesterase NP is produced as intracellular protein; its structure is unknown.

Although the process gave (*S*)-naproxen with excellent optical purity (99% ee) at an overall yield of 95%, it has not reached commercialization.

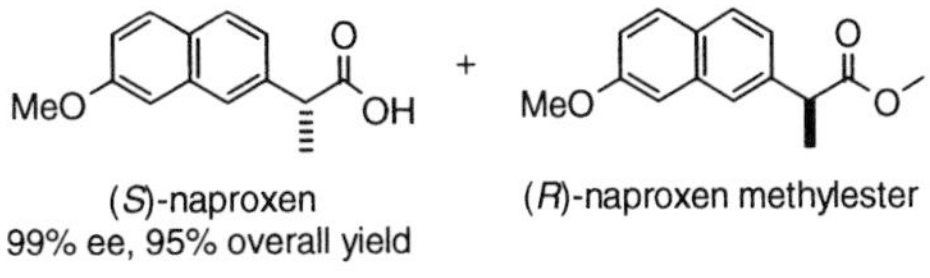

Fig. 5. Synthesis of (*S*)-naproxen by kinetic resolution of the (*R,S*)-methyl ester with carboxylesterase NP followed by chemical racemization of the (*R*)-naproxen methylester (QUAX and BROEKHUIZEN, 1994).

In contrast, Chiroscience (UK) commercialized the resolution of the ethylester of naproxen at ton scale, in which another recombinant esterase is used (BROWN et al., 1999).

2.6 Pyrethroid Precursors

2.6.1 Via Oxynitrilases

Oxynitrilases (also named hydroxynitrile lyases, HNL) comprise a group of enzymes which are capable of creating a carbon–carbon bond utilizing HCN as donor and an aldehyde (or ketone) as acceptor (BUNCH, 1998). Initially, enzymes isolated from various plant sources such as cassava, *Sorghum bicolor*, and *Hevea* sp. had been employed. In recent years, especially the groups of EFFENBERGER (Stuttgart, Germany) and GRIENGL (Graz, Austria) successfully cloned the genes encoding either (*S*)- or (*R*)-specific HNLs into heterologous hosts thus making the enzymes available at competitive prices and in sufficient amounts (WAJANT et al., 1994; EFFENBERGER, 1 994; WAJANT and EFFENBERGER, 1996; HASSLACHER et al., 1997; JOHNSON and GRIENGL, 1998).

A process based on a HNL-catalyzed reaction was developed by DSM in collaboration with the university of Graz. A prerequisite for a successful commercialization was the usage of an already existing plant, which allowed a safe handling of HCN. Currently, DSM is producing (*S*)-m-phenoxybenzaldehyde cyanohydrine on a multi-ton scale. The product is an important precursor for the synthesis of pyrethroids such as (*S*, *S*)-fenvalerate and deltamethrin (Fig. 6). The (*S*)-oxynitrilase expressed recombinantly in the yeast *Pichia pastoris* presumably originates from *Hevea brasiliensis* and is produced by Roche Diagnostics (Penzberg, Germany).

2.6.2 Via a Monooxygenase

BASF has commercialized a process in which (*R*)-2-phenoxypropionic acid is selectively hydroxylated to (*R*)-2-(4-hydroxyphenoxy)propionic acid (H-POPS). H-POPS is an intermediate in the synthesis of optically pure herbicides of the aryloxyphenoxypropionate type. Out of 7,900 strains tested, the fungus *Beauveria bassiana* was identified (Fig. 7). Considerable effort was spent for strain improvement by mutagenesis to increase productivity as well as tolerance of the microorganism towards high substrate and product concentrations. Thus productivity was increased from 0.2 g L^{-1} d^{-1} to 7 g L^{-1} d^{-1}.

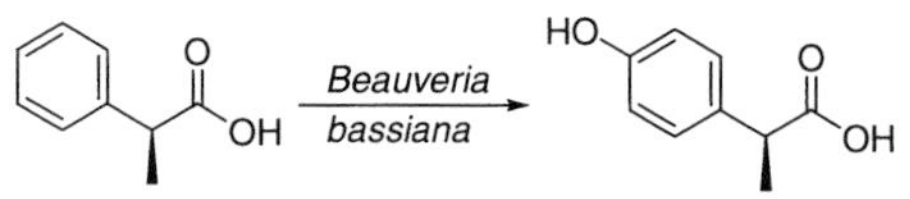

Fig. 7. Synthesis of (*R*)-2-(4-hydroxyphenoxy)propionic acid by BASF uses *Beauveria bassiana*.

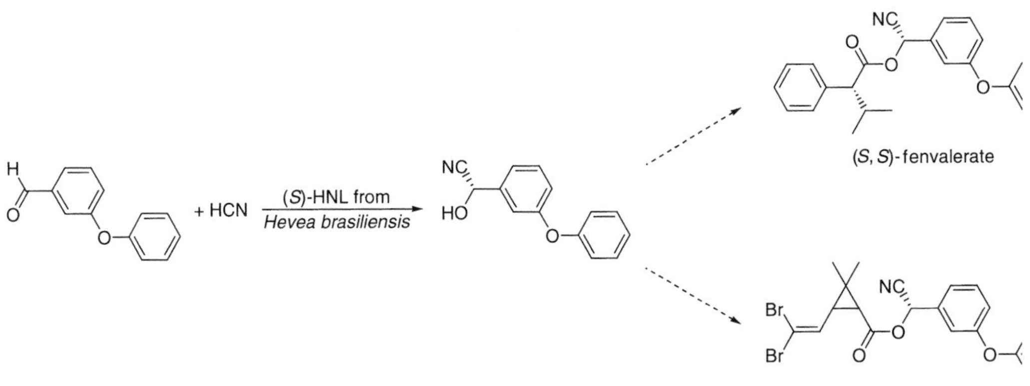

Fig. 6. Synthesis of pyrethroid precursors using a hydroxynitrilase (NHL) from *Hevea brasiliensis*.

2.6.3 Via Esterase

(+)-*trans*-(1*R*,3*R*) chrysanthemic acid is an important precursor of pyrethrin insecticides. An efficient kinetic resolution starting from the (±)-*cis-trans* ethylester was achieved using an esterase from *Arthrobacter globiformis* resulting in the sole formation of the desired enantiomer (>99% ee, at 77% conversion) (Fig. 8). The enzyme was purified and the gene was cloned in *E. coli* (NISHIZAWA et al., 1993). In a 160 g scale process reported by Sumitomo researchers, hydrolysis is performed at pH 9.5 at 50 °C. Acid produced is separated through a hollow-fiber membrane module and the esterase was very stable over four cycles of 48 h (NISHIZAWA et al., 1995). It is unclear, whether the process is currently run on an industrial scale.

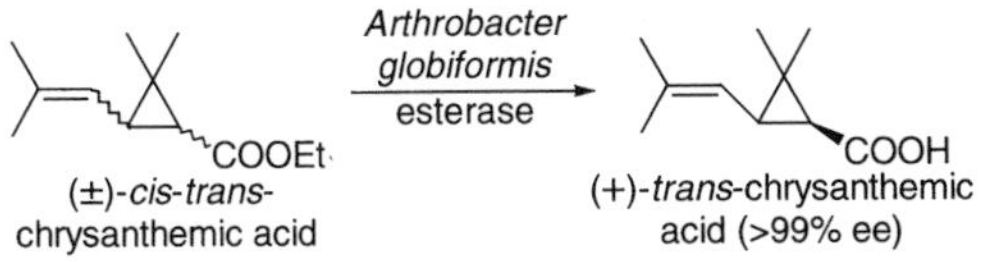

COOEt
(±)-*cis-trans*-chrysanthemic acid
Arthrobacter globiformis esterase
COOH
(+)-*trans*-chrysanthemic acid (>99% ee)

Fig. 8. Synthesis of (+)-*trans* chrysanthemic acid by kinetic resolution of the ethylester using an esterase from *Arthrobacter globiformis*.

2.7 Production of Chiral C_3 Building Blocks

Optically active C_3 compounds are versatile building blocks for organic synthesis, and the most important ones are epichlorohydrin, 3-chloro-1,2-propane diol, and glycidol. Stereoselective synthesis can be performed by chemical means (e.g., using the Jacobsen–Katsuki catalyst, Sharpless epoxidation, or D-mannitol), as well as by a range of biocatalytic routes. KASAI et al. (1998) provide an excellent overview for both approaches.

For instance, DSM-Andeno (Netherlands) produce (*R*)-glycidol butyrate using porcine pancreatic lipase (Fig. 9) by kinetic resolution, but they did not reveal details of the process (LADNER and WHITESIDES, 1984; KLOOSTERMAN et al., 1988). Several groups have since studied this reaction and its scale-up in more detail (WALTS and FOX, 1990; WU et al., 1993; VAN TOL et al., 1995a, b).

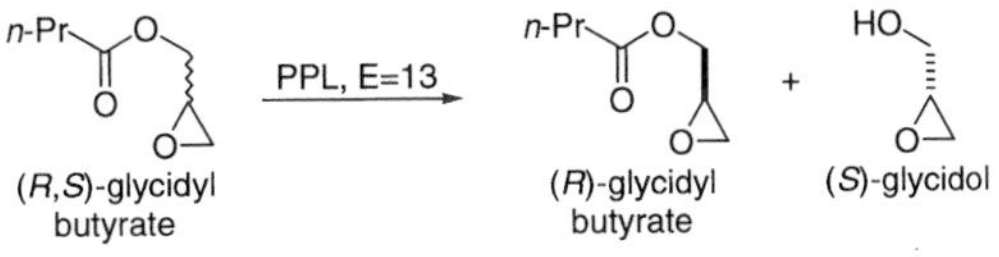

(*R,S*)-glycidyl butyrate
PPL, E=13
(*R*)-glycidyl butyrate + (*S*)-glycidol

Fig. 9. Resolution of glycidol butyrate by DSM-Andeno using porcine pancreatic lipase (PPL).

Other groups described resolution of epoxides using an epoxide hydrolase or by microbial degradation. Researchers at Daiso, Japan, discovered an (*R*)-2,3-dichloro-1-propanol assimilating bacterium, *Pseudomonas* sp. OS-K-29 and a (*S*)-selective strain from *Alcaligenes* sp. DS-K-S38 (Fig. 10). Both strains exhibited excellent selectivity and allowed for the isolation of 50% remaining substrate having either (*S*) or (*R*)-configuration (KASAI et al., 1990, 1992a, b). Although one enantiomer is lost, it is claimed that the fermentative process established in 1994 is cost-effective (KASAI et al., 1998). In a similar manner, 3-chloro-1,2-propane diol can be selectively degraded by microorganisms, and a fermentation process has been developed on an industrial scale at Daiso (SUZUKI and KASAI, 1991; SUZUKI et al., 1992, 1993). Further examples for industrial-scale biocatalysis by dehalogenases have recently been reviewed (SWANSON, 1999).

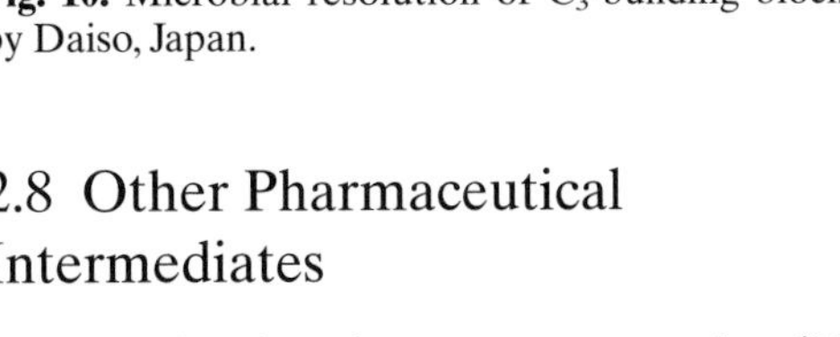

Pseudomonas sp.
100% ee, 38%
Alcaligenes sp.
>99% ee, 40-45%

Fig. 10. Microbial resolution of C_3-building blocks by Daiso, Japan.

2.8 Other Pharmaceutical Intermediates

Glaxo developed a process to resolve (1*S*, 2*S*)-*trans*-2-methoxy cyclohexanol, a secondary alcohol for the synthesis of a tricyclic β-lactam antibiotic (STEAD et al., 1996). The slow reacting enantiomer needed for synthesis is recovered in 99% ee from an acetylation of the racemate with vinyl acetate in cyclohexane. Immobilized lipase B from *Candida antarctica* (CAL-B) and lipase from *Pseudomonas fluorescens* (PFL) both showed high enantioselectivity, but CAL-B was more stable over multiple use cycles. Other workers had resolved this alcohol by hydrolysis of its esters with lipase from *Pseudomonas cepacia* (PCL), *Candida rugosa* (CRL), or pig liver acetone powder (LAUMEN et al., 1989; HÖNIG and SEUFER-WASSERTHAL, 1990; BASAVAIAH and KRISHNA, 1994), but Glaxo chose resolution by acylation of the alcohol because it yields the required slow-reacting alcohol directly (Fig. 11).

Researchers reported a number of other kilogram scale routes to pharmaceutical precursors that involve lipases. Selected examples are summarized in Fig. 12.

Other carboxylic acids are also important intermediates in the synthesis of pharmaceuticals. Researchers have published kilogram-scale resolutions by using proteases (Fig. 13).

Chiroscience (UK) also established a process for the large-scale production of a key intermediate for the synthesis of the reverse transcriptase inhibitor Abacavir (Ziagen). The process relies on a recombinant β-lactamase yielding the required (−)-lactam in >98% ee (E>400) and is operated by Glaxo-Wellcome (Fig. 14) (BROWN et al., 1999).

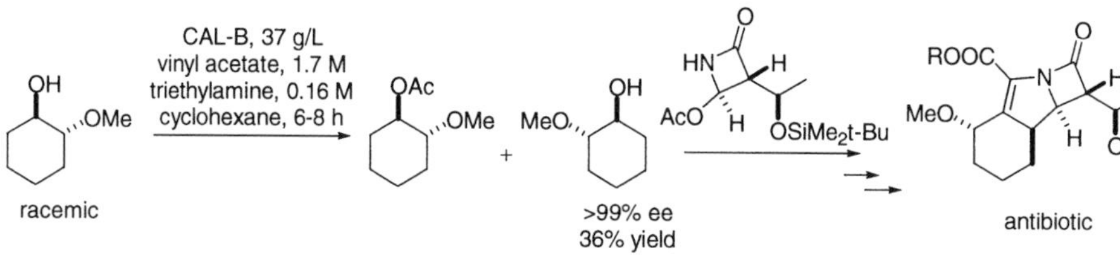

CAL-B, 37 g/L
vinyl acetate, 1.7 M
triethylamine, 0.16 M
cyclohexane, 6-8 h
racemic
>99% ee
36% yield
antibiotic

Fig. 11. Glaxo resolves a building block for antibiotic synthesis.

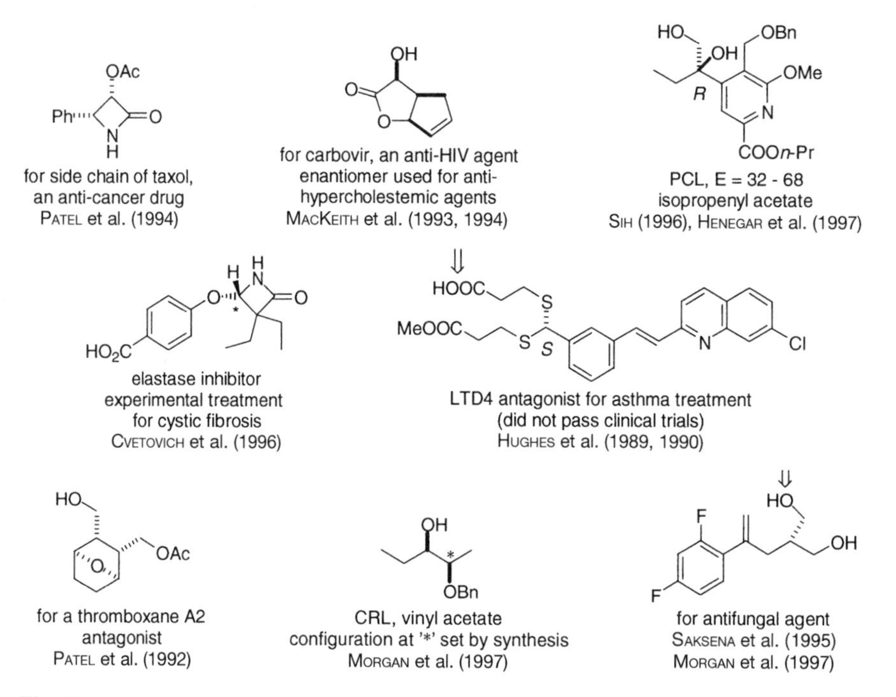

Fig. 12. Kilogram-scale routes to pharmaceutical precursors involving lipases.

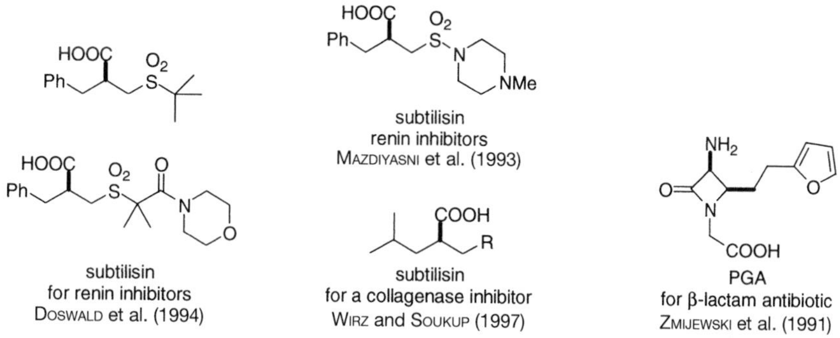

HOOC
O2
Ph
S
HOOC
O2
O
Ph
S
N
O
subtilisin
for renin inhibitors
DOSWALD et al. (1994)
HOOC
O2
Ph
S
N
NMe
subtilisin
renin inhibitors
MAZDIYASNI et al. (1993)
COOH
R
subtilisin
for a collagenase inhibitor
WIRZ and SOUKUP (1997)
NH2
O
N
O
COOH
PGA
for β-lactam antibiotic
ZMIJEWSKI et al. (1991)

Fig. 13. Kilogram-scale routes to pharmaceutical precursors involving proteases or amidases.

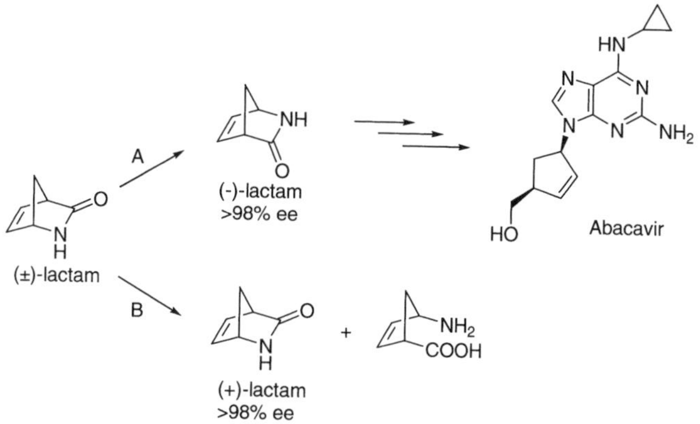

(±)-lactam
A
B
NH
O
(-)-lactam
>98% ee
O
N
H
(+)-lactam
>98% ee
+
NH2
COOH
HN
N
N
N
N
NH2
HO
Abacavir

Fig. 14. Process for the resolution of (±)-2-azabicyclo[2.2.1]hept-5-en-3-one using enantiocomplementary β-lactamases. The enzyme furnishing the (−)-lactam has been cloned and is currently used in a multi-ton scale process for the production of an Abacavir intermediate.

3 Amino Acids

3.1 Natural Amino Acids

A wide range of naturally occurring (proteinogenic) L-amino acids are currently produced by biotransformation. Especially in Japan, several L-amino acids are produced by fermentation (see Chapter 8, this volume). For instance, L-glutamic acid is produced annually at the 300,000 metric ton scale by Ajinomoto, Japan, using *Corynebacterium glutamicum*. On the other hand, L-amino acids can be produced from chemically synthesized racemic mixtures by kinetic resolution of, e.g., *N*-acetyl derivatives (see also Sect. 3.2). L-Aspartate is produced by Tanabe Seiyaku and Kyowa Hakko Kogyo (both Japan) from fumaric acid using an aspartate ammonia lyase as catalyst.

3.2 Non-Natural Amino Acids

In principle, some of the routes developed for the production of L-amino acids can also be applied for the synthesis of non-natural amino acids which consist of either natural amino acids, but in the D-configuration or non-proteinogenic amino acids (usually also in the D-configuration). The most prominent examples are D-phenylglycine and D-4-hydroxyphenylglycine, which are precursors of the semi-synthetic penicillins Ampicillin and Amoxicillin and are currently produced in a hydantoinase/carbamoylase process at a >1,000 tonnes per year scale (see also Sect. 2.3).

Most non-natural amino acids are produced commercially via acylase-, hydantoinase- and amidase-catalyzed routes (for reviews, see BOMMARIUS et al., 1992; KAMPHUIS et al., 1992). A few examples are shown in Fig. 15.

Acylase-catalyzed routes are best for the L-amino acids because acylases favor the L-enantiomer. Racemization of unreacted *N*-acyl derivative of the D-amino acid avoids wasting half of the starting material. Hydantoinase-catalyzed routes are best for producing D-amino acids because the most common hydantoinases favor the D-enantiomer. Amidase-catalyzed resolutions are best for unusual amino acids whose derivatives are not substrates for acylases or hydantoinases. For example, α-alkyl-α-amino acids are resolved by amidases as are the cyclic amino acids 2-piperidine carboxylic acid (pipecolic acid) and piperazine-2-carboxylic acid. Amidases usually favor the natural L-enantiomer.

NSC Technologies has succeeded in producing D-phenylalanine and D-tyrosine on a multiton scale. The biocatalytic route is based on the shikimic acid metabolic pathway and relies on a D-transaminase acting on phenylpyruvic acid (STINSON, 1998).

Kyowa Hakko Kogyo researchers also reported the synthesis of D-alanine using a D,L-alaninamide hydrolase found in *Arthrobacter* sp. NJ-26 (YAGASAKI and OZAKI, 1998). D-Glutamate is produced by the same company starting from cheap L-glutamate. A combination of a glutamate racemase (from *Lactobacillus brevis* ATCC8287) and an L-glutamate decarboxylase (from *E. coli* ATCC11246) allowed the synthesis of D-glutamate (99% ee) at 50% yield by a two-step fermentation. In a similar manner D-proline can be produced (YAGASAKI and OZAKI, 1998).

Researchers at DSM (The Netherlands), described a new amidase activity discovered in *Ochrobactrum anthropi* NCIMB 40321, which accepts α,α-disubstituted amino acids as substrates. The amidase gene was successfully cloned and overexpressed in a host organism, allowing the application of the enzyme in a commercial process, probably dealing with the synthesis of 4-methylthio- and 4-methylsulfonyl substituted (2*S*,3*R*)-3-phenylserines, (*S*)-(+)-cericlamine or L-methylphenylalanine (VAN DOOREN and VAN DEN TWEEL, 1992; VAN DEN TWEEL et al., 1993; KAPTEIN et al., 1994, 1995, 1998; SCHOEMAKER et al., 1996).

Chiroscience (UK) has scaled up the dynamic kinetic resolution of (*S*)-*tert*-leucine, an intermediate for the synthesis of conformationally restricted peptides and chiral auxiliaries (TURNER et al., 1995; McCague and TAYLOR, 1997). However, the only commercialized route is a process by Degussa (Hanau, Germany) utilizing a leucine dehydrogenase. The NADH consumed during the reductive amination of the α-keto acid is recycled with a formate dehydrogenase/ammonium formate system. Recycling of the cofactor allows for total turnover numbers in the range of >10,000 making the process cost-effective (BOMMARIUS et al., 1992, 1995).

3.3 Aspartame

The largest scale application (hundreds to thousands of tons) of protease-catalyzed peptide synthesis is the thermolysin catalyzed synthesis of aspartame, a low calorie sweetener (Fig. 16) (ISOWA et al., 1979; OYAMA, 1992; see Chapter 2, this volume). Precipitation of the product drives this thermodynamically controlled synthesis. The high regioselectivity of thermolysin ensures that only the α-carboxyl group in aspartate reacts. Thus, there is no need to protect the β-carboxylate. The high enantioselectivity allows Tosoh to use racemic amino acids; only the L-enantiomer reacts.

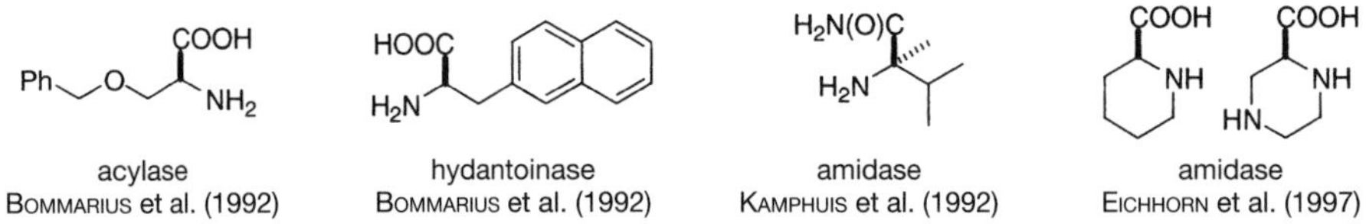

COOH
Ph O NH2
acylase
BOMMARIUS et al. (1992)

HOOC
H2N
hydantoinase
BOMMARIUS et al. (1992)

H2N(O)C
H2N
amidase
KAMPHUIS et al. (1992)

COOH NH
COOH NH HN
amidase
EICHHORN et al. (1997)

Fig. 15. Examples of non-natural amino acids produced via acylase, hydantoinase, or amidase.

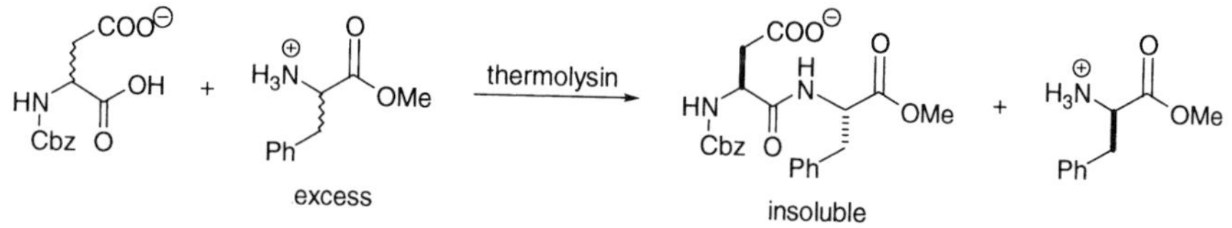

Fig. 16. Commercial process for the production of aspartame (α-L-aspartyl-L-phenylalanine methyl ester).

4 Other Applications

4.1 Acrylamide/Nicotinamide

Although nitriles can also be hydrolyzed to the corresponding carboxylic acid by strong acid or base at high temperatures, nitrile hydrolyzing enzymes have the advantages that they require mild conditions and do not produce large amounts of by-products. In addition, during hydrolysis of dinitriles, they are often regioselective so that only one nitrile group is hydrolyzed and they can be used for the synthesis of optically active substances.

The enzymatic hydrolysis of nitriles follows two different pathways (Fig. 17). Nitrilases (EC 3.5.5.1) directly catalyze the conversion of a nitrile into the corresponding acid plus ammonia. In the other pathway, a nitrile hydratase (NHase, EC 4.2.1.84; a lyase) catalyzes the hydration of a nitrile to the amide, which may be converted to the carboxylic acid and ammonia by an amidase (EC 3.5.1.4) (BUNCH, 1998).

Both pathways occur in the biosynthesis of the phytohormone indole-3-acetic acid from indole-3-acetonitrile for a nitrilase from *Alcaligenes faecalis* JM3 (KOBAYASHI et al., 1993) and for the nitrile hydratase/amidase system in *Agrobacterium tumefaciens* and *Rhizobium* sp. (KOBAYASHI et al., 1995).

Pure nitrilases and nitrile hydratases are usually unstable; so most researchers use them in whole cell preparations. Furthermore, the nitrile hydrolyzing activity must be induced first.

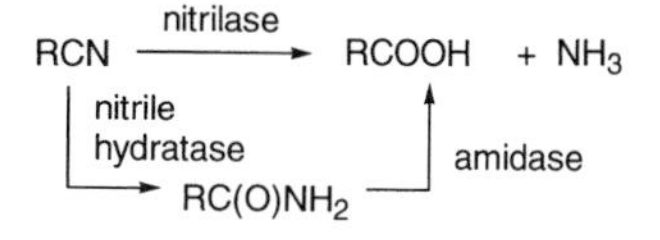

Fig. 17. Hydrolysis of nitriles follows two different pathways.

Two applications based on nitrile hydratase in *Rhodococcus rhodochrous* originally isolated by Yamada's group (NAGASAWA and YAMADA, 1995) have been commercialized. The large production of the commodity chemical acrylamide (Fig. 18) is performed by Nitto Chemical (Yokohama, Japan) on a >30,000 metric tons per year scale. Initially, strains from *Rhodococcus* sp. N-774 or *Pseudomonas chlororaphis* B23 were used, however the current process uses the 10-fold more productive strain *R. rhodochrous* J1. The productivity is >7,000 g acrylamide per g cells at a conversion of acrylonitrile of 99.97%. Formation of acrylic acid is barely detectable at the reaction temperature of 2–4 °C. In laboratory-scale experiments with resting cells up to 656 g acrylamide per liter reaction mixtures were achieved (KOBAYASHI et al., 1992).

One of the most important characteristics of the *R. rhodochrous* strain is its tolerance toward high concentrations of acrylamide (up to 50%). Induction of NHase activity is performed using urea resulting in more than 50% of high molecular weight NHase of the total soluble protein. Cobalt ions are essential to get active NHase. Besides acrylamide, a wide range of other amides can also be produced, e.g., acetamide (150 g L^{-1}), isobutyramide (100 g L^{-1}), methacrylamide (200 g L^{-1}), propionamide (560 g L^{-1}), and crotonamide (200 g L^{-1}) (KOBAYASHI et al., 1992).

Furthermore, *R. rhodochrous* J1 also accepts aromatic and arylaliphatic nitriles as substrates. For instance, also the conversion of 3-cyanopyridine to nicotinamide (a vitamin in animal feed supplementation, Fig. 18) is catalyzed (NAGASAWA et al., 1988), which was industrialized by Switzerland's Lonza on a 3,000 metric tons per year scale in their plant in Guangzhou, China. In contrast to the chemical process, no formation of the by-product nicotinic acid is observed using the nitrile hydratase system.

DuPont researchers described a route to 1,5-dimethyl-2-piperidone from 2-methylglutaronitrile (MGN). MGN is first hydrolyzed to 4-cyanopentanoic acid ammonium salt using immobilized cells of *Acidovorax facilis* 72W (ATCC55746) which contain a nitrilase. The selectivity of this biocatalyst is >98% yielding the desired monoacid as major product in concentrations between 150–200 g L^{-1}, which after chemical hydrogenation furnishes the desired product. Compared to the chemical route, the biocatalytic approach yields a single lactam product, less waste, and higher overall yields (DiCosimo et al., 1997; Gavagan et al., 1998).

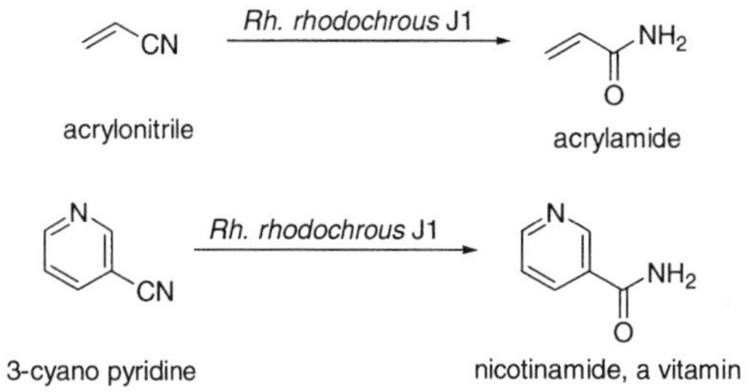

Fig. 18. Commercial production of acrylamide and nicotinamide using resting cells of *Rhodococcus rhodochrous* J1.

4.2 L-Carnitine

L-Carnitine [(*R*)-3-hydroxy-4-(trimethylammonio)butanoate] belongs to a group of vitamin-like nutrients. L-Carnitine plays an essential role in the β-oxidation of long-chain fatty acids by mitochondria and is synthesized in the human liver from lysine and methionine. Insufficient production causes muscle weakness, fatigue, and elevated levels of triglycerides in the blood. Besides its use in infant, health, sport and geriatric nutrition, it might also lead to enhanced growth rates in the animal feed sector. Because D-carnitine competes with the L-form for active transport into the eukaryotic cell, the pure L-enantiomer is required for use in the medical, nutritional and feed applications of carnitine. Several routes to L-carnitine have been described in the literature including resolution of racemic mixtures (see Zimmermann et al., 1997, and references cited therein). For instance, researchers at Daiso, Japan, reported a route starting from optically pure epichlorohydrin (Kasai and Sakaguchi, 1992; Kasai et al., 1998). However, it is unclear whether the process has been commercialized. Another access by Italians Sigma-Tau might be based on the reduction of γ-chloroacetoacetate ester to (*R*)-γ-chloro-β-hydroxy butanoate mediated by a β-ketoreductase from *Saccharomyces cerevisiae.*

Large-scale production by Lonza (Switzerland) starts from easily available butyrobetaine which is converted in a multi-enzyme reaction to L-carnitine (Fig. 19) (Zimmermann

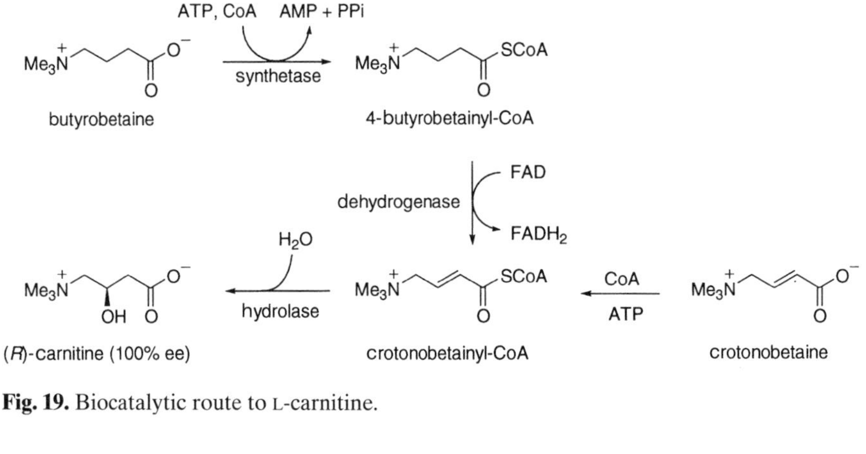

Fig. 19. Biocatalytic route to L-carnitine.

et al., 1997). The initially found most active strain (HK4) was related taxonomically to the soil microorganisms *Agrobacterium* and *Rhizobium*. For large-scale production, a strain improvement program was undertaken with the aim to increase the tolerance of the strain towards high concentrations of substrates and products. In addition, a dehydrogenase activity was irreversibly inactivated by frame shift mutation to avoid degradation via undesired metabolic pathways. The thus obtained strain (HK1349) quantitatively converts 4-butyrobetaine (and crotonobetaine) into L-carnitine (100% ee), which is excreted into the medium. Maximum productivity is in the range of 100 g L^{-1}. Although a fed-batch process gives lower productivity, this route was chosen due to an easier downstream processing and lower investment costs. Researchers at Lonza also developed a recombinant production strain; however, this one is not yet used in the process.

4.3 Lipid Modification

Already in the early 1970s, the application of lipases for the modification of their natural substrates – fats and oils – was exploited by many researchers in industry and academia. Although strong efforts were undertaken to use lipases to obtain free fatty acids by hydrolysis or to produce monoglycerides by glycerolysis, these enzymatic processes were never effective enough to compete with well established chemical methods. Current processes rather exploit 1,3-regiospecificity of lipases to produce so-called structured triglycerides (ST). ST are triglycerides modified in either the type of fatty acid or the position of the fatty acids and the most prominent examples are cocoa butter equivalents and ABA lipids (triglycerides with fatty acids A at *sn*-1 and *sn*-3 and fatty acid B at *sn*-2-position). CHRISTOPHE (1998) provides an excellent overview of nutritional properties and enzymatic synthesis of structured triglycerides.

4.3.1 Cocoa Butter Equivalents

Cocoa butter is predominantly 1,3-disaturated-2-oleyl-glyceride, where palmitic, stearic and oleic acids account for more than 95% of the total fatty acids. Cocoa butter is crystalline and melts between 25 and 35 °C imparting the desirable "mouth feel". Unilever (COLEMAN and MACRAE, 1977) and Fuji Oil (MATSUO et al., 1981) filed the first patents for the lipase-catalyzed synthesis of cocoa butter equivalents. Both companies currently manufacture cocoa butter equivalents on a multi-ton scale using 1,3-selective lipases to replace palmitic acid with stearic acid at the *sn*-1 and *sn*-3 positions (for reviews, see MACRAE, 1983; MACRAE and HAMMOND, 1985; QUINLAN and MOORE, 1993). The Unilever subsidiary Quest-Loders Croklaan (The Netherlands) uses RML and palm oil mid fraction or high oleate sunflower oil as starting materials (Fig. 20). Other suitable starting oils are rape seed (ADLERCREUTZ, 1994) or olive oils (CHANG et al., 1990).

4.3.2 Other Structured Lipids

Another structured triglyceride is Betapol, a formula additive for premature infants. Human milk contains the more easily digested OPO, but vegetable oils contain the POO isomer (OPO: 1,3-dioleyl-2-palmitoyl glyceride, POO: mixture of 1,2-dioleyl-3-palmitoyl glyceride and its enantiomer). Interesterification of tripalmitin with oleic acid using RML at low water activity yields OPO. Evaporation removes excess fatty acids and crystallization removes remaining PPP. Betapol is currently produced by a Unilever subsidiary.

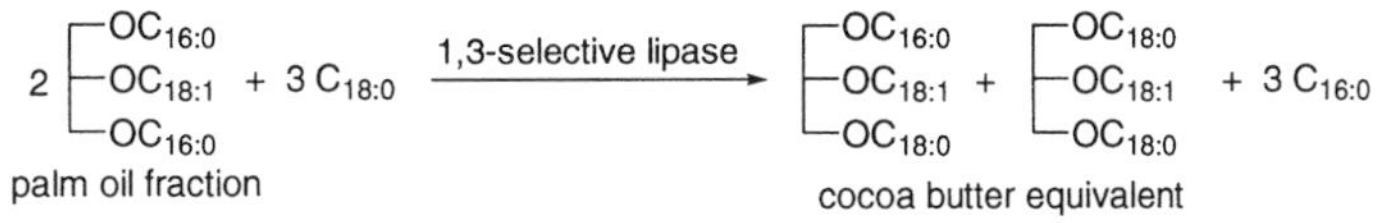

$$2\,\left[\begin{array}{l}OC_{16:0}\\ OC_{18:1}\\ OC_{16:0}\end{array}\right] + 3\,C_{18:0} \xrightarrow{\text{1,3-selective lipase}} \left[\begin{array}{l}OC_{16:0}\\ OC_{18:1}\\ OC_{18:0}\end{array}\right] + \left[\begin{array}{l}OC_{18:0}\\ OC_{18:1}\\ OC_{18:0}\end{array}\right] + 3\,C_{16:0}$$

palm oil fraction — cocoa butter equivalent

Fig. 20. Lipase-catalyzed synthesis of cocoa-butter equivalent.

Other processes under development focus on the enrichment or isolation of polyunsaturated fatty acids such as eicosapentaenoic acid (EPA) or docosahexaenoic acid (DHA) from fish oil. Diets supplemented with EPA or DHA in the triglycerides are reported to show a range of nutritional benefits.

Acknowledgements
The author thanks Dr. P. RASOR (Roche Diagnostics, Penzberg, Germany), Dr. J. PETERS (Bayer AG, Wuppertal, Germany), Dr. B. HAUER (BASF AG, Ludwigshafen, Germany), Dr. A. KIENER (Lonza, Visp, Switzerland), and Prof. R.J. KAZLAUSKAS (McGill University, Montreal, Canada) for their support during preparation of the manuscript.

5 References

ADLERCREUTZ, P. (1994), Enzyme-catalyzed lipid modification, *Biotechnol. Genet. Eng. Rev.* **12**, 231–254.

ARNOLD, F. H. (1998), Design by directed evolution, *Acc. Chem. Res.* **31**, 125–131.

BALDARO, E., FUGANTI, C., SERVI, S., TAGLIANI, A., TERRENI, M. (1992), The use of immobilized penicillin G acylase in organic synthesis, in: *Microbial Reagents in Organic Synthesis* (SERVI, S., Ed.), pp. 175–188. Dordrecht: Kluwer Academic.

BALKENHOHL, F., DITRICH, K., HAUER, B., LADNER, W. (1997), Optically active amines via lipase-catalyzed methoxyacetylation, *J. Prakt. Chem.* **339**, 381–384.

BASAVAIAH, D., KRISHNA, P. R. (1994), Pig liver acetone powder (PLAP) as biocatalyst: enantioselective synthesis of *trans*-2-alkoxycyclohexan-1-ols, *Tetrahedron* **50**, 10521–10530.

BAYER, T., WULLBRANDT, D. (1999), Enzymatische Herstellung von 7-Aminocephalosporansäure, in: *Industrielle Nutzung von Biokatalysatoren* (HEIDEN, S., BOCK, A.-K., ANTRANIKIAN, G., Eds.), pp. 187–194. Berlin: Erich-Schmidt-Verlag.

BOMMARIUS, A. S., DRAUZ, K., GROEGER, U., WANDREY, C. (1992), Membrane bioreactors for the production of enantiomerically pure α-amino acids, in: *Chirality in Industry* Vol. I (COLLINS, A. N., SHELDRAKE, G. N., CROSBY, J., Eds.), pp. 371–397. Chichester: John Wiley & Sons.

BOMMARIUS, A. S., SCHWARM, M., STINGL, K., KOTTENHAHN, M., HUTHMACHER, K., DRAUZ, K. (1995), Synthesis and use of enantiomerically pure *tert*-leucine, *Tetrahedron: Asymmetry* **6**, 2851–2888.

BORNSCHEUER, U. T. (1998), Directed evolution of enzymes, *Angew. Chem.* (*Int. Edn. Engl.*) **37**, 3105–3108.

BORNSCHEUER, U. T., KAZLAUSKAS, R. J. (1999), *Hydrolases in Organic Synthesis – Regio- and Stereoselective Biotransformations*. Weinheim: Wiley-VCH.

BROWN, R. C., TAYLOR, S. J. C., TAYLOR, I. N., KEENE, P. A., BAUER, S. (1999), Different cloning approaches for industrially useful biocatalysts, *Abstract Book, Biotrans '99 Conference, Taormina, Italy*.

BUNCH, A. (1998), Nitriles, in: *Biotechnology* 2nd Edn., Vol. 8a (REHM, H.-J., REED, G., PÜHLER, A., STADLER, P., Eds.), pp. 277–324. Weinheim: Wiley-VCH.

CHANG, M. K., ABRAHAM, G., JOHN, V. T. (1990), Production of cocoa butter-like fat from interesterification of vegetable oils, *J. Am. Oil Chem. Soc.* **67**, 832–834.

CHRISTOPHE, A. B. (Ed.) (1998), *Structural Modified Food Fats: Synthesis, Biochemistry, and Use*. Champaign, IL:AOCS Press.

COLEMAN, M. H., MACRAE, A. R. (1977), Rearrangement of fatty acid esters in fat reaction reactants, *German Patent* DE 2705608 (Unilever N.V.) (*Chem. Abstr.* **87**, 166 366).

COLLINS, A. N., SHELDRAKE, G. N., CROSBY, J. (Eds.) (1992), *Chirality in Industry I*. New York: John Wiley & Sons.

COLLINS, A. N., SHELDRAKE, G. N., CROSBY, J. (Eds.) (1997), *Chirality in Industry II*. Chichester: John Wiley & Sons.

CONLON, H. D., BAQAI, J., BAKER, K., SHEN, Y. Q., WONG, B. L. et al. (1995), Two-step immobilized enzyme conversion of cephalosporin C to 7-aminocephalosporanic acid, *Biotechnol. Bioeng.* **46**, 510–513.

CVETOVICH, R. J., CHARTRAIN, M., HARTNER, F. W., ROBERGE, C., AMATO, J. S., GRABOWSKI, E. J. (1996), An asymmetric synthesis of L-694,458, a human leukocyte elastase inhibitor, via novel enzyme resolution of β-lactam esters, *J. Org. Chem.* **61**, 6575–6580.

DICOSIMO, R., FALLON, R. D., GAVAGAN, J. E. (1997), Preparation of 5- or 6-membered lactams from aliphatic α,ω-dinitriles, *Int. Patent Appl.* WO9744318 (DuPont) (*Chem. Abstr.* **128**, 89231; Corr. of **128**, 48589).

DOSWALD, S., ESTERMANN, H., KUPFER, E., STADLER, H., WALTHER, W. et al. (1994), Large scale preparation of chiral building blocks for the P3 site of renin inhibitors, *Bioorg. Med. Chem.* **2**, 403–410.

EFFENBERGER, F. (1994), Synthesis and reactions of optically-active cyanohydrins, *Angew. Chem.* (*Int. Edn. Engl.*) **33**, 1555–1564.

EICHHORN, E., RODUIT, J.-P., SHAW, N., HEINZMANN, K., KIENER, A. (1997), Preparation of (*S*)- piperazine-2-carboxylic acid, (*R*)-piperazine-2-carboxylic acid, and (*S*)-piperidine-2-carboxylic acid by kinetic resolution of the corresponding racemic carboxamides with stereoselective amidases in whole bacterial cells, *Tetrahedron: Asymmetry* **8**, 2533–2536.

FURUI, M., FURUTANI, T., SHIBATANI, T., NAKAMOTO, Y., MORI, T. H. (1996), A membrane bioreactor combined with crystallizer for production of optically active (2*R*,3*S*)-3-(4-methoxyphenyl)glycidic acid methyl ester, *J. Ferment. Bioeng.* **81**, 21–25.

GAVAGAN, J. E., FAGER, S. K., FALLON, R. D., FOLSOM, P. W., HERKES, F. E. et al. (1998), Chemoenzymic production of lactams from aliphatic α,ω-dinitriles, *J. Org. Chem.* **63**, 4792–4801.

HASSLACHER, M., SCHALL, M., HAYN, M., BONA, R., RUMBOLD, K. et al. (1997), High-level intracellular expression of hydroxynitrile lyase from the tropical rubber tree *Hevea brasiliensis* in microbial hosts, *Protein Expr. Purif.* **11**, 61–71.

HENEGAR, K. E., ASHFORD, S. W., BAUGHMAN, T. A., SIH, J. C., GU, R. L. (1997), Practical asymmetric synthesis of (*S*)-4-ethyl-7,8-dihydro-4-hydroxy-1H-pyrano[3,4-f]indolizine-3,6,10(4H)-trione, a key intermediate for the synthesis of irinotecan and other camptothecin analogs, *J. Org. Chem.* **62**, 6588–6597.

HÖNIG, H., SEUFER-WASSERTHAL, P. (1990), A general method for the separation of enantiomeric *trans*-2-substituted cyclohexanols, *Synthesis*, 1137–1140.

HUGHES, D. L., BERGAN, J. J., AMATO, J. S., REIDER, P. J., GRABOWSKI, E. J. J. (1989), Synthesis of chiral dithioacetals: a chemoenzymic synthesis of a novel LTD4 antagonist, *J. Org. Chem.* **54**, 1787–1788.

HUGHES, D. L., BERGAN, J. J., AMATO, J. S., BHUPATHY, M., LEAZER, J. L. et al. (1990), Lipase-catalyzed asymmetric hydrolysis of esters having remote chiral/prochiral centers, *J. Org. Chem.* **55**, 6252–6259.

HULSHOF, L. A., ROSKAM, J. H. (1989), Phenylglycidate stereoisomers, conversion products thereof with e.g. 2-nitrophenol and preparation of diltiazem, *European Patent* EP0343714 (Stamicarbon) (*Chem. Abstr.* **113**, 76603).

ISOWA, Y., OHMORI, M., ICHIKAWA, T., MORI, K., NONAKA, Y. et al. (1979), The thermolysin-catalyzed condensation reactions of *N*-substituted aspartic acid and glutamic acids with phenylalanine alkyl esters, *Tetrahedron Lett.* 2611–2612.

JAEGER, K.-E., REETZ, M. T. (1998), Microbial lipases form versatile tools for biotechnology, *Trends Biotechnol.* **16**, 396–403.

JOHNSON, D. V., GRIENGL, H. (1998), Biocatalytic applications of hydroxynitrile lyases, in: *Advances in Biochemical Engineering/Biotechnology* Vol. 63 (SCHEPER, T., Ed.), pp. 31–55. Berlin: Springer-Verlag.

KAMPHUIS, J., BOESTEN, W. H. J., KAPTEN, B., HERMES, H. F. M., SONKE, T. et al. (1992), The production and uses of optically pure natural and unnatural amino acids, in: *Chirality in Industry* Vol. I (COLLINS, A. N., SHELDRAKE, G. N., CROSBY, J., Eds.), pp. 187–208. Chichester: John Wiley & Sons.

KAPTEIN, B., MOODY, H. M., BROXTERMAN, Q. B., KAMPHUIS, J. (1994), Chemo-enzymic synthesis of (*S*)-(+)-cericlamine and related enantiomerically pure 2,2-disubstituted-2-aminoethanols, *J. Chem. Soc., Perkin Trans.* **1**, 1495–1498.

KAPTEIN, B., MONACO, V., BROXTERMAN, Q. B., SCHOEMAKER, H. E., KAMPHUIS, J. (1995), Synthesis of dipeptides containing α-substituted amino acids; their use as chiral ligands in Lewis acid-catalyzed reactions, *Recl. Trav. Chim. Pays-Bas* **114**, 231–238.

KAPTEIN, B., VAN DOOREN, T. J. G. M., BOESTEN, W. H. J., SONKE, T., DUCHATEAU, A. L. L. et al. (1998), Synthesis of 4-sulfur-substituted (2*S*,3*R*)-3-phenylserines by enzymic resolution. Enantiopure precursors for Thiamphenicol and Florfenicol, *Org. Process Res. Dev.* **2**, 10–17.

KASAI, N., SAKAGUCHI, K. (1992), An efficient synthesis of (*R*)-carnitine, *Tetrahedron Lett.* **33**, 1211–1212.

KASAI, N., TSUJIMURA, K., UNOURA, K., SUZUKI, T. (1990), Degradation of 2,3-dichloro-1-propanol by a *Pseudomonas* sp., *Agric. Biol. Chem.* **54**, 3185–3190.

KASAI, N., TSUJIMURA, K., UNOURA, K., SUZUKI, T. (1992a), Isolation of (*S*)-2,3-dichloro-1-propanol assimilating bacterium, its characterization, and its use in preparation of (*R*)-2,3-dichloro-1-propanol and (*S*)-epichlorohydrin, *J. Ind. Microbiol.* **10**, 37–43.

KASAI, N., TSUJIMURA, K., UNOURA, K., SUZUKI, T. (1992b), Preparation of (*S*)-2,3-dichloro-1-propanol by *Pseudomonas* sp. and its use in the synthesis (*R*)-epichlorohydrin, *J. Ind. Microbiol.* **9**, 97–101.

KASAI, N., SUZUKI, T., FURUKAWA, Y. (1998), Chiral C3 epoxides and halohydrins: Their preparation and synthetic application, *J. Mol. Catal. B: Enzymatic* **4**, 237–252.

KAZLAUSKAS, R. J., BORNSCHEUER, U. T. (1998), Biotransformations with lipases, in: *Biotechnology* Vol. 8a (REHM, H. J., REED, G., PÜHLER, A., STADLER, P. J. W., Eds.), pp. 37-191. Weinheim: John Wiley & Sons.

KLOOSTERMAN, M., ELFERINK, V. H. M., IERSEL, J. V., ROSKAM, J. H., MEIJER, E. M. et al. (1988), Lipases in the preparation of β blockers, *Trends Biotechnol.* **6**, 251–256.

KOBAYASHI, M., NAGASAWA, T., YAMADA, H. (1992),

Enzymatic syntheses of acrylamide: a success story not yet over, *Trends Biotechnol.* **10**, 402–408.
KOBAYASHI, M., IZUI, H., NAGASAWA, T., YAMADA, H. (1993), Nitrilase in biosynthesis of the plant hormone indole-3-acetic acid from indole-3-acetonitrile: Cloning of the *Alcaligenes* gene and site-directed mutagenesis of cysteine residues, *Proc. Natl. Acad. Sci. USA* **90**, 247–251.
KOBAYASHI, M., SUZUKI, T., FUJITA, T., MASUDA, M., SHIMIZU, S. (1995), Occurrence of enzymes involved in biosynthesis of indole-3-acetic acid from indole-3-acetonitrile in plant-associated bacteria, *Agrobacterium* and *Rhizobium*, *Proc. Natl. Acad. Sci. USA* **92**, 714–718.
LADNER, W. E., WHITESIDES, G. M. (1984), Lipase-catalyzed hydrolysis as a route to esters of chiral epoxy alcohols, *J. Am. Chem. Soc.* **106**, 7250–7251.
LAUMEN, K., BREITGOFF, D., SEEMAYER, R., SCHNEIDER, M. P. (1989), Enantiomerically pure cyclohexanols and cyclohexane-1,2-diol derivatives, chiral auxiliaries and substitutes for (−)-8-phenylmenthol. A facile enzymic route, *J. Chem. Soc., Chem. Commun.* 148–150.
MACKEITH, R.A., MCCAGUE, R., OLIVO, H. F., PALMER, C. F., ROBERTS, S. M. (1993), Conversion of (−)-4-hydroxy-2-oxabicyclo[3.3.0]oct-7-en-3-one into the anti-HIV agent carbovir, *J. Chem. Soc., Perkin Trans.* **I**, 313–314.
MACKEITH, R. A., MCCAGUE, R., OLIVO, H. F., ROBERTS, S. M., TAYLOR, S. J. C., XIONG, H. (1994), Enzyme-catalyzed kinetic resolution of 4-endo-hydroxy-2- oxabicyclo[3.3.0]oct-7-en-3-one and employment of the pure enantiomers for the synthesis of antiviral and hypocholestemic agents, *Bioorg. Med. Chem.* **2**, 387–394.
MACRAE, A. R. (1983), Lipase-catalyzed interesterification of oils and fats, *J. Am. Oil Chem. Soc.* **60**, 291–294.
MACRAE, A. R., HAMMOND, R. C. (1985), Present and future applications of lipases, *Biotechnol. Genet. Eng. Rev.* **3**, 193–217.
MATSUMAE, H., FURUI, M., SHIBATANI, T. (1993), Lipase-catalyzed asymmetric hydrolysis of 3-phenylglycidic acid ester, the key intermediate in the synthesis of diltiazem hydrochloride, *J. Ferment. Bioeng.* **75**, 93–98.
MATSUMAE, H., FURUI, M., SHIBATANI, T., TOSA, T. (1994), Production of optically active 3-phenylglycidic acid ester by the lipase from *Serratia marcescens* on a hollow-fiber membrane reactor, *J. Ferment. Bioeng.* **78**, 59–63.
MATSUMOTO, K. (1992), Production of 6-APA, 7-ACA, and 7-ADCA by immobilized penicillin and cephalosporin amidases, in: *Industrial Applications of Immobilized Biocatalysts* (TANAKA, A., TOSA, T., KOBAYASHI, T., Eds.). New York: Marcel Dekker.
MATSUO, T., SAWAMURA, N., HASHIMOTO, Y., HASHIDA, W. (1981), The enzyme and method for enzymatic transesterification of lipid, *European Patent* EP 0035883 (Fuji Oil Co.) (*Chem. Abstr.* **96**, 4958).
MAZDIYASNI, H., KONOPACKI, D. B., DICKMAN, D. A., ZYDOWSKY, T. M. (1993), Enzyme-catalyzed synthesis of optically-pure β-sulfonamidopropionic acids. Useful starting materials for P-3 site modified renin inhibitors, *Tetrahedron Lett.* **34**, 435–438.
MCCAGUE, R., TAYLOR, S. J. C. (1997), Four case studies in the development of biotransformation-based processes, in: *Chirality in Industry* Vol. II (COLLINS, A. N., SHELDRAKE, G. N., CROSBY, J., Eds.), pp. 183–206. Chichester: John Wiley & Sons.
MORGAN, B., STOCKWELL, B. R., DODDS, D. R., ANDREWS, D. R., SUDHAKAR, A. R. et al. (1997), Chemoenzymatic approaches to SCH 56592, a new azole antifungal, *J. Am. Oil Chem. Soc.* **74**, 1361–1370.
NAGASAWA, T., YAMADA, H. (1995), Microbial production of commodity chemicals, *Pure Appl. Chem.* **67**, 1241–1256.
NAGASAWA, T., MATHEW, C. D., MAUGER, J., YAMADA, H. (1988), Nitrile hydratase-catalyzed production of nicotinamide from 3-cyanopyridine in *Rhodococcus rhodochrous* J1, *Appl. Environ. Microbiol.* **54**, 1766–1769.
NISHIZAWA, M., GOMI, H., KISHIMOTO, F. (1993), Purification and some properties of carboxylesterase from *Arthobacter globiformis;* stereoselective hydrolysis of ethyl chrysanthemate, *Biosci. Biotech. Biochem.* **57**, 594–598.
NISHIZAWA, M., SHIMIZU, M., OHKAWA, H., KANAOKA, M. (1995), Stereoselective production of (+)-*trans*-chrysanthemic acid by a microbial esterase: cloning, nucleotide sequence, and overexpression of the esterase gene of Arthrobacter globiformis in *Escherichia coli*, *Appl. Environ. Microbiol.* **61**, 3208–3215.
OYAMA, K. (1992), Industrial production of aspartame, in: *Chirality in Industry* Vol. I (COLLINS, A. N., SHELDRAKE, G. N., CROSBY, J., Eds.), pp. 237–247. Chichester: John Wiley & Sons.
PATEL, R. N., LIU, M., BANERJEE, A., SZARKA, L. J. (1992), Stereoselective enzymatic hydrolysis of (*exo,exo*)-7-oxabicyclo[2.2.1]heptane-2,3-dimethanol diacetate ester in a biphasic system, *Appl. Microbiol. Biotechnol.* **37**, 180–183.
PATEL, R. N., BANERJEE, A., KO, R. Y., HOWELL, J. M., LI, W.-S. et al. (1994), Enzymic preparation of (3*R*-*cis*)-3-(acetyloxy)-4-phenyl-2-azetidinone – a taxol side-chain synthon, *Biotechnol. Appl. Biochem.* **20**, 23–33.
PHYTIAN, S. J. (1998), Esterases, in: *Biotechnology* Vol. 8a (REHM, H. J., REED, G., PÜHLER, A., STADLER, P. J. W., Eds.), pp. 193–241. Weinheim: Wiley-VCH.

QUAX, W. J., BROEKHUIZEN, C. P. (1994), Development of a new *Bacillus* carboxyl esterase for use in the resolution of chiral drugs, *Appl. Microbiol. Biotechnol.* **41**, 425–431.

QUINLAN, P., MOORE, S. (1993), Modification of triglycerides by lipases: process technology and its application to the production of nutritionally improved fats, *Inform* **4**, 579–583.

REETZ, M. T., JAEGER, K.-E. (1999), Superior biocatalysts by directed evolution, *Topic Curr. Chem.* **200**, 31–57.

ROZZELL, J. D. (1999), Commercial scale biocatalysis: myths and realities, *Bioorg. Med. Chem.* **7**, 2253-2261.

SAKSENA, A. K., GIRIJAVALLABHAN, V. M., LOVEY, R. G., PIKE, R. E., WANG, H. et al. (1995), Highly stereoselective access to novel 2,2,4-trisubstituted tetrahydrofurans by halocyclization: practical chemoenzymic synthesis of SCH 51048, a broad-spectrum orally active antifungal agent, *Tetrahedron Lett.* **36**, 1787–1790.

SCHMID, R. D., VERGER, R. (1998), Lipases – interfacial enzymes with attractive applications, *Angew. Chem.* (*Int. Edn. Engl.*) **37**, 1608–1633.

SCHOEMAKER, H. E., BOESTEN, W. H. J., KAPTEIN, B., ROOS, E. C., BROXTERMAN, Q. B. et al. (1996), Enzymic catalysis in organic synthesis. Synthesis of enantiomerically pure Cα-substituted α-amino and α-hydroxy acids, *Acta Chem. Scand.* **50**, 225–233.

SHELDON, R. A. (1993), *Chirotechnology – Industrial Synthesis of Optically-Active Compounds*. New York: Marcel Dekker.

SIH, J. C. (1996), Application of immobilized lipase in production of Camptosar (CPT-11), *J. Am. Oil Chem. Soc.* **73**, 1377–1378.

STEAD, P., MARLEY, H., MAHMOUDIAN, M., WEBB, G., NOBLE, D. et al. (1996), Efficient procedures for the large-scale preparation of (1*S*,2*S*)-*trans*-2-methoxycyclohexanol, a key chiral intermediate in the synthesis of tricyclic β-lactam antibiotics, *Tetrahedron: Asymmetry* **7**, 2247–2250.

STEMMER, W. P. C. (1994), DNA shuffling by random fragmentation and reassembly: *In vitro* recombination for molecular evolution, *Proc. Natl. Acad. Sci. USA* **91**, 10747–10751.

STINSON, S. C. (1998), Counting on chiral drugs, *Chem. Eng. News* Sept. 21, 83–104.

SUZUKI, T., KASAI, N. (1991), A novel method for the generation of (*R*)- and (*S*)-3-chloro-1,2-propanediol by stereospecific dehalogenating bacteria and their use in the preparation of (*R*)- and (*S*)-glycidol, *Bioorg. Med. Chem. Lett.* **1**, 343–346.

SUZUKI, T., KASAI, N., YAMAMOTO, R., MINAMIURA, N. (1992), Isolation of a bacterium assimilating (*R*)-3-chloro-1,2-propanediol and production of (*S*)-3-chloro-1,2-propanediol using microbial resolution, *J. Ferment. Bioeng.* **73**, 443–448.

SUZUKI, T., KASAI, N., YAMAMOTO, R., MINAMIURA, N. (1993), Production of highly optically active (*R*)-3-chloro-1,2-propanediol using a bacterium assimilating the (*S*)-isomer, *Appl. Microbiol. Biotechnol.* **40**, 273–278.

SWANSON, P. E. (1999), Dehalogenases applied to industrial-scale biocatalysis, *Curr. Opin. Biotechnol.* **10**, 365–369.

TAYLOR, S. J. C., MCCAGUE, R., WISDOM, R., LEE, C., DICKSON, K. et al. (1993), Development of the biocatalytic resolution of 2-azabicyclo[2.2.1]-hept-5-en-3-one as an entry to single-enantiomer carbocyclic nucleosides, *Tetrahedron: Asymmetry* **4**, 1117–1128.

TURNER, M. (1998), Perspectives in Biotransformations, in: *Biotechnology* 2nd Edn., Vol. 8a (REHM, H.-J., REED, G., PÜHLER, A., STADLER, P., Eds.), pp. 15–17. Weinheim: Wiley-VCH.

TURNER, N. J., WINTERMAN, J. R., MCCAGUE, R., PARRATT, J. S., TAYLOR, S. J. C. (1995), Synthesis of homochiral L-(*S*)-tert-leucine via a lipase catalyzed dynamic resolution process, *Tetrahedron Lett.* **36**, 1113–1116.

VAN DEN TWEEL, W. J. J., VAN DOOREN, T. J. G. M., DE JONGE, P. H., KAPTEIN, B., DUCHATEAU, A. L. L., KAMPHUIS, J. (1993), *Ochrobactrum anthropi* NCIMB 40321: a new biocatalyst with broad-spectrum L-specific amidase activity, *Appl. Microbiol. Biotechnol.* **39**, 296–300.

VAN DOOREN, T. J. G. M., VAN DEN TWEEL, W. J. J. (1992), Enzymic preparation of optically active carboxylic acids, *Eur. Patent Appl.* EP494716 (DSM) (*Chem. Abstr.* **117**, 210758).

VAN TOL, J. B. A., JONGEJAN, J. A., DUINE, J. A. (1995a), Description of hydrolase-enantioselectivity must be based on the actual kinetic mechanism: analysis of the kinetic resolution of glycidyl (2,3-epoxy-1-propyl) butyrate by pig pancreas lipase, *Biocatal. Biotransform.* **12**, 99–117.

VAN TOL, J. B. A., KRAAYVELD, D. E., JONGEJAN, J. A., DUINE, J. A. (1995b), The catalytic performance of pig pancreas lipase in enantioselective transesterification in organic solvents, *Biocatal. Biotransform.* **12**, 119-136.

WAJANT, H., EFFENBERGER, F. (1996), Hydroxynitrile lyases of higher plants, *Biol. Chem.* **377**, 611–617.

WAJANT, H., MUNDRY, K. W., PFIZENMAIER, K. (1994), Molecular cloning of hydroxynitrile lyase from *Sorghum bicolor* (L.). Homologies to serine carboxypeptidases, *Plant. Mol. Biol.* **26**, 735–746.

WALTS, A. E., FOX, E. M. (1990), A lipase fraction for resolution of glycidyl esters to high enantiomeric excess, *US Patent* 4923810 (Genzyme) (*Chem. Abstr.* **113**, 113879).

WIRZ, B., SOUKUP, M. (1997), Enzymatic preparation of homochiral 2-isobutyl succinic acid derivatives, *Tetrahedron: Asymmetry* **8**, 187–189.

WOOLLEY, P., PETERSEN, S. B. (1994), Lipases: *Their Structure, Biochemistry, and Application*. Cambridge: Cambridge University Press.

WU, D. R., CRAMER, S. M., BELFORT, G. (1993), Kinetic resolution of racemic glycidyl butyrate using a multiphase membrane enzyme reactor: experiments and model verification, *Biotechnol. Bioeng.* **41**, 979–990.

YAGASAKI, M., OZAKI, A. (1998), Industrial biotransformations for the production of D-amino acids, *J. Mol. Catal. B: Enzymatic* **4**, 1–11.

ZIMMERMANN, T. P., ROBINS, K. T., WERLEN, J., HOEKS, F. W. J. M. M. (1997), Bio-transformation in the production of L-carnitine, in: *Chirality in Industry* Vol. II (COLLINS, A. N., SHELDRAKE, G. N., CROSBY, J., Eds.), pp. 287–305. New York: John Wiley & Sons.

ZMIJEWSKI, M. J., JR., BRIGGS, B. S., THOMPSON, A. R., WRIGHT, I. G. (1991), Enantioselective acylation of a β-lactam intermediate in the synthesis of loracarbef using penicillin G amidase, *Tetrahedron Lett.* **32**, 1621–1622.

7 Regioselective Oxidation of Aminosorbitol with *Gluconobacter oxydans*, Key Reaction in the Industrial 1-Deoxynojirimycin Synthesis

MICHAEL SCHEDEL
Wuppertal, Germany

1 Introduction

1-Deoxynojirimycin is the precursor to the potent α-glucosidase inhibitor Miglitol, its *N*-substituted analog (GERARD et al., 1987; SCOTT and TATTERSALL, 1988). Miglitol was launched as a new therapeutic drug for the treatment of non-insulin-dependent diabetes mellitus in 1998 (BERGMAN, 1999) in Europe, North America, and some other countries. The structures of 1-deoxynojirimycin and of Miglitol are shown in Fig. 1.

The industrial production of 1-deoxynojirimycin is a new example of a combined biotechnological-chemical synthesis carried out at a large scale. This synthesis integrates the selectivity offered by an enzymatic reaction into a sequence of chemical steps and results in an elegant, short and economical process (KINAST and SCHEDEL, 1981). The biotechnological key reaction is the regioselective oxidation of 1-amino-1-deoxy-D-sorbitol to 6-amino-6-deoxy-L-sorbose. Whole cells of *Gluconobacter oxydans* having a high specific activity of the membrane bound polyol dehy-

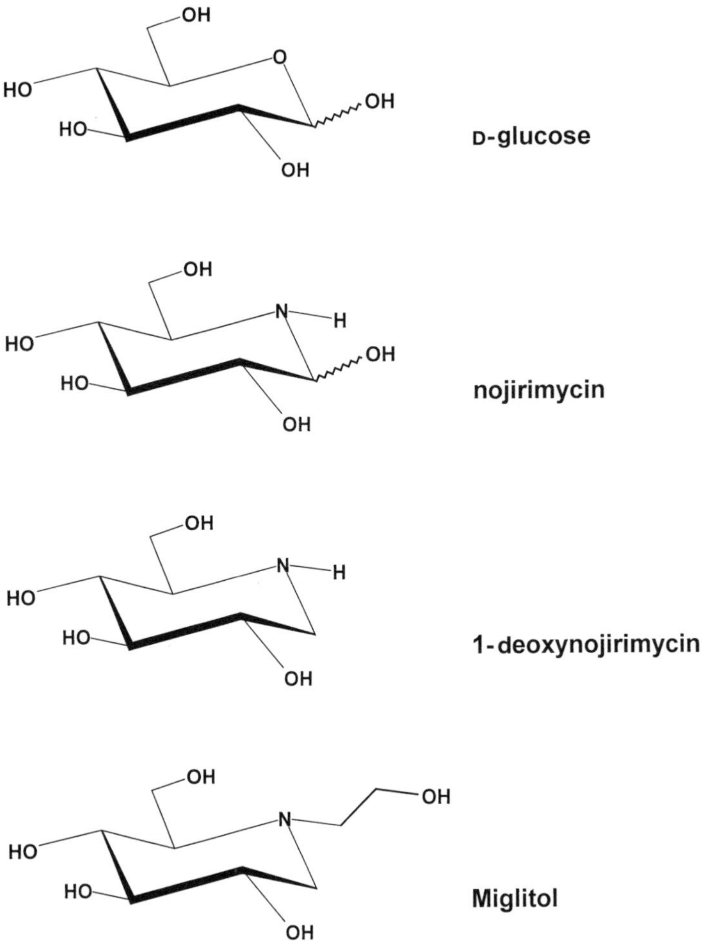

OH
HO
HO
O
OH
OH
D-glucose

OH
HO
HO
N—H
OH
OH
nojirimycin

OH
HO
HO
N—H
OH
1-deoxynojirimycin

OH
HO
HO
N
OH
OH
Miglitol

Fig. 1. The structures of D-glucose, nojirimycin, 1-deoxynojirimycin, and Miglitol.

drogenase are used as catalyst. This central biotransformation step is flanked by four chemical reactions.

The fruitful combination of biotechnology and chemistry opens up new and creative synthetic pathways. One of the best known examples is the synthesis of vitamin C (REICHSTEIN and GRÜSSNER, 1934; KULHÁNEK, 1970; BOUDRANT, 1990). In this efficient process the oxidation of D-sorbitol to L-sorbose with *Gluconobacter oxydans* (*Acetobacter suboxydans*) is part of a five-step synthesis starting from D-glucose. After its publication by REICHSTEIN it was for more than 60 years the method of choice – due to the short reaction sequence, the cheap availability of D-glucose as starting substrate, and the chemical stability of the intermediates. Various improvements have been introduced over the years.

The industrial synthesis of 1-deoxynojirimycin is similar to the vitamin C process. In both cases D-glucose is the starting material and the ability of *G. oxydans* to regioselectively oxidize straight chain polyhydric alcohols having a well defined steric configuration is employed. However, in the vitamin C synthesis D-sorbitol, a naturally occurring hexitol, is the substrate of the microbial oxidation step whereas in the 1-deoxynojirimycin case a chemically modified polyol not found in nature is used. The oxidation of such a terminally modified polyol was possible because *G. oxydans* demands a specific steric configuration at one side of the substrate to be oxidized whereas structure and size of the other part of the molecule are of little importance.

The present chapter describes the development of the combined biotechnological-chemical 1-deoxynojirimycin synthesis, compares it to other production alternatives, and summarizes relevant aspects of the oxidation of 1-amino-1-deoxy-D-sorbitol with *G. oxydans* at the production scale.

The name *Gluconobacter oxydans* is used according to the taxonomical classification described in *Bergey's Manual of Systematic Bacteriology* (DE LEY et al., 1984). In some earlier literature this bacterium was named *Gluconobacter suboxydans* or *Acetobacter suboxydans*. If such literature is cited the name used by the authors is given.

2 Nojirimycin, 1-Deoxynojirimycin, and Miglitol

1-Deoxynojirimycin is a natural compound which occurs in plants and in moths feeding on such plants; it can also be found in fermentation broths of various bacteria. Its isolation from natural sources was first described in a patent applied for by the Japanese company Nippon Shinyaku in 1976 (YAGI et al., 1976, 1977; MURAI et al., 1977); the inventors named this strong α-glucosidase inhibitor Moranoline and claimed its extraction from *Morus* plants and its use as antidiabetic agent.

Fig. 1 shows the structure of 1-deoxynojirimycin. It can be described as 1,5-dideoxy-1,5-imino-D-glucitol or as 5-hydroxymethyl-2,3,4-trihydroxypiperidine. This piperidinose shows structural similarities to D-glucose; it has the same steric configuration at four carbon atoms, however, nitrogen replaces oxygen in the ring. Ten years before l-deoxynojirimycin was found as a natural compound PAULSEN (1966) already had published its synthesis from 6-amino-6-deoxy-L-sorbose. The early work on 1-deoxynojirimycin was closely related to studies with its parent compound nojirimycin (Fig. 1) first isolated in 1965 as potent antibacterial substance from the fermentation broth of *Streptomyces roseochromogenes* R-468 and later also detected in the growth medium of other *Streptomyces* and of *Bacillus* strains (NISHIKAWA and ISHIDA, 1965; INOUYE et al., 1966, 1968; ISHIDA et al., 1967a, b; ARGOUDELIS and REUSSER, 1976; ARGOUDELIS et al., 1976; SCHMIDT et al., 1979). 1-Deoxynojirimycin can be prepared from nojirimycin by catalytic hydrogenation or reduction with sodium borohydride (INOUYE et al., 1966).

Both nojirimycin and 1-deoxynojirimycin strongly inhibit α-glucosidases isolated from the intestinal tract of animals and man (NIWA et al., 1970; REESE et al., 1971; MURAI et al., 1977; FROMMER et al., 1978a, b, 1979a; SCHMIDT et al., 1979; KODAMA et al., 1985) and immediately attracted the interest of various research groups. As lead structures they offered a significant potential for the development of new therapeutic drugs to treat diabetes. Many semi-synthetic *N*-substituted derivatives of 1-deoxynojirimycin have been prepared (JUNGE et al., 1979; MATSUMURA et al., 1979a) and Miglitol, the *N*-hydroxyethyl analog, was identified as one of the most interesting candidates showing a desired enzyme inhibitory profile (GERARD et al., 1987; SCOTT and TATTERSALL, 1988; JUNGE et al., 1996).

Miglitol was developed by Bayer as drug for the treatment of type II (non-insulin-dependent) diabetes mellitus. It was authorized in 1996 for marketing in Europe (trade name: Diastabol®) and the United States (Glyset®) and launched in 1998 in cooperation with licensing partners first in Germany and later in many other countries including the United States, Canada, and Australia. Miglitol is the third α-glucosidase inhibitor to reach the market – after Acarbose (Glucobay®, Precose®), a secondary metabolite of pseudotetrasaccharide structure produced by fermentation of *Actinoplanes* sp. (SCHMIDT et al., 1977; CLISSOLD and EDWARDS, 1988; BISCHOFF, 1994) and Voglibose (Basen®), an *N*-substituted valiolamine derivative which is on the market in Japan and some other Far Eastern countries (HORII et al., 1982; NAKAMURA et al., 1993; ODAKA and IKEDA, 1995).

3 Three Routes to Produce 1-Deoxynojirimycin

For the industrial production of Miglitol 1-deoxynojirimycin is needed as starting material. There are three principal routes to prepare 1-deoxynojirimycin (HUGHES and RUDGE, 1994; STÜTZ, 1999):

(1) Extraction from plants like the mulberry tree.
(2) Fermentation of various *Bacillus* or *Streptomyces* strains.
(3) Chemical synthesis following different synthetic strategies.

3.1 Extraction from Plants

1-Deoxynojirimycin can be extracted with water or polar solvents from the bark, the roots, and the leaves of the mulberry tree (*Morus alba, M. bombycis, M. nigra*) and also from other plants including *Jacobinia suberecta* and *J. tinctoria*, the Chinese herbs *Vespae nidus* and *Bombyx batryticatus*, and two Euphorbiaceae species, the rain forest tree *Endospermum medullosum* and the liana *Omphalea queenslandiae* (YAGI et al., 1976, 1977; MATSUMURA et al., 1980; EVANS et al., 1985; DAIGO et al., 1986; KITE et al., 1991; YAMADA et al., 1993; SIMMONDS et al., 1999). The extraction of 1-deoxynojirimycin from plant material as a route of production on the industrial scale has, however, never seriously been regarded as feasible. The content of 1-deoxynojirimycin in plants is low and the purification process would not be simple as structurally very similar substances are also present in these plants like nojirimycin and 1-deoxymannojirimycin (DAIGO et al., 1986; ASANO et al., 1994a, b).

Moreover, the sufficient and steady supply would be difficult to establish, pose an immense logistic problem, and need the availability of sufficient land.

3.2 Fermentation

The fermentative route offers an interesting large scale production alternative. A number of *Bacillus* and *Streptomyces* strains were found to synthesize 1-deoxynojirimycin in appreciable amounts and secrete it into the medium. SCHMIDT et al. (1979) and FROMMER et al. (1978a, 1979b) isolated 1-deoxynojirimycin from the culture broth of *Bacillus amyloliquefaciens, B. polymyxa*, and *B. subtilis*. FROMMER and SCHMIDT (1980) claimed a fermentation process using *Bacillus* strains; they were able to isolate 360 g of pure 1-deoxynojirimycin base after a multistep purification process starting from 360 L of fermentation broth. 1-Deoxynojirimycin production using *Streptomyces lavendulae* was reported by MATSUMURA et al. (1979b), MURAO and MIYATA (1980), in a patent of Amano Pharmaceuticals (SUMITA and MURAO, 1980), and by EZURE et al. (1985). HARDICK et al. (1991, 1992) and HAR-

DICK and HUTCHINSON (1993) investigated the biosynthesis of 1-deoxynojirimycin in *Streptomyces subrutilis* and *Bacillus subtilis* var. niger.

To produce 1-deoxynojirimycin economically by fermentation a yield approaching 50 g per liter would be necessary to reach a price of less than 100 US$ per kg. It does not seem unrealistic that this goal can be achieved as some of the wild type strains reported in the literature already produced 1-deoxynojirimycin in gram quantities. However, an intensive genetic strain development program and the optimization of the fermentation process over several years would be necessary.

3.3 Chemical Synthesis

There are many publications describing different synthetic routes to 1-deoxynojirimycin; they were reviewed by NISHIMURA (1991), HUGHES and RUDGE (1994), JUNGE et al. (1996), and LA FERLA and NICOTRA (1999). The synthesis of 1-deoxynojirimycin presents a stereochemical challenge due to the presence of four contiguous chiral centers and the same functional group occurring several times. The various synthetic routes described in the literature may be classified into three groups:

(1) Starting material is D-glucose, the nitrogen is introduced at C-1, the hydroxyl group at C-5 is oxidized to the keto group, and 1-deoxynojirimycin is obtained by reductive ring closure. An example is the original synthesis published by PAULSEN (PAULSEN, 1966, 1999; PAULSEN et al., 1967).

(2) Starting material is again D-glucose, the nitrogen is introduced at C-5 while maintaining the configuration present in D-glucose followed by reductive ring closure. This synthetic strategy was first described for nojirimycin by INOUE et al. (1968). The synthetic routes in this and the first group take advantage of the fact that already three correct stereocenters needed in 1-deoxynojirimycin are contributed by the homochiral starting material D-glucose.

(3) A number of other syntheses start from building blocks which do or do not have chiral centers like glucosylamine (BERNOTAS and GANEM, 1985), mannose or idose derivatives (FLEET, 1989; FLEET et al., 1990), tartaric acid (IIDA et al., 1987), pyroglutamic acid (IKOTA, 1989), or phenylalanine (RUDGE et al., 1994). An example of a chemo-enzymatic process has been reported by WONG et al. (1995). In their synthesis dihydroxyacetone phosphate is stereoselectively connected to 3-azido-2-hydroxypropan-al catalyzed by rabbit muscle aldolase.

All synthetic routes are complex; most of them have a relatively large number of individual reaction steps and need a substantial amount of protection group chemistry. In some of the syntheses proposed the lack of selectivity in crucial steps led to significant amounts of one or more of the 15 other possible stereoisomers. 1-Deoxynojirimycin production by chemical syntheses on an industrial scale following the above mentioned strategies was not economical in those cases considered.

4 Concept for a Combined Biotechnological-Chemical Synthesis of 1-Deoxynojirimycin

An interesting concept for a combined biotechnological-chemical synthesis was proposed by KINAST and SCHEDEL (1981) which is shown in Fig. 2. It took advantage of the ability of *Gluconobacter oxydans* to regio- and stereoselectively oxidize polyol substrates with high productivity according to the rule of BERTRAND and HUDSON (BERTRAND, 1904; HANN et al., 1938). The key reaction in the proposed three-step synthesis was the regioselective microbial oxidation of 1-amino-1-deoxy-D-sorbitol to 6-amino-6-deoxy-L-sorbose. This crucial step was flanked by well established and comparatively simple chemical reactions: the reductive amination of D-glucose to give 1-amino-1-deoxy-D-sorbitol and the stereoselective

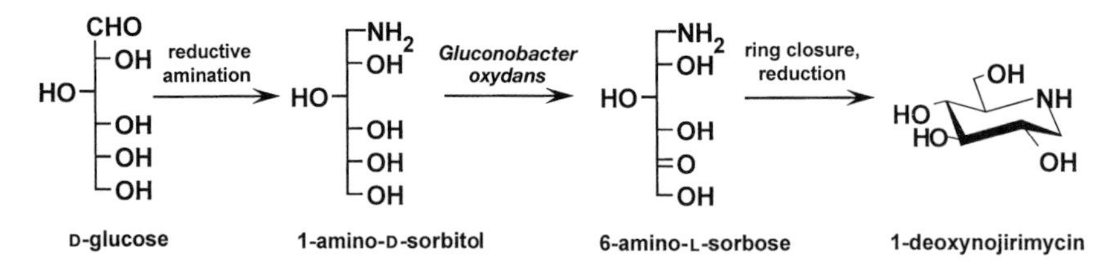

Fig. 2. Concept for the combined biotechnological-chemical synthesis of 1-deoxynojirimycin.

reductive ring closure of 6-amino-6-deoxy-L-sorbose to lead to 1-deoxynojirimycin.

The strategy of this synthesis was comparable to the original synthesis by PAULSEN (PAULSEN, 1966; PAULSEN et al., 1967): Use of D-glucose as starting material which contains at carbon atoms 2, 3, and 4 three correct stereocenters, introduction of nitrogen at carbon atom C-1, oxidation of the hydroxyl group at C-5 followed by reductive ring closure. Use of the selectivity of the polyol dehydrogenase enzyme of *G. oxydans* to oxidize 1-amino-1-deoxy-D-sorbitol made a short reaction sequence possible with no need for protection group chemistry. Regarding the final stereoselective ring closure and reduction step it was known from the studies of PAULSEN et al. (1967) that the microbial oxidation product 6-amino-6-deoxy-L-sorbose existed in solution at alkaline pH-values in the piperidinose form and could readily be reduced with appropriate reducing agents like sodium borohydride (KÖBERNICK, 1982) to give 1-deoxynojirimycin.

5 Oxidation of 1-Amino-1-Deoxy-D-Sorbitol by *Gluconobacter oxydans*

1-Amino-1-deoxy-D-sorbitol, the substrate of the microbially catalyzed key reaction, is a terminally modified polyol. For the success of the proposed 1-deoxynojirimycin synthesis it was essential that the introduction of the amino group at carbon atom C-1 did not influence the capability of *Gluconobacter oxydans* to oxidize the hydroxyl group at carbon atom C-5 according to the Bertrand–Hudson rule (BERTRAND, 1904; HANN et al., 1938). This rule states that *G. oxydans* regioselectively oxidizes the middle of three terminal hydroxyl groups if they have the D-erythro configuration. According to the literature the size and structure of the remaining part of the molecule had in the majority of cases little or no effect on the oxidation capability of *Gluconobacter* (KULHÁNEK, 1989). There are more than 30 publications demonstrating the successful oxidation of the penultimate hydroxyl group of polyols carrying the favorable Bertrand–Hudson configuration at one end of the molecule but are chemically modified or differ in carbon chain length or configuration at the other end; some relevant publications are summarized in Tab. 1. BERTRAND (1898, 1900), FULMER and UNDERKOFLER (1947), HANN et al. (1938), HANN and HUDSON (1939), ONISHI and SUZUKI (1969), MOSES and FERRIER (1962), BERTRAND and NITZBERG (1928), TILDEN (1939), MACLAY et al. (1942), ETTEL and LIEBSTER (1949), PRATT and RICHTMYER (1955), and others oxidized polyols with different carbon chain length ranging from glycerol to heptitols and octitols. Particularly the oxidation of several heptitols showed that the configuration outside the Bertrand–Hudson group was of little effect on the oxidation reaction (PRATT and RICHTMYER, 1955). It is of interest that even a cyclitol (inositol) could be oxidized (DUNNING et al., 1938; CHARGAFF and MAGASANIK, 1946). JONES and MITCHELL (1959), JONES et al. (1962), and HOUGH et al. (1959) used *Acetobacter suboxydans* for the oxidation of modified pentitols and BOSSHARD and REICHSTEIN (1935), HOUGH et al. (1959), JONES et al. (1961a, b), KULHÁNEK et al. (1977), BUDESÍNSKÝ et al. (1984), and TIWARI et al. (1986) reported the oxidation of several chemically modified hexitols. In these studies various deoxy-, deoxyamino-, deoxyhalogen- and deoxy-

Tab. 1. Oxydation of Various Polyols According to the Rule of BERTRAND and HUDSON

Polyol(s) Oxidized	References
Glycerol	BERTRAND (1898)
meso-Erythritol	BERTRAND (1900)
	WHISTLER and UNDERKOFLER (1938)
	FULMER and UNDERKOFLER (1947)
	HU et al. (1965)
D-Arabitol	HANN et al. (1938)
	ONISHI and SUZUKI (1969)
Ribitol	MOSES and FERRIER (1962)
D-Sorbitol	BERTRAND (1904)
D-Allitol	CARR et al. (1968)
D-Mannitol	CUSHING and DAVIS (1958)
D-Fructose	AVIGARD and ENGLARD (1965)
meso-, l-, d-, epi-Inositol	DUNNING et al. (1938)
	CHARGAAF and MAGASANIK (1946)
alpha-Glucoheptit	BERTRAND and NITZBERG (1928)
D-*alpha*-Mannoheptitol (perseitol)	HANN et al. (1938)
D-*alpha,beta*-Glucooctitol	
Other polyols	
D-Manno-D-gala-heptitol (perseitol)	HANN and HUDSON (1939)
	TILDEN (1939)
L-Gluco-L-gulo-heptitol (*alpha*-glucoheptitol)	MACLAY et al. (1942)
Volemitol (*alpha*-sedoheptitol)	ETTEL and LIEBSTER (1949)
	STEWART et al. (1949)
D-Gluco-D-ido-heptitol	PRATT et al. (1952)
beta-Sedoheptitol	STEWART et al. (1952)
(D-altro-D-gluco-heptitol or L-gulo-D-talo-heptitol)	
meso-Glycero-allo-heptitol	PRATT and RICHTMYER (1955)
D-Glycero-D-altro-heptitol	
D-Erythro-D-galacto-octitol	CHARLSON and RICHTMYER (1960)
2-Deoxy-D-erythro-pentitol	LINEK et al. (1979)
1-Deoxy-1-*S*-ethyl-D-arabitol	JONES and MITCHELL (1959)
L-Sorbitol-3-methyl-ether	BOSSHARD and REICHSTEIN (1935)
1-Deoxy-D-glucitol	KAUFMANN and REICHSTEIN (1967)
6-Deoxy-L-allitol	
2-Deoxy-D-sorbitol	FULMER and UNDERKOFLER (1947)
1-Deoxy-1-*S*-ethyl-D-glucitol	HOUGH et al. (1959)
1-Deoxy-1-*S*-ethyl-D-arabitol	
1-Deoxy-1-*S*-ethyl-D-galactitol	
2-Amino-2-deoxy-D-glucitol	
2-Acetamido-2-deoxy-D-glucitol	
5-Deoxy-D-ribitol	
6-Deoxy-L-galactitol	
5-O-Methyl-D-ribitol	
6-O-Acetyl-DL-galactitol	
6-Deoxy-6-*S*-ethyl-L-galactitol	
1-O-Acetyl-DL-galactitol	
2-Acetamido-2-deoxy-D-glucitol	JONES et al. (1961a)
1-Deoxy-1-N-methylacetamido-D-glucitol	JONES et al. (1961b)
2-Acetamido-1,2-dideoxy-D-glucitol	
2-Deoxy-2-fluoro-D-glucitol	KULHÁNEK et al. (1977)
3-Deoxy-3-fluoro-D-mannitol	BUDESÍNSKÝ et al. (1984)
2-Deoxy-D-arabino-hexitol	TIWARI et al. (1986)
L-Fucitol (6-deoxy-L-galactitol)	RICHTMYER et al. (1950)
D-Rhamnitol	ANDERSON and LARDY (1948)
omega-Deoxy sugar alcohols	BOLLENBACK and UNDERKOFLER (1950)

sulfuralditols were employed as substrates. Anderson and Lardy (1948), Richtmyer et al. (1950), and Bollenback and Underkofler (1950) oxidized ω-deoxy sugars and Hough et al. (1959) in addition ω-O-methyl- and ω-deoxy-acetyl sugars. In these polyols the hydroxyl group at the third to last carbon atom was oxidized. To stay in accord with the Bertrand–Hudson rule the authors regarded the terminal methyl group as equal to hydrogen, i.e., the terminal two-carbon group CH_3–CHOH was considered as a slightly extended CH_2OH group.

All the above mentioned and in Tab. 1 summarized polyols could be oxidized according to the rule of Betrand and Hudson with *G. oxydans*. In some studies the yield reported was, however, low and the fermentation time needed long (up to two or three weeks). Earlier reviews summarizing the oxidation of polyhydric alcohols by *G. oxydans* were written by Fulmer and Underkofler (1947), Janke (1960), Spencer and Gorin (1968), and Kulhánek (1989).

Examples of polyhydroxy compounds which were not oxidized by *G. oxydans* although they have the favorable Bertrand–Hudson configuration are D-glucose dimethylacetal (Iselin, 1948), D-mannose diethylmercaptal (Hann et al., 1938), the 1,1-diethyldithioacetals of D-arabinose, D-glucose, and D-galactose and the 1,1-dibenzyldithioacetal of D-arabinose (Hough et al., 1959). These results indicate that there are additional factors which are relevant for the oxidation of open chain polyols.

1-Amino-1-deoxy-D-sorbitol contains the Bertrand–Hudson-configuration and could readily be oxidized to 5-amino-5-deoxy-L-sorbose employing various *G. oxydans* strains (Kinast and Schedel, 1980a). The oxidation product 6-amino-6-deoxy-L-sorbose was, however, not stable in water at near neutral pH under the conditions of the biotransformation reaction: spontaneous ring closure occurred followed by removal of water leading to 3-hydroxy-2-hydroxymethyl-pyridine (Fig. 3). Paulsen et al. (1967) had reported that under their conditions the pyridine formation from 6-amino-6-deoxy-L-sorbose was quantitative after five days. Due to the loss of 6-amino-6-deoxy-L-sorbose the necessary high 1-deoxynojirimycin yield could not be obtained. To prevent the spontaneous intramolecular ring closure reaction the amino group had to be protected. Various protection groups could successfully be employed. In a series of patents different protection groups were claimed which could be removed either by hydrogenolytic cleavage (Kinast and Schedel, 1980b), under alkaline or acidic conditions (Kinast et al., 1982; Schröder and Stubbe, 1987), or enzymatically (Schutt, 1993). Astonishingly, many of these *N*-protected polyols were readily oxidized by *G. oxydans* even if the protection group used – like the benzyloxycarbonyl group – was large and significantly lowered the water solubility of the substrate to be employed.

In the reaction sequence of the combined biotechnological-chemical synthesis of 1-deoxynojirimycin two additional steps were thus included (Fig. 4): Introduction of a favorable protection group before the oxidation step and its removal afterwards prior to reduction. Small and hydrophilic groups like formyl-, dichloroacetyl-, or trichloroacetyl- which could be cleaved off under acidic or alkaline conditions were preferred. All further work to be described in the following sections was done with *N*-formyl-1-amino-1-deoxy-D-sorbitol as example.

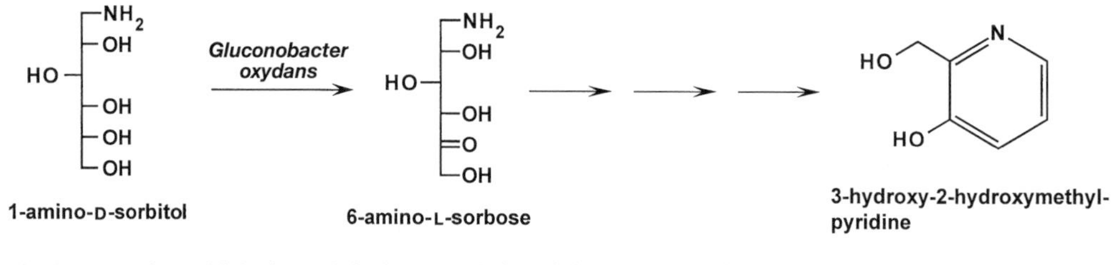

Fig. 3. Formation of 3-hydroxy-2-hydroxymethyl-pyridine from 6-amino-6-deoxy-sorbose.

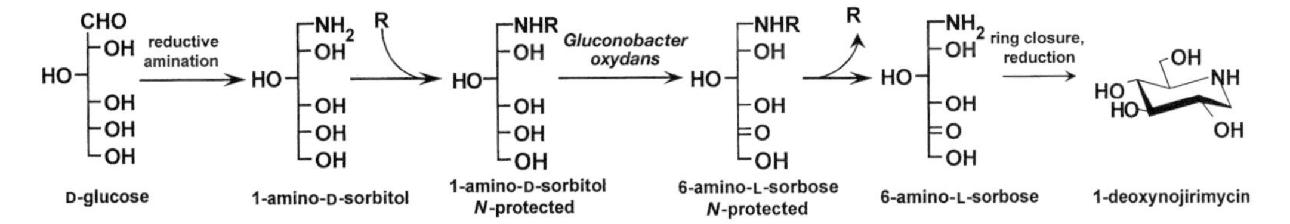

Fig. 4. The combined biotechnological-chemical synthesis of 1-deoxynojirimycin.

6 Biotransformation of *N*-Formyl-1-Amino-1-Deoxy-D-Sorbitol on the Industrial Scale

The large scale industrial production of L-sorbose as part of the vitamin C process is carried out with *Gluconobacter oxydans* cells growing on D-sorbitol under fermentation conditions (ROSENBERG et al., 1993). *N*-Formyl 1-amino-1-deoxy-D-sorbitol could, however, not be oxidized in the same way: This modified polyol did not promote growth of *G. oxydans* when added in high concentrations instead of D-sorbitol as sole polyol substrate to a yeast extract/salt medium. The same result was obtained when other protecting groups were employed. *N*-Formyl-amino-1-deoxy-D-sorbitol even inhibited growth in a concentration dependent manner when added together with D-sorbitol. The biotransformation reaction had, therefore, to be carried out as a two-step process (SCHEDEL, 1997): In the first step *G. oxydans* biomass was produced by fermentation on a "good" growth substrate, preferably D-sorbitol, separated from the broth and stored under cooling as slurry in a stirred tank. In the second step *N*-formyl-1-amino-1-deoxy-D-sorbitol which had been dissolved in water was oxidized with *G. oxydans* biomass as catalyst under "resting cell" conditions with intensive aeration and at a slightly acidic pH value. The productivity of the microbial oxidation reaction using *G. oxydans* strain DSM 2003 was very high: 250 kg of substrate per m^3 could quantitatively be oxidized in less than 16 h with a biomass concentration of about 2.5 kg cell dry weight per m^3. The specific oxidation activity was around 50 mmol of substrate oxidized per hour and g cell dry weight. A typical time course is shown in Fig. 5.

It is of interest to note that under these conditions resting *G. oxydans* cells oxidized D-sorbitol, *N*-formyl-1-amino-1-deoxy-D-sorbitol, or other polyols completely uncoupled: The electrons removed from the substrate were transferred to oxygen without conservation of metabolically utilizable energy; only heat was produced. The non-existence of a respiratory regulation was the prerequisite for this industrial biotransformation process employing resting cells suspended simply in water. The uncoupled dehydrogenation of monosaccharides by *G. oxydans* had been pointed out by KULHÁNEK (1989) and USPENSKAYA and LOITSYANSKAYA (1979).

Four main aspects of the *N*-formyl-1-amino-1-deoxy-D-sorbitol biotransformation reaction were important on the production scale under reaction conditions as described above, i.e., a *Gluconobacter* biomass concentration of about 2.5 kg m^{-3} and a substrate concentration of 250 kg m^{-3}:

(1) High oxygen demand (about 6 kg O_2 m^{-3} h^{-1}).
(2) Removal of the process heat (about 10 KW m^{-3}).
(3) Labor intensive handling of large quantities of solid substrate.
(4) Stability of the *G. oxydans* biocatalyst.

Regarding point (1) and (2) an adequately equipped stirred fermenter was necessary to meet the high oxygen demand and to effectively remove the heat which was produced by the biological reaction and introduced by the stirrer. To cope with the solid substrate handling – about 10 t of material had to be filled into the fermenter within a reasonably short period of time – an automated bigbag delivery

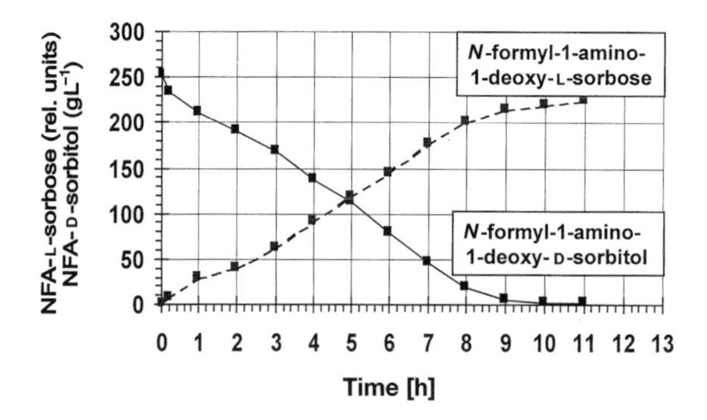

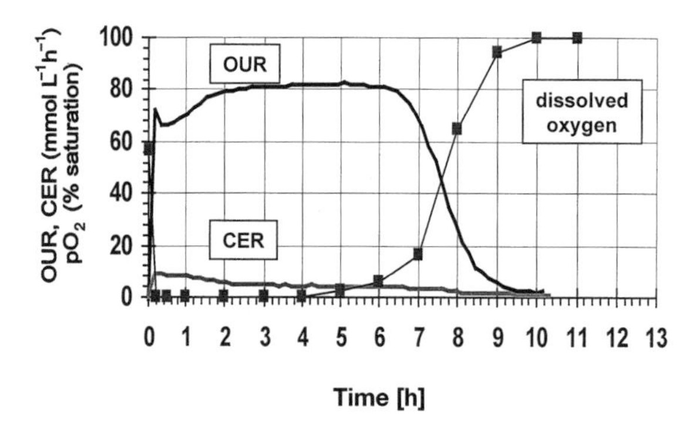

Fig. 5. Time course of the oxidation of *N*-formyl-1-amino-1-deoxy-D-sorbitol on the production scale with *Gluconobacter oxydans* DSM 2003.

system to charge the fermenter was installed. Due to instability problems after prolonged storage it was not possible to supply *N*-formyl-1-amino-1-deoxy-D-sorbitol as a concentrated solution which could be pumped from a storage tank to the medium preparation tank. The activity loss of *G. oxydans* cells under "resting cell conditions" limited their reuse and also the possibilities to develop an immobilized biocatalyst. After one biotransformation cycle *Gluconobacter* cells had lost about half of their specific activity (SCHEDEL, 1997). Successful immobilization was also hindered by the very high oxygen demand and the enormous substrate turnover rate which caused a severe transport limitation within immobilization pearls. The oxidation of D-sorbitol to L-sorbose with immobilized *Gluconobacter* cells had repeatedly been reported in the literature (SCHNARR et al., 1977; STEFANOVA et al., 1987; TRIFONOV et al., 1991; KOSSEVA et al., 1991).

7 Fermentative Production of *Gluconobacter oxydans* Cells on the Industrial Scale

The fermentative production of *Gluconobacter oxydans* cells was carried out in a yeast extract/salt medium containing D-sorbitol; other oxidizable substrates like glycerol could also be used. Fig. 6 gives a typical time course of a fermentation on the 30 m^3 scale with *G. oxydans* strain DSM 2003 on D-sorbitol. The final biomass yield was low: about 5 g cell dry weight per liter. Only a very small portion of the D-sorbitol was used as carbon source by *G. oxydans* DSM 2003 and flowed into the metabolism, about 95% of the substrate was oxidized to L-sorbose and deposited in the medium. Correspondingly, the respiratory coefficient for such a "tight" *Gluconobacter* strain was low throughout the fermentation and ranged

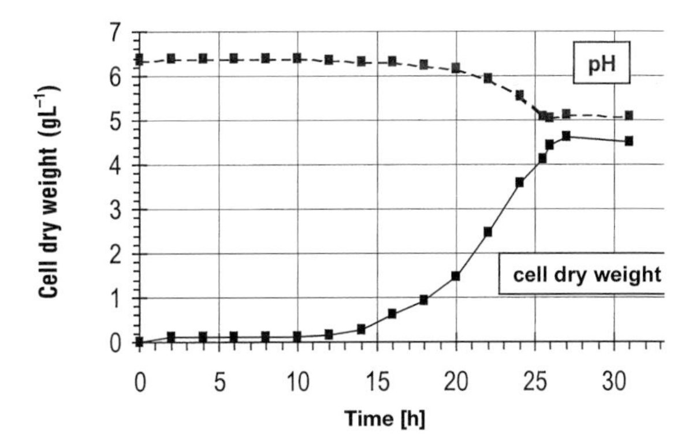

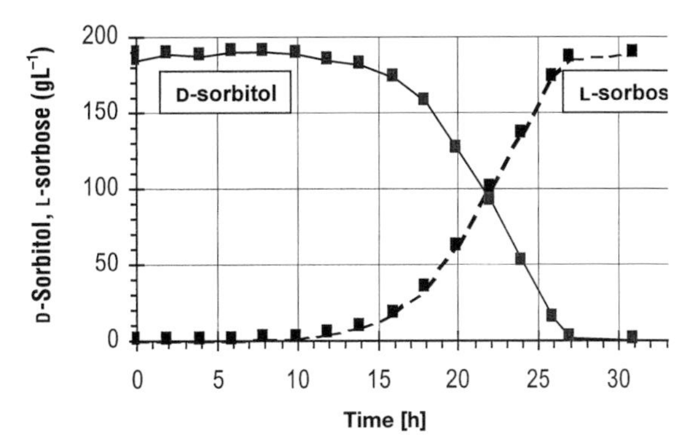

Fig. 6. Time course of the *Gluconobacter oxydans* fermentation on the production scale (strain DSM 2003).

around a value of 0.15. As reported in the literature the specific oxidation activity was highest at the entry to the stationary phase (BATZING and CLAUS, 1973; WHITE and CLAUS, 1982). Therefore, when the D-sorbitol substrate was used up and the biomass yield no longer increased, the cells were separated from the medium and stored as a cooled slurry.

The *Gluconobacter* biomass production could successfully be run at the 30 L laboratory scale as a continuous process under chemostat conditions with D-sorbitol limitation; more than 25 volumes of medium were passed through the fermenter with no loss in specific oxidative activity of the *Gluconobacter* cells obtained (SCHEDEL, 1997). There are several reports in the literature that the L-sorbose fermentation using *Gluconobacter* was carried out continuously (see, for instance: ELSWORTH et al., 1959; MÜLLER, 1966; NICOLSKAYA et al., 1984). In the pharmaceutical industry, however, continuous fermentations to be run at large scale under completely contamination free conditions are very rarely the system of choice. There are several reasons for this: Increased risk of non-sterility, possible genetic instability of the production organism during long-term fermentation necessitating extensive analytical characterization work to assure constant product quality, high investment into storage tanks upstream and downstream from the fermentation itself. The production of *Gluconobacter* biomass has, however, a good chance to become one of the very few large scale continuous fermentations in the pharmaceutical area. This is mainly due to the fact that no significant additional investment will be needed neither for the continuous preparation of medium or its storage under sterile conditions nor for the harvest and storage of large volumes of product. The medium can be continuously mixed together from three liquid streams (water, D-sorbitol syrup, concentrated basal medium containing salts and yeast extract; see Fig. 7) and passed via a continuous sterilizer into the fermenter. The microbial biomass is a small

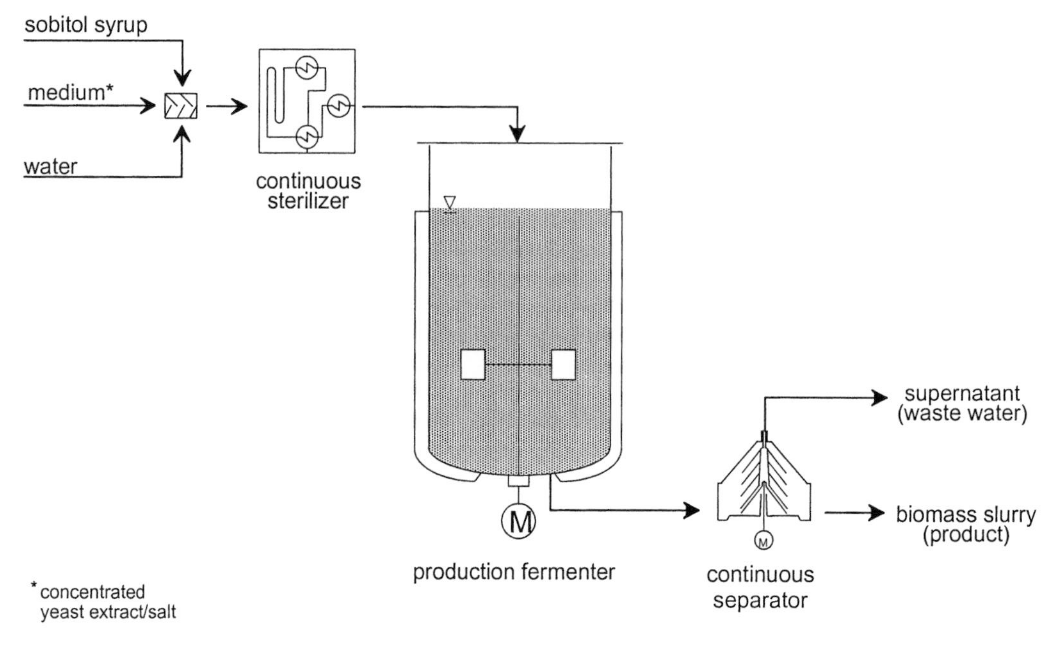

Fig. 7. Concept for the continuous fermentation of *Gluconobacter oxydans* on the production scale under chemostat conditions.

volume product which is harvested continuously with an existing separator; the culture supernatant, the large volume side product is directly passed to the wastewater plant without intermediate storage. The existing batch process plant needs, therefore, only minor amendments to set up a fully continuous *Gluconobacter* fermentation.

The enzymatic activity for the oxidation of *N*-protected 1-amino-1-deoxy-D-sorbitol is located in the membrane fraction. Kinetic studies indicated close similarity between D-sorbitol and *N*-formyl-1-amino-1-deoxy-D-sorbitol oxidation. It is likely – but not yet proved – that both substrates are oxidized by the same enzyme which belongs to either of the two types of membrane bound D-sorbitol dehydrogenases described for *G. oxydans:* The cytochrome containing three-subunit enzyme with a total molecular weight of about 140,000 Da described by SHINAGAWA et al. (1982) and CHOI et al. (1995) or the one-subunit enzyme with a molecular weight of around 80,000 Da reported by HOSHINO et al. (1996). The membrane bound polyol dehydrogenases do not need cofactors like NAD or NADP as reported for the sorbitol dehydrogenases which had been isolated from the soluble fraction by CUMMINS et al. (1957) or HAHN et al. (1989).

8 Conclusions and Outlook

The combined biotechnological-chemical synthesis of 1-deoxynojirimycin is meanwhile a well established process on the industrial scale which enables the economic production of this precursor to Miglitol. Combination of biotechnology and chemistry has led to a short and elegant synthesis with little protection group chemistry.

Regarding potential safety risks and environmental aspects the central biotransformation step has significant advantages: The microbially catalyzed polyol oxidation is run under physiological conditions and – apart from acids and bases needed for pH-control – involves no toxic, easily inflammable, or otherwise hazardous substances. The microorganism used is safe. All organic components are readily degradable.

Compared to a fully chemical synthesis, however, a significant investment is necessary into a large fermenter with state-of-the-art technology. A working volume of 30, 50, or 100 m^3 is generally not standard in chemical production. The *Gluconobacter* fermentation step has to be run under completely contamination free conditions, a challenge to the equipment and the qualification of the operators. The *Gluconobacter* fermentation produces a large volume wastewater stream with a high biological oxygen demand. If the fermentation is to be run continuously the capacity of the wastewater plant will soon set the limit to the overall process.

9 References

ANDERSON, L., LARDY, H. A. (1948), The synthesis of D-fructomethylose by biochemical oxidation, *J. Am. Chem. Soc.* **70**, 594–597.

ARGOUDELIS, A. D., REUSSER, F. (1976), Process for producing antibiotic U-51640, *US Patent* 3998698.

ARGOUDELIS, A. D., REUSSER, F., MIZSAK, S. A., BACZYNSKYJ, L. (1976), Antibiotics produced by *Streptomyces ficellus* II. Feldamycin and nojirimycin, *J. Antibiot.* **29**, 1007–1014.

ASANO, N., OSEKI, K., TOMIOKA, E., KIZU, H., MATSUI, K. (1994a), *N*-Containing sugars from *Morus alba* and their glycosidase inhibitory activities, *Carbohydr. Res.* **259**, 243–255.

ASANO, N., TOMIOKA, E., KIZU, H., MATSUI, K. (1994b), Sugars with nitrogen in the ring isolated from the leaves of *Morus bombycis, Carbohydr. Res.* **253**, 235–245.

AVIGARD, G., ENGLARD, S. (1965), 5-Keto-D-fructose. I. Chemical characterization and analytical determination of the dicarbonyl hexose produced by *Gluconobacter cerinus*, *J. Biol. Chem.* **240**, 2290–2296.

BATZING, B. L., CLAUS, G. W. (1973), Fine structural changes of *Acetobacter suboxydans* during growth in a defined medium, *J. Bacteriol.* **113**, 1455–1461.

BERGMAN, H. D. (1999), Major new drugs. Part 2, *Commun. Pharm.* **90**, 35–39.

BERNOTAS, R. C., GANEM, B. (1985), Efficient preparation of enantiomerically pure cyclic aminoalditols, total synthesis of 1-deoxynojirimycin and 1-deoxymannojirimycin, *Tetrahedron Lett.* **26**, 1123–1126.

BERTRAND, M. G. (1898), Action de la bactérie du sorbose sur les alcools plurivalents, *C. R. Acad. Sci.* **126**, 762–765.

BERTRAND, M. G. (1900), Sur l'oxidation de l'érythrite par la bactérie du sorbose; production d'un nouveau sucre: l'érythrulose, *C. R. Acad. Sci.* **130**, 1330–1333.

BERTRAND, M. G. (1904), Étude biochimique de la bactérie du sorbose, *Ann. Chim. Phys.* **3**, 181–288.

BERTRAND, G., NITZBERG, G. (1928), La fonction cétonique de l'α-glucoheptulose, *C. R. Acad. Sci.* **134**, 1172–1175.

BISCHOFF, H. (1994), Pharmacology of α-glucosidase inhibition, *Eur. J. Clin. Invest.* **24**, 3–10.

BOLLENBACK, G. N., UNDERKOFLER, L. A. (1950), The action of *Acetobacter suboxydans* upon ω-desoxy sugar alcohols, *J. Am. Chem. Soc.* **72**, 741–745.

BOSSHARD, W., REICHSTEIN, T. (1935), 1-Sorbose-4-methyläther, *Helv. Chim. Acta* **18**, 959–961.

BOUDRANT, J. (1990), Microbial processes for ascorbic acid biosynthesis: a review, *Enzyme Microb. Technol.* **12**, 322–329.

BUDESÍNSKÝ, M., CERNÝ, M., DOLEZALOVÁ, J., KULHÁNEK, M., PACÁK, J., TADRA, M. (1984), 4-Deoxy-4-fluoro-D-fructose: Preparation and structure, *Coll. Czech. Chem. Commun.* **49**, 267–274.

CARR, J. G., COGGINS, R. A., HOUGH, L., STACY, B. E., WHITING, G. C. (1968), Microbiological oxidation of allitol to L-ribo-hexulose by *Acetomonas oxydans*, *Phytochemistry* **7**, 1–4.

CHARGAFF, E., MAGASANIK, B. (1946), Oxidation of stereoisomers of the inositol group by *Acetobacter suboxydans*, J. *Biol. Chem.* **165**, 379–380.

CHARLSON, A. J., RICHTMYER, N .K. (1960), The isolation of an octulose and an octitol from natural sources: D-glycero-D-manno-octulose and D-erythro-D-galacto-octitol from the avocado and D-glycero-D-manno-octulose from *Sedum* species, *J. Am. Chem. Soc.* **82**, 3428–3434.

CHOI, E.-S., LEE, E.-H., RHEE, S.-K. (1995), Purification of a membrane-bound sorbitol dehydrogenase from *Gluconobacter suboxydans, FEMS Microbiol. Letters* **125**, 45–50.

CLISSOLD, S. P., EDWARDS, C. (1988), Acarbose, a preliminary review of its pharmacodynamic and pharmacokinetik properties, and therapeutic potential, *Drugs* **35**, 214–243.

CUMMINS, J. T., CHELDELIN, V. H., KING, T. S. (1957), Sorbitol dehydrogenases in *Acetobacter suboxydans, J. Biol. Chem.* **226**, 301–306.

CUSHING, I. B., DAVIS, R. V. (1958), Purification of fructose on ion exchange resins, *J. Am. Pharm. Assoc.* **47**, 765–768.

DAIGO, K., INAMORI, Y., TAKEMOTO, T. (1986), Studies

on the constituents of the water extract of the root of mulberry tree (*Morus bombycis* KOIDZ), *Chem. Pharm. Bull.* **34**, 2243–2246.

DE LEY, J., GILLIS, M., SWINGS, J. (1984), Family VI. Acetobacteraceae, in: *Bergey's Manual of Systematic Bacteriology,* Vol. 1 (KRIEG, N. R., HOLT, J. G., Eds.), pp. 267–278. Baltimore: William and Wilkins.

DUNNING, J. W., FULMER, E. I., GUYMON, J. F., UNDERKOFLER, L. A. (1938), The growth and chemical action of *Acetobacter suboxydans* upon *i*-inositol, *Science* **87**, 72.

ELSWORTH, R., TELLING, R. C., EAST, D. N. (1959), The investment value of continuous culture, *J. Appl. Bacteriol.* **22**, 138–152.

ETTEL, V., LIEBSTER, J. (1949), Oxydation biochimique de la volémite, *Coll. Czech. Chem. Commun.* **40**, 80–90.

EVANS, S. V., FELLOWS, L. E., SHING, T. K. M., FLEET, G. W. J. (1985), Glycosidase inhibition by plant alkaloids which are structural analogues of monosaccharides, *Phytochemistry* **24**, 1953–1955.

EZURE, Y., MARUO, S., MIYAZAKI, K., KAWAMATA, M. (1985), Moranoline (1-deoxynojirimycin) fermentation and its improvement, *Agric. Biol. Chem.* **49**, 1119–1125.

FLEET, G. W. J. (1989), Homochiral compounds from sugars, *Chem. in Britain* **25**, 287–292.

FLEET, G. W. J., CARPENTER, N. M., PETURSSON, S., RAMSDEN, N. G. (1990), Synthesis of deoxynojirimycin and of nojirimycin δ-lactam, *Tetrahedron Lett.* **31**, 409–412.

FROMMER, W., SCHMIDT, D. D. (1980), 1-Desoxynojirimycin-Herstellung, *German Patent Offenlegung* 2907190.

FROMMER, W., MÜLLER, L., SCHMIDT, D. D., PULS, W., KRAUSE, H.-P. (1978a), Inhibitoren für α-Glucosidasen, *German Patent Offenlegung* 2658561.

FROMMER, W., MÜLLER, L., SCHMIDT, D. D., PULS, W., KRAUSE, H.-P., HEBER, U. (1978b), Verfahren zur Herstellung von Saccharase-Inhibitoren, die im Verdauungstrakt wirksam werden, aus Bacillen, *DE* 2658563.

FROMMER, W., MÜLLER, L., SCHMIDT, D. D., PULS, W., KRAUSE, H.-P. (1979a), Inhibitoren für α-Glucosidasen *German Patent Offenlegung* 2726898.

FROMMER, W., MÜLLER, L., SCHMIDT, D. D., PULS, W., KRAUSE, H.-P. (1979b), Inhibitoren für Glykosid-Hydrolasen aus Bacillen, *German Patent Offenlegung* 2726899.

FULMER, E. I., UNDERKOFLER, L. A. (1947), Oxidation of polyhydric alcohols by *Acetobacter suboxydans, Iowa State Coll. J. Sci.* **21**, 251–270.

GERARD, J., HILLEBRAND, I., LEFÈBVRE, P. J. (1987), Assessment of the clinical efficacy and tolerance of two new α-glucosidase inhibitors in insulin-treated diabetics, *Int. J. Clin. Pharmacol.* **25**, 483–488.

HAHN, H., HEETHOFF, M., LIST, H.-J., GASSEN, H. G., JANY, K. D. (1989), Polyol-dehydrogenase from *Gluconobacter oxydans:* Identification, cloning and expression of the NADH-dependent sorbitol dehydrogenase II, *DECHEMA Biotechnology Conferences* **3**, pp. 345–349. Weinheim: VCH.

HANN, R. M., HUDSON, C. S. (1939), Proof of the structure and configuration of perseulose (L-galaheptulose), *J. Am. Chem. Soc.* **61**, 336–340.

HANN, R. M., TILDEN, E. B., HUDSON, C. S. (1938), The oxidation of sugar alcohols by *Acetobacter suboxydans, J. Am. Chem. Soc.* **60**, 1201–1203.

HARDICK, D. J., HUTCHINSON, D. W. (1993), The biosynthesis of 1-deoxynojirimycin in *Bacillus subtilis* var. niger, *Tetrahedron* **49**, 6707–6716.

HARDICK, D. J., HUTCHINSON, D. W., TREW, S. J., WELLINGTON, E. M. H. (1991), The biosynthesis of deoxynojirimycin and deoxymannonojirimycin in *Streptomyces subrutilis, J. Chem. Soc., Chem. Commun.* 729–730.

HARDICK, D. J., HUTCHINSON, D. W., TREW, S. J., WELLINGTON, E. M. H. (1992), Glucose is a precursor of 1-deoxynojirimycin and 1-deoxymannonojirimycin in *Streptomyces subrutilis, Tetrahedron* **48**, 6285–6296.

HORII, S., KAMEDA, Y., FUKASE, H. (1982), *N*-substituted pseudo-aminosugars, their production and use, *European Patent* 56194.

HOSHINO, T., OJIMA, S., SUGISAWA, T. (1996), D-Sorbitol dehydrogenase, *European Patent* 728840.

HOUGH, L., JONES, J. K. N., MITCHELL, D. L. (1959), The oxidation of some terminal-substituted polyhydric alcohols by *Acetobacter suboxydans, Can. J. Chem.* **37**, 725–730.

HU, C. L., MCCOMB, E. A., RENDIG, V. V. (1965), Identification of altro-heptulose and L-threitol as products of meso-erythritol oxidation by *Acetobacter suboxydans, Arch. Biochem. Biophys.* **110**, 350–353.

HUGHES, A. B., RUDGE, A. J. (1994), Deoxynojirimycin: Synthesis and biological activity, *Nat. Prod. Rep.* **11**, 135–162.

IIDA, H., YAMAZAKI, N., KIBAYASHI, C. (1987), Total synthesis of (+)-nojirimycin and (+)-1-deoxynojirimycin, *J. Org. Chem.* **52**, 3337–3342.

IKOTA, N. (1989), Synthesis of (+)-1-deoxynojirimycin from (*S*)-pyroglutamic acid, *Heterocycles* **29**, 1469–1472.

INOUYE, S., TSURUOKA, T., NIIDA, T. (1966), The structure of nojirimycin, a piperidinose sugar antibiotic, *J. Antibiot.* **19**, 288–292.

INOUYE, S., TSURUOKA, T., ITO, T., NIIDA, T. (1968), Structure and synthesis of nojirimycin, *Tetrahedron* **24**, 2125–2144.

ISELIN, B. (1948), Oxidations by *Acetobacter suboxydans, J. Biol. Chem.* **175**, 997–998.

ISHIDA, N., KUMAGAI, K., NIIDA, T., HAMAMOTO, K., SHOMURA, T. (1967a), Nojirimycin, a new antibiotic. I Taxonomy and fermentation, *J. Antibiot.* **20**, 62–65.

ISHIDA, N., KUMAGAI, K., NIIDA, T., HAMAMOTO, K., SHOMURA, T. (1967b), Nojirimycin, a new antibiotic. II Isolation, characterization and biological activity, *J. Antibiot.* **20**, 66–71.

JANKE, A. (1960), Die Essigsäuregärung, *Handbuch der Pflanzenphysiologie* Vol. 12, pp. 670–746. Berlin: Springer-Verlag.

JONES, J. K. N., MITCHELL, D. L. (1959), The synthesis of 5-deoxy-5-*S*-ethyl-D-threo-pentulose, *Can. J. Chem.* **37**, 1561–1566.

JONES, J. K. N., PERRY, M. B., TURNER, J. C. (1961a), The synthesis of acetamido-deoxy ketoses by *Acetobacter suboxydans*. Part I, *Can. J. Chem.* **39**, 965–972.

JONES, J. K. N., PERRY, M. B., TURNER, J. C. (1961b), The synthesis of acetamido-deoxy ketoses by *Acetobacter suboxydans*. Part II, *Can. J. Chem.* **39**, 2400–2410.

JONES, J. K. N., PERRY, M. B., TURNER, J. C. (1962), The synthesis of acetamido-deoxy ketoses by *Acetobacter suboxydans*. Part III, *Can. J. Chem.* **40**, 503–510.

JUNGE, B., KRAUSE, H.-P., MÜLLER, L., PULS, W. (1979), Neue Derivate von 3,4,5-Trihydroxypiperidin, Verfahren zu ihrer Herstellung und ihre Verwendung, *European Patent* 947.

JUNGE, B., MATZKE, M., STOLTEFUSS, J. (1996), Chemistry and structure–activity relationships of glucosidase inhibitors, in: *Handbook of Experimental Pharmacology:* Oral Antidiabetics (KUHLMANN, J., PULS, W., Eds.), pp. 411–482. Berlin: Springer-Verlag.

KAUFMANN, H., REICHSTEIN, T. (1967), 6-Desoxy-L-idose, 6-Desoxy-L-sorbose und 6-Desoxy-L-piscose, *Helv. Chim. Acta* **50**, 2280–2287.

KINAST, G., SCHEDEL, M. (1980a), Verfahren zur Herstellung von 6-Amino-6-desoxy-L-sorbose, *European Patent* 8031.

KINAST, G., SCHEDEL, M. (1980b), Verfahren zur Herstellung von *N*-substituierten Derivaten des 1-Desoxynojirimycins, *European Patent* 12278.

KINAST, G., SCHEDEL, M. (1981), Vierstufige 1-Desoxynojirimycin-Synthese mit einer Biotransformation als zentralem Reaktionsschritt, *Angew. Chem.* **93**, 799–800.

KINAST, G., SCHEDEL, M., KÖBERNICK, W. (1982), Verfahren zur Herstellung von *N*-substituierten Derivaten von 1-Desoxynojirimycin, *European Patent* 49858.

KITE, G. C., FELLOWS, L. E., LEES, D. C., KITCHEN, D., MONTEITH, G. B. (1991), Alkaloidal glycosidase inhibitors in nocturnal and diurnal uraniine moths and their respective foodplant genera, *Endospermum* and *Omphalea, Biochem. System. Ecol.* **19**, 441–445.

KÖBERNICK, W. (1982), Verfahren zur Herstellung von 1,5-Didesoxy-1,5-imino-D-glucitol und dessen *N*-Derivaten, *European Patent* 55431.

KODAMA, Y., TSURUOKA, T., NIWA, T., INOUYE, S. (1985), Molecular structure and glycosidase-inhibitory activity of nojirimycin bisulfite adduct, *J. Antibiot.* **38**, 116–118.

KOSSEVA, M., BESCHKOV, V., POPOV, R. (1991), Biotransformation of D-sorbitol to L-sorbose by immobilized cells of *Gluconobacter suboxydans* in a bubble column, *J. Biotechnol.* **19**, 301–308.

KULHÁNEK, M. (1970), Fermentation processes employed in vitamin C synthesis, *Adv. Appl. Microbiol.* **12**, 11–30.

KULHÁNEK, M. (1989), Microbial dehydrogenation of monosaccharides, *Adv. Appl. Microbiol.* **34**, 141–182.

KULHÁNEK, M., TADRA, M., PACÁK, J., TREJBALOVÁ, H., CERNÝ, M. (1977), 5-Deoxy-5-fluoro-L-sorbose originating from 2-deoxy-2-fluoro-D-glucitol by fermentation with *Acetomonas oxydans, Folia Microbiol.* **22**, 295–297.

LA FERLA, B., NICOTRA, F. (1999), Synthetic methods for the preparation of iminosugars, in: *Iminosugars as Glycosidase Inhibitors. Nojirimycin and Beyond* (STÜTZ, A. E., Ed.), pp. 69–92. Weinheim: Wiley-VCH.

LINEK, K., SANDTNEROVA, R., STICZAY, T., KOVACIK, V. (1979), The synthesis of 4-doxy-L-glycero-pentulose by biochemical dehydrogenation of 2-deoxy-D-erythro-pentitol, *Carbohydr. Res.* **76**, 290–294.

MACLAY, W. D., HANN, R. M., HUDSON, C. S. (1942), Some studies on L-glucoheptulose, *J. Am. Chem. Soc.* **64**, 1606–1609.

MATSUMURA, S., ENOMOTO, H., AOYAGI, Y., YOSHIKINI, Y., KURA, K., YAGI, M. (1979a), Neue *N*-substituierte Moranolinderivate, *German Patent Offenlegung* 2915037.

MATSUMURA, S., ENOMOTO, H., AOYAGI, Y., EZURE, Y., YOSHIKUNI, Y., YAGI, M. (1979b), Verfahren zur Gewinnung von Moranolin, *German Patent Offenlegung* 2850467.

MATSUMURA, S., ENOMOTO, H., AOYAGI, Y., YOSHIKUNI, Y., YAGI, M. (1980), Moranoline, *Chem. Abstr.* **93**, 66581s.

MOSES, V., FERRIER, R. J. (1962), The biochemical preparation of D-xylose and L-ribulose. Details of the action of *Acetobacter suboxydans* on D-arabitol, ribitol and other polyhydroxy compounds, *Biochem. J.* **83**, 8–14.

MÜLLER, J. (1966), Untersuchungen über die Sorbosegärung in stationärer und kontinuierlicher Kultur, *Zbl. Bakt.* **120**, 349–378.

MURAI, H., OHATA, K., ENOMOTO, H., YOSHIKUNI, Y., KONO, T., YAGI, M. (1977), 2-Hydroxymethyl-3,4,5-trihydroxypiperidin, Extraktionsverfahren zu seiner Herstellung und seine Verwendung als Arzneimittel, *German Patent Offenlegung*

2656602.

MURAO, S., MIYATA, S. (1980), Isolation and characterization of a new trehalase inhibitor, S-GI, *Agric. Biol. Chem.* **44**, 219–221.

NAKAMURA, T., TAKEBE, K., KUDOH, K., TERADA, A., TANDOH, Y. et al. (1993), Effect of an α-glucosidase inhibitor on intestinal fermentation and faecal lipids in diabetic patients, *J. Int. Med. Res.* **21**, 257–267.

NICOLSKAYA, N., BELAYKOVA, M., STATKEVITSCH, B., POMORZEVA, N., ABRAMOVA, N. (1984), Plant for continuous microbiological oxidation of D-sorbitol to L-sorbose, *Khim. Farm. Zh.* **18**, 879–892.

NISHIKAWA, T., ISHIDA, N. (1965), A new antibiotic R-468 active against drug-resistant *Shigella*, *J. Antibiot.* **18**, 132–133.

NISHIMURA, Y. (1991), The synthesis and biological activity of glucosidase inhibitors, *J. Synth. Org. Chem.* **49**, 846–857.

NIWA, T., INOUYE, S., TSURUOKA, T., KOAZE, Y., NIIDA, T. (1970), "Nojirimycin" as a potent inhibitor of glucosidase, *Agr. Biol. Chem.* **34**, 966–968.

ODAKA, H., IKEDA, H. (1995), Voglibose (AO-128): A hypoglycemic agent, *J. Takeda Res. Lab.* **54**, 21–33.

ONISHI, H., SUZUKI, T. (1969), Microbial production of xylitol from glucose, *Appl. Microblol.* **18**, 1031–1035.

PAULSEN, H. (1966), Kohlenhydrate mit Stickstoff oder Schwefel im "Halbacetal"-Ring, *Angew. Chem.* **78**, 495–510.

PAULSEN, H. (1999), The early days of monosaccharides containing nitrogen in the ring, in: *Iminosugars as Glycosidase Inhibitors. Nojirimycin and Beyond* (STÜTZ, A. E., Ed.), pp. 1–7. Weinheim, Wiley-VCH.

PAULSEN, H., SANGSTER, I., HEYNS, K. (1967), Monosaccharide mit stickstoffhaltigem Ring, XIII. Synthese und Reaktionen von Keto-piperidinosen, *Chem. Ber.* **100**, 802–815.

PRATT, J. W., RICHTMYER, N. K. (1955), D-Glycero-D-allo-heptose, L-allo-heptulose, D-talo-heptulose and related substances derived from the addition of cyanide to D-allose, *J. Am. Chem. Soc.* **77**, 6326–6328.

PRATT, J. W., RICHTMYER, N. K., HUDSON, L. C. (1952), D-Idoheptulose and 2,7-anhydro-β-D-idoheptulopyranose, *J. Am. Chem. Soc.* **74**, 2210–2214.

REESE, E. T., PARRISH, F. W., ETTLINGER, M. (1971), Nojirimycin and D-glucono-1,5-lactone as inhibitors of carbohydrases, *Carbohydr. Res.* **18**, 381–388.

REICHSTEIN, T., GRÜSSNER, A. (1934), Eine ergiebige Synthese der 1-Ascorbinsäure (C-Vitamin), *Helv. Chim. Acta* **17**, 311–328.

RICHTMYER, N. K., STEWART, L. C., HUDSON, C. S. (1950), L-Fuco-4-ketose, a new sugar produced by the action of *Acetobacter suboxydans* on L-fucitol, *J. Am. Chem. Soc.* **72**, 4934–4937.

ROSENBERG, M., SVITEL, J., ROSENBERGOVÁ, I., STURDÍK, E. (1993), Optimization of sorbose production from sorbitol by *Gluconobacter oxydans*, *Acta Biotechnol.* **35**, 269–274.

RUDGE, A. J., COLLINS, I., HOLMES, A. B., BAKER, R. (1994), An enantioselective synthesis of deoxynojirimycin, *Angew. Chem.* (*Int. Edn. Engl.*) **33**, 2320–2322.

SCHEDEL, M. (1997), Oxidation von Aminosorbit durch *Gluconobacter oxydans:* Ein wichtiger Schritt der technischen Miglitolsynthese, Abstracts of 15th Annual Meeting of Biotechnologists, p. 178. Frankfurt: DECHEMA.

SCHMIDT, D. D., FROMMER, W., JUNGE, B., MÜLLER, L., WINGENDER, W. et al. (1977), α-Glucosidase inhibitors. New complex oligosaccharides of microbial origin, *Naturwissenschaften* **64**, 535–536.

SCHMIDT, D. D., FROMMER, W., MÜLLER, L., TRUSCHEIT, E. (1979), Glucosidase-Inhibitoren aus Bazillen, *Naturwissenschaften* **66**, 584–585.

SCHNARR, G. W., SZAREK, W. A., JONES, J. K. N. (1977), Preparation and activity of immobilized *Acetobacter suboxydans* cells, *Appl. Environ. Microbiol.* **33**, 732–734.

SCHRÖDER, T., STUBBE, M. (1987), Verfahren zur Herstellung von 1-Desoxynojirimycin und dessen *N*-Derivaten, *European Patent* 240868.

SCHUTT, H. (1993), Enzymatic deacylation of acylaminosorboses, *US Patent* 5177004.

SCOTT, A. R., TATTERSALL, R. B. (1988), α-Glucosidase inhibition in the treatment of non-insulin-dependent diabetes mellitus, *Diabet. Med.* **5**, 42–46.

SHINAGAWA, E., MATSUSHITA, K., ADACHI, O., AMEYAMA, M. (1982), Purification and characterization of D-sorbitol dehydrogenase from membrane of *Gluconobacter suboxydans* var. α, *Agric. Biol. Chem.* **46**, 135–141.

SIMMONDS, M. S. J., KITE, G. C., PORTER, E. A. (1999), Taxonomic distribution of iminosugars in plants and their biological activities, in: *Iminosugars as Glycosidase Inhibitors. Nojirimycin and Beyond* (STÜTZ, A. E., Ed.), pp. 8–30. Weinheim: Wiley-VCH.

SPENCER, J. F. T., GORIN, P. A. J. (1968), Microbiological transformations of sugars and related compounds, *Progr. Ind. Microbiol.* **7**, 177–220.

STEFANOVA, S., KOSEVA, M., TEPAVICHAROVA, I., BESCHKOV, V. (1987), L-Sorbose production by cells of the strain *Gluconobacter suboxydans* entrapped in a polyacrylamide gel, *Biotechnol. Lett.* **9**, 475–477.

STEWART, L. C., RICHTMYER, N. K., HUDSON, C. S. (1949), The oxidation of volemitol by *Acetobacter suboxydans* and by *Acetobacter xylinum*, *J. Am. Chem. Soc.* **71**, 3532–3534.

STEWART, L. C., RICHTMYER, N. K., HUDSON, C. S.

(1952), L-Guloheptulose and 2,7-anhydro-β-L-guloheptulopyranose, *J. Am. Chem. Soc.* **74**, 2206–2210.

STÜTZ, A. E. (Ed.) (1999), *Iminosugars as Glycosidase Inhibitors. Nojirimycin and Beyond*, Weinheim: Wiley-VCH.

SUMITA, M., MURAO, S. (1980), Fermentative production of 1-deoxynojirimycin by cultivating *Streptomyces* microorganisms, *Japanese Patent* 55120792.

TILDEN, E. B. (1939), The preparation of perseulose by oxidation of perseitol with *Acetobacter suboxydans*, *J. Bacteriol.* **37**, 629–637.

TIWARI, K. N., DHAWALE, M. R., SZAREK, W. A., HAY, G. W., KROPINSKI, A. M. B. (1986), A synthesis of 2-deoxy-D-arabino-hexitol and its oxidation to 5-deoxy-D-threo-hexulose ("5-deoxy-D-fructose") using immobilized cells of *Gluconobacter oxydans*, *Carbohydr. Res.* **156**, 19–24.

TRIFONOV, A., STEFANOVA, S., KONSTANTINOV, H., TEPAVICHAROVA, I. (1991), Biochemical oxidation of D-sorbitol to L-sorbose by immobilized *Gluconobacter oxydans* cells, *Appl. Biochem. Biotechnol.* **28/29**, 397–405.

USPENSKAYA, S. N., LOITSYANSKAYA, M. S. (1979), Effectiveness of the utilization of glucose by *Gluconobacter oxydans*, *Mikrobiologiya* **48**, 400–405.

WHISTLER, R. L., UNDERKOFLER, L. A. (1938), The production of 1-erythrulose by the action of *Acetobacter suboxydans* upon erythritol, *J. Am. Soc.* **60**, 2507–2508.

WHITE, S. A., CLAUS, G. W. (1982), Effect of intracytoplasmic membrane development on oxidation of sorbitol and other polyols by *Gluconobacter oxydans*, *J. Bacteriol.* **150**, 934–943.

WONG, C.-H., HALCOMB, R. L., ICHIKAWA, Y., KAJIMOTO, T. (1995), Enzyme in der organischen Synthese: Das Problem der molekularen Erkennung von Kohlenhydraten (Teil 1), *Angew. Chem.* **107**, 453–474.

YAGI, M., KOUNO, T., AOYAGI, Y., MURAI, H. (1976), The structure of Moranoline, a piperidine alkaloid from *Morus* species, *Nippon Nogeikagaku Kaishi* **50**, 571–572.

YAGI, M., KUONO, T., AOYAGI, Y., MURAI, H. (1977), The structure of Moranoline, a piperidine alkaloid from *Morus* species, *Chem. Abstr.* **86**, 167851r.

YAMADA, H., OYA, I., NAGAI, T., MATSUMOTO, T., KIYOHARA, H., OMURA, S. (1993), Screening of α-glucosidase inhibitor from Chinese herbs and its application on the quality control of mulberry bark, *Chem. Abstr.* **119**, 146433r.

8 Engineering Microbial Pathways for Amino Acid Production

IAN G. FOTHERINGHAM

Mt. Prospect, IL, USA

1 Introduction

Many thousands of tons of different amino acids are produced annually for a variety of industrial applications. In particular, monosodium glutamate, phenylalanine, aspartic acid, and glycine are produced as human food additives and lysine, threonine, methionine, and tryptophan are manufactured as animal feed additives (ABE and TAKAYAMA, 1972; BATT et al., 1985; GARNER et al., 1983; SHIIO, 1986). These and other amino acids are also used as pharmaceutical intermediates and in a variety of nutritional and health care industries. In many cases bacteria are used to produce these amino acids through large scale fermentation, particularly when a high degree of enantiomeric purity is required. In some cases chemoenzymatic methods have also been successfully employed, particularly for L-phenylalanine and non-proteinogenic amino acids.

The development of strains for fermentation approaches initially consisted of classical random mutagenesis methods and screening with toxic amino acid analogs. In this way commercially viable amino acid overproducing strains were developed. However, in the 1980s this approach began to be complemented by the availability of molecular biological methods and a rapid increase in the understanding of the molecular genetics of amino acid production in particular bacterial species. The availability of plasmid vector systems, transformation protocols, and general recombinant DNA methodology enabled biochemical pathways to be altered in much more precise ways to further enhance amino acid titer and production efficiency in *Escherichia coli* and in the coryneform bacteria *Corynebacterium* and *Brevibacterium*, the most prominent amino acid producing bacteria. This chapter will center upon the most significant aspects of biochemical pathway engineering such as the identification of rate limiting enzymatic steps and the elimination of allosteric feedback inhibition of key enzymes. Considerable emphasis will be placed upon the shikimate and phenylalanine pathways of *E. coli* which are among the best characterized biochemically and genetically and have been the subject of some of the most widespread efforts in amino acid production. It will also address aspects of metabolic engineering which have been applied in the coryneform organisms due to the rapidly increasing understanding of the molecular genetics of those organisms, and their importance in amino acid production. Additionally, it will examine the increasing potential to construct completely novel biochemical pathways in organisms such as *E. coli* in order to produce non-proteinogenic or unnatural amino acids.

2 Bacterial Production of L-Phenylalanine

2.1 The Common Aromatic and L-Phenylalanine Biosynthetic Pathways

Efforts to develop L-phenylalanine overproducing organisms have been vigorously pursued by Nutrasweet Company, Ajinomoto, Kyowa Hakko Kogyo, and others. The focus has centered upon bacterial strains which have previously demonstrated the ability to overproduce other amino acids. Such organisms include principally the coryneform bacteria, *Brevibacterium flavum* (SHIIO et al., 1988) and *Corynebacterium glutamicum* (HAGINO and NAKAYAMA, 1974b; IKEDA et al., 1993) used in L-glutamic acid production. In addition *Escherichia coli* (TRIBE, 1987) has been extensively studied in L-phenylalanine manufacture due to the detailed characterization of the molecular genetics and biochemistry of its aromatic amino acid pathways and its amenability to recombinant DNA methodology. The biochemical pathway which results in the synthesis of L-phenylalanine from chorismate is identical in each of these organisms and shown in Fig. 1. The common aromatic pathway to chorismate in *E. coli* is shown in Fig. 2. In each case the L-phenylalanine biosynthetic pathway comprises three enzymatic steps from chorismic acid, the product of the common aromatic pathway. The precursors of the common aromatic pathway, phosphoenolpyruvate (PEP)

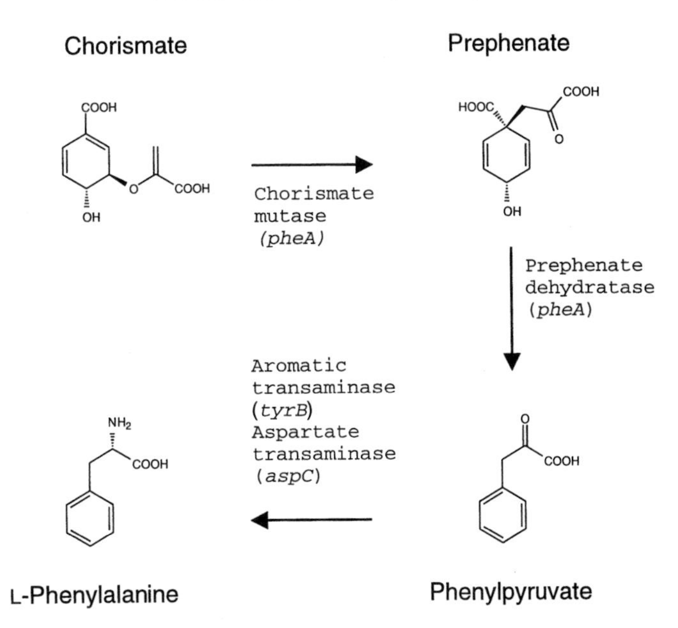

Fig. 1. The biosynthetic pathway from chorismate to L-phenylalanine in *E. coli* K12.

and erythrose-4-phosphate (E4P) derive respectively from the glycolytic and pentose phosphate pathways of sugar metabolism. Although the intermediate compounds in both pathways are identical in each of the organisms, there are differences in the organization and regulation of the genes and enzymes involved (HUDSON and DAVIDSON, 1984; OZAKI et al., 1985; SHIIO et al., 1988). Nevertheless, the principal points of pathway regulation are very similar in each of the three bacteria (HUDSON and DAVIDSON, 1984; IKEDA et al., 1993; SHIIO and SUGIMOTO, 1981b). Conversely, tyrosine biosynthesis which is also carried out in three biosynthetic steps from chorismate, proceeds through different intermediates in *E. coli* than in the coryneform organisms (HUDSON and DAVIDSON, 1984; SHIIO and SUGIMOTO, 1981b).

In *E. coli*, regulation of phenylalanine biosynthesis occurs both in the common aromatic pathway, and in the terminal phenylalanine pathway and the vast majority of efforts to deregulate phenylalanine biosynthesis have focused upon two specific rate limiting enzymatic steps. These are the steps of the aromatic and the phenylalanine pathways carried out respectively by the enzymes 3-deoxy-D-arabino-heptulosonate-7-phosphate (DAHP) synthase and prephenate dehydratase. Classical mutagenesis approaches using toxic amino acid analogs and the molecular cloning of the genes encoding these enzymes have led to very significant increases in the capability of host strains to overproduce L-phenylalanine. This has resulted from the elimination of the regulatory mechanisms controlling enzyme synthesis and specific activity. The cellular mechanisms which govern the activity of these particular enzymes are complex and illustrate many of the sophisticated means by which bacteria control gene expression and enzyme activity. Chorismate mutase, the first step in the phenylalanine specific pathway and the shikimate kinase activity of the common aromatic pathway are subject to a lesser degree of regulation and have also been characterized in detail (MILLAR et al., 1986; NELMS et al., 1992).

2.2 Classical Mutagenesis/Selection

Classical methods of strain improvement have been widely applied in the development of phenylalanine overproducing organisms (SHIIO et al., 1973; TOKORO et al., 1970; TRIBE,

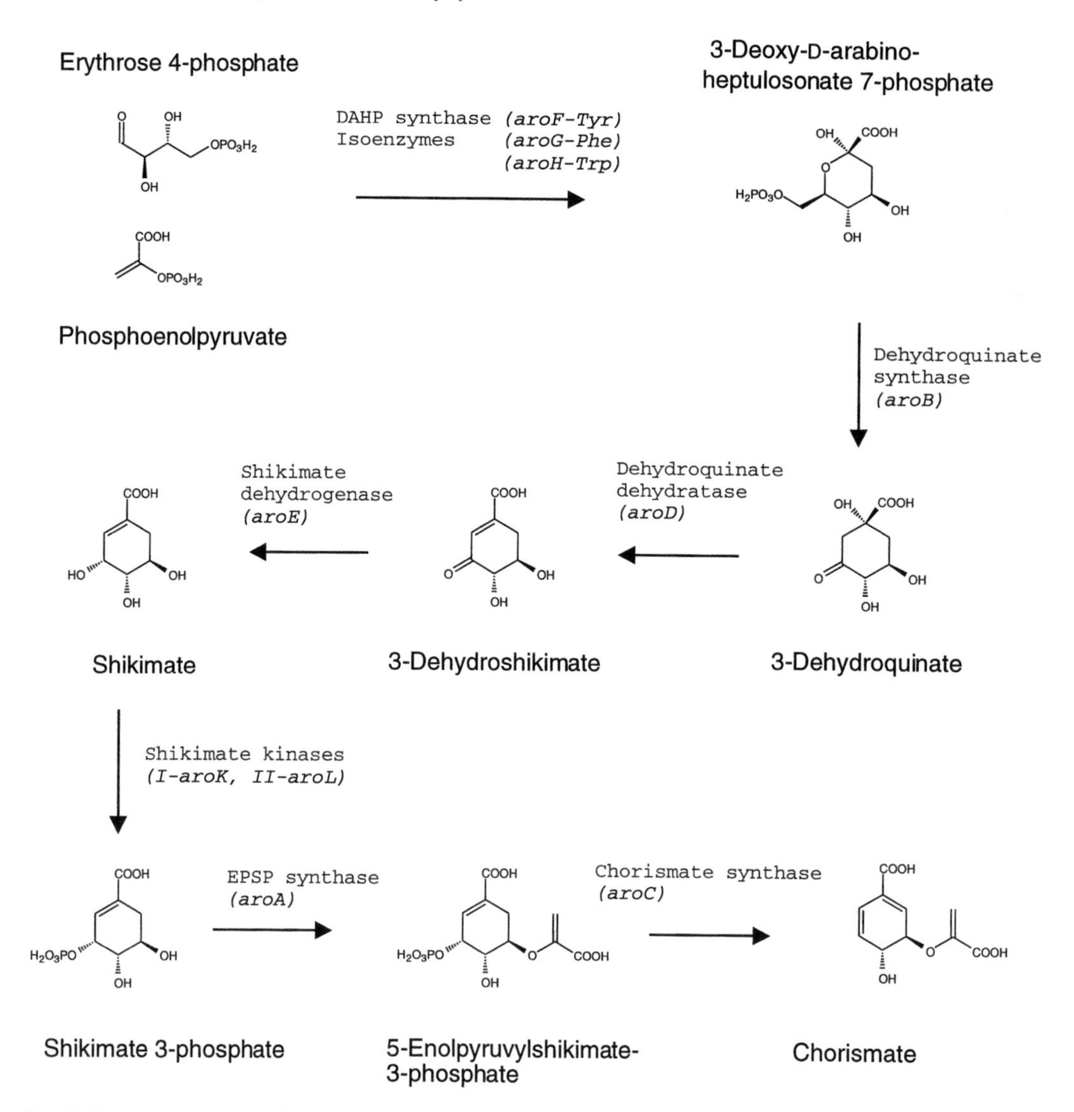

Fig. 2. The common aromatic pathway to chorismate in *E. coli* K12. The mnemonic of the genes involved are shown in parentheses below the enzymes responsible for each step.

1987; TSUCHIDA et al., 1987). Tyrosine auxotrophs have frequently been used in efforts to increase phenylalanine production through mutagenesis. These strains often already overproduce phenylalanine due to the overlapping nature of tyrosine and phenylalanine biosynthetic regulation (HAGINO and NAKAYAMA, 1974b; HWANG et al., 1985). Limiting tyrosine availability leads to partial genetic and allosteric deregulation of common biosynthetic steps (WALLACE and PITTARD, 1969). Such strains have been subjected to a variety of mutagenesis procedures to further increase the overall titer and the efficiency of phenylalanine production. In general this has involved the identification of mutants which display resistance to toxic analogs of phenylalanine or tyrosine such as β-2-thienyl-D,L-alanine and *p*-fluoro-D,L-phenylalanine (CHOI and TRIBE, 1982; HAGINO and NAKAYAMA, 1974b; TRIBE, 1987). Such mutants can be readily identified on selective plates in which the analog is present in the growth medium. The mutations responsible for the phenylalanine overproduction have frequently been located in the genes encoding the enzymatic activities DAHP synthase, chorismate mutase, and prephenate dehydratase. In turn this has prompted molecular genetic approaches to further increase phenylalanine production through the isolation and *in vitro* manipulation of these genes as described below.

2.3 Deregulation of DAHP Synthase Activity

The activity of DAHP synthase commits carbon from intermediary metabolism to the common aromatic pathway converting equimolar amounts of PEP and E4P to DAHP (HASLAM, 1974). In *E. coli* there are three isoenzymes of DAHP synthase of comparable catalytic activity encoded by the genes *aroF*, *aroG*, and *aroH* (PITTARD and GIBSON, 1970). Enzyme activity is regulated respectively by the aromatic amino acids tyrosine, phenylalanine, and tryptophan (BROWN, 1968; BROWN and SOMERVILLE, 1971; CAMAKARIS and PITTARD, 1973; IM et al., 1971; WALLACE and PITTARD, 1969). In each case regulation is mediated both by repression of gene transcription and by allosteric feedback inhibition of the enzyme, though to different degrees.

The *aroF* gene lies in an operon with *tyrA* which encodes the bifunctional protein chorismate mutase/prephenate dehydrogenase. Both genes are regulated by the TyrR repressor protein complexed with tyrosine. The *aroG* gene product accounts for 80% of the total DAHP synthase activity in wild type *E. coli* cells. The *aroG* gene is repressed by the TyrR repressor protein complexed with phenylalanine and tryptophan. Repression of *aroH* is mediated by tryptophan and the TrpR repressor protein (BROWN, 1968). The gene products of *aroF* and *aroG* are almost completely feedback inhibited respectively by low concentrations of tyrosine or phenylalanine (GARNER et al., 1983) whereas the *aroH* gene product is subject to maximally 40% feedback inhibition by tryptophan (GARNER et al., 1983). In *Brevibacterium flavum*, DAHP synthase forms a bifunctional enzyme complex with chorismate mutase and is feedback inhibited by tyrosine and phenylalanine synergistically but not by tryptophan (SHIIO et al., 1974; SUGIMOTO and SHIIO, 1980). Similarly in *Corynebacterium glutamicum* DAHP synthase is inhibited most significantly by phenylalanine and tyrosine acting in concert (HAGINO and NAKAYAMA, 1975b) but, unlike *Brevibacterium flavum*, reportedly it does not show tyrosine mediated repression of transcription (HAGINO and NAKAYAMA, 1974c; SHIIO et al., 1974).

Many analog resistant mutants of these organisms display reduced sensitivity of DAHP synthase to feedback inhibition (HAGINO and NAKAYAMA, 1974b; OZAKI et al., 1985; SHIIO and SUGIMOTO, 1979; SHIIO et al., 1974; SUGIMOTO et al., 1973; TRIBE, 1987). In *E. coli* the genes encoding the DAHP synthase isoenzymes have been characterized and sequenced (DAVIES and DAVIDSON, 1982; HUDSON and DAVIDSON, 1984; RAY et al., 1988). The mechanism of feedback inhibition of the *aroF*, *aroG*, and *aroH* isoenzymes has been studied in considerable detail (RAY et al., 1988) and variants of the *aroF* gene on plasmid vectors have been used to increase phenylalanine overproduction. Simple replacement of transcriptional control sequences with powerful constitutive or inducible promoter regions and the use of high copy number plasmids has readily enabled overproduction of the enzyme (EDWARDS et al., 1987; FOERBERG et al., 1988), and reduction of tyrosine mediated feedback inhibition has been described using resistance to the aromatic amino acid analogs β-2-thi-enyl-D,L-alanine and *p*-fluoro-D,L-phenylalanine (EDWARDS et al., 1987; TRIBE, 1987).

2.4 Deregulation of Chorismate Mutase and Prephenate Dehydratase Activity

The three enzymatic steps by which chorismate is converted to phenylalanine appear to be identical between *C. glutamicum*, *B. flavum*, and *E. coli* although only in *E. coli* have detailed reports appeared upon the characterization of the genes involved. In each case the principal regulatory step is that catalyzed by prephenate dehydratase. In *E. coli*, chorismate is converted firstly to prephenate and then to phenyl pyruvate by the action of a bifunctional enzyme, chorismate mutase/prephenate dehydratase (CMPD) encoded by the *pheA* gene (HUDSON and DAVIDSON, 1984). The final step, in which phenyl pyruvate is converted to L-phenylalanine, is carried out predominantly by the aromatic aminotransferase encoded by *tyrB* (GARNER et al., 1983). However both the aspartate aminotransferase, encoded by *aspC* and the branched chain aminotransferase encoded by *ilvE* can catalyze this reaction (FOTHERINGHAM et al., 1986). Phenylalanine biosynthesis is regulated by control of CMPD through phenylalanine mediated attenuation of *pheA* transcription (HUDSON and DAVIDSON, 1984) and by feedback inhibition of the prephenate dehydratase (PD) and chorismate mutase (CM) activities of the enzyme. Inhibition is most pronounced upon the prephenate dehydratase activity, with almost total inhibition observed at micromolar phenylalanine concentrations (DOPHEIDE et al., 1972; GETHING and DAVIDSON, 1978). Chorismate mutase activity in contrast is maximally inhibited to only 40% (DOPHEIDE et al., 1972). In *B. flavum*, prephenate dehydratase and chorismate mutase are encoded by distinct genes. Prephenate dehydratase is again the principal point of regulation with the enzyme subject to feedback inhibition, but not transcriptional repression, by phenylalanine (SHIIO and SUGIMOTO, 1976; SUGIMOTO and SHIIO, 1974). Chorismate mutase which in this organism forms a bifunctional complex with DAHP synthase is maximally inhibited by phenylalanine and tyrosine to 65%, but this is significantly diminished by the presence of very low levels of tryptophan (SHIIO and SUGIMOTO, 1981a). Expression of chorismate mutase is repressed by tyrosine (SHIIO and SUGIMOTO, 1979; SUGIMOTO and SHIIO, 1985). The final step is carried out by at least one transaminase (SHIIO et al., 1982). Similarly in *C. glutamicum*, the activities are encoded in distinct genes with prephenate dehydratase again being the more strongly feedback inhibited by phenylalanine (DE BOER and DIJKHUIZEN, 1990; HAGINO and NAKAYAMA, 1974a, 1975a). The only transcriptional repression reported is that of chorismate mutase by phenylalanine. The *C. glutamicum* genes encoding prephenate dehydratase and chorismate mutase have been isolated and cloned from analog resistant mutants of *C. glutamicum* (IKEDA and KATSUMATA, 1992; OZAKI et al., 1985) and used along with the cloned DAHP synthase gene to augment L-phenylalanine biosynthesis in overproducing strains of *C. glutamicum* (IKEDA and KATSUMATA, 1992).

Many publications and patents have described successful efforts to reduce and eliminate regulation of prephenate dehydratase activity in phenylalanine overproducing organisms (FOERBERG et al., 1988; NELMS et al., 1992; TRIBE, 1987). As with DAHP synthase the majority of the reported efforts have focused upon the *E. coli* enzyme encoded by the *pheA* gene which is transcribed convergently with the tyrosine operon on the *E. coli* chromosome (HUDSON and DAVIDSON, 1984). The detailed characterization of the *pheA* regulatory region has facilitated expression of the gene in a variety of transcriptional configurations leading to elevated expression of CMPD, and a number of mutations in *pheA* which affect phenylalanine mediated feedback inhibition have been described (BACKMAN and BALAKRISHNAN, 1988; EDWARDS et al., 1987; NELMS et al., 1992). Increased expression of the gene is readily achieved by cloning *pheA* onto multicopy plasmid vectors and deletion of the nucleotide sequences comprising the transcription attenuator illustrated in Fig. 3. Most mutations which affect the allosteric regulation of the enzyme by phenylalanine have been identified through resistance to phenylalanine analogs such as β-2-thienylalanine, but there are examples of feedback resistant mutations arising through insertional mutagenesis and gene truncation (BACKMAN and BALAKRISHNAN, 1988; EDWARDS et al., 1987). Two regions of the enzyme in particular have been shown to reduce feedback inhibition to different degrees. Mutations at position Trp338 in the peptide sequence desensitize the enzyme to levels of phenylalanine in the 2–5 mM range but are insufficient to confer resistance to higher concentrations of L-phenylalanine (BACKMAN and BALAKRISHNAN, 1988; EDWARDS et al., 1987).

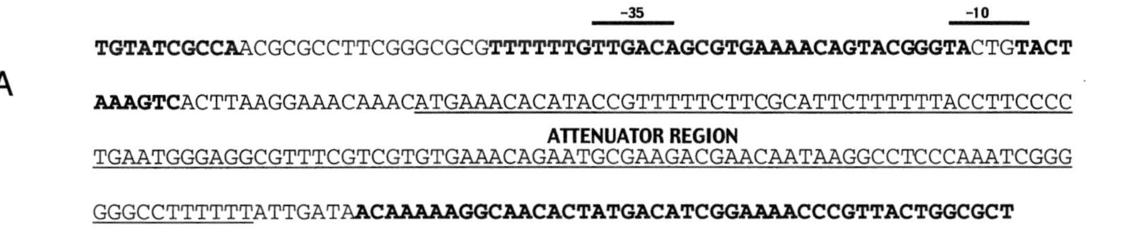

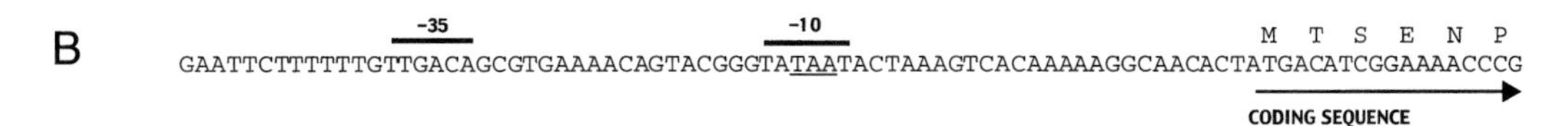

Fig. 3. A. Sequence of the wild type promoter region of the *E. coli* K12 *pheA* gene. The -35 and -10 hexamers are indicated. Bases retained in the deregulated promoter are in bold type. Sequence involved in the attenuator region is shown underlined. **B.** Sequence of the deregulated *pheA* promoter region used to produce chorismate mutase/prephenate dehydratase. Altered bases in the -10 region are shown underlined.

Mutations in the region of residues 304–310 confer almost total resistance to feedback inhibition at L-phenylalanine concentrations of at least 200 mM (NELMS et al., 1992). Feedback inhibition profiles of four such variants (JN305–JN308) are shown in Fig. 4, in comparison to the profile of wild-type enzyme (JN302). It is not clear, if the mechanism of resistance is similar in either case, but the difference is significant to commercial application as

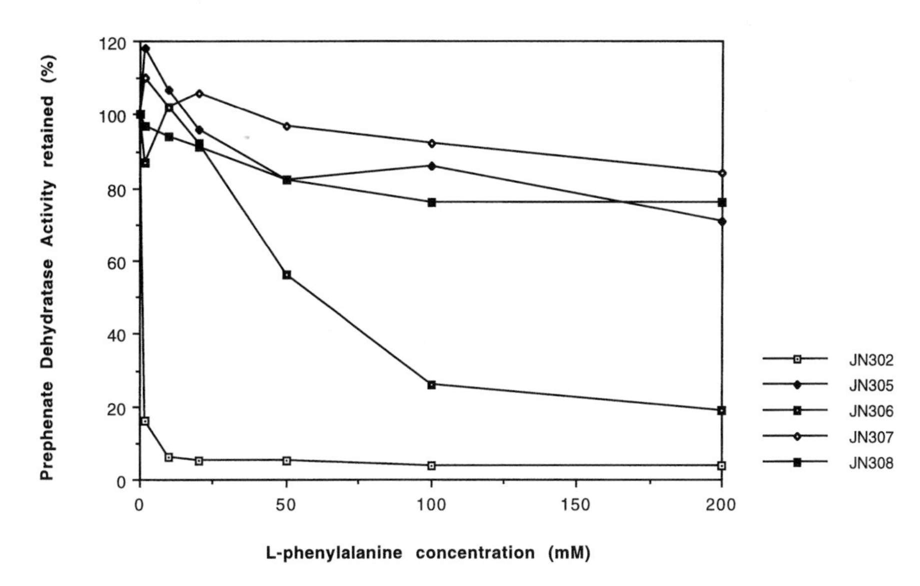

Fig. 4. L-Phenylalanine mediated feedback inhibition of wild type *E. coli* K12 prephenate dehydratase (JN302) and four feedback inhibition resistant enzyme variants (JN305–JN308). Activity is expressed as a percentage of normal wild-type enzyme activity.

overproducing organisms readily achieve extracellular concentrations of L-phenylalanine over 200 mM.

2.5 Precursor Supply

Rate limiting steps in the common aromatic and phenylalanine biosynthetic pathways are obvious targets in the development of phenylalanine overproducing organisms. However, the detailed biochemical and genetic characterization of *E. coli* has enabled additional areas of its metabolic function to be specifically manipulated to determine their effect on aromatic pathway throughput. Besides efforts to eliminate additional lesser points of aromatic pathway regulation, attempts have been made to enhance phenylalanine production by increasing the supply of aromatic pathway precursors and by facilitating exodus of L-phenylalanine from the cell. The precursors of the common aromatic pathway D-erythrose 4-phosphate and phosphoenolpyruvate are the respective products of the pentose phosphate and glycolytic pathways. Precursor supply in aromatic amino acid biosynthesis has been reviewed recently (BERRY, 1996; FROST and DRATHS, 1995). Theoretical analyses of the pathway and the cellular roles of these metabolites suggest that the production of aromatic compounds is likely to be limited by PEP availability (PATNAIK and LIAO, 1994; PATNAIK et al., 1995), since phosphoenolpyruvate is involved in a number of cellular processes including the generation of metabolic energy through the citric acid cycle (MILLER et al., 1987) and the transport of glucose into the cell by the phosphotransferase system (POSTMA, 1987). Strategies to reduce the drain of PEP by these processes have included mutation of sugar transport systems to reduce PEP dependent glucose transport (FLORES et al., 1996) and modulation of the activities of pyruvate kinase, PEP synthase, and PEP carboxylase which regulate PEP flux to pyruvate, oxaloacetate, and the citric acid cycle (BERRY, 1996; BLEDIG, 1994; MILLER et al., 1987). Similarly, the availability of E4P has been increased by altering the levels of transketolase, the enzyme responsible for E4P biosynthesis (DRATHS et al., 1992). In general these efforts have been successful in directing additional flux of PEP or E4P into the aromatic pathway, although their effect has not proved to be as predictable as the deregulation of rate limiting pathway steps and their overall impact on L-phenylalanine overproduction has not been well characterized.

2.6 Secretion of Phenylalanine from the Cell

Exodus of phenylalanine is of manifest importance in the large scale production of phenylalanine as the amino acid is typically recovered only from the extracellular medium and washed cells. Lysis of cells to recover additional phenylalanine is not generally practical and so L-phenylalanine remaining within the cells following washing is usually lost. Since cell biomass is extremely high in large scale fermentations this can represent up to 5% of total phenylalanine produced. Additionally, since L-phenylalanine is known to regulate its own biosynthesis at multiple points through feedback inhibition, attenuation, and TyrR mediated repression, methods to reduce intracellular concentrations of phenylalanine through reduced uptake or increased exodus throughout the fermentation have been sought. Studies have addressed the means by which L-phenylalanine is both taken up and excreted by bacterial cells, and whether this can be altered to increase efflux. In overproducing strains it is likely that most phenylalanine leaves the cell by passive diffusion but specific uptake and exodus pathways also exist. In *E. coli* L-phenylalanine is taken up by at least two permeases (WHIPP and PITTARD, 1977), one specific for phenylalanine encoded by *pheP*, and the other encoded by *aroP* a general aromatic amino acid permease responsible for the transport of tyrosine and tryptophan in addition to phenylalanine. The genes encoding the permeases have been cloned and sequenced and *E. coli* mutants deficient in either system have been characterized (HONORE and COLE, 1990; PI et al., 1991). The effect of an *aroP* mutation has been evaluated in L-phenylalanine overproduction (TRIBE, 1987). In addition specific export systems have been identified which though not completely characterized have been successfully used to increase exodus of L-phenylalanine from strains of *E. coli* (FOTHERINGHAM et al., 1994; MULCAHY, 1988). In one such system the Cin invertase of bacteriophage P1 has been shown to induce a metastable phenylalanine hyper-secreting phenotype upon a wild type strain of *E. coli*. The phenotype can be stabilized by transient introduction of the invertase on a temperature sensitive plasmid replicon and is sufficient to establish L-phenylalanine overproduction in the absence of any alterations to the normal biosynthetic regulation (FOTHERINGHAM et al., 1994).

The incremental gains made from the various levels at which phenylalanine biosynthesis has been addressed have led to the high efficiency production strains currently in use. Almost all large scale L-phenylalanine manufacturing processes in present operation are fermentations employing bacterial strains such as those described above in which classical strain development and/or molecular genetics have been extensively applied to bring about phenylalanine overproduction. The low substrate costs and the economics of scale associated with this approach have resulted in significant economic advantages.

3 Bacterial Production of L-Lysine and L-Threonine

3.1 Biochemical Pathway Engineering in Corynebacteria

The fermentative production of L-glutamate and amino acids of the aspartate family such as lysine and threonine, has been dominated by the use of the coryneform organisms such as *C. glutamicum*. Several strains of the corynebacteria have been used for decades to produce L-glutamic acid (ABE and TAKAYAMA, 1972; GARNER et al., 1983; YAMADA et al., 1973). These organisms have been found to secrete the amino acid at very high concentrations through changes in the cell membrane induced either by biotin limitation or by the addition of various surfactants (CLEMENT and LANEELLE, 1986; DUPERRAY et al., 1992). The enzyme glutamate dehydrogenase encoded by the *gdh* gene is principally responsible for the biosynthesis of L-glutamate in the corynebacteria from the keto acid precursor 2-ketoglutarate (TESCH et al., 1998). Although relatively few genetic studies have been published on this enzyme (BOERMANN et al., 1992), many advances in both the genetic organization and the technology to manipulate DNA in the corynebacteria have been made in the last two decades (ARCHER and SINSKEY, 1993; MARTIN et al., 1987; SANTAMARIA et al., 1987). This has enabled molecular genetic approaches to complement conventional strain development for the production of amino acids such as L-threonine and L-lysine in *Corynebacterium* (ISHIDA et al., 1993, 1989; JETTEN and SINSKEY, 1995). These advances have been thoroughly reviewed quite recently (JETTEN et al., 1993; JETTEN and SINSKEY, 1995; NAMPOOTHIRI and PANDAY, 1998) and will be summarized here with only key aspects relevant to microbial pathway engineering emphasized.

The use of *E. coli* has been a valuable tool to the researchers in the corynebacteria as many of the genes encoding the key enzymes in the biosynthesis of these amino acids were initially isolated through complementation of the corresponding mutants in *E. coli* (JETTEN and SINSKEY, 1995). Similarly, transconjugation of plasmids from *E. coli* to various *Corynebacterium* strains, facilitating gene disruption and replacement, has helped characterize the roles of specific genes, particularly in lysine and threonine biosynthesis (PÜHLER et al., 1990; SCHWARZER and PÜHLER, 1991). The recent development of *Corynebacterium* cloning vectors has further facilitated the studies of gene regulation and expression in these strains (JETTEN et al., 1994; MARTIN et al., 1987; MIWA et al., 1985).

3.2 Engineering the Pathways of L-Lysine and L-Threonine Biosynthesis

The biosynthesis of the aspartate derived amino acids is significantly impacted by the regulation of aspartokinase, the enzyme governing the flow of carbon entering the initial common pathway, shown in Fig. 5, and encoded by the *ask* gene in *C. glutamicum*. In the case of *C. glutamicum* and *B. flavum* aspartokinase is tightly regulated by concerted feedback inhibition, mediated by threonine and lysine (SHIIO and MIYAJIMA, 1969). Through classical mutagenesis/selection using the toxic lysine analog *S*-(α-aminoethyl)-D,L-cysteine, lysine overproducing strains of *C. glutamicum* were obtained which frequently carried variants of aspartokinase resistant to feedback inhibition by lysine and/or threonine (KALINOWSKI et al., 1991; TOSAKA and TAKINAMI, 1986). Sequence comparisons between the wild type and mutant *ask* gene sequences assigned mutations conferring feedback resistance to the β-subunit of the enzyme (FOLLETTIE et al., 1993; KALINOWSKI et al., 1991), although the mechanism of feedback inhibition and of resistance remains unknown. Similar mutations have been identified in the β-subunit of *C. lactofermentum* and in *B. flavum* (SHIIO et al., 1993, 1990).

By analogy to the feedback resistant mutations in prephenate dehydratase and the DAHP synthases of the *E. coli* phenylalanine and common aromatic pathways, the identification of feedback resistant variants of aspartokinase has contributed significantly to efforts to engineer lysine and threonine overproducing strains of *Corynebacterium*. Introduction of feedback resistant Ask variants to lysine overproducing strains of *C. glutamicum* and *C. lactofermentum* enhanced extracellular lysine levels, the latter in a gene dosage dependent manner (CREMER et al., 1991; JETTEN et al., 1995). Overexpression of the wild type dihydropicolinate synthase, on a multi-copy plasmid also leads to elevated lysine production (CREMER et al., 1991) as might be expected, as this is the branch point of lysine and threonine biosynthesis in *Corynebacterium*. No increase in extracellular lysine was observed with the overexpression of each of the remaining enzymes of the primary lysine biosynthetic pathway. A secondary pathway of lysine biosynthesis exists in *C. glutamicum*, which allows complementation of mutations in the *ddh* gene encoding diaminopimelate dehydrogenase, but was unable at least in the wild type state to provide high throughput of precursors to D,L diaminopimelate (CREMER et al., 1991; SCHRUMPF et al., 1991).

As a converse approach to the funneling of carbon to the lysine specific pathways by overproduction of dihydropicolinate synthase, the elimination of this activity by mutation of the *dapA* gene led to significant overproduction of L-threonine (SHIIO et al., 1989) to over 13 g L^{-1} in a strain of *B. flavum* which also carried a feedback inhibition resistant aspartokinase. This was determined to be comparable in productivity to the more typical overproducing strains of *B. flavum* carrying mutations in homoserine dehydrogenase which eliminate the normal threonine mediated feedback inhibition. This is typically selected by the now familiar mutagenesis/selection methods involving resistance to the toxic threonine analog D,L-α-amino-β-hydroxyvaleric acid. Similar levels of threonine overproduction have been reported in recombinant *E. coli* strains in which the threonine operon from a classically derived overproducing mutant was cloned on a multi-copy plasmid and re-introduced to the same host (MIWA et al., 1983). Significantly greater titers of threonine production have been reported by introducing the same cloned threonine operon into mutant strains of *B. flavum* grown with rigorous plasmid maintenance (ISHIDA et al., 1989).

The ability of the corynebacteria to overproduce these amino acids sufficiently for commercial use is partly due to the extraordinary efficiency with which this type of organism can secrete amino acids. For this reason it has been useful in the production of isoleucine (COLON et al., 1995) and other amino acids (NAMPOOTHIRI and PANDEY, 1998) without necessitating the elimination of feedback inhibition of key enzymes in the biochemical pathways involved. Nevertheless, in the examples described above the availability of detailed information on the molecular genetic control of metabolic and biosynthetic pathways is of fundamental importance in achieving optimal levels of amino acid production. Similarly the application of recombinant DNA methodology to a wider array of microbes is enhancing the potential to effectively screen for mutations which relieve feedback inhibition, frequently the most crucial aspect of deregulating biosynthesis of primary metabolites such as amino

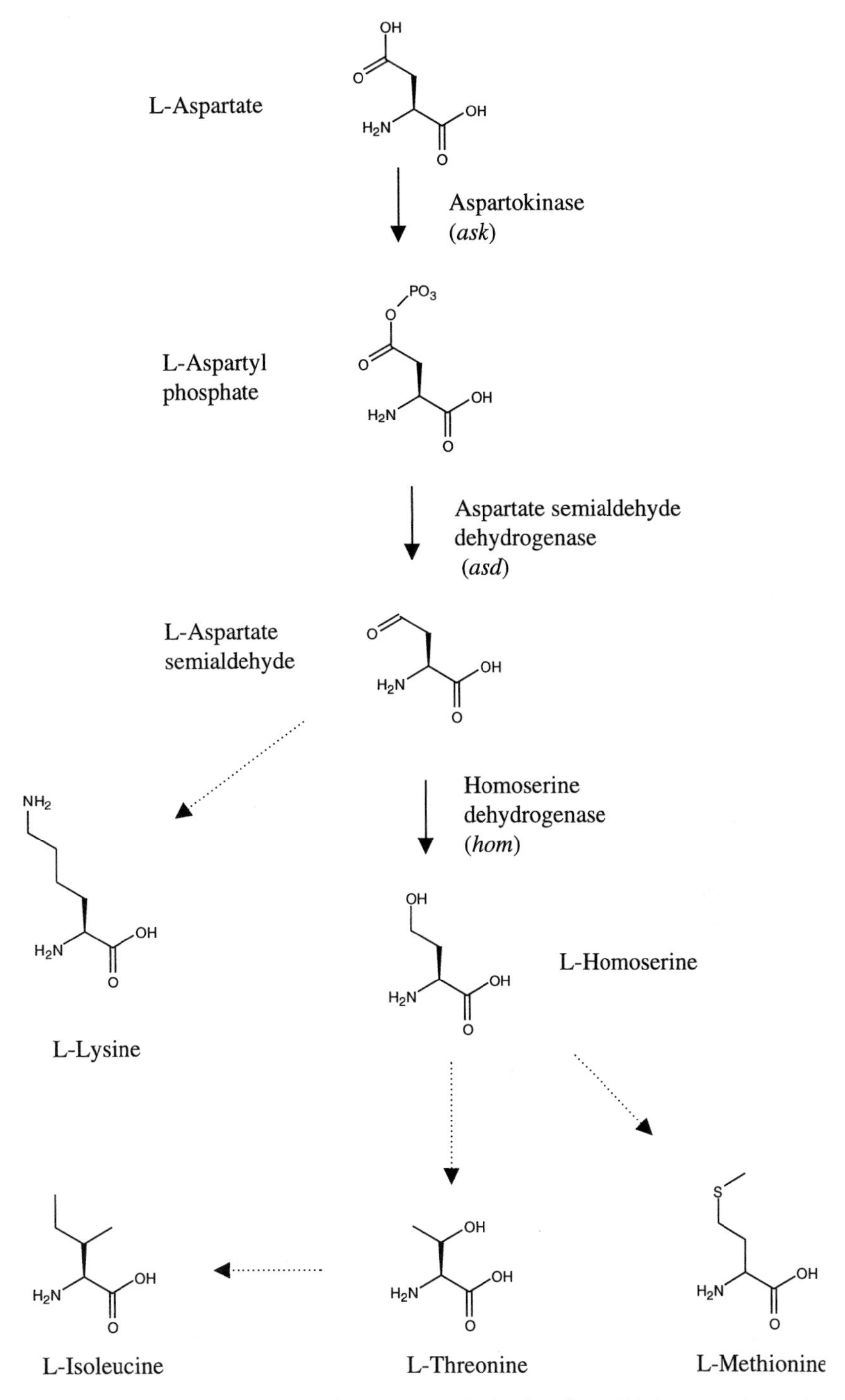

Fig. 5. The common pathway of the aspartate derived amino acids in corynebacteria. The mnemonic of the genes involved are shown in parentheses below the enzymes responsible for each step. Dotted lines indicate multiple enzymatic steps.

acids. As a results the potential to engineer microbial pathways is increasing significantly and includes opportunities to construct novel biochemical routes for the biosynthesis of nonproteinogenic amino acids.

4 Engineering Novel Biochemical Pathways for Amino Acid Production

The biosynthesis of most proteinogenic amino acids including those described above directly or indirectly involves a transamination step (Garner et al., 1983). Biochemical studies on a number of the bacterial L-amino acid transaminases (aminotransferases) responsible for this reaction have revealed that these enzymes frequently display broad substrate specificity. The aromatic transaminase of *E. coli*, e.g., is capable of synthesizing aspartate and leucine in addition to the aromatic amino acids phenylalanine and tyrosine. Similarly, the *E. coli* branched chain transaminase can synthesize phenylalanine and methionine in addition to the branched chain amino acids, isoleucine, leucine, and valine (Chihara, 1985). The relatively relaxed substrate specificity of microbial transaminases has been useful in the development of biotransformation approaches for the synthesis of non-proteinogenic L-amino acids, which are now in increasing demand as intermediates for the synthesis of peptidomimetic pharmaceuticals. Additionally the availability of a highly efficient screen has enabled the cloning of a number of genes encoding D-amino acid transaminase from a variety of microbial species. One of these genes, from *Bacillus sphaericus,* has been engineered into a synthetic pathway in *E. coli* along with an L-amino acid deaminase from *Proteus myxofaciens* and an amino acid racemase from *Salmonella typhimurium* to enable fermentative production of a number of D-amino acids from L-amino acid starting materials. Using these approaches, compounds such as D-phenylalanine, L-tertiary leucine, and both isomers of 2-aminobutyrate have been produced (Ager et al., 1997; Fotheringham et al., 1997). Some aspects of these processes are described below.

4.1 A Novel Biosynthetic Pathway to L-2-Aminobutyrate

In addition to the identification of the appropriate biosynthetic enzyme, the feasibility of all biotransformation processes depends heavily upon other criteria such as the availability of inexpensive starting materials, the reaction yield and the complexity of product recovery. In the case of transaminase processes, the reversible nature of the reaction as shown in Fig. 6 and the presence of a keto acid by-product is a concern which limits the overall yield and purity of the amino acid product, and has led to efforts to increase the conversion beyond the typical 50% yield of product (Crump and Rozzell, 1992; Taylor et al., 1998). Additionally, there are cost considerations in the large scale preparation of keto acid substrates such as 2-ketobutyrate, which are not commodity chemicals.

These issues are very readily addressed in whole cell systems as additional microbial enzymes can be used to generate substrates from inexpensive precursors and convert unwanted by-products to more easily handled compounds. For the biosynthesis of 2-aminobutyrate using the *E. coli* aromatic transaminase (Fotheringham et al., 1999), the engineered *E. coli* incorporates the cloned *E. coli* K12 *ilvA* gene encoding threonine deaminase to generate 2-ketobutyrate from the commodity amino acid L-threonine, and the cloned *alsS* gene of *Bacillus subtilis* 168 encoding acetolactate synthase which eliminates pyruvate, the keto acid

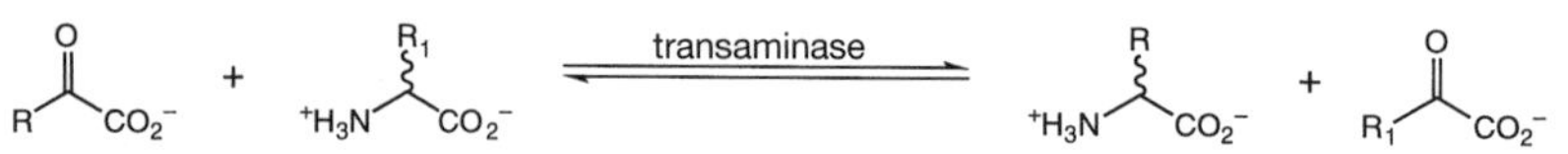

$$R\text{-}CO\text{-}CO_2^- + {}^+H_3N\text{-}CH(R_1)\text{-}CO_2^- \xrightleftharpoons{\text{transaminase}} {}^+H_3N\text{-}CH(R)\text{-}CO_2^- + R_1\text{-}CO\text{-}CO_2^-$$

Fig. 6. Transaminase reaction scheme. Transaminases occur as L-amino acid or D-amino acid specific.

by-product of the reaction, through the formation of non-reactive acetolactate. The *B. subtilis* enzyme is preferable to the acetolactate synthases of *E. coli*, which also use 2-ketobutyrate as a substrate, due to their overlapping roles in branched chain amino acid biosynthesis. The process is operated as a whole cell biotransformation in a single strain with the reaction scheme as shown in Fig. 7 producing 27 g L^{-1} of L-2-aminobutyrate. The effect of the concerted action of these three enzymes on the process of L-2-aminobutyrate biosynthesis is very significant. Process economics benefit from the use of an inexpensive amino acid such as L-threonine as the source of 2-ketobutyrate and L-aspartic acid as the amino donor. Secondly with the additional acetolactate synthase activity present the ratio of L-2-aminobutyrate to L-alanine, the major amino acid impurity increases to 22.5:1 from 2.4:1 reducing the complexity of the product recovery. Incremental improvements remain to be made in the prevention of undesired catabolism of substrates by metabolically active whole cells as the overall product yield is only 54% although the substrates are almost entirely consumed. The detailed understanding of *E. coli* metabolism and genetics increases the likelihood of eliminating such undesirable catabolic side reactions.

4.2 A Novel Biosynthetic Pathway to D-Phenylalanine

Unlike the L-amino acid transaminases (LAT), very few D-amino acid transaminases (DAT) (E.C. 2.6.1.21) have been extensively

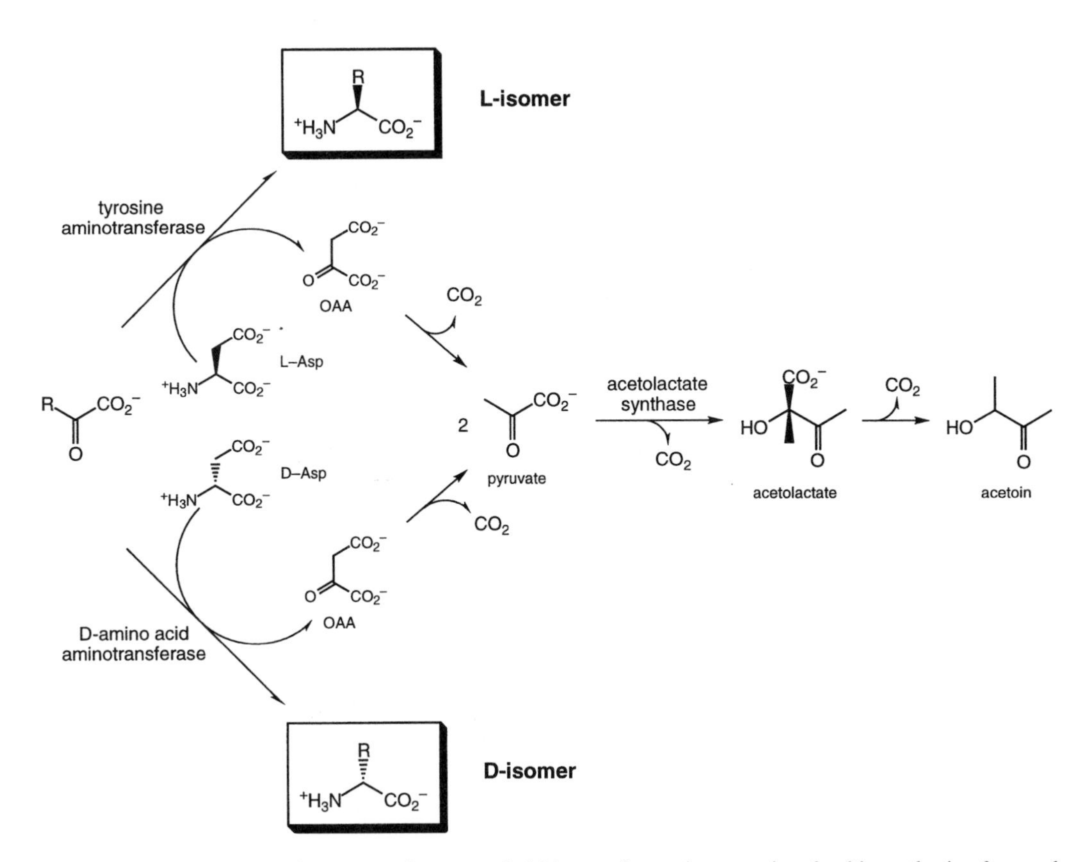

Fig. 7. Transaminase/acetolactate synthase coupled biotransformation reaction for biosynthesis of L- and D-2-aminobutyrate.

studied. One of the best understood is that from *Bacillus* sp. YM1, which has been cloned, sequenzed (TANIZAWA et al., 1989), and the crystal structure solved (SUGIO et al., 1995). Recently, the *dat* genes from *Staphylococcus haemolyticus* (PUCCI et al., 1995),), *Bacillus licheniformis* (TAYLOR and FOTHERINGHAM, 1997), and *Bacillus sphaericus* (FOTHERINGHAM et al., 1998a) have been cloned, sequenced, and overexpressed in *E. coli*. These factors, plus their broad substrate specificity make them ideal candidates for use in the production of D-amino acids.

As with the LATs, the DATs have an equilibrium constant near unity. Fortunately, many of the procedures devised to optimize product formation with the LATs are also applicable to the DATs. Product insolubility, high substrate concentrations, unstable keto acid products (e.g., oxaloacetate), and the removal of the keto acid product enzymatically (e.g., conversion of pyruvate to acetolactate) can be used to alter the equilibrium of the DAT reaction in favor of the D-amino acid product. The supply of D-amino acid donor can be problematical because most D-amino acids are either unavailable as commodity chemicals or are expensive. This problem is generally solved by initially using an L-amino acid such as L-aspartate, L-glutamate, or L-alanine, which is converted *in situ* to a racemate using an appropriate racemase. The DNA sequences of a number of racemases are available in the GenBank DNA database (NCBI, Bethesda, MD, USA) and the genes are easily obtainable by PCR methodology.

The whole cell bioconversion reaction described for L-amino acids is equally applicable to the biosynthesis of D-amino acids using the D-amino acid transaminase as shown in Fig. 7. This has been successfully applied to the production of a number of D-amino acids. These include both D-2-aminobutyrate (FOTHERINGHAM et al., 1997) and D-glutamate. Theoretically, this route is applicable to the production of other D-amino acids, as long as the keto acid cannot serve as a substrate for acetolactate synthase, and the product is not enzymatically racemized by the recombinant cell.

SODA and colleagues (GALKIN et al., 1997a, b; NAKAJIMA et al., 1988) have described a procedure which uses a thermostable DAT from *Bacillus* sp. YM1 in a coupled reaction with L-alanine dehydrogenase and alanine racemase (KURODA et al., 1990; TANIZAWA et al., 1988) such that pyruvate is recycled to D-alanine via L-alanine. The process requires an NADH regeneration system, which is supplied by the action of a cloned, thermostable formate dehydrogenase (GALKIN et al., 1995). The genes encoding each enzyme were cloned and expressed in a single *E. coli* strain. The overall procedure works efficiently for D-glutamate and D-leucine while other keto acids tested showed poor yields and/or poor enantiomeric excess. These problems were attributed to a number of causes including reaction of the L-alanine dehydrogenase with a number of keto acids tested, racemization of some of the amino acid products by the thermostable alanine racemase, and transamination of keto acids by the L-amino acid transaminases of the host. Other drawbacks include the need for relatively low substrate concentrations (by industrial standards) and the high cost of the keto acid supply.

A versatile, modular approach for the production of D-amino acids is currently being developed which addresses both D-amino acid donor and the α-keto acid acceptor supply, but does not require exogenous regeneration of co-factors. A generic depiction of this approach is shown in Fig. 8. The key step is the inclusion of the *lad* gene of *Proteus myxofaciens*, encoding the L-amino acid deaminase which is used to produce the α-keto acid *in situ* from the L-amino acid which is provided as substrate or overproduced by the host. In either case, the deamination occurs intracellularly. The D-amino acid required as amino donor is similarly generated *in situ* from the L-isomer by an appropriate cloned racemase. The two substrates are then converted by a cloned DAT to produce the corresponding D-amino acid. Because the reaction is carried out under fermentative conditions, the reaction is shifted towards 100% yield by catabolism of the α-keto product and by regeneration of the amino donor by cellular enzymes. This system can be used to produce D-amino acids with high enantiomeric excess even in the presence of the endogenous L-transaminases of the host organism. Using a combination of the approaches described, D-phenylalanine can be

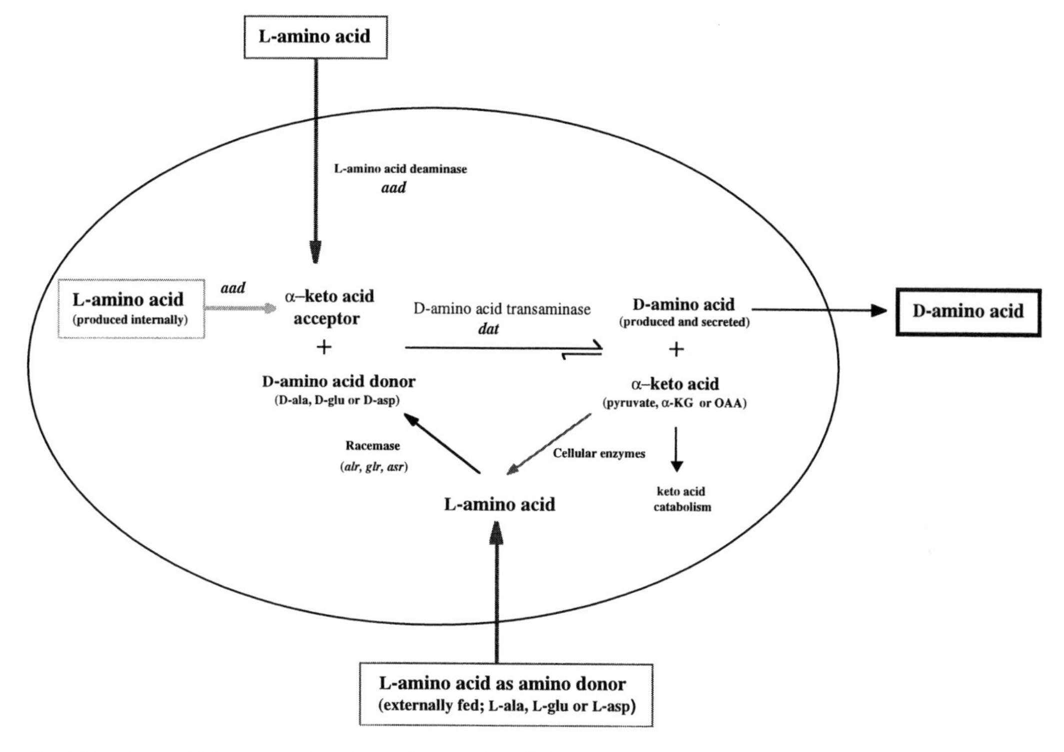

Fig. 8. Engineered biochemical pathway for whole cell biosynthesis of D-phenylalanine.

produced in high yields with 99% enantiomeric excess (FOTHERINGHAM et al., 1998b). D-Tyrosine can similarly be synthesized using a fed batch fermentation process but with improved overall yields due to the insolubility of the final product (R. F. SENKPIEL, unpublished data). The promiscuous substrate range of L-amino acid deaminase (L-AAD) and DAT facilitates the use of this type of approach to generate a wide range of D-amino acids from their L-amino acid counterparts.

5 Conclusions

A variety of bacterial species have been used for almost 40 years in the large scale fermentative production of amino acids for industrial uses. The corynebacteria have been most widely used in this application due to their propensity to overproduce and excrete very high concentrations of amino acids under specific process conditions and their early development for monosodium glutamate production. However, other organisms including *E. coli* have also been successfully employed in the production of amino acids such as phenylalanine.

The engineering of bacterial metabolic pathways to improve commercial fermentative production of amino acids has traditionally involved the use of relatively crude, non specific mutagenesis methods coupled to repeated rounds of arduous screening for resistance to toxic amino acid analogs. Although inelegant in execution, these methods nevertheless led to highly efficient and economically important organisms for large scale production of many amino acids, even though the approach becomes increasingly self-limiting as production organisms accumulate non-specific mutations.

With the increasing understanding of the biochemistry and molecular genetics of amino acid production in bacteria and the availability of recombinant DNA methodology in the 1980s the rate limiting steps of biosynthetic pathways and precursor supply were addressed through the cloning and characterization of the genes encoding the key biosynthetic enzymes. This led to a greater understanding of the molecular mechanisms involved in biosynthetic pathway regulation and the characterization of mutations introduced previously through classical methods. Deregulation of gene expression, re-introduction of specific genes on multi-copy plasmids, and the relief of end product mediated feedback inhibition has led to even greater titers of product from amino acid overproducing bacterial strains.

Consequently, many successes of amino acid overproduction in bacteria have come from the powerful hybrid approach of rational design and more traditional mutagenesis/selection methodology. As microbial DNA sequence information becomes increasingly available through genomic sequencing the possibility to combine genes from diverse organisms in a single host is providing new opportunities to further engineer specific host bacterial strains for amino acid overproduction. Accordingly, there are increasing examples of *E. coli* strains containing genes from multiple organisms to enable efficient whole cell biosynthesis of D-amino acids and non-proteinogenic L-amino acids.

As our understanding of microbial molecular biology increases and with the availability of newer mutagenesis methods such as error prone PCR and gene shuffling the capability to engineer microbial biosynthetic pathways to produce natural and unnatural amino acids will continue to increase and diversify.

Acknowledgement
I would like to thank Dr. PAUL TAYLOR and Dr. DAVID PANTALEONE for their assistance in the preparation of this chapter.

6 References

ABE, S., TAKAYAMA, K. (1972), in: *Amino Acid-Producing Microorganisms. Variety and Classification.* pp. 3–38. Tokyo: Kodansha.

AGER, D. J., FOTHERINGHAM, I. G., LANEMAN, S. A., PANTALEONE, D. P., TAYLOR, P. P. (1997), The largescale synthesis of unnatural amino acids, *Chim. Oggi* **15**, 11–14.

ARCHER, J. A. C., SINSKEY, A. J. (1993), The DNA sequence and minimal replicon of the *Corynebacterium glutamicum* plasmid pSR1: evidence of a common ancestry with plasmids from *C. diphtheriae, J. Gen. Microbiol.* **139**, 1753–1759.

BACKMAN, K. C., BALAKRISHNAN, R. (1988), Plasmids encoding *pheA* gene products resistant to phenylalanine inhibition for use in microbial manufacture of phenylalanine, *U.S. Patent* 4753883.

BATT, C. A., FOLLETTIE, M. T., SHIN, H. K., YEH, P., SINSKEY, A. J. (1985), Genetic engineering of coryneform bacteria, *Trends Biotechnol.* **3**, 305–310.

BERRY, A. (1996), Improving production of aromatic compounds in *Escherichia coli* by metabolic engineering, *Trends Biotechnol.* **14**, 250–256.

BLEDIG, S. A. (1994), Investigation of the genetic regulation affecting carbon flux through phosphoenol pyruvate in *Escherichia coli* K12, *Thesis*, University of Warwick, England.

BOERMANN, E. R., EIKMANNS, B. J., SAHM, H. (1992), Molecular analysis of the *Corynebacterium glutamicum gdh* gene encoding glutamate dehydrogenase, *Mol. Microbiol.* **6**, 317–326.

BROWN, K. D. (1968), Regulation of aromatic amino acid biosynthesis in *Escherichia coli* K12, *Genetics* **60** (Pt. 1), 31–48.

BROWN, K. D., SOMERVILLE, R. L. (1971), Repression of aromatic amino acid biosynthesis in *Escherichia coli* K12, *J. Bacteriol.* **108**, 386–399.

CAMAKARIS, H., PITTARD, J. (1973), Regulation of tyrosine and phenylalanine biosynthesis in *Escherichia coli* K12. Properties of the *tyrR* gene product, *J. Bacteriol.* **115**, 1135–1144.

CHOI, Y. J., TRIBE, D. E. (1982), Continuous production of phenylalanine using an *Escherichia coli* regulatory mutant, *Biotechnol. Lett.* **4**, 223–228.

CLEMENT, Y., LANEELLE, G. (1986), Glutamate excretion mechanism in *Corynebacterium glutamicum:* triggering by biotin starvation or by surfactant addition, *J. Gen. Microbiol.* **132**, 925–929.

COLON, G. E., NGUYEN, T. T., JETTEN, M. S. M., SINSKEY, A. J., STEPHANOPOULOS, G. (1995), Production if isoleucine by overexpression of *ilvA* in a *Corynebacterium lactofermentum* threonine producer, *Appl. Microbiol. Biotechnol.* **43**, 482–488.

CREMER, J., EGGELING, L., SAHM, H. (1991), Control

of the lysine biosynthesis sequence in *Corynebacterium glutamicum* as analyzed by overexpression of the individual corresponding genes, *Appl. Environ. Microbiol.* **57**, 1746–1752.

CRUMP, S. P., ROZZELL, J. D. (1992), in: *Biocatalytic Production of Amino Acids and Derivatives* (ROZZELL, D. J., WAGNER, F., Eds.), pp. 43–58. New York: Oxford University Press.

DAVIES, W. D., DAVIDSON, B. E. (1982), The nucleotide sequence of *aroG*, the gene for 3-deoxy-D-arabinoheptulosonate-7-phosphate synthetase (phe) in *Escherichia coli* K12, *Nucleic Acids Res.* **10**, 4045–4058.

DE BOER, L., DIJKHUIZEN, L. (1990), Microbial and enzymic processes for L-phenylalanine production, *Adv. Biochem. Eng./Biotechnol.* **41** (*Microb. Bioprod.*), 1–27.

DOPHEIDE, T. A. A., CREWTHER, P., DAVIDSON, B. E. (1972), Chorismate mutase-prephenate dehydratase from *Escherichia coli* K12. II. Kinetic properties, *J. Biol. Chem.* **247**, 4447–4452.

DRATHS, K. M., POMPLIANO, D. L., CONLEY, D. L., FROST, J. W., BERRY, A. et al. (1992), Biocatalytic synthesis of aromatics from D-glucose: the role of transketolase, *J. Am. Chem. Soc.* **114**, 3956–3962.

DUPERRAY, F., JEZEQUEL, D., GHAZI, A., LETELLIER, L., SHECHTER, E. (1992), Excretion of glutamate from *Corynebacterium glutamicum* triggered by amine surfactants, *Biochim. Biophys. Acta* **1103**, 250–258.

EDWARDS, M. R., TAYLOR, P. P., HUNTER, M. G., FOTHERINGHAM, J. G. (1987), *Europ. Patent Appl.* 229,161.

FLORES, N., XIAO, J., BERRY, A., BOLIVAR, F., VALLE, F. (1996), Pathway engineering for the production of aromatic compounds in *Escherichia coli, Nature Biotechnology* **14**, 620–623.

FOERBERG, C., ELIAESON, T., HAEGGSTROEM, L. (1988), Correlation of theoretical and experimental yields of phenylalanine from non-growing cells of a rec *Escherichia coli* strain, *J. Biotechnol.* **7**, 319–332.

FOLLETTIE, M. T., PEOPLES, O. P., AGOROPOULOU, C., SINSKEY, A. J. (1993), Gene structure and expression of the *Corynebacterium flavum* N13 *ask-asd* operon, *J. Bacteriol.* **175**, 4096–4103.

FOTHERINGHAM, I. G., DACEY, S. A., TAYLOR, P. P., SMITH, T. J., HUNTER, M. G. et al. (1986), The cloning and sequence analysis of the *aspC* and *tyrB* genes from *Escherichia coli* K12. Comparison of the primary structures of the aspartate aminotransferase and aromatic aminotransferase of *E. coli* with those of the pig aspartate aminotransferase isoenzymes, *Biochem. J.* **234**, 593–604.

FOTHERINGHAM, I. G., TON, J., HIGGINS, C. (1994), Amino acid hypersecreting *Escherichia coli* and their preparation, *US Patent* 5,354,672.

FOTHERINGHAM, I. G., PANTALEONE, D. P., TAYLOR, P. P. (1997), Biocatalytic production of unnatural amino acids, mono esters, and N-protected derivatives, *Chim. Oggi* **15**, 33–37.

FOTHERINGHAM, I. G., BLEDIG, S. A., TAYLOR, P. P. (1998a), Characterization of the genes encoding D-amino acid transaminase and glutamate racemase, two D-glutamate biosynthetic enzymes of *Bacillus sphaericus* ATCC 10208, *J. Bacteriol.* **180**, 4319–4323.

FOTHERINGHAM, I. G., TAYLOR, P. P., TON, J. L. (1998b), Preparation of D-amino acids by direct fermentative means, *US Patent* 5,728,555.

FOTHERINGHAM, I. G., GRINTER, N., PANTALEONE, D. P., SENKPIEL, R. F., TAYLOR, P. P. (1999), Engineering of a novel biochemical pathway for the biosynthesis of L-2-aminobutyric acid in *Escherichia coli* K12, *Bioorg. Med. Chem.* **7**, 2209–2213.

FROST, J. W., DRATHS, K. M. (1995), Biocatalytic syntheses of aromatics from D-glucose: renewable microbial sources of aromatic compounds, *Annu. Rev. Microbiol.* **49**, 557–579.

GALKIN, A., KULAKOVA, L., TISHKOV, V., ESAKI, N., SODA, K. (1995), Cloning of formate dehydrogenase gene from a methanol-utilizing bacterium *Mycobacterium vaccae* N10, *Appl. Microbiol. Biotechnol.* **44**, 479–483.

GALKIN, A., KULAKOVA, L., YAMAMOTO, H., TANIZAWA, K., TANAKA, H. (1997a), Conversion of α-keto acids to D-amino acids by coupling of four enzyme reactions, *J. Ferment. Bioeng.* **83**, 299–300.

GALKIN, A., KULAKOVA, L., YOSHIMURA, T., SODA, K., ESAKI, N. (1997b), Synthesis of optically active amino acids from α-keto acids with *Escherichia coli* cells expressing heterologous genes, *Appl. Environ. Microbiol.* **63**, 4651–4656.

GARNER, C. C., HERRMANN, K. M., YOSHINAGA, F., NAKAMORI, S. (1983), in: *Amino Acids: Biosynthesis and Genetic Regulation* (HERRMANN, K. M., SOMMERVILLE, R. L., Eds.), Reading, MA: Addison-Wesley.

GETHING, M. J., DAVIDSON, B. E. (1978), Chorismate mutase/prephenate dehydratase from *Escherichia coli* K12. Binding studies with the allosteric effector phenylalanine, *Eur. J. Biochem.* **86**, 165–174.

HAGINO, H., NAKAYAMA, K. (1974a), Production of aromatic amino acids by microorganisms. VI. Regulatory properties of prephenate dehydrogenase and prephenate dehydratase from *Corynebacterium glutamicum, Agric. Biol. Chem.* **38**, 2367–2376.

HAGINO, H., NAKAYAMA, K. (1974b), Production of aromatic amino acids by microorganisms. IV. L-Phenylalanine production by analog-resistant mutants of *Corynebacterium glutamicum, Agr. Biol. Chem.* **38**, 157–161.

HAGINO, H., NAKAYAMA, K. (1974c), Production of

aromatic amino acids by microorganisms. V. DAHP [3-deoxy-D-arabinoheptulosonate 7-phosphate] synthetase and its control in *Corynebacterium glutamicum, Agric. Biol. Chem.* **38**, 2125–2134.

HAGINO, H., NAKAYAMA, K. (1975a), Production of aromatic amino acids by microorganisms. VIII. Regulatory properties of chorismate mutase from *Corynebacterium glutamicum, Agric. Biol. Chem.* **39**, 331–342.

HAGINO, H., NAKAYAMA, K. (1975b), Production of aromatic amino acids by microorganisms. X. Biosynthetic control in aromatic amino acid producing mutants of *Corynebacterium glutamicum, Agric. Biol. Chem.* **39**, 351–361.

HASLAM, E. (1974), *The Shikimate Pathway*. New York: Halsted.

HONORE, N., COLE, S. T. (1990), Nucleotide sequence of the *aroP* gene encoding the general aromatic amino acid transport protein of *Escherichia coli* K-12: homology with yeast transport proteins, *Nucleic Acids Res.* **18**, 653.

HUDSON, G. S., DAVIDSON, B. E. (1984), Nucleotide sequence and transcription of the phenylalanine and tyrosine operons of *Escherichia coli* K12, *J. Mol. Biol.* **180**, 1023–1051.

HWANG, S. O., GIL, G. H., CHO, Y. J., KANG, K. R., LEE, J. H., BAE, J. C. (1985), The fermentation process for L-phenylalanine production using an auxotrophic regulatory mutant of *Escherichia coli, Appl. Microbiol. Biotechnol.* **22**, 108–113.

ICHIHARA, A. (1985), in: *Transaminases* (CHRISTEN, P., METZLER, P. D., Eds.), p. 430. New York: John Wiley & Sons.

IKEDA, M., KATSUMATA, R. (1992), Metabolic engineering to produce tyrosine or phenylalanine in a tryptophan-producing *Corynebacterium glutamicum* strain, *Appl. Environ. Microbiol.* **58**, 781–785.

IKEDA, M., OZAKI, A., KATSUMATA, R. (1993), Phenylalanine production by metabolically engineered *Corynebacterium glutamicum* with the *pheA* gene of *Escherichia coli, Appl. Microbiol. Biotechnol.* **39**, 318–323.

IM, S. W. K., DAVIDSON, H., PITTARD, J. (1971), Phenylalanine and tyrosine biosynthesis in *Escherichia coli* K-12. Mutants derepressed for 3-deoxy-D-arabinoheptulosonic acid 7-phosphate synthetase (phe), 3-deoxy-D-arabinoheptulosonic acid 7-phosphate synthetase (tyr), chorismate mutase T-prephenate dehydrogenase, and transaminase A, *J. Bacteriol.* **108**, 400–409.

ISHIDA, M., YOSHINO, E., MAKIHARA, R., SATO, K., ENEI, H., NAKAMORI, S. (1989), Improvement of an L-threonine producer derived from *Brevibacterium flavum* using threonine operon of *Escherichia coli* K-12, *Agric. Biol. Chem.* **53**, 2269–2271.

ISHIDA, M., SATO, K., HASHIGUCHI, K., ITO, H., ENEI, H., NAKAMORI, S. (1993), High fermentative production of L-threonine from acetate by a *Brevibacterium flavum* stabilized strain transformed with a recombinant plasmid carrying the *Escherichia coli thr* operon, *Biosci. Biotechnol. Biochem.* **57**, 1755–1756.

JETTEN, M. S. M., SINSKEY, A. J. (1995), Recent advances in the physiology and genetics of amino acid-producing bacteria, *Crit. Rev. Biotechnol.* **15**, 73–103.

JETTEN, M. S. M., GUBLER, M. E., MCCORMICK, M. M., COLON, G. E., FOLLETTIE, M. T., SINSKEY, A. J. (1993), Molecular organization and regulation of the biosynthetic pathway for aspartate-derived amino acids in *Corynebacterium glutamicum, Proc. Ind. Microorg.* pp. 97–104.

JETTEN, M. S. M., FOLLETTIE, M. T., SINSKEY, A. J. (1994), Metabolic engineering of *Corynebacterium glutamicum, Ann. N. Y. Acad. Sci.* **721** (Recombinant DNA Technology II), 12–29.

JETTEN, M. S. M., FOLLETTIE, M. T., SINSKEY, A. J. (1995), Effect of different levels of aspartokinase on the lysine production by *Corynebacterium lactofermentum, Appl. Microbiol. Biotechnol.* **43**, 76–82.

KALINOWSKI, J., CREMER, J., BACHMANN, B., EGGELING, L., SAHM, H., PRUEHLER, A. (1991), Genetic and biochemical analysis of the aspartokinase from *Corynebacterium glutamicum, Mol. Microbiol.* **5**, 1197–1204.

KURODA, S., TANIZAWA, K., TANAKA, H., SODA, K., SAKAMOTO, Y. (1990), Alanine dehydrogenases from two *Bacillus* species with distinct thermostabilities: molecular cloning, DNA and protein sequence determination, and structural comparison with other $NAD(P)^+$-dependent dehydrogenases, *Biochemistry* **29**, 1009–1015.

MARTIN, J. F., SANTAMARIA, R., SANDOVAL, H., DEL REAL, G., MATEOS, L. M. et al. (1987), Cloning systems in amino acid-producing corynebacteria, *Bio/Technology* **5**, 137–146.

MILLAR, G., LEWENDON, A., HUNTER, M. G., COGGINS, J. R. (1986), The cloning and expression of the *aroL* gene from *Escherichia coli* K12. Purification and complete amino acid sequence of shikimate kinase II, the *aroL*-gene product, *Biochem. J.* **237**, 427–437.

MILLER, J. E., BACKMAN, K. C., O'CONNOR, M. J., HATCH, R. T. (1987), Production of phenylalanine and organic acids by phosphoenolpyruvate carboxylase-deficient mutants of *Escherichia coli, J. Ind. Microbiol.* **2**, 143–149.

MIWA, K., TSUCHIDA, T., KURAHASHI, O., NAKAMORI, S., SANO, K., MOMOSE, H. (1983), Construction of L-threonine overproducing strains of *Escherichia coli* K-12 using recombinant DNA techniques,

Agric. Biol. Chem. **47**, 2329–2334.

MIWA, K., MATSUI, K., TERABE, M., ITO, K., ISHIDA, M. et al. (1985), Construction of novel shuttle vectors and a cosmid vector for the glutamic acid-producing bacteria *Brevibacterium lactofermentum* and *Corynebacterium glutamicum, Gene* **39**, 281–286.

MULCAHY, M. (1988), *Thesis*, University of Dundee, Scotland.

NAKAJIMA, N., TANIZAWA, K., TANAKA, H., SODA, K. (1988), Enantioselective synthesis of various D-amino acids by a multi-enzyme system, *J. Biotechnol.* **8**, 243–248.

NAMPOOTHIRI, M., PANDEY, A. (1998), Genetic tuning of coryneform bacteria for the overproduction of amino acids, *Process Biochem.* **33**, 147–161.

NELMS, J., EDWARDS, R. M., WARWICK, J., FOTHERINGHAM, J. (1992), Novel mutations in the *pheA* gene of *Escherichia coli* K-12 which result in highly feedback inhibition-resistant variants of chorismate mutase/prephenate dehydratase, *Appl. Environ. Microbiol.* **58**, 2592–2598.

OZAKI, A., KATSUMATA, R., OKA, T., FURUYA, A. (1985), Cloning of the genes concerned in phenylalanine biosynthesis in *Corynebacterium glutamicum* and its application to breeding of a phenylalanine producing strain, *Agric. Biol. Chem.* **49**, 2925–2930.

PATNAIK, R., LIAO, J. C. (1994), Engineering of *Escherichia coli* central metabolism for aromatic metabolite production with near theoretical yield, *Appl. Environ. Microbiol.* **60**, 3903–3908.

PATNAIK, R., SPITZER, R. G., LIAO, J. C. (1995), Pathway engineering for production of aromatics in *Escherichia coli:* confirmation of stoichiometric analysis by independent modulation of AroG, TktA, and Pps activities, *Biotechnol. Bioeng.* **46**, 361–370.

PI, J., WOOKEY, P. J., PITTARD, A. J. (1991), Cloning and sequencing of the *pheP* gene, which encodes the phenylalanine-specific transport system of *Escherichia coli, J. Bacteriol.* **173**, 3622–3629.

PITTARD, J., GIBSON, F. (1970), Regulation of biosynthesis of aromatic amino acids and vitamins, *Curr. Top. Cell Regul.* **2**, 29–63.

POSTMA, P. W. (1987), in: *Escherichia coli and Salmonella Cellular and Molecular Biology* (NEIDHARDT, F. C., Ed.), pp. 127–141. Washington, DC: ASM Press.

PUCCI, M. J., THANASSI, J. A., HO, H.-T., FALK, P. J., DOUGHERTY, T. J. (1995), *Staphylococcus hemolyticus* contains two D-glutamic acid biosynthetic activities, a glutamate racemase and a D-amino acid transaminase, *J. Bacteriol.* **177**, 336–342.

PÜHLER, A., KASSING, F., WINTERFELDT, A., KALINOWSKI, J., SCHAEFER, A. et al. (1990), A novel system for genetic engineering of amino acid producing corynebacteria and the analysis of AEC-resistant mutants, *Proc. ECB5* (CHRISTIANSEN, C., MUNCK, L., VILLADSEN, J., Eds.), pp. 975–978. Copenhagen: Munksgaard.

RAY, J. M., YANOFSKY, C., BAUERLE, R. (1988), Mutational analysis of the catalytic and feedback sites of the tryptophan-sensitive 3-deoxy-D-arabino-heptulosonate-7-phosphate synthase of *Escherichia coli, J. Bacteriol.* **170**, 5500–5506.

SANTAMARIA, R. I., MARTÍN, J. F., GIL, J. A. (1987), Identification of a promoter sequence in the plasmid pUL340 of *Brevibacterium lactofermentum* and construction of new cloning vectors for corynebacteria containing two selectable markers, *Gene* **56**, 199–208.

SCHRUMPF, B., SCHWARZER, A., KALINOWSKI, J., PÜHLER, A., EGGELING, L., SAHM, H. (1991), A functionally split pathway for lysine synthesis in *Corynebacterium glutamicum, J. Bacteriol.* **173**, 4510–4516.

SCHWARZER, A., PÜHLER, A. (1991), Manipulation of *Corynebacterium glutamicum* by gene disruption and replacement, *Bio/Technology* **9**, 84–87.

SHIIO, I. (1986), Tryptophan, phenylalanine, and tyrosine, *Prog. Ind. Microbiol.* **24**, (Biotechnol. Amino Acid Prod.), 188–206.

SHIIO, I., MIYAJIMA, R. (1969), Concerted inhibition and its reversal by end products as aspartate kinase in *Brevibacterium flavum, J. Biochem.* **65**, 849–859.

SHIIO, I., SUGIMOTO, S. (1976), Altered prephenate dehydratase in phenylalanine-excreting mutants of *Brevibacterium flavum, J. Biochem.* **79**, 173–183.

SHIIO, I., SUGIMOTO, S. (1979), Two components of chorismate mutase in *Brevibacterium flavum, J. Biochem.* **86**, 17–25.

SHIIO, I., SUGIMOTO, S. (1981a), Effect of enzyme concentration on regulation of dissociable chorismate mutase in *Brevibacterium flavum, J. Biochem.* **89**, 1483–1492.

SHIIO, I., SUGIMOTO, S. (1981b), Regulation at metabolic branch points of aromatic amino acid biosynthesis in *Brevibacterium flavum, Agric. Biol. Chem.* **45**, 2197–2207.

SHIIO, I., ISHII, K., YOKOZEKI, K. (1973), Production of L-tryptophan by 5-fluorotryptophan resistant mutants of *Bacillus subtilis, Agr. Biol. Chem.* **37**, 1991–1200.

SHIIO, I., SUGIMOTO, S., MIYAJIMA, R. (1974), Regulation of 3-deoxy-D-arabino-heptulosonate 7-phosphate synthetase in *Brevibacterium flavum, J. Biochem.* **75**, 987ff.

SHIIO, I., MORI, M., OZAKI, H. (1982), Amino acid aminotransferases in an amino acid-producing bacterium. *Brevibacterium flavum, Agric. Biol.*

Chem. **46**, 2967–2977.

SHIIO, I., SUGIMOTO, S., KAWAMURA, K. (1988), Breeding of phenylalanine-producing *Brevibacterium flavum* strains by removing feedback regulation of both the two key enzymes in its biosynthesis, *Agric. Biol. Chem.* **52**, 2247–2253.

SHIIO, I., TORIDE, Y., YOKOTA, A., SUGIMOTO, S., KAWAMURA, K. (1989), Threonine manufacture by dihydropicolinate synthase-deficient *Brevibacterium, Can. Patent Appl.* CAN 113:38946.

SHIIO, I., YOSHINO, H., SUGIMOTO, S. (1990), Isolation and properties of lysine-producing mutants with feedback-resistant aspartokinase derived from a *Brevibacterium flavum* strain with citrate synthase- und pyruvate kinase-defects ans feedback-resistant phosphoenolpyruvate carboxylase, *Agric. Biol. Chem.* **54**, 3275–3282.

SHIIO, I., SUGIMOTO, S., KAWAMURA, K. (1993), Isolation and properties of α-ketobutyrate-resistant lysine-producing mutants from *Brevibacterium flavum, Biosci. Biotechnol. Biochem.* **57**, 51–55.

SUGIMOTO, S., SHIIO, I. (1974), Regulation of prephenate dehydratase in *Brevibacterium flavum, J. Biochem.* **76**, 1103–1111.

SUGIMOTO, S., SHIIO, I. (1980), Purification and properties of bifunctional 3-deoxy-D-arabino-heptulosonate 7-phosphate synthetase-chorismate mutase component A from *Brevibacterium flavum, J. Biochem.* **87**, 881–890.

SUGIMOTO, S., SHIIO, I. (1985), Enzymes of common pathway for aromatic amino acid biosynthesis in *Brevibacterium flavum* and its tryptophan-producing mutants, *Agric. Biol. Chem.* **49**, 39–48.

SUGIMOTO, S., MAKAGAWA, M., TSUCHIDA, T., SHIIO, I. (1973), Regulation of aromatic amino acid biosynthesis and production of tyrosine and phenylalanine in *Brevibacterium flavum, Agr. Biol. Chem.* **37**, 2327–2336.

SUGIO, S., PETSKO, G. A., MANNING, J. M., SODA, K., RINGE, D. (1995), Crystal structure of a D-amino acid aminotransferase: how the protein controls stereoselectivity, *Biochemistry* **34**, 9661–9669.

TANIZAWA, K., OHSHIMA, A., SCHEIDEGGER, A., INAGAKI, K., TANAKA, H., SODA, K. (1988), Thermostable alanine racemase from *Bacillus stearothermophilus:* DNA and protein sequence determination and secondary structure prediction, *Biochemistry* **27**, 1311–1316.

TANIZAWA, K., ASANO, S., MASU, Y., KURAMITSU, S., KAGAMIYAMA, H. et al. (1989), The primary structure of thermostable D-amino acid aminotransferase from a thermophilic *Bacillus* species and its correlation with L-amino acid aminotransferases, *J. Biol. Chem.* **264**, 2450–2454.

TAYLOR, P. P., FOTHERINGHAM, I. G. (1997), Nucleotide sequence of the *Bacillus licheniformis* ATCC 10716 *dat* gene and comparison of the predicted amino acid sequence with those of other bacterial species, *Biochim. Biophys. Acta* **1350**, 38–40.

TAYLOR, P. P., PANTALEONE, D. P., SENKPIEL, R. F., FOTHERINGHAM, I. G. (1998), Novel biosynthetic approaches to the production of unnatural amino acids using transaminases, *Trends Biotechnol.* **16**, 412–418.

TESCH, M., EIKMANNS, B. J., DE GRAAF, A. A., SAHM, H. (1998), Ammonia assimilation in *Corynebacterium glutamicum* and a glutamate dehydrogenase-deficient mutant, *Biotechnol. Lett.* **20**, 953–957.

TOKORO, Y., OSHIMA, K., OKII, M., YAMAGUCHI, K., TANAKA, K., KINOSHITA, S. (1970), Microbial production of L-phenylalanine from *n*-alkanes, *Agr. Biol. Chem.* **34**, 1516–1521.

TOSAKA, O., TAKINAMI, K. (1986), Lysine, *Prog. Ind. Microbiol.* **24** (Biotechnol. Amino Acid Prod.), 152–172.

TRIBE, D. E. (1987), Preparation and use of novel *Escherichia coli* strains to manufacture phenylalanine, *US Patent* 4,681,852.

TSUCHIDA, T., KUBOTA, K., MORINAGA, Y., MATSUI, H., ENEI, H., YOSHINAGA, F. (1987), Production of L-phenylalanine by a mutant of *Brevibacterium lactofermentum* 2256, *Agric. Biol. Chem.* **51**, 2095–2101.

WALLACE, B. J., PITTARD, J. (1969), Regulator gene controlling enzymes concerned in tyrosine biosynthesis in *Escherichia coli, J. Bacteriol.* **97**, 1234–1241.

WHIPP, M. J., PITTARD, A. J. (1977), Regulation of aromatic amino acid transport systems in *Escherichia coli* K-12, *J. Bacteriol.* **132**, 453–461.

YAMADA, K., KINOSHITA, S., TSUNODA, T., AIDA, K. (1973), *The Microbial Production of Amino Acids.* New York: Halsted.

9 Biotechnological Production of Natural Aroma Chemicals by Fermentation Processes

JÜRGEN RABENHORST
Holzminden, Germany

1 Introduction

The demand for natural flavors by the customers of the Western world increased during the last decade. Naturalness is correlated by many people with health and other positive meanings. In Europe and the USA natural flavors are clearly defined legally.

In the EU Directive 88/388/EEC the term "natural" is allowed for a flavoring substance which is obtained "by appropriate physical processes (including distillation and solvent extraction) or enzymatic or microbiological processes from material of vegetable or animal origin either in the raw state or after processing for human consumption by traditional food preparation processes (including drying, torrefaction, and fermentation)".

In the USA the following definition of natural flavors applies: "The term *natural flavor* or *natural flavoring* means the essential oil oleoresin, essence or extractive, protein hydrolysate, distillate, or any product of roasting, heating, or enzymolysis, which contains the flavoring constituents derived from a spice, fruit juice, vegetable, or vegetable juice, edible yeast, herb, bark, bud, root, leaf, or similar plant material, meat, seafood, poultry, eggs, dairy products, or fermentation products thereof, whose significant function in food is flavoring rather than nutritional ..." (CFR, 1993).

Since the 1970s several reports appeared on the microbial and enzymatic synthesis of fragrance and aroma chemicals (for reviews, see SCHINDLER and SCHMID, 1982; and SCHARPF, JR. et al., 1986).

2 Fermentation Processes

2.1 Fermentative Production of Vanillin

Vanillin is the most important aroma chemical with an annual consumption of about 12,000 tons. It is broadly used for the flavoring of many different foods like chocolate, ice-cream, baked goods or liqueurs, and also in perfumery. Vanillin has an intensely sweet and very tenacious creamy-vanilla-like odor and a typical sweet vanillin-like taste.

Synthetic vanillin is currently produced from guaiacol and lignin (CLARK, 1990).

Natural vanilla flavor is classically obtained from the alcoholic extracts of cured vanilla pods (RANADIVE, 1994). Due to the limited availability of vanilla pods and the fluctuations of delivery based on varying harvest yields, depending on the weather, efforts were undertaken to find alternative sources.

If the price of natural vanillin was calculated on the yield in cured vanilla pods, which comprise less than 2% vanillin, it would be in the range of 2,000 U.S. $ kg^{-1}, which is extremely high compared to this nature identical vanillin obtained by chemical synthesis that only costs about 12 U.S. $ kg^{-1}. This is one of the incentives of the flavor industry for the search of natural vanillin from other sources.

2.1.1 Outline of Different Processes Applied to the Formation of Vanillin

Since the beginning of the 1990s a number of patents for the biotechnological production of vanillin have been disclosed. In some earlier studies vanillin was mentioned only as an intermediate in the microbial catabolism of different chemicals of plant origin. For example, TADASA described the degradation of eugenol with a *Corynebacterium* sp. (TADASA, 1977) and a *Pseudomonas* sp. (TADASA and KAYAHARA, 1983) and found vanillin as one of the intermediates. Many studies have been performed on the degradation of lignin. Ferulic acid has frequently been used as a simple model compound in studies for the degradation of lignin polymers. A number of workers found vanillin as an intermediate in the ferulic acid catabolism of different microorganisms (TOMS and WOOD, 1970; SUTHERLAND et al., 1983; ÖTÜK, 1985).

The first patent for the alternative production of natural vanillin was disclosed by DOLFINI et al. (1990). This was not a real biotechnological process because there turmeric oleoresin, in which curcumin is the main ingredient, was subjected to the action of heat and pressure in the presence of water by a continuous or batch process to produce vanillin (Fig. 1). This process yields only 20% of vanillin.

The first enzymatic process for the production of vanillin was disclosed by Yoshimoto et al. (1990a). They used a dioxygenase, isolated from a new *Pseudomonas* sp. TMY1009 (Yoshimoto et al., 1990b). As substrates they used isoeugenol (Fig. 2c) and 1,2-bis(4-hydroxy-3-methoxyphenyl)ethylene (Fig. 2a) or the corresponding glucoside (Fig. 2b).

The first microbial fermentation process for the production of vanillin was disclosed by Rabenhorst and Hopp (1991) with Haarmann & Reimer. They used bacteria of the genera *Klebsiella, Proteus,* and *Serratia* and eugenol or isoeugenol as substrate (Fig. 3). Only with isoeugenol reasonable vanillin concentrations of 3.8 g L^{-1} were obtained after 9 d.

Markus et al. (1993) with Quest disclosed a different enzymatic approach for the formation of vanillin. They used lipoxygenase, also known as lipoxidase (EC.1.13.11.12), an enzyme common in different plants like soy, potato, cucumber, grape, and tea. As substrates they used benzoe siam resin, in which the main component is coniferyl benzoate, or isoeugenol and eugenol. Only with isoeugenol they reached reasonable vanillin concentrations of about 10 g L^{-1} within 4 d. With benzoe siam resin and eugenol only concentrations of 2 to 4 g L^{-1} and 0.3–0.5 g L^{-1} were obtained, respectively. Due to the high price of lipoxygenase this process seems to be commercially unattractive. This is also true for all processes based on natural isoeugenol, because it is only available from some essential oils (ylang-ylang, nutmeg) as a minor component. Therefore, the price of this substrate is unacceptably high.

An interesting approach was followed by a US biotech company, ESCAgenetics Corp., who tried to produce vanillin by plant cell culture (Knuth and Sahai, 1991) in fermenters. With a cell line with an extremely high doubling time of 106 h they obtained concentrations of 11,900 mg L^{-1} vanillin under high cell density conditions (27 g cell dry weight per L; Sahai, 1994) but they seemed not to be able to scale up the process to a commercial scale, because the product never reached the market and the company does not exist anymore. A very ambitious approach was followed by Frost and coworkers, who were trying to obtain vanillin from glucose by metabolic engineering in *Escherichia coli* via the shikimic ac-

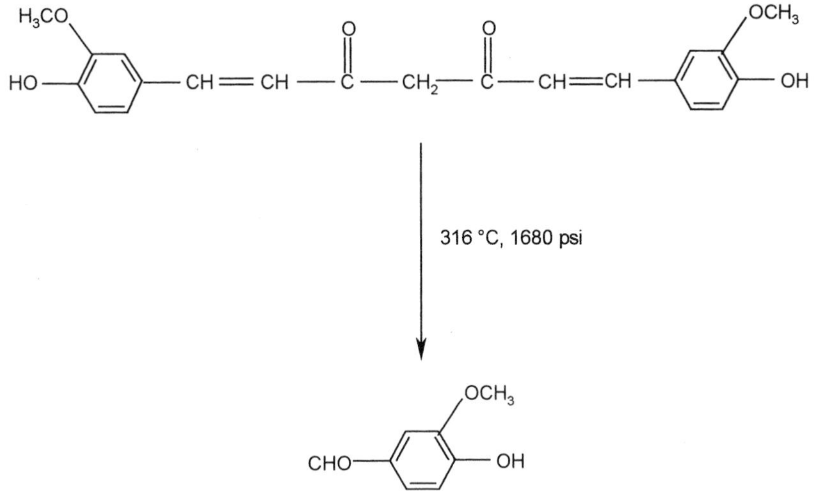

Fig. 1. Hydrolysis of curcumin according to Dolfini et al. (1990).

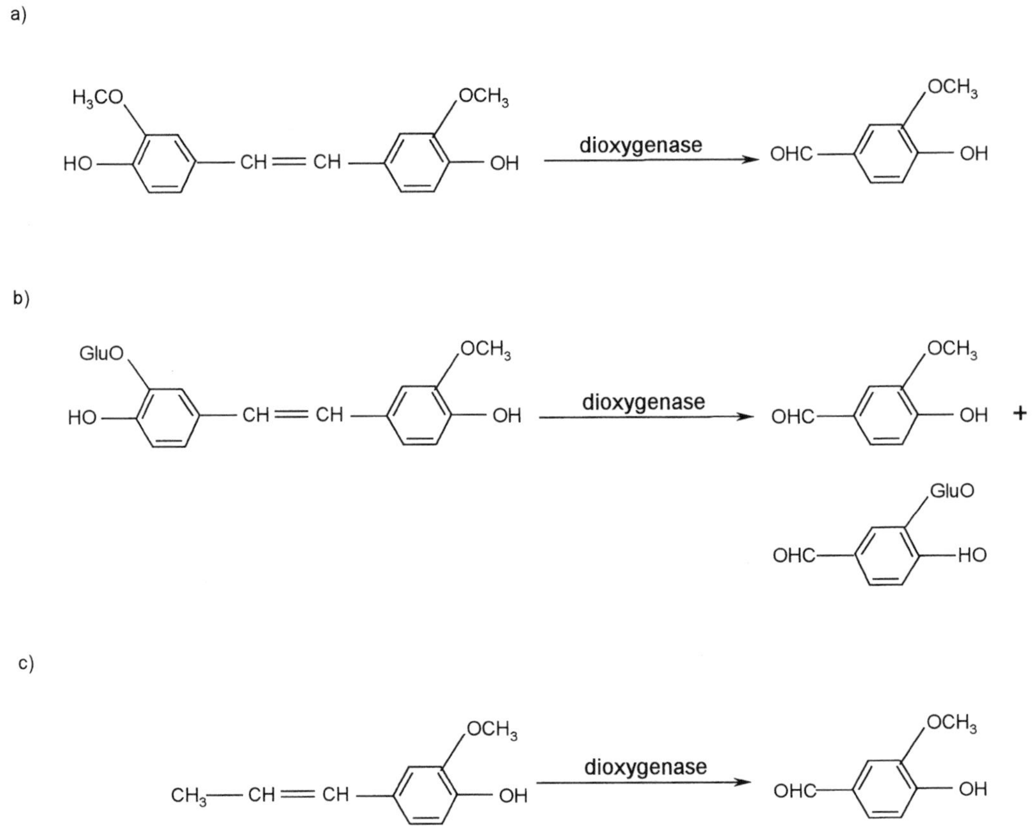

Fig. 2. Formation of vanillin with dioxygenase according to YOSHIMOTO et al. (1990a, b).

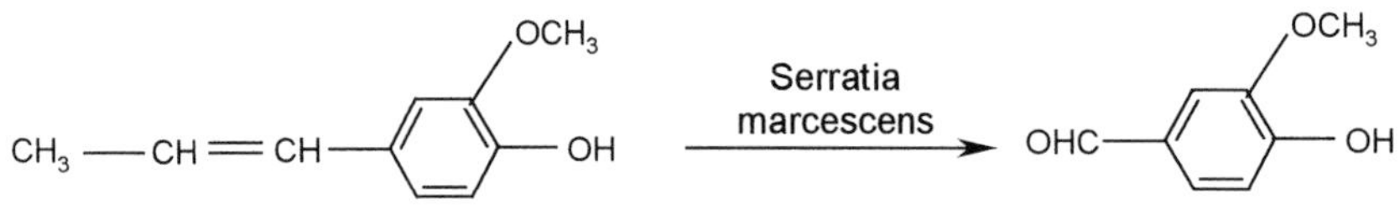

Fig. 3. Vanillin formation from isoeugenol by *Serratia marcescens* according to RABENHORST and HOPP (1990).

id pathway (LI and FROST, 1998). The main limiting steps are a lack of a sufficient catechol-O-methyltransferase activity for the formation of vanillic acid from protocatechuic acid, and a lack of selectivity for the aryl aldehyde dehydrogenase, so that about 10% of isovanillin were synthesized. Until now, obviously, only parts of the pathway have been cloned within one strain.

2.1.2 Production of Ferulic Acid as an Intermediate for Vanillin Production

Ferulic acid is an ubiquitously found phenolic cinnamic acid derivative in plants. It is linked to various carbohydrates as glycosidic conjugates, occurs as an ester or amide in various natural products, and is a part of the common plant polymer lignin (ROSAZZA et al., 1995). Free ferulic acid is not easily available in nature (STRACK, 1990).

In a patent ANTRIM and HARRIS disclosed the isolation of ferulic acid from corn hulls (ANTRIM and HARRIS, 1977). Due to the chemical alkali hydrolysis of the corn hulls to cleave the ester bonds, the resulting material cannot be considered as natural.

Another possible source is eugenol, the main component of the essential oil of the clove tree *Syzigium aromaticum* (syn. *Eugenia cariophyllus*). It is a cheap natural aroma chemical and available in sufficient amounts on the world market. It is used in many perfume and aroma compositions due to its oriental and spicy clove odor (BAUER et al., 1990). When eugenol is metabolized by different bacteria, ferulic acid is formed as an intermediate (TADASA, 1977; TADASA and KAYAHARA, 1983).

In 1993, HOPP and RABENHORST disclosed a patent application with Haarmann & Reimer for a new *Pseudomonas* sp. HR 199, isolated from soil, and a process for the production of methoxyphenols from eugenol. With this fermentation process ferulic acid concentrations of 5.8 g L^{-1} within 75 h (Fig. 4) were reached. Due to the toxicity of eugenol a repeated addition of the substrate was necessary. They obtained a high molar conversion of eugenol to ferulic acid of more than 50%.

By variation of the process parameters other interesting chemicals like coniferyl alcohol or vanillic acid could also be obtained in good yields (RABENHORST, 1996).

In the meantime the process has been further optimized for higher volumetric and molar yield and is run now successfully on a 40 m^3 scale.

Because the degradation of eugenol is extremely rare within the genus *Pseudomonas* (STEINBÜCHEL, personal communication), and the process of ferulic acid production is commercially important, the genes and enzymes of the metabolic pathway were isolated and characterized (STEINBÜCHEL et al., 1998). This pathway with the genes and enzymes, therefore, is outlined in Fig. 5.

During these studies it was found that vanillin is also an intermediate of the degradation of eugenol, but it was detected only occasionally in the culture broth in trace amounts. The reaction of ferulic acid to vanillic acid is obviously very effective, probably due to the higher reactivity and toxicity to the cells of the free aldehyde.

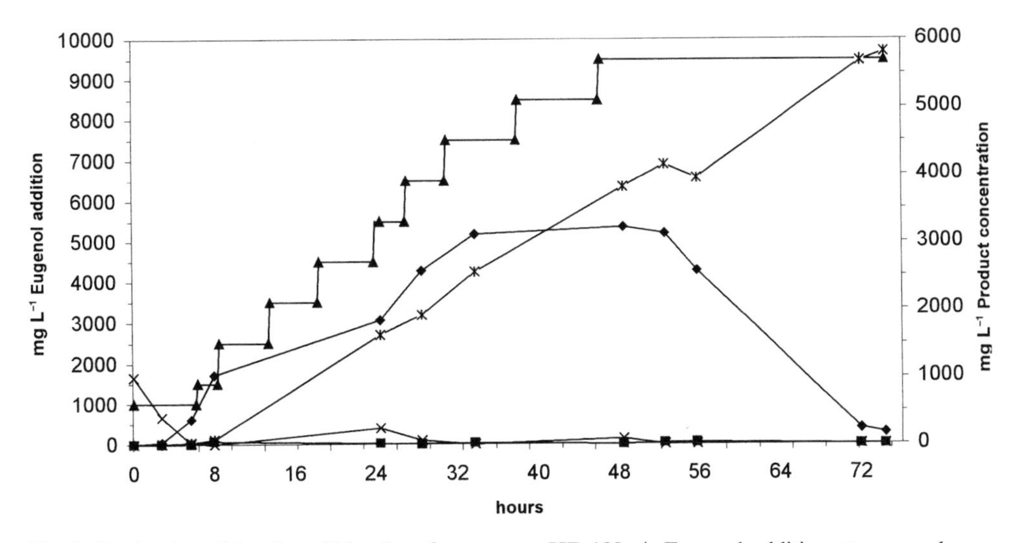

Fig. 4. Production of ferulic acid by *Pseudomonas* sp. HR 199. ▲ Eugenol addition; × eugenol concentration; ∗ ferulic acid concentration; ◆ coniferyl alcohol concentration; ■ vanillic acid concentration.

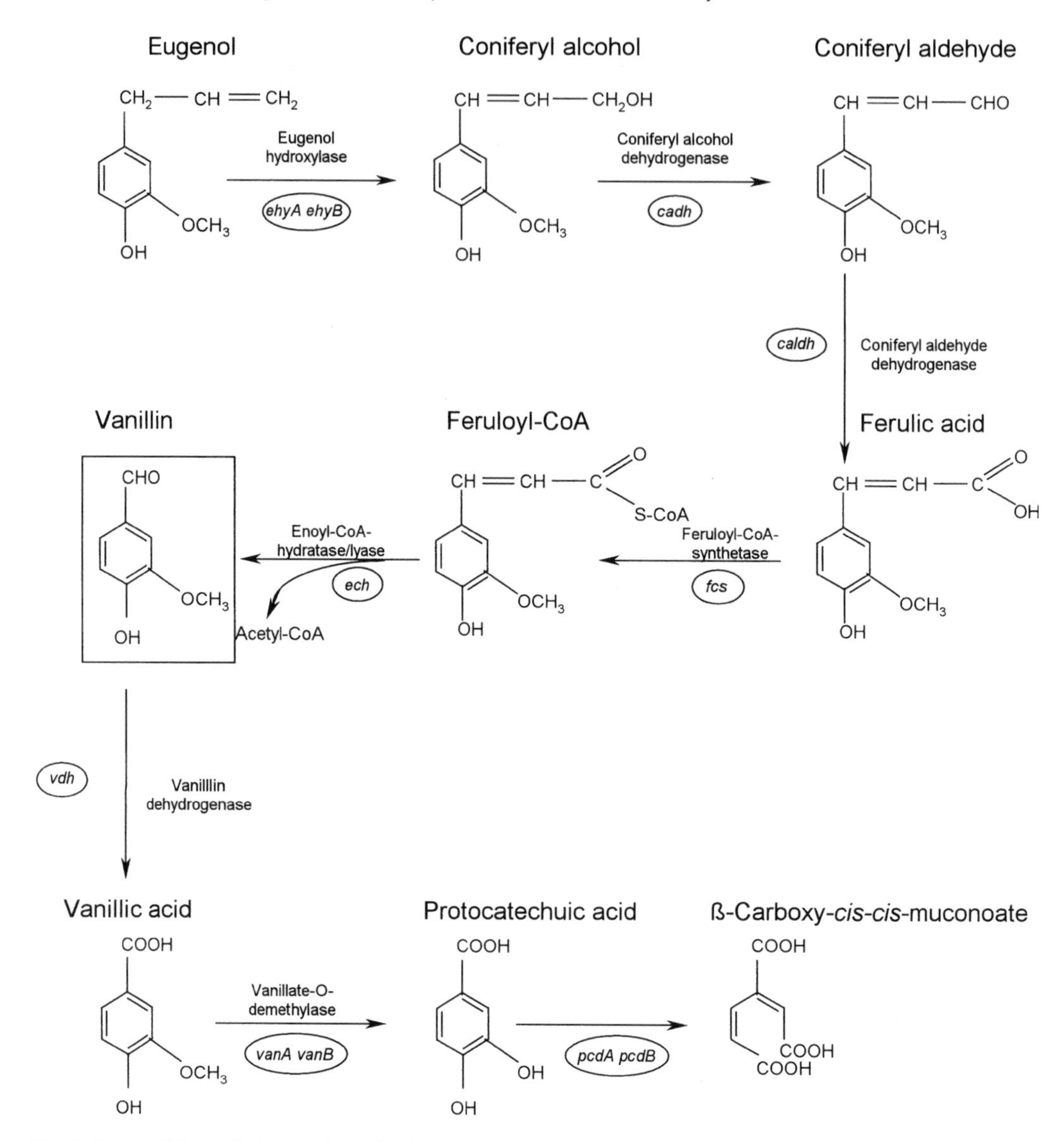

Fig. 5. Eugenol degradation pathway in *Pseudomonas* sp. HR 199 (Steinbüchel et al., 1998).

2.1.3 Production of Vanillin from Ferulic Acid

There were several attempts to use ferulic acid as a substrate for the production of natural vanillin. The chemical reaction looks very easy: a hydrolytic cleavage with the release of an acetate residue should give vanillin (Fig. 6).

In 1992, a process from Labuda et al. was disclosed for Kraft General Foods, who produced vanillin from ferulic acid with different microorganisms such as *Corynebacterium glutamicum, Aspergillus niger, Pseudomonas putida*, and *Rhodotorula glutinis* in the presence of sulfhydryl compounds such as dithiothreitol (Labuda et al., 1993). The vanillin concentra-

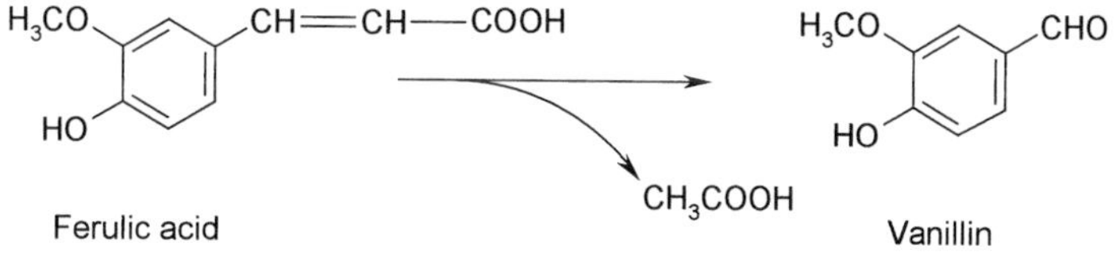

Fig. 6. Vanillin formation from ferulic acid.

tions were quite low and did not exceed 210 mg L^{-1} after 1,296 h. Due to the use of the expensive sulfhydryl compounds, the low yield, and long process times this process seems not to be commercially feasible. Two groups at INRA (Institut National de la Recherche Agronomique) together with Pernod-Ricard have been working for several years on a fungal bioconversion process for the production of vanillin from ferulic acid (GROSS et al., 1991; FALCONNIER et al., 1994; MICARD et al., 1995; LESAGE-MEESSEN et al., 1996a, b, 1997). They used a strain of *Pycnoporus cinnabarinus*, but had problems to channel the metabolic flux from ferulic acid to vanillin. They found at least three pathways for the metabolism of ferulic acid. A reductive pathway leads to coniferyl aldehyde, another one by the propenoic side chain degradation to vanillic acid, and a third one via laccase to unsoluble polymers. The formed vanillic acid is either decarboxylated to methoxyhydroquinone or reduced to vanillin and vanillyl alcohol. So the concentration of vanillin was only 64 mg L^{-1} after 6 d (FALCONNIER et al., 1994). To avoid the decarboxylation of vanillic acid they found that it was advantageous to use cellobiose as carbon source and fed additional cellobiose prior to the addition of ferulic acid. This resulted in an increase of vanillin up to 560 mg L^{-1} after 7 d. They postulated that cellobiose acts either as an easily metabolizable carbon source, required for the reductive pathway to occur, or as an inducer of the cellobiose quinone oxidoreductase, a known inhibitor of vanillic acid decarboxylation (LESAGE-MEESSEN et al., 1997). Another approach was a two-step process with the combination of two fungal strains, i.e., an *Aspergillus niger* fermentation together with the *Pycnoporus cinnabarinus* or *Phanerochaete chrysosporium*, where one strain converted ferulic acid to vanillic acid, while the second strain reduced vanillic acid to vanillin. But here vanillin concentrations were also <1 g L^{-1}.

The process with the highest yields of vanillin so far is that disclosed by Haarmann & Reimer (RABENHORST and HOPP, 1997).

In this process a filamentous growing actinomycete HR 167 was used. This strain was identified by the DSMZ, Braunschweig, Germany, as a new *Amycolatopsis* species. This result was based on the typical fatty acid pattern of *iso*/*anteiso*- plus 10-methyl-branched saturated and unsaturated fatty acids plus 2-hydroxy fatty acids, which are diagnostic for representatives of the genus *Amycolatopsis*.

Since HR 167 also displayed the typical appearance of representatives of the genus *Amycolatopsis* – beige to yellow substrate mycelium and, on some media, also a fine white aerial mycelium – the assignment of HR 167 to the genus *Amycolatopsis* was additionally supported by the morphology.

The comparison of the fatty acid patterns of the strain with the entries of the DSM fatty acid databases by principal component analysis resulted in an assignment to *Amycolatopsis mediterranei* for the new strain. However, the similarity index is too low (0.037 and 0.006) to permit definitive species assignment.

Additionally performed analyses of the 16S rDNA and comparison of the diagnostic partial sequences with the 16S rDNA database entries of *Amycolatopsis* type strains supported the results of the fatty acid analyses. In the case of the 16S rDNA sequences, high homology could also not be demonstrated with the known *Amycolatopsis* type strains. On the basis of the great differences in the 16S rDNA sequences (similarity 99.6%), it has to be stated that HR 167 is a new strain of a previously unknown *Amycolatopsis* species.

With this new strain a fed-batch fermentation was established. Within 32 h vanillin concentrations of 11.5 g L^{-1} vanillin with a molar yield of nearly 80% were obtained on the 10 L scale (Fig. 7).

This process has been successfully scaled up to the 40 m^3 production scale.

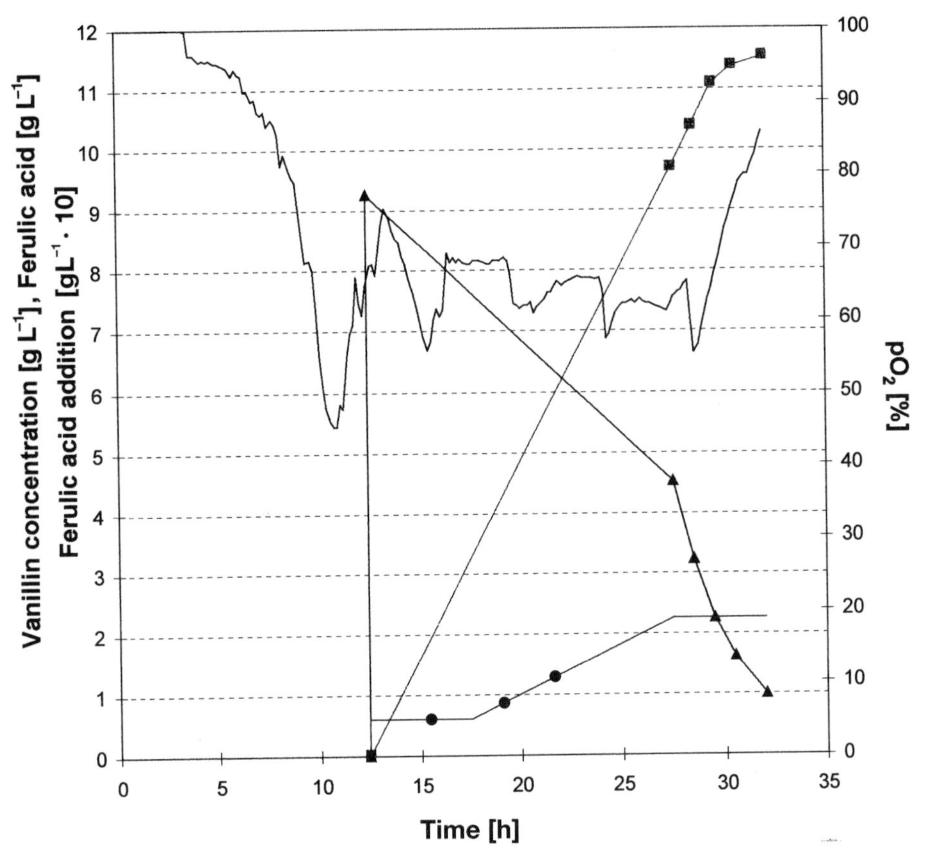

Fig. 7. Time course of vanillin production with *Amycolatopsis* sp. HR 167 in a 10 L fermenter. —■— Vanillin concentration; —▲— ferulic acid concentration; —●— ferulic acid addition; —— pO_2.

The vanillin is then isolated from the fermentation broth by well established physical methods like extraction, distillation, and crystallization.

The genes and the enzymes of this reaction were also characterized in *Pseudomonas* sp. (Fig. 5). Ferulic acid is first activated by a ferulic acid CoA synthase to ferulic acid CoA ester. This ester is then cleaved by an enoyl CoA hydratase to vanillin (NARBAD et al., 1997; GASSON et al., 1998; NARBAD and GASSON, 1998; STEINBÜCHEL et al., 1998). Vanillin is then further metabolized via vanillic acid, protocatechuic acid, and 3-carboxymuconolactone and 3-oxoadipate to succinyl-CoA, which is part of the tricarboxylic acid cycle (PRIEFERT et al., 1997; OVERHAGE et al., 1999).

2.2 Fermentative Production of Flavor-Active Lactones

A number of lactones are important flavor chemicals. Most important is γ-decalactone. It has a typical peachy flavor (Tab. 1). Many patents and articles on the formation of γ-decalactone have been published over the years.

Most of the processes use ricinoleic acid, the main fatty acid in castor oil, or esters thereof for the formation of γ-decalactone. The formation of γ-decalactone in the catabolism of ricinoleic acid by yeasts of the genus *Candida* was first observed by OKUI et al. (1963). Ricinoleic acid is degraded by four successive cycles of β-oxidation into 4-hydroxydecanoic acid,

Tab. 1. Sensoric Descriptions of Flavor Active Lactones

Name	Formula	Sensoric Description
γ-Decalactone	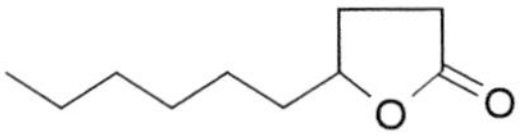	oily-peachy, extraordinarily tenacious odor; very powerful, creamy-fruity, peach-like taste in concentrations <5 ppm
γ-Dodecalactone	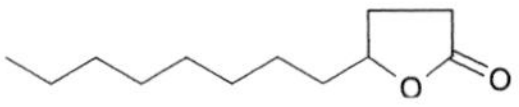	fatty-peachy, somewhat musky odor
g-Octalactone	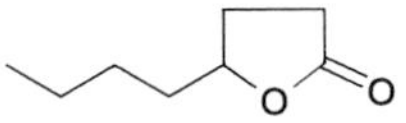	sweet-herbaceous coconut-like odor
δ-Decalactone	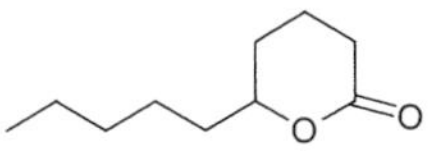	very powerful and tenacious sweet creamy, nut-like odor with a heavy fruity undertone; taste is creamy, sweet coconut-peach-milk-like <2 ppm
δ-Dodecalactone	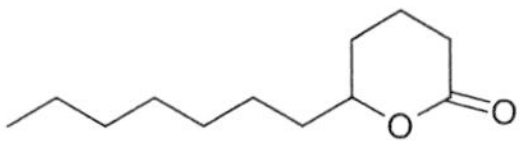	powerful fresh-fruity, oily odor, pear-peach-plum-like taste at 2–5 ppm
δ-Octalactone	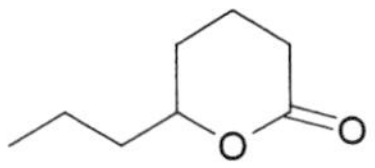	soft, lactonic, milk-like, cream-like, little fruity, peach-like, coconut milk note
6-Pentyl-2-pyrone	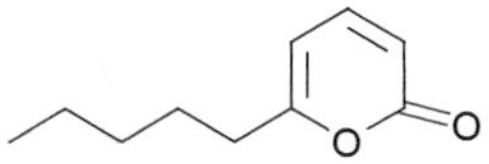	very tenacious, soft, lactonic, sweet, fatty, creamy, butter-like, fruity flavor of peach or apricot

which lactonizes at lower pH values to γ-decalactone (GATFIELD et al., 1993; Fig. 8). It is worth mentioning that the microbial lactone formation results in the same enantiomeric configuration of the lactone as it is found in peaches and other fruits.

A number of different yeasts have been used for the production of γ-decalactone.

In the first patent FARBOOD and WILLIS (1983) with IFF claimed the production of γ-decalactone with *Yarrowia lipolytica* and various other yeasts from ricinoleic acid, but the yields were less than 1 g L^{-1}. In a patent WINK et al. (1988) with Hoechst disclosed a process for the production of peach flavor with *Monilia fructicola*. After extraction of the fermentation broth and distillation they obtained a mixture of γ-octalactone (Tab. 1), γ-decalactone, and phenyl ethanol in a concentration of 0.2 g L^{-1}–1 g L^{-1}.

CHEETHAM et al. (1988) used *Sporobolomyces odorus* and *Rhodotorula glutinis*.

GATFIELD et al. (1993) reported the use of castor oil and *Candida lipolytica* for the production of γ-decalactone. They obtained γ-decalactone concentrations of 5 g L^{-1} within 2 d. In the fermentation broth they identified an additional lactone, 3-hydroxy-γ-decalactone (Fig. 9), as a by-product of γ-decalactone production. It is probably produced by the lactonization of 3,4-dihydroxy decanoic acid, which is formed as a result of the β-oxidation of 4-hydroxy decanoic acid.

The group of SPINNLER studied the γ-decalactone production with *Sporidiobolus* spp. like *Sporidiobolus salmonicolor, Sporidiobolus ruinenii, Sporidiobolus johnsonii*, and *Sporidiobolus pararoseus*. These strains are very sensitive to γ-decalactone, while the open form, i.e., 4-hydroxydecanoic acid, is less toxic (DU-

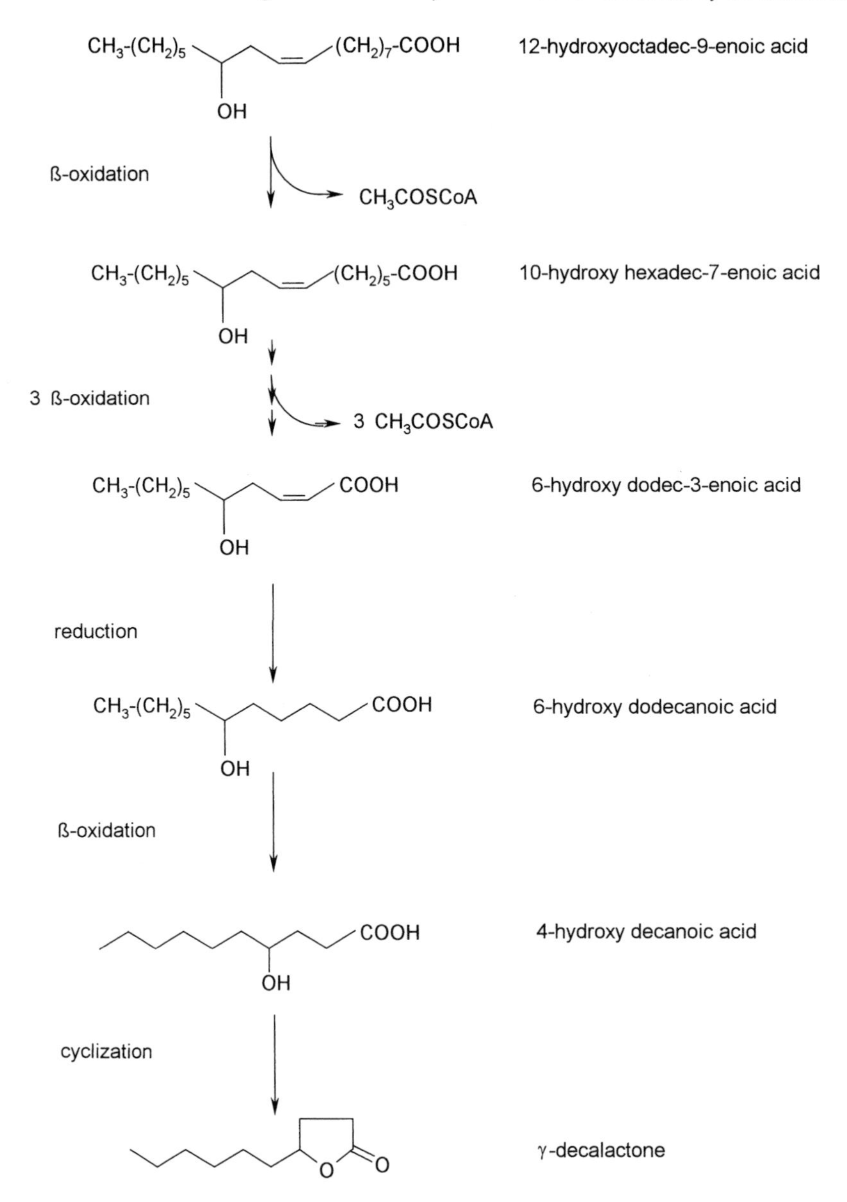

Fig. 8. β-Oxidation pathway of ricinoleic acid for the formation of γ-decalactone in *Yarrowia lipolytica* (after GATFIELD et al., 1993).

FOSSÉ et al., 1998). So the yields of γ-decalactone in fermentations with these yeasts were always very low (<1 g L^{-1}).

In another patent application CARDILLO et al. (1989) used *Aspergillus niger, Pichia etchellsii*, and *Cladosporium suaveolens* to produce γ-decalactone from castor oil. Additionally they produced 1 g L^{-1} of γ-octalactone from 10 g L^{-1} coconut oil.

NICAUD et al. (1996) used a multiple auxotrophic mutant, *Yarrowia lipolytica* PO1D, to obtain γ-decalactone in high yields. At the end

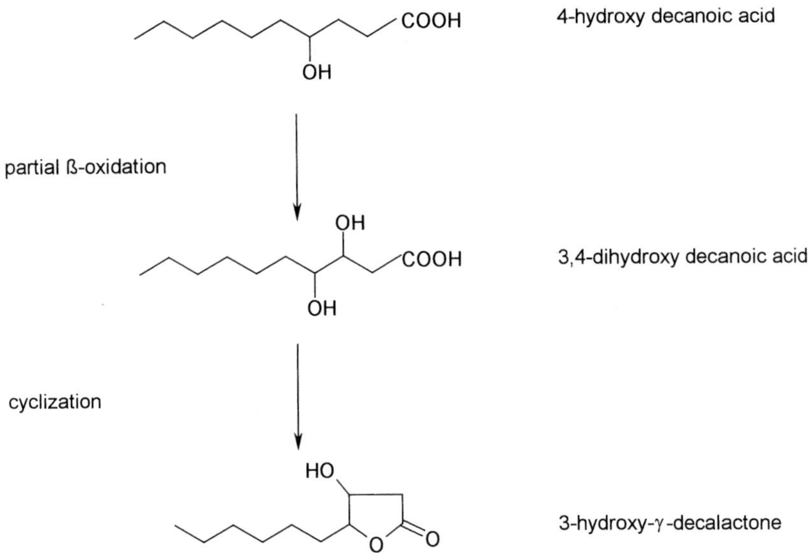

Fig. 9. Formation of 3-hydroxy-γ-decalactone.

of the growth phase they transferred the concentrated biomass into an uracil limited medium where the biotransformation took place. After 75 h the culture broth yields 9.5 g L^{-1} of the product (PAGOT et al., 1997). They used methyl ricinoleate as substrate.

A different approach has been followed by KÜMIN and MÜNCH (1997). They used *Mucor circillenoides* for the biotransformation of ethyl decanoate. After 60 h they reached 10.5 g L^{-1} γ-decalactone.

A synopsis of the different ways to γ-decalactone has been presented recently by KRINGS and BERGER (1998).

A Japanese group at T. Hasegawa Co. (GOCHO et al., 1995) studied the biotransformation of oleic acid to optically active γ-dodecalactone. They established a two-step process. First, oleic acid is oxidized to 10-hydroxystearic acid with a newly isolated gram-positive bacterial strain, bakers' yeast was then used for the biotransformation of the hydroxy acid to (*R*)-γ-dodecalactone (Fig. 10). The biotransformation yield from oleic acid to γ-dodecalactone was 22.5% on the flask scale.

A process for the production of δ-decalactone and δ-dodecalactone was established by PFW (VAN DER SCHAFT and DE LAAT, 1991; VAN DER SCHAFT et al., 1992). They used the massoi lactones 2-decen-1,5-olide and 2-dodecen-1,5-olide, which are the main components of massoi bark oil with 80% and 7%, respectively, as substrate. These unsaturated lactones were hydrogenated with baker's yeast (Fig. 11). By stepwise addition of 2-decen-5-olide they obtained about 1.2 g L^{-1} δ-decalactone

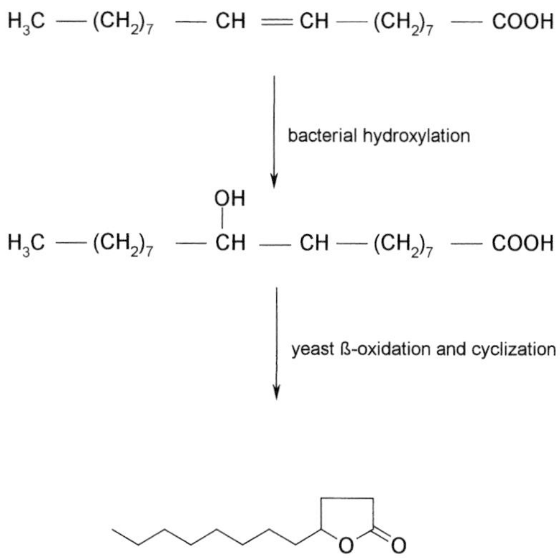

Fig. 10. Formation of γ-dodecalactone from oleic acid.

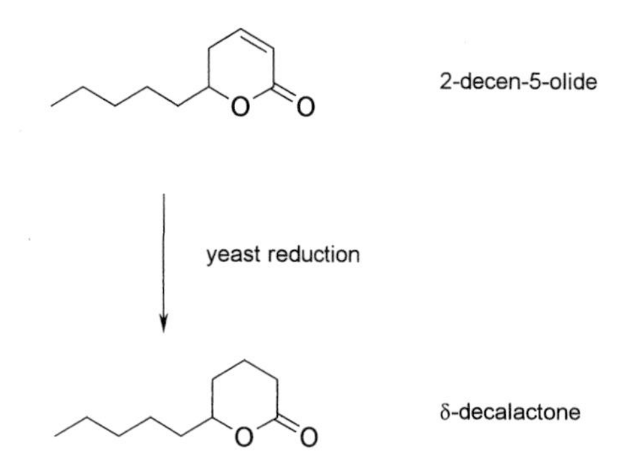

Fig. 11. Formation of δ-decalactone from massoi lactone.

within 16 h. By addition of β-cyclodextrin this process could be accelerated significantly. They also found that a number of molds performed the same reaction.

The same reaction with the use of different bacteria has also been patented by Hasegawa (GOCHO et al., 1998). The best results were obtained with *Pseudomonas putida* ATCC 33015. With this strain they were able to isolate nearly 6 g L^{-1} δ-decalactone from the fermentation broth after 48 h of incubation.

2.3 Fermentative Production of Carboxylic Acids

A number of short-chain carboxylic acids are also important as natural flavor ingredients, either for use *per se* or as substrates for the production of a broad number of flavor esters, which can be obtained by enzymatic synthesis with lipases and esterases (for a review, see GATFIELD, 1992). Acetic acid, which is probably the largest by volume and commonly used as vinegar is not mentioned here.

In Tab. 2 examples of several important carboxylic acids and their sensory descriptions are presented.

Tab. 2. Carboxylic Acids Used in Flavor Compositions

Name	Structure	Sensoric Description[a]
n-Butyric acid	$CH_3-CH_2-CH_2-COOH$	powerful, penetrating, diffusive sour odor, reminiscent of rancid butter; used in butter, cheese, nut, fruit, and other flavors
Isobutyric acid	$(H_3C)_2CH-COOH$	powerful, diffusive sour odor, slightly less repulsive, and also less buttery than *n*-butyric acid; in extreme dilution almost pleasant, fruity; the taste is, in proper dilution and with adequate sweetening, pleasant creamy-fruity, while buttery-cheesy notes dominate in the absence of sweeteners
Isovaleric acid	$(H_3C)_2CH-CH_2-COOH$	very diffusive, acid-acrid, in moderate dilution cheesy, unpleasant odor of poor tenacity; in extreme dilution the odor becomes more agreeable, herbaceous and dry; used in nut and coffee flavors due to its peculiar warm-herbaceous taste in concentrations <20 ppm
2-Methyl butyric acid	$H_3C-CH_2-CH(CH_3)-COOH$	pungent, acrid odor reminiscent of Roquefort cheese; acrid taste, which becomes quite pleasant and fruity-sour at dilutions <10 ppm; used in cheese, butter, cream, and chocolate flavors
Propionic acid	H_3C-CH_2-COOH	pungent sour odor reminiscent of sour milk, cheese, or sour butter; used in raspberry, strawberry, cognac, butter flavors

[a] Sensoric descriptions according to ARCTANDER, 1969

For the production of butyric acid two different ways are feasible. One route is the classical anaerobic fermentation with *Clostridium butyricum* (VANDAK et al., 1996). The other one is the microbial oxidation of butanol with *Gluconobacter roseus* (GATFIELD and SAND, 1988). With this process about 13 g L^{-1} butyric acid was obtained in high molar yield. The oxidation of butanol was also studied by DRUAUX et al. (1997) with *Acetobacter aceti*. After a bioconversion phase of 60 h with continuous feeding of the substrate they reached butyric acid concentrations about 39 g L^{-1}, with 86% molar yield. They also tested several other alcohols like isoamyl alcohol, 3-methylbutanol, and 2-phenylethanol, but there the molar yields were significantly lower and no product concentrations were presented.

Propionic acid can be obtained by *de novo* synthesis with *Propionibacterium shermanii* or *Propionibacterium acidipropionici* on cheap substrates like whey (CARRONDO et al., 1988; CHAMPAGNE et al., 1989). Propionic acid can also be produced by oxidation of propanol with *Gluconobacter oxydans* (SVITEL and STURDIK, 1995). This process gives nearly 44 g L^{-1} after 70 h.

The alcohol oxidation with *Gluconobacter roseus* was also applied by GATFIELD and SAND for the production of several other acids. So after addition of 20 g L^{-1} isobutanol to a 30 L fermentor 21 g L^{-1} isobutyric acid was obtained after 15 h.

The addition of 10 g L^{-1} 2-methyl butanol to the *Gluconobacter roseus* culture resulted in 7 g L^{-1} 2-methyl butyric acid after 24 h; 10 g L^{-1} isoamyl alcohol were oxidized to 11 g L^{-1} isovaleric acid.

The alcohols used as substrates for the microbial oxidations are cheaply available from fusel oil residues from ethanol fermentations of yeast.

2.4 Fermentation of Other Natural Aroma Compounds

2.4.1 Production of Flavor-Active Aldehydes and Alcohols

Aldehydes are another important group of flavor chemicals. Most important are after vanillin, which has been mentioned above, the so-called "green notes". These are *trans*-2-hexenal, hexanal, and the alcohols *cis*-3-hexen-1-ol, *trans*-2-hexen-1-ol, and hexanol.

These compounds are formed from linolenic and linoleic acid through the action of lipoxygenase and hydroperoxide lyase. This reaction typically occurs in plant tissues such as leaves after damage (HATANAKA, 1993).

As early as 1965 COLLINS and KALNINS found 2-hexenal together with acetaldehyde and 2-methylbutanal in the culture broth of the oak wilt fungus *Ceratocystis fagacearum*. A commercial process for the production of these green notes does not use microorganisms but enzymes. So soybean lipoxygenase is used to produce hydroperoxy acids which are then cleaved by lyases from plant homogenates of guava (MULLER et al., 1993), or the foliage of other plants (HOLTZ et al., 1995). The obtained aldehydes can be reduced by bakers' yeast to the corresponding alcohols. The hydroperoxide lyase of banana leaves has been

Tab. 3. Alcohol Conversion to Aldehydes by *Pichia pastoris* (after MURRAY et al., 1988; MURRAY and DUFF, 1990)

Product	Concentration [g L^{-1}]	Conversion Period [h]
Acetaldehyde	30.8	12
Propionaldehyde	28	24
Butyraldehyde	14	4
Isobutyraldehyde	10.8	8
Isovaleraldehyde	11.3	72
Hexanal	13.8	24

cloned recently (Häusler et al., 1997; Häusler and Münch, 1997) and hence a microbial system for the production of the aldehydes and alcohols can be established.

Another way to flavor-active aldehydes is the oxidation of the corresponding alcohol with an alcohol oxidase. For this process either methylotrophic yeasts or several bacteria were used.

Duff and Murray (1988) used cells of *Pichia pastoris* for the production of acetaldehyde from ethanol (Murray et al., 1989), and expanded this technology to benzaldehyde (Duff and Murray, 1989), and higher aliphatic aldehydes up to C-11 (Murray et al., 1988; Duff and Murray, 1991). The aldehyde concentrations they obtained were very high (Tab. 3).

A group at Takasago used *Candida boidinii* for the production of flavor aldehydes. They obtained isovaleraldehyde concentrations of about 40 g L^{-1} after 7 h and were able to produce also *trans*-2-hexenal and hexanal in higher amounts (Nozaki et al., 1995).

Molinari et al. (1995) used a strain of *Gluconobacter oxydans* to produce isovaleraldehyde from isoamyl alcohol. They reached 12 g L^{-1} within 8 h. By the use of an extractive bioconversion with a hollow-fiber membrane reactor they were able to increase the yield to about 35 g L^{-1} (Molinari et al., 1997).

2.4.2 Phenolic Compounds

In a review (Rosazza et al., 1995) an overview was presented of reactions of ferulic acid. A number of reactions results in flavor chemicals. Vinyl guaiacol, which has a phenolic, smoky, vanilla extract-like, roasty flavor is used in roast, vanilla, and cocoa flavorings.

Vinyl guaiacol is formed by decarboxylation of ferulic acid (Fig. 12a). A number of different organisms are known, which perform this reaction, for example, *Fusarium solani* (Nazareth and Mavinkurve, 1986), *Nocardia* sp. (Malarczyk et al., 1990), *Colletotrichum gloeosporoides* (Fuganti et al., 1995), several yeasts (Huang et al., 1994; Sutherland et al., 1995), and *Bacillus* species (Lee et al. 1998). The decarboxylase enzyme from *Bacillus pumilus* (Degrassi et al., 1995) and *Pseudomonas fluorescens* (Huang et al., 1994) have been isolated and characterized and the gene has also been cloned (Zago et al., 1995; Ago and Kikuchi, 1998).

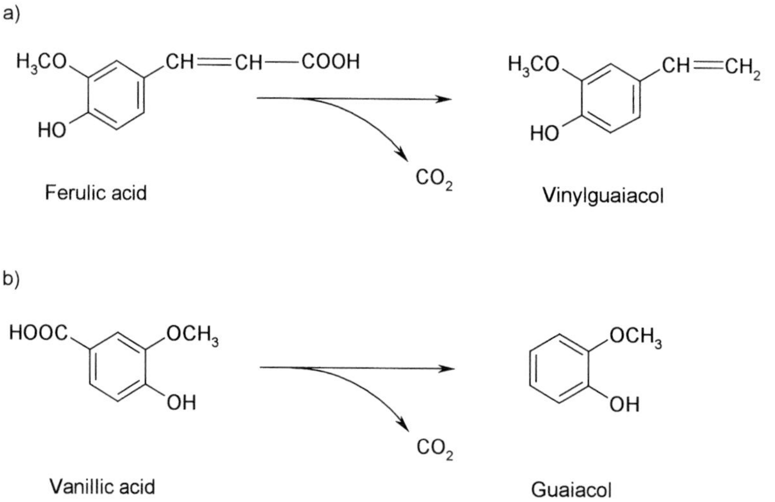

a)

Ferulic acid → Vinylguaiacol + CO_2

b)

Vanillic acid → Guaiacol + CO_2

Fig. 12. Formation of **a** vinylguaiacol and **b** guaiacol by decarboxylation of vanillic acid and ferulic acid by *Bacillus pumilus* decarboxylase.

The only process with which higher yields of vinyl guaiacol have been obtained was done with *Bacillus pumilus*. In an aqueous-organic two-phase system LEE et al. (1998) obtained 9.6 g L^{-1} vinyl guaiacol from ferulic acid within 10 h.

In an analogous reaction with vanillic acid guaiacol is obtained as product (Fig. 12b). Guaiacol has a powerful, smoke-like, somewhat medicinal odor and a warm, medicinal and somewhat sweet taste, but accompanied by a burning sensation. Guaiacol is used in flavor compositions for coffee, vanilla, whisky, tobacco, and in several fruit and spice complexes (ARCTANDER, 1969).

2.4.3 Methyl Ketones

Methyl ketones, especially 2-pentanone, 2-heptanone, 2-nonanone, and 2-undecanone, are the flavor impact compounds of blue veined cheese, which is classically produced with the mold *Penicillium roqueforti*. These typical flavors are used for flavoring of products such as salad dressings, soups, crackers, and cakes.

The methyl ketones are formed due to the action of several enzymes of the mold during cheese ripening, like lipase for the release of free fatty acids from milk triglycerides, which act as precursors for the formation of methyl ketones. The C-6, C-8, C-10, and C-12 fatty acids act as substrates for the β-ketoacyl decarboxylase, which generates the corresponding C_{n-1} methyl ketones. The biosynthesis of the methyl ketones has been reviewed by KINSELLA and HWANG (1976). Based on this basic knowledge several processes for the biotechnological production of methyl ketones have been developed (LUKSAS, 1973; WINK et al., 1988; LARROCHE et al., 1989).

3 Conclusions

For the production of natural flavor chemicals microbial fermentation is a competitive technology to obtain chemicals which cannot be isolated from plant extracts, essential oils, or other sources. It is important to realize that chemical synthesis, which can produce these simple chemicals much more cheaply, cannot compete due to the legal status of natural flavors. The increasing knowledge of the enzymatic pathways for the biosynthesis and metabolism of aroma chemicals gives new opportunities to create tailor-made genetically engineered production strains. This is actually done for the production of vanillin.

A problem remains in the low demand for extremely active flavor specialities like some pyrazines or sulfur containing chemicals, where the annual consumption is sometimes only in the area of hundreds of grams or a few kilograms. This prevents the establishment of extensive research programs for these chemicals.

The formation of flavors by the use of enzymes has not been mentioned within this chapter, but this is an equally important biotechnological method for the production of natural flavors (for reviews, see GATFIELD, 1992; BERGER, 1995; SCHREIER, 1997).

4 References

AGO, S., KIKUCHI, Y. (1998), *European Patent* 857 789.

ANTRIM, R. L., HARRIS, D. W. (1977), *US Patent* 4,038,481.

ARCTANDER, S. (1969), *Perfume and Flavor Chemicals*. Montclair, N J: Published by the author.

BAUER, K., GARBE, D., SURBURG, H. (1990), *Common Fragrance and Flavor Materials*. 2nd Edn., Weinheim: VCH.

BERGER, R. G. (1995), *Aroma Biotechnology*. Berlin: Springer-Verlag.

CARDILLO, R., FUGANTI, C., SACERDOTE, C., BARBENI, M., CABELLA, P., SQUARCIA, F. (1989), *European Patent* 356 291.

CARRONDO, M. J. T., CRESPÒ, J. P. S. G., MOURA, M. J. (1988), Production of propionic acid using xylose utilizing *Propionibacterium, Appl. Biochem. Biotechnol.* **17**, 295–312.

CHAMPAGNE, C. P., BAILLAGEON-COTE, C., GOULET, J. (1989), Whey fermentation by immobilized cells of *Propionibacterium shermanii, J. Appl. Bacteriol.* **66**, 175–184.

CHEETHAM, P. S. J., MAUME, K. A., DE ROOJI, J. F. (1988), *European Patent* 0258 993.

CLARK, G. S. (1990), Vanillin, perfum, *Flavor* **15**, 45–54.

Code of Federal Regulations (1993), *21 Food and Drugs*, Parts 100-169, revised April 1, 1993. Washington, DC: National Archives and Records Administration.

COLLINS, R. P., KALNINS, K. (1965), Carbonyl compounds produced by *Ceratocystis fagacearum, Am. J. Bot.* **52**, 751–754.

DEGRASSI, G., DE LAURETO, P. P., BRUSCHI, C. V. (1995), Purification and characterization of ferulate and *p*-coumarate decarboxylase from *Bacillus pumilus, Appl. Environ. Microbiol.* **61**, 326–332.

DOLFINI, J. E., GLINKA, J., BOSCH, A. C. (1990), *US Patent* 4,927,805.

DRUAUX, D., MANGEOT, G., ENDRIZZI, A., BELIN, J.-M. (1997), Bacterial bioconversion of primary aliphatic and aromatic alcohols into acids: effects of molecular structure and physico-chemical conditions, *J. Chem. Tech. Biotechnol.* **68**, 214–218.

DUFF, S. J. B., MURRAY, W. D. (1988), Comparison of free and immobilized *Pichia pastoris* cells for conversion of ethanol to acetaldehyde, *Biotechnol. Bioeng.* **31**, 790–795.

DUFF, S. J. B., MURRAY, W. D. (1989), Oxidation of benzyl alcohol by whole cells of *Pichia pastoris* and by alcohol oxidase in aqueous and non-aqueous reaction media, *Biotechnol. Bioeng.* **34**, 153–159.

DUFF, S. J. B., MURRAY, W. D. (1991), *US Patent* 5,010,005.

DUFOSSÉ, L., FERON, G., MAUVAIS, G., BONNARME, P., DURAND, A., SPINNLER, H.-E. (1998), Production of γ-decalactone and 4-hydroxy-decanoic acid in the genus *Sporidiobolus, J. Ferment. Bioeng.* **86**, 169–173.

FALCONNIER, B., LAPIERRE, C., LESAGE-MEESSEN, L., YONNET, G., BRUNERIE, P. et al. (1994), Vanillin as a product of ferulic acid biotransformation by the white-rot fungus *Pycnoporus cinnabarinus* I-937: Identification of metabolic pathways, *J. Biotechnol.* **37**, 123–132.

FARBOOD, M. I., WILLIS, B. J. (1983), *WO Patent* 83/01072.

FUGANTI, C., MAZZOLARI, Y., ZUCCHI, G., BARBENI, M., CISERO, M., VILLA, M. (1995), On the mode of biodegradation of cinnamic acids by *Colletotrichum gloeosporoides:* a direct acces to 4-vinyl phenol and 2-methoxy-4-vinylphenol, *Biotechnol. Lett.* **17**, 707–710.

GASSON, M. J., KITAMURA, Y., MCLAUCHLAN, W. R., NARBAD, A., PARR, A. J. et al. (1998), Metabolism of ferulic acid to vanillin. A bacterial gene of the enoyl-SCoA hydratase/isomerase superfamily encodes an enzyme for the hydration and cleavage of a hydroxy cinnamic acid SCoA thioester, *J. Biol. Chem.* **273**, 4163–4170.

GATFIELD, I. L. (1992), Bioreactors for industrial production of flavors: use of enzymes, in: *Bioformation of Flavors* (PATTERSON, R. L. S., CHARLWOOD, B. V., MACLEOD, G., WILLIAMS, A. A., Eds.), pp. 171–185. Cambridge: The Royal Society of Chemistry.

GATFIELD, I. L., SAND, T. (1988), *European Patent* 289 822.

GATFIELD, I. L., GÜNTERT, M., SOMMER, H., WERKHOFF, P. (1993), Some aspects of microbiological manufacture of flavor-active lactones with particular reference to γ-decalactone, *Chem. Mikrobiol. Technol. Lebensm.* **15**, 165–170.

GOCHO, S., RUMI, K., TSUYOSHI, K. (1998), *US Patent* 5,763,233.

GOCHO, S., TABOGAMI, N., INAGAKI, M., KAWABATA, C., KOMAI, T. (1995), Biotransformation of oleic acid to optically active γ-dodecalactone, *Biosci. Biotech. Biochem.* **59**, 1571–1572.

GROSS, B., ASTHER, M., BRUNERIE, P. (1991), *European Patent* 453 368.

HATANAKA, A. (1993), The biogeneration of green odor by green leaves, *Phytochemistry* **34**, 1201–1218.

HÄUSLER, A., MÜNCH, T. (1997), Microbial production of natural flavors, *ASM News* **63**, 551–559.

HÄUSLER, A., LERCH, K., MUHEIM, A., SILKE, N. (1997), *European Patent* 801 133.

HOLTZ, R. B., MCCULLOCH, M. J., GARGER, S. J. (1995), *WO Patent* 95/23413.

HOPP, R., RABENHORST, J. (1993), *European Patent* 583 687.

HUANG, Z., DOSTAL, L., ROSAZZA, J. P. N. (1994), Purification and characterization of a ferulic acid decarboxylase from *Pseudomonas fluorescens, J. Bacteriol.* **176**, 5912–5918.

KINSELLA, J. E., HWANG, D. H. (1976), Enzymes of *Penicillium roqueforti* involved in the biosynthesis of cheese flavor, *Crit. Rev. Food Sci. Nutr.* **8**, 191–228.

KNUTH, M., SAHAI, O. (1991), *US Patent* 5,057,424.

KRINGS, U., BERGER, R. G. (1998), Biotechnological production of flavors and fragrances, *Appl. Microbiol. Biotechnol.* **49**, 1–8.

KÜMIN, B., MÜNCH., T. (1997), *European Patent* 795-607.

LABUDA, I. M., GOERS, S. K., KEON, K. A. (1992), *US Patent* 5,128,253.

LABUDA, I. M., GOERS, S. K., KEON, K. A. (1993), Microbial bioconversion process for the production of vanillin, in: *Progress in Flavor Precursor Studies*, Proc. Int. Conf. Würzburg, Germany, September 30–October 2, 1992 (SCHREIER, P., WINTERHALTER, P., Eds.), pp. 477–482. Carol Stream, IL: Allured Publishing.

LARROCHE, C., ARPAH, M., GROS, J.-B. (1989), Methyl-ketone production by Ca-alginate/Eudragit RL entrapped spores of *Penicillium roqueforti, Enzyme Microb. Technol.* **11**, 106–112.

LEE, I.-Y., VOLM, T. G., ROSAZZA, J. P. N. (1998), Decarboxylation of ferulic acid to 4-vinyl guaiacol by *Bacillus pumilus* in aqueous-organic solvent two-phase system, *Enzyme Microb. Technol.* **23**, 261–266.

LESAGE-MEESSEN, L., DELATTRE, M., HAON, M., ASTHER, M. (1996a), *WO Patent* 96/08576.

LESAGE-MEESSEN, L., DELATTRE, M., HAON, M., THIBAULT, J. F., CECCALDI, B. C. et al. (1996b), A two-step bioconversion process for vanillin production from ferulic acid combining *Aspergillus niger* and *Pycnoporus cinnabarinus, J. Biotechnol.* **50**, 107–113.

LESAGE-MEESSEN, L., DELATTRE, M., HAON, M., THIBAULT, J. F., CECCALDI, B. C. et al. (1997), An attempt to channel the transformation of vanillic acid into vanillin by controlling methoxy hydroquinone formation in *Pycnoporus cinnabarinus* with cellobiose, *Appl. Microbiol. Biotechnol.* **47**, 393–397.

LI, K., FROST, J. W. (1998), Synthesis of vanillin from glucose, *J. Am. Chem. Soc.* **120**, 10545–10546.

LUKSAS, A. J. (1973), *US Patent* 3,720,520.

MALARCZYK, E., KOCHMANSKA-RDEST, J., APALOVIC, R., LEONOWICZ, A. (1990), Production of vinyl guaiacol, isoeugenol and other aromas from lignin-related materials by *Nocardia, Prog. Biotechnol.* **6**, 403–409.

MARKUS, P. H., PETERS, A. L. J., ROOS, R. (1993), *European Patent* 0542 348.

MICARD, V., RENARD, C., THIBAULT, J. F., LESAGE-MEESSEN, L., HAON, M. et al. (1995), Est-il possible d'obtenir de la vanilline naturelle à partir de pulpes de betterave?, *Colloq.-Inst. Natl. Rech. Agron.* **71**, 149–153.

MOLINARI, F., VILLA, R., MANZONI, M., ARAGOZZINI, F. (1995), Aldehyde production by alcohol oxidation with *Gluconobacter oxydans, Appl. Microbiol. Biotechnol.* **43**, 989–994.

MOLINARI, F., ARAGOZZINI, F., CABRAL, J. M. S., PRAZERES, D. M. F. (1997), Continuous production of isovaleraldehyde through extractive bioconversion in a hollow-fiber membrane bioreactor, *Enzyme Microb. Technol.* **20**, 604–611.

MULLER, B., GAUTIER, A., DEAN, C., KUHN, J.-C. (1993), *WO Patent* 93/24644.

MURRAY, W. D., DUFF, S. J. B. (1990), Bio-oxidation of aliphatic and aromatic high-molecular weight alcohols by *Pichia pastoris* alcohol oxidase, *Appl. Microbiol. Biotechnol.* **33**, 202–205.

MURRAY, W. D., DUFF, S. J. B., LANTHIER, P. H., ARMSTRONG, D. W., WELSH, D. W., WILLIAMS, R. E. (1988), Development of biotechnological processes for the production of natural flavors and fragrances, in: *Frontiers of Flavors*, Proc. 5th Int. Flavor Conf., Porto Karras, Chalkidiki, Greece, 1–3 July, 1987 (CHARALAMBOUS, G., Ed.), pp. 1–18. Amsterdam: Elsevier Science Publisher.

MURRAY, W. D., DUFF, S. J. B., LANTHIER, P. H. (1989), *US Patent* 4,871,669.

NARBAD, A., GASSON, M. J. (1998), Metabolism of ferulic acid via vanillin using a novel CoA-dependent pathway in a newly isolated strain of *Pseudomonas fluorescens, Microbiology* **144**, 1397–1405.

NARBAD, A., RHODES, M. J. C., GASSON, M. J., WALTON, N. J. (1997), *WO Patent* 97/35999.

NAZARETH, S., MAVINKURVE, S. (1986), Degradation of ferulic acid via 4-vinyl guaiacol by *Fusarium solani* (Mart.) Sacc., *Can. J. Microbiol.* **32**, 494–497.

NICAUD, J. M., BELIN, J. M., PAGOT, Y., ENDRIZZI, A. (1996), *French Patent* 2734-843.

NOZAKI, M., WASHIZU, Y., SUZUKI, N., KANISAWA, T. (1995), Microbial oxidation of alcohols by *Candida boidinii*, in: *Bioflavor* 95 (ETIÉVANT, P., SCHREIER, P., Eds.), pp. 255–260. Paris: INRA.

OKUI, S., UCHIYAMA, M., MIZUGAKI, M. (1963), Metabolism of hydroxy fatty acids. II. Intermediates of the oxidative breakdown of ricinoleic acid by genus *Candida, J. Biochem.* **54**, 536–540.

ÖTÜK, G. (1985), Degradation of ferulic acid by *Escherichia coli, J. Ferment. Technol.* **63**, 501–506.

OVERHAGE, J., KRESSE, A. U., PRIEFERT, H., SOMMER, H., KRAMMER, G. et al. (1999), Molecular characterization of the genes *pcaG* and *pcaH*, encoding protocatechuate 3,4-dioxygenase, which are essential for vanillin catabolism in *Pseudomonas* sp. strain HR 199, *Appl. Environ. Microbiol.* **65**, 951–960.

PAGOT, Y., ENDRIZZI, A., NICAUD, J. M., BELIN, J. M. (1997), Utilization of an auxotrophic strain of the yeast *Yarrowia lipolytica* to improve gamma-decalactone production yields, *Lett. Appl. Microbiol.* **25**, 113–116.

PRIEFERT, H., RABENHORST, J., STEINBÜCHEL, A. (1997), Molecular characterization of genes of *Pseudomonas* sp. Strain HP 199 involved in bioconversion of vanillin to protocatechuate, *J. Bacteriol.* **179**, 2595–2607.

RABENHORST, J. (1996), Production of methoxyphenol-type natural aroma chemicals by biotransformation of eugenol with a new *Pseudomonas* sp., *Appl. Microbiol. Biotechnol.* **46**, 470–474.

RABENHORST, J., HOPP, R. (1991), *European Patent* 405 197.

RABENHORST, J., HOPP, R. (1997), *German Patent* 19532 317.

RANADIVE, A. S. (1994), Vanilla – cultivation, curing, chemistry, technology and commercial products, *Dev. Food Sci.* **34**, 517–577.

ROSAZZA, J. P. N., HUANG, Z., DOSTAL, L., ROSSEAU, B. (1995), Review: Biocatalytic transformation of ferulic acid: an abundant natural product, *J. Ind. Microbiol.* **15**, 457–471.

SAHAI, O. (1994), Plant tissue culture, in: *Bioprocess Production of Flavor, Fragrance and Color Ingredients* (GABELMAN, A., Ed.), pp. 239–275. New York: John Willey & Sons.

SCHARPF, JR., L. G., SEITZ, E. W., MORRIS, J. A., FARBOOD, M. I. (1986), Generation of flavor and odor compounds through fermentation, in: *Biogeneration of Aromas*, ACS Symp. Ser. 317 (PARLIMENT, T. H., CROTEAU, R., Eds.), pp. 323–346. Washington, DC: American Chemical Society.

SCHINDLER, J., SCHMID, R. D. (1982), Fragrance or aroma chemicals – microbial synthesis and enzymatic transformation, *Proc. Biochem.* **5**, 2–8.

SCHREIER, P. (1997), Enzymes and flavor biotechnology, in: *Advances in Biochemical Engineering/Biotechnology* Vol. 55 (SCHEPER, T., Ed.), pp. 51–72. Berlin: Springer-Verlag.

STEINBÜCHEL, A., PRIEFERT, H., RABENHORST, J. (1998), *European Patent* 0 845 532.

STRACK, D. (1990), Metabolism of hydroxycinnamic acid conjugates, *Bull. Liaison – Groupe Polyphenols* **15**, 55–64.

SUTHERLAND, J. B., CRAWFORD, D. L., POEMTTO III, A. L. (1983), Metabolism of cinnamic, *p*-coumaric, and ferulic acids by *Streptomyces setonii, Can. J. Microbiol.* **29**,1253–1257.

SUTHERLAND, J. B., TANNER, A. A., MOORE, J. D., FREEMAN, J. P., DECK, J., WILLIAMS, A. J. (1995), Conversion of ferulic acid to 4-vinyl guaiacol by yeasts isolated from frozen concentrated orange juice, *J. Food Prot.* **58**, 1260–1262.

SVITEL, J., STURDIK, E. (1995), *n*-Propanol conversion to propionic acid by *Gluconobacter oxydans, Enzyme Microb. Technol.* **17**, 549–550.

TADASA, K. (1977), Degradation of eugenol by a microorganism, *Agric. Biol. Chem.* **41**, 925–929.

TADASA, K., KAYAHARA, H. (1983), Initial steps of eugenol degradation pathway of a microorganism, *Agric. Biol. Chem.* **47**, 2639–2640.

TOMS, A., WOOD, J. M. (1970), The degradation of *trans*-ferulic acid by *Pseudomonas acidovorans, Biochemistry* **9**, 337–343.

VANDAK, D., TELGARSKY, M., STURDIK, E. (1996), Influence of growth factor supplementation on butyric acid production from sucrose by *Clostridium butyricum, Folia Microbiol.* **40**, 669–672.

VAN DER SCHAFT, P. H., DE LAAT, W. T. A. M. (1991), *European Patent* 425 001.

VAN DER SCHAFT, P. H., TER BURG, N., VAN DEN BOSCH, S., COHEN., A. M. (1992), Microbial production of natural δ-decalactone and δ-dodecalactone from the corresponding α,β-unsaturated lactones in Massoi bark oil, *Appl. Microbiol. Biotechnol.* **36**, 712–716.

WINK, J., FRICKE, U., DEGER, H.-M., MIXICH, J., BAUER, D., JUSTINSKI, U. (1988), *German Patent* 3701 836.

WINK, J., VOELSKOW, H., GRABLEY, S., DEGER, H. M. (1988), *European Patent* 286 950.

YOSHIMOTO, T., SAMEJIMA, M., HANYU, N., KOMAI, T. (1990a), *Japanese Patent* 02195871.

YOSHIMOTO, T., SAMEJIMA, M., HANYU, N., KOMAI, T. (1990b), *Japanese Patent* 02200192.

ZAGO, A., DEGRASSI, G., BRUSCHI, C. V. (1995), Cloning, sequenzing, and expression in *Escherichia coli* of the *Bacillus pumilus* gene for ferulic acid decarboxylase, *Appl. Environ. Microbiol.* **61**, 4484–4486.

10 Synthetic Applications of Enzyme-Catalyzed Reactions

JANE MCGREGOR-JONES

Tempe, AZ, USA

1 Introduction

The ability of microorganisms and enzymes to act as selective and chiral catalysts has been acknowledged for many years (see Chapter 1, this volume), but it is only in more recent times that biotransformations have become accepted as routine procedures in organic synthesis. Synthetic chemists have been hesitant to learn the unfamiliar techniques required to use microorganisms and enzymes, especially since new synthetic reagents and catalysts now ease the synthesis of complex molecules that once seemed unthinkable. While nonbiological catalysts will continue to be developed and improved, the enormous range of biological catalysts provides a complementary set of reagents which have unique and competitive advantages. The relative merits of the chiral pool, asymmetric synthesis, and biotransformations as means for preparing chiral intermediates can be judged from an extensive review of the synthesis of chiral pheromones (MORI, 1989).

Enzyme catalyzed biotransformations such as the fermentation of sugar to ethanol by yeast featured in some of the earliest scriptures. Modification and optimization continues today. Countless examples of biotransformations can be found in the world around us: citric acid, obtained through fermentation, is used in the food and drink industry as a flavoring agent; the penicillins and cephalosporins are both secondary metabolites produced by fungi and are used clinically as antibacterial agents; anti-inflammatory steroids are prepared from more common naturally occurring steroids and bacterial proteases are used as additives in biological washing powders.

The aim of this chapter is to demonstrate the wide utility and applicability of biotransformation products in synthesis today, with particular reference to those that are flexible synthons and which can be used to reach common targets of general interest. Such examples of utility abound in the pharmaceutical industry where enantiomerically pure intermediates are necessary, and are frequently easily obtained through biotransformations.

The chapter has been subdivided according to the structure of the bioproduct, i.e., aliphatics, alicyclics, polycyclics, heterocycles, and aromatics. Each division is further divided according to the type of biochemical transformation used to produce the bioproduct. Only a few examples in each category are given – this chapter serves only as an overview of the usefulness of biotransformations. More exhaustive texts should be consulted for further information (DAVIES et al., 1989; DRAUZ and WALDMANN, 1995; FABER, 1995; WONG and WHITESIDES, 1994; JONES, 1986).

2 Aliphatic Biotransformation Products

2.1 Hydrolysis/Condensation Reactions

2.1.1 Pig Liver Esterase

Hydrolytic enzymes require no cofactors, but pH control is frequently essential. Pig Liver Esterase (PLE) (GREENZAID and JENCKS, 1971) catalyzes the stereoselective hydrolysis of a wide variety of esters. The enantioselective hydrolysis of substituted glutarate diesters (**1**) to half esters (**2**) (Fig. 1) can be rationalized by JONES' model (JONES, 1993; see Chapter 4, this volume).

If the undesired ester is formed the two enantiomeric series may be interconverted by selective manipulation of the ester and acid groups (e.g., (*S*)- and (*R*)-(**3b**)) or by selective hydrolysis of a methyl ester (**2c**) in the presence of a *t*-butyl ester. The latter method was exploited in a synthesis (WANG et al., 1982) of the antibiotic negamycin (**6**) (Fig. 2) (HAMADA et al., 1970; KONDO et al., 1971).

The half ester (**2c**) was also cleverly exploited in the synthesis of *trans*-carbapenems (KARADY et al., 1981; SALTZMANN et al., 1980). PLE hydrolysis of the diester (**1c**) gave half ester (**2c**) with high enantiomeric purity which was further enhanced by crystallization. Cleavage of the carbobenzyloxy (Z) group by hydrogenation and cyclization using Mukaiyama's reagent gave the β-lactam (**9**) in excellent yield (Fig. 3) (KOBAYASHI et al., 1981). Diastereoselective alkylation adjacent to the chiral center

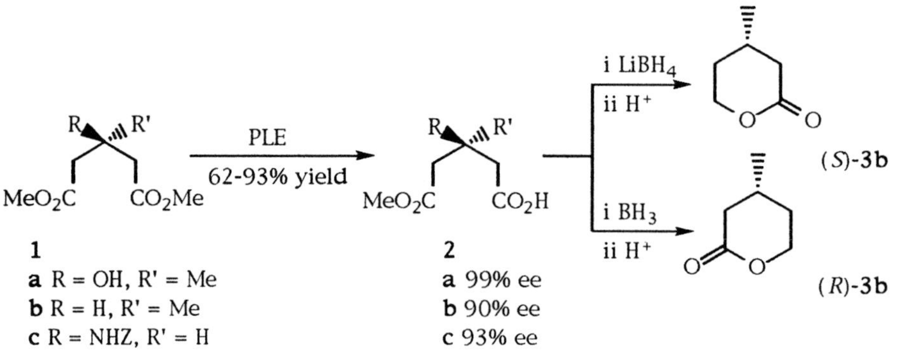

Fig. 1. Pig liver esterase hydrolysis of glutarate diesters.

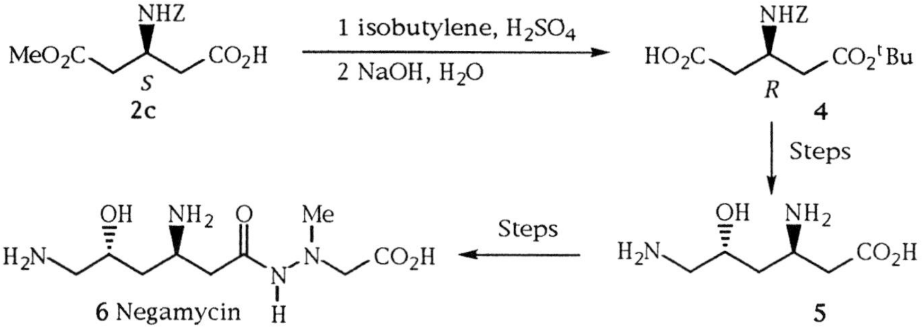

Fig. 2. The use of pig liver esterase in negamycin synthesis.

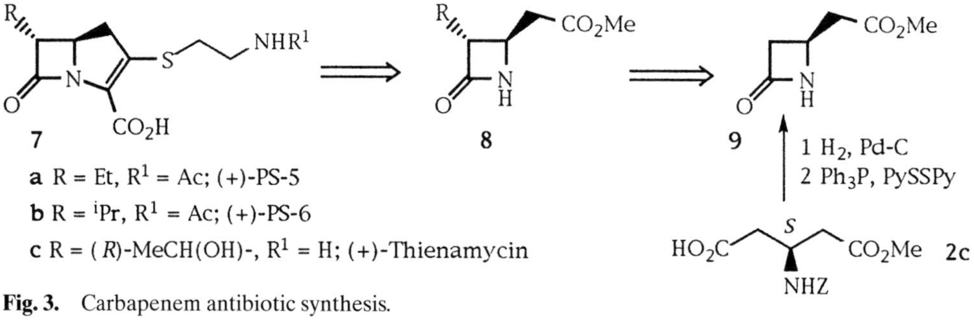

Fig. 3. Carbapenem antibiotic synthesis.

(**8**) and elaboration of the pyrroline ring completed the synthesis (OKANO et al., 1983).

2.1.2 α-Chymotrypsin

α-Chymotrypsin is a protease with a preference for L-amino acid residues that bear aromatic or hydrophobic groups, however it also has strong esterase activity. It can be used in place of PLE in the hydrolyses of C-3 substituted glutarate diesters and generally exhibits opposite enantiospecificities. The stereoselectivity is easily rationalized by overlaying the structure of the ester with that of the natural L-amino acid substrates.

2.1.3 Thermolysin Peptide Synthesis

Despite the emergence of several important methods and coupling reagents for peptide bond formation over the last 50 years, almost all involve the possibility of side reactions and racemization. Conditions in chemical syntheses must be carefully controlled to minimize or eliminate these reactions. Condensation of amino acids using proteolytic enzymes (the reverse of the usual reaction) takes place efficiently under mild conditions, opening up a useful synthesis of peptides (ISOWA et al., 1977b). It is the thermodynamic products that eventually accumulate in enzyme catalyzed processes, as with all reactions. Unfortunately the equilibrium constant for peptide formation usually lies heavily on the side of the constituent amino acids, but it can be displaced by the formation of insoluble peptides, by continuous product removal (e.g., two-phase reactions), the use of low water systems, or reactions under kinetic control.

The exploitation of the different specificities of various proteolytic enzymes provides convergent, multi-peptide bond syntheses in good yields from the component amino acids (GLASS, 1981). In some cases these reactions have been run on a multi-ton scale, for example in the formation of the aspartame precursor (**12**) (Fig. 4). Aspartame is a dipeptide synthetic sweetener, more commonly known under the brand name Nutrasweet®.

The coupling of L-phenylalanine methyl ester (**11**) with *N*-carbobenzyloxy-L-aspartic acid (**10b**) to give the dipeptide (**12**) (Fig. 4) is catalyzed by immobilized thermolysin (OYAMA et al., 1981), with a yield of 82%. The use of an immobilized enzyme (e.g., on a solid support of silica or porous ceramics) helps to drive the reaction in the required direction with high yields, as does the insolubility of the dipeptide (**12**). Thermolysin is highly specific in its catalytic activity, so that protection of the side chain carboxyl group of the aspartic acid derivative is not necessary. A commercial production route of the synthesis of aspartame (**13**) involves the use of racemic phenylalanine methyl ester *rac*-(**11**), in which the unreactive D-isomer is recovered and recycled.

Thermolysin and other proteolytic enzymes such as papain, pepsin, and nargase are commonly used in the formation of small peptides such as di-, tri-, and tetrapeptides. The synthesis of peptides, even by convergent techniques requires many steps, consequently the yields of individual steps need to be extremely high if the route is to be viable. An outstanding example is provided by the octapeptide angiotensin II (Fig. 5) (ISOWA et al., 1977a) which was prepared in an overall yield of 85%. The general principles of peptide coupling by enzymes are illustrated by synthese of leucinyl- and methionyl-enkephalins (**15**) and (**16**) (KULLMAN, 1981) and dynorphin (**17**) (KULLMAN, 1982) (Fig. 6). In these reactions selectivity is predominantly determined by the residues at the carboxylic acid terminus of the peptide chain. Thus chymotrypsin preferentially couples the carboxylic group of terminal aromatic amino acids with most other amino acids. In contrast, for papain catalyzed couplings the penultimate amino acid at the carboxylic terminus is preferably phenylalanine, but other hydrophobic

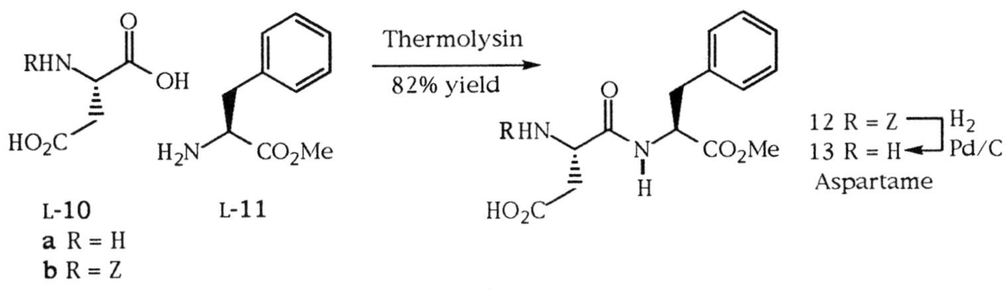

Fig. 4. Enzymatic aspartame synthesis.

H-Asn-Arg-Val-Tyr-Val-His-Pro-Phe-OH

14 Angiotensin II

Fig. 5. Angiotensin II.

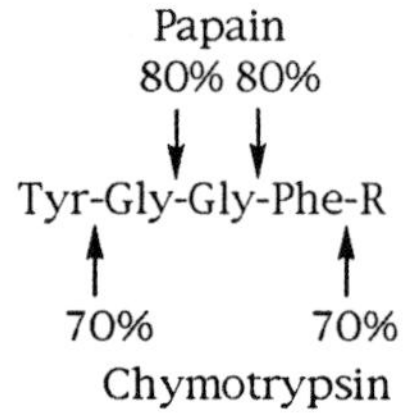

Papain
80% 80%

Tyr-Gly-Gly-Phe-R

70% 70%
Chymotrypsin

15 R = Leu 52% yield
16 R = Met 52% yield
Enkephalins

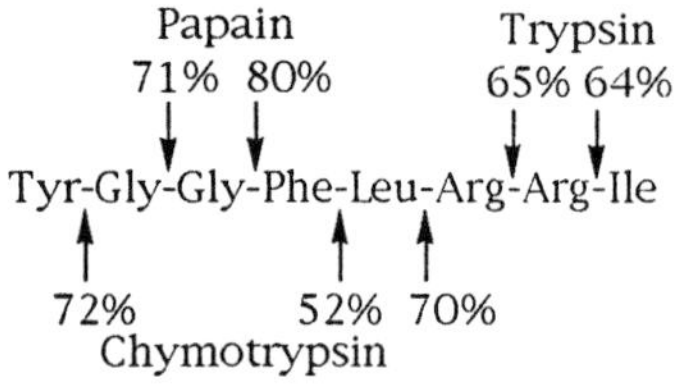

Papain 71% 80% Trypsin 65% 64%

Tyr-Gly-Gly-Phe-Leu-Arg-Arg-Ile

72% 52% 70%
Chymotrypsin

17 Dynorphin 50% yield

Fig. 6. Synthesis of leucyl and methionyl enkephalins.

amino acids such as leucine and valine are also accepted. The nature of the terminal amino acid plays only a minor role. In the examples shown (Fig. 6), the influence of the aromatic amino acid on papain catalyzed couplings enables the coupling of glycine to Tyr-Gly, and extends a further residue in the coupling of phenylalanine to Tyr-Gly-Gly. Finally, trypsin catalyzed couplings require arginine or lysine to be present at the carboxylic terminus, but do not discriminate between amino acid coupling partners. Carboxypeptidases have complementary selectivity, i.e., they discriminate the amino terminus of the peptide chain.

2.1.4 Hog Kidney Acylase

The first major application of the resolution of enantiomers by enzymes was the use of hog kidney acylase (HKA) to hydrolyze *N*-acyl amino acids (GREENSTEIN, 1954). HKA accommodates a broad range of structural types and is highly stereoselective. As with virtually all mammalian enzymes that act on amino acids it is selective for the L-enantiomer (Fig. 7). The remaining *N*-acyl D-enantiomer D-(**21**) is recycled via chemically induced racemization. This reaction is still of industrial importance today (CHIBATA et al., 1976).

Hog acylase I was used by BALDWIN (BALDWIN et al., 1976) in the first stereocontrolled synthesis of a penicillin from a peptide. Incubation of the chloroacetyl amide (**23**) with Hog kidney acylase I and separation of the unreacted amide, followed by acid hydrolysis gave the D-isodehydrovaline (**24**) in 60% yield (Fig. 8). The material obtained was identical to a sample obtained by degradation of penicillin.

In another example, MORI and IWASAWA (1980) prepared both enantiomers of *threo*-2-amino-3-methylhexanoic acid (**27**) by enzymatic resolution. These are precursors to *threo*-4-methylheptan-3-ol (**28**), a pheromone component of the smaller european elm bark beetle (Fig. 9).

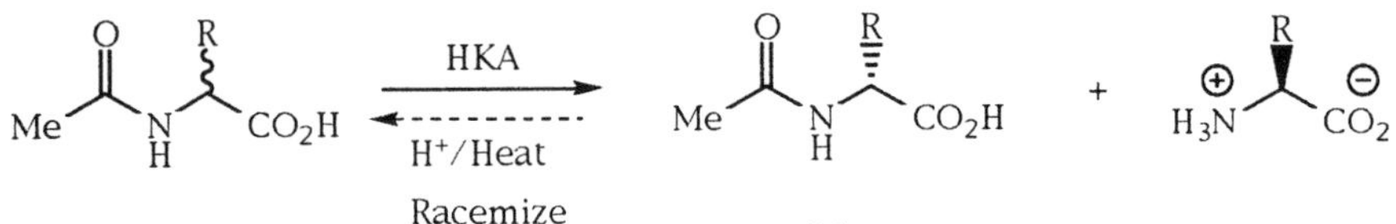

Me–C(=O)–NH–CH(R)–CO_2H (*rac*-**21**) —HKA→ / ←H$^+$/Heat, Racemize— Me–C(=O)–NH–CH(R)–CO_2H (D-**21**) + $H_3N^{\oplus}$–CH(R)–$CO_2^{\ominus}$ (L-**22**)

Fig. 7. L-Stereospecificity of acylases.

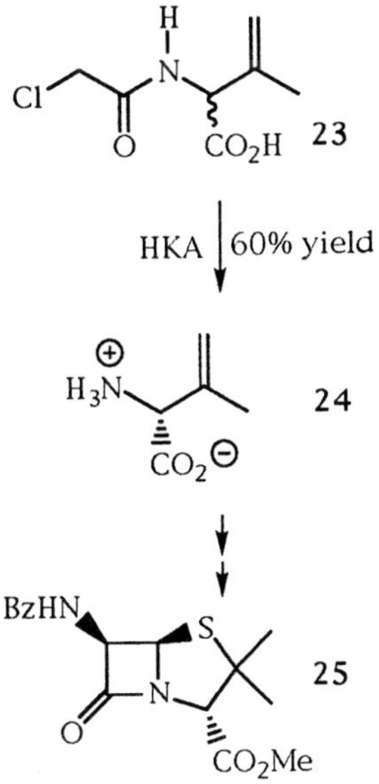

Fig. 8. Stereocontrolled synthesis of a penicillin system.

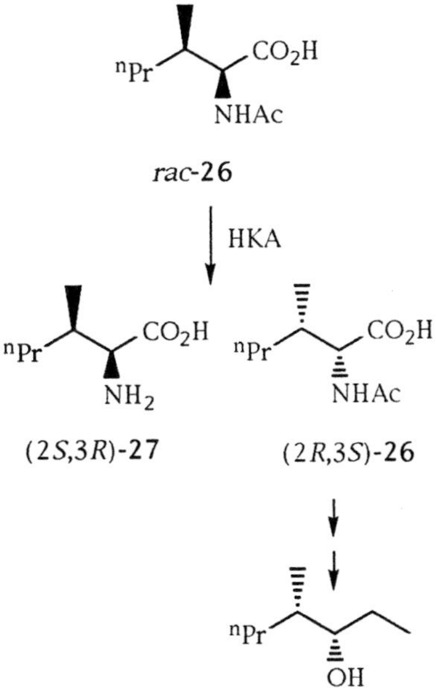

(3*S*,4*S*)-*threo* **28**

Fig. 9. Enzymatic resolution to pheromone precursors.

2.1.5 Microbial Esterase

Methyl *trans*-(2*R*,4*R*)-2,4-dimethylglutarate (**30**) is a useful chiral synthon for macrolide and polyether antibiotics (Fig. 10). One of the most convenient routes to this intermediate is by microbial esterase catalyzed hydrolysis of the racemic diester (**29**). The unreactive *trans*-diester (2*S*,4*S*)-(**29**) cannot be easily recycled to give further amounts of (2*R*,4*R*)-(**30**) because its base mediated epimerization also produces the unwanted *meso-cis*-diester (2*S*,4*R*)-(**29**) (CHEN et al., 1981). The bifunctional chiral synthon (2*R*,4*R*)-(**30**) is a potental precursor to the polyether antibiotic calcimycin (**31**) a metabolite of *Streptomyces* microorganisms (Fig. 11) (EVANS et al., 1979). Similarly, treatment of dimethyl *cis*-2,4-dimethyl glutarate (**29**) (Fig. 12) with microbial esterase results in stereospecific hydrolysis of the pro-*R* ester group in high yield (75%, 98%ee). Pig liver esterase preferentially cleaves the pro-*S* ester grouping of the *meso*-diester (**29**), but with a poor ee of 64%.

The bifunctional chiral synthon (2*R*,4*S*)-(**30**) represents a partial structural unit commonly encountered in macrolide antibiotics,

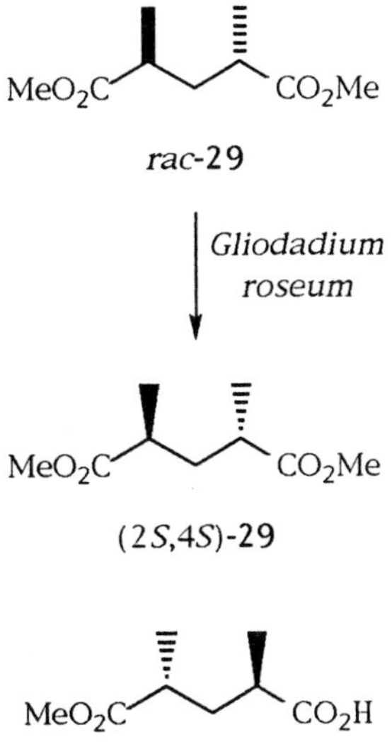

(2*R*,4*R*)- **30** ; 50% yield
up to 98%ee

Fig. 10. A useful chiral synthon for antibiotics.

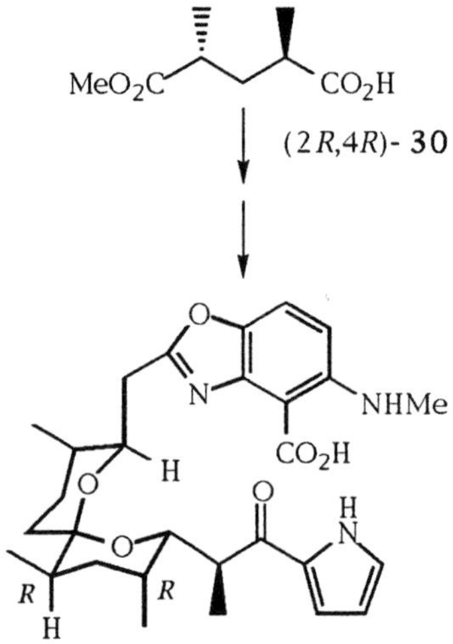

MeO_2C CO_2H
(2*R*,4*R*)- **30**

31 Calcimycin

Fig. 11. Synthesis of calcimycin.

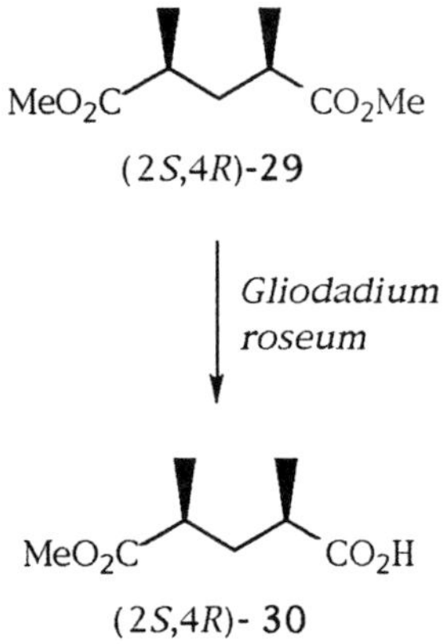

MeO_2C CO_2Me
(2*S*,4*R*)-**29**

Gliodadium roseum

MeO_2C CO_2H
(2*S*,4*R*)- **30**

Fig. 12. Stereospecific microbial esterase hydrolysis.

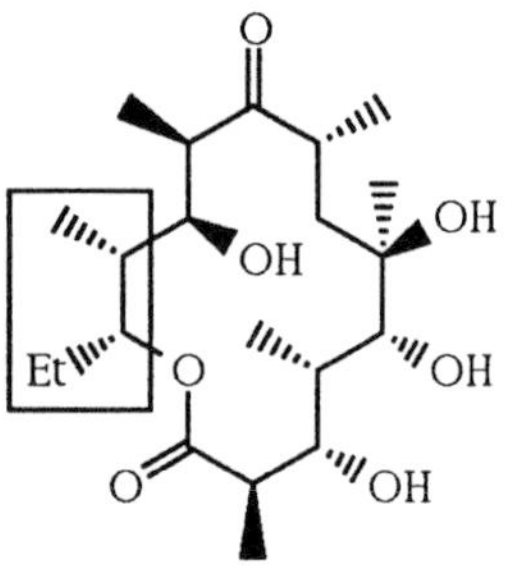

32 Erythronolide B

Fig. 13. Erythronolide B.

and was used in COREY's total synthesis of erythronolide B (**32**) (Fig. 13) (COREY et al., 1978a, b). The erythromycins, produced by the fungus *Streptomyces* erythreus, constitute one of the most important, widespread and effective of all known families of antibiotics commonly used against penicillin resistant *Staphylococcus* strains, and are the drugs of choice to treat Legionnaires' disease. Another macrolide antibiotic obtainable through the chiral synthon (2*R*,4*R*)-(**30**) is methmycin (HIRAMA et al., 1979).

Numerous polyether antibiotics are also accessible, including monensin (**35**) (Fig. 14) (COLLUM et al., 1980a), which has a stereochemically complex array of 17 asymmetric centers. The approach to monensin is based on the synthesis and coupling of three advanced optically active fragments, within one of which, the synthon (2*S*,4*R*)-(**30**) is a part.

2.1.6 Porcine Pancreatic Lipase

The Sharpless enantioselective epoxidation of allylic alcohols is one of the most reliable abiotic asymmetric reactions (KATSUKI and MARTIN, 1996; KATSUKI and SHARPLESS, 1980; BEHRENS and SHARPLESS, 1985). Porcine pancreatic lipase catalyzed hydrolysis of racemic epoxy esters *rac*-(**36**) provides a synthesis of both enantiomers and is able to accept a large variety of structures (Fig. 15) (MARPLES and ROGER-EVANS, 1989; LADNER and WHITESIDES, 1984). These hydrolyses proceed with high enantiospecificity with long alkyl ester groups R, and are simpler to perform than Sharpless epoxidation. Porcine pancreatic lipase is active at water/organic solvent interfaces, so the solubility of the substrate in water is not essential.

Fig. 14. Synthesis of monensin.

2.2 Enzyme Catalyzed Reduction Reactions

2.2.1 Yeast Alcohol Dehydrogenase

2.2.1.1 Reduction of Ketones and Aldehydes

An investigation of yeast reductions of simple ketones in the 1960's by MOSHER (MAC LEOD et al., 1964) showed that in most cases the (S)-configuration alcohols were obtained by hydride addition to the *Re*-face of the carbonyl group. The stereoselectivity can be rationalized by Prelog's model which is based on the differences in size of the groups, however these are not always congruent with Cahn-Ingold-Prelog priorities (see Chapter 9, this volume; SERVI, 1990).

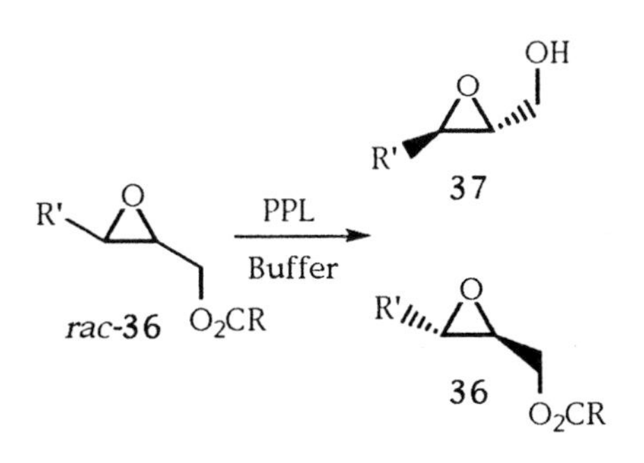

R'	R	ee %
H	Et	98
H	$^nC_3H_7$	92
H	$^nC_4H_9$	96
$^nC_3H_7$	$^nC_3H_7$	>90

Fig. 15. Porcine pancreatic lipase catalyzed hydrolysis.

2.2.1.2 Reduction of β-Keto Esters

The enantioselectivity of the reduction of β-keto acids by yeast can be modulated by changing the alkoxyl moiety of the ester. For example, comparison of reduction of the potassium salt (**38a**) with the methyl (**38b**) and butyl (**38c**) esters shows a decline in enantiomeric excess as the size of the alkoxyl moiety increases (HIRAMA et al., 1983). Moreover conversion to the carboxylic acid salt increases the aqueous solubility of the substrate and product relative to the corresponding esters and the work-up is simplified. Filtration of the reaction and extraction with organic solvents, removes the biomass and nonpolar compounds, respectively. Treatment with diazomethane and a second organic extraction yields the carboxylic acid fraction as the corresponding methyl esters. The versatile synthon (**39b**) was used in the total synthesis of compactin (**40a**) which was first isolated in 1976 from the molds *Penicillium citrinum* and *P. brevi compactum* and showed marked inhibitory activity against HMG-CoA reductase, an enzyme involved in the biosynthetic pathway to cholesterol. Mevinolin (**40b**) is even more potent and is used in treatment of patients with dangerously high cholesterol levels (Fig. 16) (HIRAMA and UEI, 1982).

The reduction of ethyl acetoacetate (**41**) (Fig. 17) with bakers' yeast to ethyl (*S*)-3-hydroxybutanoate (**42**) has been extensively investigated and reported in an *Organic Syntheses* preparation (SEEBACH et al., 1985). It was used in an extraordinary synthesis of the main component of the pheromone of the solitary bee, *Andrena wikella* (**44**). In a virtuoso demonstration of synthetic mastery, methyl acetoacetate was alkylated sequentially with two equivalents of the iodide (**43**). Thus the 11 carbon atoms of the pheromone (**44**) and a molecule of carbon dioxide were derived solely from two molecules of ethyl and one of methyl acetoacetate (MORI and TANIDA, 1981; TENGÖ et al., 1990).

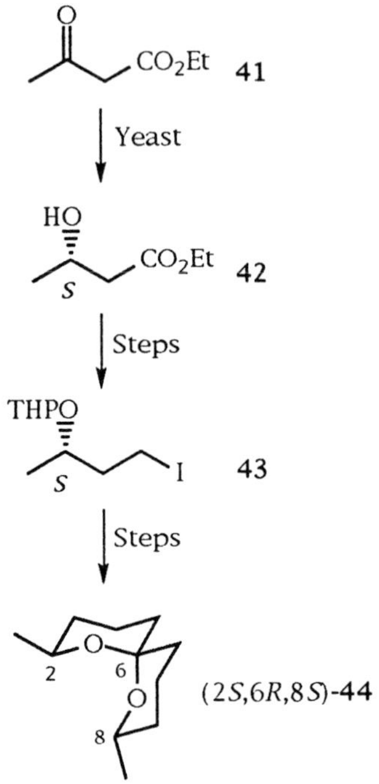
O
CO_2Et
41
Yeast
HO
CO_2Et
42
S
Steps
THPO
I
43
S
Steps
O
2
6
O
8
(2*S*,6*R*,8*S*)-44

Fig. 17. Synthesis of a pheromone component of *Andrena wilkella*.

The tri-isopropylsilyl derivative (**47**) was used in a synthesis of the important carbape-

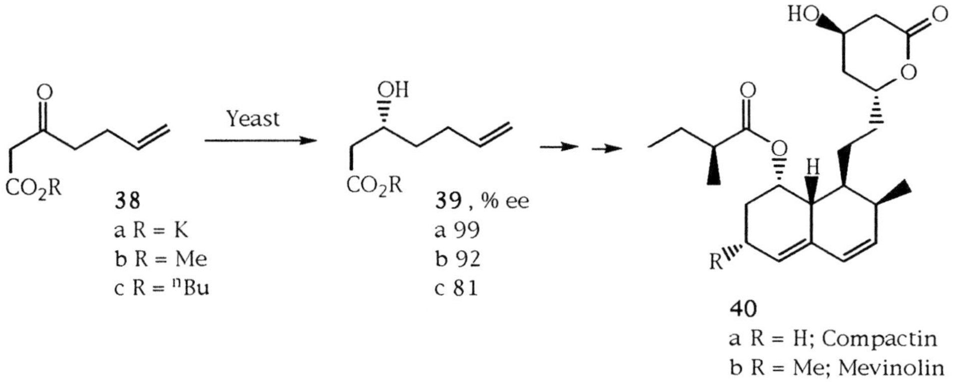
O
CO_2R
38
a R = K
b R = Me
c R = ^{n}Bu
Yeast
OH
CO_2R
39 , % ee
a 99
b 92
c 81
HO
O
O
O
O
H
R
40
a R = H; Compactin
b R = Me; Mevinolin

Fig. 16. Chiral total synthesis of compactin.

nem intermediate (**49**). The chiral center was used to direct the stereochemistry of a ketene-imine cycloaddition to generate the useful C-4 keto-substituted azetidinone (**48**) (Fig. 18) (TSCHAEW et al., 1988) in 90% overall yield. Ethyl (*S*)-3-hydroxybutanoate (**42**) was also used as a source of the remote stereocenter in griseoviridin (**50**), a member of the streptogramin family of antibiotics (Fig. 19) (MEYERS and AMOS, 1980).

Many microorganisms such as *Alcaligenes eutrophus* and *Zoogloea ramigera* I-16-M utilize (*R*)-poly-3-hydroxybutanoate as an energy reserve, which is stored as granules within the cells. The abundance of this material within the cells is extraordinary. For example, ethanolysis of 50 g of *Z. ramigera* gave 33 g of the monomer (*R*)-(**42**) (MORI, 1989).

Ethyl (*R*)-3-hydroxybutyrate (*R*)-(**42**) has been utilized in the synthesis of stegobinone (**54**) – the pheromone of the drugstore beetle (Fig. 20) (MORI and EBATA, 1986a). Diastereoselective methylation of the dianion of (*R*)-(**42**) *anti* to the alkoxide furnished the methyl group destined to be at C-3 of the pheromone (**54**) (FRATER, 1979a, b). The stereochemistry of putative C-2 was achieved by Mitsunobu inversion with 2,4-dinitrobenzoic acid to give a crystalline product. Methyl (*R*)-(–)-*β*-Hydroxyisobutyric acid (**52**) is manufactured by *β*-hydroxylation of *iso*-butyric acid with *Candida* sp., followed by esterification.

All four possible stereoisomers of 5-hydroxy-4-methyl-3-heptanone (**56**), the pheromone of the weevil of the genus *Sitophilus*, were prepared from methyl (*R*)-3-hydroxypentanoate (**55**) which was manufactured by *β*-hydroxylation of methyl pentanoate or pentanoic acid with *Candida rugosa* (Fig. 21) (MORI and EBATA, 1986b). The principal reactions were diastereoselective methylation and Mitsunobu inversion as used in the synthesis of stegobinone (**54**). Similarly diastereoselective allylation of methyl (3*R*)-3-hydroxybutanoate (*R*)-(**46**) and ethyl (3*S*)-3-hydroxybutanoate (*S*)-(**42**) (Fig. 22) furnished (*S*)- and (*R*)-lavandulol esters (**57**), respectively, which were used in the synthesis of optically active acyclic C45 and C50 carotenoids (VON KRAMER and PFANDER, 1982). (For other examples of the uses of 3-hydroxybutanoate esters see SEURLING and SEEBACH, 1977; FRATER, 1979a, b, 1980; FRATER et al., 1984.)

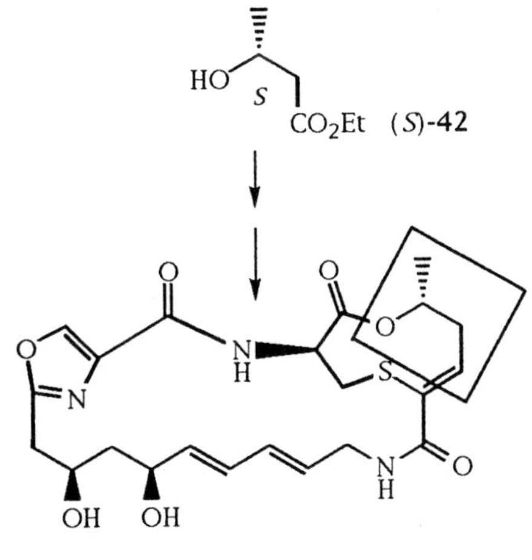

(*S*)-42

50 Griseoviridin

Fig. 19. Antibiotic synthesis using yeast.

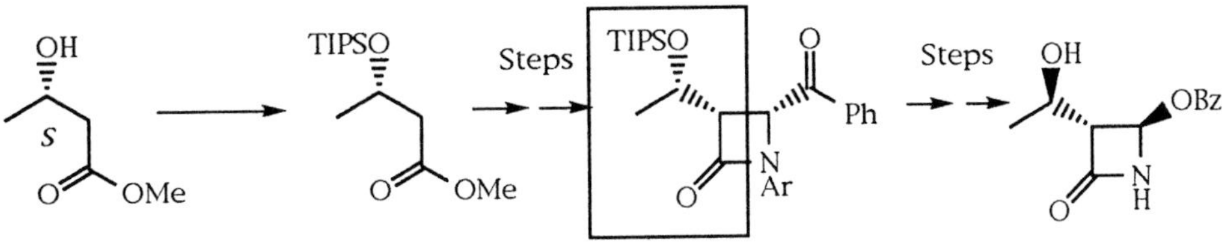

46 47 48 49

Fig. 18. *β*-Lactam intermediate synthesis.

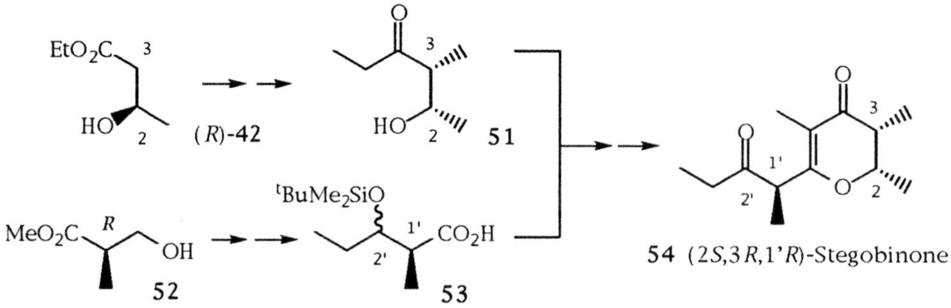

EtO_2C 3 HO 2 (*R*)-**42** → → O 3 HO 2 **51**

MeO_2C *R* OH **52** → → $^tBuMe_2SiO$ 1' 2' CO_2H **53**

→ → O O 1' 2' 3 O 2 **54** (2*S*,3*R*,1'*R*)-Stegobinone

Fig. 20. Pheromone synthesis.

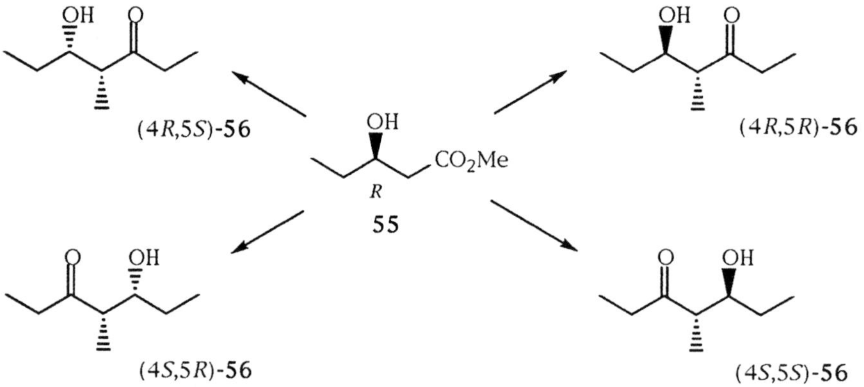

OH O (4*R*,5*S*)-**56**

OH O (4*R*,5*R*)-**56**

OH CO_2Me *R* **55**

O OH (4*S*,5*R*)-**56**

O OH (4*S*,5*S*)-**56**

Fig. 21. Stereoisomers of *Sitophilus* weevil pheromone component.

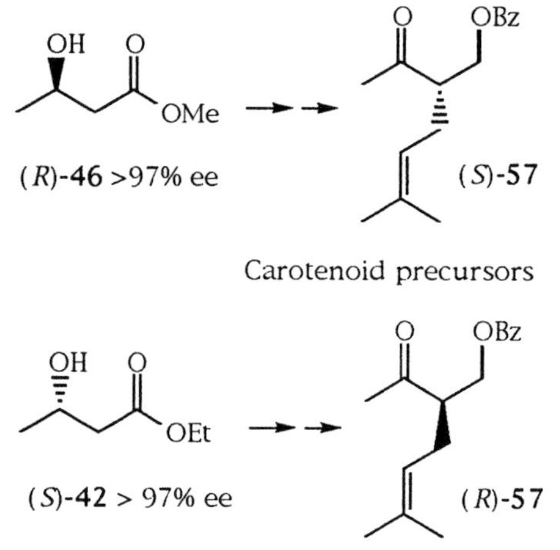

OH O OMe (*R*)-**46** >97% ee → → O OBz (*S*)-**57**

Carotenoid precursors

OH O OEt (*S*)-**42** > 97% ee → → O OBz (*R*)-**57**

Fig. 22. Routes to carotenoids.

2.3 Oxidation Reactions

2.3.1 Horse Liver Alcohol Dehydrogenase

Horse liver alcohol dehydrogenase (HLADH) is probably the most intensely studied oxidoreductase. It accommodates a wide range of substrates, but is highly stereoselective. HLADH is normally used for the reduction of aldehydes and ketones which is the thermodynamically preferred direction, however, with a suitable cofactor regenerating system, oxidations are also possible.

Oxidation of the diol (**58**) proceeds with pro-*S* enantioselectivity to give the hydroxy aldehyde (**59**). This undergoes *in situ* hemiacetal formation to give the lactol (**60**) which is further oxidized to the lactone (*S*)-(**3b**). Using the conditions indicated 5.85 g (68% yield) of the lactone was prepared in a single reaction (Fig. 23) (DRUECKHAMMER et al., 1985). It is a phytyl chain synthon (FISCHLI, 1980). Similar stereoselectivity is exhibited by the HLADH catalyzed oxidation of glycerol (**61**) to L-(−)-glyceraldehyde (**62**) (BALLY and LEUTHARDT, 1970).

2.3.2 *Pseudomonas putida*

The conversion of an unactivated carbon–hydrogen bond to a carbon–oxygen bond can be accomplished using ozone, peracids, or high oxidation state iron reagents such as the Gif system. However these reagents have low selectivity (usually for methine carbons) and do not tolerate most functional groups. The Barton reaction of steroidal 20-nitrites and Breslow's remote functionalization templates are regioselective due to proximity effects, but

HLADH
OH OH
58
Coenzymes:
NAD$^+$, FMN,
FMN reductase,
catalase, pH 9, 14d
S
CHO
OH
59
HLADH
O O
(*S*)-3b 70% yield, 87% ee
O OH
60

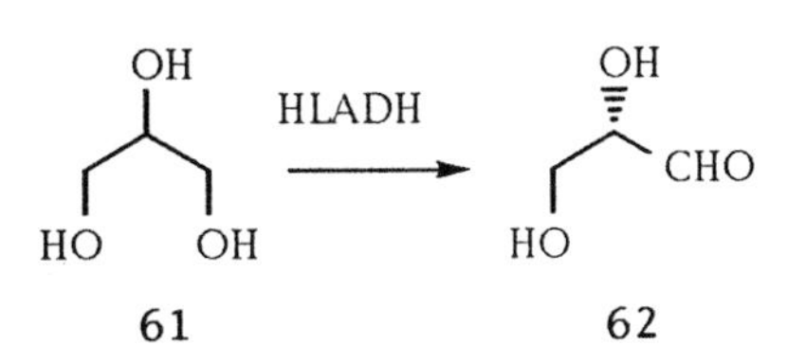

Fig. 23. Enantiotopic oxidation of prochiral *meso*-alcohols.

are not generally applicable. In contrast, many enzymes are capable of effecting such reactions regio- and stereoselectively in the presence of most functional groups (JOHNSON, 1978; DAVIES et al., 1989; DRAUZ and WALDMANN, 1995).

A major area of interest has been the β-hydroxylation of carboxylic acids and esters. (*S*)-(+)-β-hydroxyisobutyric acid (**64**) is manufactured in 48% yield by oxidation of the pro-*S*-methyl group of isobutyric acid (**63**) by *Pseudomonas putida* (Fig. 24) (GOODHUE and SCHAEFFER, 1971). It has been used as a synthon in syntheses of α-tocopherol (**66**) (vitamin E) (COHEN et al., 1976), (*R*)-muscone and, unnatural (*S*)-muscone (**67**) (BRANCA and FISCHLI, 1977).

The ansa macrocyclic maytansinoids (**69**) have antitumor activity (KUPCHAN et al., 1977), and have been the focus of many pharmacological and synthetic efforts. The "northeastern region" (**68**) synthon containing four chiral centers was prepared in gram quantities from (*S*)-(+)-β-hydroxyisobutyric acid (**64**) (Fig. 25). Both centers destined to become C-5 and C-7 were sequentially converted to aldehydes and alkylated without epimerization at C-6. The two "new" chiral centers at C-4 and C-5 were introduced by Sharpless chiral epoxidation, whereas that at C-7 resulted from diastereoselective addition to an aldehyde (MEYERS and HUDSPETH, 1981).

Further examples of the importance of (**64**) as a chiral synthon can be found throughout the literature, including the synthesis of calcimycin (**31**) (Fig. 11) (EVANS et al., 1979) and other polyether and macrolide antibiotics (JOHNSON et al., 1979; JOHNSON and KISHI, 1979).

2.3.3 Lipoxygenases

Lipoxygenases are non-heme iron containing dioxygenases which catalyze the reaction of an allyl moiety with oxygen to give a hydroperoxide. Overall the process is equivalent to an ene reaction with oxygen acting as the electrophile. The methylene group participating in the reaction must be located between two alkene groups and hence in principle lipoxygenation of arachidonic acid (**70**) (Fig. 26) could

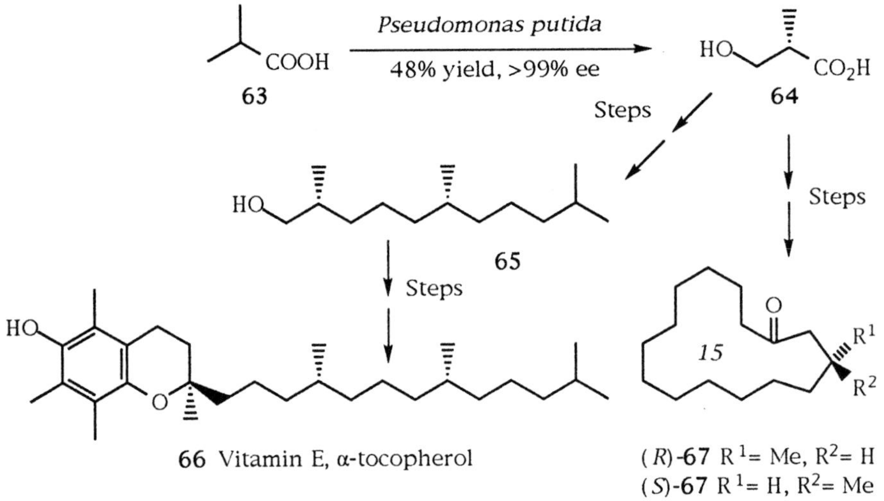

COOH

63

Pseudomonas putida

48% yield, >99% ee

HO

CO_2H

64

Steps

Steps

HO

65

Steps

HO

O

66 Vitamin E, α-tocopherol

O

15

R^1

R^2

(*R*)-67 R^1= Me, R^2= H

(*S*)-67 R^1= H, R^2= Me

Fig. 24. Enantiotopically specific hydroxylation of isobutyric acid and uses.

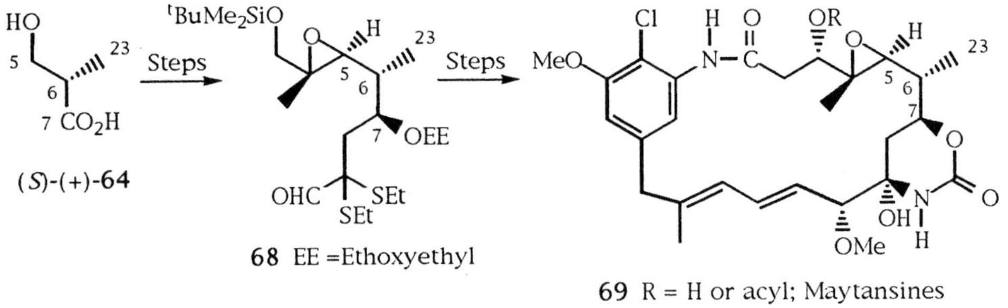

HO

5

23

6

7 CO_2H

(*S*)-(+)-64

Steps

$^tBuMe_2SiO$

O

H

23

5

6

7

OEE

OHC

SEt

SEt

68 EE =Ethoxyethyl

Steps

MeO

Cl

H

N—C

O

OR

O

H

23

5

6

7

O

N

O

OH

H

OMe

69 R = H or acyl; Maytansines

Fig. 25. A biotransformation routine to maytansinoids.

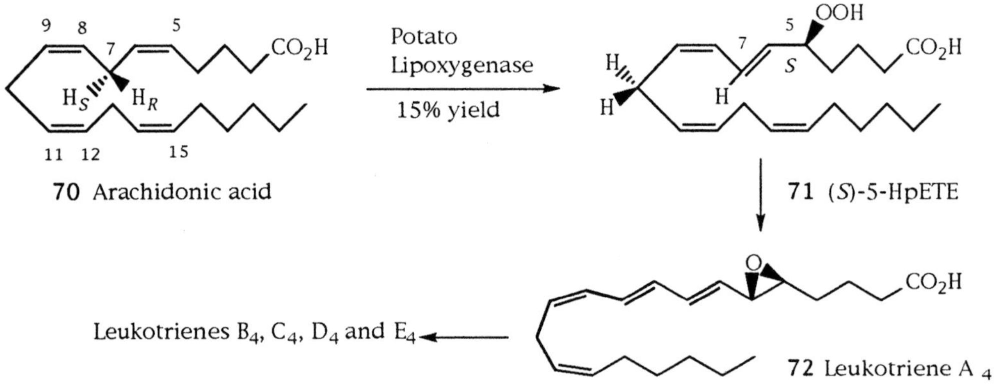

9 8 5

7

CO_2H

H_S H_R

11 12 15

70 Arachidonic acid

Potato

Lipoxygenase

15% yield

OOH

5

7

S

CO_2H

H

H

H

71 (*S*)-5-HpETE

O

CO_2H

72 Leukotriene A $_4$

Leukotrienes B_4, C_4, D_4 and E_4

Fig. 26. The leukotrienes from arachidonic acid via lipoxygenase.

occur at carbons 5, 8, 9, 11, 12, or 15. Soybean lipoxygenase catalyzes the formation of (*S*)-15-HPETE (15-hydroperoxyeicosatetraenoic acid) and potato lipoxygenase (*S*)-5-HPETE (**71**). The pro-*S* hydrogen at C-7 is lost stereospecifically in this reaction and indeed thus far all enzyme catalyzed lipoxygenations occur such that the hydroperoxide group is introduced *anti* to the eliminated hydrogen (COREY et al., 1980; COREY and LANSBURY, 1983).

2.4 Carbon–Carbon Bond Formation Reactions

2.4.1 Farnesyl Pyrophosphate Synthetase

The enzymes of terpene biosynthesis are rarely employed in biotransformations, nevertheless they have tremendous potential. Both enantiomers of 4-methyldihomofarnesol (**75**) (Fig. 27) were prepared by coupling the isopentenyl pyrophosphate homologs (**74**) with tritiated dihomogeranyl pyrophosphate (**73**) catalyzed by partially purified farnesyl diphosphate synthetase from pig liver. Cleavage of the pyrophosphate groups by alkaline phosphatase gave the 4-methyldihomofarnesols (*S*)-(**75**) (810 µg), (*R*)-(**75**) (590 µg). The radio activity of the tritium label was used to guide the purification, which otherwise would have been extremely difficult on this minute scale (KOYAMA et al., 1980, 1987; KOBAYASHI et al., 1980).

2.4.2 Aldolases

Asymmetric carbon–carbon bond formation based on catalytic aldol addition reactions remains one of the most challenging subjects in synthetic organic chemistry. Many successful nonbiological strategies have been developed (EVANS et al., 1982; HEATHCOCK, 1984), but most have drawbacks – the need for stoichiometric auxiliary reagents and the need for metal-enolate complexes for stereoselectivity are two. However, examples of preparative scale aldolase catalyzed reactions abound in the literature.

High yields of condensation products are obtained with lower primary and secondary aldehydes, but as expected a tertiary carbon atom α to the aldehyde stops the reaction (EFFENBERGER and STRAUB, 1987), and long chains or branching aliphatics lead to lower yields. The reaction tolerates a wide range of functionality (EFFENBERGER and STRAUB, 1987; O'CONNELL and ROSE, 1973). It is worth noting the mono-functionalization of the dialdehydes since the high yields obtained (94–98%) would be very difficult using classical chemistry without the use of protecting groups.

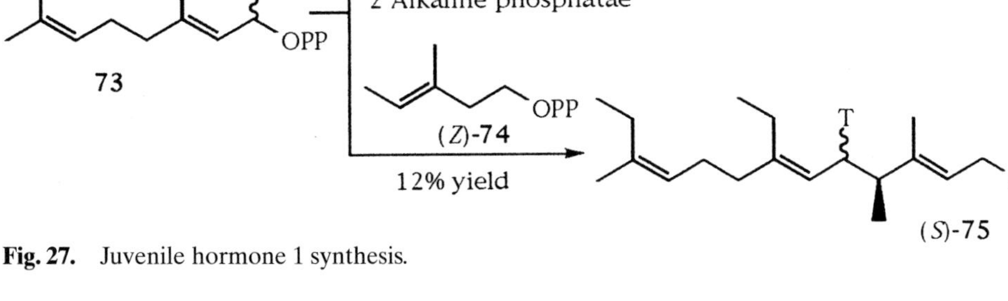

Fig. 27. Juvenile hormone 1 synthesis.

2.4.3 Acyloin Condensation

Considerable attention in fluorine chemistry has been focused on the search of chiral synthetic tools for the asymmetric synthesis of fluorinated bioactive molecules (QUISTAD et al., 1981). Yeast catalyzed reaction has been used as a means of preparing chiral trifluoromethyl compounds (KITAZUME and ISHIKAWA, 1984). Fermentation of trifluoroethanol and an α,β-unsaturated ketone (**77**) (Fig. 28) with yeast, gave trifluoroketols (**78**) in yields of 26–41%, but with an ee of over 90%. These acyloin-type reactions result in valuable chiral synthons, as they do not necessarily correspond to those easily obtained from the chiral pool of natural products.

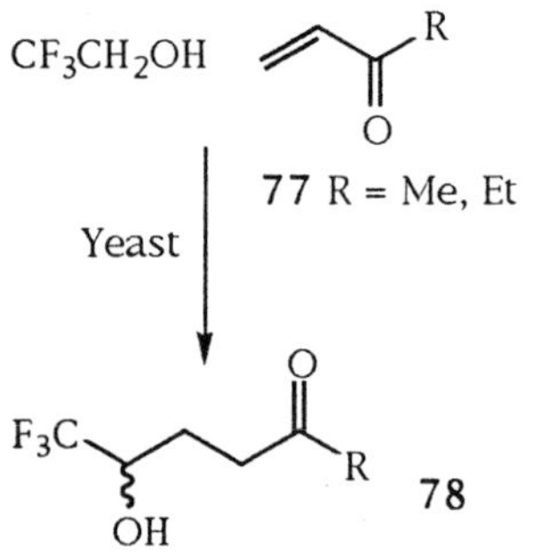

CF_3CH_2OH

77 R = Me, Et

Yeast

78

Fig. 28. Yeast catalyzed acyloin condensation.

2.4.4 Cyanohydrin Formation

Oxynitrilase enzymes, among others, catalyze the asymmetric addition of hydrogen cyanide to the carbonyl group of an aldehyde (**79**), to form a chiral cyanohydrin (**80**) (Fig. 29). These are versatile starting materials for the synthesis of several useful synthons such as α-hydroxy acids (**81**), aminoalcohols (**82**), and acyloins (**83**) (BECKER and PFEIL, 1966; see Chapter 6, this volume). Since only a single enantiomer is produced during the addition to a prochiral substrate, the availability of different enzymes that give either (*R*)- or (*S*)-cyanohydrins is important.

2.5 Multiple Enzyme Reactions

Most enzymes are specific with respect to the type of reaction they catalyze, enabling them to operate independently on their own substrate in the presence of other enzymes and their substrates. This allows multiple, sequential, synthetic transformations to be carried out in "one pot reactions" (Fig. 30).

Many amino acids can be prepared by lyase catalyzed addition of ammonia to the requisite alkene. Aspartic acid (**10a**) is produced for the manufacture of aspartame (**13**) (cf. Fig. 4) by the aspartase catalyzed addition of ammonia to fumaric acid. Using a whole cell system both this step and decarboxylation were used to prepare L-alanine (**85**) (Fig. 30) (TAKAMATSU et al., 1982). Similarly a combination of aspartase, aspartate racemase, and D-amino acid aminotransferase converts fumaric acid to D-alanine (YAMAUCHI et al., 1992).

WHITESIDES and WONG (WONG et al., 1983) have combined enzyme catalyzed and conventional steps for the preparation of unusual sugars (Fig. 30). Aldolase catalyzed condensations

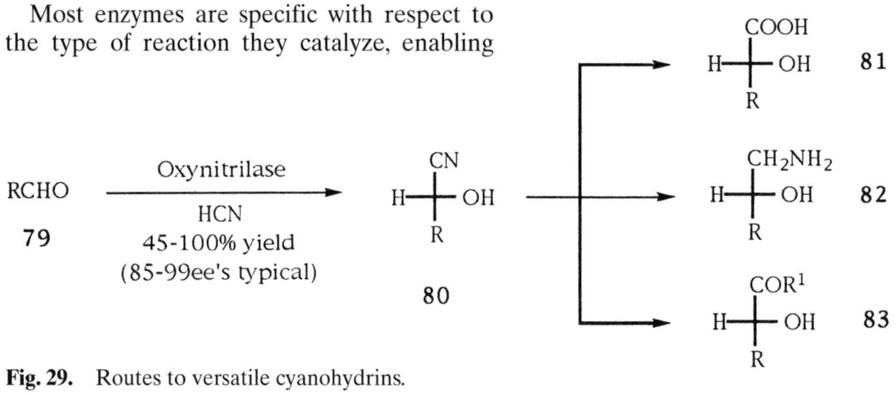

RCHO **79** → Oxynitrilase, HCN, 45-100% yield (85-99ee's typical) → H–C(CN)(R)–OH **80**

→ H–C(COOH)(R)–OH **81**

→ H–C(CH_2NH_2)(R)–OH **82**

→ H–C(COR^1)(R)–OH **83**

Fig. 29. Routes to versatile cyanohydrins.

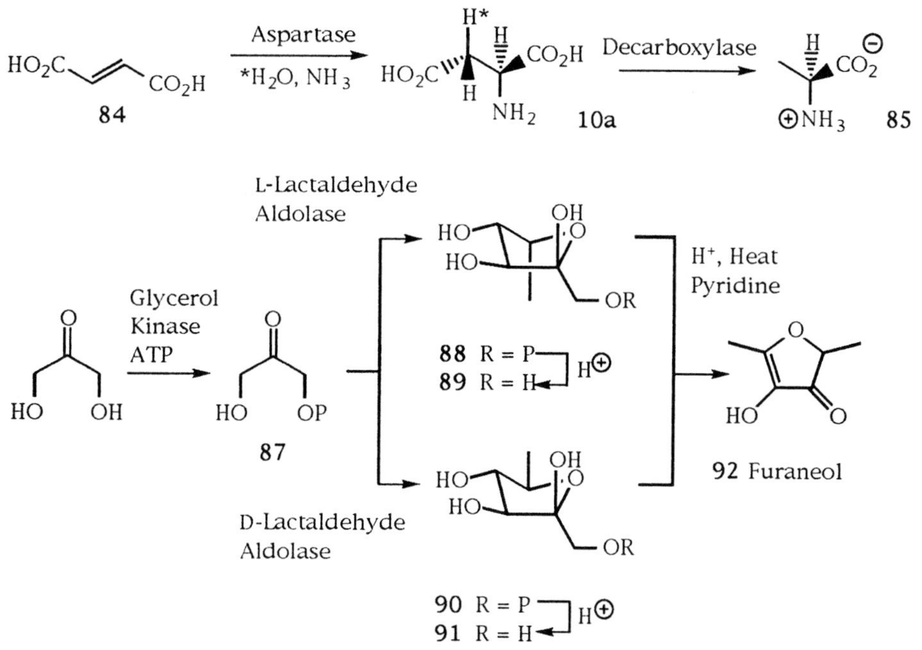

Fig. 30. Examples of multiple enzyme transformations.

of dihydroxyacetone phosphate (**87**) with L- or D-lactaldehyde lead to 6-deoxysorbose (**89**) and the blood group related monosaccharide 6-deoxyfucose (**91**), respectively. Either of these is readily converted into the flavor principle furaneol (**92**) – a caramel flavor component.

3 Alicyclic Biotransformation Products

3.1 Hydrolysis/Condensation Reactions

3.1.1 Pig Liver Esterase

A wide range of cyclic (and acyclic) *meso*-diesters are hydrolyzed by PLE. In the hydrolysis of monocyclic *cis*-1,2-diesters (**93**) a reversal of stereospecificity is observed as one goes from six-membered rings to three membered (Fig. 31) (SABBIONI et al., 1984; MOHR et al., 1983). The cyclopentyl diester marks the "crossover" point with only a 17% ee. As with the acyclic half esters, either the ester or the carboxylic acid may be reduced selectively with lithium borohydride or borane, respectively, and the products lactonized. For example, oxidative ring cleavage of the lactone (**96**) gives the dicarboxylic acid, which after esterification can be regioselectively cyclized with potassium *t*-butoxide to the cyclopentenolactone (**97**) (Dieckmann cyclization). This has been used in the synthesis of prostacyclin (**98**) (GAIS et al., 1984) and brefeldin A (**99**) (GAIS and LUKAS, 1984) (Fig. 32).

The diacetate (**100**) is hydrolyzed with high enantioselectivity by PLE or PPL (Fig. 33) to give the hydroxy ester (**101**). Moreover, the corresponding diol is acetylated by ethyl acetate under PPL catalysis to give *ent*-(**101**) (HEMMERLE and GAIS, 1987). Similarly PLE catalyzed hydrolysis of the diacetate (**102**)

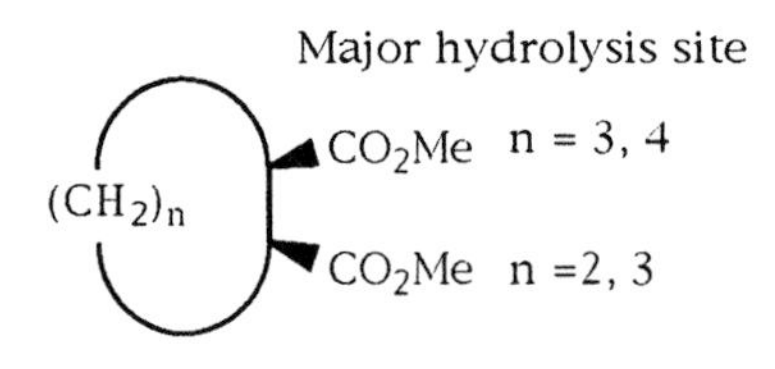

n	Yield %	%ee
1	90	>97 (1*R*,2*S*)
2	98	>97 (1*R*,2*S*)
3	98	17 (1*S*,2*R*)
4	98	>97 (1*S*,2*R*)

Fig. 31. Pig liver esterase specificity.

gives the hydroxy ester (**103**) which is a key intermediate in the 3-component synthesis of prostaglandins, (+)-biotin (**210**) (Fig. 64), and A-factor (WANG et al., 1984).

3.1.2 Lipases

3.1.2.1 *Arthrobacter* Lipase

A useful application of *Arthrobacter* lipase is the resolution of the alcohol moiety of the pyrethroid insecticides. The crude mixture of hydrolyzed (**105**) and unhydrolyzed (**106**) products was heated with mesyl chloride followed by aqueous base to selectively invert the stereochemistry of the alcohol (**105**) (Fig. 34). In this manner, only one enantiomer was obtained from racemic starting material (**104**).

3.1.2.2 Enzymic Ester Resolutions

Enantiomeric distinctions are observed in hydrolyses of esters of chiral alcohols using hydrolytic enzymes. The synthesis of L-menthol (−)-(**108**) through enzyme catalyzed hydrolysis of its racemic ester (±)-(**107**) is of commercial interest (Fig. 35) (MOROE et al., 1970).

3.1.2.3 Porcine Pancreatic Lipase

The asymmetric hydrolysis of cyclic *meso*-diacetates using PPL is complementary to the PLE catalyzed hydrolysis of the corresponding *meso*-1,2-dicarboxylates. In fact the cyclopentane derivative, obtained with low ee's using PLE (Fig. 31), is produced with 86% ee with PPL (KASEL et al., 1985; LAUMEN and SCHNEIDER, 1985). When a range of cyclic diesters (**93**) and (**94**) were subject to hydrolysis by PPL and PLE, the reactions with PPL terminated when

Fig. 32. Enzymic synthesis and uses of chiral bicyclic lactones.

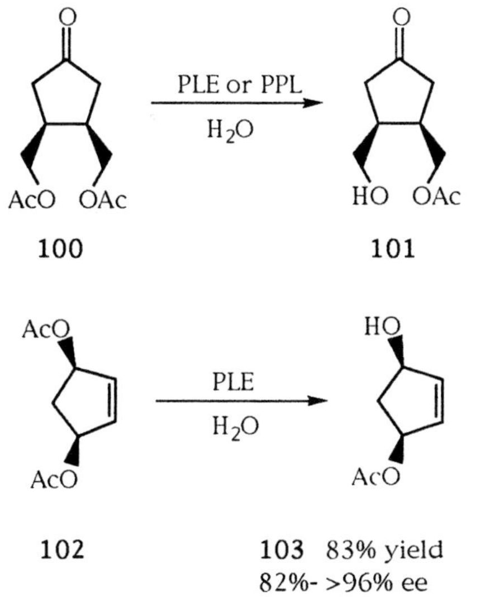

Fig. 33. Examples of the use of pig liver esterase.

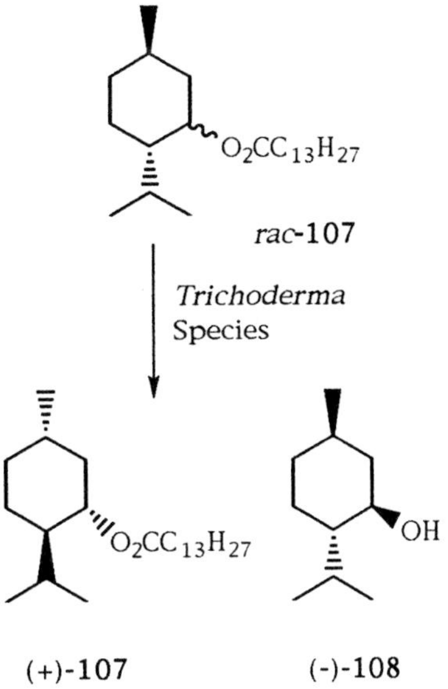

Fig. 35. Resolution of a L-menthol ester.

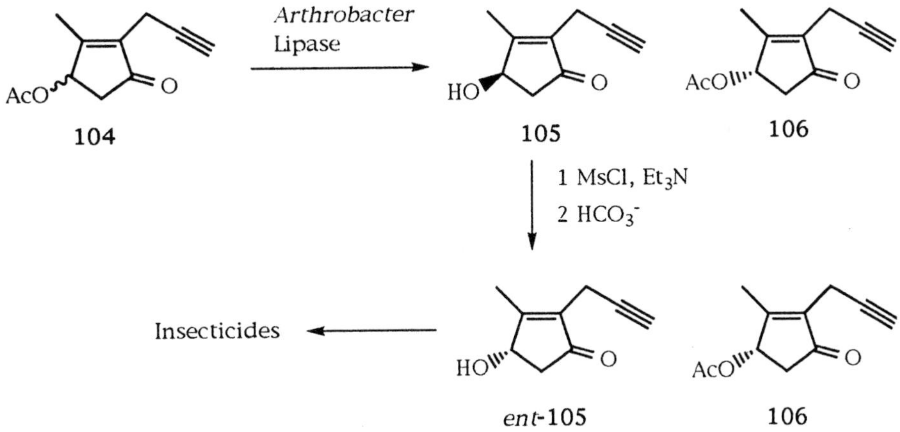

Fig. 34. Pyrethroid insecticides using lipase resolution.

only one ester group had been cleaved, whereas those catalyzed with PLE proceeded to cleave the "second" ester group unless the reaction was carefully monitored (LAUMEN and SCHNEIDER, 1985; cf. SABBIONI et al., 1984). PLE and PPL hydrolyses of esters are best run at pH 7–8 (maintained with a pH-stat) to avoid concomitant, base catalyzed hydrolysis. These mild conditions were exploited in the PPL catalyzed hydrolysis of the methyl ester of the highly sensitive prostaglandin precursors (**109**) and (**112**) to the corresponding acids (**110**) and (**113**) (Fig. 36, PORTER et al., 1979; Fig. 37, LIN et al., 1982, respectively).

3.1.3 Yeast

Although yeast are well known for reducing ketones to alcohols however they also act as esterases. Bakers' yeast was used to hydrolyze 10,11-dihydro-PGA$_1$ methyl ester (**115**) without reduction of the ketone group or the 13,14-alkene bond (Fig. 38) (SIH et al., 1972).

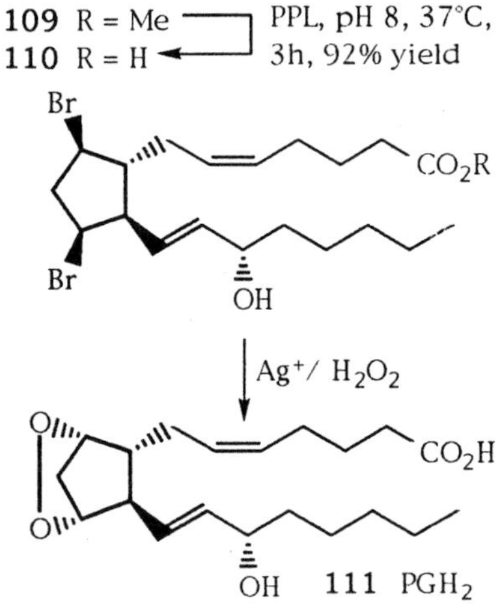

Fig. 36. Use of porcine pancreatic lipase in prostaglandin synthesis.

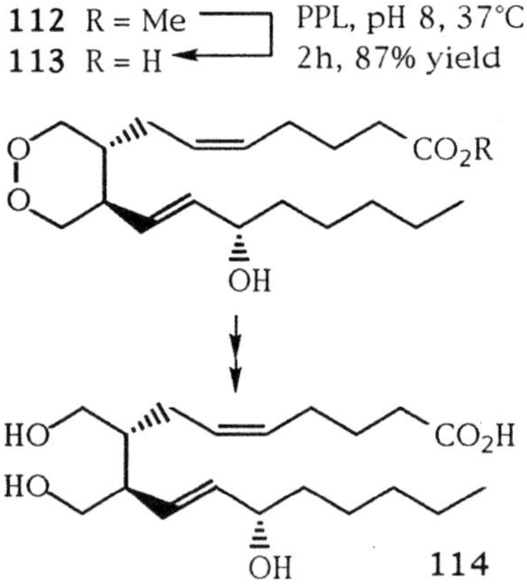

Fig. 37. Synthesis of prostaglandin endoperoxide analogs.

3.2 Enzyme Catalyzed Reduction Reactions

3.2.1 Yeast

In analogy to the yeast reduction of β-keto esters (Sect. 2.2.1.2), cyclic β-keto esters (**117**) reliably yield the (2*R*,3*S*) products (**118**) (BROOKS et al., 1984). Reductions involving cyclic substrates are often more diastereo- and enantioselective than the corresponding acyclic compound. The hydroxy esters (**118**) were converted to unsaturated ketals which underwent diastereoselective addition and alkylation to give (2*S*)-2-hydroxycyclohexane 6-alkyl esters (**119**) and (**120**) with three contiguous stereogenic centers for use in natural product synthesis (Fig. 39) (SEEBACH and HERRADON, 1987). The hydroxy ester (**118**) has been used in an approach to the southern region (**121**) of avermectin A_{26} (**122**) (Fig. 40) (BROOKS et al., 1982). Avermectins and milbemycins are potent parasitic and anthelmintic agents (BARRETT and CAPPS, 1986).

BROOKS et al. (1982, 1983, 1984) investigated the yeast reduction of a wide range of simple cyclic β-diketones. Reduction of racemic cyclopentadiones, e.g., (**123**) gave predominantly (2*S*,3*S*)-3-hydroxycyclopentanones, e.g., (**124**), whereas cyclohexadiones, e.g., (**127**) gave predominantly (2*R*,3*S*)-3-hydroxycyclohexanones, e.g., (**128**). The other enantiomer of the starting material, e.g., (**123**) and (**124**) was either recovered or reduced to the other 3-hydroxycycloalkanone diastereoisomer pair. With larg-

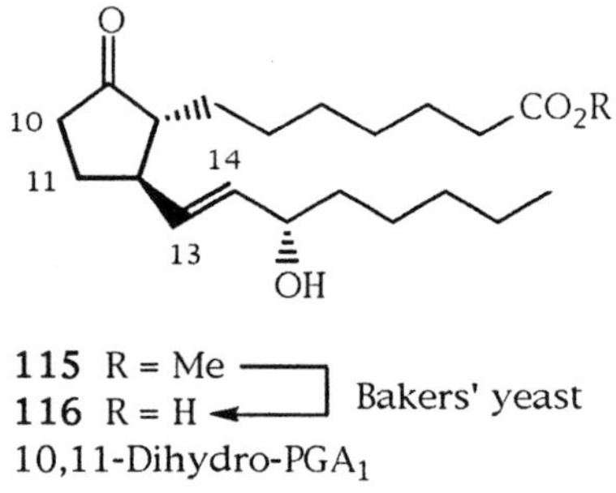

Fig. 38. Ester hydrolysis using yeast.

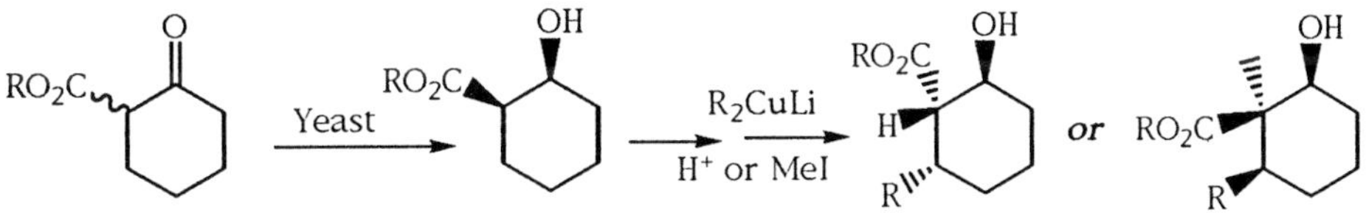

Fig. 39. Diastereoselective elaboration of β-hydroxyesters.

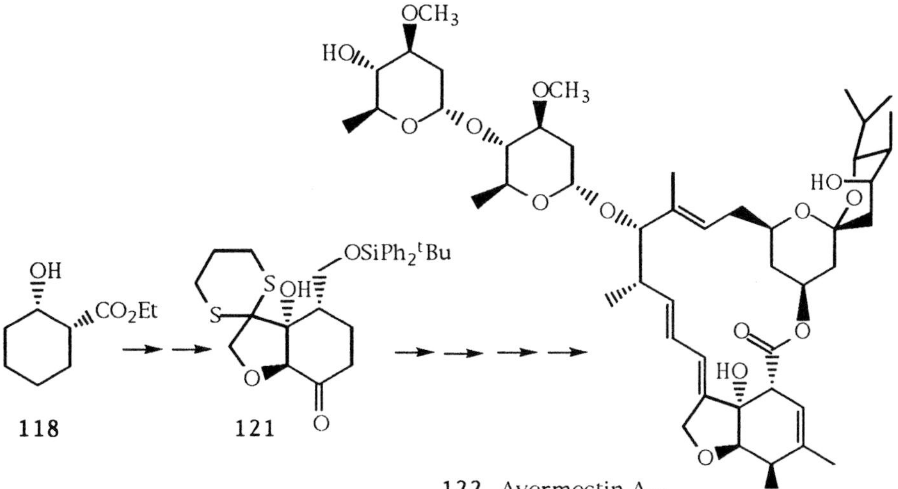

Fig. 40. The use of yeast in an approach to avermectin.

er membered rings (7–9) the stereochemical outcome was less predictable, nevertheless the enantiomeric excesses were still high, albeit the yields were lower. The alcohol (**124**) was used to prepare an intermediate in the synthesis of anguidine (**126**) (BROOKS et al., 1982, 1983). The synthesis of zoapatanol (**129**) requires a (2*S*,3*R*)-3-hydroxycyclohexanone which was prepared from both the major (2*R*,3*S*)-(**128**) and the minor (2*S*,3*S*)-diastereoisomers (Fig. 41) (BROOKS et al., 1984).

Yeast has also be used in large-scale reactions, as illustrated in the reduction of the oxoisophorone (**130**) (Fig. 42) to the carotenoid precursor (**131**), which has been performed on 13 kg of the starting enedione (LEUENBERGER et al., 1979).

3.2.2 *Rhizopus arrhizus*

Rhizopus arrhizus induced reduction of the trione (**132**) (Fig. 43) is regiospecific and enantiospecific with hydride addition being directed to the *Re*-face of the pro-*R*-carbonyl group to give the (2*R*,3*S*)-stereoisomer (**133**) (VELLUZ, 1965). The triones (**132**) can be prepared by treatment of 2-methylcyclopenta-1,3-dione with base and an α,β-unsaturated ketone. In the presence of further base they cyclize by an intramolecular aldol reaction to give bicyclic dione intermediates for steroid synthesis (Robinson annulation). The need for reduction is questionable because treatment of the triones (**132**) with L- or D-proline effects an asymmetric aldol reaction to give the bicyclic

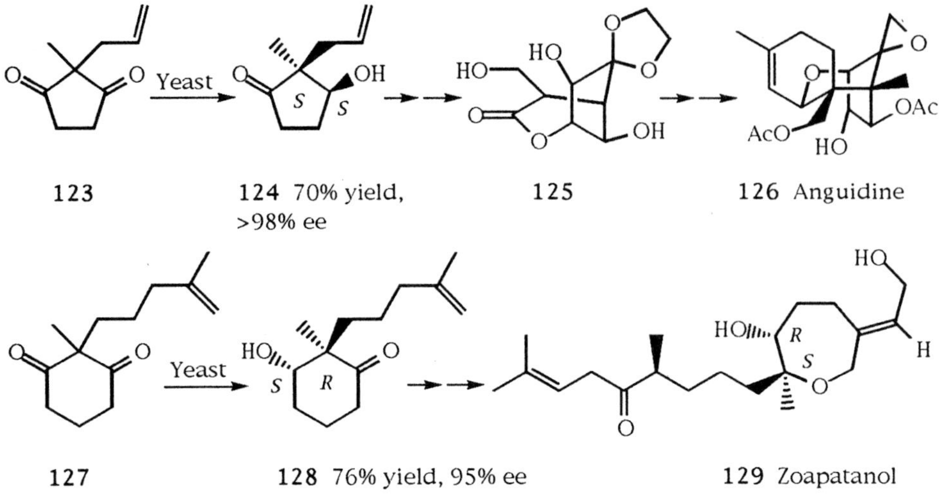

Yeast

123

124 70% yield, >98% ee

125

126 Anguidine

Yeast

127

128 76% yield, 95% ee

129 Zoapatanol

Fig. 41. Routes to anguidine and zoapatanol.

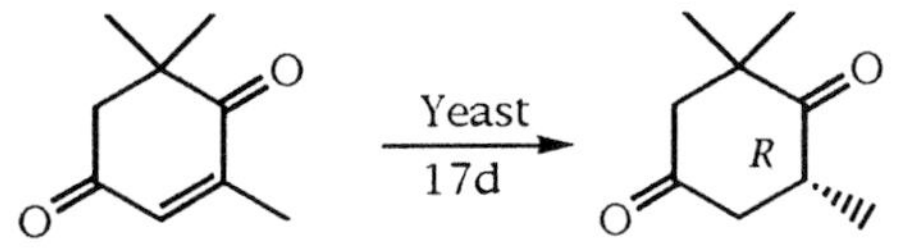

Yeast 17d

130 Oxoisophorone

131 83% yield, 100% ee

Fig. 42. Synthesis of a carotenoid precursor.

ketones as single enantiomers (Hajos-Parrish reaction).

3.3 Oxidation Reactions

3.3.1 Horse Liver Alcohol Dehydrogenase

Unless protecting groups are used, discrimination between unhindered primary and secondary alcohol functions in diols such as (**134**) and (**136**) (Fig. 44) is difficult to achieve in nonenzymic single step reactions. However, with HLADH only those hydroxyl groups that can locate at the oxidoreduction site will be oxidized, so the primary and secondary functions can be discriminated on a regional basis.

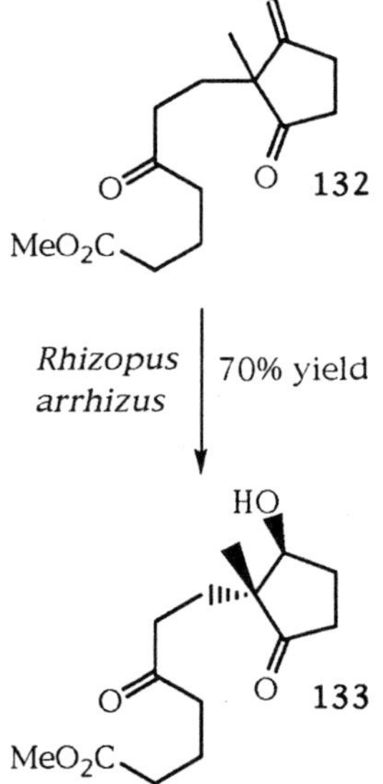

Fig. 43. Steroid intermediate from microbial reduction.

The initially formed hydroxyaldehyde (**137**) undergoes further oxidation via the hemiacetal (**138**) in another enantioselective process to yield lactone (+)-(**139**) of interest as a synthon for prostaglandin analogs. The other enantiomer of the bicyclic lactone (**139**)

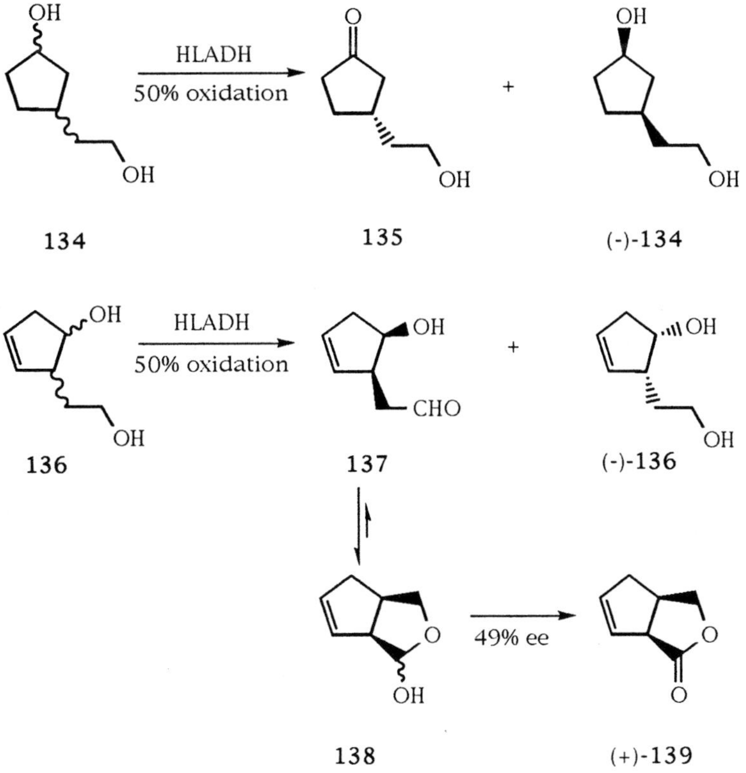

Fig. 44. Prostaglandin synthons using HLADH.

which is the precursor of natural prostaglandins, has been obtained by chemical oxidation of the enantiomer recovered from the HLADH-catalyzed oxidation of *rac*-(**136**) (IRWIN and JONES, 1977; NEWTON and ROBERTS, 1982).

3.3.2 Microbial Enzymes

The most useful enzymes for use in organic synthesis are those which accept a broad range of substrates, while being able to act stereospecifically on each. Microbial enzymes have narrower structural specificity tolerances than mammalian ones, however the larger selection of microorganisms available compensates for this. The identification of an enzyme capable of catalyzing a specific reaction on a particular substrate is performed as for any chemical transformation – by searching the literature for analogies. For example, the known conversion of cinerone (**140**) to cinerolone (**141**) (Fig. 45) (TABENKIN et al., 1969) was used to choose *Aspergillus niger* as a suitable microorganism for the stereospecific hydroxylation of the α,β-unsaturated ketone (**142**) to the prostaglandin synthon (**143**) (KUROZUMI et al., 1973).

Fragrance and flavor chemicals have an enormous world market. A considerable number of these substances are monoterpenoids, sesquiterpenoids, and similar structures. Very often the special properties of terpenoids and their derivatives depend on the absolute configurations of the molecules, so the synthesis or modification of these substances demand reactions with high stereospecificity and methods of introducing chiral centers. Biotransformations are useful in this respect (ROSAZZA, 1982; KIESLICH, 1976) with many applications having

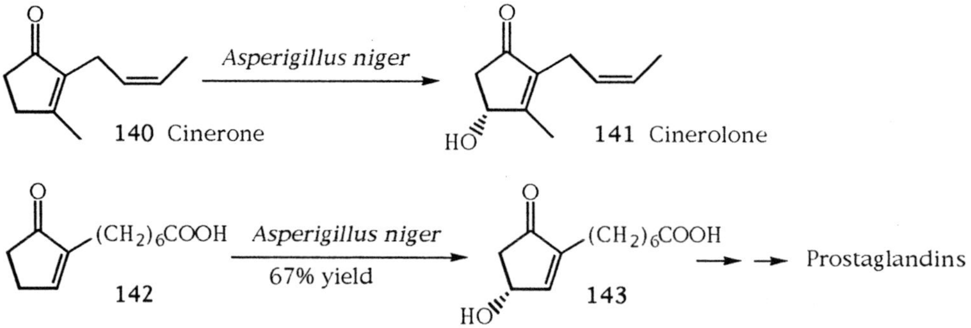

Fig. 45. Stereospecific hydroxylations.

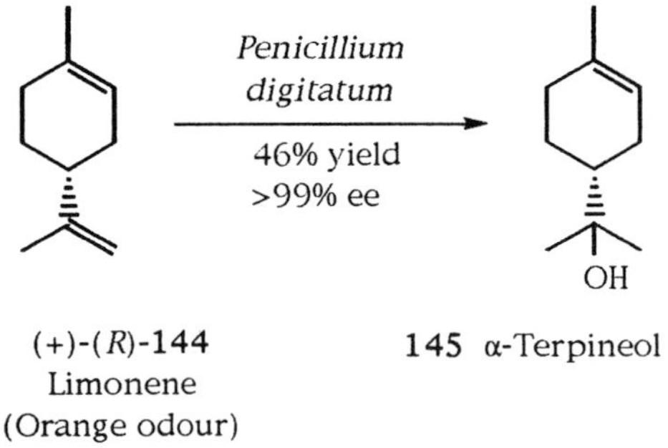

Fig. 46. Stereospecific modification of terpenoids.

been described (KRASNOBAJEW, 1984; SCHINDLER and SCHMID, 1982).

For example, limonene (**144**) (Fig. 46) is a readily available and inexpensive natural product. Specific hydration at the double bond of the isopropenyl substituent using *Penicillium digitatum* resulted in complete transformation of (+)-(*R*)-limonene (**144**) to α-terpineol (**145**) (KRAIDMAN et al., 1969). Even racemic limonene affords pure α-terpineol. With microorganisms other than *Penicillium digitatum*, limonene may undergo attack at the 1,2-double bond to form 1,2-dihydro-1,2-*trans*-diols. A number of fungi are useful in this biotransformation, with *Corynespora cassiicola* and *Diplodia gossydina* being the most appropriate strains. This provides an economical method to these glycols, useful starting materials in the synthesis of menthadienol, carvone, and other compounds.

3.3.2.1 *Pseudomonas putida*

GIBSON et al. (1970) reported the enantioselective oxidation of benzene (**146**) to *cis*-cyclohexadienediol (**147a**) (Fig. 47) and of toluene to *cis*-dihydrotoluenediol (**153**) (Fig. 49). These compounds are often referred to (incorrectly) as benzene and toluene *cis*-diol. Many arenes and aromatic heterocycles have been shown to yield diols through microbial techniques (SUGIYAMA and TRAGER, 1986).

It is important to note that these compounds cannot currently be made in quantity by conventional abiotic chemistry. Hence they offer unparalleled potential for use in natural product and other syntheses. The microbial conversion is facile and high yielding. It is then easy to make derivatives of benzene-*cis*-diol in a standard radical-type polymerization, such as the polymer (**148**) (Fig. 47), which on heating gives polyphenylene (**149**) by eliminating two molecules of H-OR (BALLARD et al., 1983). Polyphenylene itself is intractable and not easily fabricated, but this method allows production of polyphenylene films, fibers, coatings, and liquid crystals.

Benzene-*cis*-diol (**147a**) has also been used in the synthesis of both enantiomers of pinitol (**152**) (Fig. 48). Esterification of the two hydroxyl groups with benzoyl chloride, pyridine, and DMAP furnished the dibenzoate (**147c**) which underwent epoxidation *anti* to the ester groups. Treatment with acidified methanol effected ring opening of the epoxide (**150**) regiospecifically adjacent to the alkene bond to

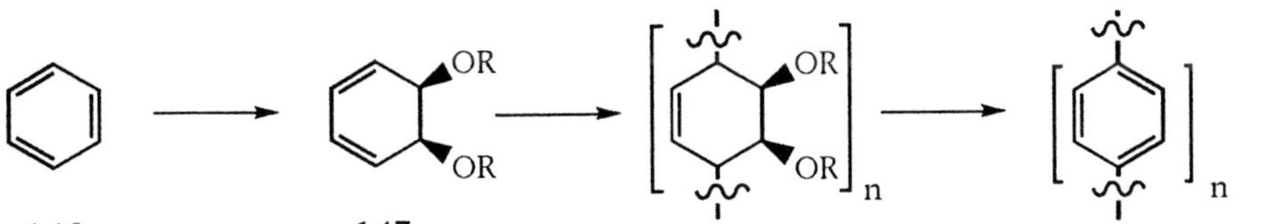

Fig. 47. The preparation of polyphenylene.

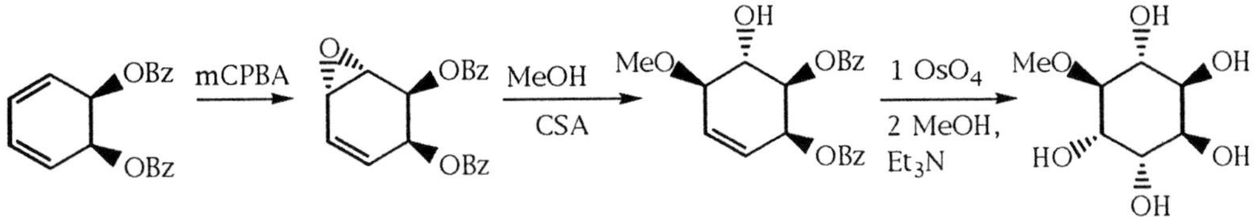

Fig. 48. Synthesis of (+)-pinitol.

give the methyl ether (**151**).The final two hydroxyl groups were then installed by osmylation (once more *anti* to the ester groups), followed by deprotection to give *rac*-pinitol *rac*-(**152**). The individual enantiomers were prepared by a parallel route involving resolution by esterification with an enantiomerically pure acid chloride (LEY and STERNFELD, 1989). Essentially identical methodology was used in the preparation of (+)- and (−)-pinitol from 1-bromobenzene-*cis*-diol (HUDLICKY et al., 1990).

The diol derived from toluene (**153**) (Fig. 49) was ketalized with 2,2-dimethoxypropane and pTsOH (85% yield) and then ozonolyzed at −60°C in ethyl acetate to give the keto aldehyde (**154**) (60–70% yield). Aldol ring closure catalyzed by alumina gave the α,β-unsaturated ketone (**155**), a known prostaglandin intermediate (HUDLICKY et al., 1988).

Hydrogenation of the toluene derived diol (**153**) was virtually nonselective, however, the diols (**156**) and (**157**) were easily separated. Treatment with periodate gave the dialdehydes (**158**) which underwent intramolecular aldol condensation to give the cyclopentenyl carboxaldehydes (**159**), which are potential synthons for bulnesene and kessane sesquiterpenes (HUDLICKY et al., 1988).

Benzene-*cis*-diol (**147a**) was also used in the synthesis of D-*myo*-inositol-1,4,5-triphosphate (**162**) (Fig. 50). The key reactions are diastereoselective epoxidation and regioselective ring opening of the epoxides, which resemble those used in the previous syntheses of pinitol (HUDLICKY et al., 1991).

Pseudomonas putida also catalyzes the oxidation of styrene, chlorobenzene and derivatives to the corresponding *cis*-diols, which were used for synthesis of natural products (HUDLICKY et al., 1989; HUDLICKY and NATCHUS, 1992; ROUDEN and HUDLICKY, 1993). The acetone ketals of D- and L-erythrose, and L-ribonolactone which are frequently used in natural product synthesis (HANESSIAN, 1983), were obtained from chlorobenzene (HUDLICKY and PRICE, 1990; HUDLICKY et al., 1989) in up to 50% yield.

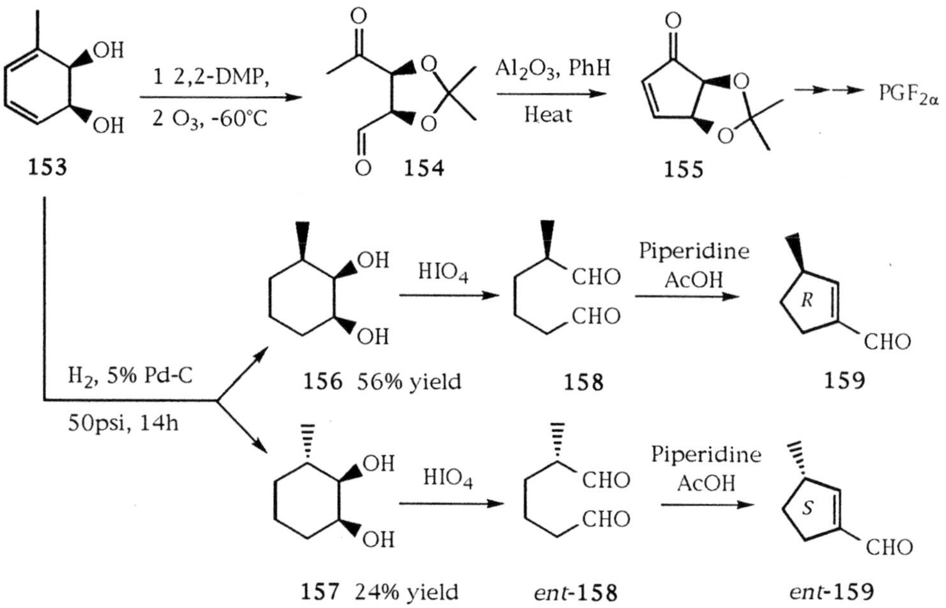

Fig. 49. Uses of toluene *cis*-glycol synthons.

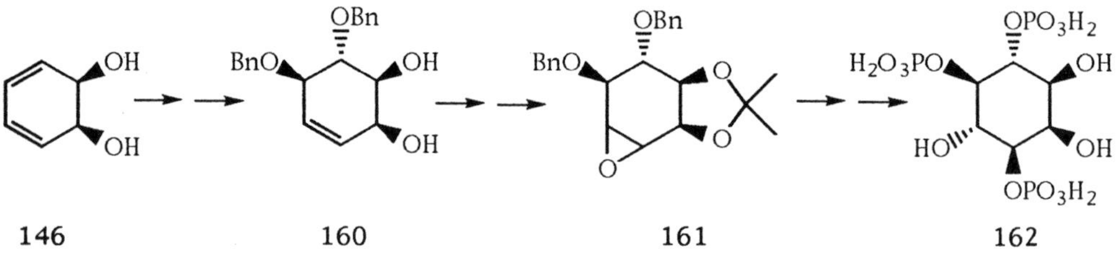

Fig. 50. Synthesis of *myo*-inositol phosphates.

4 Polycyclic Biotransformation Products

4.1 Hydrolysis/Condensation Reactions

4.1.1 Pig Liver Esterase

PLE catalyzed hydrolysis of the diester (**165**) (Fig. 51, X=O or CH_2) gives a quantitative yield of the monoester (**166**), with an enantiomeric excess of 77% which can be improved to >95% by recrystallization. Ozonolysis of the double bond yields 1,4-dialkylcyclopentanes or tetrahydrofurans which have been used in the synthesis of carbocyclic nucleotides, aristeromycin (**169**) and neplanocin A (**170**) (ARITA et al., 1983) or conventional nucleotides such as showdomycin, 6-azapseudourine, and cordycepin (ARITA et al., 1984).

4.1.2 α-Chymotrypsin

Acetate and dihydrocinnamate esters are hydrolyzed at almost identical rates using hy-

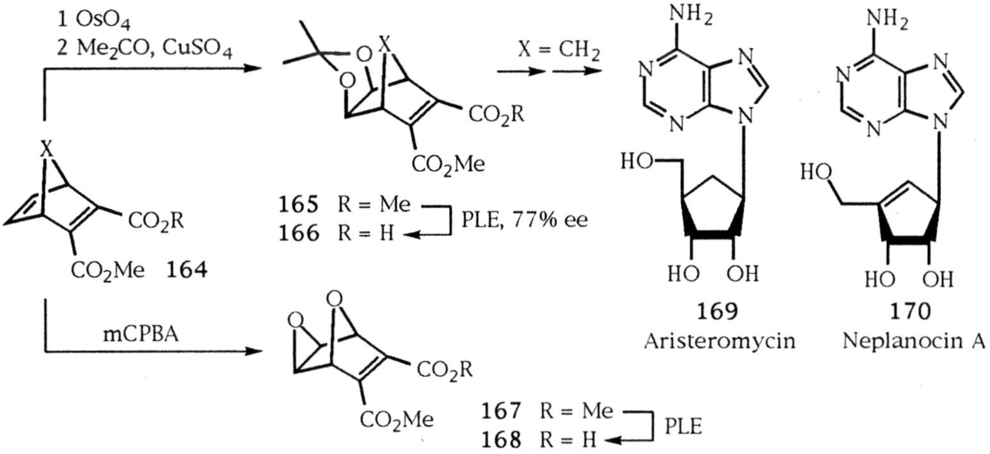

Fig. 51. Enantioselective hydrolysis and uses of tricyclic.

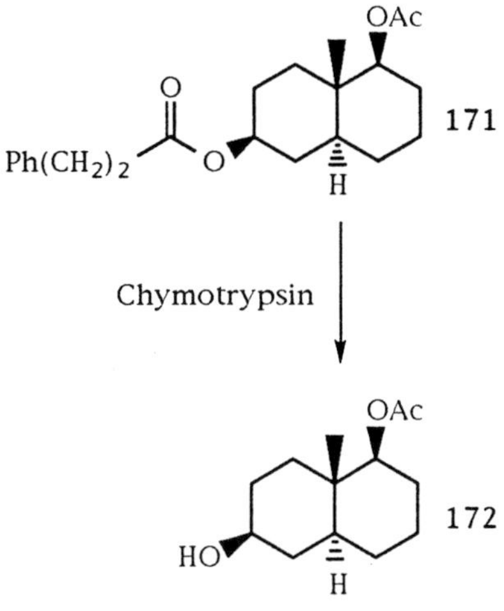

Fig. 52. Hydrolysis using α-chymotrypsin.

droxide ion (BENDER et al., 1964), however α-chymotrypsin regiospecifically cleaves dihydrocinnamate esters from simple mixed acetate-dihydrocinnamate diesters (Fig. 52) (JONES and LIN, 1972). For example, the diester (**171**) is hydrolyzed to give the 1β-acetate (**172**) in 53% yield with sodium hydroxide, with 35% of the diol also isolated. With α-chymotrypsin, the 1β-acetate is obtained in quantitative yield, although the rate of hydrolysis is slower. The "normal" substrate for chymotrypsin is the cleavage of the C-terminal amide group of aromatic amino acids in peptides. Hence in the hydrolysis of the diester (**171**) the dihydrocinnamyl group is acting as a surrogate for this group.

4.2 Enzyme Catalyzed Reduction Reactions

4.2.1 Yeast

The potent biological activity of prostaglandins has spawned enormous synthetic efforts (BINDRA and BINDRA, 1977; SCHIENMANN and ROBERTS, 1982; NEWTON and ROBERTS, 1982). The majority of the activity resides in the natural enantiomer and hence early syntheses of racemic prostaglandins initially sufficed, however, these were quickly supplanted by syntheses of enantiomerically pure materials. In one approach, the racemic bicyclic ketone (**173**) was reduced with yeast to give the diastereoisomers (**174**) and (**175**) derived from different enantiomeric series (Fig. 53). These were separated by column chromatography and then reoxidized to the ketone and converted to the bromohydrins (**176**) in one step. Each of these was then converted to natural prostaglandins, e.g., (**177**) by different routes (NEWTON and ROBERTS, 1980, 1982; ROBERTS, 1985).

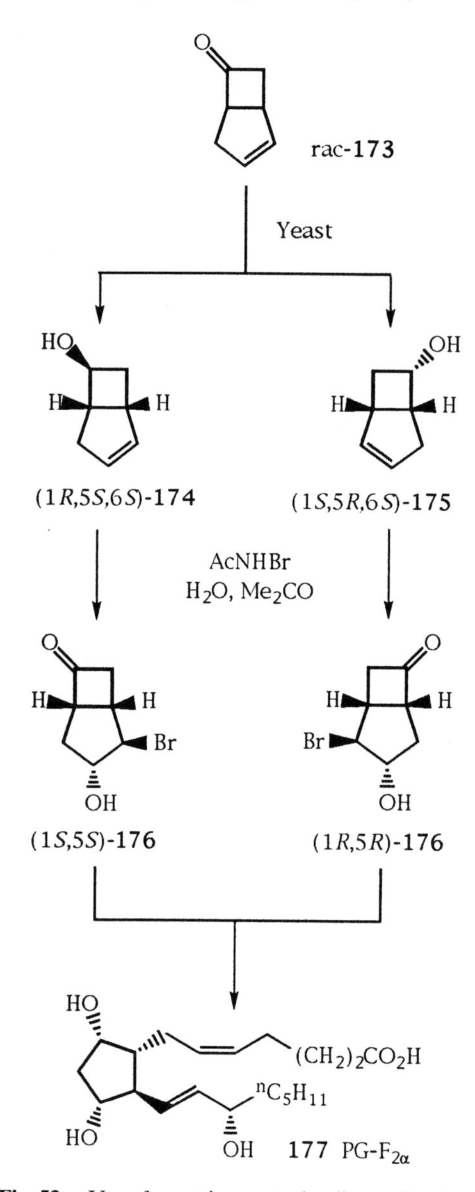

Fig. 53. Use of yeast in prostaglandin synthesis.

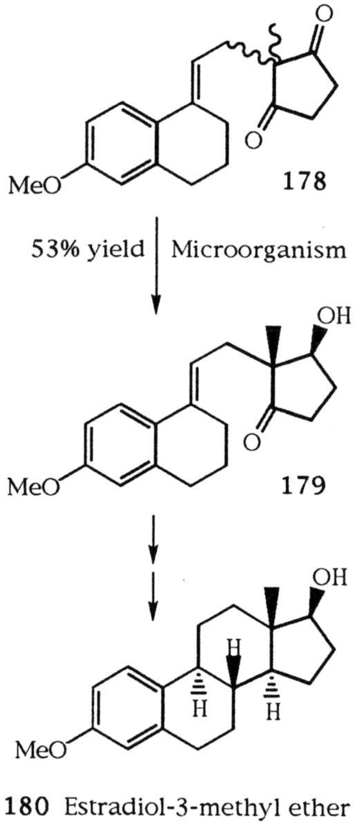

Fig. 54. Microorganism reductions in steroid synthesis.

4.2.2 Microorganism Reductions

As an extension to the reduction of cyclic β-keto esters (Sect. 3.2.1) compounds with a second, or third, ring are also reduced in the same manner with high selectivity (BROOKS et al., 1987). Early interest in the reduction of cyclic β-diketones centered around the conversion of diones such as (**178**) (Fig. 54) into chiral intermediates, such as the hydroxy ketone (**179**) (KOSMOL et al., 1967) using microorganisms such as *Bacillus* spp. and *Saccharomyces* spp. The intermediates could then be used in the total synthesis of steroids, such as estradiol-3-methylether (**180**) (DAI and ZHOU, 1985; REHM and REED, 1984).

Hydroxysteroid dehydrogenases (HSDH) catalyze the regiospecific reduction of ketones (and oxidation of hydroxy groups) of steroids. Enzymes selective for reduction or oxidation at positions 3, 7, 12, and 20 (Fig. 55) have been reported (BONARA et al., 1986). In most cases HSDHs have high regio- and diastereoselectivity which cannot be achieved with conventional abiotic reagents. For example 12-dehydrocholic acid (**181**) undergoes reduction at

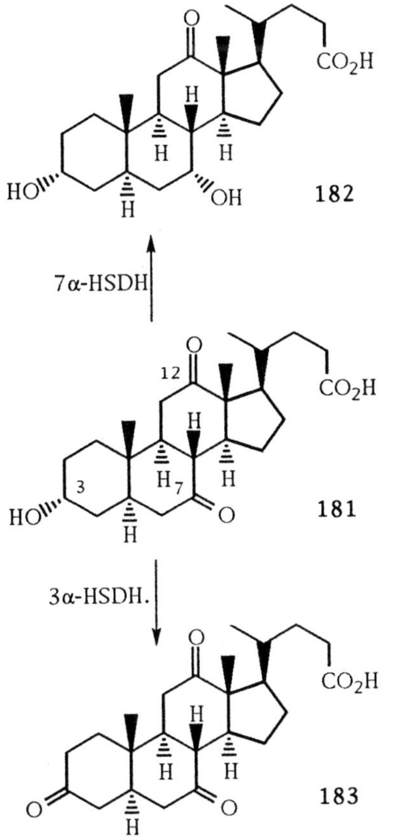

Fig. 55. Regiospecific ketone reduction.

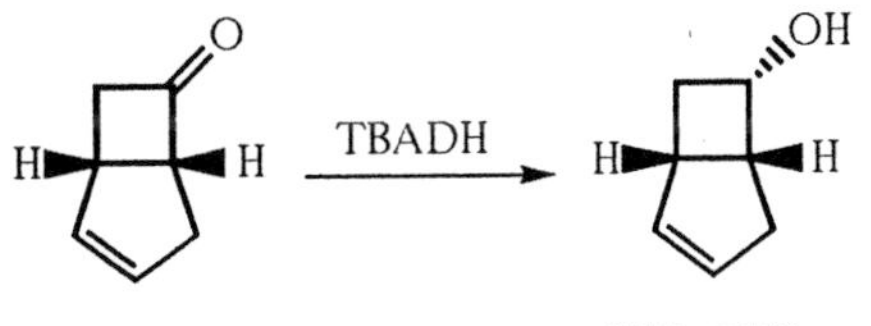

Fig. 56. Preparation of optically active synthons for prostaglandin synthesis.

the 7-position (**182**) and oxidation at the 3-position (**183**) in 97–99% yield without affecting the 12-keto group (CARREA et al., 1984). It should be noted that HSDH is a functional description usually derived from substrate screening. In many cases the biologically significant substrates for these enzymes are unknown and are unlikely to be steroids. However, the ability to reduce large hydrophobic substrates suggests that these enzymes will have broad applicability.

Thermoanaerobium brockii is a thermophilic organism which thrives best at high temperatures. Its enzymes have adapted to this regime and hence may be sufficiently robust for industrial applications. The alcohol dehydrogenase from *Thermoanaerobium brockii* (TBADH) reduces methyl ketones most efficiently and hence *iso*-propanol or acetone are frequently used as the donor or acceptor for coupled redox reactions. However bicyclic ketones, e.g., (**173**) (Fig. 56) are also efficiently reduced with *Re*-face delivery of hydride. In contrast to the corresponding yeast reduction (Fig. 53) the other enantiomer of the ketone (**173**) is only reduced. This is an advantage because it is easier to separate a ketone (**173**) from an alcohol (**175**) than to separate two alcohols (**174**) and (**175**) (ROBERTS, 1985).

Fermentative reduction of the racemic diketone (**184**) with *Aureobasidium pullulans* is stereoselective for the *Re*-faces of both carbonyl groups (**184**) (Fig. 57). The *trans*-1,4-diol product (**185**) was isolated in 33% yield accompanied by a *cis*-1,4-diol derived from *ent*-(**184**) and other diastereoisomers. The *trans*-1,4-diol product (**185**) has a C_2-axis of symmetry (like the diketone (**184**)) and hence additions to the alkene give a single stereoisomer. This was exploited in a synthesis of compactin (**40a**) (Fig. 16) (WANG et al., 1981).

4.2.3 Horse Liver Alcohol Dehydrogenase

HLADH reduces all simple cyclic ketones from four- to nine-membered, and parallels the rate of reduction by sodium borohydride in isopropanol (VAN OSSELAER et al., 1978). It excels over traditional chemical methods in combining regio- and enantiomeric specificities (JONES, 1985). HLADH catalyzes the reduction of many *cis*- and *trans*-decalindiones. These highly symmetrical compounds are potentially useful chiral synthons. The reductions are specific for pro-*R* ketones only (**186**), to

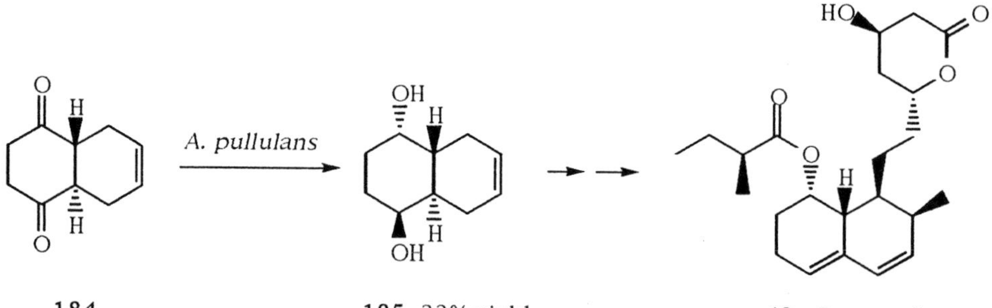

184 — *A. pullulans* → 185 33% yield → → 40a Compactin

Fig. 57. The first route to compactin.

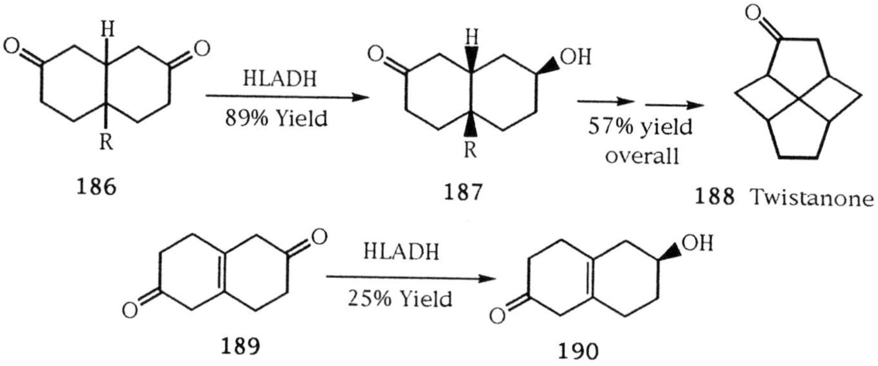

186 — HLADH, 89% Yield → 187 → → 57% yield overall → 188 Twistanone

189 — HLADH, 25% Yield → 190

Fig. 58. HLADH-catalyzed reductions.

give enantiomerically pure keto alcohols (**187**) (Fig. 58). Even with more symmetrical diones, for example, (**189**), which lacks a ring junction prochiral center, the reductions are stereospecific. The decalin (**187**, R=H) has been converted to (+)-4-twistanone (**188**) (DODDS and JONES, 1982, 1988).

These representative hydroxy ketones are not readily available using traditional chemical methods, and they are clearly broadly useful synthons for steroidal, terpenoid and related structures.

NAKAZAKI and NAEMURA (NAKAZAKI et al., 1983; NAEMURA et al., 1984, 1986) have shown that HLADH accepts a variety of cage shaped molecules as substrate, and reduces them with enantioselectivity. Such rigid hydrocarbons and derivatives are of special interest as test cases for chiroptic theories.

4.3 Oxidation Reactions

4.3.1 Microorganism Oxidation

Except in the area of steroid research, little work on the use of oxygenases to functionalize nonactivated carbon centers has been done. These provide useful synthons for the preparation of medicinally important compounds. Sixty-one cultures were screened for their capacity to hydroxylate cyclohexylcyclohexane (**191**) (Fig. 59) (DAVIES et al., 1986). Two species (*Cunninghamella blakesleeana* and *Geotrichum lacrispora*) gave the diequatorial diol (**192**) as the major product, for use in the synthesis of the carboxylic acid (**193**), an analog of the chemotactic agent, leukotriene B_3 (**194**).

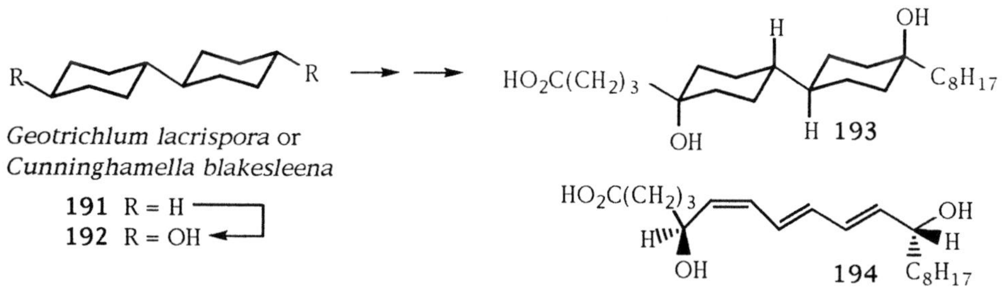
R
R
Geotrichlum lacrispora or
Cunninghamella blakesleena
191 R = H
192 R = OH
HO2C(CH2)3
H
OH
C8H17
OH
H 193
HO2C(CH2)3
OH
H
194 C8H17

Fig. 59. Preparation of leukotriene analogs.

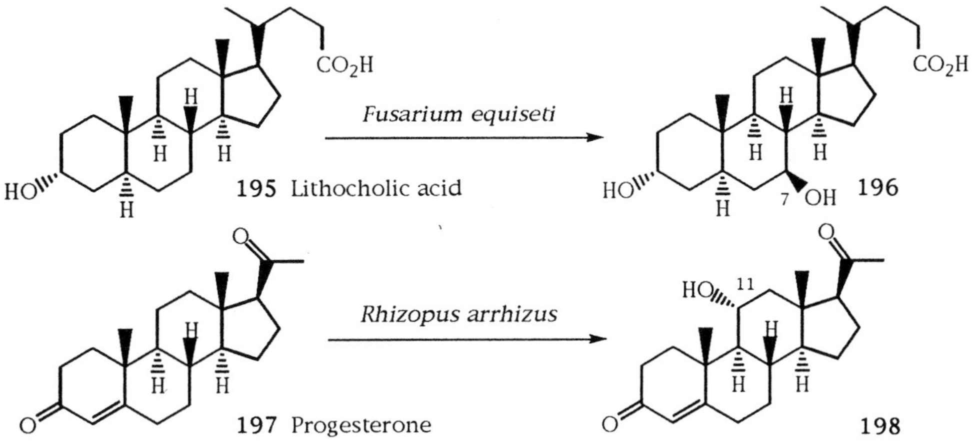
CO2H
H
H
H
HO
H
195 Lithocholic acid
Fusarium equiseti
CO2H
HO
7 OH
196
Rhizopus arrhizus
197 Progesterone
HO 11
198

Fig. 60. Transformations of progesterone and lithocholic acid.

4.3.2 Hydroxylation of Steroids

The hydroxylation of nonactivated centers in hydrocarbons is a very useful biotransformation, since it has very few counterparts in traditional organic synthesis. Intense research on the stereoselective hydroxylation of alkanes began in the late 1940s in the steroid field, when progesterone (**197**) was converted to 11α-hydroxyprogesterone (**198**) (Fig. 60), thus halving the 37 steps needed using conventional chemistry, and making (**198**) available for therapy at a reasonable cost. This regioselective oxidation became commercially important for the manufacture of cortisone, since 11β-hydroxy configuration is required for the optimum biological activity of hydrocortisone and the corticosteroids, but difficult to obtain using traditional chemistry. Nowadays, virtually any center in a steroid can be regioselectively hydroxylated by choosing the appropriate microorganism (DAVIES et al., 1989). The highly selective hydroxylation of lithocholic acid (**195**) in the 7β-position was achieved using *Fusarium equiseti* (SAWADA et al., 1982). The product ursodeoxycholic acid (**196**) (Fig. 60) is capable of dissolving cholesterol and can therefore be used in the therapy of gallstones. There is a great deal of research activity in this area due to the immense industrial importance, most notably by KIESLICH, JONES, HOLLAND, and CRABB.

4.4 Carbon–Carbon Bond Formation Reactions

4.4.1 Yeast Cyclases

Some enzymes involved in the biosynthesis of steroids have recently been used in organic synthesis. 2,3-Oxidosqualene-lanosterol cyclase, from bakers' yeast (*Saccharomyces cerevisiae*) catalyzes the synthesis of a number of lanosterol analogs from the corresponding 2,3-oxidosqualene derivatives (MEDINA and KYLER, 1988). In the cyclization of the vinyl derivative (**199**) (Fig. 61), the cyclization cascade is followed by hydrogen and methyl migrations as usual, but in addition the vinyl group also undergoes a 1,2-shift from C-8 to C-14 (**200**). The structure was confirmed by conversion to the alcohol (**201**) (MEDINA et al., 1989) which is a natural product and an inhibitor of HMG-CoA reductase (GRUNDY, 1988).

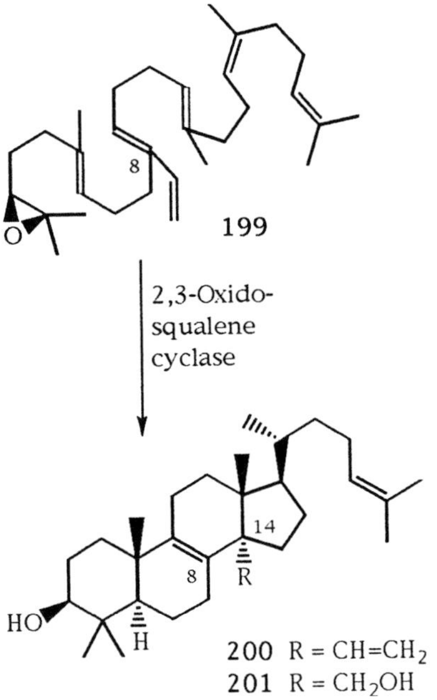

199

2,3-Oxido-squalene cyclase

200 R = $CH{=}CH_2$
201 R = CH_2OH

Fig. 61. Use of lanosterol cyclase.

4.4.2 Reactions Involving Acetyl CoA

Coenzyme A (CoA) thioesters are involved in the biosynthesis of steroids, terpenoids, macrolides, fatty acids, and other substrates (BILLHARDT et al., 1989; PATEL et al., 1986). However, these enzymes can only be used practically in organic synthesis if the CoA thioester is recycled, due to its high cost (BILLHARDT et al., 1989).

4.5 Exploiting Combinations of Enzyme Specificity

4.5.1 Combinations of Dehydrogenases

The degree of stereochemical control achievable with nonenzymic methods in asymmetric synthesis has made great improvements in recent years. Still, no chemical chiral reagent can yet approach the abilities of enzymes to combine several different specificities in a single step reaction.

Reduction of the racemic ketone (**202**) by yeast is *Re*-face selective, consequently both enantiomers are reduced to (*S*)-alcohols (Fig. 62). The desired alcohol (1′*S*,7*R*)-(**203**) was reoxidized to the ketone (7*R*)-(**202**) and utilized in the synthesis of 4-demethoxydaunorubicin (**205**), a member of the adriamycin family of anthracycline antibiotics. The unwanted (7*S*)-epimer was similarly reoxidized to the ketone (7*S*)-(**202**), racemized with pTsOH in acetic acid at 110 °C (TERASHIMA and TAMATO, 1982), and resubjected to yeast reduction.

4.6 Multiple Enzyme Reactions

Interest in multienzyme synthesis has intensified, as it has enormous potential, for the manufacture of complex pharmaceuticals. For example, the readily accessible Reichstein's compound S (**206**) undergoes 11β-hydroxylation with *Curvularia lunata*, but with other microorganisms dehydrogenation also occurs to give prednisolone (**207**) (Fig. 63) (MOSBACH et al., 1978).

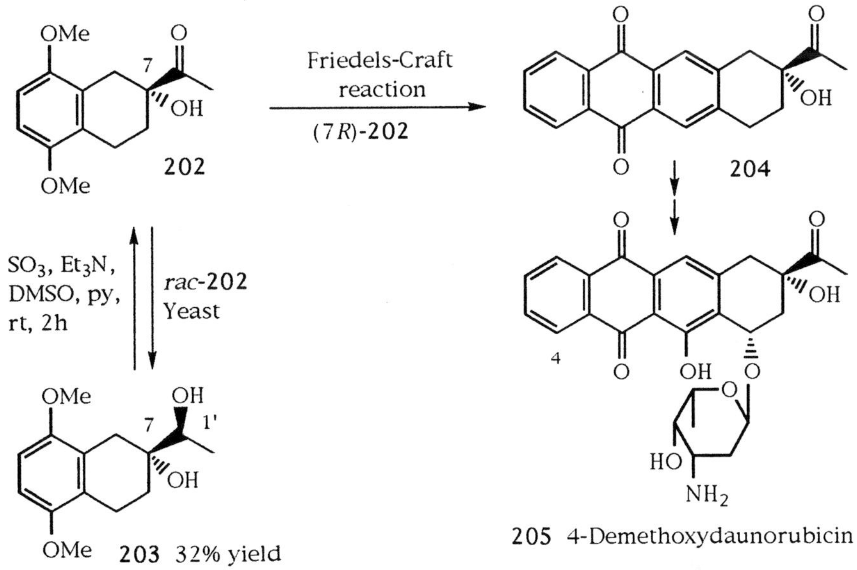

Friedels-Craft reaction

(7*R*)-202

202

204

SO_3, Et_3N, DMSO, py, rt, 2h

rac-202 Yeast

203 32% yield

205 4-Demethoxydaunorubicin

Fig. 62. Use of alcohol dehydrogenases.

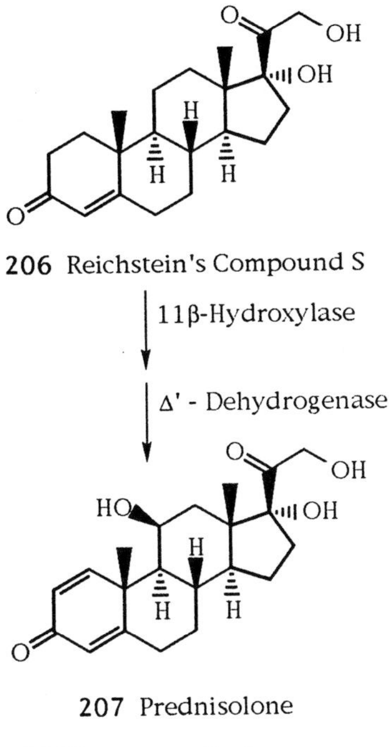

206 Reichstein's Compound S

11β-Hydroxylase

Δ' - Dehydrogenase

207 Prednisolone

Fig. 63. Multienzyme synthesis.

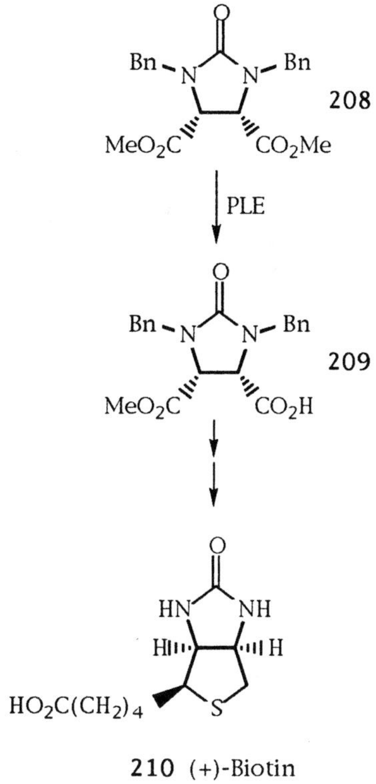

208

PLE

209

210 (+)-Biotin

Fig. 64. Enantioselective hydrolysis using pig liver esterase.

5 Heterocyclic Biotransformation Products

5.1 Hydrolysis/Condensation Reactions

5.1.1 Pig Liver Esterase

The general hydrolysis of *meso*-diesters has already been discussed in Sect. 3.1.1. The example shown in Fig. 64 shows the enantioselective hydrolysis of a *meso*-diester (**208**) which affords an intermediate for the synthesis of (+)-biotin (**210**) (IRIUCHIJIMA et al., 1978). Biotin functions as a cocarboxylase in a number of biochemical reactions. Initial experiments showed that the pro-(*S*) group of the bis-ester (**208**) was selectively cleaved, although the half ester (**209**) was only obtained in 38% ee. In contrast, the bis-propyl ester affords the corresponding monoester in 85% yield, and with 75% ee. The diacetate of the diol corresponding to (**208**) has been hydrolyzed by PPL to the mono-alcohol, by preferentially cleaving the

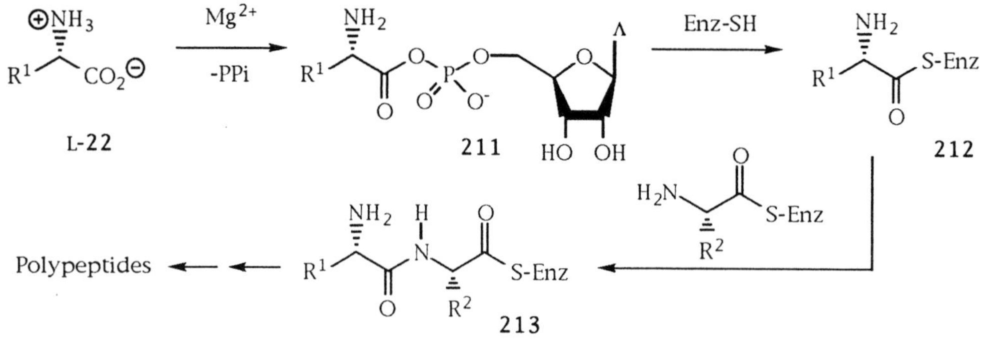

Fig. 65. Peptide synthesis.

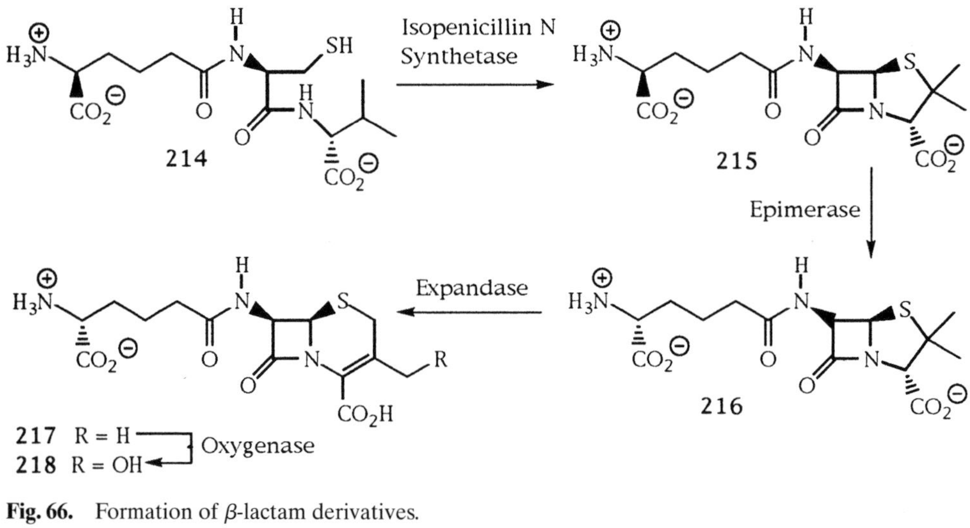

Fig. 66. Formation of β-lactam derivatives.

pro-(R) acetoxy group, in 70% yield and 92% ee. This provides an alternative route to (+)-biotin (WANG and SIH, 1984).

5.1.2 δ-(L-α-Aminoadipyl)-L-cysteinyl-D-valine (ACV) Synthetase

In the biosynthesis of non-ribosomal peptides (e.g., peptide antibiotics) the carboxyl group of amino acids is activated by esterification with ATP with displacement of pyrophosphate to give aminoacyl adenylate (**211**). This is transferred to an enzyme bound thioester, which then reacts with the amino group of another aminoacyl thioester to form a peptide bond (Fig. 65) (LIPMANN, 1973).

The peptide chain is thus extended from N- to C-terminus on a multienzyme template. An example of this is seen in the biosynthesis of δ-(L-α-aminoadipyl)-L-cysteinyl-D-valine (ACV) (**214**) catalyzed by ACV synthetase (from *Streptomyces clavuligerus*) (Fig. 66) (BANKO et al., 1987). The enzyme accepts many non-protein amino acids as well as hydroxy acids. Since ACV is a precursor of penicillins and cephalosporin, the enzymatically formed ACV analogs may be converted to the corresponding β-lactam derivatives. ACV (**214**) is cyclized to isopenicillin N (**215**) in a remarkable reaction which occurs in a single step with retention of configuration at both carbon centers participating in the cyclization. Epimerization gives penicillin N (**216**), which undergoes ring expansion to give desacetoxycephalosporin C (**217**) and allylic hydroxylation to desacetylcephalosporin C (**218**) (BANKO, 1987).

5.1.3 Lipases

5.1.3.1 *Pseudomonas* Lipases

The lipases isolated from different *Pseudomonas* species (PSL) are highly selective, especially for the hydrolysis of the esters of secondary alcohols, and the corresponding reverse reactions (BOLAND et al., 1991). The low cost and high selectivity and stability of PSL make it a very useful reagent for organic synthesis (Fig. 67) (KAN et al., 1985a, b).

All commercially available *Pseudomonas* sp. lipases possess a stereochemical preference for the (R)-configuration at the reaction center of secondary alcohols. *Pseudomonas* lipase (P-3C) was found to be highly selective ($E > 100$) for the hydrolysis of 3-hydroxy-4-phenyl β-lactam derivatives, in an approach to the synthesis of the C-13 side chain derivative of taxol (BRIENA et al., 1993). The selectivity remained high for the hydrolysis of the 3-acetoxy derivatives, with the ring nitrogen free or protected.

5.1.3.2 *Mucor* Species Lipases

The lipases from *Mucor* species such as *M. miehei* and *M. javanicus* have been used in synthesis (CHAN et al., 1988). Both enzymes seem to possess a stereochemical preference similar to that of *Pseudomonas* lipase.

M. miehei lipase was used in the resolution of methyl-*trans*-β-phenylglycidate (**223**) via transesterification in hexane:*iso*-butyl alcohol (1:1) (Fig. 68) (YEE et al., 1992). The unreacted substrate (2R,3S)-(**223**) and product (2S,3R)-(**224**) were obtained in 95% ee. Both

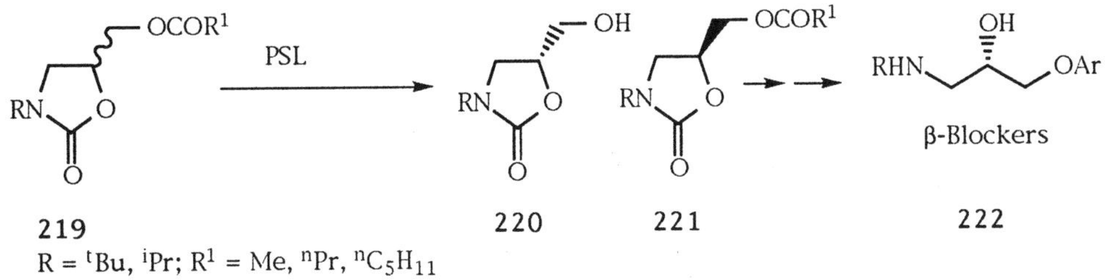

OCOR^1 PSL OH OCOR^1 OH
RN O RN O RN O RHN OAr
O O O β-Blockers
219 220 221 222
R = ^{t}Bu, ^{i}Pr; R^1 = Me, ^{n}Pr, ^{n}C$_5$H$_{11}$

Fig. 67. Synthesis of β-blockers.

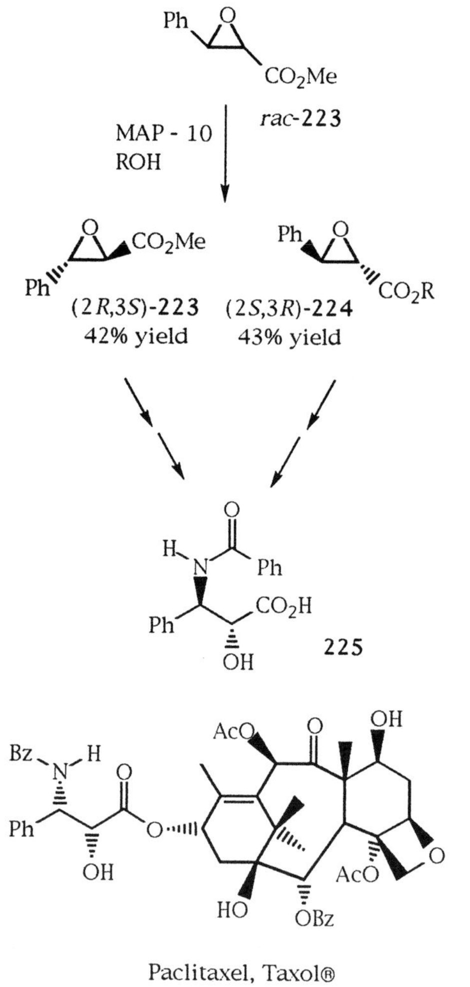

Ph CO2Me rac-223 MAP - 10 ROH (2R,3S)-223 42% yield (2S,3R)-224 43% yield CO2R 225 Paclitaxel, Taxol®

Fig. 68. Use of lipases in taxol synthesis.

enantiomers were converted to *N*-benzoyl-(2*R*,3*S*)-3-phenyl isoserine (**225**), the C-13 side chain of the antitumor agent taxol. The hindered *iso*-butyl alcohol was used to avoid the reverse transesterification.

A kinetic resolution for the preparation of the anti-inflammatory agent ketorolac (**227**) was achieved with *M. miehei* lipase catalyzed hydrolysis of the racemic methyl ester (Fig. 69) (FULLING and SIH, 1987). Since the ester is easily racemized at pH. 9.7, both enantiomers are eventually converted to the (*R*)-product. Enantiocomplementary to the process is the use of protease N and some other microbial proteases to produce the (*S*)-enantiomer (**227**), which is reported to be more potent in animal studies.

5.1.4 Penicillin Acylase

It has been shown that penicillin catalyzed hydrolysis of the phenylacetamido group of penicillin G (**228**) (Fig. 70), does not affect the sensitive β-lactam ring (BRIENA et al., 1993). This is now an important part of the procedure for the industrial production of 6-aminopenicillanic acid (**229**) (LAGERLÖF et al., 1976).

5.1.4.1 Transacylations Using Penicillin Acylase

Transacylation reactions can be induced, as represented in the penicillin and cephalosporin fields (Fig. 71). This methodology is advantageous in that no protecting groups are required. The protease from *Xanthomonas citri* condenses 6-aminopenicillanic acid (**229**) and D-phenylglycine methyl ester (**230a**) or its *p*-hydroxy derivative (**230b**) to give high yields of ampicillin (**231a**) and amoxicillin (**231b**), respectively (Fig. 71) (KATO et al., 1980). Similarly 6-aminocephalosporanic acid (**232**) is converted into cephalexin (**233a**) (CHOI et al., 1981).

Trypsin mediated exchange of threonine (as its ester derivatives) for the terminal alanine residue of porcine insulin is the basis of a commercial process for the production of insulin for human use.

5.2 Oxidation Reactions

5.2.1 Horse Liver Alcohol Dehydrogenase

As has been noted previously (Sects. 2.3.1 and 3.3.1) a broad range of *meso*-diols can be enantiospecifically oxidized using HLADH to lactones via the corresponding hemiacetals

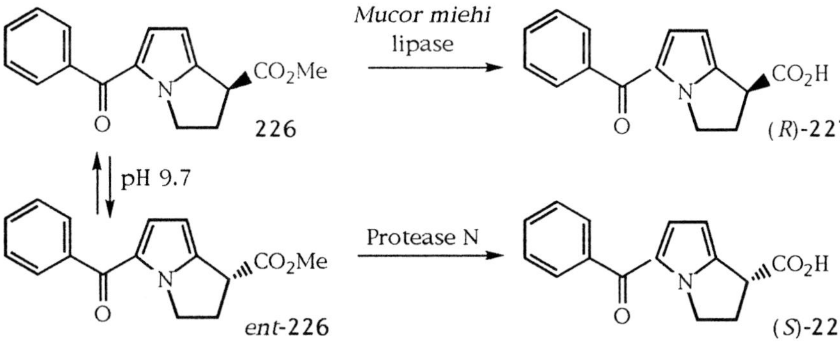

Fig. 69. Preparation of ketorolac.

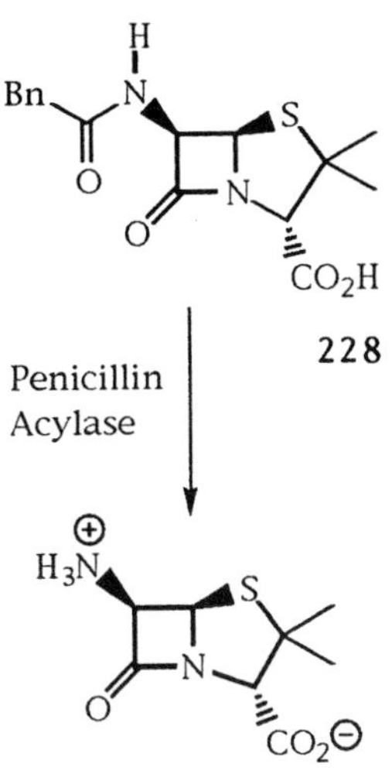

Fig. 70. Hydrolysis of penicillin G.

(Fig. 72) (JAKOVIC et al., 1982). Formation of the lactol or lactone boosts the enantiomeric excess by protecting the less reactive hydroxyl group from oxidation. In support of this assertion, oxidation of the corresponding *trans*-compounds gives essentially racemic products. The lactone (**235**) was produced on a large scale (JAKOVIC et al., 1982) in high yield and excellent enantiomeric excess. It provided both chiral centers of the iridoid aglycone methyl sweroside (**236**) (HUTCHINSON and IKEDA, 1984). Equally good results were obtained in the oxidation of the cyclobutane diol (**237**) which was used in a synthesis of grandisol (**239**) a major component of male boll weevil sex pheromone (JONES et al., 1982). Similarly the cyclopropyl lactone (**241**) (Fig. 72) was readily transformed into (+)-(1*R*,2*S*)-*cis*-methylchrysanthanenate (**242**) providing an attractive route to the pyrethroids (JONES, 1985). Further examples of such biotransformations in synthesis can be found in macrolide (COLLUM et al., 1980b) and prostaglandin synthesis.

5.2.2 Dihydrofolate Reductase

Dihydrofolate reductase catalyzes the *in vivo* interconversion of folate and tetrahydrofolate by NADPH. The reduction of dihydrofolic acid to chiral tetrahydrofolic acid was investigated using enzymic and nonenzymic means (REES et al., 1986). The results of the enzymic route far exceeded the chemical methods, providing pure stable tetrahydrofolate derivatives whereas traditional chemical routes gave minimal enantiomeric excesses. The technique was applied to the synthesis of enantiopure 5-formyltetrahydrofolate for use in cancer "rescue" therapy for patients undergoing cancer chemotherapy with methotetrexate.

5.2.3 Yeast

Yeast biotransformation of the furanyl acrolein (**243**) provided two intermediates

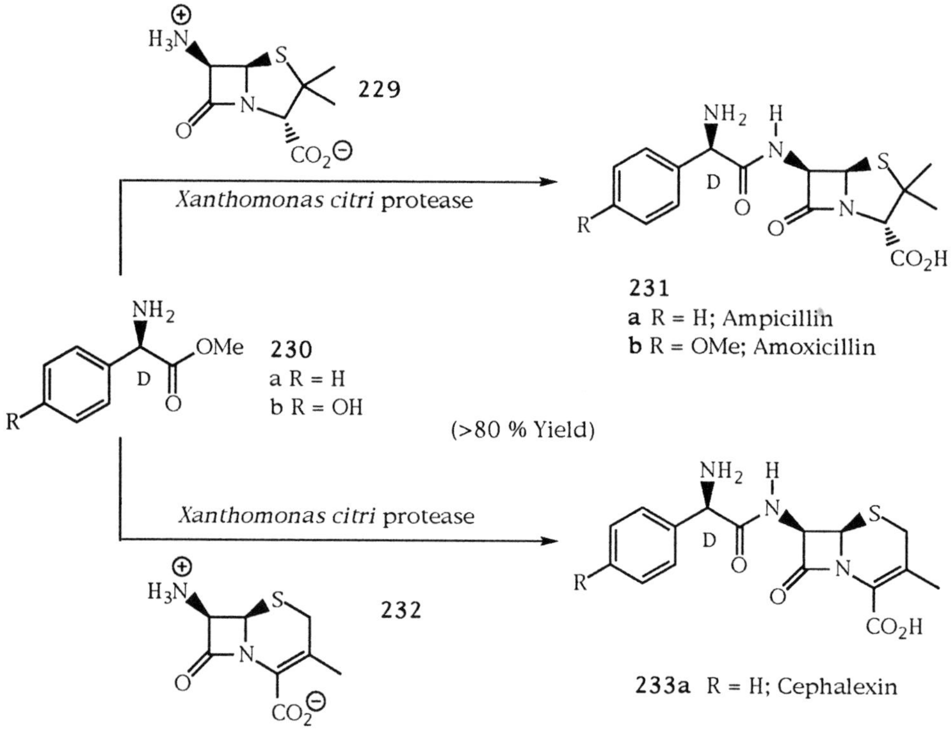

Fig. 71. Transacylation reactions.

(**244**) and (**246**) for the synthesis of α-tocopherol (**66**) (Fig. 73). Fermentation supplemented with pyruvate provided the diol (**244**) by acyloin condensation, whereas standard reductive fermentation effected reduction of both the alkene bond and the aldehyde group to give the alcohol (**246**). Cleavage of the furan ring provided the functionality to conjoin the fragments (FUGANTI and GRASSELLI, 1985a).

5.3 Carbon–Carbon Bond Formation Reactions

5.3.1 Rabbit Muscle Aldolase

Rabbit muscle aldolase (RAMA) has been used in the synthesis of numerous oxygen heterocycles (BEDNARSKI et al., 1989). The enzyme accepts a wide range of aldehydes, but is virtually specific for dihydroxyacetone phosphate (**248**) as the nucleophilic component (Fig. 74). The stereochemistry at the carbons engaged in the newly formed bond is always *syn*-3,4-(3*S*) which is equivalent to D-*threo*. RAMA catalyzed aldol reaction of *N*-acetylaspartate β-semialdehyde (**249**) gave the diol (**250**) with a maximum of 40% conversion (37% yield). Reduction of the 2-keto group with sodium or tetramethylammonium triacetoxyborohydride gave predominantly the desired 2,3-*anti*-diol. The *N*-acetyl group was cleaved with 6 N hydrochloric acid, to give the amine which was converted to a ketone (**251**) (13% overall yield) by transamination with sodium glyoxylate. Surprisingly enzyme catalyzed transamination with a range of enzymes was less effective than the abiotic reaction (TURNER and WHITESIDES, 1989).

Products from RAMA catalyzed aldol reactions have been used in efficient approaches to C-glycosides (**254**) and cyclitols (**255**) (Fig. 75) (SCHMID and WHITESIDES, 1990) and the bee-

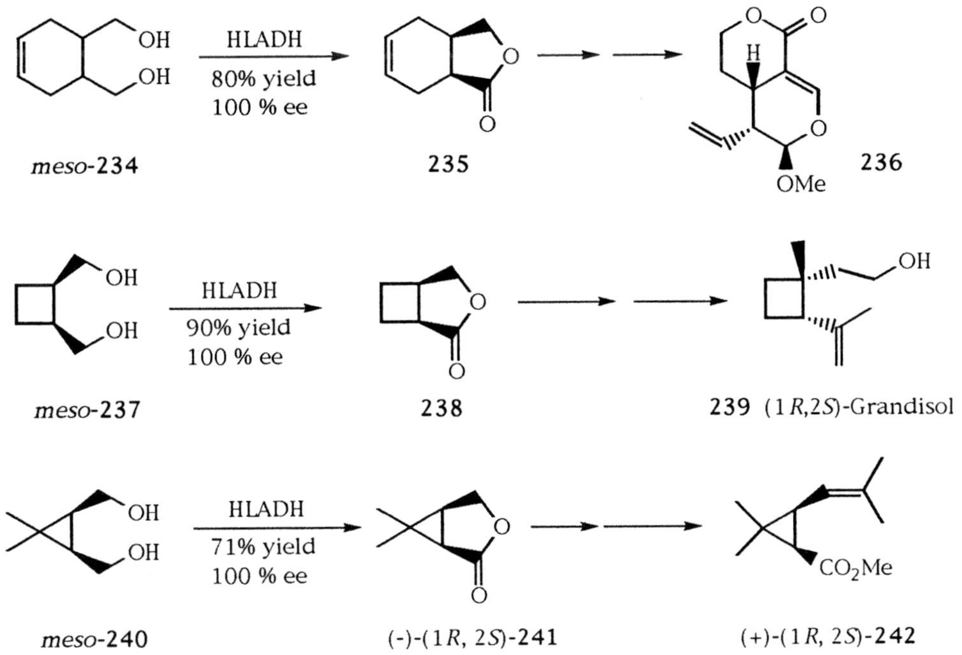

HLADH 80% yield 100 % ee

meso-234 235 236

HLADH 90% yield 100 % ee

meso-237 238 239 (1*R*,2*S*)-Grandisol

HLADH 71% yield 100 % ee

meso-240 (-)-(1*R*, 2*S*)-241 (+)-(1*R*, 2*S*)-242

Fig. 72. Use of lactone biotransformation products.

tle pheromone (+)-*exo*-brevicomin (**259**) (SCHULTZ et al., 1990).

5.4 Nucleotide Chemistry

5.4.1 Phosphorylation

Many structurally specific phosphate hydrolyzing enzymes are known, and have been widely used in organic synthesis. Of particular synthetic importance is their use in selective phosphate bond formation without the need for protecting groups, for example, in the selective mono- and pyrophosphorylations of monosaccharide moieties (SABINA et al., 1984; LADNER and WHITESIDES, 1985).

5.4.2 Gene Synthesis

Enzyme mediated phosphorylation is of great value in the nucleic acid field. In polynucleotide synthesis enzyme catalyzed phosphate bond formations provide solutions to the problems of controlled couplings of oligonucleotide intermediates. Gene synthesis relies heavily on this technique, as seen in the structural genes for yeast analine tRNA and *E. coli* tyrosine suppressor tRNA (KHORANA, 1976).

5.4.3 Base Exchange

Enzyme catalyzed base exchange has proved to be of value in other areas of nucleotide chemistry, such as in the preparation of possible antiviral agents (MORISAWA et al., 1980). The antiviral or antitumor activity of 9-β-D-arabinofuranosylpurines has generated interest in such nucleotides, and although they are obtainable by a sugar–base coupling reaction there remain definite practical limitations, such as low yields through a laborious and time consuming process.

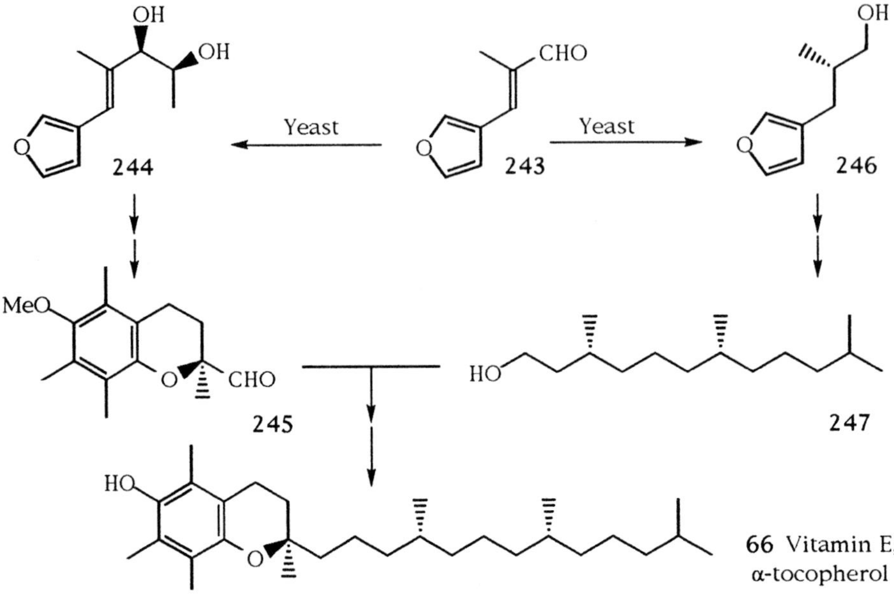

Fig. 73. Synthesis of natural α-tocopherol.

6 Aromatic Biotransformation Products

6.1 Hydrolysis/Condensation Reactions

6.1.1 α-Chymotrypsin

α-Chymotrypsin is a versatile enzyme capable of showing enantiomeric specificity on a broad range of racemic ester substrates (Fig. 76). The use of chymotrypsin as an enzymic catalyst is advantageous in that its stereospecificity is predictable using a simple model. The reactive ester stereoisomers are those whose chiralities parallel those of the natural L-amino acids, and the unreactive enantiomers can all be recycled via chemical racemization.

6.2 Oxidation Reactions

6.2.1 Hydroxylation

Selective hydroxylation of aromatic compounds is a difficult task in preparative organic chemistry, particularly when the compounds to be hydroxylated (or their products) are optically active. In such cases the reaction should be conducted rapidly and mildly to prevent racemization and decomposition. KLIBANOV showed that under certain conditions horseradish peroxidase (HRPO) can be used for fast, convenient, and selective hydroxylations in yields of up to 75%. L-DOPA (**263**) was produced from L-tyrosine (**262**), and L-epinephrine (adrenaline) (**265**) from L-(−)-phenylephrine (**264**) using HRPO catalyzed hydroxylation (Fig. 77) (KLIBANOV et al., 1981).

The preparation of 2-arylpropionic acids by hydroxylation and oxidation of cumenes with *Cordyceps* sp. (SUGIYAMA and TRAGER, 1986) is possible, where the cumene is hydroxylated not on the benzylic carbon as expected, but se-

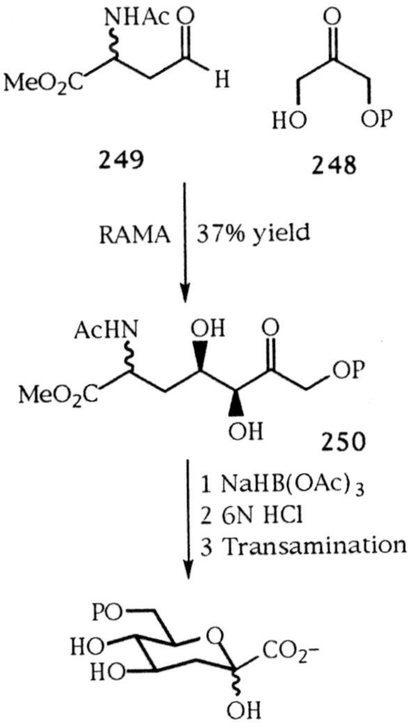

NHAc O MeO2C H 249 · HO OP 248 · RAMA 37% yield · AcHN OH O MeO2C OP OH 250 · 1 NaHB(OAc)3 2 6N HCl 3 Transamination · PO HO HO O CO2- OH

251 3-Deoxy-D-*arabino*-heptulosonic acid

Fig. 74. Synthesis of 3-deoxy-D-arabino-heptulosonic acid using RAMA.

lectively at the pro-(R) methyl group. For example, the naphthalene derivative (**266**) is transformed by *Cunninghamella militaris* into (S)-naproxen (**267**) in 98% ee (Fig. 78) (PHILLIPS et al., 1986).

The hydroxylation of alkaloids often modifies their biological activity and makes available compounds that are otherwise difficult to synthesize. For example, acronycine (**268**), an antitumor alkaloid, is hydroxylated in the 9-position in 30% yield by *Cunninghamella echinulata* (Fig. 78) (BETTS et al., 1974).

6.2.2 Oxidative Degradation

Hydroxylation of a substrate by an organism is often a prelude to its utilization as an energy source, which may involve complete conversion to CO_2. Processes which effect substantial

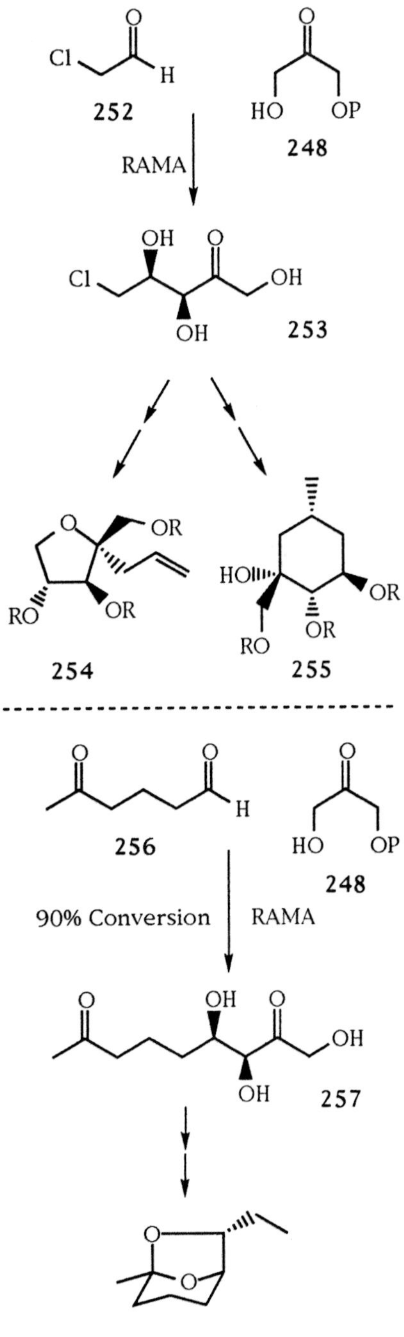

Cl O H 252 · O HO OP 248 · RAMA · OH O Cl OH OH 253 · O OR RO OR 254 · HO OR OR RO 255

O O H 256 · O HO OP 248 · 90% Conversion RAMA · O OH O OH OH 257 · O O

259 (+)-*exo*-Brevicomin

Fig. 75. The use of RAMA in carbocycle and pheromone synthesis.

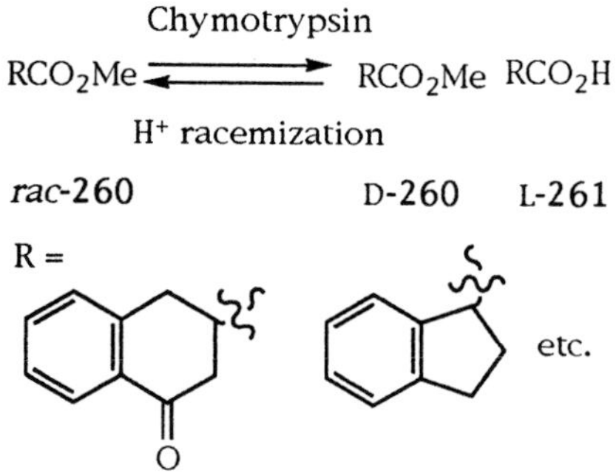

Chymotrypsin
RCO_2Me
RCO_2Me RCO_2H
H^+ racemization
rac-260
D-260 L-261
R =
etc.
O

Fig. 76. Aromatic chiral acids and esters.

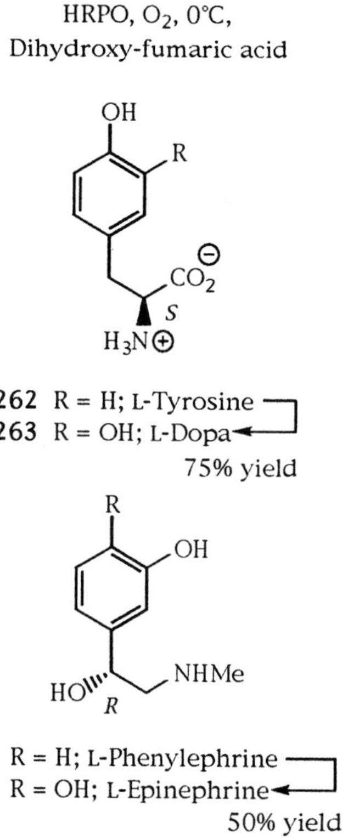

HRPO, O_2, 0°C,
Dihydroxy-fumaric acid
OH
R
$CO_2^{\ominus}$
S
$H_3N^{\oplus}$
262 R = H; L-Tyrosine
263 R = OH; L-Dopa
75% yield
R
OH
NHMe
HO
R
264 R = H; L-Phenylephrine
265 R = OH; L-Epinephrine
50% yield

Fig. 77. Enzymic hydroxylation in drug synthesis.

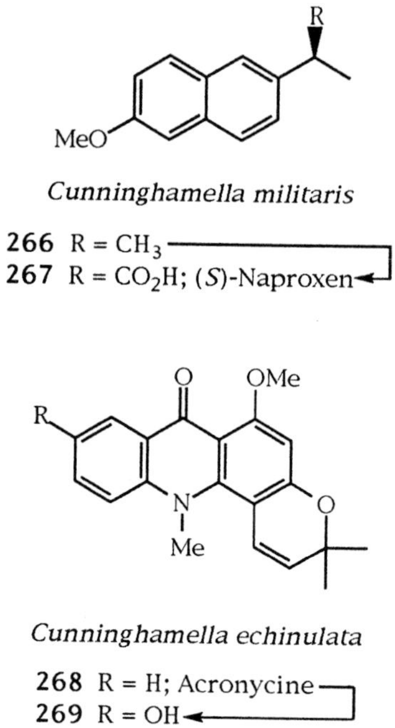

R
MeO
Cunninghamella militaris
266 R = CH_3
267 R = CO_2H; (S)-Naproxen
O
OMe
R
N
Me
O
Cunninghamella echinulata
268 R = H; Acronycine
269 R = OH

Fig. 78. Hydroxylation of cumenes and preparation of an antitumor alkaloid.

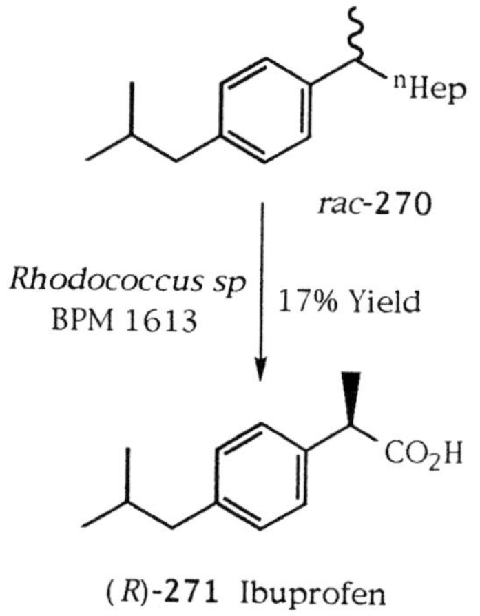

^{n}Hep
rac-270
Rhodococcus sp
BPM 1613
17% Yield
CO_2H
(R)-271 Ibuprofen

Fig. 79. Ibuprofen from stereoselective degradation.

chemical modification but which do not lead to complete degradation are particularly valuable. Treatment of *rac*-(**270**) with a *Rhodococcus* sp., results in complete degradation of one enantiomer, however oxidation of the other enantiomer terminates at the carboxylic acid stage to give the anti-inflammatory drug ibuprofen (*R*)-(**271**) (Fig. 79) (SUGAI and MORI, 1984).

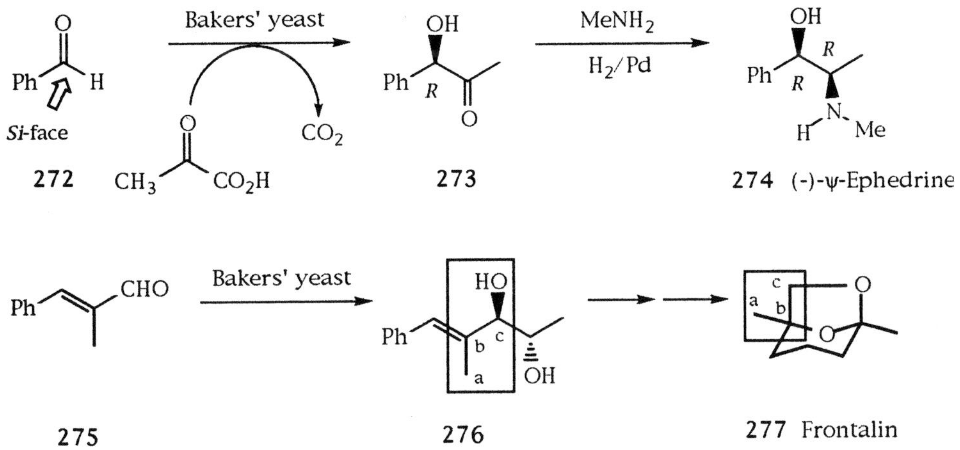

Fig. 80. The synthesis of (−)-ephedrine and frontalin.

6.3 Carbon–Carbon Bond Formation Reactions

6.3.1 Acyloin Condensation

The formation of phenyl acetyl carbinol (**273**) (Fig. 80) from benzaldehyde (**272**) by fermenting yeast was first observed in 1921 (CROUT et al., 1991).

The enzyme system involves pyruvic acid, decarboxylation of which provides the C_2 unit (an acyl anion equivalent) which is transferred to the *Si*-face of the aldehyde to form an (3*R*)-α-hydroxy ketone (acyloin) (**273**).

There is remarkable tolerance by the enzyme system with respect to changes in the structure of the aldehyde. α,β-unsaturated aliphatic and aromatic aldehydes undergo alkylation and reduction of the acyloin to give (2*S*,3*R*)-diols (**276**) (FRONZA et al., 1982). The yields of chiral diols are poor (10–35%), but this is offset by the ease of reaction and the cheapness of starting reagents used. SERVI and FUGANTI (FUGANTI and SERVI, 1988; FUGANTI, 1986; FUGANTI and GRASSELLI, 1985b) have exploited these compounds for numerous syntheses of chiral natural products. An interesting example is provided by the synthesis of (−)-frontalin (**277**) which is a pheromone of several bark beetles (Fig. 80). The two chiral centers created by the yeast reduction are used to induce stereoselective addition of a Grignard reagent and subsequently both are destroyed by a periodate cleavage. Thus only three carbons and no chiral centers from the biotransformation product are carried through to (−)-frontalin (**277**) (Fig. 80) (FUGANTI et al., 1983).

7 References

ARITA, M., ADACHI, K., ITO, Y., SAWAI, H., OHNO, M. (1983), Enantioselective synthesis of the carbocyclic nucleoside (−)-aristeromycin and (−)-neplanocin A by a chemicoenzymatic approach, *J. Am. Chem. Soc.* **105**, 4049–4055.

ARITA, M., ADACHI, K., ITO, Y., SAWAI, H., OHNO, M., SHIBATA, K. (1984), Synthetic studies on biologically active natural products by a chemicoenzymatic approach, *Tetrahedron* **40**, 145–152.

BALDWIN, J. E., CHRISTIE, M. A., HABER, S. B., KRUSE, L. I. (1976), Stereospecific synthesis of penicillins. Conversion from a peptide precursor, *J. Am. Chem. Soc.* **98**, 3045–3047.

BALLARD, D. G., COURTIS, A., SHIRLEY, I. M., TAYLOR, S. C. (1983), A biotech route to polyphenylene, *J. Chem. Soc., Chem. Commun.*, 954–955.

BALLY, C., LEUTHARDT, F. (1970), Die Stereospezifität der Alkoholdehydrogenase, *Helv. Chim. Acta* **53**, 732–738.

BANKO, G., DEMAIN, A. L., WOLFE, S. (1987), ACV synthetase: A multifunctional enzyme with a broad substrate specificity for the synthesis of penicillin and cephalosporin precursors, *J. Am. Chem. Soc.* **109**, 2858–2860.

BARRETT, A. G. M., CAPPS, N. K. (1986), Synthetic approaches to the avermectins : Studies on the hexahydrobenzofuran unit, *Tetrahedron Lett.* **27**, 5571–5574.

BECKER, W., PFEIL, E. (1966), Continuous synthesis of optically active α-hydroxynitriles, *J. Am. Chem. Soc.* **88**, 4299–4300.

BEDNARSKI, M. D., SIMON, E. S., BISCHOFBERGER, N., FESSNER, W.-F., KIM, M.-K. et al. (1989), Rabbit muscle aldolase as a catalyst in organic synthesis, *J. Am. Chem. Soc.* **111**, 627, 635.

BEHRENS, K. B., SHARPLESS, K. B. (1985), Selective transformations of 2,3-epoxy alcohols and related derivatives – strategies for nucleophilic attack at carbon-3 or carbon-2, *J. Org. Chem.* **50**, 5687–5696.

BENDER, M. L., KEDZY, F. J., GUNTER, C. R. (1964), The anatomy of an enzymatic catalysis. α-Chymotrypsin, *J. Am. Chem. Soc.* **86**, 3714–3721.

BETTS, R. E., WALTERS, D. E., ROSAZZA, J. P. (1974), Microbial transformations of antitumor compounds. 1. Conversion of acronycins to 9-hydroxy-acronycine by *Cunninghamella echinulata*, *J. Med. Chem.* **17**, 599–602.

BILLHARDT, U. M., STEIN, P., WHITESIDES, G. M. (1989), Enzymatic methods for the preparation of acetyl CoA and analogs, *Bioorg. Chem.* **17**, 1–12.

BINDRA, J. S., BINDRA, R. (1977), *Prostaglandin Synthesis*. New York: Academic Press.

BOLAND, W., FROBL, C., LORENZ, M. (1991), Esterolytic and lipolytic enzymes in organic synthesis, *Synthesis,* 1049–1072.

BONARA, R., CARREA, G., PASTA, P., RIVA, S. (1986), Preparative scale regio- and stereospecific oxidoreduction of cholic acid and dehydrocholic acid catalyzed by hydroxysteroid dehydrogenases, *J. Org. Chem.* **51**, 2902–2906.

BRANCA, Q., FISCHLI, A. (1977), Eine chiral-ökonomische Totalsynthese von (R)- und (S)-Muskon via Epoxysulfoncyclofragmentierung, *Helv. Chim. Acta.* **60**, 925–944.

BRIENA, R., CRICH, J. Z., SIH, C. J. (1993), Chemoenzymatic synthesis of the C-13 side chain of taxol: Optically active 3-hydroxy-4-phenyl β-lactam derivatives, *J. Org. Chem.* **58**, 1068–1075.

BROOKS, D. W., GROTHAUS, P. G., PALMER, J. T. (1982), Synthetic studies of trichothecenes, an enantioselective synthesis of a C-ring, precursor of anguidine, *Tetrahedron Lett.* **23**, 4187–4190.

BROOKS, D. W., GROTHAUS, P. G., MAZDIYASNI, H. (1983), Total synthesis of the trichothecene mycotoxin anguidine, *J. Am. Chem. Soc.* **105**, 4472–4473.

BROOKS, D. W., MAZDIYASNI, H., CHAKRABARTI, S. (1984), Chiral cyclohexanoid synthetic precursors via asymmetric microbial reduction of prochiral cyclohexanediones, *Tetrahedron Lett.* **25**, 1241–1244.

BROOKS, D. W., WILSON, M., WEBB, M. (1987), Different enzymatic reactions of an enantiomeric pair: Simultaneous dual kinetic resolution of a keto ester by bakers' yeast, *J. Org. Chem.* **52**, 2244–2248.

CARREA, G., BONARA, R., CREMONESI, P., LODI, R. (1984), Enzymatic preparation of 12-ketochenodeoxycholic acid with NADP regeneration, *Biotechnol. Bioeng.* **26**, 560–563.

CHAN, C., COX, P. B., ROBERTS, S. M. (1988), Convergent stereocontrolled synthesis of 13-hydroxy-9Z,11E-octadecadienoic acid (13-HODE), *J. Chem. Soc., Chem. Commun.*, 971–972.

CHEN, C.-S., FUJIMOTO, Y., SIH, C. J. (1981), Bifunctional chiral synthons via microbiological methods. 1. Optically active 2,4-dimethylglutaric acid monomethyl esters, *J. Am. Chem. Soc.* **103**, 3580–3582.

CHIBATA, I., TOSA, T., SATO, T., MORI, T. (1976), Production of L-amino acids by aminoacylase adsorbed on DEAE-Sephadex, *Methods Enzymol.* **44**, 746–759.

CHOI, W. G., LEE, S. B., RYU, D. D. Y. (1981), Cephalexin synthesis by partially purified and immobilized enzymes, *Biotechnol. Bioeng.* **23**, 361–371.

COHEN, N., EICHEL, W. F., LOPRESTI, R. J., NEUKOM, C., SAUCY, G. (1976), Synthetic studies on (2R, 4$'S$, 8$'R$)-α-tocopherol. An approach utilizing side-chain synthons of microbiological origin, *J. Org. Chem.* **41**, 3505–3511.

COLLUM, D. B., MCDONALD III, J. H., STILL, W. C. (1980a), The polyether antibiotic monensin. 1. Strategy and degradations, *J. Am. Chem. Soc.* **102**, 2117–2118.

COLLUM, D. B., MCDONALD III, J. H., STILL, W. C. (1980b), The polyether antibiotic monensin. 2. Preparation of intermediates, *J. Am. Chem. Soc.* **102**, 2118–2120.

COREY, E. J., TRYBULSKI, E. J., MELVIN, L. S., NICOLAOU, K. C., SECRIST, J. et al. (1978a), Total synthesis of erythromycins. 3. Stereoselective routes to intermediates corresponding to C(1) to C(9) and C(10) to C(13) fragments of erythronolide B, *J. Am. Chem. Soc.* **100**, 4618–4620.

COREY, E. J., KIM, S., YOO, S., NICOLAOU, K. C., MELVIN, L. S. et al. (1978b), Total synthesis of erythromycins. 4. Total synthesis of erythronolide B, *J. Am. Chem. Soc.* **100**, 4620–4622.

COREY, E. J., ALBRIGHT, J. O., BARTON, A. E., HASHIMOTO, S. (1980), Chemical and enzymic syntheses of 5-HPETE, a key biological precursor of slow-reacting substance of anaphylaxis (SRS), and 5-HETE, *J. Am. Chem. Soc.* **102**, 1435–1436.

COREY, E. J., LANSBURY, P. T. (1983), Stereochemical course of 5-lipoxygenation of arachidonate by rat basophil leukemic cell (RBL-1) and potato enzymes, *J. Am. Chem. Soc.* **105**, 4093–4094.

CROUT, D. H. G., DALTON, H., HUTCHINSON, D. W., MIYAGOSHI, M. (1991), Studies on pyruvate decarboxylase acyloin formation from aliphatic, aromatic and heterocyclic aldehydes, *J. Chem. Soc., Perkin. Trans. I*, 1329–1334.

DAI, W. M., ZHOU, W. S. (1985), New synthesis of two optically active steroid C D ring synthons by microbial asymmetric reduction, *Tetrahedron* **41**, 4475–4482.

DAVIES, H. G., DAWSON, M. J., LAWRENCE, G. C., MAYALL, J., NOBLE, D. et al. (1986), Microbial hydroxylation of cyclohexylcyclohexane; Synthesis of an analog of leukotriene B_3, *Tetrahedron Lett.* **27**, 1089–1092.

DAVIES, H. G., GREEN, R. H., KELLY, D. R., ROBERTS, S. M. (1989), Biotransformations in preparative organic chemistry. London: Academic Press.

DODDS, D. R., JONES, J. B. (1982), Selective and stereospecific enzyme-catalyzed reductions of *cis*- and *trans*-decalindiones to enantiomerically pure hydroxy-ketones; an efficient access to (+)-4-twistanone, *J. Chem. Soc., Chem. Commun.*, 1080–1081.

DODDS, D. R., JONES, J. B. (1988), Enzymes in organic synthesis. 38. Preparations of enantiomerically pure chiral hydroxydecalones via stereospecific horse liver alcohol dehydrogenase-catalyzed reductions of decalindiones, *J. Am. Chem. Soc.* **110**, 577–583.

DRAUZ, K., WALDMANN, H. (1995), *Enzyme Catalysis in Organic Synthesis, A Comprehensive Handbook*. Weinheim: VCH.

DRUECKHAMMER, D. G., RIDDLE, V. W., WONG, C.-H. (1985), FMN reductase catalyzed regeneration of NAD(P) for use in enzymatic synthesis, *J. Org. Chem.* **50**, 5387–5389.

EFFENBERGER, F., STRAUB, A. (1987), A novel, convenient preparation of dihydroxyacetone phosphate and its use in Enzymatic aldol reactions, *Tetrahedron Lett.* **28**, 1641–1644.

EVANS, D. A., SACKS, C. E., KLESCHICK, W. A., TABER, T. R. (1979), Polyether antibiotics synthesis and absolute configuration of the ionophore A-23187, *J. Am. Chem. Soc.* **101**, 6789–6791.

EVANS, D. A., NELSON, J. V., TABER, T. R. (1982), Stereoselective aldol condensation, *Top. Stereochem.* **13**, 1–115.

FABER, K. (1995), *Biotransformations in Organic Chemistry*, 2nd Edn. Berlin, New York: Springer-Verlag.

FISCHLI, A. (1980), *Modern Synthetic Methods*, Vol. 2 (SCHEFFOLD, R., Ed.). Frankfurt: Salle-Sauerländer.

FRATER, G. (1979a), Über die Stereospezifität der α-Alkylierung von β-Hydroxycarbonsäureestern, *Helv. Chim. Acta.* **62**, 2825–2828.

FRATER, G. (1979b), Stereospezifische Synthese von (+)-(3*R*,4*R*)-4-methyl-3-Heptanol. Das Enantiomer eines Pheromons des kleinen Ulmensplintkäfers, *Helv. Chim. Acta.* **62**, 2829–2832.

FRATER, G. (1980), Über die Stereoselektivität der α-Alkylierung von (1*R*,2*S*)-(+)-*cis*-2-hydroxycyclohexancarbonsäureethylester, *Helv. Chim. Acta.* **63**, 1383–1390.

FRATER, G., MÜLLER, U., GÜNTHER, W. (1984), The stereoselective α-alkylation of chiral β-hydroxy esters and applications thereof, *Tetrahedron* **40**, 1269–1277.

FRONZA, G., FUGANTI, C., MAJORI, L., PEDROCCHI-FANTONI, G., SPREAFICO, F. (1982), Synthesis of enantiomerically pure forms of *N*-acyl derivatives of C-methyl analogs of the aminodeoxy sugar L-acosamine from non-carbohydrate precursors, *J. Org. Chem.* **47**, 3289–3296.

FUGANTI, C. (1986), Bakers' yeast-mediated preparation of carbohydrate-like chiral synthons, in: *Enzymes as Catalysts in Organic Synthesis* (SCHNEIDER, M. P., Ed.) NATO ASI Series, Vol. 178. Dordrecht: Reidel-Gruyter.

FUGANTI, C., GRASSELLI, P. (1985a), Enzymes in Organic synthesis, in: *CIBA Foundation Symposium III*, pp. 112–127. London: Pitman.

FUGANTI, C., GRASSELLI, P. (1985b), Stereochemistry and synthetic applications of products of fermentation of α,β-unsaturated aromatic aldehydes by bakers' yeast, in: *Enzymes as Catalysts in Organic Synthesis* (CLARK, S., PORTER, R., Eds.). London: Pitman.

FUGANTI, C., SERVI, S. (1988), *Bioflavour '87* (SCHEIER, P., Ed.). Berlin: de Gruyter.

FUGANTI, C., GRASSELLI, P., SERVI, S. (1983), Synthesis of (−)-frontalin from the (2*S*,3*R*)-diol prepared from α-methylcinnamaldehyde and fermenting bakers' yeast, *J. Chem. Soc., Perkin Trans. I*, 241–244.

FULLING, G., SIH, C. J. (1987), Enzymatic second order asymmetric hydrolysis of ketorolac esters: *in situ* racemization, *J. Am. Chem. Soc.* **109**, 2845–2846.

GAIS, H.-J., LUKAS, K. L. (1984), Enantioselective and enantioconvergent syntheses of building blocks for the total synthesis of cyclopentanoid natural products, *Angew. Chem.* (Int. Edn. Engl.) **23**, 142–143.

GAIS, H.-J., LIED, T., LUKAS, K. L. (1984), Asymmetric synthesis of a novel enantiomerically pure prostaglandin building block, *Angew. Chem.* (Int. Edn. Engl.) **23**, 511–512.

GIBSON, D. T., HENSLEY, M., YOSHIOKA, H., MABRY, T. J. (1970), Formation of (+)-*cis*-2,3-dihydroxy-1-methylcyclohexa-4,6-diene from toluene by *Pseudomonas putida*, *Biochemistry* **9**, 1626–1630.

GLASS, J. D. (1981), Enzymes as reagents in the synthesis of peptides, *Enzyme Microb. Technol.* **3**, 1–8.

GOODHUE, C. T., SCHAEFFER, J. R. (1971), Preparation of (L)-(+)-β-hydroxyisobutyric acid by bacterial oxidation of isobutyric acid, *Biotechnol. Bioeng.* **13**, 203–214.

GREENSTEIN, J. P. (1954), The resolution of racemic α-hydroxy amino acids, *Adv. Protein Chem.* **9**, 121–202.

GREENZAID, P., JENCKS, W. P. (1971), Pig liver esterase reactions with alcohols, structure–reactivity correlations and the acyl-enzyme intermediate, *Biochemistry* **10**, 1210–1222.

GRUNDY, S. M. (1988), HMG-CoA reductase inhibitors for treatment of hypercholesterolemia, *N. Eng. J. Med.* **319**, 24–33.

HAMADA, M., TAKEUCHI, T., KONDO, S., IKEDA, Y., NAGANAWA, H. et al. (1970), A new antibiotic, negamycin, *J. Antibiot.* **23**, 170–171.

HANESSIAN, S. (1983), Total synthesis of natural products: The 'chiron' approach. Oxford: Pergamon Press.

HEATHCOCK, C. H. (1984), The aldol addition reaction, *Asymm. Synth.* **3**, 111–212.

HEMMERLE, H., GAIS, H.-J. (1987), Asymmetric hydrolysis and esterification catalyzed by esterases from porcine pancreas in the synthesis of both enantiomers of cyclopentanoid building blocks, *Tetrahedron Lett.* **28**, 3471–3474.

HIRAMA, M., UEI, M. (1982), Chiral total synthesis of compactin, *J. Am. Chem. Soc.* **104**, 4251–4253.

HIRAMA, M., GARVEY, P. S., LU, L. D. L., MASAMUNE, S. (1979), Use of the *E*-vinyloxyborane derived from *S*-phenyl propanethioate for stereospecific aldol-type condensation. A simplified synthesis of the Prelog-Djerassi lactonic acid, *Tetrahedron Lett.* **41**, 3937–3940.

HIRAMA, M., SHIMIZU, T., IWASHITA, M. (1983), Enantiospecific syntheses of trifunctional (*R*)-3-hydroxy esters by bakers' yeast reduction, *J. Chem. Soc., Chem. Commun.*, 599–600.

HUDLICKY, T., NATCHUS, M. (1992), Chemoenzymatic enantiocontrolled synthesis of (−)-specionin, *J. Org. Chem.* **57**, 4740–4746.

HUDLICKY, T., PRICE, J. D. (1990), Microbial oxidation of chloroaromatics in the enantioselective synthesis of carbohydrates – L-ribonic γ-lactone, *Synlett*, 159–160.

HUDLICKY, T., LUNA, H., BARBIERI, G., KWART, L. D. (1988), Enantioselective synthesis through microbial oxidation of arenes. Efficient preparation of terpene and prostanoid synthons, *J. Am. Chem. Soc.* **110**, 4735–4741.

HUDLICKY, T., LUNA, H., PRICE, J. D., RULIN, F. (1989), An enantiodivergent approach to D- and L-erythrose via microbial oxidation of chlorobenzene, *Tetrahedron Lett.* **30**, 4053–4054.

HUDLICKY, T., PRICE, J. D., RULIN, F., TSUNODA, T. (1990), Efficient and enantiodivergent synthesis of (+)- and (−)-pinitol, *J. Am. Chem. Soc.* **112**, 9439–9440.

HUDLICKY, T., LUNA, H., OLIVO, H. F., ANDERSEN, C., NUGENT, T., PRICE, J. D. (1991), Biocatalysis as the strategy of choice in the exhaustive enantiomerically controlled synthesis of conduritols, *J. Chem. Soc., Perkin Trans. I*, 2907–2917.

HUTCHINSON, C. R., IKEDA, T. (1984), A general, enantiospecific synthesis of cyclopentanoid monoterpenes (iridoids), The total synthesis of (−)-1-*O*-methylsweroside aglucone, *J. Org. Chem.* **49**, 2837–2838.

IRIUCHIJIMA, S., HASEGAWA, K., TSUCHIHASHI, G. (1978), Structure of a new phenolic glycoside, chesnatin, from chestnut galls, *Agric. Biol. Chem.* **46**, 1907–1910.

IRWIN, A. J., JONES, J. B. (1977), Asymmetric synthesis via enantiotopically selective horse liver alcohol dehydrogenase-catalyzed oxidations of diols containing a prochiral center, *J. Am. Chem. Soc.* **99**, 556–561.

ISOWA, Y., OHMORI, M., SATA, M., MORI, K. (1977a), The enzymatic synthesis of protected valine-5-angiotensin II amide-1, *Bull. Chem. Soc. Jpn.* **50**, 2766–2772.

ISOWA, Y., OHMORI, M., ICHIKAWA, T., KURITA, H., SATO, M., MORI, K. (1977b), the synthesis of peptides by means of proteolytic enzymes, *Bull. Chem. Soc. Jpn.* **50**, 2762–2765.

JAKOVIC, I. J., GOODBRAND, H. B., LOK, P. K., JONES, J. B. (1982), Enzymes in organic synthesis. 24. Preparations of enantiomerically pure chiral lactones via stereospecific horse liver alcohol dehydroagenase-catalyzed oxidations of monocyclic *meso*-diols, *J. Am. Chem. Soc.* **104**, 4659–4665.

JOHNSON, R. A. (1978), Oxidation in organic chemistry, Part C (TRAHANOVSKY, W. S., Ed.). New York: Academic Press.

JOHNSON, M. R., KISHI, Y. (1979), Cooperative effect by α-hydroxy and ether oxygen in peroxidation with a peracid, *Tetrahedron Lett.* **45**, 4347–4350.

JOHNSON, M. R., NAKATA, T., KISHI, Y. (1979), Stereo- and regioselective methods for the synthesis of three consecutive asymmetric units in many natural products, *Tetrahedron Lett.* **45**, 4343–4346.

JONES, J. B. (1985), Enzymes in organic synthesis, in: *CIBA Foundation Symposium III* (CLARK, S., PORTER, R., Eds.), pp. 3–21. London: Pitman.

JONES, J. B. (1986), Enzymes in organic synthesis, *Tetrahedron* **42**, 3351–3403.

JONES, J. B. (1993), Probing the specificity of synthetically useful enzymes, *Aldrichimica Acta* **26**, 105–112.

JONES, J. B., LIN, Y. Y. (1972), Evaluation of some of the factors involved in the selective hydrolysis of aromatic ester protecting groups by α-chymotrypsin, *Can. J. Chem.* **50**, 2053–2058.

JONES, J. B., FINCH, M. A. W., JAKOVIC, I. J. (1982), Enzymes in organic synthesis. 26. Synthesis of enantiomerically pure grandisol from an enzyme-generated chiral synthon, *Can. J. Chem.* **60**, 2007–2011.

KAN, K., MIYAMA, A., HAMAGUCHI, S., OHASHI, T., WATANABE, K. (1985a), Synthesis of (*S*)-β-blocker from (*S*)-5-hydroxymethyl-3-*tert*-butyl-2-oxazolidinone or (*S*)-5-hydroxymethyl-3-isopropyl-2-oxazolidinone, *Agric. Biol. Chem.* **49**, 207– 210.

KAN, K., MIYAMA, A., HAMAGUCHI, S., OHASHI, T., WATANABE, K. (1985b), Stereochemical inversion of (*R*)-5-hydroxymethyl-3-*tert*-butyl-2-oxazolidinone or (*R*)-5-hydroxymethyl-3-isopropyl-2-oxazolidinone to the corresponding (*S*)-isomer, *Agric. Biol. Chem.* **49**, 1669–1674.

KARADY, S., AMATO, J. S., REAMER, R. A., WEINSTOCK, L. M. (1981), Stereospecific conversion of penicillin to thienamicin, *J. Am. Chem. Soc.* **103**, 6765–6767.

KASEL, W., HULTIN, P. G., JONES, J. B. (1985), Preparations of chiral hydroxy ester synthons via stereoselective porcine pancreatic lipase-catalyzed hydrolyses of *meso*-diesters, *J. Chem. Soc., Chem. Commun.*, 1563–1564.

KATO, K., KAWAHARA, K., TAKAHASHI, T., IGARASI, S. (1980), Enzymatic synthesis of amoxicillin by the cell bound α-amino acid ester hydrolase of *Xanthomonas citri*, *Agric. Biol. Chem.* **44**, 821–825.

KATSUKI, T., MARTIN, V. S. (1996). Catalytic asymmetric epoxidation of allylic alcohols, *Org. React.* **48**, 1–130.

KATSUKI, T., SHARPLESS, K. B. (1980), The first practical method for asymmetric epoxidation, *J. Am. Chem. Soc.* **102**, 5974–5976.

KHORANA, H. G. (1976), Total synthesis of the structural gene for the precursor of a tyrosine suppressor transfer RNA from *Escherichia coli*, *J. Biol. Chem.* **251**, 565–586.

KIESLICH, K. (1976), Microbial transformations of non-steroid cyclic compounds. Stuttgart: Thieme.

KITAZUME, T., ISHIKOWA, N. (1984), Introduction of center of chirality into fluorocompounds by microbial transformation of 2,2,2-trifluoroethanol, *Chem. Lett.*, 1815–1818.

KLIBANOV, A. M., BERMAN, Z., ALBERTI, B. N. (1981), Preparative hydroxylation of aromatic compounds catalyzed by peroxidase, *J. Am. Chem. Soc.* **103**, 6263–6264.

KOBAYASHI, M., KOYAMA, T., OGURA, K., SETO, S., RITTER, F. J., BRUGGERMANN-ROTGANS, I. E. M. (1980), Bioorganic synthesis and absolute configuration of faranal, *J. Am. Chem. Soc.* **102**, 6602–6604.

KOBAYASHI, S., IIMORI, T., IZAWA, T., OHNO, M. (1981), $Ph_3P-(PyS)_2-CH_3CN$ as an excellent condensing system for β-lactam formation from β-amino acids, *J. Am. Chem. Soc.* **103**, 2406–2408.

KONDO, S., SHIBAHARA, S., TAKAHASHI, S., MAEDA, K., UMEZAWA, H., OHNO, M. (1971), Negamycin, a novel hydrazide antibiotic, *J. Am. Chem. Soc.* **93**, 6305–6306.

KOSMOL, H., KIESLICH, K., VOSSING, R., KOCH, H.-J., PETZOLD, K., GIBIAN, H. (1967), Mikrobiologische stereospezifische Reduktion von 3-Methoxy-8.14-seco-1.3.5(10).9-ostratraen-14.17-dion, *Liebigs Ann. Chem.* **701**, 198–205.

KOYAMA, T., SAITO, A., OGURA, K., SETO, S. (1980), Substrate specificity to farnesyl pyrophosphate synthetase, application to asymmetric synthesis, *J. Am. Chem. Soc.* **102**, 3614–3618.

KOYAMA, T., OGURA, K., BAKER, F. C., JAMIESON, G. C., SCHOOLERY, D. A. (1987), Synthesis and absolute configuration of 4-methyl juvenile hormone 1 (4-MeJH 1) by a biogenetic approach: A combination of enzymatic synthesis and biotransformation, *J. Am. Chem. Soc.* **109**, 2853–2854.

KRAIDMAN, G., MUKHERJEE, B. B., HILL, J. D. (1969), Cyclic vicinal glycols as microbial transformation products of some monocyclic terpene hydrocarbons, *Bacteriol. Proc.* **63**.

KRASNOBAJEW, V. (1984), Microbial transformations of terpenoids, in: *Biotechnology* 1st Edn., Vol. 6a. *Biotransformations* (REHM, H.-J., REED, G., Eds.). Weinheim: Verlag Chemie.

KULLMAN, W. (1981), Protease-mediated peptide bond formation, *J. Biol. Chem.* **256**, 1301–1304.

KULLMAN, W. (1982), Enzymatic synthesis of dynorphin (1–8), *J. Org. Chem.* **47**, 5300–5303.

KUPCHAN, S. M., KOMODO, Y., BRANFMAN, A. R., SNEDEN, A. T., COURT, W. A. et al. (1977), The maytansinoids. Isolation, structural elucidation and chemical interrelation of novel ansa macrolides, *J. Org. Chem.* **42**, 2349–2357.

KUROZUMI, S., TORA, T., OSHIMOTO, S. (1973), Preparation of 2-(4-hydroxy-1-oxocyclopent-2-ene)-heptanoic acid, an important prostaglandin synthon, *Tetrahedron Lett.* **49**, 4959–4960.

LADNER, W. E., WHITESIDES, G. M. (1984), Lipase-catalyzed hydrolysis as a route to esters of chiral epoxy alcohols, *J. Am. Chem. Soc.* **106**, 7250–7251.

LADNER, W. E., WHITESIDES, G. M. (1985), Enzymatic synthesis of deoxy ATP using RNA as starting material, *J. Org. Chem.* **50**, 1076–1079.

LAGERLÖF, E., NATHORST-WESTFELOT, L., EKSTROM, B., SJOBERG, B. (1976), Production of 6-aminopenicillanic acid with immobilized *Escherichia coli*, *Methods Enzymol.* **44**, 759–768.

LAUMEN, K., SCHNEIDER, M. (1985), Enantioselective hydrolysis of *cis*-1,2-diacetoxycycloalkane dimethanols: Enzymatic preparations of chiral building blocks from prochiral *meso* substrates,

Tetrahedron Lett. **26**, 2073–2076.

LEUENBERGER, H. G., BOGUTH, W., BARNER, R., SCHMID, M., ZELL, R. (1979), Large-scale reduction of oxoisophorone in the synthesis of carotenoids, *Helv. Chem. Acta* **62**, 455–465.

LEY, S. V., STERNFELD, F. (1989), Microbial oxidation in synthesis: Preparation of (+)- and (−)-pinitol from benzene, *Tetrahedron* **45**, 3463–3476.

LIN, C. H., ALEXANDER, D. L., CHICHESTER, C. G., GORMAN, R. R., JOHNSON, R. A. (1982), 10-nor-9,11-secoprostaglandins. Synthesis, structure and biology of endoperoxide analogs, *J. Am. Chem. Soc.* **104**, 1621–1628.

LIPMANN, F. (1973), Nonribosomal polypeptide synthesis on polyenzyme templates, *Acc. Chem. Res.* **6**, 361–367.

MACLEOD, R., PROSSER, H., FIKENTSCHER, L., LANYI, J., MOSHER, H. S. (1964), Asymmetric reductions. XII. Stereoselective ketone reductions by fermenting yeast, *Biochemistry* **3**, 838–846.

MARPLES, B. A., ROGER-EVANS, M. (1989), Enantioselective lipase-catalyzed hydrolysis of esters of epoxy secondary alcohols: An alternative to Sharpless oxidation, *Tetrahedron Lett.* **30**, 261–264.

MEDINA, J. C., KYLER, K. S. (1988), Enzymatic cyclization of hydroxylated surrogate squalenoids with bakers' yeast, *J. Am. Chem. Soc.* **110**, 4818–4821.

MEDINA, J. C., GUAJARDO, R., KYLER, K. S. (1989), Vinyl group rearrangement in the enzymatic cyclization of squalenoids: Synthesis of 30-oxysterols, *J. Am. Chem. Soc.* **111**, 2310–2311.

MEYERS, A. I., AMOS, R. A. (1980), Studies directed toward the total synthesis of streptogramin antibiotics. Enantiospecific approach to the nine-membered macrocycle of grisoviridin, *J. Am. Chem. Soc.* **102**, 870–872.

MEYERS, A. I., HUDSPETH, J. P. (1981), Enantioselective synthesis of C_3–C_{10} fragment (north east zone) of maytanisoids with four chiral centers (4*S*,5*S*,6*R*,7*S*), *Tetrahedron Lett.* **22**, 3925–3928.

MOHR, P., WAESPE-SARCEVIC, N., TAMM, C., GAWRONSKA, K., GAWRONSKI, J. K. (1983), A Study of stereoselective hydrolysis of symmetrical diesters with pig liver esterase, *Helv. Chim. Acta.* **66**, 2501–2511.

MORI, K. (1989), Synthesis of optically active pheromones, *Tetrahedron* **45**, 3233–3298.

MORI, K., EBATA, T. (1986a), Synthesis of (2*S*,3*R*,1′*R*)-stegobinone, the pheromone of the drugstore beetle with stereocontrol at C-2 and C-2′, *Tetrahedron* **42**, 4413–4420.

MORI, K., EBATA, T. (1986b), Synthesis of all of the four possible stereoisomers of 5-hydroxy-4-methyl-3-heptanone (sitophilure). The aggregation pheromone of the rice weevil and the maize weevil, *Tetrahedron* **42**, 4421–4426.

MORI, K., IWASAWA, H. (1980), Preparation of both enantiomers of *threo*-2-amino-3-methyl-hexanoic acid by enzymatic resolution and their conversion to optically active forms of *threo*-4-methylheptan-3-ol, a pheromone component of the smaller European elm bark beetle, *Tetrahedron* **36**, 2209–2213.

MORI, K., TANIDA, K. (1981), Synthesis of three stereoisomeric forms of 2,8-dimethyl-1.7-dioxaspiro(5.5)undecane, the main component of the cephalic secretion of *Andrena wilkella*, *Heterocycles* **15**, 1171–1179.

MORISAWA, H., UTAGAWA, T., MIJOSHI, T., YOSHINAGA, F., YAMAZAKI, A., MITSUGI, K. (1980), A new method for the synthesis of some 9-β-D-arabinofuranosylpurines by a combination of chemical and enzymatic reactions, *Tetrahedron Lett.* **21**, 479–482.

MOROE, T., HATTORI, S., KOMATSU, A., YAMAGUCHI, Y. (1970), *Chem. Abstr.* **73**, 33900d.

MOSBACH, K., OHLSON, S., LARSSON, P. D. (1978), Steroid transfer by activated living immobilized *Arthrobacter simplex* cells, *Biotechnol. Bioeng.* **20**, 1267–1284.

NAEMURA, K., KATOH, T., CHIKAMATSU, H., NAKAZAKI, M. (1984), The preparation and chirotopical properties of cage-shaped pentacyclic hydrocarbons, *Chem. Lett.*, 1371–1374.

NAEMURA, K., FUJII, T., CHIKAMATSU, H. (1986), Selective and stereospecific horse liver alcohol dehydrogenase-catalyzed reduction of cage-shaped *meso*-diketones. An efficient access to optically active D_3-trishomocubane derivatives, *Chem. Lett.*, 923–926.

NAKAZAKI, M., CHIKAMATSU, H., FUJII, T., SASAKI, Y., AO, Y. (1983), Stereochemistry of horse liver alcohol dehydrogenase-mediated oxidoreduction of 2-brendanone type cage-shaped tricyclic ketones and the related stereoisomeric alcohols, *J. Org. Chem.* **48**, 4337–4345.

NEWTON, R. F., ROBERTS, S. M. (1980), Steric controlled prostaglandin synthesis involving bicyclic and tricyclic intermediates, *Tetrahedron* **36**, 2163–2196.

NEWTON, R. F., ROBERTS, S. M. (1982), Prostaglandin and thromboxanes: An introductory text. London: Butterworths.

O'CONNELL, E. L., ROSE, I. A. (1973), Affinity labeling of phosphoglucose isomerase by 1,2-anhydrohexitol-6-phosphates, *J. Biol. Chem.* **248**, 2225–2231.

OKANO, K., IZAWA, T., OHNO, M. (1983), A general approach of *trans*-carbapenem antibiotics. Enantioselective synthesis of key intermediates for (+)-PS-5, (+)-PS-6 and (+)-thienamycin, *Tetrahedron Lett.* **24**, 217–220.

OYAMA, K., NISHIMURA, S., NONAKA, Y., KIHARA, Y., HASHIMOTO, T. (1981), Synthesis of an aspartame precursor by immobilized thermolysin in an organic solvent, *J. Org. Chem.* **46**, 5241–5242.

PATEL, S. S., CONLON, H. D., WALT, D. R. (1986), Enzyme-catalyzed synthesis of L-acetylcarnitine and citric acid using acetyl coenzyme A recycling, *J. Org. Chem.* **51**, 2842–2844.

PHILLIPS, G. T., MATCHAM, G. W. J., BERTOLA, M. A., MARX, A. F., KOGER, H. S. (1986), *Eur. Patent Appl.* EP 205,21517.

PORTER, N. A., BYERS, J. D., HOLDEN, K. M., MENZEL, D. B. (1979), Synthesis of prostaglandin H_2, *J. Am. Chem. Soc.* **101**, 4319–4322.

QUISTAD, G. B., CERF, D. C., SCHOOLEY, D. A., STAAL, G. B. (1981), Fluoromenalonate acts as an inhibitor of insect juvenile hormone biosynthesis, *Nature* **289**, 176–177.

REHM, H.-J., REED, G. (Eds.) (1984), *Biotechnology* 1st. Edn., Vol. 6a *Biotransformations* (KIESLICH, K., Ed.). Weinheim: Verlag Chemie.

REES, L., VALENTE, E., SUCKLING, C. J., WOOD, H. C. S. (1986), Asymmetric reduction of dihydrofolate using dihydrofolate reductases and chiral boron-containing compounds, *Tetrahedron* **42**, 117–136.

ROBERTS, S. M. (1985), Enzymes in organic synthesis, in: *CIBA Foundation Symposium III*, pp. 31–39. London: Pitman.

ROSAZZA, J. P. (1982), *Microbial Transformations of Bioactive Compounds*, Vol. 1. Boca Raton, FL: CRC Press.

ROUDEN, J., HUDLICKY, T. (1993), Total synthesis of (+)-kifunensine, a potent glycosidase inhibitor, *J. Chem. Soc., Perkin Trans. I*, 1095–1097.

SABBIONI, G., SHEA, M. L., JONES, J. B. (1984), Preparations of bicyclic chiral lactone synthons via stereospecific pig liver esterase-catalyzed hydrolyses of *meso*-diesters. Ring-sized induced reversal of stereospecificity, *J. Chem. Soc., Chem. Commun.*, 236–238.

SABINA, R. L., HOLMES, E. W., BECKER, M. A. (1984), The enzymatic synthesis of 5-amino-4-imidazolecarbocamide riboside triphosphate (ZTP), *Science* **223**, 1193–1195.

SALTZMANN, T. N., RATCLIFFE, R., CHRISTENSEN, B. G., BOUFFARD, F. A. (1980), A stereocontrolled synthesis of (+)-thienamycin, *J. Am. Chem. Soc.* **102**, 6161–6163.

SAWADA, S., KULPRECHA, N., NILUBOL, N., YOSHIDA, S., KINOSHITA, S., TAGUCHI, H. (1982), Microbial production of ursodeoxycholic acid from lithiocholic acid by *Fusarium equiseti* M41, *Appl. Environ. Microbiol.* **44**, 1249–1252.

SCHIENMANN, F., ROBERTS, S. M. (1982), *Recent Synthetic Routes to Prostaglandins and Thromboxanes.* London: Academic Press.

SCHINDLER, J., SCHMID, R. D. (1982), Fragrance or aroma chemicals – microbial synthesis and enzymatic transformation – a review, *Proc. Biochem.* **2**, October 2–8.

SCHMID, W., WHITESIDES, G. M. (1990), A new approach to cyclitols based on rabbit muscle aldolase (RAMA), *J. Am. Chem. Soc.* **112**, 9670–9671.

SCHULTZ, M., WALDMANN, H., VOGT, W., KUNZ, H. (1990), Stereospecific C—C bond formation with rabbit muscle aldolase – a chemicoenzymatic synthesis of (+)-*exo*-brevicomin, *Tetrahedron Lett.* **31**, 867–868.

SEEBACH, D., HERRADON, B. (1987), Diastereoselective elaboration of the carbon skeleton of β-hydroxy esters from yeast reductions, *Tetrahedron Lett.* **28**, 3791–3794.

SEEBACH, D., SUTTER, R. H., WEBER, R. H., ZUGER, M. F. (1985) Ethyl (*S*)-3-hydroxybutanoate, *Org. Synth.* **63**, 1–7.

SERVI, S. (1990), Bakers' yeast as a reagent in organic synthesis, *Synthesis*, 1–25.

SEURLING, B., SEEBACH, D. (1977), Synthese von vier chiralen, elektrophilen C_3- und C_4-Synthesebausteinen aus Hydroxycarbonsäuren, *Helv. Chim. Acta* **60**, 1175–1181.

SIH, C. J., SALOMAN, R. G., PRICE, P., SOOD, R., PERLUZZOTTI, G. (1972), Total synthesis of prostaglandins III. 11-Deoxyprostaglandins, *Tetrahedron Lett.* **24**, 2435–2437.

SUGAI, T., MORI, K. (1984), Preparation of (*R*)-ibuprofen and related carbocyclic acids using microbial oxidation of an aromatic hydrocarbon by *Rhodococcus* sp. BPM 1613, *Agric. Biol. Chem.* **48**, 2501–2504.

SUGIYAMA, K., TRAGER, W. F. (1986), Prochiral selectivity and intramolecular isotope effects in the cytochrome P-450 catalyzed ω-hydroxylation of cumene, *Biochemistry* **25**, 7336–7343.

TABENKIN, B., LEMAHIEN, R. A., BERGER, J., KIERSTEAD, R. W. (1969), Microbiological hydroxylation of cinerone to cinerolone, *Appl. Microbiol.* **17**, 714–717.

TAKAMATSU, S., UMEMURA, I., TOSA, T., SATO, T., YAMAMOTO, K., CHIBATA, C. (1982), Production of L-alanine from ammonium fumarate using two types of immobilized microbial cells, *Eur. J. Appl. Microbiol. Biotechnol.* **15**, 142–152.

TENGÖ, J., ÅGREN, L., BAUR, B., ISAKSSON, R., LILJEFORS, T. et al. (1990), *Andrena wilkella* male bees discriminate enantiomers of cephalic secretion components, *J. Chem. Ecol.* **16**, 429–441.

TERASHIMA, S., TAMOTO, K. (1982), An efficient synthesis of optically pure anthracycline intermediates by the novel use of microbial reduction, *Tetrahedron Lett.* **23**, 3715–3718.

TSCHAEW, D. M., FUENTES, L. M., LYNCH, J. E., LASWELL, W. L., VOLANTE, R. P., SHINKAI, I. (1988), An efficient synthesis of 4-benzoyloxyazetidinone: An important carbapenem nucleus, *Tetrahedron Lett.* **29**, 2779–2782.

TURNER, N. J., WHITESIDES, G. M. (1989), A combined chemical-enzymatic synthesis of 3-deoxy-δ-arabino-heptulosonic acid 7-phosphate, *J. Am. Chem. Soc.* **111**, 624–627.

VAN OSSELAER, T. A., LEMIERE, G. L., LEPOIVRE, J. A., ALDERWEIRELOT, F. C. (1978), Enzymatic *in vitro* reduction of ketones, *Bull. Soc. Chim. Belg.* **87**, 153–154.

VON KRAMER, A., PFANDER, H. (1982), C_{45}- und C_{50}-Carotinoide: Synthese von (*R*)- und (*S*)-Lavandulol, *Helv. Chim. Acta.* **65**, 293–301.

VELLUZ, L. (1965), Recent advances in the total synthesis of steroids, *Angew. Chem.* (Int. Edn. Engl.) **4**, 181–200.

WANG, Y.-F., SIH, C. J. (1984), Bifunctional chiral synthons via biochemical methods 4. Chiral precursors to (+)-biotin and (−)-A-factor, *Tetrahedron Lett.* **25**, 4999–5002.

WANG, N. Y., HSU, C. T., SIH, C. J. (1981), Total synthesis of (+)-compactin (ML-236B), *J. Am. Chem. Soc.* **103**, 6538–6539.

WANG, Y.-F., IZAWA, T., KOBAYASHI, S., OHNO, M. (1982), Stereocontrolled synthesis of (+)-negamycin from an acyclic homoallyamine by 1,3-asymmetric induction, *J. Am. Chem. Soc.* **104**, 6465–6466.

WANG, Y.-F., CHEN, C.-S., GIRDAUKAS, G., SIH, C. J. (1984), Bifunctional chiral synthons via biochemical methods. 3. Optical purity enhancement in enzymic asymmetric catalysis, *J. Am. Chem. Soc.* **106**, 3695–3697.

WONG, C.-H., WHITESIDES, G. M. (1994), *Enzymes in Synthetic Organic Chemistry*, Tetrahedron Organic Chemistry Series Vol. 12. Trowbridge: Pergamon.

WONG, C. H., MAZENOID, F. P., WHITESIDES, G. M. (1983), Chemical and enzymatic syntheses of 6-deoxyhexoses. Conversion to 2,5-dimethyl-4-hydroxy-2,3-dihydrofuran-3-one (furaneol) and analogs, *J. Org. Chem.* **48**, 3493–3497.

YAMAUCHI, T., CHOI, S. Y., OKADA, H., YOHDA, M., KUMAGAI, H. et al. (1992), Properties of aspartate racemase, a pyridoxal 5′-phosphate-independent amino acid racemase, *J. Biol. Chem.* **267**, 18361–18365.

YEE, C., BLYTHE, T. A., MCNABB, T. J., WALTS, A. E. (1992), Biocatalytic resolution of tertiary α-substituted carbocyclic acid esters: Efficient preparation of a quaternary asymmetric carbon center, *J. Org. Chem.* **57**, 3525–3527.

The Future of Biotransformations

11 Catalytic Antibodies

George Michael Blackburn
Arnaud Garçon
Sheffield, UK

1 Introduction

This review seeks to deal with most of the important developments in the field of catalytic antibodies since the initial successes were announced over a dozen years ago (POLLACK et al., 1986; TRAMONTANO et al., 1986). The development of catalytic antibodies has required contributions from a number of scientific disciplines, which traditionally have not worked in concert. Thus, while this chapter describes catalytic antibodies, their reactions, and their mechanisms from a biotransformations viewpoint, it will also provide a review that demands only a basic biochemical knowledge of antibody structure, function, and production. Sufficient details of these matters have been supplied to meet the needs of expert and non-expert readers alike. In Sect. 11, there is a glossary of most of the immunological terms used in this review, in language familiar to chemists. While this survey is not fully comprehensive, it seeks to focus on the most significant parts of a subject which, in a little over a decade, has achieved much more than most critics expected at the outset. A moderately complete survey of the literature is presented in the form of an appendix (Sect. 12), which lists over 100 examples of reactions catalyzed by antibodies, the haptens employed, and their kinetic parameters.

1.1 Antibody Structure and Function

One of the most important biological defense mechanisms for higher organisms is the immune response. It relies on the rapid generation of structurally novel proteins that can identify and bind tightly to foreign substances that would otherwise be damaging to the parent organism. These proteins are called immunoglobulins and constitute a protein superfamily. In their simplest form they are made up of four polypeptide chains: one pair of identical short chains and one pair of identical long chains, interconnected by disulfide bridges. The two light and two identical heavy chains contain repeated homologous sequences of about 110 amino acids which fold individually into similar structural domains, essentially a bilayer of antiparallel β-pleated sheets. This gives the structure of an IgG immunoglobulin molecule whose core is formed from twelve, similar structural domains: eight from the two heavy chains and four from the two light chains (Fig. 1) (BURTON, 1990). Notwithstanding this apparent homogeneity, the *N*-terminal regions of antibody light and heavy chains vary greatly in the variety and number of their constituent amino acids and thereby provide binding regions of great diversity called hypervariable regions. The variety of proteins so generated approaches 10^{10} in higher mammals.

The essential property of the immune system is its ability to respond to single or multiple foreign molecular species (antigens) through rapid diversification of the sequences of these hypervariable regions by processes involving mutation, gene splicing, and RNA splicing. This initially provides a vast number of different antibodies which subsequently are selected and amplified in favor of those structures with the strongest affinity for one particular antigen.

1.2 The Search for a New Class of Biocatalyst

Fifty years ago, LINUS PAULING clearly set out his theory that enzymes achieve catalysis because of their complementarity to the transition state for the reaction being catalyzed (PAULING, 1948). With hindsight, this concept could be seen as a logical extension of the new transition state theory that had been recently developed to explain chemical catalysis (EVANS and POLANYI, 1935; EYRING, 1935). The basic proposition was that the rate of a reaction is related to the difference in Gibbs free energy (ΔG_0) between the ground state of reactant(s) and the transition state for that reaction. For catalysis to be manifest, either the energy of the transition state has to be lowered (transition state stabilization) or the energy of the substrate has to be elevated (substrate destabilization). PAULING applied this concept to enzyme catalysis by stating that an enzyme preferentially binds to and thereby stabilizes the transition state for a reaction relative to

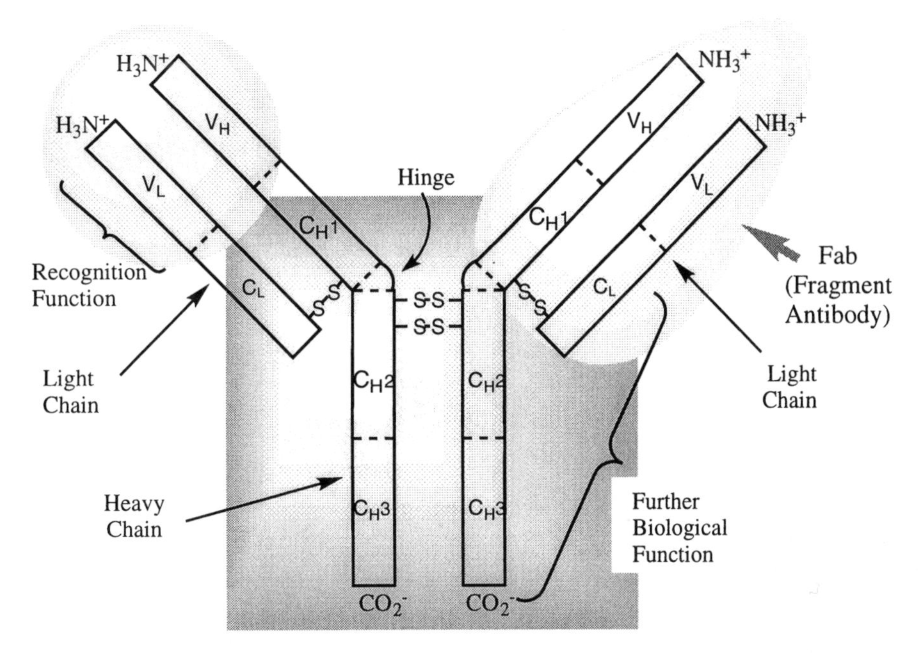

Fig. 1. Scheme to show the structure of the peptide components of an IgG immunoglobulin: the two light (L) and two heavy (H) polypeptide chains; the disulfide bridges (–S–S–) connecting them; four variable regions of the light (V_L) and heavy (V_H) chains; and the eight "constant" regions of the light (C_L) and heavy (C_{H^1}, C_{H^2}, C_{H^3}) chains (shaded rectangle). The hypervariable regions that achieve antigen recognition and binding are located within six polypeptide loops, three in the V_L and three in the V_H sections (shaded circle, top left). These can be excised by protease cleavage to give a fragment antibody, Fab (shaded lobe, top right).

the ground state of substrate(s) (Fig. 2). This has become a classical theory in enzymology and is widely used to explain the way in which such biocatalysts are able to enhance specific processes with rate accelerations of up to 10^{17} over background (ALBERY and KNOWLES, 1976, 1977; ALBERY, 1993 for a review).

PAULING apparently did not bring ideas about antibodies into his concept of enzyme catalysis, although there is a tantalizing photograph in a volume of PAULING's lectures ca. 1948 which shows on a single blackboard a cartoon of an energy profile diagram for the lowering of a transition state energy profile and also reference to an immunoglobulin (PAULING, 1947). And so it fell to BILL JENCKS in his magisterial work 1969 on catalysis to bring together the opportunity for synthesis of an enzyme by the use of antibodies that had been engineered by manipulation of the immune system (JENCKS, 1969).

"One way to do this (i.e., *synthesize an enzyme*) is to prepare an antibody to a haptenic group which resembles the transition state of a given reaction".[1]

The practical achievement of this goal was held up for 18 years, primarily because of the great difficulty in isolation and purification of single-species proteins from the immune repertoire. During that time, many attempts to demonstrate catalysis by inhomogeneous (i.e., polyclonal) mixtures of antibodies were made and failed (e.g., RASO and STOLLAR, 1975;

[1] In making this statement, JENCKS was apparently not aware of PAULING's idea (JENCKS, personal communication).

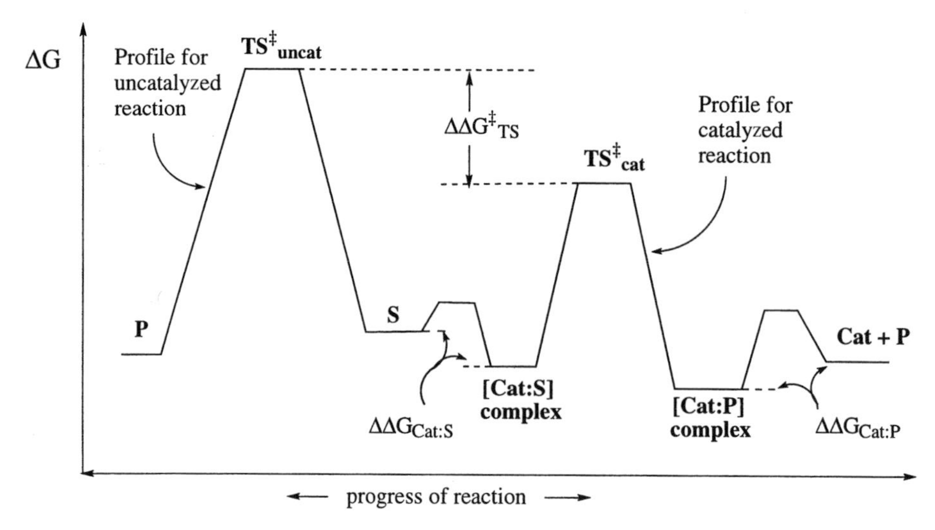

Fig. 2. Catalysis is achieved by lowering the free energy of activation for a process, i.e., the catalyst must bind more strongly to the transition state ($TS^{\ddagger}$) of the reaction than to either reactants or products. Thus: $\Delta\Delta G^{\ddagger} \gg \Delta\Delta G_{Cat:S}$ and $\Delta\Delta G_{Cat:P}$.

SUMMERS, 1983). The problem was resolved in 1976 by KÖHLER and MILSTEIN's development of hybridoma technology, which has made it possible both to screen rapidly the "complete" immune repertoire and to produce relatively large amounts of one specific monoclonal antibody species *in vitro* (KÖHLER and MILSTEIN, 1975; KÖHLER et al., 1976).

While transition states have been discussed in terms of their free energies, there have been relatively few attempts to describe their structure at atomic resolution for most catalyzed reactions. Transition states are high energy species, often involving incompletely formed bonds, and this makes their specification very difficult. In some cases these transient species have been studied using laser femtochemistry (ZEWAIL and BERNSTEIN, 1988; ZEWAIL, 1997), and predictions of some of their geometries have been made using molecular orbital calculations (HOUK et al., 1995). Intermediates along the reaction coordinate are also often of very short lifetime, though some of their structures have been studied under stabilising conditions while their existence and general nature can often be established using spectroscopic techniques or trapping experiments (MARCH, 1992a).

The Hammond postulate predicts that if a high energy intermediate occurs along a reaction pathway, it will resemble the transition state nearest to it in energy (HAMMOND, 1955). Conversely, if the transition state is flanked by two such intermediates, the one of higher energy will provide a closer approximation to the transition state structure. This assumption provides a strong basis for the use of mimics of unstable reaction intermediates as transition state analogs (BARTLETT and LAMDEN, 1986; ALBERG et al., 1992).

1.3 Early Examples of Catalytic Antibodies

In 1986, RICHARD LERNER and PETER SCHULTZ independently reported antibody catalysis of the hydrolysis of aryl esters and of carbonates respectively (POLLACK et al., 1986; TRAMONTANO et al., 1986). Such reactions are well-known to involve the formation and breakdown of an unstable tetrahedral intermediate (**2**), and so this can be deemed to be closely related to the transition state ($TS^{\ddagger}$) of the reaction (Fig. 3).

Transition states of this tetrahedral nature have now been effectively mimicked by a range of stable analogs, including phosphonic acids, phosphonate esters, α-difluoroketones, and hydroxymethylene functional groups (JACOBS, 1991). LERNER's group elicited antibodies to a tetrahedral anionic phosphonate hapten[2] (**3**) (AE[3] 2.9) while SCHULTZ's group isolated a protein with high affinity for *p*-nitrophenyl cholyl phosphate (**5**) (Fig. 4, AE 3.2).

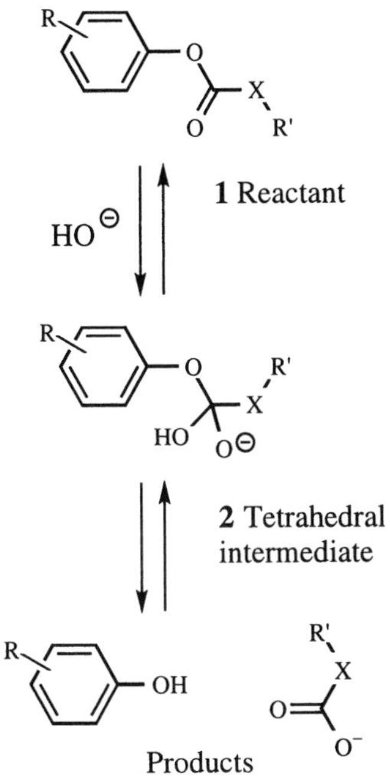

1 Reactant

2 Tetrahedral intermediate

Products

Fig. 3. The hydrolysis of an aryl ester (**1**) ($X=CH_2$) or a carbonate (**1**) ($X=O$) proceeds through a tetrahedral intermediate (**2**) which is a close model of the transition state for the reaction. It differs substantially in geometry and charge from both reactants and products.

[2] It might be helpful to the reader to indicate that the pyridine-2-6-dicarboxylate component of (**3**) was designed for a further purpose, neither used nor needed for the activity described in the present scheme.

[3] AE = Appendix Entry

1.4 Methods for Generating Catalytic Antibodies

It is appropriate at this stage in the review to consider the stages in production of a catalytic antibody and to put in focus the relative roles of chemistry, biochemistry, immunology, and molecular biology. Nothing less than the full integration of these cognate sciences is needed for the fullest realization of the most difficult objectives in the field of catalytic antibodies. In broad terms, the top section of the flow diagram for abzyme production (Fig. 5) involves chemistry, the right hand side is immunology, the bottom sector is biochemistry, and molecular biology completes the core of the scheme.

Chemistry
At the outset, chemistry dominates the selection of the process to be investigated. The targeted reaction should meet most if not all of following criteria:

(1) have a slow but measurable spontaneous rate under ambient conditions;
(2) be well analyzed in mechanistic terms;
(3) be as simple as possible in number of reaction steps;
(4) be easy to monitor;
(5) lead to the design of a synthetically accessible transition state analog (TSA) of adequate stability.

As we shall see later, many catalytic antibodies achieve rate accelerations in the range 10^3 to 10^6. It follows that for a very slow reaction, e.g., the alkaline hydrolysis of a phosphate diester with k_{OH} ca. $10^{-11}\ M^{-1}\ s^{-1}$, direct observation of the reaction is going to be experimentally problematic. Given that concentrations of catalytic antibodies employed are usually in the 1–10 μM range, it has been far more realistic to target the hydrolysis of an aliphatic ester, with k_{OH} ca. $0.1\ M^{-1}\ s^{-1}$ under ambient conditions.

The need for a good understanding of the mechanism of the reaction is well illustrated by the case of amide hydrolysis. Many early enterprises sought to employ TSAs that were based on a stable anionic tetrahedral intermediate, as had been successful for ester hydrolysis. Such approaches identified catalytic anti-

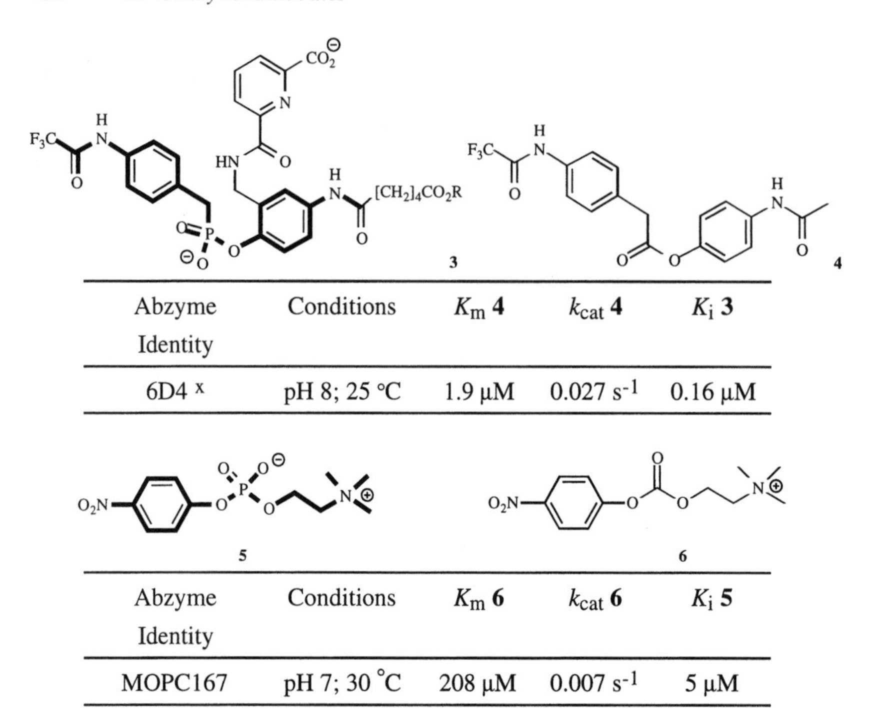

Abzyme Identity	Conditions	K_m **4**	k_{cat} **4**	K_i **3**
6D4 x	pH 8; 25 °C	1.9 μM	0.027 s^{-1}	0.16 μM

Abzyme Identity	Conditions	K_m **6**	k_{cat} **6**	K_i **5**
MOPC167	pH 7; 30 °C	208 μM	0.007 s^{-1}	5 μM

Fig. 4. LERNER's group used phosphonate (**3**) as the hapten to raise an antibody which was capable of hydrolyzing the ester (**4**). SCHULTZ found that naturally occurring antibodies using phosphate (**5**) as their antigen could hydrolyze the corresponding *p*-nitrophenyl choline carbonate (**6**). (Parts of haptens (**3**) and (**5**) required for antibody recognition have been emphasized in bold).

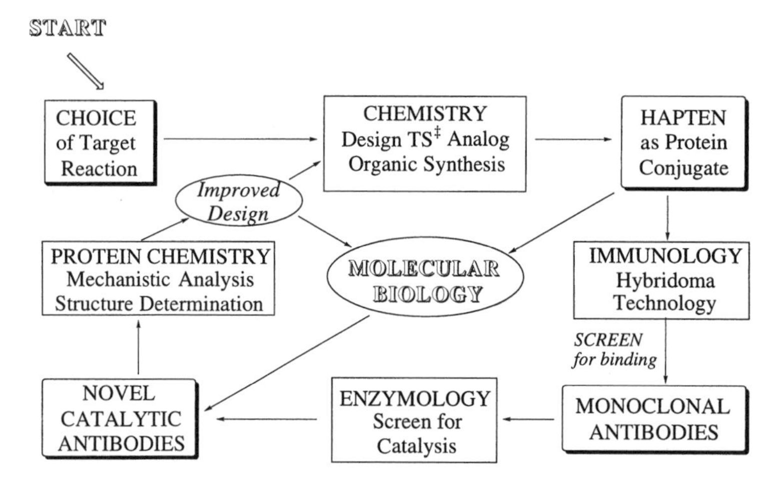

Fig. 5. Stages in the production of a catalytic antibody.

bodies capable of ester hydrolysis but not of amide cleavage! There is good evidence that for aliphatic amides, breakdown of the tetrahedral intermediate is the rate determining step. This step is catalyzed by protonation of the nitrogen leaving group and hence this feature must be part of the TSA design.

The importance of minimizing the number of covalent steps in the process to be catalyzed is rather obvious. Single- and two-step processes dominate the abzyme scene. However, there is substantial evidence that some acyl transfer reactions involve covalent antibody intermediates and so must proceed by up to four covalent steps. Nonetheless, such antibodies were not elicited by intentional design but rather discovered as a consequence of efficient screening for reactivity.

Direct monitoring of the catalyzed reaction has most usually been carried out in real time by light absorption of fluorescent emission analysis and some initial progress has been made with light emission detection. The low quantities of abzymes usually available at the screening stage put a premium on the sensitivity of such methods. However, some work has been carried out of necessity using indirect analysis, e.g., by HPLC or NMR.

Finally, this area of research might well have supported a *Journal of Unsuccessful Abzymes*. It is common experience in the field that some three out of four enterprises fail, and for no obvious reason. It is therefore imperative that chemical synthesis of a TSA should not be the rate determining step of an abzyme project. The average performance target is to achieve hapten synthesis within a year: one or two examples have employed TSAs that could be found in a chemical catalog, the most synthetically-demanding cases have perforce employed multi-step routes of considerable sophistication (e.g., AE 13.2). Lastly, the TSA has to survive *in vivo* for at least two days to elicit the necessary antigenic response.

Immunology

The interface of chemistry and immunology requires conjugation of multiple copies of the TSA to a carrier protein for production of antibodies by standard monoclonal technology (KÖHLER and MILSTEIN, 1976). One such conjugate is used for mouse immunization and a second one for ELISA screening purposes. The carrier proteins selected for this purpose are bovine serum albumin (BSA, RMM 67,000), keyhole limpet hemocyanin (KLH, RMM $4 \cdot 10^6$), and chicken ovalbumin (RMM 32,000). All of these are basic proteins of high immunogenicity and with multiple surface lysine residues that are widely used as sites for covalent attachment of hapten. Successful antibody production can take some three months and should deliver from 20 to 200 monoclonal antibody lines for screening, preferably of IgG isotype.

Screening in early work sought to identify high affinity of the antibody for the TSA, using a process known as ELISA. This search can now be performed more quantitatively by BIAcore analysis, based on surface plasmon resonance methodology (LÖFÅS and JOHNSSON, 1990). A subsequent development is the catELISA assay (TAWFIK et al., 1993) which searches for product formation and hence the identification of abzymes that can generate product.

Methods of this nature are adequate for *screening* sets of hybridomas but not for selection from much larger libraries of antibodies. So, most recently, selection methods employing suicide substrates (Sect. 6.2) (JANDA et al., 1997) or DNA amplification methodology (FENNIRI et al., 1995) have been brought into the repertoire of techniques for the direct identification of antibodies that can turnover their substrate. However, the time-consuming screening of hybridomas remains the mainstay of abzyme identification.

Biochemistry

A family of 100 hybridoma antibodies can typically provide 20 tight binders and these need to be assayed for catalysis. At this stage in the production of an abzyme, the benefit of a sensitive, direct screen for product formation comes into its own. Following identification of a successful catalyst, the antibody-producing cell line is usually recloned to ensure purity and stabilization of the clone, then protein is produced in larger amount (ca. 10 mg) and used for determination of the kinetics and mechanism of the catalyzed process by classical biochemistry. Digestion of such protein with trypsin or papain provides fragment antibodies (Fabs), that contain only the attenuated upper limbs of the intact IgG (Fig. 1). It is these components that have been crystallized, in many cases with the substrate analog, product, or TSA bound in the combining site, and their structures determined by X-ray diffraction.

Molecular Biology
Only a few abzymes have reached the stage where mutagenesis is being employed in order to improve their performance (MILLER et al., 1997). Likewise, HILVERT is the first to have reached the stage of redesign of the hapten to attempt the production of antibodies with enhanced performance (KAST et al., 1996). So, the circle of production has now been completed for at least one example, and chemistry can start again with a revised synthetic target.

2 Approaches to Hapten Design

One can now recognize a variety of strategies in addition to the earliest ones deployed for hapten design. Some of these were presented originally as discrete solutions of the problem of abzyme generation, but it is now recognized that they need not be mutually exclusive either in design or in application. Indeed, more recent work often brings two or more of them together interactively. They can be classified broadly into five categories for the purposes of analysis of their principal design elements. The sequence of presentation of these here is in part related to the chronology of their appearance on the abzyme scene.

(1) Transition state analogs.
(2) Bait and switch.
(3) Entropy traps.
(4) Desolvation.
(5) Supplementation of chemical functionality.

2.1 Transition State Analogs

As has clearly been shown by the majority of all published work on catalytic antibodies, the original guided methodology, i.e., the design of stable transition state analogs (TSAs) for use as haptens to induce the generation of catalytic antibodies, has served as the bedrock of abzyme research. Most work has been directed at hydrolytic reactions of acyl species, perhaps because of the broad knowledge of the nature of reaction mechanisms for such reactions and the wide experience of deploying phosphoryl species as stable mimics of unstable tetrahedral intermediates. More than 80 examples of hydrolytic antibodies have been reported, including the 47 examples of acyl group transfer to water (entries under 1–5 of the Appendix).

Most such acyl transfer reactions involve stepwise addition of the nucleophile followed by expulsion of the leaving group with a transient, high energy, tetrahedral intermediate separating these processes. The fastest of such reactions generally involve good leaving groups and the addition of the nucleophile is the rate determining step. This broad conclusion from much detailed kinetic analysis has been endorsed by computation for the hydrolysis of methyl acetate (TERAISHI et al., 1994). This places the energy for product formation from an anionic tetrahedral intermediate some 7.6 kcal mol^{-1} lower than for its reversion to reactants. So, for the generation of antibodies for the hydrolysis of aryl esters, alkyl esters, carbonates, and activated anilides, the design of hapten has focused on facilitating nucleophilic attack, and with considerable success.

The tetrahedral intermediates used for this purpose initially deployed phosphorus (V) systems, relying on the strong polarization of the P=O bond (arguably more accurately represented as P^+–O^-). The range has included many of the possible species containing an ionized P–OH group (Fig. 6). One particularly good feature of such systems is that the P^+–O^- bond is intermediate in length (1.521 Å) between the C–O^- bond calculated for an anionic tetrahedral intermediate (0.2–0.3 Å shorter) and for the C···O breaking bond in the transition state (some 0.6 Å longer) (TERAISHI et al., 1994). Other tetrahedral systems used have included sulfonamides (SHEN, 1995) and sulfones (BENEDETTI et al., 1996), secondary alcohols (SHOKAT et al., 1990), and α-fluoroketone hydrates (KITAZUME et al., 1994).

It is clear that phosphorus-based transition states have had the greatest success, as shown by the many entries under 1–5 of the Appendix. This may be a direct result of their anionic or partial anionic character, a feature not generally available for the other species illustrated in Fig. 6, though α-difluorosulfonamides might reasonably also share this feature as a result of their enhanced acidity.

Not surprisingly, most of the catalytic antibody binding sites examined in structural detail have been found to contain a basic residue that provides a coulombic interaction with these TSAs, for which the prototype is the natural antibody McPC603 to phosphoryl choline, where the phosphate anion is stabilized by coulombic interaction with Arg^{H52} (PADLAN et al., 1985). In particular, X-ray structures analyzed by FUJII (FUJII et al., 1995) have shown that the protonated His^{H27d} in catalytic antibodies 6D9, 4B5, 8D11, and 9C10 (AE 1.8) is capable of forming a hydrogen bond to the oxyanion in the transition state for ester hydrolysis.

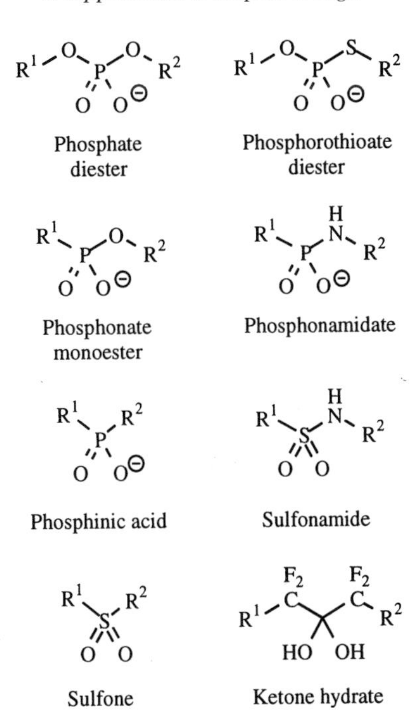

Fig. 6. Transition state analogs for acyl cleavage reaction via tetrahedral transitions states.

In similar vein, KNOSSOW has identified His^{H35} located proximate to the oxyanion of p-nitrophenyl methanephosphonate in the crystalline binary complex of antibody CNJ206 and TSA, a system designed to hydrolyze p-nitrophenyl acetate (c.f. AE 2.7) (CHARBONNIER et al., 1995). A third example is seen in SCHULTZ's structure of antibody 48G7, which hydrolyzes methyl p-nitrophenyl carbonate (AE 3.1c). The hapten p-nitrophenyl 4-carboxybutane-phosphonate is proximate to Arg^{L96} and also forms hydrogen bonds to His^{H35}, Tyr^{H33}, and Tyr^{L94} (Fig. 7) (PATTEN et al., 1996).

Clearly, the oxyanion hole is now as significant a feature of the binding site of such acyl transfer abzymes as it is already for esterases and peptidases – and not without good reason.

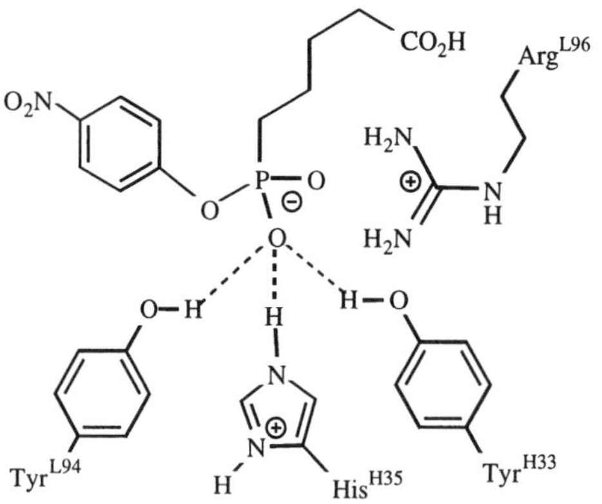

Fig. 7. Binding site details for antibody 48G7 complexed with hapten *p*-nitrophenyl 4-carboxybutanephosphonate (PATTEN et al., 1996). NB Amino acid residues in antibodies are identified by their presence in the light (L) or heavy (H) chains with a number denoting their sequence position from the *N*-terminus of the chain.

KNOSSOW has analyzed the structures of three esterase-like catalytic antibodies, each elicited in response to the same phosphonate TSA hapten (CHARBONNIER et al., 1997). Catalysis for all three is accounted for by transition state stabilization and in each case there is an oxyanion hole involving a tyrosine residue. This strongly suggests that evolution of immunoglobulins for binding to a single TSA hapten followed by selection from a large hybridoma repertoire by screening for catalysis leads to antibodies with structural convergence. Furthermore, the juxtaposition of X-ray structures of the unliganded esterase mAb D2.3 and its complexes with a substrate analog and with one of the products provides a complete description of the reaction pathway. D2.3 acts at high pH by attack of hydroxide on the substrate with preferential stabilization of the oxyanion anionic tetrahedral intermediate, involving one tyrosine and one arginine residue. Water readily diffuses to the reaction center through a canal that is buried in the protein structure (GIGANT et al., 1997). Such a clear picture of catalysis now opens the way for site-directed mutagenesis to improve the performance of this antibody.

2.2 Bait and Switch

Charge–charge complementarity is an important feature involved in the specific and tight binding of antibodies to their respective antigens. It is the amino acid sequence and conformation of the hypervariable (or complementarity determining regions, CDRs) in the antibody combining site which determine the interactions between antigen and antibody. This has been exploited in a strategy dubbed "bait and switch" for the induction of antibody catalysts which perform β-elimination reactions (SHOKAT et al., 1989; THORN et al., 1995), acyl-transfer processes (JANDA et al., 1990b, 1991b; SUGA et al., 1994a; LI and JANDA, 1995), *cis–trans* alkene isomerizations (JACKSON and SCHULTZ, 1991), and dehydration reactions (UNO and SCHULTZ, 1992).

The bait and switch methodology deploys a hapten to act as a "bait". This bait is a modified substrate that incorporates ionic functions intended to represent the coulombic distribution expected in the transition state. It is thereby designed to induce complementary, oppositely charged residues in the combining site of antibodies produced by the response of the immune system to this hapten. The catalytic ability of these antibodies is then sought by a subsequent "switch" to the real substrate and screening for product formation, as described above.

The nature of the combining site of an antibody responding to charged haptens was first elucidated by GROSSBERG and PRESSMAN (GROSSBERG and PRESSMAN, 1960), who used a cationic hapten containing a *p*-azophenyltrimethylammonium ion to make antibodies with a combining site carboxyl group, essential for substrate binding (as shown by diazoacetamide treatment).

The first example of "bait and switch" for catalytic antibodies was provided by SHOKAT (SHOKAT et al., 1989), whose antibody 43D4-3D12 raised to hapten (**7**) was able to catalyze the β-elimination of (**8**) to give the *trans*-enone (**9**) with a rate acceleration of $8.8 \cdot 10^4$ above background (Fig. 8; AE 8.2). Subsequent analysis has identified a carboxylate residue, Glu^{46H} as the catalytic function induced by the cationic charge in (**7**) (Fig. 6) (SHOKAT et al., 1994).

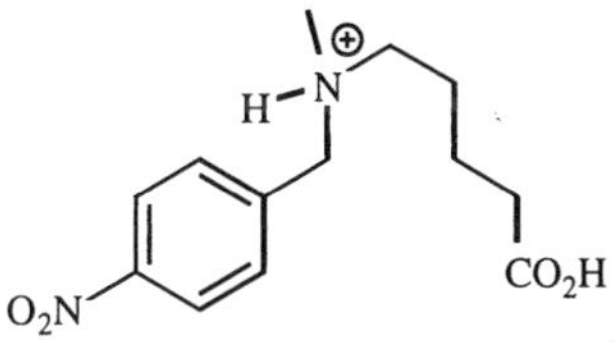

Abzyme Identity		Conditions
43D4-3D12		pH 6; 37 °C
K_m **8**	k_{cat} **8**	K_i **7**
182 μM	0.003 s^{-1}	0.29 μM

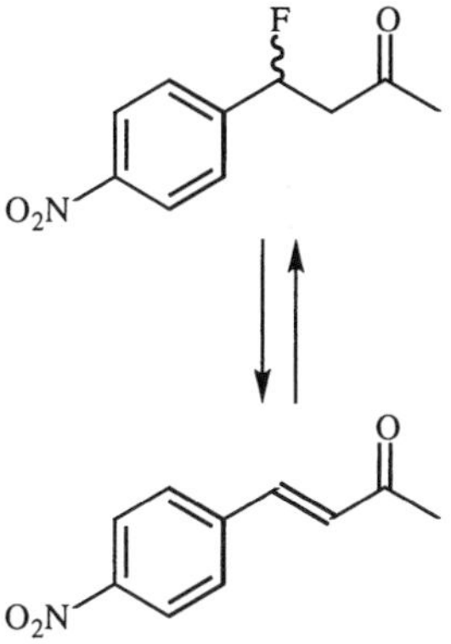

Fig. 8. Using the "bait and switch" principle, hapten (**7**) elicited an antibody, 43D4-3D12, which catalyzed the *β*-elimination of (**8**) to a *trans*-enone (**9**). The carboxyl function in (**7**) is necessary for its attachment to the carrier protein.

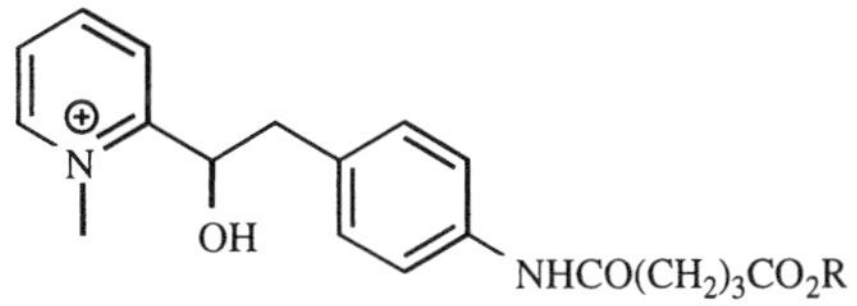

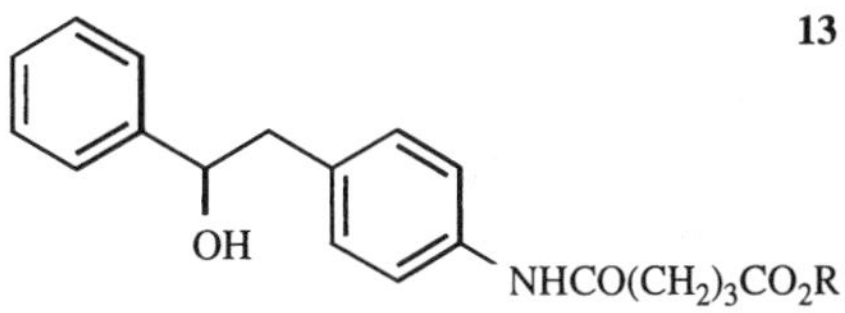

R = succinimidyl

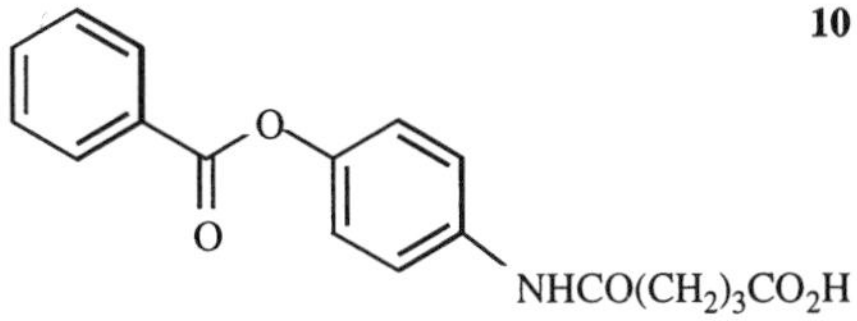

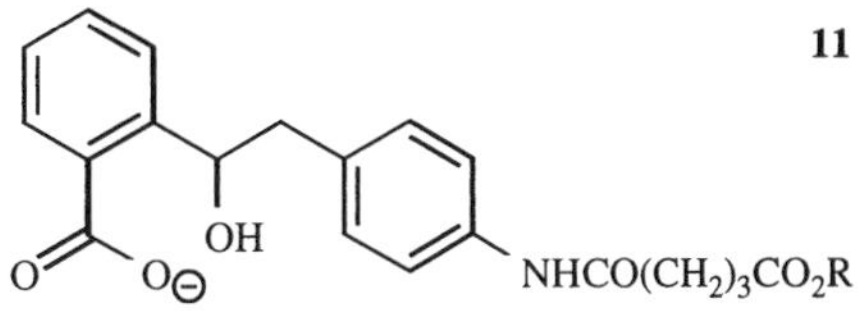

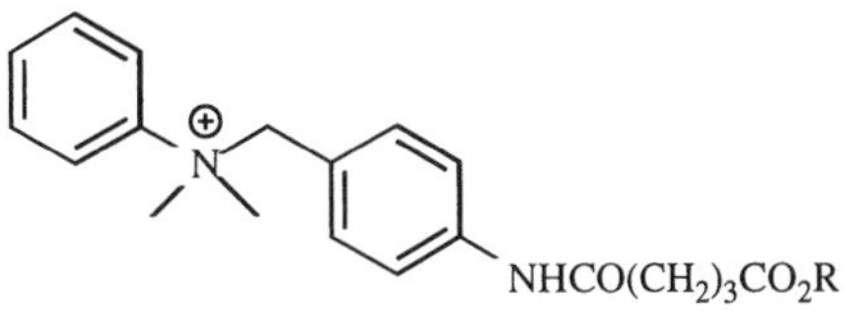

Fig. 9. The original hapten (**10**) demonstrated the utility of the "bait and switch" strategy in the generation of antibodies to hydrolyze the ester substrate (**11**). Three haptens (**12**), (**13**), (**14**) were designed to examine further the effectiveness of point charges in amino acid induction. Both charged haptens (**12**), (**13**) produced antibodies that catalyzed the hydrolysis of (**11**) whereas the neutral hapten (**14**) generated antibodies which bound the substrate unproductively.

A similar "bait and switch" approach has been exploited for acyl-transfer reactions (JANDA et al., 1990b, 1991b). The design of hapten (**10**) incorporates both a transition state mimic and the cationic pyridinium moiety, designed to induce the presence of a potential general acid/base or nucleophilic amino acid residue in the combining site, able to assist in catalysis of the hydrolysis of substrate (**11**) (Fig. 9, AE 2.6).

Some 30% of all of the monoclonal antibodies obtained using hapten (**10**) were catalytic, and so the work was expanded to survey three other antigens based on the original TSA design (JANDA et al., 1991b). The carboxylate anion in (**12**) was designed to induce a cationic combining site residue, while the quaternary ammonium species (**13**) combines both tetrahedral mimicry and positive charge in the same locus. Finally, the hydroxyl group in (**14**) was designed to explore the effects of a neutral antigen.

Three important conclusions arose from this work.

(1) A charged functionality is crucial for catalysis.
(2) Catalytic antibodies are produced from targeting different regions of the binding site with positive and negative haptens (although more were generated in the case of the cationic hapten used originally).
(3) The combination of charge plus mimicry of the transition state is required to induce hydrolytic esterases.

Esterolytic antibodies have also been produced by SUGA using an alternative "bait and switch" strategy (AE 2.1) (SUGA et al., 1994a). A 1,2-aminoalcohol function was designed for generating not only esterases but also amidases. In order to elicit an anionic combining site for covalent catalysis, three haptens were synthesized, one contained a protonated amine (**15**) and two featured trimethylammonium cations (**16**), (**17**) (Fig. 10). The outcome was interpreted as suggesting that haptens containing a trimethylammonium group were too sterically demanding, so that the induced anionic amino acid residues in the antibody binding pocket were too distant to provide nucleophilic attack at the carbonyl carbon of substrate (**18**). An alternative explanation may be that coulombic interactions lacking any hydrogen bonding capability will not be sufficiently short-range for the purpose intended.

The use of secondary hydroxyl groups in the haptens (**15**) and (**16**) was designed to mimic the tetrahedral geometry of the transition state (as in JANDA's work), while the third hapten (**17**) replaced the neutral OH with an anionic phosphate group, designed to elicit a cationic combining site residue to stabilize the transition state oxyanion. However, this function in (**17**) may have proved too large to induce a catalytic residue close enough to the developing oxyanion, since weaker catalysis was observed relative to haptens (**15**) and (**16**) ($k_{cat}/k_{uncat} = 2.4 \cdot 10^3$, $3.3 \cdot 10^3$, and $\sim 1 \cdot 10^3$ for (**15**), (**16**), and (**17**), respectively) (Fig. 10).

In order to achieve catalysis employing both acid and basic functions, an alternative zwitterionic hapten was proposed in which the anion-

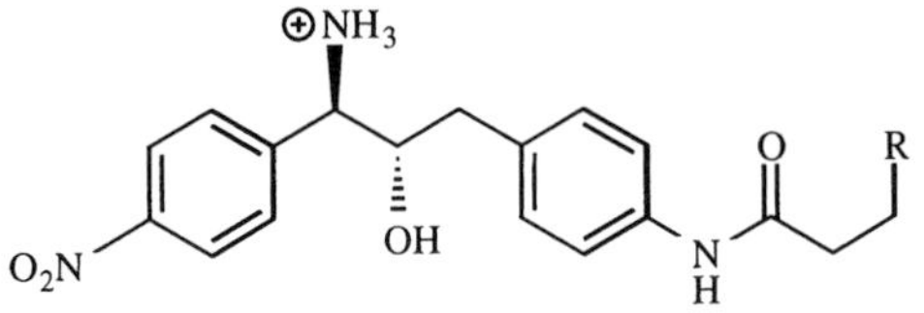

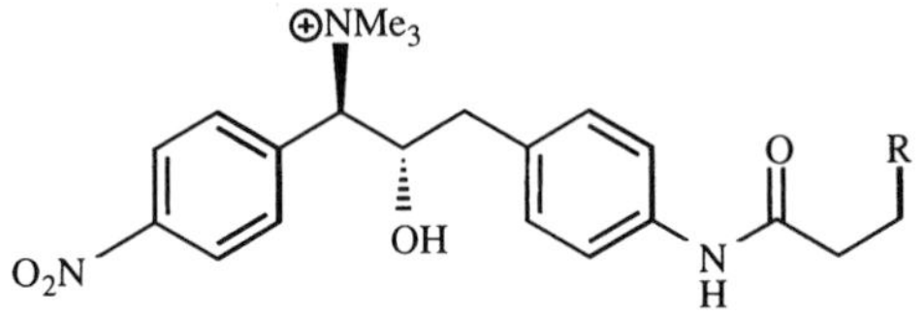

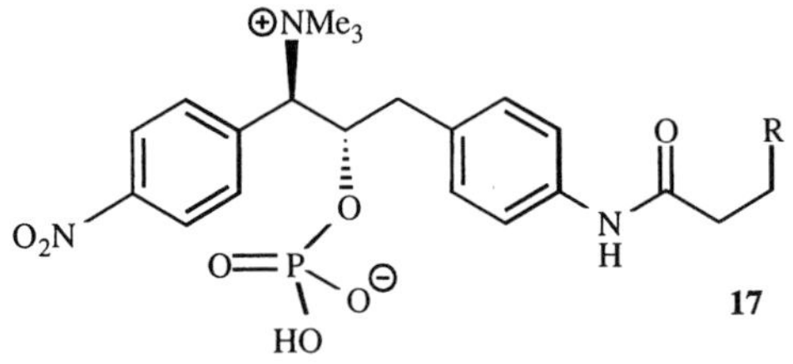

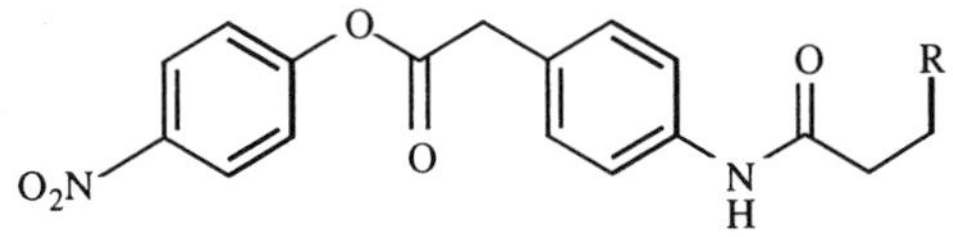

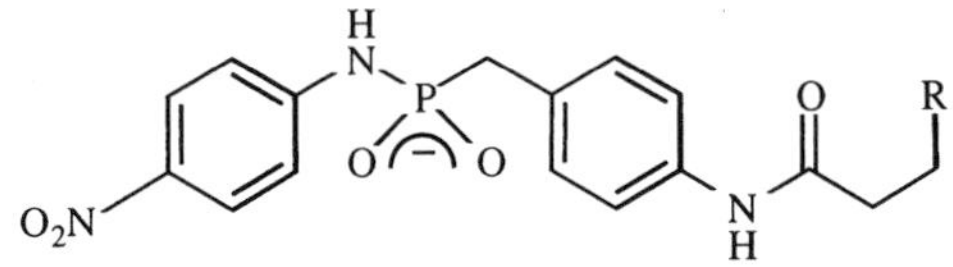

Fig. 10. Three haptens (**15**), (**16**), (**17**) containing 1,2-aminoalcohol functionality were investigated as alternatives for esterase and amidase induction. Half of the antibodies raised against hapten (**15**) were shown to catalyze the hydrolysis of ester (**18**), thereby establishing the necessity for a compact haptenic structure. Hapten (**19**) along with (**16**) was employed in a heterologous immunization program to elicit both a general and acid base function in the antibody binding site.

ic phosphoryl core is incorporated alongside the cationic trimethylammonium moiety (cf. **17**) (SUGA et al., 1994b). The difficulty in synthesizing such a target hapten can be overcome by stimulating the immune system first with the cationic and then with the anionic point charges using the two structurally related haptens (**16**) and (**19**), respectively. Such a sequential strategy has been dubbed "heterologous immunization" (Fig. 10) and this results of this strategy were compared with those from the individual use of haptens (**16**) and (**19**) in a "homologous immunization" routine. Of 48 clones produced as a result of the homologous protocols, 7 were found to be catalytic, giving rate enhancements up to $3 \cdot 10^3$. By contrast, 19 of the 50 clones obtained using the heterologous strategy displayed catalysis, the best being up to two orders of magnitude better.

A final example of the bait and switch strategy (THORN et al., 1995) focuses on the base-promoted decomposition of substituted benzisoxazole (**20**) to give cyanophenol (**21**) (Fig 11, AE 8.4). A cationic hapten (**22**) was used to mimic the transition state geometry of all reacting bonds. It was anticipated that if the benzimidazole hapten (**22**) induced the presence of a carboxylate in the binding site, it would be ideally positioned to make a hydrogen bond to the N-3 proton of the substrate. The resultant abzymes would thus have general base capability for abstracting the H-3 in the substrate.

Two monoclonals, 34E4 and 35F10, were found to catalyze the reaction with a rate acceleration greater than 10^8, while the presence of a carboxylate-containing binding site residue was confirmed by pH-rate profiles and covalent modification by a carbodiimide, which reduced catalysis by 84%.

The bait and switch tactic clearly illustrates that antibodies are capable of a coulombic response that is potentially orthogonal to the use of transition state analogs in engendering catalysis. By variations in the hapten employed, it is possible to fashion antibody combining sites that contain individual residues to deliver intricate mechanisms of catalysis.

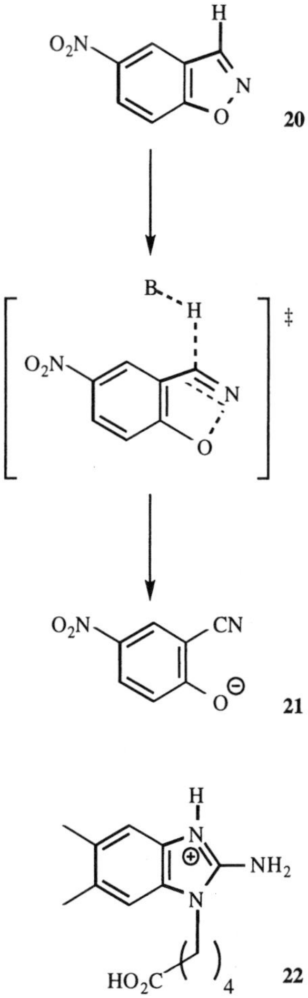

O_2N H N O **20**

B H ‡ O_2N N O

O_2N CN $O^{\ominus}$ **21**

H N ⊕ NH_2 N HO_2C 4 **22**

Fig. 11. The use of a cationic hapten (**22**) mimics the transition state of the base-promoted decomposition of substituted benzisoxazole (**20**) to cyanophenol (**21**) and also acts as a "bait" to induce the presence of an anion in the combining site that may act as a general base.

2.3 Entropic Trapping

Rotational Entropy

An important component of enzyme catalysis is the control of translational and rotational entropy in the transition state (PAGE and JENCKS, 1971). This is well exemplified for unimolecular processes by the enzyme chorismate mutase, which catalyzes the rearrangement of chorismic acid (**23**) into prephenic acid (**24**)

(Fig. 12). This reaction proceeds through a cyclic transition state having a pseudo-diaxial conformation (**25**) (ADDADI et al., 1983). With this analysis, BARTLETT designed and synthesized a transition state analog (**26**) which proved to be a powerful inhibitor for the enzyme (BARTLETT and JOHNSON, 1985). X-ray structures of mutases from *Escherichia coli* (LEE, et al., 1995), *Bacillus subtilis* (CHOOK et al., 1993, 1994), and *Saccharomyces cerevisiae* (XUE and LIPSCOMB, 1995) complexed to (**26**) show completely different protein architectures although the bacterial enzymes have similar values of k_{cat}/k_{uncat} ($3 \cdot 10^6$) and of K_i for (**26**). It appears that these enzymes exert their catalysis through a combination of conformational control and enthalpic lowering. Supporting this, HILLIER has carried out a hybrid quantum mechanical/molecular mechanics calculation on the *B. subtilis* complex with substrate (**23**). He concluded that interactions between protein and substrate are maximal close to the transition state (**25**) and lead to a lowering of the energy barrier greater than is needed to produce the observed rate acceleration (DAVIDSON and HILLIER, 1994).

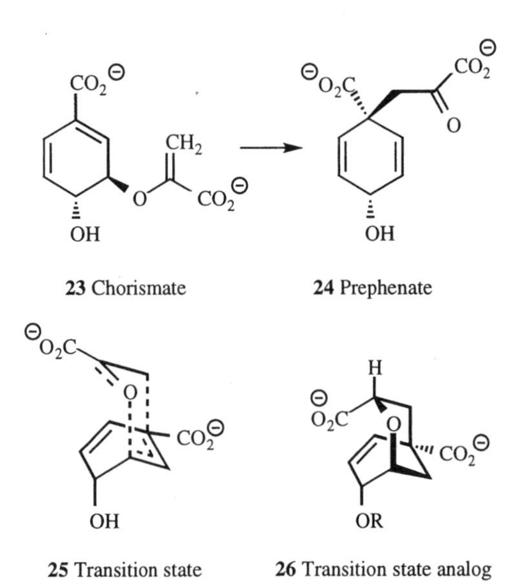

Fig. 12. Chorismate mutase.

SCHULTZ employed TSA (**26**) as a hapten to generate antibodies to catalyze this same isomerization reaction (**23**)→(**24**) (Fig. 12) (JACKSON et al., 1988). His kinetic analysis of one purified antibody revealed that it increases the entropy of activation of the reaction by 12 cal mol^{-1} K^{-1} (Tab. 1, Antibody 11F1-2E11, AE 13.2), and gives a rate enhancement of 10^4. He suggested that this TSA induces a complementary combining site in the abzyme that constrains the reactants into the correct conformation for the [3,3]-sigmatropic reaction and designated this strategy as an "entropic trap".

HILVERT's group used the same hapten (**26**) with a different spacer to generate an antibody catalyst which has very different thermodynamic parameters. It has a high entropy of activation but an enthalpy lower than that of the wild type enzyme (Tab. 1, Antibody 1F7, AE 13.2) (HILVERT et al., 1988; HILVERT and NARED, 1988). WILSON has determined an X-ray crystal structure for the Fab' fragment of this antibody in a binary complex with its TSA (HAYNES, et al., 1994) which shows that amino acid residues in the active site of the antibody catalyst faithfully complement the components of the conformationally ordered transition state analog (Fig. 13) while a trapped wa-

Tab. 1. Kinetic and Thermodynamic Parameters for the Spontaneous, Enzyme Catalyzed and Antibody Catalyzed Conversion of Chorismic Acid (**23**) into Prephenic Acid (**24**)

Catalyst	Relative Rate	$\Delta G^{\ddagger}$ [kcal mol^{-1}]	$\Delta H^{\ddagger}$ [kcal mol^{-1}]	$\Delta S^{\ddagger}$ [cal mol^{-1} K^{-1}]	K_m **23**	k_{cat} **23**	K_i **26**
Spontaneous[a]	1	24.2	20.5	−12.9			
Chorismate mutase[b]	$3 \cdot 10^6$	15.9	15.9	0	45 μM	1.35 s^{-1}	75 nM
Antibody 1F7[c]	250	21.3	15.0	−22	49 μM	0.023 min^{-1}	600 nM
11F1-2E11[d]	10,000	18.7	18.3	− 1.2	260 μM	0.27 min^{-1}	9 μM

[a] At 25 °C; [b] *E. coli* enzyme at 25 °C; [c] pH 7.5, 14 °C; [d] pH 7.0, 10 °C.

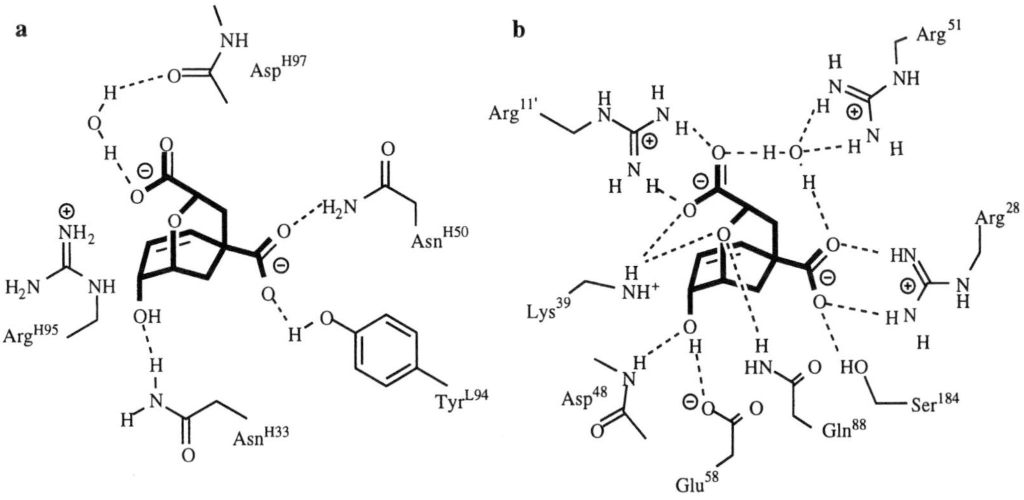

Fig. 13. Schematic diagrams of X-ray crystal structures show the hydrogen bonding (dashed lines) and electrostatic interactions between the transition state analog (**26**) (in grey) with relevant side chains of **a** antibody 1F7 (HAYNES et al., 1994) and **b** the active site of the *E. coli* enzyme (LEE et al., 1995).

ter molecule is probably responsible for the adverse entropy of activation. Thus it appears that antibodies have emulated enzymes in finding contrasting solutions to the same catalytic problem.

Further examples of catalytic antibodies that are presumed to control rotational entropy are AZ-28, that catalyzes an oxy-Cope [3.3]-sigmatropic rearrangement (AE 13.1) (BRAISTED and SCHULTZ, 1994; ULRICH et al., 1996) and 2E4 which catalyzes a peptide bond isomerization (AE 13.3) (GIBBS, et al., 1992b; LIOTTA, et al., 1995). Perhaps the area for the greatest opportunity for abzymes to achieve control of rotational entropy is in the area of cationic cyclization reactions (LI, et al., 1997). The achievements of the LERNER group in this area (AE 15.1–15.4) will be discussed later in this article (Sect. 5.3).

Translational Entropy

The classic example of a reaction that demands control of translational entropy is surely the Diels–Alder cycloaddition. It is accelerated by high pressure and by solutions 8M in LiCl (BLOKZIJL and ENGBERTS, 1994; CIOBANU and MATSUMOTO, 1997; DELL, 1997) and proceeds through an entropically disfavored, highly ordered transition state, showing large activation entropies in the range of -30 to -40 cal mol^{-1} K^{-1} (SAUER, 1966).

While it is one of the most important and versatile transformations available to organic chemists, there is no unequivocal example of a biological counterpart. However, two cell free fractions from *Alternaria solani* have been purified (KATAYAMA et al., 1998) which are capable of converting prosolanapyrone II to solanapyrones A and D (OIKAWA et al., 1999). This reaction can be rationalized as an oxidation followed by an intramolecular Diels–Alder reaction (OIKAWA et al., 1998, 1997). In addition there are many natural products, such as himgravine (BALDWIN et al., 1995) and the manzamines (BALDWIN et al., 1998) which are plausibly formed biosynthetically by intramolecular Diels–Alder reactions (ICHIHARA and OIKAWA, 1998), albeit no enzyme has been implicated in these latter processes so far. Hence, attempts to generate antibodies which could catalyze this reaction were seen as an important target.

The major task in producing a "Diels–Alderase" antibody lies in the choice of a suitable haptenic structure because the transition state for the reaction resembles product more closely than reactants (Fig. 14). The reaction product itself is an inappropriate hapten because it is likely to result in severe product inhibition of the catalyst, thereby preventing turnover.

Tetrachlorothiophene dioxide (TCTD) (**27**) reacts with *N*-ethylmaleimide (**28**) to give an unstable, tricyclic intermediate (**29**) that extrudes SO_2 spontaneously to give a dihydrophthalimide as the bicyclic adduct (**30**) (RAASCH, 1980). This led to the design of hapten as a bridged dichlorotricycloazadecene derivative (**31**) which closely mimics the high energy intermediate (**29**), while being sufficiently different from the product (**30**) to avoid the possibility of inhibition by end-product (HILVERT, et al., 1989).

Several antibodies raised to the hapten (**31**) accelerated the Diels–Alder cycloaddition between (**27**) and (**28**) (Fig. 14). The most efficient of these, 1E9, performs multiple turnovers showing that product inhibition has been largely avoided. Comparison of k_{cat} with the second order rate constant for the uncatalyzed reaction ($k_{uncat} = 0.04\ M^{-1}\ min^{-1}$, 25 °C) gives an effective molarity, EM,[4] of 110 M (AE 17.1) (HILVERT et al., 1989). This value is several orders of magnitude larger than any attainable concentration of substrates in aqueous solution, therefore the antibody binding site confers a significant entropic advantage over the bimolecular Diels–Alder reaction.

A number of further examples of Diels–Alder catalytic antibodies have been described (AE 17.2–17.5) and they must necessarily benefit from the same entropic advantage over spontaneous reactions, albeit without HILVERT's clever approach to avoiding product inhibition. Their success in achieving control of regio- and stereochemistry will be discussed later (Sect. 5.1).

Of greater long-term significance is the control of translational entropy for antibody catalyzed synthetic purposes. BENKOVIC's description of an antibody ligase capable of joining an activated amino acid (e.g., **32**) to a second amino acid to give a dipeptide and to a dipeptide

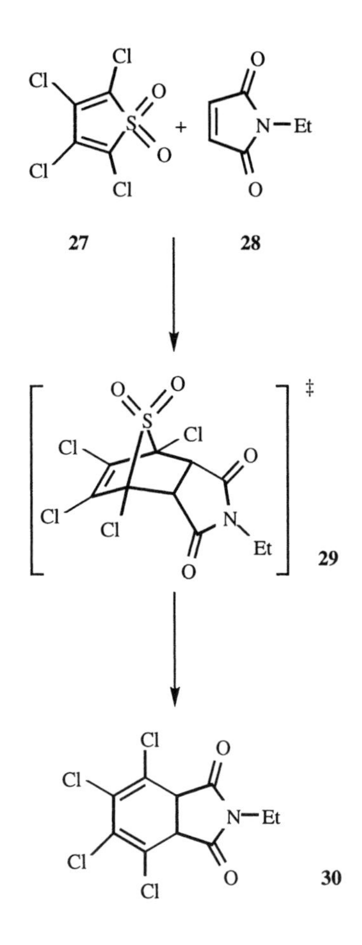

Fig. 14. The Diels–Alder cycloaddition of TCTD (**27**) and *N*-ethylphthalimide (**28**) proceeds through an unstable intermediate (**29**) which extrudes SO_2 spontaneously to give the dihydrophthalimide adduct (**30**). Hapten (**31**) was designed as a stable mimic of (**29**) that would be sufficiently different from product (**30**) to avoid product inhibition of the antibody catalyst.

[4] The EM is equivalent to the concentration of substrate that would be needed in the uncatalyzed reaction to achieve the same rate as achieved by the antibody ternary complex (KIRBY, 1980).

(e.g., **33**) to give a tripeptide with only low product inhibition is particularly significant (Fig. 15, AE 18.4) (SMITHRUD et al., 1997). Antibody 16G3 can achieve 92% conversion of substrates for tripeptide formation and 70% for tetrapeptide synthesis within an assay time of 20 min. A concentration of 20 μM antibody can produce a 1.8 mM solution of a dipeptide in 2 h. The very good regiocontrol of the catalyzed process is shown by the 80:1 ratio for formation of the programmed peptide (**34**) compared to the unprogrammed product (**35**) whereas the uncatalyzed reaction gives a 1:1 ratio.

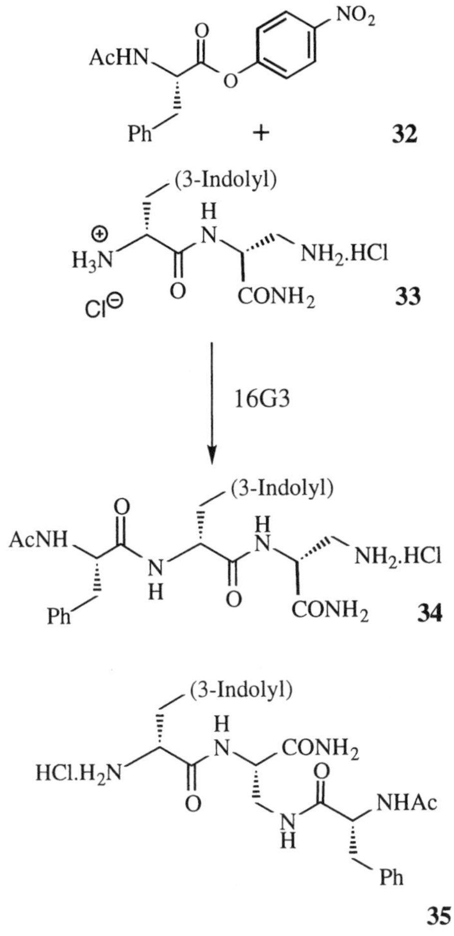

Fig. 15. Antibody 16G3 catalyzed peptide bond formation predominantly yields the peptide (**34**) (**34**:**35**, 80:1), whereas the uncatalyzed reaction gives a 1:1 ratio of the two peptides.

2.4 Substrate Desolvation

A classic example of rate acceleration by desolvation is the Kemp decarboxylation of 6-nitro-3-carboxybenzisoxazole (**36**). The change from water to a less polar environment can effect a 10^7-fold rate acceleration which has been ascribed to a combination of substrate destabilization by loss of hydrogen bonding to solvent and transition state stabilization in a dipolar aprotic solvent (KEMP et al., 1975). Both HILVERT and KIRBY have sought to generate abzymes for this process (AE 9.1) (LEWIS et al., 1991; SERGEEVA et al., 1996) HILVERT generated several antibodies using TSA (**37**) and the best, 25E10, gave a rate acceleration of $23.2 \cdot 10^3$ for decarboxylation of (**36**), comparable to rate accelerations found in other mixed solvent systems but much less than for hexamethylphosphoric triamide (10^8, Fig. 16). In particular, it is of some concern that the K_m for this antibody is as high as 25 mM, which reflects the tenuous relationship between the hapten design and the substrate/transition state structure. Unfortunately, apparently better designed TSAs, e.g. (**38**) (SERGEEVA et al., 1996), fared worse in outcome, possibly through the absence of a countercation in the binding site. This may offer an opportunity for protein engineering to induce the presence of an *N,N,N*-trimethyl-lysine residue in the active site to provide a non-hydrogen bonding salt pair.

Selenoxide *syn*-eliminations are another type of reaction favored by less polar solvents (REICH, 1979). The planar 5-membered, pericyclic transition state for *syn*-elimination of (**40**) was mimicked by the racemic proline-based *cis*-hapten (**39**) to give 28 monoclonal antibodies (AE 8.5) (ZHOU et al., 1997). Abzyme SZ-*cis*-42F7 converted substrate (**40**) exclusively into *trans*-anethole (**41**) with an enhancement ratio (ER) of 62 (R = Me, X = NO_2) and with a low K_m of 33 μM. Abzyme SZ-*cis*-39C11 gave a good acceleration, k_{cat} 0.036 min^{-1}, k_{cat}/K_m 2,400 M^{-1} min^{-1} (substrate **40**, R = H = X) comparable to the rate in 1,2-dichloroethane solution. Unexpectedly, the catalytic benefit appears to be mainly enthalpic for both the antibody and for the solvent switch, as shown by the data in Tab. 2.

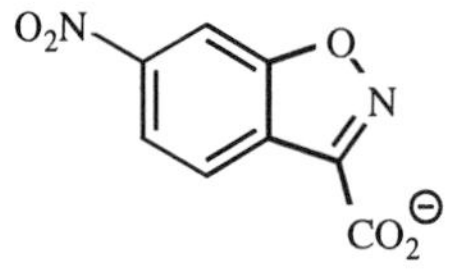

36

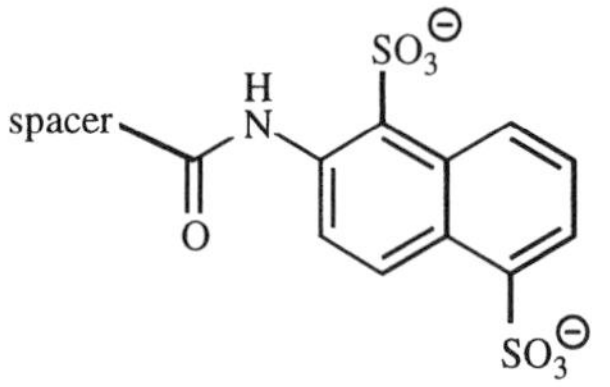

37

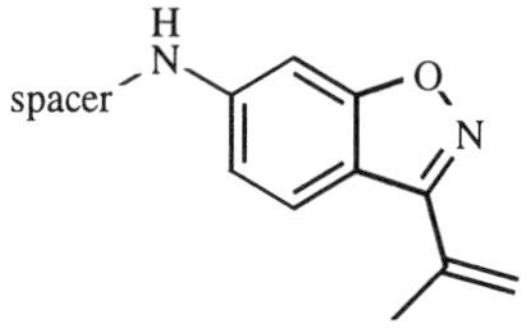

38

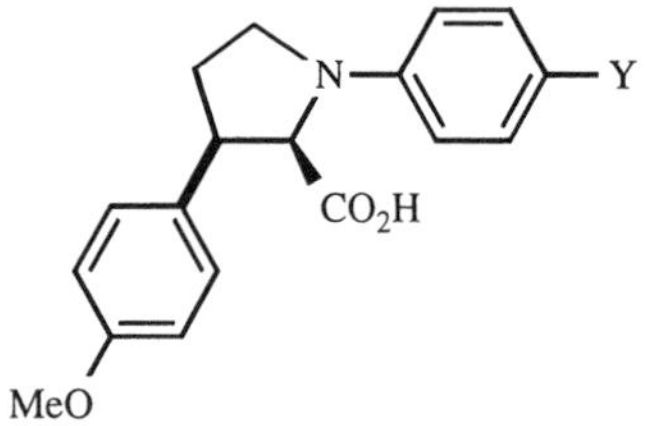

39

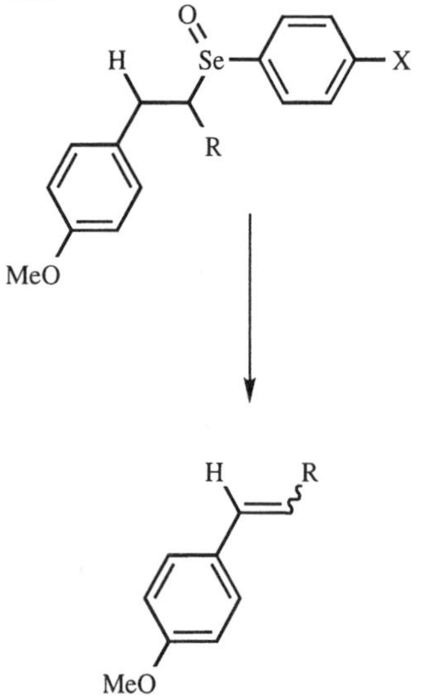

40

41

Fig. 16. TSAs (**37**), (**38**) for Kemp decarboxylation of 6-nitro-3-carboxybenzisoxazole (**36**) and for selenoxide elimination (**39**).

2.5 Supplementation of Chemical Functionality

Several antibodies have been modified to incorporate natural or synthetic groups to aid catalysis (POLLACK et al., 1988). POLLACK and SCHULTZ reported the first example of a semisynthetic abzyme by the introduction of an imidazole residue into the catalytic site through selective modification of the thiol-containing antibody MOPC315 (POLLACK and SCHULTZ, 1989). This yielded a chemical mutant capable of hydrolyzing coumarin ester (**48**) with k_{cat} 0.052 min^{-1} at pH 7.0, 24 °C. Incorporation of the nucleophilic group alone was previously shown to accelerate hydrolysis of the ester by a factor of 10^4 over controls (POLLACK et al., 1988).

The process of modification is shown in Fig. 17. Lys-52 (**42**) is first derivatized with 4-thiobutanal and then an imidazole is bonded through a disulfide bridge into the active site (**47**). This can now act as a general base/nucleophile in the hydrolysis of (**48**), as was established first by the pH-rate profile and then by complete deactivation of the antibody by diethyl pyrocarbonate (an imidazole-specific inactivating reagent).

The first success in sequence-specific peptide cleavage by an antibody was claimed by IVERSON (IVERSON and LERNER, 1989) using hapten (**49**) which contains an inert Co(III)-(trien) complexed to the secondary amino acid of a tetrapeptide (Fig. 18). This approach aimed towards the elucidation of monoclonal antibodies with a binding site that could simultaneously accommodate a substrate molecule and a kinetically labile complex such as Zn(II)(trien) or Fe(III)(trien), designed to provide catalysis. Much early work by BUCKINGHAM and SARGESON had shown that such cobalt complexes are catalytic for amide hydrolysis via polarization of the carbonyl group, through nucleophilic attack of metal-bound hydroxide, or by a combination of both processes (SUTTON and BUCKINGHAM, 1987; HENDRY and SARGESON, 1990).

Peptidolytic monoclonal antibody 287F11 was selected for further analysis. At pH 6.5, cleavage of substrate (**50**) was observed with a variety of metal complexes. The Zn(II)(trien)

Tab. 2. Parameters for the *syn*-Elimination of Selenoxide (**39**) (R = X = H) in Water, DCM, and Catalyzed by Antibody SZ-*cis*-39C11

Catalyst	ΔG‡	ΔH‡ [kcal mol^{-1}]	ΔS‡ [cal mol^{-1} K^{-1}]	k_{cat}/K_m [M^{-1} min^{-1}]	k_{cat} (k_{obs})[a] [min^{-1}]	ER
Water	26.3±0.15	26.3±0.15	+0.014±0.47		1.6 E-5[b]	
SZ-*cis*-39C11	22.2±1.2	19.7±1.2	−7.8±4.1	2,400	3.5 E-2	2,200
(DCM)	21.8±0.5	20.3±0.5	−4.8±1.7		4.4 E-2	2,750

[a] 25 °C; [b] here and elsewhere exponents are expressed as, e.g., E-5 ... $\cdot 10^{-5}$.

complex was the most efficient with 400 turnovers per antibody combining site and a turnover number of $6 \cdot 10^{-4}$ s^{-1} (Fig. 18). While this approach is undoubtedly ingenious, there are some doubts about its actual performance. The site of cleavage of the substrate is not, as would be expected from the design of the hapten, between the *N*-terminal phenylalanine and glycine, but rather it is between glycine and the internal phenylalanine.

In an equivalent approach, MAHY has generated a peroxidase-like catalytic antibody which binds an iron(III) *ortho*-carboxy substituted tetraarylporphyrin, achieving oxidation rate up to 540 min^{-1} with ABTS as a cosubstrate (DE LAUZON et al., 1999).

A major achievement in augmenting the chemical potential of antibodies has been in the area of redox processes. Many examples now exist of stereoselective reductions, particularly recruiting sodium cyanoborohydride (AE under 22). A growing number of oxidation reactions can now be catalyzed by abzymes, with augmentation from oxidants such as hydrogen peroxide and sodium periodate (AE under 21).

3 Spontaneous Features of Antibody Catalysis

Despite the attention given to the programmed relationship of hapten design and consequent antibody catalytic activity, there are now many examples where the detailed examination of catalysis reveals mechanistic features that were not evidently design features of the system at the outset. Such discoveries are clearly a strength rather than a weakness of the abzyme field, and two of these out-turns are described below.

3.1 Spontaneous Covalent Catalysis

The nucleophilic activity of serine in the hydrolysis of esters and amides by many enzymes is one of the classic features of covalent catalysis by enzymes. So it was perhaps inevitable that an antibody capable of catalyzing the hydrolysis of a phenyl ester should emerge having the same property. SCANLAN has provided just that example with evidence from kinetic and X-ray structural analysis to establish that the hydrolysis of phenyl (*R*)-*N*-formylnorleucine (**51**) proceeds via an acyl antibody intermediate with abzyme 17E8 (AE 2.3) (ZHOU et al., 1994; Fig. 19). The antibody reaction has a bell-shaped profile corresponding to ionizable groups of pK_a 10.0 and 9.1. Based on X-ray analysis, they appear to be respectively Lys^{H97} and probably Tyr^{H101}. This system is deemed to activate Ser^{H99} as part of a catalytic diad with His^{H35} (**52**). In addition to the kinetic and structural evidence for this claim, a trapping experiment with hydroxylamine generated a mixture of amino acid and amino hydroxamic acid products from substrate (**51**) in the presence of antibody.

In a similar vein, antibody NPN43C9 appears to employ a catalytic histidine, His^{L91}, as a nucleophilic catalyst in the hydrolysis of a *p*-nitrophenyl phenylacetate ester (AE 2.8) (GIBBS et al., 1992a; CHEN, et al., 1993).

3.2 Spontaneous Metal Ion Catalysis

JANDA and LERNER tried to show that a metal ion or coordination complex need not be included within the hapten used for the induction of abzymes in order that they could (1) bind a metallo-complex and thereby (2) provide a suitable environment for catalysis (WADE et al., 1993). The pyridine ester (**54**) was screened as a substrate for 23 antibodies raised against the hapten (**53**) (Fig. 20). Antibody 84A3 showed catalytic activity in the hydrolysis of (**54**) only in the presence of zinc, with a rate enhancement of 12,860 over the spontaneous rate and 1,230 over that seen in the presence of zinc alone. Other metals, such as Cd^{2+}, Co^{2+}, Ni^{2+}, were inactive. The affinity of 84A3 for the substrate was high (3.5 μM) whereas the affinity for zinc in the presence of substrate was much lower (840 μM). This is far weaker than any affinity of real potential for the recruitment of metal ion activity into the catalytic antibody repertoire (plasma $[Zn^{2+}]$ is 17.2 μM). However, we can anticipate that application of site-directed mutagenesis and computer assisted design will be targeted on this problem with real expectation of success.

LERNER's group has shown that the catalytic performance of antibody 38C2 in catalyzing an aldol reaction is improved in the presence of one equivalent of Pd(II) (FINN et al., 1998). 38C2 binds Pd specifically ($K_d \sim 1$ μM) and an allosteric effect is observed. This study suggests that the metal does not bind into the catalytic site, but induces a conformational change promoting a closer binding to the substrate.

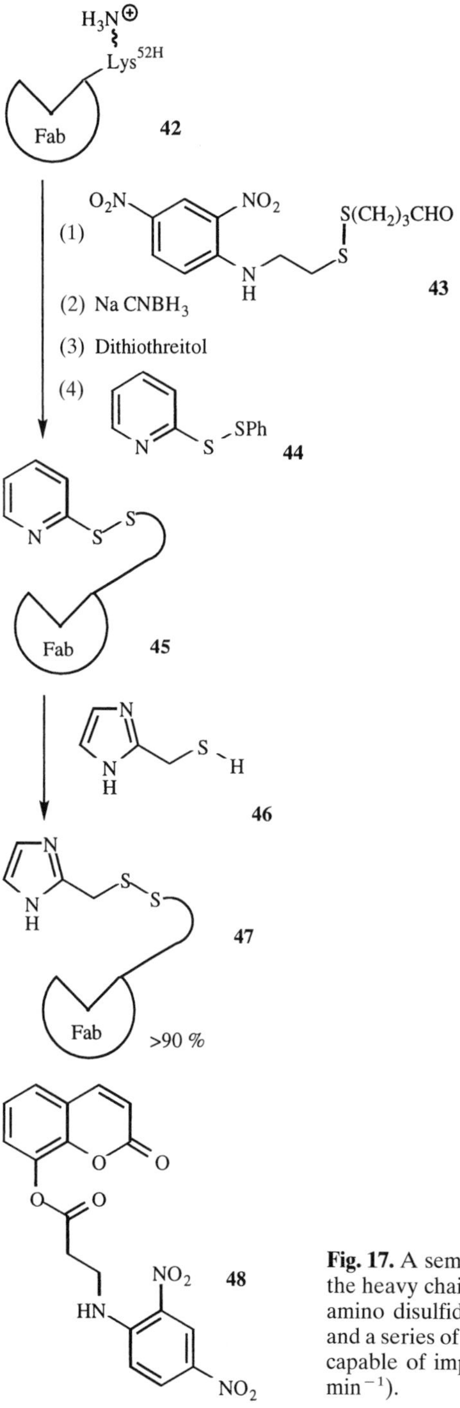

$H_3N^{\oplus}$ Lys[52H] Fab **42**

(1) O_2N ... NO_2 ... N H ... S ... $S(CH_2)_3CHO$ **43**

(2) Na $CNBH_3$

(3) Dithiothreitol

(4) N S SPh **44**

N S S Fab **45**

N N H S H **46**

N N H S S Fab **47** >90 %

O O O O HN NO_2 NO_2 **48**

Fig. 17. A semi-synthetic abzyme. Selective derivatization of lysine-52 in the heavy chain of MOPC315 (**42**) gives an imine which is reduced to an amino disulfide with sodium cyanoborohydride. Reduction to the thiol and a series of disulfide exchanges gives an imidazole derivatized abzyme capable of improved hydrolysis of the coumarin ester (**48**) (k_{cat} – 0.052 min^{-1}).

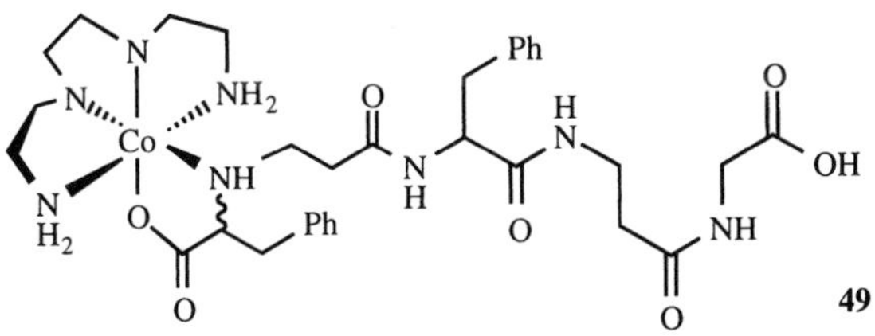

Fig. 18. A metal complex **49** used as hapten to raise antibodies capable of incorporating metal cofactors to facilitate the cleavage of **50** at the position indicated.

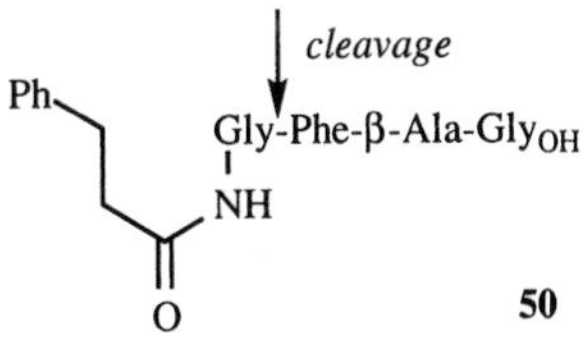

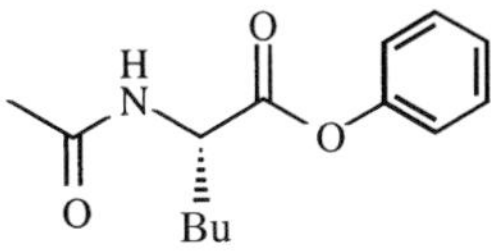

Fig. 19. The hydrolysis of (*R*)-*N*-formylnorleucine (**51**) proceeds via an acyl antibody intermediate with abzyme 17E8 (**52**). (ZHOU et al., 1994).

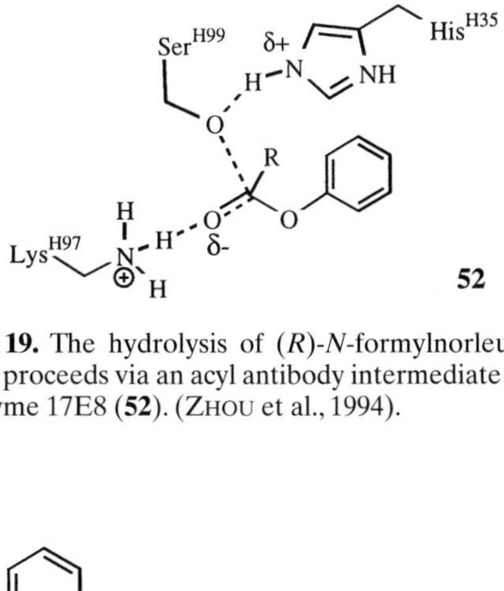

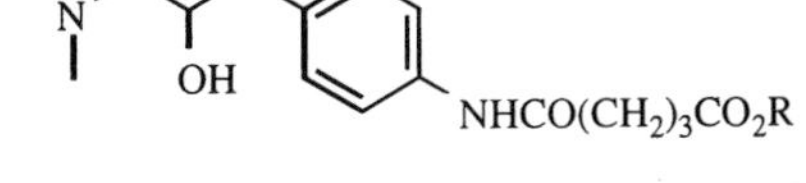

Fig. 20. The pyridine ester (**54**) was screened as a substrate for 23 antibodies raised against the hapten (**53**).

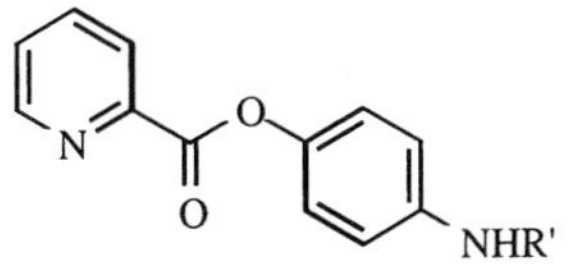

In view of the great general importance of metalloproteinases, it seems inevitable that further work will be directed at this key area either by designed or opportunistic incorporation of metal ions into the catalytic apparatus of abzymes.

4 How Good are Catalytic Antibodies?

In the first years of abzyme research, acyl group transfer reactions have appeared in a majority of examples. Many of these endeavors have been based on mimicry of a high energy, tetrahedral intermediate that lies along such reaction pathways (Sect. 2.1) and which, though not truly a "transition state analog", provides a realistic target for production of a stable TSA. Many of these haptens were based on four coordinate phosphoryl centers.

In 1991, JACOBS analyzed 18 examples of antibody catalysis of acyl transfer reactions as a test of the PAULING concept, i.e., delivering catalysis by $TS^{\ddagger}$ stabilization. The range of examples included the hydrolysis of both aryl and alkyl esters as well as aryl carbonates. In some cases more than one reaction was catalyzed by the same antibody, in others the same reaction was catalyzed by different antibodies (JACOBS, 1991).

Much earlier, WOLFENDEN and THOMPSON (WESTERICK and WOLFENDEN, 1972; THOMPSON, 1973), established a criterion for enzyme inhibitors working as TSAs. They proposed that such activity should be reflected by a linear relationship between the inhibition constant for the enzyme K_i and its inverse second order rate constant, K_m/k_{cat}, for pairs of inhibitors and substrates that differ in structure only at the TSA/substrate locus. That has been well validated, *inter alia*, for phosphonate inhibitors of thermolysin (BARTLETT and MARLOWE, 1983) and pepsin (BARTLETT and GIANGIORDANO, 1996). In order to apply such a criterion to a range of catalytic antibodies, JACOBS assumed firstly that the spontaneous hydrolysis reaction proceeds via the same $TS^{\ddagger}$ as that for the antibody mediated reaction and secondly that all corrective factors due to medium effects are constant. By treating the hydrolyses reactions as a pseudo-first-order process, one can derive a simple relationship with approximations of K_{TS} and K_S to provide a mathematical statement in terms of K_i, K_m, k_{cat}, and k_{uncat} (Fig. 21) (WOLFENDEN, 1969; JENCKS, 1975; BENKOVIC et al., 1988; JACOBS, 1991).

A log–log plot using K_i, K_m, k_{cat}, and k_{uncat} data from the 18 separate cases of antibody catalysis exhibited a linear correlation over four orders of magnitude and with a gradient of 0.86 (Fig. 22)[5]. Considering the assumptions made, this value is sufficiently close to unity to suggest that the antibodies do stabilize the transition state for their respective reactions. However, even the highest k_{cat}/k_{uncat} value of 10^6 in this series (TRAMONTANO et al., 1988) barely compares with enhancement ratios seen for weaker enzyme catalysts (LIENHARD, 1973).

The fact that many values of K_m/K_i fall below the curve (Fig. 22) suggested that interactions between the antibody and the substrate are largely passive in terms of potential catalytic benefit. This conclusion exposes a limitation in the design of haptens, if solely based on the transition state concept. It is well known that enzymes utilize a wide range of devices to achieve catalysis as well as dynamic interactions to guide substrate towards the transition state, which is then selectively stabilized. However, as has been illustrated above, the original concept of transition state stabilization has been augmented by a range of further approaches in the generation of catalytic antibodies and with considerable success.

[5] It may also be worth mentioning here that many early estimates of K_d for the affinity of the antibody to their TSA were upper limits, being based on inhibition kinetics using concentrations of antibody that were significantly higher than the true K_i being determined.

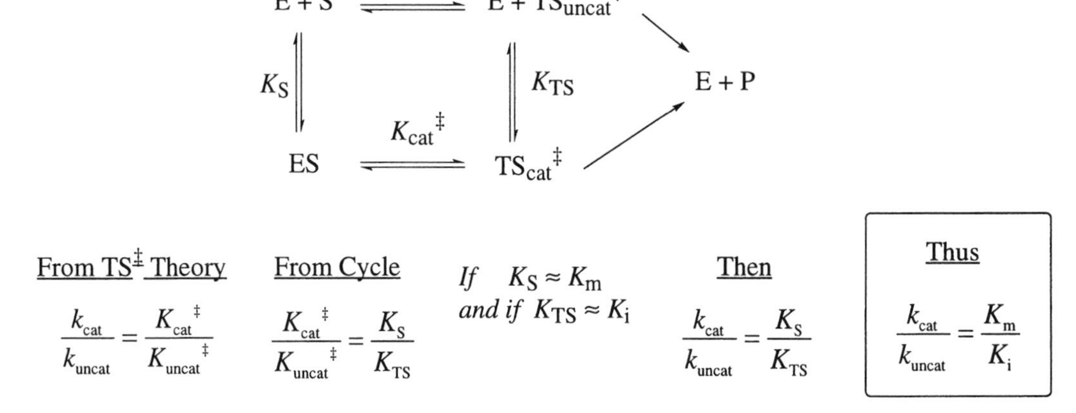

Fig. 21. A thermodynamic cycle linked to transition state theory gives an equation relating the enhancement ratio for a biocatalyzed process to the ratio of equilibrium constants for the complex between the biocatalyst and (i) substrate and (ii) the transition state for the reaction. These two values can be estimated as K_m and K_i for the TSA, respectively.

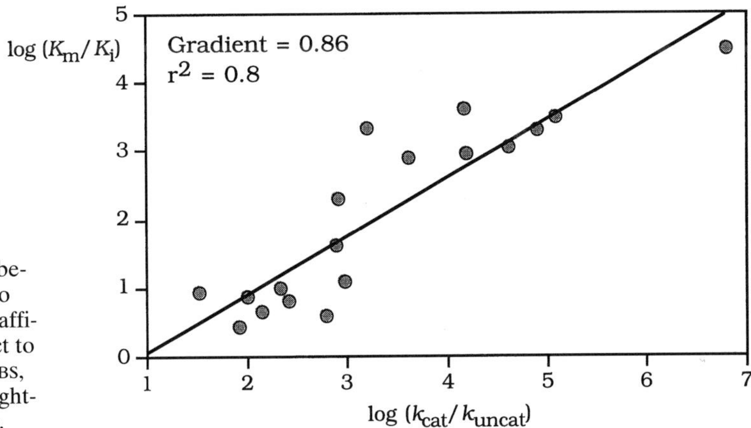

Fig. 22. JACOBS' correlation between the enhancement ratio (k_{cat}/k_{uncat}) and the relative affinity for the TSA with respect to the substrate (K_m/K_i) (JACOBS, 1991). The slope is an unweighted linear regression analysis.

A second use of this type of analysis has been presented by STEWART and BENKOVIC. They showed that the observed rate accelerations for some 60 antibody catalyzed processes can be predicted from the ratio of equilibrium binding constants to the catalytic antibodies for the reaction substrate, K_m, and for the TSA used to raise the antibody, K_i. In particular, this approach supports a rationalization of product selectivity shown by many antibody catalysts for disfavored reactions (Sect. 5.2) and predictions of the extent of rate accelerations that may ultimately be achieved by abzymes. They also used the analysis to underline some differences between mechanism of catalysis by enzymes and abzymes (STEWART and BENKOVIC, 1995). It is interesting to note that the data plotted (Fig. 23) show a high degree of scatter

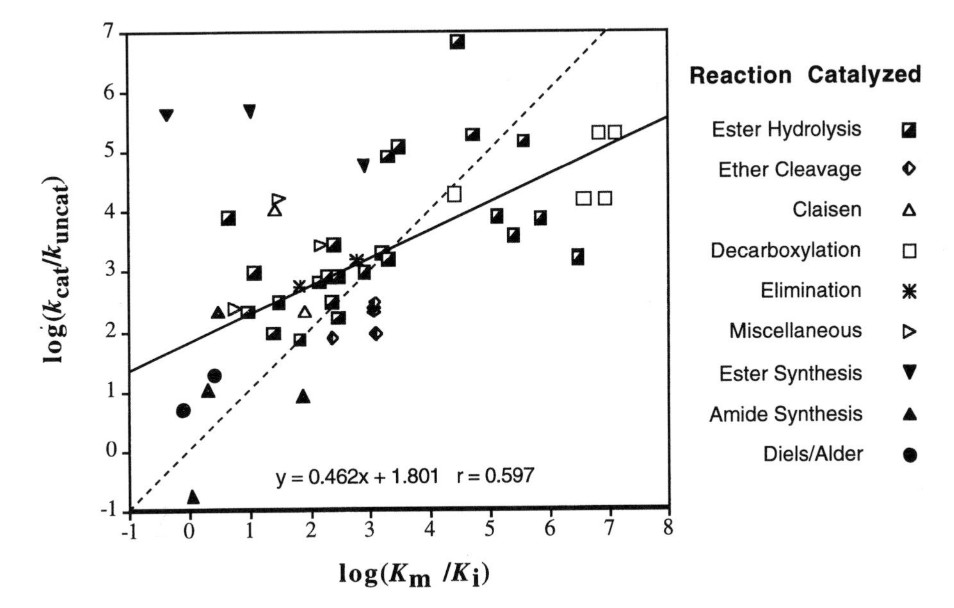

Fig. 23. The Stewart-Benkovic plot of rate enhancement vs relative binding of substrate and TSA for 60 abzyme catalyzed reactions (STEWART and BENKOVIC, 1995). The theoretical unit slope (----) diverges from the calculated linear regression slope (—) for these data (its equation is shown).

with a correlation coefficient for the linear fit of only 0.6 and with a slope of 0.46 very different from the "theoretical slope" of unity. Perhaps of greatest significance are the many positive deviations from the general pattern. These appear to show that antibody catalysis can achieve more than is predicted from catalysis through transition state stabilization alone.

5 Changing the Regio- and Stereochemistry of Reactions

The control of kinetic vs thermodynamic product formation can often be achieved by suitable modificaton of reaction conditions. A far more difficult task is to enhance the formation of a disfavored minor product to the detriment of a favored major product, especially when the transition states for the two processes share most features in common. This goal has been met by antibodies with considerable success, both for reaction pathways differing in regioselectivity and also for ones differing in stereoselectivity. In both situations, control of entropy in the transition state must hold the key.

5.1 Diels–Alder Cycloadditions

In the Diels–Alder reaction between an unsymmetrical diene and dienophile up to eight stereoisomers can be formed (MARCH, 1992b). It is known that the regioselectivity of the Diels–Alder reaction can be biased so that only the four *ortho* adducts are produced (Fig. 24) through increasing the electron-withdrawing character of the substituent on the dienophile (DANISHEFSKY and HERSHENSON, 1979). However, stereochemical control of the Diels–Alder reaction to yield the disfavored *exo*-products in enantiomerically pure form has proved to be very difficult. GOUVERNEUR and co-workers were interested in controlling the outcome of the reaction between diene (**55**) and *N*,*N*-dimethylacrylamide (**54**) (Fig. 25) (GOUVERNEUR et al., 1993).

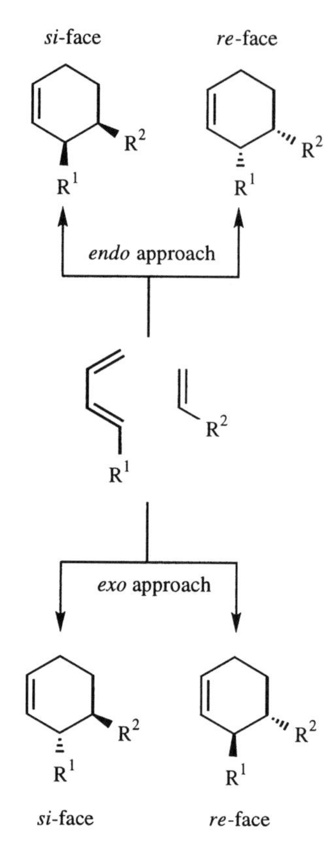

Fig. 24. Enantio- and diastereoselectivity of the Diels–Alder reaction for *ortho*-approach.

They had shown experimentally that the uncatalyzed reaction gave only two stereoisomers: the *ortho-endo* (*cis*) (**57**) and the *ortho-exo* (*trans*) (**58**) adducts in an 85:15 mixture. This experimental observation was underpinned by *ab initio* transition state modeling for the reaction of *N*-butadienylcarbamic acid (**59**) with acrylamide (**60**), which showed that the relative activation energies of the *ortho-endo* and *ortho-exo* transition states were of considerably lower energy than the *meta-endo* and *meta-exo* transition structures (not illustrated) (Tab. 3). The design of hapten was crucial for the generation of catalytic antibodies to deliver full regio- and diastereoselectivity. Transition state analogs were therefore devised to incorporate features compatible with either the disfavored *endo* (**63**) or favored *exo* (**65**) transition states (Fig. 26) (AE 17.5).

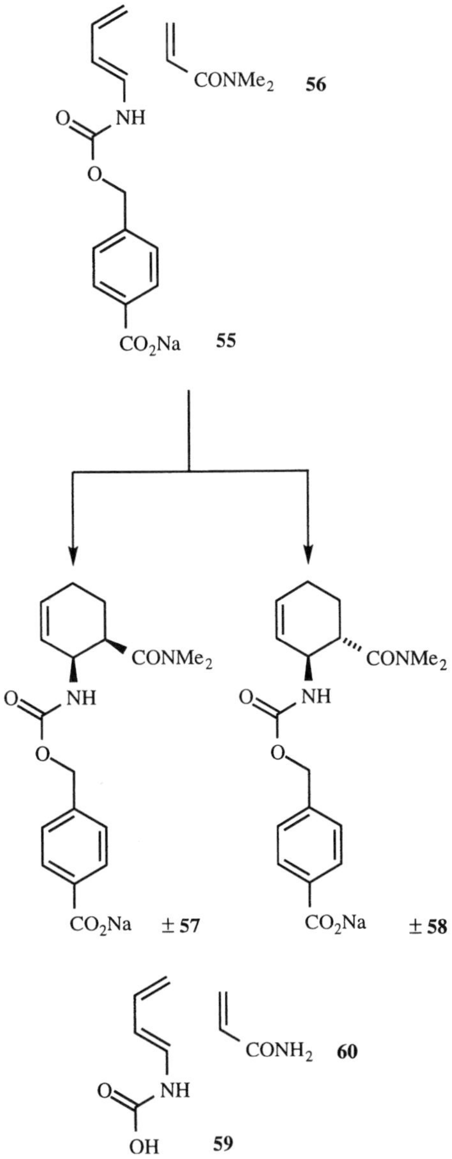

HILVERT developed a new strategy to minimize product binding to the abzyme (HILVERT et al., 1989). Based on the knowledge that the transition state for Diels–Alder processes is very product-like, he designed haptens (**64**) and (**66**) to mimic a high energy, boat conformation for each product, thereby ensuring avoidance of product inhibition.

Two of the monoclonal antibodies produced, 7D4 and 22C8, proved to be completely stereoselective, catalyzing the *endo* and the *exo* Diels–Alder reactions, with a k_{cat} of $3.44 \cdot 10^{-3}$ and $3.17 \cdot 10^{-3}$ min^{-1}, respectively at 25 °C. That the turnover numbers are low was attributed in part to limitations in transition state representation: modeling studies had shown that the transition states for both the *exo* and *endo* processes were asynchronous whereas both TSAs (**64**) and (**66**) were based on synchronous transition states (GOUVERNEUR et al., 1993).

In a further experiment, compounds (**67**) and (**68**) were perceived as freely rotating haptens for application as TSAs for the same Diels–Alder addition (Fig. 27). As expected, each proved capable of inducing both *endo*- and *exo*-adduct forming abzymes. It can be noted that (**67**) produced more "*exo*-catalysts" (6 out of 7) whereas (**68**) favored the production of "*endo*-catalysts" (7 out of 8) though it is difficult to draw any conclusion from this observation (AE 17.5) (YLI-KAUHALUOMA et al., 1995; HEINE et al., 1998).

Fig. 25. The Diels–Alder cycloaddition between the diene (**55**) and the dienophile (**56**) yields two diastereoisomers (**57**) and (**58**). Attenuated substrate analogs (**59**) and (**60**) were used in molecular orbital calculations of this reaction.

5.2 Disfavored Regio- and Stereoselectivity

Reversal of Kinetic Control in a Ring Closure Reaction

When several different outcomes are possible in a reaction, the final product distribution reflects the relative free energies of each transition state when the reaction is under kinetic control (SCHULTZ and LERNER, 1993). BALDWIN's rules predict that for acid catalyzed ring closure of the hydroxyepoxide (**70**) the tetrahydrofuran product (**69**) arising from 5-*exo-tet* attack will be preferred (Fig. 28) (BALDWIN, 1976). By raising antibodies to the

Tab. 3. Calculated Activation Energies of the Transition Structures Relative to Reactants for the Reaction of *N*-(1-Butadienyl)-carbamic Acid (**59**) with Acrylamide (**60**)

Transition State Geometry	Calculated Activation Energy [kcal mol^{-1}]	
	RHF/3-21G	6-31G·/3-21G
ortho-endo	27.3	40.80
ortho-exo	28.84	42.70
meta-endo	29.82	42.88
meta-exo	30.95	43.94

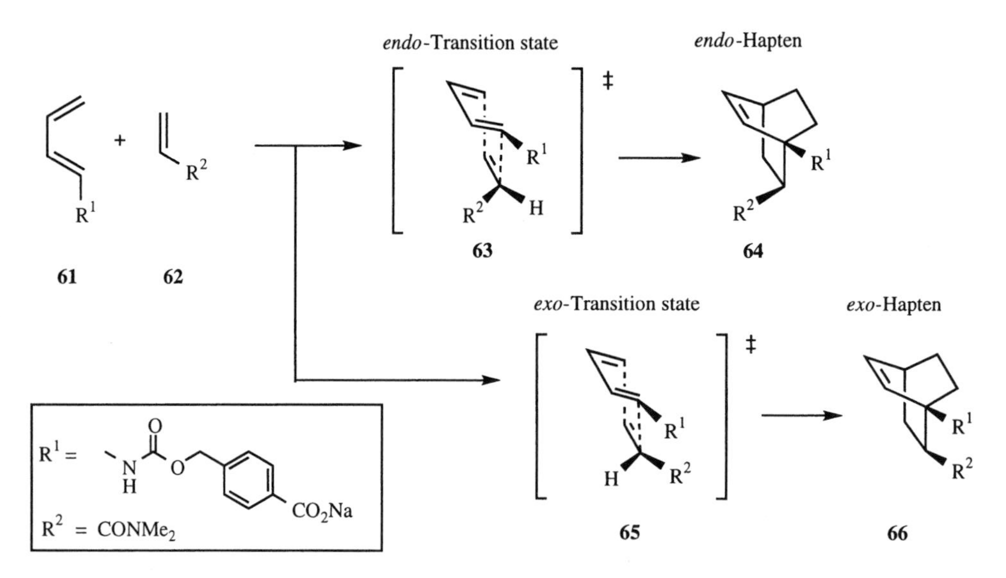

Fig. 26. Haptens (**64**) and (**66**) were designed as analogs of the favored *endo*-(**63**) or disfavored *exo*-(**65**) transition states, respectively.

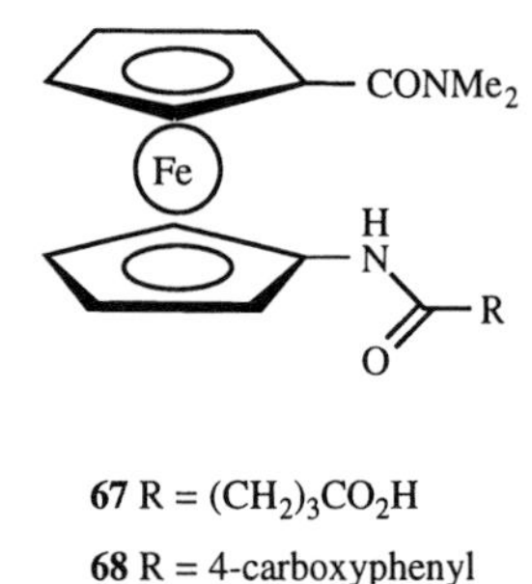

Fig. 27. An alternative strategy for eliciting Diels-Alderase antibodies has employed the freely rotating ferrocenes (**67**) and (**68**) as TSAs.

charged hapten (**72**), Janda and co-workers generated an abzyme which accelerated 6-*exo* attack of the racemic epoxide to yield exclusively the disfavored tetrahydropyran product (**71**) and in an enantiomerically pure form (AE 14.1) (Janda et al., 1993).

This work reveals that an antibody can selectively deliver a single regio- and stereochemically defined product for a reaction in which multiple, alternative transition states are accessible and can also selectively lower the energy of the higher of two alternative transition states (Na and Houk, 1996).

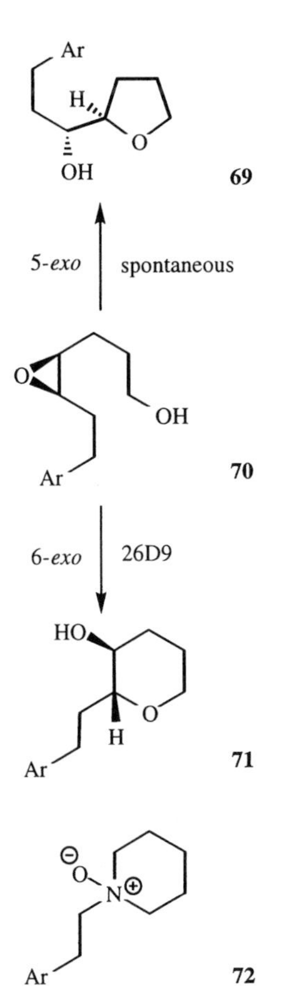

Fig. 28. The monoclonal antibody 26D9, generated to the *N*-oxide hapten (**72**), catalyzed the 6-*exo-tet* ring closure of (**70**) regioselectively to yield the *disfavored* tetrahydropyran product (**71**). This is a formal violation of BALDWIN's rules which predict a spontaneous 5-*exo-tet* process to generate tetrahydrofuran derivative (**69**).

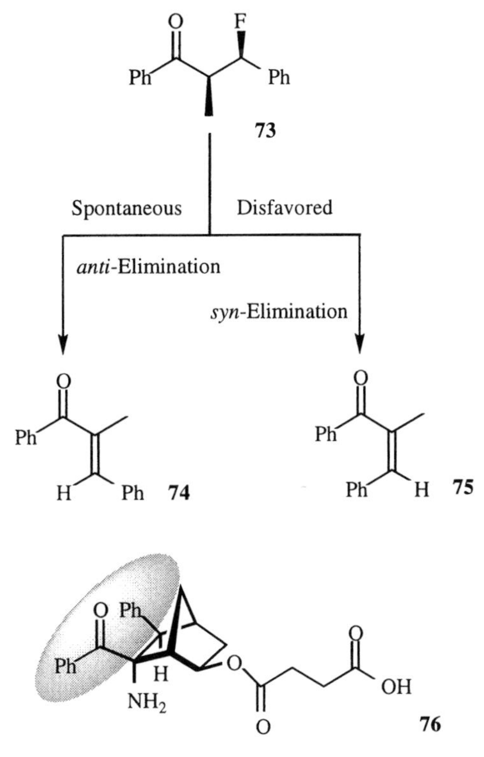

Fig. 29. The elimination of HF from the β-fluoroketone (**73**) is catalyzed by antibody 1D4, elicited to hapten (**76**), to form the disfavored (Z)-α,β-unsaturated ketone (**75**). This contrasts the spontaneous process in which an *anti*-elimination reaction yields the (E)-α,β-unsaturated ketone (**74**). The *syn*-eclipsed conformation of (**76**) is shaded.

Syn-Elimination of β-Fluoroketones

The base-catalyzed β-elimination of HF from the ketone (**73**) normally gives the thermodynamically more stable (E)-alkene (**74**) via a staggered conformation in the transition state (Fig. 29). Hapten (**76**) was designed to enforce the *syn*-coplanar conformation of the phenyl and benzoyl functions in the transition state and so catalyze the disfavored *syn*-elimination of (**73**) to give the (Z)-α,β-unsaturated ketone (**75**) (Fig. 29). Preliminary estimations of the energy difference between the favored and disfavored processes are close to 5 kcal mol^{-1} (CRAVATT et al., 1994), though this value is exceeded in the antibody catalyzed rerouting of carbamate hydrolysis from E1cB to $B_{Ac}2$ (Sect. 8.2, AE 4.3) (WENTWORTH et al., 1997). Antibody 1D4, raised to hapten (**76**) and used in 15% DMSO at pH 9.0, gave exclusively the (Z)-alkene (**75**) with K_m 212 μM and k_{cat} $2.95 \cdot 10^{-3}$ min^{-1}. Under the same conditions, k_{obs} is $2.48 \cdot 10^{-4}$ min^{-1} for formation of (**74**) and immeasurably slow for the (undetectable) formation of (**75**) (AE 8.1).

5.3 Carbocation Cyclizations

The cationic cyclization of polyenes giving multi-ring carbocyclic compounds with many sterically defined centers is one of the more remarkable examples of regioselective and stereoselective enzyme control which has provided a major challenge for biomimetic chemistry (JOHNSON, 1968). This system provides an excellent opportunity for the application of regio- and stereocontrol by catalytic antibodies.

LI and co-workers have discussed the use of catalytic antibodies to control the reactivity of carbocations (LI et al., 1997). At an entry level, antibody 4C6 was raised to hapten (**77**) and used to convert acyclic olefinic sulfonate ester (**78**) into cyclic alcohol (**79**) (98%) with very little cyclohexene (2%) produced (Fig. 30, AE 15.1) (LI, et al., 1994).

Moving closer to a cationic transition state mimic, HASSERODT and co-workers used the amidinium ion (**80**) as a TSA for cyclization of the arenesulfonate ester (**81**) (Hasserodt et al., 1996). One antibody raised to this hapten, HA1-17G8, catalyzed the conversion of substrate (**81**) into a mixture of the 1,6-dimethylcyclohexene (**82**) and 2-methylene-1-methylcyclohexane (**83**) (Fig. 31) (AE 15.3). By contrast, the uncatalyzed cyclization of (**81**) formed a mixture of 1,2-dimethyl-cyclohexanols and a little 1,2-dimethylcyclohexene. Evidently, the antibody both excludes water from the transition state and also controls the loss of a proton following cyclization.

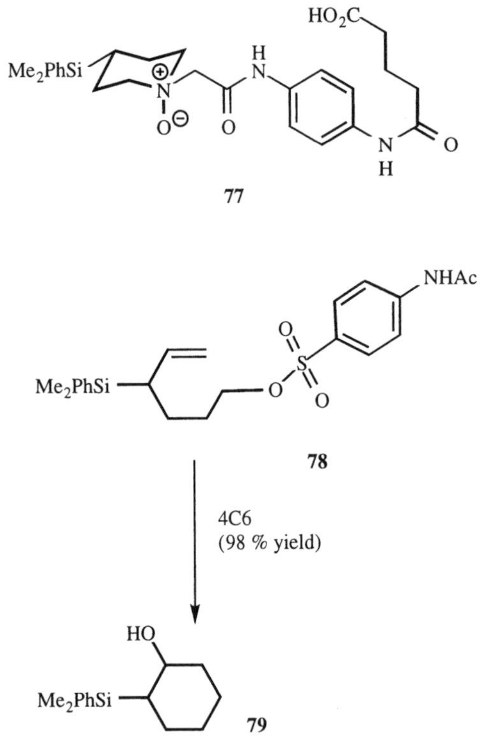

Me_2PhSi HO$_2$C N H N H O O$^{\ominus}$ O **77**

NHAc Me_2PhSi O S O O **78**

4C6 (98 % yield)

HO Me_2PhSi **79**

Fig. 30. The *N*-oxide hapten (**77**) was used to elicit mAb 4C6 which catalyzed the cyclization of the allylsilane (**78**) to form the cyclohexanol (**79**).

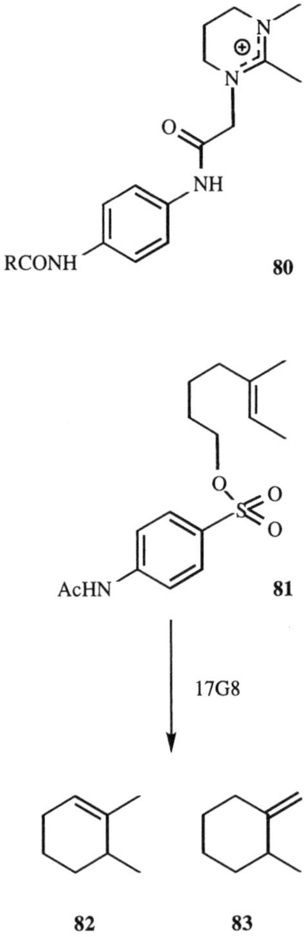

N N O NH RCONH **80**

O S O O AcHN **81**

17G8

82 **83**

Fig. 31. Antibody HA1-17G8 raised against TSA (**80**) catalyzed the cyclization of (**81**) to give (**82**) and (**83**).

While cyclopentanes have also been produced by antibody catalyzed cyclization (AE 15.2) (LI et al., 1996), much the most striking example of cationic cyclization by antibodies is the formation of the decalins (**85**), (**86**), and (**87**) (Fig. 32). The *trans*-decalin epoxide (**88**) ($t_{1/2}$ 100 h at 37 °C) was employed as a mixture of two enantiomeric pairs of diastereoisomers as a TSA to raise antibodies, among which was produced HA5-19A4 as the best catalyst for cyclization of substrate (**84**) (AE 15.4) (HASSERODT et al., 1997).

Sufficient substrate (**84**) was transformed to give 10 mg of mixed products. The olefinic fraction (70%) was predominantly a mixture of the three decalins (**85**), (**86**), and (**87**) (2:3:1 ratio) and formed along with a mixture of cyclohexanol diastereoisomers (30%). Moreover, the decalins were formed with enantiomeric excesses of 53, 53, and 80%, respectively. It is significant that the (*Z*)-isomer of (**84**) is not a substrate for this antibody.

The ionization of the arenesulfonate is the first step catalyzed by the antibody which generates a carbocation as part of a process that shows an ER of 2,300 with K_m 320 μM. The resulting cation can then either cyclize to decalins in a concerted process (as in transition state **89**) or in two stepwise cyclizations. The formation of significant amounts of cyclohexanols seems to favor the latter of possibilities. Most interestingly, inhibition studies suggest that the isomer of the haptenic mixture that elicited this antibody has structure (**88**), therefore, locating the leaving group in an axial position. This is contrary to the STORK-ESCHENMOSER concept of equatorial leaving group and presents a challenge for future examination (STORK and BURGSTAHLER, 1955; ESCHENMOSER et al., 1955).

It is an exciting prospect that catalysts of this nature may lead to artificial enzymes capable of processing natural and unnatural polyisoprenoids to generate various useful terpenes.

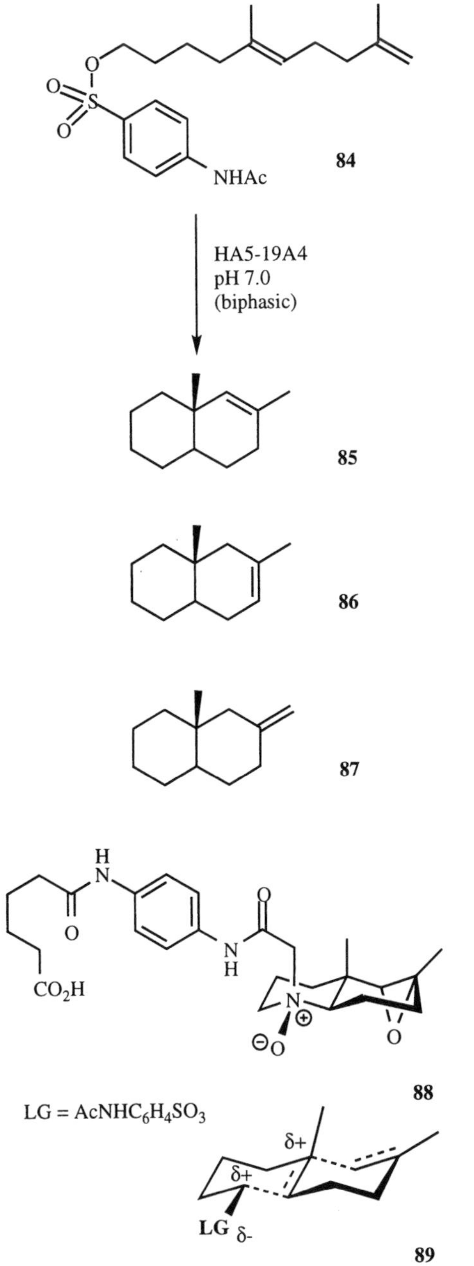

O
O
S
O
NHAc
84
HA5-19A4
pH 7.0
(biphasic)
85
86
87
H
N
O
O
N
H
CO_2H
N
⊕
⊖O
O
88
LG = $AcNHC_6H_4SO_3$
δ+
δ+
LG δ-
89

Fig. 32. Formation of isomeric decalins (**85**)–(**87**) by cyclization of a terpenoid alcohol catalyzed by antibody HA5-19A4 raised to hapten (**88**). The transition state (**89**) has the leaving group in the equatorial position, as favored by the STORK-ESCHENMOSER hypothesis.

6 Difficult Processes

As scientists studying catalytic antibodies have gained more confidence in the power of abzymes, their attention has turned from reactions having moderate to good feasibility to more demanding ones. Their work has on the one hand tackled more adventurous stereochemical problems and on the other hand they are seeking to catalyze reactions whose spontaneous rates are very slow indeed. Examples of both of these areas are discussed in this section.

6.1 Resolution of Diastereoisomers

Antibodies generally show very good selectivity into the recognition of their antigens and discriminate against regio- or stereoisomers of them. This results from a combination of the inherent chirality of proteins and the refined response of the immune system (PLAYFAIR, 1992). In extension, this character suggests that a catalytic antibody should be capable of similar discrimination in its choice of substrate and the transition state it can stabilize, as determined by the hapten used for its induction. As already shown above, the murine immune system can respond to a single member of a mixture of stereoisomers used for immunization (Sect. 5.3). Such discrimination has been exemplified in antibody catalyzed enantioselective ester hydrolysis (JANDA et al., 1989; POLLACK et al., 1989; SCHULTZ, 1989) and transesterification reactions (WIRSCHING et al., 1991; JACOBSEN et al., 1992).

One study has made use of abzyme stereoselectivity to resolve the four stereoisomers (*R,R′*, *S,S′*, *R,S′*, and *S,R′*) of 4-benzyloxy-3-fluoro-3-methylbutan-2-ol (**94**)–(**97**) through the antibody mediated hydrolysis of a diastereoisomeric mixture of their phenacetyl esters (**90**)–(**93**) (KITAZUME et al., 1991b). Antibodies were raised separately to each of four phosphonate diastereoisomers (**98**)–(**101**), corresponding to the four possible transition states for the hydrolysis of the four diastereoisomeric esters (Fig. 33) (AE 1.12). Each antibody operated on a mixture of equal parts of the four diastereoisomers as substrate to give each alcohol in ca. 23% yield, with >97% ee/de, and leaving the three other stereoisomers unchanged. Following successive incubation with the four antibodies in turn, the mixture of diastereoisomers could effectively be separated completely (Tab. 4). In a similar vein, KITAZUME also resolved the enantiomers of 1,1,1-trifluorodecan-2-ol with 98.5% ee (AE 1.11) (KITAZUME et al., 1991a).

6.2 Cleavage of Acetals and Glycosides

The synthesis, modification, and degradation of carbohydrates by antibody catalyzed transformations are subjects of active investigation. Preliminary studies have reported antibody hydrolysis of model glycoside substrates (YU et al., 1994) while the regio- and stereoselective deprotection of acylated carbohydrates has been achieved using catalytic antibodies

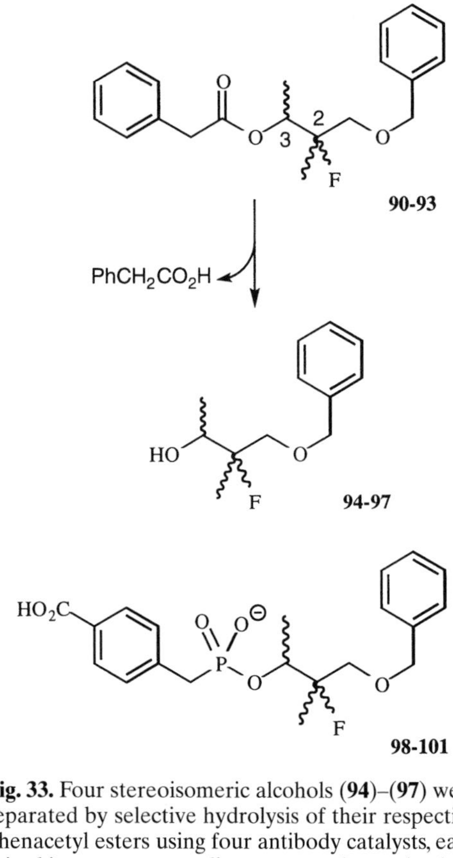

Fig. 33. Four stereoisomeric alcohols (**94**)–(**97**) were separated by selective hydrolysis of their respective phenacetyl esters using four antibody catalysts, each raised in response to a discrete stereoisomeric phosphonate hapten (**98**)–(**101**).

Tab. 4. Kinetic Parameters for those Antibodies Raised against Phosphonates (**98**)–(**101**) which Effect the Resolution of the Fluorinated Alcohols (94)–(97). The Configuration of the Diastereoisomerically Pure Product from Each Antibody Catalyzed Process was Shown to Correspond to that of the Antibody-Inducing Hapten

Product Alcohol	Hapten	Configuration	k_{cat}[a] [min^{-1}]	K_m [μM]	Product ee/de [%]
(**94**)	(**98**)	2*R*, 3*R* (+)	0.88	390	99.0
(**95**)	(**99**)	2*S*, 3*S* (−)	0.91	400	98.5
(**96**)	(**100**)	2*R*, 3*S* (+)	0.94	410	98.5
(**97**)	(**101**)	2*S*, 3*R* (−)	0.86	380	98.0

[a] At 25 °C.

with moderate rate enhancements. Antibodies raised to the TSA (**102**) were screened for their hydrolytic properties of the diester (**103**). One antibody, 17E11, used in a 20% concentration with respect to (**102**), effected hydrolysis exclusively at C-4. This process was fast enough to make spontaneous acyl migration from C-3 to C-4 of no significance: i.e., no C^3-OH product was detected. [This migration reaction is generally fast compared to chemical deacylation (Fig. 34. AE 1.7) (IWABUCHI et al., 1994)]. In this context, the use of an antibody to cleave a trityl ether by an S_N1 process may have further applications (AE 7.1) (IVERSON et al., 1990). Also, the objective of utilizing abzymes in the regioselective removal of a specified protecting group has been extended to show that an antibody esterase can have broad substrate tolerance (AE 1.18) (LI et al., 1995b).

The research towards the demanding glycosylic bond cleavage is progressing well, albeit slowly. As a first step, REYMOND has described an antibody capable of catalyzing the acetal hydrolysis of a phenoxytetrahydropyran, although it is slow, with k_{cat} $7.8 \cdot 10^{-5}$ s^{-1} at 24 °C, and has a modest ER of 70 (AE 7.4B) (REYMOND et al., 1991).

A classic approach to the problem has been to raise antibodies against TSAs related by design to well-known inhibitors of glycosidases. Piperidino and pyrrolidino cations have high affinity for pyranosidases and furanosidases (WINCHESTER and FLEET, 1992; WINCHESTER et al., 1993) and can also be envisaged as components of a "bait and switch" approach to antibody production. Thus, SCHULTZ has described the hydrolysis of the 3-indolyl acetal (**105**) by antibody AA71.17 raised to transition state analog (**104**) (Fig. 35) (AE 7.2) (YU, et al., 1994). Antibody AA71.17 has a useful K_m of 320 μM but is rather slow in turnover, k_{cat} 0.015 min^{-1}. By contrast, two other haptens

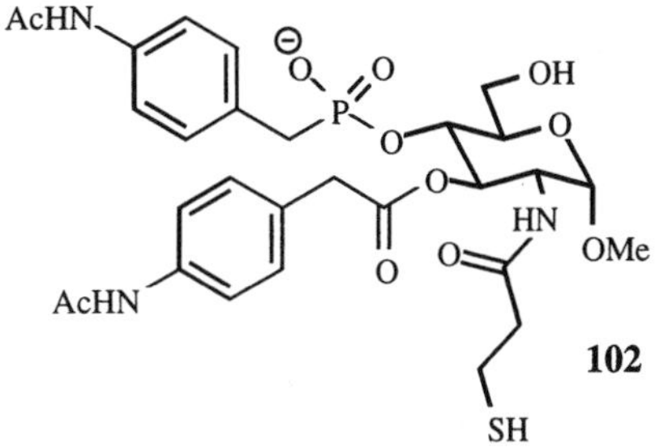

AcHN
OH
O
HN
OMe
AcHN
SH
102

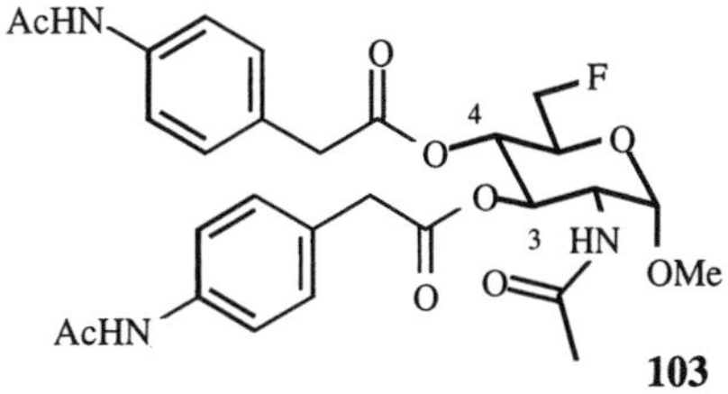

AcHN
F
4
3 HN
OMe
AcHN
103

Abzyme Identity		Conditions
17E11		pH 8.2; 20 °C
K_m **103**	k_{cat} **103**	K_i **102**
6.6 μM	0.182 min^{-1}	0.026 μM

Fig. 34. Antibody 17E11 raised against the TSA (**102**) was screened for its ability to hydrolyze diester (**102**) and, used in a 20% concentration with respect to (**102**), effected hydrolysis exclusively at C-4.

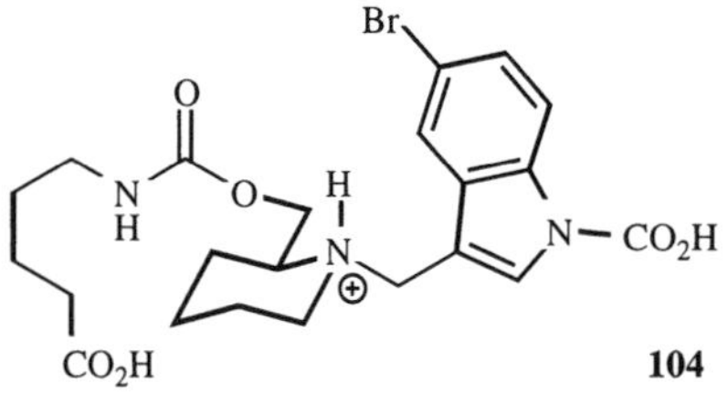

Fig. 35. Antibody AA71.17 raised against hapten (**104**) catalyzes the hydrolysis of the aryl acetal (**105**) (Yu et al., 1994).

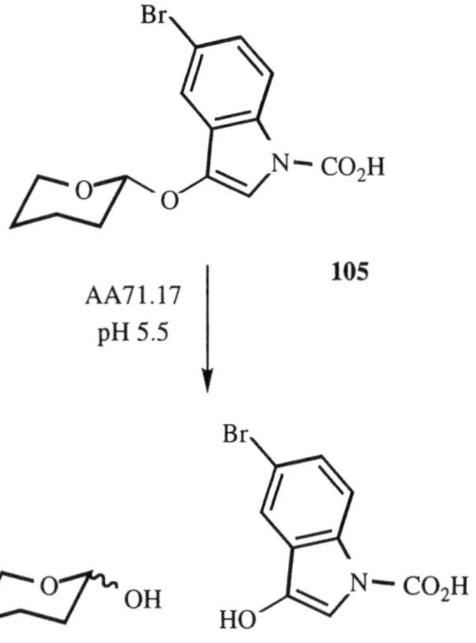

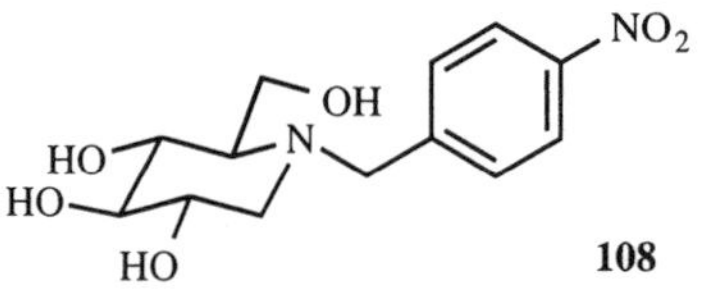

Fig. 36. Antibody 4f4f raised to (**108**) catalyzes the hydrolysis of (**109**) (Yu et al., 1998).

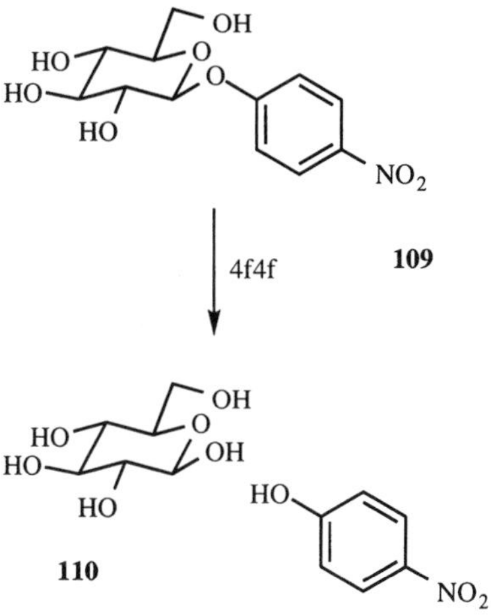

based on a guanidino and a dihydropyran inhibitor did not elicit any antibodies showing glycosidase activity.

Further studies led the same team to generate glycosidase antibodies by *in vitro* immunization. They raised antibodies to the chair-like transition state analog (**108**) derived from 1-deoxynojirimycin, a classic glycosidase inhibitor. They isolated antibody 4f4f which shows k_{cat} $2.8 \cdot 10^{-3}$ min^{-1} and K_m 22 μM, respectively, for the hydrolysis of substrate (**109**) (Yu et al., 1998) (Fig. 36).

Janda has described the production of a galactopyranosidase antibody in response to hapten (**112**) (Fig. 37). This was designed to accommodate several features of the transition state for glycoside hydrolysis: notably a flattened half-chair conformation and substantial sp^2 character at the anomeric position. Some 100 clones were isolated in response to immunization with (**112**) and used to generate a cDNA library for display on the surface of phage (AE 7.3) (Janda et al., 1997). Rather than proceed to the normal screening for turnover, Janda then created a suicide substrate system to trap the catalytic species.

Halazy had earlier shown that phenols with *ortho-* or *para*-difluoromethyl substituents spontaneously eliminate HF to form quinonemethides that are powerful electrophiles and that this activity can be used to trap glycosidases (Halazy et al., 1992). So, glycosylic bond cleavage in (**113**) results in formation of the quinonemethide (**114**) that covalently traps the antibody catalyst. By suitable engineering a bacteriophage system, Janda was able to screen a large library of Fab fragment antibodies and select for catalysis. Fab 1B catalyzed the hydrolysis of *p*-nitrophenyl β-galactopyranoside with k_{cat} 0.007 min^{-1} and K_m

Fig. 37. Fragment antibody Fab1B is selected by suicide selection with substrate (**113**) from a library of antibodies generated to hapten (**112**). The suicide intermediate is the *o*-quinonemethide (**114**).

530 μM, corresponding to a rate enhancement of $7 \cdot 10^4$. Moreover, this activity was inhibited by hapten (**112**) with K_i 15 μM. By contrast, the best catalytic antibody, 1F4, generated from hapten (**112**) by classical hybridoma screening showed k_{cat} 10^{-5} min^{-1} and K_m 330 μM, a rate enhancement of only 100.

Clearly, this work offers an exciting method for screening for antibodies that can lead to suicide product trapping and also appears to offer a general approach to antibodies with glycosidase activity.

6.3 Phosphate Ester Cleavage

The mechanisms of phosphate ester cleavage vary significantly between monoesters, diesters, and triesters (THATCHER and KLUGER, 1989). Each of these is a target for antibody catalyzed hydrolysis and progress has been reported for all three cases.

Phosphate Monoesters

The achievement of this reaction is particularly important in light of the fact that phosphoryl transfers involving tyrosine, serine, or threonine play crucial roles in signal transduction pathways that control many aspects of cellular physiology. The production of an abzyme would provide a crucial tool for the investigation and manipulation of such processes.

SCHULTZ's group employed an α-hydroxyphosphonate hapten (**115**) and subsequently isolated 20 cell lines of which 5 catalyzed the hydrolysis of the model substrate *p*-nitrophenol phosphate (**116**) above background (Fig. 38)

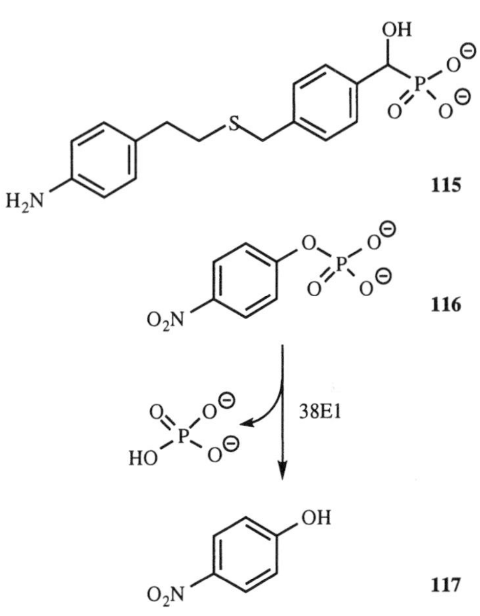

Abzyme Identity		Conditions
38E1		pH 9.0
K_m **116**	k_{cat}[a] **116**	K_i **115**
155 μM	0.0012 min^{-1}	34 μM

[a] At 30 °C

Fig. 38. Antibody 38E1, generated from a α-hydroxyphosphonate hapten (**115**), catalyzed the hydrolysis of *p*-nitrophenyl phosphate (**116**) to *p*-nitrophenol (**117**).

(SCANLAN et al., 1991). Antibody 38E1 was characterized in more detail and kinetic parameters were afforded by LINEWEAVER-BURKE analysis. This antibody exhibited 11 turnovers per binding site with no change in V_{max}, and thus acted as a true catalyst. Moreover, examination of substrate specificity showed that catalysis was entirely selective for *para*-substituted species (AE 6.6).

Phosphodiester Cleavage
The phosphodiester bond being one of the most stable chemical linkages in nature, its cleavage is an obvious and challenging target for antibody catalysis. In an attempt to model a metal-independent mechanism, a nucleotide analog (**118**) featuring an *O*-phosphorylated hydroxylamine moiety was chosen by SAKURAI and co-workers (SAKURAI et al., 1996) (Fig. 39).

This hapten design aims to mimic the geometry and spatial constraints in the phosphate linkage, to retain the stereoelectronic configuration of the phosphorus atom, and to act as a simple model of a dinucleotide. For that purpose, the retention of the phosphate backbone seeks to facilitate the formation of an oxyanion hole in which the electrophilicity of the phosphorus center is increased in the bound substrate, while the positive charge on the hapten is designed to induce an anionic amino acid in the abzyme binding site, either to act as a general base to activate a nucleophilic water molecule, or as a nucleophile operating directly at the phosphorus center. More details are awaited from this work.

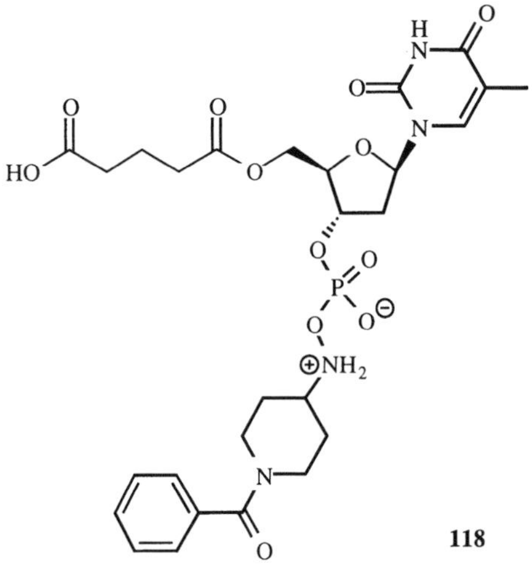

118

Fig. 39. Hapten for generating catalytic antibodies for phosphodiester cleavage (SAKURAI et al., 1996).

The classic case of assisted hydrolysis of phosphate diesters is neighboring group participation by a vicinal hydroxyl group, specifically the phosphate ester of a 1,2-diol, which provides a rate acceleration greatly in excess of 10^6 (WESTHEIMER, 1968). While vanadate complexes of 1,2-diols have been explored as pentacoordinate species for inhibiting enzymes, they are toxic and too labile for use in murine immunization (CRANS, et al., 1991). JANDA has solved this problem by using of pentacoordinate oxorhenium chelates (WEINER et al., 1997). By employing hapten (**119**) as separate diastereoisomers, 50 monoclonal antibodies were generated and screened for their ability to hydrolyze uridine 3′-(*p*-nitrophenyl phosphate) (**120a**) (Fig. 40) (AE 6.4). The most active of the three antibodies showing catalysis, 2G12, had k_{cat} $1.53 \cdot 10^{-3}$ s^{-1} at 25 °C and K_m 240 μM giving k_{cat}/k_{uncat} 312. A more favorable expression for protein catalysts working at substrate concentrations below K_m is $k_{cat}/(K_m \cdot k_{uncat})$ (RADZICKA and WOLFENDEN, 1995), which is $1.3 \cdot 10^6$ M^{-1} and compares favorably to the value for RNase A of 10^{10} M^{-1} for the same substrate. The TSA *anti*-(**119**) proved to be a powerful inhibitor for 2G12 with K_i estimated at 400 nM.

Evidently, this is a system with scope alike for improvement in design and for broader application. It is clearly one of the most successful examples of antibody catalysis of a difficult reaction.

More recently, the same team has used a bait and switch strategy in order to improve this system. A nucleophilic residue activating the 2′-hydroxyl is induced by a cationic trimethylammonium group, while the 3′-hydroxyl is substituted by a *p*-nitrophenyl phosphoester in a substrate-like shape (**120b**). Two antibodies were isolated of which MATT.F-1 was found to obey classic Michaelis-Menten kinetics with the values $k_{cat} = 0.44$ min^{-1} and $K_m = 104$ μM. This antibody is 5 times more efficient than the one studied before (WENTWORTH et al., 1998).

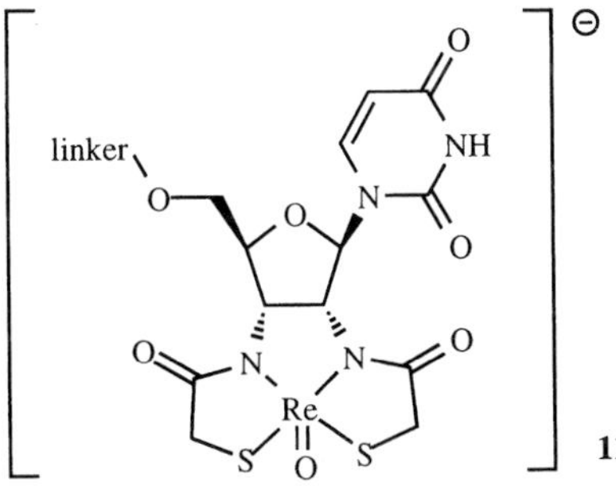

Fig. 40. Hapten *anti*-(**119**) was used to generate 25 mAbs from which 2G12 proved to catalyze the hydrolysis of the phosphate diester (**120a**). (**120b**) was later used as a hapten in a bait and switch strategy to obtain antibodies hydrolyzing (**120a**).

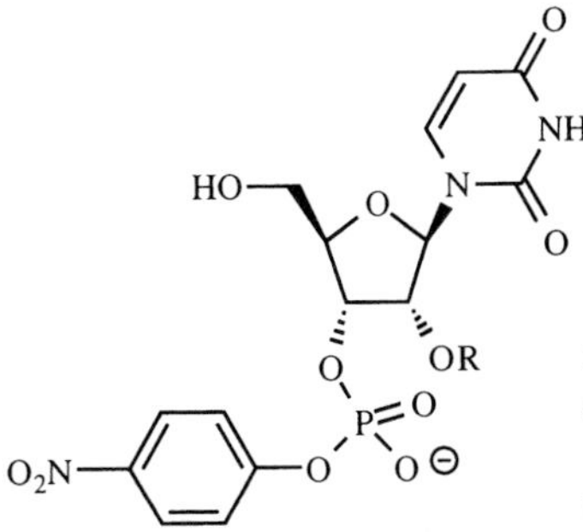

Phosphotriester Hydrolysis

Catalysis of this reaction was first exhibited by antibodies raised by ROSENBLUM and co-workers (ROSENBLUM et al., 1995). More recently, LAVEY and JANDA have explored the generation of abzymes able to catalyze the breakdown of poisonous agrochemicals (LAVEY and JANDA, 1996a). 25 Monoclonal antibodies were raised against the *N*-oxide hapten (**121**) of which two were found to be catalytic. The hapten was designed to generate antibodies for the hydrolysis of triester (**122**) using the "bait and switch" strategy: cationic charge on the nitrogen atom targeted to induce anionic amino acids to act as general base catalysts; partial negative charge on oxygen to favor the selection of antibody residues capable of stabilizing negative charge in the transition state (Fig. 41).

Antibody 15C5 was able to catalyze the hydrolysis of the phosphate triester (**122**) with a value of $k_{cat} = 2.65 \cdot 10^{-3}\ min^{-1}$ at 25 °C. Later, a second antibody from the same immunization program was found to hydrolyze the acetylcholinesterase inhibitor Paraoxon (**123**) with $k_{cat} = 1.95 \cdot 10^{-3}\ min^{-1}$ at 25 °C (AE 6.2) (LAVEY and JANDA, 1996b). Antibody 3H5 showed Michaelis-Menten kinetics and was strongly inhibited by the hapten (**121**). It exhibited a linear dependence of the rate of hydrolysis on hydroxide ion suggesting that 3H5 effects catalysis by transition state stabilization rather than by general acid/base catalysis.

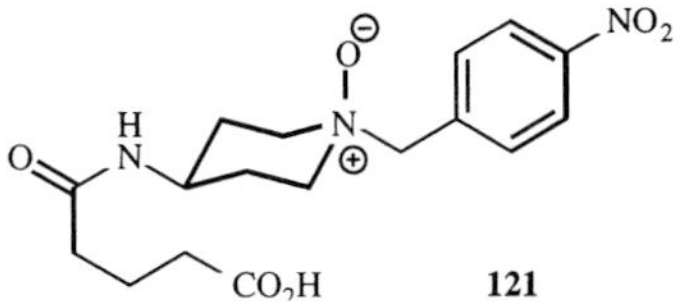

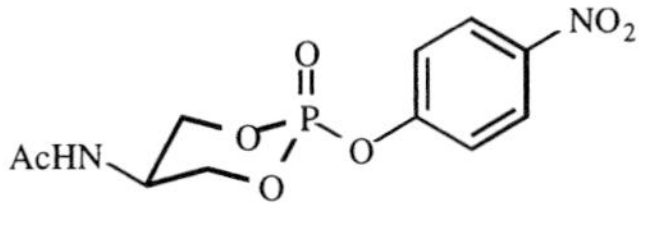

Abzyme Identity	Substrate	Conditions	K_m	$10^3 \cdot k_{cat}$ [a]	K_i **121**
15C5	**122**	pH 8.1	87 µM	2.7 min^{-1}	–
3H5	**123**	pH 9.15	5.05 mM	2.0 min^{-1}	0.98 µM

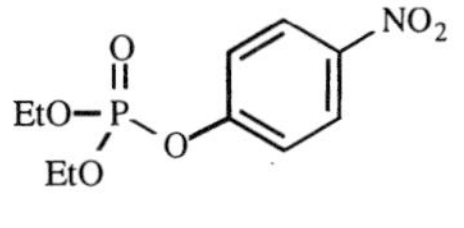

[a] Measured at pH 8.25 and 25 °C

Fig. 41. The *N*-oxide (**121**) was used as hapten to raise mAbs to catalyze the hydrolysis of both triesters (**122**) and (**123**).

Phosphate ester hydrolysis is one of the most demanding of reactions for catalyst engineering. The progress made so far with catalytic antibodies is extremely promising and appears to be competitive with studies using metal complexes if only because they can deliver *turnover* while metal complexes have for the most part to solve the problem of tight product binding.

6.4 Amide Hydrolysis

While ester, carbonate, carbamate, and anilide hydrolyses have been effectively catalyzed by antibodies, the difficult tasks of catalyzing the hydrolysis of an aliphatic amide or a urea remain largely unsolved. Much of this problem comes from the fact that breakdown of a $TI^{\pm}$ is the rate determining step, as established by much kinetic analysis and, more recently, by computation. TERAISHI has computed that C–N bond cleavage for the TI^{-} for hydrolysis of *N*-methylacetamide (or aminolysis of acetic acid) lies some 18 kcal mol^{-1} above that for C–O(H) bond cleavage (TERAISHI et al., 1994). Clearly, protonation of nitrogen is an essential step in the breakdown process of such intermediates and for anilides that is hardly a practical proposition under ambient conditions. To date only three investigations of this problem have shown any success.

A group at IGEN raised antibodies to the dialkylphosphinic acid (**124**) (Fig. 42). These were screened for their ability to hydrolyze four alkyl esters and four primary amides at pH 5.0, 7.0, and 9.0. Just one out of 68 antibodies, 13D11, hydrolyzed the C-terminal carboxamide with stereospecificity for the (*R*)-substrate (**125a**) (rendered visible by the attachment of a dansyl fluorophore to support HPLC analysis of the course of the reaction) (AE 5.1) (MARTIN et al., 1994). At pH 9.0 and 37 °C, mAb 13D11 showed K_m 432 μM and k_{cat} $1.65 \cdot 10^{-7}$ s^{-1}, corresponding to a half-life of 42 d. This activity was fully inhibited by hapten (**124**) with K_i 14 μM. Unexpectedly, the dansyl group proved to be an essential component of the substrate. Even more surprisingly, the antibody did not catalyze the hydrolysis of the corresponding methyl ester (**125b**).

More recently, JANDA and co-workers have produced catalytic antibodies for hydrolysis of a similar primary amide (GAO et al., 1998). They raised antibodies to the trigonal boronic acid hapten (**126**) in order to catalyze the hydrolysis of the primary amide (**127a**) and methyl ester (**127b**) (Fig. 43). Using antibodies obtained by standard hybridoma techniques, they built a combinatorial library and selecting with a boronic acid probe and phage display, they were able to isolate Fab-BL25 catalyzing the hydrolysis of the L-enantiomer of (**127a**). This Fab fragment showed Michaelis-Menten kinetics (K_m 150 μM and k_{cat} 0.003 min^{-1}) with a half-life for the substrate of 3.9 h, more than two orders of magnitude better than for 13D11. It is interesting to note that no catalytic antibodies were obtained by the standard immunization route.

Whereas most amide substrates for catalytic antibodies have been activated by the use of aromatic amines (AE 5.3, 5.4, 5.5), BLACKBURN and co-workers chose to explore hydrolysis of an aliphatic amide activated through halogenation in the acyl moiety. Chloram-

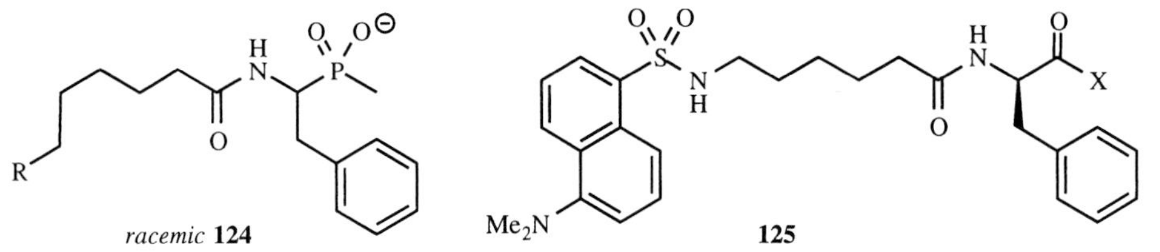

racemic **124**

125
a X = NH_2
b X = OMe

Fig. 42. One out of 68 antibodies raised to the dialkylphosphinic acid hapten (**124**), hydrolyzed the carboxamide (**125a**), but not the ester (**125b**).

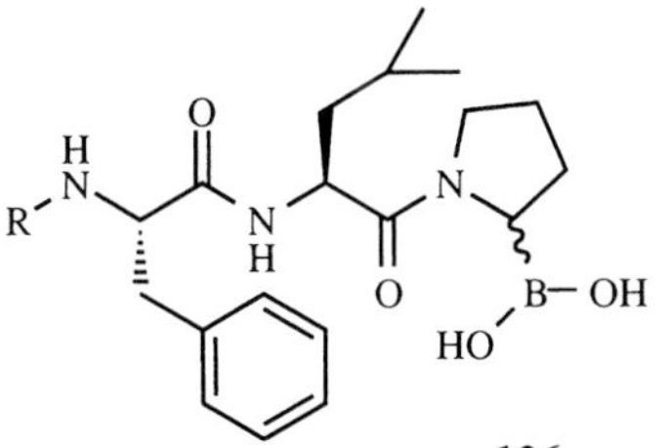

126

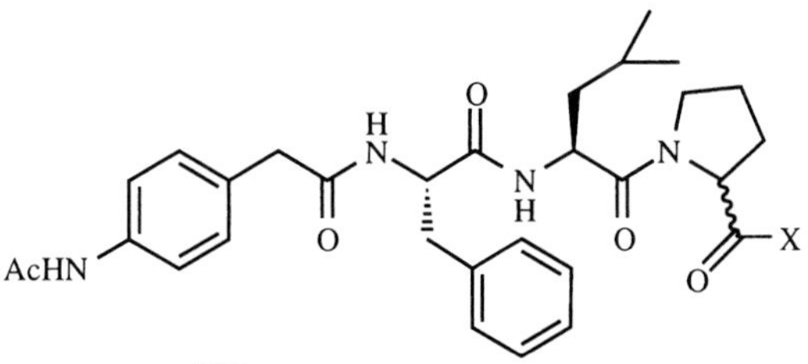

127
a X = NH_2
b X = OMe

Fig. 43. Antibodies raised to the trigonal boronic acid hapten (**126**) catalyzed the hydrolysis of the L-enantiomer of the primary amide (**127a**).

phenicol (**128**) was selected as substrate on account of its dichloroacetamide function and the tetrahedral intermediate for hydrolysis was mimicked by the neutral sulfonamide (**129**) and the zwitterionic "stretched transition state analog" aminophosphinic acid (**130**) (Fig. 44). Antibodies produced to each of these haptens proved too weak to hydrolyze chloramphenicol (**128**) at a rate sufficiently above background (k_{OH} $1.3 \cdot 10^{-3}$ M^{-1} s^{-1}) for further study. However, the switch to a more reactive amide substrate, trifluoramphenicol (**131**) (k_w $6 \cdot 10^{-7}$ s^{-1}, k_{OH} $6.3 \cdot 10^{-2}$ M^{-1} s^{-1} at 37 °C), enabled useful data to be obtained for antibody 2B5. This showed Michaelis-Menten kinetics with k_{cat} $2 \cdot 10^{-6}$ s^{-1} and K_m 640 µM at pH 7.0, 37 °C. The use of $k_{cat}/(K_m \cdot k_{uncat})$ gave an ER of 5,200. The relatively high K_m may be a consequence of switching the dichloroacetamide moiety in the hapten for the trifluoroacetamide group in the substrate and so could be improved by appropriate modification to the hapten. The low rate of turnover achieved clearly indicates the difficult task ahead for antibody cleavage of a peptide based on tetrahedral intermediate mimicry alone (SHEN, 1995; DATTA et al., 1996).

A new strategy to achieve cleavage of an amide bond has used an external cofactor (ERSOY et al., 1998). Antibodies were raised to an arylphosphonate diester. In this case the arylpropylamide is attacked by any of three nucleophile cofactors catalyzing a N→O acyl transfer. This generates an ester with a high spontaneous rate of cleavage. In this way, aryl amide hydrolysis was achieved with k_{cat} of $1.3–13.3 \cdot 10^{-5}$ min^{-1} and K_m of 77–370 µM for the substrate and 14–136 µM for the three cofactors.

By contrast, the reverse reaction, that of amide synthesis, has proved to be a good target for antibody catalysis and a range of different enterprises have been successful (AE 18.1–18.4). It would appear here that little more is needed than a good leaving group and satisfactory design of a TSA based on an anionic tetrahedral intermediate (BENKOVIC et al., 1988; JANDA et al., 1988a; HIRSCHMANN et al. 1994; JACOBSEN and SCHULTZ, 1994).

7 Reactive Immunization

An alternative approach for the induction of catalysis in antibody binding sites is a strategy named "reactive immunization" (WIRSCHING et al., 1995). This system uses haptens of intermediate chemical stability as antigens. After the first stimulation of the mouse B cells to generate antibodies, one of the products of *in vivo* chemical transformation of the original hapten is then designed to act as a second immunogen to stimulate further maturation of antibodies that will be better able to catalyze the desired reaction. The system seems well-designed to achieve the benefits of a neutral and a charged hapten within the same family of monoclonal antibodies.

An organophosphate diester (**132**), was chosen as the primary reactive immunogen. Fol-

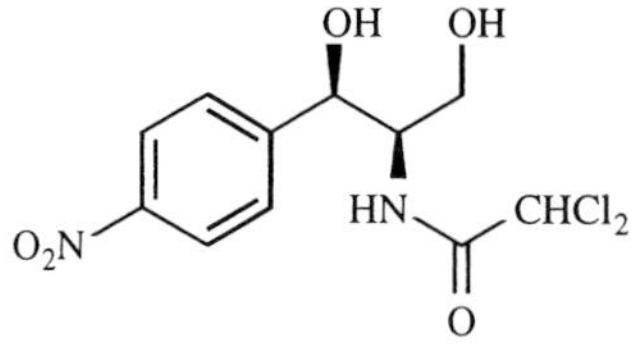

128
Chloramphenicol

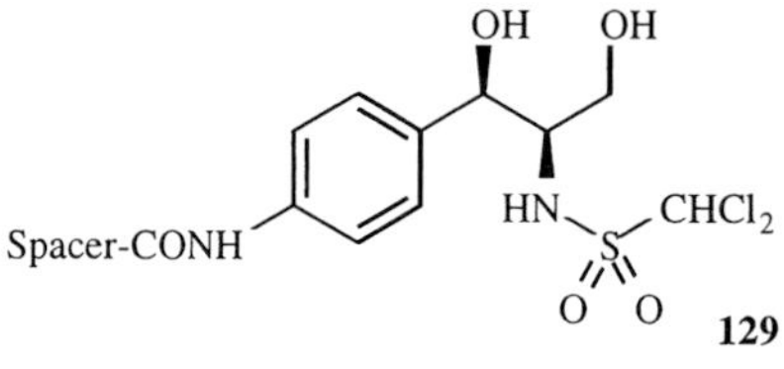

129

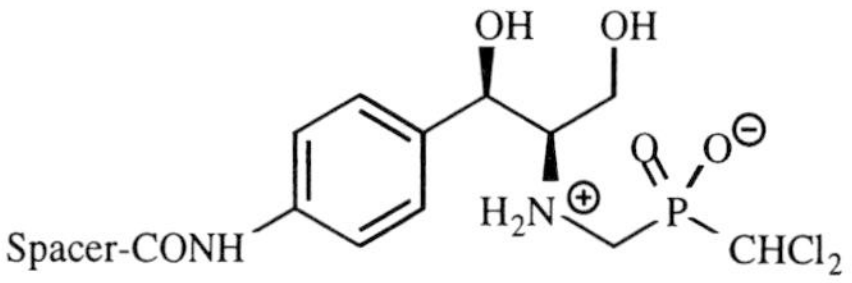

130

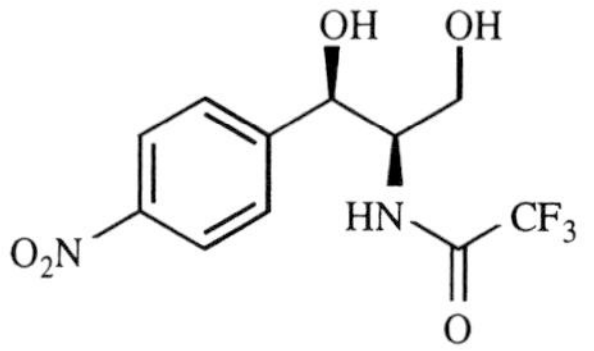

131
Trifluoramphenicol

Fig. 44. Antibodies produced to each of haptens (**129**), (**130**) proved too weak to hydrolyze chloramphenicol (**128**) at a rate sufficiently above background. However, the more reactive substrate, trifluoramphenicol (**131**) enabled useful data to be obtained for antibody 2B5 (Shen, 1995; Datta et al., 1996).

REACTIVE IMMUNOGEN

Ar = 4-(methylsulfonyl)phenyl

a R = represents position of linker to which carrier protein is attached
b R = H in free hapten

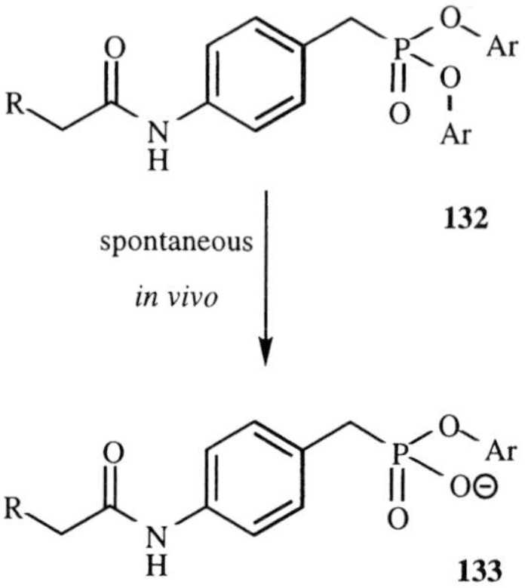

132

spontaneous *in vivo*

133

STABLE IMMUNOGEN - TSA

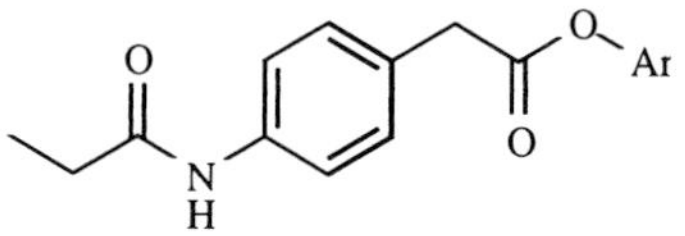

134

SUBSTRATE

Fig. 45. Antibodies raised simultaneously against the reactive (**132**) and stable immunogen (**133**) shown above were capable of efficient "turnover" of the related aryl ester substrate (**134**) (Ab 49H4: $K_m = 300\ \mu M$; $k_{cat} = 31\ min^{-1}$ at 22 °C).

lowing spontaneous hydrolysis *in vivo*, it becomes a stable monoester transition state analog (**133**), which in turn gives a new challenge to the immune system (Fig. 45) (AE 2.14). Cross-reactivity has been demonstrated as being an advantage in this process since heterologous immunization with both diaryl ester (**132**) and the corresponding monoaryl ester gave cross-reacting serum with enhanced affinity for the monoaryl TSA. This further promotes the induction of amino acids capable of acting as nucleophiles or general acid/base catalysts in the active site. In practice, reactive immunization with (**132a**) generated 19 monoclonal antibodies, 11 of which were able to hydrolyze substrate (**132b**). The most efficient abzyme, SPO49H4, was analyzed kinetically using radioactive substrates. It was established that 49H4 had undergone reactive immunization, since it was able to hydrolyze the aryl carboxylate aryl (**134**) very effectively with $K_m = 300\ \mu M$; $k_{cat} = 31\ min^{-1}$.

A similar approach has been used by LERNER and BARBAS to induce catalytic antibodies mimicking Type I aldolases. The reaction scheme is shown below: the aim here was to induce an enamine moiety which can achieve catalysis through lowering the entropy for bimolecular reaction between ketone substrate and aldol acceptor. Compound (**135**) is a 1,3-diketone which acts as a trap for the "critical lysine", forming the vinylogous amide (**138**), which can be monitored by spectrophotometry at 318 nm (Fig. 46) (AE 16.2) (LERNER and BARBAS, 1996). Screening for this catalytic intermediate by incubation with hapten facilitated the detection of two monoclonal antibodies with k_{cat} $2.28 \cdot 10^{-7}$ M^{-1} min^{-1}. Furthermore, $k_{cat}/(K_m \cdot k_{uncat})$ is close to 10^9, making these antibodies nearly efficient as the naturally occurring fructose 1,6-bisphosphate aldolase. Studies on the stoichiometry of the reaction by titration of antibody with acetylacetone indicated two binding sites to be involved in the reaction.

The stereochemical control shown by these antibodies was remarkable, and so it has been suggested that it might be possible to program an antibody to accept *any* aldol donor and acceptor. In the event, the antibodies generated in this program accepted a broad range of substrates including acetone, fluoroacetone, 2-butanone, 3-pentanone, and dihydroxyacetone.

Finally, the process of reactive immunization leads to the concept of using mechanism-based inhibitors as haptens, actively promoting the desired mechanism by contrast to their conventional use as irreversible enzyme inhibitors.

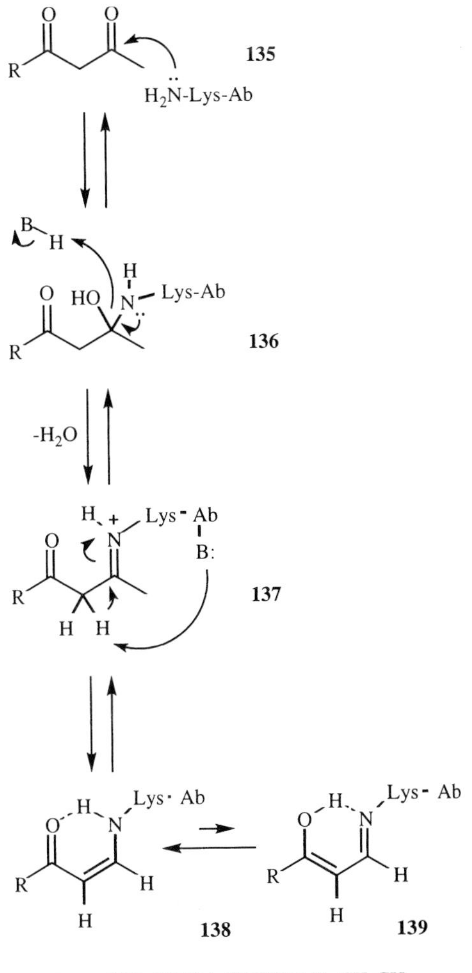

$R = p\text{-}(HO_2C(CH_2)_3CONH)C_6H_4\text{-}CH_2CH_2\text{-}$

Fig. 46. The mechanism by which an essential Lys residue in the antibody combining site is trapped using the 1,3-diketone (**135**) to form the covalently linked vinylogous amide (**138**).

8 Potential Medical Applications

8.1 Detoxification

The idea that abzymes might be used therapeutically to degrade harmful chemicals in man potentially offers a new route to treatment for victims of drug overdose. LANDRY's group has generated antibodies to catalyze the hydrolysis of the benzoyl ester of cocaine (**140**) yielding the ecgonine methyl ester (**141**) and benzoic acid, products which retain none of the stimulant or reinforcing properties of the parent drugs. The transition state analog used in this experiment was the stable phosphonate monoester (**142**) which led to a range of antibodies of which 3B9 and 15A10 were the most effective (Fig. 47) (AE 1.3) LANDRY et al., 1993).

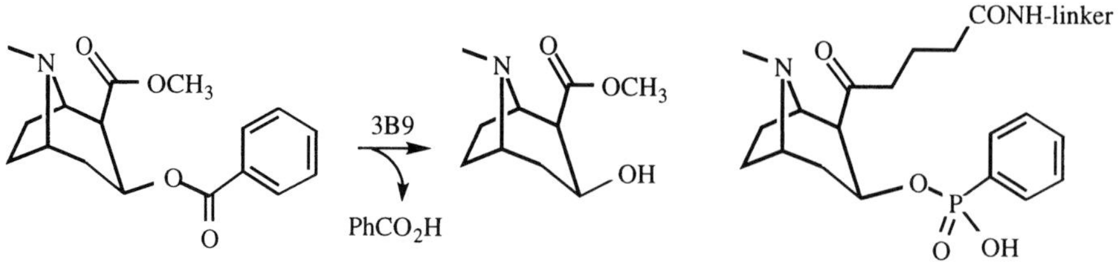

Abzyme Identity	Conditions	K_m	k_{cat} [a]	K_i **142**
3B9	pH 7.7	490 μM	0.11 min^{-1}	<2 μM
15A10	pH 8.0	220 μM	2.3 min^{-1}	–

[a] Temperature not defined

Fig. 47. Hapten (**142**) was used to raise an antibody 3B9 capable of the hydrolysis of cocaine (**140**) to the alcohol (**141**) thereby effecting cocaine detoxification.

In a more recent study, the same group examined the efficiency of antibody 15A10 *in vivo* (METS et al., 1998). They studied the effect of mAb 15A10 on seizure and lethality in a rat model of overdose and its effect on cocaine induced reinforcement in a rat model of addiction. They showed how this mAb while effectively increasing the degradation rate of the drug, protects against its lethal effect and blocks its reinforcing effects. The authors suggest the use of a humanized mAb to transpose this system to humans.

8.2 Prodrug Activation

Many therapeutic agents are administered in a chemically modified form to improve features such as their solubility characteristics, ease of administration, and bioavailability (BOWMAN and RAND, 1988). Such a "prodrug" must be designed to break down *in vivo* releasing the active drug, sometimes at a particular stage of metabolism or in a particular organ. The limitation that this imposes on the choice of masking function could be overcome by the use of an antibody catalyst for activating the prodrug which could, in principle, be concentrated at a specified locus in the body. Such selectivity could have implications in targeted therapies.

Antibody mediated prodrug activation was first illustrated by FUJII who raised antibodies against phosphonate (**144**) to hydrolyze a prodrug of chloramphenicol (**145**). Antibody 6D9 was shown to operate on substrate (**145**) to release the antibiotic (**128**) with an ER of $1.8 \cdot 10^3$ (AE 1.8) (Fig. 48) (MIYASHITA et al., 1993). Furthermore, FUJII showed unequivocally that antibody catalyzed prodrug activation is viable by demonstrating inhibition of the growth of *B. subtilis* by means of the ester (**145**) only when antibody mAb 6D9 was present in the cell culture medium. No product inhibition was observed by chloramphenicol at 10 mM supporting the multiple turnover effect seen in the growth inhibition assay.

CAMPBELL and co-workers also succeeded with this type of strategy by producing antibody 49.AG.659.12 against a phosphonate TSA (**147**), designed to promote release of the anti-cancer drug, 5-fluorodeoxyuridine from a D-valyl ester prodrug (**148**) (Fig. 49, AE 1.10, CAMPBELL, et al., 1994). This catalyst was able to bring about inhibition of the growth of *E. coli* by the release of the cytotoxic agent, 5-fluorodeoxyuridine *in vitro*.

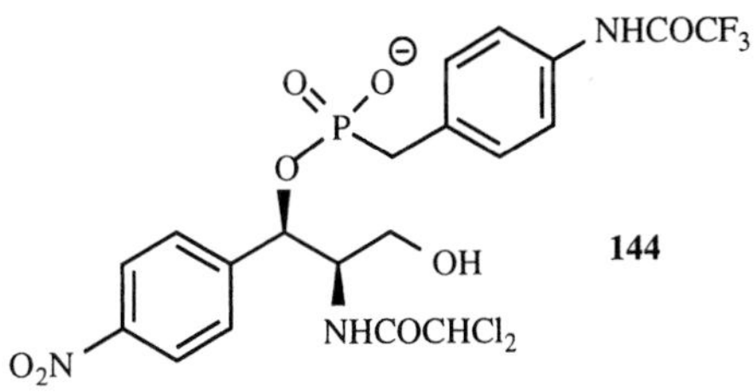

144

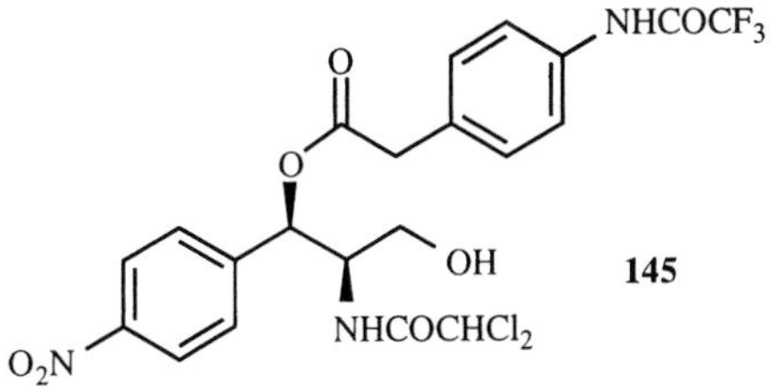

145

6D9

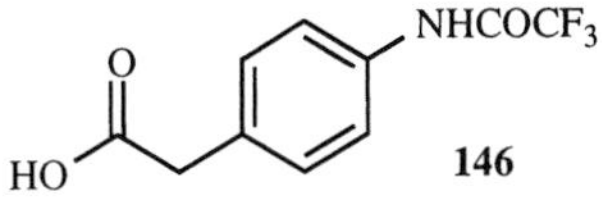

146

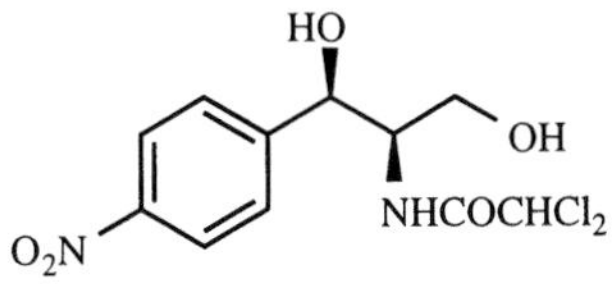

128
Chloramphenicol

Abzyme Identity		Conditions
6D9		pH 8.0; 30 °C
K_m **145**	k_{cat} **145**	K_i **144**
64 μM	0.133 min^{-1}	0.06 μM

Fig. 48. The monoclonal, 6D9, raised against phosphonate (**144**) catalyzed the hydrolysis of one possible regio-isomer (**145**) of a phenacetyl ester prodrug derived from chloramphenicol (**128**).

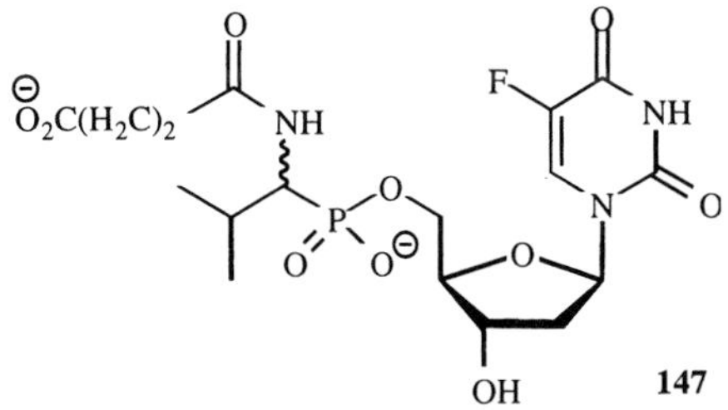

147

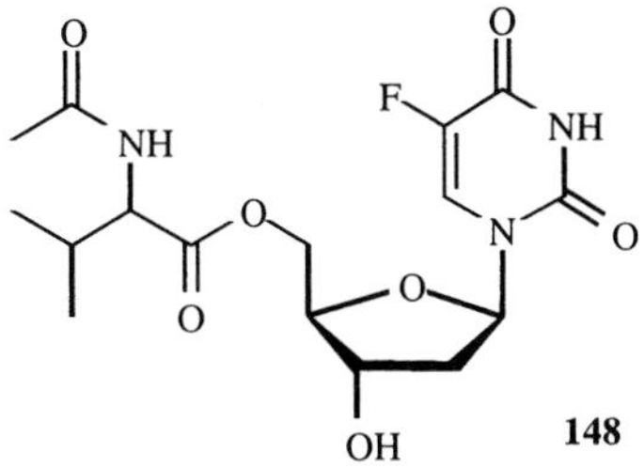

148

Abzyme Identity		Conditions
49.AG.659.12		pH 8.0; 37 °C
K_m **148**	k_{cat} **148**	K_i **147**
218 μM	0.03 min^{-1}	0.27 μM

Fig. 49. Prodrug (**148**) is an acylated derivative of the anticancer drug 5-fluorodeoxyuridine. Antibody 49.AG.659.12, raised against phosphonate (**147**) was found to activate the prodrug (**148**) *in vitro*, thereby inhibiting the growth of *E. coli*. Kinetic parameters are listed.

Much the most developed example of prodrug activation comes from our own laboratory. The cytotoxicity of nitrogen mustards is dependent on substitution on the nitrogen atom: electron-withdrawing substituents deactivate and electron-releasing substituents activate a bifunctional mustard. Thus, antibody catalyzed cleavage of a carbamate ester of a phenolic mustard can enhance its cytotoxicity to establish the carbamate as a viable prodrug for cancer chemotherapy (BLAKEY, 1992; BLAKEY et al., 1995). So the target for prodrug activation is defined as an aryl carbamate whose nitrogen substituent is either an aryl (**149a**) or alkyl (**149b**) moiety (Fig. 50).

Aryl carbamates are known to cleave by an E1cB process with a high dependency on the pK_a of the leaving phenol ($\rho^- = 2.5$). By contrast, aryl *N*-methylcarbamates are hydrolyzed by a $B_{Ac}2$ process with a much lower dependency on leaving group ($\rho^0 = 0.8$) (WILLIAMS and DOUGLAS, 1972a, b). Given the electron-releasing nature of the nitrogen mustard function (σ ca. -0.5), the kinetic advantage of antibody hydrolysis via the $B_{Ac}2$ pathway coupled to the proved ability of antibodies to stabilize tetrahedral transition states led to the formulation of TSAs (**152a**) and (**152b**). Siting the linker in the locus of the nitrogen mustard was designed (1) to minimize potential alkylation of the antibody by the mustard function and (2) to support mechanistic investigations by variation of the *p*-substituent of the aryl carbamate with little or no change in K_m, both features that were realized in the outcome. Both of these TSAs generated large numbers of hybridomas including many catalysts capable of carbamate hydrolysis.

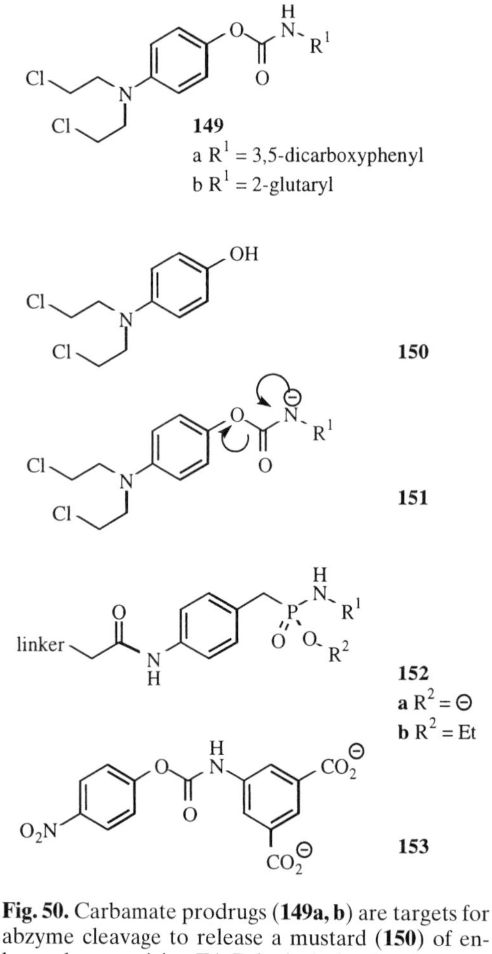

149
a R^1 = 3,5-dicarboxyphenyl
b R^1 = 2-glutaryl

150

151

152
a $R^2 = \ominus$
b R^2 = Et

153

Fig. 50. Carbamate prodrugs (**149a, b**) are targets for abzyme cleavage to release a mustard (**150**) of enhanced cytotoxicity. E1cB hydrolysis of aryl carbamates involves the anion (**151**). TSAs (**152a**) and (**152b**) were used to generate antibodies to catalyze a $B_{Ac}2$ mechanism for hydrolysis whose kinetic behavior was evaluated with ester (**153**).

A mechanistic analysis of antibody DF8-D5 showed it to cleave *p*-nitrophenyl carbamate (**153**) with k_{cat} 0.3 s^{-1}, K_m 120 μM, and k_{cat}/k_{uncat} 300 at 14 °C (AE 4.3) (WENTWORTH et al., 1997). This is some tenfold more active than a carbamatase antibody generated by SCHULTZ to a *p*-nitrophenyl phosphonate TSA but with a similar ER (AE 4.1) (VAN VRANKEN et al., 1994). Most significantly, variations in the *p*-substituent in substrates for DF8-5 hydrolysis identified a Hammett ρ^0 value[6] of 0.526 to establish the $B_{Ac}2$ nature of the reaction. For the *p*-methoxyphenyl carbamate substrate ($\sigma^0 = -0.3$) the apparent ER is $1.2 \cdot 10^6$. Given that there is a 10^8 difference in rate for the E1cB and $B_{Ac}2$ processes for the *p*-nitrophenyl carbamate (**153**), the data show that antibody DF8-D5 has promoted the disfavored $B_{Ac}2$ process relative to the spontaneous E1cB cleavage by some 13 kcal mol^{-1}. Lastly, it is noteworthy that DF8-D5 was raised against the phosphonamidate ethyl ester (**149a**) as hapten, which raises the possibility that it may be an unexpected product of reactive immunization (Sect. 7).

The medical potential of such carbamatases depends on their ability to deliver sufficient cytotoxic agent to kill cells. Antibody EA11-D7, raised against TSA (**152b**) proved able to hydrolyze the prodrug (**149b**) with K_m 201 μM and k_{cat} 1.88 min^{-1} at 37 °C (AE 4.2) (WENTWORTH et al., 1996). *Ex vivo* studies with this abzyme and human colonic carcinoma (LoVo) cells led to a marked reduction in cell viability relative to controls. This cytotoxic activity was

[6] A ρ/σ^- plot would have an even flatter slope.

reproduced exactly by the Fab derived from EA11-D7 and was fully inhibited by a stoichiometric amount of the TSA (**152b**). Using EA11-D7 at 0.64 μM, some 70% of cells were killed in a 1 h incubation with prodrug (**149b**) and the antibody transformed a net 4.18 μM of prodrug delivering more than $2 \cdot IC_{50}$ of the cytotoxic agent (**150**). This performance is, however, well behind that of the bacterial carboxypeptidase CPG2 used by Zeneca in their ADEPT system (BAGSHAWE, 1990; BLAKEY et al., 1995), being 10^3 slower than the enzyme and $4 \cdot 10^4$ inferior in selectivity ratio. Nonetheless, it is the first abzyme system to show genuine medical potential and should stimulate further work in this area.

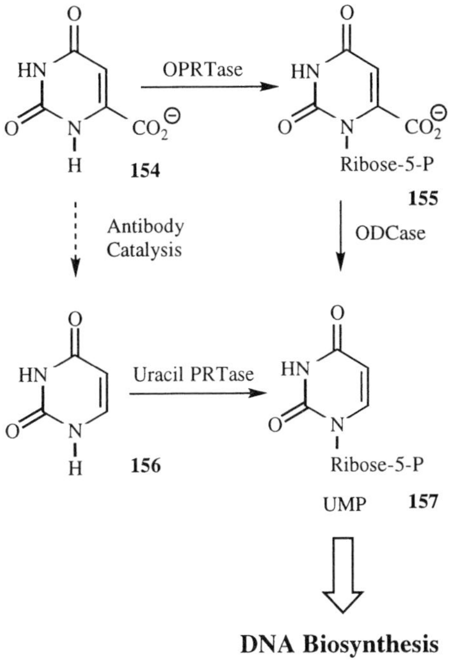

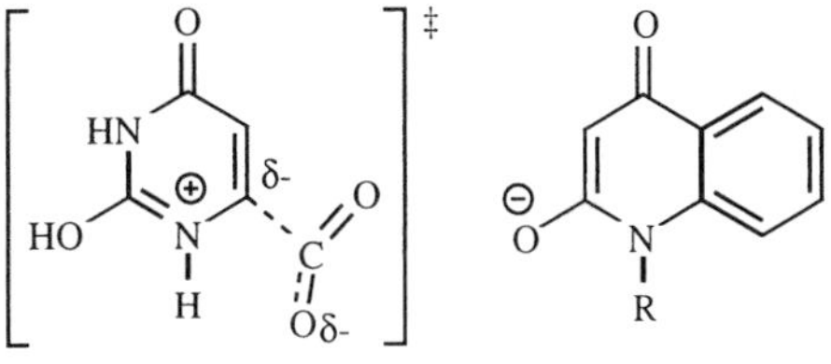

8.3 Abzyme Screening via Cell Growth

BENKOVIC proposed a new method of selecting Fabs from the whole immunological repertoire in order to facilitate a metabolic process (SMILEY and BENKOVIC, 1994). A cDNA library for antibodies was raised against hapten (**159**) and then expressed in an *E. coli* strain devoid of any native orotic acid decarboxylase activity (Fig. 51). The bacteria were then established in a pyrimidine-free environment where only those bacteria grew which expressed an antibody able to provide pyrimidines essential for DNA synthesis, and hence bacterial growth. Six colonies expressing an active Fab fragment were found viable in a screen of 16,000 transformants (AE 9.2). The remarkable feature of this system is that OCDase, which catalyzes the decarboxylation of orotidylic acid to UMP (Fig. 51), is thought to be at the top end of performance by any enzyme in accelerating this decarboxylation by some 10^{17} (RADZICKA and WOLFENDEN, 1995).

Such an example illustrates the ability of abzymes to implant cell viability in the face of a damaged or deleted gene for a crucial metabolic process. The medical opportunities for applications of such catalysis are clear.

There is sufficient encouragement in these examples to show that out of all the prospects for the future development of catalytic antibodies, those in the field of medical applications, where selectivity in transformation of unusual substrates may be of greater importance than sheer velocity of turnover of substrate, may well rank highest.

Fig. 51. Pathways for uridylate biosynthesis. PRTase: phosphoribosyl transferase; ODCase: OMP decarboxylase; OMP: orotidine 5′-phosphate; UMP: uridine 5′-phosphate. Mutants lacking enzymes PRTase or ODCase can complete a route to UMP provided by an antibody orotate decarboxylase in conjunction with the naturally occurring uracil PRTase. Decarboxylation of orotic acid (**154**) is thought to proceed through the transition state (**158**), for which the hapten (**159**) was developed (SMILEY and BENKOVIC, 1994).

9 Industrial Future of Abzymes

In view of the widespread interest in biocatalysis, it is not surprising that in little over a dozen years after catalytic antibodies were first reported, a vast literature has developed documenting them. The field of research workers is international, with groups from four continents showing activity in the area. The potential of "designer enzymes" is already becoming a reality in relation to both the chemical industry and the pharmaceutical field.

Over 70 different chemical reactions, ranging from hydrolyses to carbon–carbon bond forming reactions, have been catalyzed by abzymes and their application to general synthetic organic chemistry seems promising. Typical Michaelis constants (K_m) lie in the range 10 μM to 1 mM and binding selectivity for the TSA over the substrate is in the range 10- to 10^5-fold. It therefore appears that antibodies have fulfilled most expectations that they would be capable of substrate discrimination comparable to that of enzymes but over a wider range of substrate types than foreseen, and especially effective when programmed to a designated substrate. The range of reactions that may be catalyzed by antibodies appears to be limited only by a sufficient knowledge of the transition state for any given reaction combined with synthetic accessibility to a stable TSA or to a suitable "suicide" substrate.

On the other hand, antibodies are generally able to accelerate reactions by at most 10^7 times the rate of the uncatalyzed process. It has to be said that scientists at large are looking for a major step forward in abzymes properties to achieve rate accelerations up to 10^9 that would establish abzymes as a feature of synthetically useful biotransformations. At the same time, it is essential to demonstrate that product inhibition is not an obstacle to the scaled-up use of abzymes.

In relation to syntheses able to deliver usable amounts of product, LERNER has shown that stereoselective reactions can be performed on a gram scale, as in the enantioselective hydrolysis of a 2-benzylpropenyl methyl ether to the corresponding (S)-2-benzylpropanol of high ee (AE 7.4A) (REYMOND et al., 1994). In addition, JANDA has described an automated method of transposing antibody catalyzed transformations of organic molecules onto the multigram scale by employment of a biphasic system. The viability of this system was demonstrated by an epoxide ring-closing antibody, 26D9, to transform 2.2 g of substrate, corresponding to a turnover of 127 molecules per catalytic site in each batch process. This proves that the catalytic antibody does not experience inhibition by product (AE 14.1) (SHEVLIN et al., 1994). It would appear that an improvement in abzyme performance of little more than two orders of magnitude is needed before catalytic antibodies can be efficiently put to work in bioreactors and participate in kilogram scale production.

Lastly, two technical features of antibody production may be valuable for the future production of cheaper abzymes of commercial value. First, the use of polyclonal catalysts, primarily from sheep, has had a tough early passage but now appears to be established for a wide range to transformations (GALLACHER et al., 1990, 1992; STEPHENS and IVERSON, 1993; TUBUL et al., 1994; BASMADJIAN et al., 1995; WALLACE and IVERSON, 1996). While these catalysts may not lend themselves to detailed examination by physical organic chemistry, they should be able to deliver catalysts at a much lower cost. Secondly, as science becomes more environmentally friendly and animal experimentation is more tightly regulated, approaches to screening antibodies with *in vitro* libraries may become a more important component of this domain of investigation. THOMAS has made a beginning with *in vitro* immunization and shown that useful catalysis can be identified (STAHL et al., 1995). While there are some limitations in this system, notably the relatively low substrate affinity of antibodies generated in this way, it is capable of refinement and may become a useful component of future abzyme selection systems.

10 Conclusions

On an evolutionary time scale, abzyme research is just reaching adolescence (THOMAS,

1996), yet already over 80 different antibody catalyzed chemical reactions have been listed during its first decade of life. The details uncovered concerning the mechanisms of abzyme catalyzed reactions have been richer than expected. A wide diversity of purposely designed catalysts has been explored, with the potential impact on the field of medicine and fine chemicals production being implicated. However, the immaturity of antibody catalysis has been exposed by its poor efficiency, which in spite of intense research efforts to improve all aspects of abzyme generation, continues to hinder wide-scale acknowledgement of its contribution to biocatalysis, particularly from under the shadow of powerful enzymes.

There now exists sufficient literature about catalytic antibodies not only in terms of their kinetic behavior (Appendix, Sect. 12) but also through structural information derived from X-ray crystallographic data (GOLINELLI-PIMPANEAU et al., 1994; HAYNES et al., 1994; ZHOU et al., 1994) and 3-D modeling of protein sequences (ROBERTS et al., 1994), that it has become possible to speculate on a more general basis concerning the scope, limitations, and realistic future of the field (STEWARD and BENKOVIC, 1995; KIRBY, 1996). In terms of transition state stabilization, catalytic antibodies have been shown to recognize features of the putative transition state structure encoded by their haptens with affinity constants in the nanomolar region whereas it has been estimated that enzymes can achieve transition state complementarity with association constants of the order of 10^{-24} M to deliver rate accelerations of up to 10^{17} (RADZICKA and WOLFENDEN, 1995). The whole subject of binding energy and catalysis has been authoritatively and critically reviewed by MADER and BARTLETT, with especial focus on the relationship between transition state analogs and catalytic antibodies (MADER and BARTLETT, 1997). Enzymes have evolved to interact with every species along the reaction pathways that they catalyze whereas our manipulation of the immune system is still relatively simplistic, using a single hapten to stimulate a full, often multistep, reaction sequence of catalysis.

The serendipity that may be involved in the isolation of an efficient antibody catalyst is now well appreciated while recent studies have shown that non-specific binding proteins such as BSA may display catalysis approaching the level of abzymes, albeit without any substrate selectivity (HOLLFELDER et al., 1996). All of this serves to emphasize the fact that protein recognition of discrete high energy reaction intermediates need not translate into efficient protein catalysis. However, the improvements in hapten design and antibody generation strategies described above are being used to highlight more intricate catalytic features. Charged and nucleophilic active site residues, substrate distortion (DATTA et al., 1996; YLI-KAUHALUOMA et al., 1996), desolvation, and proximity effects have all been now identified as components of antibody mediated catalysis. Using structural information available for an ever increasing number of catalytic antibodies, manipulation of the antibody combining site is now attainable using procedures such as chemical modification (POLLACK and SCHULTZ, 1989; SCHULTZ, 1989) and site-directed mutagenesis (JACKSON et al., 1991; STEWART et al., 1994; KAST et al., 1996) to pinpoint or to improve the action of abzymes. The semi-rational design of antibody catalysts using a combination of such techniques is also supporting the systematic dissection of these primitive protein catalytic systems so as to provide valuable information concerning the origin and significance of catalytic mechanisms employed in enzymes.

If the ultimate worth of antibody catalysts is to be more than academic, then the key must be found in their programmability. Here, the capabilities of abzymes such as their promotion of disfavored processes and selectivity for substrates and transformations for which there are no known enzymes may offer prospects more significant than the further pursuit of performance levels comparable to those of enzymes.

11 Glossary

Abzyme: An alternative name for a catalytic antibody (derived from **Ab**-en**zyme**).

Affinity labeling: A method for identifying amino acid residues located in the binding

site(s) of a peptide. The protein is treated with a ligand which binds to the binding site and reacts with proximal amino acid residues to form a covalent derivative. Upon hydrolysis of the protein, individual amino acids or short peptide fragments linked to the ligand are separated and identified.
Antibodies: Proteins of the immunoglobulin superfamily, carrying *antigen* binding sites that bind non-covalently to the corresponding *epitope.* They are produced by B lymphocytes and are secreted from plasma cells in response to antigen stimulation.
Antigen: A molecule, usually peptide, protein, or polysaccharide, that elicits an immune response when introduced into the tissues of an animal.
B cells: (Also known as B lymphocytes). Derived from the bone marrow, these cells are mediators of humoral immunity in response to *antigen*, where they differentiate into antibody forming plasma cells and B memory cells.
Bait and switch: A strategy whereby the charge–charge complementarity between antibody and hapten is exploited. By immunizing with haptens containing charges directed at key points of the reaction transition state, complementary charged residues are induced in the active site which are then used in catalysis of the substrate.
BSA: Bovine Serum Albumin. Extracted from cattle serum and used as a *carrier* protein.
Carrier protein: Macromolecule to which a hapten is conjugated, thereby enabling the hapten to stimulate the immune response.
catELISA: Similar to an ELISA, except that the assay detects catalysis as opposed to simple binding between hapten and antibody. The substrate for a reaction is bound to the surface of the microtiter plate, and putative catalytic *antibodies* are applied. Any product molecules formed are then detected by the addition of anti-product antibodies, usually in the form of a polyclonal mixture raised in rabbits. The ELISA is then completed in the usual way, with an anti-rabbit "second antibody" conjugated to an enzyme, and the formation of colored product upon addition of the substrate for this enzyme. The intensity of this color is then indicative of the amount of product formed, and thus catalytic antibodies are selected directly.
Conjugate: In immunological terms this usually refers to the product obtained from the covalent coupling of a protein (e.g., a *carrier* protein) with either a hapten, a label such as fluorescein, or an enzyme.
Conjugation: The process of covalently bonding (multiple) copies of a hapten to a *carrier* protein, usually by means of a linker/spacer to distance the hapten from the surface of the carrier protein by a chain of about six atoms.
ELISA: (Enzyme Linked Immunosorbent Assay), An immunoassay in which *antibody* or *antigen* is detected. To detect antibody, antigen is first adsorbed onto the surface of microtiter plates, after which the test sample is applied. Any unbound (non-antigen specific) material is washed away, and remaining antibody–antigen complexes are detected by an antiimmunoglobulin *conjugated* to an enzyme. When the substrate for this enzyme is applied, a colored product is formed which can be measured spectrophotometrically. The intensity of the colored product is proportional to the concentration of antibody bound.
Enhancement ratio, ER: Quantified as k_{cat}/k_{uncat}, is used to express the catalytic power of a biocatalyst. It is a comparison between the catalyzed reaction occurring at its optimal rate and the background rate.
Entropic trap: A strategy aimed at improving the efficiency of catalytic *antibodies*, via the incorporation of a molecular constraint in the transition state analog that gives the hapten a higher-energy conformation than the reaction product.
Epitope: The region of an *antigen* to which antibody binds specifically. This is also known as the antigenic determinant.
Fab: The fragment obtained by papain hydrolysis of immunoglobulins. The fragment has a molecular weight of ~45 kDa and consists of one light chain linked to the *N*-terminal half of its corresponding heavy chain. A Fab contains one *antigen* binding site (as opposed to bivalent *antibodies*), and can combine with antigen as a univalent antibody.
Fab': The fragment obtained by pepsin digestion of immunoglobulins, followed by reduction of the interchain disulfide bond between the two heavy chains at the hinge region. The resulting fragment is similar to a Fab fragment in that it can bind with *antigen* univalently, but it has the extra hinge region of the heavy chain.

Hapten: Substance that can interact with antibody but cannot elicit an immune response unless it is conjugated to a *carrier* protein before its introduction into the tissues of an animal. Haptens are mostly small molecules of less than 1 kDa. For the generation of a catalytic antibody, a *TSA* is attached to a spacer molecule to give a hapten of which multiple copies can be linked to a *carrier* protein.
Hybridoma: Cell produced by the fusion of antibody producing plasma cells with myeloma/carcinoma cells. The resultant hybrids have then the capacity to produce antibody (as determined by the properties of the plasma cells), and can be grown in continuous culture indefinitely due to the immortality of the myeloma fusion partner. This technique enabled the first continuous supply of monoclonal *antibodies* to be produced.
IgG: The major immunoglobulin in human serum. There are four subclasses of IgG: IgG1, IgG2, IgG3, and IgG4, yet this number varies in different species. All are able to cross the placenta, and the first three subclasses fix complement by the classical pathway. The molecular weight of human IgG is 150 kDa and the normal serum concentration in man is 16 mg mL^{-1}.
Immunoglobulin: Member of a family of proteins containing heavy and light chains joined together by interchain disulfide bonds. The members are divided into classes and subclasses, with most mammals having five classes (IgM, IgG, IgA, IgD, and IgE).
k_{cat}: The rate constant for the formation of product from a particular substrate. k_{cat} is obtained by dividing the Michaelis-Menten parameter, V_{max}, by the total enzyme concentration. It is also called the "maximum turnover number", because it represents the maximum number of substrate molecules converted per active site per unit time (FERSHT, 1985).
k_{cat}/K_m: See specificity constant.
KLH: Keyhole limpet haemocyanin, used for its excellent antigenic properties. It is used as a *carrier* protein in order to bestow immunogenicity to small haptens.
K_m: Is the Michaelis-Menten constant, which is defined as the substrate concentration at which the biocatalyst is working at half its maximum rate (V_{max}). In practice, K_m gives a measure of the binding affinity between the substrate and biocatalyst; the smaller the value, the tighter the binding in the complex (FERSHT, 1985).
Library: A collection of antibodies, usually Fab or scFv fragments, in the range of 10^6 to 10^{10} and displayed on the surface of bacteriophage whose DNA gene contains a DNA sequence capable of expression as the antibody protein. Thus, identification of a single member of the library by selection can be used to generate multiple copies of the phage and sizeable amounts of the antibody protein.
Monoclonal antibody, mAb: Describes an antibody derived from a single clone of cells or a clonally obtained cell line. Its common use denotes an antibody secreted by a hybridoma cell line. Monoclonal *antibodies* are used very widely in the study of antigens, and as diagnostics.
Polyclonal antibodies: Antibodies derived from a mixture of cells, hence containing various populations of antibodies with different amino acid sequences. Are of limited use in that they will not all bind to the same *epitopes* following immunization with a hapten/*carrier* protein *conjugate*. They are also difficult to purify and characterize, yet have been used with success in the catELISA system.
Positive clones: A phrase usually used to describe those hybridoma clones which bind reasonably well to their respective hapten in an enzyme linked immunosorbent assay, thereby eliminating non-specific *antibodies* raised to different *epitopes* of the hapten/carrier *conjugate.*
Residues: General term for the unit of a polymer, that is, the portion of a sugar, amino acid or nucleotide that is added as part of the polymer chain during polymerization.
Site-directed mutagenesis: A change in the nucleotide sequence of DNA at particular nucleotide residues and/or complete codons. This is usually done in order to change the corresponding amino acid residue in the protein encoded by the nucleotide sequence so as to introduce or remove functionality from the protein.
Single Chain Antibody (scFv): Comprises a V_L linked to a V_H chain via a polypeptide linker. It is thus a univalent functioning antibody containing both of the variable regions of the parent antibody.

Somatic hypermutation: Mutations occurring in the variable region genes of the light and heavy chains during the formation of memory B cells. Those B cells whose affinity is increased by such mutations are positively selected by interaction with *antigen*, and this leads to an increase in the average affinity of the *antibodies* produced.
Specificity constant: Is defined as k_{cat}/K_m. It is a pseudo-second-order rate constant which, in theory, would be the actual rate constant if formation of the enzyme–substrate complex was the rate determining step (FERSHT, 1985).
TSA: Transition state analog; frequently a stable analog of an unstable, high energy, reaction intermediate that is close to related energy barriers in a multi-step reaction.

12 Appendix

12.1 Catalog of Antibody Catalyzed Processes

A. HYDROLYTIC AND DISSOCIATIVE PROCESSES

1. Aliphatic Ester Hydrolysis

Reaction / Conditions	Hapten / Comments / K_i	K_m [/µM]	k_{cat} [/min^{-1}]	k_{cat}/k_{uncat}	Entry AE
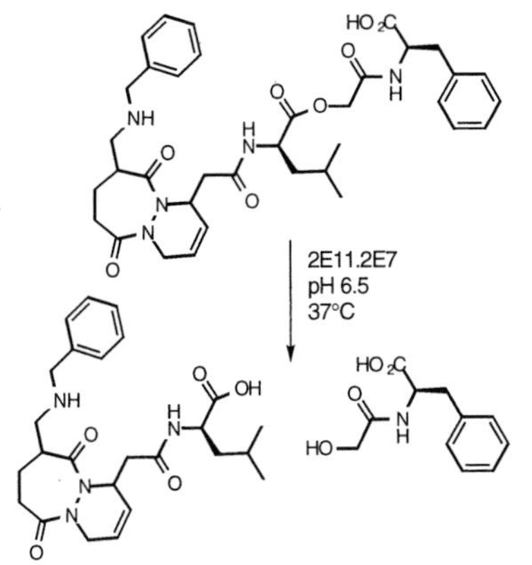 HO_2C O NH O H N O O N N O O H N 2E11.2E7 pH 6.5 37°C NH O OH O H N O N N O O HO_2C O HO N H	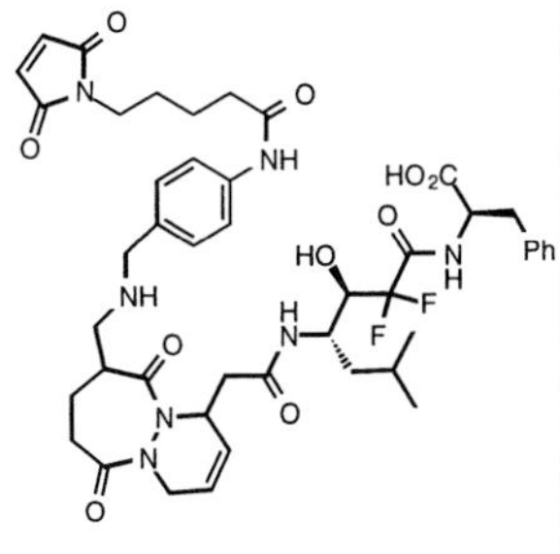 O O N O O NH HO_2C O HO N H Ph F F NH H N O O N N O O *TSA* K_i < 1.0 E+0 µM	4.4 E+3	8.0 E+0	nr	**1.1**
NHCbz R O O NO_2 NH_2 O NH R = 3G2 | 7G12 pH 8.0, 25°C NHCbz R O O NO_2 NHCbz R O O NO_2 NHCbz R OH O 4-NO_2BnOH 94% *ee* NHCbz R OH O 4-NO_2BnOH 96% *ee*	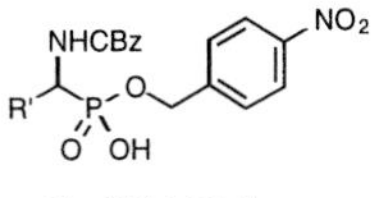 NHCBz R' P O O OH NO_2 R' = $(CH_2)_4CO_2H$ 7G12 3G2 *TSA* 7G12: K_i 1.9 E-2 µM 3G2: K_i 4.7 E-2 µM	 1.3 E+1 5.4 E+0	 7.0 E-2 3.3 E-2	 3.7 E+3 1.7 E+3	**1.2**
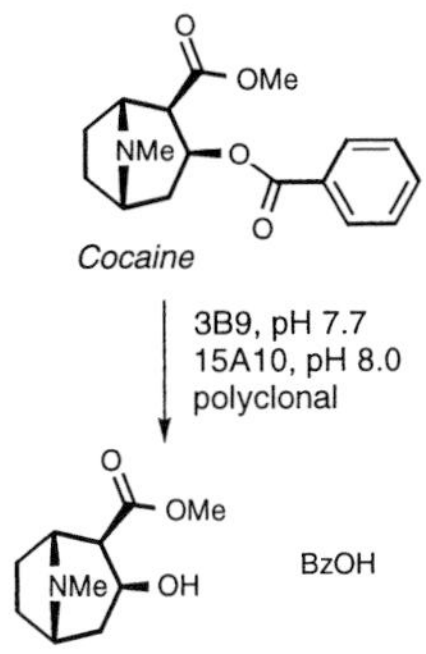 O OMe NMe O O *Cocaine* 3B9, pH 7.7 15A10, pH 8.0 polyclonal O OMe NMe OH BzOH	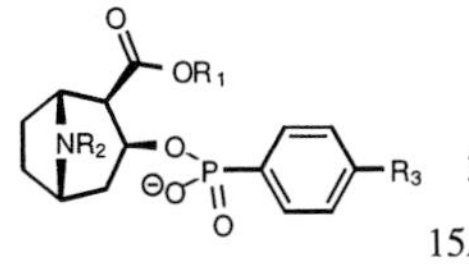 O OR_1 NR_2 O P O O R_3 3B9 15A10 $R_1 = (CH_2)_3NHCO(CH_2)_2CO_2H$, R_2 = Me, R_3 = H *TSA* 3B9: K_i < 2.0E+0 µM Vaccine Immunogen Polyclonal R_1 = Me, R_2 = Me, R_3 = NH-DT R_1 = Me, R_2 = DT, R_3 = H R_1 = DT. R_2 = Me, R_3 = H	4.9 E+2 2.2 E+2 nr	1.1 E-1 2.3 E+0 nr	5.4 E+2 2.3 E+4 nr	**1.3**

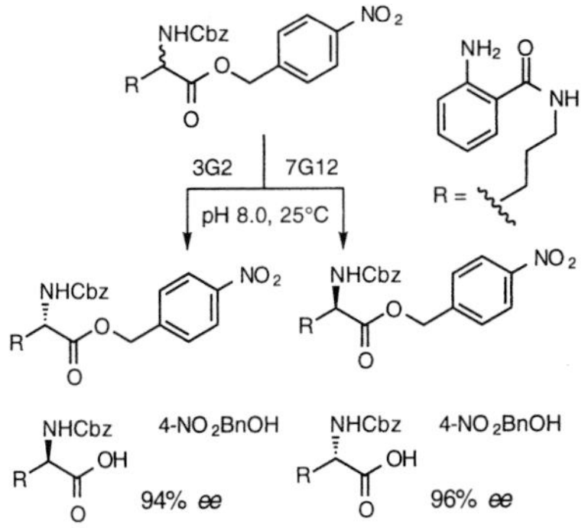

Reaction	TSA				
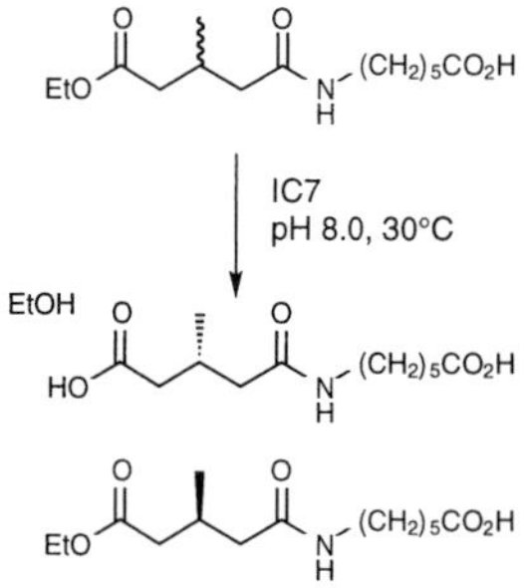 O, O, EtO, N, H, (CH2)5CO2H; IC7, pH 8.0, 30°C; EtOH; O, O, HO, N, H, (CH2)5CO2H; O, O, EtO, N, H, (CH2)5CO2H	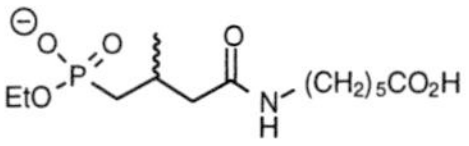 ⊖O, O, P, EtO, O, N, H, (CH2)5CO2H; *TSA*; K_i 4.0 E+0 μM	2.9 E+2	2.0 E+0	nr	**1.4**
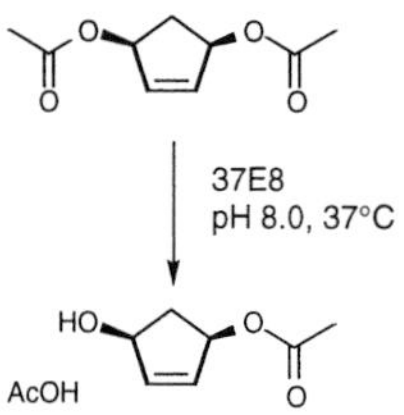 O, O, O, O; 37E8, pH 8.0, 37°C; HO, O, O; AcOH	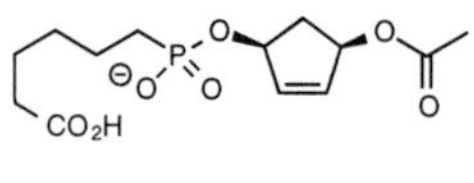 P, O, O, ⊖O, O, O, O; CO2H; *TSA*; K_i 7.0 E+0 μM	1.8 E+2	7.0 E-3	8.8 E+1	**1.5**
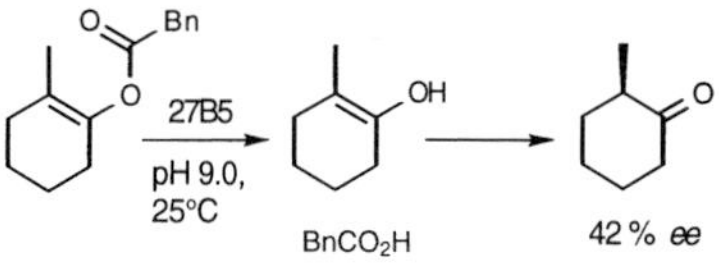 O, Bn, O; 27B5, pH 9.0, 25°C; OH; BnCO2H; O; 42 % *ee*	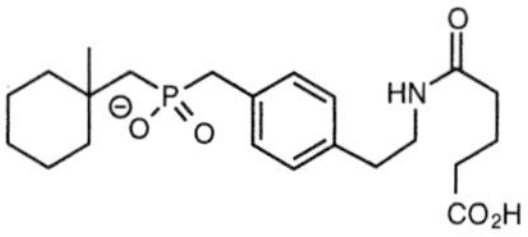 ⊖O, P, O; HN, O; CO2H; *TSA*; K_i 4.3 E+0 μM	9.9 E+2	1.0 E-2	3.0 E+2	**1.6**
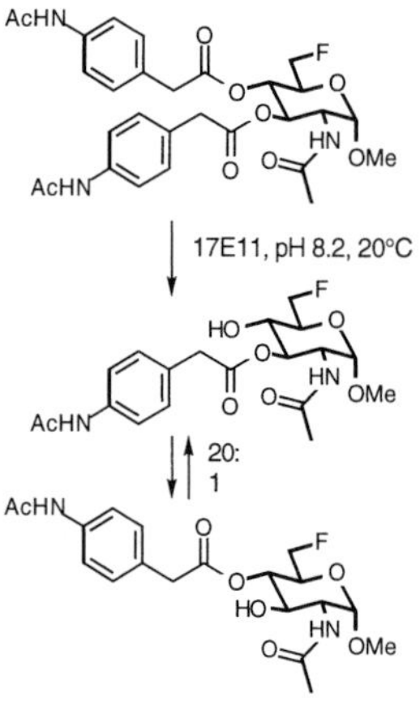 AcHN, O, O, F, O, O, HN, OMe, O, O, AcHN; 17E11, pH 8.2, 20°C; HO, F, O, O, HN, OMe, O, O, AcHN; 20: 1; AcHN, O, O, F, O, HO, HN, OMe, O	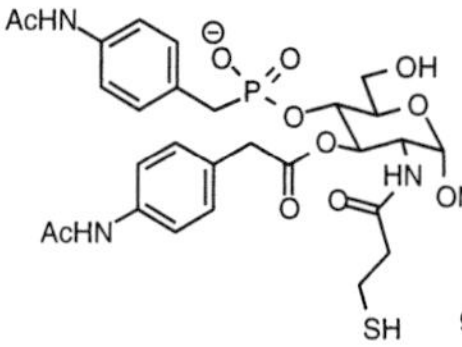 AcHN, ⊖O, O, P, O, OH, O, O, HN, OMe, O, O, AcHN; SH; 90; *TSA*; K_i 2.6 E-2 μM	6.6 E+0	1.8 E-1	2.7 E+3	**1.7**

Reaction	TSA				
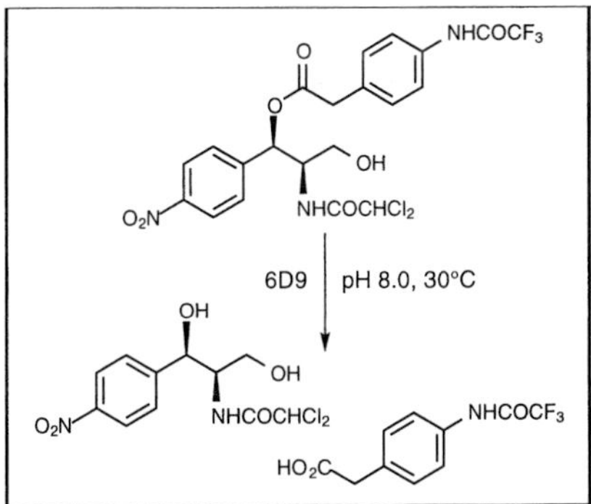 6D9, pH 8.0, 30°C	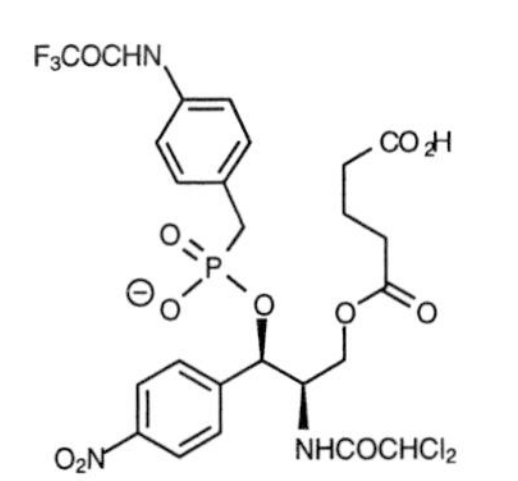 TSA, K_i 6.0 E-2 μM	6.4 E+1	1.3 E-1	1.8 E+3	**1.8**
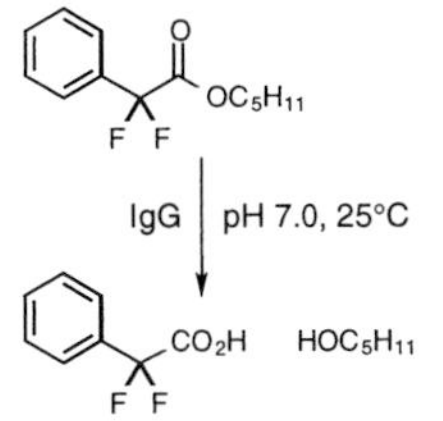 IgG, pH 7.0, 25°C	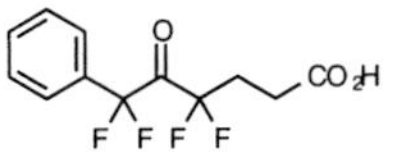 TSA	3.1 E+2	6.7 E-1	9.1 E+3	**1.9**
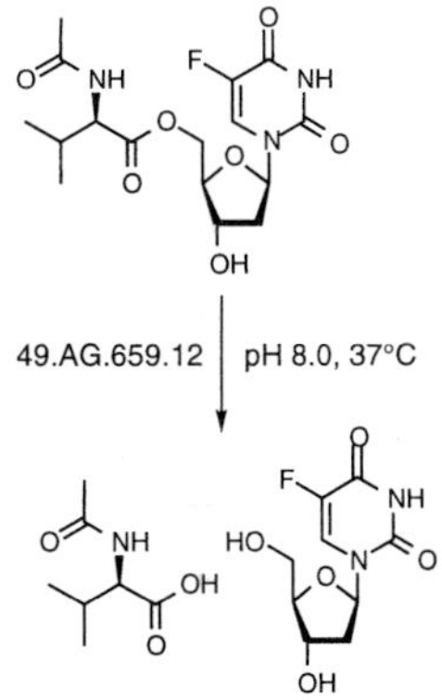 49.AG.659.12, pH 8.0, 37°C	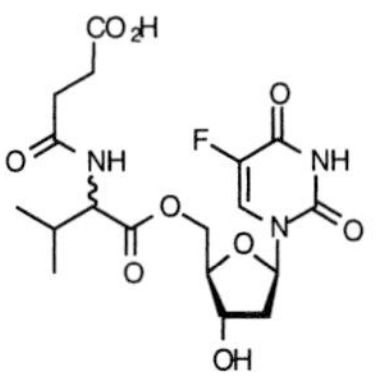 TSA, K_i 2.7 E-1 μM	2.2 E+2	3.0 E-2	9.7 E+2	**1.10**
A / B, pH 7.3, 26°C; A: 98.5 % *ee*; B: > 98 % *ee*	A	4.3 E+2	8.9 E-1	nr	**1.11**
	B; Enantiomers immunized separately; TSA	3.9 E+2	8.6 E-1	nr	

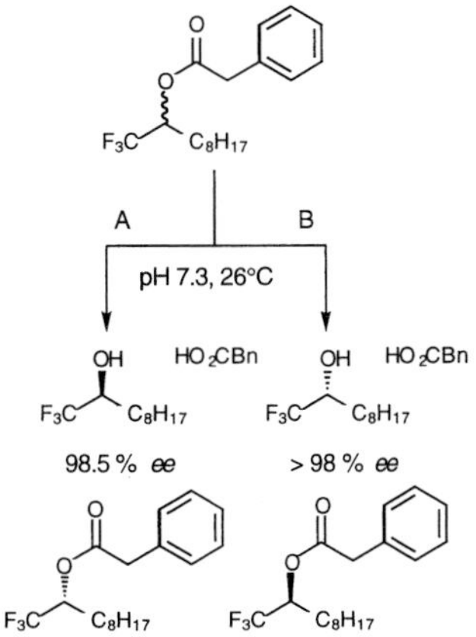

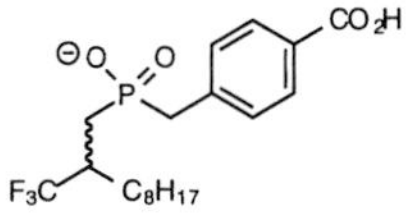

Reaction	TSA					
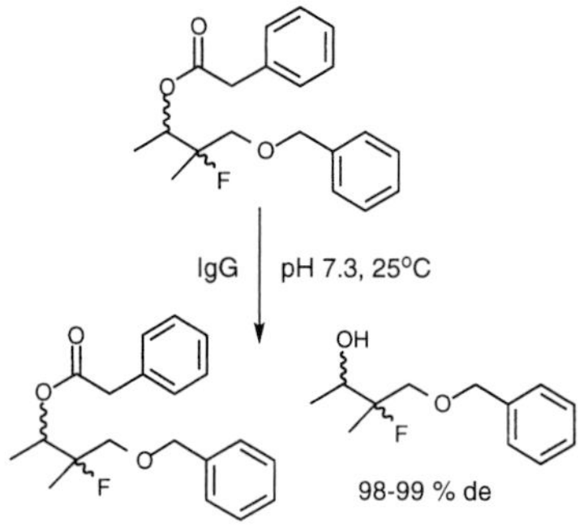 IgG, pH 7.3, 25°C; 98-99 % de	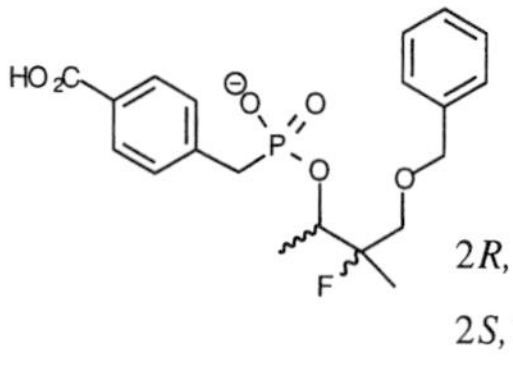 TSA	2*R*, 3*R*	3.9 E+2	8.8 E-1	nr	**1.12**
		2*S*, 3*S*	4.0 E+2	9.1 E-1	nr	
		2*R*, 3*S*	4.1 E+2	9.4 E-1	nr	
		2*S*, 3*R*	3.8 E+2	8.6 E-1	nr	
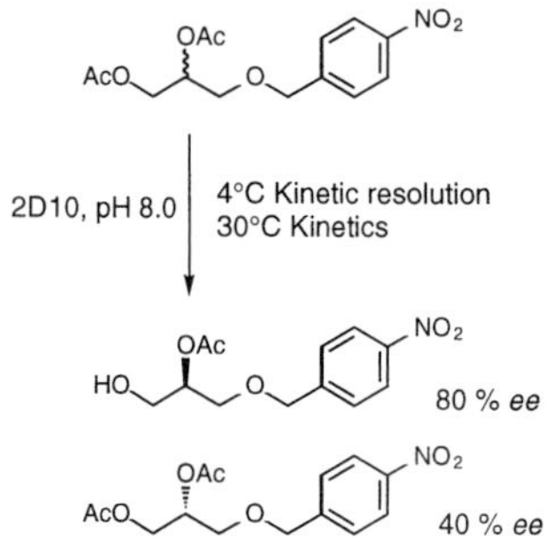 2D10, pH 8.0; 4°C Kinetic resolution; 30°C Kinetics; 80 % *ee*; 40 % *ee*	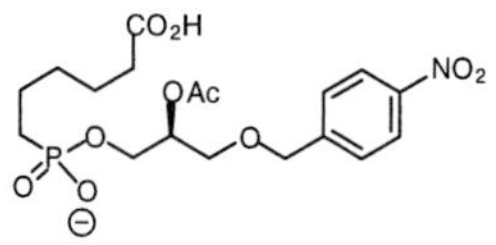 *TSA*; K_i 2.8 E+0 μM		1.3 E+3	2.0 E+0	2.4 E+2	**1.13**
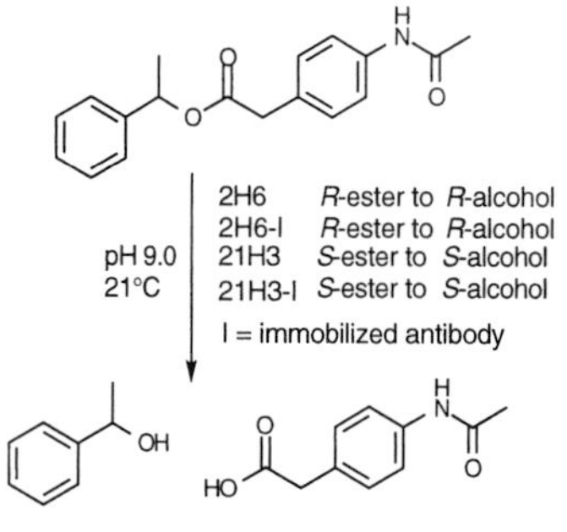 pH 9.0, 21°C; 2H6 *R*-ester to *R*-alcohol; 2H6-I *R*-ester to *R*-alcohol; 21H3 *S*-ester to *S*-alcohol; 21H3-I *S*-ester to *S*-alcohol; I = immobilized antibody	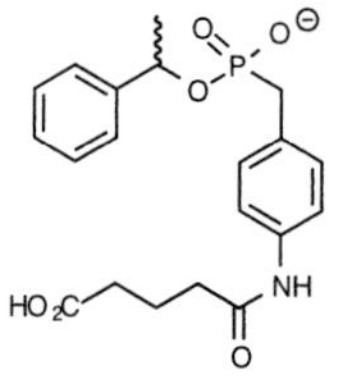 *TSA*; 2H6: K_i 2.0 E+0 μM; 21H3: K_i 1.9 E-1 μM	2H6	4.0 E+3	4.6 E+0	8.3 E+4	**1.14**
		2H6-I	2.2 E+3	4.0 E+0	7.2 E+4	
		21H3	3.9 E+2	9.0 E-2	1.6 E+3	
		21H3-I	2.0 E+2	6.0 E-2	1.1 E+3	
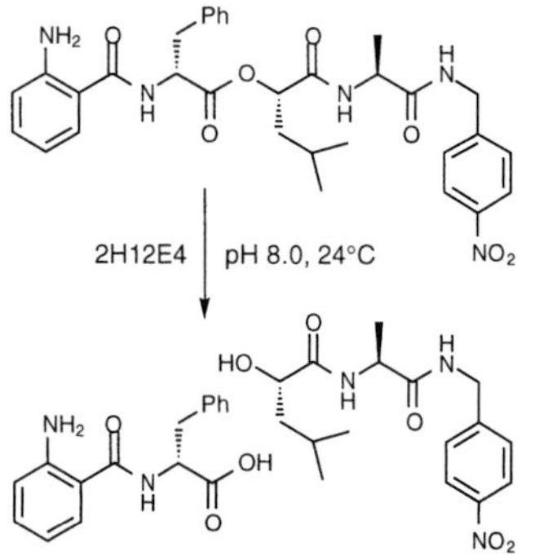 2H12E4, pH 8.0, 24°C	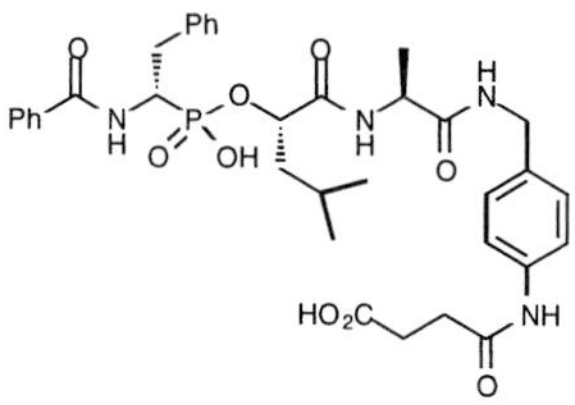 *TSA*; K_i 2.4 E+0 μM		1.5 E+1	1.9 E-2	2.7 E+2	**1.15**

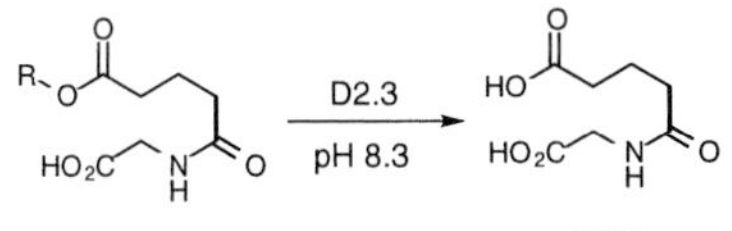

Reaction	Hapten / Method					Entry
R-O, O, HO₂C, N, H, O → HO, O, HO₂C, N, H, O; D2.3, pH 8.3; **A** R = 4-Nitrobenzyl; **B** R = 4-Nitrophenyl; ROH	O₂N, O, P, O, O⊖, HO₂C, N, H, O; *TSA*	**A**	2.8 E+2	7.4 E+0	2.6 E+5	**1.16**
		B	3.3 E+1	3.4 E+0	7.2 E+3	
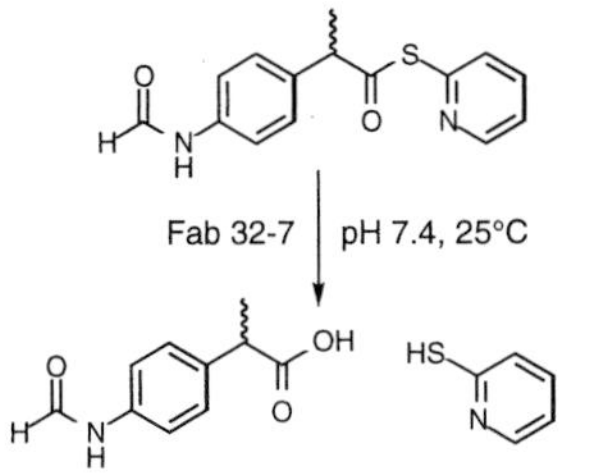 H, O, N, H, O, S, N; Fab 32-7, pH 7.4, 25°C; H, O, N, H, OH, O, HS, N	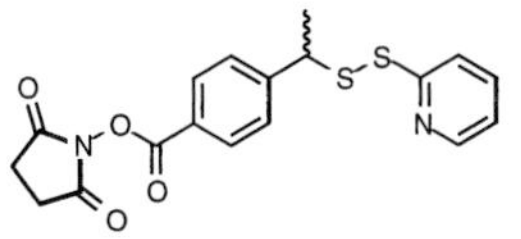 O, N, O, O, O, S, S, N; *in vitro Chemical Selection*		1.0 E+2	3.0 E-2	3.0 E+1	**1.17**
Esterase with broad substrate tolerance; 3-examples shown below; O₂N, O, O, R; **A** R = allyl; **B** R = ethyl; **C** R = *m*-nitrobenzyl; 27H9, pH 9.0, 23°C (biphasic); O₂N, OH, O; ROH	O₂N, P, O, O, O⊖, N, CO₂H, O; *TSA*; K_i (substrate A) 1.2 E+1 μM	**A**	3.4 E+3	6.2 E-2	2.9 E+4	**1.18**
		B	2.8 E+3	1.1 E-2	8.8 E+3	
		C	1.5 E+3	4.7 E-1	1.4 E+6	
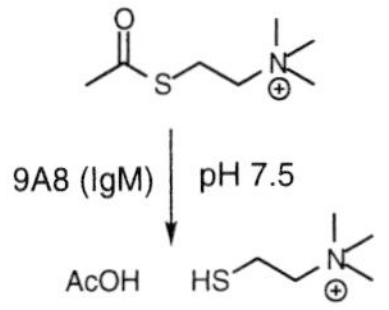 O, S, N⊕; 9A8 (IgM), pH 7.5; AcOH, HS, N⊕	ANTI-IDIOTYPIC CATALYSTS; Antibodies elicited against mAbE-2, an anti- acetylcholinesterase antibody		6.0 E+2	4.9 E+3	4.2 E+8	**1.19**

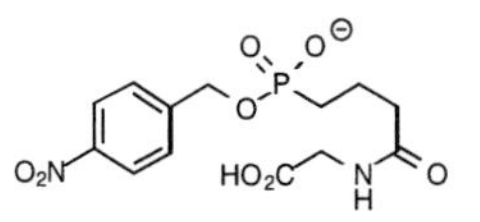

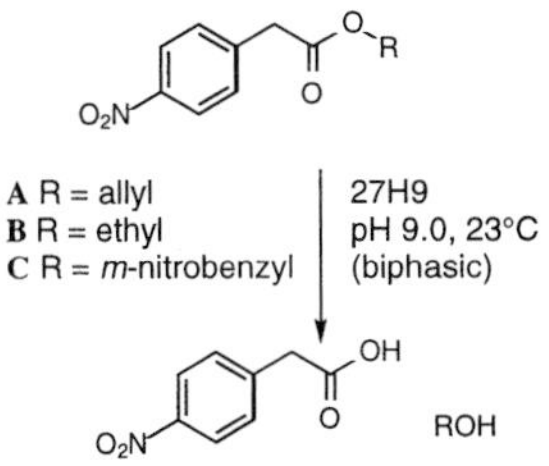

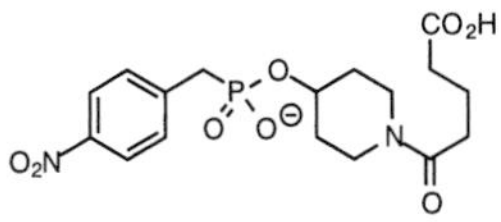

2. Aryl Ester Hydrolysis

Reaction	Hapten / Method				Entry
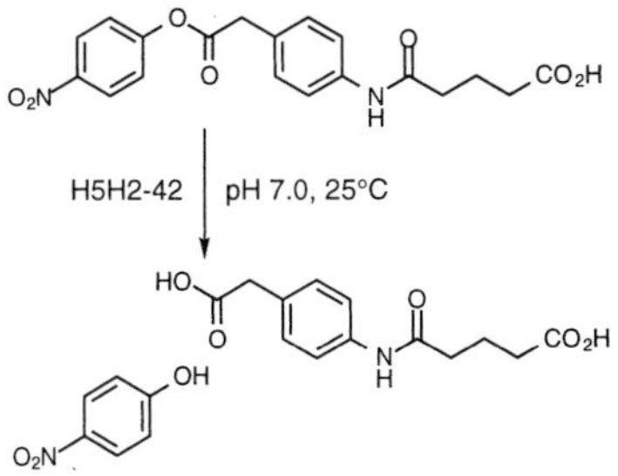 O₂N, O, O, N, H, O, CO₂H; H5H2-42, pH 7.0, 25°C; HO, O, N, H, O, CO₂H; O₂N, OH	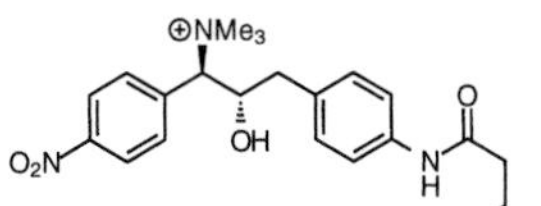 ⊕NMe₃, O₂N, OH, N, H, O, CO₂H; K_i 3.0 E+2 μM; 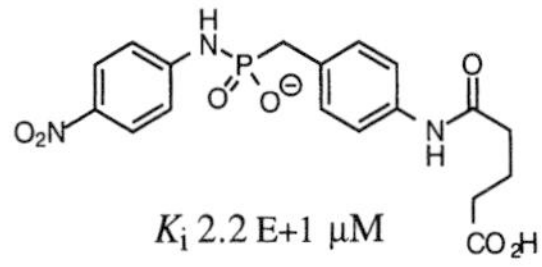 O₂N, H, N, P, O, O⊖, N, H, O, CO₂H; K_i 2.2 E+1 μM; *Heterologous Immunization*	2.4 E+2	1.3 E+1	6.8 E+4	**2.1**

Reaction	TSA					
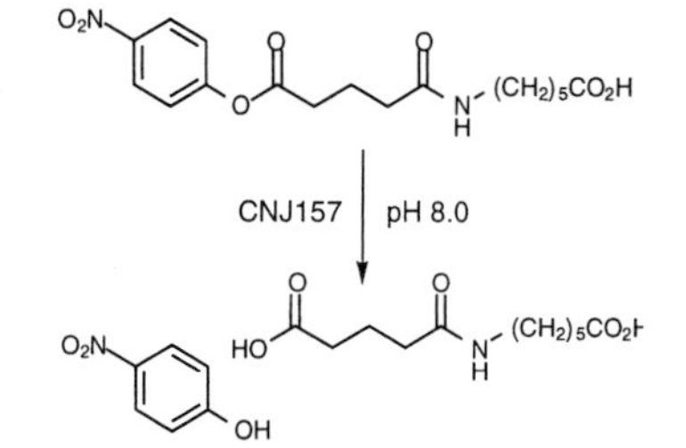 O_2N-C₆H₄-O-CO-(CH₂)₃-CO-NH-$(CH_2)_5CO_2H$ → CNJ157, pH 8.0 → O_2N-C₆H₄-OH + HO-CO-(CH₂)₃-CO-NH-$(CH_2)_5CO_2H$	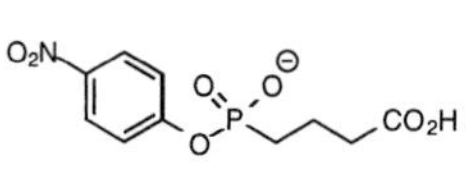 *TSA* K_i 3.4 E+3 μM		1.1 E+2	2.4 E+0	9.7 E+3	**2.2**
17E8, pH 8.7 and 9.5	*TSA* K_i 5.0 E-1 μM	pH 8.7	2.6 E+2	1.0 E+2	1.3 E+4	**2.3**
		pH 9.5	nr little variance with pH	2.2 E+2	2.2 E+4	
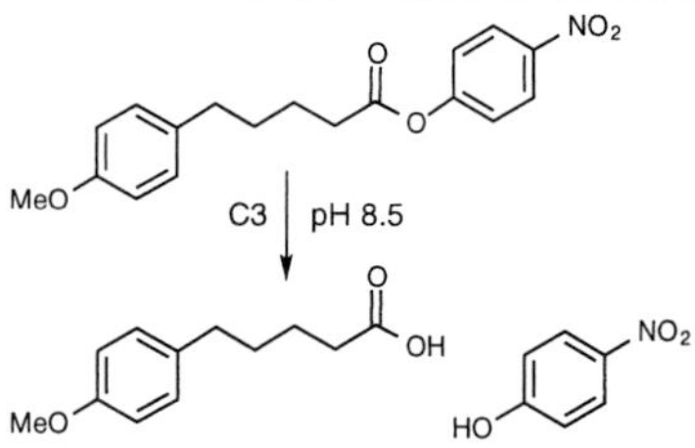 C3, pH 8.5	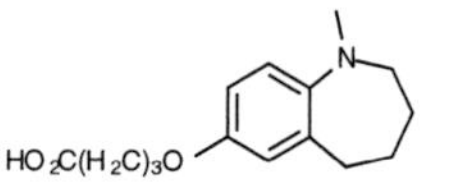 $HO_2C(H_2C)_3O$- *TSA* K_d 2.4 E-2 μM		4.0 E+2	1.4 E+2	3.5 E+6	**2.4**
→ AcOH; **A** 20G9, pH 8.8, 25°C; **Bi** 20G9, pH 8.5, 35°C; **Bii** 20G9 (reverse micelles) - Wo 23	*TSA* **A**, K_i 2.2 E-3 μM; **Bi**, K_i 3.9 E-2 μM	**A**	3.6 E+1	5.4 E-1	6.9 E+1	**2.5**
		Bi	1.6 E+2	1.9 E+1	1.7 E+4	
		Bii	5.7 E+2	3.9 E+0	nr	

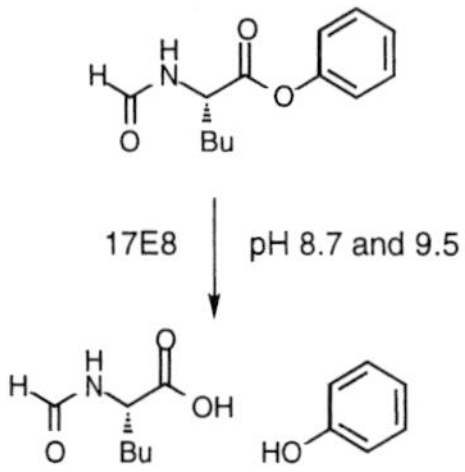

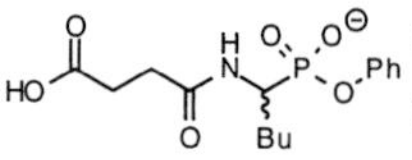

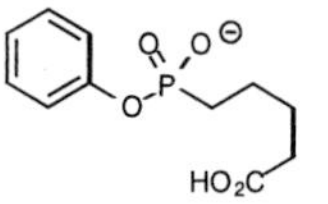

Reaction	Antibody / hapten				
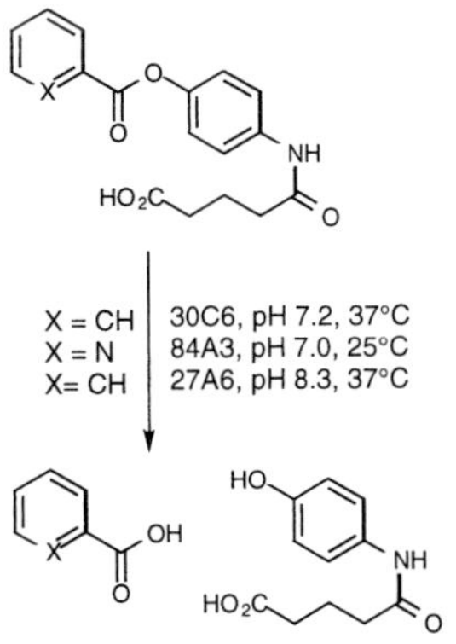 X O O NH HO_2C O X = CH: 30C6, pH 7.2, 37°C X = N: 84A3, pH 7.0, 25°C X= CH: 27A6, pH 8.3, 37°C X OH O HO NH HO_2C O	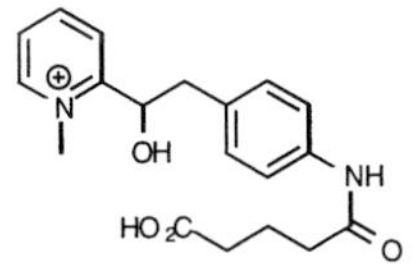 ⊕ N OH NH HO_2C O 30C6	1.1 E+3	5.0 E-3 (app)	1.0 E+6	**2.6**
	84A3 (zinc dependent) 30C6: K_i 8.3E+1 µM	3.5 E+0	2.7 E+0	1.2 E+3	
	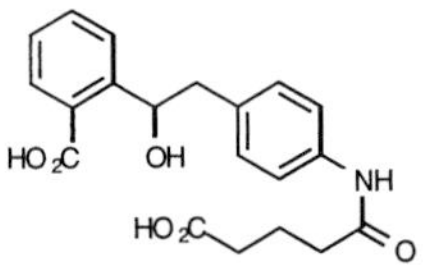 HO_2C OH NH HO_2C O 27A6 27A6: K_i 6.0 E+0 µM *Bait and Switch (BS)*	2.4 E+2	2.0 E-3 (app)	nr	
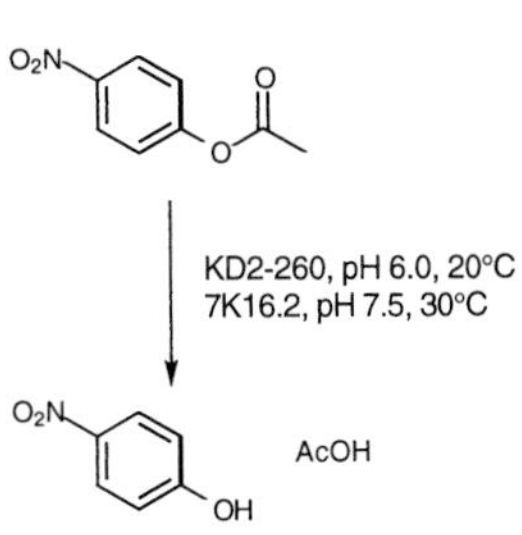 O_2N O O KD2-260, pH 6.0, 20°C 7K16.2, pH 7.5, 30°C O_2N OH AcOH	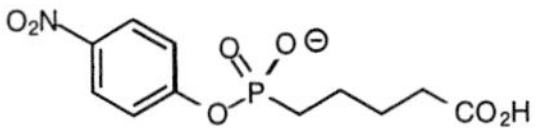 O_2N O O ⊖ O P CO_2H KD2-260 KD2-260: $K_{i\,(30°C)}$ 1.2 E-1 µM	4.9 E+0	2.5 E+0	3.1 E+3	**2.7**
	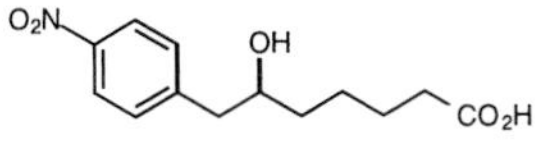 O_2N OH CO_2H 7K16.2 7K16.2: K_i 1.4 E+2 µM *TSA*	3.7 E+3	7.2 E-1	2.3 E+3	
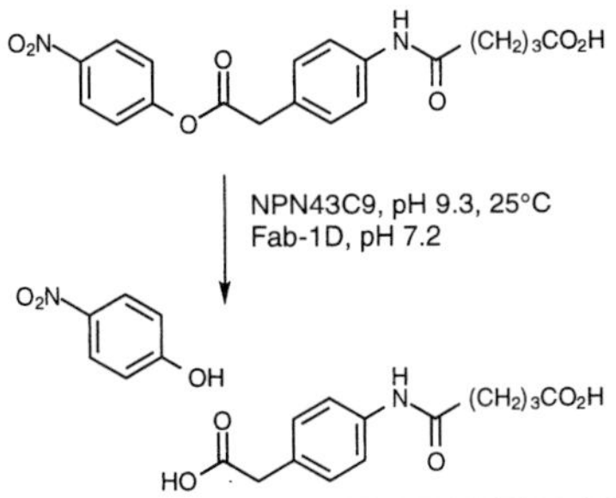 O_2N O O H N $(CH_2)_3CO_2H$ O NPN43C9, pH 9.3, 25°C Fab-1D, pH 7.2 O_2N OH HO O H N $(CH_2)_3CO_2H$ O	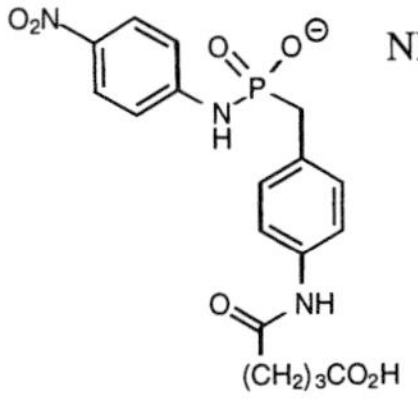 O_2N O O⊖ P N H NPN43C9	5.3 E+1	1.5 E+3 (estimate based on pH rate profile)	2.7 E+4	**2.8**
	O NH $(CH_2)_3CO_2H$ Fab-1D *TSA* NPN43C9: $K_{d\,(pH\,7)}$ 1.0 E+0 µM	1.1 E+2	2.5 E-1	nr	
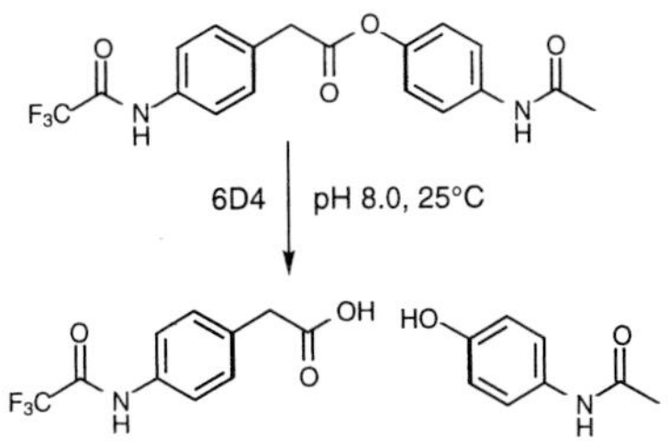 F_3C O N H O O O N H 6D4 pH 8.0, 25°C F_3C O N H OH O HO O N H	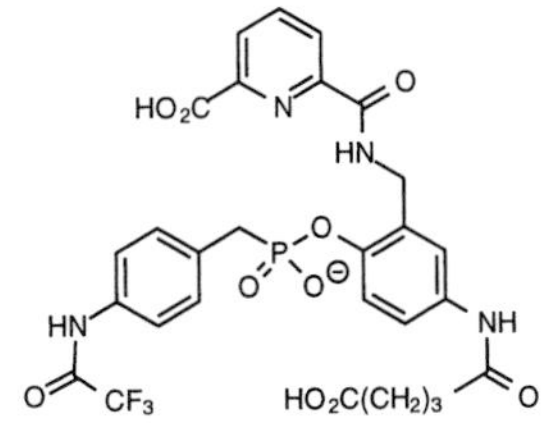 HO_2C N O HN P O O O⊖ HN NH O CF_3 $HO_2C(CH_2)_3$ O 6D4: K_i 1.6 E-1 µM *TSA*	1.9 E+0	1.6 E+0	9.6 E+2	**2.9**

Reaction	Hapten				
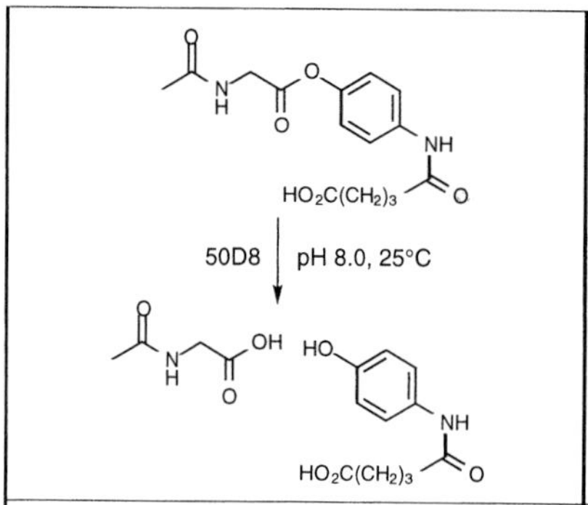 $HO_2C(CH_2)_3$; 50D8, pH 8.0, 25°C; $HO_2C(CH_2)_3$	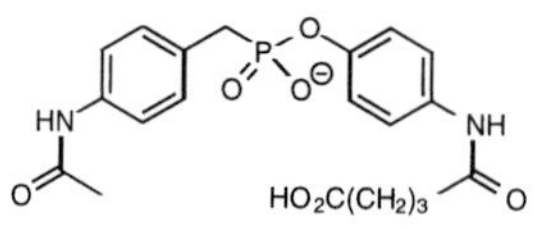 $HO_2C(CH_2)_3$; 50D8: K_i 5.0 E-2 μM; *TSA*	1.5 E+3	1.2 E+3	6.3 E+6	**2.10**
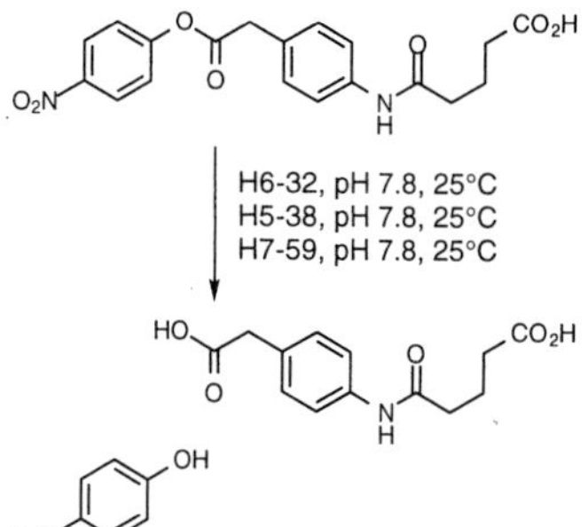 O_2N; CO_2H; H6-32, pH 7.8, 25°C; H5-38, pH 7.8, 25°C; H7-59, pH 7.8, 25°C; HO; CO_2H; O_2N; OH	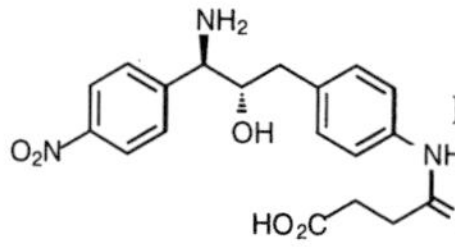 NH_2; O_2N; OH; H6-32; NH; HO_2C; H6-32: K_i 3.6 E+2 μM	8.5 E+2	7.1 E-1	2.4 E+3	**2.11**
	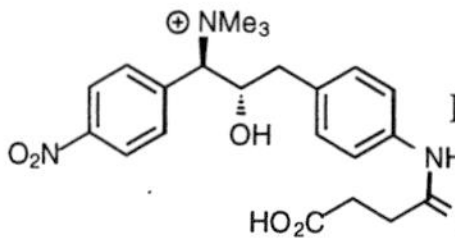 ⊕ NMe_3; O_2N; OH; H5-38; NH; HO_2C; H5-38: K_i 5.0 E+0 μM	8.7 E+2	1.0 E+0	3.3 E+3	
	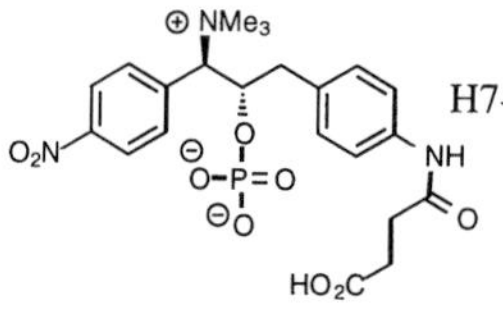 ⊕ NMe_3; O_2N; ⊖ O–P=O ⊖O; H7-59; NH; HO_2C; H7-59: K_i 2.3 E+2 μM	1.1 E+3	4.9 E-1	1.6 E+3	
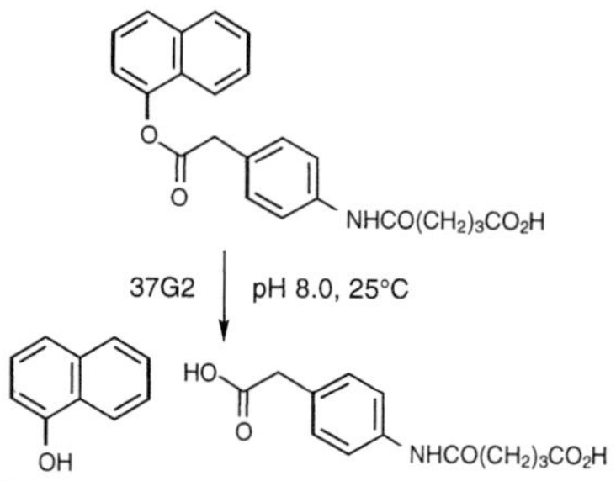 $NHCO(CH_2)_3CO_2H$; 37G2, pH 8.0, 25°C; OH; HO; $NHCO(CH_2)_3CO_2H$	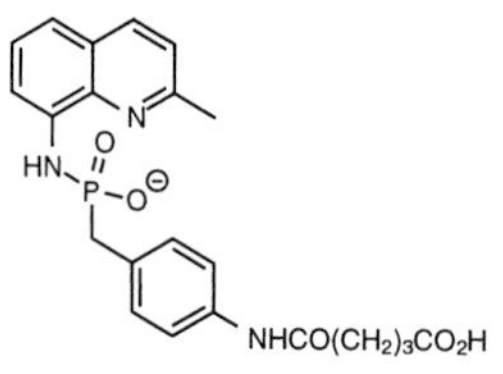 HN; P; $O^{\ominus}$; $NHCO(CH_2)_3CO_2H$; *TSA*; K_i 5.0 E-1 μM	7.7 E+2	2.0 E-1	2.0 E+3	**2.12**
Lactone Formation; 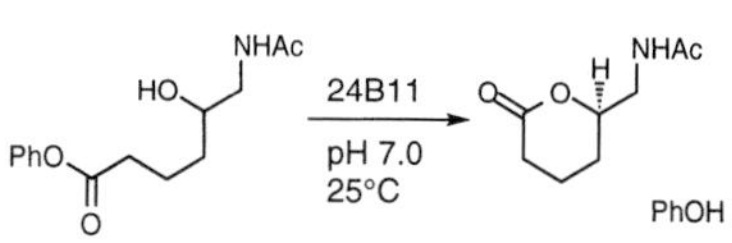 PhO; HO; NHAc; 24B11, pH 7.0, 25°C; NHAc; PhOH	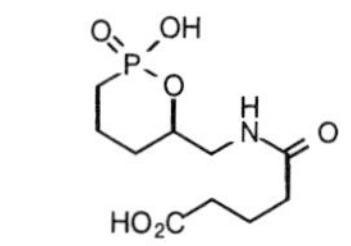 OH; P; O; H; N; HO_2C; *TSA*, K_i 2.5 E-1 μM	7.6 E+1	5.0 E-1	1.7 E+2	**2.13**

Reaction	Hapten				
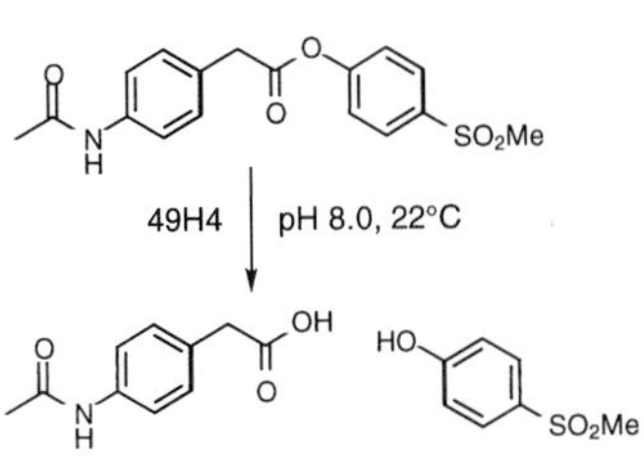 49H4, pH 8.0, 22°C; products: acid + HO-C_6H_4-SO_2Me	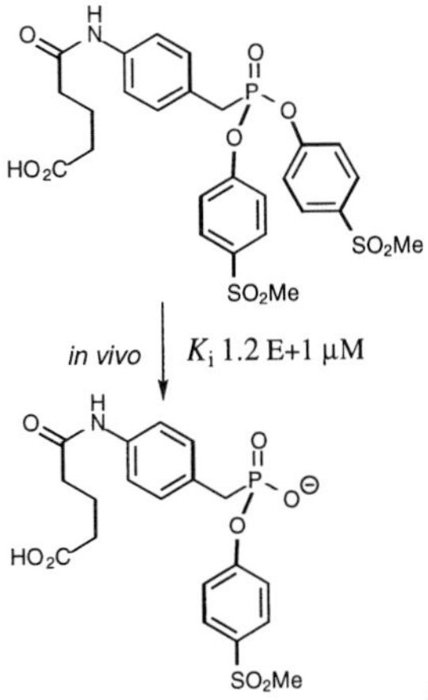 in vivo, K_i 1.2 E+1 μM; *R*; *Reactive Immunization (RI)*	3.0 E+2	3.1 E+1	6.7 E+3	**2.14**
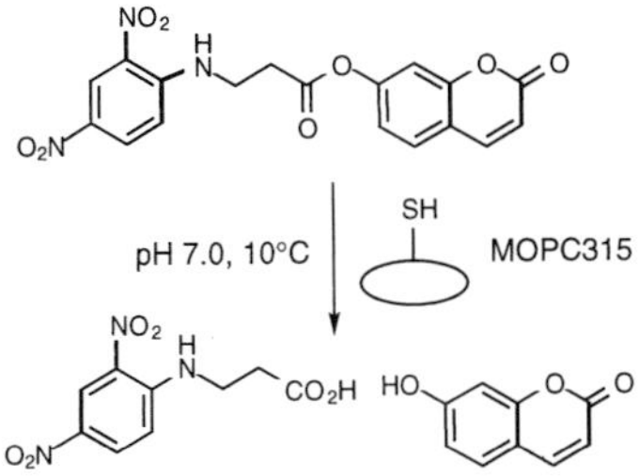 pH 7.0, 10°C; SH; MOPC315; products: CO_2H, HO	Semisynthetic Antibodies Nucleophilic thiol groups were introduced into a 2,4-dinitrophenyl ligand specific antibody binding site by chemical modification K_i (DNP-Gly) 2.5 E-1 μM	1.2 E+0	8.7 E-1	6.0 E+4	**2.15**

3. Carbonate Hydrolysis

Reaction	Hapten					
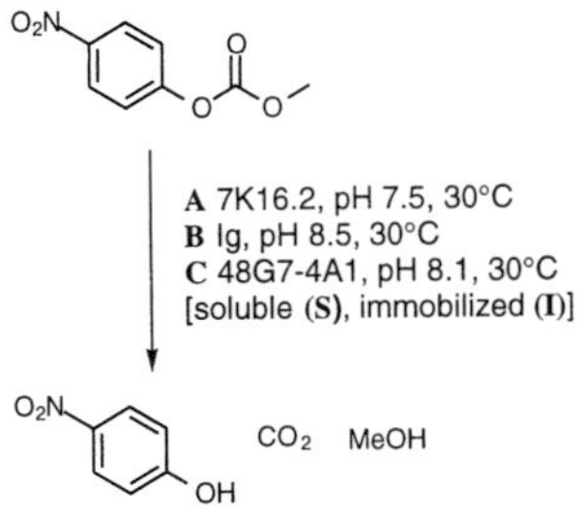 **A** 7K16.2, pH 7.5, 30°C **B** Ig, pH 8.5, 30°C **C** 48G7-4A1, pH 8.1, 30°C [soluble (**S**), immobilized (**I**)] products: CO_2, MeOH	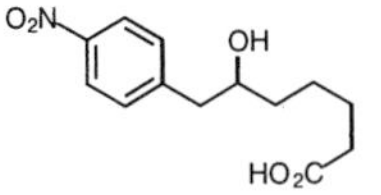 OH, HO_2C	**A**	3.3 E+3	3.1 E-1	9.3 E+2	**3.1**
	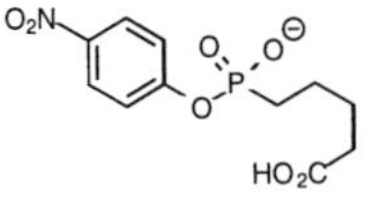 O_2N, HO_2C	**B**	6.6 E+2	1.4 E+0	8.1 E+2	
		C(S)	4.3 E+2	4.0 E+1	2.3 E+4	
	B: K_i 3.3 E+0 μM; *TSA*	**C(I)**	6.8 E+2	2.3 E+1	1.3 E+4	
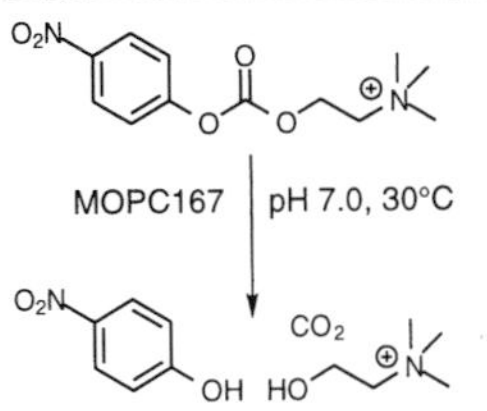 MOPC167, pH 7.0, 30°C; products: OH, CO_2, HO	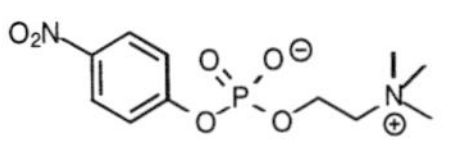 *TSA*; K_i 5.0 E+0 μM		2.1 E+2	4.0 E-1	7.7 E+2	**3.2**

	TSA					
O_2N-C$_6$H$_4$-O-C(=O)-O-CH$_3$ → (A 7K16.2, pH 7.5, 30°C; B Ig, pH 8.5, 30°C; C 48G7-4A1, pH 8.1, 30°C [soluble (S), immobilized (I)]) → O_2N-C$_6$H$_4$-OH + CO_2 + MeOH	A	3.3 E+3	3.1 E-1	9.3 E+2	3.1
	B	6.6 E+2	1.4 E+0	8.1 E+2	
	C(S)	4.3 E+2	4.0 E+1	2.3 E+4	
	C(I)	6.8 E+2	2.3 E+1	1.3 E+4	
	B: K_i 3.3 E+0 μM; *TSA*				
MOPC167, pH 7.0, 30°C; products: O_2N-C$_6$H$_4$-OH, CO_2, HO-CH$_2$CH$_2$-$N^{\oplus}$(CH$_3$)$_3$	*TSA*; K_i 5.0 E+0 μM	2.1 E+2	4.0 E-1	7.7 E+2	3.2
Polyclonal Immunization (Sheep no. 270), pH 8.0, 25°C; *in vitro* Immunization (IVCAT2-6), pH 8.0, 30°C; products: O_2N-C$_6$H$_4$-OH, CO_2, HO-C$_6$H$_4$-CH$_2$C(=O)NHBu	Sheep no. 270	3.3 E+0	1.7 E+2 (based on 1% active protein)	1.5 E+4	3.3
	IVCAT2-6	9.8 E+2	7.2 E+1	5.5 E+3	
	Sheep no. 270, R = OH; IVCAT2-6, R = $NH(CH_2)_3CH_3$; *TSA*; Sheep no. 270: K_i 9.0 E-3 μM; IVCAT2-6: K_i 2.0 E+2 μM				
Polyclonal; products: phenol (OH), CO_2, HO-CH$_2$-C$_6$H$_4$-C(=O)NH-(CH$_2$)$_3$-NH-CH$_2$-C$_6$H$_3$(SO_3H)$_2$ (HO$_3$S, SO$_3$H)	*TSA*; K_d (app) 6.9 E+0 μM	8.9 E+1 (app)	2.1 E-1 (app)	4.3 E+3 (app)	3.4

4. Carbamate Ester Hydrolysis

Reaction	Hapten				
O_2N, O, O, N, H, CO_2H; 33B4F11 pH 7.0, 25°C; O_2N, OH, H_2N, CO_2H, CO_2	O_2N, O, O, $O^{\ominus}$, P, HO_2C; *TSA* K_i 1.0 E-1 μM	5.5 E+0	1.5 E+0	2.6 E+2	**4.1**
Cl, Cl, N, O, O, N, H, CO_2H, CO_2H; EA11-D7 pH 7.0, 37°C; Cl, Cl, N, OH, H_2N, CO_2H, CO_2H, CO_2	CO_2H, EtO, O, P, N, H, CO_2H, HN, O, CO_2H; *TSA* K_d 2.0 E+8 M	2.0 E+2	1.9 E+0	nr	**4.2**
Y, O, O, N, H, CO_2H, CO_2H; Y = NO_2, Br, F, OMe; DF8-D5 pH 6.5, 14°C; Y, OH, CO_2, H_2N, CO_2H, CO_2H	CO_2H, EtO, O, P, N, H, CO_2H, HN, O, CO_2H; *TSA*; Y: NO_2	1.2 E+2	1.8 E+1	3.0 E+2	**4.3**
	Br	8.0 E+1	6.0 E+0	1.0 E+4	
	F	4.1 E+1	7.2 E+0	4.0 E+4	
	MeO	5.8 E+1	4.9 E+0	1.2 E+6	

5. Amide Hydrolysis

Reaction	Hapten				
MeHN, O, S, O, N, H, O, N, H, Bn, O, NH_2; 13D11 pH 9.5, 37°C; MeHN, O, S, O, N, H, O, N, H, Bn, O, OH, NH_3	O, N, O, O, N, H, O, P, $O^{\ominus}$; *TSA* K_i 1.4 E+1 μM	4.3 E+2	9.9 E-6	1.3 E+2	**5.1**

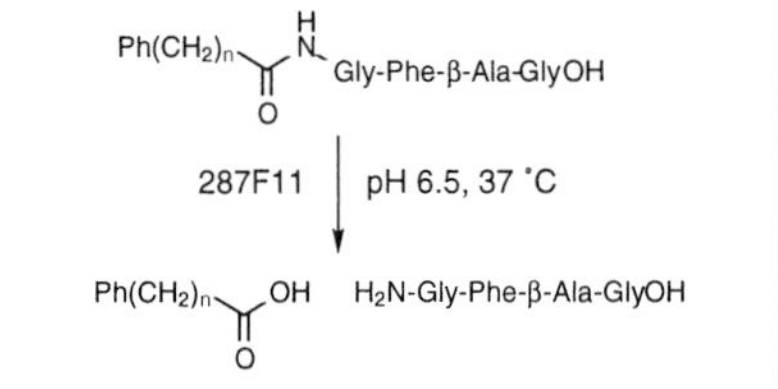 $Ph(CH_2)_n$–C(=O)–N(H)–Gly-Phe-β-Ala-GlyOH 287F11 ↓ pH 6.5, 37 °C $Ph(CH_2)_n$–C(=O)OH H_2N-Gly-Phe-β-Ala-GlyOH	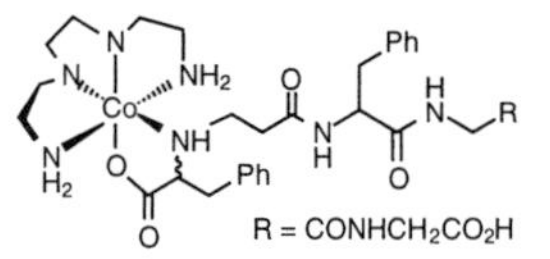 N, N, N, Co, NH_2, NH, H_2N, O, O, Ph, O, N, H, O, Ph, H, N, R R = $CONHCH_2CO_2H$ *Metal complex cofactor*	nr	3.6 E-2	2 E+5	**5.2**
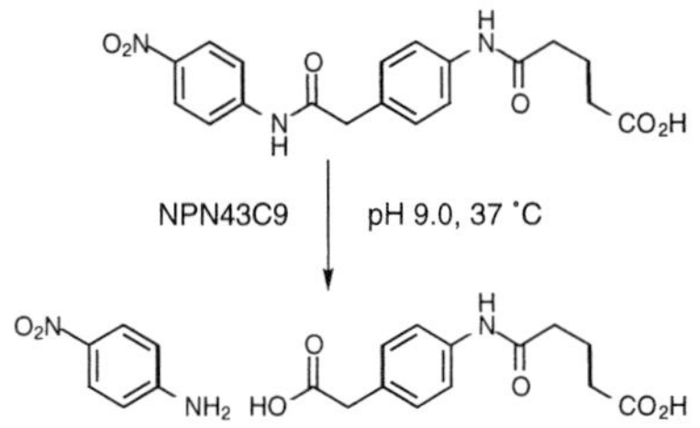 O_2N, O, N, H, H, N, O, CO_2H NPN43C9 ↓ pH 9.0, 37 °C O_2N, NH_2, HO, O, H, N, O, CO_2H	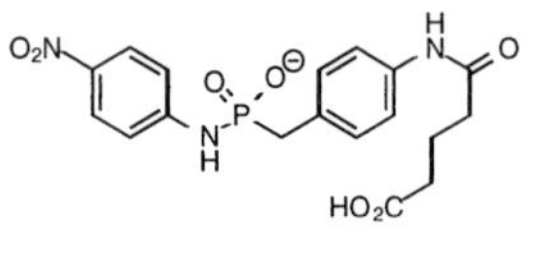 O_2N, N, H, O, P, O⊖, H, N, O, HO_2C *TSA* K_i 1.0 E+1 μM	5.6 E+2	8.0 E-2	2.5 E+5	**5.3**
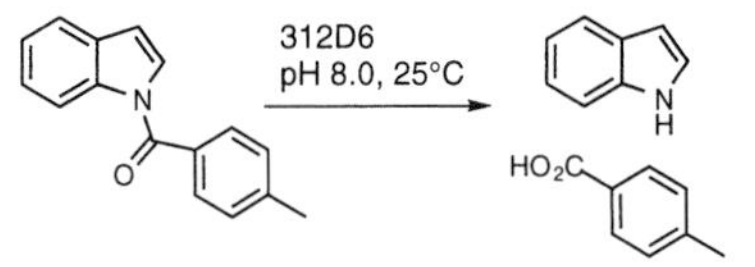 N, O → (312D6, pH 8.0, 25°C) → N, H + HO_2C	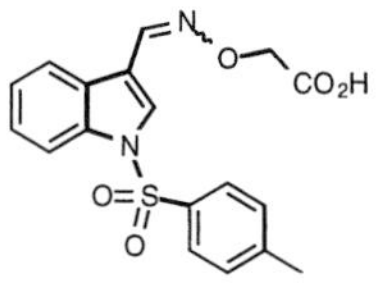 N, O, CO_2H, N, O=S, O *TSA*	3.6 E+1	4.5 E-4	7.5 E+2	**5.4**
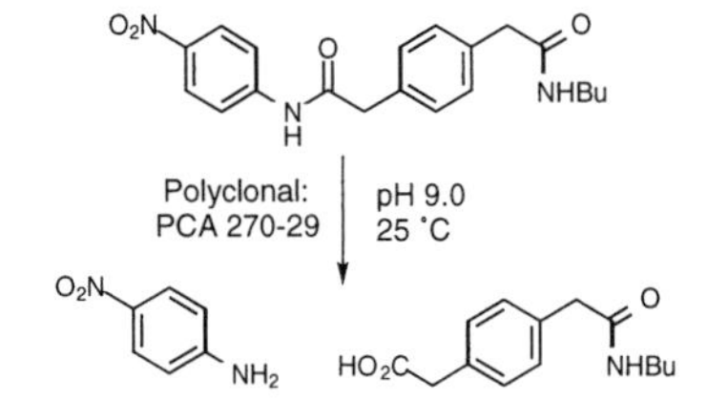 O_2N, O, N, H, O, NHBu Polyclonal: PCA 270-29 ↓ pH 9.0, 25 °C O_2N, NH_2, HO_2C, O, NHBu	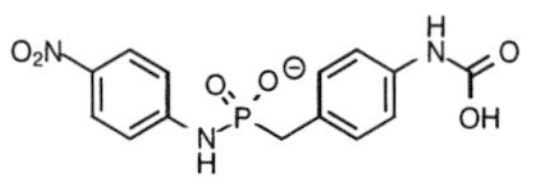 O_2N, N, H, O, P, O⊖, H, N, O, OH *TSA*	5.4 E+0	3.6 E-1 (based on 1% active protein)	1.1 E+3	**5.5**
Gln^{16}–Met^{17} VIP Fab ↓ pH 8.5, 38 °C VIP (1-16) VIP (17-28)	Autoantibodies Human serum IgG fraction was found to hydrolyze vasoactive intestinal polypeptide (VIP). Unknown immunogen	3.8 E-2	1.6 E+1	nr	**5.6**
Tg ↓ Polyclonal Tg-Ab Tg-fragments	Autoantibodies Human serum IgG fraction was found to hydrolyze thyroglobulin (Tg). Unknown immunogen	3.9 E-2	3.9 E-3	nr	**5.7**
Boc-EAR-MCA ↓ BJP-B6 Boc-EAR MCA	Autoantibodies Bence Jones proteins (BJPs) (monoclonal antibody light chains) isolated from the urine of multiple myeloma patients, were found to hydrolyze peptide methylcoumarin amide peptide-MCA substrates	1.5 E+1	3.3 E-2	nr	**5.8**

6. *Phosphate Ester Hydrolysis*

Reaction	Hapten					
O, P, O, O, O_2N, O, A, NO_2, O-P, O, O, B; Tx1-4C6, pH 8.5, 30°C; $O^{\ominus}$, O-P, O, O; HO, NO_2	$O^{\ominus}$, O, Si, N, ⊕, NH, NH, HO_2C, O *Electrostatic TS Stabilization* K_d (fluorescence quench) 6.7 E-1 μM	A	1.8 E+1	1.9 E-3	3.6 E+2	**6.1**
		B	3.6 E+1	1.0 E-2	9.8 E+2	
1. *N*-Acyl Serinol Triester Hydrolysis: O_2N, O, AcHN, O-P, O, O; 15C5, pH 8.1, 25°C; O_2N, OH, OH, AcHN, O-P, O, O 2. Paraoxon Hydrolysis: O_2N, O, O, P, OEt, OEt; 3H5, pH 9.2, 25°C; O_2N, OH, $O^{\ominus}$, O, P, OEt, OEt	CO_2H, H, N, O, $O^{\ominus}$, N, ⊕, NO_2 *Electrostatic TS Stabilization and BS* 3H5: K_i (pH 9.3) 9.8 E-1 μM	15C5	8.7 E+1	2.7 E-3	1.3 E+2	**6.2**
		3H5	5.1 E+3	2.0 E-3	3.5 E+2	
Phosphonofluoridate Hydrolysis: O, F, P, O; IIA12, pH 7.0; O, ⊖O, P, O; HF	O, O, P, MeO, O, O, Protein—N, H *TSA*		3.3 E+2	4.0 E+0	5.5 E+3	**6.3**
O, NH, HO, O, N, O, O, OH, O, P, O, O_2N, O, O⊖; pH 6.0, 25°C, 2G12; O, NH, HO, O, N, O, O_2N, OH, O, O, P, ⊖O, O	HO_2C, ⊖, O, O, NH, O, O, N, O, O, N, Re, N, O, S, O, S *TSA* K_i <4.0 E-1 μM		2.4E+2	9.2 E-2	3.1 E+2	**6.4**

Reaction	Immunogen				Entry
Plasmid DNA (pUC18) → Nicked DNA; Fab fragment from an IgG purified from human sera, pH 7.5, 30°C	Autoantibodies Human serum IgG fraction (Fab) was found to hydrolyze DNA. Unknown immunogen	4.3 E+1	1.4 E+1	nr	**6.5**
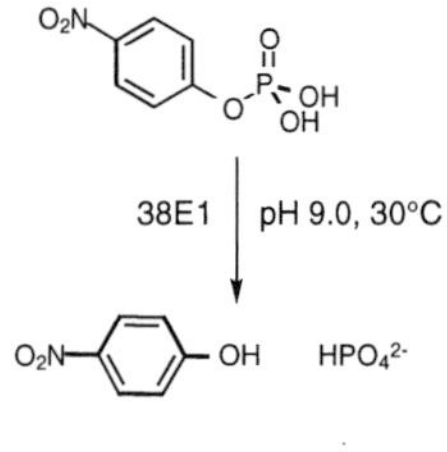 O_2N-C6H4-O-P(O)(OH)OH → O_2N-C6H4-OH + HPO_4^{2-}; 38E1, pH 9.0, 30°C	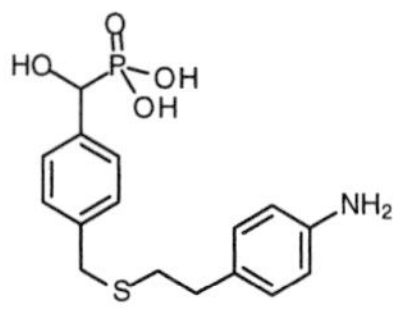 K_i 3.4 E+1 μM	1.6 E+2	1.2 E-3	8.0 E+3	**6.6**

7. Miscellaneous Hydrolyses

Reaction	Immunogen				Entry
Ether Hydrolysis 37C4, pH 6.0; polyclonal, pH 7.2 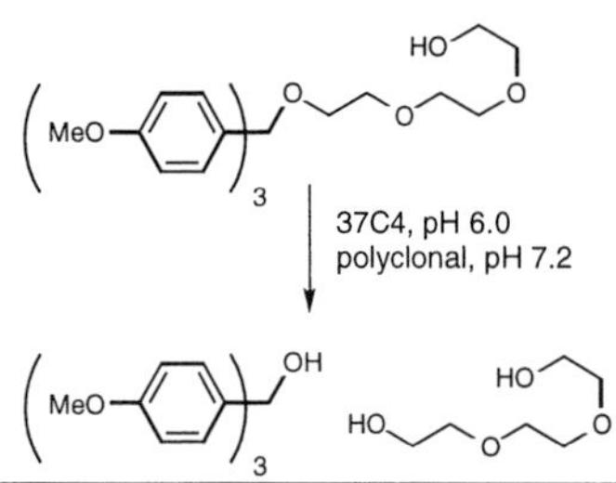	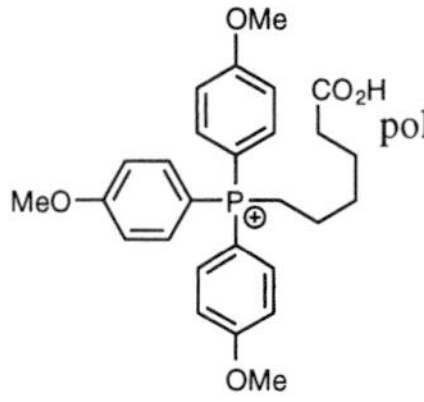 37C4 polyclonal *TSA* 37C4: K_d 2.5 E-2 μM	3.1 E+1 3.1 E+1	1.0 E-1 2.0 E-2 (based on 12 % active protein)	2.7 E+2 1.3 E+2	**7.1**
Glycoside Hydrolysis 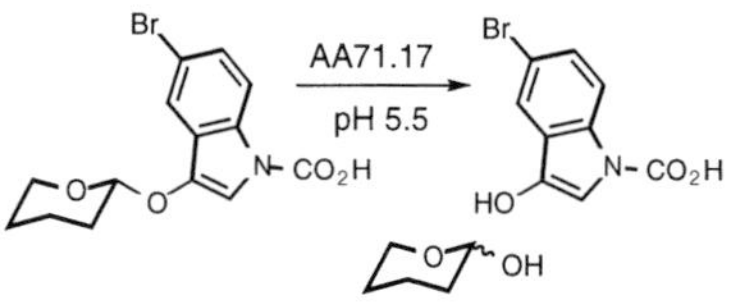 AA71.17, pH 5.5	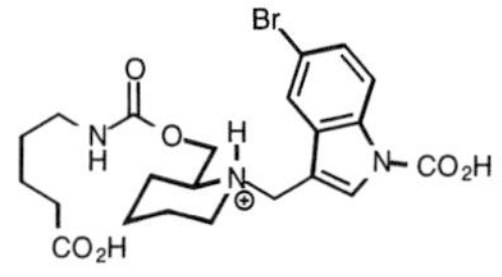 *TSA/ BS* K_i 3.5 E+1 μM	3.2 E+2	1.5 E-2	nr	**7.2**
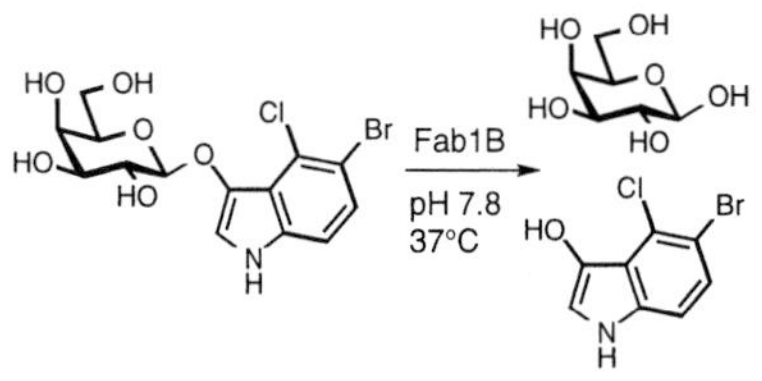 Fab1B, pH 7.8, 37°C	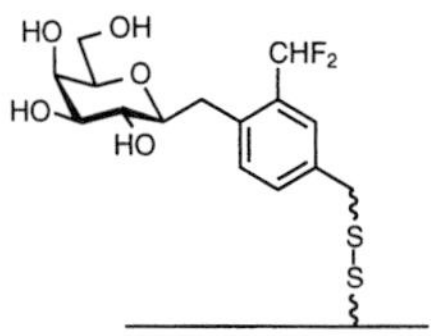 **in vitro** *Chemical Selection* K_i 1.5 E+1 μM	5.3 E+2	7.0 E-3	7.0 E+4	**7.3**

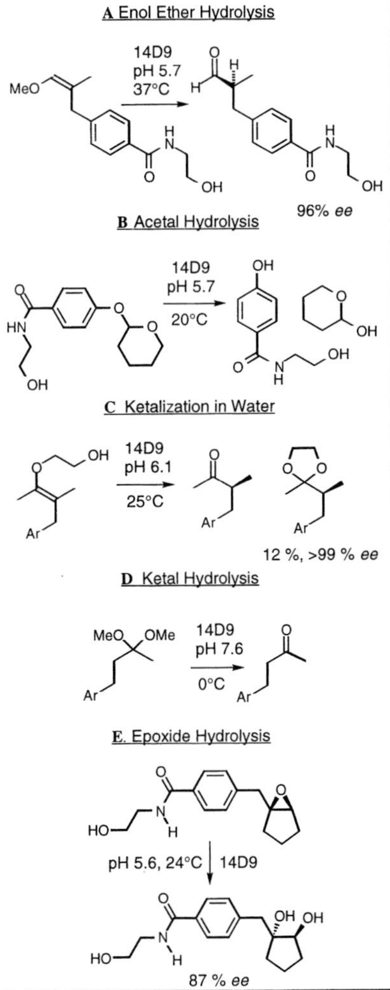

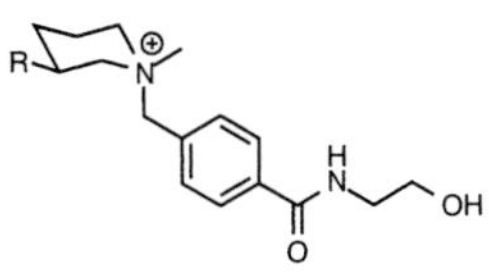

A Enol Ether Hydrolysis MeO… → 14D9, pH 5.7, 37°C → … 96% *ee* **B** Acetal Hydrolysis 14D9, pH 5.7, 20°C **C** Ketalization in Water 14D9, pH 6.1, 25°C 12 %, >99 % *ee* **D** Ketal Hydrolysis 14D9, pH 7.6, 0°C **E**. Epoxide Hydrolysis pH 5.6, 24°C, 14D9 87 % *ee*	R = $CH_2NHCO(CH_2)_3CO_2H$ *BS* **A**: K_d 1.0 E-2 μM	**A** **B** **C** **D** **E**	3.4 E+2 1.0 E+2 5.0 E+1 2.3 E+2 2.5 E+1	5.7 E-3 4.7 E-3 7.2 E-2 1.0 E-2 1.5 E-3	2.5 E+3 7.0 E+1 6.0 E+2 4.3 E+2 4.4 E+2	**7.4**

8. *Eliminations*

Disfavored *syn*-Elimination 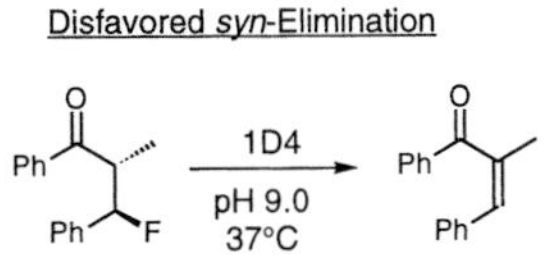 1D4, pH 9.0, 37°C	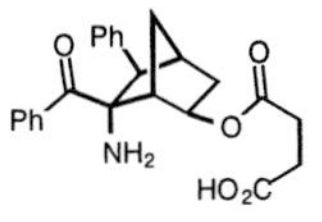 *BS and Entropic Trap*	2.1 E+2	3.0 E-3	nr	**8.1**

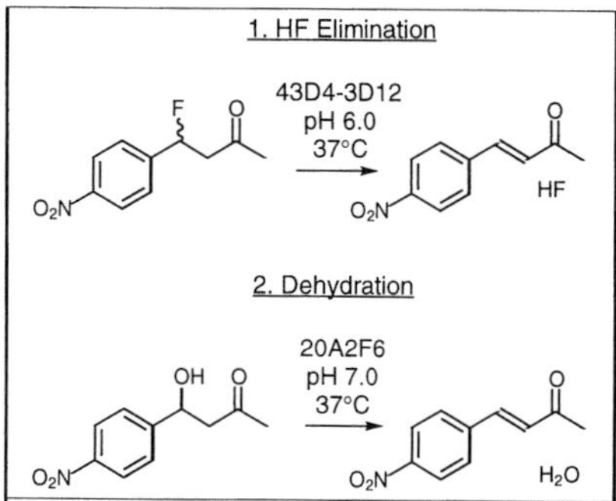1. HF Elimination F O 43D4-3D12 pH 6.0 37°C O O2N O2N HF 2. Dehydration OH O 20A2F6 pH 7.0 37°C O O2N O2N H2O	43D4-3D21 20A2F6 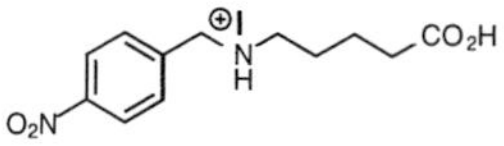N H CO2H O2N *BS* 43D4-3D12: K_i 2.9 E-1 μM 20A2F6: K_i 1.6 E+1 μM	1.8 E+2 1.1 E+3	1.9 E-1 3.5 E-4	8.8 E+4 1.2 E+3	**8.2**
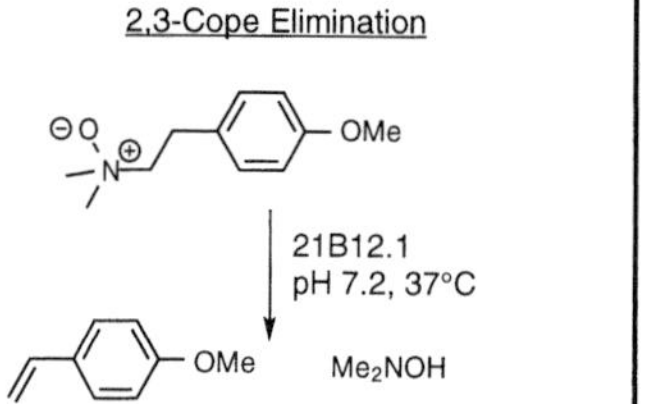2,3-Cope Elimination O N OMe 21B12.1 pH 7.2, 37°C OMe Me2NOH	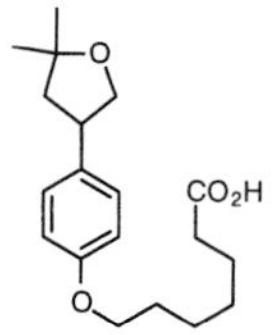O CO2H O *TSA* K_i 2.0 E-1 μM	2.4 E+2	2.4 E-5	9.1 E+2	**8.3**
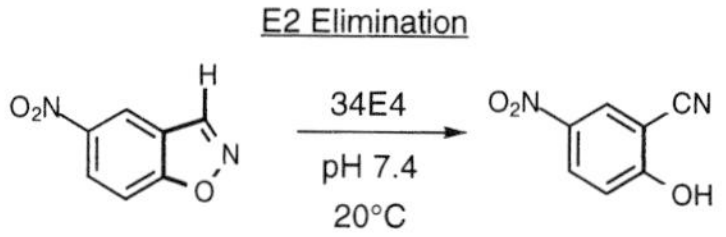E2 Elimination O2N H N O 34E4 pH 7.4 20°C O2N CN OH	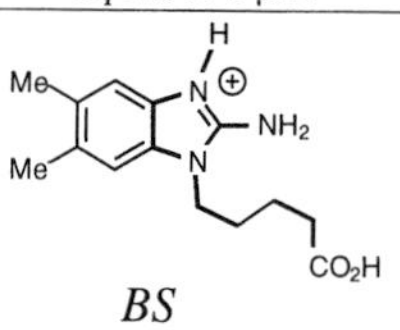H Me Me N NH2 N CO2H *BS*	1.2 E+2	4.0 E+1	2.1 E+4	**8.4**
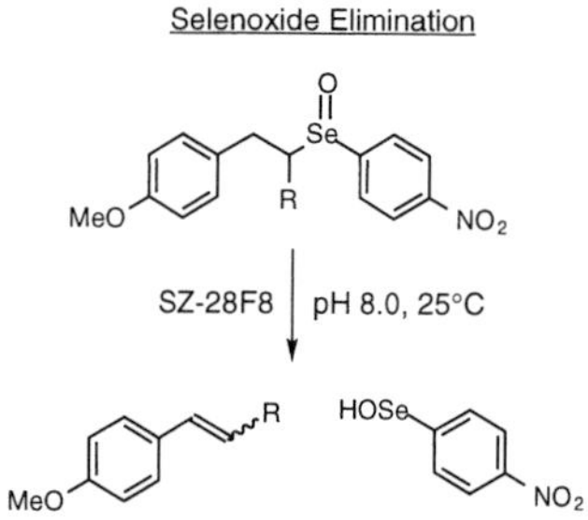Selenoxide Elimination O Se R MeO NO2 SZ-28F8 pH 8.0, 25°C R HOSe MeO NO2	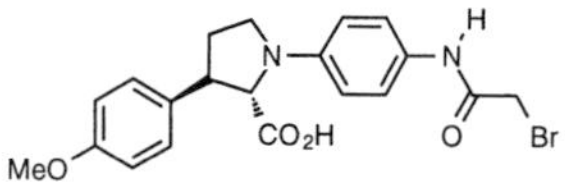N H N CO2H O Br MeO *TSA* K_i 8.2 E-2 μM	1.5 E+0	1.8 E-1	1.6 E+2	**8.5**

9. *Decarboxylations*

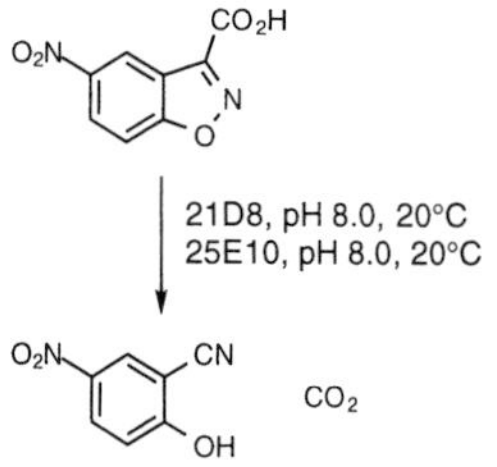CO2H O2N N O 21D8, pH 8.0, 20°C 25E10, pH 8.0, 20°C O2N CN CO2 OH	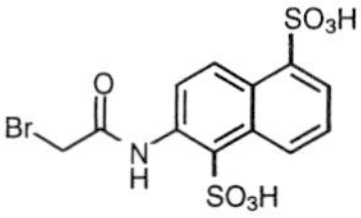SO3H Br O N H SO3H 21D8 25E10 *Medium Effect* 21D8: K_i 6.8 E-3 μM 25E10: K_i 2.4 E-3 μM	1.7 E+2 2.6 E+2	1.7 E+1 2.3 E+1	1.9 E+4 2.3 E+4	**9.1**

Reaction	Hapten				
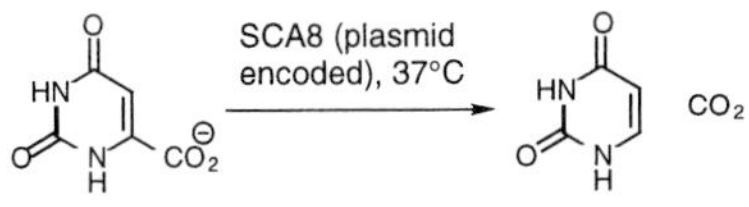 SCA8 (plasmid encoded), 37°C; CO_2	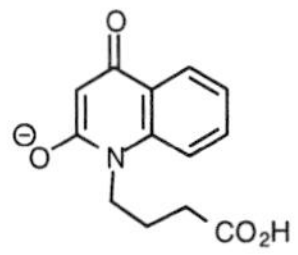	nr	2.7 E-4	1.0 E+8	**9.2**
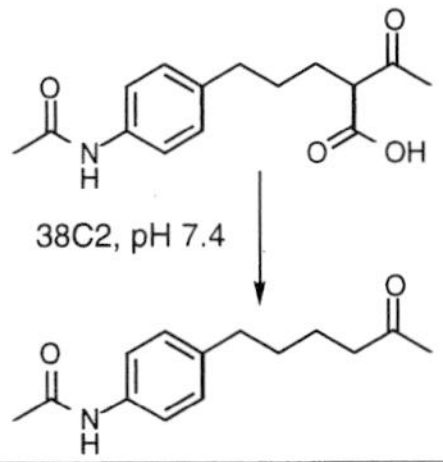 38C2, pH 7.4	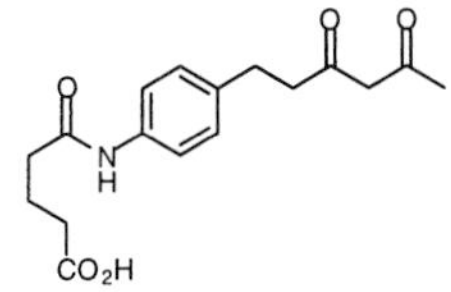 *Reactive Immunization*	9.5 E+2	1.6 E-1	1.5 E+4	**9.3**
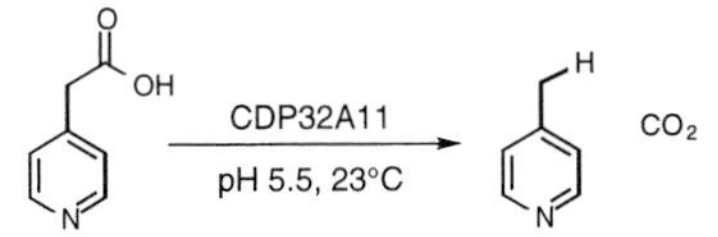 CDP32A11, pH 5.5, 23°C; CO_2	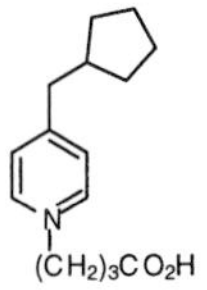 $(CH_2)_3CO_2H$ *Medium Effect* K_i 1.0 E-2 μM	1.4 E+5	2.8 E-2	1.9 E+5	**9.4**

10. Cycloreversions

Reaction	Hapten					
Retro Diels-Alder Reaction 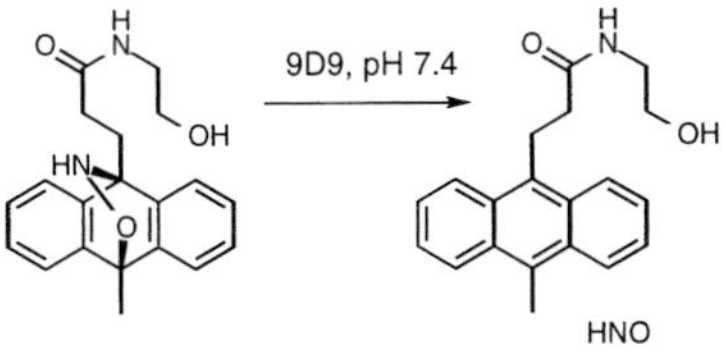 9D9, pH 7.4; HNO	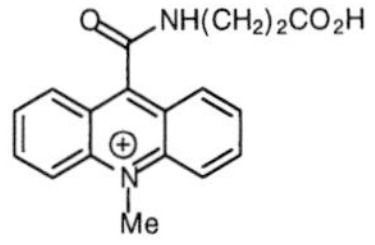 $NH(CH_2)_2CO_2H$ *Heterologous Immunization* (with hydroxylated form) $K_{i(pH\ 9.0)}$ 9.0 E-1 μM		1.3 E+2	7.3 E-2	2.3 E+2	**10.1**
[2+2] 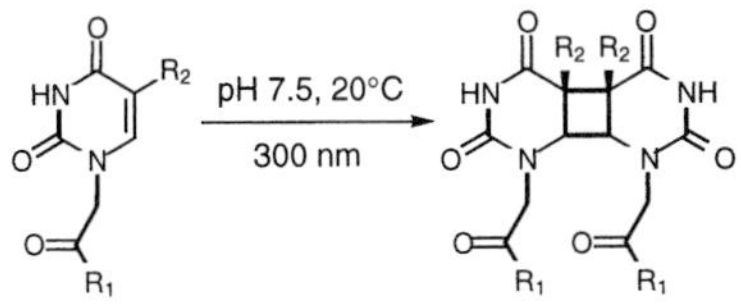 pH 7.5, 20°C, 300 nm **A** 15F1-3B1, R_1 = OH, R_2 = H, **B** UD4C3.5, R_1 = $NHCH_2CO_2Me$, R_2 = H	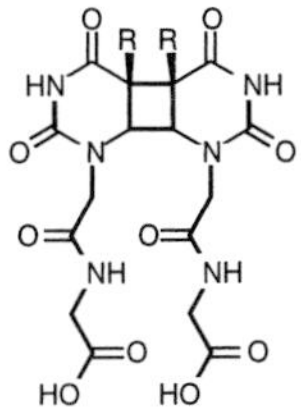 R = Me *cis, syn* R = H *trans, syn* *BS* 15F1-B1: K_i < 1.0 E+0 μM UD4C3.5: K_d (fluorescence quench) 5.4 E-2 μM	**A** **B**	6.5 E+0 2.8 E+2	1.2 E+0 4.7 E-1	2.2 E+2 3.8 E+2	**10.2**

11. Retro-Henry and Retro-Aldol Reactions

Reaction / Conditions	Hapten / Comments / K_i			Entry AE
Retro-Henry Addition 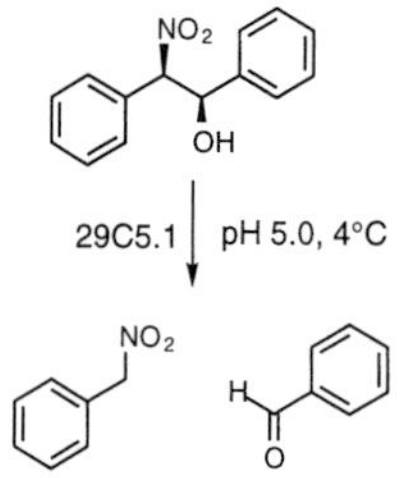 29C5.1 pH 5.0, 4°C	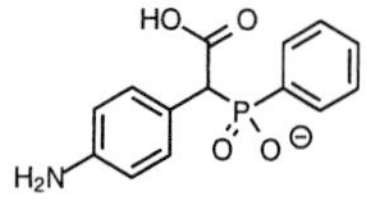 *TSA* K_i (app) 2.6 E+0 μM	k_{cat}/K_m 1.3 E+2 $M^{-1}min^{-1}$ (k_{cat} and K_m not measured separately)	k_{imid} 2.5 E-4 M^{-1} min^{-1}	**11.1**
(*S*)-Selective retro-aldol and (*R*)-selective elimination 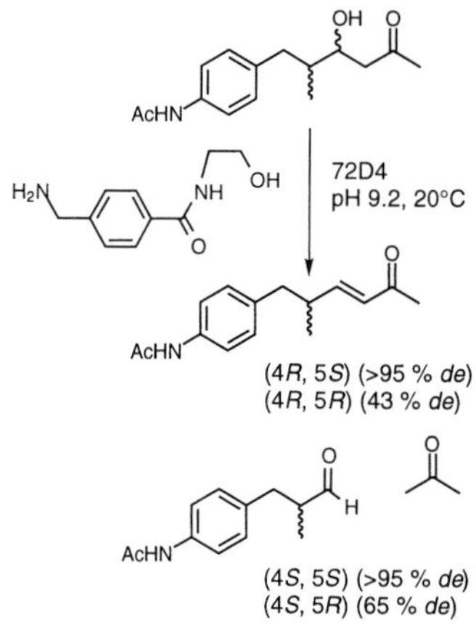 72D4 pH 9.2, 20°C (4*R*, 5*S*) (>95 % *de*) (4*R*, 5*R*) (43 % *de*) (4*S*, 5*S*) (>95 % *de*) (4*S*, 5*R*) (65 % *de*)	(*R*)-Selective elimination (0.8 mM Amine) (4*R*, 5*S*) (4*R*, 5*R*) 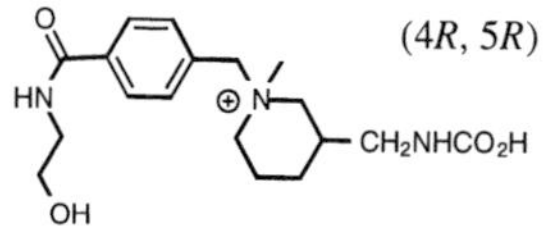 CH_2NHCO_2H (*S*)-Selective retro-aldol (0.8 mM Amine) (4*S*, 5*S*) (4*S*, 5*R*)	$k_{cat}/K_m M^{-1}min^{-1}$ (app) 1.1 E-1 2.2 E-2 7.8 E-2 2.2 E-1	nr nr	**11.2**

B. INTRAMOLECULAR PROCESSES

12. Isomerizations

Reaction / Conditions	Hapten / Comments / K_i	K_m [/μM]	k_{cat} [/min^{-1}]	k_{cat}/k_{uncat}	Entry AE
Peptidyl-prolyl *cis-trans* isomerisation 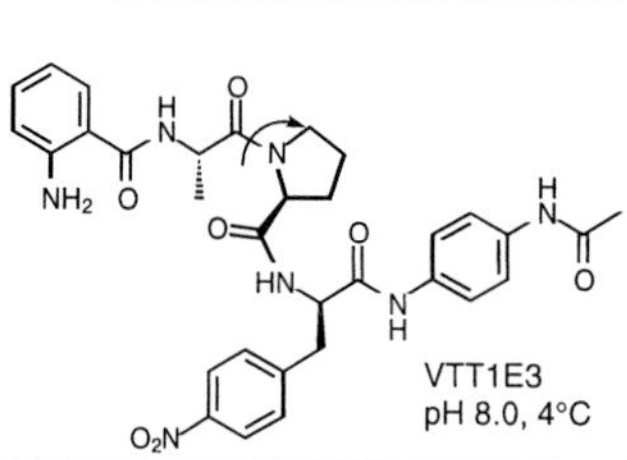 VTT1E3 pH 8.0, 4°C	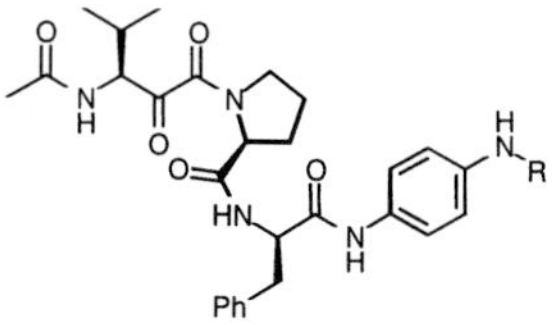 R = $CO(CH_2)_3COOH$ *TSA*, K_i 1.0 E+1 μM	1.0 E+2	6.6 E+0	2.7 E+1	**12.1**
cis-trans Isomerization 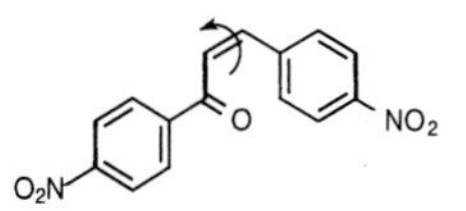 DYJ10-4 pH 7.5 25°C	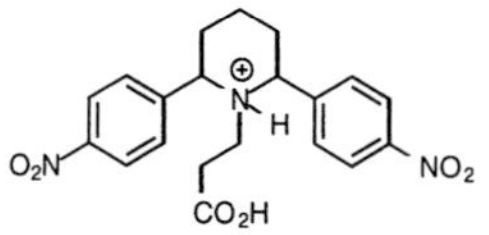 *BS*, K_i 6.7 E+0 μM	2.2 E+2	4.8 E+0	1.5 E+4	**12.2**

Reaction	TSA				
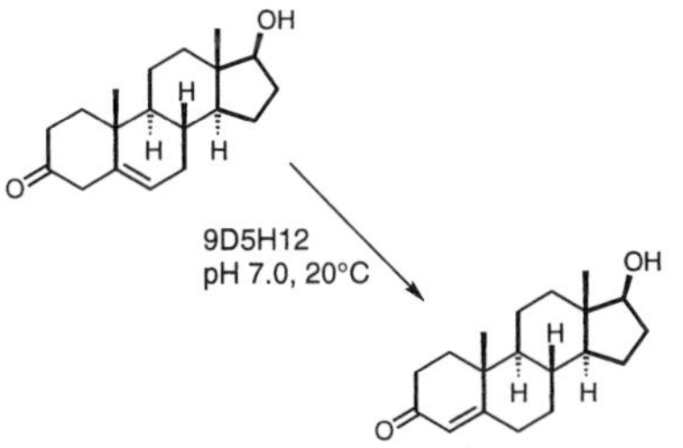 9D5H12 pH 7.0, 20°C	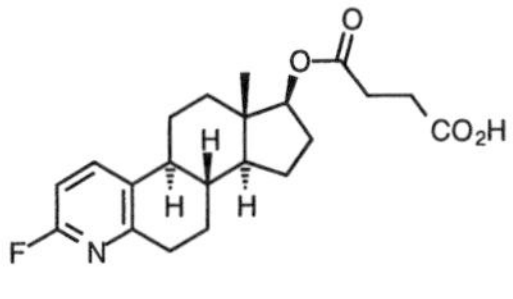 *TSA* K_i 1.3 E+2 μM	1.6 E+2	2.3 E+0	8.3 E+2	**12.3**
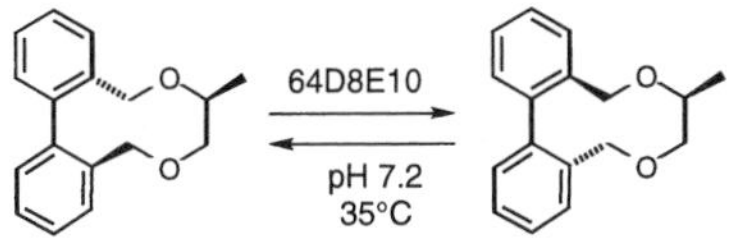 64D8E10 pH 7.2 35°C	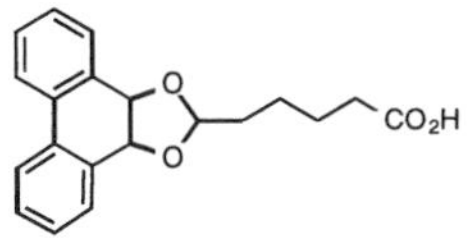 *TSA* K_d (BIAcore) 2.1 E-1 μM	4.2 E+2	2.6 E-3	2.9 E+3	**12.4**

13. Rearrangements

Reaction	TSA				
Cope Rearrangement 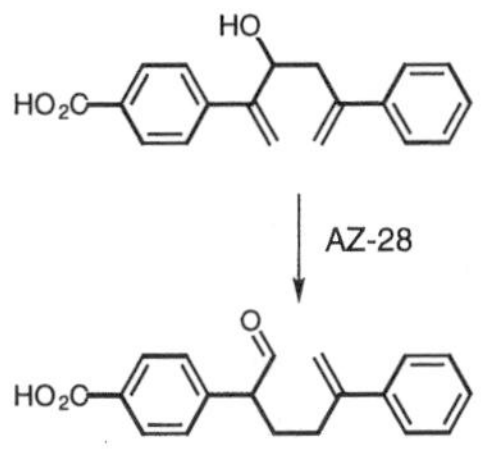 AZ-28	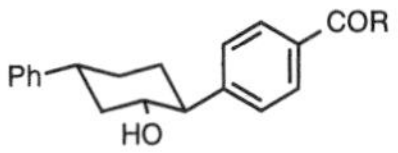 R = $NH(CH_2)_2O(CH_2)_2NH_2$ *TSA* K_i 3.0 E-2-1.6 E-1 μM	9.7 E+1 (app) 4.9 E+1 (cor)	2.6 E-2	5.3 E+3	**13.1**
Claisen Rearrangement 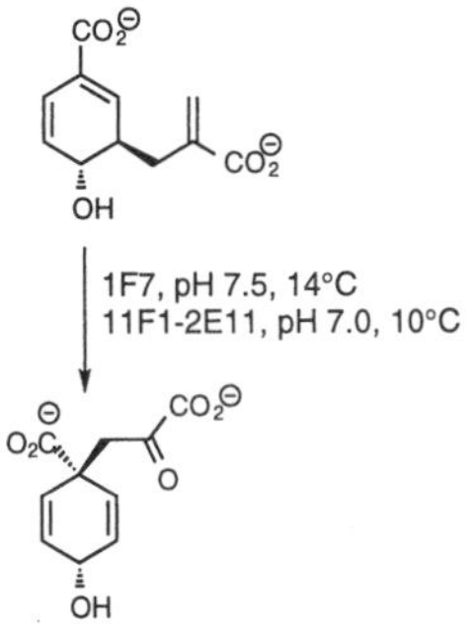 1F7, pH 7.5, 14°C 11F1-2E11, pH 7.0, 10°C	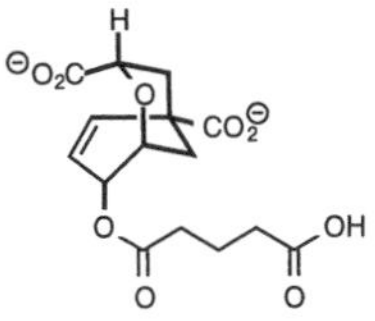 1F7 1F7: K_i 6.0 E-1 μM	4.9 E+1	2.3 E-2	2.5 E+2	**13.2**
	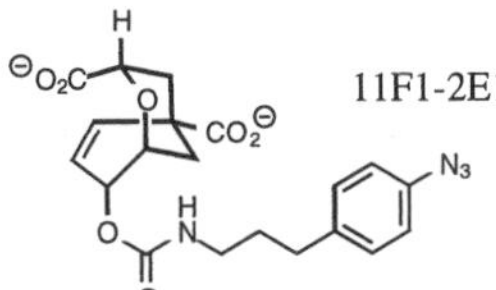 11F1-2E11 11F1-2E11: K_i 9.0 E+0 μM	2.6 E+2	2.7 E+0	1.0 E+4	

Reaction	TSA					
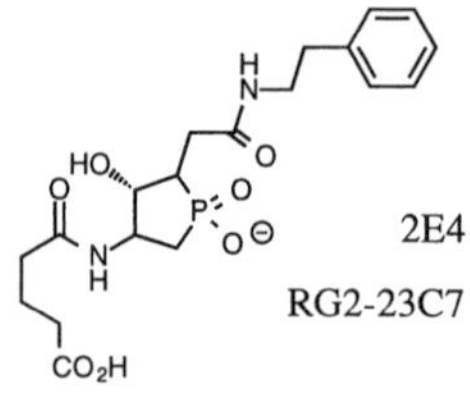 Peptide Bond Rearrangement and Succinimide Hydrolysis 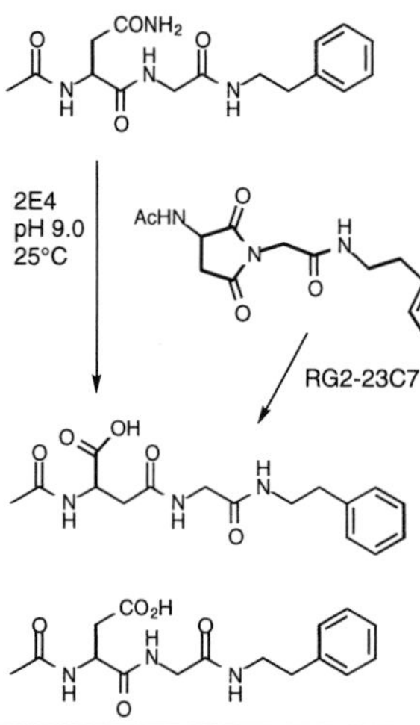 2E4, pH 9.0, 25°C AcHN RG2-23C7 CONH2, OH, CO2H	TSA 2B4: K_i 1.0 E-1 μM HO, NH, CO_2H	2E4 RG2-23C7	1.9 E+2 8.3 E-1	7.2 E-3 3.6 E+1	7.0 E+ nr	**13.3**
1,2-Rearrangement 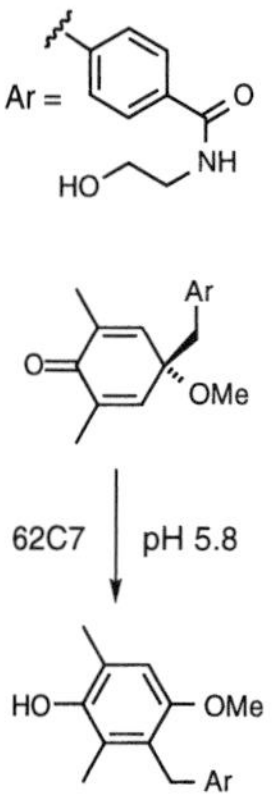 Ar = HO, NH Ar, OMe 62C7, pH 5.8 HO, OMe, Ar	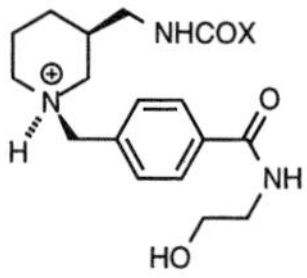 NHCOX NH, HO TSA X = Linker		6.7 E+2	7.3 E-5	8.0 E+1	**13.4**
Ring Opening of a Dinitrospiropyran 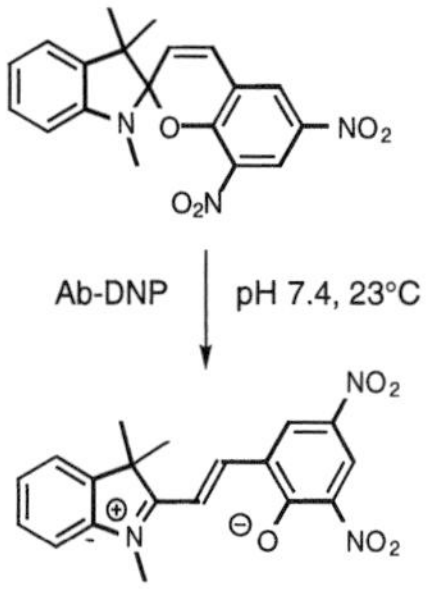 NO_2, O_2N Ab-DNP, pH 7.4, 23°C NO_2, NO_2	OH O_2N, NO_2		1.7 E+5	1.8 E+1	1.9 E+4	**13.5**

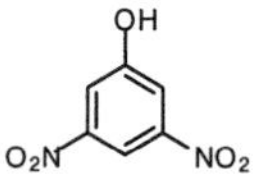

14. Epoxide Opening

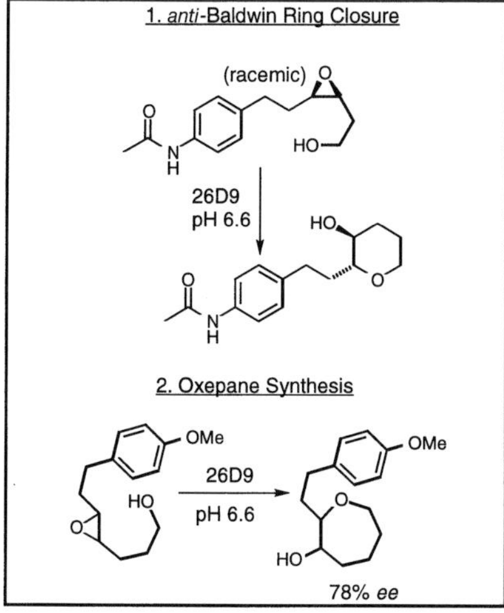 1. *anti*-Baldwin Ring Closure (racemic) O HO N H O 26D9 pH 6.6 HO O N H O 2. Oxepane Synthesis OMe HO O 26D9 pH 6.6 OMe O HO 78% *ee*	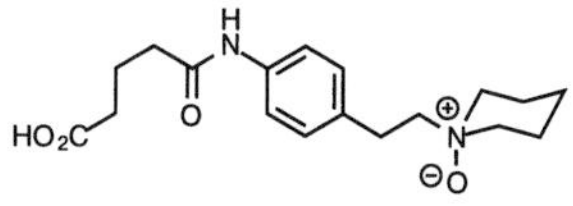 H N HO_2C O ⊕ N ⊖O *TSA* 1. 2.	3.6 E+2 2.0 E+2	9.0 E-1 9.0 E-1	nr nr	**14.1**

15. Cationic Cyclization

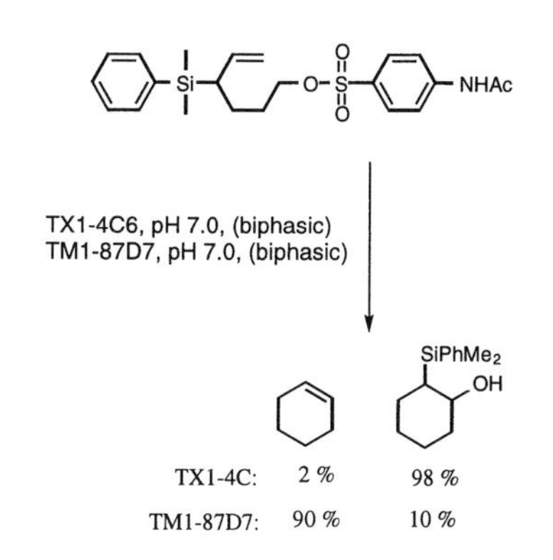 Si O-S(O)(O) NHAc TX1-4C6, pH 7.0, (biphasic) TM1-87D7, pH 7.0, (biphasic) $SiPhMe_2$ OH TX1-4C: 2 % 98 % TM1-87D7: 90 % 10 %	TXI-4C6 TMI-87D7 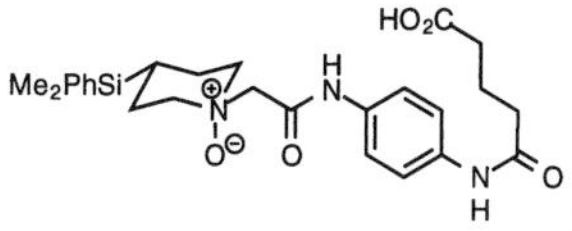 HO_2C Me_2PhSi ⊕N O⊖ O H N N H O *TSA* TX1-4C6: K_i 1.0 E+0 μM TM1-87D7: K_i 1.4 E+0 μM	2.3 E+2 2.5 E+1	2.0 E-2 2.0 E-2	nr nr	**15.1**
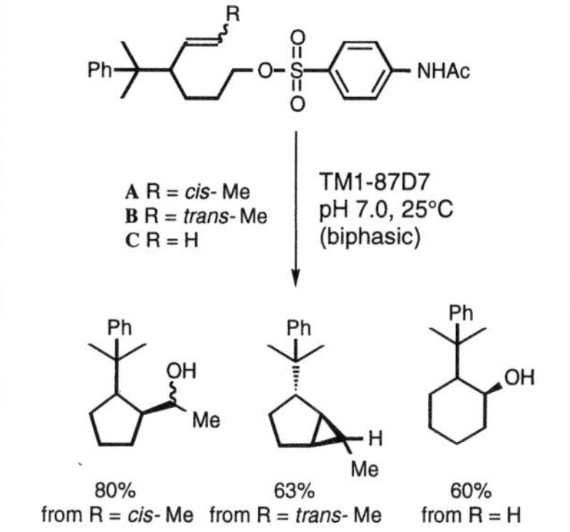 R Ph O-S(O)(O) NHAc **A** R = *cis*- Me **B** R = *trans*- Me **C** R = H TM1-87D7 pH 7.0, 25°C (biphasic) Ph OH Me 80% from R = *cis*- Me Ph H Me 63% from R = *trans*- Me Ph OH 60% from R = H	**A** **B** **C** 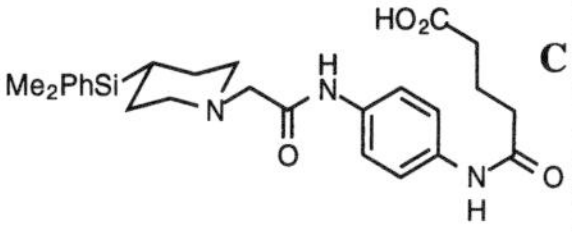 HO_2C Me_2PhSi N O H N N H O *TSA* **A**: K_i 1.0 E+0 μM **B**: K_i 1.0 E+0 μM **C**: K_i 1.0 E+0 μM	5.8 E+1 1.0 E+2 3.1 E+1	1.3 E-2 2.1 E-2 1.0 E-2	nr nr nr	**15.2**

Reaction / Conditions	Hapten / Comments (K_i)	K_m [μM]	k_{cat} [min^{-1}]	k_{cat}/k_{uncat} [M]	Entry AE
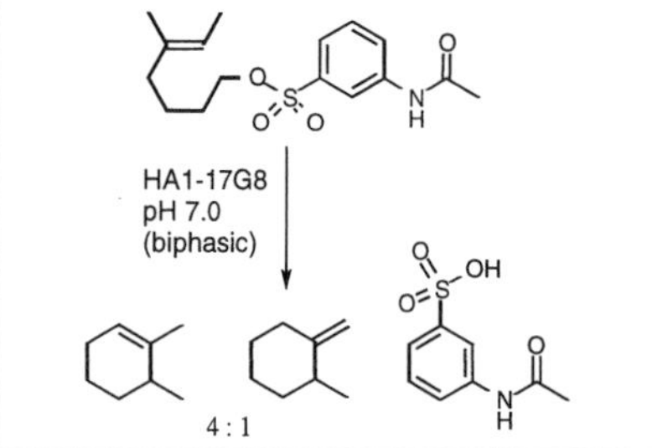 HA1-17G8 pH 7.0 (biphasic); 4 : 1	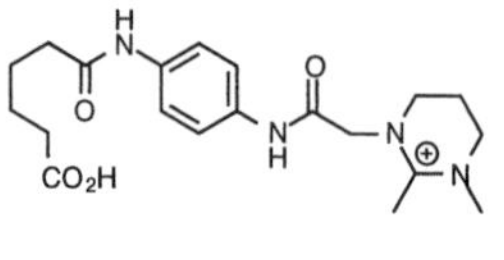 TSA; IC_{50} 1.2 E+0 μM	3.5 E+1	2.5 E-2	7.0 E+1	**15.3**
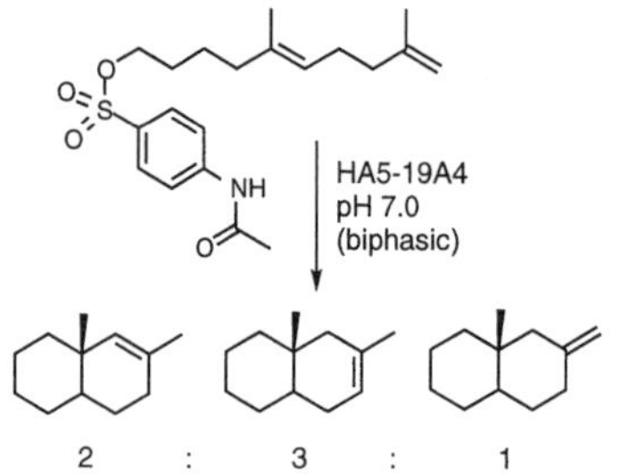 HA5-19A4 pH 7.0 (biphasic); 2 : 3 : 1	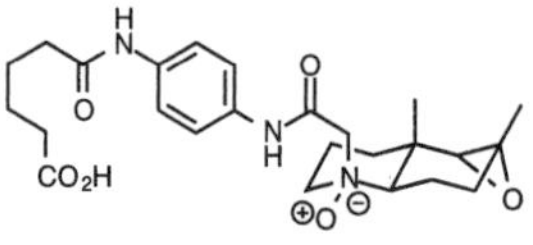 TSA; K_i 1.4 E+0 μM	3.2 E+2	2.1 E-2	2.3 E+3	**15.4**

C. BIMOLECULAR ASSOCIATIVE AND SUBSTITUTION PROCESSES

16. Aldol Reactions

Reaction / Conditions	Hapten / Comments (K_i)	K_m [μM]	k_{cat} [min^{-1}]	k_{cat}/k_{uncat} [M]	Entry AE
Aldol and Disfavored Elimination; 78H6 pH 7.5	BS; **A**	3.6 E+2	1.4 E-4	2.0 E+5	**16.1**
	B	4.7 E+2	4.9 E-4	3.6 E+4	
Aldol and Retroaldol Reaction With A Range of Aldehydes and Ketones; 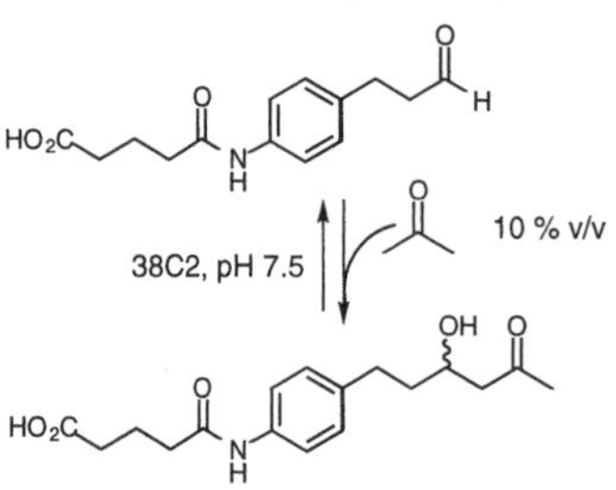 38C2, pH 7.5; 10 % v/v	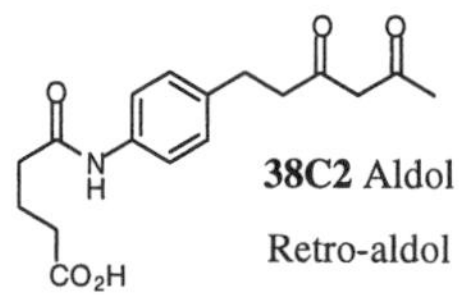 Reactive Immunization; **38C2** Aldol	1.7 E+1	6.7 E-3	2.9 E+4	**16.2**
	Retro-aldol	5.4 E+1	4.4 E-3	nr	

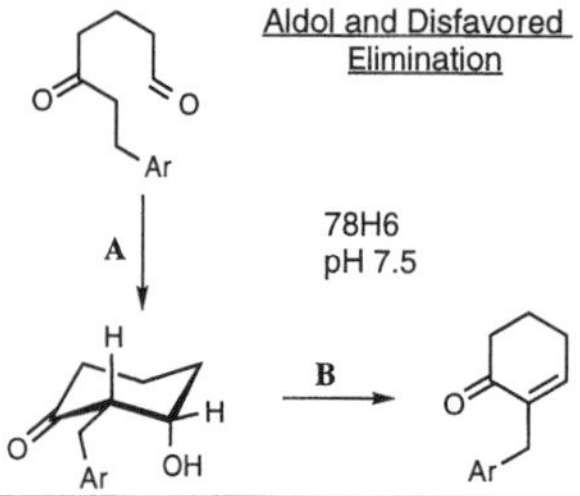

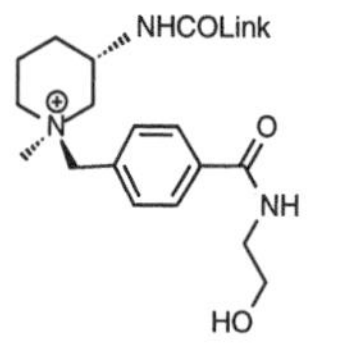

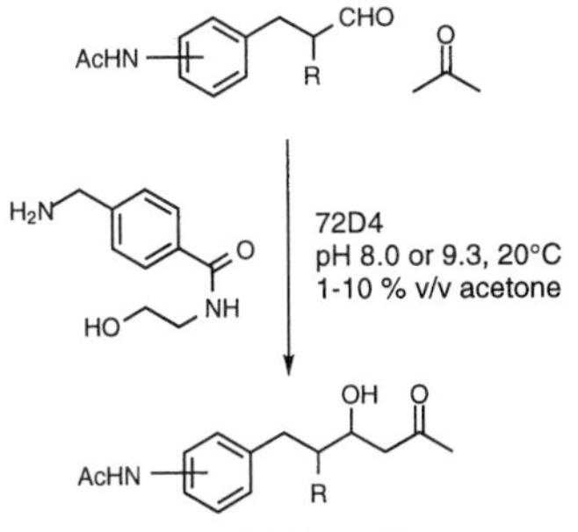 AcHN, CHO, R; H$_2$N, O, NH, HO; 72D4, pH 8.0 or 9.3, 20°C, 1-10 % v/v acetone; AcHN, OH, O, R; **A** *p*-AcNH, R = Me; **B** *m*-AcNH, R = H	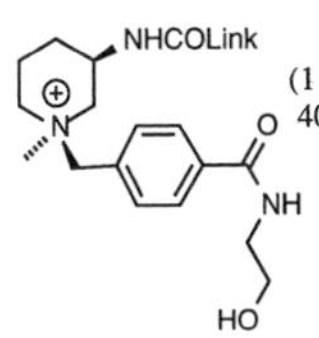 NHCOLink, O, NH, HO; **A** (1 % v/v acetone, 400 μM, amine); **B**	**A** 1.8 E+3 (app) **B** 4.9 E+3 (app)	**A** 1.8 E-4 (app) **B** 9.6 E-5 (app)	**A** 1.0 E+2 (app) **B** 1.7 E+2 (app)	**16.3**

17. Diels-Alder Cycloaddition

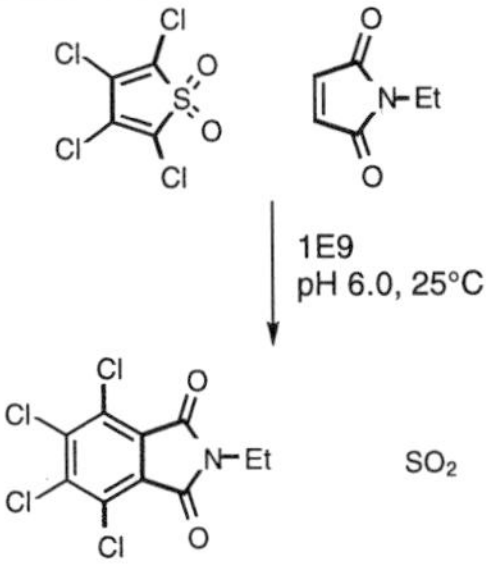 Cl, S, O, O, Cl, Cl, Cl; O, N-Et, O; 1E9, pH 6.0, 25°C; Cl, Cl, Cl, Cl, O, N-Et, O; SO$_2$	Dienophile ([Diene] 0.61 mM); 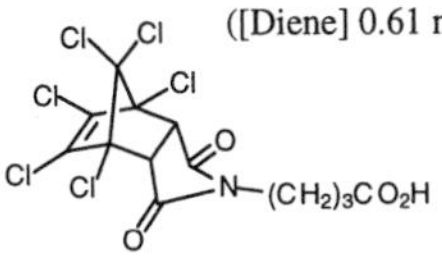 Cl, Cl, Cl, Cl, Cl, Cl, O, N-(CH$_2$)$_3$CO$_2$H, O; *TSA, Entropic Trap*	2.1 E+4 (app)	4.3 E+0 (app)	1.1 E+2 (app)	**17.1**
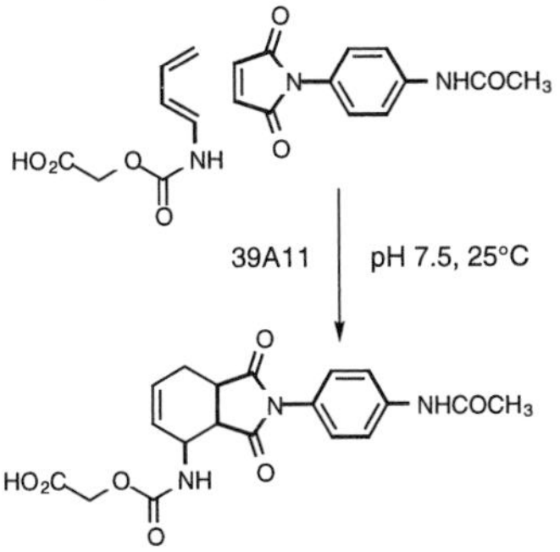 HO$_2$C, O, NH, O; O, N, NHCOCH$_3$, O; 39A11, pH 7.5, 25°C; O, N, NHCOCH$_3$, HO$_2$C, O, NH, O, O	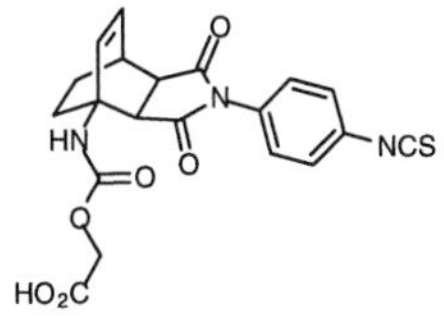 Diene Dienophile O, N, NCS, HN, O, O, HO$_2$C; *TSA*, K_i 1.3 E-1 μM	Diene 1.1 E+3 Dienophile 7.4 E+2	4.0 E+1	3.5 E-1	**17.2**
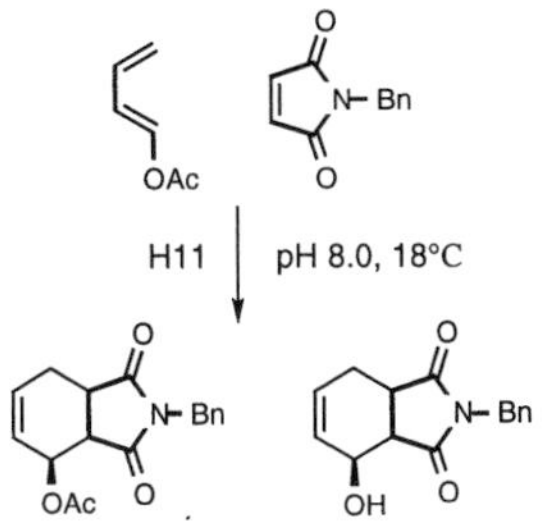 OAc; O, N-Bn, O; H11, pH 8.0, 18°C; O, N-Bn, OAc, O; O, N-Bn, OH, O	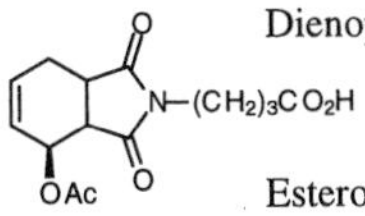 Diene Dienophile O, N-(CH$_2$)$_3$CO$_2$H, OAc, O Esterolysis *TSA*	Diene nr Dienophile 8.3 E+3 Esterolysis 1.1 E+3	Diene nr Dienophile 3.3 E+1 Esterolysis 5.5 E-2	Diene nr Dienophile 1.8 E+1 Esterolysis nr	**17.3**

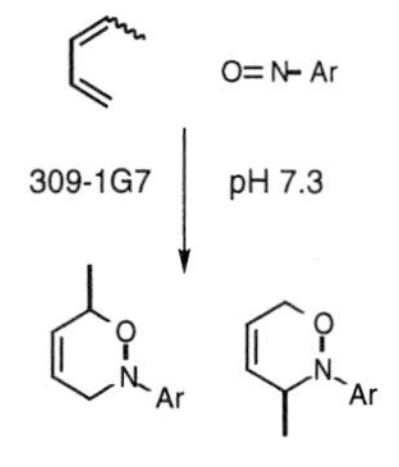

Reaction	TSA / Substrate				
309-1G7, pH 7.3; O=N–Ar; Ar = p-$C_6H_4CONHPr$	Dienophile ([*trans* Diene] 5.0 mM)	3.1 E+3 (app)	2.0 E+1 (app)	1.2 E+3 (app)	**17.4**
NHCOR; R = $(CH_2)_5CO_2H$	Dienophile ([*cis* Diene] 5.0 mM)	3.9 E+3 (app)	1.1 E+1 (app)	2.6 E+3 (app)	
	TSA				
$CON(CH_3)_2$; NH; HO_2C; 7D4, pH 7.4, 37°C; 22C8, pH 7.4, 37°C; 4D5, pH 7.4, 37°C; 13G5, pH 7.4, 37°C; $CON(CH_3)_2$; NH; CO_2H	**7D4** (*endo*) Diene	9.6 E+2	3.4 E-3	4.8 E+0	**17.5**
	Dienophile; $CON(CH_3)_2$; HN; $XOC(H_2C)_3$	1.7 E+3			
	22C8 (*exo*) Diene	7.0 E+2	3.2 E-3	1.8 E+1	
	Dienophile; HN; $CON(CH_3)_2$; $XOC(H_2C)_3$	7.5 E+3			
	4D5 (*endo*) Diene	1.6 E+3	3.5 E-3	4.9 E+0	
	Dienophile; $CONMe_2$; Fe; $NHCO(CH_3)_2CO_2H$	5.9 E+3			
	13G5 (*exo*) Diene	2.7 E+3	1.2 E-3	6.9 E+1	
	Dienophile	1.0 E+4			
	TSA				

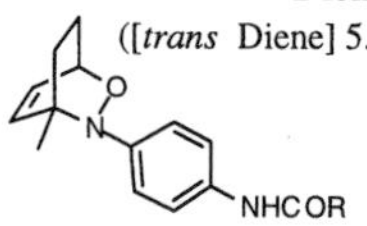

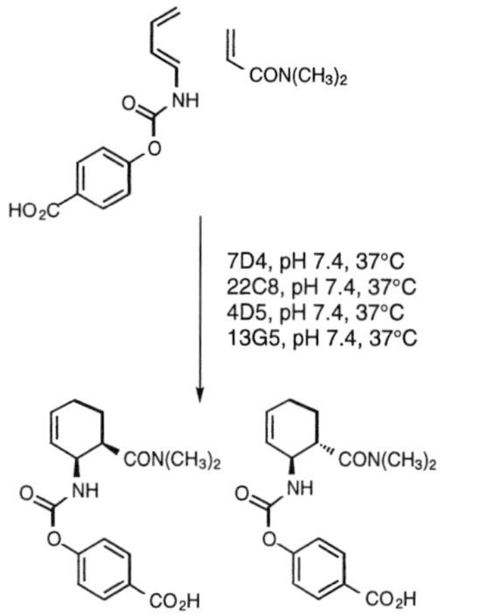

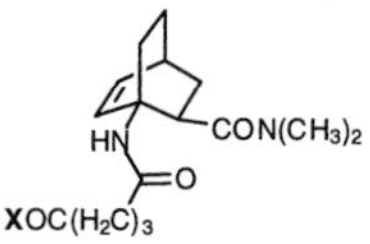

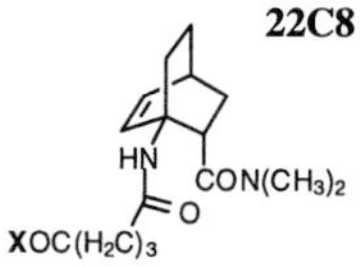

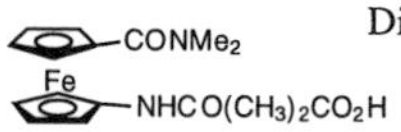

18. Acyl Transfer Reactions

Reaction	TSA / Substrate				
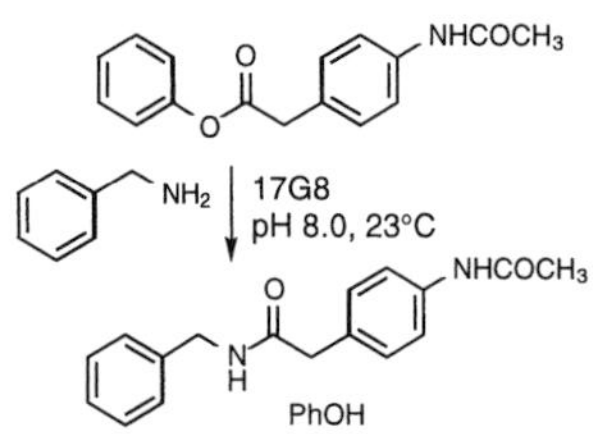 Amide Formation; $NHCOCH_3$; NH_2; 17G8, pH 8.0, 23°C; $NHCOCH_3$; PhOH	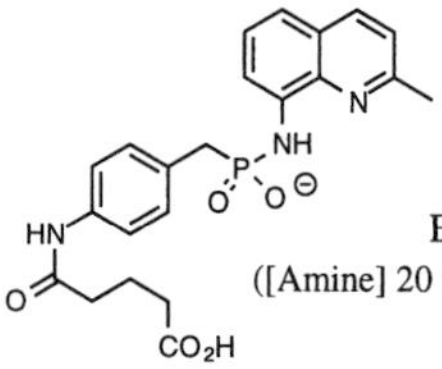 NH; HN; CO_2H; Ester ([Amine] 20 mM); *TSA*, IC_{50} 1.0 E+1 μM	2.2 E+3 (app)	2.3 E-2 (app)	1.1 E+1 (app)	**18.1**
NH_2; H_2N; 24B11, pH 7.0; OH	OH; Lactone	4.9 E+3	6.6 E-2	1.6 E+1	**18.2**
	Aniline; HO_2C; *TSA*, K_i 7.5 E-2 μM	1.2 E+3			

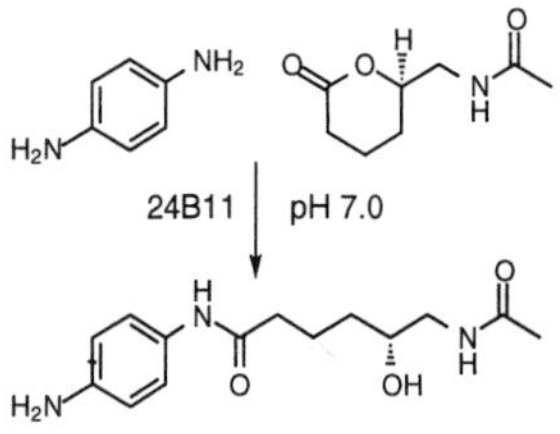

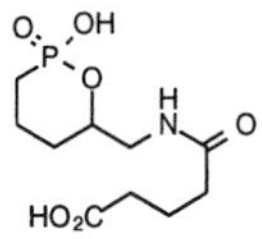

Reaction	TSA					Entry
Peptide Bond Formation; 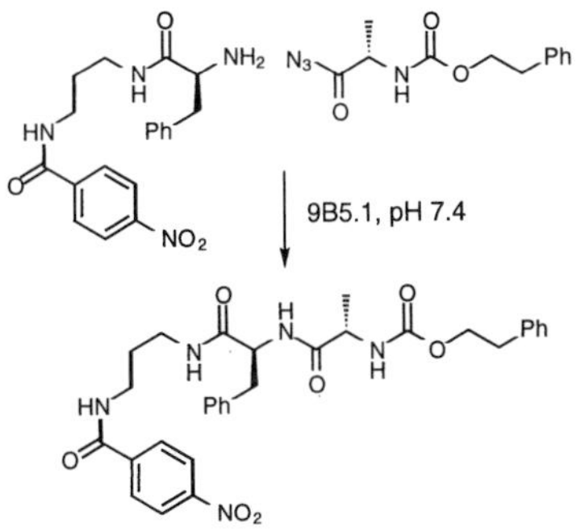 HN, H, N, O, NH_2, Ph, N_3, NO_2; 9B5.1, pH 7.4	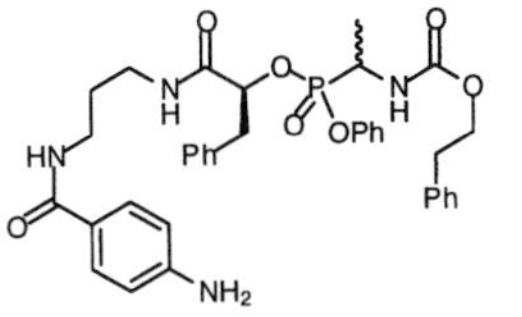 HN, H, N, O, P, OPh, Ph, NH_2; TSA, K_d 1.9E-2 μM	Azide	1.5 E+1	5.9 E-2	1.0 E+4	**18.3**
		Amine	1.5 E+3			
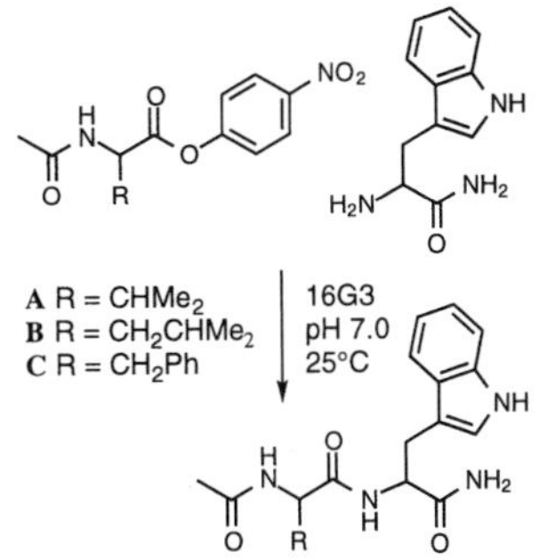 H, N, O, R, NO_2, NH, H_2N, NH_2; **A** R = $CHMe_2$; **B** R = CH_2CHMe_2; **C** R = CH_2Ph; 16G3, pH 7.0, 25°C	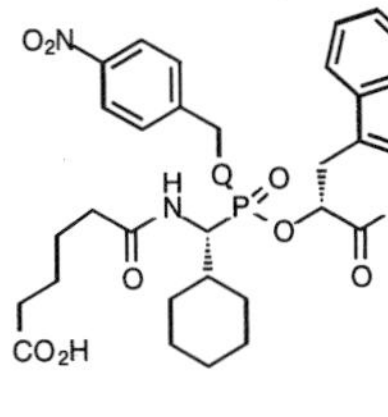 O_2N, Q, H, N, P, O, NH, NH_2, CO_2H; TSA	**A** L-Ester	4.0 E+3	1.3 E+1	1.9 E+2	**18.4**
		L- Amine	1.6 E+4			
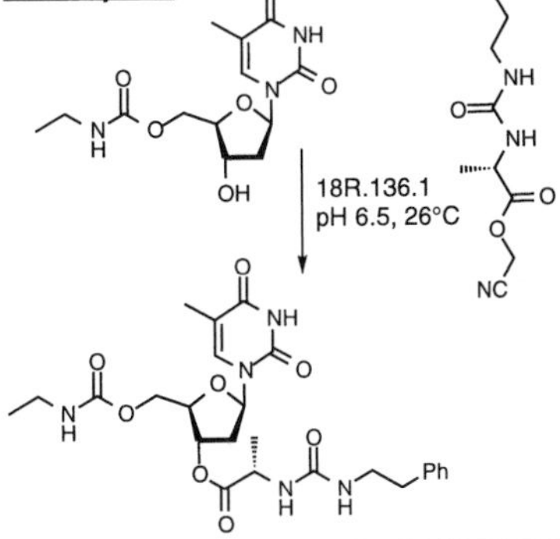 Aminoacylation; O, NH, N, H, OH, Ph, NC; 18R.136.1, pH 6.5, 26°C	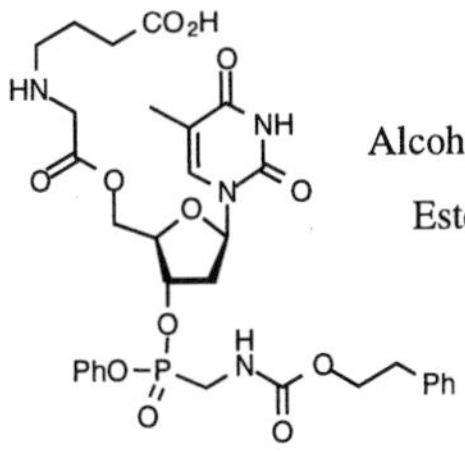 CO_2H, HN, O, NH, N, PhO-P, H, Ph; TSA, K_d 2.4 E-4 μM	Alcohol	7.7 E+2	1.4 E+1	5.5 E+4	**18.5**
		Ester	2.6 E+2			
Transesterification; OH, O, H, N; **A** 21H3, pH 9.0; **B** 21H3. pH 8.5 , 23°C (96 % octane); CH_3CHO	O, P, H, N, HO_2C; TSA; 21H3: K_i (app) 2.0 E+0 μM	**A** Ester	3.0 E+3	2.1 E+1	nr	**18.6**
		alcohol	7.3 E+3			
		B Ester	1.1 E+2	3.0 E+0	nr	
		alcohol	2.3 E+3			

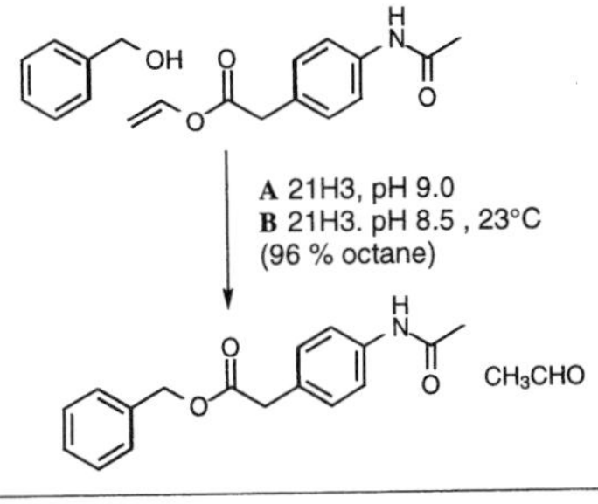

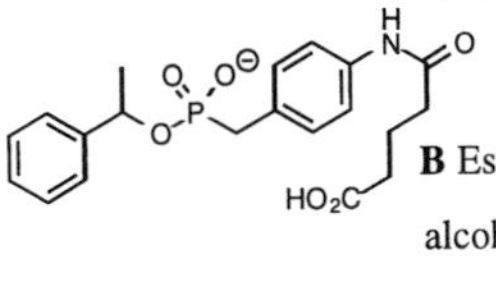

Reaction	Substrate / hapten				
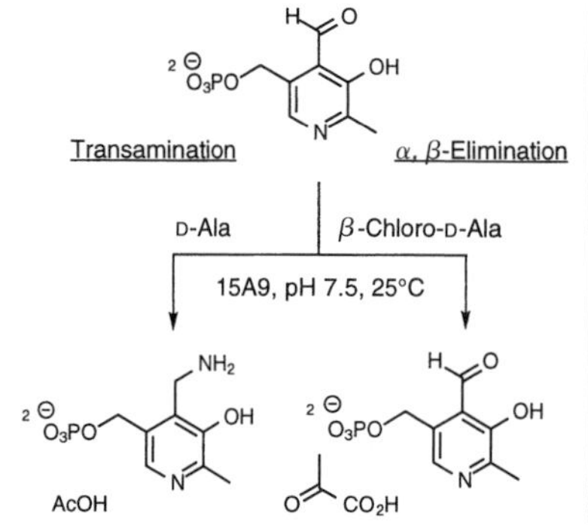 H O; $^{2\ominus}O_3PO$; OH; N; Transamination; α, β-Elimination; D-Ala; β-Chloro-D-Ala; 15A9, pH 7.5, 25°C; NH_2; $^{2\ominus}O_3PO$; OH; N; AcOH; H O; $^{2\ominus}O_3PO$; OH; N; O; CO_2H	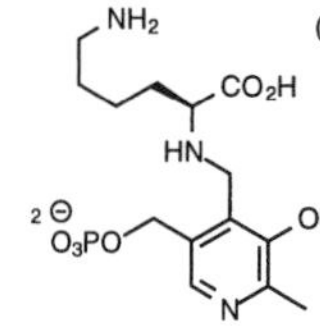 Transamination (100 μM PLP); NH_2; CO_2H; HN; $^{2\ominus}O_3PO$; OH; N	2.5 E+3 (app)	4.2 E-1 (app)	nr	**18.7**
	Elimination (100 μM PLP)	10 E+3 (app)	5.0 E+1 (app)	nr	

19. Amination Reactions

Reaction	Substrate / hapten				
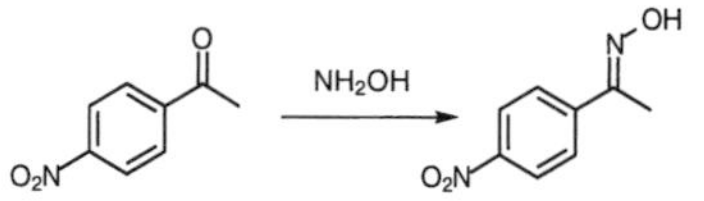 Oxime Formation; O; O_2N; NH_2OH; N OH; O_2N; 20AF2F6, pH 7.3, 25°C, *syn* :*anti* 9:1; 43D4-3D12, pH 6.5, 25°C, *syn* :*anti* 1:9	20AF2F6 Ketone ($[NH_2OH]$ 20 mM); 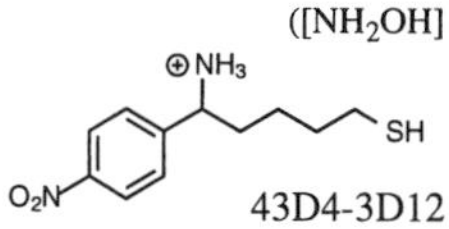 $\oplus NH_3$; SH; O_2N	2.7 E+3 (app)	1.1 E+1 (app)	1.7 E+4 (app)	**19.1**
	43D4-3D12 Ketone ($[NH_2OH]$ 20 mM); 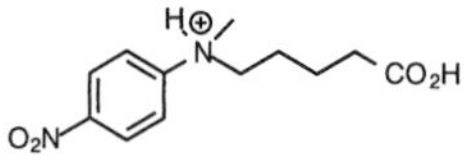 H⊕; N; CO_2H; O_2N; *TSA*	9.4 E+2 (app)	6.7 E+0 (app)	2.9 E+3 (app)	
Aldimine Formation; 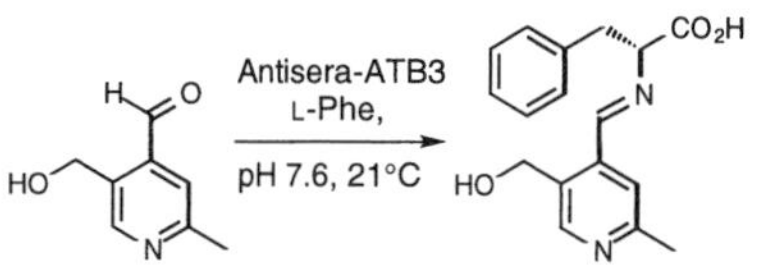 H O; HO; N; Antisera-ATB3, L-Phe, pH 7.6, 21°C; CO_2H; N; HO; N	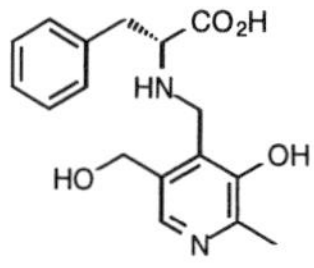 CO_2H; HN; HO; OH; N; Pyridoxal	3.9 E+3	1.5 E+1	nr	**19.2**
	L-Phe	1.6 E+2			
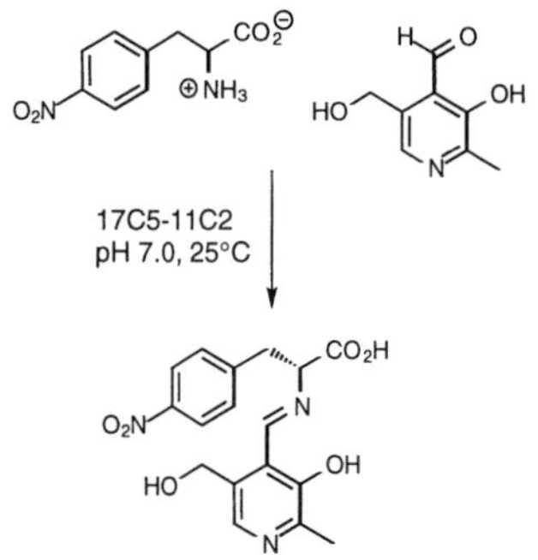 O_2N; $CO_2^{\ominus}$; $\oplus NH_3$; H O; HO; OH; N; 17C5-11C2, pH 7.0, 25°C; O_2N; CO_2H; N; HO; OH; N	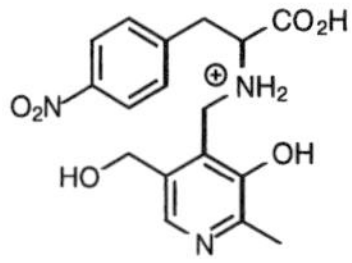 O_2N; CO_2H; $\oplus NH_2$; HO; OH; N; Amino acid	1.2 E+2	1.8 E+1	2.1 E-1	**19.3**
	Pyridoxal	7.1 E+2			
	L-amino acid: K_d 6.0 E-3 μM; D-amino acid: K_d 1.7 E-2 μM				

20. Miscellaneous

Reaction / Conditions	Hapten / Comments (K_i)	K_m [μM]	k_{cat} [min^{-1}]	k_{cat}/k_{uncat} [M]	Entry AE
Nucleophilic Substitution 16B5 NaI, 37°C (biphasic)	Sulfonate ([NaI] 0.15M)	1.3 E+2 (app)	2.8 E-5 (app)	5.8 E+2 (app)	**20.1**
	NaI ([Sulfonate] 0.75mM) *TSA* K_i *ca.* 1.0 E+1 μM	1.5 E+5 (app)	2.8 E-2 (app)	5.8 E+2 (app)	
Conjugate Addition 5G4, NaCN pH 7.0, 37°C	Enone	6.4 E+1	2.1 E-2	3.0 E-2	**20.2**
	CN^- *TSA* K_i 1.1 E+1 μM	1.4 E+2			
Porphyrin Metalation Porphyrin + M^{2+} 7G12-A10-G1-A12 pH 8.0, 26°C Porphyrin-M^{2+}	Zn^{2+}	4.9 E+1	5.2 E-4	2.6 E+3	**20.3**
	Cu^{2+}	5.0 E+1	8.4 E-5	1.7 E+3	

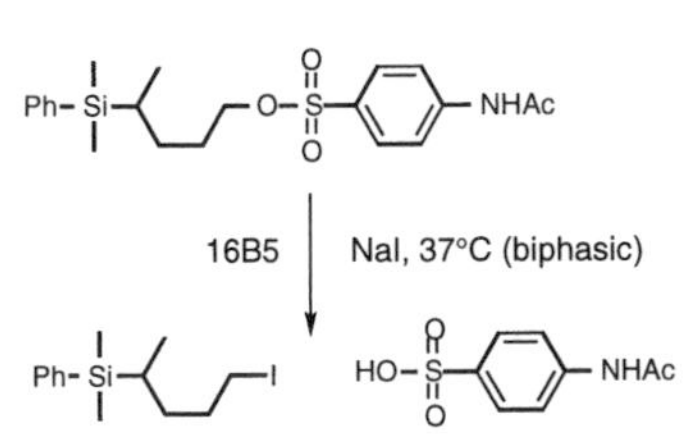

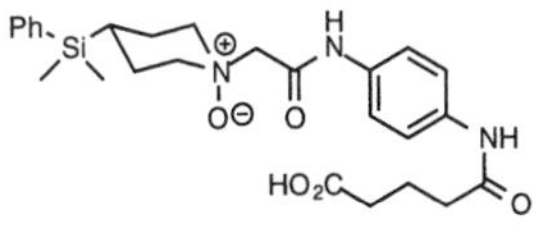

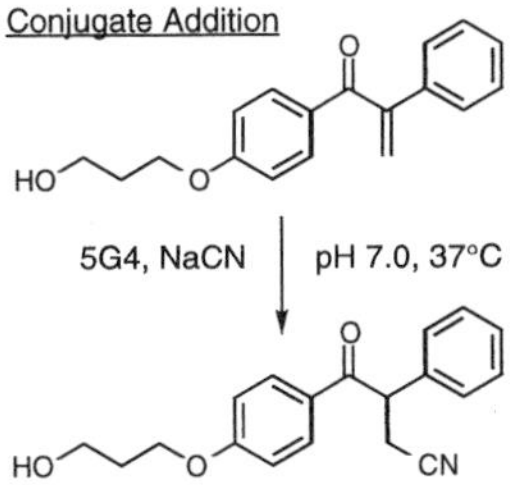

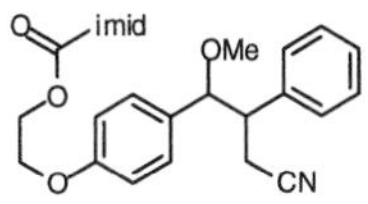

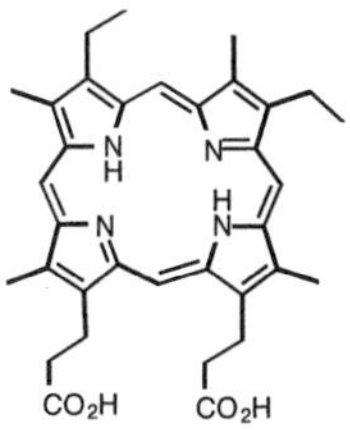

D. REDOX REACTIONS

21. Oxidations

Reaction / Conditions	Hapten / Comments (K_i)	K_m [μM]	k_{cat} [min^{-1}]	k_{cat}/k_{uncat} [M]	Entry AE
$NaIO_4$ 28B4.2 pH 5.5, 23°C $NaIO_3$	Sulfide	4.3 E+1	8.2 E+0	9.4 E+6	**21.1**
	$NaIO_4$ *TSA* K_d 5.2 E-2 μM	2.5 E+2			

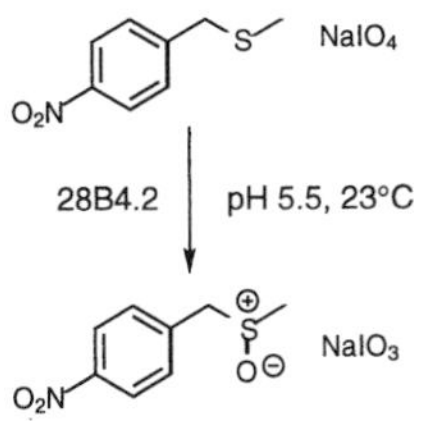

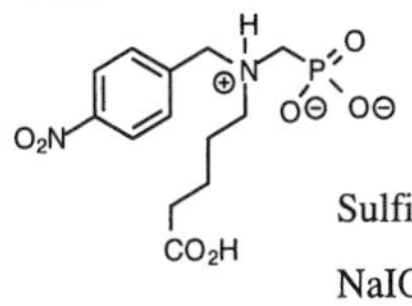

R, R, R, R, N, Mn^{3+}, N, N, N; PhIO, reverse micelles; PhI	R, R, R, R, NH, N, N, HN; *Metal cofactor complex*	nr	nr	30-60% rate enhancement	**21.2**
Ar; 20B11, CH_3CN; H_2O_2, pH 6.6; O; Ar; 66 % *ee*	Linker; ⊕ N; Ar; *TSA*	2.6 E+2 (app)	8.4 E-4 (app)	6.0 E+1 (app)	**21.3**
OH, OH; 7G12-A10-G1-A12 pH 8.0, 10°C; Fe(III), H_2O_2, mesoporphyrin; O, O	NH, N, N, N; CO_2H, CO_2H; *Metal cofactor complex*	2.4 E+4	4.0 E+2	2.4 E+1	**21.4**

22. Reductions

O_2N; O; H N; O; NaBH$_3$CN (1 mM); A5 pH 5.0, 22°C; O_2N; OH; H N; O; *de* > 99 %	O_2N; O; O ⊖; P; O; H_2NOC; *TSA*; K_d 6.1 E-1 μM	1.2 E+3 (app)	1.0 E-1 (app)	2.9 E+2 (app)	**22.1**
N, N; H_2N; ⊕ N; ⊕ ⊖ NH_3Cl; Cl⊖; Safranine T [O]; mAb, flavin, dithionite; Safranine T [R]	O; O; CO_2H; N, N, O; NH; N; O; *Cofactor complex*; K_d 8.0 E-3 μM	nr	nr	nr	**22.2**

HO, O, N, O⊖ ; 66D2, pH 5.8, 25°C ; HO, O, N, O	NCS, CO_2H, HO, O, O ; Sulfite Resazurin	3.0 E+3 6.0 E-1	1.2 E+0	6.0 E+1	**22.3**
O_2N, R, O ; R = Et R = isopropyl R = Bn ; 37B39.3 $NaCNBH_3$ pH 5.0, 22°C ; O_2N, R, OH ; R = Et, 99 % *S*	O_2N, N, O⊖, CO_2H ; R = Et 50 mM $NaBH_3CN$ 0.15 mM $NaBH_3CN$ *TSA* K_d 3.3 E-2 µM	5.2 E+1 (app) 5.7 E+4 (app)	9.7 E-2 (app) 1.7 E-1 (app)	*k*uncat 1.1 E-3 min^{-1} M^{-1}	**22.4**

12.2 Key to Bibliography

A Hydrolytic and Dissociative Processes

1. *Aliphatic Ester Hydrolysis*
1.1 Shen et al., 1992; **1.2** Tanaka et al., 1996; **1.3** 3B9 Landry et al., 1993, 15A10 Yang et al., 1996, polyclonal Basmadjian et al., 1995; **1.4** Nakatani et al., 1993; **1.5** Ikeda et al., 1991; **1.6** Fujii et al., 1991; **1.7** Iwabuchi et al., 1994; **1.8** Miyashita et al., 1993; **1.9** Kitazume et al., 1994; **1.10** Campbell et al., 1994; **1.11** Kitazume et al., 1991a; **1.12** Kitazume et al., 1991b; **1.13** Ikeda and Achiwa, 1997; **1.14** Janda et al., 1989, 1990a; **1.15** Pollack et al., 1989; **1.16** A Tawfik et al., 1993, B Tawfik et al., 1997); **1.17** Janda et al., 1994; **1.18** Li et al., 1995b; **1.19** Izadydar et al., 1993.

2. *Aryl Ester Hydrolysis*
2.1 Suga et al., 1994b; **2.2** Tawfik et al., 1990; **2.3** pH 8.7 Guo et al., 1994, pH 9.5 Zhou et al., 1994; **2.4** Khalaf et al., 1992; **2.5** A Martin et al., 1991, B Durfor et al., 1988; **2.6** 30C6 Janda et al., 1990b, 84A3 Wade et al., 1993, 27A6 Janda et al., 1991b; **2.7** KD2-260 Ohkubo et al., 1993, 7K16.2 Shokat et al., 1990; **2.8** NPN43C9 Gibbs et al., 1992a, Fab-1D Chen et al., 1993; **2.9** Tramontano et al., 1986; **2.10** Tramontano et al., 1988; **2.11** Suga et al., 1994a; **2.12** Janda et al., 1991a; **2.13** Napper et al., 1987; **2.14** Wirsching et al., 1995; **2.15** Pollack et al., 1988.

3. *Carbonate Hydrolysis*
3.1 A Shokat et al., 1990, B Jacobs et al., 1987, C Spitznagel et al., 1993; **3.2** Pollack et al., 1986; **3.3** "Sheep no. 270" Gallacher et al., 1991, IV-CAT 2-6 Stahl et al., 1995; **3.4** Wallace and Iverson, 1996.

4. *Carbamate Ester Hydrolysis*
4.1 Van Vranken et al., 1994; **4.2** Wentworth et al., 1996; **4.3** Wentworth et al., 1997.

5. *Amide Hydrolysis*
5.1 Martin et al., 1994; **5.2** Iverson and Lerner, 1989; **5.3** Janda et al., 1988b; **5.4** Benedetti et al., 1996; **5.5** Gallacher et al., 1992; **5.6** Paul et al., 1989; **5.7** Li et al., 1995a; **5.8** Paul et al., 1995.

6. *Phosphate Ester Hydrolysis*
6.1 Rosenblum et al., 1995; **6.2** 15C5 Lavey and Janda, 1996a, 3H5 Lavey and Janda, 1996b; **6.3** Brimfield et al., 1993; **6.4** Weiner et al., 1997; **6.5** Shuster et al., 1992, Gololobov et al., 1995; **6.6** Scanlan et al., 1991.

7. *Miscellaneous Hydrolyses*
7.1 37C4 Iverson et al., 1990, polyclonal Stephens and Iverson, 1993; **7.2** Yu et al., 1994; **7.3** Janda et al., 1997; **7.4** A Reymond et al., 1992, 1993, 1994, B Reymond et al., 1991, C Shabat et al., 1995, D Sinha et al., 1993b, E Sinha et al., 1993a.

8. *Eliminations*
8.1 Cravatt et al., 1994; **8.2** 43D4-3D21, Shokat et al., 1989, 20A2F6 Uno and Schultz, 1992; **8.3** Yoon et al., 1996; **8.4** Thorn et al., 1995; **8.5** Zhou et al., 1997.

9. *Decarboxylations*
9.1 21D8 Lewis et al., 1991, 25E10 Tarasow et al., 1994; **9.2** Smiley and Benkovic, 1994; **9.3** Björnestedt et al., 1996; **9.4** Ashley et al., 1993.

10. *Cycloreversions*
10.1 Bahr et al., 1996; **10.2** A Cochran et al., 1988, B Jacobsen et al., 1995.

11. *Retro-Henry and Retro-Aldol Reactions*
11.1 Flanagan et al., 1996; **11.2** Reymond, 1995.

B Intramolecular Processes

12. *Isomerizations*
12.1 Yli-Kauhaluoma et al., 1996; **12.2** Jackson and Schultz, 1991; **12.3** Khettal et al., 1994; **12.4** Uno et al., 1996.

13. *Rearrangements*
13.1 Braisted and Schultz, 1994, Ulrich et al., 1996; **13.2** 1F7 Hilvert et al., 1988, Hilvert and Nared, 1988, 11F1-2E11 Jackson et al., 1988; **13.3** 2B4 Gibbs et al., 1992b, RG2-23C7 Liotta et al., 1993; **13.4** Chen et al., 1994; **13.5** Willner et al., 1994.

14. *Epoxide Opening*
14.1 1. Janda et al., 1993, Shevlin et al., 1994, 2. Janda et al., 1995.

15. *Cationic Cyclization*
15.1 TX1-4C6 Li et al., 1994, TMI 87D7 Li et al., 1995c; **15.2** Li et al., 1996; **15.3** Hasserodt et al., 1996; **15.4** Hasserodt et al., 1997.

C Bimolecular Associative and Substitution Reactions

16. *Aldol Reactions*
16.1 Koch et al., 1995; **16.2** Wagner et al., 1995; **16.3** A Reymond and Chen, 1995a, B Reymond and Chen, 1995b.

17. *Diels-Alder Cycloaddition*
17.1 Hilvert et al., 1989; **17.2** Braisted and Schultz, 1990; **17.3** Suckling et al., 1993; **17.4** *trans* Meekel et al., 1995, *cis* Resmini et al., 1996; **17.5** 7D4 and 22C8 Gouverneur et al., 1993, 4D5 and 13G5 Yli-Kauhaluoma et al., 1995.

18. *Acyl Transfer Reactions*
18.1 Janda et al., 1988a; **18.2** Benkovic et al., 1988; **18.3** Jacobsen and Schultz, 1994; **18.4** Hirschmann et al., 1994, Smithrud et al., 1997; **18.5** Jacobsen et al., 1992; **18.6** A Wirsching et al., 1991, B Ashley and Janda, 1992; **18.7** Gramatikova and Christen, 1996.

19. *Amination Reactions*
19.1 Uno et al., 1994; **19.2** Tubul et al., 1994; **19.3** Cochran et al., 1991.

20. *Miscellaneous*
20.1 Li et al., 1995d; **20.2** Cook et al., 1995; **20.3** Cochran and Schultz, 1990a, for a second example see Kawamura-Konishi et al., 1996.

D Redox Reactions

21. *Oxidations*
21.1 Hsieh et al., 1994; **21.2** Keinan et al., 1990; **21.3** Koch et al., 1994; **21.4** Cochran and Schultz, 1990b.

22. *Reductions*
22.1 Nakayama and Schultz, 1992; **22.2** Shokat et al., 1988; **22.3** Janjic and Tramontano, 1989; **22.4** Hsieh et al., 1993.

13 References

ADDADI, L., JAFFI, E. K., KNOWLES, J. R. (1983), Secondary tritium isotope effects as probes of the enzymic and nonenzymic conversion of chorismate to prephenate, *Biochemistry* **22**, 4494–4501.

ALBERG, D. G., LAUHON, C. T., NYFELER, R., FASSLER, A., BARTLETT, P. A. (1992), Inhibition of EPSP synthase by analogues of the tetrahedral intermediate and of EPSP, *J. Am. Chem. Soc.* **114**, 3535–3546.

ALBERY, W. J. (1993), Transition state theory revisited, *Adv. Phys. Org. Chem.* **28**, 139–170.

ALBERY, J., KNOWLES, J. R. (1976), Evolution of enzyme function and development of catalytic efficiency, *Biochemistry* **15**, 5631–5640.

ALBERY, J., KNOWLES, J. R. (1977), Efficiency and evolution of enzyme catalysis, *Angew. Chem.* (*Int. Edn. Engl.*) **16**, 285–293.

ASHLEY, J. A., JANDA, K. D. (1992), Antibody catalysis in low water content media, *J. Org. Chem.* **57**, 6691–6693.

ASHLEY, J. A., LO, C.-H. L., MCELHANEY, G. P., WIRSCHING, P., JANDA, K. D. (1993), A catalytic antibody model for PLP-dependent decarboxylases, *J. Am. Chem. Soc.* **115**, 2515–2516.

BAGSHAWE, K. D. (1990), Antibody-directed enzyme/prodrug therapy (ADEPT), *Biochem. Soc. Trans.* **18**, 750–752.

BAHR, N., GÜLLER, R., REYMOND, J.-L., LERNER, R. A. (1996), A nitroxyl synthase catalytic antibody, *J. Am. Chem. Soc.* **118**, 3550–3555.

BALDWIN, J. E. (1976), Approach vector analysis: A stereochemical approach to reactivity, *J. Chem. Soc. Chem. Commun.* 738–741.

BALDWIN, J. E., CHESWORTH, R., PARKER, J. S., RUSSELL, A. T. (1995), Studies towards a postulated biomimetic Diels–Alder reaction for the synthesis of himgravine, *Tetrahedron Lett.* **36**, 9551–9554.

BALDWIN, J. E., CLARIDGE, T. D. W., CULSHAW, A. J., HEUPEL, F. A., LEE, V. et al. (1998), Investigations into the manzamine alkaloid biosynthesis hypothesis, *Angew. Chem.* (*Int. Edn. Engl.*) **37**, 2661–2663.

BARTLETT, P. A., GIANGIORDANO, M. A. (1996), Transition-state analogy of phosphonic acid peptide inhibitors of pepsin, *J. Org. Chem.* **61**, 3433–3438.

BARTLETT, P. A., JOHNSON, C. R. (1985), An inhibitor of chorismate mutase resembling the transition-state conformation, *J. Am. Chem. Soc.* **107**, 7792–7793.

BARTLETT, P. A., LAMDEN, L. A. (1986), Inhibition of chymotrypsin by phosphonate and phosphonamidate peptide analogs, *Bioorg. Chem.* **14**, 356–377.

BARTLETT, P. A., MARLOWE, C. K. (1983), Phosphonamidates as transition state analogue inhibitors of thermolysin, *Biochemistry* **22**, 4618–4624.

BASMADJIAN, G. P., SINGH, S., SASTRODJOJO, B., SMITH, B. T., AVOR, K. S. et al. (1995), Generation of polyclonal catalytic antibodies against cocaine using transition state analogs of cocaine conjugated to diptheria toxoid, *Chem. Pharm. Bull.* **43**, 1902–1911.

BENEDETTI, F., BERTI, F., COLOMBATTI, A., EBERT, C., LINDA, P., TONIZZO, F. (1996), anti-Sulfonamide antibodies catalyse the hydrolysis of a heterocyclic amide, *Chem. Commun.* 1417–1418.

BENKOVIC, S. J., NAPPER, A. D., LERNER, R. A. (1988), Catalysis of a stereospecific bimolecular amide synthesis by an antibody, *Proc. Natl. Acad. Sci. USA* **85**, 5355–5358.

BJÖRNESTEDT, R., ZHONG, G., LERNER, R. A., BARBAS III, C. F. (1996), Copying nature's mechanism for the decarboxylation of β-keto acids into catalytic antibodies by reactive immunization, *J. Am. Chem. Soc.* **118**, 11720–11724.

BLAKEY, D. C. (1992), Drug targeting with monoclonal antibodies – a review, *Acta Oncologica*, **31**, 91–97.

BLAKEY, D. C., BURKE, P. J., DAVIES, D. H., DOWELL, R. I., MELTON, R. G. et al. (1995), Antibody-directed enzyme prodrug therapy (ADEPT) for treatment of major solid tumour disease, *Biochem. Soc. Trans.* **23**, 1047–1050.

BLOKZIJL, W., ENGBERTS, J. B. F. N. (1994), Enforced hydrophobic interactions and hydrogen-bonding in the acceleration of Diels–Alder reactions in water, *ACS Symp. Series* **568**, 303–317.

BOWMAN, W. C., RAND, M. J. (1988), *Textbook of Pharmacology*. Oxford: Blackwell Scientific Publications.

BRAISTED, A. C., SCHULTZ, P. G. (1990), An Antibody-Catalyzed bimolecular Diels–Alder reaction, *J. Am. Chem. Soc.* **112**, 7430–7431.

BRAISTED, A. C., SCHULTZ, P. G. (1994), An antibody-catalyzed oxy-Cope rearrangement, *J. Am. Chem. Soc.* **116**, 2211–2212.

BRIMFIELD, A. A., LENZ, D. E., MAXWELL, D. M., BROOMFIELD, C. A. (1993), Catalytic antibodies hydrolysing organophosphorus esters, *Chemico-Biological Interactions – England* **87**, 95–102.

BURTON, D. R. (1990), Antibody: The flexible adaptor molecule, *Trends Biochem. Sci.* **15**, 64–69.

CAMPBELL, D. A., GONG, B., KOCHERSPERGER, L. M., YONKOVICH, S., GALLOP, M. A., SCHULTZ, P. G. (1994), Antibody-catalyzed prodrug activation, *J. Am. Chem. Soc.* **116**, 2165–2166.

CHARBONNIER, J.-B., CARPENTER, E., GIGANT, B., GOLINELLI-PIMPANEAU, B., ESHHAR, Z. et al. (1995), Crystal structure of the complex of a catalytic antibody Fab fragment with a transition-state analog – structural similarities in esterase-like catalytic antibodies, *Proc. Natl. Acad. Sci. USA* **92**, 11721–11726.

CHARBONNIER, J.-P., GOLINELLI-PIMPANEAU, B., GIGANT, B., TAWFIK, D. S., CHAP, R. et al. (1997), Structural convergence in the active sites of a family of catalytic antibodies, *Science* **275**, 1140–1142.

CHEN, Y.-C. J., DANON, T., SASTRY, L., MUBARAKI, M., JANDA, K. D., LERNER, R. A. (1993), Catalytic antibodies from combinatorial libraries, *J. Am. Chem. Soc.* **115**, 357–358.

CHEN, Y., REYMOND, J.-L., LERNER, R. A. (1994), An antibody-catalyzed 1,2-rearrangement of carbon–carbon bonds, *Angew. Chem. (Int. Edn. Engl.)* **33**, 1607–1609.

CHOOK, Y. M., KE, H. M., LIPSCOMB, W. N. (1993), Crystal sturctures of the monofunctional chorismate mutase from *Bacillus subtilis* and its complex with a transition state analog, *Proc. Natl. Acad. Sci. USA* **90**, 8600–8603.

CHOOK, V. M., GRAY, J. V., KE, H. M. L., LIPSCOMB, W. N. (1994), The monofunctional chorismate mutase from *Bacillus subtilis* – structure determination of chorismate mutase and its complexes with a transition-state analog and prephenate and implications for the mechanism of enzymatic reaction, *J. Mol. Biol.* **240**, 476–500.

CIOBANU, M., MATSUMOTO, K. (1997), Recent advances in organic synthesis under high pressure, *Liebigs Ann. Chem.* **4**, 623–635.

COCHRAN, A. G., SCHULTZ, P. G. (1990a), Antibody-catalysed porphyrin metallation, *Science*, **249**, 781– 783.

COCHRAN, A. G., SCHULTZ, P. G. (1990b), Peroxidase activity of an antibody-heme complex, *J. Am. Chem. Soc.* **112**, 9414–9415.

COCHRAN, A. G., SUGASAWARA, R., SCHULTZ, P. G. (1988), Photosensitized cleavage of a thymine dimer by an antibody, *J. Am. Chem. Soc.* **110**, 7888–7890.

COCHRAN, A. G., PHAM, T., SUGASAWARA, R., SCHULTZ, P. G. (1991), Antibody-catalyzed bimolecular imine formation, *J. Am. Chem. Soc.* **113**, 6670–6672.

COOK, C. E., ALLEN, D. A., MILLER, D. B., WHISNANT, C. C. (1995), Antibody-Catalyzed Michael reaction of cyanide with an α,β-unsaturated ketone, *J. Am. Chem. Soc.* **117**, 7269–7270.

CRANS, D. C., FELTY, R. A., MILLER, M. M. (1991), Cyclic vanadium(V) alkoxide: An analogue of the ribonuclease inhibitors, *J. Am. Chem. Soc.* **113**, 265–269.

CRAVATT, B. F., ASHLEY, J. A., JANDA, K. D., BOGER, D. L., LERNER, R. A. (1994), Crossing extreme mechanistic barriers by antibody catalysis: *Syn* elimination to a *cis* olefin, *J. Am. Chem. Soc.* **116**, 6013–6014.

DANISHEFSKY, S., HERSHENSON, F. M. (1979), Regiospecific synthesis of isogabaculine, *J. Org. Chem.* **44**, 1180–1181.

DATTA, A., PARTRIDGE, L. J., BLACKBURN, G. M. (1996), European network on antibody catalysis 1993–1995. *Brussels, Luxembourg*, CSC-EC-EA-EC, 36.

DAVIDSON, M. M., HILLIER, I. H. (1994), Claisen rearrangement of chorismic acid and related analogues: An *ab initio* molecular orbital study, *J. Chem. Soc. Perkin Trans. II*, 1415–1417.

DE LAUZON, S., DESFOSSES, B., MANSUY, D., MAHY, J.-P. (1999), Studies of the reactivity of artificial peroxidase-like hemoproteins based on antibodies elicited against a specifically designed *ortho*-carboxy substituted tetraarylporphyrin, *FEBS Lett.* **443**, 229–234.

DELL, C. P. (1997), Cycloadditions in synthesis, *Contemp. Org. Synth.* **4**, 87–117.

DURFOR, C. N., BOLIN, R. J., SUGASAWARA, R. J., MASSEY, R. J., JACOBS, J. W., SCHULTZ, P. G. (1988), Antibody catalysis in reverse micelles, *J. Am. Chem. Soc.* **110**, 8713–8714.

ERSOY, O., FLECK, R., SINSKEY, A., MASUMUNE, S. (1998), Antibody catalyzed cleavage of an amide bond using an external nucleophilic cofactor, *J. Am. Chem. Soc.* **120**, 817–818.

ESCHENMOSER, A., RUZICKA, L., JEGER, O., ARIGONI, D. (1955), Eine stereochemische Interpretation der biogenetischen Isoprenregel bei den Triterpenen, *Helvet. Chim. Acta* **38**, 1890–1904.

EVANS, M. G., POLANYI, M. (1935), Some applications of the transition state method to the calculation of reaction velocities, especially in solution, *Trans. Faraday Soc.* **31**, 875–894.

EYRING, H. (1935), The activated complex and the absolute rate of chemical reactions, *Chem. Rev.* **17**, 65–77.

FENNIRI, H., JANDA, K. D., LERNER, R. A. (1995), Encoded reaction cassette for the highly sensitive detection of the making and breaking of chemical bonds, *Proc. Natl. Acad. Sci. USA* **92**, 2278–2282.

FERSHT, A. F. (1985), *Enzyme Structure and Mechanism*, pp. 98–154. New York: Freeman.

FINN, M. G., LERNER, R. A., BARBAS III, C. F. (1998), Cofactor-induced refinement of catalytic antibody activity: A metal-specific allosteric effect, *J. Am. Chem. Soc.* **120**, 2963–2964.

FLANAGAN, M. E., JACOBSEN, J. R., SWEET, E., SCHULTZ, P. G. (1996), Antibody-catalyzed retro-aldol reaction, *J. Am. Chem. Soc.* **118**, 6078–6079.

FUJII, I., LERNER, R. A., JANDA, K. D. (1991), Enantiofacial protonation by catalytic antibodies, *J. Am. Chem. Soc.* **113**, 8528–8529.

FUJII, I., TANAKA, F., MIYASHITA, H., TANIMURA, R., KINOSHITA, K. (1995), Correlation between antigen-combining-site structures and functions within a panel of catalytic antibodies generated against a single transition state analog, *J. Am. Chem. Soc.* **117**, 6199–6209.

GALLACHER, G., JACKSON, C. S., TOPHAM, C. M., SEARCEY, M., TURNER, B. C. et al. (1990), Polyclonal-antibody-catalysed hydrolysis of an aryl nitrophenyl carbonate, *Biochem. Soc. Trans.* **18**, 600–601.

GALLACHER, G., JACKSON, C. S., SEARCEY, M., BADMAN, G. T., GOEL, R. et al. (1991), A polyclonal antibody preparation with Michaelian catalytic properties, *Biochem. J.* **279**, 871–881.

GALLACHER, G., SEARCEY, M., JACKSON, C. S., BROCKLEHURST, K. (1992), Polyclonal antibody-catalysed amide hydrolysis, *Biochem. J.* **284**, 675–680.

GAO, C., LAVEY, B. J., LO, C. H. L., DATTA, A., WENTWORTH JR., P., JANDA, K. D. (1998), Direct selection for catalysis from combinatorial antibody libraries using a boronic acid probe: primary amide bond hydrolysis, *J. Am. Chem. Soc.* **120**, 2211–2217.

GIBBS, R. A., BENKOVIC, P. A., JANDA, K. D., LERNER, R. A., BENKOVIC, S. J. (1992a), Substituent effects on an antibody catalyzed hydrolysis of phenyl esters: Further evidence for an acyl-antibody intermediate, *J. Am. Chem. Soc.* **114**, 3528–3534.

GIBBS, R. A., TAYLOR, S., BENKOVIC, S. J. (1992b), Antibody-catalysed rearrangement of the peptide bond, *Science* **258**, 803–805.

GIGANT, B., CHARBONNIER, J.-B., ESHHAR, Z., GREEN, B. S., KNOSSOW, M. (1997), X-ray structures of a hydrolytic antibody and of complexes elucidate catalytic pathway from substrate binding and transition state stabilization through water attack and product release, *Proc. Natl. Acad. Sci. USA* **94**, 7857–7861.

GOLINELLI-PIMPANEAU, B., GIGANT, B., BIZEBARD, T., NAVAZA, J., SALUDJIAN, P. et al. (1994), Crystal structure of a catalytic antibody with esterase-like activity, *Structure* **2**, 175–183.

GOLOLOBOV, G. V., CHERNOVA, E. A., SCHOUROV, D. V., SMIRNOV, I. V., KUDELINA, I. A., GABIBOV, A. G. (1995), Cleavage of supercoiled plasmid DNA by autoantibody Fab fragment: Application of the flow linear dichroism technique, *Proc. Natl. Acad. Sci. USA* **92**, 254–257.

GOUVERNEUR, V. E., HOUK, K. N., DE PASCUAL-TERESA, B., BENO, B., JANDA, K. D., LERNER, R. A. (1993), Control of the *exo* and *endo* pathways of the Diels–Alder reaction by antibody catalysis, *Science* **262**, 204–208.

GRAMATIKOVA, S. I., CHRISTEN, P. (1996), Pyridoxal 5′-phosphate-dependent catalytic antibody, *J. Biol. Chem.* **271**, 30583–30586.

GROSSBERG, A. L., B PRESSMAN, D. (1960), Nature of the combining site of antibody against a hapten bearing a positive charge, *J. Am. Chem. Soc.* **82**, 5478–5482.

GUO, J., HUANG, W., SCANLAN, T. S. (1994), Kinetic and mechanistic characterization of an efficient hydrolytic antibody: Evidence for the formation of an acyl intermediate, *J. Am. Chem. Soc.* **116**, 6062–6069.

HALAZY, S., BERGES, V., ERHARD, A., DANZIN, C. (1992), Difluoromethylphenyl glycosides as suicide substrates for glycosidases, *Bioorg. Chem.* **18**, 330–335.

HAMMOND, G. S. (1955), A correlation of reaction rates, *J. Am. Chem. Soc.* **77**, 334–338.

HASSERODT, J., JANDA, K. D., LERNER, R. A. (1996), Antibody catalyzed terpenoid cyclization, *J. Am. Chem. Soc.* **118**, 11654–11655.

HASSERODT, J., JANDA, K. D., LERNER, R. A. (1997), Formation of bridge-methylated decalins by antibody-catalyzed tandem cationic cyclization, *J. Am. Chem. Soc.* **119**, 5993–5998.

HAYNES, M. R., STURA, E. A., HILVERT, D., WILSON, I. A. (1994), Routes to catalysis: Structure of a catalytic antibody and comparison with its natural counterpart, *Science* **263**, 646–652.

HEINE, A., STURA, E. A., YLI-KAUHALUOMA, J. T., GAO, C. C., DENG, Q. L. et al. (1998), An antibody *exo*-Diels-Alderase inhibitor complex at 1.95Å resolution, *Science* **279**, No. 5358, 1934–1940.

HENDRY, P., SARGESON, A. M. (1990), Metal-ion promoted reactions of phosphate derivatives, *Progr. Inorg. Chem.* **38**, 201–258.

HILVERT, D., NARED, K. D. (1988), Stereospecific Claisen rearrangement catalyzed by an antibody, *J. Am. Chem. Soc.* **110**, 5593–5594.

HILVERT, D., CARPENTER, S. H., NARED, K. D., AUDITOR, M.-T. M. (1988), Catalysis of concerted reactions by antibodies: The Claisen rearrangement, *Proc. Natl. Acad. Sci. USA* **85**, 4953–4955.

HILVERT, D., HILL, K. W., NARED, K. D., AUDITOR, M.-T. M. (1989), Antibody catalysis of a Diels–Alder reaction, *J. Am. Chem. Soc.* **111**, 9261–9262.

HIRSCHMANN, R., SMITH III, A. B., TAYLOR, C. M., BENKOVIC, P. A., TAYLOR, S. D. et al. (1994), Peptide synthesis catalyzed by an antibody containing a binding site for variable amino acids, *Science* **265**, 234–237.

HOLLFELDER, F., KIRBY, A. J., TAWFIK, D. S. (1996), "Off-the-Shelf" proteins that rival tailor-made antibodies as catalysts, *Nature* **353**, 60–63.

HOUK, K. N., GONZALES, J., LI, Y. (1995), Pericyclic reaction transition states: passions and punctilios, 1935–1995, *Acc. Chem. Res.* **28**, 81–90.

HSIEH, L. C., YONKOVICH, S., KOCHERSPERGER, L., SCHULTZ, P. G. (1993), Controlling chemical reactivity with antibodies, *Science* **260**, 337–339.

HSIEH, L. C., STEPHANS, J. C., SCHULTZ, P. G. (1994), An efficient antibody-catalyzed oxygenation reaction, *J. Am. Chem. Soc.* **116**, 2167–2168.

ICHIHARA, A., OIKAWA, H. (1998), Diels–Alder type natural products – Structures and biosynthesis, *Curr. Org. Chem.* **2**, 365–394.

IKEDA, K., ACHIWA, K. (1997), Antibody-mediated regio- and enantioselective resolution of a glycerol derivative, *Bioorg. Med. Chem. Lett.* **7**, 225–228.

IKEDA, S., WEINHOUSE, M. I., JANDA, K. D., LERNER, R. A. (1991), Asymmetric induction via a catalytic antibody, *J. Am. Chem. Soc.* **113**, 7763–7764.

IVERSON, B. L., LERNER, R. A. (1989), Sequence-specific peptide cleavage catalyzed by an antibody, *Science* **243**, 1184–1189.

IVERSON, B. L., CAMERON, K. E., JAHANGIRI, G. K., PASTERNAK, D. S. (1990), Selective cleavage of trityl protecting groups catalyzed by an antibody, *J. Am. Chem. Soc.* **112**, 5320–5323.

IWABUCHI, Y., MIYASHITA, H., TANIMURA, R., KINOSHITA, K., KIKUCHI, M., FUJII, I. (1994), Regio- and stereoselective deprotection of acylated carbohydrates via catalytic antibodies, *J. Am. Chem. Soc.* **116**, 771–772.

IZADYAR, L., FRIBOULET, A., REMY, M. H., ROSETO, A., THOMAS, D. (1993), Monoclonal anti-idiotypic antibodies as functional internal images of enzyme active sites: Production of catalytic antibody with a cholinesterase activity, *Proc. Natl. Acad. Sci. USA* **90**, 8876–8880.

JACKSON, D. Y., SCHULTZ, P. G. (1991), An antibody-catalyzed *cis–trans* isomerization reaction, *J. Am. Chem. Soc.* **113**, 2319–2321.

JACKSON, D. Y., JACOBS, J. W., SUGASAWARA, R., REICH, S. H., BARTLETT, P. A., SCHULTZ, P. G. (1988), An antibody-catalyzed Claisen rearrangement, *J. Am. Chem. Soc.* **110**, 4841–4842.

JACKSON, D. Y., PRUDENT, J. R., BALDWIN, E. P., SCHULTZ, P. G. (1991), A mutagenesis study of a catalytic antibody, *Proc. Natl. Acad. Sci. USA* **88**, 58–62.

JACOBS, J. W. (1991), New perspectives on catalytic antibodies, *Bio. Technology* **9**, 258–262.

JACOBS, J., SCHULTZ, P. G., SUGASAWARA, R., POWELL, M. (1987), Catalytic antibodies, *J. Am. Chem. Soc.* **109**, 2174–2176.

JACOBSEN, J. R., SCHULTZ, P. G. (1994), Antibody catalysis of peptide bond formation, *Proc. Natl. Acad. Sci. USA* **91**, 5888–5892.

JACOBSEN, J. R., PRUDENT, J. R., KOCHERSPERGER, L., YONKOVICH, S., SCHULTZ, P. G. (1992), An efficient antibody-catalyzed aminoacylation reaction, *Science* **256**, 365–367.

JACOBSEN, J. R., COCHRAN, A. G., STEPHANS, J. C., KING, D. S., SCHULTZ, P. G. (1995), Mechanistic studies of antibody-catalyzed pyrimidine dimer photocleavage, *J. Am. Chem. Soc.* **117**, 5453–5461.

JANDA, K. D., LERNER, R. A., TRAMONTANO, A. (1988a), Antibody catalysis of bimolecular amide formation, *J. Am. Chem. Soc.* **110**, 4835–4837.

JANDA, K. D., SCHLOEDER, D., BENKOVIC, S. J., LERNER, R. A. (1988b), Induction of an antibody that catalyzes the hydrolysis of an amide bond, *Science* **241**, 1188–1191.

JANDA, K. D., BENKOVIC, S. J., LERNER, R. A. (1989), Catalytic antibodies with lipase activity and *R* or *S* substrate selectivity, *Science* **244**, 437–440.

JANDA, K. D., ASHLEY, J. A., JONES, T. M., MCLEOD, D. A., SCHLOEDER, D. M., WEINHOUSE, M. I. (1990a), Immobilized catalytic antibodies in aqueous and organic solvents, *J. Am. Chem. Soc.* **112**, 8886–8888.

JANDA, K. D., WEINHOUSE, M. I., SCHLOEDER, D. M., LERNER, R. A., BENKOVIC, S. J. (1990b), Bait and switch strategy for obtaining catalytic antibodies with acyl-transfer capabilities, *J. Am. Chem. Soc.* **112**, 1274–1275.

JANDA, K. D., BENKOVIC, S. J., MCLEOD, D. A., SCHLOEDER, D. M., LERNER, R. A. (1991a), Substrate attenuation: An approach to improve antibody catalysis, *Tetrahedron* **47**, 2503–2506.

JANDA, K. D., WEINHOUSE, M. I., DANON, T., PACELLI, K. A., SCHLOEDER, D. M. (1991b), Antibody bait and switch catalysis: A survey of antigens capable of inducing abzymes with acyl-transfer properties, *J. Am. Chem. Soc.* **113**, 5427–5434.

JANDA, K. D., SHEVLIN, C. G., LERNER, R. A. (1993), Antibody catalysis of a disfavored chemical transformation, *Science* **259**, 490–493.

JANDA, K. D., LO, C.-H. L., LI, T., BARBAS III, C. F., WIRSCHING, P., LERNER, R. A. (1994), Direct selection for a catalytic mechanism from combinatorial antibody libraries, *Proc. Natl. Acad. Sci. USA* **91**, 2532–2536.

JANDA, K. D., SHEVLIN, C. G., LERNER, R. A. (1995), Oxepane synthesis along a disfavored pathway: The rerouting of a chemical reaction using a catalytic antibody, *J. Am. Chem. Soc.* **117**, 2659–2660.

JANDA, K. D., LO, L.-C., LO, C.-H. L., SIM, M.-M., WANG, R. et al. (1997), Chemical selection for catalysis in combinatorial antibody libraries, *Science* **275**, 945–948.

JANJIC, N., TRAMONTANO, A. (1989), Antibody-catalyzed redox reaction, *J. Am. Chem. Soc.* **111**, 9109–9110.

JENCKS, W. P. (1969), Mechanisms for catalysis and carbonyl and acyl-group reactions, in: *Catalysis in Chemistry and Enzymology*, pp. 3–6 and 523–537. New York: McGraw Hill.

JENCKS, W. P. (1975), Binding energy, specificity and enzyme catalysis: the circe effect, *Adv. Enzymol. Relat. Areas Mol. Biol.* **43**, 219–410.

JOHNSON, W. S. (1968), Non-enzymic biogenetic-like

olefinic cyclisations, *Acc. Chem. Res.* **1**, 1–8.

KAST, P., ASIF-ULLAH, M., JIANG, N., HILVERT, D. (1996), Exploring the active site of chorismate mutase by combinatorial mutagenesis and selection: The importance of electrostatic catalysis, *Proc. Natl. Acad. Sci. USA* **93**, 5043–5048.

KATAYAMA, K., KOBAYASHI, T., OIKAWA, H., HOMA, M., ICHIHARA, A. (1998), Enzymatic activity and partial purification of solanapyrone synthase: first enzyme catalysing Diels–Alder reaction, *Biochim. Biophys. Acta* **1384**, 387–395.

KAWAMURA-KONISHI, Y., HOSOMI, N., NEYA, S., SUGANO, S., FUNASAKI, N., SUZUKI, H. (1996), Kinetic characterization of antibody-catalyzed insertion of a metal ion into porphyrin, *J. Biochem.* **119**, 857–862.

KEINAN, E., SINHA, S. C., SINHA-BAGCHI, A., BENORY, E., GHOZI, M. C. et al. (1990), Towards antibody-mediated metallo-porphyrin chemistry, *Pure Appl. Chem.* **62**, 2013–2019.

KEMP, D. S., COX, D. D., PAUL, K. G. (1975), The physical organic chemistry of benzisoxazoles. IV. The origins and catalytic nature of the solvent rate acceleration for the decarboxylation of 3-carboxybenzisoxazoles, *J. Am. Chem. Soc.* **97**, 7312–7318.

KHALAF, A. I., PROCTOR, G. R., SUCKLING, C. J., BENCE, L. H., IRVINE, J. I., STIMSON, W. H. (1992), Remarkably efficient hydrolysis of a 4-nitrophenyl ester by a catalytic antibody raised to an ammonium hapten, *J. Chem. Soc. Perkin Trans. I,* 1475–1481.

KHETTAL, B., DE LAUZON, S., DESFOSSES, B., USHIDA, S., MARQUET, A. (1994), A catalytic antibody isomerizing a Δ^5-3-ketosteroid, *C. R. Acad. Sci. Paris, Life Sciences* **317**, 381–385.

KIRBY, A. J. (1980), Effective molarities for intramolecular reactions, *Adv. Phys. Org. Chem.* **17**, 183–278.

KIRBY, A. J. (1996), The potential of catalyic antibodies, *Acta Chim. Scand.* **50**, 203–210.

KITAZUME, T., LIN, J. T., YAMAMOTO, T., YAMAZAKI, T. (1991a), Antibody-catalyzed double stereoselection in fluorinated materials, *J. Am. Chem. Soc.* **113**, 8573–8575.

KITAZUME, T., LIN, J. T., TAKEDA, M., YAMAZAKI, T. (1991b), Stereoselective synthesis of fluorinated materials by an antibody, *J. Am. Chem. Soc.* **113**, 2123–2126.

KITAZUME, T., TSUKAMOTO, T., YOSHIMURA, K. (1994), A catalytic antibody elicited by a hapten of tetrahedral carbon-type, *J. Chem. Soc. Chem. Commun.* 1355–1356.

KOCH, A., REYMOND, J.-L., LERNER, R. A. (1994), Antibody-catalyzed activation of unfunctionalized olefins for highly enantioselective asymmetric epoxidation, *J. Am. Chem. Soc.* **116**, 803–804.

KOCH, T., REYMOND, J.-L., LERNER, R. A. (1995), Antibody catalysis of multistep reactions: An aldol addition followed by a disfavored elimination, *J. Am. Chem. Soc.* **117**, 9383–9387.

KÖHLER, G., MILSTEIN, C. (1975), Continuous cultures of fused cells secreting antibody of predefined specificity, *Nature* **256**, 495–497.

KÖHLER, G., MILSTEIN, C. (1976), Derivation of specific antibody-producing tissue culture and tumour lines by cell fusion, *Eur. J. Immunol.* **6**, 511–519.

KÖHLER, G., HOWE, S. C., MILSTEIN, C. (1976), Fusion between immunoglobulin-secreting and nonsecreting myeloma cell lines, *Eur. J. Immunol.* **6**, 292–295.

LANDRY, D. W., ZHAO, K., YANG, G. X.-Q., GLICKMAN, M., GEORGIADIS, M. (1993), Antibody-catalyzed degradation of cocaine, *Science* **259**, 1899–1901.

LAVEY, B. J., JANDA, K. D. (1996a), Antibody catalyzed hydrolysis of a phosphotriester, *Bioorg. Med. Chem. Lett.* **6**, 1523–1524.

LAVEY, B. J., JANDA, K. D. (1996b), Catalytic antibody mediated hydrolysis of Paraoxon, *J. Org. Chem.* **61**, 7633–7636.

LEE, A. Y., STEWART, J. D., CLARDY, J., GANEM, B. (1995), New insight into the catalytic mechanism of chorismate mutases from structural studies, *Chem. Biol.* **2**, 195–203.

LERNER, R. A., BARBAS III, C. F. (1996), Using the process of reactive immunisation to induce catalytic antibodies with complex mechanisms: Aldolases, *Acta Chim. Scand.* **50**, 672–678.

LEWIS, C., KRÄMER, T., ROBINSON, S., HILVERT, D. (1991), Medium effects in antibody-catalyzed reactions, *Science* **253**, 1019–1022.

LI, T., JANDA, K. D. (1995), Synthesis of a bifunctional hapten designed to mimic both transition state and to induce "Bait and Switch" catalysis, *Bioorg. Med. Chem. Lett.* **5**, 2001–2004.

LI, T., JANDA, K. D., ASHLEY, J. A., LERNER, R. A. (1994), Antibody-catalyzed cationic cyclization, *Science* **264**, 1289–1293.

LI, L., PAUL, S., TYUTYULKOVA, S., KAZATCHKINE, M. D., KAVERI, S. (1995a), Catalytic activity of anti-thyroglobulin antibodies, *J. Immunol.* 3328–3332.

LI, T., HILTON, S., JANDA, K. D. (1995b), The potential application of catalytic antibodies to protecting group removal: Catalytic antibodies with broad substrate tolerance, *J. Am. Chem. Soc.* **117**, 2123–2127.

LI, T., HILTON, S., JANDA, K. D. (1995c), Remarkable ability of different antibody catalysts to control and diversify the product outcome of cationic cyclization reactions, *J. Am. Chem. Soc.* **117**, 3308–3309.

LI, T., JANDA, K. D., HILTON, S., LERNER, R. A.

(1995d), Antibody catalyzed nucleophilic substitution reaction at a primary carbon that appear to proceed by an ionisation mechanism, *J. Am. Chem. Soc.* **117**, 2367–2368.

LI, T., JANDA, K. D., LERNER, R. A. (1996), Cationic cyclopropanation by antibody catalysis, *Nature* **379**, 326–327.

LI, T., LERNER, R. A., JANDA, K. D. (1997), Antibody-catalyzed cationic reactions: Rerouting of chemical transformations via antibody catalysis, *Acc. Chem. Res.* **30**, 115–121.

LIENHARD, G. E. (1973), Enzymatic catalysis and transition-state theory, *Science* **180**, 149–154.

LIOTTA, L. J., BENKOVIC, P. A., MILLER, G. P., BENKOVIC, S. J. (1993), A catalytic antibody for imide hydrolysis featuring a bifunctional transition-state mimic, *J. Am. Chem. Soc.* **115**, 350–351.

LIOTTA, L. J., GIBBS, R. A., TAYLOR, S. D., BENKOVIC, P. A., BENKOVIC, S. J. (1995), Antibody-catalyzed rearrangement of a peptide bond: mechanistic and kinetic investigations, *J. Am. Chem. Soc.* **117**, 4729–4741.

LÖFÅS, S., JOHNSSON, B. (1990), A novel hydrogel matrix on gold surfaces in surface plasmon resonance sensors for fast and efficient covalent immobilisation of ligands, *J. Chem. Soc. Chem. Commun.* 1526–1528.

MADER, M. M., BARTLETT, P. A. (1997), Binding energy and catalysis: the implications for transition state analogs and catalytic antibodies, *Chem. Rev.* **97**, 1281–1301.

MARCH, J. (1992a), Determination of the presence of an intermediate, in: *Advanced Organic Chemistry: Reactions Mechanisms and Structure*, pp. 217–219. New York: John Wiley & Sons.

MARCH, J. (1992b), Addition to carbon–carbon multiple bonds, in: *Advanced Organic Chemistry: Reactions Mechanisms and Structure*, pp. 839–852. New York: John Wiley & Sons.

MARTIN, M. T., NAPPER, A. D., SCHULTZ, P. G., REES, A. R. (1991), Mechanistic studies of a tyrosine-dependent catalytic antibody, *Biochemistry* **30**, 9757–9761.

MARTIN, M. T., ANGELES, T. S., SUGASAWARA, R., AMAN, N. I., NAPPER, A. D. et al. (1994), Antibody-catalyzed hydrolysis of an unsubstituted amide, *J. Am. Chem. Soc.* **116**, 6508–6512.

MEEKEL, A. A. P., RESMINI, M., PANDIT, U. K. (1995), First example of an antibody-catalyzed hetero-Diels–Alder reaction, *J. Chem. Soc. Chem. Commun.* 571–572.

METS, B., WINGER, G., CABRERA, C., SEO, S., JAMDAR, S. et al. (1998), A catalytic antibody against cocaine prevents cocaine's reinforcing and toxic effects in rats, *Proc. Natl. Acad. Sci. USA* **95**, 10176.

MILLER, G. P., POSNER, B. A., BENKOVIC, S. J. (1997), Expanding the 43C9 class of catalytic antibodies using a chain-shuffling approach, *Bioorg. Med. Chem.* **5**, 581–590.

MIYASHITA, H., KARAKI, Y., KIKUCHI, M., FUJII, I. (1993), Prodrug activation via catalytic antibodies, *Proc. Natl. Acad. Sci. USA* **90**, 5337–5340.

NA, J., HOUK, K. N. (1996), Predicting antibody catalyst selectivity from optinum binding of catalytic groups to a hapten, *J. Am. Chem. Soc.* **118**, 9204–9205.

NAKATANI, T., HIRATAKE, J., SHINZAKI, A., UMESHITA, R., SUZUKI, T. et al. (1993), A mode of inhibition of an esterolytic antibody, *Tetrahedron Lett.* **34**, 4945–4948.

NAKAYAMA, G. R., SCHULTZ, P. G. (1992), Stereospecific antibody-catalyzed reduction of an α-keto amide, *J. Am. Chem. Soc.* **114**, 780–781.

NAPPER, A. D., BENKOVIC, S. J., TRAMONTANO, A., LERNER, R. A. (1987), A stereospecific cyclisation catalysed by an antibody, *Science* **237**, 1041–1043.

OHKUBO, K., URATA, Y., SERI, K., ISHIDA, H., SAGAWA, T. et al. (1993), The catalytic activities of IgG and IgM monoclonal antibodies for the hydrolysis of *p*-nitrophenyl acetate, *Chem. Lett.* **6**, 1075–1078.

OIKAWA, H., YAGI, K., WATANABE, K., HONMA, M., ICHIHARA, A. (1997), Biosynthesis of macrophomic acid: plausible involvement of intramolecular Diels–Alder reaction, *J. Chem. Soc., Chem. Commun.* 97–98.

OIKAWA, H., KOBAYASHI, T., KATAYAMA, K., SUZUKI, Y., ICHIHARA, A. (1998), Total synthesis of (–)-solanapyrone via enzymatic Diels–Alder reaction of prosolanapyrone, *J. Org. Chem.* **63**, 8748–8756.

OIKAWA, H., SUZUKI, Y., KATAYAMA, K., NAYA, A., SAKANO, C., ICHIHARA, A. (1999), Involvement of the Diels–Alder reaction in the biosynthesis of secondary natural products: the late stages of the biosynthesis of the phytotoxins, solanapyrones, *J. Chem. Soc., Perkin Trans. I*, 1225–1232.

PADLAN, E. A., COHEN, G. H., DAVIES, D. R. (1985), On the specificity of antibody 3-antigen interactions – phosphocholine binding to MCPC603 and the correlation of 3-dimensional structure and sequence data, *Annales de l'Institut Pasteur – Immunology* **C136**, 271.

PAGE, M. I., JENCKS, W. P. (1971), Entropic contributions to rate accelerations in enzymic and intramolecular reactions and the chelate effect, *Proc. Natl. Acad. Sci. USA* **68**, 1678–1683.

PATTEN, P. A., GRAY, N. S., YANG, P. L., MARKS, C. B., WEDEMAYER, G. J. et al. (1996), The immunological evolution of catalysis, *Science* **271**, 1086–1091.

PAUL, S., VOLLE, D. J., BEACH, C. M., JOHNSON, D. R., POWELL, M. J., MASSEY, R. J. (1989), Catalytic hydrolysis of vasoactive intestinal peptide by human autoantibody, *Science* **244**, 1158–1162.

PAUL, S., LI, L., KALAGA, R., WILKINS-STEVENS, P., STEVENS, F. J., SOLOMON, A. (1995), Natural catalytic antibodies: Peptide hydrolyzing activities of Bence Jones proteins and VL fragment, *J. Biol. Chem.* **270**, 15257–15261.

PAULING, L. (1947), Chemical achievement and hope for the future, *Silliman Lecture*, Yale University Press.

PAULING, L. (1948), Nature of forces between large molecules of biological interest, *Nature* **161**, 707–709.

PLAYFAIR, J. H. L. (1992), *Immunology at a Glance*. Oxford: Blackwell Scientific.

POLLACK, S. J., SCHULTZ, P. G. (1989), A semisynthetic catalytic antibody, *J. Am. Chem. Soc.* **111**, 1929–1931.

POLLACK, S. J., JACOBS, J. W., SCHULTZ, P. G. (1986), Selective chemical catalysis by an antibody, *Science* **234**, 1570–1573.

POLLACK, S. J., NAKAYAMA, G. R., SCHULTZ, P. G. (1988), Introduction of nucleophiles and spectroscopic probes into antibody combining sites, *Science* **242**, 1038–1040.

POLLACK, S. J., HSIUN, P., SCHULTZ, P. G. (1989), Stereospecific hydrolysis of alkyl esters by antibodies, *J. Am. Chem. Soc.* **111**, 5961–5962.

RAASCH, M. S. (1980), Annelations with tetrachlorothiophene 1,1-dioxide, *J. Org. Chem.* **45**, 856–867.

RADZICKA, A., WOLFENDEN, R. (1995), A proficient enzyme, *Science* **267**, 90–93.

RASO, V., STOLLAR, B. D. (1975), The antibody–enzyme analogy. Characterisation of antibodies to phosphopyridoxyltyrosine derivatives, *Biochemistry* **14**, 584–591.

REICH, H. J. (1979), Functional group manipulation using organoselenium reagents, *Acc. Chem. Res.* **12**, 22–30.

RESMINI, M., MEEKEL, A. A. P., PANDIT, U. K. (1996), Catalytic antibodies: Regio- and enantioselectivity in a hetero Diels–Alder reaction, *Pure Appl. Chem.* **68**, 2025–2028.

REYMOND, J.-L. (1995), (*S*)-Selective retroaldol reaction and the (*R*)-selective β-elimination with an aldolase catalytic antibody, *Angew. Chem.* (*Int. Edn. Engl.*) **34**, 2285–2287.

REYMOND, J.-L., CHEN, Y. (1995a), Catalytic, enantioselective aldol reaction using antibodies against a quaternary ammonium ion with a primary amine cofactor, *Tetrahedron Lett.* **36**, 2575– 2578.

REYMOND, J.-L., CHEN, Y. (1995b), Catalytic, enantioselective aldol reaction with an artificial aldolase assembled from a primary amine and an antibody, *J. Org. Chem.* **60**, 6970–6979.

REYMOND, J.-L., JANDA, K. D., LERNER, R. A. (1991), Antibody catalysis of glycosidic bond hydrolysis, *Angew. Chem.* (*Int. Edn. Engl.*) **30**, 1711–1713.

REYMOND, J., JANDA, K. D., LERNER, R. A. (1992), Highly enantioselective protonation catalyzed by an antibody, *J. Am. Chem. Soc.* **114**, 2257–2258.

REYMOND, J.-L., JAHANGIRI, G. K., STOUDT, C., LERNER, R. A. (1993), Antibody catalyzed hydrolysis of enol ethers, *J. Am. Chem. Soc.* **115**, 3909–3917.

REYMOND, J.-L., REBER, J.-L., LERNER, R. A. (1994), Enantioselective, multigram-scale synthesis with a catalytic antibody, *Angew. Chem.* (*Int. Edn. Engl.*) **33**, 475–477.

ROBERTS, V. A., STEWART, J., BENKOVIC, S. J., GETZOFF, E. D. (1994), Catalytic antibody model and mutagenesis implicate arginine in transition-state stabilization, *J. Mol. Biol.* **235**, 1098–1116.

ROSENBLUM, J. S., LO, L.-C., LI, T., JANDA, K. D., LERNER, R. A. (1995), Antibody-catalyzed phosphate triester hydrolysis, *Angew. Chem.* (*Int. Edn. Engl.*) **34**, 2275–2277.

SAKURAI, M., WIRSCHING, P., JANDA, K. D. (1996), Synthesis of a nucleoside hapten with a [P(O)–O–N] linkage to elicit catalytic antibodies with phosphodiesterase activity, *Bioorg. Med. Chem. Lett.* **6**, 1055–1060.

SAUER, J. (1966), Diels–Alder reactions: new preparative aspects, *Angew. Chem.* (*Int. Edn. Engl.*) **5**, 211–220.

SCANLAN, T. S., PRUDENT, J. R., SCHULTZ, P. G. (1991), Antibody-catalyzed hydrolysis of phosphate monoesters, *J. Am. Chem. Soc.* **113**, 9397–9398.

SCHULTZ, P. G. (1989), Catalytic antibodies, *Acc. Chem. Res.* **22**, 287–294.

SCHULTZ, P. G., LERNER, R. A. (1993), Antibody catalysis of difficult chemical transformations, *Acc. Chem. Res.* **26**, 391–395.

SERGEEVA, M. V., YOMTOVA, V., PARKINSON, A., OVERGAAUW, M., POMP, R. et al. (1996), Hapten design for antibody-catalyzed decarboxylation and ring-opening of benzisoxazoles, *Isr. J. Chem.* **36**, 177–183.

SHABAT, D., ITZHAKY, H., REYMOND, J.-L., KEINAN, E. (1995), Antibody catalysis of a reaction otherwise disfavoured in water, *Nature* **374**, 143–146.

SHEN, J.-Q. (1995), Catalytic antibodies for the hydrolysis of chloramphenicol, *Thesis,* Sheffield University.

SHEN, R., PRIEBE, C., PATEL, C., RUBO, L., SU, T. et al. (1992), An approach for the generation of secondary structure specific abzymes, *Tetrahedron Lett.* **33**, 3417–3420.

SHEVLIN, C. G., HILTON, S., JANDA, K. D. (1994), Automation of antibody catalysis: A practical methodology for the use of catalytic antibodies in organic synthesis, *Bioorg. Med. Chem. Lett.* **4**, 297–302.

SHOKAT, K. M., LEUMANN, C. J., SUGASAWARA, R., SCHULTZ, P. G. (1988), An antibody-mediated redox reaction, *Angew. Chem.* (*Int. Edn. Engl.*) **27**, 1172–1175.

SHOKAT, K. M., LEUMANN, C. J., SUGASAWARA, R., SCHULTZ, P. G. (1989), A new strategy for the generation of catalytic antibodies, *Nature* **338**, 269–271.

SHOKAT, K. M., KO, M. K., SCANLAN, T. S., KOCHERSPERGER, L., YONKOVICH, S. et al. (1990), Catalytic antibodies: A new class of transition-state analogues used to elicit hydrolytic antibodies, *Angew. Chem.* (*Int. Edn. Engl.*) **29**, 1296–1303.

SHOKAT, K., UNO, T., SCHULTZ, P. G. (1994), Mechanistic studies of an antibody-catalyzed elimination reaction, *J. Am. Chem. Soc.* **116**, 2261–2270.

SHUSTER, A. M., GOLOLOBOV, G. V., KVASHUK, O. A., BOGOMOLOVA, A. E., SMIRNOV, I. V., GABIBOV, A. G. (1992), DNA hydrolysing antibodies, *Science* **256**, 665–667.

SINHA, S. C., KEINAN, E., REYMOND, J.-L. (1993a), Antibody-catalyzed enantioselective epoxide hydrolysis, *J. Am. Chem. Soc.* **115**, 4893–4894.

SINHA, S. C., KEINAN, E., REYMOND, J.-L. (1993b), Antibody-catalyzed reversal of chemoselectivity, *Proc. Natl. Acad. Sci. USA* **90**, 11910–11913.

SMILEY, J. A., BENKOVIC, S. J. (1994), Selection of catalytic antibodies for a biosynthetic reaction from a combinatorial cDNA library by complementation of an auxotrophic *Escherichia coli:* Antibodies for orotate decarboxylation, *Proc. Natl. Acad. Sci. USA* **91**, 8319–8323.

SMITHRUD, D. B., BENKOVIC, P. A., BENKOVIC, S. J., TAYLOR, C. M., YAGER, K. M. et al. (1997), Investigations of an antibody ligase, *J. Am. Chem. Soc.* **119**, 278–282.

SPITZNAGEL, T. M., JACOBS, J. W., CLARK, D. S. (1993), Random and site-specific immobilisation of catalytic antibodies, *Enzyme Microb. Technol.* **15**, 916–921.

STAHL, M., GOLDIE, B., BOURNE, S. P., THOMAS, N. R. (1995), Catalytic antibodies generated by *in vitro* immunization, *J. Am. Chem. Soc.* **117**, 5164–5165.

STEPHENS, D. B., IVERSON, B. L. (1993), Catalytic polyclonal antibodies, *Biochem. Biophys. Res. Commun.* **192**, 1439–1444.

STEWART, J. D., BENKOVIC, S. J. (1995), Transition-state stabilization as a measure of the efficiency of antibody catalysis, *Nature* **375**, 388–391.

STEWART, J. D., ROBERTS, V. A., THOMAS, N. R., GETZOFF, E. D., BENKOVIC, S. J. (1994), Site-directed mutagenesis of a catalytic antibody: An arginine and a histidine residue play key roles, *Biochemistry* **33**, 1994–2003.

STORK, G., BURGSTAHLER, W. A. (1955), Stereochemistry of polyene cyclization, *J. Am. Chem. Soc.* **77**, 5068–5077.

SUCKLING, C. J., TEDFORD, C., BENCE, L. M., IRVINE, J. I., STIMSON, W. H. (1993), An antibody with dual catalytic activity, *J. Chem. Soc. Perkin Trans. I*, 1925–1929.

SUGA, H., EESOY, O., TSUMURAYA, T., LEE, J., SINSKEY, A. J., MASAMUNE, S. (1994a), Esterolytic antibodies induced to haptens with a 1,2-amino alcohol functionality, *J. Am. Chem. Soc.* **116**, 487–494.

SUGA, H., ERSOY, O., WILLIAMS, S. F., TSUMURAYA, T., MARGOLIES,, M. N. et al. (1994b), Catalytic antibodies generated via heterologous immunization, *J. Am. Chem. Soc.* **116**, 6025–6026.

SUMMERS, R. (1983), Catalytic principles of enzyme chemistry, *Thesis,* Harvard University.

SUTTON, P. A., BUCKINGHAM, D. A. (1987), Cobalt (III)-promoted hydrolysis of amino acid esters and peptides and the synthesis of small peptides, *Acc. Chem. Res.* **20**, 357–364.

TANAKA, F., KINOSHITA, K., TANIMURA, R., FUJII, I. (1996), Relaxing substrate specificity in antibody-catalyzed reactions: Enantioselective hydrolysis of *N*-Cbz-amino acid esters, *J. Am. Chem. Soc.* **118**, 2332–2339.

TARASOW, T. M., LEWIS, C., HILVERT, D. (1994), Investigation of medium effects in a family of decarboxylase antibodies, *J. Am. Chem. Soc.* **116**, 7959–7963.

TAWFIK, D. S., ZEMEL, R. R., ARAD-YELLIN, R., GREEN, B. S., ESHHAR, Z. (1990), A simple method of selecting catalytic monoclonal antibodies that exhibit turnover and specificity, *Biochemistry* **29**, 9916–9921.

TAWFIK, D. S., GREEN, B. S., CHAP, R., SELA, M., ESHHAR, Z. (1993), catELISA: A facile route to catalytic antibodies, *Proc. Natl. Acad. Sci. USA* **90**, 373–377.

TAWFIK, D. S., LINDNER, A. B., CHAP, R., ESHHAR, Z., GREEN, B. S. (1997), Efficient and selective *p*-nitrophenyl-ester-hydrolyzing antibodies elicited by a *p*-nitrobenzyl hapten, *Eur. J. Biochem.* **244**, 619–626.

TERAISHI, K., SAITO, M., FUJII, NAKAMURA, H. (1994), Design of the hapten for the induction of antibodies catalysing aldol reaction, *J. Mol. Graphics* **12**, 282–285.

THATCHER, G. R. J., KLUGER, R. (1989), Mechanism and catalysis of nucleophilic substitution in phosphate esters, *Adv. Phys. Org. Chem.* **25**, 99–265.

THOMAS, N. R. (1996), Catalytic antibodies – reaching adolescence?, *Natural Product Reports* **13**, 479–512.

THOMPSON, R. C. (1973), Use of peptide aldehydes to generate transition-state analogs of elastase, *Biochemistry* **12**, 47–51.

THORN, S. N., DANIELS, R. G., AUDITOR, M.-T. M., HILVERT, D. (1995), Large rate accelerations in antibody catalysis by strategic use of haptenic charge, *Nature* **373**, 228–230.

TRAMONTANO, A., JANDA, K. D., LERNER, R. A. (1986), Catalytic antibodies, *Science* **234**, 1566–1570.

TRAMONTANO, A., AMMANN, A. A., LERNER, R. A. (1988), Antibody catalysis approaching the activity of enzymes, *J Am. Chem. Soc.* **110**, 2282–2286.

TUBUL, A., BRUN, P., MICHEL, R., GHARIB, B., DE REGGI, M. (1994), Polyclonal antibody-catalyzed aldimine formation, *Tetrahedron Lett.* **35**, 5865–5868.

ULRICH, H. D., DRIGGERS, E. M. G., SCHULTZ, P. G. (1996), Antibody catalysis of pericyclic reactions, *Acta Chim. Scand.* **50**, 328–332.

UNO, T., SCHULTZ, P. G. (1992), An antibody-catalyzed dehydration reaction, *J. Am. Chem. Soc.* **114**, 6573–6574.

UNO, T., GONG, B., SCHULTZ, P. G. (1994), Stereoselective antibody-catalyzed oxime formation, *J. Am. Chem. Soc.* **116**, 1145–1146.

UNO, T., KU, J., PRUDENT, J. R., HUANG, A., SCHULTZ, P. G. (1996), An antibody-catalyzed isomeriziation reaction, *J. Am. Chem. Soc.* **118**, 3811–3817.

VAN VRANKEN, D. L., PANOMITROS, D., SCHULTZ, P. G. (1994), Catalysis of carbamate hydrolysis by an antibody, *Tetrahedron Lett.* **35**, 3873–3876.

WADE, W. S., ASHLEY, J. A., JAHANGIRI, G. T., MCELHANEY, G., JANDA, K. D., LERNER, R. A. (1993), A highly specific metal-activated catalytic antibody, *J. Am. Chem. Soc.* **115**, 4906–4907.

WAGNER, J., LERNER, R. A., BARBAS III, C. F. (1995), Efficient aldolase catalytic antibodies that use the enamine mechanism of the natural enzyme, *Science* **270**, 1797–1800.

WALLACE, M. B., IVERSON, B. L. (1996), The influence of hapten size and hydrophobicity on the catalytic activity of elicited polyclonal antibodies, *J. Am. Chem. Soc.* **118**, 251–252.

WEINER, D. P., WIEMANN, T., WOLFE, M. M., WENTWORTH JR., P., JANDA, K. D. (1997), A pentacoordinate oxorhenium(V) metallochelate elicits antibody catalysts for phosphodiester cleavage, *J. Am. Chem. Soc.* **119**, 4088–4089.

WENTWORTH JR., P., DATTA, A., BLAKEY, D., BOYLE, T., PARTRIDGE, L. J., BLACKBURN, G. M. (1996), Towards antibody directed abzyme prodrug therapy, ADAPT: Carbamate prodrug activation by a catalytic antibody and its *in vitro* application to human tumor cell-killing, *Proc. Natl. Acad. Sci. USA* **93**, 799–803.

WENTWORTH JR., P., DATTA, A., SMITH, S., MARSHALL, A., PARTRIDGE, L. J., BLACKBURN, G. M. (1997), Antibody catalysis of BAc2 aryl carbamate ester hydrolysis: A highly disfavored chemical process, *J. Am. Chem. Soc.* **119**, 2315–2316.

WENTWORTH JR., P., LIU, Y., WENTWORTH, D. A., FAN, P., FOLEY, M. J., JANDA, K. D. (1998), A bait and switch hapten strategy generates catalytic antibodies for phosphodiester hydrolysis, *Proc. Natl. Acad. Sci. USA* **95**, 5971.

WESTERICK, J. O., WOLFENDEN, R. (1972), Aldehydes as inhibitors of papain, *J. Biol. Chem.* **247**, 8195–8197.

WESTHEIMER, F. H. (1968), Pseudorotation in the hydrolysis of phosphate esters, *Acc. Chem. Res.* **1**, 70–78.

WILLIAMS, A., DOUGLAS, K. T. (1972a), E1cB mechanisms. Part II. Base hydrolysis of substituted phenyl phosphoramidates, *J. Chem. Soc. Perkin Trans. II*, 1455–1459.

WILLIAMS, A., DOUGLAS, K. T. (1972b), E1cB mechanisms. Part III. Effect of ionisation of the amido-NH on the reaction of hydroxide ion with methyl 3- and 4-benzamidobenzoates, *J. Chem. Soc. Perkin Trans. II*, 2112–2115.

WILLNER, I., BLONDER, R., DAGAN, A. (1994), Reversible optical-recording by a dinitrophenol antibody-catalyzed ring-opening of 6,8-dinitrospiropyran, *J. Am. Chem. Soc.* **116**, 3121–3122.

WINCHESTER, B., FLEET, G. W. (1992), Amino-sugar glycosidase inhibitors – versatile tools for glycobiologists, *Glycobiology* **2**, 199–210.

WINCHESTER, B., ALDAHER, S., CARPENTER, N. C., DIBELLO, I. C., CHOI, S. S. et al. (1993), The structural basis of the inhibition of human *alpha*-mannosidases by azafuranose analogs of mannose, *Biochemical J.* **290**, 742–749.

WIRSCHING, P., ASHLEY, J. A., BENKOVIC, S. J., JANDA, K. D., LERNER, R. A. (1991), An unexpectedly efficient catalytic antibody operating by ping-pong and induced fit mechanisms, *Science* **252**, 680–685.

WIRSCHING, P., ASHLEY, J. A., LO, C.-H. L., JANDA, K. D., LERNER, R. A. (1995), Reactive immunization, *Science* **270**, 1775–1782.

WOLFENDEN, R. (1969), Transition state analogues for enzyme catalysis, *Nature* **223**, 704–705.

XUE, Y. F., LIPSCOMB, W. N. (1995), Location of the active-site of allosteric chorismate mutase from *Saccharomyces cervisiae*, and comments on the catalytic and regulatory mechanisms, *Proc. Natl. Acad. Sci. USA* **92**, 10595–10598.

YANG, G., CHUN, J., ARAKAWA-URAMOTO, H., WANG, X., GAWINOWICZ, M. A. et al. (1996), Anti-cocaine catalytic antibodies: A synthetic approach to improved antibody diversity, *J. Am. Chem. Soc.* **118**, 5881–5890.

YLI-KAUHALUOMA, J. T., ASHLEY, J. A., LO, C.-H., TUCKER, L., WOLFE, M. M., JANDA, K. D. (1995), Anti-metallocene antibodies: A new approach to enantioselective catalysis of the Diels–Alder reaction, *J. Am. Chem. Soc.* **117**, 7041–7047.

YLI-KAUHALUOMA, J. T., ASHLEY, J. A., LO, C.-H. L., COAKLEY, J., WIRSCHING, P., JANDA, K. D. (1996), Catalytic antibodies with peptidyl-prolyl *cis-trans* isomerase activity, *J. Am. Chem. Soc.* **118**, 5496–5497.

YOON, S. S., OEI, Y., SWEET, E., SCHULTZ, P. G. (1996), An antibody-catalyzed [2,3]-elimination reaction,

J. Am. Chem. Soc. **118**, 11686–11687.
YU, J., HSIEH, L. C., KOCHERSPERGER, L., YONKOVICH, S., STEPHANS, J. C. et al. (1994), Progress towards an antibody glycosidase, *Angew. Chem. (Int. Edn. Engl.)* **33**, 339–341.
YU, J., CHOI, S. Y., MOON, K.-D., CHUNG, H.-H., YOUN, H. J. et al. (1998), A glycosidase antibody elicited against a chair-like transition state analog by *in vitro* immunization, *Proc. Natl. Acad. Sci. USA* **95**, 2880.
ZEWAIL, A. H. (1997), Femtochemistry, *Adv. Chem. Phys.* **101**, 892 (and other articles in this issue).
ZEWAIL, A. H., BERNSTEIN, R. B. (1988), Special Report – Real time laser femtochemistry – viewing the transition from reagents to products, *Chem. Eng. News* **66**, 24–43.
ZHOU, G. W., GUO, J., HUANG, W., FLETTERICK, R. J., SCANLAN, T. S. (1994), Crystal structure of a catalytic antibody with a serine protease active site, *Science* **265**, 1059–1064.
ZHOU, Z. S., JIANG, N., HILVERT, D. (1997), An antibody-catalyzed selenoxide elimination, *J. Am. Chem. Soc.* **119**, 3623–3624.

12 Synthetic Enzymes – Artificial Peptides in Stereoselective Synthesis

PAUL A. BENTLEY

New York, NY 10027, USA

1 Introduction

Low molecular weight peptides may have functioned as the earliest natural catalysts prior to the evolution of their sophisticated enzyme relatives (EHLER and ORGEL, 1976; FERRIS et al., 1996; LEE et al., 1996). Now they are being *revitalized* for use in the laboratory. Synthetic applications of polypeptides (EBRAHIM and WILLS, 1997), and polypeptides plus other polymers (PU, 1998) have been reviewed recently.

This chapter will review the applications of synthetic peptides to stereoselective synthesis, acquainting the reader with the history and latest developments, as well as speculating on the future directions of this intriguing area.

2 Cyclic Dipeptides

One of the greatest challenges for synthetic chemists is to design small molecules which have the catalytic selectivity of enzymes. However the complex and ordered architecture of an enzyme active site, which is seemingly a prerequisite for selectivity is not easily mimicked by small acyclic molecules. Consequently chemists have resorted to cyclic structures which are constrained to be highly ordered and which *prima facie* have properties which can be predicted with some certainty. This was INOUE's premise for studying cyclic dipeptides (diketopiperazines). After initial studies with very limited success (OKU et al., 1979, 1982), he was rewarded in 1981 when the dipeptide, *cyclo*-[(*S*)-Phe-(*S*)-His][1] (**3**) in benzene catalyzed the addition of hydrogen cyanide to benzaldehyde (**1a**) giving the (*R*)-cyanohydrin (**2a**, mandelonitrile) in 90% ee at 40% conversion (OKU and INOUE, 1981). This reaction was subsequently optimized to 97% ee in 97% yield by running the reaction in toluene at −20 °C with two equivalents of hydrogen cyanide (Fig. 1 and Tab. 1, Entry 5) (TANAKA et al., 1990). An important feature of this reaction is that the catalyst must be present in the form of a gel for enantioselectivity to be achieved.

Despite wide ranging studies, variation of the original ligand's structure (**3**) (Tab. 1) has achieved very limited success. Replacement of (*S*)-phenylalanine with (*S*)-leucine gives the opposite enantiomer of the cyanohydrin (*ent*-**2a**) (Tab. 1, Entry 4). If the phenyl ring of phenylalanine is changed to naphthalene or anthracene (Tab. 1, Entries 11 and 12) enantio-

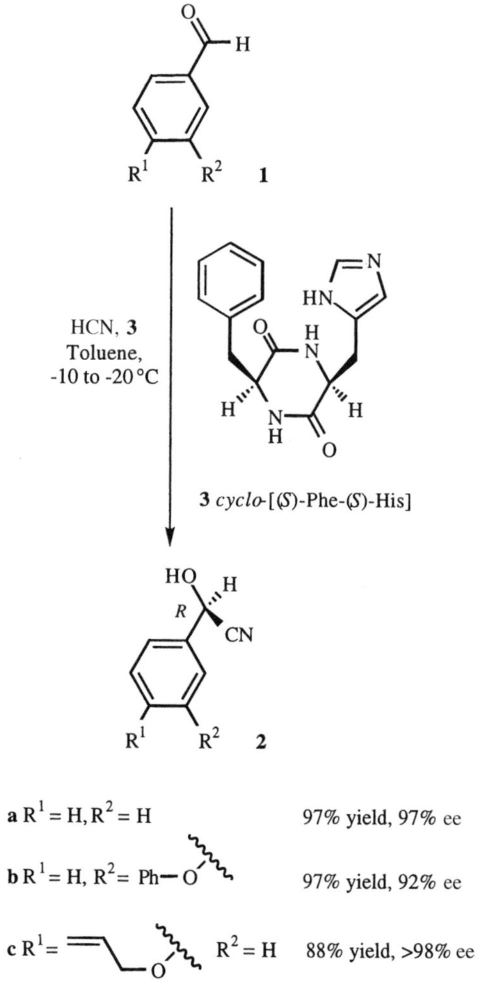

HCN, **3**
Toluene,
-10 to -20 °C

3 *cyclo*-[(*S*)-Phe-(*S*)-His]

a R^1 = H, R^2 = H — 97% yield, 97% ee

b R^1 = H, R^2 = Ph–O– — 97% yield, 92% ee

c R^1 = allyl–O–, R^2 = H — 88% yield, >98% ee

Fig. 1. Optimal substrate and conditions for cyclic dipeptide hydrocyanation.

[1] For 18 of the 20 amino acids involved in DNA encoded peptide biosynthesis the Cahn–Ingold–Prelog designation (*S*) is equivalent to the Fischer–Rosanoff designation L, which is the predominant naturally occurring enantiomer. The exceptions are glycine (achiral) and cysteine.

Tab. 1. Cyclic Dipeptides Used in Stereoselective Hydrocyanation of Benzaldehyde and Derivative **1b**

Entry	Catalyst	R¹	R²	Product	ee [%]	Y/C [%]	Time [h] (Reference)
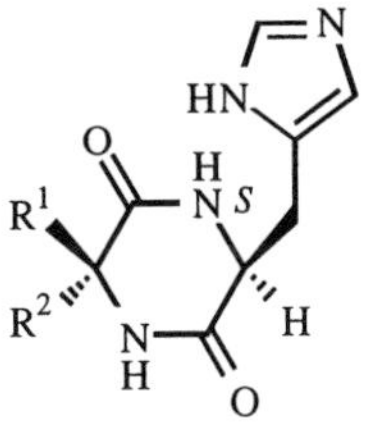	**4** *cyclo*-[X-(*S*)-His]						
1	**4** *cyclo*-[Gly-(*S*)-His]	H	H	(*R*)-**2a**	3.3	70[c]	44 (1)
2	**4** *cyclo*-[(*S*)-Ala-(*S*)-His]	Me	H	(*R*)-**2b**	26	21[y]	20 (2)
				(*R*)-**2a**	9.9	50[c]	47 (1)
3	**4** *cyclo*-[(*R*)-Ala-(*S*)-His]	H	Me	(*R*)-**2a**	7.5	90[c]	47 (1)
4	**4** *cyclo*-[(*S*)-Leu-(*S*)-His]	^{i}Bu	H	(*S*)-**2a**	55	85[y]	5 (3)
5	**4** *cyclo*-[(*S*)-Phe-(*S*)-His] ≡ **3**	Bn	H	(*R*)-**2a**	90	40[c]	0.5 (4)
				(*R*)-**2a**	10.1	90[c]	42 (1)
				(*R*)-**2a**	71	75[y]	17 (2)
				(*R*)-**2b**	97	97[c]	8 (5)
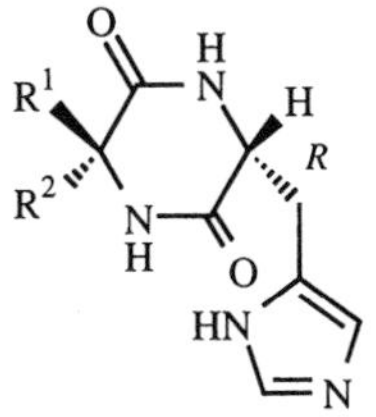	**5** *cyclo*-[X-(*R*)-His]						
6	**5** *cyclo*-[(*R*)-Phe-(*R*)-His]	H	Bn	(*S*)-**2a**	93	95[c]	8 (5)
				(*S*)-**2b**	88	93[y]	18.5 (2)
7	**4** *cyclo*-[(*S*)-Phe-(*R*)-His]	Bn	H	(*S*)-**2b**	11	75[y]	21 (2)
8	**5** *cyclo*-[(*S*)-PhGly-(*S*)-His]	Ph	H	**2b**	0	30[y]	20 (2)
9	**4** *cyclo*-[(*S*)-Tyr-(*S*)-His]	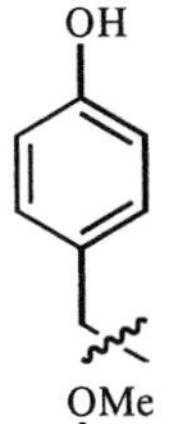 OH	H	(*R*)-**2b**	21	61[y]	24 (2)
10	**4** *cyclo*-[(*S*)-MeOTyr-(*S*)-His]	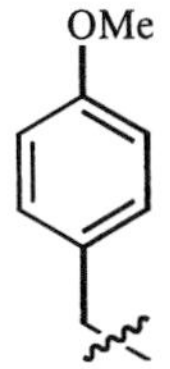 OMe	H	(*R*)-**2b**	28	74[y]	42 (2)

Tab. 1. Continued

Entry	Catalyst	R¹	R²	Product	ee [%]	Y/C [%]	Time [h] (Reference)
11	**4** *cyclo*-[(*S*)-1-naphthylAla-(*S*)-His]	R¹ = [1-naphthylmethyl]	R² = H	(*R*)-**2a**	0	<10[c]	– (6)
12	**4** *cyclo*-[(*S*)-1-anthracenylAla-(*S*)-His]	R¹ = [anthracenylmethyl]	R² = H	(*R*)-**2a**	0	<10[c]	– (6)
13	**4** *cyclo*-[(*S*)-His-(*S*)-His]	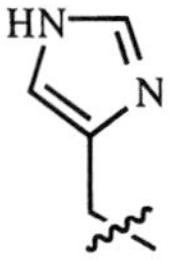 [imidazolylmethyl]	H	(*R*)-**2b** (*R*)-**2a**	2 2.5	90[y] 50[c]	42 (2) 23 (1)
14	**4** *cyclo*-[(*R*)-His-(*S*)-His]	H	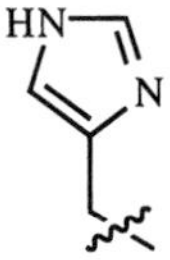 [imidazolylmethyl]	(*S*)-**2b**	10	50[y]	21(2)
15	**4** *cyclo*-[(*S*)-3-thienyl-Ala-(*S*)-His]	 [3-thienylmethyl]	H	**2a**	0	<20[c]	– (6)
16	**4** *cyclo*-[(*S*)-2-thienylAla-(*S*)-His]	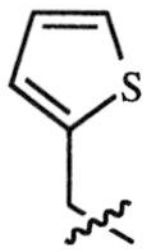 [2-thienylmethyl]	H	(*R*)-**2a**	72	40[c]	– (6)
17	**4** *cyclo*-[(*S*)-2-(5-Me-thienyl)Ala-(*S*)-His]	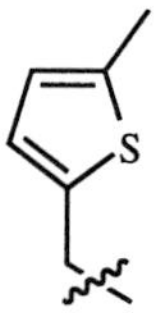 [5-methyl-2-thienylmethyl]	H	**2a**	0	<20[c]	– (6)
18	**4** *cyclo*-[(*S*)-2-(5-Br-thienyl)Ala-(*S*)-His]	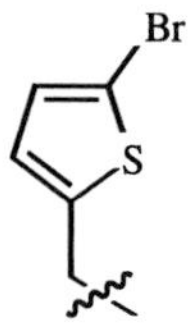 [5-bromo-2-thienylmethyl] (Br, S)	H	**2a**	0	<20[c]	– (6)

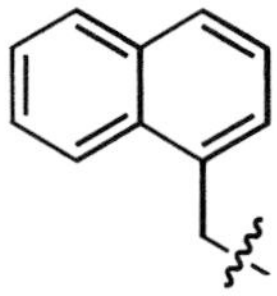

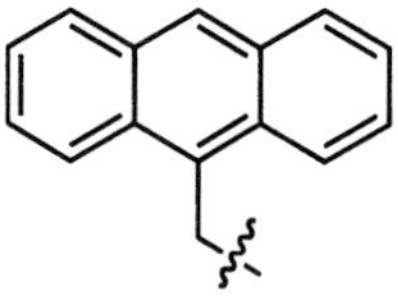

Tab. 1. Continued

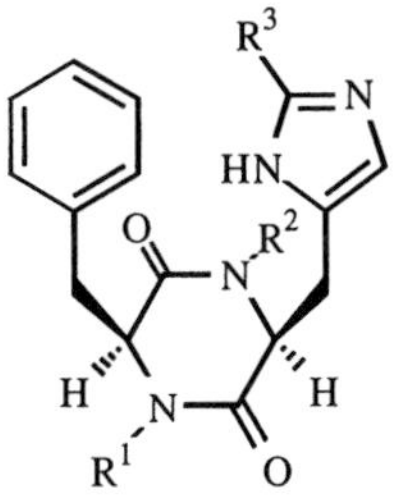

Entry	Catalyst	R^1	R^2	Product	ee [%]	Y/C [%]	Time [h] (Reference)
	6						
19	**6** *cyclo*-[(*S*)-(*N*-Me)-Phe-(*S*)-His]; R^3 = H	Me	H	**2a**	0	<10[c]	– (7)
20	**6** *cyclo*-[(*S*)-Phe-(*S*)-(*N*-Me)-His]; R^3 = H	H	Me	**2a**	0	<10[c]	– (7)
21	**6** *cyclo*-[(*S*)-Phe-(*S*)-(2-Me)-His]; R^3 = Me	H	H	(*R*)-**2a**	22	10[c]	– (7)
	7						
22	**7** *cyclo*-[(*S*)-(β,β-diMe)-Phe-(*S*)-His]	Me	Me	(*R*)-**2a**	60	20[c]	– (7)
23	**7** *cyclo*-[(*S*)-(β-Ph)-Phe-(*S*)-His]	Ph	H	(*R*)-**2a**	36	20[c]	– (7)
24	**4** *cyclo*-[(*S*)-(α-Me)-Phe-(*S*)-His]	Bn	Me	(*R*)-**2a** (*R*)-**2a**	15 99	50[c] 98[y]	– (7) – (8)
25	**4** *cyclo*-[(*R*)-(α-Me)-Phe-(*S*)-His]	Me	Bn	(*R*)-**2a**	32	90[y]	– (8)
26	**8** (5*R*)-5-(4-imidazolylmethyl-3-*N*-phenyl-2,4-imidazolidinedione			(*S*)-**2a**	37	90[c]	1 (9)
	8						

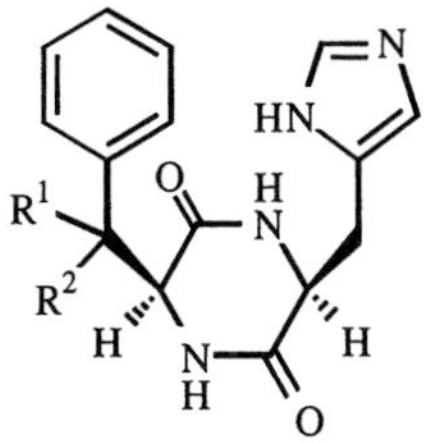

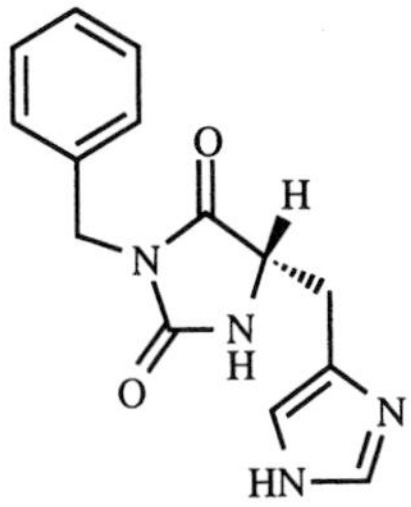

Y; [y] = yield; C, [c] = conversion. References: 1 Oku et al., 1982; 2 Jackson et al., 1988; 3 Mori et al., 1989; 4 Oku and Inoue, 1981; 5 Tanaka et al., 1990; 6 Noe et al., 1996; 7 Noe et al., 1997; 8 Hulst et al., 1997; 9 Danda, 1991.

selectivity is completely lost, possibly due to the disruption of supramolecular assembly. Substitution of the amidic protons believed to be involved in intermolecular hydrogen bonding (see Fig. 4) by a methyl group also leads to a loss of any stereoselectivity (Tab. 1, Entries 19 and 20). Incorporation of (*S*)-α-methylphenylalanine (Tab. 1, Entry 24) in place of (*S*)-phenylalanine (Tab. 1, Entry 5) is the only modification which retains high enantioselectivity. Replacement of the six membered ring, by a five membered ring (**8**), reduces enantioselectivity, however the nature of the enantiodiscrimination achieved [(*S*)-cyanohydrin (*ent*-**2a**), (*R*)-histidinyl residue] suggests that the same mode of discrimination is operating as in the six membered ring case (Tab. 1, Entries 6, 7, and 26).

The expansion of substrate range (MATTHEWS et al., 1988; KIM and JACKSON, 1994) (Tab. 2) has led to a system of great practical value (for a review see NORTH, 1993) which has been applied to the preparation of natural products, drugs and their analogs (Fig. 2) (BROWN et al., 1994). However several limitations have been revealed. Aromatic aldehydes bearing electron donating substituents undergo hydrocyanation with high enantioselectivity, but enantioselectivity is reduced if electron withdrawing substituents are present. The enantioselectivity of hydrocyanation of aliphatic aldehydes (which mostly react rapidly) is only low to moderate, and that of conjugated aldehydes (<31% ee) and ketones (<19% ee) is even lower.

The addition of cyanide to imines (**9**) is the enantiodiscriminating step in the Strecker synthesis of α-amino nitriles (**10**), which are precursors of α-amino acids (Fig. 3). Catalysis of this reaction by *cyclo*-[(*S*)-Phe-(*S*)-His] (**3**) was unsuccessful, which was attributed to failure of the imidazole group to accelerate proton transfer to the imine. Replacement of the histidine residue in the cyclic dipeptide with (*S*)-α-amino-γ-guanidinobutyric acid (which

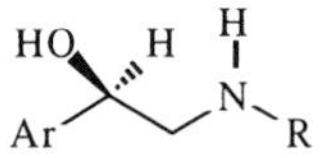

HO H H N Ar R

Ar = *p*-OMe-C_6H_4, R = PhCO; Tembamide
Ar = *p*-OMe-C_6H_4, R = (*E*) Ph-CH=CH-; Aegeline
Ar = *m*-OPh-C_6H_4, R = ^{t}Bu; Salbutamol derivative
Ar = *p*-OH-C_6H_4, R = CH_2Bz-*m*,*p*-OMe; Denopamine

Fig. 2. Natural products and drugs produced using cyclic dipeptide catalyzed hydrocyanation as the key step (BROWN et al., 1994).

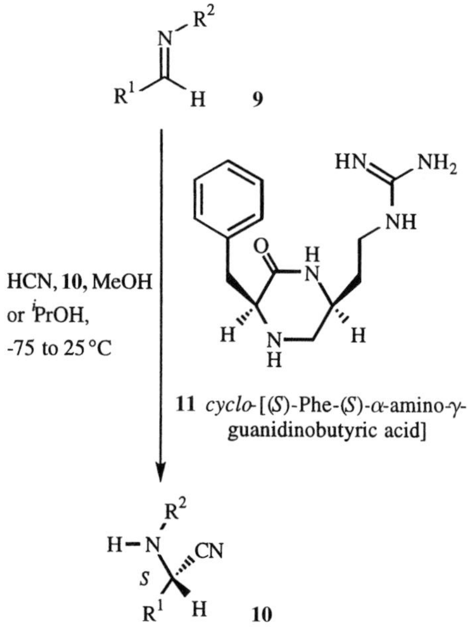

9

HCN, **10**, MeOH or ^{i}PrOH, -75 to 25 °C

11 *cyclo*-[(*S*)-Phe-(*S*)-α-amino-γ-guanidinobutyric acid]

10

$R^1 = R^2$ = Aromatic; 82-98% yield, 10->99% ee
R^1 = Aliphatic, R^2 = Aromatic; 88% yield, 75% ee
R^1 = Aromatic, R^2 = Aliphatic, 71-81% yield, 10-17% ee

Fig. 3. Cyclic dipeptide catalyzed enantioselective Strecker reaction (IYER et al., 1996).

Tab. 2. Substrate Range and Enantioselectivity for Asymmetric Hydrocyanation of Aldehydes Catalyzed by *cyclo*-[(*S*)-Phe-(*S*)-His)] (**3**)

Functionality (No. of Examples)	0–30% ee	31–60% ee	61–90% ee	91–100% ee
Aromatic aldehydes (43)	9%	28%	46%	16%
Aliphatic aldehydes (18)	50%	44%	5%	–

has a more basic side chain group) yielded an extremely efficient catalyst (**11**). The results with a range of substrates show some interesting parallels with hydrocyanation. Aromatic imines with electron donating substituents give α-amino nitriles with high enantiomeric excesses, whereas those bearing electron withdrawing substituents, heteroaromatic and aliphatic functionality give low enantiomeric excesses. Unlike hydrocyanation the Strecker reaction is conducted as a homogeneous reaction in methanol, moreover the cyclic dipeptide (**11**) gave the opposite enantiomer as the major product when compared with the equivalent hydrocyanation process, suggesting the involvement of different mechanistic factors (IYER et al., 1996).

Cyclic dipeptide hydrocyanation is one of the few examples of reactions which exhibit autoinduction (BOLM et al., 1996). In the hydrocyanation of *m*-phenoxy-benzaldehyde (**1b**, Fig. 1) catalyzed by *cyclo*-[(*R*)-Phe-(*R*)-His] (*ent*-**3**) a gradual increase in enantioselectivity for formation of the (*S*)-cyanohydrin (*ent*-**2b**) was observed as the reaction progressed. When a small amount of the enantiopure product of this reaction, the (*S*)-cyanohydrin (*ent*-**2b**), was included prior to addition of hydrogen cyanide, the optical purity of the product remained at circa 96% ee throughout the course of the reaction. The presence of an initial trace of (*R*)-cyanohydrin (**2b**) gave a slightly lower enantioselectivity, but a similar rate of increase in enantioselectivity for formation of the (*S*)-cyanohydrin (*ent*-**2b**), over the duration of the reaction when compared with no pre-addition (DANDA et al., 1991; cf. KOGUT et al., 1998). Moreover, pre-addition of achiral 2-cyano-propan-2-ol (acetone cyanohydrin) increases the enantioselectivity of furfural hydrocyanation from 53 to 73% ee. Pre-addition of (*S*)-1-phenyl-1-ethanol causes a similar enhancement of enantioselectivity (72% ee), but the (*R*)-enantiomer had only a modest effect (58% ee), which was identical to that of methanol (58% ee) (KOGUT et al., 1998). These phenomena are broadly consistent with a catalyst in which hydrogen bonding plays a major role and, as might be expected, an excess of the alcohol causes a drop in enantioselectivity, attributable to the disruption of hydrogen bonds.

As noted above, the dipeptide catalyst must be present as a gel in solution for enantioselective hydrocyanation to occur. In early experiments the gel tended to dissolve as the reaction proceeded with erosion of the enantioselectivity. Switching from benzene to toluene largely solved this problem, and initial gel formation is promoted by a range of prescriptions for pretreatment of the dipeptide to give amorphous non-crystalline material. This is achieved by rapid crystallization of the dipeptide from methanol alone or by addition of ether (TANAKA et al., 1990; DANDA et al., 1991), spray drying (DONG and PETTY, 1985), interaction with supercritical CO_2 (SHVO et al., 1996), or lyophilization (KOGUT et al., 1998). IR observations indicate that during this activation process intermolecular hydrogen bonds (Fig. 4) are broken, while intramolecular hydrogen bonds are formed (JACKSON et al., 1992). However, the importance of intermolecular hydrogen bonding for good enantiodiscrimination is clearly underscored by the poor results which have been obtained when the cyclic dipeptide has been immobilized (KIM and JACKSON, 1992). The importance of hydrogen bonding is also clear from the fact that methanol destroys chiral control in the reaction.

The mechanism by which cyclic dipeptides provide such high enantiodiscrimination has

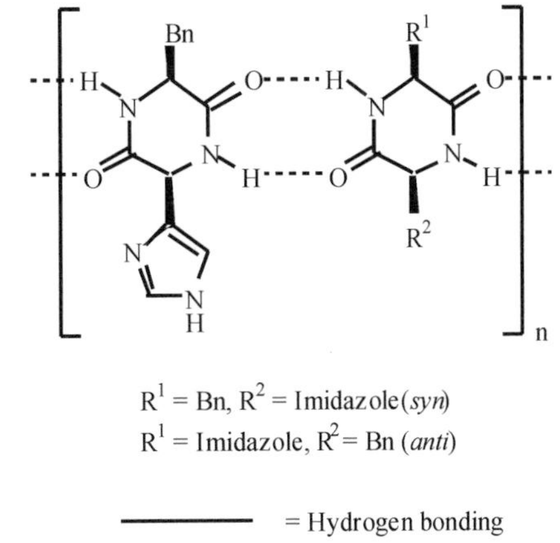

R^1 = Bn, R^2 = Imidazole(*syn*)
R^1 = Imidazole, R^2 = Bn (*anti*)

——— = Hydrogen bonding

Fig. 4. Intermolecular hydrogen bonding of cyclic dipeptides used for enantioselective hydrocyanation (SHVO et al., 1996; KOGUT et al., 1998).

14

---- = Hydrogen bonding

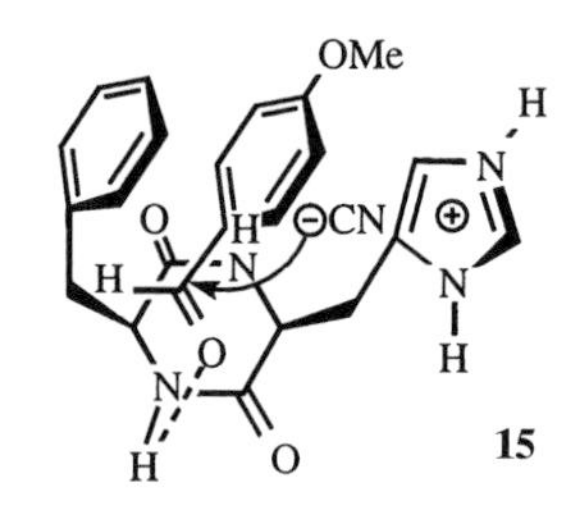

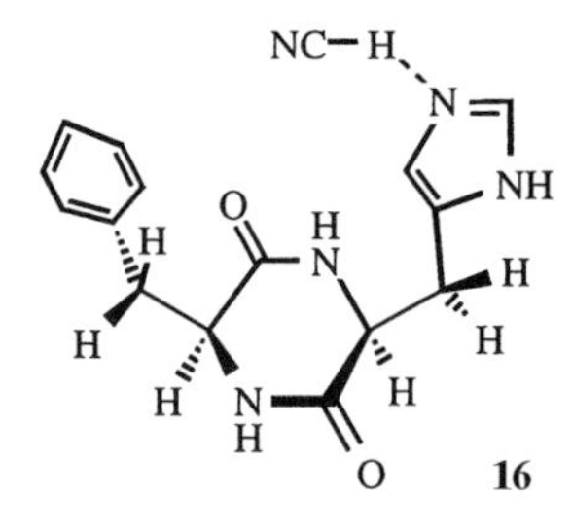

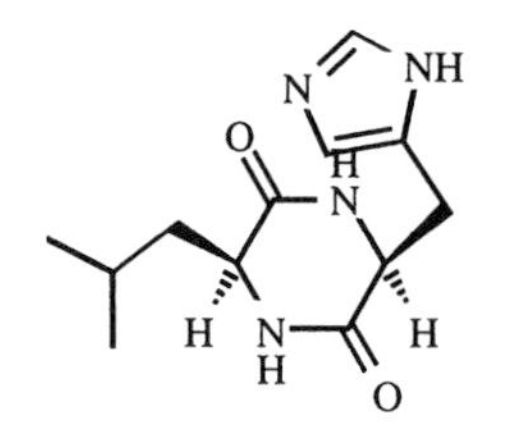

17 *cyclo*-[(*S*)-Leu-(*S*)-His]

Fig. 5. Imidazole/cyanide salt mechanism for cyclic dipeptide hydrocyanation and possible conformation of the cyclic dipeptides (Tanaka et al., 1990; Kim and Jackson, 1994).

been the subject of considerable investigation. If it is assumed that catalysis involves augmented direct addition of cyanide to the aldehyde (**1**, Fig. 1) by *cyclo*-[(*S*)-Phe-(*S*)-His] (**3**), then the (*R*)-cyanohydrins (e.g., **2**) must be formed by addition of cyanide to the *Si*-face of the carbonyl group. Localization of the hydrogen cyanide most plausibly occurs by coordination to the basic imidazole ring of histidine. The only issues remaining are then the position of the aldehyde and the means by which it is orientated and/or activated to undergo nucleophilic attack.

One proposal is that the transition state resembles (**14**) (Fig. 5), where the imidazole ring is rotationally restricted by hydrogen bonding to the carbonyl group of histidine (Kim and Jackson, 1994), while the aldehydic oxygen is hydrogen bonded to the amide group of histidine (Tanaka et al., 1990). π-Stacking may also exist between ligand and substrate, the overall effect being to create a U-shape molecule. Cyanide becomes associated with the imidazole ring by ionic interaction. An alternative arrangement is supported by NMR data and has the aldehydic oxygen hydrogen bonded to the phenylalanine amide group (**15**) (Hulst et al., 1997). NMR and molecular modeling studies suggest that the pre-transition state model (**16**) does not adopt a U-shape and that HCN forms a covalent bond to imidazole (Callant et al., 1992).

Interestingly *cyclo*-[(*S*)-Leu-(*S*)-His] (**17**) gives the *ent*-cyanohydrin (*ent*-**2**) when compared to the product from *cyclo*-[(*S*)-Phe-(*S*)-His] (**3**) (Hogg et al., 1994). *Cyclo*-[(*S*)-Leu-(*S*)-His] (**17**) adopts a conformation in solution in which the imidazole ring is folded over the diketopiperazine ring, whereas the predominant conformer for *cyclo*-[(*S*)-Phe-(*S*)-His] (**3**, Fig. 6) has the phenyl ring folded over the diketopiperazine ring (Hogg et al., 1994). This may account for the difference in the substrate profile of the two catalysts. *Cyclo*-[(*S*)-Leu-(*S*)-His] (**17**) performs consistently better for aliphatic aldehydes compared to aromatic aldehydes (Mori et al., 1989), whereas the converse is true for *cyclo*-[(*S*)-Phe-(*S*)-His] (**3**).

A radically different mechanism was proposed based on the observation that *cyclo*-[(*S*)-Phe-(*S*)-His] (**3**) in the presence of dioxygen catalyzes the oxidation of benzaldehyde to benzoic acid. In the absence of catalyst this reaction proceeds via an aldehyde hydrate, whereas the catalyst can accelerate this reac-

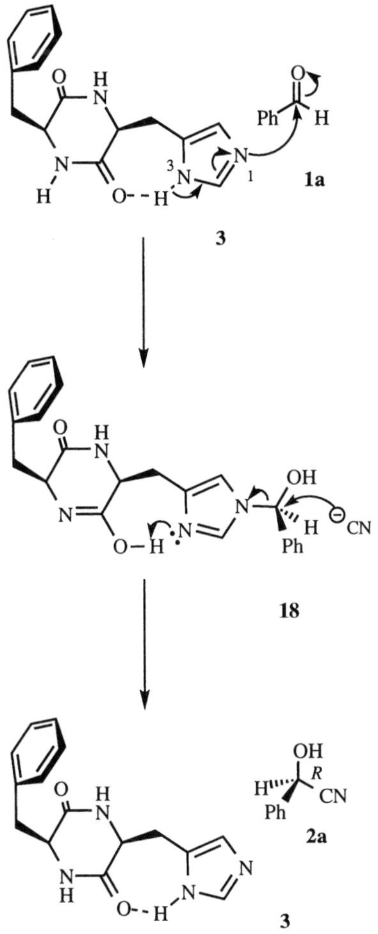

Fig. 6. Aminol mechanism for cyclic peptide catalyzed hydrocyanation.

tion by attack of benzaldehyde (**1a**) by the imidazole moiety to give an aminol intermediate (**18**). This could also serve as a substrate for nucleophilic displacement by cyanide to give the cyanohydrin (**2a**) (HOGG et al., 1993) (Fig. 6). The conformation of *cyclo*-[(*S*)-Phe-(*S*)-His] (**3**) has been determined in considerable detail by solution state NMR (HOGG and NORTH, 1993), molecular modeling (NORTH, 1992), and solid state NMR studies (APPERLEY et al., 1995). These studies indicate that a strong hydrogen bond exists between the hydrogen on *N*-3 of the imidazole ring and the amidic carbonyl group of the histidine residue. This of course could play a major role in enhancing the nucleophilicity of *N*-1 as required for the aminol mechanism. Whether this involves full proton transfer to give the enolic form of the amide (Fig. 6) or a protonated amide or some other pathway is conjecture at present.

Moreover hydrocyanation has second order kinetics with respect to the cyclic dipeptide suggesting two molecules of cyclic dipeptide are involved in the rate determining step. It is plausible that the aldehyde is bound to one imidazole group and the hydrogen cyanide to another imidazole group with the catalytic event occurring in the void between two molecules of dipeptide (SHVO et al., 1996). If this is the case, and catalysis involves multiple dipeptide units, it will be ironic, because the inital premise for using cyclic dipeptides was that their structural rigidity would mimic the hydrogen bonding arrays of long acyclic polypeptides. Have the catalysts evolved (by selection for enantioselectivity) towards structures which resemble acyclic polypeptides?

3 Poly-Leucine

The first breakthrough in the use of synthetic peptides in stereoselective synthesis was achieved by JULIÁ (JULIÁ et al., 1980; COLONNA et al., 1987). Using poly-L-alanine to catalyze the epoxidation of chalcone (**19**) under triphasic conditions (polypeptide/organic solvent/aqueous phase) the (2*R*,3*S*)-epoxide (**20**) was obtained in up to 93% ee (Fig. 7). By using poly-D-alanine the *ent* epoxide (**21**) was prepared with comparable enantioselectivity. In collaboration with COLONNA the reaction was found to be applicable to a fair range of α,β-unsaturated ketones (Tab. 3), gave optimal results in toluene and carbon tetrachloride and while the rate of the reaction was dependent on the substrate : catalyst ratio, the enantiomeric excess obtained was largely unaffected. These latter two features indicate enzyme-like catalysis and a low background uncatalyzed epoxidation rate (JULIÁ et al., 1982). The

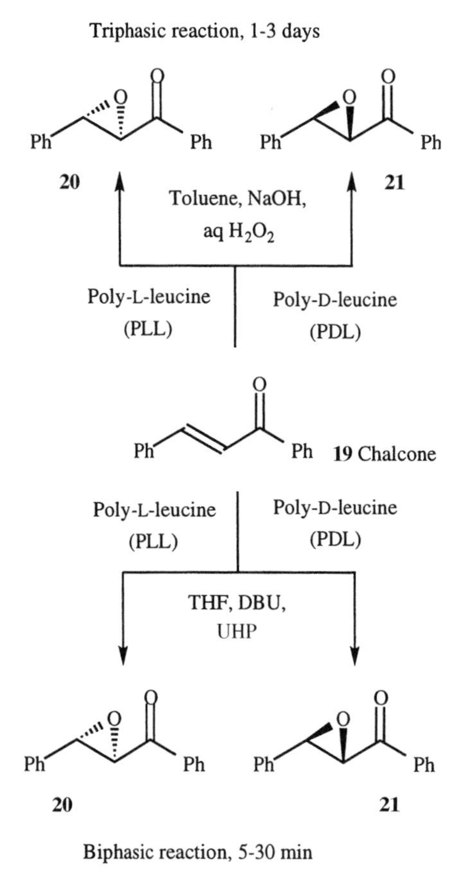

Fig. 7. Biphasic and triphasic poly-leucine stereoselective epoxidation.

so-called "poly-alanine" used in this work was prepared by polymerization of the *N*-carboxyanhydride of L-alanine initiated by *n*-butylamine, thus the polymer bears a *C*-terminal *N*-butyl amide group. However the catalytic activity is unaffected by the nature of this group and even if the *C*-terminus is used as a site for immobilization to polystyrene, enantioselective catalytic activity is retained (BANFI et al., 1984). Poly-L-alanine oligomers with an average of 5, 7, 10 and 30 residues were tested as catalysts, with only 10- and 30-mers making viable peptides for epoxidation, although the 30-mer was considered marginally optimal (JULIÁ et al., 1980, 1982). More than 15 oligopeptides were prepared and assayed for enantioselective epoxidation of chalcone (**18**), but the only examples with practical levels of enantioselection were poly-leucine and poly-isoleucine (COLONNA et al., 1983).

The major disadvantage of triphasic reactions is that the reactions are mostly slow (1–3 d) which results in partial degradation of the poly-alanine and partial loss of activity (JULIÁ et al., 1982, 1983; COLONNA et al., 1983, 1984). Despite the slow rate of the reaction it was exploited successfully in the synthesis of flavonoids (cf. **29**, Fig. 8) (BEZUIDENHOUDT et al., 1987; AUGUSTYN et al., 1990; VAN RENSBURG et al., 1996), although superior results were subsequently achieved with biphasic conditions (NEL et al., 1998).

The next developments in the use of polypeptides as epoxidation catalysts came about as a consequence of a large scale synthesis of

Tab. 3. The Epoxidation of "Chalcone-Type" Substrates (**19** and **19a–d**) Using Poly-L-Alanine Under Triphasic Conditions (JULIÁ et al., 1980, 1982, 1983; COLONNA et al., 1983, 1984)

R^1 / R^2 — **19** and **19a-d** → (Toluene, H_2O_2, aq NaOH / Poly-L-alanine, triphasic conditions) → R^1 / R^2 — **20** and **20a-d**

Entry	Substrate	R¹	R²	Time [h]	ee [%]	Yield [%]
1	**19**	Ph	Ph	37	93	75
2	**19a**	^{i}Pr	Ph	42	50	68
3	**19b**	Ph	^{i}Pr	52	60	49
4	**19c**	Ph	^{t}Bu	49	100	24
5	**19d**	Ph	Naphthyl	44	80	77

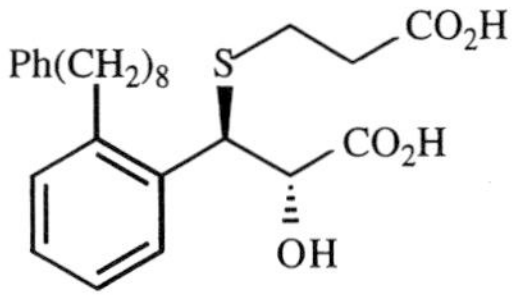

22 SK&F 104353

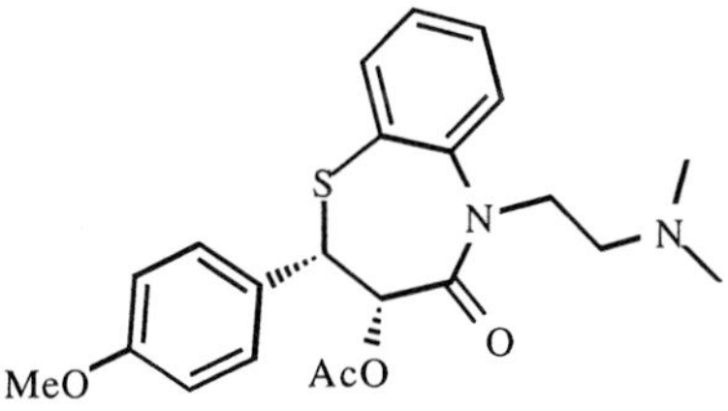

23 Diltiazem

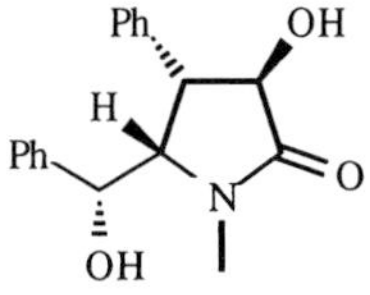

24 Clausenamide

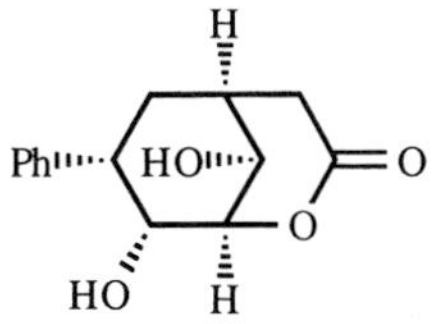

25 (+)-Goniopypyrone

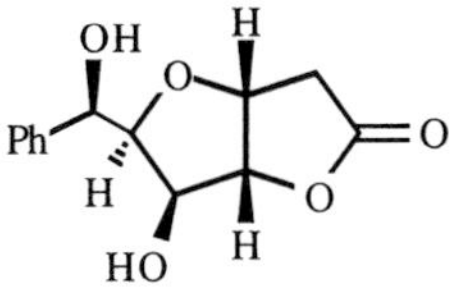

26 (+)-Goniofufurone

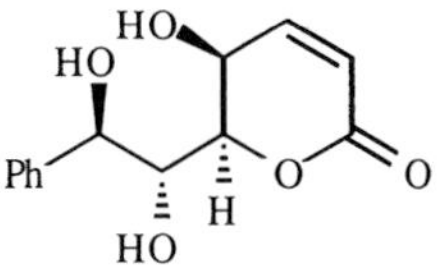

27 (+)-Goniotriol

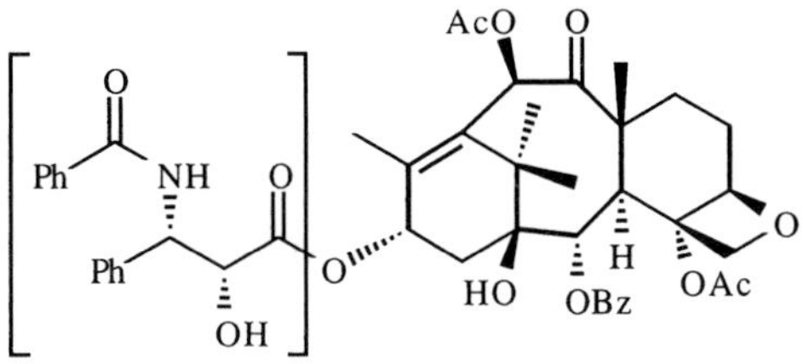

28 Taxol^{TM}[Side chain]

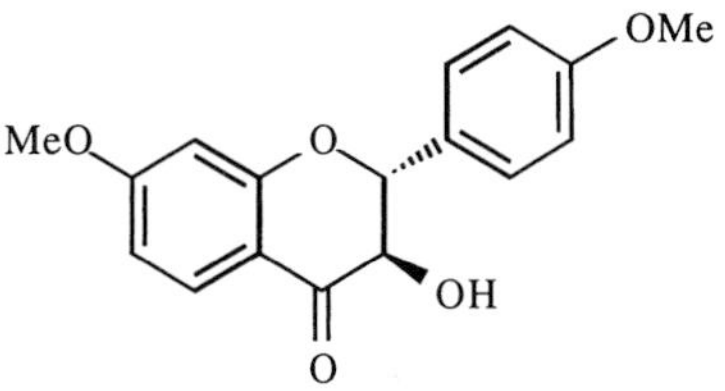

29 Dihydroflavonols

Fig. 8. Natural products and drugs prepared using poly-leucine catalyzed epoxidation.

SK&F 104353 (**22**, Fig. 8). A selection of commercial homo-polypeptides were investigated and poly-L-leucine was found to give the most consistent results. The catalyst was prepared by placing the *N*-carboxyanhydride of L-leucine into trays to a depth of 6–8 cm in a chamber at 70–75% relative humidity. Polymerization was initiated by water from the air. Preswelling the peptide for 6 h under the reaction conditions prior to the addition of the substrate and the use of hexane as solvent led to higher enantioselectivity in shorter times, with the additional bonus that the recovered catalyst retained full activity over 6 cycles (BAURES et al., 1990; FLISAK et al., 1993).

ITSUNO prepared immobilized poly-alanine and poly-leucine by initiating polymerization of the *N*-carboxyanhydrides with aminomethyl groups on the surface of cross-linked microporous polystyrene. This provided a more robust catalyst, affording a higher degree of recyclability (12 cycles) and reproducibility (ITSUNO et al., 1990). Surprisingly, even peptides with as few as four residues had substantial enantioselective catalytic activity (83% ee), whereas the corresponding non-immobilized peptides gave only poor enantiomeric excesses and yields.

In the last few years, the Roberts group at Exeter and latterly Liverpool has further expanded the scope of poly-leucine catalyzed epoxidation (for a review see LASTERRA-SÁNCHEZ and ROBERTS, 1997). Initially the triphasic conditions were extended to an increased range of (*E*)-α,β-unsaturated ketones (Tab. 4) (LASTERRA-SÁNCHEZ and ROBERTS, 1995; LASTERRA-SÁNCHEZ et al., 1996; KROUTIL et al., 1996a,b) and subsequently to (*Z*)-disubstituted, digeminally substituted, and trisubstituted α,β-unsaturated ketones (Tab. 5) (RAY and ROBERTS, 1999; BENTLEY, 1999).

Phase transfer reagents were added to increase the rate of the reaction, without loss of enantioselectivity, suggesting the phase transfer properties of poly-leucine were not implicit in its enantioselectivity. These conditions were used in the epoxidation of phenyl-β-styryl sulfone, but only achieved 21% ee. This disappointing result can be rationalized by increased steric hindrance of the sulfone oxygens or weaker hydrogen bonds possible through the sulfone oxygens (compared with the ketone) to the amide of the peptide backbone. The substrate would be less tightly bound into the chiral environment enhancing the role of the background reaction. This could indicate key hydrogen bonding of the substrate to poly-leucine via the ketone in enones (BENTLEY, 1999). Alternative oxidation systems to hydrogen peroxide/sodium hydroxide were found including sodium percarbonate (LASTERRA-SÁNCHEZ et al., 1996) and sodium perborate/sodium hydroxide (KROUTIL et al., 1996a; SAVIZKY et al., 1998). These methods have some advantages, but the reaction times are still prolonged (>24 h). The discovery of biphasic conditions, in which the urea–hydrogen peroxide complex (UHP) (HEANEY, 1993) is used with 1,8-diazobicyclo[5.4.0]undec-7-ene (DBU) (YADAV and KAPOOR, 1995) as the base in THF led to a dramatic decrease in reaction time (5–30 min; BENTLEY et al., 1997a). The biphasic conditions were successfully applied in the synthesis of diltiazem (**23**) (ADGER et al., 1997), clausenamide (**24**) (CAPPI et al., 1998), (+)-goniopypyrone (**25**), (+)-goniofufurone (**26**), and (+)-goniotriol (**27**) (CHEN and ROBERTS, 1999), Taxol™ side chain (**28**) (ADGER et al., 1997), and dihydroflavonols (**29**) (NEL et al., 1998) (Fig. 8).

The standard THF/DBU/UHP biphasic reaction protocol has been significantly improved in terms of enantioselectivity, with limited loss of reaction rate. This has been achieved by dilution of the reagents and increasing the amount of poly-leucine relative to the substrate, while changing the solvent to *iso*-propyl acetate (Tab. 5). These conditions were used to expand the substrate range to encompass trisubstituted and digeminally substituted enones (**32**) (BENTLEY and ROBERTS, unpublished data). Substrate (**32**) offered an interesting contrast to prior results (e.g., **30f**); these "tethered chalcones" (**30** and **32**) could

Tab. 4. The Epoxidation of Structurally Varied Substrates (**19e–j**) Using Poly-L-Leucine Under Triphasic Conditions (LASTERRA-SÁNCHEZ and ROBERTS, 1995; LASTERRA-SÁNCHEZ et al., 1996; KROUTIL et al., 1996a,b)

19e-j → **20e-j** (Toluene, H_2O_2, aq NaOH; Poly-L-leucine, triphasic conditions)

Entry	Substrate	R^1	R^2	Time [h]	ee [%]	Yield [%]
1	**19e**	Cyclopropyl	Ph	18	77	85
2	**19f**	Ph-C≡C	Ph	96	90	57
3	**19g**	Ph	$(SCH_2)_2C=CH$	115	>94	51
4	**19h**	PhCO	Ph	–	76	76
5	**19i**	PhCH=CH	2-Naphthyl	72	>96	78
6	**19j**	Ph	PhCH=CH	–	>98	–

Tab. 5. The Epoxidation of Digeminal and Trisubstituted Alkenes (**30a–h**, **32**) Catalyzed by Immobilized Poly-L-Leucine Under Biphasic Conditions. **Protocol:** Substrate (**30a–h**, **32**) (0.24 mmol), ^{i}PrOAc (1.6 mL), activated immobilized PLL (200 mg), urea–hydrogen peroxide complex (UHP, 11 mg) and 1,8-diazobicyclo[5.4.0]undec-7-ene (DBU, 30 mg). UHP and DBU were added every 12 h until reaction was complete (Bentley, 1999)

30a-h → **31a-h**; **32** → **33**

^{i}PrOAc, UHP, DBU

Immobilized poly-L-leucine, biphasic conditions

Substrate	R	n	Time [h]	ee [%]	Yield [%]
30a	Ph	2	90	84.5	76
30b	p-Br-C_6H_4	2	72	82.5	81
30c	p-NO_2-C_6H_4	2	78	96.5	85
30d	Me	2	60	92.0	66
30e	^{t}Bu	2	192	83.0	63
30f	Ph	1	48	88.0	72
30g	Ph	3	168	59.0	74
30h[a]	H	2	7	94.0	64
32	–	–	72	0.0	67

[a] Reagent additions made after 0.5, 1, 3 and 6 h

give an indication as to the orientation that the (*E*)-disubstituted α,β-unsaturated ketone adopts so as to be processed by poly-leucine.

Although initial studies found the range of bases to be of utility with UHP very limited, further work demonstrated potassium carbonate to provide comparable enantioselectivity to DBU, ultimately leading to sodium percarbonate's (Bentley, 1999) replacement of UHP and DBU. Development of these conditions has provided distinct improvements (Ray and Roberts, 1999; Allen et al., 1999) in the reaction rate of this economically significant alternative biphasic reaction.

Poly-leucine is prepared from leucine *N*-carboxyanhydride (NCA) (**34**) (Fig. 9) (Kricheldorf, 1987), which in turn is prepared from leucine and triphosgene (Daly and Poché, 1988). There is a strong correlation between the purity of NCA as assessed by melting point (should be within 2 °C of literature values; LeuNCA: 76–78 °C) and the quality of poly-leucine as a catalyst. Good quality NCA is obtained by controlling the temperature of the reaction and limiting the exposure of NCA to moisture at all stages of the process. An extra step of stirring poly-leucine in aqueous NaOH and toluene followed by washing with selected solvents is required to provide superior material (**35**) for biphasic epoxidation reactions, when compared to unactivated poly-leucine (Allen et al., 1998; Bentley et al., 1998; Nugent, unpublished data).

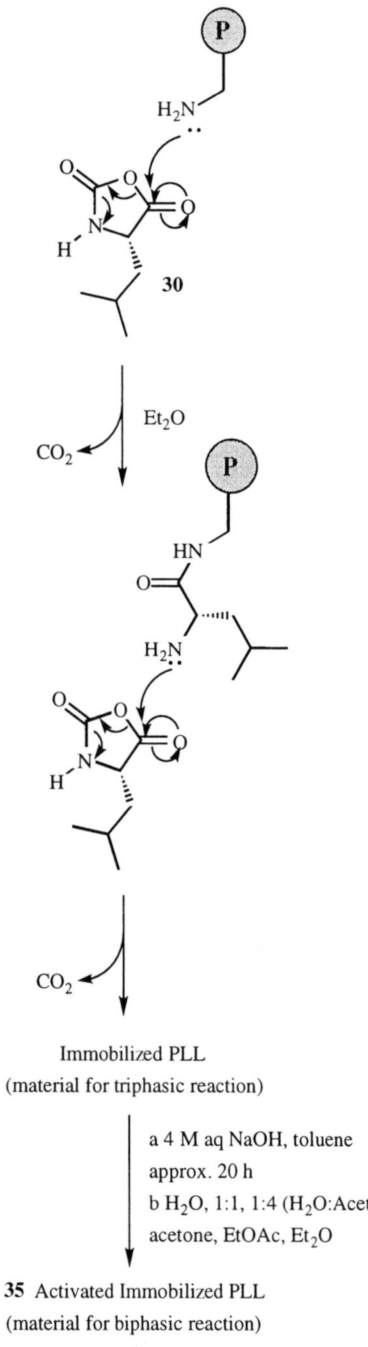

30

Et_2O

CO_2

CO_2

Immobilized PLL
(material for triphasic reaction)

a 4 M aq NaOH, toluene
approx. 20 h
b H_2O, 1:1, 1:4 (H_2O:Acetone),
acetone, EtOAc, Et_2O

35 Activated Immobilized PLL
(material for biphasic reaction)

Fig. 9. Preparation of poly-L-leucine. P indicates polymer backbone.

The means by which poly-leucine achieves its impressive chiral discrimination has recently begun to be addressed by the preparation of polypeptides of precisely defined lengths, containing both D- and L-residues attached via a spacer to a polystyrene resin. The homo-decapeptide of L-leucine (P1, Fig. 10) catalyzes the epoxidation of chalcone to give the expected epoxide (**20**) with a moderate enantiomeric excess, whereas the 5D/5L diastereoisomer (P2) gives essentially racemic epoxide (**20**). The 20-mer homologs (P3–P8) behave differently. The 7D/13L (P4), and 5D/15L (P5) diastereoisomers give predominantly the enantiomeric epoxide (**21**), apparently under control by the D-leucine residues in the *N*-terminal region (BENTLEY et al., 1998). It is clear from the 1L/5D/14L diastereoisomer (P6) giving predominantly the epoxide (**21**), under control by the D-leucine residues, 2L/5D/13L (P7) giving virtually racemic epoxide and good selectivity for the epoxide (**20**) being restored with the 3L/5D/12L diastereoisomer (P8) that the terminal trimer is the most important determinant of enantioselectivity (BENTLEY, unpublished data). Moreover, the terminal amino group can be replaced by *N*,*N*-dimethylamino- (BENTLEY et al., 1998) or other substituents (-JULIÁ et al., 1982) without a substantial effect on enantioselectivity. Cleavage of the peptide from the polymer does not change the stereocontrol of the *N*-terminal region, eliminating its comparatively less hindered environment as an explanation for these observation.

The *C*-terminal region also plays a significant role in determining enantioselectivity, potentially providing the correct scaffold allowing the *N*-terminal region to be presented in an effective orientation. An immobilized 18-mer of L-leucine is a highly enantioselective epoxidation catalyst (Fig. 11, P1), whereas the corresponding *tert*-leucine analog (P4) is non-enantioselective. As might be anticipated, polypeptides with 9 *tert*-leucine or a mixture of D,L-leucine residues in the *C*-terminal region (P2 and P5) and 9 L-leucine residues at the *N*-terminus are also selective catalysts, but interestingly the 9′L/9L catalyst (P3) is also fairly enantioselective (BENTLEY et al., 1997b).

In summary, the length of poly-leucine must be at least 10 residues for good enantioselectivity, with only 18–20 residues providing opti-

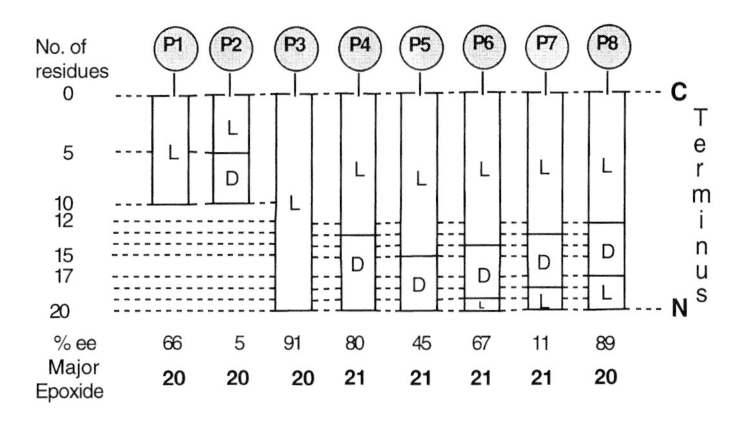

Fig. 10. The determination of the role of amino acids at various positions in poly-leucine using D- and L-leucine (shown as D and L, respectively).

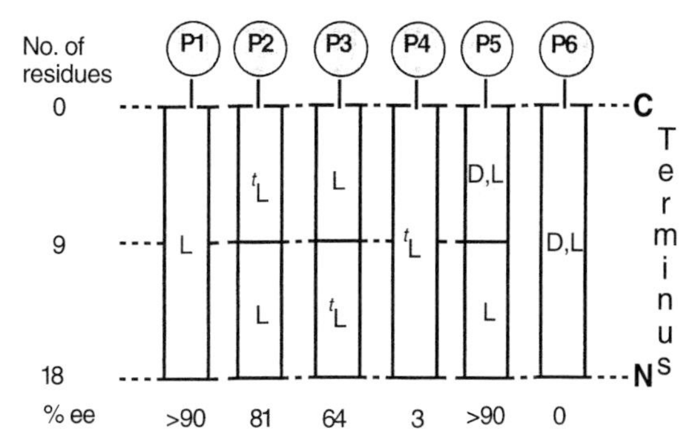

Fig. 11. The determination of the role of amino acids in various regions of poly-leucine, using L-leucine, racemic leucine, and L-^{t}leucine [shown as L, (D,L), and ^{t}L, respectively].

mum stereoselectivity and a higher rate (FLOOD, unpublished data). Shorter polypeptides are more enantioselective if immobilized. Three to five residues at the *N*-terminus are the most important determinant of enantioselectivity, but this can be modified profoundly by five to fifteen *C*-terminal residues, although the stereochemistry of these residues does not seem to be important. The terminal amino group does not play a key role in determining enantioselectivity.

Other examples of reactions for which polypeptides can be used as enantioselective catalysts are rather sparse but they do give important directions for expansion of poly-leucine technology. It has been shown that palladium binds to the peptide backbone (FRANCIS et al., 1996). Poly-L-leucine has been used in an enantioselective palladium catalyzed carbonylation to give the (*R*)-lactone (**37**) (Fig. 12) and gave better chiral induction than a range of chiral alcohols (e.g., menthol, 1,1-bi-2-naphthol, and diethyl tartrate) and phosphines (BINAP and CHIRAPHOS). Poly-D-alanine gave the (*S*)-lactone (*ent*-**37**), but in only 8% enantiomeric excess. It has been speculated that the reaction goes via the complex (**38**), although no explanation was proffered for invoking the unusual protonated amide (ALPER and HAMEL, 1990).

3.1 Other Applications of Oligopeptides

Stereoselective electrochemical reduction and oxidation reactions have been carried out using electrodes coated with a range of homopolypeptides. Citraconic acid (**39**) (ABE et al., 1983; NONAKA et al., 1983; ABE and NONAKA, 1983) and 4-methylcoumarin (**41**) (ABE et al., 1983) were reduced on poly-L-valine coated graphite electrodes with moderate success (Fig. 13). Amazingly, the authors reported that both poly-D- and poly-L-valine coated elec-

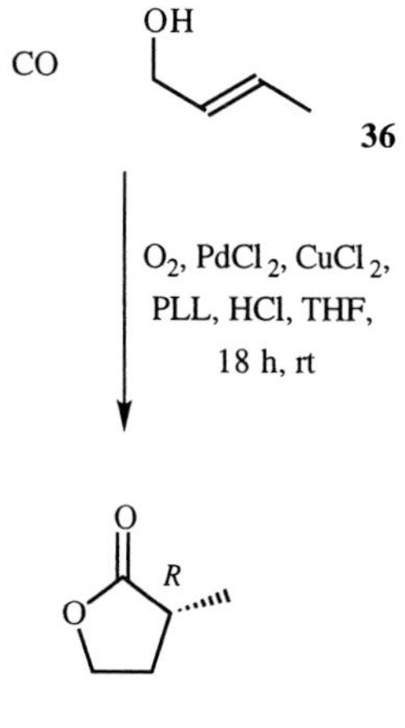

CO **36**

O_2, $PdCl_2$, $CuCl_2$, PLL, HCl, THF, 18 h, rt

37 49% yield, 61% ee

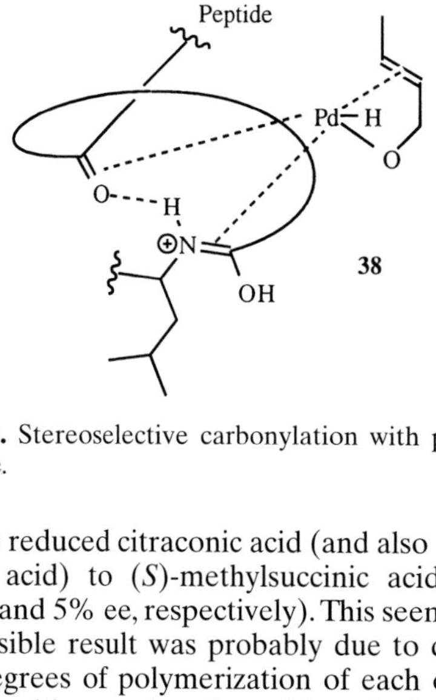

Peptide

38

Fig. 12. Stereoselective carbonylation with poly-L-leucine.

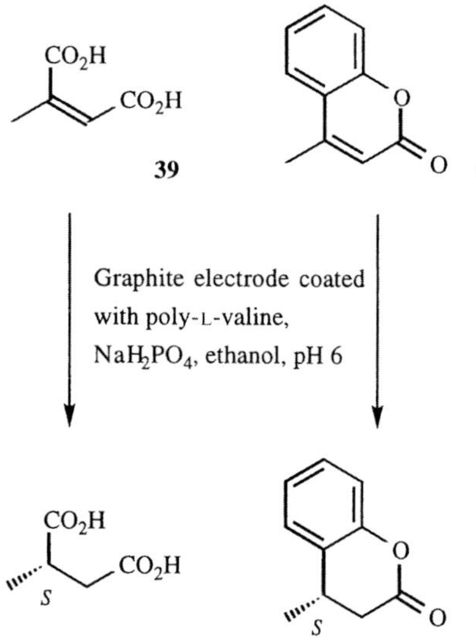

39 **41**

Graphite electrode coated with poly-L-valine, NaH_2PO_4, ethanol, pH 6

40 25% ee **42** 43% ee

Fig. 13. Stereoselective reduction with poly-L-valine.

Tab. 6. Electrochemical Enantioselective Oxidation of Sulfides to Sulfoxides with Poly-L-Valine Coated Electrodes (KOMORI and NONAKA, 1983, 1984)

43a-f → Pt electrode coated with poly-pyrrole and poly-L-valine, ^{n}Bu$_4$BF$_4$, aq MeCN → **44a-f**

Entry	Substrate	R	ee [%]	Yield [%]
1	**43a**	Me	1	–
2	**43b**	^{n}Bu	18	–
3	**43c**	^{i}Bu	44	69
4	**43d**	^{i}Pr	77	56
5	**43e**	^{t}Bu	93	45
6	**43f**	^{c}C$_6$H$_{11}$	54	–

trodes reduced citraconic acid (and also mesaconic acid) to (*S*)-methylsuccinic acid (**40**) (25% and 5% ee, respectively). This seemingly impossible result was probably due to different degrees of polymerization of each of the polypeptide samples, since although they had opposite optical rotations these were not identical in magnitude. More fruitful studies involved the oxidation of phenyl sulfides (**43**) to the corresponding sulfoxides (**44**). Partially reusable platinum electrodes were prepared by covalently coating with poly-pyrrole and then dipping in a solution of poly-L-valine. Although the enantioselectivity of the electrodes diminished with each cycle of use, the activity could be almost wholly restored by redipping in the solution of poly-L-valine. In some cases good enantioselectivity was observed (**43e**) (Tab. 6) (KOMORI and NONAKA, 1983, 1984).

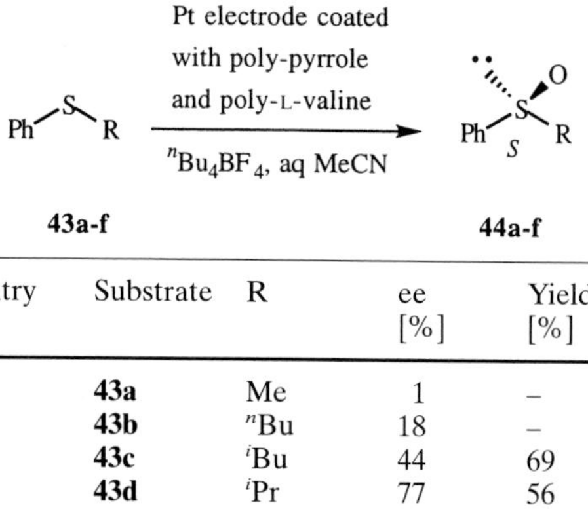

The selective epoxidation of the terminal alkene bond of the hexaene squalene, is one of the "landmark" achievements of organic synthesis. The result was rationalized as due to the polyene chain forming a coiled conformation in polar media which minimizes contact of the non-polar chain with the solvent and shields the internal alkene bonds from reaction with the epoxidation reagent (VAN TAMELEN, 1968). This idea was extended by preparing homochiral peptide derivatives of (*E*,*E*)-farnesic acid with the anticipation that the peptide chain would induce a helical conformation which would render epoxidation both regio- and enantioselective.

Initial studies using hexa-L-phenylalanyl methylester farnesate (**45a**; Fig. 14) gave poor results, possibly because the material had low solubility and aggregation to give β-pleated sheet structures occurred. Insertion of α-amino-*iso*-butyric acid (Aib) residues into the peptide chain increased solubility and consequently improved the results obtained. The best results were achieved with the *N*-terminal farnesoyl derivative of the nonapeptide; [L-Phe-L-Phe-Aib]$_3$OBz (**45b**). Bromohydroxylation followed by base treatment gave the (*R*)-epoxide (**46b**) but with only 25% chiral induction, whereas epoxidation with *m*-chloroperoxybenzoic acid gave the (*S*)-enantiomer (*ent*-**46b**) (12% ee) plus *bis*-epoxide. The enantiocomplementary results with bromohydroxylation/base and peracid are consistent with approach to the less hindered face of an alkene bond in a right-handed α-helix (BUDT et al., 1986).

45a,b

a NBS, glyme:water (5:1), 0 °C;
b K_2CO_3, MeOH, 0 °C

R

46
a R = (L-Phe)$_6$-OMe
b R = (L-Phe-L-Phe-Aib)$_3$-OBz; 25% ee

Fig. 14. Regio- and enantioselective epoxidation of farnesamide *N*-oligopeptides (BUDT et al., 1986).

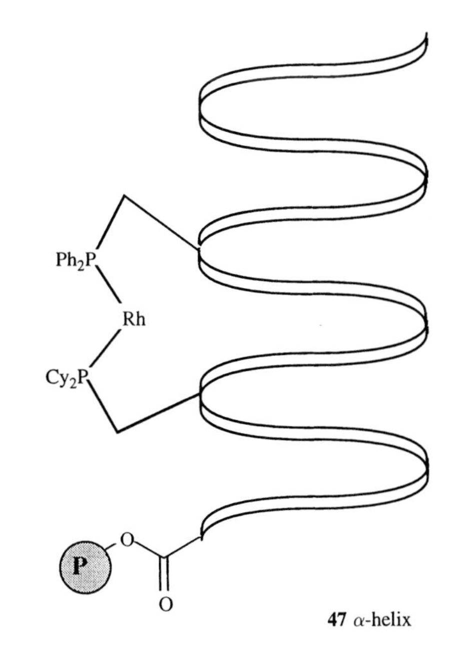

47 α-helix

48 Pps

49 Cps

H_2, 1200 psi, **47**
4-9% ee

50 **51**

Fig. 15. Stereoselective hydrogenation using a rhodium polypeptide complex (Ac-Ala-Aib-Ala-Pps-Ala-Ala-Aib-Cps-Ala-Ala-Aib-Ala **47**).

Hydrogenation reactions have been carried out using rhodium and a derivatized oligomer of alanine (**47**) (Fig. 15) with incorporation of Aib to enhance secondary structure. Although initially poor enantiomeric excess was obtained (GILBERTSON et al., 1996), an intriguing combinatorial strategy was adopted from this approach (GILBERTSON and WANG, 1996). A 63-member library was created, its generic features were Ac-Ala-Aib-Ala-1-2-3-4-5-Ala-Aib-Ala-NH_2. The peptides either had two Pps (cf. **48**), two Cps (cf. **49**) or one of each. In some cases these residues were at positions 1 and 5, in others they were adjacent to each other. The remaining positions were taken with combinations of Ala, Aib, Phe, Val, His, and Ile with the objective of placing the latter four residues in proximity to the metal. When the catalysts were used in the hydrogenation of (**50**) at 400 psi the highest enantioselectivity (**51**, 18% ee), was achieved with Cps-Ala-Ala-Aib-Cps. A general trend found was that the better catalysts had Ala at positions 2 and 3 with another amino acid at 4. It shows some of the potential for accurate rational modification of single amino acids in an oligomer so as to design oligopeptides utilizing variation in side chains in the assembly of a chiral scaffold.

A noteworthy failure is an attempted stereoselective epoxidation of unfunctionalized olefins using a metalloporphyrin strapped with a peptide (Ac-Ala-Cys-Glu-Gln-Leu-Leu-Lys-Glu-Leu-Leu-Gln-Lys-Cys-Ala-NH_2) through the cysteine sulfur atoms (GEIER and SASAKI, 1997). Despite no enantiodiscrimination being observed, the study demonstrates a thorough rationalization of the design process (GEIER and SASAKI, 1999; GEIER et al., 1999).

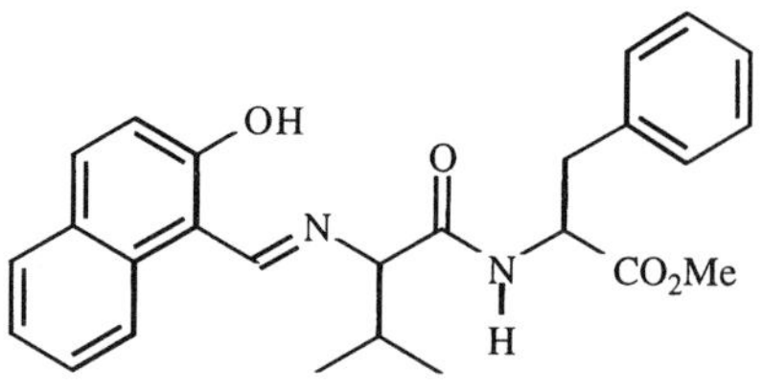

52
a Nap-(*S*)-Val-(*S*)-Phe
b Nap-(*R*)-Val-(*S*)-Phe
c Nap-(*R*)-Val-(*R*)-Phe

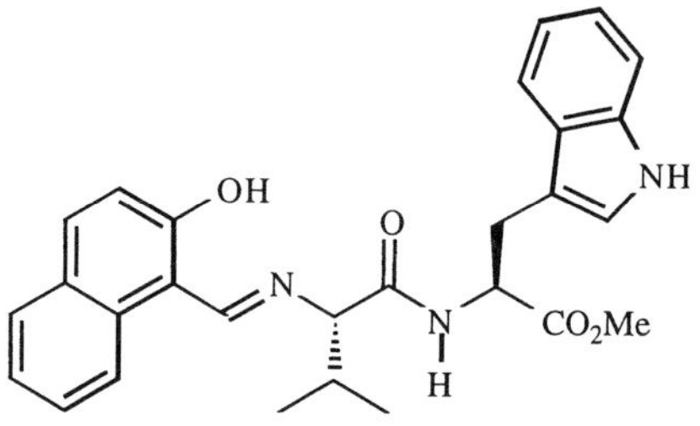

53 Nap-(*S*)-Val-(*S*)-Trp

Fig. 16. Dipeptide ligands for titanium alkoxide catalyzed enantioselective hydrocyanation of aldehydes (MORI et al., 1991; ABE et al., 1992; NITTA et al., 1992).

4 Use of Linear Di-, Tri- and Tetrapeptides

INOUE's considerable contribution to this whole field has also embraced linear dipeptides attached to a Schiff base (**52**, **53**) (Fig. 16), which in the presence of titanium(IV), catalyze enantioselective hydrocyanation (Tab. 7). Good enantioselectivity has been obtained with a diverse range of aldehydes (**54a–o**) (MORI et al., 1991; ABE et al., 1992), with lower temperatures providing higher enantiodiscrimination. While Ile, Leu, Ala, and phenyl glycine have also been studied as part of the catalyst, Val, Phe, and Trp appear to provide the best results (NITTA et al., 1992).

An elegant development of this methodology is the titanium(IV)-catalyzed addition of trimethylsilyl cyanide (TMSCN) to *meso*-epoxides using a combinatorially optimized ligand (interesting examples in the series: ligand **56** I–IV and i–iv) for each substrate type (**57a–c** and **59**) (Tab. 8, Entries 1, 2; 7, 8) (COLE et al., 1996; SHIMIZU et al., 1997).

The appendage of glycine as R^3 has provided increases of enantioselectivities of up to 20% ee compared to R^3=OMe with different ligands, for three substrates (Tab. 8, Entries 3, 4; 10, 11; 12, 13). Interestingly it appears the chirality of specific amino acids do not entirely determine which product is the major enantiomer in the reaction; the functionality of the side chain is at least partially responsible (Tab. 8, Entries 4, 5, and 6) (SHIMIZU et al., 1997).

Extension of this work has produced an even more successful enantioselective addi-

Tab. 7. Enantioselective Hydrocyanation of Aldehydes Using Titanium Dipeptide Complex Catalysts

Ligands **52** or **53**, toluene, HCN, $Ti(OEt)_4$ or $Ti(O^iPr)_4$

54a-o → **(R)-55a-o** + **(S)-55a-o**

54a ≡ 1a **54b** **54c** OPh **54d** **54e** $^nH_{13}C_6$ **54f**

Ph **54g** $^nH_{11}C_5$ **54h** ^nPr **54i** **54j** **54k**

54l ^nPr Et **54m** **54n** ^nBu **54o**

Entry	Ligand	Substrate	Temp [°C]	Time [h]	Product ee [%]	Conv. [%]
1	**52a**	**54a**	−40	4	(*R*)-**55a**, 86	85
2	**52b**	**54a**	−20	4	(*S*)-**55a**, 38	84
3	**52c**	**54a**	−40	7	(*S*)-**55a**, 84	86
4	**53**	**54a**	−40	3	(*R*)-**55a**, 88	88
5	**53**	**54b**	−40	7.5	(*R*)-**55b**, 90	88
6	**53**	**54c**	−40	7.5	(*R*)-**55c**, 86	74
7	**52c**	**54d**	−40, −20	11, 12	(*S*)-**55d**, 86	85
8	**52a**	**54e**	−40	1.5	(*R*)-**55e**, 54	99
9	**52a**	**54f**	−40	1.5	(*R*)-**55f**, 74	99
10	**52a**	**54g**	−40	18	(*R*)-**55g**, 81	82
11	**52a**	**54h**	−60	119	(*R*)-**55h**, 89	83
12	**52a**	**54i**	−20	22	(*R*)-**55i**, 85	85
13	**52a**	**54j**	−60	71	**55j**, 70	74
14	**52a**	**54k**	−60	20	**55k**, 37	22
15	**52a**	**54l**	−60	46	**55l**, 72	90
16	**52a**	**54m**	−60	143	**55m**, 60	28
17	**52a**	**54n**	−60	143	**55n**, 60	78
18	**52a**	**54o**	−40	69	**55o**, 68	57

tion of cyanide to imines (**62**) (Tab. 9). Again, the peptide/Schiff base (**61**) used was optimized along with the metal salt, solvent, and cyanide donor, allowing access to the α-amino nitriles (**63**) with excellent enantioselectivity. Initially conversion proved disappointing

Tab. 8. Titanium Dipeptide Complex Catalyzed Enantioselective *meso*-Epoxide Ring Opening with Trimethylsilyl Cyanide (Cole et al., 1996; Shimizu et al., 1997)

57 → 58: $Ti(O^iPr)_4$, toluene, 4 °C, TMSCN (ligand 56)

59 → 60: $Ti(O^iPr)_4$, toluene, 4 °C, TMSCN

R^1 = I, II, III, IV

R^2 = i, ii, iii, iv

Entry	Ligand **56** R^1	R^2	R^3	Substrate	n	Product **58** or **60** ee [%]	Yield [%]
1	I	ii	OMe	**57a**	1	64	56
2	II	iii	OEt	**57a**	1	75	57
3	II	ii	OMe	**57a**	1	63	–
4	II	ii	Gly	**57a**	1	83	72
5	II	i	Gly	**57a**	1	10	–
6	II	iv	Gly	**57a**	1	*ent*-**58a**, 58	–
7	I	ii	OMe	**57b**	2	86	65
8	II	iii	OEt	**57b**	2	70	65
9	II	iii	Gly	**57c**	3	52	–
10	III	ii	OMe	**57c**	3	69	–
11	III	ii	Gly	**57c**	3	84	79
12	IV	ii	OMe	**59**	–	58	–
13	IV	ii	Gly	**59**	–	78	69

(<39% after 48 h), but gradual addition of *iso*-propanol provided the product in higher yield and improved enantioselectivity with some substrates. It is thought that the alcohol causes slow *in situ* release of hydrogen cyanide from the TMSCN (Krueger et al., 1999). The use of the diphenylmethyl protecting group enables the nitrile group and the amino protecting group to be cleaved simultaneously with 6 N hydrochloric acid to give the corresponding amino acid without racemization. A similar type of catalyst and approach can be seen with another stereoselective Strecker reaction (Sigman and Jacobsen, 1998) and epoxidation (Francis and Jacobsen, 1999).

Recently Miller and coworkers have developed tetrapeptide catalysts for the acylation of alcohols incorporating a 3-(1-imidazolyl)-L-alanine residue (**64a** shown as an *N*-acetylated intermediate; Fig. 17). The challenge in

Tab. 9. Titanium Dipeptide Complex Catalyzed Enantioselective Addition of Cyanide to Imines (Strecker Reaction)

61 (0.1 equiv.)

$Ti(O^iPr)_4$ (0.10 equiv.), toluene, TMSCN (2 equiv.), ^iPrOH (1.5 equiv.) added over 20 h, 4 °C

62a-g → **63a-g**

62, 63
a $R^3 = R^4 = H$
b $R^3 = Cl, R^4 = H$
c $R^3 = Br, R^4 = H$
d $R^3 = H, R^4 = OMe$

e **f** **g** (^tBu)

Entry	Ligand **61** R^1	R^2	Substrate	Product **63a–g** ee [%]	Yield [%]
1	OMe	H	**62a**	97	82
2	Cl	Cl	**62b**	93	85
3	Cl	Cl	**62c**	94	93
4	Cl	Cl	**62d**	94	99
5	OMe	H	**62e**	93	80
6	OMe	H	**62f**	90	87
7	Br	Br	**62g**	85	97

creating an "active site" with a short peptide is to induce it to fold back upon itself, rather than form an extended structure. This is achieved in this case by the combination of proline and Aib (Aib is a seemingly popular choice for synthetic peptides) to give a β-turn. The α-methylbenzamide group was incorporated to provide π-stacking with the acylimidazolium ion. The catalyst (**64a**) showed good enantioselectivity in the acylation of a range of alcohols (e.g., **65**) bearing an adjacent amide group, which presumably orientates the alcohol in the hydrogen bonding array of the catalyst. The importance of hydrogen bonding was further demonstrated by a solvent optimization study. The highest enantioselectivity was achieved in non-polar, non-Lewis base solvents, whereas protic solvents resulted in totally non-selective acylation. The role of the imidazole group in the catalyst (**64a**) was demonstrated by replacing it with a phenyl group (i.e., a phenylalanine residue), to give an analog (**64b**) which did not act as an acylation catalyst (Miller et al., 1998).

Further investigations studied the replacement of the *C*-terminal α-methylbenzamide

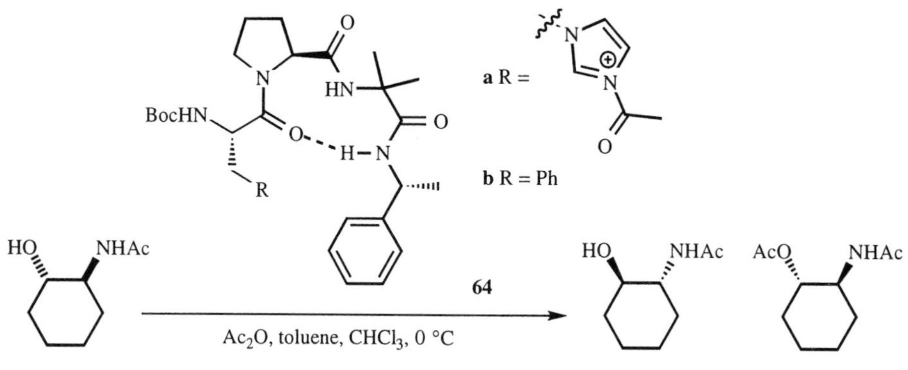

Ac_2O, toluene, $CHCl_3$, 0 °C

rac-**65** → (2*R*,3*R*)-**65** (2*S*,3*S*)-**66** 84% ee

64: **a** R = (acylimidazolium); **b** R = Ph

Fig. 17. Tripeptide benzamide catalyzed enantioselective acylation (MILLER et al., 1998).

group with phenylalanine, valine, or glycine to give the tetrapeptide (**67**) (Tab. 10) and replacement of the L-proline residue by a D-proline residue (**68**) (Tab. 11).

The diastereomeric catalysts (**67**) and (**68**) give enantiocomplementary products with the latter being more enantioselective in all cases. Evidently the chirality of the proline residue dictates which enantiomer undergoes acylation, but the enantioselectivity is also modulated mainly by histidine, but to some extent by the *C*-terminal residue. For the L-proline based catalyst (**67**), D-amino acids at the *C*-terminus increase enantioselectivity, while the converse is true for the D-proline based catalyst (**68**). Evidently, the *u*-relationship is advantageous for selective catalysis over the *l*-form. NMR and IR data were used to determine that the

Tab. 10. L-Proline Containing Tetrapeptide Catalysts for the Acylation of Alcohols (COPELAND et al., 1998)

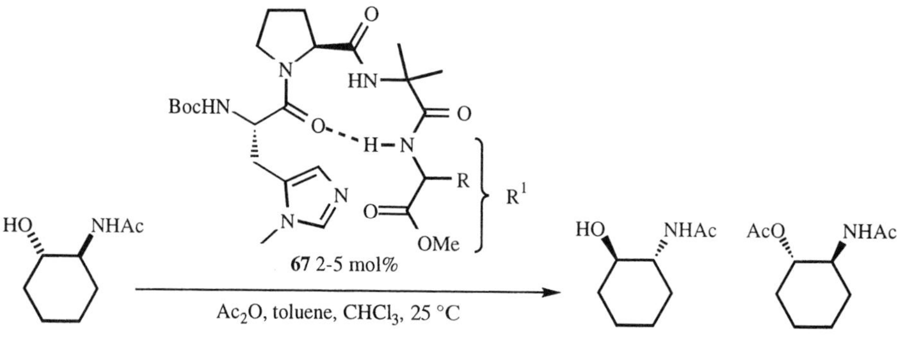

rac-**65** → (2*R*,3*R*)-**65** (2*S*,3*S*)-**66**

67 2-5 mol%

Ac_2O, toluene, $CHCl_3$, 25 °C

Entry	Catalyst **67** R[1]	*S*	(2*R*,3*R*)-**65** ee [%]	(2*S*,3*S*)-**66** ee [%]	Conv. [%]
1	L-Phe	3.0	44	34	56
2	D-Phe	5.7	89	36	71
3	L-Val	3.4	54	35	61
4	D-Val	4.3	65	39	63
5	Gly	3.5	50	38	57

Tab. 11. D-Proline Containing Tetrapeptide Catalysts for the Acylation of Alcohols (COPELAND et al., 1998)

HO NHAc *rac*-65 → **68** 2-5 mol%, Ac_2O, toluene, $CHCl_3$, 25 °C → (2*S*,3*S*)-**65** + (2*R*,3*R*)-**66**

Entry	Catalyst **68** R[1]	*S*	(2*S*,3*S*)-**65** ee [%]	(2*R*,3*R*)-**66** ee [%]	Conv. [%]
1	L-Phe	28	98	73	58
2	D-Phe	14	89	66	57
3	L-Val	21	99	63	61
4	D-Val	9.2	88	55	62
5	Gly	14	97	57	63

L-proline based catalyst (**67**) gives a type II *β*-turn with one hydrogen bond and the D-proline based catalyst (**68**) gives a type II′ *β*-turn with two hydrogen bonds. This probably gives a more rigid structure, which is the origin of the greater enantioselectivity of the latter catalyst (COPELAND et al., 1998).

5 Summary and Future Directions

While successful examples of peptides being used in stereoselective synthesis are currently limited in number, the few positive examples demonstrate a significant capacity for their viable application. The trend towards combinatorially designed ligands, which allow adaptation for individual substrates, increases the potential importance. To this end peptides uniquely offer a cheap, simple, enantiopure (with both enantiomers available), diverse range of subunits (amino acids) with the potential for metal binding and hydrogen bonding of reagents and substrates (BURGER and STILL, 1995; FRANCIS et al., 1996; SEVERIN et al., 1998). The chemistry for preparation of the ligand (peptide) is very well understood with facility for solid phase synthesis and good comprehension of their structure. Such compatibility offers the exciting possibility of multidomain peptides facilitating multiple stereoselective reactions tailored to specific requirements. Further exploitation of the versatility of synthetic peptides as catalysts in stereoselective synthesis is inevitable.

Since the preparation of this chapter there have been interesting studies using combinatorial variation of peptides to optimize serine-protease mimics (DE MUYNCK et al., 2000) and in the search to find means of controlling cyclizations of epoxy-alcohols to provide the energetically disfavored product (CHIOSIS, 2000). In addition to this, diphenylglycine has displayed an interesting capacity for the enantioselective inclusion of sulfixides (AKAZOME et al., 2000).

6 References

ABE, S., NONAKA, T. (1983), Electroorganic reactions on organic electrodes. Part 5. Durable chiral graphite electrodes modified chemically with poly(L-valine) and poly(*N*-acryloyl-L-valine methyl ester), *Chem. Lett.*, 1541–1542.

ABE, S., NONAKA, T., FUCHIGAMI, T. (1983), Electroorganic reactions on organic electrodes. 1. Asymmetric reduction of prochiral activated olefins on poly-L-valine-coated graphite, *J. Am. Chem. Soc.* **105**, 3630–3632.

ABE, H., NITTA, H., MORI, A., INOUE, S. (1992), A facile synthesis of optically active γ-cyanoallylic alcohols using asymmetric hydrocyanation of α,β-alkenyl aldehydes followed by stereospecific [3.3] sigmatropic chirality transfer of the cyanohydrin acetates, *Chem Lett.*, 2443–2446.

ADGER, B. M., BARKLEY, J. V., BERGERON, S., CAPPI, M. W., FLOWERDEW, B. E. et al. (1997), Improved procedure for Juliá–Colonna asymmetric epoxidation of α,β-unsaturated ketones: total synthesis of diltiazem and Taxol™ side-chain, *J. Chem. Soc., Perkin Trans.* 1, 3501–3507.

AKAZOME, M., UENO, Y., OOISO, H., OGURA, K. (2000), Enantioselective inclusion of methyl phenyl sulfoxides and benzyl methyl sulfoxides by (*R*)-phenylglycyl-(*R*)-phenylglycyl and the crystal structures of the inclusion cavities, *J. Org. Chem.* **65**, 68–76.

ALLEN, J. V., ROBERTS, S. M., WILLIAMSON, N. M. (1998), Polyamino acids as man-made catalysts, *Adv. Biochem. Eng. Biotechnol.* **63**, 126–143.

ALLEN, J. V., DRAUZ, K.-H., FLOOD, R. W., ROBERTS, S. M., SKIDMORE, J. (1999), Polyamino acid catalyzed asymmetric epoxidation: sodium percarbonate as a source of base and oxidant, *Tetrahedron Lett.* **40**, 5417–5420.

ALPER, H., HAMEL, N. (1990), Poly-L-leucine as an added chiral ligand for the palladium catalyzed carbonylation of allylic alcohols, *J. Chem. Soc., Chem. Commun.*, 135–136.

APPERLEY, D., NORTH, M., STOKOE, R. B. (1995), The catalytically active conformation of *cyclo*-[(*S*)-His-(*S*)-Phe] as determined by solid state NMR, *Tetrahedron: Asymmetry* **6**, 1869–1872.

AUGUSTYN, J. A. N., BEZUIDENHOUDT, B. C. B., FERREIRA, D. (1990), Enantioselective synthesis of flavonoids. Part 1. Poly-oxygenated chalcone epoxides, *Tetrahedron* **46**, 2651–2660.

BANFI, S., COLONNA, S., MOLINARI, H., JULIÁ, S., GUIXER, J. (1984), Asymmetric epoxidation of electron-poor olefins. 5. Influence on stereoselectivity of the structure of poly-α-aminoacids used as catalysts, *Tetrahedron* **40**, 5207–5211.

BAURES, P. W., EGGLESTON, D. S., FLISAK, J. R., GOMBATZ, K. et al. (1990), An efficient asymmetric synthesis of substituted phenyl glycidic esters, *Tetrahedron Lett.* **31**, 6501–6504.

BENTLEY, P. A. (1999), The use of poly-leucine in stereoselective synthesis, *Thesis*, University of Liverpool.

BENTLEY, P. A., BERGERON, S., CAPPI, M. W., HIBBS, D. E. et al. (1997a), Asymmetric epoxidation of enones employing polymeric α-amino acids in non-aqueous media, *J. Chem. Soc., Chem. Commun.*, 739–740.

BENTLEY, P. A., KROUTIL, W., LITTLECHILD, J. A., ROBERTS, S. M. (1997b), Preparation of polyamino acid catalysts for use in Juliá asymmetric epoxidation, *Chirality* **9**, 198–202.

BENTLEY, P. A., CAPPI, M. W., FLOOD, R. W., ROBERTS, S. M., SMITH, J. A. (1998), Towards a mechanistic insight into the Juliá–Colonna asymmetric epoxidation of α,β-unsaturated ketones using discrete lenghts of poly-leucine, *Tetrahedron Lett.* **39**, 9297–9300.

BEZUIDENHOUDT, B. C. B., SWANEPOEL, A., AUGUSTYN, J. A. N., FERREIRA, D. (1987), The first enantioselective synthesis of poly-oxygenated α-hydroxydihydro-chalcones, *Tetrahedron Lett.* **28**, 4857–4860.

BOLM, C., BIENEWALD, F., SEGER, A. (1996), Asymmetric autocatalysis with amplification of chirality, *Angew. Chem. (Int. Edn. Engl.)* **35**, 1657–1659.

BROWN, R. F. C., DONOHUE, A. C., JACKSON, W. R., MCCARTHY, T. D. (1994), Synthetic applications of optically active cyanohydrins. Enantioselective syntheses of the hydroxyamides Tembamide and Aegeline, the cardiac drug Denopamine, and some analogs of the bronchodilator Salbutamol, *Tetrahedron* **50**, 13739–13752.

BUDT, K.-H., VATELE, J.-M., KISHI, Y. (1986), Terminal epoxidation of farnesate attached to helical peptides, *J. Am. Chem. Soc.* **108**, 6080–6082.

BURGER, M. T., STILL, W. C. (1995), Synthetic ionophores. Encoded combinatorial libraries of cyclen-based receptors for Cu^{2+} and Co^{2+}, *J. Org. Chem.* **60**, 7382–7383.

CALLANT, D., COUSSENS, B., MATEN, T. V. D., DE VRIES, J. G., DE VRIES, K. (1992), NMR measurements and semi-empirical calculations in a first approach to elucidate the mechanism of enantioselective cyanohydrin formation catalyzed by *cyclo*-(*S*)-Phe-(*S*)-His, *Tetrahedron: Asymmetry* **3**, 401–414.

CAPPI, M. W., CHEN, W. P., FLOOD, R. W., LIAO, Y. W., ROBERTS, S. M. et al. (1998), New procedures for the Juliá–Colonna asymmetric epoxidation: synthesis of (+)-Clausenamide, *J. Chem. Soc., Chem. Commun.*, 1159–1160.

CHEN, W.-P., ROBERTS, S. M. (1999), Juliá–Colonna asymmetric epoxidation of furyl styryl ketone as a route to intermediates to naturally occurring styryl lactones, *J. Chem. Soc., Perkin Trans.* 1, 103–105.

CHIOSIS, G. (2000), Study into the cyclization of an epoxy-alcohol to the energetically disfavored product by peptides from non-biased combinatorial libaries, *Tetrahedron Lett.* **41**, 801–806.

COLE, B. M., SHIMIZU, K. D., KRUEGER, C. A., HARRITY, J. P. A., SNAPPER, M. L., HOVEYDA, A. H. (1996), Discovery of chiral catalysts through ligand diversity: Ti-catalyzed enantioselective addition of TMSCN to *meso*-epoxides, *Angew. Chem. (Int. Edn. Engl.)* **35**, 1668–1671.

COLONNA, S., MOLINARI, H., BANFI, S., JULIÁ, S., MASANA, J., ALVAREZ, A. (1983), Synthetic enzymes 4. Highly enantioselective epoxidation by means of polyaminoacids in a triphasic system: influence of structural variations within the catalyst, *Tetrahedron* **39**, 1635–1641.

COLONNA, S., JULIÁ, S., MOLINARI, H., BANFI, S. (1984), Catalytic asymmetric epoxidation by means of polyaminoacids in a triphase system, *Heterocycles* **21**, 548.

COLONNA, S., MANFREDI, A., SPADONI, M. (1987), Asymmetric syntheses catalyzed by natural and synthetic peptides, in: *Organic Synthesis: Modern Trends for Organic Synthesis* (CHIZHOV, O., Ed.), pp. 275–284. Oxford: Blackwell.

COPELAND, G. T., JARVO, E. R., MILLER, S. J. (1998), Minimal acylase-like peptides. Conformational control of absolute stereospecificity, *J. Org. Chem.* **63**, 6784–6785.

DALY, W. H., POCHÉ, D. (1988), The preparation of *N*-carboxyanhydrides of α-amino acids using bis(trichloromethyl)carbonate, *Tetrahedron Lett.* **29**, 5859–5862.

DANDA, H. (1991), Asymmetric hydrocyanation of benzaldehyde catalyzed by (5*R*)-5-(4-imidazolylmethyl)-2,4-imidazolidinedione, *Bull. Chem. Soc. Jpn.* **64**, 3743–3745.

DANDA, H., NISHIKAWA, H., OTAKA, K. (1991), Enantioselective autoinduction in the asymmetric hydrocyanation of 3-phenoxybenzaldehyde catalyzed by *cyclo*-[(*R*)-phenylalanyl-(*R*)-histidyl], *J. Org. Chem.* **56**, 6740–6741.

DE MUYNCK, H., MADDER, A., FARCY, N., DE CLERCQ, P. J., PÉREZ-PAYÁN, M. N. et al. (2000), Application of combinatorial procedures in the search for serine-protease-like activity with focus on the acyl transfer step, *Angew. Chem.* (*Int. Edn.*) **39**, 145–148.

DONG, W., PETTY, W. (1985), *U.S. Patent* **102**, 4,554.

EBRAHIM, S., WILLS, M. (1997), Synthetic applications of polymeric *a*-amino acids, *Tetrahedron: Asymmetry* **8**, 3163–3173.

EHLER, K. W., ORGEL, L. E. (1976), *N,N′*-carbonyldiimidazole induced peptide formation in aqueous solution, *Biochim. Biophys. Acta* **434**, 233–243.

FERRIS, J. P., HILL, A. R., LUI, R., ORGEL, L. E. (1996), Synthesis of long prebiotic oligomers on mineral surfaces, *Nature* **381**, 59–61.

FLISAK, J. R., GOMBATZ, K. J., HOLMES, M. M., JARMAS, A. A., LANTOS, I. et al. (1993), A practical, enantioselective synthesis of SK&F 104353, *J. Org. Chem.* **58**, 6247–6254.

FRANCIS, M. B., JACOBSEN, E. N. (1999), Discovery of novel catalysts for alkene epoxidation from metal-binding combinatorial libraries, *Angew. Chem. (Int. Edn. Engl.)* **38**, 937–941.

FRANCIS, M. B., FINNEY, N. S., JACOBSEN, E. N. (1996), Combinatorial approach to the discovery of novel coordination complexes, *J. Am. Chem. Soc.* **118**, 8983–8984.

GEIER, G. R. I., SASAKI, T. (1997), The design, synthesis and characterization of a porphyrin–peptide conjugate, *Tetrahedron Lett.* **38**, 3821–3824.

GEIER, G. R. I., SASAKI, T. (1999), Catalytic Oxidation of alkene with a surface-bound metalloporphyrin–peptide conjugate, *Tetrahedron* **55**, 1859–1870.

GEIER, G. R. I., LYBRAND, T. P., SASAKI, T. (1999), On the absence of stereoselectivity in the catalytic oxidation of alkenes with a surface-bound metalloporphyrin–peptide conjugate, *Tetrahedron* **55**, 1871–1880.

GILBERTSON, S. R., WANG, X. (1996), The combinatorial synthesis of chiral phosphine ligands, *Tetrahedron Lett.* **37**, 6475–6478.

GILBERTSON, S. R., WANG, X., HOGE, G. S., KLUG, C. A., SCHAEFER, J. (1996), Synthesis of phosphine–rhodium complexes attached to a standard peptide synthesis resin, *Organometallics* **15**, 4678–4680.

HEANEY, H. (1993), Oxidation reactions using magnesium monoperphtalate and urea hydrogen peroxide, *Aldrichim. Acta* **26**, 35–45.

HOGG, D. J. P., NORTH, M. (1993), Mechanistic studies on the asymmetric addition of HCN to aldehydes catalyzed by *cyclo*-[(*S*)-His-(*S*)-Phe], *Tetrahedron* **49**, 1079–1090.

HOGG, D. J. P., NORTH, M., STOKOE, R. B., TEASDALE, W. G. (1993), The oxidation of benzaldehyde to benzoic acid catalyzed by *cyclo*-[(*S*)-His-(*S*)-Phe], and its implications for the catalytic asymmetric addition of HCN to aldehydes, *Tetrahedron: Asymmetry* **4**, 1553–1558.

HOGG, D. J. P., NORTH, M., STOKOE, R. B. (1994), A conformational comparison of *cyclo*-[(*S*)-His-(*S*)-Leu] and *cyclo*-[(*S*)-His-(*S*)-Phe], catalysts for asymmetric addition, *Tetrahedron* **50**, 7933–7946.

HULST, R., BROXTERMAN, Q. B., KAMPHUIS, J., FORMAGGIO, F., CRISMA, M. et al. (1997), Catalytic enantioselective addition of hydrogen cyanide to benzaldehyde and *p*-methoxybenzaldehyde using *cyclo*-His-(α-Me)Phe as catalyst, *Tetrahedron: Asymmetry* **8**, 1987–1999.

ITSUNO, S., SAKAKURA, M., ITO, K. (1990), Polymer-supported poly(amino acids) as new asymmetric epoxidation catalyst of α,β-unsaturated ketones, *J. Org. Chem.* **55**, 6047–6049.

IYER, M. S., GIGSTAD, K. M., NAMDEV, N. D., LIPTON, M. (1996), Asymmetric catalysis of the Strecker amino acid synthesis by a cyclic dipeptide, *J. Am. Chem. Soc.* **118**, 4910–4911.

JACKSON, W. R., JAYATILAKE, G. S., MATTHEWS, B. R., WILSHIRE, C. (1988), Evaluation of some cyclic dipeptides as catalysts for the asymmetric hydrocyanation of aldehydes, *Aust. J. Chem.* **41**, 203–213.

JACKSON, W. R., JACOBS, H. A., KIM, H. J. (1992), Spectroscopic characterization of the active form of the "Inoue" dipeptides (*cyclo*-[(*S*)-Phe-(*S*)-His-]), *Aust. J. Chem.* **45**, 2073–2076.

JULIÁ, S., MASANA, J., VEGA, J. C. (1980), Synthetic enzymes – highly stereoselective epoxidation of chalcone in a triphasic toluene–water–poly[(*S*)-alanine] system, *Angew. Chem. (Int. Edn. Engl.)* **19**, 929–931.

JULIÁ, S., GUIXER, J., MASANA, J., ROCAS, J., COLONNA, S. et al. (1982), Synthetic enzymes. 2. Catalytic asymmetric epoxidation by means of polyaminoacids in a triphase system, *J. Chem. Soc., Perkin Trans.* 1, 1317–1324.

JULIÁ, S., MASANA, J., ROCAS, J., COLONNA, S., ANNUNZIATA, R., MOLINARI, H. (1983), Synthetic enzymes. 3. Highly stereoselective epoxidation of chalcones in a triphasic toluene–water–poly[(*S*)-alanine] system, *Anales De Quimica Serie C-Quimica Organica Y Bioquimica* **79**, 102–104.

KIM, H. J., JACKSON, W. R. (1992), Polymer attached cyclic dipeptidedes as catalysts for enantioselective cyanohydrin formation, *Tetrahedron: Asymmetry* **3**, 1421–1430.

KIM, H. J., JACKSON, W. R. (1994), Substituent effects on the enantioselective hydrocyanation of aryl aldehyde, *Tetrahedron: Asymmetry* **5**, 1541–1548.

KOGUT, E. F., THOEN, J. C., LIPTON, M. A. (1998), Examination and enhancement of enantioselective autoinduction in cyanohydrin formation by *cyclo*-[(*R*)-His-(*R*)-Phe], *J. Org. Chem.* **63**, 4604–4610.

KOMORI, T., NONAKA, T. (1983), Electroorganic reactions on organic electrodes. 3. Electrochemical asymmetric oxidation of phenyl cyclohexyl sulfide on poly(L-valine)-coated platinum electrodes, *J. Am. Chem. Soc.* **105**, 5690–5691.

KOMORI, T., NONAKA, T. (1984), Electroorganic reactions on organic electrodes. 6. Electrochemical asymmetric oxidation of unsymmetric sulfides to the corresponding chiral sulfoxides on poly(amino acids)-coated electrodes, *J. Am. Chem. Soc.* **106**, 2656–2659.

KRICHELDORF, H. R. (1987), *α-Aminoacid-N-Carboxyanhydrides and Related Heterocycles.* Berlin: Springer-Verlag.

KROUTIL, W., LASTERRA-SÁNCHEZ, M. E., MADDRELL, S. J., MAYON, P., MORGAN, P. et al. (1996a), Development of the Juliá asymmetric epoxidation reaction. Part 2. Application of the oxidation to alkyl enones, enediones and unsaturated keto esters, *J. Chem. Soc., Perkin Trans.* 1, 2837–2844.

KROUTIL, W., MAYON, P., LASTERRA-SÁNCHEZ, M. E., MADDRELL, S. J., ROBERTS, S. M. et al. (1996b), Unexpected asymmetric epoxidation reactions catalyzed by polyleucine-based systems, *J. Chem. Soc., Chem. Commun.*, 845–846.

KRUEGER, C. A., KUNTZ, K. W., DZIERBA, C. D., WIRSCHUM, W. G., GLEASON, J. D. et al. (1999), Ti-catalyzed enantioselective addition of cyanide to imines. A practical synthesis of optically pure α-amino acids, *J. Am. Chem. Soc.* **121**, 4284–4285.

LASTERRA-SÁNCHEZ, M. E., ROBERTS, S. M. (1995), Enantiocomplementary asymmetric epoxidation of selected enones using poly-L-leucine and poly-D-leucine, *J. Chem. Soc., Perkin Trans.* 1, 1467–1468.

LASTERRA-SÁNCHEZ, M. E., ROBERTS, S. M. (1997), An important niche in synthetic organic chemistry for some biomimetic oxidation reactions catalyzed by polyamino acids, *Curr. Org. Chem.* **1**, 187–196.

LASTERRA-SÁNCHEZ, M. E., FELFER, U., MAYON, P., ROBERTS, S. M., THORNTON, S. R., TODD, C. J. (1996), Development of the Juliá asymmetric epoxidation reaction. Part 1. Application of the oxidation to enones other than chalcones, *J. Chem. Soc., Perkin Trans.* 1, 343–348.

LEE, D. H., GRANJA, J. R., MARTINEZ, J. A., SEVERIN, K., GHADIRI, M. R. (1996), A self-replicating peptide, *Nature* **382**, 525–528.

MATTHEWS, B. R., JACKSON, W. R., JAYATILAKE, G. S., WILSHIRE, C., JACOBS, H. A. (1988), Asymmetric hydrocyanation of a range of aromatic and aliphatic aldehydes, *Aust. J. Chem.* **41**, 1697–1709.

MILLER, S. J., COPELAND, G. T., PAPAIOANNOU, N., HORSTMANN, T. E., RUEL, E. M. (1998), Kinetic resolution of alcohols catalyzed by tripeptides containing the *N*-alkylimidazole substructure, *J. Am. Chem. Soc.* **120**, 1629–1630.

MORI, A., IKEDA, Y., KINOSHITA, K., INOUE, S. (1989), Cyclo-((*S*)-leucyl-(*S*)-histidyl). A catalyst for asymmetric addition of hydrogen cyanide to aldehydes, *Chem. Lett.*, 2119–2122.

MORI, A., NITTA, H., KUDO, M., INOUE, S. (1991), Peptide–metal complex as an asymmetric catalyst. A catalytic enantioselective cyanohydrin synthesis, *Tetrahedron Lett.* **32**, 4333–4336.

NEL, R. J. J., VAN HEERDEN, P. S., VAN RENSBURG, H., FERREIRA, D. (1998), Enantioselective synthesis of flavonoids. Part 5. Poly-oxygenated β-hydroxydihydrochalcones, *Tetrahedron Lett.* **39**, 5623–5626.

NITTA, H., YU, D., KUDO, M., MORI, A., INOUE, S. (1992), Peptide–titanium complex as catalyst for asymmetric addition of hydrogen cyanide to aldehyde, *J. Am. Chem. Soc.* **114**, 7969–7975.

NOE, C. R., WEIGAND, A., PIRKER, S. (1996), Studies on cyclic dipeptides, I: Aryl modifications of *cyclo*-[Phe-His], *Monatshefte für Chemie* **127**, 1081–1097.

NOE, C. R., WEIGAND, A., PIRKER, S., LIEPERT, P. (1997), Studies on cyclic dipeptides, II. Methylated modifications of *cyclo*-[Phe-His], *Monatshefte für Chemie* **128**, 301–316.

NONAKA, T., ABE, S., FUCHIGAMI, T. (1983), Electroorganic reactions on organic electrodes. Part 2. Electrochemical asymmetric reduction of citraconic and mesaconic acids on optically active poly(amino acid)-coated electrodes, *Bull. Chem. Soc. Jpn.* **56**, 2778–2783.

NORTH, M. (1992), A conformational study on *cyclo*-[(*S*)-phenylalanyl-(*S*)-histidyl] by molecular modeling and NMR techniques, *Tetrahedron* **48**, 5509–5522.

NORTH, M. (1993), Catalytic asymmetric cyanohydrin synthesis, *Synlett*, 807–820.

OKU, J.-I., INOUE, S. (1981), Asymmetric cyanohydrin synthesis catalyzed by a synthetic cyclic dipeptide, *J. Chem. Soc., Chem. Commun.*, 229–230.

OKU, J.-I., ITO, N., INOUE, S. (1979), Asymmetric cyanohydrin synthesis catalyzed by synthetic dipeptides, 1, *Makromol. Chem.* **180**, 1089–1091.

OKU, J.-I., ITO, N., INOUE, S. (1982), Asymmetric cyanohydrin synthesis catalyzed by synthetic dipeptides, 2, *Makromol. Chem.* **183**, 579–586.

PU, L. (1998), Recent developments in asymmetric catalysis using synthetic polymers with main chain chirality, *Tetrahedron: Asymmetry* **9**, 1457–1477.

RAY, P. C., ROBERTS, S. M. (1999), Overcoming diastereoselection using poly amino acids as chiral epoxidation catalysts, *Tetrahedron Lett.* **40**, 1779–1782.

SAVIZKY, R. M., SUZUKI, N., BOVÉ, J. L. (1998), The use of sonochemistry in the asymmetric epoxidation of substituted chalcones with sodium perborate tetrahydrate, *Tetrahedron: Asymmetry* **9**, 3967–3969.

SEVERIN, K., BERGS, R., BECK, W. (1998), Bioorganometallic chemistry–transition metal complexes with α-amino acids and peptides, *Angew. Chem. (Int. Edn. Engl.)* **37**, 1634–1654.

SHIMIZU, K. D., COLE, B. M., KRUEGER, C. A., KUNTZ, K. W., SNAPPER, M. L., HOVEYDA, A. H. (1997), Search for chiral catalysts through ligand diversity: substrate-specific catalysts and ligand screening on solid phase, *Angew. Chem. (Int. Edn. Engl.)* **36**, 1704–1707.

SHVO, Y., GAL, M., BECKER, Y., ELGAVI, A. (1996), Asymmetric hydrocyanation of aldehydes with cyclo-dipeptides: a new mechanistic approach, *Tetrahedron: Asymmetry* **7**, 911–924.

SIGMAN, M. S., JACOBSEN, E. N. (1998), Schiff base catalysts for the asymmetric Strecker reaction identified and optimized form of parallel synthetic libraries, *J. Am. Chem. Soc.* **120**, 4901–4902.

TANAKA, K., MORI, A., INOUE, S. (1990), The cyclic dipeptide *cyclo*-[(*S*)-phenylalanyl-(*S*)-histidyl] as a catalyst for asymmetric addition of hydrogen cyanide to aldehydes, *J. Org. Chem.* **55**, 181–185.

VAN RENSBURG, H., VAN HEERDEN, P. S., BEZUIDENHOUDT, B. C. B., FERREIRA, D. (1996), The first enantioselective synthesis of *trans*- and *cis*-dihydroflavonols, *J. Chem. Soc., Chem. Commun.*, 2747–2748.

VAN TAMELEN, E. E. (1968), Bioorganic chemistry: Sterols and acyclic terpene epoxides, *Acc. Chem. Res.* **1**, 111–120.

YADAV, V. K., KAPOOR, K. K. (1997), 1,8-Diazobicyclo[5.4.0]undec-7-ene: a remarkable base in the epoxidation of α,β-unsaturated lactones and other enones with anhydrous *t*-BuOOH, *Tetrahedron Lett.* **51**, 8573–8584.

Index

B

C

D

E

I

Q

R

U

V

Y

Z